Interaction of Mechanics and Mathematics

The Interaction of Mechanics and Mathematics (IMM) series publishes advanced textbooks and introductory scientific monographs devoted to modern research in the wide area of mechanics. The authors are distinguished specialists with international reputation in their field of expertise. The books are intended to serve as modern guides in their fields and anticipated to be accessible to advanced graduate students. IMM books are planned to be comprehensive reviews developed to the cutting edge of their respective field and to list the major references.

More information about this series at http://www.springer.com/series/5395

Martin Kružík · Tomáš Roubíček

Mathematical Methods in Continuum Mechanics of Solids

Martin Kružík
Institute of Information Theory
and Automation
Czech Academy of Sciences
Prague, Czech Republic

Faculty of Civil Engineering
Czech Technical University
Prague, Czech Republic

Tomáš Roubíček
Mathematical Institute, Faculty
of Mathematics and Physics
Charles University
Prague, Czech Republic

Institute of Thermomechanics
Czech Academy of Sciences
Prague, Czech Republic

ISSN 1860-6245 ISSN 1860-6253 (electronic)
Interaction of Mechanics and Mathematics
ISBN 978-3-030-02064-4 ISBN 978-3-030-02065-1 (eBook)
https://doi.org/10.1007/978-3-030-02065-1

Library of Congress Control Number: 2018960190

This Springer imprint is published by the registered company Springer Nature Switzerland AG
The registered company address is: Gewerbestrasse 11, 6330 Cham, Switzerland

Preface

> Mechanics is the paradise of mathematical sciences, because with that one comes to the mathematical fruit.
>
> LEONARDO DI SER PIERO DA VINCI (1452–1519)

> Pure mathematicians sometimes are satisfied with showing that the non-existence of a solution implies a logical contradiction, while engineers might consider a numerical result as the only reasonable goal. Such one sided views seem to reflect human limitations rather than objective values.[1]
>
> RICHARD COURANT (1888–1972)

> To those who do not know mathematics it is difficult to get across a real feeling as to the beauty, the deepest beauty, of nature... If you want to learn about nature, to appreciate nature, it is necessary to understand the language that she speaks in. ... It's an appreciation of the mathematical beauty of nature, of how she works inside.
>
> RICHARD PHILLIPS FEYNMAN (1918–1988)

> It is certain that continuum mechanics, which emerged during the 1950th, would not have experienced such rapid growth without amazing efficiency of the mathematical methods that were so well adapted to the problems formulated in the context of theoretical mechanics. [217]
>
> PAUL GERMAIN (1920–2009)

Mechanics have accompanied the development of mankind since ancient times and ultimately called for rational thinking, primarily based on more or less rigorous mathematical tools. And conversely, in this way, mechanics have served historically as a constant and probably main inspiration for mathematics. Such *Interaction between Mechanics and Mathematics*, being reflected also by the name of this Springer series, accelerated the development of modern mathematical tools in twentieth century when mathematics becomes relatively well applicable also to various nonlinear and coupled problems in mechanics and thermomechanics.

A very particular interaction has been developed between mechanics of continuum media and the theory of partial differential equations. Important concepts have been created on both sides, and a lot of mechanicians have become fairly applied mathematicians and vice versa.

In spite of such occasional and well-connecting "bridges", there are barriers between continuum mechanics and applied mathematics, caused probably partly by a

[1] See [129].

traditional way of education on specialized departments with very diverse professional accents. In this way, mechanical engineers and computational physicists do not support their models even by a very basic qualitative analysis and very typically use models whose solutions may even not exist under some generic circumstances. Usually, they also use approximations whose numerical stability and convergence are not granted and even sometimes there is a numerical evidence that they do not converge to any solution of a continuous model (in some cases because such solutions even may not exist). And quite often, prefabricated software packages are used without a solid knowledge what they really calculate, and such unrealistic trust and overuse of digital technology sometimes result in the deterioration of cognitive abilities of young generations.[2] A large portion of computational simulations performed in engineering and in physics, nowadays under euphemistic labels like "computational modeling", "numerical modeling", or "computational analysis", have unintentionally moved rather to a position that can be called, with possibly a little exaggeration, a "computer-assisted science fiction". On the other hand, this is also partly due to mathematicians because they often slide into very academic models, which have a little or no relevance for real-world demands, and into very particular results which, when in addition expressed in complicated mathematical language, are not understandable even for mathematically oriented mechanicians and physicists.

Viewed from the optimistic perspective, these (both historically developed and newly arising) barriers yield even more challenges both to design more applicable mathematics and to apply it to even more interesting (often of a "multi-character")[3] problems in continuum (thermo)mechanics. It is certainly not possible to smear out these barriers by only a single book. Anyhow, this book aims at contributing at least a bit to make these barriers smaller. It tries to present selected basic mechanical concepts in the context of their (at least to some extent rigorous) mathematical handling, typically focused on existence of solutions of particular models possibly together with some additional attributes as smoothness or uniqueness, outlining their approximations that would suggest computationally implementable algorithms. To make it readable for engineers or computational physicists, advanced analytical tools and results are suppressed to a minimal reasonable level, most of them being only briefly exposed in the four Appendices (A–D) without proofs, and we intentionally avoid really "exotic" concepts like non-metrizable topologies, measures which are only finitely additive, or convergence in terms of nets instead of conventional sequences. We also reduce advanced analytical tools handling set-valued and nonsmooth mappings, except those which arise from convex analysis.

The primal focus is on static boundary-value problems arising in mechanics of solids at large or at small strains (Part I) and their various evolution variants (Part II). As the title already suggests, we intentionally exclude fluids, although some spots

[2] A certain general parallel and a possible origin of this phenomenon has been articulated by M. Spitzer [491].

[3] To advertise complexity of models from various aspects, the popular adjective "multi" is used in literature in connection of multi-scale, multi-component, multi-continuum, multi-phase, multi-field, multi-physics, multi-disciplinary, multi-functional, multi-ferroic, and, of course, multi-dimensional.

as Sects. 3.6, 5.7, 6.6, 7.6, or 8.6 have a slight relevance to fluid mechanics, too. This still represents a very ambitious plot, henceforth some (otherwise important) areas are omitted: in particular contact mechanics, i.e. phenomena like friction, adhesion, or wear will not be addressed. Also, homogenization methods for composite materials and various dimensional reductions of three-dimensional continua to two-dimensional plates, shells, or membranes, or one-dimensional beams, trusses, or rods are omitted, too. Numerical approximation as a wide and important area of computational mechanics is presented only in its very minimal extent. At this point, specialized monographs at pp. 473–478 as a further reading are advisable.

The book can serve as an advanced textbook and introductory scientific monograph for graduate or Ph.D. students in programs such as mathematical modeling, applied mathematics, computational continuum physics, or mechanical engineering. Henceforth, also some exercises (sometimes with solutions outlined on pp. 557–574) are involved, too. Besides, we believe that experts actively working in theoretical or computational continuum mechanics and thermomechanics of solids will find useful material here.

The book reflects both our experience with graduate classes within the program "Mathematical modeling" at Charles University in Prague taught during 2005–2018 and some other occasional teaching activities in this area,[4] reflecting also our own research[5] during the past several decades. Particular spots of the book benefit from our own computational activity in this area of continuum mechanics during many decades and from our collaboration with experts in mechanical engineering. The presented computer simulations have been provided by Barbora Benešová, José Reinoso, Jan Valdman, and Roman Vodička to whom we thus express our truly deep thanks. We are also deeply indebted to Katharina Brazda and Riccarda Rossi for careful reading of some chapters.

Eventually, we truly appreciate a constructive attitude of the Springer publisher allowing for printing this book essentially from our own pdf-file (only with reflecting a partial language corrections made in India[6]).

Prague, July 2018 *Martin Kružík & Tomáš Roubíček*

[4] In particular, it concerns regular courses taught by M.K. at Technical University München in 2006-8 and at University of Würzburg in 2016, and by T.R. at University Vienna in 2017. Also, it concerns short intensive courses by T.R. about damage and plasticity at SISSA, Trieste, in 2015 and at Humboldt University Berlin in 2016.

[5] At this occasion, we would like to acknowledge the support by the Czech Science Foundation under grants 16-03823S "Homogenization and multi-scale computational modeling of flow and nonlinear interactions in porous smart structures", 16-34894L "Variational structures in thermomechanics of solids", 17-04301S "Advanced mathematical methods for dissipative evolutionary systems", 18-03834S "Localization phenomena in shape memory alloys: experiments & modeling", "Large strain challenges in materials science", and 19-04956S "Dynamic and nonlinear behaviour of smart structures; modelling and optimization".

[6] After reducing the production procedure in India to minimum, the proofs and final corrections of our own files were thus accomplished in January 2019.

Contents

Part II EVOLUTION PROBLEMS

Part I

STATIC PROBLEMS

The static[7] problems addressed in Part I have own importance in applications where the outer conditions are constant in time so that the mechanical systems have enough time to relax and to stay in a steady state. Simultaneously they serve as a preliminary tool for evolution situations treated in Part II. We confine ourselves to so-called *hyperelastic materials* whose important attribute is the existence of a *stored energy*, reflecting the idea that, microscopically, materials are composed of atoms and the energy of interatomic links related to mutual positions of atoms together with the position with respect to outer fields (e.g. the gravitational field) forms in sum the mentioned macroscopic stored energy. The kinetic energy of vibrations of the atoms is macroscopically interpreted as heat and, although interesting static anisothermal situations do exist, we will treat heat-transfer problems later in Part II, similarly as situations where the local composition of material is varied and then the energy of interatomic links also contains a chemical energy. (Of course, the energy in nuclei of atoms is irrelevant in the context of continuum mechanics and thermomechanics.)

In some sense the governing *principle* for steady-state problems is the *minimization of the stored energy*, although in a lot of situations this minimization should be only local in some sense.[8] This relatively simple minimization principle allows for a lot of results that sometimes are even difficult to adopt to evolutionary situations. Hence, static problems have also their own theoretical importance, both historically for the development of several parts of mathematics, in particular Calculus of Variations, and for motivation of temporary mathematics by formulating sometimes very nontrivial problems, some of them still open, cf. also pp. 479–482.

[7] To avoid possible confusion, let us explain that the short adjective "static" is mostly used in this book for thermodynamically equilibrated steady states, in contrast to general steady states which are not evolving in time but need not be in thermodynamical equilibrium, cf. Section 5.7.

[8] Think about a truss with the zero displacement condition on two opposite bases. The identity deformation is certainly a minimizer, but also one can rotate the bases by 360 degrees in opposite directions to get another (stressed) local minimizer. And even a countable number of different local minimizers exist by iterating the rotation argument.

Chapter 1
Description of deformable stressed bodies

> The second half of the 20th century was witness of the vehement research in the field theories of mechanical systems. The renaissance of these continuum theories begun during World War II and was motivated by practical needs in the description of rubber, napalm and many other complicated materials with essentially nonlinear behavior.
>
> KRZYSZTOF WILMANSKI (1940–2012)

One of our main objectives will be the description of a mechanical response of materials. A key ingredient here is a concept of deformations which identifies a new "shape" of the specimen. Often one can identify an original (reference) configuration of the material. While such configuration is suitable for mathematical considerations because it is fixed, physically relevant is the deformed configuration because there one can measure forces caused by deformations. These forces will manifest themselves in terms of stresses. The Cauchy stress vector reflecting internal material forces compensating for the external ones is introduced axiomatically and its existence is fundamental for continuum mechanics of solids. In order to map stress fields from the deformed to the reference configurations, we introduce the so-called Piola transform which formally allows for the same form of equilibrium equations in both configurations.

1.1 Kinematics

Throughout this book, $d \in \mathbb{N}$ will denote the space dimension (usually considered 2 or 3) of the Euclidean space $\mathbb{R}^d$. We assume that $\mathbb{R}^d$ is equipped with a right-handed orthonormal basis $\{\mathbf{e}_1, \ldots, \mathbf{e}_d\}$ such that $(\mathbf{e}_1 \times \mathbf{e}_2) \cdot \mathbf{e}_3 = 1$ if $d = 3$ and that $\mathbf{e}_1$ can be rotated to $\mathbf{e}_2$ by a ninety-degree counter-clockwise rotation if $d = 2$. The set of $d \times d$ matrices will be denoted by $\mathbb{R}^{d \times d}$.

In what follows, $\Omega \subset \mathbb{R}^d$ will be a bounded *domain* (i.e., an open and connected set) with sufficiently regular boundary $\Gamma := \partial\Omega$; the specific regularity will be specified at particular places. Its closure, denoted by $\bar{\Omega}$, represents the body before it is deformed, so we call $\bar{\Omega}$ the *reference configuration*. We wish to emphasize that the choice of the reference configuration is arbitrary and thus it is in some sense only fictitious, although sometimes it is a natural (usually stress-free) initial *configuration* related with the *material*. A point $x \in \bar{\Omega}$ is therefore called a *material point*.

M. Kružík and T. Roubíček, *Mathematical Methods in Continuum Mechanics of Solids*, Interaction of Mechanics and Mathematics,
https://doi.org/10.1007/978-3-030-02065-1_1

We refer to Appendix A for definitions of various sets of matrices and vectors and operations defined on them used later in this chapter.

1.1.1 Deformation mappings

A *deformation* of $\bar{\Omega}$ is defined through a mapping $y : \bar{\Omega} \to \mathbb{R}^d$ that is smooth enough, injective (possibly except on Γ), i.e., $y|_\Omega$ is invertible (one-to-one), and $\det \nabla y > 0$ in Ω, i.e., y is *orientation-preserving*. Indeed, choosing $x \in \Omega$ arbitrary[1] and defining $F := \nabla y(x)$ we get that $\{F\mathrm{e}_1, \ldots, F\mathrm{e}_d\}$ is again a right-handed basis of $\mathbb{R}^d$. This means that $(F\mathrm{e}_1 \times F\mathrm{e}_2) \cdot (F\mathrm{e}_3) > 0$ for $d := 3$. If $d := 2$ the same must hold if we embed $F \in \mathbb{R}^{2\times2}$ into $\mathbb{R}^{3\times3}$ by setting $F_{3i} = F_{i3} := \delta_{3i}$ for $1 \le i \le 3$ and for δ_{3i} the *Kronecker's symbol*, cf. (A.2.5). Moreover, $y(\bar{\Omega})$ denotes the deformed configuration and we write[2]

$$x^y := y(x) \text{ and } \bar{\Omega}^y := y(\bar{\Omega}). \tag{1.1.1}$$

Actually, various other notations can be found in the literature, too. For example, $x = y(X)$ for $X \in \bar{\Omega}$ is often used. The inverse of y, $y^{-1} : y(\bar{\Omega}) \to \bar{\Omega}$, is called the *reference mapping*. Hence, perhaps except the boundary of $\bar{\Omega}$, we can use both x as well as x^y to describe equivalently the deformation of the body. The description in terms of the *material coordinates* $x \in \bar{\Omega}$ is called a *Lagrangian description*. On the other hand, the so-called *Eulerian description* is the one in terms of x^y, it refers to the *actual deformed configuration* and it uses *spatial coordinates*. The former description, i.e., the Lagrangian one is usually used in continuum mechanics of solids and we will also follow this approach. The Eulerian description is then more often used in fluid mechanics. We call $v \in \mathbb{R}^d$ *a material vector* if there are two points $x_1, x_2 \in \bar{\Omega}$ such that $v = x_2 - x_1$. On the other hand, we call v^y *a spatial vector* if $v^y = y(x_2) - y(x_1)$ for some $x_1, x_2 \in \bar{\Omega}$. Thus, material vectors are associated with the ambient space surrounding the reference configuration $\bar{\Omega}$ while spatial vectors are associated with the ambient space of the deformed configuration $\bar{\Omega}^y$.

The *deformation gradient* F is defined as

$$F := \nabla y \tag{1.1.2}$$

and the orientation preservation implies that $\det F > 0$ in Ω if y is smooth enough. Written componentwise (see (A.3.25)), for every $i, j \in \{1, \ldots, d\}$, we will ocasionally use the notation

[1] Later we will require this property only for almost every $x \in \Omega$ with respect to the d-dimensional Lebesgue measure. This will correspond to searching for deformations in Sobolev spaces.

[2] This notation already appeared in [114].

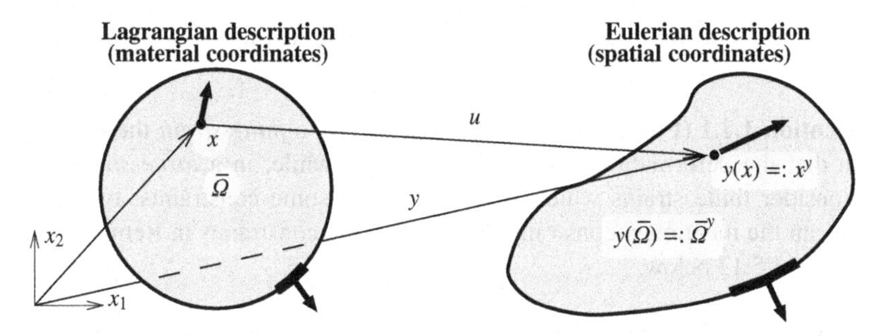

Fig. 1.1 A reference configuration $\bar{\Omega}$ (left) and the deformed one as its image under a continuous deformation y (right), indicating also how vectors connected with deforming material are transferred, cf. (1.1.1), and how the unit outward normal together with the surface area is transformed, cf. (1.1.18) and (1.1.21).

$$F_{ij} = \partial_{x_j} y_i . \tag{1.1.3}$$

The *polar decomposition theorem* asserts that if $F \in \mathrm{GL}^+(d)$ then there is a rotation $R \in \mathrm{SO}(d)$ and positive definite symmetric matrices $U \in \mathrm{GL}^+(d)$, and $V \in \mathrm{GL}^+(d)$ such that

$$F = RU = VR. \tag{1.1.4}$$

In continuum mechanics, U in (1.1.4) is called *stretch* and it easily follows that

$$U^2 = F^\top F \quad \text{and} \quad V^2 = FF^\top . \tag{1.1.5}$$

The *displacement* $u : \bar{\Omega} \to \mathbb{R}^d$ is defined for all $x \in \bar{\Omega}$ as

$$u(x) := y(x) - x \tag{1.1.6}$$

and then the *displacement gradient* reads

$$\nabla u(x) = \nabla y(x) - \mathbb{I}, \tag{1.1.7}$$

where $\mathbb{I}$ is the identity matrix in $\mathbb{R}^{d\times d}$, i.e., a matrix satisfying $\mathbb{I}v = v$ for every $v \in \mathbb{R}^d$.

As we will see in many situations, available mathematical tools and approaches are not yet well developed to preserve various nonconvex algebraic constraints on the deformation gradient F which come from mechanical considerations. The requirement $\det F > 0$ can serve as an example. In order to distinguish situations in which we do prescribe some constraints from others we make the following convention:[3]

Convention 1.1.1 (Finite strain/Large strain). By a *finite strain* theory, we will mean that the deformation gradient is a.e. finite, while, in a *large strain* theory, we consider finite strains which are subjected to some constraints; typically we will mean the mentioned constraint $\det F > 0$ or the constraints in Remarks 3.4.14, 4.5.12, or 4.5.13 below.

Convention 1.1.2 (Mechanics versus engineering). Terminology in continuum mechanics of solids varies a lot, the main difference being probably between mechanics and engineering. What we call deformation and denote by y (being sometimes denoted alternatively as φ or f or χ in literature), in engineering often means $F = \nabla y$, being called gradient of displacement, in contrast to the displacement as used here, i.e. $u = y -$ identity. Here we tried to use rather mechanical terminology and careful identification of notation when reading particular literature is generally advisable.[4]

The deformation is called *homogeneous* if F is constant in $\bar{\Omega}$. Correspondingly, $F \in \mathbb{R}^{d\times d}$ and $y(x) := Fx + c$ for some $c \in R^d$ and all $x \in \bar{\Omega}$. This means that if $x_1, x_2 \in \bar{\Omega}$ then $y(x_1) - y(x_2) = F(x_1 - x_2)$. As $x_1 - x_2$ is a vector in the reference (material) configuration and $y(x_1) - y(x_2)$ is a vector in the deformed (spatial) configuration, we see that F maps material vectors to spatial ones. Hence, F^{-1} maps spatial vectors to material ones. Assume now, that $v^y \in \mathbb{R}^d$ is a spatial vector and y is a homogeneous deformation $y(x) := Fx + c$. Then

$$v^y \cdot (y(x_1) - y(x_2)) = v^y \cdot F(x_1 - x_2) = F^\top v^y \cdot (x_1 - x_2).$$

Therefore, $F^\top v^y$ is a material vector and $F^\top$ maps spatial vectors to material vectors. Obviously, $F^{-\top}$ brings material vectors to spatial ones. If a deformation is homogeneous and $F = R$ for some $R \in \mathrm{SO}(d)$ then we call it a *rigid* deformation.

We define a *vectorial pullback* which maps a spatial vector $v^y \in \mathbb{R}^d$ to a material vector $\mathscr{P}_F v^y$. Examples are

$$\mathscr{P}_F v^y := F^\top v^y \quad \text{or} \tag{1.1.8}$$

$$\mathscr{P}_F v^y := F^{-1} v^y. \tag{1.1.9}$$

In both cases, $\mathscr{P}_F v^y$, $i = 1, 2$ are material vectors. As F is invertible we can also define the inverse operation called *a vectorial pushforward* $\mathscr{P}^{-1}$ which maps material vectors on spatial ones. Examples include $\mathscr{P}_F^{-1} v = F^{-\top} v$ and $\mathscr{P}_F^{-1} v = Fv$. These will be used later to transform vectorial quantities between the reference and the deformed configurations. To summarize, we have

[3] Usually, however, "the finite-strain theory" and "the large-strain theory" are considered identical in literature.

[4] The deformation y as used here is sometimes called a motion or a placement and then ∇y is called "motion gradient" or "displacement gradient", cf. [145, 340, 342, 518].

$$F, F^{-\top} : \text{material vectors} \to \text{spatial vectors}, \qquad \text{(pushforward)} \qquad (1.1.10a)$$

$$F^{\top}, F^{-1} : \text{spatial vectors} \to \text{material vectors}. \qquad \text{(pullback)} \qquad (1.1.10b)$$

These four tensors[5] are called *mixed tensor fields* because they map material vectors to spatial ones and vice versa. A tensor which maps material vectors to material vectors or spatial vectors to spatial vectors is called *a material tensor* and *a spatial tensor*, respectively. In view of (1.1.10), we get that, for instance, $F^{\top}F$ is a material tensor, while $FF^{\top}$ is a spatial tensor.

Let us now assume that $G^y \in \mathbb{R}^{d\times d}$ is a spatial tensor, i.e., it maps spatial vectors to spatial vectors. A *(tensorial) pullback* $\mathscr{P}$ is a linear transformation of G^y which maps material vectors to material vectors, i.e., it is a material tensor. Hence, tensorial pullback maps spatial tensors to material ones. Examples include

$$\mathscr{P}_F G^y := F^{\top} G^y F, \qquad (1.1.11a)$$

where F is a deformation gradient. Other choices are

$$\mathscr{P}_F G^y := F^{-1} G^y F^{-\top}, \quad \text{or} \qquad (1.1.11b)$$

$$\mathscr{P}_F G^y := F^{-1} G^y F, \quad \text{or also} \qquad (1.1.11c)$$

$$\mathscr{P}_F G^y := F^{\top} G^y F^{-\top}. \qquad (1.1.11d)$$

On the other hand, the deformation gradient F can be also used to transform material tensors to spatial ones. This operation is called *a (tensorial) pushforward* $\mathscr{P}^{-1}$ and it can be seen as the inverse operation to the (tensorial) pullback. This also suggests the chosen notation. Let $G \in \mathbb{R}^{d\times d}$ be a material tensor field. Then, for $\mathscr{P}$ from (1.1.11a), we have[6]

$$\mathscr{P}_F^{-1} G := F^{-\top} G F^{-1},$$

while if (1.1.11b) is considered, then the corresponding pushforward is just

$$\mathscr{P}_F^{-1} G := F G F^{\top}.$$

Actually, in particular models, the pullbacks (1.1.11) can be premultiplied by various scalar factors which then occurs also in (1.1.12), cf. also (9.1.15) and (9.2.25)–(9.2.26) below.

For a general deformation, we get by the Taylor expansion (assuming enough smoothness of y) that, for all $x_1, x_2 \in \bar{\Omega}$, it holds

$$y(x_2) - y(x_1) = F(x_1)(x_2 - x_1) + o(|x_2 - x_1|).$$

---

[5] In what follows, tensor (or tensor field) is a linear transformation from $\mathbb{R}^d$ to $\mathbb{R}^d$.

[6] Indeed, a short calculation shows that, for $\mathscr{P}_F$ defined in (1.1.11a) and for all spatial tensors G^y and for all material tensors G, it holds $\mathscr{P}_F^{-1}(\mathscr{P}_F G^y) = F^{-\top} F^{\top} G^y F F^{-1} = G^y$ and $\mathscr{P}_F(\mathscr{P}_F^{-1} G) = F^{\top} F^{-\top} G F^{-1} F = G$.

This implies that every smooth deformation behaves locally as a homogeneous one and that (1.1.10) holds pointwise also for non-constant F.

Proposition 1.1.3 (Properties of deformations [114]). *Let $\Omega\subset\mathbb{R}^d$ be an open bounded set and let $y\in C(\bar\Omega;\mathbb{R}^d)$ be a mapping whose restriction on Ω is injective. Then*

$$y(\bar\Omega)=\overline{y(\Omega)},\qquad y(\Omega)\subset\operatorname{int}y(\bar\Omega),\quad \textit{and}\quad y(\partial\Omega)\supset\partial y(\bar\Omega). \tag{1.1.13}$$

If, additionally[7], $\operatorname{int}\bar\Omega=\Omega$ *and* $y\in C(\bar\Omega;\mathbb{R}^d)$ *is injective then*

$$y(\Omega)=\operatorname{int}y(\bar\Omega)\quad \textit{and}\quad y(\partial\Omega)=\partial y(\Omega)=\partial y(\bar\Omega). \tag{1.1.14}$$

Proof. Take $x^y\in y(\bar\Omega)$. Then there is $x\in\bar\Omega$ such that $y(x)=x^y$. Let $\lim_{k\to\infty}x_k=x$, $\{x_k\}\subset\Omega$. Due to continuity of y, we have $y(x)=\lim_{k\to\infty}y(x_k)$. Thus $y(\bar\Omega)\subset\overline{y(\Omega)}$. Since $\bar\Omega$ is compact, so is $y(\bar\Omega)$. Hence, $y(\Omega)\subset y(\bar\Omega)$ implies $\overline{y(\Omega)}\subset\overline{y(\bar\Omega)}=y(\bar\Omega)$. Therefore, $y(\bar\Omega)=\overline{y(\Omega)}$.

We see that $y(\Omega)$ is open due to Theorem A.1.2, and it is contained in $y(\bar\Omega)$. Hence, we get $y(\Omega)\subset\operatorname{int}y(\bar\Omega)$. Moreover, $y(\bar\Omega)=\operatorname{int}y(\bar\Omega)\cup\partial y(\bar\Omega)$ and $\operatorname{int}y(\bar\Omega)\cap\partial y(\bar\Omega)=\emptyset$. On the other hand,

$$y(\bar\Omega)=y(\Omega\cup\partial\Omega)=y(\Omega)\cup y(\partial\Omega)\quad\text{and}\quad y(\Omega)\subset\operatorname{int}y(\bar\Omega),$$

so that $y(\partial\Omega)\supset\partial y(\bar\Omega)$. This proves the first part of the proposition.

To prove the second part of the proposition, take $x^y\in\operatorname{int}y(\bar\Omega)$ and assume that $x^y\notin y(\Omega)$. A continuous mapping $y:\bar\Omega\to y(\bar\Omega)$ is bijective and $\bar\Omega$ is compact, so that $y^{-1}:y(\bar\Omega)\to\bar\Omega$ is also continuous. By Theorem A.1.2, $y^{-1}(\operatorname{int}y(\bar\Omega))$ is an open subset of $\bar\Omega$ that contains $y^{-1}(x^y)=x$. As $x=y^{-1}(x^y)\notin\Omega$ we have the existence of an open subset of $\bar\Omega$ which strictly contains Ω, a contradiction.

Further, $\overline{y(\Omega)}=y(\Omega)\cup\partial y(\Omega)$. Since $y:\bar\Omega\to y(\bar\Omega)$ is injective, we have $\partial y(\Omega)=y(\partial\Omega)$. As $y(\Omega)=\operatorname{int}y(\bar\Omega)$ we also have

$$\overline{y(\Omega)}=y(\Omega)\cup\partial y(\bar\Omega)\quad\text{and}\quad y(\Omega)\cap\partial y(\bar\Omega)=\emptyset,$$

and it yields $\partial y(\bar\Omega)=\partial y(\Omega)$. □

Example 1.1.4 (A domain for which $\operatorname{int}\bar\Omega\neq\Omega$*).* Consider $\Omega:=\{x\in\mathbb{R}^d:\ |x|\le1\}\setminus\{x=(0,x_2,...,x_d)\in\mathbb{R}^d:\ x_2\ge0\}$ for $d\ge2$ it is a connected open set (domain) for which $\operatorname{int}\bar\Omega=\operatorname{int}\{x\in\mathbb{R}^d:\ |x|\le1\}=\{x\in\mathbb{R}^d:\ |x|<1\}\supsetneq\Omega$. See also Fig. 3.2-middle for another example. On the other hand, if $\Omega\subset\mathbb{R}^d$ is a bounded Lipschitz domain then $\operatorname{int}\bar\Omega=\Omega$.

Remark 1.1.5 (Variational characterization of the polar decomposition). It is proved in [336] that if $F\in\mathrm{GL}^+(d)$ then the rotation $R\in\mathrm{SO}(d)$ which appeared in (1.1.4) satisfies $R=\min_{Q\in\mathrm{SO}(d)}|F-Q|$.

[7] Here "int" denotes the interior.

Exercise 1.1.6. Decide whether $\operatorname{Cof} F$ (for F the deformation gradient) is a material, spatial, or a mixed tensor. Find the tensorial pushforward for the pullback $\mathscr{P}_F G^y = F^{-1} G^y \operatorname{Cof} F$ for every spatial tensor $G^y \in \mathbb{R}^{d\times d}$.

Exercise 1.1.7. Prove formula (A.2.22) in Appendix A.

Exercise 1.1.8. Assume that $G \in \mathrm{GL}(d)$. Show that $G^\top G$ and $GG^\top$ are positive definite.

1.1.2 Piola transform

As mentioned before, there are two configurations of the specimen: the reference one and the deformed one. While measurements are usually done on the latter one, for theoretical consideration the reference configuration is more suitable. Namely, it is a fixed domain in the space. In what follows, we will describe a suitable way how to transform various quantities between these two configurations of the material.

The Piola transform establishes a correspondence between tensor fields defined in the deformed and reference configurations, respectively. If $T^y(x^y)$ denotes a tensor field over $y(\bar{\Omega})$ then we define for $x^y = y(x)$ a matrix-valued map $T : \bar{\Omega} \to \mathbb{R}^{d\times d}$ by

$$T(x) := (\det \nabla y(x)) T^y(x^y)(\nabla y(x))^{-\top} = T^y(x^y) \operatorname{Cof} \nabla y(x). \tag{1.1.15}$$

In the following theorem we assume that all involved maps are smooth enough so that all operations make sense.

Theorem 1.1.9 (Properties of Piola's transform). *Let $\Omega \subset \mathbb{R}^d$ be a bounded domain, $y \in C^2(\bar{\Omega};\mathbb{R}^d)$ be injective, and $T : \bar{\Omega} \to \mathbb{R}^{d\times d}$ be the Piola transform of $T^y \in C^1(y(\bar{\Omega});\mathbb{R}^{d\times d})$. Then*

$$\operatorname{div} T(x) = (\det \nabla y(x)) \operatorname{div}^y T^y(x^y) \qquad \forall x^y = y(x), \quad x \in \bar{\Omega}, \tag{1.1.16}$$

$$\int_{\partial\omega} T(x)\vec{n}\, \mathrm{d}S = \int_{\partial\omega^y} T^y(x^y)\vec{n}^y\, \mathrm{d}S^y \qquad \forall \omega \subset \bar{\Omega}, \tag{1.1.17}$$

where ω is an arbitrary domain with a smooth boundary and $\vec{n}$ and $\vec{n}^y$ are unit outer normals to $\partial\omega$ and $\partial\omega^y$, respectively. In particular, the area elements $\mathrm{d}S$ and $\mathrm{d}S^y$ at the points $x \in \Gamma := \partial\Omega$ and $x^y = y(x) \in \partial(\bar{\Omega})$ are related by applying (1.1.17) to $T^y := \mathbb{I}$, i.e.,

$$\det \nabla y(x) \big| [\nabla y(x)^{-\top}]\vec{n} \big| \mathrm{d}S = \big| [\operatorname{Cof} \nabla y(x)]\vec{n} \big| \mathrm{d}S = |\vec{n}^y| \mathrm{d}S^y = \mathrm{d}S^y. \tag{1.1.18}$$

Lemma 1.1.10 (Piola's identity). *If $y \in C^2(\bar{\Omega};\mathbb{R}^d)$ then it holds for every $x \in \bar{\Omega}$ that*

$$\operatorname{div}(\operatorname{Cof} \nabla y(x)) = 0. \tag{1.1.19}$$

Proof. It is obvious for $d = 2$ in view of (A.2.21) and (A.3.29). If $d = 3$ we have due to (A.2.20) and (A.3.29) that

$$(\mathrm{Cof}\,\nabla y)_{ij} = \partial_{x_{j+1}} y_{i+1} \partial_{x_{j+2}} y_{i+2} - \partial_{x_{j+2}} y_{i+1} \partial_{x_{j+1}} y_{i+2}$$

counting the indices modulo 3, i.e., $4 \mapsto 1$ and $5 \mapsto 2$. Hence, it holds $\sum_{j=1}^{3} \partial_{x_j} (\mathrm{Cof}\,\nabla y)_{ij} = 0$. □

Proof of Theorem 1.1.9. We have using the Einstein's summation convention that

$$\begin{aligned} T_{ij}(x) &= (\det \nabla y(x)) T^y_{ik}(y(x))(\nabla y(x)^{-\top})_{kj} \quad \text{and} \\ \partial_{x_j} T_{ij} &= (\det \nabla y(x)) \partial_{x_j} [T^y_{ik}(y(x))](\nabla y(x)^{-\top})_{kj} \\ &\quad + T^y_{ik}(y(x)) \partial_{x_j} [(\det \nabla y(x)(\nabla y(x)^{-\top})_{kj}] \\ &= (\det \nabla y(x)) \partial_{x_j} [T^y_{ik}(y(x))](\nabla y(x)^{-\top})_{kj} , \end{aligned}$$

where the last equality is by the Piola identity (1.1.19). Exploiting the chain rule we get $\partial_{x_j}[T^y_{ik}(y(x))] = \partial_{y_l} T^y_{ik}(y(x) \partial_{x_j} y_l(x)$. Hence,

$$(\det \nabla y(x)) \partial_{x_j} [T^y_{ik}(y(x))](\nabla y(x)^{-\top})_{kj} = (\det \nabla y(x)) \partial_{y_l} [T^y_{ik}(y(x))] \partial_{x_j} [y_l(x)](\nabla y(x)^{-1})_{jk}.$$

This proves (1.1.16).

In order to show (1.1.17), we calculate using Theorems B.3.8 and B.3.11 for $\omega \subset \bar{\Omega}$ smooth that

$$\begin{aligned} \int_{\partial\omega} T(x)\vec{n}\,\mathrm{d}S &= \int_{\omega} \mathrm{div}\, T(x)\,\mathrm{d}x = \int_{\omega} (\det \nabla y(x)) \mathrm{div}^y T^y(y(x))\,\mathrm{d}x \\ &= \int_{y(\omega)} \mathrm{div}^y T^y(x^y)\,\mathrm{d}x^y = \int_{\partial y(\omega)} T^y(x^y)\vec{n}^y\,\mathrm{d}S^y , \end{aligned}$$

which implies (1.1.17). Applying (1.1.17) to $T^y = \mathbb{I}$ we get

$$\int_{\partial\omega} \mathrm{Cof}\,\nabla y(x)\vec{n}\,\mathrm{d}S = \int_{\partial\omega^y} \vec{n}^y\,\mathrm{d}S^y \tag{1.1.20}$$

which is (1.1.18) because $|\vec{n}^y| = |\vec{n}| = 1$. □

Remark 1.1.11 (Normal vector in the deformed configuration). We see that if $x \in \Gamma$ and $x^y \in \partial y(\Omega)$ then

$$\vec{n}^y(x^y) = \frac{\mathrm{Cof}\,\nabla y(x)\vec{n}}{|\mathrm{Cof}\,\nabla y(x)\vec{n}|} = \frac{(\nabla y(x))^{-\top}\vec{n}}{|(\nabla y(x))^{-\top}\vec{n}|} , \tag{1.1.21}$$

i.e., this formula says how to calculate normal vectors to the deformed boundary. Let us notice also that we implicitly assume in (1.1.21) that $\det \nabla y(x) > 0$.

1.1.3 Volume, surface, and length in deformed configurations

Let $\mathrm{d}x$ denotes a volume element around a point x of the reference configuration. The volume element $\mathrm{d}x^y$ in the deformed configuration is formally given by (see Theorem B.3.10) the formula

$$\mathrm{d}x^y = \det\nabla y(x)\,\mathrm{d}x. \tag{1.1.22}$$

If $\omega \subset \bar{\Omega}$ then (under the assumptions of Theorem B.3.11) $\mathrm{meas}_d(\omega) = \int_\omega \mathrm{d}x$ and $\mathrm{meas}_d(\omega^y) = \int_{\omega^y} \mathrm{d}x^y = \int_\omega \det\nabla y(x)\,\mathrm{d}x$.

As we have seen before in (1.1.18), the surface element in the deformed configuration is measured by means of the cofactor of the gradient. Indeed, if Ω and y is as in Theorem 1.1.9 and if $\omega \subset \Gamma$ then $y(\omega) = \omega^y \subset \partial y(\Omega)$ and

$$\mathrm{meas}_{d-1}(\omega^y) = \int_{y(\omega)} \mathrm{d}S^y = \int_\omega \left|\mathrm{Cof}\,\nabla y(x)\vec{n}\right| \mathrm{d}S\,,$$

i.e., we write

$$\mathrm{d}S^y = \left|\mathrm{Cof}\,\nabla y(x)\vec{n}\right| \mathrm{d}S\,. \tag{1.1.23}$$

Finally, if y is differentiable at $x \in \Omega$ then we write, for all points $x + \tilde{x} \in \bar{\Omega}$ for some $\tilde{x} \in \Omega$,

$$y(\tilde{x}) = y(x) + \nabla y(x)(\tilde{x} - x) + o(|\tilde{x} - x|)\,. \tag{1.1.24}$$

Hence, the squared distance of $y(\tilde{x})$ and $y(x)$ is

$$|y(\tilde{x}) - y(x)|^2 = (\tilde{x} - x)^\top[(\nabla y(x))^\top \nabla y(x)](\tilde{x} - x) + o(|\tilde{x} - x|^2)\,. \tag{1.1.25}$$

Recalling notation (1.1.2), the symmetric tensor

$$C := F^\top F = (\nabla y)^\top \nabla y \tag{1.1.26}$$

is called the *right Cauchy-Green strain tensor*. Roughly speaking, this tensor measures the change of the (Euclidean) distance of two nearby points after the deformation. Namely, it follows from (1.1.25) that neglecting $o(|\tilde{x} - x|^2)$, it holds for the right Cauchy-Green strain tensor evaluated at $\nabla y(x)$, i.e., $C(x) := (\nabla y(x))^\top \nabla y(x)$, that

$$\left|y(\tilde{x}) - y(x)\right| \approx \sqrt{(\tilde{x}-x)^\top \cdot [C(x)](\tilde{x}-x)}\,. \tag{1.1.27}$$

This again implies that $C(x)$ must be positive-definite for all $x \in \Omega$; cf. also Exercise 1.1.8. We can also use it to measure how much the deformation differs from the rigid one. Indeed, if for some $c \in \mathbb{R}^d$ and $R \in \mathrm{SO}(d)$ we define $y(x) := c + Rx$ for all $x \in \bar{\Omega}$ then $\nabla y = R$ and $C = \mathbb{I}$. Let us notice that if (1.1.27) held exactly, positive definiteness of C would be equivalent to injectivity of deformations. This is, however, not the case, see Figure 3.2 and the related example. Thus, the *Green-Lagrange* (sometimes also called *Green-St. Venant*) *strain tensor*

$$E := \frac{1}{2}(C - \mathbb{I}) \tag{1.1.28}$$

indicates how much the current deformation y differs from the rigid one. Indeed, we can easily see this by applying the following deep result due to Friesecke, James, and Müller [206], known as a *rigidity estimate* and Exercise 1.1.15 below. In the following theorem, $\mathrm{dist}(F, \mathrm{SO}(d)) := \min_{R \in \mathrm{SO}(d)} |F - R|$.

Theorem 1.1.12 (Rigidity estimate). *Let $d \geq 2$, let $1 < p < +\infty$, and let Ω be a bounded Lipschitz domain. Then there is a constant $K > 0$ such that for every $y \in W^{1,p}(\Omega;\mathbb{R}^d)$ there exists a fixed rotation matrix $R \in \mathrm{SO}(d)$ with the property*

$$\|\nabla y - R\|_{L^p(\Omega;\mathbb{R}^{d\times d})} \leq K \big\|\mathrm{dist}(\nabla y, \mathrm{SO}(d))\big\|_{L^p(\Omega)}. \tag{1.1.29}$$

Let us notice that in view of the polar decomposition $F = RU$, where $R \in \mathrm{SO}(d)$ and $U = U^\top$ is positive definite, we also have

$$E = \frac{1}{2}(U^2 - \mathbb{I}). \tag{1.1.30}$$

If we write $F = \mathbb{I} + D$ where D is a placeholder for the displacement gradient ∇u (1.1.7) then we see that

$$E = \frac{1}{2}(D + D^\top + D^\top D). \tag{1.1.31}$$

The right Cauchy-Green strain tensor is also sometimes called the *right Cauchy-Green deformation tensor.* Many other tensors measuring deformations appear in nonlinear elasticity. For example the *Finger tensor* is the inverse of C, i.e., $C^{-1} = F^{-1}F^{-\top}$ or the *Cauchy deformation tensor* is given as $F^{-\top}F^{-1}$. It is sometimes also called the *Piola tensor.*

Remark 1.1.13 (Seth-Hill family of strain measures). Various strain measures are used in physics and engineering. A set of strain tensor of the form

$$E^{(m)} := \frac{1}{2m}(C^m - \mathbb{I}) \tag{1.1.32}$$

for different values of m is called the *Seth-Hill family* of strain measures. For example, if $m = 1$ we get the Green-Lagrange strain tensor, if $m := 1/2$ then $E^{(1/2)} := C^{1/2} - \mathbb{I}$ is called the *Biot strain tensor*, etc.

Exercise 1.1.14. Assume that $F, \hat{F} \in \mathbb{R}^{d\times d}$ are matrices with positive determinants and such that $F^\top F = \hat{F}^\top \hat{F}$. Show that this is equivalent to $F = R\hat{F}$ for some $R \in \mathrm{SO}(d)$.

Exercise 1.1.15. Consider $y : \Omega \to \mathbb{R}^d$ smooth enough and such that $\det \nabla y > 0$ and $(\nabla y)^\top \nabla y = \mathbb{I}$ in Ω. Show that there is $R \in \mathrm{SO}(d)$ satisfying $\nabla y = R$ in Ω, i.e., that the deformation is rigid.

1.2 Forces and stresses

Up to now, we have presented deformations and their various quantified descriptions. In this section, we are going to deal with their cause, i.e., with forces and stresses. Forces and stresses are the reason for the appearance of deformations and strains, and vice versa. The effect of forces applied on the specimen is postulated by the Axiom of the stress principle which goes back to Cauchy and Euler.

1.2.1 Applied forces and Cauchy stress tensor

We will consider two types of applied forces:

a/ *applied body forces* defined through the density $f^y : \bar{\Omega}^y \to \mathbb{R}^d$ per unit volume in the deformed configuration,

b/ *applied surface forces* defined by $g^y : \Gamma_{\mathrm{N}}^y \to \mathbb{R}^d$ on a meas_{d-1}-measurable subset $\Gamma_{\mathrm{N}}^y \subset \Gamma^y$. Then, g^y is the density per unit area in the deformed configuration.

Note that f^y is measured in $\mathrm{N/m^3}$ while the physical unit of g^y is Pascal, i.e., Pa = $\mathrm{N/m^2} = \mathrm{J/m^3}$.

Now we are ready to postulate the existence of internal forces in the deformed specimen.

Axiom (Stress principle of Euler and Cauchy): Consider a body occupying a fixed deformed configuration $\bar{\Omega}^y$ and subjected to applied forces represented by densities $f^y : \bar{\Omega}^y \to \mathbb{R}^d$ and $g^y : \Gamma_{\mathrm{N}}^y \to \mathbb{R}^d$. Let further $S^{d-1} \subset \mathbb{R}^d$ denote the unit sphere centered at the origin. We assume that there exists a vector field

$$t^y : \bar{\Omega}^y \times S^{d-1} \to \mathbb{R}^d, \tag{1.2.1a}$$

called *Cauchy's stress vector*, such that:

i/ For any subdomain $\omega^y \subset \bar{\Omega}^y$ and any point $x^y \in \Gamma_{\mathrm{N}}^y \cap \partial\omega^y$ where the joint outer unit normal vector $\vec{n}^y$ exists, it holds

$$t^y(x^y, \vec{n}^y) = g^y(x^y). \tag{1.2.1b}$$

ii/ *Axiom of balance of forces.* For any subdomain $\omega^y \subset \bar{\Omega}^y$, it holds

$$\int_{\omega^y} f^y(x^y)\mathrm{d}x^y + \int_{\partial\omega^y} t^y(x^y,\vec{n}^y)\mathrm{d}S^y = 0. \tag{1.2.1c}$$

Again, $\vec{n}^y$ denotes the outer unit normal to $\partial\omega^y$.

iii/ *Axiom of balance of momenta.* For any subdomain $\omega^y \subset \bar{\Omega}^y$ with the outer unit normal $\vec{n}^y$, it holds

$$\int_{\omega^y} x^y \times f^y(x^y)\mathrm{d}x^y + \int_{\partial\omega^y} x^y \times t^y(x^y,\vec{n}^y)\mathrm{d}S^y = 0. \tag{1.2.1d}$$

Remark 1.2.1 (Equilibrium equations). This axiom asserts that there are elementary forces $t^y(x^y,\vec{n}^y)\mathrm{d}S^y$ along boundaries of any subdomain of $\bar{\Omega}^y$. These forces depend on ω^y only through the outer unit normal to $\partial\omega^y$. Moreover, the deformed configuration $\bar{\Omega}^y$ is in the static equilibrium by ii/ and iii/.

Theorem 1.2.2 (Cauchy's theorem). *Let $\Omega^y \subset \mathbb{R}^d$ be open. Let the applied force density $f^y : \bar{\Omega}^y \to \mathbb{R}^d$ be continuous and let $t^y(\cdot,\vec{n}) \in C^1(\bar{\Omega}^y;\mathbb{R}^d)$ for every $\vec{n} \in S^{d-1}$ and $t^y(x^y,\cdot) \in C(S^{d-1};\mathbb{R}^d)$ for any $x^y \in \bar{\Omega}^y$. Then the Axiom "Stress principle of Euler and Cauchy" implies the existence of a symmetric tensor $T^y : \bar{\Omega}^y \to \mathbb{R}^{d\times d}$ belonging to $C^1(\bar{\Omega}^y;\mathbb{R}^{d\times d})$ such that:*

$$t^y(x^y,\vec{n}) = T^y(x^y)\vec{n} \qquad \forall x^y \in \bar{\Omega}^y \ \ \forall \vec{n} \in S^{d-1}, \tag{1.2.2a}$$

$$-\mathrm{div}^y T^y(x^y) = f^y(x^y) \qquad \forall x^y \in \Omega^y, \tag{1.2.2b}$$

$$T^y(x^y)\vec{n}^y = g^y(x^y) \qquad \forall x^y \in \Gamma^y_{\mathrm{N}}, \tag{1.2.2c}$$

where $\vec{n}^y$ is the unit outer normal vector to $\Gamma^y_{\mathrm{N}} \subset \Gamma^y$, a part of the boundary where surface forces are prescribed.

Proof. We prove the theorem only for $d = 3$ while the case $d = 2$ is left to the reader. Let $\{\mathbf{e}_i\}_{i=1,2,3}$ denote a orthonormal basis. Consider a point $x^y \in \Omega^y$ and a tetrahedron T_h with three faces parallel to the coordinate planes and $h = \mathrm{dist}([v_1 v_2 v_3], x^y)$; here v_1, v_2, v_3, and x^y are vertices of the tetrahedron, cf. Figure 1.2, and $[v_1 v_2 v_3]$ denotes the triangle with the specified vertices. Let us notice that this tetrahedron is

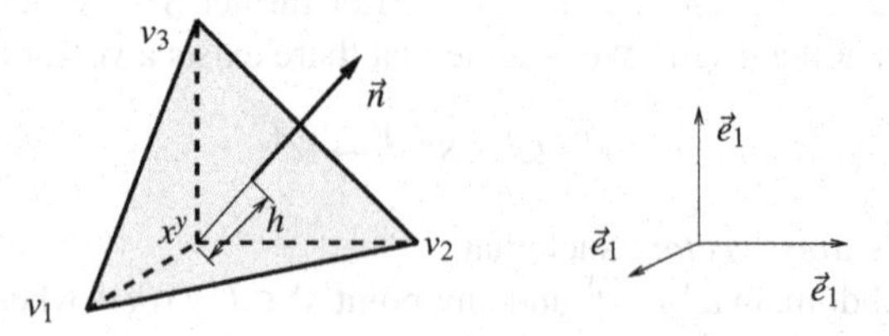

Fig. 1.2 The tetrahedron T_h used in the proof of Cauchy's Theorem 1.2.2.

a Lipschitz domain, so that the outer normal vector exists almost everywhere in the sense of the two-dimensional Lebesgue measure. We suppose that the tetrahedron is contained in Ω^y, which is an open set. Note also that components of the normal unit vector $\vec{n} := (n_1, n_2, n_3)$ to the plane $v_1v_2v_3$ fulfill $n_i > 0$, $i = 1,2,3$.

Let us further denote by $\text{meas}_2[v_1v_2v_3]$ the area of the triangle $[v_1v_2v_3]$. An analogous notation is used for other faces. The volume of the tetrahedron is proportional to $h\,\text{meas}_2[v_1v_2v_3]$ and $\text{meas}_2[v_2x^yv_3] = n_1\text{meas}_2[v_1v_2v_3]$ and similarly for the area of other faces. Finally, realize that by the action-reaction principle $t^y(x^y,\vec{m}) = -t^y(x^y,-\vec{m})$ for any $x^y \in \Omega^y$ and any $\vec{m} \in S^2$.

We calculate the force balance (1.2.1c) on the tetrahedron T_h:

$$\int_{T_h} f^y(a^y)\,\mathrm{d}a^y + \int_{\partial T_h} t^y(a^y,\vec{n}^y)\,\mathrm{d}S^y = 0.$$

Further, for any $i = 1,2,3$, it holds

$$\int_{\partial T_h} t_i^y(a^y,\vec{n})\,\mathrm{d}S^y = \int_{[v_1v_2v_3]} t_i^y(a^y,\vec{n})\,\mathrm{d}S^y - \int_{[v_2x^yv_3]} t_i^y(a^y,\mathbf{e}_1)\,\mathrm{d}S^y$$
$$- \int_{[v_1x^yv_3]} t_i^y(a^y,\mathbf{e}_2)\,\mathrm{d}S^y - \int_{[v_1x^yv_2]} t_i^y(a^y,\mathbf{e}_3)\,\mathrm{d}S^y.$$

Hence,

$$\frac{1}{\text{meas}_2[v_1v_2v_3]}\int_{[v_1v_2v_3]} t_i^y(a^y,\vec{n})\,\mathrm{d}S^y$$
$$= \frac{n_1}{\text{meas}_2[v_2x^yv_3]}\int_{[v_2x^yv_3]} t_i^y(a^y,\mathbf{e}_1)\,\mathrm{d}S^y + \frac{n_2}{\text{meas}_2[v_1x^yv_3]}\int_{[v_1x^yv_3]} t_i^y(a^y,\mathbf{e}_2)\,\mathrm{d}S^y$$
$$+ \frac{n_3}{\text{meas}_2[v_1x^yv_2]}\int_{v_1x^yv_2} t_i^y(a^y,\mathbf{e}_3)\,\mathrm{d}S^y - \frac{1}{\text{meas}_2[v_1v_2v_3]}\int_{T_h} f(a^y)\,\mathrm{d}a^y.$$

We have by continuity

$$\lim_{h\to 0}\frac{n_1}{\text{meas}_2[v_2x^yv_3]}\int_{[v_2x^yv_3]} t_i^y(a^y,\mathbf{e}_1)\,\mathrm{d}S^y = t_i^y(x^y,\mathbf{e}_1)n_1$$

and similarly for other components, too. Further,

$$\lim_{h\to 0}\frac{1}{\text{meas}_2[v_1v_2v_3]}\left|\int_{T_h} f(a^y)\,\mathrm{d}a^y\right| = 0,$$

because $\|f^y\|_{C(T_h;\mathbb{R}^d)}$ is bounded independently of $h > 0$ by our assumption. Altogether, for $i = 1,2,3$, we have $t_i^y(x^y,\vec{n}) = \sum_{i=1}^d t_i(x^y,\mathbf{e}_j)n_j$ or, equivalently,

$$t^y(x^y,\vec{n}) = \sum_{i=1}^d t(x^y,\mathbf{e}_j)n_j. \tag{1.2.3}$$

As $t_i^y(x^y,\mathbf{e}_i) = -t_i^y(x^y,-\mathbf{e}_i)$ it follows that (1.2.3) holds even if $n_j \le 0$ for some j. We define $T_{ij}^y : \bar{\Omega}^y \to \mathbb{R}$ by $t^y(x^y,\mathbf{e}_j) = \sum_i T_{ij}^y(x^y)\mathbf{e}_i$, so that $t^y(x^y,\vec{n}) = \sum_{i,j} T_{ij}^y(x^y)\mathbf{e}_i n_j$.

Hence, $t_i^y(x^y,\vec{n}) = \sum_j T_{ij}^y(x^y)n_j$ for all $x^y \in \bar{\Omega}^y$ and all $n \in S^2$, or, in other words,

$$t^y(x^y,\vec{n}) = T^y(x^y)\vec{n}. \tag{1.2.4}$$

The tensor $T^y = (T_{ij}^y)_{i,j=1}^d$ is called the *Cauchy stress tensor*. Let us notice, in particular, that it shows a linear dependence of t^y on the normal $\vec{n}$ and one can even show that it is really a 2nd-order tensor, see e.g. [388].

We use the axiom of force balance and the Green theorem (B.3.14) to infer that

$$\begin{aligned}0 = \int_{\omega^y} f^y(x^y)\,\mathrm{d}x^y + \int_{\partial\omega^y} t^y(x^y,\vec{n}^y)\,\mathrm{d}S^y &= \int_{\omega^y} f^y(x^y)\,\mathrm{d}x^y + \int_{\partial\omega^y} T^y(x^y)n^y\,\mathrm{d}S^y \\ &= \int_{\omega^y} f^y(x^y)\,\mathrm{d}x^y + \int_{\omega^y} \mathrm{div}^y T^y(x^y)\,\mathrm{d}x^y,\end{aligned}$$

which shows (1.2.2b) because $\omega^y \subset \Omega^y$ was arbitrary.

In view of the momentum balance we have, when using the *Levi-Civita permutation symbol* ε_{ijk} and *Kronecker's delta* δ_{ij} together with the summation convention, see (A.2.3) and (A.2.5) for definitions:

$$\begin{aligned}0 = \int_{\omega^y} \varepsilon_{ijk}x_j^y f_k^y(x^y)\,\mathrm{d}x^y + \int_{\partial\omega^y} x_j^y T_{km}^y(x^y)n_m^y\,\mathrm{d}S^y = \int_{\omega^y} \varepsilon_{ijk}x_j^y f_k^y(x^y)\,\mathrm{d}x^y \\ + \int_{\omega^y} \varepsilon_{ijk}\partial_{x_m^y}\left(x_j^y T_{km}^y(x^y)\right)\mathrm{d}x^y = \int_{\omega^y} \varepsilon_{ijk}x_j^y \underbrace{\left(f_k^y(x^y) + \partial_{x_m^y}T_{km}^y(x^y)\right)}_{=0\text{ by (1.2.2b)}}\mathrm{d}x^y \\ + \int_{\omega^y} \varepsilon_{ijk}T_{km}^y(x^y)\delta_{jm} = \int_{\omega^y} \varepsilon_{ijk}T_{kj}^y(x^y)\,\mathrm{d}x^y,\end{aligned}$$

which implies symmetry of T^y. Finally, (1.2.2c) follows from (1.2.1b). □

Let us discuss three important examples of T^y as in [114]. Let us first take $p > 0$ and put $T^y(x^y) = -pI$, so that $t^y(x^y,\vec{n}^y) = -p\vec{n}^y$. This defines the pressure load on Ω^y; cf. Figure 1.3a.

Secondly, let $T^y(x^y) = \tau e\otimes e$ where $\tau > 0$, $|e| = 1$; recall the notation $(a\otimes b)_{ij} = a_i b_j$, $i,j = 1,2,3$. Then $t^y(x^y,\vec{n}^y) = T^y(x^y)n^y = \tau(e\cdot n^y)e$; cf. Figure 1.3b and it is called pure tension.

Finally, take $\sigma > 0$ and unit mutually perpendicular vectors $e, \tilde{e}$ and put $T^y(x^y) = \sigma(e\otimes\tilde{e} + \tilde{e}\otimes e)$. This yields $t^y(x^y,\vec{n}^y) = \sigma((e\cdot n)\tilde{e} + (\tilde{e}\cdot\vec{n})e)$; cf. Figure 1.3c and we call it pure shear.

The *Axiom of material frame-indifference*[8] states that, if a deformation y is composed with another deformation z of $\bar{\Omega}^y$ where $z(x) := Ry(x)$ for all $x \in \bar{\Omega}$, i.e., $x^z = Rx^y$ and some rotation $R \in \mathrm{SO}(d)$, then, for all $x \in \bar{\Omega}$ and any $n \in S^{d-1}$, it holds that

[8] It is sometimes also called frame-invariance.

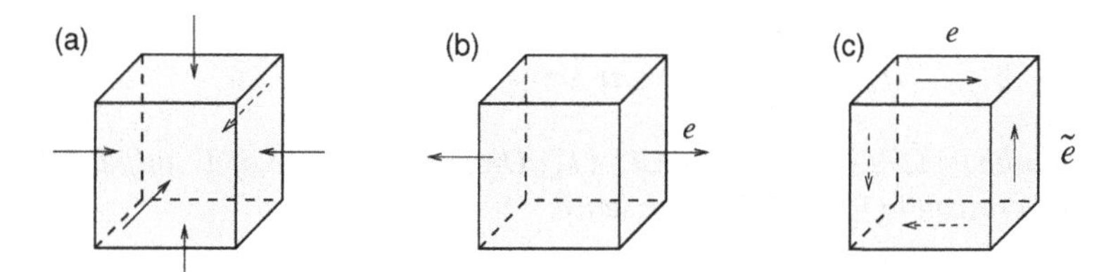

Fig. 1.3 Three basic types of load (strains):
(a) isotropic compression with the pressure $p > 0$,
(b) tension at the direction e, and
(c) pure shear.

$$t^z(x^z, Rn) = Rt^y(x^y, n) \,. \tag{1.2.5}$$

Let us notice that we can write

$$t^z(x^z, Rn) = T^z(x^z)Rn = Rt^y(x^y, n) = RT^y(x^y)n \,.$$

Let $m \in S^{d-1}$ be such that $Rn = m$. Then we immediately get that $T^z(x^z) = RT^y(x^y)R^\top$ for any rotation R.

Exercise 1.2.3. Prove Theorem 1.2.2 for $d = 2$.

1.2.2 Principle of virtual work and Piola-Kirchhoff stresses

We are now about to formulate equilibrium equations in a deformed configuration. This configuration is, however, not known a-priori. Therefore, later in Sect. 1.2.4, we will formulate the corresponding boundary value problem also in the reference configuration. We will see that it is very convenient to map the Cauchy stress tensor to the reference configuration by means of the Piola transform. Namely, the resulting equations have the same form in both configurations.

Theorem 1.2.4 (Principle of virtual work in the deformed configuration). *A weak solution T^y (in the sense of Sect. C.1.1) of the following equations*

$$-\mathrm{div}^y T^y = f^y \qquad \text{in } \Omega^y \,, \tag{1.2.6a}$$

$$T^y n^y = g^y \qquad \text{on } \Gamma_\mathrm{N}^y \tag{1.2.6b}$$

formally[9] *satisfies the integral identity (see* (B.3.14)*)*

[9] The adverb "formally" here wants to emphasize that we do not specify assumptions needed to apply the Green formula (like some minimal regularity of Ω^y or integrability of the stress and force densities) mainly because they would not be realistic even for very regular bodies Ω under very regular loadings.

$$\int_{\Omega^y} T^y : \nabla^y v \, dx^y = \int_{\Omega^y} f^y \cdot v \, dx^y + \int_{\Gamma_N^y} g^y \cdot v \, dS^y \tag{1.2.7}$$

for all smooth $v : \bar{\Omega}^y \to \mathbb{R}^d$, $v := 0$ *on* $\Gamma^y \setminus \Gamma_N^y$. *On the other hand, if* T^y *fulfills* (1.2.7) *then it also satisfies* (1.2.6) *in a weak sense.*

Heuristic proof. We apply the following version of Green's theorem for v as in the statement of the theorem.

$$\int_{\Omega^y} \mathrm{div}^y T^y \cdot v \, dx = -\int_{\Omega^y} T^y : \nabla^y v \, dx^y + \int_{\Gamma_N^y} T^y n^y \cdot v \, dS^y. \tag{1.2.8}$$

Thus,

$$0 = \int_{\Omega^y} (\mathrm{div}^y T^y + f^y) \cdot v \, dx^y = \int_{\Omega^y} (f^y \cdot v - T^y : \nabla^y v) \, dx^y + \int_{\Gamma_N^y} T^y n^y \cdot v \, dS^y,$$

which shows that (1.2.6) implies (1.2.7). Conversely, take v with $v := 0$ on Γ and check that (1.2.7) implies (1.2.6a). Then (1.2.6b) easily follows from (1.2.8) and (1.2.6a). □

The formulation (1.2.6) equipped with the condition $T^y = T^{y\top}$ (which is automatically satisfied by Theorem 1.2.2) is called *equilibrium equations in a deformed configuration.*

The problem is that (1.2.6) is formulated in the deformed configuration, which is apriori not known and is a part of a sought solution. Hence, it is desirable to transform the equilibrium equations to the reference configuration. We define the *1st Piola-Kirchhoff stress tensor* $S : \bar{\Omega} \to \mathbb{R}^{d \times d}$, as the Piola transform of the Cauchy stress tensor T^y, i.e.[10],

$$S(x) := T^y(x^y) \mathrm{Cof} \, \nabla y(x), \qquad x^y = y(x), \quad x \in \bar{\Omega}. \tag{1.2.9}$$

It follows from the properties of the Piola transform that

$$\mathrm{div} \, S(x) = (\det \nabla y(x)) \mathrm{div}^y T^y(x^y). \tag{1.2.10}$$

Let us notice that S is not symmetric in general. The symmetric tensor $S_K : \bar{\Omega} \to \mathbb{R}^{d \times d}$ defined by

$$S_K(x) := S(x)(\nabla y(x))^\top = (\det \nabla y(x)) T^y(x^y) \tag{1.2.11}$$

is called the *Kirchhoff stress tensor*. Finally, we define the *2nd Piola-Kirchhoff stress tensor* $\Sigma : \bar{\Omega} \to \mathbb{R}^{d \times d}$ by the formula

[10] Sometimes we write $[S(\nabla y)](x)$ to indicate the dependence on ∇y.

$$\Sigma(x) := (\nabla y(x))^{-1}S(x) = (\nabla y(x))^{-1}T^y(x^y)\mathrm{Cof}\,\nabla y(x), \tag{1.2.12}$$

which is clearly symmetric.

1.2.3 Applied forces in reference configurations

Next we rewrite force densities from $\bar{\Omega}^y$ to $\bar{\Omega}$. Having $f^y : \bar{\Omega}^y \to \mathbb{R}^d$ a body force density (per volume) we look for $f : \Omega \to \mathbb{R}^d$ such that, for every subdomain $\omega \subset \Omega$, it holds

$$\int_\omega f(x)\,\mathrm{d}x = \int_{\omega^y} f^y(x^y)\,\mathrm{d}x^y,$$

i.e., such that the total force acting on an arbitrary smooth subset ω of the specimen in the reference configuration and on the corresponding set $\omega^y \subset \Omega^y$ is the same. Hence, applying the change of variables formula (Thm. B.3.11) we arrive at

$$f(x) := f^y(x^y)\det\nabla y(x), \quad x^y = y(x). \tag{1.2.13}$$

Then f is the (volume) density of body forces in the reference configuration.

If $\varrho : \Omega \to \mathbb{R}$ and $\varrho^y : \Omega^y \to \mathbb{R}$ are *mass densities* in the reference and deformed configurations, respectively, we have

$$\varrho(x) = \varrho^y(x^y)\det\nabla y(x), \quad x^y = y(x). \tag{1.2.14}$$

Let us notice that again $\int_\Omega \varrho(x)\,\mathrm{d}x = \int_{\Omega^y} \varrho^y(x^y)\,\mathrm{d}x^y$, i.e., the total mass of the body is conserved, i.e. independent of the deformed configuration.

Similarly we proceed with surface forces. We look for $g : \Gamma_\mathrm{N} \to \mathbb{R}^d$, $y(\Gamma_\mathrm{N}) =: \Gamma_\mathrm{N}^y$ such that for every $\gamma \subset \Gamma_\mathrm{N}$ and its image $\gamma^y := y(\gamma) \subset \Gamma_\mathrm{N}^y$

$$\int_\gamma g(x)\,\mathrm{d}S = \int_{\gamma^y} g^y(x^y)\,\mathrm{d}S^y.$$

Here $\Gamma_\mathrm{N} \subset \Gamma$ is a part of the boundary where we apply forces. Thus, using the properties of Piola's transform, we have that

$$g(x) = g^y(x^y)\big|(\mathrm{Cof}\,\nabla y(x))\vec{n}(x)\big|, \quad x^y = y(x), \quad x \in \Gamma_\mathrm{N}, \tag{1.2.15}$$

is the density of surface forces in the reference configuration.

An applied body *force* is called a *dead load* if its associated density in the reference configuration is independent of the deformation y. A simple example is a homogeneous gravitational field $f(x) = (0,0,-\mathrm{const.}\ \varrho(x))$, $x \in \Omega$. Then $f^y(x^y) = (0,0,-\mathrm{const.}\ \varrho^y(x^y))$ for all $x^y \in \Omega^y$. Likewise, an applied surface force is a dead load if its associated density in the reference configuration is independent of the deformation y. We would like to emphasize that applied forces are only very rarely dead loads in reality and that it is usually a mathematical simplification.

Consider an applied surface force being a pressure load. In this situation,

$$g^y(x^y) = -p\vec{n}^y(x^y), \quad x^y \in \Gamma_{\rm N}^y \text{ and } p \geq 0. \tag{1.2.16}$$

Clearly, if $p \neq 0$ then, in general, the pressure load is different for different deformations.

In order to fix ideas, we will suppose that the applied force densities are of the form

$$f(x) = \tilde{f}(x, y(x), \nabla y(x)), \qquad x \in \Omega, \text{ and} \tag{1.2.17a}$$

$$g(x) = \tilde{g}(x, y(x), \nabla y(x)), \qquad x \in \Gamma_{\rm N}, \tag{1.2.17b}$$

where $\tilde{f} : \bar{\Omega} \times \mathbb{R}^d \times \mathbb{R}^{d\times d} \to \mathbb{R}^d$ and $\tilde{g} : \Gamma_{\rm N} \times \mathbb{R}^d \times {\rm GL}^+(d) \to \mathbb{R}^d$ are given.

Having a deformation gradient $F : \bar{\Omega} \to \mathbb{R}^{d\times d}$ and a material vector field $v : \bar{\Omega} \to \mathbb{R}^d$ we can use the inverse of $\mathscr{P}_F$ from (1.1.8) to define a spatial field as a vectorial pushforward of v as

$$v^y := F^{-\top} v. \tag{1.2.18}$$

In this case, we say that v^y is a *normal convection* of v. The name of this transformation is motivated by the formula (1.1.21) because v transforms as the normal vector to Γ. We can also transform v by means of $\mathscr{P}_F^{-1}$ for $\mathscr{P}_F$ defined in (1.1.9). This creates a *tangential convection* and

$$v^y := Fv. \tag{1.2.19}$$

Let us notice that, if v and $\tilde{v} : \bar{\Omega} \to \mathbb{R}^d$ are orthogonal, then also $F^{-\top}v$ and $F\tilde{v}$ are perpendicular,too. This explains the adjective "tangential" for the transformation (1.2.19).

Let us discuss the two proposed convections more in detail. Assume that $y : \mathbb{R}^d \to \mathbb{R}^d$ is a homogeneous deformation, i.e., $y(x) := Fx$ for some $F \in {\rm GL}^+(d)$ and all $x \in \mathbb{R}^d$. Take $0 \neq a \in \mathbb{R}^d$ and define a hyperplane $H := \{x \in \mathbb{R}^d;\, a \cdot x = 0\}$. Then $H^y := y(H) = \{x^y \in \mathbb{R}^d;\, (F^{-\top}a) \cdot x^y = 0\}$ is the "deformed"[11] hyperplane, i.e., the hyperplane H after the deformation. It is again a hyperplane because y is a homogeneous deformation. We see that the normal to H^y is $F^{-\top}a \in \mathbb{R}^d$. We also observe that if $x \cdot a = 0$ then $Fx \cdot F^{-\top}a = 0$, it means that Fx is a vector in the plane H^y, or, in other words, it is a tangent vector to it.

Now take an arbitrary injective and smooth deformation $y : \bar{\Omega} \to \mathbb{R}^d$. If $v^y : \bar{\Omega}^y \to \mathbb{R}^d$ is a vector field in the deformed configuration and we want to evaluate its flux through the boundary of a smooth subdomain $\omega^y \subset \bar{\Omega}^y$, we write in view of (1.1.20)

[11] The quotation marks are here because it is still a hyperplane although after the deformation by y.

$$\int_{\partial\omega^y} v^y(x^y)\cdot\vec{n}^y(x^y)\,\mathrm{d}S^y = \int_{\partial\omega} v^y(y(x))\cdot\mathrm{Cof}\,\nabla y(x)\vec{n}(x)\,\mathrm{d}S$$
$$= \int_{\partial\omega} (\mathrm{Cof}\,\nabla y)^\top v^y(y(x))\cdot\vec{n}(x)\,\mathrm{d}S = \int_{\partial\omega} v(x)\cdot\vec{n}(x)\,\mathrm{d}S\,,$$

where, for every $x\in\bar{\Omega}$, we define $v(x) := (\mathrm{Cof}\,\nabla y)^\top v^y(y(x))$. Saying otherwise, writing $F := \nabla y$ we get, for every $x\in\bar{\Omega}$, that

$$\det F(x)v^y(y(x)) = F(x)v(x)\,. \tag{1.2.20}$$

Altogether we see, that $(\det F)v^y$ is a tangential convection of v according to (1.2.19). We also refer to (9.1.9) for further details.

In order to understand better the normal convection, consider a smooth curve $c^y(\xi):\xi\mapsto x^y(\xi)$ in the deformed configuration $\bar{\Omega}^y$ where $\xi_1\le\xi\le\xi_2$ is a parametrization of this curve. The tangent vector to c^y is given as $(x^y)' := \mathrm{d}x^y/\mathrm{d}\xi$. If again $v^y:\bar{\Omega}^y\to\mathbb{R}^d$ is a vector field in the deformed configuration and we want to calculate its integral along c^y we proceed as follows

$$\int_{c^y} v^y(x^y)\mathrm{d}x^y := \int_{\xi_0}^{\xi_1} v^y(x^y(\xi))\cdot(x^y(\xi))'\mathrm{d}\xi = \int_{\xi_0}^{\xi_1} v^y(y(x(\xi)))\cdot\nabla y(x(\xi))x'(\xi)\,\mathrm{d}\xi$$
$$= \int_{\xi_0}^{\xi_1} v(x(\xi))\cdot x'(\xi)\,\mathrm{d}\xi = \int_c v(x)\,\mathrm{d}x\,, \tag{1.2.21}$$

where for every $x\in\bar{\Omega}$ we have $v(x) := (\nabla y(x))^\top v^y(y(x))$, which is a vector in the reference configuration. Again, if $F := \nabla y$ then $v^y = F^{-\top}v$, i.e., v^y is a normal convection of v in this case by (1.2.18). Here $c:\xi\mapsto x(\xi) = y^{-1}(x^y(\xi))$ is the preimage of c_y, i.e., a curve in the reference configuration.

Exercise 1.2.5 (Transformation of a dead load). Consider an elastic bar fixed on its left-hand side by a Dirichlet condition and exposed to a homogeneous gravitational field so that the reference configuration (see Figure 1.4-left) is bent by the deformation (see Figure 1.4-right). The vertical "downward" gravitational force at one particular point at the right-hand side is depicted in the actual deformed configuration. Decide how this force is seen in the reference configuration: option A (again downward) or option B (nearly horizontally with the same direction towards the boundary of the specimen)? Give a reason. How it is in a nonhomogeneous gravitational field? Compare it with Figure 4.5 in Section 4.5.4.

1.2.4 Principles of virtual work in the reference configuration

We have the following boundary value problems in the reference configuration.

Theorem 1.2.6 (Equilibrium equations for the 1st Piola-Kirchhoff tensor). *A weak solution (in the sense of Sect. C.1.1) S of the following equations*

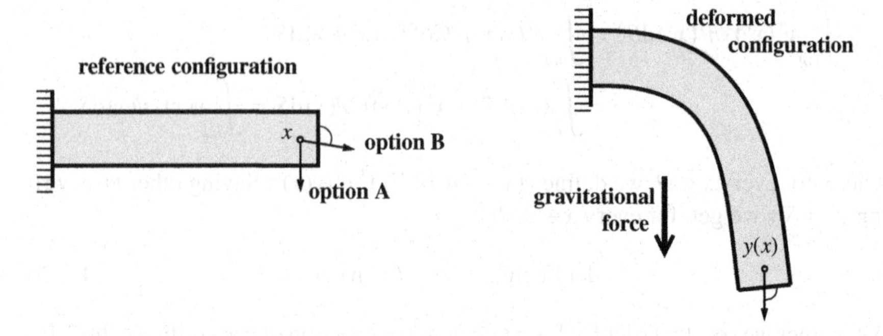

Fig. 1.4 An example how the dead (e.g. gravitational) load transforms from the actual into a reference configuration.

$$-\mathrm{div}\, S = f \qquad \text{in } \Omega, \text{ and} \tag{1.2.22a}$$
$$S\vec{n} = g \qquad \text{on } \Gamma_{\mathrm{N}} \tag{1.2.22b}$$

also formally satisfies the integral identity (see (B.3.14)*)*

$$\int_\Omega S:\nabla v\,\mathrm{d}x = \int_\Omega f\cdot v\,\mathrm{d}x + \int_{\Gamma_{\mathrm{N}}} g\cdot v\,\mathrm{d}S \tag{1.2.23}$$

for all smooth $v:\bar{\Omega}\to\mathbb{R}^d$, $v=0$ *on* $\Gamma\setminus\Gamma_{\mathrm{N}}$.

Heuristic proof. The proof follows from (1.2.6) and definitions of f, g, and S. □

In view of (1.2.12) we have the equilibrium problem for the second Piola-Kirchhoff stress tensor. Let us notice that $S = \nabla y\Sigma$ which relates the following theorem with Theorem 1.2.6

Theorem 1.2.7 (Equilibrium equations for the 2nd Piola-Kirchhoff tensor). *A weak solution (in the sense of Sect. C.1.1)* Σ *of the following equations*

$$-\mathrm{div}(\nabla y\Sigma) = f \qquad \text{in } \Omega \text{ and} \tag{1.2.24a}$$
$$\nabla y\Sigma\vec{n} = g \qquad \text{on } \Gamma_{\mathrm{N}} \tag{1.2.24b}$$

also formally fulfills (see (B.3.14)*) the integral identity*

$$\int_\Omega \nabla y\Sigma:\nabla v\,\mathrm{d}x = \int_\Omega f\cdot v\,\mathrm{d}x + \int_{\Gamma_{\mathrm{N}}} g\cdot v\,\mathrm{d}S \tag{1.2.25}$$

for all smooth $v:\bar{\Omega}\to\mathbb{R}^d$, $v=0$ *on* $\Gamma\setminus\Gamma_{\mathrm{N}}$.

The problems (1.2.22) and (1.2.24) are called *equilibrium equations in the reference configurations* and the corresponding integral identities are referred to as *principles of virtual work in the reference configuration.*

Applied forces f and g from (1.2.17) are *conservative* if there is $\mathcal{F} : \{y : \bar{\Omega} \to \mathbb{R}^d\} \to \mathbb{R}$, such that the Gâteaux differential $\mathcal{F}'$ of $\mathcal{F}$ reads as[12]

$$\begin{aligned}\mathcal{F}'(y)v &= \int_\Omega \tilde{f}(x, y(x), \nabla y(x))\cdot v(x)\,\mathrm{d}x + \int_{\Gamma_\mathrm{N}} \tilde{g}(x, y(x), \nabla y(x))\cdot v(x)\,\mathrm{d}S \\ &= \int_\Omega f(x)\cdot v(x)\,\mathrm{d}x + \int_{\Gamma_\mathrm{N}} g(x)\cdot v(x)\,\mathrm{d}S \end{aligned} \tag{1.2.26}$$

for all smooth $v : \bar{\Omega} \to \mathbb{R}^d$ vanishing on $\Gamma_\mathrm{D} \subset \Gamma$. This is a part of the boundary where we prescribe *Dirichlet boundary conditions* of admissible deformations, while $\Gamma_\mathrm{N} = \Gamma \setminus \Gamma_\mathrm{D}$ serves for prescribing *Neuman or Robin conditions.*

Proposition 1.2.8 (Pressure load is conservative). *Let $p : \mathbb{R}^d \to [0; +\infty)$ be smooth and let $y : \bar{\Omega} \to \bar{\Omega}^y$ be a given deformation. Then the pressure force $g^y(x^y) := -p(x^y)\vec{n}^y(x^y)$ for all $x^y \in \Gamma^y$ mapped to the reference configuration is conservative.*

Proof. When combining Remark 1.1.11 with (1.2.15), we obtain that $g(x) = \tilde{g}(x, y(x), \nabla y(x)) = -p(y(x))[\mathrm{Cof}\,\nabla y(x)]\vec{n}(x)$. Consider $\Gamma_\mathrm{N} := \Gamma$ and the functional

$$\mathcal{F}(y) = -\int_\Omega p(y(x))\det\nabla y(x)\,\mathrm{d}x. \tag{1.2.27}$$

Then it holds for a smooth map $v : \bar{\Omega} \to \mathbb{R}^d$ that

$$\begin{aligned}\mathcal{F}'(y)v = \int_\Omega ((\mathrm{Cof}\,\nabla y(x))\nabla p(y(x)) - [\nabla^y p](y(x))\det\nabla y(x))\cdot v(x)\,\mathrm{d}x \\ - \int_\Gamma p(y(x))\,\mathrm{Cof}\,\nabla y(x))\vec{n}(x)\cdot v(x)\,\mathrm{d}S \\ = -\int_\Gamma p(y(x))[\mathrm{Cof}\,\nabla y(x))\vec{n}(x)]\cdot v(x)\,\mathrm{d}S\end{aligned}$$

because

$$\begin{aligned}\int_\Omega ((\mathrm{Cof}\,\nabla y(x))\nabla p(y(x)))\cdot v(x)\,\mathrm{d}x &= \int_\Omega \sum_{i,j=1}^d (\mathrm{Cof}\,\nabla y)_{ij} \sum_{k=1}^d p'(y(x))_k (\nabla y)^\top_{jk} v_i(x)\,\mathrm{d}x \\ &= \int_\Omega [\nabla^y p](y(x))\cdot v(x)\det\nabla y(x))\,\mathrm{d}x\end{aligned}$$

where we used the identity $(\mathrm{Cof}\,F)F^\top = F^\top\,\mathrm{Cof}\,F = (\det F)\mathbb{I}$ valid for all invertible matrices $F \in \mathbb{R}^{d\times d}$. □

Conservative forces play an important role in the variational formulation of elasticity problems in Chapter 2.

[12] Here we assume that the deformation y is smooth enough (say, $y \in C^1(\bar{\Omega}; \mathbb{R}^d)$), so that the right-hand side of (1.2.26) is well-defined.

Chapter 2
Elastic materials

Ut tensio, sic vis (= As extension, so force)

ROBERT HOOKE (1635–1703)

We must choose a law expressing the stress-strain relationships which is sufficiently simple to allow of considerable mathematical development and which at the same time expresses the known behavior of as wide a range of highly elastic materials as possible. ... It is necessary ... to strike a compromise between mathematical tractability, breadth of applicability and exactitude of applicability.[1]

RONALD S. RIVLIN (1915–2005)

Looking at the force balance (1.2.6), we see that we have only d equations for $d(d{+}3)/2$ unknowns (namely d components of y and $d(d{+}1)/2$ components of T^y). Therefore, we complete (1.2.6) by material constitutive relations describing particular materials as far as its elastic properties concerns. Such constitutive relations should involve the stress tensor and the deformation gradient and may be in general implicit, cf. [427]. Without excluding too many applications, we confine ourselves to the case when the stress is explicitly determined by the strain.

2.1 Response functions

Materials are called elastic (or *Cauchy elastic*) if the Cauchy stress is determined only by the current deformation. More precisely, we say that a material is elastic if there exists a mapping

$$\tilde{T}^D : \bar{\Omega} \times \mathrm{GL}^+(d) \to \mathbb{R}^{d\times d}_{\mathrm{sym}}, \tag{2.1.1}$$

called a response function for the Cauchy stress, such that

$$T^y(x^y) = \tilde{T}^D(x, \nabla y(x)) \tag{2.1.2}$$

[1] See [46].

M. Kružík and T. Roubíček, *Mathematical Methods in Continuum Mechanics of Solids*, Interaction of Mechanics and Mathematics,
https://doi.org/10.1007/978-3-030-02065-1_2

for every $x\in\bar{\Omega}$. The relation (2.1.2) is called the constitutive equation of the material. The function $\tilde{T}^D$ does not depend on y and the upper index D refers to the fact that it defines the stress tensor in the deformed configuration. The response function obviously depends on the chosen reference configuration. Although the stress in a Cauchy elastic material depends only on the state of deformation, the work done by stresses may also depend on the path of deformation; cf. [519]. Therefore a Cauchy elastic material, in general, has a non-conservative structure. We refer e.g. to [430] for examples of non-conservative Cauchy elastic materials.

Similarly, keeping in mind relationship between the Cauchy and both Piola-Kirchhoff stress tensors, we can find response functions for the latter ones

$$\tilde{S}(x,F) = (\det F)\tilde{T}^D(x,F)F^{-\top} \quad \text{and} \quad \tilde{\Sigma}(x,F) = (\det F)F^{-1}\tilde{T}^D(x,F)F^{-\top}$$

with $F\in\mathrm{GL}^+(d)$ and a point $x\in\bar{\Omega}$. Then the Piola-Kirchhoff stress tensors read, for all $x\in\bar{\Omega}$, respectively as

$$S(x) = \tilde{S}(x,\nabla y(x)) \quad \text{and} \quad \Sigma(x) = \tilde{\Sigma}(x,\nabla y(x)). \tag{2.1.3}$$

An "elastic material" is a theoretical construction and we cannot really prove that a particular piece of matter is elastic, i.e., that it fulfills our definition. We can only suggest a response function and compare our predictions with experiments. A material in the reference configuration is called homogeneous if its response function does not depend on x, otherwise it is called inhomogeneous. Homogeneity is related to a particular reference configuration, while the response function $\tilde{T}^D$ is related to a particular deformed configuration. There are elasticity theories relating $T^y(x^y)$ to the gradient ∇y in the whole Ω (the so-called nonlocal elasticity) or taking higher order gradients into considerations (the so-called nonsimple materials cf. Section 2.5 below).

The Axiom of material frame-indifference asserts that for all $x\in\bar{\Omega}$, $R\in\mathrm{SO}(d)$ and any $F\in\mathrm{GL}^+(d)$ we have $\tilde{T}^D(x,RF) = R\tilde{T}^D(x,F)R^\top$. Consequently, the response function $\tilde{S}$ of the first Piola-Kirchhoff stress tensor satisfies $R^\top\tilde{S}(x,RF) = \tilde{S}(x,F)$. Indeed,

$$\tilde{S}(x,RF) = \det(RF)\tilde{T}^D(x,RF)RF^{-\top} = \det(RF)R\tilde{T}^D(x,F)R^\top RF^{-\top} = R\tilde{S}(x,F). \tag{2.1.4}$$

Analogously, we have for the response function of the second Piola-Kirchhoff stress tensor that for every $R\in\mathrm{SO}(d)$

$$\tilde{\Sigma}(x,RF) = \tilde{\Sigma}(x,F). \tag{2.1.5}$$

The polar decomposition (A.2.27) then implies that

$$\tilde{\Sigma}(x,F) = \tilde{\Sigma}_1(x,F^\top F) \tag{2.1.6}$$

for some function $\tilde{\Sigma}_1$.

It is an interesting question whether one can consider linear constitutive laws valid in finite elasticity. Clearly, an affirmative answer would allow for simpler constitutive models. Unfortunately, a negative answer to this question was given by Fosdick and Serrin in [197]. Let $U := \{D \in \mathbb{R}^{d\times d}; \det(\mathbb{I}+D) > 0\}$. Then we have for the first Piola-Kirchhoff stress tensor $S(\mathbb{I}+D) = \bar{S}(D)$, which defines $\bar{S}$ in a neighborhood of the origin. In view of (2.1.4) we have for any $R \in \mathrm{SO}(d)$ that

$$S(R+RD) = \bar{S}(RD+R-\mathbb{I}) = RS(\mathbb{I}+D) = R\bar{S}(D).$$

Assume that $\bar{S}$ is linear, i.e., $\bar{S}_{ij}(D) = a_{ijkl}D_{kl}$ for $i,j,k,l = 1,\ldots,d$. Hence, setting $D := 0$ we get that $\bar{S}(R-\mathbb{I}) = 0$ for every $R \in \mathrm{SO}(d)$. We take $R_\mu := \exp(\mu A)$ for $A \in \mathbb{R}^{d\times d}$ skew symmetric and $\mu \in \mathbb{R}$. Here the matrix exponential is defined as $\exp(\mu A) := \mathbb{I} + \sum_{i=1}^{\infty} \frac{\mu^i A^i}{i!}$. Therefore, it yields

$$\frac{\mathrm{d}^k}{\mathrm{d}\mu^k}\bar{S}(R_\mu - \mathbb{I}) = \bar{S}\left(\frac{\mathrm{d}^k}{\mathrm{d}\mu^k}R_\mu\right) = 0.$$

Setting $\mu = 0$ and $k = 1,2$ we get $\bar{S}(A) = \bar{S}(A^2) = 0$. As $A \in \mathbb{R}^{d\times d}$ is skew we have $A^2 = b\otimes b - |b|^2\mathbb{I}$, where b is the axial vector of A; cf. Exercise 2.1.1. Putting $b := \mathbf{e}_i$ for $i = 1,\ldots,d$ we have $\bar{S}(\sum_{i=1}^{d} \mathbf{e}_i \otimes \mathbf{e}_i - d\mathbb{I}) = 0$. We get from the linearity of $\bar{S}$ that $\bar{S}((1-d)\mathbb{I}) = 0$ which implies that $\bar{S}(b\otimes b) = 0$ for every vector b. Consequently, since any symmetric matrix $\tilde{A}$ can be written as $\tilde{A} = \lambda_i v_i \otimes v_i$ where $\lambda_i \in \mathbb{R}$ and $v_i \in \mathbb{R}^d$ are eigenvalues and eigenvectors of $\tilde{A}$, respectively we see that $\bar{S}(\tilde{A}) = 0$. This, together with $\bar{S}(A) = 0$ for any skew matrix A yields $\bar{S} = 0$. This means that there is not a nonzero linear function assigning to a displacement gradient D the first Piola-Kirchhoff stress tensor.

Exercise 2.1.1. Show that if $A \in \mathbb{R}^{d\times d}$ is skew symmetric then $A^2 = b\otimes b - |b|^2\mathbb{I}$ where $b \in \mathbb{R}^d$ is the axial vector of A, i.e., $Aa = b\times a$ for all $a \in \mathbb{R}^d$. Here $|b|^2 = b\cdot b$.

2.2 Isotropic materials

The intuitive idea of isotropy is that at a given point of our material its response is the same in all directions, i.e., isotropy is directional uniformity. As an example, we can consider polycrystalline materials, metals, or dough. Isotropic materials are easy to shape. On the other hand, wood or sedimented soil/rocks are examples of anisotropic materials because they behavior is different along and across fibers or layers, respectively. We now give a mathematical meaning of isotropy. Consider a deformation $y : \bar{\Omega} \to y(\bar{\Omega})$. Then we have by (2.1.2) that

$$T^y(x^y) = \tilde{T}^D(x, \nabla y(x)).$$

Let us take $x_0 \in \bar{\Omega}$ and rotate $\bar{\Omega}$ around this point by a rotation $R^\top \in \mathrm{SO}(d)$, i.e., define $v(z) := x_0 + R^\top(z - x_0)$ for all $z \in \bar{\Omega}$. Let us further consider $\tilde{y} := y \circ v^{-1} : v(\bar{\Omega}) \to y(\bar{\Omega})$, so that $\tilde{y}(\tilde{x}) = y(x_0 + R(\tilde{x} - x_0))$ if $\tilde{x} \in v(\bar{\Omega})$. However, $x_0^y = x_0^{\tilde{y}}$ and therefore

$$T^y(x_0^y) = \tilde{T}^D(x_0, \nabla y(x_0)) = T^{\tilde{y}}(x_0^{\tilde{y}}) = \tilde{T}^D(x_0, \nabla\tilde{y}(x_0)) = \tilde{T}^D(x_0, \nabla y(x_0)R). \quad (2.2.1)$$

Hence, we say that a material is isotropic at a point $x_0 \in \bar{\Omega}$ if the response function for the Cauchy stress satisfies for all $F \in \mathrm{GL}^+(d)$ and all $R \in \mathrm{SO}(d)$ that

$$\tilde{T}^D(x_0, F) = \tilde{T}^D(x_0, FR).$$

Using the response functions for S and Σ we get analogously for all F, R as before that

$$\tilde{S}(x_0, FR) = \tilde{S}(x_0, F)R \quad \text{and} \quad \tilde{\Sigma}(x_0, FR) = R^\top \tilde{\Sigma}(x_0, F)R.$$

Isotropy is related to a particular reference configuration. For example, if we pre-stress a sample along one coordinate axis, it may have different elastic properties in various directions.

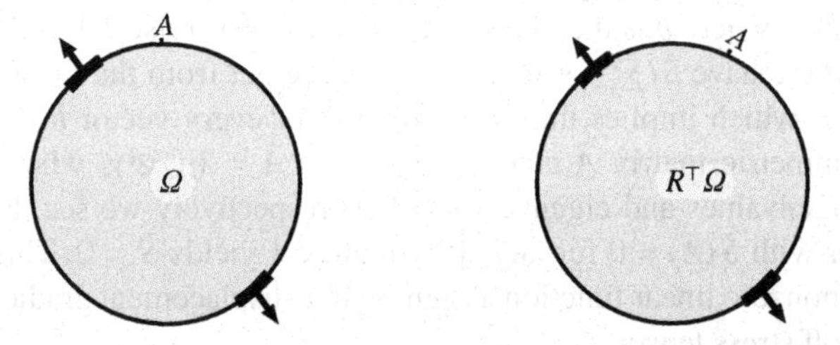

Fig. 2.1 Ω (left) and $R^\top\Omega$ (right) are experimentally indistinguishable states of the isotropic material. The arrows denote applied loading.

2.3 Hyperelastic materials

An elastic *material* is *hyperelastic* (or sometimes called Green-elastic) if there is a *stored energy* function $\varphi : \bar{\Omega} \times \mathrm{GL}^+(d) \to [0; +\infty)$ such that, for all $x \in \bar{\Omega}$ and all $F \in \mathrm{GL}^+(d)$, it holds

$$\tilde{S}(x, F) = \partial_F \varphi(x, F). \quad (2.3.1)$$

As before, a hyperelastic material is a model and its relevancy cannot be proven. However, it emphasizes reversibility of deformations and the idea that energy can be stored in the material and used later *without losses* to do work. In this sense, hyperelastic materials are conservative. The first models of hyperelastic materials were developed by R. Rivlin and M. Mooney.

Upcoming calculations follow [114] and show restrictions put on φ by the Axiom of material frame-indifference. Let us notice that we have from (2.1.4) that $\tilde{T}^D(x,RF) = R\tilde{T}^D(x,F)R^\top$ for an arbitrary proper rotation R. This means in view of (2.1.4) that

$$\forall R \in \mathrm{SO}(d): \quad R^\top \partial_F \varphi(x,RF) = \partial_F \varphi(x,F). \tag{2.3.2}$$

Let us fix a rotation R and denote $\varphi_R(x,F) := \varphi(x,RF)$. Then we get by the Taylor formula for $\tilde{F} \in \mathrm{GL}^+(d)$ such that $\det(F+\tilde{F}) > 0$ that

$$\begin{aligned}\varphi_R(x,F+\tilde{F}) &= \varphi(x,RF+R\tilde{F}) = \varphi(x,RF) + \partial_F \varphi(x,RF){:}R\tilde{F} + o(|\tilde{F}|)\\ &= \varphi_R(x,F) + R^\top \partial_F \varphi(x,RF){:}\tilde{F} + o(|\tilde{F}|)\\ &= \varphi_R(x,F) + \partial_F \varphi_R(x,F){:}\tilde{F} + o(|\tilde{F}|).\end{aligned} \tag{2.3.3}$$

Therefore, in view of (2.3.2) and (2.3.3), it holds

$$\partial_F \varphi(x,F) = \partial_F \varphi_R(x,F).$$

In other words, for all $F \in \mathrm{GL}^+(d)$, it holds $\partial_F(\varphi(x,F) - \varphi(x,RF)) = 0$. As $\mathrm{GL}^+(d)$ is a connected set[2], we infer that there is a constant K_R (depending on R) such that $\varphi(x,RF) = \varphi(x,F) + K_R$. Testing this equality for $F := \mathbb{I}$, $F := R$, $F := R^2$, etc., we get that $\varphi(x,R^n) = \varphi(x,\mathbb{I}) + nK_R$. Thus, if $K_R \neq 0$ then inevitably $\lim_{n\to+\infty} |\varphi(x,R^n)| = +\infty$. However, the set $\{R^n\}_{n\in\mathbb{N}}$ is compact and $\varphi(x,\cdot)$ is differentiable and therefore $|\varphi(x,\cdot)|$ must be continuous for almost every $x \in \Omega$. Consequently, $\lim_{n\to+\infty} |\varphi(x,R^n)| < +\infty$. For this reason, $K_R = 0$. Altogether, *frame indifference of hyperelastic materials* means that, for all rotations R and all $F \in \mathrm{GL}^+(d)$, it holds

$$\varphi(x,RF) = \varphi(x,F). \tag{2.3.4}$$

As we can always find a decomposition $F = RU$ where $R \in \mathrm{SO}(d)$ and $U = U^\top \in \mathrm{GL}^+(d)$ with $U^2 = F^\top F$, it is clear that, for some function $\hat{\varphi} : \bar{\Omega} \times \{A = A^\top; A \in \mathrm{GL}^+(d)\} \to \mathbb{R}$, it holds

[2] Indeed, notice that if $F \in \mathrm{GL}^+(d)$ then there is an upper triangular matrix $\tilde{F}$ and $R \in \mathrm{SO}(d)$ such that $F = R\tilde{F}$. Moreover, the diagonal components of $\tilde{F}$ can be taken all positive (and then the decomposition of F is unique). Hence, $\det F = \det \tilde{F} = \prod_{i=1}^d \tilde{F}_{ii}$. Let for $t \in [0,1]$ $t \mapsto \tilde{F}_t$ be defined in such a way that $\tilde{F}_t$ has the same diagonal as $\tilde{F}$ but its off-diagonal elements are $(1-t)$ multiples of off-diagonal elements of $\tilde{F}$. Therefore, $\tilde{F}_1 = \mathrm{diag}(\tilde{F}_{11},\dots,\tilde{F}_{dd})$. Now we extend the mapping $t \to \tilde{F}_t$ to the interval $[1,2]$ in the following way: If $t \in [1,2]$ then $\tilde{F}_t := \mathrm{diag}((1-\tilde{F}_{11})t + 2\tilde{F}_{11} - 1,\dots,(1-\tilde{F}_{dd})t + 2\tilde{F}_{dd} - 1)$. In particular, $\tilde{F}_2 = \mathbb{I}$ and the path $\{t \in [0,2] \mapsto \tilde{F}_t\} \subset \mathrm{GL}^+(d)$ and it is continuous. This means that $t \mapsto F_t := R\tilde{F}_t$ makes a continuous path between F and R. As R is a rotation it can be joined with the identity by a continuous path as it can be readily seen from the expression of R in terms of axial rotation angles (called Euler's decomposition). Altogether, we see that F is connected with the identity. Consequently, $\mathrm{GL}^+(d)$ is connected.

$$\varphi(\cdot,F)=\hat{\varphi}(\cdot,C) \quad \text{with } C=F^\top F. \tag{2.3.5}$$

A question immediately arises if the concept of hyperelastic materials which defines the first Piola-Kirchhoff stress tensor in terms of a potential φ automatically ensures that the corresponding Cauchy stress tensor is symmetric. We recall that, by (1.2.9), the relation between the 1st Piola-Kirchhoff stress tensor S and the Cauchy stress tensor T^y is

$$S(x)=T^y(y(x))\operatorname{Cof}\nabla y(x) \tag{2.3.6}$$

for every $x\in\bar{\Omega}$. In view of (A.2.20) on p. 491, the symmetry of T^y yields

$$\frac{S(x)}{\det\nabla y(x)}\nabla y(x)^\top=\nabla y(x)\frac{S(x)^\top}{\det\nabla y(x)}. \tag{2.3.7}$$

To find this out we apply the inverse of the Piola transform defined in (1.1.15). Thus, symmetry of the Cauchy stress implies that for all $F\in\mathrm{GL}^+(d)$ it must hold that

$$\partial_F(\varphi(\cdot,F))F^\top=F(\partial_F(\varphi(\cdot,F)))^\top. \tag{2.3.8}$$

Taking into account (2.3.5), however, we see that

$$\tilde{S}(\cdot,F)=\partial_F\varphi(\cdot,F)=2F\partial_C\hat{\varphi}(\cdot,C)=2F(\partial_C\hat{\varphi}(\cdot,C))^\top, \tag{2.3.9}$$

where the last equality follows from the symmetry of $C=F^\top F$. Consequently, (2.3.8) holds if (2.3.4) does.

The tensor on the left-hand side of (2.3.8) is the response function of the Kirchhoff stress tensor, $\tilde{S}_K$, defined in (1.2.11), i.e., involving (2.3.1) we have for every $x\in\Omega$ that

$$\tilde{S}_K(x,F):=\partial_F(\varphi(\cdot,F))F^\top=\tilde{S}(\cdot,F)F^\top. \tag{2.3.10}$$

Let us notice also that the response function for the second Piola-Kirchhoff stress tensor Σ is easily computed from the stored-energy density. Namely, we have from (2.3.9) that

$$\tilde{\Sigma}(\cdot,F)=F^{-1}\tilde{S}(\cdot,F)=2\partial_C\tilde{\varphi}(\cdot,C). \tag{2.3.11}$$

By (2.3.5) and Exercise 2.3.7, we see that the stored energy density φ of an hyperelastic isotropic material satisfies

$$\forall F\in\mathbb{R}^{d\times d}\ \ \forall R,Q\in\mathrm{SO}(d):\quad \varphi(\cdot,F)=\varphi(\cdot,RFQ) \tag{2.3.12}$$

and thus, $\varphi(x,F) = \hat{\varphi}(x,C) = \hat{\varphi}(x,Q^\top CQ)$ is frame-indifferent and isotropic. Hence, it can be expressed in terms of *principal stretches*, i.e., for some function $\bar{\varphi}$ we get

$$\varphi(\cdot,F) = \bar{\varphi}(\cdot,\lambda_1,\ldots,\lambda_d), \tag{2.3.13}$$

where $\lambda_1,\ldots,\lambda_d$ are eigenvalues of the right Cauchy-Green tensor $C = F^\top F$ or alternatively of the *left Cauchy-Green strain* tensor

$$B = FF^\top. \tag{2.3.14}$$

In the case of hyperelastic material and if the applied forces are conservative, a solution of the equilibrium equations is *formally equivalent* to finding a critical point of the functional

$$\mathcal{E}(y) := \int_\Omega \varphi(x,\nabla y(x))\,\mathrm{d}x - \mathcal{F}(y). \tag{2.3.15}$$

By the formal equivalence we mean that partial differential equations (usually called *Euler-Lagrange equations*, see (A.3.4)) together with boundary conditions describing critical points of $\mathcal{E}$ are derived under such qualification of solutions y which is not much realistic; for instance the regularity of y allowing us to exchange limits and integrals is generically not satisfied.

The functional $y \mapsto \int_\Omega \varphi(x,\nabla y(x))\,\mathrm{d}x$ is called the *stored energy*[3], while $\mathcal{E}$ is called the *total energy* of the specimen. As φ is a density of energy stored in the elastic material, it is natural to require that

$$\varphi(x,F)\begin{cases} \to +\infty & \text{if } \det F \to 0_+, \\ = +\infty & \text{if } \det F \le 0, \end{cases} \tag{2.3.16}$$

because compression of the material to zero volume should require the infinite amount of energy and (local) interpenetration is merely forbidden. Additionally, we extended $\varphi(x,\cdot)$ by $+\infty$ to the set of matrices with non-positive determinants (without changing the notation). This prevents admissible deformations from changing the orientation and compressing the material to a point. Moreover, it makes, together with (2.3.16), $\varphi(x,\cdot) : \mathbb{R}^{d\times d} \to \mathbb{R}\cup\{+\infty\}$ continuous.

Theorem 2.3.1 (Formal Euler-Lagrange equations for $\mathcal{E}$). *Let there be given a hyperelastic material subjected to conservative applied body and surface forces in the sense of* (1.2.26), *and subjected to Dirichlet boundary conditions on $\Gamma_\mathrm{D} \subset \Gamma$. Let also $\mathcal{A} := \big\{\tilde{y} \in C^2(\bar{\Omega};\mathbb{R}^d);\ \det\nabla y > 0$ in $\bar{\Omega}$ and $y = y_\mathrm{D}$ on $\Gamma_\mathrm{D}\big\}$. Then $y \in \mathcal{A}$ is a solu-*

[3] Some authors call it also strain energy.

tion[4] *to the boundary-value problem*

$$-\mathrm{div}\,\partial_F\varphi(x,\nabla y(x)) = \tilde{f}(x,y(x),\nabla y(x)) \quad \textit{for } x\in\Omega, \tag{2.3.17a}$$

$$\partial_F\varphi(x,\nabla y(x))\vec{n}(x) = \tilde{g}(x,y(x),\nabla y(x)) \quad \textit{for } x\in\Gamma_{\mathrm{N}},\ \textit{and} \tag{2.3.17b}$$

$$y(x) = y_{\mathrm{D}}(x) \quad \textit{for } x\in\Gamma_{\mathrm{D}} \tag{2.3.17c}$$

with $\vec{n}(x)$ *denoting the outer unit normal to* $\Gamma_{\mathrm{N}}\subset\Gamma$ *at* $x\in\Gamma_{\mathrm{N}}$ *if and only if it satisfies*

$$\mathcal{E}'(y)v = 0 \tag{2.3.18}$$

for all $v\in C^2(\bar{\Omega};\mathbb{R}^d)$ *such that* $v=0$ *on* Γ_{D}. *In particular, if* $y\in\mathcal{A}$ *and* $\mathcal{E}(y) = \min_{\tilde{y}\in\mathcal{A}}\mathcal{E}(\tilde{y})$, *then* y *satisfies (2.3.17).*

Heuristic proof. Let us take $v\in C^2(\bar{\Omega};\mathbb{R}^d)$ such that $v=0$ on Γ_{D} and $y\in\mathcal{A}$. Define $h(t) := \mathcal{E}(y+tv)$ for $t\in\mathbb{R}$. Notice that $y+tv = y_{\mathrm{D}}$ on Γ_{D} and $y+tv\in\mathcal{A}$ for every t small enough. If $y\in\mathcal{A}$ satisfies (2.3.18), then

$$\begin{aligned}
0 &= \int_\Omega \partial_F\varphi(x,\nabla y(x)){:}\nabla v(x)\,\mathrm{d}x - \mathcal{F}'(y)v \\
&= -\int_\Omega \mathrm{div}\,\partial_F\varphi(x,\nabla y(x))v(x)\,\mathrm{d}x + \int_{\Gamma_{\mathrm{N}}}[\partial_F\varphi(x,\nabla y(x))\vec{n}(x)]\cdot v(x)\,\mathrm{d}S \\
&\quad - \int_\Omega \tilde{f}(x,y(x),\nabla y(x))\cdot v(x)\,\mathrm{d}x - \int_{\Gamma_{\mathrm{N}}}\tilde{g}(x,y(x),\nabla y(x))\cdot v(x)\,\mathrm{d}S \\
&= \int_\Omega [-\mathrm{div}\,\partial_F\varphi(x,\nabla y(x)) - \tilde{f}(x,y(x),\nabla y(x))]v(x)\,\mathrm{d}x \\
&\quad - \int_{\Gamma_{\mathrm{N}}}[\partial_F\varphi(x,\nabla y(x))\vec{n}(x) - \tilde{g}(x,y(x),\nabla y(x))]\cdot v(x)\,\mathrm{d}S\,.
\end{aligned} \tag{2.3.19}$$

The arbitrariness of v yields (2.3.17). On the other hand, if y satisfies (2.3.17) then (2.3.19) holds, i.e., (2.3.18) is satisfied, too.

The converse argumentation holds, too. Moreover, if y is a minimizer of $\mathcal{E}$, then (2.3.18) certainly holds. □

We again wish to stress that (2.3.18) need not express necessary conditions for minimizers of $\mathcal{E}$. First of all, when deriving them we must evaluate the functional at $y+tv$ and we have no information if the determinant $\det(\nabla y + t\nabla v) > 0$. Moreover, let us recall the directional derivative

$$\mathrm{D}\mathcal{E}(u,v) = [\mathcal{E}'(y)]v = \lim_{t\to 0}\frac{\mathcal{E}(y+tv) - \mathcal{E}(y)}{t}, \tag{2.3.20}$$

cf. the definition (A.3.1) on p. 494. The derivation of Euler-Lagrange equations implicitly assumes that we can exchange the limit and the integral in (2.3.20). This is,

[4] Such a solution, having the full number of derivatives satisfying the equations in (2.3.17) pointwise, is called a classical solution. Of course, it is also assumed that $\varphi(x,F)$, $f(x,y,F)$, $g(x,y,F)$, and $y_{\mathrm{D}}(x)$ are defined for all (not only almost all) x.

however, not always possible if $y \notin W^{1,\infty}(\Omega;\mathbb{R}^d)$. This problem appears already in one-dimensional variational problems as observed by Ball and Mizel in [38]. As we cannot assume this regularity of y, in general, we resort to finding a minimizer of $\mathcal{E}$ instead of searching for a weak solution to (2.3.17). Every minimizer of $\mathcal{E}$ is then called a *variational solution* to (2.3.17).

Nevertheless, as shown by J.M. Ball in [29], in some cases one can prove that minimizers of $\mathcal{E}$ satisfy equilibrium equations with the Kirchhoff stress tensor (1.2.11). We will follow the reasoning from [29]. Let us assume that φ does not depend on x and that there is $K > 0$ such that, for all $F \in \mathrm{GL}^+(d)$, we get

$$|\partial_F\varphi(F)F^\top| \le K(\varphi(F)+1). \tag{2.3.21}$$

We have due to symmetry of the Kirchhoff stress tensor that

$$\begin{aligned}|\partial_F\varphi(F)F^\top|^2 &= (\partial_F\varphi(F)F^\top):(F\partial_F\varphi(F)^\top)\\ &= (F^\top\partial_F\varphi(F)):(F^\top\partial_F\varphi(F))^\top \le |F^\top\partial_F\varphi(F)|^2,\end{aligned} \tag{2.3.22}$$

hence, (2.3.21) holds, for example, if for all $F \in \mathrm{GL}^+(d)$ it holds

$$|F^\top\partial_F\varphi(F)| \le K(\varphi(F)+1). \tag{2.3.23}$$

Lemma 2.3.2 (J.M. Ball [29]). *Let us assume that $\varphi : \mathrm{GL}^+(d) \to [0;+\infty)$ satisfies (2.3.21). Then there is $\gamma > 0$ such that for all $\tilde{F} \in \mathrm{GL}^+(d)$ satisfying $|\tilde{F} - \mathbb{I}| < \gamma$ it holds that for all $F \in \mathrm{GL}^+(d)$*

$$|\partial_F\varphi(\tilde{F}F)F^\top| \le 3K(\varphi(F)+1). \tag{2.3.24}$$

After this preparatory lemma we can prove the result formulated and proved already in [29]. We define, for some $y_\mathrm{D} \in W^{1,1}(\Omega;\mathbb{R}^d)$, the following set of admissible deformations

$$\mathcal{A} := \{y \in W^{1,1}(\Omega;\mathbb{R}^d) :\ \det\nabla y > 0 \text{ a.e.},\ y = y_\mathrm{D} \text{ on } \Gamma_\mathrm{D}\}. \tag{2.3.25}$$

Theorem 2.3.3 (Necessary conditions for minimizers [29]). *Let $\mathcal{F} = 0$ in (2.3.15) and let $y \in \mathcal{A}$ be a minimizer of $\mathcal{E}$ in $\mathcal{A}$ where $\mathcal{A}$ is given in (2.3.25). Let further (2.3.21) hold for $\varphi : \mathbb{R}^{d\times d} \to [0;+\infty]$ such that $\varphi \in C^1(\mathrm{GL}^+(d))$ for almost all $x \in \Omega$. Then*

$$\int_\Omega \partial_F\varphi(\nabla y(x))\nabla y(x)^\top:\nabla\tilde{y}(x)\,\mathrm{d}x = 0 \tag{2.3.26}$$

for all $\tilde{y} \in C^1(\mathbb{R}^d;\mathbb{R}^d)$ bounded and such that also $\nabla\tilde{y}$ is bounded on $\mathbb{R}^d$, and, moreover, $\tilde{y} \circ y = 0$ on Γ_D.

Proof. Define $y_t := y + t\tilde{y} \circ y$ for $|t|$ small enough. Notice that

$$\nabla y_t = \nabla y + t\nabla\tilde{y}(y)\nabla y = (\mathbb{I} + t\nabla\tilde{y}(y))\nabla y.$$

In particular, $\det\nabla y_t > 0$ in Ω if $|t|$ is small enough and $\|y-y_t\|_{W^{1,1}(\Omega;\mathbb{R}^d)} \to 0$ if $t\to 0$. We get

$$0 = \frac{\mathcal{E}(y_t)-\mathcal{E}(y)}{t} = \frac{1}{t}\int_\Omega\int_0^1 \frac{\mathrm{d}}{\mathrm{d}s}\varphi((\mathbb{I}+st\nabla\tilde{y}(y(x)))\nabla y(x))\,\mathrm{d}s\,\mathrm{d}x$$

$$= \int_\Omega\int_0^1 \partial_F\varphi((\mathbb{I}+st\nabla\tilde{y}(y(x)))\nabla y(x)):[\nabla\tilde{y}(y(x))\nabla y(x)]\,\mathrm{d}s\,\mathrm{d}x. \qquad (2.3.27)$$

The inner (x-dependent) integrand is, however, bounded by $x\mapsto 3K(\varphi(\nabla y(x)) + 1)\sup_{z\in\mathbb{R}^d}|\nabla\tilde{y}(z)|$ which is an integrable function. The passage for $t\to 0$ and the Lebesgue dominated convergence theorem finish the proof. □

We assume that there are positive constants $\varepsilon, p, q, r > 0$ such that, for each $x\in\bar{\Omega}$ and all $F\in\mathbb{R}^{d\times d}$,

$$\varphi(x,F) \ge \begin{cases} \varepsilon(|F|^p + |\mathrm{Cof}\,F|^q + (\det F)^r) & \text{if } \det F > 0,\\ +\infty & \text{otherwise.}\end{cases} \qquad (2.3.28)$$

This explicitly assumes that large deformation gradients and changes of area and of volume between the reference and the deformed configurations inevitably contribute to the energy stored in the material. Of course, this is not realistic because real materials sooner or later break if large-strain measures are applied. However, (2.3.28) is an important assumption from the mathematical point of view because it will ensure coercivity; c.f. (3.4.5).

Proposition 2.3.4 (Non-existence of a convex stored energy). *If $d\ge 2$, there is no function $\varphi:\bar{\Omega}\times\mathbb{R}^{d\times d}\to\mathbb{R}\cup\{+\infty\}$ such that $\varphi(x,\cdot)$ is convex and finite on* $\mathrm{GL}^+(d)$ *for $x\in\bar{\Omega}$ and satisfies* (2.3.16).

Proof. We will argue by contradiction. Take $x\in\bar{\Omega}$ fixed. Since $\mathrm{GL}^+(d)$ is not convex, there is $\mu\in(0,1)$ and $F,\tilde{F}\in\mathrm{GL}^+(d)$ such that $\mu F+(1-\mu)\tilde{F}\notin\mathrm{GL}^+(d)$. Moreover, if $\varphi(x;\cdot)$ were convex then

$$\sup_{0\le\mu\le 1}\varphi(x,\mu F+(1-\mu)\tilde{F}) \le \max(\varphi(x,F),\varphi(x,\tilde{F})) < +\infty.$$

Further, due to continuity of the determinant there is $\lambda_0\in(0,\mu]$ such that $\det(\lambda F+(1-\lambda)\tilde{F}) > 0$ for $\lambda\in[0,\lambda_0)$ and $\det(\lambda_0 F+(1-\lambda_0)\tilde{F}) = 0$. But this means that $\lim_{\lambda\to\lambda_0-}\varphi(x,\lambda F+(1-\lambda)\tilde{F}) = +\infty$, a contradiction. □

Convexity is an important ingredient if one proves existence of minima to variational problems. The following example shows that a minimum of an integral functional with a nonconvex term in the "gradient variable" does not necessarily exist.

Example 2.3.5 (Non-existence of a minimizer). Consider the following minimization problem with a nonconvex functional φ:

$$\text{minimize } J(y) := \int_0^1 y^2(x) + \Big(\Big(\frac{\mathrm{d}y}{\mathrm{d}x}\Big)^2(x) - 1\Big)^2 \,\mathrm{d}x, \qquad y \in W^{1,4}(0,1). \tag{2.3.29}$$

We can see that $J > 0$. Take $\bar{y} : \mathbb{R} \to \mathbb{R}$ the one-periodic extension of $y : [0,1] \to \mathbb{R}$ given by

$$y(x) := \begin{cases} x & \text{if } 0 \le x \le \frac{1}{2}, \\ -x+1 & \text{if } \frac{1}{2} \le x \le 1. \end{cases}$$

Then e.g. the sequence $y_k(x) := (\bar{y}(kx) + 1)/k$ illustrated in Fig. 2.2 belongs to $W^{1,4}(0,1)$ and $J(y_k) = \mathscr{O}(1/k^2) \to 0$ as $k \to \infty$[5]. Thus $\{y_k\}_{k\in\mathbb{N}}$ is a minimizing sequence since obviously $J \ge 0$, and therefore $\inf J = 0$. On the other hand, there is no y for which $J(y) = 0$. Indeed, then both $\int_0^1 ((\frac{\mathrm{d}y}{\mathrm{d}x})^2 - 1)^2 \mathrm{d}x$ and $\int_0^1 y^2 \mathrm{d}x$ simultaneously would have to be zero, so that $y = 0$, but then also $\frac{\mathrm{d}y}{\mathrm{d}x} = 0$, which however contradicts $\int_0^1 ((\frac{\mathrm{d}y}{\mathrm{d}x})^2 - 1)^2 \mathrm{d}x = 0$.

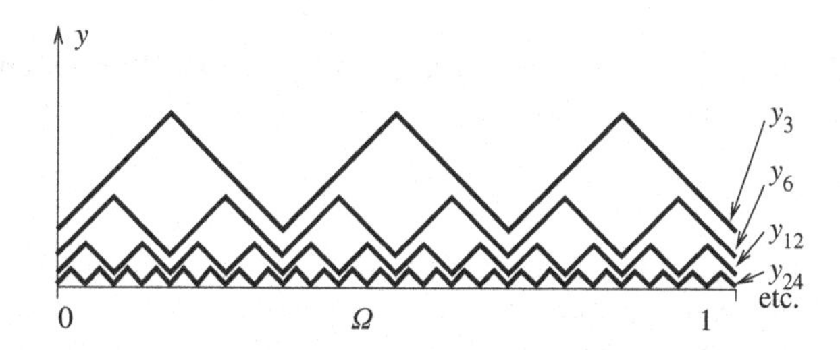

Fig. 2.2 A minimizing sequence for J from (2.3.29). The corresponding gradients exhibit faster and faster spatial oscillations. This explains also occurrence of the microstructure in some materials, cf. Fig. 4.6 on p. 146.

Remark 2.3.6 (Influence of boundary conditions). Existence/non-existence of solutions to minimization problems very much depends on boundary conditions. Indeed, consider

$$\left.\begin{aligned} &\text{minimize } J(y) = \int_0^1 \Big(\Big(\frac{\mathrm{d}y}{\mathrm{d}x}\Big)^2(x) - 1\Big)^2 + y^2(x)\,\mathrm{d}x, \\ &\text{subject to } y \in W^{1,4}(0,1),\ y(0) = y(1) = 0,\ -1 \le \frac{\mathrm{d}y}{\mathrm{d}x} \le 1. \end{aligned}\right\} \tag{2.3.30}$$

A slight modification of the minimizing sequence introduced in Example 2.3.5 shows that there is no minimizer of J but again $\inf J = 0$. On the other hand, changing the boundary condition at the right-end point suitably, the resulting problem

$$\left.\begin{aligned} &\text{minimize } J(y) = \int_0^1 \Big(\Big(\frac{\mathrm{d}y}{\mathrm{d}x}\Big)^2(x) - 1\Big)^2 + y^2(x)\,\mathrm{d}x, \\ &\text{subject to } y \in W^{1,4}(0,1),\ y(0) = 0,\ y(1) = 1,\ -1 \le \frac{\mathrm{d}y}{\mathrm{d}x} \le 1. \end{aligned}\right\} \tag{2.3.31}$$

[5] Recall the Bachmann-Landau notation meaning that $\limsup_{k\to\infty} k^2 J(y_k) < +\infty$.

clearly has a unique solution $y(x) = x$ for all $x \in (0,1)$. This example shows that convexity of J is not necessary for the existence of minimizers.

Exercise 2.3.7. Show that for an isotropic hyperelastic material $\varphi(x,F) = \varphi(x,FQ)$ for all $x \in \bar{\Omega}$, $Q \in \mathrm{SO}(d)$ and all $F \in \mathrm{GL}^+(d)$.

Exercise 2.3.8. Formulate necessary condition for a minimizer of $\mathcal{E}$ as in Theorem 2.3.3 but for $\mathcal{F}$ a linear functional.

2.4 Examples of hyperelastic materials

Here we mention a few important examples of hyperelastic materials. To ease our notation, except for the St.Venant-Kirchhoff material, we only define the densities for matrices of positive determinant. Otherwise, the densities are implicitly extended by infinity.

Geometrically nonlinear Lamé material (St.Venant-Kirchhoff material). The response function of the 1st Piola-Kirchhoff stress tensor is

$$\tilde{S}(F) := \lambda(\mathrm{tr}\,E)\mathbb{I} + 2GE, \tag{2.4.1}$$

where $E = (F^\top F - \mathbb{I})/2$, and $\lambda \geq -dG/2$ and $G > 0$, the latter one is called the *shear modulus*. The shear modulus G is often also denoted as μ, and called together with λ *Lamé constants.* [6] On the other hand, the *bulk modulus* K measures the resistance of the material to hydrostatic pressure, and is defined as $K := \lambda + 2/d\mu$.

Often, elasticity of such (isotropic) materials is described in terms of the *Young modulus* E_{Young} and the *Poisson ratio* ν, having the meaning of the stress arising by unit uni-axial elongation and the relative shrinkage of the cross-section diameter under such loading, when nonlinear effects due to large strain are neglected. The bulk modulus K and the Lamé constant G can then be expressed in terms of E_{Young} and ν we have

$$K := \frac{E_{\mathrm{Young}}}{3(1-2\nu)} \quad \text{and} \quad G := \frac{E_{\mathrm{Young}}}{2(1+\nu)}. \tag{2.4.2}$$

The physical dimension of λ, K, G, and the Young modulus E_{Young} is Pascal ($=\mathrm{J/m^3}=\mathrm{N/m^2}$) while the Poisson ratio is dimensional-less. The energy corresponding to the stress/strain response (2.4.1) is, for $\det F > 0$, expressed as

$$\varphi(F) := \frac{\lambda}{2}(\mathrm{tr}\,E)^2 + G|E|^2 \quad \text{with} \quad E = \frac{1}{2}(F^\top F - \mathbb{I}). \tag{2.4.3}$$

[6] Simultaneously, as used also in this book, μ very standardly denotes the chemical potential, henceforth we used the notation G instead of μ for this second Lamé constant.

Equivalently, up to an (unimportant) additive constant, it can be expressed in terms of the right Cauchy-Green tensor C instead of the Green-St. Venant tensor E as

$$\varphi(F) = -\frac{3\lambda+2G}{4}\mathrm{tr}\,C + \frac{\lambda+2G}{8}\mathrm{tr}\,C^2 + \frac{\lambda}{4}\mathrm{tr}\,\mathrm{Cof}\,C \quad \text{with} \quad C = F^\top F. \tag{2.4.4}$$

We can also consider a non-isotropic material described by a Saint Venant-Kirchhoff energy density of the form

$$\varphi_{\mathrm{SVK}}(F) := \frac{1}{8}(C-\mathbb{I}){:}\mathbb{C}(C-\mathbb{I}) \quad \text{with} \quad C = F^\top F. \tag{2.4.5}$$

Here $\mathbb{C}$ is the fourth-order tensor of elastic constants of the material and it is positive definite, so that $\varphi_{\mathrm{SVK}} \geq 0$ and $\varphi_{\mathrm{SVK}}(F) = 0$ if and only if F is a rotation matrix.

Compressible Mooney-Rivlin material [374, 442]. This material has a stored energy of the form

$$\varphi(x,F) := a(x)|F|^2 + b(x)|\mathrm{Cof}\,F|^2 + \gamma(\det F), \tag{2.4.6}$$

where $a,b > 0$ and $\gamma(\delta) = c_1\delta^2 - c_2\log\delta$, $c_1,c_2 > 0$. It can be shown [114, Thm. 4.10.2] that for $d = 3$

$$\varphi(F) := \frac{\lambda}{2}(\mathrm{tr}\,E)^2 + G|E|^2 + R(E) \quad \text{with} \quad E = \frac{1}{2}(C-\mathbb{I}), \tag{2.4.7}$$

where $R(E) = \mathscr{O}(|E|^3)$ for $E \to 0$, λ and G are the constants as in (2.4.3). Indeed, it is a matter of a tedious computation to show that, given λ and G, the following equations must be fulfilled by constants a,b,c_1, and c_2. In particular, $c_2 := (\lambda+2G)/2$, $2a+2b = G$, and $4b+4c_1 = \lambda$.

Compressible neo-Hookean material. This material, proposed in [442], has a stored energy of the form

$$\varphi(x,F) := a(x)|F|^2 + \gamma(\det F), \tag{2.4.8}$$

with the constants as for compressible Mooney-Rivlin materials.

Ogden material [403]. This material has a stored energy of the form (recall that $C = F^\top F$)

$$\varphi(x,F) := \sum\nolimits_{i=1}^{M} a_i(x)\mathrm{tr}\,C^{\gamma_i/2} + \sum\nolimits_{i=1}^{N} b_i(x)\mathrm{tr}\,(\mathrm{Cof}\,C)^{\delta_i/2} + \gamma(\det F), \tag{2.4.9}$$

and $a_i, b_i > 0$, $\lim_{\delta\to 0_+}\gamma(\delta) = +\infty$ with some $\gamma : (0,+\infty) \to (0,+\infty)$ convex.

Remark 2.4.1 (St.Venant-Kirchhoff material modified). One can think about a commutative *multiplicative decomposition* of the deformation gradient $F \in \mathrm{GL}^+(d)$ to the *isochoric and volumetric* parts as

$$F = F_{\rm ISO} F_{\rm VOL} = F_{\rm VOL} F_{\rm ISO} \quad \text{with} \quad F_{\rm VOL} = (\det F)^{1/d}\mathbb{I}. \tag{2.4.10}$$

Noteworthy, $F_{\rm VOL}$ records the mere volume variation while $F_{\rm ISO}$ is the corresponding volume-preserving "complement" because obviously $\det F_{\rm ISO} = \det(F_{\rm VOL}^{-1}F) = (\det F_{\rm VOL}^{-1})\det F = (\det F)^{-1}\det F = 1$. This is an analogue to the orthogonal deviatoric/spherical decomposition of the small-strain tensor, cf. (A.2.10). As an alternative to the St.Venant-Kirchhoff model (2.4.3), one can then devise the energy expressed in terms of the bulk and shear moduli as

$$\varphi(F) := \frac{d}{2}K\big|F_{\rm VOL} - \mathbb{I}\big|^2 + G\big|(\det F)^{1/d}U - \mathbb{I}\big|^2 \quad \text{where } U = \sqrt{F^\top F}; \tag{2.4.11}$$

note that U is the stretch tensor from the polar decomposition of F; cf. (1.1.4). Such φ is frame indifferent and, in contrast to (2.4.3), has an at most quadratic growth. This is a counterpart of the linear Lamé material at small strains, cf. (6.7.9) below. When still augmented by a term of the type $1/|F_{\rm VOL}|^q$ and a higher-gradient term, we can avoid local interpenetration, cf. Theorem 2.5.3.

2.5 Nonsimple hyperelastic materials

The concept of materials whose energy depends not only on basic quantities (as the deformation gradient F) but also on their gradients has been developing for long time, since the work by R.A. Toupin [516, 517] and R.D. Mindlin [369], under various names like *nonsimple materials* as e.g. in [203, 416, 482], *strain gradient materials* as e.g. in [307], or *multipolar materials*, see e.g. [228] (in particular fluids),[7] alternatively also referred as the concept of hyper- or couple-stresses [370, 418, 516]. Thermodynamical consistency of nonsimple materials has been studied much later in [105, 169]. A formal justification relies on a further expansion in (1.1.24) so that for the difference of images via y of points $x, \tilde{x} \in \Omega$ can be, for $1 \le i \le d$, estimated as[8]

$$y_i(\tilde{x}) = y_i(x) + \nabla y_i(x)(\tilde{x} - x) + \frac{1}{2}\nabla^2 y_i{:}(\tilde{x} - x)\otimes(\tilde{x} - x) \\ + \frac{1}{6}\nabla^3 y_i \,\vdots\, (\tilde{x} - x)\otimes(\tilde{x} - x)\otimes(\tilde{x} - x) + \cdots. \tag{2.5.1}$$

It also modifies the Cauchy Axiom because the stress vector may depend also on derivatives of the normal along the surface. In this book, we will consider only 2nd-grade nonsimple materials, i.e., second-order deformation gradients (= first-order

[7] In fluid mechanics, the gradient-dependence concept was pioneered by D.J. Korteweg [284] for describing capillarity effects.

[8] If $a, b, c \in \mathbb{R}^d$ then $a \otimes b \otimes c \in \mathbb{R}^{d \times d \times d}$ and $(a \otimes b \otimes c)_{ijk} := a_i b_j c_k$ for all $i, j, k \in \{1, \ldots, d\}$.

strain gradients) are involved. Engineering/physical literature often speaks about *strain gradient elasticity (material)* [302, 411]. In dynamical situations, this may offer a suitable tool to model a dispersion, cf. Remark 6.3.6 on p. 207. The main mathematical advantage of nonsimple materials is that higher-order deformation gradients bring additional regularity of deformations and, possibly, also compactness of the set of admissible deformations in a stronger topology. Moreover, the stored energy can be even convex in the highest derivatives of the deformation which is helpful in proving existence of minimizers.

The downside of this approach is that there are not many physically justified models of nonsimple materials and material constants are rarely available. Also, the stress principle of Euler and Cauchy must be revisited because now the Cauchy stress vector in (1.2.1a) does not depend only on the normal vector to the boundary of a subdomain of Ω^y but also on the curvature of this boundary, cf. Remark 1.2.1.

Remark 2.5.1 (Nonlocal-material and gradient-theory concepts). An worth historical reminiscence is due to G.A. Maugin [345] who wrote "....a step consists in giving up the standard Cauchy argument used to introduce the notion of stress, and then introducing a notion of nonlocality, weak or strong according to the author [339]. This yields the notions of gradient theories (a so-called *weak* nonlocality) and *truly nonlocal*theories[9] involving interactions at a rather long distance that replace strict contact interactions (see Eringen [176], for this aspect). Concerning this last point, stretching a little the historical perspective, the idea that the mechanical response at a material point may depend on a larger spatial domain than the immediate neighborhood of the considered point may be traced back to Duhem [165]. Within the framework of gradient theories, simple materials are described within a first-order gradient theory of displacement. A pioneering step towards higher-order gradient theories is the consideration, in a variational formulation, of the second gradient of the elastic displacement by Le Roux [310]."

Moreover, a derivation of a nonlocal model with higher order deformation gradients in one dimension can be found, for example, in [520]. Nonlocal models of beams are studied e.g. in [174].

2.5.1 Second-grade nonsimple materials

We assume that the stored energy density has an enhanced form

$$\varphi_{\rm E} = \varphi_{\rm E}(x, F, G) \tag{2.5.2}$$

with $F \in \mathbb{R}^{d\times d}$ being again a placeholder for ∇y while $G \in \mathbb{R}^{d\times d\times d}$ is a placeholder for the gradient of F, i.e. the Hessian $\nabla^2 y$ of y. This means that $\varphi_{\rm E} : \bar{\Omega} \times \mathbb{R}^{d\times d} \times$

[9] For the terminology weak vs. strong (truly) nonlocal we refer to D. Rogula [447] or also Z. Bažant and M. Jirásek [50].

$\mathbb{R}^{d\times d\times d} \to \mathbb{R}\cup\{+\infty\}$ such that the first Piola-Kirchhoff stress tensor S is determined by a smooth deformation $y : \bar{\Omega} \to \mathbb{R}^d$ is given as

$$S(x) = \partial_F \varphi_{\rm E}(x, \nabla y(x), \nabla^2 y(x)) - {\rm Div}\,\mathfrak{H}. \tag{2.5.3}$$

Here, $\mathfrak{H} := \partial_G \varphi_{\rm E}$ is a so-called *hyperstress*, cf. [196, 418], i.e., if $y : \bar{\Omega} \to \mathbb{R}^d$ is a smooth deformation then the hyperstress is defined in the reference configuration for $x \in \bar{\Omega}$ by

$$\mathfrak{H}(x) := \partial_G \varphi_{\rm E}(x, \nabla y(x), \nabla^2 y(x)). \tag{2.5.4}$$

More in detail, (2.5.3)–(2.5.4) reads componentwise for $i, j \in \{1,\ldots,d\}$ as

$$S_{ij}(x) := \partial_{F_{ij}} \varphi_{\rm E}(x, \nabla y(x), \nabla^2 y(x)) - \underbrace{\sum_{k=1}^{d} \partial_{x_k} \mathfrak{H}_{ijk}}_{=:\,[{\rm Div}\,\mathfrak{H}]_{ij}} \ \text{ with } \ \mathfrak{H}_{ijk} = \partial_{G_{ijk}} \varphi_{\rm E}(x, \nabla y(x), \nabla^2 y(x)).$$

We now define an energy

$$\mathcal{E}(y) := \int_\Omega \varphi_{\rm E}(x, \nabla y(x), \nabla^2 y(x)) - f(x)\cdot y(x)\,{\rm d}x - \int_{\Gamma_{\rm N}} g(x)\cdot y(x) + g_1(x)\cdot \partial_{\vec{n}} y(x)\,{\rm d}S\,, \tag{2.5.5}$$

where $g_1 : \Gamma_{\rm N} \to \mathbb{R}^d$ is the surface density of (hypertraction) forces balancing the hyperstress $\partial_G \varphi_{\rm E}(x, \nabla y(x), \nabla^2 y(x))$. The other terms containing f and g are volume and surface forces which appeared already in Theorem 2.3.1, for instance. Here we, however, assume for simplicity that f, g, and g_1 depend only on $x \in \Omega$ and $x \in \Gamma_{\rm N}$, respectively, and are independent of deformations. We also implicitly assume that g and g_1 are extended by zero to the whole Γ.

Let us suppose that y used in (2.5.5) for $f = g = g_1 = 0$ is followed by a rigid body rotation $x^y \mapsto Rx^y$ where $R \in {\rm SO}(d)$ and $x^y \in \Omega^y$. The composition of the two deformations is $\hat{y} : \bar{\Omega} \to \mathbb{R}^d$ given as $\hat{y}(x) := Ry(x)$ for every $x \in \bar{\Omega}$. Since $\hat{y}$ differs from y only by a rigid body rotation and also all applied forces are rotated, it must hold by the *Axiom of material frame-indifference* that $\mathcal{E}(\hat{y}) = \mathcal{E}(y)$. Consequently,

$$\begin{aligned}\int_\Omega \varphi_{\rm E}(x, \nabla\hat{y}(x), \nabla^2\hat{y}(x))\,{\rm d}x &= \int_\Omega \varphi_{\rm E}(x, R\nabla y(x), R\nabla^2 y(x))\,{\rm d}x \\ &= \int_\Omega \varphi_{\rm E}(x, \nabla y(x), \nabla^2 y(x))\,{\rm d}x. \end{aligned}\tag{2.5.6}$$

Localizing (2.5.6) at Lebesgue points and taking into account that $R \in {\rm SO}(d)$ was arbitrary, we arrive at the conclusion that, for almost all $x \in \Omega$, for all $R \in {\rm SO}(d)$, for all $F \in \mathbb{R}^{d\times d}$, and all $G \in \mathbb{R}^{d\times d\times d}$, it holds

$$\varphi_{\rm E}(x,RF,RG)=\varphi_{\rm E}(x,F,G)\,. \tag{2.5.7}$$

Here we implicitly extended $\varphi_{\rm E}(x,\cdot,G)$ by $+\infty$ to $(d\times d)$-matrices in $\mathbb{R}^{d\times d}$ with non-positive determinants. If $y:\bar\Omega\to\mathbb{R}^d$ is smooth we write for $x\in\Omega$ that $\nabla y(x)=R(x)U(x)$ where $R(x)\in{\rm SO}(d)$ and $U(x)\in\mathbb{R}^{d\times d}$ come from the polar decomposition of $\nabla y(x)$; see (1.1.4). Formula (2.5.7) says that if $y:\bar\Omega\to\mathbb{R}^d$ is smooth we can write for almost every $x\in\Omega$ and $R^\top(x)\in{\rm SO}(d)$

$$\varphi_{\rm E}(x,R^\top(x)R(x)U(x),R^\top(x)[\nabla(RU)](x))=\varphi_{\rm E}(x,\nabla y(x),\nabla^2y(x))\,. \tag{2.5.8}$$

The term $R^\top\nabla R$ which appears on the left-hand side of (2.5.8) can be expressed in terms of U and ∇U. Indeed, if $d=3$ and taking into account that $0={\rm curl}\,\nabla y={\rm curl}(RU)$ we calculate, exploiting the Levi-Civita symbol ε from (A.2.3), and also (A.3.32), that

$$0=({\rm curl}\,\nabla y)_{ij}=\sum_{k,l=1}^{3}\Big(\varepsilon_{jkl}\sum_{m=1}^{3}\partial_{x_k}(R_{im})U_{ml}+\varepsilon_{jkl}\sum_{m=1}^{3}R_{im}\partial_{x_k}U_{ml}\Big). \tag{2.5.9}$$

Multiplying (2.5.9) by $R^\top_{ni}$ and summing over i yields

$$0=\sum_{k,l=1}^{3}\Big(\varepsilon_{jkl}\sum_{i,m=1}^{3}R_{in}\partial_{x_k}(R_{im})U_{ml}+\varepsilon_{jkl}\partial_{x_k}U_{nl}\Big). \tag{2.5.10}$$

Tedious computations show (see [150, Formula (2.9)] and [194]) that

$$(R^\top\nabla R)_{nmk}=\sum_{i=1}^{3}R_{in}\partial_{x_k}R_{im}=\sum_{j=1}^{3}\varepsilon_{nmj}\Lambda_{jk} \tag{2.5.11}$$

where all indices run from one to three and[10]

$$\Lambda=\frac{1}{\det U}\Big(U({\rm curl}\;U)^\top U-\frac{1}{2}[U\!:\!{\rm curl}\;U]U\Big). \tag{2.5.12}$$

If $d=2$ we first embed ∇y, R, and U from the polar decomposition of ∇y to GL(3) by defining $R_{3i}=R_{i3}=(\nabla y)_{i3}=(\nabla y)_{3i}=U_{3i}=U_{i3}=\delta_{3i}$ where δ_{3i} is the Kronecker's symbol and $1\le i\le 3$. Then we continue as in the case $d=3$.

Hence, (2.5.7) implies that, for almost all $x\in\Omega$, it holds

$$\varphi_{\rm E}(x,\nabla y(x),\nabla^2y(x))=\hat\varphi_{\rm E}(x,C(x),\nabla C(x))\,, \tag{2.5.13}$$

where $C(x)=(\nabla y(x))^\top\nabla y(x)=U^2(x)$ is the right Cauchy-Green strain tensor defined in (1.1.26). Indeed, simple algebraic manipulations show that $\nabla U=\nabla C(\sqrt{C})^{-1}/2$ and $U=\sqrt{C}$.

[10] See (A.3.30) for a definition of curl U.

Similarly, as in the case of simple materials, we have the following formal derivation of Euler-Lagrange equations for minimizers of $\mathcal{E}$. Again, the approach is far from being rigorous because, in particular, we should compose deformations rather than to add them together. In contrast to the simple-material situation, here the smoothness of $\Gamma = \partial\Omega$ is important because the mean curvature of the boundary enters the equations. We also refer to (1.2.15) where the surface force density depends on the deformation gradient. This dependence is even more pronounced in models of nonsimple materials; see also (C.1.7).

Theorem 2.5.2 (Formal Euler-Lagrange equations for $\mathcal{E}$ from (2.5.5)**).** *Let $\Omega \subset \mathbb{R}^d$ be a bounded domain with a smooth boundary. Let $\Gamma :== \Gamma_{\rm N} \cup \Gamma_{\rm D}$ where $\Gamma_{\rm D}$ is closed and disjoint from $\Gamma_{\rm N}$ and let $\mathcal{A} := \{y \in C^4(\bar{\Omega};\mathbb{R}^d);\ \det\nabla y > 0$ in $\bar{\Omega}$ and $y = y_{\rm D}$ on $\Gamma_{\rm D}\}$. Then, any $y \in \mathcal{A}$ satisfying*

$$-\operatorname{div} S(x) = f(x) \qquad \text{for } x\in\Omega, \tag{2.5.14a}$$

$$S(x)\vec{n}(x) - \operatorname{div}_S(\mathfrak{H}(x)\vec{n}(x)) - 2\kappa(x)[\mathfrak{H}(x)\vec{n}(x)]\vec{n}(x) = g(x) \qquad \text{for } x\in\Gamma_{\rm N}, \tag{2.5.14b}$$

$$\mathfrak{H}(x)(\vec{n}(x)\otimes\vec{n}(x)) = g_1(x) \qquad \text{for } x\in\Gamma, \tag{2.5.14c}$$

$$y(x) = y_{\rm D}(x) \qquad \text{for } x\in\Gamma_{\rm D}, \tag{2.5.14d}$$

with S from (2.5.3) and $\mathfrak{H}$ from (2.5.4), satisfies also

$$\mathcal{E}'(y)v = 0 \tag{2.5.15}$$

for all smooth mappings $v : \bar{\Omega} \to \mathbb{R}^d$ vanishing on $\Gamma_{\rm D}$. Here $\vec{n}(x)$ is the outer unit normal to $\Gamma_{\rm N}$ at x and $\kappa = \frac{1}{2}\operatorname{div}_S\vec{n} : \Gamma_{\rm N} \to \mathbb{R}$ is the mean curvature of $\Gamma_{\rm N}$. Conversely, if $y \in \mathcal{A}$ satisfies (2.5.15) for all $v : \bar{\Omega} \to \mathbb{R}^d$ smooth and vanishing on $\Gamma_{\rm D}$, then (2.5.14) is satisfied, too. In particular, if $y \in \mathcal{A}$ and $\mathcal{E}(y) = \min_{\tilde{y}\in\mathcal{A}} \mathcal{E}(\tilde{y})$, then (2.5.15) and thus also (2.5.14) are satisfied.

Heuristic proof. Let us take $v \in C^4(\bar{\Omega};\mathbb{R}^d)$ such that $v(x) = 0$ for every $x \in \Gamma_{\rm D}$ and put $h(t) := \mathcal{E}(y + tv)$. If (2.5.15) holds, then $h'(0) = 0$ and consequently (see also (C.1.4b))

$$\begin{aligned}
0 &= \int_\Omega \partial_F\varphi_{\rm E}(x,\nabla y(x),\nabla^2 y(x)){:}\nabla v + \partial_G\varphi_{\rm E}(x,\nabla y(x),\nabla^2 y(x))\,\vdots\,\nabla^2 v\,{\rm d}x \\
&\quad - \int_\Omega f(x)\cdot v(x)\,{\rm d}x - \int_{\Gamma_{\rm N}} (g(x)\cdot y(x) + g_1(x)\cdot\partial_{\vec{n}}v(x))\,{\rm d}S \\
&= -\int_\Omega \operatorname{div}(\partial_F\varphi_{\rm E}(x,\nabla y(x),\nabla^2 y(x)))\cdot v\,{\rm d}x + \int_{\Gamma_{\rm N}} \partial_F\varphi_{\rm E}(x,\nabla y(x),\nabla^2 y(x))\vec{n}\cdot v\,{\rm d}S \\
&\quad - \int_\Omega \operatorname{div}(\partial_G\varphi_{\rm E}(x,\nabla y(x),\nabla^2 y(x))){:}\nabla v\,{\rm d}x + \int_\Gamma \partial_G\varphi_{\rm E}(x,\nabla y(x),\nabla^2 y(x))\vec{n}{:}\nabla v\,{\rm d}S \\
&\quad - \int_\Omega f(x)\cdot v(x)\,{\rm d}x - \int_{\Gamma_{\rm N}} (g(x)\cdot y(x) + g_1(x)\cdot\partial_{\vec{n}}v(x))\,{\rm d}S \\
&= -\int_\Omega \operatorname{div}(\partial_F\varphi_{\rm E}(x,\nabla y(x),\nabla^2 y(x)) - \operatorname{div}(\partial_G\varphi_{\rm E}(x,\nabla y(x),\nabla^2 y(x))))\cdot v(x)\,{\rm d}x
\end{aligned}$$

$$
\begin{aligned}
&+\int_{\Gamma_{\rm N}}[\partial_F\varphi_{\rm E}(x,\nabla y(x),\nabla^2 y(x))-{\rm div}(\partial_G\varphi_{\rm E}(x,\nabla y(x),\nabla^2 y(x)))]\vec{n}\cdot v\,{\rm d}S\\
&-\int_\Omega f(x)\cdot v(x)\,{\rm d}x-\int_{\Gamma_{\rm N}}(g(x)\cdot\hat{y}(x)+g_1(x)\cdot\partial_{\vec{n}}v(x))\,{\rm d}S\\
&+\int_\Gamma\partial_G\varphi_{\rm E}(x,\nabla y(x),\nabla^2 y(x))\vec{n}{:}\nabla v\,{\rm d}S\,. \qquad (2.5.16)
\end{aligned}
$$

The last term in (2.5.16) can be further manipulated using the surface Green formula, cf. (B.3.26), as follows

$$
\begin{aligned}
&\int_\Gamma\partial_G\varphi_{\rm E}(x,\nabla y(x),\nabla^2 y(x))\vec{n}{:}\nabla v\,{\rm d}S=\int_\Gamma\partial_G\varphi(x,\nabla y(x),\nabla^2 y(x)){:}(\vec{n}\otimes\vec{n})\cdot\partial_{\vec{n}}v(x)\\
&+\int_\Gamma\partial_G\varphi_{\rm E}(x,\nabla y(x),\nabla^2 y(x))\vec{n}{:}\nabla_{\rm S}v(x)\,{\rm d}S=\int_\Gamma\partial_G\varphi_{\rm E}(x,\nabla y(x),\nabla^2 y(x)){:}(\vec{n}\otimes\vec{n})\cdot\partial_{\vec{n}}v(x)\\
&-\int_{\Gamma_{\rm N}}\Big({\rm div}_{\rm S}(\partial_G\varphi_{\rm E}(x,\nabla y(x),\nabla^2 y(x))\vec{n})+2\kappa[\partial_G\varphi_{\rm E}(x,\nabla y(x),\nabla^2 y(x))\vec{n}]\vec{n}\Big)\cdot v(x)\,{\rm d}S\,.
\end{aligned}
\qquad (2.5.17)
$$

Putting together corresponding integrands and arbitrariness of v yield the result.

The converse argumentation holds, too. Moreover, if y is a minimizer of $\mathcal{E}$, then (2.5.15) certainly holds. □

The next theorem shows that the second deformation gradient together with suitable coercivity conditions on the determinant of the inverse gradient ensure positivity of the determinant of ∇y everywhere in $\bar{\Omega}$. As we will also have $y\in W^{1,d}(\Omega;\mathbb{R}^d)$ it implies in view of [190, Thm. 6.1] that y is *locally almost invertible* in Ω in the sense that for almost every $x_0\in\Omega$ there is $r>0$, an open set $\omega\subset\Omega$, and $z:B(y(x_0);r)\to\omega$ such that $z(y(x))=x$ for almost every $x\in\omega$.

Theorem 2.5.3 (Local invertibility, T.J. Healey and S. Krömer [245]). *Let $p>d$, $q>pd/(p-d)$, and*

$$
\mathcal{A}_c:=\Big\{y\in W^{2,p}(\Omega;\mathbb{R}^d):\ \|y\|_{W^{2,p}(\Omega;\mathbb{R}^d)}+\int_\Omega\frac{1}{(\det\nabla y(x))^q}\,{\rm d}x\le c,\ \det\nabla y>0\text{ a.e.in }\Omega\Big\}
$$

with c given so that $\mathcal{A}_c\ne\emptyset$. Then:

$$
\exists\varepsilon=\varepsilon(p,q,c,d)>0\ \ \forall y\in\mathcal{A}_c:\qquad \det\nabla y\ge\varepsilon\ \ \text{on }\bar{\Omega} \qquad (2.5.18)
$$

and y is locally almost invertible.

Proof. We have that $y\in W^{2,p}(\Omega;\mathbb{R}^d)$ with $p>d$, i.e., $\nabla y\in C^{0,\lambda}(\bar{\Omega};\mathbb{R}^{d\times d})$ for $\lambda=(p-d)/p$. As $\det\nabla y>0$ almost everywhere in Ω, we first show that $\det\nabla y>0$ on Γ and then that $\det\nabla y>0$ everywhere in Ω. This will imply that $\det\nabla y>0$ on $\bar{\Omega}$ and continuity of $x\mapsto\det\nabla y(x)$ for $x\in\bar{\Omega}$ implies the claim. Assume that there is $x_0\in\Gamma$

such that $\det\nabla y(x_0) = 0$. We can, without loss of generality, assume that $x_0 := 0$. Further, we estimate for every $x \in \Omega$

$$0 < \det\nabla y(x) = \det\nabla y(x) - \det\nabla y(0) \leq K|\nabla y(x) - \nabla y(0)| \leq \ell|x|^\lambda, \qquad (2.5.19)$$

where $K = K(\|\nabla y\|_{C(\bar\Omega;\mathbb{R}^{d\times d})})$ and $\ell > 0$ comes from the Hölder estimate of ∇y; cf. (A.1.3). The above inequality combines λ-Hölder continuity of ∇y and local Lipschitz continuity of $A \mapsto \det A$ for $A \in \mathbb{R}^{d\times d}$. Altogether, we see that if $|x| \leq r$ then $(\det\nabla y(x))^{-q} \geq \ell^{-q} r^{-\lambda q}$. We have for $r > 0$ small enough

$$\int_\Omega \frac{1}{(\det\nabla y(x))^q}\,\mathrm{d}x \geq \int_{B(0,r)\cap\Omega} \frac{1}{(\det\nabla y(x))^q}\,\mathrm{d}x \geq \frac{\tilde\ell r^d}{\ell^q r^{\lambda q}}, \qquad (2.5.20)$$

where $\tilde\ell r^d \leq \mathrm{meas}_d(B(0,r)\cap\Omega)$ and $\tilde\ell > 0$ is independent of r. Note that this is possible because Ω is Lipschitz and therefore it has the *cone property*, see Definition B.3.1. Passing to the limit for $r \to 0$ in (2.5.20), we see that $\int_\Omega (\det\nabla y(x))^{-q}\,\mathrm{d}x = +\infty$ because $\lambda q > d$. This, however, contradicts our assumptions and, consequently, $\det\nabla y > 0$ everywhere in Γ. An analogous argument shows that $\det\nabla y > 0$ everywhere in Ω. Altogether $\det\nabla y > 0$ in $\bar\Omega$.

It remains to show that the bound is uniform for every $y \in \mathcal{A}_c$. Assume that it is not the case, i.e., that for every $k \in \mathbb{N}$ there is $y_k \in \mathcal{A}_c$ and $x_k \in \Omega$ such that $\det\nabla y_k(x_k) < 1/k$. The uniform (with the same Hölder constant $\ell > 0$ for all $k \in \mathbb{N}$) Hölder continuity of $x \mapsto \det\nabla y_k(x)$ implies that $|\det\nabla y_k(x) - \det\nabla y_k(x_k)| \leq \ell|x - x_k|^\lambda$ and therefore $0 < \det\nabla y_k(x) \leq 1/k + \ell|x - x_k|^\lambda$. Take $0 < r_k$ so small that $B(x_k, r_k) \subset \Omega$ for all $k \in \mathbb{N}$. Then

$$\int_\Omega (\det\nabla y_k(x))^{-q}\,\mathrm{d}x \geq \int_{B(x_k,r_k)} (\det\nabla y_k(x))^{-q}\,\mathrm{d}x \geq \int_{B(x_k,r_k)} \Big(\frac{k}{1+\ell k r_k^\lambda}\Big)^q\,\mathrm{d}x.$$

The right-hand side is, however arbitrarily large if k is large enough and r_k is sufficiently small. The statement is proved because this contradicts $y_k \in \mathcal{A}_c$. □

It is clear from the proof of Theorem 2.5.3 that the assumptions can be weakened, namely only $\det\nabla y$ is to be controlled instead of the full deformation gradient ∇y:

Theorem 2.5.4 (Local invertibility under weaker assumptions). *Let $p > d$, $q > pd/(p-d)$, and*

$$\mathcal{A}_c := \Big\{ y \in W^{1,d}(\Omega;\mathbb{R}^d) : \|\det\nabla y\|_{W^{1,p}(\Omega)} + \int_\Omega \frac{1}{(\det\nabla y(x))^q}\mathrm{d}x \leq c,\ \det\nabla y > 0 \text{ a.e. in } \Omega \Big\}$$

with c given so that $\mathcal{A}_c \neq \emptyset$. Then:

$$\exists \varepsilon = \varepsilon(p,q,c,d) > 0\ \ \forall y \in \mathcal{A}_c: \qquad \det\nabla y \geq \varepsilon \ \text{ on } \bar\Omega \qquad (2.5.21)$$

and y is locally almost invertible.

Let us notice, however, that the energy density function of the Mooney-Rivlin material (2.4.6) grows more slowly to $+\infty$ near a noninvertible matrix than it is required in Theorems 2.5.3 and 2.5.4.

Remark 2.5.5 (Nonsimple isotropic materials). Similarly as in Section 2.2 we discuss restrictions on the material response if we additionally assume that the material is isotropic; see also [384]. Let $\Omega \subset \mathbb{R}^d$ be a ball centered at the origin. Take $R \in \mathrm{SO}(d)$ arbitrary. The idea of isotropy means that if we perform a mechanical experiment on Ω or on $R\Omega$ the results should be equivalent. In particular, the energy content stored in the specimen by applying a deformation $y : \bar{\Omega} \to \mathbb{R}^d$ or $x \mapsto y(Rx) : \bar{\Omega} \to \mathbb{R}^d$ must be the same. Hence,

$$\int_\Omega \varphi_{\mathrm{E}}(x, \nabla y(x), \nabla^2 y(x))\,\mathrm{d}x = \int_{R^{-1}\Omega} \varphi_{\mathrm{E}}(Rx, \nabla[y(Rx)], \nabla^2[y(Rx)])\,\mathrm{d}x$$
$$= \int_\Omega \varphi_{\mathrm{E}}(x, \nabla y(x)R, \nabla^2 y(x)RR)\,\mathrm{d}x. \tag{2.5.22}$$

Assuming further that (2.5.22) can be localized and taking into account that $R \in \mathrm{SO}(d)$ was arbitrary, we arrive at the conclusion that, for all $R \in \mathrm{SO}(d)$, all $F \in \mathbb{R}^{d\times d}$, all $G \in \mathbb{R}^{d\times d\times d}$, and almost all $x \in \Omega$ it holds

$$\varphi_{\mathrm{E}}(x, FR, GRR) = \varphi_{\mathrm{E}}(x, F, G). \tag{2.5.23}$$

Here we implicitly extended $\varphi_{\mathrm{E}}(x, \cdot, G)$ by $+\infty$ to $(d{\times}d)$-matrices in $\mathbb{R}^{d\times d}$ with non-positive determinants.

Example 2.5.6 (Linear 2nd-grade term). A frame-indifferent stored energy, i.e. complying with (2.5.7), is

$$\varphi_{\mathrm{E}}(F,G) := \varphi(F) + \frac{1}{2} \sum_{i,j,k,l,m=1}^{d} \mathbb{H}_{jklm} G_{ijk} G_{ilm}$$
$$= \varphi(F) + \frac{1}{2} \sum_{i=1}^{d} \mathbb{H} G_i : G_i \qquad \Big(\text{with } G_i := (G_{ikl})_{k,l=1}^d\Big) \tag{2.5.24a}$$
$$= \varphi(F) + \frac{1}{2} \mathbb{H} G \,\vdots\, G \qquad \Big(\text{with } [\mathbb{H}G]_{ijk} := \sum_{l,m=1}^{d} \mathbb{H}_{jklm} G_{ilm}\Big) \tag{2.5.24b}$$

where $\varphi \geq 0$ is frame indifferent, i.e., $\varphi(F) = \varphi(RF)$ for every $F \in \mathbb{R}^{d\times d}$ and every $R \in \mathrm{SO}(d)$. Introducing the notation

$$\mathrm{SLin}(\mathbb{R}^{d\times d}_{\mathrm{sym}}) := \{\mathbb{C} \in \mathbb{R}^{(d\times d)\times(d\times d)} : \ \mathbb{C}_{ijkl} = \mathbb{C}_{klij} = \mathbb{C}_{jikl}\}, \tag{2.5.25}$$

we also assume that $\mathbb{H} \in \mathrm{SLin}(\mathbb{R}^{d\times d}_{\mathrm{sym}})$ and positive definite tensor, i.e., $\sum_{k,l,m,n=1}^d \mathbb{H}_{klmn} A_{kl} A_{mn} > 0$ for every $0 \neq A \in \mathbb{R}^{d\times d}_{\mathrm{sym}}$; note that the Hessians of y_i, i.e.

$G_i = [\partial^2_{x_j x_k} y_i]_{j,k=1}^d$ are symmetric, hence valued in $\mathbb{R}^{d\times d}_{\rm sym}$. Here we used two quite standard conventions, the former (2.5.24a) using a symmetrical notation while the later (2.5.24b) using the analog of the standard matrix-like notation of the type "Ax" with understanding $\mathbb{H}$ as an operator[11] $\mathbb{R}^{d\times d}_{\rm sym} \to \mathbb{R}^{d\times d}_{\rm sym}$. Note that, with any $G\in\mathbb{R}^{d\times d\times d}$ with $G_i \in \mathbb{R}^{d\times d}_{\rm sym}$ for $i = 1,...,d$, we have

$$\forall R\in{\rm SO}(d): \qquad \sum_{i,j,k,l,m,n,p=1}^{d} R_{ij}G_{jkl}\mathbb{H}_{klmn}R_{ip}G_{pmn} = \sum_{i,k,l,m,n=1}^{d} G_{ikl}\mathbb{H}_{klmn}G_{imn}\,, \tag{2.5.26}$$

which implies the frame indifference. For example, taking $\mathbb{H}_{klmn} := \delta_{kl}\delta_{mn}$ yields $\sum_{i,k,l,m,n=1}^d \mathbb{H}_{klmn}\frac{\partial^2 y_i}{\partial x_k\partial x_l}\frac{\partial^2 y_i}{\partial x_m\partial x_n} = \sum_{i=1}^d |\Delta y_i|^2$ while $\mathbb{H}_{klmn} = \delta_{km}\delta_{ln}$ yields $\sum_{i,k,l,m,n=1}^d \mathbb{H}_{klmn}\frac{\partial^2 y_i}{\partial x_k\partial x_l}\frac{\partial^2 y_i}{\partial x_m\partial x_n} = \sum_{i=1}^d |\nabla^2 y_i|^2$; here δ denotes Kronecker's symbol.

Exercise 2.5.7. Express the stored energy density defined in (2.5.24) in terms of the Cauchy-Green tensor C and its gradient.

Exercise 2.5.8. Replace the $\mathcal{A}_c$ in Theorem 2.5.4 by

$$\Big\{y\in W^{1,d}(\Omega;\mathbb{R}^d):\ \|\det\nabla y\|_{W^{1,r}(\Omega)} + \int_\Omega h(\det\nabla y(x))\,{\rm d}x \le c,\ \det\nabla y > 0 \text{ a.e.in } \Omega\Big\},$$

where $h : (0,+\infty)\to(0,+\infty)$ is a decreasing continuous function. Specify how h should behave near the origin so that the conclusion of Theorem 2.5.4 still holds?

2.5.2 *Nonlocal simple or nonsimple materials*

Alternative models exhibiting compactifying properties can be based on the compact embedding $H^{1+\gamma}(\Omega) \subset W^{1,p}(\Omega)$ or, for the purpose of Theorem 2.5.3, on $H^{2+\gamma}(\Omega) \subset W^{2,p}(\Omega)$. Note that we have confined the discussion to the Hilbert-type Sobolev-Slobodetskiĭ spaces $H^{1+\gamma}$ and $H^{2+\gamma}$, so that the quadratic energy functionals can be used to result in linear terms in corresponding Euler-Lagrange equations, being particularly useful in evolutionary problems in Part II. We refer the mentioned cases as nonlocal simple and nonlocal nonsimple materials, respectively. The concept of nonlocal materials may advantageously serve to model dispersion of elastic waves, cf. Remark 6.3.7 on p. 208, and it is even physically motivated by dispersion, see e.g. [219].

[11] See (A.2.33) for the definition of $\vdots$. Notice also that this we can require without any loss of generality because $G_i\otimes G_i$ is already symmetric in kl and mn for every i admissible. Here $(G_i\otimes G_i)_{klmn} := G_{ikl}G_{imn}$ for all indices ranging $\{1,\dots,d\}$.

In the former case, we may rely on the stored energy augmented by the nonlocal term but still involving only the first deformation gradient, i.e. the simple-material concept. More specifically, the extended stored energy is then considered as[12]

$$y \mapsto \int_\Omega \varphi(\nabla y)\,\mathrm{d}x + \mathscr{H}(\nabla y) \quad \text{with}$$
$$\mathscr{H}(F) := \frac{1}{4}\int_{\Omega\times\Omega} \mathbb{K}(x,\tilde{x})(F(x)-F(\tilde{x})):(F(x)-F(\tilde{x}))\,\mathrm{d}x\mathrm{d}\tilde{x} \tag{2.5.27}$$

with a suitable singular symmetric positive-semidefinite kernel $\mathbb{K}:\Omega\times\Omega\to \mathrm{SLin}(\mathbb{R}^{d\times d}_{\mathrm{sym}})$, including also $\mathbb{K}(x,\tilde{x}) = \mathbb{K}(\tilde{x},x)$. For $\mathscr{H}(F) < +\infty$, $(x,\tilde{x}) \mapsto \mathbb{K}(x,\tilde{x})(F(x)-F(\tilde{x})):(F(x)-F(\tilde{x})) \in L^1(\Omega\times\Omega)$ and, by Fubini's theorem, it means that (2.5.27) can be written as

$$y \mapsto \int_\Omega\bigg(\varphi(\nabla y(x)) + \frac{1}{4}\int_\Omega \mathbb{K}(x,\tilde{x})(\nabla y(x)-\nabla y(\tilde{x})):(\nabla y(x)-\nabla y(\tilde{x}))\,\mathrm{d}\tilde{x}\bigg)\mathrm{d}x.$$

It again allows for treating non-quasiconvex φ like the 2nd-grade nonsimple concept from Sect. 2.5.1.

Models with scalar-valued kernel, i.e. in our notation $\mathbb{K}_{ijkl}(x,\tilde{x}) = \frac{1}{4}\kappa(x,\tilde{x})\delta_{ik}\delta_{jl}$ for some $\kappa:\Omega\times\Omega\to[0,+\infty)$, have been suggested by R.C. Rogers [446] and for small-strain elasticity by A.C. Eringen [176, Chap.6]. An "academical-type" example is the kernel $\mathbb{K}_{ijkl}(x,\tilde{x}) = \frac{1}{4}\delta_{ik}\delta_{jl}/|x-\tilde{x}|^{d+2\gamma}$ with $0<\gamma<1$, which gives the standard Gagliardo seminorm in the *Sobolev-Slobodetskiĭ space* $H^{1+\gamma}(\Omega;\mathbb{R}^d)$, cf. (B.3.23) on p. 519. Let us emphasize that, considering the Frobenius matrix norm $|\cdot|$, the functional (2.5.27) is frame indifferent if φ is so. Indeed, it follows from that the rotation by a matrix $R\in\mathrm{SO}(d)$ preserves distances used in the nonlocal term, namely:

$$|R\nabla y(x) - R\nabla y(\tilde{x})|^2 = |\nabla y(x) - \nabla y(\tilde{x})|^2.$$

For a general kernel $\mathbb{K}$, we will assume that, for some $\exists\varepsilon>0$, it holds

$$\forall x,\tilde{x}\in\Omega,\ F\in\mathbb{R}^{d\times d}: \quad \Big(\frac{\varepsilon|F|^2}{|x-\tilde{x}|^{d+2\gamma}} - \frac{1}{\varepsilon}\Big)^+ \le \mathbb{K}(x,\tilde{x})F{:}F \le \frac{|F|^2}{\varepsilon|x-\tilde{x}|^{d+2\gamma}}. \tag{2.5.28}$$

Here $(\cdot)^+ = \max(0,\cdot)$ so that we admitted (perhaps more realistically) that this nonlocal term can be effective only in short-range interactions with only a bounded radius ρ if $\mathbb{K}(x,\tilde{x}) = 0$ for $|x-\tilde{x}|>\rho$. If $-\frac{1}{\varepsilon}$ were omitted, the quadratic functional $\mathscr{H}$ from (2.5.27) is equivalent to the Gagliardo seminorm $|\cdot|_{\gamma,2}$ on $H^\gamma(\Omega;\mathbb{R}^{d\times d})$, while otherwise it induces the same topology because only the asymptotic behavior in the neighborhood of the diagonal in $\bar{\Omega}\times\bar{\Omega}$ is important.

The Gâteaux derivative $\mathscr{H}'$ of the quadratic functional $\mathscr{H}$ from (2.5.27) determines a linear functional on $H^\gamma(\Omega;\mathbb{R}^{d\times d})$, i.e. an element of $H^\gamma(\Omega;\mathbb{R}^{d\times d})^*$. In general

[12] See (A.2.30) and (A.2.31) for the definition the dot product if 4th-order tensors are involved. Here, we used the convention (2.5.24b).

$H^\gamma(\Omega) \not\subset C(\Omega)$, so that $H^\gamma(\Omega)^*$ contains distributions on Ω not necessarily living in $L^1(\Omega)$. Yet here we have automatically a certain regularity:

Lemma 2.5.9 (Nonlocal stress). *Let $\mathbb{K} : \Omega\times\Omega \to \mathrm{SLin}(\mathbb{R}^{d\times d}_{\rm sym})$ be symmetric in the sense $\mathbb{K}(x,\tilde{x}) = \mathbb{K}(\tilde{x},x)$ for almost all $(x,\tilde{x}) \in \Omega\times\Omega$ and assume that it satisfies (2.5.28). Then $\mathscr{H} : H^\gamma(\Omega;\mathbb{R}^{d\times d}) \to \mathbb{R}$ is Gâteaux differentiable and $\mathscr{H}'(F) \in L^1(\Omega)$ for $F \in H^\gamma(\Omega;\mathbb{R}^{d\times d})$, having a representation*

$$[\mathfrak{S}(F)](x) = \int_\Omega \mathbb{K}(x,\tilde{x})(F(x) - F(\tilde{x}))\,\mathrm{d}\tilde{x}. \tag{2.5.29}$$

Proof. This follows from the calculations of the Gâteaux derivative

$$\begin{aligned}
\langle \mathscr{H}'(F), \tilde{F}\rangle &\overset{\substack{\text{“scalar pro-}\\ \text{duct on } H^\gamma\text{”}}}{=} \frac{1}{2}\int_{\Omega\times\Omega} \mathbb{K}(x,\tilde{x})(F(x) - F(\tilde{x})){:}(\tilde{F}(x) - \tilde{F}(\tilde{x}))\,\mathrm{d}\tilde{x}\mathrm{d}x \\
&\overset{\substack{\text{symmetry}\\ \text{of } \mathbb{K}}}{=} \int_{\Omega\times\Omega} \mathbb{K}(x,\tilde{x})(F(x) - F(\tilde{x})){:}\tilde{F}(x)\,\mathrm{d}\tilde{x}\mathrm{d}x \\
&\overset{\substack{\text{Fubini}\\ \text{theorem}}}{=} \int_\Omega \underbrace{\Big(\int_\Omega \mathbb{K}(x,\tilde{x})(F(x) - F(\tilde{x}))\,\mathrm{d}\tilde{x}\Big)}_{= [\mathfrak{S}(F)](x)}{:}\tilde{F}(x)\,\mathrm{d}x,
\end{aligned} \tag{2.5.30}$$

where F here denotes the field $\Omega \to \mathbb{R}^{d\times d}$. For the scalar product in the case of the Riesz kernel see (B.3.24). The usage of the Fubini theorem B.2.12 is legitimate whenever $(x,\tilde{x}) \mapsto \mathbb{K}(x,\tilde{x})(F(x) - F(\tilde{x})){:}(\tilde{F}(x) - \tilde{F}(\tilde{x})) \in L^1(\Omega\times\Omega)$. In particular, it yields (2.5.29) with $\mathfrak{S}(F) \in L^1(\Omega;\mathbb{R}^{d\times d})$. Simultaneously, $\mathfrak{S}(F)$ is a representation of a functional from $H^\gamma(\Omega;\mathbb{R}^{d\times d})^*$. $\square$

The advantage of this nonlocal regularization is that it leads to boundary-value problems like in (1.2.22) only with the stress augmented by a nonlocal term, i.e.

$$[S(\nabla y)](x) = \varphi'(\nabla y(x)) + \int_\Omega \mathbb{K}(x,\tilde{x})(\nabla y(x) - \nabla y(\tilde{x}))\,\mathrm{d}\tilde{x}. \tag{2.5.31}$$

One can think about a generalization, using the kernel F-dependent, i.e. $\mathfrak{K} : \Omega\times\Omega\times\mathbb{R}^{d\times d} \to \mathrm{SLin}(\mathbb{R}^{d\times d}_{\rm sym})$, and then $\mathscr{H}$ in (2.5.27) is generalized as

$$\mathscr{H}(F) := \frac{1}{4}\int_{\Omega\times\Omega} \mathfrak{K}(x,\tilde{x},F(x){-}F(\tilde{x}))(F(x){-}F(\tilde{x})){:}(F(x){-}F(\tilde{x}))\,\mathrm{d}x\mathrm{d}\tilde{x}. \tag{2.5.32}$$

This possibly nonquadratic variant would lead to a nonlinear nonlocal simple material. This can control y in $W^{\gamma,p}(\Omega;\mathbb{R}^d)$. In the Healey-Krömer's Theorem 2.5.3, we need in fact the compact embedding $W^{2,p}(\Omega) \subset C^{1,\alpha}(\Omega)$. Now, it would suffice to have $W^{\gamma,p}(\Omega) \subset C^\alpha(\Omega)$ compactly. It allows us to modify the Theorem 2.5.3:

Corollary 2.5.10 (Local invertibility in nonlinear nonlocal materials). *Consider the functional* $\mathcal{E}: y \mapsto \int_\Omega \varphi(\nabla y)\,\mathrm{d}x + \mathscr{H}(\nabla y) : W^{1+\gamma,p}(\Omega;\mathbb{R}^d) \to \mathbb{R}\cup\{+\infty\}$ *for some* $d/p < \gamma < 1$ *which is coercive on* $W^{1+\gamma,p}(\Omega;\mathbb{R}^d) \to \mathbb{R}\cup\{+\infty\}$ *and with* φ *satisfying* $\varphi(F) \geq 1/(\det F)^q$ *for* $q > pd/(\gamma p - d)$ *and every* $F \in \mathrm{GL}^+(d)$. *Then for every* $c \in \mathbb{R}$ *such that there is* $y \in W^{1+\gamma,p}(\Omega;\mathbb{R}^d)$ *with* $\mathcal{E}(y) \leq c$, *there is* $\varepsilon > 0$ *satisfying*

$$\det(\nabla y) \geq \varepsilon \quad \text{on } \bar{\Omega} \tag{2.5.33}$$

and y *is locally almost invertible.*

Proof. As $y \in W^{1+\gamma,p}(\Omega;\mathbb{R}^d)$ it follows that $\nabla y \in W^{\gamma,p}(\Omega;\mathbb{R}^{d\times d})$. Theorem B.3.16 then implies that $\nabla y \in C^{0,\alpha}(\bar{\Omega};\mathbb{R}^{d\times d})$ where $\alpha = \gamma - d/p$. Then $q > pd/(\gamma p - d)$ allows us to finish the proof as in the case of Theorem 2.5.3. □

In the nonlocal nonsimple case, the stored energy (2.5.24) can be modified in the spirit of (2.5.27). In this way, $\varphi(\nabla y)$ can be augmented by the nonlocal integral term acting on $\nabla^2 y$ instead of ∇y again with $0 < \gamma < 1$ and using the hyperelastic-moduli symmetric positive-semidefinite kernel $\mathfrak{K}: \Omega\times\Omega \to \mathrm{SLin}(\mathbb{R}^{d\times d}_{\mathrm{sym}})$:

$$y \mapsto \int_\Omega \varphi(\nabla y)\,\mathrm{d}x + \mathscr{H}(\nabla^2 y) = \int_\Omega \bigg(\varphi(\nabla y(x)) + \frac{1}{4}\int_\Omega \mathfrak{K}(x,\tilde{x})(\nabla^2 y(x) - \nabla^2 y(\tilde{x}))\,\vdots\,(\nabla^2 y(x) - \nabla^2 y(\tilde{x}))\,\mathrm{d}\tilde{x}\bigg)\mathrm{d}x. \tag{2.5.34}$$

Relying on the assumed symmetry of $\mathfrak{K}$ including also $\mathfrak{K}(x,\tilde{x}) = \mathfrak{K}(\tilde{x},x)$, we write the 1st-order conditions for critical points of (2.5.34) in the weak form

$$\begin{aligned} 0 &= \int_\Omega \bigg(\partial\varphi(\nabla y(x)):\nabla\tilde{y}(x) + \frac{1}{2}\int_\Omega \mathfrak{K}(x,\tilde{x})(\nabla^2 y(x) - \nabla^2 y(\tilde{x}))\,\vdots\,(\nabla^2\tilde{y}(x) - \nabla^2\tilde{y}(\tilde{x}))\,\mathrm{d}\tilde{x}\bigg)\mathrm{d}x \\ &= \int_\Omega \bigg(\partial\varphi(\nabla y):\nabla\tilde{y} + \int_\Omega \mathfrak{K}(x,\tilde{x})(\nabla^2 y(x) - \nabla^2 y(\tilde{x}))\mathrm{d}\tilde{x}\,\vdots\,\nabla^2\tilde{y}(x)\bigg)\mathrm{d}x \end{aligned} \tag{2.5.35}$$

for any $\tilde{y} \in H^{2+\gamma}(\Omega;\mathbb{R}^d)$, where for the latter equality we used the Fubini theorem B.2.12. Like in Lemma 2.5.9 applied now on the G- instead of F-field, it reveals that (2.5.34) gives rise to the *nonlocal hyperstress* $\mathfrak{H} = \mathfrak{H}(\nabla^2 y)$ given by

$$\mathfrak{H}(x) = \int_\Omega \mathfrak{K}(x,\tilde{x})(\nabla^2 y(x) - \nabla^2 y(\tilde{x}))\,\mathrm{d}\tilde{x}; \tag{2.5.36}$$

here again the convention (2.5.24b) and (A.2.32) has been used.

Like in (2.5.28), we will assume that, for some $\varepsilon > 0$, it holds

$$\forall x,\tilde{x}\in\Omega,\ A\in\mathbb{R}^{d\times d}_{\mathrm{sym}}: \quad \left(\frac{\varepsilon|A|^2}{|x-\tilde{x}|^{d+2\gamma}} - \frac{1}{\varepsilon}\right)^+ \leq \mathfrak{K}(x,\tilde{x})A:A \leq \frac{|A|^2}{\varepsilon|x-\tilde{x}|^{d+2\gamma}}; \tag{2.5.37}$$

here, A is a placeholder for G_i as used in (2.5.24a), cf. also (A.2.33).

Corollary 2.5.11 (Local invertibility for nonlocal nonsimple materials). *Consider the functional $\mathcal{E}(y) := \int_\Omega \varphi(\nabla y)\,\mathrm{d}x + \mathscr{H}(\nabla^2 y)$ as in (2.5.34) with (2.5.37) satisfied for some $\gamma > d/2-1$ and with φ satisfying $\varphi(F) \geq 1/(\det F)^q$ for $q > 2d/(2\gamma{+}2{-}d)$. Then, for any $c \in \mathbb{R}$, such that there is $y \in H^{2+\gamma}(\Omega;\mathbb{R}^d)$ with $\mathcal{E}(y) < c$ there is $\varepsilon > 0$ satisfying* (2.5.33) *and y is locally almost invertible.*

Sketch of the proof. We have the embedding[13] $H^{2+\gamma}(\Omega) \subset W^{2,p}(\Omega)$ if $p < 2d/(d{-}2\gamma)$. To facilitate usage of Theorem 2.5.3, we need also $p > d$. Therefore, we need $2d/(d{-}2\gamma) > d$ to allow for an existence of p satisfying both these bounds. This results in the condition $\gamma > d/2-1$. The exponent p does not explicitly occur in the statement of this corollary and plays an auxiliary role only. The strategy is to use p as big as possible, i.e. $p \nearrow 2d/(d{-}2\gamma)$, because this allows for $q \geq pd/(p{-}d)$ as low as possible. After some algebra, it reveals the limit value $2d/(2\gamma{+}2{-}d)$ for q, as specified in the statement of this Corollary. □

Note that for two-dimensional problems we need only $\gamma > 0$ to apply this Corollary, which is "nearly" the conventional linear 2nd-order nonsimple material model working with $\gamma = 0$. Yet, independently of the dimension, if γ is very close to the bound $\gamma > d/2-1$, then $q \to +\infty$ so that the growth of φ for $\det F \to 0_+$ must be very steep.

An analog for Theorem 2.5.4 uses $\mathscr{H}$ with a positive-definite matrix-valued kernel $K : \Omega\times\Omega \to \mathbb{R}^{d\times d}_{\mathrm{sym}}$ with $K(x,\tilde{x}) = K(\tilde{x},x)$. Realizing the calculus $\nabla(\det F(x)) = (\det F(x))' {:} \nabla F(x) = \mathrm{Cof}\, F(x) {:} \nabla F(x)$, cf. (A.2.22), we can consider the functional[14]

$$\mathcal{E}(y) := \int_\Omega \varphi(\nabla y)\,\mathrm{d}x + \mathscr{H}(\mathrm{Cof}\,\nabla y {:} \nabla^2 y) \quad \text{with}$$
$$\mathscr{H}(v) = \frac{1}{4}\int_{\Omega\times\Omega} (v(x){-}v(\tilde{x}))\cdot K(x,\tilde{x})(v(x){-}v(\tilde{x}))\,\mathrm{d}x\mathrm{d}\tilde{x}. \qquad (2.5.38)$$

Corollary 2.5.12 (Bound on the determinant under still weaker assumptions). *Let $\mathcal{E}$ be from (2.5.38) with K satisfying (2.5.28) with $\gamma > d/2-1$ and with $\mathbb{R}^{d\times d}$ and $\mathbb{K}$ replaced by $\mathbb{R}^d$ and K, respectively, and let φ satisfy again $\varphi(F) \geq 1/(\det F)^q$ for $q > 2d/(2\gamma{+}2{-}d)$. Then for any $c \in \mathbb{R}$ and any $y \in W^{1,d}(\Omega;\mathbb{R}^d)$ with $\det\nabla y \in H^{1+\gamma}(\Omega)$ such that $\mathcal{E}(y) < c$ there is $\varepsilon > 0$ such that (2.5.33) holds and y is locally almost invertible.*

Exercise 2.5.13. Derive the weak formulation of the Euler-Lagrange equation resulting from minimizing the energy (2.5.38) and identify the stress.

---

[13] Intuitively, this embedding can be seen when applying the embedding $H^\gamma(\Omega) \subset L^p(\Omega)$ on $\nabla^2 y$ and, using the Sobolev-embedding theorem B.3.4 on p. 515 (which is formulated for $\gamma = 1,2,...$) also for $0 < \gamma < 1$.

[14] The functional (2.5.38) yields the hyperstress $\mathfrak{H} = \mathfrak{H}(\nabla y, \nabla^2 y)$ given by $\mathfrak{H}_{ijk}(x) = [\mathrm{Cof}\,\nabla y(x)]_{ij} \int_\Omega K_{lk}(x,\tilde{x})(\partial_{x_l}\det\nabla y(x) - \partial_{x_l}\det\nabla y(\tilde{x}))\,\mathrm{d}\tilde{x}$ for $i,j,k \in \{1,\ldots,d\}$. Now the $\mathscr{H}$-term contributes also to the stress tensor, computation of this contribution being rather tedious however and we leave it as an exercise for the interested reader.

Chapter 3
Polyconvex materials: existence of energy-minimizing deformations

Nothing takes place in the world whose meaning is not that of some maximum or minimum.

LEONHARD PAUL EULER (1707–1783)

As already mentioned below Theorem 2.3.1, the derivation of Euler-Lagrange equations for the elasticity functional $\mathcal{E}$ is only formal and we cannot rely on the fact that minimizers of $\mathcal{E}$ satisfy these equations. Interestingly, this problem already appears in one-dimensional examples due to J.M. Ball and V.J. Mizel [39] who showed that the problem

$$\left.\begin{array}{l} \text{minimize} \quad J(y) := \displaystyle\int_0^1 (x^2 - y^3(x))^2 (y'(x))^{14} + \varepsilon (y'(x))^2 \,\mathrm{d}x \\ \text{subject to} \quad y(0) = 0 \quad \text{and} \quad y(1) = k \end{array}\right\} \tag{3.0.1}$$

for $\varepsilon = 0$ and $0 < k \leq 1$ has the (unique) solution $y(x) = \min(x^{2/3}, k)$ for $x \in [0,1]$ exhibiting a singularity/nonsmoothness of y at zero which persists even for $\varepsilon > 0$ small enough. Hence, the minimizer is not a solution to the corresponding Euler-Lagrange equation. Consequently, we resort to finding minimizers of elastic-energy functionals in place of solving Euler-Lagrange equations to obtain equilibrium configurations of the elastic body.[1]

The aim of this chapter is to show existence of minimizers for functionals measuring the elastic energy of a body. We will distinguish a few situations. First, in Section 3.4, we show the existence for pure displacement and displacement-traction problems, i.e., when we prescribe Dirichlet boundary conditions and surface and body forces acting on the specimen. We also include the so-called Ciarlet-Nečas condition ensuring almost-everywhere invertibility of minimizers.

The main tool will be the *direct method* of the calculus of variations, an approach, proposed by David Hilbert around the year 1900, to show (in a non-constructive way but without any auxiliary approximation) the existence of a solution to a minimization problem.

---

[1] We refer to [23] for an approach ensuring uniqueness of an elastic equilibrium.

M. Kružík and T. Roubíček, *Mathematical Methods in Continuum Mechanics of Solids*, Interaction of Mechanics and Mathematics,
https://doi.org/10.1007/978-3-030-02065-1_3

3.1 Polyconvex stored energy functions

In 1977, John Ball in his seminal paper [26] defined a new notion of convexity called polyconvexity. As we will see later, this notion allows us to prove existence of minimizers of our energy functional. If $F \in \mathbb{R}^{d\times d}$ we denote $\mathsf{M}(F)$ a vector of all *minors* of the matrix F, i.e., $\mathsf{M}(F) := (F, \det F)$ if $d = 2$ or $\mathsf{M}(F) = (F, \mathrm{Cof}\, F, \det F)$ for $d = 3$. For a general $d \in \mathbb{N}$, the number of components of $\mathsf{M}(F)$ is $\binom{2d}{d} - 1$.

Definition 3.1.1 (J.M. Ball's polyconvexity). [2] We say that $\varphi : \mathbb{R}^{d\times d} \to \mathbb{R}\cup\{+\infty\}$ is *polyconvex* if there exists a convex and continuous function $\tilde{\varphi} : \mathbb{R}^{\binom{2d}{d}-1} \to \mathbb{R}\cup\{+\infty\}$ such that

$$\varphi(F) = \tilde{\varphi}(\mathsf{M}(F)). \tag{3.1.1}$$

An elastic material obeying a polyconvex stored energy function will be called polyconvex.

It is clear that convex functions are polyconvex. On the other hand, $F \mapsto \det F$, $F \in \mathbb{R}^{d\times d}$ is an example of a function which is not convex but it is polyconvex if $d \geq 2$. Hence, polyconvexity really generalizes the notion of convexity. See [135] for polyconvex functions defined on general, possibly non-square, matrices.

Now we derive an interesting property of polyconvex functions, namely the so-called rank-one convexity which plays a crucial role in the calculus of variations and mathematical elasticity.

Take $A \in \mathbb{R}^{d\times d}$ and $a, b \in \mathbb{R}^d$. Consider a function $\alpha : \mathbb{R} \to \mathbb{R}$, $\alpha(t) := \det(A + ta\otimes b)$. We claim that for all $t \in \mathbb{R}$: $\alpha''(0) = 0$. First notice that if $r, s \in C^2(\mathbb{R})$ then $(rs)'' = r''s + 2r's' + rs''$. We can write

$$\alpha(t) = \det(A + ta\otimes b) = \sum_{i=1}^{d} (A + ta\otimes b)_{i1}[\mathrm{Cof}(A + ta\otimes b)]_{i1}. \tag{3.1.2}$$

Let us fix $i \in \{1,2,3\}$ and set $r(t) := (A + ta\otimes b)_{i1}$ and $s(t) := [\mathrm{Cof}(A + ta\otimes b)]_{i1}$. We immediately see that $r''(t) = 0$, $s''(t) = \mathrm{Cof}(ta\otimes b)_{i1} = 0$ because the rank of $a\otimes b$ is at most one, so that every subdeterminant of the order two must be inevitably zero. Finally, we calculate that $r'(0)s'(0) = \sum_{i=1}^{d} (a\otimes b)_{i1} \frac{\mathrm{d}}{\mathrm{d}t}[\mathrm{Cof}(A + ta\otimes b)]_{i1}|_{t=0} = 0$ in view of (A.2.34) for $d = 3$. If $d = 2$ the calculation is trivial. Altogether, we get that α is affine. Consequently, if $A, B \in \mathbb{R}^{d\times d}$ fulfill $\mathrm{rank}(A - B) \leq 1$ (or equivalently that $\exists a, b \in \mathbb{R}^d$: $A - B = a\otimes b$) then for all $0 \leq \lambda \leq 1$

$$\det(\lambda A + (1-\lambda)B) = \lambda \det A + (1-\lambda)\det B. \tag{3.1.3}$$

[2] Interestingly, already Morrey observed in [376, Thm. 5.3] that if we take $\tilde{\varphi}$ a positively one-homogeneous convex function of minors then $y \mapsto \int_\Omega \tilde{\varphi}(\mathsf{M}(\nabla y(x)))\,\mathrm{d}x$ is weak* lower semicontinuous on $W^{1,\infty}(\Omega;\mathbb{R}^d)$.

An analogous result holds for the cofactor because it is a matrix of 2×2 subdeterminants, i.e.,

$$\mathrm{Cof}(\lambda A + (1-\lambda)B) = \lambda \mathrm{Cof}\, A + (1-\lambda)\mathrm{Cof}\, B\,. \tag{3.1.4}$$

Assuming that $\varphi : \mathbb{R}^{d\times d} \to \mathbb{R}\cup\{+\infty\}$ is polyconvex we get, for every $0 \le \lambda \le 1$ and every $A, B \in \mathbb{R}^{d\times d}$ such that $\mathrm{rank}(A-B) \le 1$ that

$$\varphi(\lambda A + (1-\lambda)B) \le \lambda\varphi(A) + (1-\lambda)\varphi(B)\,. \tag{3.1.5}$$

This property is called *rank-one convexity* of φ. We just showed that polyconvexity implies rank-one convexity.

Definition 3.1.2 (Rank-one convexity). We say that $\varphi : \mathbb{R}^{d\times d} \to \mathbb{R}\cup\{+\infty\}$ is rank-one convex if (3.1.5) holds for every $0 \le \lambda \le 1$ and every $A, B \in \mathbb{R}^{d\times d}$ such that $\mathrm{rank}(A-B) \le 1$.

Rank-one convexity plays an important role in the vectorial calculus of variations. Although it looks very similar to usual convexity it has some surprising properties. In particular, there is a significant difference between between rank-one convex functions which can take the value $+\infty$ and the finite ones. Indeed, assume that $\varphi : \mathbb{R}^{d\times d} \to [0;+\infty]$ is such that

$$\varphi(F) := \begin{cases} 0 & \text{if } F \in \{A_1, A_2, A_3, A_4\}, \\ +\infty & \text{otherwise,} \end{cases} \tag{3.1.6}$$

where $A_i, A_j \in \mathbb{R}^{d\times d}$ are not pairwise *rank-one connected*, i.e., $\mathrm{rank}(A_i - A_j) > 1$ whenever $i \ne j$ and $1 \le i \le 4$ and $1 \le j \le 4$. Let us notice that $\varphi(F) = 0$ if and only if $F \in \{A_1, A_2, A_3, A_4\}$ and otherwise $\varphi(F) > 0$. It is easy to verify that φ is rank-one convex. The question which we can ask is the following: having four matrices $A_1, \ldots, A_4 \in \mathbb{R}^{d\times d}$ which are not pairwise rank-one connected, is it always possible to construct a finite rank-one convex function $\varphi : \mathbb{R}^{d\times d} \to [0;+\infty)$ such that $\varphi(F) = 0$ if and only if $F \in \{A_1, A_2, A_3, A_4\}$? Surprisingly, the answer is negative, as it was already observed in [22] and later e.g. in [108, 135, 338, 381, 504].

Example 3.1.3 (T_4 configuration). Consider four diagonal matrices in the plane: $A_1 := \mathrm{diag}(-1,-3)$, $A_2 := \mathrm{diag}(-3,1)$, $A_3 = -A_1$ and $A_4 = -A_2$. We identify diagonal matrices with points (x_1, x_2) in the plane via the map $A := \mathrm{diag}(x_1, x_2)$. Two diagonal matrices are rank-one connected only if they are on the same horizontal or a vertical line. Besides matrices A_i for $1 \le i \le 4$ we have four auxiliary diagonal matrices B, C, D, E in the corners of the square in Figure 3.1. We easily verify the following inequalities exploiting rank-one convexity of φ:

$$\varphi(B) \le \frac{1}{2}\varphi(A_1) + \frac{1}{2}\varphi(C)\,,$$

$$\frac{1}{2}\varphi(C) \le \frac{1}{4}\varphi(A_2) + \frac{1}{4}\varphi(D)\,,$$

$$\frac{1}{4}\varphi(D) \le \frac{1}{8}\varphi(A_3) + \frac{1}{8}\varphi(E),$$

$$\frac{1}{8}\varphi(E) \le \frac{1}{16}\varphi(A_4) + \frac{1}{16}\varphi(B).$$

This together implies that

$$\frac{15}{16}\varphi(B) \le \frac{1}{2}\varphi(A_1) + \frac{1}{4}\varphi(A_2) + \frac{1}{8}\varphi(A_3) + \frac{1}{16}\varphi(A_4). \tag{3.1.7}$$

In other words, $\varphi(B) = 0$. Analogously, we get that $\varphi(C) = \varphi(D) = \varphi(E) = 0$. Consequently, any nonnegative rank-one convex function $\varphi : \mathbb{R}^{2\times 2} \to \mathbb{R}$ which is zero at A_i for $1 \le i \le 4$ defined above must be inevitably zero also at the closure of the square BCDE as well as on the closed segments A_1B, A_2C, A_3D, and A_4D; cf. Figure 3.1.

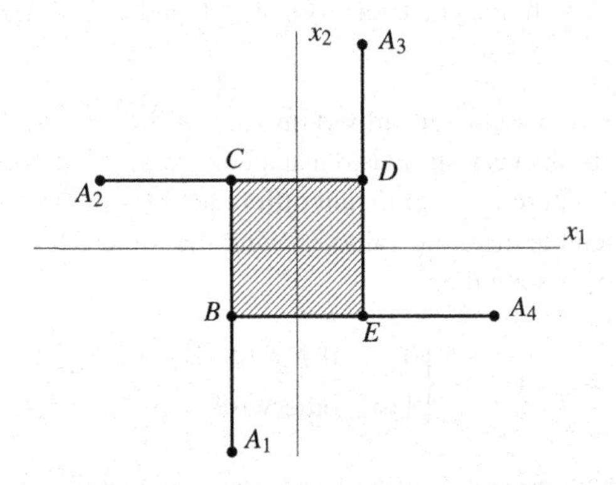

Fig. 3.1 The T_4 configuration of 2×2 diagonal matrices.

As matrices of rank at most one can be identified with the set $\{a\otimes b;\, a,b \in \mathbb{R}^d\}$ we see that $\varphi : \mathbb{R}^{d\times d}$ is rank-one convex if and only if $t \mapsto \varphi(A + ta\otimes b)$ is convex for every $A \in \mathbb{R}^{d\times d}$ and every $a, b \in \mathbb{R}^d$ as a function of a real variable t. Due to arbitrariness of A, a, and b this can be rephrased as convexity of $t \mapsto \varphi(A + ta\otimes b)$ at the point $t = 0$. If $\varphi \in C^2(\mathbb{R}^{d\times d})$, its rank-one convexity can be equivalently expressed as the so-called *Legendre-Hadamard condition*

$$\forall A \in \mathbb{R}^{d\times d},\ a,b \in \mathbb{R}^d :\ \partial^2_{A_{ij}A_{kl}}\varphi(A)a_ib_ja_kb_l \ge 0. \tag{3.1.8}$$

We end this section by noticing that the St.Venant-Kirchhoff energy density (2.4.3) is not polyconvex. Interestingly, it is used in many engineering calculations mainly because it allows for the explicit involvement of standard elastic moduli.

Proposition 3.1.4. *If $d \ge 2$, φ given by (2.4.3) is not polyconvex.*

Proof. The contradiction argument we are going to follow can be found e.g. in [135]. We show that φ is not even rank-one convex, so it cannot be polyconvex.

We perform the proof for $d = 3$, the modification for $d = 2$ being straightforward. Take $\varepsilon > 0$ and two families of matrices $F_\varepsilon = \varepsilon\mathbb{I}$ and $\tilde{F}_\varepsilon = \varepsilon\,\mathrm{diag}(1,1,3)$. We observe that $F_\varepsilon - \tilde{F}_\varepsilon$ is a rank-one matrix for $\varepsilon > 0$. For $\varepsilon > 0$ but small enough

$$\varphi\Big(\frac{1}{2}(F_\varepsilon + \tilde{F}_\varepsilon)\Big) > \frac{1}{2}\varphi(F_\varepsilon) + \frac{1}{2}\varphi(\tilde{F}_\varepsilon), \tag{3.1.9}$$

i.e., φ is not even rank-one convex. Hence, it cannot be polyconvex. □

Exercise 3.1.5. Express (3.1.8) for $A \mapsto \det A$ and for $A \mapsto \mathrm{Cof}\,A$. Use (3.1.8) to characterize rank-one convexity of the function $\varphi : \mathbb{R}^{d\times d} \to \mathbb{R}$ given as $\varphi(F) := \mathbb{A}_{ijkl}F_{ij}F_{kl}$ for some $\mathbb{A} \in \mathbb{R}^{(d\times d)^2}$.

3.2 Weak continuity of minors

Mappings $y \mapsto \mathsf{M}(\nabla y)$ with $\mathsf{M}(\nabla y)$ denoting all minors of ∇y for $y \in W^{1,p}(\Omega;\mathbb{R}^d)$ play an important role in the variational calculus because of the surprising property which is studied in Theorems 3.2.1 and 3.2.2 and Exercise 3.2.6. If $d = 2$ the only nonlinear minor of a matrix F is the determinant, $\det F$, because the cofactor of F has (up to the minus sign in front of the off-diagonal elements) the same entries as F itself, see (A.2.21). If $d = 3$ then the nonlinear minors are only $\mathrm{Cof}\,F$ and $\det F$. The following two classical results are formulated for $d = 3$, the two-dimensional case is left as an exercise to the interested reader.

Theorem 3.2.1 (Weak convergence of cofactors [114]). *Let $\Omega \subset \mathbb{R}^3$ be a bounded Lipschitz domain. Then for all $p \geq 2$ the mapping $y : W^{1,p}(\Omega;\mathbb{R}^3) \to \mathrm{Cof}\,\nabla y \in L^{p/2}(\Omega;\mathbb{R}^{3\times3})$ is well defined and continuous. Further, let $y_k \to y$ weakly in $W^{1,p}(\Omega;\mathbb{R}^3)$ and $\mathrm{Cof}\,\nabla y_k \to A$ weakly in $L^q(\Omega;\mathbb{R}^{3\times3})$ for some $q \geq 1$. Then $A = \mathrm{Cof}\,\nabla y$.*

Proof. The good sense and continuity of the mapping in question follows by Hölder's inequality. Take $y \in C^2(\bar{\Omega};\mathbb{R}^3)$ then

$$(\mathrm{Cof}\,\nabla y)_{ij} = \partial_{x_{i+2}}\big(y_{j+2}\partial_{x_{i+1}}y_{j+1}\big) - \partial_{x_{i+1}}\big(y_{j+2}\partial_{x_{i+2}}y_{j+1}\big),$$

(here we do not sum over repeated indices and we count them modulo three).

Taking, $v \in C_0^\infty(\Omega)$ we get (no summation)

$$\int_\Omega (\mathrm{Cof}\,\nabla y)_{ij} v\,\mathrm{d}x = -\int_\Omega y_{j+2}\partial_{x_{i+1}}y_{j+1}\partial_{x_{i+2}}v\,\mathrm{d}x + \int_\Omega y_{j+2}\partial_{x_{i+2}}y_{j+1}\partial_{x_{i+1}}v\,\mathrm{d}x. \tag{3.2.1}$$

Both sides of the above identity are continuous in $C^2(\bar{\Omega};\mathbb{R}^3)$ equipped with the $W^{1,p}(\Omega;\mathbb{R}^3)$-norm if v is fixed. Indeed, e.g.

$$\left|\int_\Omega (\mathrm{Cof}\,\nabla y)_{ij} v\,\mathrm{d}x\right| \le \left\|(\mathrm{Cof}\,\nabla y)_{ij}\right\|_{L^1(\Omega;\mathbb{R}^{3\times3})}\left\|v\right\|_{L^\infty(\Omega)} \le K(v)\left\|y\right\|_{W^{1,p}(\Omega;\mathbb{R}^3)},$$

where $K(v) > 0$ is a constant depending on v. We recall that $C^2(\bar{\Omega};\mathbb{R}^3)$ is dense in $W^{1,p}(\Omega;\mathbb{R}^3)$. Thus, (3.2.1) remains true in $W^{1,p}(\Omega;\mathbb{R}^3)$, $p \ge 2$. Due to the compact embedding of $W^{1,p}(\Omega;\mathbb{R}^3)$ to $L^r(\Omega;\mathbb{R}^3)$ if $1 \le r < p^{*}$[3] we can take $r < p^*$ and simultaneously $r^{-1} + p^{-1} \le 1$. Then we have that $y_k \to y$ strongly in $L^r(\Omega;\mathbb{R}^3)$ and hence, for example,

$$\int_\Omega \underbrace{[y_k]_{j+2}}_{\text{strongly}} \underbrace{\partial_{x_{i+1}}[y_k]_{j+1}}_{\text{weakly}} \partial_{x_{i+2}} v\,\mathrm{d}x \to \int_\Omega y_{j+2}\partial_{x_{i+1}} y_{j+1}\partial_{x_{i+2}} v\,\mathrm{d}x.$$

In other words, observing (3.2.1), we get

$$\lim_{k\to\infty}\int_\Omega ((\mathrm{Cof}\,\nabla y_k)_{ij} - (\mathrm{Cof}\,\nabla y)_{ij})v\,\mathrm{d}x = 0$$

and by our assumption $A = \mathrm{Cof}\,\nabla y$. □

Theorem 3.2.2 (Weak convergence of determinants [114]). *Let $\Omega \subset \mathbb{R}^3$ be a bounded Lipschitz domain. For any $p \ge 2$ and any $q \ge p/(p-1)$ the mapping $W^{1,p}(\Omega;\mathbb{R}^3)\times L^q(\Omega;\mathbb{R}^{3\times3}) \to L^s(\Omega)$, $1/s = 1/p + 1/q$, given by (summation over j)*

$$(y, \mathrm{Cof}\,\nabla y) \mapsto \det\nabla y = \partial_{x_j} y_1 (\mathrm{Cof}\,\nabla y)_{1j}$$

is well defined and continuous. Moreover, if $y_k \to y$ weakly in $W^{1,p}(\Omega;\mathbb{R}^3)$, $\mathrm{Cof}\,\nabla y_k \to A$ in $L^q(\Omega;\mathbb{R}^{3\times3})$ and $\det\nabla y_k \to \delta$ in $L^r(\Omega)$, $r \ge 1$ then $A = \mathrm{Cof}\,\nabla y$ and $\delta = \det\nabla y$.

Proof. We follow [114]. The convergence result for cofactors follows from Theorem 3.2.1. The continuity of the mapping follows again by Hölder's inequality. Using the Piola identity (cf. Lemma 1.1.15) we have that for, $y \in C^2(\bar{\Omega};\mathbb{R}^3)$, it holds that

$$\partial_{x_j} y_1 (\mathrm{Cof}\,\nabla y)_{1j} = \det\nabla y.$$

Thus, for any $v \in C_0^\infty(\Omega)$, we have

$$\int_\Omega \partial_{x_j} y_1 (\mathrm{Cof}\,\nabla y)_{1j} v\,\mathrm{d}x = -\int_\Omega y_1 (\mathrm{Cof}\,\nabla y)_{1j}\partial_{x_j} v\,\mathrm{d}x. \tag{3.2.2}$$

If $p \ge 3$ we proceed similarly as in the proof of Theorem 3.2.1 because $y \mapsto \int_\Omega \partial_{x_j} y_1 (\mathrm{Cof}\,\nabla y)_{1j} v\,\mathrm{d}x$ is continuous with respect to the norm of $W^{1,p}(\Omega;\mathbb{R}^3)$. It remains to prove the case $p \in [2,3)$ where the continuity does not hold.

Notice that the bilinear form $W^{1,p}(\Omega;\mathbb{R}^3)\times L^{p'}(\Omega;\mathbb{R}^{3\times3}) \to \mathbb{R}$ defined through

$$(y, A) \mapsto \int_\Omega \partial_{x_j} y_1 A_{1j} v\,\mathrm{d}x$$

[3] Recall that $p^* = 3p/(3-p)$ if $p < 3$, or $p^* < +\infty$ if $p \ge 3$.

is continuous if $p' = p/(p-1)$. However,

$$\int_\Omega \partial_{x_j} y_1 A_{1j} v\,\mathrm{d}x = -\int_\Omega y_1 A_{1j} \partial_{x_j} v\,\mathrm{d}x \tag{3.2.3}$$

does not generally hold unless $\partial_{x_j} A_{1j} = 0$ for smooth A_{1j}. But this is true for the cofactor as $\operatorname{div}\operatorname{Cof}\nabla y = 0$ if $y \in C^2(\bar\Omega;\mathbb{R}^3)$. Therefore, $\int_\Omega (\operatorname{Cof}\nabla y)_{1j}\partial_{x_j} v\,\mathrm{d}x = 0$ for any $v \in C_0^\infty(\Omega)$. Similarly as before we see that for every $y \in C^2(\bar\Omega;\mathbb{R}^3)$

$$y \mapsto \int_\Omega (\operatorname{Cof}\nabla y)_{1j}\partial_{x_j} v\,\mathrm{d}x$$

is continuous with respect to the $W^{1,p}(\Omega;\mathbb{R}^3)$-norm if $p \ge 2$. Subsequently, the density argument implies that $\int_\Omega (\operatorname{Cof}\nabla y)_{1j}\partial_{x_j} v\,\mathrm{d}x = 0$ for any $y \in W^{1,p}(\Omega;\mathbb{R}^3)$ and any $v \in C_0^\infty(\Omega)$.

Having $w \in L^{p'}(\Omega;\mathbb{R}^3)$ satisfying

$$\int_\Omega w_j \partial_{x_j} v\,\mathrm{d}x = 0 \tag{3.2.4}$$

for all $v \in C_0^\infty(\Omega)$ we get that, for every $z \in W^{1,p}(\Omega)$, it holds

$$-\int_\Omega z w_j \partial_{x_j} v\,\mathrm{d}x = \int_\Omega (\partial_{x_j} z) w_j v\,\mathrm{d}x. \tag{3.2.5}$$

For fixed w and v, the above relation is linear and continuous in z, so that it is sufficient to consider $z \in C^\infty(\bar\Omega)$ because of the density argument. Then $zv \in C_0^\infty$ and we see that (3.2.5) is implied by (3.2.4). Putting $z := y_1$ and $w = (\operatorname{Cof}\nabla y)_{1j}$ we get that (3.2.3) holds for $A := \operatorname{Cof}\nabla y$. Using the same (strong,weak)-convergence argument as in Thm. 3.2.1 we have, for $1/\tilde r + 1/p' \le 1$, that

$$\int_\Omega [y_k]_1 (\operatorname{Cof}\nabla y_k)_{1j}\partial_{x_j} v\,\mathrm{d}x \to \int_\Omega y_1 (\operatorname{Cof}\nabla y)_{1j}\partial_{x_j} v\,\mathrm{d}x.$$

This holds if $\tilde r < p^* = 3p/(3-p)$, i.e. in our case now $2 \le p < 3$. Hence $\lim_{k\to\infty}\int_\Omega \det\nabla y_k v\,\mathrm{d}x = \int_\Omega \det\nabla y v\,\mathrm{d}x$ for all $v \in C_0^\infty(\Omega)$.[4] □

A special case of Theorem 3.2.2 and Exercise 3.2.6 is the following corollary.

Corollary 3.2.3. *Let $\Omega \subset \mathbb{R}^d$ be a bounded Lipschitz domain. Let $p > d$ and let $y_k \to y$ weakly in $W^{1,p}(\Omega;\mathbb{R}^d)$ as $k \to \infty$. Then* $\det\nabla y_k \to \det\nabla y$ *weakly in $L^{p/d}(\Omega)$.*

---

[4] We showed that weak convergence of $y_k \to y$ in $W^{1,p}(\Omega;\mathbb{R}^3)$, $p \ge 2$ and of $\operatorname{Cof}\nabla y_k \to \operatorname{Cof}\nabla y$ in $L^q(\Omega;\mathbb{R}^{3\times 3})$ results in the convergence of $\det\nabla y_k$ to $\det\nabla y$ in the sense of distributions. This is an example of the so-called compensated compactness studied by F. Murat and L. Tartar; cf. [385, 503] for a survey and references therein. Namely, notice that if $y \in W^{1,p}(\Omega;\mathbb{R}^3)$ is smooth then $\det\nabla y = \frac{1}{3}(\nabla y)_{ij}(\operatorname{Cof}\nabla y)_{ij}$, $\operatorname{div}\operatorname{Cof}\nabla y = 0$ due to the Piola identity (1.1.19) and $\operatorname{curl}\nabla y = 0$.

The conclusion of Corollary 3.2.3 is not valid if $p = d$. To see this, we define $\omega := \{x \in B(0,1) : x_d < 0\}$ and find $y \in W_0^{1,d}(B(0,1);\mathbb{R}^d)$ such that $\int_\omega \det \nabla y(x)\,\mathrm{d}x < 0$, and extend it by zero to the whole $\mathbb{R}^d$. Let us define $y_k(x) := y(kx)$ for $k \in \mathbb{N}$. This implies that $\nabla y_k = k\nabla y(kx)$ and $y_k \to 0$ weakly in $W^{1,d}(B(0,1);\mathbb{R}^d)$. This convergence is even in measure. However, we have

$$\int_\omega \det \nabla y_k(x)\,\mathrm{d}x = k^d \int_\omega \det \nabla y(kx)\,\mathrm{d}x$$
$$= \int_{\{z \in B(0,k):\, z_d < 0\}} \det \nabla y(z)\,\mathrm{d}z = \int_\omega \det \nabla y(x)\,\mathrm{d}x < 0. \tag{3.2.6}$$

Consequently, $\lim_{k\to\infty} \int_\omega \det \nabla y_k(x)\,\mathrm{d}x < 0$. On the other hand, as $\int_{B(0,1)} \det \nabla y(x)\,\mathrm{d}x = 0$ because $y = 0$ on $\partial B(0,1)$ in the sense of traces, we immediately get

$$\lim_{k\to\infty} \int_{B(0,1)\setminus\omega} \det \nabla y_k(x)\,\mathrm{d}x = -\int_\omega \det \nabla y(x)\,\mathrm{d}x > 0.$$

Altogether, it yields $\lim_{k\to\infty} \int_{B(0,1)} \det \nabla y_k(x)\,\mathrm{d}x = 0$.

We have just shown that for a particular sequence and a particular domain we can get all possible relations between the limit of the sequence and the value at the weak limit. In general, the sequential weak continuity of $y \mapsto \int_\omega \det \nabla y(x)\,\mathrm{d}x$ may break down for $p = d$ *only* if $\{|\nabla y_k|^d\}_{k\in\mathbb{N}} \subset L^1(\omega)$ concentrates[5] at the boundary $\partial\omega$. Another example can be found in [333] where the author constructs a counterexample showing that that $y \mapsto \int_\Omega \det \nabla y(x)\,\mathrm{d}x$ is not weakly continuous even along a sequence of diffeomorphisms.

However, if we assume a pointwise constraint $\det \nabla y_k \geq 0$ almost everywhere in Ω for all $k \in \mathbb{N}$ the situation is much different. The surprising observation that non-negativity of $\det \nabla y_k$ improves compactness of $\{\det \nabla y_k\} \subset L^1(\Omega)$ is due to S. Müller [378, 379]. He even proved that if $y \in W^{1,d}(\Omega;\mathbb{R}^d)$ and $\det \nabla y \geq 0$ almost everywhere in Ω then $(\det \nabla y)\ln(2 + \det \nabla y) \in L^1(\omega)$ for every compact $\omega \subset \Omega$. Moreover, if $\det \nabla y \geq 0$ almost everywhere in Ω then

$$\left\|(\det \nabla y)\ln(2 + \det \nabla y)\right\|_{L^1(\omega)} \leq C(\omega, \|y\|_{W^{1,d}(\Omega;\mathbb{R}^d)}) \tag{3.2.7}$$

for some constant $C(\omega, \|y\|_{W^{1,d}(\Omega;\mathbb{R}^d)}) > 0$ depending only on ω and the Sobolev norm of y. Employing parametrized measures, we obtain the following two propositions which show that superlinear integrability of sequences of nonnegative determinants.

Proposition 3.2.4 (Superlinear integrability of determinant I). *Let $\Omega \subset \mathbb{R}^d$ be a bounded Lipschitz domain. Let $\{y_k\}_{k\in\mathbb{N}} \subset W^{1,d}(\Omega;\mathbb{R}^d)$ be such that $y_k \to y$ weakly in $W^{1,d}(\Omega;\mathbb{R}^d)$ as $k \to \infty$. If $y_k = y$ on Γ or y_k has periodic boundary conditions for all $k \in \mathbb{N}$ then for all $g \in C(\bar{\Omega})$*

[5] It means that $\{|\nabla y_k|^d\}$ weakly* converges to a measure whose support intersects with $\partial\omega$.

$$\lim_{k\to\infty}\int_\Omega g(x)\det\nabla y_k(x)\,\mathrm{d}x = \int_\Omega g(x)\det\nabla y(x)\,\mathrm{d}x. \tag{3.2.8}$$

If additionally for all $k\in\mathbb{N}$ either $\det\nabla y_k\ge 0$ or $\det\nabla y_k\le 0$ almost everywhere in Ω then (3.2.8) holds for all $g\in L^\infty(\Omega)$, i.e., $\det\nabla y_k\to\det\nabla y$ weakly in $L^1(\Omega)$.

Proof. Take a (non relabeled) subsequence of $\{\nabla y_k\}$ which generates $(\sigma,\tilde\nu)\in\mathcal{GDM}^p_{\mathrm{S}}(\Omega;\mathbb{R}^{d\times d})$ for $p:=d$. As $F\mapsto\det F$ is positively d-homogeneous we can write (D.2.3) for $v(F):=\det F$ and $z_k:=\nabla y_k$ for all $k\in\mathbb{N}$ to get for all $g\in C(\bar\Omega)$

$$\begin{aligned}\lim_{k\to\infty}\int_\Omega g(x)\det\nabla y_k(x)\,\mathrm{d}x &= \int_{\bar\Omega}\int_{\beta_{\mathrm{S}}\mathbb{R}^{d\times d}}\frac{\det F}{1+|F|^d}\tilde\nu_x(\mathrm{d}F)g(x)\sigma(\mathrm{d}x)\\ &= \int_\Omega g(x)\det\nabla y(x)\,\mathrm{d}x.\end{aligned} \tag{3.2.9}$$

The last equality follows from the fact that $F\mapsto\pm\det F$ is polyconvex and therefore (D.2.5b) and (D.2.5c) hold as equalities. As the result holds for an arbitrary subsequence of $\{\nabla y_k\}$ we get (3.2.8). If $\det\nabla y_k\ge 0$ or $\det\nabla y_k\le 0$ we get the L^1-weak convergence in view of Lemma D.2.4. □

Proposition 3.2.5 (Superlinear integrability of determinant II). *Let $\Omega\subset\mathbb{R}^d$ be a bounded Lipschitz domain. Let $\{y_k\}_{k\in\mathbb{N}}\subset W^{1,d}(\Omega;\mathbb{R}^d)$ be such that $y_k\to y$ weakly in $W^{1,d}(\Omega;\mathbb{R}^d)$ as $k\to\infty$. Then for every $\varepsilon>0$ there is a measurable set $\Omega\supset\Omega_\varepsilon\supset\{x\in\Omega:\mathrm{dist}(x,\Gamma)>\varepsilon\}$ such that for all $g\in C(\bar\Omega_\varepsilon)$*

$$\lim_{k\to\infty}\int_{\Omega_\varepsilon} g(x)\det\nabla y_k(x)\,\mathrm{d}x = \int_{\Omega_\varepsilon} g(x)\det\nabla y(x)\,\mathrm{d}x. \tag{3.2.10}$$

If additionally for all $k\in\mathbb{N}$ either $\det\nabla y_k\ge 0$ or $\det\nabla y_k\le 0$ almost everywhere in Ω_ε then (3.2.10) holds for all $g\in L^\infty(\Omega_\varepsilon)$, i.e., $\det\nabla y_k\to\det\nabla y$ weakly in $L^1(\Omega_\varepsilon)$.

Proof. We apply Theorem D.2.2 and choose Ω_ε in such a way that $\sigma(\Gamma_\varepsilon)=0$ where σ is the measure defined in (3.2.9). □

Exercise 3.2.6. Prove a version of Theorem 3.2.2 for $d=2$. Let $\Omega\subset\mathbb{R}^2$ be a bounded Lipschitz domain. If for any $p\ge 2$, $y_k\to y$ weakly in $W^{1,p}(\Omega;\mathbb{R}^2)$, and if $\det\nabla y_k\to\delta$ weakly in $L^r(\Omega)$ for some $r\ge 1$ then $\delta=\det\nabla y$.

Exercise 3.2.7. Use (3.2.7) and prove that if $\{y_k\}_{k\in\mathbb{N}}\subset W^{1,d}(\Omega;\mathbb{R}^d)$ and $\det\nabla y_k\ge 0$ almost everywhere in Ω for all $k\in\mathbb{N}$ then $\det\nabla y_k\to\det\nabla y$ weakly in $L^1(\omega)$ for every compact set $\omega\subset\Omega$.

3.3 Lower semicontinuity

Here we prove the following statement on (sequential) lower semicontinuity of integral functionals. It appeared in [172] for finite integrands and it was extended to our needs in [31, Thm. 5.4]. We also refer to [192, Cor. 7.9] for even a more general version. The integrand ζ can also take infinite values and the lower semicontinuity is observed simultaneously with respect to "in measure" and weak convergence. Let us notice that convexity of ζ is required only in the argument in which we plug a weakly converging sequence. Altogether, the resulting functional in not necessarily convex.

Theorem 3.3.1 (Weak lower semicontinuity [31, 172]). *Let* $\zeta : \Omega\times\mathbb{R}^s\times\mathbb{R}^\sigma \to \mathbb{R}\cup\{+\infty\}$ *satisfy the following properties:*
(i) $\zeta(\cdot,z,v) : \Omega \to \mathbb{R}\cup\{+\infty\}$ *is measurable for all* $(z,v)\in\mathbb{R}^s\times\mathbb{R}^\sigma$,
(ii) $\zeta(x,\cdot,\cdot) : \mathbb{R}^s\times\mathbb{R}^\sigma \to \mathbb{R}\cup\{+\infty\}$ *is continuous for almost every* $x\in\Omega$,
(iii) $\zeta(x,z,\cdot) : \mathbb{R}^\sigma \to \mathbb{R}\cup\{+\infty\}$ *is convex. Assume further that for all* $(z,v)\in\mathbb{R}^s\times\mathbb{R}^\sigma$ $\zeta(\cdot,z,v)\ge\psi$ *for some* $\psi\in L^1(\Omega)$.
Furthermore, let $z_k \to z$ *almost everywhere in* Ω *and let* $v_k \to v$ *weakly in* $L^1(\Omega;\mathbb{R}^\sigma)$. *Then*

$$\int_\Omega \zeta(x,z(x),v(x))\,\mathrm{d}x \le \liminf_{k\to\infty}\int_\Omega \zeta(x,z_k(x),v_k(x))\,\mathrm{d}x\,.$$

Proof. We can, without loss of generality, suppose that $\zeta\ge 0$ for otherwise we work with $\zeta-\psi$. Moreover, we can extract a (non-relabeled) subsequence realizing lim inf. Hence, assume that $\lim_{k\to\infty}\int_\Omega\zeta(x,z_k(x),v_k(x))\,\mathrm{d}x = \alpha\in\mathbb{R}$. Namely, if $\alpha=+\infty$ then there is nothing to prove. Denote $g_k(x) := \zeta(x,z_k(x),v_k(x) - \zeta(x,z(x),v_k(x)))$. We show that g_k converges to zero in measure as $k\to\infty$. Indeed, if it is not the case there would be $\varepsilon>0$ and $\delta>0$ such that for a (non-relabeled) subsequence $\mathrm{meas}_d(\Omega_k)\ge\delta$ where $\Omega_k := \{x\in\Omega : |g_k(x)|\ge\varepsilon\}$. We can assume that $\|v_k\|_{L^1(\Omega;\mathbb{R}^\sigma)}\le K$ and also that $\int_\Omega\zeta(x,z_k(x),v_k(x))\,\mathrm{d}x\le K$ for some $K>0$. Consequently, $\mathrm{meas}_d(\omega_k)<\delta/2$ where $\omega_k := \{x\in\Omega : |v_k(x)|\ge 4K/\delta$ or $\zeta(x,z(x),v(x))\ge 4K/\delta\}$. Therefore $\mathrm{meas}_d(\Omega_k\setminus\omega_k) > \delta/2$. By the Arzela-Young Lemma 3.3.3 below, $\bigcap_{k=1}^\infty(\Omega_k\setminus\omega_k)\ne\emptyset$. If $x\in\Omega$ belongs to this set we see $|g_k(x)|\ge\varepsilon$ and $|v_k(x)|\le 4K/\delta$. Extracting a convergent subsequence from $\{v_k(x)\}_{k\in\mathbb{N}}$ leads to the contradiction with continuity of $\zeta(x,\cdot,\cdot)$. Altogether, $g_k\to 0$ in measure and some subsequence converges to zero almost everywhere in Ω. We apply Mazur's lemma A.1.7 to $\{v_k\}$. This means that there are coefficients of convex combinations $\{\lambda_k^j\}$ such that $w^k = \sum_{j=k}^{N(k)}\lambda_k^j v_k$ and $w_k\to w$ in $L^1(\Omega;\mathbb{R}^\sigma)$. Thus,

$$\zeta(x,z(x),w_k(x)) + \sum_j \lambda_k^j g_j(x) \le \sum_j \lambda_k^j\zeta(x,z_j(x),v_j(x))\,.$$

Passing to the limit for $k\to\infty$, integrating over Ω, and applying Fatou's lemma B.2.3 yields the result. □

Lemma 3.3.2 (Measure of the union and the intersection).
(i) *Let* $\Omega_{k-1}\subset\Omega_k\subset\mathbb{R}^d$, *for* $k\in\mathbb{N}$ *be measurable sets. Then* $\mathrm{meas}_d(\bigcup_{i=1}^\infty\Omega_k) = \lim_{k\to\infty}\mathrm{meas}_d(\Omega_k)$.
(ii) *Let* $\Omega_k\subset\Omega_{k-1}\subset\mathbb{R}^d$, *for* $k\in\mathbb{N}$ *be measurable sets such that* $\mathrm{meas}_d(\Omega_1)<+\infty$. *Then* $\mathrm{meas}_d(\bigcap_{i=1}^\infty\Omega_k) = \lim_{k\to\infty}\mathrm{meas}_d(\Omega_k)$.

Proof. Denote $\omega_k := \Omega_k \setminus \Omega_{k-1}$ for $k > 1$ and $\omega_1 := \Omega_1$. Notice that $\omega_k \bigcap \omega_j = \emptyset$ for $j \neq k$ and that $\Omega_i = \bigcup_{k=1}^{i} \omega_k$. Consequently, $\bigcup_{k=1}^{\infty} \Omega_k = \bigcup_{k=1}^{\infty} \omega_k$. Thus, due to σ-additivity of measures we have

$$\operatorname{meas}_d\Big(\bigcup_{i=1}^{\infty} \Omega_k\Big) = \operatorname{meas}_d\Big(\bigcup_{i=1}^{\infty} \omega_k\Big) = \lim_{i\to+\infty} \sum_{k=1}^{i} \operatorname{meas}_d(\omega_k) = \lim_{i\to+\infty} \operatorname{meas}_d(\Omega_i).$$

This proves (i).

To prove (ii), denote $\omega_k := \Omega_1 \setminus \Omega_k$. Then $\omega_{k-1} \subset \omega_k$ and $\operatorname{meas}_d(\Omega_k) = \operatorname{meas}_d(\Omega_1) - \operatorname{meas}_d(\omega_k)$. By the first statement, we have $\lim_{k\to\infty} \operatorname{meas}_d(\omega_k) = \operatorname{meas}_d(\bigcup_{k=1}^{\infty} \omega_k)$. However, $\bigcup_{k=1}^{\infty} \omega_k = \bigcup_{k=1}^{\infty} (\Omega_1 \setminus \Omega_k) = \Omega_1 \setminus \bigcap_{k=1}^{\infty} \Omega_k$. Altogether,

$$\operatorname{meas}_d\Big(\bigcap_{k=1}^{\infty} \Omega_k\Big) = \lim_{k\to\infty} \operatorname{meas}_d(\Omega_1) - \lim_{k\to\infty} \operatorname{meas}_d(\omega_k) = \lim_{k\to\infty} \operatorname{meas}_d(\Omega_k).$$

□

The following result is sometimes called the Arzela-Young lemma; see for example [528].

Lemma 3.3.3 (Arzela-Young). *Let $\Omega \subset \mathbb{R}^d$ be measurable,* $\operatorname{meas}_d(\Omega) < +\infty$. *Assume that $\{\Omega_k\}_{k\in\mathbb{N}}$ is a sequence of subsets of Ω such that* $\operatorname{meas}_d(\Omega_k) \geq \varepsilon > 0$ *for some ε. Then there exists a (non relabeled) subsequence of $\{\Omega_k\}_{k\in\mathbb{N}}$ such that $\bigcap_{k=1}^{\infty} \Omega_k \neq \emptyset$. Moreover, the Lebesgue measure of the set of points belonging to infinitely many sets Ω_k is at least ε.*

Proof. Define $\chi(x) := \lim_{m\to\infty} \sum_{k=1}^{m} \chi_{\Omega_k}(x)$ where χ_{Ω_k} is the characteristic function of Ω_k in Ω. If the claim of the lemma were not true we would have that $\chi(x) < +\infty$ for all $x \in \Omega$. The Lusin theorem B.2.7 on p. 511 asserts that there is $\omega \subset \Omega$ such that $\operatorname{meas}_d(\Omega \setminus \omega) < \varepsilon/2$ such that χ is continuous on ω. Consequently, $\chi(x) < K < +\infty$ for all $x \in \omega$. Hence, $\operatorname{meas}_d(\Omega_k \bigcap \omega) > \varepsilon/2$. We calculate

$$\begin{aligned} K\operatorname{meas}_d(\omega) &\geq \int_\omega \chi(x)\,\mathrm{d}x = \lim_{m\to\infty} \sum_{k=1}^{m} \int_\omega \chi_{\Omega_k\cap\omega}(x)\,\mathrm{d}x \\ &= \lim_{m\to\infty} \sum_{k=1}^{m} \operatorname{meas}_d(\Omega_k \cap \omega) > \lim_{m\to\infty} \frac{m\varepsilon}{2} = +\infty, \end{aligned}$$

which gives a contradiction[6]. □

3.4 Existence results in hyperelasticity

The concept of hyperelastic materials enables us to use the basic tool from the variational calculus, namely the direct method for the underlying energy functional, to show existence of at least one critical point as a global minimizer.

---

[6] Another proof was proposed by S. Luckhaus (as mentioned in [172]). Denote, for $k \in \mathbb{N}$, $\omega_k := \Omega_k \bigcup \Omega_{k+1} \ldots$. Then $\omega_{k+1} \subset \omega_k$ for all k. Moreover, $\operatorname{meas}_d(\omega_k) \geq \varepsilon$. Consequently, we get that the set $\operatorname{meas}_d(\{x \in \Omega : \chi(x) = +\infty\}) \geq \varepsilon$ due to Lemma 3.3.2(ii).

3.4.1 Polyconvex materials

We show existence of minimizers for elastic energy functional $\mathcal{E}$ given by (2.3.15), i.e.

$$\mathcal{E}(y) = \int_\Omega \varphi(x, \nabla y(x)) - \mathcal{F}(y), \tag{3.4.1}$$

involving now a polyconvex energy density $\varphi(x.\cdot)$. First, we define for $p > 1$ and $d = 2$ or $d = 3$ the following constant

$$q_{pd} := \begin{cases} p/(p-1) & \text{if } d = 3, \\ p & \text{if } d = 2. \end{cases} \tag{3.4.2}$$

The functional $-\mathcal{F}$ expresses the work done by conservative force on the specimen. We will assume that $-\mathcal{F}$ is weakly lower semicontinuous on $W^{1,p}(\Omega;\mathbb{R}^d)$. We adopt this assumption because external loading is a part of a physical experiment which we control. In literature, it is often considered that $\mathcal{F}$ is a bounded linear functional on $W^{1,p}(\Omega;\mathbb{R}^d)$ which is usually a mathematical simplification which does not hold in reality; cf. Proposition 1.2.8, for instance.

Theorem 3.4.1 (Existence result for polyconvex materials). *Let $d := 2$ or $d := 3$, $\Omega \subset \mathbb{R}^d$ be a bounded Lipschitz domain and let $\varphi : \bar{\Omega} \times \mathbb{R}^{d \times d} \to \mathbb{R} \cup \{+\infty\}$ be a stored energy function with the following properties:*

(a) Polyconvexity: let there be a Carathéodory integrand $\tilde{\varphi} : \bar{\Omega} \times \mathbb{R}^{\binom{2d}{d}-1} \to \mathbb{R} \cup \{+\infty\}$ such that $\tilde{\varphi}(x, \cdot)$ is convex for a.a. $x \in \Omega$ and

$$\forall_{\text{a.a.}} x \in \Omega \ \forall F \in \mathbb{R}^{d \times d} : \quad \varphi(x, F) = \tilde{\varphi}(x, \mathsf{M}(F)) \, ; \tag{3.4.3}$$

(b) Let, for a.a. $x \in \Omega$, the blow-up condition (2.3.16) and the frame-indifference (2.3.4) *hold and*

$$\exists \varepsilon > 0, \ p \geq 2, \ q \geq q_{pd}, \ r > 1 \ \forall_{\text{a.a.}} x \in \Omega \ \forall F \in \mathbb{R}^{d \times d} : \quad \text{(2.3.28) } \textit{holds.}$$

(c) Let further $\Gamma = \Gamma_{\rm D} \cup \Gamma_{\rm N}$ be a measurable partition of $\Gamma = \partial\Omega$ with the $\mathrm{meas}_{d-1}(\Gamma_{\rm D}) > 0$*, let $y_{\rm D} \in W^{1,p}(\Omega;\mathbb{R}^d)$ be given, and let*

$$\mathcal{A} := \Big\{ y \in W^{1,p}(\Omega;\mathbb{R}^d); \ \mathrm{Cof}\, \nabla y \in L^q(\Omega;\mathbb{R}^{d \times d}), \\ \det \nabla y \in L^r(\Omega), \ y = y_{\rm D} \ \textit{on } \Gamma_{\rm D}, \ \det \nabla y > 0 \ \textit{a.e.} \Big\} \neq \emptyset.$$

(d) Let further the functional

$$y \mapsto \mathcal{F}(y) : W^{1,p}(\Omega;\mathbb{R}^d) \to \mathbb{R} \tag{3.4.4}$$

be such that $-\mathcal{F}$ *is weakly lower semicontinuous and such that* $\mathcal{F}(y) \leq \tilde{K}(\|y\|^s_{W^{1,p}(\Omega;\mathbb{R}^d)} + 1)$ *for some* $\tilde{K} > 0$ *and* $1 \leq s < p$.

(e) Eventually, let there be $y \in \mathcal{A}$ *such that* $\mathcal{E}(y) < +\infty$.

Then the infimum of $\mathcal{E}$ *defined in (2.3.15) on* $\mathcal{A}$ *is attained, i.e. the minimum of* $\mathcal{E}$ *on* $\mathcal{A}$ *exists.*

Proof. Note that $x \mapsto \tilde{\varphi}(x, \mathsf{M}(\nabla y(x)))$ is measurable because φ is a Carathéodory integrand. Using (b) we get

$$\mathcal{E}(y) \geq \varepsilon \int_\Omega |\nabla y|^p + |\mathrm{Cof}\,\nabla y|^q + (\det \nabla y)^r \,\mathrm{d}x - \tilde{K}\|y\|^s_{W^{1,p}(\Omega;\mathbb{R}^d)} - \tilde{K}. \tag{3.4.5}$$

Applying the Poincaré inequality (see Theorem B.3.15) we conclude that there is a constant $\varepsilon > 0$ and

$$\mathcal{E}(y) \geq \varepsilon\Big(\|y\|^p_{W^{1,p}(\Omega;\mathbb{R}^d)} + \|\mathrm{Cof}\,\nabla y\|^q_{L^q(\Omega;\mathbb{R}^{d\times d})} + \|\det \nabla y\|^r_{L^r(\Omega)}\Big) - \hat{K}$$

for some $\hat{K} > 0$ and all $y \in \mathcal{A}$. Let $\{y_k\} \subset \mathcal{A}$ be a minimizing sequence of $\mathcal{E}$, i.e.,

$$\lim_{k\to\infty} \mathcal{E}(y_k) = \inf_{\mathcal{A}} \mathcal{E} < +\infty.$$

By (3.4.5) the sequence $\{(y_k, \mathrm{Cof}\,\nabla y_k, \det \nabla y_k)\}_{k\in\mathbb{N}}$ is bounded in the reflexive Banach space $W^{1,p}(\Omega;\mathbb{R}^d) \times L^q(\Omega;\mathbb{R}^{d\times d}) \times L^r(\Omega)$. Hence it has a subsequence weakly converging to $(y, A, \delta) \in W^{1,p}(\Omega;\mathbb{R}^d) \times L^q(\Omega;\mathbb{R}^{d\times d}) \times L^r(\Omega)$ and by Theorems 3.2.1 and 3.2.2 $A = \mathrm{Cof}\,\nabla y$ and $\delta = \det \nabla y$. To sum up, there is a minimizing sequence $\{y_k\}$ s.t. $y_k \to y$ weakly in $W^{1,p}(\Omega;\mathbb{R}^d)$, $\mathrm{Cof}\,\nabla y_k \to \mathrm{Cof}\,\nabla y$ weakly in $L^q(\Omega;\mathbb{R}^{d\times d})$ and $\det \nabla y_k \to \det \nabla y$ weakly in $L^r(\Omega)$. Moreover, $y = y_\mathrm{D}$ on Γ_D due to Theorem B.3.6.

Now we apply Theorem 3.3.1 with $v_k := \mathsf{M}(\nabla y_k)$, $z_k := y_k$, $v := \mathsf{M}(\nabla y)$, $z := y$, and $\xi(x, z, v) := \varphi(x, \mathsf{M}(\nabla y))$. This shows that $\mathcal{E}$ is (sequentially) weakly lower semicontinuous. It remains to prove that $y \in \mathcal{A}$ which amounts in verifying that $\det \nabla y > 0$ almost everywhere in Ω. This, however, follows from the fact that $\mathcal{E}(y) = \liminf_{k\to\infty} \mathcal{E}(y_k) < +\infty$ because $\{y_k\} \subset \mathcal{A}$ is a minimizing sequence and from the assumption (b). □

3.4.2 Ensuring injectivity almost everywhere

Local invertibility of a deformation $y \in C^1(\bar{\Omega};\mathbb{R}^d)$ is ensured by the condition $\det \nabla y > 0$ in $\bar{\Omega}$. On the other hand, local invertibility does not entail the global one. Indeed, as already observed in [114], if we consider $\bar{\Omega}$ a rod with a rectangular cross-section contained in the open half-space $x_1 > 0$, e.g. $\Omega = (1,2)\times(0,4\pi)\times(1,2)$, and the mapping $y : \bar{\Omega} \to \mathbb{R}^d$ given by

$$y(x_1, x_2, x_3) = (x_1 \cos x_2, x_1 \sin x_2, x_3),$$

then y is not injective. Let us notice that

$$\nabla y(x) = \begin{pmatrix} \cos x_2 & -x_1 \sin x_2 & 0 \\ \sin x_2 & x_1 \cos x_2 & 0 \\ 0 & 0 & 1 \end{pmatrix}$$

and therefore $\det \nabla y = x_1 > 0$ but the injectivity is lost. Indeed, we have $y(x_1,x_2,x_3) = y(x_1, 2\pi + x_2, x_3)$ if $0 < x_2 < 2\pi$. Altogether, we witness self-penetration of the material. The right Cauchy-Green strain tensor (1.1.26) equals

$$C(x) := (\nabla y(x))^\top \nabla y(x) = \begin{pmatrix} 1 & 0 & 0 \\ 0 & x_1^2 & 0 \\ 0 & 0 & 1 \end{pmatrix},$$

which is necessarily a positive definite matrix. This, in particular, shows that we cannot replace "≈" by "=" in (1.1.27) and that terms neglected in (1.1.25) can be significant.

In the following theorem, which already appeared in [114], the matrix norm is considered to be the operator norm subordinated to the Frobenius vector norm. It means that $|A| = \sup_{|x|=1} |Ax|$ for an arbitrary matrix $A \in \mathbb{R}^{d\times d}$. This result gives a sufficient condition on the norm of the displacement to ensure global injectivity.

Theorem 3.4.2 (Injectivity under a small-strain condition [114]). *Let $y = \mathbb{I} + u : \Omega \subset \mathbb{R}^d \to \mathbb{R}^d$ be a mapping differentiable at a point $x \in \Omega$. Then if $|\nabla u(x)| < 1$ we have $\det \nabla y(x) > 0$. Moreover, if Ω is convex then any mapping $y = \mathrm{id} + u \in C^1(\bar{\Omega}; \mathbb{R}^d)$ satisfying $\sup_{x\in\bar{\Omega}} |\nabla u(x)| < 1$ is injective.*

Proof. Let $x \in \Omega$ be a point at which $|\nabla u(x)| < 1$. Then $\det(\mathbb{I} + t\nabla u(x)) \neq 0$ if $0 \le t \le 1$.[7] On the other hand, the function $\delta : [0,1] \to \mathbb{R}$, $\delta(t) = \det(\mathbb{I} + t\nabla u(x))$ is continuous and therefore $\delta([0,1])$ is a closed interval in reals. As $\delta([0,1])$ contains $1 = \delta(0)$ but not 0 we infer that $\det(\mathbb{I} + \nabla u(x)) = \delta(1) > 0$. This proves the first statement.

As in the second assertion we suppose that Ω is convex, so is $\bar{\Omega}$. Thus, take $x_1, x_2 \in \bar{\Omega}$ and apply the mean-value theorem to y. We get

$$|y(x_1) - y(x_2) - (x_1 - x_2)| = |u(x_1) - u(x_2)| \le \sup_{x \in]x_1, x_2[} |\nabla u(x)||x_1 - x_2|.$$

Hence $|y(x_1) - y(x_2) - (x_1 - x_2)| < |x_1 - x_2|$ if $x_1 \neq x_2$ and therefore $y(x_1) \neq y(x_2)$. □

When speaking about elasticity, we understand that bodies deform reversibly under stress. This means that, upon releasing all loads, a body should deform back to its original (reference) configuration. This suggests that the inverse deformation y^{-1} should have the same quality as y. The following result due to J.M. Ball [27].

Theorem 3.4.3 (Inverse deformation y^{-1}, [27]). *Let $\Omega \subset \mathbb{R}^d$ be a bounded Lipschitz domain. Let $y_\mathrm{D} : \bar{\Omega} \to \mathbb{R}^d$ be continuous in $\bar{\Omega}$ and one-to-one in Ω such that*

[7] Indeed, for otherwise, there is $\mathbb{R}^d \ni v \neq 0$ such that $\nabla u(x)v = -v$. Consequently, $|\nabla u(x)| \ge 1$ which contradicts our assumption.

$y_{\rm D}(\Omega)$ is also bounded and Lipschitz continuous. Let $y \in W^{1,p}(\Omega;\mathbb{R}^d)$ for some $p > d$, $y(x) = y_{\rm D}(x)$ for all $x \in \Gamma$, and let $\det\nabla y > 0$ *a.e. in Ω. Finally, assume that, for some $q > d$, it holds*

$$\int_\Omega \big|(\nabla y(x))^{-1}\big|^q \det\nabla y(x)\,\mathrm{d}x < +\infty\,. \tag{3.4.6}$$

Then $y(\bar\Omega) = y_{\rm D}(\bar\Omega)$ and y is a homeomorphism of Ω onto $y_{\rm D}(\Omega)$. Moreover, the inverse map $y^{-1} \in W^{1,q}(y_{\rm D}(\Omega);\mathbb{R}^d)$ and $\nabla^y y^{-1}(x^y) = (\nabla y(x))^{-1}$ for a.a. $x \in \Omega$.

The drawback of Theorem 3.4.3 is that we need to prescribe Dirichlet boundary conditions on the whole boundary Γ which is usually unrealistic in lab experiments. The following statement ensures the injectivity in Ω; cf. [119]. Formula (3.4.7) is known as the Ciarlet-Nečas condition and it allows for a self-contact, i.e., the deformed body can touch itself. In the following theorem, $N(x^y,y,\Omega)$ denotes the Banach indicatrix, see (B.3.15).

Theorem 3.4.4 (Ciarlet-Nečas condition, [119]). *Let Ω be a bounded Lipschitz domain in $\mathbb{R}^d$, let $p > d$, and $y \in W^{1,p}(\Omega;\mathbb{R}^d)$ be such that* $\det\nabla y > 0$ *almost everywhere in Ω and*

$$\int_\Omega \det\nabla y(x)\,\mathrm{d}x \le \mathrm{meas}_d(y(\Omega))\,. \tag{3.4.7}$$

Then $N(x^y,y,\Omega) = 1$ for almost every $x^y \in \Omega^y$, i.e., for almost every $x^y \in \Omega^y$ there is only one $x \in \Omega$ satisfying $y(x) = x^y$.

Proof. Theorem B.3.10, the assumption $\det\nabla y > 0$ in Ω, and (3.4.7) imply that

$$\int_\Omega \det\nabla y(x)\,\mathrm{d}x = \int_{y(\Omega)} N(x^y,y,\Omega)\,\mathrm{d}x^y \le \mathrm{meas}_d(y(\Omega)) = \int_{y(\Omega)} 1\,\mathrm{d}x^y\,.$$

In view of the fact that it always holds $N(x^y,y,\Omega) \ge 1$, the above calculation shows that necessarily $N(x^y,y,\Omega) = 1$ for almost every $x^y \in y(\Omega) =: \Omega^y$. □

In fact, equality in (3.4.7) has the same strength. The illustration of a situation when the inequality in (3.4.7) is violated is in Figure 3.2.

We are going to show that the injectivity condition can be imposed on admissible deformations and an existence result similar to Theorem 3.4.1 still holds. First, however, let us discuss the notion of *injectivity almost everywhere.*

Definition 3.4.5 (Injectivity almost everywhere). We say that $y : \Omega \to \mathbb{R}^d$ is injective almost everywhere in a bounded domain $\Omega \subset \mathbb{R}^d$ if there is $\omega \subset \Omega$ such that $\mathrm{meas}_d(\omega) = 0$ and $y(x_1) \ne y(x_2)$ for every $x_1, x_2 \in \Omega \setminus \omega$ satisfying $x_1 \ne x_2$.

If the map y in Definition 3.4.5 satisfies Lusin's condition N (see Definition B.3.9) then $\mathrm{meas}_d(y(\Omega)) = \mathrm{meas}_d(y(\Omega\setminus\omega))$ and $y : \Omega\setminus\omega \to y(\Omega\setminus\omega)$ is injective and the domain and the range have both the full measure.

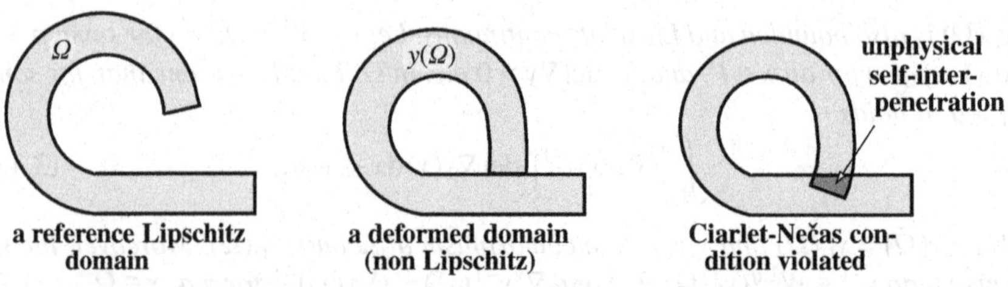

Fig. 3.2 A reference configuration Ω as a Lipschitz domain (left) deformed up to a self-contact so that Ω is no longer a Lipschitz domain but the Ciarlet-Nečas condition (3.4.7) still holds (middle). Further, already unphysical deformation would lead to an overlap and to violation of (3.4.7) because the dark-gray part is counted twice into the left-hand side integral in (3.4.7) (right).

Theorem 3.4.6 (Injectivity almost everywhere). *Let all assumptions of Theorem 3.4.1 hold and $p > d$. Let*

$$\mathcal{A}_{\rm inj} := \mathcal{A} \cap \{y \in W^{1,p}(\Omega;\mathbb{R}^d);\ (3.4.7)\ holds\ \}.$$

Then there is a minimizer of $\mathcal{E}$ on $\mathcal{A}_{\rm inj}$ which is injective almost everywhere in Ω.

Proof. We follow [119] and we only show that y obtained in the proof of Theorem 3.4.1 satisfies injectivity. As $p > d$ we know by the embedding theorem that $y \in C(\bar{\Omega};\mathbb{R}^d)$. So, a subsequence $\{y_k\}$ of the minimizing sequence converges uniformly to y. Moreover, y satisfies Lusin's condition N; see Theorem B.3.10. As Ω is a Lipschitz domain, we get $\mathrm{meas}_d(\Gamma) = 0$ and due to Lusin's condition N also $\mathrm{meas}_d(y(\Gamma)) = 0$. Consequently, $\mathrm{meas}_d(y(\Omega)) = \mathrm{meas}_d(y(\bar{\Omega}))$. We also have that $y(\bar{\Omega})$ is compact and therefore measurable, so that there is for any $\varepsilon > 0$ an open set O_ε such that $y(\bar{\Omega}) \subset O_\varepsilon$ and $\mathrm{meas}_d(O_\varepsilon \setminus y(\bar{\Omega})) < \varepsilon$. We claim that there is a number $\delta(\varepsilon) > 0$ such that

$$\bigcup_{x\in y(\bar{\Omega})} B(x,\delta(\varepsilon)) \subset O_\varepsilon .$$

For if not, there is $\varepsilon > 0$ and sequences $\{x_k\} \in y(\bar{\Omega})$, $\{\tilde{x}_k\} \notin O_\varepsilon$ and $\delta_k \to 0$ if $k \to \infty$ such that $|\tilde{x}_k - x_k| < \delta_k$. By compactness we may suppose that $x_k \to x \in y(\bar{\Omega})$ and we would have also $\tilde{x}_k \to x$ but this means that $x \in \mathbb{R}^d \setminus O_\varepsilon$ but it is not possible because $y(\bar{\Omega}) \subset O_\varepsilon$. Therefore,

$$\bigcup_{x\in y(\bar{\Omega})} B(x,\delta(\varepsilon)) \subset O_\varepsilon$$

for some $\delta(\varepsilon) > 0$ and there is k_0 such that $y_k(\bar{\Omega}) \subset O_\varepsilon$ if $k \geq k_0$ because y_k converges uniformly. As $y_k \in \mathcal{A}$ we have, for $k \geq k_0$, that

$$\int_\Omega \det \nabla y_k(x)\,\mathrm{d}x \leq \mathrm{meas}_d(y_k(\bar{\Omega})) \leq \mathrm{meas}_d(O_\varepsilon).$$

By the weak convergence of the determinants, we also have

$$\int_{\Omega} \det \nabla y(x)\,\mathrm{d}x = \lim_{k\to\infty} \int_{\Omega} \det \nabla y_k(x)\,\mathrm{d}x \le \mathrm{meas}_d(O_\varepsilon)\,.$$

Yet, $\mathrm{meas}_d(O_\varepsilon) = \mathrm{meas}_d(y(\bar{\Omega})) + \mathrm{meas}_d(O_\varepsilon \setminus y(\bar{\Omega}))$ and the arbitrariness of $\varepsilon > 0$ yields

$$\int_{\Omega} \det \nabla y(x)\,\mathrm{d}x \le \mathrm{meas}_d(y(\bar{\Omega})) = \mathrm{meas}_d(y(\Omega))\,.$$

Altogether, using Theorem B.3.10, we have

$$\mathrm{meas}_d(y(\Omega)) = \int_{y(\Omega)} \mathrm{d}x^y \le \int_{y(\Omega)} N(x^y, y, y(\Omega))\mathrm{d}x^y = \int_{\Omega} \det \nabla y(x)\,\mathrm{d}x \le \mathrm{meas}_d(y(\Omega)),$$

and thus $N(x^y, y, y(\Omega)) = 1$ for almost every $x^y \in y(\Omega)$. Let $\omega \subset y(\Omega)$ be a null set such that $N(x^y, y, \omega) > 1$. Since, y satisfies Lusin's condition N^{-1}, cf. Theorem B.3.13, we get that $\mathrm{meas}_d(\{x \in \Omega; y(x) \in \omega\}) = 0$, which concludes the proof. □

3.4.3 Ensuring injectivity everywhere

Maps that are injective almost everywhere still include rather nonphysical situations. We refer to [258] for some examples.This can be prevented if the deformation is injective everywhere. As we will see this can be achieved, for example, if the deformation is an open and discrete mapping. We are now going to state a few definitions and results which will help us to the improve Theorem 3.4.6.

Definition 3.4.7 (Open/discrete mapping). Let $\Omega \subset \mathbb{R}^d$ be a domain. We say that the mapping $y : \Omega \to \mathbb{R}^d$ is open if $y(\omega)$ is open for each open set $\omega \subset \Omega$. The mapping y is called discrete if the pre-image of each point x^y, i.e., $y^{-1}(x^y)$ is a discrete set (this means a set of isolated points) in Ω.

Theorem 3.4.8 (Injectivity everywhere I.). *Let all the assumptions of Theorem 3.4.6 hold. Assume that the minimizer of $\mathcal{E}$ is an open mapping. Then it is injective everywhere in Ω.*

Proof. Let y be a minimizer obtained in Theorem 3.4.6, i.e., it is injective almost everywhere in Ω. We prove that it is injective everywhere. Assume that it is not the case, i.e. that there are two different points $x_1 \in \Omega$ and $x_2 \in \Omega$ such that $y(x_1) = y(x_2) =: a$. As Ω is open and y is an open map then $y(\Omega)$ is open. Hence, there is $\varepsilon > 0$ such that $B(a;\varepsilon) \subset y(\Omega)$. Continuity of the inverse implies that $y^{-1}(B(a;\varepsilon)) \subset \Omega$ is open and it contains x_1 and x_2. Therefore there are two open disjoint neighborhoods U, V such that $x_1 \in U$, $x_2 \in V$ and $y(U) \cap y(V) \ni a$. As $y(U)$ and $y(V)$ are both open their intersection is also open and therefore $\mathrm{meas}_d(y(U) \cap y(V)) > 0$, i.e. y cannot be almost everywhere injective. □

Notice that if the assumptions of Theorem 3.4.8 are satisfied then the conclusions of Proposition 1.1.3 hold, too.

We will now assume that the admissible set contains only such deformations which have a finite distortion in the sense:

Definition 3.4.9 (Mappings of finite distortion). Let $\Omega \subset \mathbb{R}^d$ be an open bounded domain and $y \in W^{1,p}(\Omega;\mathbb{R}^d)$, $p > 1$, be such that $\det \nabla y > 0$ almost everywhere in Ω and that $\det \nabla y \in L^1(\Omega)$. We say that y has a finite distortion if the function $K_y : \Omega \to [0;+\infty]$ (called distortion[8]) defined as

$$K_y(x) := \begin{cases} |\nabla y(x)|^d / \det \nabla y(x) & \text{if } \det \nabla y(x) > 0, \\ 1 & \text{otherwise} \end{cases} \tag{3.4.8}$$

is such that $K_y(x) < +\infty$ for almost every $x \in \Omega$.

The following result is a consequence of [248, Thm. 3.4].

Lemma 3.4.10 (Openness and discreteness of maps with integrable distortion). *Let $\Omega \subset \mathbb{R}^d$ be a bounded domain and $y \in W^{1,p}(\Omega;\mathbb{R}^d)$ for some $p > d$ such that* $\det \nabla y > 0$ *almost everywhere in Ω. Assume that the distortion $K_y \in L^q(\Omega)$ for some $q > 2$ if $d = 3$ or $q = 1$ if $d = 2$. Then y is either constant or both open and discrete.*

Putting together Theorem 3.4.8, Lemma 3.4.10, and the Ciarlet-Nečas condition (3.4.7), we get the following result.

Theorem 3.4.11 (Maps injective everywhere). *Let $\Omega \subset \mathbb{R}^d$ be a bounded Lipschitz domain, let $y \in W^{1,p}(\Omega;\mathbb{R}^d)$ for some $p > d$ be such that* $\det \nabla y > 0$ *almost everywhere in Ω and let* (3.4.7) *hold. If its distortion K_y belongs to $L^q(\Omega)$ for some $q > 2$ if $d = 3$ or $q = 1$ if $d = 2$. then y is injective everywhere in Ω.*

Proof. Following the arguments of the proof of Theorem 3.4.6 we show that y is injective almost everywhere. Lemma 3.4.10 implies that y is either constant or both open and discrete. However, it cannot be constant because it is almost everywhere injective. Hence, it is open. Assume that that there are two different points $x_1 \in \Omega$ and $x_2 \in \Omega$ such that $y(x_1) = y(x_2) =: a$. As Ω is open and y is an open map then $y(\Omega)$ is open. Hence, there is $\varepsilon > 0$ such that $B(a;\varepsilon) \subset y(\Omega)$. Continuity of the inverse implies that $y^{-1}(B(a;\varepsilon)) \subset \Omega$ is open and it contains x_1 and x_2. Therefore there are two open disjoint neighborhoods U, V such that $x_1 \in U$, $x_2 \in V$ and $y(U) \cap y(V) \ni a$. As $y(U)$ and $y(V)$ are both open their intersection is also open and therefore $\mathrm{meas}_d(y(U) \cap y(V)) > 0$, i.e. y cannot be almost everywhere injective. □

Corollary 3.4.12 (Injectivity everywhere II.). *Let all the assumptions of Theorem 3.4.6 hold and let, for $q > 2$ if $d = 3$ or $q = 1$ if $d = 2$, it holds for almost every $x \in \Omega$ that*

$$\exists \varepsilon > 0: \quad \varphi(x,F) \geq \varepsilon \frac{|F|^{dq}}{(\det F)^q}. \tag{3.4.9}$$

Then every minimizer of $\mathcal{E}$ from (3.4.1) is injective everywhere in Ω.

[8] Sometimes K_y is called the "optimal" distortion while distortion (not determined uniquely) is a mapping $k_y : \Omega \to [0;+\infty]$ which is a.e. finite and satisfies $|\nabla y(x)|^d \leq k_y(x) \det \nabla y(x)$ for a.a. $x \in \Omega$; see [248].

Remark 3.4.13 (Geometric meaning of the distortion). Consider the open ball $\Omega := \{x \in \mathbb{R}^d;\ \sum_{i=1}^d x_i^2 < 1\}$ and a homogeneous deformation (stretch) $y : \Omega \to \mathbb{R}^d$ defined for all $x \in \Omega$ as $y(x) := Fx$ where $F := \mathrm{diag}(F_{11}, \dots, F_{dd})$ with $F_{ii} > 0$ for all $1 \le i \le d$. It is easy to see that $y(\Omega)$ is an open ellipsoid given by

$$\Big\{x^y \in \mathbb{R}^d;\ \sum\nolimits_{i=1}^d \frac{(x_i^y)^2}{F_{ii}^2} < 1\Big\}. \tag{3.4.10}$$

Hence, $K_y = \sum\nolimits_{i=1}^d (F_{ii}^2)^{d/2} / \prod\nolimits_{i=1}^d F_{ii}$ measures the eccentricity of the ellipsoid. Indeed, we have $|F|^d \ge d^{d/2}\det F$, so that the distortion is the smallest if all axes of the ellipsoid are the same, i.e., if F_{ii} is independent of $1 \le i \le d$ and $y(\Omega)$ is again a ball. This reasoning is most pronounced if we take $F_{ii} \ne 1$ and $F_{jj} = 1$ for all $j \ne i$. See also Figure 3.3.

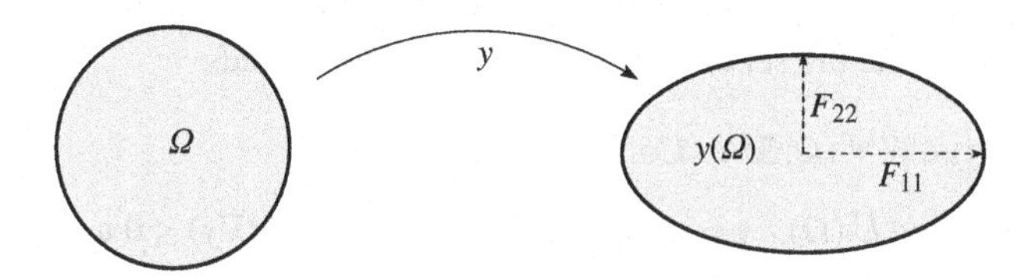

Fig. 3.3 Schematic drawing of a homogeneous deformation mapping of a ball to an ellipsoid if $d = 2$ with the semi-major axis of the length F_{11} and the semi-minor axis of the length F_{22}.

Remark 3.4.14 (Locking materials). A concept motivated by *material locking*, has been introduced by Prager in [421] see also [118, 151, 223, 413, 480] for newer results. According to Prager's classification of elastic materials, a material is called elastically *hard* if its elastic constants increase with the increasing strain. Perfectly (or ideally) locking materials extrapolate this property by assuming that the material gets locked (i.e. stiff or rigid), once some strain measure reaches a prescribed value. Prager [421] introduced a locking constraint in the form $L(\nabla y) \le 0$ almost everywhere in Ω where

$$L(F) := \Big|\frac{1}{2}(F + F^\top) + (1 - \frac{2}{d}(\mathrm{tr}\,F))\mathbb{I}\Big|^2 - \varepsilon,$$

where $\mathbb{I}$ is the identity matrix, "tr" denotes the trace, and $\varepsilon > 0$ is a material parameter. This function is, however, not suitable for nonlinear elasticity because it is not frame-indifferent, i.e., it does not satisfy $L(F) = L(RF)$ for some $R \in \mathrm{SO}(d)$ and $F \in \mathbb{R}^{d\times d}$. Ciarlet and Nečas [118] removed this issue by setting

$$L(F) := \Big|E - \frac{\mathrm{tr}\,E}{d}\mathbb{I}\Big|^2 - \varepsilon, \tag{3.4.11}$$

where $E := (F^\top F - \mathbb{I})/2$. Note that the pointwise character of locking constraints allows us to control locally the strain appearing in the material. Altogether, locking constraints can serve as indicators whether or not we are authorized to use merely elastic description of the material behavior. The concept of polyconvex materials and the corresponding existence theory allow for locking constraints, as well if one assumes that $L : \mathbb{R}^{d\times d} \to \mathbb{R}$ is polyconvex. Details are left to the reader in Exercise 3.4.16 below. Unfortunately, (3.4.11) is not polyconvex and therefore it does not fit into the present theory. Nevertheless, Theorem 3.1 in [118] shows that, for $d = 3$, there is a polyconvex function $\tilde{L} : \mathrm{GL}(3)^+ \to \mathbb{R}$ such that $\tilde{L}(F) = L(F) + o(|E|^2)$ for E close to the origin, i.e, if F is near a rotation matrix.

Exercise 3.4.15. Reformulate Theorems 3.4.1 as well as 3.4.6 to show the existence of minimizers for incompressible materials, i.e., with the condition $\det\nabla y = 1$ almost everywhere in Ω.

Exercise 3.4.16 (Locking models). Assume that all assumptions of Theorem 3.4.6 hold and that there is a locking constraint prescribed by means of a polyconvex function L. More precisely, we assume that there is a polyconvex function $L : \mathrm{GL}^+(d) \to \mathbb{R}$ and the set of admissible deformations reads

$$\mathcal{A} := \Big\{ y \in W^{1,p}(\Omega;\mathbb{R}^d);\ \ \mathrm{Cof}\,\nabla y \in L^q(\Omega;\mathbb{R}^{d\times d}),$$
$$\det\nabla y \in L^r(\Omega),\ \ y = y_{\mathrm{D}} \text{ on } \Gamma_{\mathrm{D}},\ \det\nabla y > 0 \text{ a.e.},\ L(\nabla y) \le 0 \text{ a.e.} \Big\} \ne \emptyset.$$

Show existence of a minimizer to the energy functional $\mathcal{E}$ from (3.4.1) on $\mathcal{A}$.

Exercise 3.4.17. Show that the function $(a,b) \mapsto a^m/b^n$ is convex on $(0,+\infty)\times(0,+\infty)$ if $m \ge n+1 \ge 1$ and that it is is increasing in a if b is kept fixed. Then $F \mapsto |F|^m/(\det F)^n$ is polyconvex for $F \in \mathrm{GL}^+(d)$ and $m \ge n+1 \ge 1$. Hint: Show that the second differential of $(a,b) \mapsto a^m/b^n$ is positive semidefinite.

Exercise 3.4.18. Assume that $y \in W^{1,p}(\Omega;\mathbb{R}^d)$ for some $p > d$ and let $\det\nabla y > 0$ almost everywhere. Find conditions on p and the integrability of $1/\det\nabla y$ such that the distortion satisfies assumptions of Lemma 3.4.10.

3.5 Magnetoelasticity

Magnetoelasticity describes the mechanical behavior of elastic solids under magnetic effects. These effect have an microscopical origin by electron spins and may manifest differently depending whether there is a collective tendency of these spins begin collinear or opposite, leading to ferro- or antiferro magnetism, respectively. On some conditions typically very dependent on temperature, the magnetic properties in magnetic materials can even be suppressed and magnetic materials then become paramagnetic; we then speak about *ferro/para-magnetic phase transformation*.

Here we have in mind rather magnetically soft[9] ferromagnetic materials which are mechanically not rigid. The magnetoelastic coupling is based on the presence of small magnetic domains in the material manifested by a magnetization vector $m : \Omega \to \mathbb{R}^d$. In the absence of an external magnetic field these magnetic domains are randomly oriented but when exposed to an external magnetic field they become aligned along the field and their rotations induce a deformation of the specimen. As the intensity of the magnetic field is increased, more and more magnetic domains orientate themselves so that their principal axes of anisotropy are collinear with the magnetic field in each region and finally saturation is reached. We refer to e.g. [95, 154, 155, 259, 261] for a discussion on the foundations of magnetoelasticity.

The mathematical modeling of magnetoelasticity is a vibrant area of research, triggered by the interest on so-called *multifunctional materials*.[10] Following the mentioned references, in this section we will choose the "mixed" Lagrangian/Eulerian description, considering the deformation gradient in the reference configuration but the magnetization in the actual configuration. We shall here be concerned with the total energy $\mathcal{E}$ defined as

$$\begin{aligned}\mathcal{E}(y,m) := &\int_\Omega \varphi(\nabla y(x), [m\circ y](x))\,\mathrm{d}x + \frac{\kappa}{2}\int_{\Omega^y} |\nabla m(x)|^2\,\mathrm{d}x \\ &- \int_\Omega h_{\mathrm{ext}}(y(x))\cdot m(y(x))\,\mathrm{d}x + \frac{\mu_0}{2}\int_{\mathbb{R}^d} |\nabla \phi_m(x)|^2\,\mathrm{d}x,\end{aligned} \tag{3.5.1}$$

Here φ stands for the magneto-elastic energy density, the second term is the so-called *exchange energy* considered in the actual configuration (cf. [471]) and is related to the typical size of ferromagnetic texture. The term involving external magnetic field, h_{ext}, is the work done by this external field and the term is also called the *Zeeman* energy. The last term represents the magnetostatic energy, μ_0 is the vacuum permeability, and ϕ_m is the *magnetostatic potential* induced by m. In particular, ϕ_m is a solution to the equation

$$\mathrm{div}\,(\mu_0 \nabla \phi_m - \chi_{\Omega^y} m) = 0 \ \text{ in } \ \mathbb{R}^d, \tag{3.5.2}$$

where χ_{Ω^y} is the characteristic function of the deformed configuration Ω^y. We shall consider $\mathcal{E}$ under the constraints

$$\det \nabla y = 1 \quad \text{and} \quad |m| = 1 \quad \text{a.e. on } \Omega, \tag{3.5.3}$$

which corresponds to incompressibility and magnetic saturation (with the saturation magnetization being put to 1 for notational simplicity). Both these constraints are a

[9] The attribute " magnetically soft" means that the so-called coercive force is negligible and possible reorientation of magnetization is very easy and dissipation-less. Such materials are intentionally used e.g. for transformers where any energy losses are undesirable.

[10] Among these one has to mention rare-earth alloys such as Terfenol and Galfenol. All these materials exhibit so-called giant magnetostrictive behavior as reversible strains as large as 10% can be reached by the imposition of relatively moderate magnetic fields. This strong magnetoelastic coupling makes them relevant in a wealth of innovative applications including sensors and actuators.

certain idealization. The latter one is sometimes called a Heisenberg constraint and is relevant rather for ferromagnets under absolute-zero temperature.[11]

The mentioned incompressibility constraint reads $\det\nabla y = 1$ almost everywhere in Ω. In particular, this entails invertibility of y almost everywhere in Ω through the Ciarlet-Nečas condition (3.4.7) which in our situation reads $\mathrm{meas}_d(\Omega^y) = \mathrm{meas}_d(\Omega)$ because $\det\nabla y = 1$ almost everywhere. Hence, we have that

$$\mathrm{meas}_d(\Omega^y) = \int_{\Omega^y} 1\,\mathrm{d}x^y = \int_\Omega \det\nabla y(x)\,\mathrm{d}x = \mathrm{meas}_d(\Omega)\,.$$

We shall define the sets

$$\mathcal{A} := \{y\in W^{1,p}(\Omega;\mathbb{R}^d);\ \ \det\nabla y = 1 \text{ in } \Omega,\ y = y_{\rm D} \text{ on } \Gamma_{\rm D},\ \mathrm{meas}_d(\Omega^y) = \mathrm{meas}_d(\Omega)\}$$
$$\mathcal{A}^y := \{m\in H^1(\Omega^y;\mathbb{R}^d);\ \ |m| = 1 \text{ in } \Omega\}.$$

Note that, as $p > d$, the set $\mathcal{A}$ is sequentially closed with respect to the weak topology of $W^{1,p}(\Omega;\mathbb{R}^d)$. This indeed follows from the sequential continuity of the map $y \mapsto \det\nabla y$ from $W^{1,p}(\Omega;\mathbb{R}^d)$ to $L^{p/d}(\Omega)$ (both equipped with the weak convergence), the weak closeness of the Ciarlet-Nečas condition (3.4.7), and from the compactness properties of the trace operator.

We say that $\{(y_k,m_k)\}_{k\in\mathbb{N}}$ $\mathcal{A}\times\mathcal{A}^y$-converges to $(y,m) \in \mathcal{A}\times\mathcal{A}^y$ as $k\to\infty$ if the following three conditions hold

$$y_k \rightharpoonup y \qquad \text{in } W^{1,p}(\Omega;\mathbb{R}^d), \tag{3.5.4a}$$
$$\chi_{y_k(\Omega)}m_k \to \chi_{y(\Omega)}m \qquad \text{in } L^2(\mathbb{R}^d;\mathbb{R}^d), \tag{3.5.4b}$$
$$\chi_{y_k(\Omega)}\nabla m_k \rightharpoonup \chi_{y(\Omega)}\nabla m \qquad \text{in } L^2(\mathbb{R}^d;\mathbb{R}^{d\times d}). \tag{3.5.4c}$$

Eventually, we say that a sequence $\{(y_k,m_k)\}_{k\in\mathbb{N}} \subset \mathcal{A}\times\mathcal{A}^y$ is $\mathcal{A}\times\mathcal{A}^y$-bounded if

$$\sup_{k\in\mathbb{N}}\Big(\|y_k\|_{W^{1,p}(\Omega;\mathbb{R}^d)} + \|\nabla m_k\|_{L^2(y_k(\Omega);\mathbb{R}^{d\times d})}\Big) < +\infty\,.$$

The following result is an immediate consequence of the linearity of the Maxwell equation (3.5.2):

Lemma 3.5.1. *Let $\chi_{y_k(\Omega)}m_k \to \chi_{y(\Omega)}m$ in $L^2(\mathbb{R}^d;\mathbb{R}^d)$ and let $\phi_{m_k}\in H^1(\mathbb{R}^d)$ be the solution of* (3.5.2) *corresponding to m_k. Then $\phi_{m_k} \rightharpoonup \phi_m$ in $H^1(\mathbb{R}^d)$ where ϕ_m is the solution of* (3.5.2) *corresponding to $\chi_{y(\Omega)}m$.*

Let us further assume:

$$\exists\varepsilon>0\ \forall F\in\mathbb{R}^{d\times d},\ m\in\mathbb{R}^d:\quad \varphi(F,m) \ge \varepsilon|F|^p\,, \tag{3.5.5a}$$
$$\forall R\in\mathrm{SO}(d),\ F\in\mathbb{R}^{d\times d}:\quad \varphi(RF,Rm) = \varphi(F,m)\,, \tag{3.5.5b}$$

[11] Actually, the Heisenberg constraint can be far violated in temperatures close to the point when ferromagnetic materials become antiferro- or paramagnetic, which is called the Curie temperature.

$$\forall F \in \mathbb{R}^{d\times d},\ m \in \mathbb{R}^d: \quad \varphi(F,m) = \varphi(F,-m), \tag{3.5.5c}$$

$$\forall m \in \mathbb{R}^d: \quad \varphi(\cdot;m) \text{ is polyconvex}. \tag{3.5.5d}$$

Theorem 3.5.2 (Existence of minimizers [297]). *The energy $\mathcal{E}$ is lower semicontinuous and coercive with respect to $\mathcal{A}\times\mathcal{A}^y$-convergence. In particular, it attains a minimum on $\mathcal{A}\times\mathcal{A}^y$.*

Sketch of the proof. Owing to the coercivity assumption (3.5.5a), one immediately gets that $\mathcal{E}$ sublevels are $\mathcal{A}\times\mathcal{A}^y$-bounded, hence $\mathcal{A}\times\mathcal{A}^y$-sequentially compact.

The magnetoelastic term in $\mathcal{E}$ is weakly lower semicontinuous because of the assumptions (3.5.5) on φ. The exchange energy term in $\mathcal{E}$ is quadratic hence weakly lower semicontinuous. The magnetostatic term is weakly lower semicontinuous by Lemma 3.5.1. The existence of a minimizer then follows from the direct method. Details can be found in [297]. □

An analogous result without the incompressibility constraint was recently proved in [45].

Remark 3.5.3 (Frame indifference). The condition $\varphi(RF,Rm) = \varphi(F,m)$ for all $R\in$ SO(d), cf. (3.5.5b), is automatically guaranteed by considering a "material" stored energy $\varphi_{\rm M}: \times\mathbb{R}^{d\times d}\times\mathbb{R}^d \to \mathbb{R}$ and putting $\varphi_{\rm M}(F^\top F, F^\top m) =: \varphi(F,m)$, cf. [260, Formula (3.15)]. Then, relying on $R^\top R = \mathbb{I}$ for any $R\in{\rm SO}(d)$, indeed

$$\varphi(RF,Rm) = \varphi_{\rm M}(F^\top R^\top RF, F^\top R^\top Rm) = \varphi_{\rm M}(F^\top F, F^\top m) = \varphi(F,m).$$

Alternatively, as in [467], the frame indifference can be ensured by considering $\varphi_{\rm M}$ dependent on $(F^\top F, F^{-1}m)$ because, relying again on $R^\top R = \mathbb{I}$ and now also on $R^{-1}R = \mathbb{I}$, for any $R\in{\rm SO}(d)$ then

$$\varphi(RF,Rm) = \varphi_{\rm M}(F^\top R^\top RF, F^{-1}R^{-1}Rm) = \varphi_{\rm M}(F^\top F, F^{-1}m) = \varphi(F,m).$$

Remark 3.5.4 (Exchange energy nonlocal). In analog with the nonlocal simple materials as in Sect. 2.5.2, one can replace ∇m by a nonlocal term acting on m, i.e. (3.5.1) modifies as

$$\begin{aligned}\mathcal{E}(y,m) = \int_\Omega \varphi(\nabla y(x), m^y(x)) - h_{\rm ext}(x)\cdot m^y(x)\,{\rm d}x + \frac{\mu_0}{2}\int_{\mathbb{R}^d} |\nabla\phi_m(x)|^2\,{\rm d}x \\ + \int_{\Omega\times\Omega} (m^y(x)-m^y(\tilde{x}))\cdot\mathbb{K}(x,\tilde{x})\cdot(m^y(x)-m^y(\tilde{x}))\,{\rm d}x{\rm d}\tilde{x}.\end{aligned} \tag{3.5.6}$$

When the kernel $\mathbb{K}: \Omega\times\Omega \to \mathbb{R}^{d\times d}_{\rm sym}$ is singular, satisfying (2.5.28) with $m\in\mathbb{R}^d$ in place of $F\in\mathbb{R}^{d\times d}$, the model controls m in $H^\gamma(\Omega;\mathbb{R}^d)$ and enjoys similar compactifying properties as the original model (3.5.1) controlling m in $H^1(\Omega;\mathbb{R}^d)$. Let us notice that $\nabla m = F(F^{-1}\nabla m)$ with $F^{-1}\nabla m \in L^2(\Omega)$. For rigid ferromagnets, this model has been suggested by R.C. Rogers [444, 445].

3.6 Poroelastic and swollen materials

Some elastic materials allow for a penetration of very small atoms into a solid atomic grid in crystalline solids or into spaces between big macromolecules. It occurs in some metals with the interstitial solute being hydrogen; such metals then undergo a so-called *metal-hydride phase transformation* where the hydrid phase has considerably bigger lattice parameters than the metal phase, as schematically depicted in Figure 3.4-left. The latter mentioned mechanism occurs in polymers allowing for a diffusion of a specific solvent. The solvent which diffuses thorough the elastic body influences considerably the volume, again schematically depicted in Figure 3.4-middle. In both cases, it leads to a macroscopical mechanical phenomenon of a considerable variation of volume, a so-called *swelling*. In turn, this swelling influences stress/strain distribution and therefore backward the diffusion process itself.

Other, microscopically different mechanism occurs in macroscopically elastic solid materials that allow for a penetration by another material of a fluidic character. Microscopically it is due to various pores or voids which are mutually connected. It is manifested macroscopically as a homogeneous mixture of an elastic solid body and fluid which can diffuse throughout the volume. Examples are poroelastic rocks or concrete, or porous polymers filled with water, depicted schematically in Figure 3.4-right. In all cases, the diffusing fluidic media (gasses, solvents, or liquids) are also referred to as a *diffusants*. Apart the mentioned swelling, interaction of solids with diffusants may also be manifested by *squeezing*.

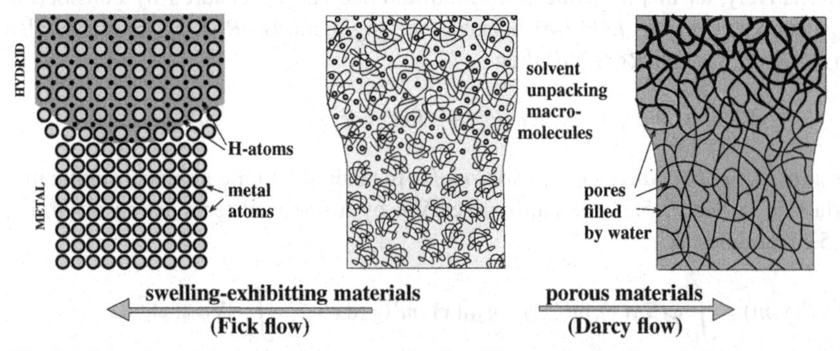

Fig. 3.4 Several variations of volume expansion by diffusant flow with underlying mechanisms ranging from an atomic via a molecular to a mesoscopical level:
Left: hydrogen diffusing inside a metallic atomic grid (metal/hydrid transformation).
Middle: macromolecules in polymers unpacked by a diffusing solvent.
Right: water flowing through pores of a (poro)elastic rock or a concrete

The solid-diffusant interaction is surely a complicated multi-scale problem and a big amount of phenomenology is usually applied to build a simplified model. A wide menagerie of models can thus be obtained, cf. [428] for a survey. Typically, small velocity of the diffusant is assumed. Later, in Sect. 5.7, we also assume small strains. In any case, one should distinguish between the general *steady-state* (sometimes also called *stationary*[12]) situations and purely *static* situations. The former one means that all fields including the specific dissipation rates do not depend on time, while the latter means in addition that that the dissipation rate is zero.

3.6.1 Static versus steady-state problems

In order to explain our ideas on static problems, we introduce a material model in which elasticity depends on an internal variable c denoting a concentration of some diffusant. More specifically, we are concerned with the following boundary value problem:

$$-\mathrm{div}\,\partial_F\tilde{\varphi}(\nabla y(x),c(x)) = \tilde{f}(x,y(x)) \qquad \text{for } x\in\Omega, \tag{3.6.1a}$$

$$\partial_F\tilde{\varphi}(\nabla y(x),c(x))\vec{n}(x) = \tilde{g}(x,\nabla y(x)) \qquad \text{for } x\in\Gamma_{\mathrm{N}}, \tag{3.6.1b}$$

$$y(x) = y_{\mathrm{D}}(x) \qquad \text{for } x\in\Gamma_{\mathrm{D}}. \tag{3.6.1c}$$

Here, $\tilde{\varphi}$ is a (spatially homogeneous) stored energy density such that the first Piola-Kirchhoff stress tensor S reads

$$S(x) := \partial_F\tilde{\varphi}(\nabla y(x),c(x)).$$

The thermodynamic driving force for diffusive processes is a so-called *chemical potential*, denoted by μ and defined as a partial derivative of $\tilde{\varphi}$ with respect to c, i.e.,

$$\mu(x) := \partial_c\tilde{\varphi}(\nabla y(x),c(x)). \tag{3.6.2}$$

Then the diffusive flux is governed by the (*generalized*) *Fick law*:

$$j = -\mathbb{M}\nabla\mu, \tag{3.6.3}$$

where $\mathbb{M} = \mathbb{M}(F,c)$ is a mobility matrix. It may cover both the *conventional Fick law* [187] and the *Darcy law* [142] which say that the flux j is driven by the (negative) gradient of concentration and pressure, respectively, cf. Sect. 5.7.1 below. For the possible dependence of $\mathbb{M}$ on the deformation gradient we refer to (9.6.3) on p. 461, although in our static case here the particular dependence of $\mathbb{M}$ is not important as far as it keeps $\mathbb{M}$ positive definite. The diffusive flux satisfies the continuity equation $\mathrm{div}\,j = 0$, i.e.

$$\mathrm{div}(\mathbb{M}(\nabla y,c)\nabla\mu) = 0. \tag{3.6.4}$$

[12] See e.g. [149, Ch. 5].

It is to be accompanied by suitable boundary conditions, e.g.

$$\mathbb{M}(\nabla y, c)\nabla\mu\cdot\vec{n} + \alpha\mu = \alpha\mu_{\rm ext} \quad \text{on } \Gamma, \tag{3.6.5}$$

where $\mu_{\rm ext} = \mu_{\rm ext}(x)$ is the chemical potential of the external environment and $\alpha = \alpha(x) \geq 0$ is the phenomenological coefficient of permeability of the boundary.

In the static situations we are addressing here, all dissipation processes vanishes, i.e. $j = 0$. In particular, all diffusive processes vanish. Here (3.6.3) with $j = 0$ and $\mathbb{M} > 0$ give $\nabla\mu = 0$ on Ω. When assuming Ω connected, this further leads to that μ is constant. Let us denote this constant by $\bar{\mu}$. Also (3.6.5) reduces to $\alpha(\bar{\mu} - \mu_{\rm ext}) = 0$ on Γ. Therefore, solvability of such problem essentially requires either the system to be in equilibrium with the external environment or to be isolated. The former option is rather trivial because (3.6.5) with $\alpha > 0$ then fixes $\mu = \bar{\mu}$. Then, in some cases, one can even eliminate c. From $\bar{\mu} = \partial_c\varphi(F, c)$, we can then find the concentration $c = [\partial_c\varphi(F,\cdot)]^{-1}(\bar{\mu})$ as a function of $F = \nabla y$ provided $\partial_c\varphi(F,\cdot)$ is invertible (i.e. increasing). This last condition means $\varphi(F,\cdot)$ convex, and then we can even write a bit more specifically

$$c = \partial_\mu\varphi^*(F, \bar{\mu}) \tag{3.6.6}$$

with $\varphi^*(F,\cdot)$ denoting the convex conjugate function of $\varphi(F,\cdot)$, cf. the definition (A.3.16) on p. 498. Note that, even if $\varphi(F,\cdot)$ is not smooth as in (3.6.9) below, $\varphi^*(F,\cdot)$ is indeed single-valued if the natural requirement $\partial^2_{cc}\varphi > 0$ holds. The concentration c can thus be completely eliminated.

The latter option (i.e. the boundary permeability coefficient $\alpha = 0$) is more interesting. If the profile $\nabla y = \nabla y(x)$ were known, since $\varphi(F,\cdot)$ is convex and the overall amount of diffusant $\int_\Omega c\,\mathrm{d}x = C_{\rm total}$ depends monotonically on $\bar{\mu}$ due to (3.6.6), it would allow us to specify $\bar{\mu}$ if $C_{\rm total}$ is given. Yet, the deformation y is a part of solution itself so that, unfortunately, it does not seem possible to fix $\bar{\mu}$ just from knowing $C_{\rm total}$ or also if $\varphi(F,\cdot)$ is not convex. In any case, in this isolated situation, it is natural to prescribe the total amount of diffusant

$$\int_\Omega c\,\mathrm{d}x = C_{\rm total} \qquad \text{with} \quad C_{\rm total} \geq 0 \quad \text{given.} \tag{3.6.7}$$

Two basic alternatives that can be distinguished in the isotropic continuum are that either the strain is directly influenced by a concentration c of the solvent which diffuses through the material or the diffusant has its own pressure which influences the overall stress and eventually indirectly the strain, too. These two options microscopically reflect that the diffusant does not directly contribute to the pressure but can change the reference configuration (as in Fig. 3.4-left) or that the diffusant flows though pores under its pressure (as in Fig. 3.4-right). These two modeling options are reflected in rather different mathematics, treated respectively in the two following sections.

3.6.2 Equilibrium configuration for poroelastic materials

Poroelastic materials are solid materials permeated by a mutually interconnected network of pores (voids) to be filled with a fluid. Here we consider only situations that the pores are fully filled, which is called a saturated flow.

In static problems, we assume that the amount of fluid contained in the material is fixed. Here we assume that the solid *material* is *hyperelastic* described a polyconvex stored energy density which depends on concentration c of the fluid. Thus, we denote $\varphi(\cdot,c):\mathbb{R}^{d\times d}\to\mathbb{R}\cup\{+\infty\}$ the hyperelastic energy density of the material depending on the concentration c. Following early ideas of [191, 309] where the notions of *cross-quasiconvexity* and joint convexity/quasiconvexity were considered. In analogy with polyconvexity as a sufficient condition for quasiconvexity, we introduce the following property, called cross-polyconvexity, which implies cross-quasiconvexity:

Definition 3.6.1 (Cross-polyconvexity). We say that $\varphi:\mathbb{R}^{d\times d}\times\mathbb{R}\to\mathbb{R}\cup\{+\infty\}$ is cross-polyconvex if

$$\forall F\in\mathbb{R}^{d\times d},\ c\in\mathbb{R}:\qquad \varphi(F,c)=\tilde{\varphi}(\mathsf{M}(F),c)$$

for some convex and continuous function $\tilde{\varphi}:\mathbb{R}^{\binom{2d}{d}-1}\times\mathbb{R}\to\mathbb{R}\cup\{+\infty\}$.

The amount of fluid in the body is fixed and we denote its value $C_{\rm total}>0$. Additionally, we require that admissible deformations of the material are orientation preserving and injective almost everywhere in Ω. The last property will be ensured by the Ciarlet-Nečas condition (3.4.7). We also assume that the elastic body is fixed on a part of its boundary by a Dirichlet condition. Recalling the Ciarlet-Nečas condition (3.4.7), we are left with the following problem:

$$\left.\begin{array}{ll}\text{Minimize} & J(y,c):=\displaystyle\int_\Omega\varphi(\nabla y,c)-f\cdot y\,\mathrm{d}x\\ \text{subject to} & \displaystyle\int_\Omega\det\nabla y\,\mathrm{d}x\le\mathrm{meas}_d(y(\Omega))\ \text{ and }\ \int_\Omega c\,\mathrm{d}x=C_{\rm total},\\ & \det\nabla y>0\ \text{ and }\ c\ge0\ \text{ a.e. on }\Omega,\\ & y\in W^{1,p}_{\Gamma_{\rm D}}(\Omega;\mathbb{R}^d)\ \text{ and }\ c\in L^r(\Omega).\end{array}\right\}\tag{3.6.8}$$

Here $W^{1,p}_{\Gamma_{\rm D}}(\Omega;\mathbb{R}^d):=\{y\in W^{1,p}(\Omega;\mathbb{R}^d):\ y=y_{\rm D}\text{ on }\Gamma_{\rm D}\}$. The direct method of the calculus of variations provides us with the following existence result.

Proposition 3.6.2 (Existence of minimizers). *Let us assume that*
(α) *the specific energy φ is cross-polyconvex,*
(β) $\varphi(F,c)\ge\varepsilon(|F|^p+\det F^{-1}+|c|^r-1)$ *for some* $\varepsilon>0$, $p>d$, *and* $r>1$,
(γ) $f\in L^{p^{*\prime}}(\Omega;\mathbb{R}^d)$.
Then, if (3.6.8) is feasible (i.e. its constraints allow for at least one admissible pair (y,c) such that $J(y,c)<+\infty$), then there exists its solution.

Proof. Let $\{(y_k,c_k)\}_{k\in\mathbb{N}}$ be a minimizing sequence of J. In particular, we can assume that $J(y_k,c_k)<+\infty$ for all $k\in\mathbb{N}$. The growth conditions (β) together with Poincaré's inequality imply that $\|y_k\|_{W^{1,p}(\Omega;\mathbb{R}^d)}+\|c_k\|_{L^r(\Omega)}<+\infty$. As $p>1$ as well as $r>1$ there are (not relabeled) subsequences, $y\in W^{1,p}_{\Gamma_{\rm D}}(\Omega;\mathbb{R}^d)$, and $c\in L^r(\Omega)$ such that $y_k\to y$ weakly $c_k\to c$, where both convergences are weak. The fact that $y\in W^{1,p}_{\Gamma_{\rm D}}(\Omega;\mathbb{R}^d)$ follows also from compactness of the trace operator. Moreover, $\int_\Omega c\,{\rm d}x=C_{\rm total}$ and $c\ge 0$ due to properties of weak convergence in $L^r(\Omega)$. Cross-polyconvexity ensures that $\liminf_{k\to\infty}J(y_k,c_k)\ge J(y,c)$ by the arguments used already in the proof of Theorem 3.4.1. Consequently, $J(y,c)<+\infty$ and, by the growth condition (β) in cooperation with the constraint $\det\nabla y_k>0$ in (3.6.8), also $\det\nabla y>0$ almost everywhere in Ω. Almost-everywhere injectivity of y follows by the same arguments as in the proof of Theorem 3.4.6. □

Remark 3.6.3 (Constancy of chemical potential). From the optimality conditions for a solution (y,c) to (3.6.8), in particular involving the partial differential with respect to c, one can read formally (if φ is suitably smooth and the constraint $c\ge 0$ is not active) that there is a Lagrange multiplier $\bar\mu\in\mathbb{R}$ to the constraint $\int_\Omega c\,{\rm d}x-C_{\rm total}=0$ and $\bar\mu=\partial_c\varphi(\nabla y,c)$ on Ω. This multiplier is in the position of chemical potential. Cf. also Sect. (5.7.2) below.

Example 3.6.4 (Biot's poroelastic model [69]). As an example of an integrand expressing the direct influence of the diffusant concentration on the volume, we can consider

$$\varphi(F,c)=\varphi_{\rm S}(F)+\frac{1}{2}M(c-c_0\beta(1-\det F))^2+\begin{cases}c(\ln(c/c_0)-1) & \text{for } c>0,\\ 0 & \text{for } c=0,\\ +\infty & \text{for } c<0.\end{cases}\qquad(3.6.9)$$

This integrand is cross-polyconvex provided the "solid" part $\varphi_{\rm S}$ is polyconvex. Here $c_0>0$ is an equilibrium concentration which minimizes $\varphi(\mathbb{I},\cdot)$ and M allowing an interpretation as the compressibility modulus of the fluid; actually, $M>0$ and $\beta>0$ are the so-called *Biot modulus* and the *Biot coefficient* used in conventional models of porous media.

Exercise 3.6.5. Prove that $(F,c)\mapsto c^m/(\det F)^{m-1}+\delta_{(0,+\infty)}(\det F)+\delta_{[0,+\infty)}(c)$ is cross-polyconvex if $m\ge 1$.

3.6.3 Equilibrium configuration for swelling-exhibiting materials

More involved models allow for modification of the reference configuration according some specific flow rules, which is particularly of interest in evolution situations in Part II describing phenomena like creep, swelling, or plasticity. Nevertheless, we can also formulate a static version of the problem. In order to involve an inelastic response of the material we introduce the Kröner-Lee-Liu *multiplicative decomposition* [288, 312] of the deformation gradient.

$$F = F_{\rm EL} F_{\rm IN} \quad \text{with the deformation gradient } F = \nabla y \tag{3.6.10}$$

and with y the deformation as in Chapters 1-2, where $F_{\rm EL}$ denotes the elastic strain. Now, instead of F, the tensor $F_{\rm EL}$ enters the stored energy.

A first and relatively simple example consists in considering $F_{\rm IN} = F_{\rm IN}(c)$ as dependent on one scalar variable with the interpretation of a concentration of a constituent that may possibly move in the solid elastic matrix through microscopical pores (or interstitial spaces in atomic grid, depending on particular applications). The dependence $F_{\rm IN}$ on c expresses the phenomenon called *swelling*.

Rather for a notational simplicity, we assume the swelling isotropic and put $F_{\rm IN}(c) := \mathbb{I}/b(c)$. Moreover, we impose an additive ansatz $\varphi(F,c) = \varphi_{\rm E}(F_{\rm EL}) + \varphi_{\rm C}(c)$ with $\varphi_{\rm E}$ the "elastic" and $\varphi_{\rm C}$ the "chemical" contributions to the overall stored energy. This leads to the stored energy

$$\varphi(F,c) = \varphi_{\rm E}(F_{\rm EL}) + \varphi_{\rm C}(c) = \varphi_{\rm E}(b(c)F) + \varphi_{\rm C}(c). \tag{3.6.11}$$

Taking the same set of admissible deformations and concentrations as in (3.6.8), we can state the following problem:

$$\left.\begin{array}{ll}\text{Minimize} & J(y,c) := \displaystyle\int_\Omega \varphi_{\rm E}(b(c)\nabla y) + \varphi_{\rm C}(c) + \frac{\kappa}{r}|\nabla c|^r - f\cdot y\,{\rm d}x \\ \text{subject to} & \displaystyle\int_\Omega \det\nabla y(x)\,{\rm d}x \le {\rm meas}_d(y(\Omega)) \ \text{ and } \ \int_\Omega c\,{\rm d}x = C_{\rm total}, \\ & \det\nabla y > 0 \ \text{ and } \ c \ge 0 \ \text{ a.e. on } \Omega, \\ & y \in W^{1,p}_{\Gamma_{\rm D}}(\Omega;\mathbb{R}^d) \ \text{ and } \ c \in W^{1,r}(\Omega). \end{array}\right\} \tag{3.6.12}$$

Let us notice that c enters φ in a more complicated way than in (3.6.9) and is, except rather trivial non-realistic cases, not compatible with the cross-polyconvexity concept. This forces us to involve also gradients of concentrations into the energy functional with $\kappa > 0$ a (presumably small) "capillarity" coefficient having (beside a physical interpretation) a regularizing effect from our mathematical viewpoint. We have the following existence result.

Proposition 3.6.6 (Existence of minimizers). *Let us assume that*
(α) *the ansatz* (3.6.11) *holds with* $\varphi_{\rm E} : \mathbb{R}^{d\times d} \to \mathbb{R}\cup\{+\infty\}$ *polyconvex and with* $\varphi_{\rm C} : [0,+\infty) \to \mathbb{R}\cup\{+\infty\}$ *lower semicontinuous,*
(β) $b : [0;+\infty) \to [0;+\infty)$ *is continuous and bounded with* $\inf b > 0$,
(γ) $\varphi_{\rm E}(F) \ge \varepsilon(|F|^q + \det F^{-1} - 1)$ *for some* $\varepsilon > 0$ *and* $q > p > d$,
(δ) $f \in L^{p^{*\prime}}(\Omega;\mathbb{R}^d)$.
Then, if (3.6.12) is feasible (i.e. its constraints allow for at least one admissible pair (y,c) *such that* $J(y,c) < +\infty$*), then there exists its solution.*

Proof. Let $\{(y_k, c_k)\}_{k\in\mathbb{N}}$ be a minimizing sequence such that $\lim_{k\to\infty} J(y_k, c_k) = \inf J$. Then $+\infty > \inf J > -\infty$ because of (γ) and because of the assumption on the existence of an admissible competitor with a finite value of J. Condition (γ) additionally implies that $\sup_{k\in\mathbb{N}} \|b(c_k)\nabla y_k\|_{L^q(\Omega;\mathbb{R}^{d\times d})} < +\infty$ and the form of the energy functional also yields that $\sup_{k\in\mathbb{N}} \|\nabla c_k\|_{L^r(\Omega;\mathbb{R}^d)} < +\infty$. Introducing a submultiplicative norm $|\cdot|$ on $\mathbb{R}^{d\times d}$, we have, for matrices $A, B \in \mathbb{R}^{d\times d}$, that $|A| \le |AB^{-1}||B|$ if B is invertible. Consequently,

$$|AB^{-1}| \ge \frac{|A|}{|B|} \ge \beta\delta^{\beta/(\beta-1)}|A|^{1/\beta} - (\beta-1)\delta|B|^{1/\beta-1}$$

for arbitrary $\beta > 1$ and $\delta > 0$. Take $\beta := q/p$ and calculate, for $1/s := 1/p - 1/q$ and all $k \in \mathbb{N}$, that

$$\begin{aligned}\|\nabla y_k b(c_k)\|^q_{L^q(\Omega;\mathbb{R}^{d\times d})} &\ge \|\nabla y_k\|^q_{L^p(\Omega;\mathbb{R}^{d\times d})}/\|b^{-1}(c_k)\mathbb{I}\|^q_{L^s(\Omega;\mathbb{R}^{d\times d})} \\ &\ge \beta\delta^{1/(\beta-1)}\|\nabla y_k\|^p_{L^p(\Omega;\mathbb{R}^{d\times d})} - (\beta-1)\delta\|b^{-1}(c_k)\mathbb{I}\|^s_{L^s(\Omega;\mathbb{R}^{d\times d})}.\end{aligned}$$

Hence for $\delta > 0$ small enough we get that $\sup_{k\in\mathbb{N}} \|\nabla y_k\|_{L^p(\Omega;\mathbb{R}^{d\times d})} < +\infty$. Altogether, we can extract a (non-relabeled) subsequences such that $y_k \rightharpoonup y$ in $W^{1,p}(\Omega;\mathbb{R}^d)$ and $c_k \rightharpoonup c$ in $W^{1,r}(\Omega)$; here we used also the boundedness of $\{c_k\}_{k\in\mathbb{N}}$ in $L^1(\Omega)$ due to the constraints in (3.6.12). Moreover, $\int_\Omega c\,\mathrm{d}x = C_{\mathrm{total}}$ and $c \ge 0$ hold for this limit, too.

Additionally, $b(c_k)\nabla y_k \rightharpoonup b(c)\nabla y$ and $\mathrm{Cof}(b(c_k)\nabla y_k) \rightharpoonup \mathrm{Cof}(b(c)\nabla y_k)$ in $L^1(\Omega;\mathbb{R}^{d\times d})$ and $\det(b(c_k)\nabla y_k) \rightharpoonup \det(b(c)\nabla y_k)$ in $L^1(\Omega)$ as $k\to\infty$. This allows us to show weak lower semicontinuity of the stored energy $(y,c) \mapsto \int_\Omega \varphi_{\mathrm{sw}}(b(c)\nabla y, c)$. Finally, the assumed polyconvexity of φ_{E} and lower semicontinuity of φ_{C} imply that J is weakly lower semicontinuous, so that we have $+\infty > \liminf_{k\to\infty} J(y_k, c_k) \ge J(y,c)$. This shows that $J(y,c) < +\infty$. Let us notice, that $b > 0$ almost everywhere in Ω otherwise $\det(b\nabla y) = 0$ which would cause $J(y,c) = +\infty$ due to (δ). By the same argument, using the growth condition (γ) in cooperation with the constraint $\det\nabla y_k > 0$ in (3.6.8), $\det\nabla y > 0$ almost everywhere in Ω. Injectivity of y almost everywhere in Ω again follows from the Ciarlet-Nečas conditions as in the proof of Theorem 3.4.6. □

Exercise 3.6.7. Show in detail that there are almost everywhere injective components y of minimizers to (3.6.12) if $p > d$. Find a condition such that y is injective everywhere.

3.7 Brief excursion into numerical approximation

Numerical approximation of the variational problem minimizing the functional $\mathcal{E}(y) = \int_\Omega \varphi(x, \nabla y(x))\,\mathrm{d}x$ is not entirely trivial. A straightforward idea consists in restricting the problem from the infinite-dimensional Sobolev space $W^{1,p}(\Omega;\mathbb{R}^d)$ with

p determined by the polynomial-type coercivity of $\varphi(t,\cdot)$ on a finite-dimensional subspace. Such a finite-dimensional minimization problem can in principle be implemented on computers, although even this is not entirely easy if the problem has many degrees of freedom and is nonconvex. This is the simplest idea behind numerical approximation, called the *Ritz method* [441] or, rather in a more general non-variational context, also the *Galerkin method* [212].

3.7.1 Basics in the Finite-Element Method (FEM)

A specific construction of these finite-dimensional subspaces uses a discretization of the domain Ω by division smaller subdomains and then using polynomials of a prescribed degree on each such a subdomain. This is called a *finite-element method*, abbreviated often as FEM. Although here we will only need to approximate maps of $W^{1,p}(\Omega;\mathbb{R}^d)$ we survey the general situation for $W^{k,p}(\Omega;\mathbb{R}^d)$ with $k \geq 1$ and exploit it fully in Section 4.5.3. It is called a *conformal* FEM if the created finite-dimensional spaces are subsets of the original space, i.e. here $W^{k,p}(\Omega)$ with $k = 0, 1, 2, \ldots$ The convergence is then achieved by refining the discretization of Ω (most often) or increasing the order of used polynomials, or both.

For problems in this book, only $k = 0, 1$ is relevant or sometimes, for 2nd-grade nonsimple materials, also $k = 2$. Simplest examples of FEM use only values of functions (not their derivatives). In the one-dimensional case ($d = 1$), the construction is rather trivial. The domain Ω is an interval to be divided into segments and the finite-dimensional subspace of $W^{k,p}(\Omega)$ consists of the piecewise (or we say element-wise) constant functions ($k = 0$), or element-wise affine C^0-functions ($k = 1$) or element-wise quadratic C^1-functions ($k = 2$) are used, cf. Figure 3.5.

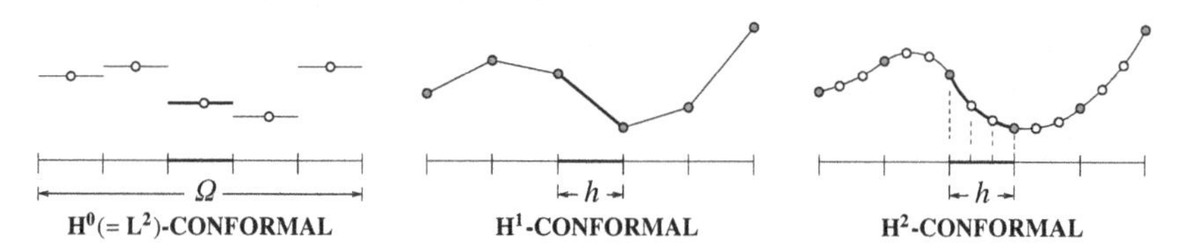

Fig. 3.5 Finite-element approximation in one-dimensional domain Ω with the mesh size h: P0 with values at barycenters s variables (left), P1 Lagrange element with nodal values as variables (middle) and P3 Lagrange element with values also at some extra-nodal points to allow for derivatives at nodal points continuous (right).

The multidimensional generalization of this construction is trivial for P0-elements. For Pk with $k \geq 1$, there are many variants to construct element-wise polynomial C^k-functions. One basic option consider each element as a simplex. P1-elements then takes only nodal values while H^2-conformal Lagrangian P3-elements consider also values at extra-nodal points on edges or inside the elements in order to make the derivatives continuous across the edges or sides. Another basic construction considers tensorial product, thus the one dimensional segment becomes a square in two dimensions and cube in three dimensions and the polynomial functions are taken in the form $v(x_1,...,x_d) = \prod_{i=1}^{d} v_i(x_i)$ with v_i being the n-degree polynomial on the one-dimensional interval. In the one-dimensional case, these constructions become trivial and naturally coincides, cf. Figure 3.6. Also P0 and Q0 (not depicted in Figure 3.6) coincide with each other. Both constructions can be combined to dis-

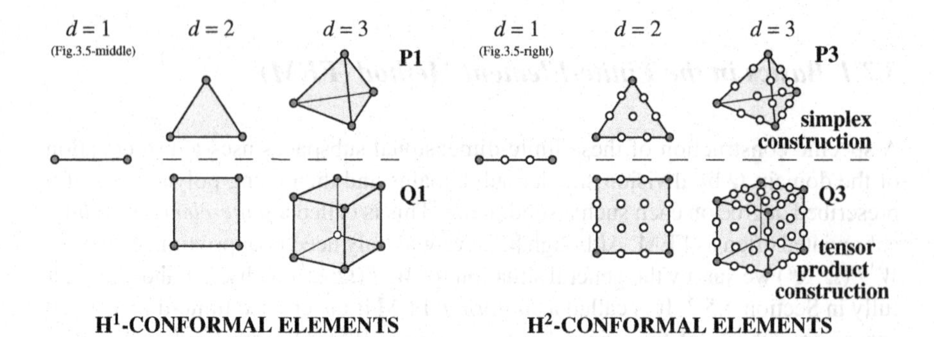

Fig. 3.6 Lagrangian finite-element approximation in multi-dimensional cases: the simplex construction in the upper row and the tensorial-product construction in the lower row.

cretize domains with a complicated shape, or possibly deformed for non-polyhedral domains.

Other construction of finite elements is based on a Hermite interpolation instead of the Lagrange one, i.e. it considers not only values but also derivatives. It opens a great menagerie of finite elements of higher order and in particular those which are H^2-conformal, cf. 88 two-dimensional H^2-conformal variants in [256].

For further purpose, without going into technical details behind the theory of the finite-element method, we denote by $\mathcal{P}_h^{(n)}$ the projector from $W^{k,p}(\Omega)$ into the Pn-finite element space on the triangulation $\mathcal{T}_h$, and we rely on the (standard) approximation property[13]

$$\lim_{h\to 0}\big\|v-\mathcal{P}_h^{(k)}v\big\|_{W^{k,p}(\Omega)} = 0 \qquad \forall v\in W^{k,p}(\Omega), \text{ and} \tag{3.7.1a}$$

$$\big\|v-\mathcal{P}_h^{(k)}v\big\|_{W^{l,p}(\Omega)} \le Ch^{k-l}\big\|v\big\|_{W^{k,p}(\Omega)} \quad \text{for } l = 0,...,k \tag{3.7.1b}$$

with some $C_{k,p}$ independent of h. The same estimates are at disposal also for the Qn-finite elements.

Remark 3.7.1 (Engineering implementation of nonsimple materials). In engineering, the nonsimple materials are rather computational implemented as a system of lower-order equations. E.g. splitting $\Delta^2 y$ is considered as Δz with $z = \Delta y$ and then on can use only H^1-conformal finite elements. This is however not fully compatible with our theoretical investigations based on the Galerkin method.

[13] Cf. e.g. [87, Thm. 4.4.20].

3.7.2 Lavrentiev phenomenon

In fact, these spaces constructed by polynomials are not only subspaces of $W^{1,p}(\Omega;\mathbb{R}^d)$ but even of $W^{1,\infty}(\Omega;\mathbb{R}^d)$. This obvious fact makes a difficulty if $\inf_{W^{1,p}(\Omega;\mathbb{R}^d)}\Phi < \inf_{W^{1,\infty}(\Omega;\mathbb{R}^d)}\Phi$, which may occur in problems in nonlinear elasticity due to the fast growth of W reflecting the local nonpenetration property (2.3.28) and which is then referred to as a Lavrentiev phenomenon. In such cases, the direct usage of FEM is impossible and one must somehow regularize the problem.

Originally, the Lavrentiev phenomenon was demonstrated on an innocent-looking one-dimensional counterexample for non-existence of a solution in the situation where rather the coercivity is weakened while the growth is preserved; more specifically the underlying Sobolev space is $W^{1,6}([0,1])$ and $\Phi\ge 0$ is continuous and weakly lower semicontinuous given

$$\Phi(u)=\int_0^1 (u^3-x)^2(u'(x))^6\,\mathrm{d}x \qquad \text{subject to} \quad u(0)=0,\ \ u(1)=1. \tag{3.7.2}$$

The minimum of (3.7.2) is obviously 0, being realized on $u(x)=x^{1/3}$. Such $u\in W^{1,1}([0,1])$ however does not belong to $W^{1,6}([0,1])$ because $u'(x)^6 = 3^{-6}x^{-4}$ is not integrable due to its singularity at $x=0$. Thus (3.7.2) attains the minimum on $W^{1,p}([0,1])$ with $1\le p<3/2$ although Φ is not (weakly lower semi-) continuous and even not finite on this space, and thus abstract Theorem A.3.3 cannot be used. A surprising and not entirely obvious phenomenon is that the infimum (3.7.2) on $W^{1,6}([0,1])$ is positive, i.e. greater than the infimum on $W^{1,p}([0,1])$ with $p<3/2$; this effect was first observed in [305], cf. also e.g. [101, Sect. 4.3.]. Note that $W^{1,6}([0,1])$ is dense in $W^{1,p}([0,1])$ but one cannot rely on $\Phi(u_k)\to\Phi(u)$ if $u_k\to u$ in $W^{1,p}([0,1])$ for $p<6$; it can even happen that $\Phi(u)=0$ while $\Phi(u_k)\to+\infty$ for $u_k\to u$, a *repulsive effect*, cf. [107, Sect. 7.3]. Here $\varphi(x,u,\cdot)$ is not uniformly convex, yet the Lavrentiev phenomenon can occur even for uniformly convex φ's, cf. [39].

An interesting observation is that polyconvex materials allow for an energy-controlled stress $|\varphi'(F)F^\top|\le C(1+\varphi(F))$ even though the so-called Kirchhoff stress $\varphi'(F)F^\top$ itself does not need to be bounded. This can be used e.g. in sensitivity analysis and to obtain modified Euler-Lagrange equations to overcome the possible failure of (C.1.3) for such materials, cf. [29]. It is worth noting that even such spatially homogeneous, frame-indifferent, and polyconvex materials can exhibit the *Lavrentiev phenomenon*, cf. [198].

3.7.3 Polyconvex materials and basic FEM approximation

Converging numerical approximation of variational problems involving physically relevant material models might be quite nontrivial. We consider here also the *Ciarlet-Nečas condition* (3.4.7), and the overall energy

$$\mathcal{E}(y) = \begin{cases} \displaystyle\int_\Omega \varphi(\nabla y)\,\mathrm{d}x - \ell(y) & \text{if } y|_{\Gamma_\mathrm{D}} = y_\mathrm{D} \text{ and } \displaystyle\int_\Omega \det(\nabla y)\,\mathrm{d}x \le \mathrm{meas}_d(y(\Omega)), \\ +\infty & \text{otherwise.} \end{cases} \tag{3.7.3}$$

We consider the polyconvex ansatz (3.1.1), i.e. $\varphi(F) = \tilde{\varphi}(F, \mathrm{Cof}\, F, \det F)$, in a bit special form[14]

$$\tilde{\varphi}(F, \mathrm{Cof}\, F, \det F) = \phi_0(F, \mathrm{Cof}\, F) + \gamma(\det F) \tag{3.7.4}$$

which covers, in particular, (2.4.6), (2.4.8), and (2.4.9). To ensure non-self-penetration, together with (3.4.7) we impose the local non-penetration $\gamma(\det F) = +\infty$ if $\det F \le 0$.

To overcome the Lavrentiev phenomenon as in Section 3.7.2 which would destroy convergence of a direct discretization of a minimization of $\mathcal{E}$ on $W^{1,p}(\Omega;\mathbb{R}^d)$, we first make an regularization of $\mathcal{E}$, considering a parameter $\varepsilon > 0$ and a suitable approximation φ_ε of φ. Applying the Yosida approximation (cf. (A.3.20) on p. 499) directly to the convex function ϕ would keep polyconvexity but might destroy the $W^{1,p}$-coercivity if $p > 2$. Thus we apply the *Yosida approximation* rather only to γ, so that altogether we consider

$$\varphi_\varepsilon(F) := \phi_0(F, \mathrm{Cof}\, F) + \gamma_\varepsilon(\det F) \quad \text{with} \quad \gamma_\varepsilon(J) := \inf_{\tilde{J} \in \mathbb{R}} \Big(\gamma(\tilde{J}) + \frac{(\tilde{J} - J)^2}{2\varepsilon}\Big). \tag{3.7.5}$$

We also penalize the global constraint involved in (3.7.3). Altogether, we thus obtain

$$\mathcal{E}_\varepsilon(y) := \begin{cases} \displaystyle\int_\Omega \varphi_\varepsilon(\nabla y)\,\mathrm{d}x - \ell(y) & \\ \displaystyle\quad + \frac{1}{\varepsilon}\Big(\int_\Omega \det(\nabla y)\,\mathrm{d}x - \mathrm{meas}_d(y(\Omega))\Big)^+ & \text{and if } y|_{\Gamma_\mathrm{D}} = y_\mathrm{D}, \\ +\infty & \text{otherwise.} \end{cases} \tag{3.7.6}$$

with $(\cdot)^+$ denoting the positive part.[15] Now we can make a finite-element discretization, considering a mesh parameter $h > 0$ and a corresponding finite-element subspace $V_h \subset W^{1,\infty}(\Omega;\mathbb{R}^d)$. Thus the approximated problem consists in minimization of $\mathcal{E}_\varepsilon$ on V_h.

To make a limit passage in the Ciarlet-Nečas condition in (3.7.6), the following continuity will be essential:

Lemma 3.7.2 (J. Malý [329]). *Let $\Omega \subset \mathbb{R}^d$ be an open set,* $\mathrm{meas}_d(\Omega) < +\infty$, *and let $y_j \to y$ in $W^{1,p}(\Omega;\mathbb{R}^d)$ weakly with some $p > d$. Then* $\mathrm{meas}_d(y_j(\Omega)) \to \mathrm{meas}_d(y(\Omega))$.

Proof. By the Sobolev embedding theorem B.3.2, $y_j \to y$ uniformly. If $K \subset \Omega$, using the area formula by Marcus and Mizel [334] and the Hölder inequality we estimate

[14] Note that if $d = 2$ then $\mathrm{Cof}\, F$ can be omitted because the set of its entries coincides (up to the sign at off-diagonal elements) with entries of F; cf. also (A.2.21).

[15] We recall that $f^+ := (|f| + f)/2$.

$$\mathrm{meas}_d(y_j(\Omega\backslash K)) \le \int_{\Omega\backslash K} |\det \nabla y_j|\,\mathrm{d}x \le \mathrm{meas}_d(\Omega\backslash K)^{1-1/p}\|\nabla y_j\|_{L^p(\Omega;\mathbb{R}^{d\times d})}.$$

It follows that given $\varepsilon > 0$ we can find a compact set $K \subset \Omega$ such that $\mathrm{meas}_d(y_j(\Omega \setminus K)) < \varepsilon$ with $j = 1,2,\ldots$. Let U be an open set containing $y(K)$ such that $\mathrm{meas}_d(U \setminus y(K)) < \varepsilon$. By the uniform convergence $y_j \to y$, there exists $k \in \mathbb{N}$ such that $y_j(K) \subset U$, $j \ge k$. For $j \ge k$ we have

$$\begin{aligned}\mathrm{meas}_d(y_j(\Omega)) &\le \mathrm{meas}_d(y_j(K)) + \mathrm{meas}_d(y_j(\Omega \setminus K))\\ &\le \mathrm{meas}_d(U) + \varepsilon \le \mathrm{meas}_d(y(\Omega)) + 2\varepsilon,\end{aligned}$$

so that

$$\mathrm{meas}_d(y(\Omega)) \ge \limsup_{j\to+\infty} \mathrm{meas}_d(y_j(\Omega)).$$

For the lower semicontinuity, recall that $W^{1,p}$-functions have the so-called Luzin (N)-property if $p > d$, i.e. sets of measure zero are mapped to sets of measure zero, see e.g. [439, Thm II.2.3]. By [337, Thm. 3.10], almost all points $z \in y(\Omega)$ are stable, this means that there exists $\delta > 0$ such that $z \in \tilde{y}(\Omega)$ for each continuous mapping $\tilde{y} : \Omega \to \mathbb{R}^d$ with $\|\tilde{y} - y\|_{C(\bar{\Omega};\mathbb{R}^d)} < \delta$. By the uniform convergence $y_j \to y$, almost all points of $\tilde{y}(\Omega)$ belong to the set $\bigcup_{k=1}^{\infty} \bigcap_{j=k}^{\infty} y_j(\Omega)$. It follows that

$$\mathrm{meas}_d(y(\Omega)) \le \liminf_{j\to+\infty} \mathrm{meas}_d(y_j(\Omega)). \qquad \square$$

We will further assume the coercivity

$$\exists c > 0,\ \ p > 1\ \ \forall F \in \mathbb{R}^{d\times d}: \quad \varphi(F) \ge c|F|^p. \tag{3.7.7}$$

Existence of a minimizer, let us denote it by $y_{\varepsilon h}$, of $\mathcal{E}_\varepsilon$ on V_h is simply just by a direct method, which works quite trivially on the finite-dimensional subspaces V_h; actually, a compatibility of the Dirichlet condition y_{D} with the finite-element discretization is needed. The a-priori bounds in $W^{1,p}(\Omega;\mathbb{R}^d)$ follow from coercivity (3.7.7) of φ which is not affected by the regularization parameter ε. Fixing $\varepsilon > 0$, the weak convergence for $h \to 0$ of $y_{\varepsilon h}$ (in terms of subsequences) to minimizers of $\mathcal{E}_\varepsilon$ on $W^{1,p}(\Omega;\mathbb{R}^d)$ is then obtained by the Γ-convergence $\mathcal{E}_\varepsilon + \delta_{V_h} \to \mathcal{E}_\varepsilon$; here coercivity, polyconvexity of φ_ε, as well as Lemma 3.7.2 is used[16]. See also Example A.3.20. Eventually, denoting by y_ε a minimizer of $\mathcal{E}_\varepsilon$ on $W^{1,p}(\Omega;\mathbb{R}^d)$, we pass the regularization ε to zero, having $y_\varepsilon \to y$ weakly in $W^{1,p}(\Omega;\mathbb{R}^d)$ in terms of subsequences, and any y obtained as a limit solves the original problem, i.e. minimizes $\mathcal{E}$ on $W^{1,p}(\Omega;\mathbb{R}^d)$. At the last point, we can rely on Γ-convergence $\mathcal{E}_\varepsilon \xrightarrow{\Gamma} \mathcal{E}$ which follows from Theorem A.3.17 and from the properties of the Yosida approximation, see (A.3.20). A diagonalization argument then leads to the conclusion that

[16] Here we also apply the formula

$$|(\alpha-\beta)^+ - (\gamma-\delta)^+| \le |\alpha-\gamma| + |\beta-\delta|$$

valid for every $\alpha,\beta,\gamma,\delta \in \mathbb{R}$. We also recall that if $a \in \mathbb{R}$ then $a^+ := \max(a,0) = (a+|a|)/2$.

$\lim_{\varepsilon\to 0}\lim_{h\to 0} y_{\varepsilon h} = y$ weak in $W^{1,p}(\Omega;\mathbb{R}^d)$ and y minimizes $\mathcal{E}$ defined in (3.7.3). Altogether we have proved the following proposition.

Proposition 3.7.3 (Convergence of the numerical approximation[17]). *Let $\Omega\subset\mathbb{R}^d$ be a bounded Lipschitz domain, let $\varphi:\mathbb{R}^{d\times d}\to\mathbb{R}\cup\{+\infty\}$ be polyconvex, and let (3.7.4),(3.7.5), and (3.7.7) hold with $p>d$. Assume further that $\ell\in W^{1,p}(\Omega;\mathbb{R}^d)^*$ and that $V_h:=\mathcal{P}_h^{(1)}W^{1,p}(\Omega;\mathbb{R}^d)$ where $\mathcal{P}_h^{(1)}$ is defined in Section 3.7.1. Then there are minimizers, $y_{\varepsilon h}\in V_h\subset W^{1,p}(\Omega;\mathbb{R}^d)$ of $\mathcal{E}_\varepsilon$ defined in (3.7.6) for every $\varepsilon>0$ and $h>0$. Moreover, there is a (non-relabeled) subsequence such that $\lim_{\varepsilon\to 0}\lim_{h\to 0} y_{\varepsilon h}=y$ weak in $W^{1,p}(\Omega;\mathbb{R}^d)$ and y minimizes $\mathcal{E}$ defined in (3.7.3).*

The above approximation allows only for a successive limit $h\to 0$ and then $\varepsilon\to 0$. The joint limit $(h,\varepsilon)\to(0,0)$ needs some (unspecified) stability criterion.

Actually, the same convergence results holds if a smooth penalty would be used in (3.7.6) instead of the nonsmooth function $(\cdot)^+$.

Remark 3.7.4 (Alternatives for fighting with Lavrentiev's phenomenon). Another discretization overcoming the Lavrentiev phenomenon may use regularization of the discrete problem not having a direct counterpart in the continuous setting (e.g. introducing an auxiliary variable η and the constraint $\eta=\nabla y$ which is penalized) or confines on convex problems, cf. [24, 35, 107, 315, 392].

[17] See also [361] even for a quasistatic variant.

Chapter 4
General hyperelastic materials: existence/nonexistence results

> Because of the success of the 'direct methods' in the Calculus of Variations, many writers have shown that certain integrals are lower semicontinuous. ... 'quasi-convexity' ... is both necessary and sufficient for the lower semicontinuity...
>
> CHARLES BRADFIELD MORREY JR. (1907–1984)

As we have seen before, polyconvexity ensures existence of a minimizer to the energy functional. Moreover, polyconvex functions are relatively easy to construct and the theory developed allows us to incorporate important physical requirements into the model. On the other hand, polyconvexity is only a sufficient condition for weak lower semicontinuity and one can ask what is a necessary condition. In this chapter, we touch these problems, show a condition called (Morrey's) *quasiconvexity* which is the sought necessary and sufficient condition ensuring lower semicontinuity in the weak topology. Moreover, some material models, however, do not allow for polyconvex bulk energy density. Beside St.Venant-Kirchhoff material (2.4.3), another prominent example are materials exhibiting clear tendency for creating a microstructure due to purely mechanical reasons as shape memory material, for instance. These materials posses non-quasiconvex stored energy density but if we include an interface energy term for each phase of the material we obtain existence results, as well.

4.1 Necessary and sufficient conditions for weak lower semicontinuity

We have seen that polyconvexity is a sufficient condition to ensure sequential weak lower semicontinuity of the energy functional $\mathcal{E}$. However, polyconvexity is far from being a necessary condition. The right condition to look at in this respect is the so-called *quasiconvexity*[1].

Let us first start with an example. Consider $\varOmega \subset \mathbb{R}^d$ an open bounded domain and Q a cube inside $\varOmega$ with faces parallel to Cartesian coordinate axes. Consider

[1] Sometimes it is called Morrey's [375] quasiconvexity in order to distinguish it from "quasiconvexity" in optimization theory which refers merely to convexity of sub-level sets.

M. Kružík and T. Roubíček, *Mathematical Methods in Continuum Mechanics of Solids*, Interaction of Mechanics and Mathematics,
https://doi.org/10.1007/978-3-030-02065-1_4

$\tilde{y} \in W_0^{1,\infty}(Q;\mathbb{R}^d)$ and extend it to the whole $\mathbb{R}^d$ periodically without changing its name. Let $\tilde{y}_k(x) := k^{-1}\tilde{y}(kx)$ for all $k \in \mathbb{N}$. Take $A \in \mathbb{R}^{d\times d}$ and define $y_k \in W^{1,\infty}(\Omega;\mathbb{R}^d)$ by

$$y_k(x) := \begin{cases} Ax & \text{if } x \in \Omega \setminus Q, \\ Ax + \tilde{y}_k(x) & \text{if } x \in Q. \end{cases}$$

Let us assume that $\mathcal{E}(y) := \int_\Omega \varphi(\nabla y(x))\,dx$ is weakly* lower semicontinuous on $W^{1,\infty}(\Omega;\mathbb{R}^d)$. Then $y_k \to y$ weakly*, $y(x) = Ax$ for $x \in \Omega$, and

$$\begin{aligned} \liminf_{k\to\infty} \mathcal{E}(y_k) &= \mathrm{meas}_d(\Omega \setminus Q)\varphi(A) + \lim_{k\to\infty} \frac{1}{k^d} \int_{kQ} \varphi(A + \nabla\tilde{y}(x))\,dx \\ &\geq \varphi(A)\mathrm{meas}_d(\Omega) = \mathcal{E}(y). \end{aligned} \tag{4.1.1}$$

Hence, we see that if $\mathcal{E}$ is sequentially weakly* lower semicontinuous then

$$\forall A \in \mathbb{R}^{d\times d}\ \forall \tilde{y} \in W_0^{1,\infty}(Q;\mathbb{R}^d): \quad \mathrm{meas}_d(Q)\varphi(A) \leq \int_Q \varphi(A + \nabla\tilde{y}(x))\,dx, \tag{4.1.2}$$

because A and y were arbitrary. This condition is called (Morrey's) *quasiconvexity* . As we will see later, it is also sufficient for weak* lower semicontinuity on $W^{1,\infty}(\Omega;\mathbb{R}^d)$. If $\Omega \subset \mathbb{R}^d$ is bounded then $W^{1,\infty}(\Omega;\mathbb{R}^d) \subset W^{1,p}(\Omega;\mathbb{R}^d)$ for every $1 \leq p < +\infty$. Moreover, it is easy to see that if $y_k \to y$ weakly* in $W^{1,\infty}(\Omega;\mathbb{R}^d)$ for $k \to \infty$ then $y_k \to y$ weakly in $W^{1,p}(\Omega;\mathbb{R}^d)$. Consequently, quasiconvexity is also a necessary condition for weak lower semicontinuity of $\mathcal{E}$ on $W^{1,p}(\Omega;\mathbb{R}^d)$ for $1 \leq p \leq +\infty$. We have the following definition of this notion.

Definition 4.1.1 (Quasiconvexity). Let $\Omega \in \mathbb{R}^d$ be a bounded Lipschitz domain. We say that $\varphi : \mathbb{R}^{d\times d} \to \mathbb{R} \cup \{+\infty\}$ is quasiconvex if for any $A \in \mathbb{R}^{d\times d}$ and every $y \in W_0^{1,\infty}(\Omega;\mathbb{R}^d)$ it holds that

$$\varphi(A) \leq \frac{1}{\mathrm{meas}_d(\Omega)} \int_\Omega \varphi(A + \nabla y(x))\,dx \tag{4.1.3}$$

whenever the integral on the right side of (4.1.3) exists. If (4.1.3) holds only for a particular $A \in \mathbb{R}^{d\times d}$, we say that φ is quasiconvex at A.

The notion of quasiconvexity is independent of a particular Lipschitz domain used in the definition; cf. e.g. [135]. Hence, Ω in Definition 4.1.1 can be replaced by an arbitrary bounded Lipschitz domain $\Omega_1 \subset \mathbb{R}^d$ and the set of quasiconvex functions $\mathbb{R}^{d\times d} \to \mathbb{R} \cup \{+\infty\}$ remains the same. The proof by Meyers [349] works only for φ finite but it is very easy. Indeed, if (4.1.3) holds for Ω and $\Omega_1 \subset \mathbb{R}^d$ is another bounded Lipschitz domain then $x_0 + \varepsilon\Omega_1 \subset \Omega$ for some $x_0 \in \mathbb{R}^d$ and $\varepsilon > 0$. Every $y \in W_0^{1,\infty}(x_0 + \varepsilon\Omega_1;\mathbb{R}^d)$ can be extended by zero to the whole Ω so that we calculate

$$\int_{x_0+\varepsilon\Omega_1} \varphi\Big(A + \nabla y\Big(\frac{x-x_0}{\varepsilon}\Big)\Big)dx + \mathrm{meas}_d(\Omega \setminus (x_0{+}\varepsilon\Omega_1))\,\varphi(A) \geq \mathrm{meas}_d(\Omega)\varphi(A),$$

from which it follows that

$$\varphi(A) \leq \frac{1}{\mathrm{meas}_d(\Omega_1)} \int_{\Omega_1} \varphi(A + \nabla y(x))\,\mathrm{d}x \tag{4.1.4}$$

The proof was modified by Ball and Murat [40] for integrands taking infinite values. Indeed, assume that (4.1.3) holds for Ω and take $\Omega_1 \subset \mathbb{R}^d$, another bounded Lipschitz domain. We take the family

$$\{\Theta_j\}_j = \left\{x \in a + \epsilon\bar{\Omega}_1 \subset \Omega;\ a \in \mathbb{R}^d,\ \epsilon > 0\right\}$$

which is a Vitali covering of Ω; cf. Definition B.4.16. By Theorem B.4.17 there exists a countable collection $\{x \in a_i + \epsilon_i\bar{\Omega}_1\}_{i\in\mathbb{N}}$, $\epsilon_i > 0$ and $a_i \in \mathbb{R}^d$ for all i, of pairwise disjoint sets and

$$\Omega = \bigcup_i \{x \in a_i + \epsilon_i\bar{\Omega}_1\} \bigcup \omega, \quad \mathrm{meas}_d(\omega) = 0.$$

We see that $\sum_i \epsilon_i^d = \mathrm{meas}_d(\Omega)/\mathrm{meas}_d(\Omega_1)$. For $y \in W_0^{1,\infty}(\Omega_1;\mathbb{R}^d)$, we define

$$\tilde{y}(x) = \begin{cases} \epsilon_i y\left(\frac{x-a_i}{\epsilon_i}\right) & \text{if } x \in a_i + \epsilon_i\Omega_1 \\ 0 & \text{otherwise}. \end{cases}$$

We easily see that $\tilde{y} \in W_0^{1,\infty}(\Omega;\mathbb{R}^d)$ and therefore it is an admissible test map for (4.1.3). Consequently,

$$\begin{aligned} \mathrm{meas}_d(\Omega)\varphi(A) &\leq \int_\Omega \varphi(A + \tilde{y}(x))\,\mathrm{d}x = \sum_i \int_{a_i+\varepsilon_i\Omega_1} \varphi\Big(A + \nabla y\Big(\frac{x-a_i}{\epsilon_i}\Big)\Big)\mathrm{d}x \\ &= \sum_i \epsilon_i^d \int_{\Omega_1} \varphi(A + \nabla y(x))\,\mathrm{d}x = \frac{\mathrm{meas}_d(\Omega)}{\mathrm{meas}_d(\Omega_1)} \int_{\Omega_1} \varphi(A + \nabla y(x))\,\mathrm{d}x, \end{aligned}$$

which implies the result.

There is an equivalent formulation of quasiconvexity which uses periodic test mappings instead of those with the zero trace. The proof of the next proposition exploits a simple fact that locally bounded functions $\mathbb{R}^{d\times d} \to \mathbb{R}$ are bounded on every compact subset of $\mathbb{R}^{d\times d}$. It easily follows by constructing a finite covering of the compact set.

Proposition 4.1.2 (Quasiconvexity with periodic test functions). *Let $\varphi : \mathbb{R}^{d\times d} \to \mathbb{R}$ be quasiconvex and continuous*[2]*. Let $\tilde{y} \in W^{1,\infty}(\mathbb{R}^d;\mathbb{R}^d)$ be $(0,1)^d$-periodic. Then, for all $A \in \mathbb{R}^{d\times d}$, it holds*

$$\varphi(A) \leq \int_{(0,1)^d} \varphi(A + \nabla\tilde{y}(x))\,\mathrm{d}x. \tag{4.1.5}$$

[2] We will see in the proof of Thm. 4.1.3 that finite functions satisfying (4.1.5) are continuous.

Proof. Define $\tilde{y}_k(x) := k^{-1}\tilde{y}(kx)$. Notice that $\tilde{y}_k \to 0$ weakly* in $W^{1,\infty}(\mathbb{R}^d;\mathbb{R}^d)$. Consider for $\ell \in \mathbb{N}$ large enough following cut-off functions $\eta_\ell \in C^1_0((0,1)^d)$, $|\nabla\eta_\ell| \le K\ell$ for some $K > 0$ independent of ℓ:

$$\eta_\ell(x) := \begin{cases} 1 & \text{if } \operatorname{dist}(x, \partial(0,1)^d) \ge \ell^{-1} \\ 0 & \text{if } x \in \partial(0,1)^d. \end{cases}$$

Consequently, $\eta_\ell\tilde{y}_k \in W^{1,\infty}_0((0,1)^d;\mathbb{R}^d)$ and, by (4.1.3), we have $\varphi(A) \le \int_{(0,1)^d} \varphi(A + \nabla\eta_\ell(x) \otimes \tilde{y}_k(x) + \eta_\ell(x)\nabla\tilde{y}_k(x))\,\mathrm{d}x$. Then we observe

$$\varphi(A) \le \int_{(0,1)^d} \varphi(A+\nabla\eta_\ell(x)\otimes\tilde{y}_k(x)+\eta_\ell(x)\nabla\tilde{y}_k(x))\,\mathrm{d}x = \int_{(\ell^{-1},1-\ell^{-1})^d} \varphi(A+\nabla\tilde{y}_k(x))\,\mathrm{d}x$$
$$+ \int_{(0,1)^d\setminus(\ell^{-1},1-\ell^{-1})^d} \varphi(A+\nabla\eta_\ell(x)\otimes\tilde{y}_k(x)+\eta_\ell(x)\nabla\tilde{y}_k(x))\,\mathrm{d}x.$$

The last term, however, converges to zero if $\ell \to +\infty$ and $k := \ell$. Namely, the integrand is bounded and the measure of the set over which we integrate vanishes. Moreover, if $k = \ell$ then

$$\int_{(\ell^{-1},1-\ell^{-1})^d} \varphi(A+\nabla\tilde{y}_k(x))\,\mathrm{d}x = k^{-d}\int_{(1,k-1)^d} \varphi(A+\nabla\tilde{y}(z))\,\mathrm{d}z$$
$$\to \int_{(0,1)^d} \varphi(A+\nabla\tilde{y}(x))\,\mathrm{d}x$$

as $k \to \infty$ because $\tilde{y}$ is $(0,1)^d$-periodic. The claim is proved. □

Quasiconvexity is very difficult or currently impossible to verify. However, there is a simpler necessary condition, namely rank-one convexity; cf. Definition 3.1.2.

Theorem 4.1.3 (Quasiconvexity implies rank-one convexity). *Assume that $\varphi : \mathbb{R}^{d\times d} \to \mathbb{R}$ satisfies (4.1.5) for every $A \in \mathbb{R}^{d\times d}$ and every $\tilde{y} \in W^{1,\infty}(\mathbb{R}^d;\mathbb{R}^d)$ which is $(0,1)^d$-periodic. Then φ is continuous and rank-one convex.*

Proof. Take $a \in \mathbb{R}^d$ arbitrary and $b \in \mathbb{R}^d$ with integer components. Let $0 \le \lambda \le 1$ and let $\chi : \mathbb{R} \to \mathbb{R}$ be a periodic extension of the function $\chi_0 : [0,1] \to \mathbb{R}$ defined as

$$\chi_0(x) := \begin{cases} 1-\lambda & \text{if } 0 \le x \le \lambda, \\ -\lambda & \text{if } \lambda \le x \le 1. \end{cases}$$

Define $\tilde{y}(x) := a\int_0^{x\cdot b} \chi(s)\,\mathrm{d}s$ for $x \in \mathbb{R}^d$. Notice that $\tilde{y} \in W^{1,\infty}(\mathbb{R}^d;\mathbb{R}^d)$ and that it is $(0,1)^d$-periodic. Indeed, we must show that $\tilde{y}(x) = \tilde{y}(x+z)$ for every $x \in \mathbb{R}^d$ and every $z \in \mathbb{R}^d$ with integer components. We have

$$\tilde{y}(x+z) = a\int_0^{(x+z)\cdot b} \chi(s)\,\mathrm{d}s = a\int_0^{z\cdot b} \chi(s)\,\mathrm{d}s + a\int_{z\cdot b}^{(x+z)\cdot b} \chi(s)\,\mathrm{d}s.$$

The first term on the right hand side equals zero because $z\cdot b$ is an integer while the second term can be written as

$$a\int_{z\cdot b}^{(x+z)\cdot b}\chi(s)\,\mathrm{d}s = a\int_0^{x\cdot b}\chi(\tilde{s}+z\cdot b)\,\mathrm{d}\tilde{s} = a\int_0^{x\cdot b}\chi(\tilde{s})\,\mathrm{d}\tilde{s} = y(x),$$

because χ is 1-periodic. Also, $\nabla\tilde{y}(x) = \chi(x\cdot b)a\otimes b$. The periodicity of $\tilde{y}$ implies $0 = \int_{(0,1)^d}\nabla\tilde{y}(x)\,\mathrm{d}x = 0$. This means that $\mathrm{meas}_d(\{x\in(0,1)^d;\ \nabla\tilde{y}(x) = (1-\lambda)a\otimes b\}) = \lambda$ and, similarly, $\mathrm{meas}_d(\{x\in(0,1)^d;\ \nabla\tilde{y}(x) = -\lambda a\otimes b\}) = 1-\lambda$. Consequently, if $A\in\mathbb{R}^{d\times d}$ and $\varphi:\mathbb{R}^{d\times d}\to\mathbb{R}$ is quasiconvex, we have

$$\varphi(A) \le \int_{(0,1)^d}\varphi(A+\nabla\tilde{y}(x))\,\mathrm{d}x = \lambda\varphi(A+(1-\lambda)a\otimes b) + (1-\lambda)\varphi(A-\lambda a\otimes b), \quad (4.1.6)$$

which, in particular, shows convexity of φ along Cartesian coordinates in $\mathbb{R}^{d\times d}$ (this property is usually called separate convexity) and therefore continuity of φ. Now consider $b\in\mathbb{R}^d$ arbitrary but with rational components, i.e., assume that $b := (\beta_1/\beta,\beta_2/\beta,\ldots)$ for β,β_i integers for all $1\le i\le d$, and $\beta\ne 0$. As $a\otimes b = (a/\beta)\otimes(\beta_1,\beta_2,\ldots)$ for every $a\in\mathbb{R}^d$ we see that

$$\varphi(A) \le \lambda\varphi(A+(1-\lambda)a\otimes b) + (1-\lambda)\varphi(A-\lambda a\otimes b) \quad (4.1.7)$$

holds also if $b\in\mathbb{R}^d$ has rational components. The general result follows by continuity of φ and by the density of rational numbers in reals. □

In many situations, we would like to use rather $y\in W_0^{1,p}(\Omega;\mathbb{R}^d)$ instead of $y\in W_0^{1,\infty}(\Omega;\mathbb{R}^d)$ in Definition 4.1.1. Before discussing this, we need the following elementary result, which can be found in a more general form e.g. in [135, Ch. 4, Lemma 2.2]:

Lemma 4.1.4 (p-Lipschitz property). *Let $\varphi:\mathbb{R}^{d\times d}\to\mathbb{R}$ be rank-one convex with $|\varphi(F)|\le K(1+|F|^p)$, $K>0$, for all $F\in\mathbb{R}^{d\times d}$ and some $1\le p<+\infty$. Then there is a constant $\tilde{K}\ge 0$ such that for every $F_1,F_2\in\mathbb{R}^{d\times d}$ it holds*

$$|\varphi(F_1)-\varphi(F_2)| \le \tilde{K}(1+|F_1|^{p-1}+|F_2|^{p-1})|F_1-F_2|. \quad (4.1.8)$$

Proposition 4.1.5 (Quasiconvexity with $W_0^{1,p}$-test functions). *Let $\varphi:\mathbb{R}^{d\times d}\to\mathbb{R}$ be continuous and such that $|\varphi(A)|\le K(1+|A|^p)$ for all $A\in\mathbb{R}^{d\times d}$ and some $K>0$ independent of A and some $1\le p<+\infty$. If φ is quasiconvex then* (4.1.3) *holds also if $y\in W_0^{1,p}(\Omega;\mathbb{R}^d)$.*

Proof. Take $y\in W_0^{1,p}(\Omega;\mathbb{R}^d)$ and a sequence $\{y_k\}_{k\in\mathbb{N}}\subset W_0^{1,\infty}(\Omega;\mathbb{R}^d)$ such that $\tilde{y}_k\to\tilde{y}$ strongly in $W^{1,p}(\Omega;\mathbb{R}^d)$. We get using (4.1.8) that

$$\int_\Omega\varphi(F+\nabla y(x))\,\mathrm{d}x = \lim_{k\to\infty}\int_\Omega\varphi(F+\nabla y_k(x))\,\mathrm{d}x. \quad (4.1.9)$$

As φ is quasiconvex, we also have

$$\lim_{k\to\infty}\int_\Omega \varphi(F+\nabla y_k(x))\,\mathrm{d}x \geq \mathrm{meas}_d(\Omega)\varphi(F),$$

which finishes the proof. □

It is an easy exercise to show that rank-one convex (and therefore also quasiconvex) functions are continuous. We leave a proof to the reader.

If we admit that $\varphi : \mathbb{R}^{d\times d} \to \mathbb{R}\cup\{+\infty\}$ it is no longer true that quasiconvexity implies rank-one convexity, see [40].

Proposition 4.1.6 (Quasiconvex functions taking value $+\infty$). *Let $d \geq 1$. There is $\varphi : \mathbb{R}^{d\times d} \to \mathbb{R}\cup\{+\infty\}$ quasiconvex which is not rank-one convex.*

Proof. Consider two nonzero vectors $a, b \in \mathbb{R}^d$ and define

$$\varphi(F) := \begin{cases} 0 & \text{if } F = 0 \text{ or } F = a\otimes b, \\ +\infty & \text{otherwise.} \end{cases}$$

Obviously, φ is not rank-one convex but a simple calculation shows that (4.1.3) holds. □

On the other hand, it has been open for about forty years if the opposite implication holds or not. In 1992, V. Šverák [500] showed that there is a rank-one convex function $\mathbb{R}^{d\times d} \to \mathbb{R}$ which is not quasiconvex if $d \geq 3$. The case $d = 2$, still remains open despite of a huge effort of many outstanding mathematicians. Let us explain the idea of Šverák's counterexample. We will first work in the space of matrices $\mathbb{R}^{3\times 2}$ and then we will extend the result to higher dimensions. First of all if $\varphi : \mathbb{R}^{3\times 2} \to \mathbb{R}$ is quasiconvex then

$$\varphi(A) \leq \int_{(0,1)^2} \varphi(A+\nabla\tilde{y}(x))\,\mathrm{d}x$$

for any $\tilde{y} \in W^{1,\infty}(\mathbb{R}^2;\mathbb{R}^3)$, $(0,1)^2$-periodic. This immediately follows from the proof of Proposition 4.1.2 because the proof does not use the fact that the the domain and the range of φ have the same dimensions. We take $A := 0$ and consider $y : (0,1)^2 \to \mathbb{R}^3$ given by the following formula

$$y(x) := \frac{1}{2\pi}(\sin 2\pi x_1, \sin 2\pi x_2, \sin 2\pi(x_1+x_2)). \tag{4.1.10}$$

We see that $\nabla y \subset L$, where

$$L := \left\{ \begin{pmatrix} r & 0 \\ 0 & s \\ t & t \end{pmatrix} ;\ r,s,t\in\mathbb{R} \right\}. \tag{4.1.11}$$

Moreover, the only rank-one matrices in L are multiples of the following three ones:

$$\begin{pmatrix} 1 & 0 \\ 0 & 0 \\ 0 & 0 \end{pmatrix}, \quad \begin{pmatrix} 0 & 0 \\ 0 & 1 \\ 0 & 0 \end{pmatrix}, \quad \text{and} \quad \begin{pmatrix} 0 & 0 \\ 0 & 0 \\ 1 & 1 \end{pmatrix}.$$

Let us define $\varphi : L \to \mathbb{R}$ by

$$\varphi\left(\begin{pmatrix} r & 0 \\ 0 & s \\ t & t \end{pmatrix}\right) = -rst.$$

Clearly, φ is rank-one convex on L but, for y from (4.1.10), it holds

$$\int_{(0,1)^3} \varphi(\nabla y(x))\,\mathrm{d}x < \varphi(0) = 0.$$

Indeed, we have

$$\begin{aligned} -\int_{(0,1)^2} & \cos 2\pi x_1 \cos 2\pi x_2 \cos 2\pi(x_1 + x_2)\,\mathrm{d}x = \\ = -\int_{(0,1)^2} & \cos^2 2\pi x_1 \cos^2 2\pi x_2 + \cos 2\pi x_1 \cos 2\pi x_1 \sin 2\pi x_2 \,\mathrm{d}x < 0. \end{aligned} \tag{4.1.12}$$

Then φ is modified and extended to $\mathbb{R}^{3\times 2}$. We must show that for every $\varepsilon > 0$ there is $k > 0$ such that $\varphi_{\varepsilon,k} : \mathbb{R}^{3\times 2} \to \mathbb{R}$ defined as

$$\varphi_{\varepsilon,k}(A) := \varphi(PA) + \varepsilon(|A|^2 + |A|^4) + k|A - PA|^2 \tag{4.1.13}$$

is rank-one convex. Here $P : \mathbb{R}^{3\times 2} \to L$ defines a projector on L. For $A \in \mathbb{R}^{3\times 2}$ such that $A := (A_{ij})$, we take

$$PA := \frac{1}{2}\begin{pmatrix} 2A_{11} & 0 \\ 0 & 2A_{22} \\ A_{21} + A_{22} & A_{21} + A_{22} \end{pmatrix}.$$

Indeed, assume for a contradiction that there is $\varepsilon_0 > 0$ such that $\varphi_{\varepsilon_0,k}$ is not rank-one convex for arbitrary $k > 0$. It just means that for every $k > 0$ we find two unit vectors $a_k \in \mathbb{R}^3$ and $b_k \in \mathbb{R}^2$, i.e., $|a_k| = |b_k| = 1$ and $B_k \in \mathbb{R}^{3\times 2}$ such that $t \mapsto \varphi_{\varepsilon_0,k}(B_k + ta_k \otimes b_k)$ is not convex. As $\varphi_{\varepsilon_0,k}$ is a polynomial, it just means that

$$\partial^2 \varphi_{\varepsilon_0,k} \partial F^2(B_k) : a_k \otimes b_k : a_k \otimes b_k < 0 \tag{4.1.14}$$

due to the Legendre-Hadamard condition (3.1.8). In other words, for all $k > 0$

$$\begin{aligned} \partial^2 \varphi(PB_k) : P(a_k \otimes b_k) : P(a_k \otimes b_k) + 2\varepsilon_0 |a_k \otimes b_k|^2 + \varepsilon_0(4|B_k|^2 |a_k \otimes b_k|^2 \\ + 8|B_k : (a_k \otimes b_k)|^2 + k|a_k \otimes b_k - P(a_k \otimes b_k)|^2 < 0. \end{aligned}$$

As the left-hand side is coercive in B_k we get that $|B_k| \le K$ for some $K > 0$. So, we can assume that (possibly for a non-relabeled subsequences) $B_k \to A$, $a_k \to a$ and $b_k \to b$ for some $A \in \mathbb{R}^{3\times 3}$ and $a, b \in \mathbb{R}^3$. We also get that $P(a \otimes b) = a \otimes b$ which contradicts rank-one affinity of φ on L. Hence, for every $\varepsilon > 0$ there is $k(\varepsilon) > 0$ such that $\varphi_{\varepsilon,k(\varepsilon)}$ is rank-one convex. If $\varepsilon > 0$ is small enough we still have

$$\int_{(0,1)^2} \varphi_{\varepsilon,k}(\nabla y(x))\,dx < \varphi_{\varepsilon,k}(0) = 0$$

because $P\nabla y = \nabla y$ and ∇y is pointwise bounded. Hence, $\varphi_{\varepsilon,k}$ is rank-one convex but not quasiconvex (at the origin). Altogether, we have the following theorem.

Theorem 4.1.7 (Rank-one convex, non-quasiconvex function, V. Šverák [500]). *Let $d \geq 3$. Then there exists a rank-one convex function $\varphi : \mathbb{R}^{d\times d} \to \mathbb{R}$ which is not quasiconvex.*

Proof. We define a projector $\tilde{P} : \mathbb{R}^{d\times d} \to \mathbb{R}^{3\times 2}$ such that $\tilde{P}A := (A_{ij})$ for $1 \leq i \leq 3$ and $1 \leq j \leq 2$ and consider $A \mapsto \varphi_{\varepsilon,k}(\tilde{P}A)$ for $\varepsilon > 0$ and $k > 0$ such that $\varphi_{\varepsilon,k} : \mathbb{R}^{3\times 2} \to \mathbb{R}$ is quasiconvex. □

Šverák's example has been essentially the only example of a rank-one convex non-quasiconvex function for many years. Another completely different set of rank-one convex but non-quasiconvex functions has recently been discovered by Grabovsky [225].

One could think that defining $\tilde{\varphi}_{\varepsilon,k} : \mathbb{R}^{2\times 3} \to \mathbb{R}$ as $\tilde{\varphi}_{\varepsilon,k}(A) := \varphi_{\varepsilon,k}(A^\top)$ for every $A \in \mathbb{R}^{2\times 3}$ provides an example of a rank-one convex function which is not quasiconvex on $\mathbb{R}^{2\times 3}$. Unfortunately, it is not the case as shown in [382].

Theorem 4.1.8 (Quasiconvexity and transposition, S. Müller [382]). *For every $\varepsilon > 0$ there is $k > 0$ such that $\bar{\varphi}_{\varepsilon,k} : \mathbb{R}^{2\times 3} \to \mathbb{R}$ is quasiconvex.*

Here we only prove that $\bar{\varphi} : \mathbb{R}^{2\times 3} \to \mathbb{R} \cup \{+\infty\}$ defined as

$$\bar{\varphi}(A) := \begin{cases} -A_{11}A_{22}(A_{13}+A_{23})/2 & \text{if } A^\top \in L, \\ +\infty & \text{otherwise} \end{cases} \tag{4.1.15}$$

is quasiconvex. This result was first proven in [289].

Lemma 4.1.9 (Gradients in the subspace). *Let $y \in W_0^{1,\infty}((0,1)^3;\mathbb{R}^2)$ and $\nabla y \in L^\top$. Then there exist $\alpha, \beta, \gamma \in W_0^{1,\infty}((0,1);\mathbb{R})$ such that*

$$y_1(x) = \alpha(x_1) + \gamma(x_3), \quad y_2(x) = \beta(x_2) + \gamma(x_3).$$

Proof. By the definition of P we have

$$\frac{\partial y_1}{\partial x_2} = \frac{\partial y_2}{\partial x_1} = \frac{\partial (y_2 - y_1)}{\partial x_3} = 0 \quad \text{for a.a. } x \in (0,1)^3,$$

whence $y_1(x) = a(x_1, x_3)$ and $y_2(x) = b(x_2, x_3)$, and also

$$\frac{\partial a}{\partial x_3}(x_1, x_3) = \frac{\partial b}{\partial x_3}(x_2, x_3);$$

the left hand side must be independent of x_1 and the right hand side independent of x_2, therefore both are functions of x_3 alone. Thus $a(x_1, x_3) = \alpha(x_1) + \gamma(x_3)$ and

$b(x_2,x_3) = \beta(x_2) + \gamma(x_3)$; the summability and periodicity properties of the three functions are straightforward. □

Proposition 4.1.10 (Quasiconvexity of $\bar{\varphi}$). *Let $\bar{\varphi}$ be given by* (4.1.15). *Then, for any $y \in W_0^{1,\infty}((0,1)^3;\mathbb{R}^2)$ and $A \in \mathbb{R}^{2\times 3}$, it holds*

$$\bar{\varphi}(A) \leq \int_{(0,1)^3} \bar{\varphi}(A + \nabla y(x))\,\mathrm{d}x. \tag{4.1.16}$$

Proof. Let $A \in L^\top$ and $y \in W_0^{1,\infty}((0,1)^3;\mathbb{R}^2)$ with $\nabla y \in L^\top$ be arbitrary. We have by Lemma 4.1.9 we obtain

$$\begin{aligned}\int_{(0,1)^3} \bar{\varphi}(A + \nabla y(x)))\,\mathrm{d}x &= -\frac{1}{2}\int_{(0,1)^3} (A_{11} + \alpha'(x_1))(A_{22} + \beta'(x_2))(A_{13} + 2\gamma'(x_3))\,\mathrm{d}x \\ &\quad - \frac{1}{2}\int_0^1 A_{11}\,\mathrm{d}x_1 \int_0^1 A_{22}\,\mathrm{d}x_2 \int_0^1 A_{13}\,\mathrm{d}x_3 = \bar{\varphi}(A).\end{aligned}$$

□

Although it is does not follow generally (see Proposition 4.1.6), $\bar{\varphi}$ is rank-one convex because it is finite on the linear subspace $L^\top := \{A^\top;\ A \in L\}$.

Quasiconvexity of the integrand φ is a necessary condition for weak* lower lower semicontinuity of $y \mapsto \int_\Omega \varphi(\nabla y(x))\,\mathrm{d}x$ on $W^{1,\infty}(\Omega;\mathbb{R}^d)$. Adapting slightly Theorem 3.4.1, we easily observe that polyconvexity of φ is sufficient for weak* lower semicontinuity of $y \mapsto \int_\Omega \varphi(\nabla y(x))\,\mathrm{d}x$ on $W^{1,\infty}(\Omega;\mathbb{R}^d)$. Consequently, we have the following theorem.

Theorem 4.1.11 (Polyconvexity implies quasiconvexity). *Let $\varphi : \mathbb{R}^{d\times d} \to \mathbb{R}$ be polyconvex. Then it is also quasiconvex.*

In the next proposition, we show that quasiconvexity is also sufficient for sequential weak lower semicontinuity of integral functionals. Our proof will exploit Young measures introduced in Appendix D. Results for more general integrands can be found e.g. in [2, 135].

Proposition 4.1.12 (Quasiconvexity implies weak lower semicontinuity). *Assume that $\varphi : \mathbb{R}^{d\times d} \to [0;+\infty)$ is quasiconvex and such that $-1/K \leq \varphi \leq K(1+|\cdot|^p)$ for some $K > 0$ and $1 < p < +\infty$. Let $\{y_k\}_{k\in\mathbb{N}} \subset W^{1,p}(\Omega;\mathbb{R}^d)$ be such that $y_k \to y$ weakly in $W^{1,p}(\Omega;\mathbb{R}^d)$ as $k \to \infty$. Then*

$$\liminf_{k\to\infty} \int_\Omega \varphi(\nabla y_k(x))\,\mathrm{d}x \geq \int_\Omega \varphi(\nabla y(x))\,\mathrm{d}x, \tag{4.1.17}$$

i.e., $y \mapsto \int_\Omega \varphi(\nabla y(x))\,\mathrm{d}x$ is sequentially weak lower semicontinuous on $W^{1,p}(\Omega;\mathbb{R}^d)$.

Proof. We can consider a (non-relabeled) subsequence of $\{y_k\}$ such that "lim inf" on the left-hand side of (4.1.17) is already "lim". If necessary, we can extract another subsequence such that $\{\nabla y_k\}_{k\in\mathbb{N}}$ generates a Young measure ν; cf. Theorem D.1.1 for $N := d\times d$. In view of (D.1.3) applied to $\varphi + 1/K \geq 0$ we get

$$\lim_{k\to\infty}\int_\Omega \varphi(\nabla y_k(x))\,\mathrm{d}x \geq \int_\Omega\int_{\mathbb{R}^{d\times d}} \varphi(F)\nu_x(\mathrm{d}F)\,\mathrm{d}x. \tag{4.1.18}$$

However, ν is generated by the sequence of gradients (bounded in $L^p(\Omega;\mathbb{R}^{d\times d})$), i.e., it is a gradient Young measure. We therefore apply Theorem D.1.2 and get, due to (D.1.4b), that

$$\int_\Omega\int_{\mathbb{R}^{d\times d}} \varphi(F)\nu_x(\mathrm{d}F)\,\mathrm{d}x \geq \int_\Omega \varphi(\nabla y(x))\,\mathrm{d}x.$$

□

Analogously, we get the following sequential weak* lower semicontinuity result. Here the lower bound on φ can be neglected.

Proposition 4.1.13 (Quasiconvexity implies weak* lower semicontinuity). *Let us assume that $\varphi : \mathbb{R}^{d\times d} \to \mathbb{R}$ is quasiconvex. Let further $\{y_k\}_{k\in\mathbb{N}} \subset W^{1,\infty}(\Omega;\mathbb{R}^d)$ be such that $y_k \to y$ weakly in $W^{1,\infty}(\Omega;\mathbb{R}^d)$ as $k\to\infty$. Then*

$$\liminf_{k\to\infty}\int_\Omega \varphi(\nabla y_k(x))\,\mathrm{d}x \geq \int_\Omega \varphi(\nabla y(x))\,\mathrm{d}x, \tag{4.1.19}$$

i.e., $y \mapsto \int_\Omega \varphi(\nabla y(x))\,\mathrm{d}x$ is sequentially weak lower semicontinuous on $W^{1,\infty}(\Omega;\mathbb{R}^d)$.*

Proof. The proof follows the same reasoning as the proof of Proposition 4.1.12. Realize that Theorem D.1.1 used here with $N := d\times d$ holds true also if the integrand is bounded from below by a constant which is not necessarily zero. Namely, as $\sup_{k\in\mathbb{N}} \|\nabla y_k\|_{L^\infty(\Omega;\mathbb{R}^{d\times d})} < K$ we can replace φ by $\varphi - \min_{F\in\overline{B(0,K)}}\varphi(F)$. This will show that

$$\liminf_{k\to\infty}\int_\Omega \varphi(\nabla y_k(x))\,\mathrm{d}x \geq \int_\Omega \varphi(\nabla y(x))\,\mathrm{d}x. \tag{4.1.20}$$

The proof is finished again by applying Theorem D.1.2. □

Sequential weak lower semicontinuity can be then used to show existence of minimizers of functionals with a quasiconvex stored energy density.

Theorem 4.1.14 (Existence of minimizers and quasiconvexity). *Let $\Omega \subset \mathbb{R}^d$ be a bounded Lipschitz domain. Let $\varphi : \mathbb{R}^{d\times d} \to \mathbb{R}$ be quasiconvex, $0 \leq \varphi \leq K + K|\cdot|^p$ for some $0 < \tilde{K} < K$ and $1 < p < +\infty$. Let $y_\mathrm{D} \in W^{1,p}(\Omega;\mathbb{R}^d)$. Then the functional $y \mapsto \int_\Omega \varphi(\nabla y(x))\,\mathrm{d}x$ attains its minimum on the set*

$$\mathcal{A} := \{y \in W^{1,p}(\Omega;\mathbb{R}^d);\ y = y_\mathrm{D} \text{ on } \Gamma_\mathrm{D}\}.$$

Proof. It is a standard application of the direct method of the calculus of variations. Coercivity of φ implies that every minimizing sequence is bounded in the norm of $W^{1,p}(\Omega;\mathbb{R}^d)$. Application of Proposition 4.1.12 and weak closeness of $\mathcal{A}$ finishes the proof. □

Remark 4.1.15 (Rank-one convexity vs. quasiconvexity). The function $\varphi_{\varepsilon,k}$ given in (4.1.13) can be easily modified to provide an example of a rank-one convex function which is not quasiconvex $\mathbb{R}^{d_1\times d_2}$ for $d_1 \geq 3$ and $d_2 \geq 2$. The case $d_1 = 2$ and $d_2 \geq 2$ is open.

Remark 4.1.16 (A function changing convexity properties). If $|\cdot|$ denotes the Frobenius norm, Alibert and Dacorogna [6] showed that the function $\varphi_\gamma : \mathbb{R}^{2\times 2} \to \mathbb{R}$ defined as

$$\varphi_\gamma(F) := |F|^2(|F|^2 - 2\gamma \det F) \text{ is } \begin{cases} \text{convex} & \text{if and only if } |\gamma| \leq 2\sqrt{2}/3, \\ \text{polyconvex} & \text{if and only if } |\gamma| \leq 1, \\ \text{quasiconvex} & \text{if and only if } |\gamma| \leq 1+\varepsilon, \\ \text{rank-one convex} & \text{if and only if } |\gamma| \leq 2/\sqrt{3}. \end{cases}$$

The precise value of ε is not known and it can be only shown that $0 < \varepsilon \leq 2/\sqrt{3} - 1$. On the other hand, φ_γ is isotropic, i.e., $\varphi_\gamma(F) = \varphi_\gamma(RFQ)$ for every $F \in \mathbb{R}^{2\times 2}$ and all rotation matrices $R, Q \in \mathrm{SO}(2)$.

Remark 4.1.17 (Quasiconvexity and transposition). We emphasize that it is still an open problem if given $\varepsilon > 0$ the function $\widetilde{\varphi}_{\varepsilon,k}$ from Theorem 4.1.8 becomes quasiconvex for the same values of k for which it is also rank-one convex.

Remark 4.1.18 ($W^{1,p}$-quasiconvexity). The idea of using $y \in W_0^{1,p}(\Omega;\mathbb{R}^d)$ for $1 \leq p < +\infty$ as test functions in (4.1.3) instead of $y \in W_0^{1,\infty}(\Omega;\mathbb{R}^d)$ was first introduced by Ball and Murat in [40] and it is called there *$W^{1,p}$-quasiconvexity*. Obviously, $W^{1,p}$-quasiconvexity is generally a stronger property than $W^{1,q}$-quasiconvexity for $1 \leq p < q \leq +\infty$ because the set of test maps y is larger. Reasoning as in the proof of Proposition 4.1.5, one can show that if $\varphi : \mathbb{R}^{d\times d} \to \mathbb{R}$ satisfies (4.1.8) then, for every $A \in \mathbb{R}^{d\times d}$, it holds

$$\inf_{y\in W_0^{1,\infty}(\Omega;\mathbb{R}^d)} \int_\Omega \varphi(A + \nabla y(x))\,\mathrm{d}x = \inf_{y\in W_0^{1,1}(\Omega;\mathbb{R}^d)} \int_\Omega \varphi(A + \nabla y(x))\,\mathrm{d}x. \tag{4.1.21}$$

For a general function φ, however, these two infima are not necessarily the same,which leads to the so-called *Lavrentiev phenomenon* observed also in nonlinear elasticity [198] and which causes crucial problems in numerical approximations of variational problems see [107, 361, 392], for instance.

Exercise 4.1.19. Show that a rank-one convex function $\varphi : \mathbb{R}^{d\times d} \to \mathbb{R}$ is continuous.

4.2 Concentrations and quasiconvexity at the boundary

Consider a bounded sequence $\{y_k\}_{k\in\mathbb{N}} \subset W^{1,p}(\Omega;\mathbb{R}^d)$. Clearly, $\{|\nabla y_k|^p\}_{k\in\mathbb{N}} \subset L^1(\Omega)$. Assume that $|\nabla y_k|^p \to \sigma$ weakly* in $\mathcal{M}(\bar{\Omega})$ for $k \to \infty$. We say that $\{|\nabla y_k|^p\}_{k\in\mathbb{N}}$ concentrates at the boundary if $\sigma(\Gamma) > 0$. Concentrations of $\{|\nabla y_k|^p\}_{k\in\mathbb{N}}$ at the boundary of Ω may influence the sequential weak lower semicontinuity of integral functionals $\mathcal{E} : W^{1,p}(\Omega;\mathbb{R}^d) \to \mathbb{R}$: $\mathcal{E}(y) = \int_\Omega \varphi(\nabla y(x))\,\mathrm{d}x$ where $\varphi : \mathbb{R}^{d\times d} \to \mathbb{R}$ is continuous and such that $|\varphi| \le K(1+|\cdot|^p)$ for some constant $K > 0$. This was already observed in the discussion below Corollary 3.2.3 in the case $\varphi(F) := \det F$ for all $F \in \mathbb{R}^{d\times d}$. Indeed, consider $y \in W_0^{1,p}(B(x_0,1);\mathbb{R}^d)$, where $B(x_0,1)$ is the unit ball in $\mathbb{R}^d$ centered at $0 \in \Gamma$, and extend it by zero to the whole R^d. Let us define for $x \in \mathbb{R}^d$ and $y_k(x) = k^{n/p-1}y(kx)$ for $k \in \mathbb{N}$, i.e., $y_k \to 0$ weakly in $B(x_0,1)$ and assume that Ω is a smooth domain $\Omega \in \mathbb{R}^d$ such that, $\vec{n}$ is the outer unit normal to Γ at the origin and let there be $x \in \Omega$ such that $\vec{n}\cdot x < 0$. Moreover, take a function φ to be positively p-homogeneous, i.e., $\varphi(\alpha F) = \alpha^p \varphi(F)$ for all $\alpha \ge 0$. Then if $\mathcal{E}$ is weakly lower semicontinuous then

$$
\begin{aligned}
0 = \mathcal{E}(0) &\le \liminf_{k\to\infty} \int_\Omega \varphi(\nabla y_k(x))\,\mathrm{d}x = \liminf_{k\to\infty} \int_{B(x_0;1)\cap\Omega} \varphi(\nabla y_k(x))\,\mathrm{d}x \\
&= \liminf_{k\to\infty} \int_{B(x_0;1)\cap\Omega} k^d \varphi(\nabla(y(kx))\,\mathrm{d}x = \int_{B(x_0,1)\cap\{x\in\mathbb{R}^d;\ \vec{n}x<0\}} \varphi(\nabla(y(x))\,\mathrm{d}x.
\end{aligned}
$$

Thus, we can see that

$$
0 \le \int_{B(x_0,1)\cap\{x\in\mathbb{R}^d;\ \vec{n}x<0\}} \varphi(\nabla y(x))\,\mathrm{d}x
$$

for all $y \in W_0^{1,p}(B(x_0;1);\mathbb{R}^d)$ forms a necessary condition for weak lower semicontinuity of $\mathcal{E}$. This condition is stronger than quasiconvexity of φ at zero because, in general, $y \ne 0$ on the boundary of the half-ball $B(x_0,1)\cap\{x\in\mathbb{R}^d;\ \vec{n}\cdot x < 0\}$ while quasiconvexity would require $y = 0$ on the boundary.

In fact, it can be shown that if $\varphi : \mathbb{R}^{d\times d} \to \mathbb{R}$ is positively d-homogeneous then the precise condition ensuring weak lower semicontinuity of *quasiconvexity at the boundary*, the notion introduced by J.M. Ball and J. Marsden [37]. We refer the interested reader to [292] and [269] for more details. In order to introduce quasiconvexity at the boundary we must first define *standard boundary domains*. Roughly speaking, it is a Lipschitz domain with a flat piece of the boundary, where the outer normal vector to the boundary does not change; cf. Figure 4.1.

Definition 4.2.1 (Standard boundary domain). Let $\vec{n} \in \mathbb{R}^d$ be a unit vector and let $D_{\vec{n}}$ be a bounded Lipschitz domain. We say that $D_{\vec{n}}$ is a standard boundary domain with the normal $\vec{n}$ if there is $a \in \mathbb{R}^d$ such that $D_{\vec{n}} \subset H_{a,\vec{n}} := \{x\in\mathbb{R}^d;\ \vec{n}\cdot x < a\}$ and the $(d-1)$-dimensional interior $\Gamma_{\vec{n}}$ of $\partial D_{\vec{n}} \cap \partial H_{a,\vec{n}} \ne \emptyset$.

We are now ready to define the quasiconvexity at the boundary:

Definition 4.2.2 (Quasiconvexity at boundary [37]). Let $\vec{n} \in \mathbb{R}^d$ be a unit vector. A function $\varphi : \mathbb{R}^{d\times d} \to \mathbb{R}$ is called quasiconvex at the boundary at $F \in \mathbb{R}^{d\times d}$ with respect to $\vec{n}$ if there is $q \in \mathbb{R}^d$ such that, for all $\tilde{y} \in W_{\Gamma_{\vec{n}}}^{1,\infty}(D_{\vec{n}};\mathbb{R}^d) := \{y \in W^{1,\infty}(D_{\vec{n}};\mathbb{R}^d);\ y = 0 \text{ on } \partial D_{\vec{n}} \setminus \Gamma_{\vec{n}}\}$, it holds

$$
\int_{\Gamma_{\vec{n}}} q\cdot\tilde{y}(x)\,\mathrm{d}S + \varphi(F)\mathrm{meas}_d(D_{\vec{n}}) \le \int_{D_{\vec{n}}} \varphi(F+\nabla\tilde{y}(x))\,\mathrm{d}x \tag{4.2.1}
$$

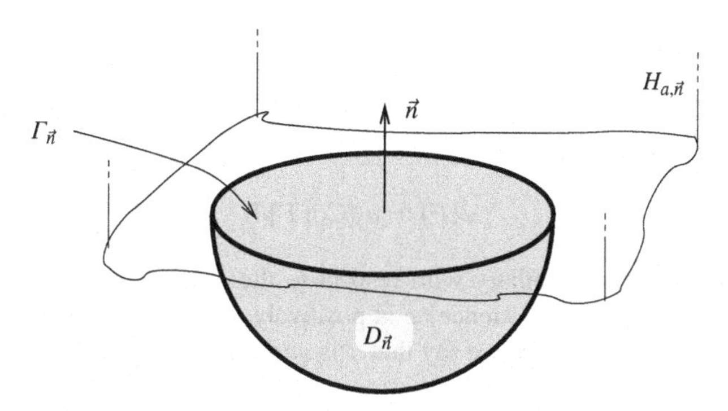

Fig. 4.1 An example of a standard boundary domain $D_{\vec{n}}$ with a half-space $H_{a,\vec{n}}$.

with $\Gamma_{\vec{n}}$ and $D_{\vec{n}}$ from Definition 4.2.1.

An immediate generalization of the previous quasiconvexity is the following:

Definition 4.2.3 ($W^{1,p}$-quasiconvexity at boundary). Let $1 \le p \le +\infty$ and $\vec{n} \in \mathbb{R}^d$ be a unit vector. A function $\varphi : \mathbb{R}^{d\times d} \to \mathbb{R}$, $|\varphi| \le K(1+|\cdot|^p)$for some $K > 0$ is called $W^{1,p}$-quasiconvex at the boundary at $F \in \mathbb{R}^{d\times d}$ with respect to $\vec{n}$ if there is $q \in \mathbb{R}^d$ such that, for all $\tilde{y} \in W^{1,p}_{\Gamma_{\vec{n}}}(D_{\vec{n}};\mathbb{R}^d) := \{y \in W^{1,p}(D_{\vec{n}};\mathbb{R}^d);\ \ y = 0 \text{ on } \partial D_{\vec{n}} \setminus \Gamma_{\vec{n}}\}$, it holds

$$\int_{\Gamma_{\vec{n}}} q\cdot\tilde{y}(x)\,\mathrm{d}S + \varphi(F)\mathrm{meas}_d(D_{\vec{n}}) \le \int_{D_{\vec{n}}} \varphi(F + \nabla\tilde{y}(x))\,\mathrm{d}x. \tag{4.2.2}$$

The following lemma shows that Definitions 4.2.2 and 4.2.3 are equivalent to each other for a class of functions whose modulus grows as the p-th power.

Lemma 4.2.4. *Let $\varphi : \mathbb{R}^{d\times d} \to \mathbb{R}$ be continuous and such that $|\varphi(A)| \le K(1+|A|^p)$ for all $A \in \mathbb{R}^{d\times d}$ and some $K > 0$ independent of A and some $1 \le p < +\infty$. If φ is quasiconvex at the boundary at $(F,\vec{n})$, it is $W^{1,p}$-quasiconvex at the boundary at $(F,\vec{n})$.*

Proof. Take $\tilde{y} \in W^{1,p}_{\Gamma_{\vec{n}}}(D_{\vec{n}};\mathbb{R}^d)$ and a sequence $\{\tilde{y}_k\}_{k\in\mathbb{N}} \subset W^{1,\infty}_{\Gamma_{\vec{n}}}(D_{\vec{n}};\mathbb{R}^d)$ such that $\tilde{y}_k \to \tilde{y}$ strongly in $W^{1,p}(D_{\vec{n}};\mathbb{R}^d)$. We get using (4.1.8) that

$$\int_{D_{\vec{n}}} \varphi(F + \nabla\tilde{y}(x))\,\mathrm{d}x = \lim_{k\to\infty} \int_{D_{\vec{n}}} \varphi(F + \nabla\tilde{y}_k(x))\,\mathrm{d}x. \tag{4.2.3}$$

As v is quasiconvex at the boundary at $(s_0,\vec{n})$ we have

$$\begin{aligned}\lim_{k\to\infty} \int_{D_{\vec{n}}} \varphi(F + \nabla\tilde{y}_k(x))\,\mathrm{d}x &\ge \mathrm{meas}_d(D_{\vec{n}})\varphi(F) + \lim_{k\to\infty} \int_{\Gamma_{\vec{n}}} q\cdot\tilde{y}_k(x)\,\mathrm{d}S \\ &= \mathrm{meas}_d(D_{\vec{n}})\varphi(F) + \int_{\Gamma_{\vec{n}}} q\cdot\tilde{y}(x)\,\mathrm{d}S\,,\end{aligned}$$

which finishes the proof in view of (4.2.3). □

Example 4.2.5. If $d = 3$ then it is shown in [483, Prop. 17.2.4] that the function $\varphi : \mathbb{R}^{d\times d} \to \mathbb{R}$ given for all F by

$$\varphi(F) := a\cdot[\mathrm{Cof}\, F]\vec{n}$$

is quasiconvex at the boundary with respect to the unit normal $\vec{n} \in \mathbb{R}^d$. Here $a \in \mathbb{R}^d$ is an arbitrary constant. Hence, φ is positively 2-homogeneous. As $-\varphi$ is also quasiconvex at the boundary, we say that φ is *quasiaffine at the boundary*.

The following result follows from Theorem D.2.3 and it shows interesting properties of quasiaffine functions. Here we take $d = 3$ because cofactor is a linear matrix function if $d = 2$, and, consequently, the result is obvious.

Proposition 4.2.6 (Superlinear integrability of cofactors [269]). *Let $\Omega \subset \mathbb{R}^3$ be a bounded smooth domain with a C^1-boundary. Let $\vec{n}$ be continuously extended from Γ to $\bar{\Omega}$, and let $a \in C(\bar{\Omega};\mathbb{R}^3)$. Let further $\{\tilde{y}_k\}_{k\in\mathbb{N}} \subset H^1(\Omega;\mathbb{R}^3)$ be such that $\tilde{y}_k \to \tilde{y}$ weakly in $H^1(\Omega;\mathbb{R}^3)$ as $k \to \infty$. Then, for all $g \in C(\bar{\Omega})$, it holds*

$$\lim_{k\to\infty} \int_\Omega g(x)a(x)\cdot[\mathrm{Cof}\,\nabla\tilde{y}_k(x)]\vec{n}(x)\,\mathrm{d}x = \int_\Omega g(x)a(x)\cdot[\mathrm{Cof}\,\nabla\tilde{y}(x)]\vec{n}(x)\,\mathrm{d}x\,. \quad (4.2.4)$$

If additionally, for all $k \in \mathbb{N}$, $a\cdot[\mathrm{Cof}\,\nabla\tilde{y}_k]\vec{n}\mathrm{d}x \ge 0$ almost everywhere in Ω (or, for all $k \in \mathbb{N}$, $a\cdot[\mathrm{Cof}\,\nabla\tilde{y}_k]\vec{n}\mathrm{d}x \le 0$ almost everywhere in Ω) then (4.2.4) holds for all $g \in L^\infty(\Omega)$, i.e., $a\cdot[\mathrm{Cof}\,\nabla\tilde{y}_k]\vec{n} \to a\cdot[\mathrm{Cof}\,\nabla\tilde{y}]\vec{n}$ weakly in $L^1(\Omega)$. In particular, there is $\zeta : \mathbb{R} \to [0;+\infty)$ with $\lim_{r\to+\infty} \zeta(r)/r = +\infty$ such that $\sup_{k\in\mathbb{N}} \int_\Omega \zeta(a\cdot[\mathrm{Cof}\,\nabla\tilde{y}_k]\vec{n}) < +\infty$.

Remark 4.2.7 (Properties of quasiconvexity at the boundary).

(i) If φ is differentiable at F then $q := \partial_F\varphi(F)\vec{n}$ is given uniquely; cf. [492]. In [37], only the case $q = 0$ is discussed.

(ii) It is clear that, if φ is quasiconvex at the boundary at $(F, \vec{n})$, it is also quasiconvex at F, i.e., (4.1.3) holds.

(iii) If (4.2.2) holds for one standard boundary domain it holds for other standard boundary domains, too.

(iv) If $p > 1$ and φ is positively p-homogeneous, i.e. $\varphi(\lambda s) = \lambda^p\varphi(F)$ for all $F \in \mathbb{R}^{d\times d}$ and all $\lambda \ge 0$, and $W^{1,p}$-quasiconvex at the boundary at $(0, \vec{n})$ then $q = 0$ in (4.2.2). Indeed, we have $\varphi(0) = 0$ and suppose that $\int_{D_{\vec{n}}} \varphi(\nabla\tilde{y}(x))\,\mathrm{d}x < 0$ for some $\tilde{y} \in W^{1,\infty}_{\Gamma_{\vec{n}}}(D_{\vec{n}};\mathbb{R}^d)$ and that $q \neq 0$. By (4.2.2), we must have for all $\lambda > 0$

$$0 \le \lambda^p \int_{D_{\vec{n}}} \varphi(\nabla\tilde{y}(x))\,\mathrm{d}x - \lambda \int_{\Gamma_{\vec{n}}} q\cdot\tilde{y}(x)\,\mathrm{d}S\,.$$

However, it is not possible for $\lambda > 0$ large enough and therefore for all $\tilde{y} \in W^{1,\infty}_{\Gamma_{\vec{n}}}(D_{\vec{n}};\mathbb{R}^d)$ it holds that $\int_{D_{\vec{n}}} \varphi(\nabla\tilde{y}(x))\,\mathrm{d}x \ge 0$. Thus, we take $q = 0$.

4.3 Necessary conditions for energy-minimizing configurations

It turns out that quasiconvexity and quasiconvexity at the boundary form necessary conditions for local minimizers of variational problems in elasticity. Indeed, assume that the elastic-energy functional has the following form:

$$\mathcal{E}(y) := \int_{\Omega} \varphi(x, \nabla y(x))\,\mathrm{d}x - \int_{\Omega} f(x)\cdot y(x)\,\mathrm{d}x - \int_{\Gamma_{\mathrm{N}}} g(x)\cdot y(x)\,\mathrm{d}S\,. \tag{4.3.1}$$

Here, $\varphi : \Omega\times\mathbb{R}^{d\times d} \to \mathbb{R}\cup\{+\infty\}$ is the stored energy density of a body $\Omega\subset\mathbb{R}^d$, $f : \Omega\to\mathbb{R}^d$ again denotes volume density of body forces and $g : \Gamma_{\mathrm{N}} \to \mathbb{R}^d$ is the density of surface forces. As before, we assume that Ω is a bounded Lipschitz domain, so that $\mathrm{meas}_d(\Gamma) = 0$. Let $y \in W^{1,1}(\Omega;\mathbb{R}^d)$ be such that (4.3.1) is well defined, i.e., all integrals exist and no two of them are infinite with opposite signs. The following definition is due to Ball and Marsden [37].

Definition 4.3.1 (Local minimizers). Let $y \in W^{1,1}(\Omega;\mathbb{R}^d)$ be such that $\mathcal{E}(y)$ is well-defined and finite and let $x_0 \in \bar{\Omega}$. We say that y is a local minimizer of $\mathcal{E}$ at x_0 in $W^{k,p}(\Omega;\mathbb{R}^d)\cap C(\bar{\Omega};\mathbb{R}^d)$ if there are numbers $\varepsilon > 0$ and $\delta > 0$ such that for every $\bar{y} \in W^{1,1}(\Omega;\mathbb{R}^d)$ satisfying $y-\bar{y} \in C^\infty(\bar{\Omega};\mathbb{R}^d)$, $y(x) = \bar{y}(x)$ for all $x \in \bar{\Omega}\setminus\overline{B(x_0;\delta)}$, and $\|\bar{y}-y\|_{W^{k,p}(\Omega;\mathbb{R}^d)} + \|\bar{y}-y\|_{C(\bar{\Omega};\mathbb{R}^d)} < \varepsilon$ the value $\mathcal{E}(\bar{y})$ is well defined, and $\mathcal{E}(\bar{y}) \geq \mathcal{E}(y)$.

Let us notice that "local" in Definition 4.3.1 means not only that $\tilde{y}$ and y are close in the Sobolev norm but also that they only differ in a neighborhood of x_0.

Local minimizers which are smooth in a neighborhood of x_0 necessarily satisfy conditions stated in the following theorem.

Theorem 4.3.2 (Necessary conditions for local minimizers [37]). *Let $1 \leq p < +\infty$, let $k \in \mathbb{N}$ satisfy $k < 1 + d/p$, let $x_0 \in \bar{\Omega}$, and let Γ_{N} be a C^∞ part of the boundary of Ω. Suppose that $y \in W^{1,1}(\Omega;\mathbb{R}^d)$ is a local minimizers of $\mathcal{E}$ from (4.3.1) at x_0 in $W^{k,p}(\Omega;\mathbb{R}^d)\cap C(\bar{\Omega};\mathbb{R}^d)$ and assume that $y \in C^1(B(x_0;r)\cap\bar{\Omega};\mathbb{R}^d)$ for some $r > 0$. Moreover, let the surface load g be continuous. Then if $x_0 \in \Omega$ the following inequality holds for every open bounded Lipschitz domain $\tilde{\Omega}\subset\mathbb{R}^d$, $\tilde{y} \in C^1_0(\tilde{\Omega};\mathbb{R}^d)$, and such that $x \mapsto \varphi(x_0, \nabla y(x_0) + \nabla\tilde{y}(x))$ is uniformly bounded in $\tilde{\Omega}$:*

$$\int_{\tilde{\Omega}} \varphi(x_0, \nabla y(x_0) + \nabla\tilde{y}(x))\,\mathrm{d}x \geq \mathrm{meas}_d(\tilde{\Omega})\varphi(x_0, \nabla y(x_0))\,. \tag{4.3.2}$$

On the other hand, if $x \in \Gamma_{\mathrm{N}}$ and $\vec{n}$ is the outer unit normal to Γ_{N} at x_0 then

$$\int_{D_{\vec{n}}} \varphi(x_0, \nabla y(x_0)+\nabla\tilde{y}(x))\,\mathrm{d}x - \int_{\Gamma_{\vec{n}}} g(x_0)\cdot\tilde{y}(x)\,\mathrm{d}S \geq \mathrm{meas}_d(D_{\vec{n}})\varphi(x_0, \nabla y(x_0)) \tag{4.3.3}$$

for every standard boundary domain $D_{\vec{n}} \subset \mathbb{R}^d$ and all $\tilde{y} \in C^1(D_{\vec{n}};\mathbb{R}^d)$ vanishing in a neighborhood of $\partial D_{\vec{n}}\setminus\Gamma_{\vec{n}}$.

Proof. We follow the proof from [37]. Assume first that $\tilde{y} \in C^\infty_0(\tilde{\Omega};\mathbb{R}^d)$. Take $\varepsilon > 0$ so small that $x_0 + \varepsilon\tilde{\Omega} \subset \Omega$ and define

$$y_\varepsilon(x) := \begin{cases} y(x) + \varepsilon\tilde{y}((x-x_0)/\varepsilon) & \text{if } x \in x_0 + \varepsilon\tilde{\Omega}, \\ y(x) & \text{otherwise.} \end{cases}$$

Notice that $y - y_\varepsilon \in C^\infty(\bar{\Omega};\mathbb{R}^d)$ and $y(x) = y_\varepsilon(x)$ if $|x - x_0| \geq c\varepsilon$ for some constant $c > 0$. Moreover, $\|y - y_\varepsilon\|_{C(\bar{\Omega};\mathbb{R}^d)} \leq \varepsilon\|\tilde{y}\|_{C(\bar{\Omega};\mathbb{R}^d)}$ and again for some $c > 0$ we calculate

$$\begin{aligned}\|y - y_\varepsilon\|_{W^{k,p}(\Omega;\mathbb{R}^d)} &= \Big(\sum_{|\beta|\leq k} \int_\Omega |\varepsilon D^\beta\tilde{y}((x-x_0)/\varepsilon)|^p \,\mathrm{d}x\Big)^{1/p} \\ &= \Big(\sum_{|\beta|\leq k} \varepsilon^{(1-|\beta|)p+d} \int_{\tilde{\Omega}} |D^\beta\tilde{y}(z)|^p \,\mathrm{d}z\Big)^{1/p} \leq c\varepsilon^{1-r+d/p}\|\tilde{y}\|_{W^{k,p}(\tilde{\Omega};\mathbb{R}^d)}.\end{aligned}$$

Altogether, if $\varepsilon > 0$ is small enough y_ε can be used as a test map in the place of $\bar{y}$ in Definition 4.3.1 and we have from the hypothesis that y is a local minimizer

$$\begin{aligned}0 \leq \mathcal{E}(y_\varepsilon) - \mathcal{E}(y) &= \int_{x_0+\varepsilon\tilde{\Omega}} \varphi(x, \nabla y_\varepsilon(x)) - f(x)\cdot y_\varepsilon(x) - \varphi(x, \nabla y_\varepsilon(x)) + f(x)\cdot y(x)\,\mathrm{d}x \\ &= \int_{\tilde{\Omega}} \varepsilon^d\Big(\varphi(x_0{+}\varepsilon z, \nabla y(x_0{+}\varepsilon z){+}\nabla\tilde{y}(z)) - f(x_0{+}\varepsilon z)\cdot\varepsilon\tilde{y}(z) - \varphi(x_0{+}\varepsilon z, \nabla y(x_0{+}\varepsilon z))\Big)\mathrm{d}z.\end{aligned}$$

Divide the previous inequality by $\varepsilon^d > 0$. As we assume that y is smoothly differentiable in a neighborhood of x_0 and that $x \mapsto \varphi(x_0, \nabla y(x_0) + \nabla\tilde{y}(x))$ is uniformly bounded in $\tilde{\Omega}$ we can apply the Lebesgue dominated-convergence theorem (cf. Theorem B.2.2) which altogether yields (4.3.2). If $\tilde{y} \in C_0^1(\tilde{\Omega};\mathbb{R}^d)$ we approximate it by a sequence $\{\tilde{y}_k\}_{k\in\mathbb{N}} \subset C_0^\infty(\tilde{\Omega};\mathbb{R}^d)$ such that $\tilde{y}_k \to \tilde{y}$ in $C_0^1(\tilde{\Omega};\mathbb{R}^d)$ as $k \to \infty$ and the same conclusion follows.

It remains to prove (4.3.3). We can assume without any loss of generality that $x_0 = 0$ and that $\vec{n} = \mathbf{e}_d = (0,\ldots,1)$ by shifting and rotating Ω if necessary. The subspace $H_{a,\vec{n}}$ from Definition 4.2.1 can be then taken with $a = 0$, i.e., $H_{0,\mathbf{e}_d} := \{x \in \mathbb{R}^d;\, x_d < 0\}$. Let $N \subset \mathbb{R}^d$ be a neighborhood of zero such that

$$\Gamma_{\mathrm{N}} \cap N := \{x \in N;\, x_d = h(x')\} \qquad \text{and} \qquad \Omega \cap N := \{x \in N;\, x_d < h(x')\},$$

where $x' := (x_1,\ldots,x_{d-1})$, and $h \in C^\infty(\mathbb{R}^{d-1})$ with $h(0) = 0$ and $\nabla h(0) = 0$. Let $\tilde{y} \in C^\infty(\bar{D}_{\mathbf{e}_d};\mathbb{R}^d)$vanishing in a neighborhood of $\partial D_{\mathbf{e}_d} \setminus \Gamma_{\mathbf{e}_d}$ be such that $x \mapsto \varphi(x_0, \nabla y(x_0) + \nabla\tilde{y}(x))$ is uniformly bounded in $D_{\mathbf{e}_d}$. For $x \in \bar{\Omega}$ define

$$y_\varepsilon(x) := \begin{cases} y(x) + \varepsilon\tilde{y}((x - h'(x')\mathbf{e}_d)/\varepsilon) & \text{if } x{\in}N \text{ and } x - h(x')\mathbf{e}_d \in \varepsilon D_{\mathbf{e}_d}, \\ y(x) & \text{otherwise.} \end{cases}$$

We now extend $\tilde{y}$ by zero to the whole subspace $H_{0,\mathbf{e}_d}$ (without changing its name). Let $z_\varepsilon \in C^\infty(\mathbb{R}^d;\mathbb{R}^d)$ be defined as

$$z_\varepsilon(x) := \frac{x - h(x')\mathbf{e}_d}{\varepsilon}.$$

It is an invertible map and $z_\varepsilon^{-1}(z) = \varepsilon z + h(\varepsilon z')\mathbf{e}_d$. We can also easily calculate the gradients of z_ε and its inverse, namely,

$$\nabla z_\varepsilon(x) = (\mathbb{I} - \mathbf{e}_d \otimes \nabla h(x'))/\varepsilon \quad \text{and} \quad \nabla z_\varepsilon^{-1}(z) = \varepsilon(\mathbb{I} + \mathbf{e}_d \otimes \nabla h(\varepsilon z')).$$

Notice that $y - y_\varepsilon \in C^\infty(\bar{\Omega};\mathbb{R}^d)$. Given $\delta > 0$, we can choose ε¿0 so small that $y(x) = y_\varepsilon(x)$ if $|x| > \delta$ and $x \in \bar{\Omega}$. Also, it holds that $\|y - y_\varepsilon\|_{C(\bar{\Omega};\mathbb{R}^d)} = \varepsilon\|\tilde{y}\|_{C(\bar{\Omega};\mathbb{R}^d)}$, so it can be made arbitrarily small. Note that $z_\varepsilon(\bar{D}_{\mathbf{e}_d}) \subset \Omega \cap N$ is a set on which y_ε differs from y. Hence,

$$\begin{aligned}
\|y - y_\varepsilon\|_{W^{k,p}(\Omega;\mathbb{R}^d)} &= \Big(\sum_{|\beta|\le k} \int_{z_\varepsilon(D_{\mathbf{e}_d})} \varepsilon^p |D^\beta(\tilde{y}(y_\varepsilon(x))|^p \, \mathrm{d}x\Big)^{1/p} \\
&\le c\Big(\sum_{|\beta|\le k} \int_{z_\varepsilon(D_{\mathbf{e}_d})} \varepsilon^{p(1-|\beta|)} |D^\beta(\tilde{y})(y_\varepsilon(x))|^p \, \mathrm{d}x\Big)^{1/p} \\
&= c\Big(\sum_{|\beta|\le k} \int_{D_{\mathbf{e}_d}} \varepsilon^{p(1-|\beta|)+d} |D^\beta \tilde{y}(x)|^p \det(\mathbb{I} + \mathbf{e}_d \otimes \nabla h(\varepsilon x'))\, \mathrm{d}x\Big)^{1/p} \le c\varepsilon^{1-k+d/p}
\end{aligned}$$

for some constant $c > 0$. Hence, also $\|y - y_\varepsilon\|_{W^{k,p}(\Omega;\mathbb{R}^d)}$ can be made arbitrarily small depending on ε. Again, we conclude that y_ε is an admissible comparison function for a local minimizer y. Therefore,

$$\begin{aligned}
0 \le \mathcal{E}(y_\varepsilon) - \mathcal{E}(y) = \int_{z_\varepsilon(D_{\mathbf{e}_d})} & \varphi(x, \nabla y_\varepsilon(x)) - \varphi(x, \nabla y(x)) - f(x) \cdot (y_\varepsilon(x) - y(x))\, \mathrm{d}x \\
& - \int_{z_\varepsilon(\Gamma_{\mathbf{e}_d})} g(x) \cdot (y_\varepsilon(x) - y(x))\, \mathrm{d}S \\
= \varepsilon^d \int_{D_{\mathbf{e}_d}} & (\varphi(z_\varepsilon(x), \nabla y(z_\varepsilon(x)) + \nabla\tilde{y}(x)(\mathbb{I} + \mathbf{e}_d \otimes h(\varepsilon x')) \\
& - \varphi(z_\varepsilon(x), \nabla y(z_\varepsilon(x))) - f((z_\varepsilon(x)) \cdot \varepsilon\tilde{y}(x)) \det(\mathbb{I} + \mathbf{e}_d \otimes h(\varepsilon x'))\, \mathrm{d}x \\
- \varepsilon^d \int_{\Gamma_{\mathbf{e}_d}} & g(z_\varepsilon(x)) \cdot \tilde{y}(x)(1 + |\nabla h(\varepsilon x')|^2)^{1/2}\, \mathrm{d}S\,.
\end{aligned}$$

Division by ε^d and the passage to the limit for $\varepsilon \to 0$ in the above inequality exploiting also Lebesgue dominated-convergence theorem (see Theorem B.2.2) yields

$$\varphi(0, \nabla y(0))\mathrm{meas}_d(D_{\mathbf{e}_d}) + \int_{\Gamma_{\mathbf{e}_d}} g(0) \cdot \tilde{y}(x)\, \mathrm{d}S \le \int_{D_{\mathbf{e}_d}} \varphi(0, \nabla y(0) + \nabla\tilde{y}(x))\, \mathrm{d}x,$$

which is just (4.3.3). If $\tilde{y} \in C^1(D_{\mathbf{e}_d};\mathbb{R}^d)$ only we again approximate it first by smooth maps. The proof is finished. □

Let us notice that no, convexity (even generalized) properties are required on $\varphi(x,\cdot)$ to hold in Theorem 4.3.2. Hence, Theorem 4.3.2 states necessary optimality conditions for smooth local minimizers.

4.4 Relaxation of integral functionals by Young measures

If $\mathcal{E}$ is not weakly lower semicontinuous on $W^{1,p}(\Omega;\mathbb{R}^d)$ the direct method fails to show the existence of minimizers. Nevertheless, if a minimizing sequence $\{y_k\}_{k\in\mathbb{N}} \subset W^{1,p}(\Omega;\mathbb{R}^d)$ is bounded we can still search for an effective description of $\liminf_{k\to\infty} \mathcal{E}(y_k)$. Namely, we need to evaluate $\liminf_{k\to\infty} \mathcal{E}(y_k)$. In order to do so, we will exploit *Young measures*, the tool which goes back to L.C. Young [537] and was further developed by many authors, see e.g. [28, 271, 272]. As there are already a few books discussing this topic in a great detail, cf. [410, 452], for instance, we only review basic facts about Young measures in Section D.1. The interested reader, however, finds there all statements needed to follow our presentation.

As we already know even from simple one-dimensional examples (see Example 2.3.5, for instance) non-convex and non-quasiconvex stored energy densities may result is non-existence of minimizers of variational problems. In such a case we only have minimizing sequences which record behavior of the functional along them. Typically, minimizing sequences oscillate faster and faster, so that the energy functional reaches its infimum. On the other hand, the limit of this oscillatory sequence has usually greater energy because the functional is not lower semicontinuous in the underlying (mostly) weak topology. We again refer to Example 2.3.5 for understanding the difficulties. The idea of *relaxation* is to overcome this problem by introducing another functional and a minimization problem. More precisely, assume that we have a set $\mathcal{A}$ and $\mathcal{E} : \mathcal{A} \to \mathbb{R} \cup \{+\infty\}$ and the problem

$$\text{minimize } \mathcal{E} \text{ over } \mathcal{A}. \tag{4.4.1}$$

Then we look for a set $\mathfrak{A}$ and another functional $\hat{\mathcal{E}} : \mathfrak{A} \to \mathbb{R} \cup \{+\infty\}$ such that the following three conditions are satisfied:

(i) $\inf_{\mathcal{A}} \mathcal{E} = \min_{\mathfrak{A}} \hat{\mathcal{E}}$,
(ii) every converging minimizing sequence $\{y_k\}_{k\in\mathbb{N}} \subset \mathcal{A}$ converges to a minimizer of $\hat{\mathcal{E}}$ on $\mathfrak{A}$, and
(iii) every minimizer of $\hat{\mathcal{E}}$ on $\mathfrak{A}$ is the limit of some minimizing sequence of $\mathcal{E}$ in $\mathcal{A}$.

The convergence in (ii) and the convergence in which the limit in (iii) is taken is the same in both instances and it must be specified for each problem extraneously. Condition (i) implicitly means that $\min_{\mathfrak{A}} \hat{\mathcal{E}}$ exists. Conditions (ii) and (iii) define a "one-to-one"[3] correspondence between minimizing sequences of $\mathcal{E}$ and minimizers of $\hat{\mathcal{E}}$. If already $\mathcal{E}$ possesses a minimum on $\mathcal{A}$ then we can trivially set $\mathfrak{A} := \mathcal{A}$ and $\hat{\mathcal{E}} := \mathcal{E}$ and the above mentioned topology is the one on $\mathcal{A}$.

[3] Quotation marks are used because, obviously, there are many minimizing sequences converging to the same minimizer. Hence, we can define equivalence classes where two sequences are in the same class if they converge to the same minimizer. The one-to-one correspondence mentioned is then between these classes and minimizers.

Definition 4.4.1 (Relaxation). Assume that $\mathcal{A}$, $\mathcal{E} : \mathcal{A} \to \mathbb{R} \cup \{+\infty\}$, $\mathfrak{A}$, and $\hat{\mathcal{E}} : \mathfrak{A} \to \mathbb{R} \cup \{+\infty\}$ are given. If conditions (i), (ii), and (iii) hold simultaneously, wee say that $\hat{\mathcal{E}}$ is a relaxation of $\mathcal{E}$ with respect to the convergence considered in (ii), and (iii).

We call the problem

$$\text{minimize } \hat{\mathcal{E}} \text{ over } \mathfrak{A} \tag{4.4.2}$$

the *relaxed problem* corresponding to (4.4.1), or simply the relaxation of (4.4.1) and $\hat{\mathcal{E}}$ the *relaxed functional* corresponding to $\mathcal{E}$.

We will show that relaxed problems can be conveniently formulated in terms of Young measures or in terms of the *quasiconvex envelope* of the stored energy density. On the other hand, we will see that there are crucial drawbacks if we want to apply the relaxation theory on energy functionals in elasticity and there are many difficult open problems which need to be solved. We also refer to [60] for a review article on this topic.

4.4.1 $W^{1,p}$-relaxation

Before we start to describe relaxation procedure we define a suitable notion o convergence, called here *Y-convergence* which is the convergence needed for (ii) and (iii).

Definition 4.4.2 (Y-convergence). Assume that $\{y_k\}_{k\in\mathbb{N}} \subset W^{1,p}(\Omega;\mathbb{R}^d)$. Then $y_k \overset{Y}{\to} (y,\nu) \in L^p(\Omega;\mathbb{R}^d) \times \mathcal{GY}^p(\Omega;\mathbb{R}^{d\times d})$ weakly as $k \to \infty$ if $y_k \to y$ in $L^p(\Omega;\mathbb{R}^d)$ and $\{\nabla y_k\}_{k\in\mathbb{N}}$ generates ν in the sense of (D.1.2). This convergence is called the Y-convergence.

Let $\Omega \subset \mathbb{R}^d$ be a bounded Lipschitz domain. Let us assume that

$$\varphi : \Omega \times \mathbb{R}^{d\times d} \to \mathbb{R} \text{ is a Carathéodory integrand} \tag{4.4.3}$$

and such that for some constant $K > 0$ all $F \in \mathbb{R}^{d\times d}$ and for some $1 < p < +\infty$ it holds that

$$\frac{1}{K}|F|^p - K \le \varphi(x,F) \le K(1 + |F|^p). \tag{4.4.4}$$

Let further $\mathcal{F} : W^{1,p}(\Omega;\mathbb{R}^d) \to \mathbb{R}$ be a weakly continuous functional describing the work of external forces and assume that for some $\tilde{K} > 0$ and every $y \in W^{1,p}(\Omega;\mathbb{R}^d)$ it holds that

$$\mathcal{F}(y) \le \tilde{K}(\|y\|^{\tilde{p}}_{W^{1,p}(\Omega;\mathbb{R}^d)} + 1) \tag{4.4.5}$$

for some $0 \le \tilde{p} < p$. We define $\mathcal{E} : W^{1,p}(\Omega;\mathbb{R}^d) \to \mathbb{R}$ as

$$\mathcal{E}(y) := \int_\Omega \varphi(x, \nabla y(x))\,dx - \mathcal{F}(y). \tag{4.4.6}$$

Given $y_D \in W^{1,p}(\Omega;\mathbb{R}^d)$ which will represent a Dirichlet boundary condition on $\Gamma_D \subset \Gamma$ such that $\text{meas}_{d-1}(\Gamma_D) > 0$ we define the set of admissible deformations as

$$\mathcal{A} := \{y \in W^{1,p}(\Omega;\mathbb{R}^d);\ y = y_D \text{ on } \Gamma_D\}. \tag{4.4.7}$$

We formulate the following problem for $\mathcal{E}$ from (4.4.6) and for the set $\mathcal{A}$ given in (4.4.7):

$$\left.\begin{array}{l}\text{minimize } \mathcal{E}(y),\\ \text{subject to } y \in \mathcal{A}.\end{array}\right\} \tag{4.4.8}$$

If φ is not quasiconvex then $\mathcal{E}$ is not sequentially weakly lower semicontinuous on $W^{1,p}(\Omega;\mathbb{R}^d)$ and the direct method of the calculus of variations cannot be applied to show the existence of minimizers. More importantly, the problem (4.4.8) does not necessarily have a solution. Nevertheless, growth conditions (4.4.4) and (4.4.5) together with the Dirichlet datum y_D allow us to bound uniformly a minimizing sequence of $\mathcal{E}$. In other words, if $\{y_k\}_{k\in\mathbb{N}} \subset \mathcal{A}$ is such that $\lim_{k\to\infty} \mathcal{E}(y_k) = \inf_{\mathcal{A}} \mathcal{E}$ then $\sup_{k\in\mathbb{N}} \|y_k\|_{W^{1,p}(\Omega;\mathbb{R}^d)} < +\infty$. Therefore, we can extract a (non-relabeled) subsequence of $\{y_k\}_{k\in\mathbb{N}} \subset \mathcal{A}$ which Y-converges to $(y, \nu) \in \mathcal{A} \times \mathcal{GY}^p(\Omega;\mathbb{R}^{d\times d})$. Indeed, $y \in \mathcal{A}$ because, due to Theorem B.3.6, $y = y_D$ on Γ_D. In view of Theorem D.1.1 and formula (D.1.3) (used here for $N := d^2$) we get that

$$\inf_{\mathcal{A}} \mathcal{E} = \lim_{k\to\infty} \mathcal{E}(y_k) \geq \int_\Omega \int_{\mathbb{R}^{d\times d}} \varphi(x, F)\nu_x(dF)\,dx - \mathcal{F}(y). \tag{4.4.9}$$

Note that $\lim_{k\to\infty} \mathcal{F}(y_k) = \mathcal{F}(y)$ due to weak continuity of $\mathcal{F}$. We know from TheoremD.1.1 that the equality holds in (4.4.9) if and only if $\{\varphi(\cdot, \nabla y_k)\}_{k\in\mathbb{N}}$ is uniformly integrable, which equivalently means (see (4.4.4)) that $\{|\nabla y_k|^p\}_{k\in\mathbb{N}}$ is so. On the other hand, due to Decomposition Lemma D.1.3 we can, without loss of generality, assume that $\{|\nabla y_k|^p\}_{k\in\mathbb{N}}$ is uniformly integrable . Indeed, the right-hand side of (4.4.9) provides a lower bound for $\inf_{\mathcal{A}} \mathcal{E}$ which is attained if $\{|\nabla y_k|^p\}_{k\in\mathbb{N}}$ is uniformly integrable, or, saying otherwise, if $\{\nabla y_k\}_{k\in\mathbb{N}}$ is p-equiintegrable; cf. Theorem D.1.1 for $N := d\times d$. Consequently, any minimizing sequence inevitably has this property. Altogether, this shows that

$$\inf_{\mathcal{A}} \mathcal{E} = \int_\Omega \int_{\mathbb{R}^{d\times d}} \varphi(x, F)\nu_x(dF)\,dx - \mathcal{F}(y). \tag{4.4.10}$$

Assume now that there is $(\tilde{y}, \mu) \in \mathcal{A} \times \mathcal{GY}^p(\Omega;\mathbb{R}^{d\times d})$ such that

$$\int_\Omega \int_{\mathbb{R}^{d\times d}} \varphi(x, F)\nu_x(dF)\,dx - \mathcal{F}(y) > \int_\Omega \int_{\mathbb{R}^{d\times d}} \varphi(x, F)\mu_x(dF)\,dx - \mathcal{F}(\tilde{y}). \tag{4.4.11}$$

But this means that there is a sequence $\{\tilde{y}_k\}_{k\in\mathbb{N}} \subset \mathcal{A}$ such that $\{\nabla \tilde{y}_k\}_{k\in\mathbb{N}}$ generates μ. Moreover, we can assume that $\{|\nabla \tilde{y}_k|^p\}_{k\in\mathbb{N}}$ is uniformly integrable and thus we see that $\{\tilde{y}_k\}_{k\in\mathbb{N}}$ Y-converges to $(\tilde{y},\mu)$ for $k\to\infty$. This, together with (4.4.10) and (4.4.11), however means that for k large enough $\mathcal{E}(\tilde{y}_k)<\inf_{\mathcal{A}}\mathcal{E}$ which is not possible. Let us now define for $(y,\nu)\in\mathcal{A}\times\mathcal{GY}^p(\Omega;\mathbb{R}^{d\times d})$

$$\bar{\mathcal{E}}(y,\nu):=\int_\Omega\int_{\mathbb{R}^{d\times d}}\varphi(x,F)\nu_x(\mathrm{d}F)\,\mathrm{d}x-\mathcal{F}(y) \tag{4.4.12}$$

and

$$\bar{\mathcal{A}}:=\{(y,\nu)\in\mathcal{A}\times\mathcal{GY}^p(\Omega;\mathbb{R}^{d\times d});\ \nabla y=\bar{\nu}\}. \tag{4.4.13}$$

Here $\bar{\nu}(x)=\int_{\mathbb{R}^{d\times d}}F\nu_x(\mathrm{d}F)$ for almost every $x\in\Omega$ is the first moment of the Young measure ν as defined in (D.1.4a).

The above considerations really show that $\bar{\mathcal{E}}$ from (4.4.12) together with $\mathfrak{A}$ from (4.4.13) and Y-convergence is the relaxed problem to (4.4.8). More precisely, we set

$$\left.\begin{array}{l}\text{minimize } \bar{\mathcal{E}}_{(}y,\nu),\\ \text{subject to } (y,\nu)\in\bar{\mathcal{A}}.\end{array}\right\} \tag{4.4.14}$$

We just have proved the following theorem:

Theorem 4.4.3 (Relaxation). *Let $\Omega\subset\mathbb{R}^d$ be a bounded Lipschitz domain, let* (4.4.3), (4.4.4),(4.4.7), *and* (4.4.5) *hold. Then the problem* (4.4.14) *is a relaxed problem to* (4.4.8) *with respect to the Y-convergence in the sense of Definition 4.4.1, i.e. $\hat{\mathcal{E}}:=\bar{\mathcal{E}}$ and $\mathfrak{A}:=\bar{\mathcal{A}}$.*

We can see that the relaxation theorem (4.4.3) is just a formal way how to express limits of $\mathcal{E}$ along minimizing sequences and does not shed much light on the original problem (4.4.8). It is therefore important to understand properties of gradient Young measures. Their characterization is a celebrated result by D. Kinderlehrer and P. Pedregal [271, 272]. The explicit description is also stated in Theorem D.1.2 in the appendix. Nevertheless, condition (D.1.4b) is exteremely difficult/impossible to verify because it is a Jensen-type inequality involving quasiconvex functions.

To make the problem more tractable we can come up is an important subset of gradient Young measures called *laminates*. This notion was introduced by Pedregal in [409]. In this case, we simply assume that (D.1.4b) holds for every rank-one convex function ψ (with the given growth). As quasiconvexity implies rank-one convexity (see Theorem 4.1.3), laminates form a subset of gradient Young measures. On the other hand, we can require (D.1.4b) for all polyconvex functions ψ (with the given growth). Since polyconvexity implies quasiconvexity (see Theorem 4.1.11), such Young measures generally form a superset of the set of gradient Young measures called in [47, 291] *polyconvex measures*. It is, however, not obvious that the inclusions are strict in general. Let us first show that there is a gradient Young measure $\nu\in\mathcal{GY}^p(\Omega;\mathbb{R}^{d\times d})$ for $d=3$ and every $1\le p\le+\infty$ which is not a laminate. This construction appeared in [291].

Let us take functions $g,h:[0,4]\to\mathbb{R}$ as

$$g(t) = \begin{cases} -3t & \text{if } 0 \le t \le 1, \\ t-4 & \text{if } 1 \le t \le 4, \end{cases} \qquad \text{and} \qquad h(t) = \begin{cases} t & \text{if } 0 \le t \le 2, \\ -t+4 & \text{if } 2 \le t \le 4, \end{cases}$$

extend both functions periodically onto the whole $\mathbb{R}$, and define a $(0,4)^2$-periodic deformation $\tilde{y} : (0,4)^2 \to \mathbb{R}^d$ as

$$\tilde{y}(x) = \begin{pmatrix} g(x_1) \\ g(x_2) \\ h(x_1+x_2) \end{pmatrix}.$$

Further set

$$B = \begin{pmatrix} -1 & 0 \\ 0 & -1 \\ -1 & -1 \end{pmatrix}.$$

The matrix $B + \nabla\tilde{y}$ takes on $(0,4)^2$ seven different values $A_1, \ldots, A_7$; cf. Figure 4.2. We have

$$A_1 = \begin{pmatrix} -4 & 0 \\ 0 & -4 \\ 0 & 0 \end{pmatrix}, \quad A_2 = \begin{pmatrix} -4 & 0 \\ 0 & 0 \\ 0 & 0 \end{pmatrix}, \quad A_3 = \begin{pmatrix} -4 & 0 \\ 0 & 0 \\ -2 & -2 \end{pmatrix}, \quad A_4 = \begin{pmatrix} 0 & 0 \\ 0 & -4 \\ 0 & 0 \end{pmatrix}, \tag{4.4.15a}$$

$$A_5 = \begin{pmatrix} 0 & 0 \\ 0 & -4 \\ -2 & -2 \end{pmatrix}, \qquad A_6 = \begin{pmatrix} 0 & 0 \\ 0 & 0 \\ -2 & -2 \end{pmatrix}, \qquad A_7 = \begin{pmatrix} 0 & 0 \\ 0 & 0 \\ 0 & 0 \end{pmatrix}. \tag{4.4.15b}$$

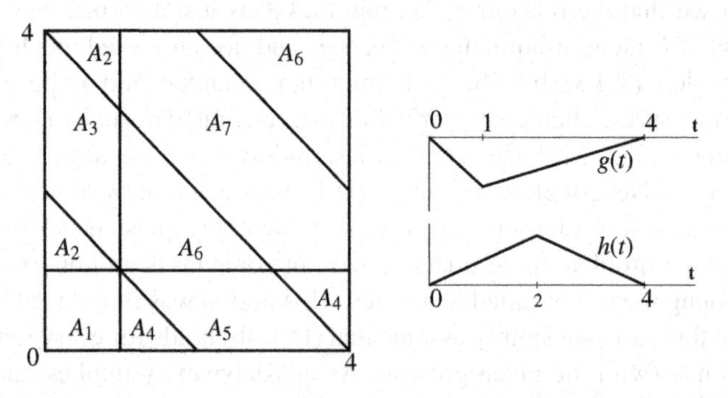

Fig. 4.2 The seven values of $B + \nabla\tilde{y}$ and signs of $g'(x_1)$, $g'(x_2)$, and $h'(x_1+x_2)$, respectively.

By construction, if $\varphi : \mathbb{R}^{3\times 2} \to \mathbb{R}$ is quasiconvex then inevitably

$$\varphi(B) \le \frac{1}{64} \int_{(0,4)^3} \varphi(B + \nabla\tilde{y}(x))\, dx = \frac{1}{16}(\varphi(A_1) + \varphi(A_2) + \varphi(A_4)) + \frac{1}{8}(\varphi(A_3) + \varphi(A_5)) + \frac{1}{4}\varphi(A_6) + 20\varphi(A_7) = \int_{\mathbb{R}^{d\times d}} \varphi(F)\nu(dF). \tag{4.4.16}$$

Hence, $\varphi(B) \le \int_{\mathbb{R}^{d\times d}} \varphi(F)\nu(dF)$, where

$$\nu := \frac{1}{16}(\delta_{A_1} + \delta_{A_2} + \delta_{A_4}) + \frac{1}{8}(\delta_{A_3} + \delta_{A_5}) + \frac{1}{4}\delta_{A_6} + \frac{5}{16}\delta_{A_7}. \tag{4.4.17}$$

Let us notice that $B = \bar{\nu}$, i.e.,

$$B = \frac{1}{16}(A_1 + A_2 + A_4) + \frac{1}{8}(A_3 + A_5) + \frac{1}{4}A_6 + \frac{5}{16}A_7. \tag{4.4.18}$$

On the other hand, taking the rank-one convex function $\varphi_{\varepsilon,k}$ from (4.1.13) we easily see that for ε small enough but still positive we have $\varphi_{\varepsilon,k}(B) > \int_{\mathbb{R}^{3\times 2}} \varphi_{\varepsilon,k}(F)\nu(\mathrm{d}F)$. This means that $\nu \in \mathcal{GY}^p(\Omega;\mathbb{R}^{3\times 2})$ from (4.4.17) is not a laminate.

At the same time, we can exploit Theorem 4.1.8 to show that

$$\nu^\top := \frac{1}{16}(\delta_{A_1^\top} + \delta_{A_2^\top} + \delta_{A_4^\top}) + \frac{1}{8}(\delta_{A_3^\top} + \delta_{A_5^\top}) + \frac{1}{4}\delta_{A_6^\top} + \frac{5}{16}\delta_{A_7^\top} \tag{4.4.19}$$

is not a gradient Young measure. Indeed, we get for a quasiconvex function $\tilde{\varphi}_{\varepsilon,k}$ given in Theorem 4.1.8 that

$$\tilde{\varphi}_{\varepsilon,k}(B^\top) > \int_{\mathbb{R}^{2\times 3}} \tilde{\varphi}_{\varepsilon,k}(F)\nu^\top(\mathrm{d}F), \tag{4.4.20}$$

which follows easily from (4.4.19). If we extend all matrices A_i, for $1 \le i \le 7$ and the matrix B to matrices in $\mathbb{R}^{3\times 3}$ by adding zeros as the third column and extending $\varphi_{\varepsilon,k}$ to a map from $\mathbb{R}^{3\times 3} \to \mathbb{R}$ as in Theorem 4.1.7 as well as $\tilde{\varphi}_{\varepsilon,k}$ we get get the same inequality as (4.4.20). This shows that $\nu^\top \notin \mathcal{GY}^p(\Omega;\mathbb{R}^{d\times d})$ for $d = 3$ and $1 \le p \le +\infty$. On the other hand, it is a matter of a simple calculation which is left as an exercise to the reader that

$$\det B = \int_{\mathbb{R}^{3\times 3}} \det F \nu^\top(\mathrm{d}F) \quad \text{and} \quad \mathrm{Cof}\, B = \int_{\mathbb{R}^{3\times 3}} \mathrm{Cof}\, F \nu^\top(\mathrm{d}F), \tag{4.4.21}$$

i.e., $\nu^\top$ is not a gradient Young measure but it satisfies the Jensen inequality (D.1.4b) (as the equality) when tested by the determinant and the cofactor. If $d = 2$ it is not known whether laminates form a strict subset of gradient Young measures. However, gradient Young measures are strictly contained in polyconvex measures. For the proof, we refer to V. Šverák [499] who showed that the function $\varphi : \mathbb{R}^{2\times 2} \to \mathbb{R} \cup \{+\infty\}$ defined as

$$\varphi(F) := \begin{cases} \det F & \text{if } F \text{ is symmetric and positive definite,} \\ 0 & \text{if } F \text{ is symmetric but not positive definite,} \\ +\infty & \text{otherwise} \end{cases} \tag{4.4.22}$$

is quasiconvex. Take

$$A_1 := \begin{pmatrix} -1 & 0 \\ 0 & -1/2 \end{pmatrix}, \quad A_2 := \begin{pmatrix} 0 & 0 \\ 0 & 1/2 \end{pmatrix}, \quad A_3 := \begin{pmatrix} 1 & 0 \\ 0 & 0 \end{pmatrix}, \quad B := \begin{pmatrix} 1/3 & 0 \\ 0 & 1/6 \end{pmatrix}, \tag{4.4.23}$$

and the measure

$$\nu := \frac{1}{9}\delta_{A_1} + \frac{4}{9}\delta_{A_2} + \frac{4}{9}\delta_{A_3}. \tag{4.4.24}$$

We get that $B = \bar{\nu}$ and that $\det B = \int_{\mathbb{R}^{2\times 2}} \det F \nu_x(\mathrm{d}F)$ so that ν is a polyconvex measure. However, $\varphi(B) > \int_{\mathbb{R}^{2\times 2}} \varphi(F)\nu_x(\mathrm{d}F)$ for the quasiconvex function φ introduced in (4.4.22)[4]. Consequently, ν in (4.4.24) is a polyconvex measure but not a gradient Young measure. Altogether, we can only speak about the inner (by laminates) and the outer (by polyconvex measures) estimate of the set $\mathcal{GY}^p(\Omega;\mathbb{R}^{d\times d})$ but we can hardly speak about the inner and the outer approximations.

The provided relaxation in Theorem 4.4.3 utilizes (gradient) Young measures. While this is a useful tool often used in the calculus of variations, it is still more common in some applications to use a relaxation by means of a so-called quasiconvex envelope of $\varphi(x,\cdot)$ and we will denote it $\mathrm{Q}\varphi(x,\cdot)$. This is the largest quasiconvex function below $\varphi(x,\cdot)$ where $x \in \Omega$. Let us, for simplicity, omit the dependence of φ on $x \in \Omega$ and assume that $\varphi : \Omega \times \mathbb{R}^{d\times d} \to \mathbb{R}$ is a Carathéodory integrand.

We define for $x_0 \in \Omega$

$$\mathrm{Q}\varphi(x_0,\cdot) := \sup\{\psi : \mathbb{R}^{d\times d} \to \mathbb{R} \text{ quasiconvex}; \psi \le \varphi(x_0,\cdot)\}. \tag{4.4.25}$$

If φ satisfies, for some $1 < p < +\infty$, almost every $x \in \Omega$, and $K > 0$,

$$\frac{1}{K}|F|^p - K \le \varphi(x,F) \le K(1+|F|^p). \tag{4.4.26}$$

Then we immediately see that

$$\frac{1}{K}|F|^p - K \le \mathrm{Q}\varphi(x,F) \le K(1+|F|^p) \tag{4.4.27}$$

because the lower bound on φ in (4.4.26) is a convex function of F and therefore it is also quasiconvex. Formula (4.4.25) is rather implicit but, if φ is Borel measurable and locally bounded, then it was proved in [135] that, for every $F \in \mathbb{R}^{d\times d}$, every $p \le q \le +\infty$, and almost every $x_0 \in \Omega$, it holds

$$\mathrm{Q}\varphi(x_0,F) = \inf_{y \in W_0^{1,q}(\Omega;\mathbb{R}^d)} \frac{1}{\mathrm{meas}_d(\Omega)} \int_\Omega \varphi(x_0, F + \nabla y(x))\,\mathrm{d}x, \tag{4.4.28}$$

where Ω is a bounded Lipschitz domain in $\mathbb{R}^d$. Similarly as the definition of quasiconvexity, the left-hand side of (4.4.28) is independent of the Lipschitz bounded domain Ω, so that one can take, e.g., $\Omega := (0,1)^d$ for definiteness. Again, this for-

[4] A careful reader might object that (4.4.22) is not an admissible test function in (D.1.4b) because it does not satisfy the required growth condition. However, the proof of Theorem D.1.2 can be easily modified also for mappings with symmetric gradients. This modification exploits that symmetric matrices form a convex set.

mula is of the limited use because to calculate the infimum on the right-hand side requires to solve a variational problem. Moreover, to make the definition sound, we must show that the right-hand side of (4.4.28) does not depend on q in the following sense.

Proposition 4.4.4 (Well-posedness of Qφ). *Assume that* (4.4.26) *holds. Then the right-hand side of* (4.4.28) *is the same for every* $q \geq p$.

Proof. Let $Q^{\infty}\varphi(x_0, F)$ be the right-hand side of (4.4.28) for $q := +\infty$ and $Q^p\varphi(x_0, F)$ the same quantity for $q := p$. Take $y \in W_0^{1,p}(\Omega;\mathbb{R}^d)$ and $\{y_k\}_{k\in\mathbb{N}} \subset W_0^{1,\infty}(\Omega;\mathbb{R}^d)$ such that $y_k \to y$ in $W^{1,p}(\Omega;\mathbb{R}^d)$ for $k \to \infty$. We have

$$\begin{aligned}0 &\leq \int_\Omega K(1+|F+\nabla y_k(x)|^p) - \varphi(x_0, F+\nabla y_k(x))\,\mathrm{d}x\\ &\leq \int_\Omega K(1+|F+\nabla y_k(x)|^p) - Q^{\infty}\varphi(x_0, F)\,\mathrm{d}x.\end{aligned}$$

Passing to the limit for $k \to \infty$ (possibly in a subsequence) and invoking the Fatou lemma B.2.3 yields that

$$Q^{\infty}\varphi(x_0, F) \leq \frac{1}{\mathrm{meas}_d(\Omega)} \int_\Omega \varphi(x_0, F+\nabla y(x))\,\mathrm{d}x.$$

Arbitrariness of y implies that $Q^{\infty}\varphi(x_0, F) \leq Q^p\varphi(x_0, F)$. The opposite inequality is obvious because $W_0^{1,\infty}(\Omega;\mathbb{R}^d) \subset W_0^{1,p}(\Omega;\mathbb{R}^d)$, hence the claim is proved. □

Let us notice that (4.4.28) can be rewritten as

$$\mathrm{Q}\varphi(x_0, F) = \inf_{y\in W_F^{1,p}(\Omega;\mathbb{R}^d)} \frac{1}{\mathrm{meas}_d(\Omega)} \int_\Omega \varphi(x_0, \nabla y(x))\,\mathrm{d}x, \tag{4.4.29}$$

where[5]

$$W_F^{1,p}(\Omega;\mathbb{R}^d) := \{\tilde{y} \in W^{1,p}(\Omega;\mathbb{R}^d);\ \tilde{y}(x) = Fx \text{ if } x \in \Gamma\}. \tag{4.4.30}$$

Since Ω in (4.4.29) and (4.4.30) can be replaced by any bounded Lipschitz domain in $\mathbb{R}^d$ and (4.4.29) is independent of p in the sense of Proposition 4.4.4 we can equivalently rewrite (4.4.29) for $\Omega := B(0,1) \subset \mathbb{R}^d$, i.e. the unit ball in R^d centered at the origin, as

$$\mathrm{Q}\varphi(x_0, F) = \inf_{y\in W_F^{1,\infty}(B(0,1);\mathbb{R}^d)} \frac{1}{\mathrm{meas}_d(B(0,1))} \int_{B(0,1)} \varphi(x_0, \nabla y(x))\,\mathrm{d}x, \tag{4.4.31}$$

[5] Unless $F = 0$, $W_F^{1,p}(\Omega;\mathbb{R}^d)$ is *not* a linear function space.

Consider $\{y_k\}_{k\in\mathbb{N}} \subset W^{1,p}_F(\Omega;\mathbb{R}^d)$ a minimizing sequence for $Q\varphi(x_0,F)$ defined in (4.4.29). This minimizing sequence is inevitably bounded in $W^{1,p}(\Omega;\mathbb{R}^d)$ due to (4.4.26) and Poincaré-type inequality (Theorem B.3.15) because of the Dirichlet boundary conditions. Hence, by Theorem D.1.1with $N := d\times d$, there is a Young measure $\nu \in \mathcal{GY}^p(\Omega;\mathbb{R}^{d\times d})$ generated by (a subsequence of) $\{\nabla y_k\}_{k\in\mathbb{N}}$ such that

$$Q\varphi(x_0,F) = \frac{1}{\mathrm{meas}_d(\Omega)}\int_\Omega\int_{\mathbb{R}^{d\times d}}\varphi(x_0,A)\nu_x(\mathrm{d}A)\,\mathrm{d}x. \tag{4.4.32}$$

Indeed, we can, without loss of generality, assume that $\{|\nabla y_k|^p\}_{k\in\mathbb{N}}$ is equiintegrable in $L^1(\Omega)$ because of Lemma D.1.3. In view of Proposition D.1.5 there is a homogeneous $\mu \in \mathcal{GY}^p(\Omega;\mathbb{R}^{d\times d})$ (depending on x_0 and F, i.e., $\mu = \mu(x_0,F)$) such that

$$Q\varphi(x_0,F) = \int_{R^{d\times d}}\varphi(A)\mu(\mathrm{d}A), \quad \bar{\mu} = F. \tag{4.4.33}$$

The Jensen inequality (D.1.4b) and the obvious relation $Q\varphi \le \varphi$ yields

$$Q\varphi(x_0,F) = \int_{R^{d\times d}}\varphi(x_0,A)\mu(\mathrm{d}A) \ge \int_{R^{d\times d}}Q\varphi(x_0,A)\mu(\mathrm{d}A) \ge Q\varphi(x_0,F), \tag{4.4.34}$$

hence

$$\int_{R^{d\times d}}\varphi(x_0,A)\mu(\mathrm{d}A) = \int_{R^{d\times d}}Q\varphi(x_0,A)\mu(\mathrm{d}A) = Q\varphi(x_0,F). \tag{4.4.35}$$

It implies that μ is supported on the set $\{A \in \mathbb{R}^{d\times d};\ \varphi(x_0,A) = Q\varphi(x_0,A)\}$. Let $y \in W^{1,p}(\Omega;\mathbb{R}^d)$ and we apply (4.4.35) for $F := \nabla y(x)$ and $x_0 := x$ for almost every $x \in \Omega$, denoting $\mu_x := \mu(x,\nabla y(x))$. We see that

$$\int_{R^{d\times d}}\varphi(x,A)\mu_x(\mathrm{d}A) = \int_{R^{d\times d}}Q\varphi(x,A)\mu_x(\mathrm{d}A) = Q\varphi(x,\nabla y(x)). \tag{4.4.36}$$

We also see that $x \mapsto \int_{R^{d\times d}}\varphi(A)\mu_x(\mathrm{d}A)$ as well as $x \mapsto \int_{R^{d\times d}}Q\varphi(A)\mu_x(\mathrm{d}A)$ is measurable because $Q\varphi(\cdot,\nabla y)$ is so due to continuity of $Q\varphi$ in its second variable. However, generally it is not obvious that $\{\mu_x\}_{x\in\Omega}$ is a Young measure because the weak* measurability of $x \mapsto \mu_x$ does not follow from the above considerations. We will show it for φ independent of $x \in \Omega$. For more general case see [501]. To prove it, we start with an auxiliary lemma.

Lemma 4.4.5. *Let $y \in W^{1,p}(\Omega;\mathbb{R}^d)$ be given, let φ (and also $Q\varphi$) be independent of $x \in \Omega$, and let (4.4.26) hold. Then there is a bounded sequence $\{\tilde{y}_k\}_{k\in\mathbb{N}} \subset W^{1,p}(\Omega;\mathbb{R}^d)$ such that $\tilde{y}_k \rightharpoonup y$ weakly in $W^{1,p}(\Omega;\mathbb{R}^d)$ as $k\to\infty$ and*

$$\lim_{k\to\infty}\int_\Omega\psi(\nabla\tilde{y}_k(x))\,\mathrm{d}x = \int_\Omega\int_{\mathbb{R}^{d\times d}}\psi(A)\mu_x(\mathrm{d}A)\,\mathrm{d}x \tag{4.4.37}$$

holds for $\psi \in \{\varphi, Q\varphi\}$.

Sketch of the proof. The sequence can be constructed by the same way as in [272, Proof of Thm. 1.1]. Take $r_k : \Omega \setminus \omega \to \mathbb{R}$ and using Theorem B.4.5 find $a_{ik} \in \Omega \setminus \omega$, $\epsilon_{ik} \leq r_k(a_{ik})$ such that

$$\lim_{k\to\infty} \sum_i V(a_{ik})\mathrm{meas}_d(a_{ik} + \epsilon_{ik}\Omega) = \int_\Omega V(x)\,\mathrm{d}x,$$

where

$$V(x) := \int_\Omega \psi(A)\mu_x(\mathrm{d}A).$$

Here we apply Theorem B.4.5 with $g := 1$ and $f_1(x) := \int_\Omega \varphi(A)\mu_x(\mathrm{d}A)$ and $f_2(x) := \int_\Omega Q\varphi(A)\mu_x(\mathrm{d}A)$ for almost all $x \in \Omega$.

We know that that $\mu_{a_{ik}}$ is a homogeneous gradient Young measure living in $\mathcal{GY}^p(\Omega;\mathbb{R}^{d\times d})$ and we call $\{\nabla y_j^{ik}\}_{j\in\mathbb{N}}$ its generating sequence. It means that

$$\lim_{j\to\infty} \int_\Omega \psi(x, \nabla y_j^{ik}(x))g(x)\,\mathrm{d}x = \int_\Omega \int_{\mathbb{R}^{d\times d}} \psi(A)\mu_{a_{ik}}(\mathrm{d}A)\,\mathrm{d}x. \tag{4.4.38}$$

By Theorem B.4.5, we have

$$\bar{\Omega} = \bigcup_i \{x \in a_{ik} + \epsilon_{ik}\bar{\Omega}\} \bigcup N_k, \quad \mathrm{meas}_d(N_k) = 0.$$

We define a sequence of smooth cut-off functions $\{\eta_\ell\}_{\ell\in\mathbb{N}}$ such that

$$\eta_\ell(x) = \begin{cases} 0 & \text{in } \Omega_\ell, \\ 1 & \text{on } \Gamma \end{cases}$$

and $|\nabla\eta_\ell| \leq C\ell$ for some $C > 0$. Further, take a sequence $\{y_k^\ell\}_{k,\ell\in\mathbb{N}} \subset W^{1,p}(\Omega;\mathbb{R}^m)$ defined by

$$y_k^\ell(x) = \begin{cases} \Big[y(a_{ik}) + \epsilon_{ik} y_j^{ik}\Big(\frac{x-a_{ik}}{\epsilon_{ik}}\Big)\Big]\Big(1 - \eta_\ell\Big(\frac{x-a_{ik}}{\epsilon_{ik}}\Big)\Big) + y(x)\eta_\ell\Big(\frac{x-a_{ik}}{\epsilon_{ik}}\Big) & \text{if } x \in a_{ik} + \epsilon_{ik}\Omega, \\ y(x) & \text{otherwise.} \end{cases}$$

A suitable diagonal subsequence of $\{\nabla y_k^\ell\}_{\ell,k\in\mathbb{N}}$ is the sought sequence $\{\nabla\tilde{y}_k\}_{k\in\mathbb{N}}$. Consequently, $\tilde{y}_k \rightharpoonup y$ weakly in $W^{1,p}(\Omega;\mathbb{R}^d)$ for k approaching infinity. □

Having $\{\tilde{y}_k\}_{k\in\mathbb{N}}$ bounded in $W^{1,p}(\Omega;\mathbb{R}^d)$ just constructed we know that it contains a (non-relabeled) subsequence $\{\nabla\tilde{y}_k\}_{k\in\mathbb{N}}$ which generates a gradient Young measure $\nu \in \mathcal{GY}^p(\Omega;\mathbb{R}^{d\times d})$. This allows us to prove the following result.

Theorem 4.4.6 (Young measures and quasiconvexification). *Let $\Omega \subset \mathbb{R}^d$ be a bounded Lipschitz domain. Let $\varphi : \mathbb{R}^{d\times d} \to \mathbb{R}$ be continuous and satisfy* (4.4.26). *Then for every $y \in W^{1,p}(\Omega;\mathbb{R}^d)$ there is $\nu \in \mathcal{GY}^p(\Omega;\mathbb{R}^{d\times d})$ such that the first moment of ν (as in* (D.1.4a)*) equals $\bar{\nu} = \nabla y$ almost everywhere in Ω and*

$$\int_\Omega \int_{R^{d\times d}} \varphi(x,A)\nu_x(\mathrm{d}A)\,\mathrm{d}x = \int_\Omega \int_{R^{d\times d}} Q\varphi(x,A)\nu_x(\mathrm{d}A) = \int_\Omega Q\varphi(x,\nabla y(x))\,\mathrm{d}x. \tag{4.4.39}$$

Moreover, if $\{\nabla\tilde{y}_k\}_{k\in\mathbb{N}}$ *generates* ν, *then*

$$\lim_{k\to\infty} \int_\Omega \varphi(x,\nabla\tilde{y}_k(x))\,\mathrm{d}x = \int_\Omega Q\varphi(x,\nabla y(x))\,\mathrm{d}x. \tag{4.4.40}$$

Proof. The measure ν is constructed as mentioned above and it follows from Lemma 4.4.5 that

$$\int_\Omega \int_{R^{d\times d}} \varphi(A)\nu_x(\mathrm{d}A)\,\mathrm{d}x = \int_\Omega \int_{R^{d\times d}} \varphi(A)\mu_x(\mathrm{d}A)\,\mathrm{d}x = \int_\Omega Q\varphi(\nabla y(x))\,\mathrm{d}x \quad \text{and}$$

$$\int_\Omega \int_{R^{d\times d}} Q\varphi(A)\nu_x(\mathrm{d}A)\,\mathrm{d}x = \int_\Omega \int_{R^{d\times d}} Q\varphi(A)\mu_x(\mathrm{d}A)\,\mathrm{d}x = \int_\Omega Q\varphi(\nabla y(x))\,\mathrm{d}x.$$

□

As we have already mentioned above, Theorem 4.4.6 holds even if φ explicitly depends on $x \in \Omega$, more precisely, if (4.4.4) holds. Theorem 4.4.6 allows us to state a different relaxation of (4.4.8) without Young measures. Instead, we employ the quasiconvex envelope of φ.

Let (4.4.3), (4.4.4), (4.4.5), and (4.4.7) hold where $\mathcal{F} : W^{1,p}(\Omega;\mathbb{R}^d) \to \mathbb{R}$ is weakly continuous. Let us define

$$\mathcal{E}_Q(y) := \int_\Omega Q\varphi(x,\nabla y(x))\,\mathrm{d}x - \mathcal{F}(y) \tag{4.4.41}$$

and the problem

$$\left.\begin{array}{l}\text{minimize } \mathcal{E}_Q(y),\\ \text{subject to } y \in \mathcal{A}.\end{array}\right\} \tag{4.4.42}$$

We have the following result.

Theorem 4.4.7 (Relaxation with the quasiconvex envelope). *Let* $\Omega \subset \mathbb{R}^d$ *be a bounded Lipschitz domain, let* (4.4.3), (4.4.4), (4.4.7), *and* (4.4.5) *hold. Then the problem* (4.4.42) *is a relaxed problem to* (4.4.8) *with respect to the the weak convergence on* $W^{1,p}(\Omega;\mathbb{R}^d)$ *in the sense of Definition 4.4.1. Hence* $\hat{\mathcal{E}} := \mathcal{E}_Q$ *and* $\mathfrak{A} := \mathcal{A}$.

Proof. If $\{y_k\}_{k\in\mathbb{N}} \subset \mathcal{A}$ is a minimizing sequence of $\mathcal{E}$ then it is uniformly bounded in $W^{1,p}(\Omega;\mathbb{R}^d)$ and a (not relabeled) subsequence $\{\nabla y_k\} \subset L^p(\Omega;\mathbb{R}^{d\times d})$ generates a Young measure $\nu \in \mathcal{GY}^p(\Omega;\mathbb{R}^{d\times d})$. Moreover $y_k \rightharpoonup y$ weakly in $W^{1,p}(\Omega;\mathbb{R}^d)$ as $k \to \infty$. We can also assume that $\{|\nabla y_k|^p\}_{k\in\mathbb{N}}$ is equiintegrable due to Lemma D.1.3 and Theorem D.1.1 used with $N := d\times d$. Consequently,

$$\lim_{k\to\infty} \mathcal{E}(y_k) \geq \liminf_{k\to\infty} \mathcal{E}_Q(y_k) = \int_\Omega \int_{\mathbb{R}^{d\times d}} Q\varphi(x,A)\nu_x(\mathrm{d}A)\,\mathrm{d}x - \mathcal{F}(y) \tag{4.4.43}$$

The right-hand side, however, must be equal to $\mathcal{E}_Q(y)$ which follows easily from Theorem 4.4.6. On the other hand, if y is a minimizer to $\mathcal{E}_Q$ on $\mathcal{A}$ then Theorem D.1.1 implies that there is a gradient Young measure and a generating sequence which is inevitably minimizing for $\mathcal{E}$. Notice that the prescribed Dirichlet boundary condition on Γ_D does not cause any difficulties because by Proposition D.1.6 there is always a generating sequence of gradients such that the underlying deformations share the trace with the weak limit. □

Remark 4.4.8 (Smoothness of $Q\varphi$). We refer to [34] for a proof that if $\varphi : \mathbb{R}^{d\times d} \to \mathbb{R}$ is differentiable and such that $\liminf_{|F|\to+\infty} \varphi(F)/|F|^p = +\infty$ and $\limsup_{|F|\to+\infty} \varphi(F)/|F|^{p+1} < +\infty$ for some $p \ge 0$ then $Q\varphi$ has a continuous gradient.

4.4.2 $W^{1,\infty}$-relaxation

In this section, we prove a relaxation result for functionals involving a continuous stored energy density function φ and a locking constraint $L(\nabla y) \le 0$; cf. also Remark 3.4.14. The considered *locking function* will be often the form

$$L(F) := |F| - \varsigma \le 0, \tag{4.4.44}$$

which means that we assume that the deformation gradient is restricted to the closed ball $\overline{B(0;\varsigma)}$ for some $\varsigma > 0$. Analogous material constraints appeared already in the theory of strain-limiting materials, see for example [99]. If we want to apply the relaxation approach by means of Young measures as in the previous section we face the problem that if the gradient Young measure is supported on a bounded set then its generating sequence is generally supported on a large set. This may cause the violation of the constraint (4.4.44). This situation is demonstrated in Proposition D.1.4. It states that if $\nu \in \mathcal{GY}^p(\Omega;\mathbb{R}^{d\times d})$ is supported on a compact convex set $\mathcal{S} \subset \mathbb{R}^{d\times d}$ then there is a generating sequence of ν which is supported in an arbitrarily small neighborhood of $\mathcal{S}$. In what follows we exploit this result and also the fact that we want to pass to a limit along the sequence just with the integrand representing the stored energy density. This leads to the idea to rescale suitably the generating sequence and this allows us to prove a relaxation result. In order to do so, we will replace Dirichlet boundary conditions by the following penalty term which takes into account (in a simplified way) elastic properties of the hard device to which the specimen is clamped. More precisely, we consider a term

$$\alpha \|y - y_D\|_{L^2(\Gamma_D;\mathbb{R}^d)}, \tag{4.4.45}$$

where $\alpha > 0$ and y_D is a desired value of the trace of y on $\Gamma_D \subset \Gamma$. Obviously, we could come up with more sophisticated terms reflecting non-isotropy, for instance, or some other material features.

Sometimes it is needed to replace the Frobenius norm in (4.4.44) by a different norm on $\mathbb{R}^{d\times d}$. This leads us to the following definition of a *strictly convex set.*

Definition 4.4.9 (Strictly convex set). A convex closed set $\mathcal{S}$ is called strictly convex if for every $A_1, A_2 \in \partial\mathcal{S}$ and every $0 < \lambda < 1$ it holds that $\lambda A_1 + (1-\lambda)A_2 \notin \partial\mathcal{S}$.

As to the functional $\mathcal{F} : W^{1,\infty}(\Omega;\mathbb{R}^d) \to \mathbb{R}$ describing the work of external force we assume that there is $C_\ell > 0$ such that for every $y, \tilde{y} \in W^{1,\infty}(\Omega;\mathbb{R}^d)$ we have

$$|\mathcal{F}(y) - \mathcal{F}(\tilde{y})| \le C_\ell \|y - \tilde{y}\|_{L^\infty(\Omega;\mathbb{R}^d)}, \tag{4.4.46}$$

which is a Lipschitz continuity of $\mathcal{F}$[6]. In more detail, we seek to solve

$$\left.\begin{array}{l} \text{minimize } \mathcal{E}(y) := \displaystyle\int_\Omega \varphi(\nabla y(x))\,\mathrm{d}x - \mathcal{F}(y) + \alpha\big\|y - y_\mathrm{D}\big\|_{L^2(\Gamma_\mathrm{D};\mathbb{R}^d)} \\ \text{subject to } y \in W^{1,\infty}(\Omega;\mathbb{R}^d) \text{ and } |\nabla y| \le \varsigma \text{ a.e. in } \Omega. \end{array}\right\} \tag{4.4.47}$$

The energy contribution (4.4.45) together with the locking constraint yields boundedness of the minimizing sequence of $\mathcal{E}$ in $W^{1,\infty}(\Omega;\mathbb{R}^d)$ owing to the Poincaré inequality. Let us notice also that the last term in $\mathcal{E}$ is continuous with respect to the weak* convergence in $W^{1,\infty}(\Omega;\mathbb{R}^d)$.

We can easily infer from Proposition 4.1.13 that if $\varsigma = +\infty$, and φ is quasiconvex then $\mathcal{E}$ is weakly* lower semicontinuous on $W^{1,\infty}(\Omega;\mathbb{R}^d)$.Nevertheless, due to the involved locking constraint and the fact that φ may not even be defined outside $\overline{B(0,\varsigma)}$, this does not directly mean that (4.4.47) possesses a solution. This issue was settled by Kinderlehrer and Pedregal [271] who, by suitable rescaling, indeed showed that (4.4.47) is solvable provided φ is quasiconvex on its domain.

On the other hand, if φ is not quasiconvex, solutions to (4.4.47) might not exist. In this case, some physically relevant quantities about the minimizing sequence can be drawn from studying minimizers of the relaxed problem. In the following we provide a relaxation of the functional $\mathcal{E}$ by means of Young measures.

We define $\bar{\mathcal{E}} : W^{1,\infty}(\Omega;\mathbb{R}^d) \times \mathcal{GY}^\infty(\Omega;\mathbb{R}^{d\times d}) \to \mathbb{R}$ and the following minimization problem via:

$$\left.\begin{array}{l} \text{minimize } \bar{\mathcal{E}}(y_\nu, \nu) := \displaystyle\int_\Omega \int_{\mathbb{R}^{d\times d}} \varphi(A)\nu_x(\mathrm{d}A)\,\mathrm{d}x - \mathcal{F}(y) + \alpha\big\|y_\nu - y_\mathrm{D}\big\|_{L^2(\Gamma_\mathrm{D};\mathbb{R}^d)} \\ \text{subject to } y_\nu \in W^{1,\infty}(\Omega;\mathbb{R}^d), \;\; \nu \in \mathcal{GY}^\infty(\Omega;\mathbb{R}^{d\times d}), \\ \qquad\qquad \operatorname{supp}\nu \subset \overline{B(0,\varsigma)}, \;\; \nabla y_\nu = \bar{\nu}. \end{array}\right\} \tag{4.4.48}$$

Proposition 4.4.10 (Relaxation of $\mathcal{E}$, cf. [61]). *Let $\mathcal{F}$ satisfy* (4.4.46), $\varphi \in C(\overline{B(0,\varsigma)})$ *for some $\varsigma > 0$, let $\alpha > 0$, and let $y_\mathrm{D} \in W^{1,\infty}(\Omega;\mathbb{R}^d)$. Then the infimum of $\mathcal{E}$ in* (4.4.47) *is the same as the minimum of $\bar{\mathcal{E}}$ in* (4.4.48)*. Moreover, every minimizing sequence of* (4.4.47) *contains a subsequence which Y-converges to a minimizer of* (4.4.48)*. On the other hand, for every minimizer (y_ν, ν) of* (4.4.48) *there exists a minimizing sequence $\{y_k\}_{k\in\mathbb{N}}$ of* (4.4.47) *such that*

[6] Note, however, that the difference on the right-hand side is measured in the L^∞-norm and not in the $W^{1,\infty}$-norm.

$$\mathcal{E}(y_k) \to \bar{\mathcal{E}}(y_\nu, \nu).$$

Proof. We first prove that every minimizing sequence of (4.4.47) (or at least a subsequence thereof) Y-converges to a minimizer of (4.4.48) and that the values of the infimum in (4.4.47) and the minimum of (4.4.48) agree.

To this end, take $\{y_k\}_{k\in\mathbb{N}}$ a minimizing sequence for (4.4.47). This sequence has to belong to the set

$$\mathcal{A}_\varsigma := \{y \in W^{1,\infty}(\Omega;\mathbb{R}^d);\ \|\nabla y\|_{L^\infty(\Omega;\mathbb{R}^{d\times d})} \le \varsigma\} \tag{4.4.49}$$

and, by definition, $\{\mathcal{E}(y_k)\}_{k\in\mathbb{N}}$ converges to $\inf_{\mathcal{A}_\varsigma}\mathcal{E}$. Inevitably, $\nabla y_k(x) \in \overline{B(0,\varsigma)}$ for almost all $x \in \Omega$, so that $L(\nabla y_k) \le 0$, with L as in (4.4.44). Moreover, as $\alpha > 0$ the Poincaré inequality implies that $\{y_k\}_{k\in\mathbb{N}}$ is uniformly bounded in $W^{1,\infty}(\Omega;\mathbb{R}^d)$. Therefore, there is a Young measure $\nu \in \mathcal{GY}^\infty(\Omega;\mathbb{R}^{d\times d})$ and a function $y_\nu \in W^{1,\infty}(\Omega;\mathbb{R}^d)$ such that $y_k \overset{Y}{\rightharpoonup} (y_\nu, \nu)$ as $k \to \infty$ (at least in terms of a non-relabeled subsequence). Moreover, ν is supported in

$$\bigcup_{\text{a.a. } x\in\Omega} \operatorname{supp}\nu_x \subset \bigcup_{\text{a.a. } x\in\Omega} \bigcap_{n=1}^{\infty} \overline{\{\nabla y_k(x),\, k \ge n\}} \subset \overline{B(0,\varsigma)}$$

(see e.g. [25]) and for the first moment of ν we have that $\bar{\nu} = \nabla y_\nu$. Now, by the representation theorem on Young measures (Thm. D.1.1), $\mathcal{E}(y_k) \to \bar{\mathcal{E}}(y_\nu, \nu)$ and, consequently, it holds that

$$\inf_{\mathcal{A}_\varsigma} \mathcal{E} = \bar{\mathcal{E}}(y_\nu, \nu).$$

We now prove that (y_ν, ν) is indeed a minimizer of $\bar{\mathcal{E}}$. Suppose, by contradiction, that this was not the case. Then there had to exist another gradient Young measure $\mu \in \mathcal{GY}^\infty(\Omega;\mathbb{R}^{d\times d})$ and a corresponding $y_\mu \in W^{1,\infty}(\Omega;\mathbb{R}^d)$ such that $\nabla y_\mu = \bar{\mu}$, μ is supported on $\overline{B(0,\varsigma)}$ and

$$\bar{\mathcal{E}}(y_\mu, \mu) < \bar{\mathcal{E}}(y_\nu, \nu).$$

We show that this is not possible. Indeed, for every $0 < \varepsilon < 1$, we find a generating sequence $\{y_k^\varepsilon\}_{k\in\mathbb{N}} \subset W^{1,\infty}(\Omega;\mathbb{R}^d)$ for μ, that is $y_k^\varepsilon \overset{Y}{\rightharpoonup} (y_\mu, \mu)$ as $k \to \infty$, such that $\sup_{k\in\mathbb{N}} \|\nabla y_k^\varepsilon\|_{L^\infty(\Omega;\mathbb{R}^{d\times d})} \le \varsigma + \varepsilon$, see Proposition D.1.4. By the fundamental theorem on Young measures,

$$\lim_{k\to\infty} \mathcal{E}(y_k^\varepsilon) = \bar{\mathcal{E}}(y_\mu, \mu) < \bar{\mathcal{E}}(y_\nu, \nu) = \inf_{\mathcal{A}_\varsigma} \mathcal{E}. \tag{4.4.50}$$

At this point, we abused the notation a bit for the sake of better readability of the proof. Indeed, $\mathcal{E}(y_k^\varepsilon)$ might not be well-defined because $\nabla y_k^\varepsilon(x)$ might not be contained in the ball $\overline{B(0,\varsigma)}$ for almost all $x \in \Omega$; however, the energy density φ is defined, originally, just on this ball. Nevertheless, relying on the Tietze theorem A.1.1, we can extend φ from the ball $\overline{B(0,\varsigma)}$ in a continuous way to the whole space. We will denote this extension by φ again and the functional into which it enters again by $\mathcal{E}$.

By (4.4.50), there is $k_0 = k_0(\varepsilon) \in \mathbb{N}$ such that $\mathcal{E}(y^\varepsilon_{k_0}) \leq \inf_{\mathcal{A}_\varsigma} \mathcal{E} + \delta$ for some $\delta > 0$ and $\|y^\varepsilon_{k_0} - y_\mu\|_{L^2(\Gamma_\mathrm{D};\mathbb{R}^d)} \leq 1$ as well as $\|y^\varepsilon_{k_0} - y_\mu\|_{L^\infty(\Omega;\mathbb{R}^d)} \leq 1$ (due to the strong convergence of $\{y^\varepsilon_k\}$ to y_μ for $k \to \infty$). Now, we can apply a trick, similar to the one used in [271, Prop. 7.1], and multiply $y^\varepsilon_{k_0}$ by $\varsigma/(\varsigma+\varepsilon)$ so that the values of the gradient belong to $\overline{B(0,\varsigma)}$. With this rescaling, we obtain

$$
\begin{aligned}
\left|\mathcal{E}(\varsigma/(\varsigma+\varepsilon)y^\varepsilon_{k_0}) - \mathcal{E}(y^\varepsilon_{k_0})\right| &\leq \left|\int_\Omega \varphi(\varsigma/(\varsigma+\varepsilon)\nabla y^\varepsilon_{k_0}(x))\mathrm{d}x - \int_\Omega \varphi(\nabla y^\varepsilon_{k_0}(x))\mathrm{d}x\right| \\
&\quad + C_\ell\|\varsigma/(\varsigma+\varepsilon)y^\varepsilon_{k_0} - y^\varepsilon_{k_0}\|_{L^\infty(\Omega;\mathbb{R}^d)} \\
&\quad + \left|\alpha\|\varsigma/(\varsigma+\varepsilon)y^\varepsilon_{k_0} - y_\mathrm{D}\|_{L^2(\Gamma_\mathrm{D};\mathbb{R}^d)} - \alpha\|y^\varepsilon_{k_0} - y_\mathrm{D}\|_{L^2(\Gamma_\mathrm{D};\mathbb{R}^d)}\right| \\
&\leq \int_\Omega \vartheta(\varepsilon/(\varsigma+\varepsilon)|\nabla y^\varepsilon_{k_0}(x)|)\,\mathrm{d}x \\
&\quad + C_\ell\|\varsigma/(\varsigma+\varepsilon)y^\varepsilon_{k_0} - y^\varepsilon_{k_0}\|_{L^\infty(\Omega;\mathbb{R}^d)} + \alpha\|\varsigma/(\varsigma+\varepsilon)y^\varepsilon_{k_0} - y^\varepsilon_{k_0}\|_{L^2(\Gamma_\mathrm{D};\mathbb{R}^d)},
\end{aligned}
$$

where $\vartheta : [0,\infty) \to [0,\infty)$ is the nondecreasing modulus of uniform continuity of φ on $\overline{B(0,\varsigma+1)}$, which satisfies $\lim_{s\to 0}\vartheta(s) = 0$.[7] Since $\nabla y^\varepsilon_{k_0}$ is uniformly bounded by $\varsigma+\varepsilon$, we have

$$
\begin{aligned}
\left|\mathcal{E}(\varsigma/(\varsigma+\varepsilon)y^\varepsilon_{k_0}) - \mathcal{E}(y^\varepsilon_{k_0})\right| &\leq \mathrm{meas}_d(\Omega)\vartheta(\varepsilon) + \frac{(C_\ell+\alpha)\varepsilon}{\varsigma+\varepsilon}\left(\|y^\varepsilon_{k_0}\|_{L^2(\Gamma_\mathrm{D};\mathbb{R}^d)} + \|y^\varepsilon_{k_0}\|_{L^\infty(\Omega;\mathbb{R}^d)}\right) \\
&\leq \mathrm{meas}_d(\Omega)\vartheta(\varepsilon) + \frac{(C_\ell+\alpha)\varepsilon}{\varsigma+\varepsilon}\left(\|y_\mu\|_{L^2(\Gamma_\mathrm{D};\mathbb{R}^d)} + \|y_\mu\|_{L^\infty(\Omega;\mathbb{R}^d)} + 2\right). \quad (4.4.51)
\end{aligned}
$$

The right-hand side in (4.4.51) tends to zero as $\varepsilon \to 0$; therefore, for $\varepsilon > 0$ small enough, it is smaller than $\delta/2$ and thus

$$
\mathcal{E}(\varsigma/(\varsigma+\varepsilon)y^\varepsilon_{k_0}) \leq \inf_{\mathcal{A}_\varsigma} \mathcal{E} - \delta/2.
$$

This closes our contradiction argument because $\varsigma/(\varsigma+\varepsilon)y^\varepsilon_{k_0} \in \mathcal{A}_\varsigma$, so that we showed that every minimizing sequence of (4.4.47) generates a minimizer of (4.4.48).

To finish the proof, we need to show that for any minimizer $(y_\nu, \nu) \in W^{1,\infty}(\Omega;\mathbb{R}^d) \times \mathcal{G}\mathcal{Y}^\infty(\Omega;\mathbb{R}^{d\times d})$ of (4.4.48) we can construct a sequence $\{y_k\}_{k\in\mathbb{N}} \subset W^{1,\infty}(\Omega;\mathbb{R}^d)$ that is a minimizing sequence of (4.4.47) and satisfies

$$
\mathcal{E}(y_k) \to \bar{\mathcal{E}}(y_\nu, \nu).
$$

The strategy is similar to above: indeed, for any $\varepsilon > 0$ we find a generating sequence $\{y^\varepsilon_k\}_{k\in\mathbb{N}} \subset W^{1,\infty}(\Omega;\mathbb{R}^d)$ for ν, that is $y^\varepsilon_k \stackrel{Y}{\to} (y_\nu, \nu)$ as $k \to \infty$, such that $\sup_{k\in\mathbb{N}}\|\nabla y^\varepsilon_k\|_{L^\infty(\Omega;\mathbb{R}^{d\times d})} \leq \varsigma + \varepsilon$. Let us rescale this sequence by $\frac{\varsigma}{\varsigma+\varepsilon}$ and consider only k large enough to obtain a sequence $\{\bar{y}^\varepsilon_k\}_{k\in\mathbb{N}}$ that is contained in $\mathcal{A}_\varsigma$ and satisfies $\|\bar{y}^\varepsilon_k - y_\nu\|_{L^2(\Gamma_\mathrm{D};\mathbb{R}^d)} \leq 1$ as well as $\|\bar{y}^\varepsilon_k - y_\nu\|_{L^\infty(\Omega;\mathbb{R}^d)} \leq 1$ for all $k \in \mathbb{N}$.

Choosing a subsequence of k's if necessary, we can assure that

[7] This meas that $|\varphi(A) - \varphi(B)| \leq \vartheta(|A - B|)$ if $A, B \in \overline{B(0,\varsigma+1)}$

$$\left|\mathcal{E}(y_k^{\varepsilon(k)}) - \bar{\mathcal{E}}(y_\nu,\nu)\right| \le \frac{1}{k},$$

for any arbitrary ε fixed. Moreover, owing to (4.4.51), we can choose $\varepsilon = \varepsilon(k)$ in such a way that

$$\left|\mathcal{E}(\bar{y}_k^{\varepsilon(k)}) - \mathcal{E}(y_k^{\varepsilon(k)})\right| \le \frac{1}{k},$$

so that

$$\left|\mathcal{E}(\bar{y}_k^{\varepsilon(k)}) - \bar{\mathcal{E}}(y_\nu,\nu)\right| \le \frac{2}{k}.$$

Thus, we can construct a sequence $\{\bar{y}_k\}_{k\in\mathbb{N}}$ by setting $\bar{y}_k = \bar{y}_k^{\varepsilon(k)}$ that lies in $\mathcal{A}_\varsigma$ and satisfies

$$\mathcal{E}(\bar{y}_k) \to \bar{\mathcal{E}}(y_\nu,\nu) \quad \text{as } k \to \infty.$$

Finally, since $\bar{\mathcal{E}}(y_\nu,\nu) = \inf_{\mathcal{A}_\varsigma}\mathcal{E}$ the constructed sequence is indeed a minimizing sequence of $\mathcal{E}$. □

Proposition 4.4.10 can be seen as a relaxation result for *locking materials* introduced by W. Prager in [421], see also e.g. [118, 151, 223, 413, 508]. These are hyperelastic materials for which the strain tensor is constrained to stay in some convex set which in our case is just $\overline{B(0,\varsigma)}$.

4.4.3 Remarks on relaxation results with determinant constraints

We have seen that Young measures and quasiconvexification can be quite useful in relaxation of variational problems. On the other hand, there are many unsolved problems related to applications in continuum mechanics of solids. In particular, (4.4.4) is not compatible with the assumptions on the stored energy density. Let us notice that (4.4.4) means that the stored energy density cannot attain the infinite values. Generally, relaxation result do not allow us to impose the orientation-preservation constraint on the deformation. It can only be checked aposteriori which, of course, can possibly be done in numerical examples but hardly in the abstract existence results. This means that general relaxation procedure cannot guarantee that obtained minimizers are physically acceptable. Therefore, relaxation results generically belong to the finite strain rather than large strain theory according to Convention 1.1.1. Let us notice, for example, that a simple Saint-Venant Kirchhoff energy density $\varphi(F) := |F^\top F - \mathbb{I}|^2$ which has the quartic growth at infinity is minimized on orthogonal matrices and obviously $Q\varphi$ is finite on matrices with zero or negative determinants, too. Consequently, one must give up requirements on injectivity and orientation-preservation of deformations. Nevertheless, there are attempts to find relaxations for functionals satisfying (at least partly) restrictions coming from the elasticity theory. These results can be then addressed as large strain relaxation theory. Here we shortly mention a few of them. The first one is the following statement due to O. Anza Hafsa and J.-P. Mandallena [15]. The model excludes the zero determinant of deformation gradients but still allows for negative determinants. In the this case, the quasiconvex envelope has polynomial growth at infinity which is heavily used in the proof.

Theorem 4.4.11 (Relaxation for invertible gradients [15]). *Let $d = 3$, $1 < p < +\infty$, and $\Omega \subset \mathbb{R}^d$ be a bounded Lipschitz domain. Let $\varphi : \mathbb{R}^{d\times d} \to [0;+\infty]$ be Borel measurable and such that*
(a) $\varepsilon|F|^p \le \varphi(F)$ for some $\varepsilon > 0$ and all $F \in \mathbb{R}^{d\times d}$,
(b) $\varphi(F) \to +\infty$ if $|\det F| \to 0$ and that for every $\varepsilon > 0$ there is $K_\varepsilon > 0$ such that $\varphi(F) \le K_\varepsilon(1 + |F|^p)$ whenever $|\det F| \ge \varepsilon$,
(c) $\varphi(F) = \varphi(RFQ)$ for every $F \in \mathbb{R}^{d\times d}$ and all rotations $R, Q \in \mathrm{SO}(d)$.
Let $y_\mathrm{D} : \Gamma \to \mathbb{R}^d$ be piecewise affine (with finitely many pieces) and given. Assume additionally, that the set $\mathcal{A} := \{y \in W^{1,p}(\Omega;\mathbb{R}^d);\ y = y_\mathrm{D}$ on $\Gamma\} \neq \emptyset$. Consider the functional $\mathcal{E}_\mathrm{Q} : W^{1,p}(\Omega;\mathbb{R}^d) \to \mathbb{R}$ given by

$$\mathcal{E}_\mathrm{Q}(y) := \int_\Omega \mathrm{Q}\varphi(\nabla y(x)), \tag{4.4.52}$$

where $\mathrm{Q}\varphi$ is defined as in (4.4.25)*. Then the problem*

$$\left.\begin{array}{l} \text{minimize } \mathcal{E}_\mathrm{Q}(y), \\ \text{subject to } y \in \mathcal{A} \end{array}\right\} \tag{4.4.53}$$

is the relaxed problem to

$$\left.\begin{array}{l} \text{minimize } \mathcal{E}(y) := \int_\Omega \varphi(\nabla y(x))\,\mathrm{d}x, \\ \text{subject to } y \in \mathcal{A} \end{array}\right\} \tag{4.4.54}$$

with respect to the weak convergence on $W^{1,p}(\Omega;\mathbb{R}^d)$.

Another relaxation result was proven by S. Conti and G. Dolzmann [126]. It properly treats the determinant constraint, however, it requires that the quasiconvex envelope is even polyconvex. This assumption is, however, generally not satisfied and also difficult to prove. Nevertheless, the growth and coercivity conditions on φ are physically justified. Indeed, we assume that there is $\gamma : (0, +\infty) \to [0, +\infty)$ convex satisfying

$$\lim_{\delta\to 0} \gamma(\delta) = +\infty \tag{4.4.55}$$

for all $a, b > 0$ and some $K > 0$

$$\gamma(ab) \le K(1 + \gamma(a))(1 + \gamma(b)) \tag{4.4.56}$$

and such that, for all $F \in \mathrm{GL}^+(d)$, it holds

$$\frac{1}{K}(|F|^p + \gamma(\det F) - K^2) \le \varphi(F) \le K(|F|^p + \gamma(\det F) + 1). \tag{4.4.57}$$

As to the loading functional, it is assumed in [126] that there is $\mathcal{F} \in \mathcal{C}(\mathbb{R})$ such that

$$|\mathcal{F}| \leq K(1+|\cdot|)^{\tilde{p}} \text{ for some } 0 \leq \tilde{p} < p \text{ and } K > 0. \tag{4.4.58}$$

Theorem 4.4.12 (Relaxation with positive determinant [126]). *Let $\Omega \subset \mathbb{R}^d$ be a bounded Lipschitz domain and $p \geq d > 1$. Let $\varphi : \mathbb{R}^{d\times d} \to \mathbb{R} \cup \{+\infty\}$ be a function continuous on* $\mathrm{GL}^+(d)$ *satisfying* (4.4.57) *for some convex $\gamma : (0;+\infty) \to [0;+\infty)$ obeying* (4.4.55) *and* (4.4.56)*. We further assume that $\varphi(F) = +\infty$ if $F \in \mathbb{R}^{d\times d} \setminus$* $\mathrm{GL}^+(d)$*. Let $y_D \in W^{1,p}(\Omega;\mathbb{R}^d)$ be given such that* $\det \nabla y_{\mathrm{D}} > 0$ *almost everywhere in Ω. Let*

$$\mathcal{A} := \{y \in W^{1,p}(\Omega;\mathbb{R}^d);\ \det \nabla y > 0 \text{ a.e. in } \Omega,\ y = y_D \text{ on } \Gamma_{\mathrm{D}}\}. \tag{4.4.59}$$

Assume, finally, that $\mathrm{Q}\varphi$ *calculated as in* (4.4.31) *is polyconvex. Then the problem* (4.4.42) *is a relaxed problem to* (4.4.8) *with respect to the the weak convergence on $W^{1,p}(\Omega;\mathbb{R}^d)$ in the sense of Definition 4.4.1. Hence $\hat{\mathcal{E}} := \mathcal{E}_{\mathrm{Q}}$ and $\mathfrak{U} := \mathcal{A}$.*

K. Koumatos, F. Rindler, and E. Wiedemann [285] defined, for $1 < p < d$, $W^{1,p}$-orientation preserving quasiconvexity of a function $\varphi : \mathbb{R}^{d\times d} \to \mathbb{R} \cup \{+\infty\}$, locally bounded on $\mathrm{GL}^+(d)$, as $W^{1,p}$-quasiconvexity tested by orientation-preserving maps. More precisely, φ is orientation-preserving $W^{1,p}$-quasiconvex if for every $F \in \mathrm{GL}^+(d)$ and every $y \in W^{1,p}_F(B(0,1);\mathbb{R}^d)$ such that $\det \nabla y > 0$ almost everywhere in $B(0,1)$ it holds that

$$\varphi(F) \leq \frac{1}{\mathrm{meas}_d(B(0,1))} \int_{B(0,1)} \varphi(\nabla y(x))\,\mathrm{d}x. \tag{4.4.60}$$

Interestingly, they showed in [285] that there is no $W^{1,p}$-orientation preserving quasiconvex function which satisfies (4.4.57) if $1 < p < d$ and if γ is convex and satisfies (4.4.55) and $\limsup_{\delta\to\infty} \gamma(\delta)/\delta^{p/d} < +\infty$. In spite of the effort of many researchers, relaxation and weak lower semicontinuity results are still not available for physically relevant stored energy density functions. We also refer to [29] for more details.

4.4.4 Measure-valued extension of Euler-Lagrange equations

The programme of designing of suitable extensions of variational problems and, in general, of partial differential equations has been articulated by David Hilbert [250, Problem 20] and was pursued during the whole 20th century. This led to development of modern theory of variational calculus and partial differential equations based on Sobolev spaces and weak solutions. In 30ties L.C. Young [537] devised parametrized measures[8] which has then been used to extension variational problems

[8] In fact, the original work [537] uses functionals on suitable test functions because Measure Theory was developed only later.

(as presented in previous sections) and later of various partial differential equations where it is known under the name *measure-valued solutions*.

The extension (relaxation) of variational problems and especially of underlying partial differential equations has to be made carefully. Often, various concepts of measure-valued solutions have been used to facilitate mathematics without any physical or mechanical interpretation. The following two (rather vague) attributes have been articulated in [464] to avoid such junk concepts:

A1: The measure-valued solutions should coincide with the standard weak solutions to the investigated problem whenever data involved in the problem apparently prevent any rapid oscillations.

A2: Also, the measure-valued solution should be unique whenever data do not give any apparent reason for a non-uniqueness.

In the case of variational problems, the role of the related differential equation is played by the Euler-Lagrange equation. A temptation to construct a measure-valued solution by the direct extension of the Euler-Lagrange (2.3.17a) (with a bit modified boundary condition $\partial_F\varphi(\nabla y)\vec{n} = g(y)$ on Γ) would lead to[9]

$$\operatorname{div} \lambda(x) + \tilde{f}(x,y(x)) = 0 \ \text{ with } \lambda(x) = \int_{R^{d\times d}} \partial_F\varphi(x,F)\nu_x(\mathrm{d}F) \tag{4.4.61a}$$

$$\text{and } \nabla y(x) = \int_{R^{d\times d}} F\nu_x(\mathrm{d}F) \quad \text{for } x\in\Omega, \text{ and} \tag{4.4.61b}$$

$$\lambda(x)\cdot\vec{n}(x) = g(x,y(x)) \qquad \text{for } x\in\Gamma_{\mathrm{N}} \tag{4.4.61c}$$

Written in the weak formulation, (4.4.61) reads as

$$\int_\Omega\Big(\int_{R^{d\times d}} \partial_F\varphi(x,F){:}\nabla\tilde{y}(x)\nu_x(\mathrm{d}F) + \tilde{f}(x,y(x))\cdot\tilde{y}(x)\Big)\mathrm{d}x \\ + \int_{\Gamma_{\mathrm{N}}}\int_{R^{d\times d}} \tilde{g}(x,F)\tilde{y}(x)\nu_x(\mathrm{d}F)\,\mathrm{d}x = 0, \tag{4.4.62}$$

which should hold for any $\tilde{y} \in W^{1,\infty}(\Omega;\mathbb{R}^d)$ vanishing on Γ_{D} together with (4.4.61b).

It is useful to have in mind the following simple example which works even in the one-dimensional case $d = 1$:[10]

$$\varphi(x,F) := \begin{cases} \frac{1}{2c}|F|^2 + \frac{c}{2} & \text{if } |F| \le c, \\ |F| & \text{if } c < |F| \le \frac{1}{c}, \\ \frac{c}{2}|F|^2 + \frac{1}{2c} & \text{if } |F| \ge \frac{1}{c}, \end{cases}$$

where $0 < c < 1$ is a given (small) constant, $\tilde{f} = 0$, and $g(y) = y$. Note that $\varphi(x,\cdot)$ is convex and smooth, and has a quadratic growth. The variational problem (based on

[9] The idea of Young-measure extension of differential equations is due to R.J. DiPerna [158]. In the evolution variant, for such Young-measure extension like (4.4.61) see e.g. [487]. See also [464] for many other examples.

[10] Cf. also [452, Remark 5.3.8] where it was formulated in the scalar but multidimensional case.

minimization of the energy $\int_\Omega \varphi(\nabla y) + \hat{F}(y)\,\mathrm{d}x + \int_\Gamma \hat{G}(y)\,\mathrm{d}x$ with $\hat{F}(x,\cdot)$ and $\hat{G}(x,\cdot)$ denoting respectively primitive functions of $\tilde{f}(x,\cdot)$ and $g(x,\cdot)$ has obviously the unique solution, namely $y = 0$. Nevertheless, any $y \in W_0^{1,\infty}(\Omega)$ with $\|y\|_{W^{1,\infty}(\Omega)} \leq (1-c^2)/(2c)$ admits a completion by an appropriate Young measure ν so that the pair (u,ν) is the measure-valued solution to (2.3.17a), i.e. it satisfies the extended Euler-Lagrange equation (4.4.62). Indeed, it suffices to take

$$\nu_x := \frac{1}{2}\delta_{\nabla y(x)+\tilde{F}(x)} + \frac{1}{2}\delta_{\nabla y(x)-\tilde{F}(x)} \quad \text{with} \quad \tilde{F}(x) := \begin{cases} \dfrac{(1+c^2)\nabla y(x)}{2c|\nabla y(x)|} & \text{if } \nabla y(x) \neq 0, \\ 0 & \text{if } \nabla y(x) = 0. \end{cases}$$

The above mentioned desirable attributes A1-A2 are thus drastically violated.

4.5 Existence results for nonsimple materials

Here we show that considering nonsimple materials allows for a stronger statement about the existence of minimizers to variational problems in nonlinear elasticity and such minimizers have physically desirable properties. We deal with two main cases. The first one will be the existence of minimizers for second-grade materials, i.e., materilas whose elastic properties depend on the strain gradient, or saying otherwise, on the second gradient of the deformation.

4.5.1 Second-grade nonsimple materials

We recall that our energy functional of a second-grade material reads (see (2.5.5))

$$\mathcal{E}(y) := \int_\Omega \varphi_{\mathrm{E}}(x, \nabla y(x), \nabla^2 y(x)) - f(x)\cdot y(x)\,\mathrm{d}x - \int_{\Gamma_{\mathrm{N}}} g(x)\cdot y(x) + g_1(x)\cdot\partial_{\vec{n}} y(x)\,\mathrm{d}S\,,$$

where $\tilde{g}$, $\tilde{g}_1 : \Gamma_{\mathrm{N}} \to \mathbb{R}^d$ is the surface density of forces and hypertraction forces, respectively. As before $\tilde{f}$ denotes the volume density of body forces. Thus, we set

$$\ell(y) := \int_\Omega f(x)\cdot y(x)\,\mathrm{d}x + \int_{\Gamma_{\mathrm{N}}} g(x)\cdot y(x) + g_1(x)\cdot\partial_{\vec{n}} y(x)\,\mathrm{d}S\,, \tag{4.5.1}$$

which is a functional describing work of external forces expended on the specimen.

We will assume that the stored energy density $\varphi_{\mathrm{E}} : \bar{\Omega} \times \mathbb{R}^{d\times d} \times \mathbb{R}^{d\times d\times d} \to \mathbb{R} \cup \{+\infty\}$ is such that it is a Carathéodory integrand in $[x,(F,G)]$ and that it satisfies the following growth conditions for some constant $\varepsilon > 0$:

$$\varphi_{\mathrm{E}}(x,F,G) \geq \begin{cases} \varepsilon(|G|^p + |F|^p) - 1/\varepsilon & \text{for } \det F > 0, \\ +\infty & \text{otherwise.} \end{cases} \tag{4.5.2}$$

Theorem 4.5.1 (Existence results for nonsimple materials I). *Let $\Omega \subset \mathbb{R}^d$ be a bounded Lipschitz domain. Let the following hold:*
(a) the function $\varphi_E(\cdot, F, G)$ is measurable for all $(F,G) \in \mathrm{GL}^+(d) \times \mathbb{R}^{d\times d\times d}$ and $\varphi_E(x,\cdot,\cdot)$ is continuous for almost all $x \in \Omega$,
(b) $\varphi_E(x,F,\cdot)$ is convex for almost all $x \in \Omega$ and all $F \in \mathbb{R}^{d\times d}$.
(c) for a.a. $x \in \Omega$, $\varphi_E(x,F,G) \to +\infty$ if $\det F \to 0_+$.
(d) There is $p > d/2$ such that, for a.a. $x \in \Omega$ and all $F \in \mathbb{R}^{d\times d}$, $G \in \mathbb{R}^{d\times d\times d}$, (4.5.2) *holds.*
(e) Let further $\Gamma = \Gamma_D \cup \Gamma_N$ be a $\mathcal{L}^{d-1}$-measurable partition of $\Gamma = \partial\Omega$ with the area of $\Gamma_D > 0$ and let $y_D \in W^{2,p}(\Omega;\mathbb{R}^d)$ be given, and let there be $y \in \mathcal{A}$ such that $\mathcal{E}(y) < +\infty$ with

$$\mathcal{A} := \left\{y \in W^{2,p}(\Omega;\mathbb{R}^d) : \text{(3.4.7) holds},\ y = y_D \text{ on } \Gamma_D,\ \det\nabla y > 0 \text{ a.e. on } \Omega\right\} \neq \emptyset.$$

(f) Let $f \in L^{p^{**\prime}}(\Omega;\mathbb{R}^d)$, $g \in L^{p^{*\#\prime}}(\Gamma_N;\mathbb{R}^d)$, and $g_1 \in L^{p^{\#\prime}}(\Gamma_N;\mathbb{R}^d)$.
Assume that there is $\tilde{y} \in \mathcal{A}$ such that $\mathcal{E}(\tilde{y}) < +\infty$. Then the infimum of $\mathcal{E}$ on $\mathcal{A}$ is attained. Moreover, every minimizer is injective almost everywhere in Ω.

Proof. Note that $x \mapsto \varphi_E(x, \nabla y(x), \nabla^2 y(x))$ is measurable because φ_E is a Carathéodory function. Using (d) and linearity of ℓ we get for some $\tilde{K} > 0$

$$\mathcal{E}(y) \geq \varepsilon \int_\Omega |\nabla y|^p + |\nabla^2 y|^p \, dx - \varepsilon^{-1}\mathrm{meas}_d(\Omega) - \tilde{K}\|y\|_{W^{2,p}(\Omega;\mathbb{R}^d)}. \tag{4.5.3}$$

Applying the Poincaré inequality we conclude that there is a constant $\varepsilon > 0$ and

$$\mathcal{E}(y) \geq \varepsilon\|y\|^p_{W^{2,p}(\Omega;\mathbb{R}^d)} - 1/\varepsilon$$

for all $y \in \mathcal{A}$. By (f),

$$\ell \in W^{2,p}(\Omega;\mathbb{R}^d)^*. \tag{4.5.4}$$

with ℓ from (4.5.1). Let $\{y_k\} \subset \mathcal{A}$ be a minimizing sequence of $\mathcal{E}$, i.e.,

$$\lim_{k\to\infty} \mathcal{E}(y_k) = \inf_{\mathcal{A}} \mathcal{E} < +\infty.$$

Then there is a minimizing sequence $\{y_k\}$ s.t. $y_k \to y$ weakly in $W^{2,p}(\Omega;\mathbb{R}^d)$. Consequently, $y_k \to y$ strongly in $W^{1,p}(\Omega;\mathbb{R}^d)$.

Now we apply Theorem 3.3.1 with $v_k := \nabla^2 y_k$, $z_k := (y_k, \nabla y_k)$, $v := \nabla^2 y$, $z := (y, \nabla y)$, and $\xi := \varphi_E - f$. This shows that $\mathcal{E}$ is (sequentially) weakly lower semicontinuous. In order to prove that $y \in \mathcal{A}$ we verify that $\det\nabla y > 0$ almost everywhere in Ω. This, however, follows from the fact that $\mathcal{E}(y) = \liminf_{k\to\infty} \mathcal{E}(y_k) < +\infty$ because $\{y_k\} \subset \mathcal{A}$ is a minimizing sequence and from the assumption. To show almost everywhere injectivity of minimizers if $p > d/2$ we realize due to Corollary B.3.4 that in this case $y \in C(\bar{\Omega};\mathbb{R}^d)$ and we proceed as in the proof of Theorem 3.4.6. □

Strengthening the assumptions in such a way that Theorem 2.5.3 applies we get the following result:

Theorem 4.5.2 (Existence results for nonsimple materials II). *Let $\Omega \subset \mathbb{R}^d$ be a bounded Lipschitz domain. Let the following hold:*

(a) the function $\varphi_{\mathrm{E}}(\cdot, F, G)$ is measurable for all $(F,G) \in \mathrm{GL}^+(d) \times \mathbb{R}^{d\times d\times d}$ and $\varphi_{\mathrm{E}}(x,\cdot,\cdot)$ is continuous for almost all $x \in \Omega$,

(b) $\varphi_{\mathrm{E}}(x,F,\cdot)$ is convex for almost all $x \in \Omega$ and all $F \in \mathbb{R}^{d\times d}$.

(c) There are constants $\varepsilon > 0$, $p > d$, and $q \geq pd/(p-d)$ such that for a.a. $x \in \Omega$ and all $F \in \mathbb{R}^{d\times d}$, $G \in \mathbb{R}^{d\times d\times d}$

$$\varphi_{\mathrm{E}}(x,F,G) \geq \begin{cases} \varepsilon(|G|^p + |F|^p + (\det F)^{-q}) - 1/\varepsilon & \text{if} \det F > 0, \\ +\infty & \text{otherwise.} \end{cases} \tag{4.5.5}$$

(d) Let further $\Gamma = \Gamma_{\mathrm{D}} \cup \Gamma_{\mathrm{N}}$ be a $\mathcal{L}^{d-1}$-measurable partition of $\Gamma = \partial\Omega$ with the area of $\Gamma_{\mathrm{D}} > 0$ and let $y_{\mathrm{D}} \in W^{2,p}(\Omega;\mathbb{R}^d)$ be given and let there be $y \in \mathcal{A}$ such that $\mathcal{E}(y) < +\infty$ with

$$\mathcal{A} := \Big\{ y \in W^{2,p}(\Omega;\mathbb{R}^d) : \ (3.4.7) \text{ holds}, \ y = y_{\mathrm{D}} \text{ on } \Gamma_{\mathrm{D}}, \ \det\nabla y > 0 \text{ a.e. on } \Omega \Big\} \neq \emptyset.$$

(e) Let finally $f \in L^{p^{**\prime}}(\Omega;\mathbb{R}^d)$, $g \in L^{p^{*\#\prime}}(\Gamma_{\mathrm{N}};\mathbb{R}^d)$, and $g_1 \in L^{p^{\#\prime}}(\Gamma_{\mathrm{N}};\mathbb{R}^d)$.

Then the infimum of $\mathcal{E}$ on $\mathcal{A}$ is attained. Moreover, every minimizer $y \in \mathcal{A}$ is injective everywhere in Ω and there is $\epsilon > 0$ such that $\det\nabla y \geq \epsilon \in \bar{\Omega}$. Moreover, $y^{-1} \in W^{1,\infty}(y(\Omega);\mathbb{R}^d)$ and $\nabla y^{-1}(x^y) = (\nabla y(y^{-1}(x^y)))^{-1}$.

Proof. The proof is mostly analogous to the one of Theorem 4.5.1. As $p > d$ and (4.5.5) holds we can apply Theorem 2.5.3 to get a positive lower bound on the determinant of a minimizer. Notice also that $y \in C^1(\bar{\Omega};\mathbb{R}^d)$ hence Theorem 3.4.4 applies. We will concentrate on proving regularity of the inverse of the minimizer, y^{-1} following [36]. As $\det\nabla y \geq \epsilon > 0$ in $\bar{\Omega}$ and $\nabla y \in W^{1,\infty}(\Omega;\mathbb{R}^{d\times d})$ we get that $(\nabla y(y^{-1}(x^y)))^{-1} \in L^\infty(\Omega;\mathbb{R}^{d\times d})$. Let us denote, for simplicity, $z := y^{-1}$. Applications of Piola's identity (1.1.19) and Theorem B.3.11 yield, for every $v \in C_0^\infty(y(\Omega))$, that

$$\begin{aligned}
&\int_{y(\Omega)} \partial_{x_j^y} z_i(x^y) v(x^y)\,\mathrm{d}x^y = -\int_{y(\Omega)} z_i(x^y)\partial_{x_j^y} v(x^y)\,\mathrm{d}x^y \\
&= -\int_\Omega x_i \partial_{x_j^y} v(y(x)) \det\nabla y(x)\,\mathrm{d}x = -\int_\Omega x_i \partial_{x_k} v(y(x))((\nabla y(x))^{-\top})_{jk} \det\nabla y(x)\,\mathrm{d}x \\
&= \int_\Omega \delta_{ik} v(y(x))(\mathrm{Cof}\,\nabla y(x))_{jk}\,\mathrm{d}x = \int_\Omega \frac{(\mathrm{Cof}\,\nabla y(x))_{ji}}{\det\nabla y(x)} v(y(x)) \det\nabla y(x)\,\mathrm{d}x \\
&= \int_{y(\Omega)} \frac{(\mathrm{Cof}\,\nabla y(z(x^y)))^\top_{ji}}{\det\nabla y(z(x^y))} v(x^y)\,\mathrm{d}x^y = \int_{y(\Omega)} [(\nabla y(z(x^y)))^{-1}]_{ij} v(x^y)\,\mathrm{d}x^y.
\end{aligned}$$

□

Convexity of $\varphi_{\mathrm{E}}(x,F,\cdot)$ can be replaced by polyconvexity defined for functions depending on higher-order gradients; cf. [31] where this scenario is undertaken.

4.5.2 Gradient-polyconvex materials

We can also weaken our assumptions on the deformations. Namely, to prove the existence of a minimizer for a model with a non-polyconvex energy density it is enough to control, besides the deformation gradient, only the gradient of the cofactor of the deformation gradient in a suitable Lebesgue space. More generally, it suffices to bound gradients of nonlinear minors along minimizing sequences but not the second gradient of the deformation. We will assume that the stored-energy density is a convex function of the gradients of the cofactor and of the determinant of the deformation. In analogy with polyconvexity, we call such stored energies gradient-polyconvex. With a slight abuse of notation, we will call materials with gradient polyconvex stored energy density *gradient-polyconvex materials.*

Definition 4.5.3 (Gradient polyconvexity [61]). Let $\tilde{\varphi} : \bar{\Omega}\times\mathbb{R}^{d\times d}\times\mathbb{R}^{d\times d\times d}\times\mathbb{R}^d \to \mathbb{R}\cup\{+\infty\}$ be measurable in $x \in \bar{\Omega}$ and continuous in all other variables. The functional

$$y \mapsto \int_\Omega \tilde{\varphi}(x, \nabla y(x), \nabla[\mathrm{Cof}\,\nabla y(x)], \nabla[\det \nabla y(x)])\mathrm{d}x, \tag{4.5.6}$$

defined for any measurable function $y : \Omega \to \mathbb{R}^d$ for which the weak derivatives ∇y, $\nabla[\mathrm{Cof}\,\nabla y]$, $\nabla[\det \nabla y]$ exist and the right-hand side of (4.5.6) is integrable is called *gradient polyconvex on a domain* $\Omega \subset \mathbb{R}^d$ (or *gradient polyconvex* for short) if the function $\tilde{\varphi}(x, F, \cdot, \cdot)$ is convex for every $F \in \mathbb{R}^{d\times d}$ and almost every $x \in \Omega$.

Definition 4.5.4 (Gradient polyconvex material [61]). If $\Omega \subset \mathbb{R}^d$ is a bounded domain and $\tilde{\varphi} : \bar{\Omega}\times\mathbb{R}^{d\times d}\times\mathbb{R}^{d\times d\times d}\times\mathbb{R}^d \to \mathbb{R}\cup\{+\infty\}$ defines a gradient-polyconvex functional in the sense of Def. 4.5.3 then the mapping

$$y \mapsto \int_\Omega \tilde{\varphi}(x, \nabla y(x)\,, \nabla \mathrm{Cof}[\nabla y(x)]\,, \nabla[\det \nabla y(x)])\,\mathrm{d}x \tag{4.5.7}$$

then represents the energy stored in the material if deformed by $y : \Omega \to \mathbb{R}^d$. We call the material *gradient polyconvex* if its stored energy is given by (4.5.7).

The following example shows that gradient-polyconvex materials admit a larger class of deformations than second-grade materials. As $\mathrm{Cof}\,F$ has the same set of entries (up to signs) as $F \in \mathbb{R}^{2\times 2}$ this can only be true if $d \geq 3$. Therefore, gradient-polyconvex materials are 2nd-grade materials if $d = 2$. For this reason, we will formulate the theorems below only for $d = 3$.

Example 4.5.5 (Deformation $y \notin W^{2,1}$ with $\det \nabla y$ and $\mathrm{Cof}\,\nabla y$ in $W^{1,\infty}$). Let us explicitly show that requiring for a deformation $y : \Omega \to \mathbb{R}^3$ to satisfy $\det \nabla y \in W^{1,r}(\Omega)$ and $\mathrm{Cof}\,\nabla y \in W^{1,q}(\Omega; \mathbb{R}^{3\times 3})$ is indeed a weaker requirement than $y \in W^{2,1}(\Omega; \mathbb{R}^3)$ for any $r, q \geq 1$. To see this, let us take the cube $\Omega = (0,1)^3$ and the following deformation (see also Figure 4.3)

$$y(x_1, x_2, x_3) := \Big(x_1\,, x_1^\alpha x_2 + \frac{1-x_1}{2}\,, x_1 x_3 + \frac{1-x_1}{2}\Big) \tag{4.5.8}$$

for some $1 > \alpha > 0$ so that

$$\nabla y(x_1,x_2,x_3) = \frac{1}{2}\begin{pmatrix} 2 & 0 & 0 \\ 2\alpha x_1^{\alpha-1}x_2 - 1 & 2x_1^{\alpha} & 0 \\ 2x_3 - 1 & 0 & 2x_1 \end{pmatrix}.$$

It follows that $\det\nabla y(x_1,x_2,x_3) = x_1^{\alpha+1} > 0$ for $x_1 \neq 0$ and

$$\mathrm{Cof}\,\nabla y(x_1,x_2,x_3) = \frac{1}{2}\begin{pmatrix} 2x_1^{\alpha+1} & -2\alpha x_1^{\alpha}x_2 + x_1 & -2x_1^{\alpha}x_3 + x_1^{\alpha} \\ 0 & 2x_1 & 0 \\ 0 & 0 & 2x_1^{\alpha} \end{pmatrix}.$$

Let us notice that $\det\nabla y \in W^{1,\infty}(\Omega)$, $\mathrm{Cof}\,\nabla y \in W^{1,\infty}(\Omega;\mathbb{R}^{3\times3})$, $(\det\nabla y)^{-1/(2(1+\alpha))} \in$

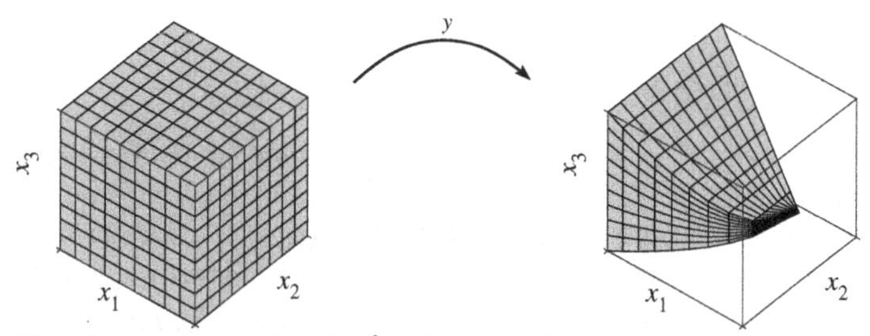

Fig. 4.3 The reference domain $\Omega = (0,1)^3$ (left) and the deformed domain $y(\Omega)$ (right) with Lipschitzian determinant and cofactor of deformation gradient for $y \notin W^{2,1}(\Omega)$ from (4.5.8) with $\alpha := 1/2$. Visualization by Jan Valdman (Czech Academy of Sciences).

$L^1(\Omega)$ but we see that the Hessian is not integrable, i.e. $\nabla^2 y \notin L^1(\Omega;\mathbb{R}^{3\times3\times3})$, which means that $y \notin W^{2,1}(\Omega;\mathbb{R}^3)$. On the other hand, $y \in W^{1,p}(\Omega;\mathbb{R}^3) \cap L^\infty(\Omega;\mathbb{R}^3)$ for every $1 \le p < 1/(1-\alpha)$.

We now state a simple algebraic lemma.

Lemma 4.5.6. *Let $F \in \mathbb{R}^{3\times3}$ be invertible. Then*

$$|(\mathrm{Cof}\,F)^{-1}| \le \frac{3}{2}|F^{-1}|^2. \tag{4.5.9}$$

If $F \in \mathbb{R}^{2\times2}$ is invertible, then $|\mathrm{Cof}\,F^{-1}| = |F^{-1}|$.

Proof. Let us recall that the cofactor of an invertible matrix $F \in \mathbb{R}^{3\times3}$ is defined as $\mathrm{Cof}\,F := (\det F)F^{-\top}$, cf. (A.2.17). Consequently, $\mathrm{Cof}\,F \in \mathbb{R}^{3\times3}$ and it consists of all nine 2×2 subdeterminants of F. If $A \in \mathbb{R}^{2\times2}$ and $|A|$ denotes the Frobenius norm of A, then the Hadamard inequality (A.2.23) implies $|\det A| \le \frac{1}{2}|A|^2$. Applying it to all nine 2×2 submatrices of $F \in \mathbb{R}^{3\times3}$ we get $|\mathrm{Cof}\,F| \le \frac{3}{2}|F|^2$. Since $\mathrm{Cof}\,F^{-1} = F^\top/\det F =$

$(\mathrm{Cof}\,F)^{-1}$, we obtain (4.5.9). If $d = 2$ then $|\mathrm{Cof}\,F^{-1}| = |(\mathrm{Cof}\,F)^{-1}| = |F^{-1}|$ and the result follows, as well. $\square$

We have the following existence results for gradient-polyconvex materials.

Theorem 4.5.7 (Existence result for gradient-polyconvex materials I [61]). *Let $\Omega\subset\mathbb{R}^3$ be a bounded Lipschitz domain representing a gradient polyconvex material with $\tilde{\varphi} : \bar{\Omega}\times\mathbb{R}^{3\times3}\times\mathbb{R}^{3\times3\times3}\times\mathbb{R}^3 \to \mathbb{R}\cup\{+\infty\}$ and assume that there are constants $\varepsilon > 0$, $p \geq 2$, $q \geq q_{pd}$ for q_{pd} given by (3.4.2), $r > 1$, $s > 0$ such that, for a.a. $x\in\Omega$, all $F\in\mathbb{R}^{3\times3}$, all $\Delta_1\in\mathbb{R}^{3\times3\times3}$, and $\Delta_2\in\mathbb{R}^3$*

$$\tilde{\varphi}(x,F,\Delta_1,\Delta_2) \geq \begin{cases} \varepsilon(|F|^p + |\mathrm{Cof}\,F|^q + (\det F)^r \\ \quad +(\det F)^{-s} + |\Delta_1|^q + |\Delta_2|^r) & \text{if } \det F > 0, \\ +\infty & \text{otherwise.} \end{cases} \tag{4.5.10}$$

Let further $\Gamma = \Gamma_{\mathrm{D}}\cup\Gamma_{\mathrm{N}}$ be a measurable partition of $\Gamma := \partial\Omega$ with $\mathrm{meas}_2(\Gamma_{\mathrm{D}}) > 0$ and let $y_{\mathrm{D}} \in W^{1,p}(\Omega;\mathbb{R}^3)$ be given. Let

$$\mathcal{A} := \Big\{y\in W^{1,p}(\Omega;\mathbb{R}^3);\ \ \mathrm{Cof}\,\nabla y\in W^{1,q}(\Omega;\mathbb{R}^{3\times3}),$$
$$\det\nabla y\in W^{1,r}(\Omega),\ \ y = y_{\mathrm{D}} \text{ on } \Gamma_{\mathrm{D}},\ \ \det\nabla y > 0 \text{ a.e.}\Big\}$$

be nonempty, and let further the functional $\mathcal{F} : W^{1,p}(\Omega;\mathbb{R}^3) \to \mathbb{R}$ describing work of external loads be upper semicontinuous. and such that, for all $y \in \mathcal{A}$, it holds that $\mathcal{F}(y) \leq K_1(\|y\|^{\tilde{p}}_{W^{1,p}(\Omega;\mathbb{R}^3)} + 1)$ for some $K_1 > 0$ and $0 \leq \tilde{p} < p$. Let for all $y \in \mathcal{A}$

$$\mathcal{E}(y) := \int_\Omega \varphi(x, \nabla y(x), \nabla[\mathrm{Cof}\,\nabla y(x)], \nabla[\det\nabla y(x)])\,\mathrm{d}x - \mathcal{F}(y). \tag{4.5.11}$$

Eventually, let there be $y \in \mathcal{A}$ such that $\mathcal{E}(y) < +\infty$. Then the infimum of $\mathcal{E}$ on $\mathcal{A}$ is attained. Moreover, if $r > 3$, $s > \frac{3r}{r-3}$ and $y \in \mathcal{A}$ is a minimizer then there is $\varepsilon > 0$ such that $\det\nabla y \geq \varepsilon$ *in $\bar{\Omega}$.*

Proof. Note that $x \mapsto \tilde{\varphi}(x, \nabla y(x), \nabla\,\mathrm{Cof}\,\nabla y(x), \nabla\det\nabla y(x))$ is measurable because $\tilde{\varphi}(x,\cdot,\cdot,\cdot)$ is continuous and $\tilde{\varphi}(\cdot, F, \Delta_1, \Delta_2)$ is measurable. Using (4.5.10) we get for every $y \in \mathcal{A}$ that

$$\mathcal{E}(y) \geq \varepsilon\int_\Omega |\nabla y|^p\,\mathrm{d}x + \|\mathrm{Cof}\,\nabla y\|^q_{W^{1,q}(\Omega;\mathbb{R}^{3\times3})} + \|\det\nabla y\|^r_{W^{1,r}(\Omega)} - K_1\|y\|^{\tilde{p}}_{W^{1,p}(\Omega;\mathbb{R}^3)}.$$

Applying the Poincaré inequality (Theorem B.3.15) we conclude that there is $\tilde{K} > 0$ such that

$$\mathcal{E}(y) \geq \varepsilon\Big(\|y\|^p_{W^{1,p}(\Omega;\mathbb{R}^3)} + \|\mathrm{Cof}\,\nabla y_k\|^q_{W^{1,q}(\Omega;\mathbb{R}^{3\times3})}$$
$$+ \|(\det\nabla y)^{-s}\|_{L^1(\Omega)} + \|\det\nabla y\|^r_{W^{1,r}(\Omega)}\Big) - \tilde{K} \tag{4.5.12}$$

for all $y \in \mathcal{A}$. Let $\{y_k\} \subset \mathcal{A}$ be a minimizing sequence of $\mathcal{E}$, i.e.,

$$\lim_{k\to\infty} \mathcal{E}(y_k) = \inf_{\mathcal{A}} \mathcal{E} < +\infty .$$

By (4.5.12) the sequence $\{(y_k, \operatorname{Cof}\nabla y_k, \det\nabla y_k)\}_{k\in\mathbb{N}}$ is uniformly bounded in the reflexive Banach space $W^{1,p}(\Omega;\mathbb{R}^3)\times W^{1,q}(\Omega;\mathbb{R}^{3\times3})\times W^{1,r}(\Omega)$. Hence it has a subsequence weakly converging to $(y,G,\delta) \in W^{1,p}(\Omega;\mathbb{R}^3)\times W^{1,q}(\Omega;\mathbb{R}^{3\times3})\times W^{1,r}(\Omega)$. Due to Theorem 3.2.2 we easily see that $G = \operatorname{Cof}\nabla y$ and $\delta = \det\nabla y$. Hence, there is a subsequence (not relabeled) such that $\det\nabla y_k \to \det\nabla y$ and $\operatorname{Cof}\nabla y_k \to \operatorname{Cof}\nabla y$ pointwise almost everywhere in Ω for $k\to\infty$ because, by the compact embedding, $\operatorname{Cof}\nabla y_k \to \operatorname{Cof}\nabla y$ in $L^q(\Omega;\mathbb{R}^{3\times3})$ and $\det\nabla y_k \to \det\nabla y$ in $L^r(\Omega)$ for $k\to\infty$. Now, we prove that $y\in\mathcal{A}$ which amounts in verifying that $\det\nabla y > 0$ almost everywhere in Ω. We follow the reasoning as in the proof of Theorem 3.4.1. If $s > \frac{3r}{r-3}$ the argument showing a uniform positive lower bound for the determinant follows from the proof of Theorem 2.5.4. We have

$$(\nabla y_k(x))^{-1} = \frac{(\operatorname{Cof}\nabla y_k(x))^\top}{\det\nabla y_k(x)} \tag{4.5.13}$$

and thus, for almost all $x\in\Omega$, it holds $(\nabla y_k(x))^{-1} \to (\nabla y(x))^{-1}$. Further, the cofactor of a matrix is invertible whenever the matrix itself is invertible and we have by Lemma 4.5.6 that for almost all $x\in\Omega$

$$\sup_{k\in\mathbb{N}}|\nabla y_k(x)| = \sup_{k\in\mathbb{N}}\det\nabla y_k(x)|(\operatorname{Cof}(\nabla y_k(x))^{-\top}| \le \sup_{k\in\mathbb{N}}\frac{3}{2}\det\nabla y_k(x)|(\nabla y_k(x))^{-1}|^2 < \infty .$$

This means that there is (an x-dependent) subsequence of $\{\nabla y_k(x)\}_{k\in\mathbb{N}}$ converging pointwise to $\nabla y(x)$ for almost every $x\in\Omega$. As all converging subsequences have the same limit we deduce that the whole sequence converges to $\nabla y(x)$ for almost every $x\in\Omega$. Consequently, for almost all $x\in\Omega$ and $k\to\infty$, we obtain

$$\begin{aligned}\nabla y_k(x) &= ((\operatorname{Cof}(\nabla y_k(x))^{-1})^\top)\det\nabla y_k(x)\\ &\to ((\operatorname{Cof}(\nabla y(x))^{-1})^\top)\det\nabla y(x) = \nabla y(x).\end{aligned}$$

This implies, in view of Theorem B.2.4, that $y_k \to y$ strongly in $W^{1,\tilde p}(\Omega;\mathbb{R}^3)$ for every $1\le\tilde p<p$.

To show weak lower semicontinuity of

$$y \mapsto \int_\Omega \tilde\varphi(x,\nabla y(x),\nabla[\operatorname{Cof}\nabla y],\nabla[\det\nabla y])\,\mathrm{d}x - \mathcal{F}(y)$$

we apply Theorem 3.3.1. Notice that $\tilde\varphi(x,F,\cdot,\cdot)$ is convex for almost all $x\in\Omega$ and all $F\in\mathbb{R}^{d\times d}$. □

To get mere existence of a minimizer we can weaken the assumptions of Theorem 4.5.7. In particular, $\nabla[\det\nabla y]$ can be omitted from the list of arguments of $\tilde\varphi$ in (4.5.6).

Theorem 4.5.8 (Existence results for gradient-polyconvex materials II). *Let $\Omega \subset \mathbb{R}^3$ be a bounded Lipschitz domain representing a gradient polyconvex material with $\tilde{\varphi} : \bar{\Omega} \times \mathbb{R}^{3\times3} \times \mathbb{R}^{3\times3\times3} \times \mathbb{R}^3 \to \mathbb{R} \cup \{+\infty\}$ and assume that there are constants $\varepsilon > 0$, $p \geq 2$, $q \geq q_{pd}$ for q_{pd} given by (3.4.2), $r > 1$, $s > 0$ such that, for a.a. $x \in \Omega$, all $F \in \mathbb{R}^{3\times3}$, and all $\Delta_1 \in \mathbb{R}^{3\times3\times3}$*

$$\tilde{\varphi}(x,F,\Delta_1) \geq \begin{cases} \varepsilon\big(|F|^p + |\mathrm{Cof}\, F|^q + (\det F)^r + (\det F)^{-s} + |\Delta_1|^q\big) & \text{if } \det F > 0, \\ +\infty & \text{otherwise.} \end{cases} \tag{4.5.14}$$

Let us assume that $\Gamma = \Gamma_\mathrm{D} \cup \Gamma_\mathrm{N}$ is a measurable partition of $\Gamma := \partial\Omega$ with $\mathrm{meas}_2(\Gamma_\mathrm{D}) > 0$ and let $y_\mathrm{D} \in W^{1,p}(\Omega;\mathbb{R}^3)$ be given. Let

$$\mathcal{A} := \big\{ y \in W^{1,p}(\Omega;\mathbb{R}^3);\ \mathrm{Cof}\,\nabla y \in W^{1,q}(\Omega;\mathbb{R}^{3\times3}), \\ \det\nabla y \in L^r(\Omega),\ \ y = y_\mathrm{D} \text{ on } \Gamma_\mathrm{D},\ \ \det\nabla y > 0 \text{ a.e.} \big\}$$

be nonempty, and let further the functional $\mathcal{F} : W^{1,p}(\Omega;\mathbb{R}^3) \to \mathbb{R}$ describing work of external loads be upper semicontinuous. and such that, for all $y \in \mathcal{A}$, it holds that $\mathcal{F}(y) \leq K_1(\|y\|^{\tilde{p}}_{W^{1,p}(\Omega;\mathbb{R}^3)} + 1)$ for some $K_1 > 0$ and $0 \leq \tilde{p} < p$. Let for all $y \in \mathcal{A}$

$$\mathcal{E}(y) := \int_\Omega \varphi(x, \nabla y(x), \nabla[\mathrm{Cof}\,\nabla y(x)])\,\mathrm{d}x - \mathcal{F}(y). \tag{4.5.15}$$

Eventually, let there be $y \in \mathcal{A}$ such that $\mathcal{E}(y) < +\infty$. Then the infimum of $\mathcal{E}$ on $\mathcal{A}$ is attained. Moreover, if $q > 3$ and $s > 6q/(q-3)$ then there is $\varepsilon > 0$ such that for every minimizer $\tilde{y} \in \mathcal{A}$ of $\mathcal{E}$ it holds that $\det\nabla\tilde{y} \geq \varepsilon$ in $\bar{\Omega}$.

Sketch of the proof. The proof proceeds almost the same way as the proof of Theorem 4.5.7. A new idea must be employed to get pointwise convergence $\det\nabla y_k \to \det\nabla y$ almost everywhere in Ω because by the assumptions $\{\det\nabla y_k\} \subset L^r(\Omega)$ only. However, having the pointwise convergence $\mathrm{Cof}\,\nabla y_k \to \mathrm{Cof}\,\nabla y$ almost everywhere in Ω at our disposal we realize that, in view of (A.2.19), for any $F \in \mathrm{GL}^+(3)$

$$\det(\mathrm{Cof}\, F) = (\det F)^2, \tag{4.5.16}$$

which yields the pointwise convergence of $\det\nabla y_k \to \det\nabla y$ almost everywhere in Ω. Now we prove the lower bound on the determinant. If $q > 3$ Theorem B.3.2 implies that $\mathrm{Cof}\,\nabla y \in C^{0,\alpha}(\bar{\Omega})$ where $\alpha = (q-3)/q < 1$. We first show that $\det\nabla y > 0$ on Γ. Assume that there is $x_0 \in \Gamma$ such that $\det\nabla y(x_0) = 0$. We can, without loss of generality, assume that $x_0 := 0$ and estimate for all $x \in \bar{\Omega}$ and in view of (4.5.16)

$$0 \leq |\mathrm{Cof}\,\nabla y(x) - \mathrm{Cof}\,\nabla y(0)| \leq C|x|^\alpha,$$

where $C > 0$. Taking into account (4.1.8) applied to determinant, and the fact that $\mathrm{Cof}\,\nabla y$ is uniformly bounded on Ω, we have for some $K, \tilde{d} > 0$ that

$$0 \le (\det \nabla y(x))^2 = |\det \operatorname{Cof} \nabla y(x) - \det \operatorname{Cof} \nabla y(0)| \\ \le \tilde{d} |\operatorname{Cof} \nabla y(x) - \operatorname{Cof} \nabla y(0)| \le K|x|^\alpha . \tag{4.5.17}$$

Altogether, we see that, if $|x| \le t$, then

$$\frac{1}{(\det \nabla y(x))^s} \ge \frac{1}{K^{st/2} t^{\alpha s/2}} .$$

We have for $t > 0$ small enough

$$\int_\Omega \frac{1}{(\det \nabla y(x))^s} \, \mathrm{d}x \ge \int_{B(0,t)\cap\Omega} \frac{1}{(\det^2 \nabla y(x))^{s/2}} \, \mathrm{d}x \ge \frac{\hat{K} t^3}{K^{s/2} t^{\alpha s/2}} , \tag{4.5.18}$$

where $\hat{K} t^3 \le \operatorname{meas}_3(B(0,t) \cap \Omega)$ and $\hat{K} > 0$ is independent of t for t small enough. Note that this is possible because Ω is Lipschitz and therefore it has the cone property. Passing to the limit for $t \to 0$ in (4.5.18) we see that $\int_\Omega (\det \nabla y(x))^{-s} \, \mathrm{d}x = \infty$ because $\alpha s/2 = ((q-3)/q)s/2 > 3$. This, however, contradicts our assumptions and, consequently, $\det \nabla y > 0$ everywhere in Γ. If we assume that there is $x_0 \in \Omega$ such that $\det \nabla y(x_0) = 0$ the same reasoning as before brings us again to a contradiction, too. It is even easier because $B(x_0, t) \subset \Omega$ for $t > 0$ small enough. As $\bar{\Omega}$ is compact and $x \mapsto \det \nabla y(x)$ is continuous it is clear that there is $\varepsilon > 0$ such that $\det \nabla y \ge \varepsilon$ in $\bar{\Omega}$. This finishes the proof if there is a finite number of minimizers.

Assume then that there are infinitely many minimizers of $\mathcal{E}$ and assume that such $\varepsilon > 0$ does not exist, i.e., that for every $k \in \mathbb{N}$ there existed $y_k \in \mathcal{A}$ such that $\mathcal{E}(y_k) = \inf_\mathcal{A} \mathcal{E}$ and $x_k \in \bar{\Omega}$ such that $\det^2 \nabla y_k(x_k) < 1/k$. We can even assume that $x_k \in \Omega$ because of the continuity of $x \mapsto \det^2 \nabla y_k(x)$. The uniform (in k) Hölder continuity of $x \mapsto \det^2 \nabla y_k(x)$ which comes from (4.5.10) and from the fact that $\mathcal{E}(y_k) = \inf_\mathcal{A} \mathcal{E}$ implies that $|\det^2 \nabla y_k(x) - \det^2 \nabla y_k(x_k)| \le K|x - x_k|^\alpha$ (with K independent of k) and therefore $0 \le \det^2 \nabla y_k(x) \le 1/k + K|x - x_k|^\alpha$. Take $r_k > 0$ so small that $B(x_k, r_k) \subset \Omega$ for all $k \in \mathbb{N}$. Then

$$\inf_\mathcal{A} \mathcal{E} \ge \int_\Omega (\det^2 \nabla y_k(x))^{-s/2} \, \mathrm{d}x \\ \ge \int_{B(x_k,r_k)} (\det^2 \nabla y_k(x))^{-s/2} \, \mathrm{d}x \ge \int_{B(x_k,r_k)} \left(\frac{k}{1 + Kkr_k^\alpha} \right)^{s/2} \mathrm{d}x .$$

The right-hand side is, however, arbitrarily large for k suitably large and r_k suitably small because $\alpha s > 6$. This contradicts $\inf_\mathcal{A} \mathcal{E} < \infty$. The proof is finished. □

If $y \in W^{1,p}(\Omega; \mathbb{R}^d)$ for some $p > d$ is injective and $\det \nabla y > 0$ almost everywhere in Ω then Theorem B.3.10 implies that for every open ball $B(x_0, r) \subset \Omega$

$$\operatorname{meas}_d(y(B(x_0, r))) = \int_{B(x_0,r)} \det \nabla y(x) \, \mathrm{d}x .$$

Hence, it yields for every Lebesgue point $x_0 \in \Omega$ of $\det \nabla y$ that

$$\det \nabla y(x_0) = \lim_{r \to 0} \frac{\mathrm{meas}_d(y(B(x_0, r)))}{\mathrm{meas}_d(B(x_0, r))}. \tag{4.5.19}$$

In particular, in gradient-polyconvex materials, the limit on the right-hand side of (4.5.19) is either weakly differentiable or at least pointwise converging as a function of $x_0 \in \Omega$.

Let $\varphi_{\mathrm{SVK}} : \mathbb{R}^{3\times 3} \to \mathbb{R}$ be a stored energy density of an *anisotropic St. Venant-Kirchhoff material* (cf. (2.4.5)), i.e.,

$$\varphi_{\mathrm{SVK}}(F) := \frac{1}{8}\mathbb{C}(F^\top F - \mathbb{I}){:}(F^\top F - \mathbb{I}), \tag{4.5.20}$$

where $\mathbb{C}$ is the fourth-order and positive definite tensor of elastic constants, $\mathbb{I}$ is the identity matrix, and ":" denotes the scalar product between matrices. Therefore, $\varphi_{\mathrm{SVK}}(F) \geq K|F|^4$ for some $K > 0$ and all matrices F. If $G : \Omega \to \mathbb{R}^{3\times 3}$ is as in Definition 4.5.3 then

$$\varphi(G) := \begin{cases} \varphi_{\mathrm{SVK}}(G) + \alpha(|\nabla \mathrm{Cof}\, G|^q + (\det G)^{-s}) & \text{if } \det G > 0, \\ +\infty & \text{otherwise} \end{cases} \tag{4.5.21}$$

for some $\alpha > 0$, $s > 0$, $q = 2$, and $r > 1$ is gradient polyconvex and Theorem 4.5.8 can be readily applied with $p = 4$. We emphasize that φ is widely used in engineering/computational community because it allows for the implementation of all elastic constants. On the other hand, φ_{SVK} is not quasiconvex which means that existence of minimizers cannot be guaranteed. Additionally, φ_{SVK} stays locally bounded even in the vicinity of non-invertible matrices and thus allows for non-realistic behavior. The function φ cures the mentioned drawbacks and (4.5.21) offers a mechanically relevant alternative to the Saint Venant-Kirchhoff model of material behavior.

Let us finally mention the following strong compactness statement which might be of an independent interests and which is a generalization of the results used in the proof of Theorem 4.5.8.

Theorem 4.5.9 (Strong compactness). *Let $\Omega \subset \mathbb{R}^d$, $d \geq 2$, be a Lipschitz bounded domain and let $\{y_k\}_{k\in\mathbb{N}} \subset W^{1,p}(\Omega;\mathbb{R}^d)$ for $p > d$ be such that, for some $s > 0$, it holds*

$$\sup_{k\in\mathbb{N}} \Big(\|y_k\|_{W^{1,p}(\Omega;\mathbb{R}^d)} + \|\mathrm{Cof}\, \nabla y_k\|_{\mathrm{BV}(\Omega;\mathbb{R}^{d\times d})} + \big\| |\det \nabla y_k|^{-s} \big\|_{L^1(\Omega)} \Big) < +\infty.$$

Then there is a (not relabeled) subsequence and $y \in W^{1,p}(\Omega;\mathbb{R}^d)$ such that for $k \to \infty$ we have the following convergence results: $y_k \to y$ in $W^{1,\tilde{p}}(\Omega;\mathbb{R}^d)$ for every $1 \leq \tilde{p} < p$, $\det \nabla y_k \to \det \nabla y$ in $L^r(\Omega)$ for every $1 \leq r < p/d$, $\mathrm{Cof}\, \nabla y_k \to \mathrm{Cof}\, \nabla y$ in $L^q(\Omega;\mathbb{R}^{d\times d})$ for every $1 \leq q < p/(d-1)$, and $|\det \nabla y_k|^{-\tilde{s}} \to |\det \nabla y|^{-\tilde{s}}$ in $L^1(\Omega)$ for every $0 \leq \tilde{s} < s$.

Proof. Reflexivity of $W^{1,p}(\Omega;\mathbb{R}^d)$ and (4.5.22) imply the existence of a subsequence of $\{y_k\}$ which we do not relabel and such that $y_k \to y$ weakly in $W^{1,p}(\Omega;\mathbb{R}^d)$. Moreover, by the compact embeddings of $\mathrm{BV}(\Omega;\mathbb{R}^{d\times d})$ to $L^1(\Omega;\mathbb{R}^{d\times d})$ we extract a further subsequence satisfying $\mathrm{Cof}\, \nabla y_k \to G$ in $L^1(\Omega;\mathbb{R}^{d\times d})$ for some function $G \in$

$L^1(\Omega;\mathbb{R}^{d\times d})$. Weak continuity of $y \mapsto \mathrm{Cof}\,\nabla y : W^{1,p}(\Omega;\mathbb{R}^d) \to L^{p/(d-1)}(\Omega;\mathbb{R}^{d\times d})$ implies that $G = \mathrm{Cof}\,\nabla y$. The strong convergence in $L^1(\Omega;\mathbb{R}^d)$ yields that we can extract a further subsequence ensuring which implies due to A.2.19 that $\det\nabla y_k \to \det\nabla y$ a.e. in Ω. By the Fatou lemma and the assumption (4.5.22)

$$+\infty > \liminf_{k\to\infty} \int_\Omega \frac{1}{|\det\nabla y_k(x)|^s}\,\mathrm{d}x \geq \int_\Omega \frac{1}{|\det\nabla y(x)|^s}\,\mathrm{d}x,$$

which shows that $\det\nabla y \neq 0$ almost everywhere in Ω. Reasoning analogous to the one in the proof of Theorem 4.5.8 results in the almost everywhere convergence $\nabla y_k \to \nabla y$ for $k \to \infty$. The Vitali convergence theorem (see Theorem B.2.4) then implies that $\nabla y_k \to \nabla y$ in $L^{\tilde{p}}(\Omega;\mathbb{R}^{d\times d})$ for every $1 \leq \tilde{p} < p$. This shows the strong convergence of $y_k \to y$ in $W^{1,\tilde{p}}(\Omega;\mathbb{R}^d)$. The same argument gives the other strong convergences of the determinant and the cofactor. □

Remark 4.5.10 (Regularity of $(\det\nabla y)^2$*).* Using (A.2.19) and Theorem B.3.5, we obtain that under the assumptions of Theorem 4.5.8 every $y \in \mathcal{A}$ satisfies $\det^2\nabla y := (\det\nabla y)^2 = \det(\mathrm{Cof}\,\nabla y) \in W^{1,r}(\Omega)$, where

$$r := \begin{cases} 3q/(9-2q) & \text{if } 9/5 \leq q < 3, \\ q & \text{if } q > 3. \end{cases}$$

Remark 4.5.11 (Regularity of of states with $\mathcal{E}(y) < +\infty$*).* Assume that $\mathrm{Cof}\,\nabla y \in W^{1,q}(\Omega;\mathbb{R}^{3\times 3})$ for $q > 3$, and that for some $\varepsilon > 0$ $\det\nabla y \geq \varepsilon$ in $\bar{\Omega}$. In view of Remark 4.5.10 we see that $\det^2\nabla y \in W^{1,q}(\Omega)$. Let us take for $z \in \mathbb{R}$ the function $h(z) := \min(1/\sqrt{|z|}, 1/\sqrt{\varepsilon})$. Then h is Lipschitz and $h(\det^2\nabla y) = (\det\nabla y)^{-1} \in W^{1,q}(\Omega)$ by Theorem B.3.5. Since

$$(\nabla y)^{-1} = \frac{(\mathrm{Cof}\,\nabla y)^\top}{\det\nabla y}$$

we get that $(\nabla y)^{-1} \in W^{1,q}(\Omega;\mathbb{R}^{3\times 3})$.[11] Consequently, we get that $(\mathrm{Cof}(\nabla y)^{-1})^\top \in W^{1,q}(\Omega;\mathbb{R}^{3\times 3})$. This again follows from Theorem B.3.5 and the fact that the cofactor matrix collects all 2×2 subdeterminants. Taking the Lipschitz function $h : \mathbb{R} \to \mathbb{R}$ defined now as $h(z) := \max(\sqrt{|z|}, \sqrt{\varepsilon})$, we get that $h(\det^2\nabla y) = \det\nabla y \in W^{1,q}(\Omega)$. Hence,

$$\nabla y = \det\nabla y(\mathrm{Cof}(\nabla y)^{-1})^\top \in W^{1,q}(\Omega;\mathbb{R}^{3\times 3}).$$

This means that, if $q > 3$ and $s > 6q/(q-3)$ in Theorem 4.5.8, then every minimizer of the functional $\mathcal{E}$ in (4.5.15) belongs to $W^{2,\min(p^*,q)}(\Omega;\mathbb{R}^3)$.

Remark 4.5.12 (Locking constraint and gradient polyconvexity). Theorem 4.5.9 implies that if $L : \mathrm{GL}^+(d) \to \mathbb{R}$ is lower semicontinuous then a locking constraint $L(\nabla y) \leq 0$ can be easily added to the minimization in Theorems 4.5.7 and 4.5.8 provided the set $\mathcal{A}$ contains deformations satisfying the locking constraint. Indeed, if $\{y_k\}_{k\in\mathbb{N}} \subset \mathcal{A}$ is a minimizing sequence then Theorem 4.5.9 implies that (at least

[11] As $q > 3$, $W^{1,q}(\Omega)$ is the Banach algebra for $\Omega \subset \mathbb{R}^3$ a bounded Lipschitz domain. In particular, this means here $F_1, F_2 \in W^{1,q}(\Omega;\mathbb{R}^{3\times 3})$ implies that also $F_1F_2 \in W^{1,q}(\Omega;\mathbb{R}^{3\times 3})$.

for a subsequence) $\nabla y_k \to \nabla y$ almost everywhere in Ω. Consequently, if $L(\nabla y_k) \le 0$ almost everywhere in Ω then lower semicontinuity of L implies that

$$L(\nabla y(x)) \le \liminf_{k\to\infty} L(\nabla y_k(x)) \le 0$$

for almost every $x \in \Omega$, i.e., y is admissible. In particular, L from (3.4.11) can be used.

Remark 4.5.13 (Strain-limiting materials). Analogously to locking constraints one can also consider the so-called *strain-limiting materials*; see e.g. [99]. For such materials, some measure of strain (a norm of the right Cauchy-Green strain tensor, for example) is bounded independently of the applied stress. This can be easily incorporated to the theory of gradient-polyconvex materials, too.

Remark 4.5.14 (Transfer of regularity). Gradient-polyconvex materials enable us to control regularity of the first Piola-Kirchhoff stress tensor by means of smoothness of the Cauchy stress. Assume that the Cauchy stress tensor $T^y : \bar{\Omega}^y \to \mathbb{R}^{d\times d}$ is Lipschitz continuous, for instance. If $\operatorname{Cof}\nabla y : \bar{\Omega} \to \mathbb{R}^{d\times d}$ is Lipschitz continuous, too then the first Piola-Kirchhoff stress tensor inherits the Lipschitz continuity from T^y; cf. (1.2.9). In a similar fashion, one can transfer Hölder continuity of T^y to S via Hölder continuity of $x \mapsto \operatorname{Cof}\nabla y$.

Remark 4.5.15 (Relationship to 2nd-grade materials). Assume that $y \in W^{2,p}(\Omega;\mathbb{R}^d)$. Then, for $1 \le i \le d$, it holds

$$\partial_{x_i} \det\nabla y = (\operatorname{Cof}\nabla y)_{jk}\partial^2_{x_k x_i} y_j \tag{4.5.22}$$

and similarly

$$\partial_{x_i}(\operatorname{Cof}\nabla y)_{jk} = \mathcal{L}_{jklm}(\nabla y)\partial^2_{x_m x_i} y_l\,, \tag{4.5.23}$$

where, for every $F \in \mathbb{R}^{d\times d}$, we denoted by $\mathcal{L}_{jklm}(F) := \partial_{F_{lm}}(\operatorname{Cof} F)_{jk}$ a polynomial of the degree at most $d-2$ in F.

Exercise 4.5.16. Prove Theorem 4.5.8 if we additionally require injectivity of deformations, i.e., the Ciarlet-Nečas condition (3.4.7).

Exercise 4.5.17. Let all the assumptions of Theorem 4.5.7 hold with $p > (d-1)d$. Let (3.4.7) hold for all $y \in \mathcal{A}$. Show that all minimizer of $\mathcal{E}$ from (4.5.11) are injective everywhere in Ω.

Exercise 4.5.18. Assume that $y_k \to y$ weakly in $W^{1,p}(\Omega;\mathbb{R}^d)$ if $k \to \infty$ and let $p > d$ for $d = 2$ or $d = 3$. Let for every $k \in \mathbb{N}$ it holds that $\det\nabla y_k \ne 0$ almost everywhere in Ω and $\det\nabla y \ne 0$ everywhere in Ω. If

$$\sup_{k\in\mathbb{N}} \|\nabla \operatorname{Cof}\nabla y_k\|_{L^1(\Omega;\mathbb{R}^{d\times d\times d})} + \|\nabla\det\nabla y_k\|_{L^1(\Omega;\mathbb{R}^d)} < +\infty$$

then there is a (not relabeled) subsequence such that $\nabla y_k \to \nabla y$ almost everywhere in Ω if $k \to \infty$.

Exercise 4.5.19. Write down Theorem 4.5.9 for $d = 2$ and find an alternative proof by means of Sobolev embedding theorems.

Exercise 4.5.20. Formulate a minimization problem for a gradient-polyconvex material whose energy density does not depend on $\nabla \mathrm{Cof}\,\nabla y$. Find a a deformation y which does not belong to $W^{2,1}(\Omega;\mathbb{R}^d)$ but is such that $\det\nabla y \in W^{1,\infty}(\Omega)$.

Exercise 4.5.21. Assume that $d = 2$, or $d = 3$, $p > d$, and $q > 1$ such that $pq/(q(d-1)+p) > 1$. Show that the functional $y \mapsto \mathcal{E}(y)$ given as

$$\begin{aligned}\mathcal{E}(y) := \int_\Omega \varphi_1(\nabla y(x))\,\mathrm{d}x + \varphi_2((\nabla y(x))^{-1})\,\mathrm{d}x \\ + \int_\Omega \det(\nabla y(x))^{-q}\,\mathrm{d}x + \int_\Omega |\nabla \det\nabla y(x))|^{p/d}\mathrm{d}x,\end{aligned} \tag{4.5.24}$$

for $0 \le \varphi_i : \mathbb{R}^{d\times d} \to \mathbb{R}$ convex if $i = 1,2$ and $\varphi_1(F) \ge \varepsilon|F|^p$ for all $F \in \mathbb{R}^{d\times d}$ and $\varepsilon > 0$, attains a minimizer on the set

$$\mathcal{A} := \{y \in W^{1,p}(\Omega;\mathbb{R}^d);\ \det\nabla y > 0,\ y(x) = x \text{ on } \Gamma\}. \tag{4.5.25}$$

Exercise 4.5.22. Assume that $\Omega \subset \mathbb{R}^3$ is a bounded Lipschitz domain, $q > 3$, and that $F : \Omega \to \mathbb{R}^{3\times 3}$ is such that $\det F \ge \varepsilon > 0$. Show that $\mathrm{Cof}\,F \in W^{1,q}(\Omega;\mathbb{R}^{3\times 3})$ is equivalent to $F \in W^{1,q}(\Omega;\mathbb{R}^{3\times 3})$.

4.5.3 Excursion into higher-order FEM approximations

To have a chance for some explicit convergence criterion in contrast what we had in Section 3.7.3 and to allow for non-polyconvex materials, we will modify the model (3.7.3) in the spirit of the theory of the second-grade nonsimple materials, cf. Section 2.5. We again consider (2.5.24).

As we now use the concept of nonsimple materials, we have "compactness of deformation gradients" and we can handle non-polyconvex materials as e.g. the St. Venant-Kirchhoff materials combined with a $\det F$-dependent term respecting the Ciarlet-Nečas condition. We thus consider the following ansatz:

$$\varphi(F) := \varphi_1(F) + \gamma(\det F) \quad \text{with } \varphi_1 \text{ continuous, satisfying} \tag{4.5.26a}$$

$$\exists p \ge 2d,\ C, c > 0: \quad c|F|^p \le \varphi_1(F) \le C(1+|F|^p), \quad \text{and} \tag{4.5.26b}$$

$$\gamma : (0,+\infty) \to [0,+\infty] \text{ lower semicontinuous, convex, proper, and } C^1 \text{ on its domain.} \tag{4.5.26c}$$

As we will have an L^∞-control on the deformation gradient (and its determinant, too), instead of the Yosida regularization (3.7.5) we can now more advantageously perform the regularization of φ only for small values of $\det F$. More specifically, we define for $\varepsilon < 1$

$$\varphi_\varepsilon(F) := \varphi_1(F) + \gamma_\varepsilon(\det F)\qquad\text{with }\ \gamma_\varepsilon(J) := \begin{cases}\gamma(J) & \text{if } J \ge \varepsilon,\\ \gamma(\varepsilon) + \gamma'(\varepsilon)(\varepsilon - J) & \text{if } J < \varepsilon,\end{cases} \tag{4.5.27}$$

cf. Figure 4.4 for an illustration. Note that γ_ε is a C^1-function with a bounded deriva-

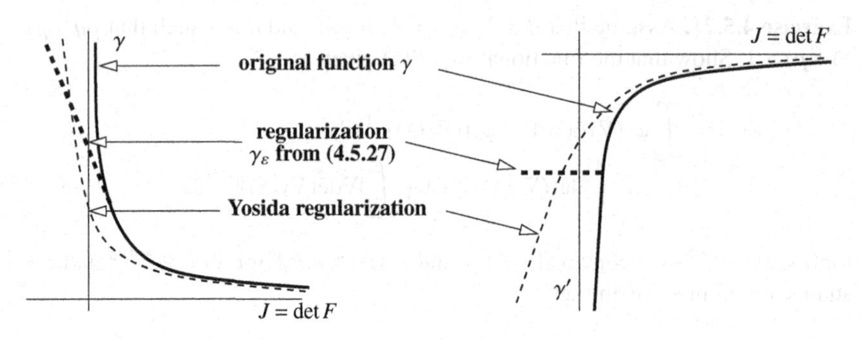

Fig. 4.4 The C^1-regularization (4.5.27) and its comparison with the Yosida regularization (3.7.5) of the original convex function γ (left) together with its derivative (right).

tive (hence in particular Lipschitz continuous) and $\gamma_\varepsilon'(J) = \gamma'(J)$ provided $J > \varepsilon$ and, relying on the convexity of γ and the monotonicity of γ', the coercivity of φ is not deteriorated in φ_ε if φ_1 is coercive. It is realistic to assume that φ_ε is p-Lipschitz continuous (see (4.1.8)) in a specific way, namely

$$\exists \ell > 0\ \forall F, \tilde F \in \mathbb{R}^{d\times d}:\quad \big|\varphi_\varepsilon(F) - \varphi_\varepsilon(\tilde F)\big| \le \ell\varepsilon^{-1}(1+|F|^{p-1}+|\tilde F|^{p-1})\big|F - \tilde F\big|. \tag{4.5.28}$$

If we assume that there is $\mathcal{F} : W^{1,p}(\Omega;\mathbb{R}^d) \to \mathbb{R}$ a weakly lower semicontinuous and Lipschitz continuous functional expressing the work of external loads and $\mathbb{H}$ is a positive definite matrix in $(\mathbb{R}^d)^4$ as in (2.5.24), the overall energy is

$$\mathcal{E}^{\rm SG}(y) := \begin{cases} \displaystyle\int_\Omega \varphi(\nabla y) + \frac{1}{2}\mathbb{H}\nabla^2 y \,\vdots\, \nabla^2 y \,\mathrm{d}x - \mathcal{F}(y) & \text{if } \displaystyle\int_\Omega \det(\nabla y)\,\mathrm{d}x \le \mathrm{meas}_d(y(\Omega))\\ & \text{and if } y|_{\Gamma_{\rm D}} = y_{\rm D},\\ +\infty & \text{otherwise.}\end{cases} \tag{4.5.29}$$

while its regularization is

$$\mathcal{E}_\varepsilon^{\rm SG}(y) := \begin{cases} \displaystyle\int_\Omega \varphi_\varepsilon(\nabla y) + \frac{1}{2}\mathbb{H}\nabla^2 y \vdots \nabla^2 y \,\mathrm{d}x - \mathcal{F}(y) \\ \displaystyle\quad + \frac{1}{\varepsilon}\Big(\int_\Omega \det(\nabla y)\,\mathrm{d}x - \mathrm{meas}_d(y(\Omega))\Big)^+ & \text{if } y|_{\Gamma_{\rm D}} = y_{\rm D}, \qquad (4.5.30) \\ +\infty & \text{otherwise.} \qquad (4.5.31) \end{cases}$$

Further, we make a discretization. Here we must use also higher-order finite elements, say P2 or Q2, relying again on (3.7.1). We define:

$$\mathcal{E}_{\varepsilon h}^{\rm SG}(y) := \begin{cases} \mathcal{E}_\varepsilon^{\rm SG}(y) & \text{if } y \in V_h, \\ +\infty & \text{otherwise,} \end{cases} \tag{4.5.32}$$

where $V_h \subset W^{2,\infty}(\Omega;\mathbb{R}^d)$ denotes the linear finite-dimensional finite-element space using a discretization of Ω with the mesh size h.

Lemma 4.5.23 (Lipschitz continuity of $\mathcal{E}_\varepsilon^{\rm SG}$). *Let the assumptions (3.7.7b-d) and* (4.5.26) *hold and let $2^* \ge p > d$ with 2^* denoting the Sobolev exponent $2d/(d-2)$. Moreover, let* (4.5.28) *hold. Then the functional $\mathcal{E}_\varepsilon^{\rm SG} : W^{1,p}(\Omega;\mathbb{R}^d) \cap H^2(\Omega;\mathbb{R}^d)) \to \mathbb{R}$ is Lipschitz continuous on (some) bounded sets; more specifically, for all $\rho \in [0,+\infty)$ there is some $L_\rho \in [0,+\infty)$ such that it holds*

$$\begin{aligned} &\forall y,\tilde y \in (W^{1,p}(\Omega;\mathbb{R}^d) \cap H^2(\Omega;\mathbb{R}^d)) : \\ &\quad \left.\begin{aligned} &\mathcal{E}_\varepsilon^{\rm SG}(y) \le \rho, \\ &\|y\|_{W^{1,p}(\Omega;\mathbb{R}^d)\cap H^2(\Omega;\mathbb{R}^d)} \le \rho, \\ &\|\tilde y\|_{W^{1,p}(\Omega;\mathbb{R}^d)\cap H^2(\Omega;\mathbb{R}^d)} \le \rho, \end{aligned}\right\} \Rightarrow \big|\mathcal{E}_\varepsilon^{\rm SG}(y) - \mathcal{E}_\varepsilon^{\rm SG}(\tilde y)\big| \le L_\rho\Big(\frac{1}{\varepsilon}\big\|y-\tilde y\big\|_{W^{1,p}(\Omega;\mathbb{R}^d)} \\ &\qquad\qquad\qquad\qquad\qquad\qquad + \big\|y-\tilde y\big\|_{H^2(\Omega;\mathbb{R}^d)}\Big) \\ &\qquad\qquad\qquad\qquad\qquad\qquad + \big|\mathrm{meas}_d(y(\Omega)) - \mathrm{meas}_d(\tilde y(\Omega))\big|. \end{aligned} \tag{4.5.33}$$

Proof. We can first estimate the difference in the elastic part by using (4.5.28) and the Hölder inequality:

$$\begin{aligned} \Big|\int_\Omega \varphi_\varepsilon(\nabla y)\,\mathrm{d}x - \int_\Omega \varphi_\varepsilon(\nabla\tilde y)\,\mathrm{d}x\Big| &\le \int_\Omega \big|\varphi_\varepsilon(\nabla y) - \varphi_\varepsilon(\nabla\tilde y)\big|\,\mathrm{d}x \\ &\le \int_\Omega \ell\varepsilon^{-1}\big|\nabla y - \nabla\tilde y\big|\big(1 + |\nabla y|^{p-1} + |\nabla\tilde y|^{p-1}\big)\,\mathrm{d}x \\ &\le \ell_p\varepsilon^{-1}\big\|\nabla y - \nabla\tilde y\big\|_{L^p(\Omega;\mathbb{R}^{d\times d})}\Big(1 + \big\|\nabla y\big\|_{L^p(\Omega;\mathbb{R}^{d\times d})}^{p-1} + \big\|\nabla\tilde y\big\|_{L^p(\Omega;\mathbb{R}^{d\times d})}^{p-1}\Big) \end{aligned} \tag{4.5.34}$$

with ℓ_p depending on ℓ from (4.5.28) and on p, too. Further we estimate the higher-order term as:

$$\begin{aligned} &\Big|\int_\Omega \mathbb{H}\nabla^2 y \vdots (\nabla^2 y) - \mathbb{H}\nabla^2\tilde y \vdots (\nabla^2\tilde y)\,\mathrm{d}x\Big| \\ &\qquad = \Big|\int_\Omega \mathbb{H}(\nabla^2 y - \nabla^2\tilde y) \vdots (\nabla^2 y + \nabla^2\tilde y)\,\mathrm{d}x\Big| \le C_\rho\big\|y-\tilde y\big\|_{H^2(\Omega;\mathbb{R}^d)} \end{aligned}$$

with C_ρ depending on ρ from (4.5.33) and on $|\mathbb{H}|$, too. The contribution from the penalization term can be estimated as[12]

$$\begin{aligned}\Big|\Big(\int_\Omega \det(\nabla y)\,\mathrm{d}x - \mathrm{meas}_d(y(\Omega))\Big)^+ - \Big(\int_\Omega \det(\nabla\tilde y)\,\mathrm{d}x - \mathrm{meas}_d(\tilde y(\Omega))\Big)^+\Big|\\ \le \int_\Omega |\det(\nabla y) - \det\nabla\tilde y|\,\mathrm{d}x + \big|\mathrm{meas}_d(y(\Omega)) - \mathrm{meas}_d(\tilde y(\Omega))\big|\\ \le C_\rho\big\|\nabla y - \nabla\tilde y\big\|_{L^p(\Omega;\mathbb{R}^{d\times d})} + \big|\mathrm{meas}_d(y(\Omega)) - \mathrm{meas}_d(\tilde y(\Omega))\big|\\ \le C_\rho\big\|y - \tilde y\big\|_{W^{1,p}(\Omega;\mathbb{R}^d)} + \big|\mathrm{meas}_d(y(\Omega)) - \mathrm{meas}_d(\tilde y(\Omega))\big|\end{aligned}$$

with C_ρ dependent on the radius of a ball in $W^{1,p}(\Omega;\mathbb{R}^d)$ where y and $\tilde y$ live. It follows from (4.1.8) applied to the determinant. □

Proposition 4.5.24 (Γ-convergence of $\mathcal{E}^{\rm SG}_{\varepsilon h}$). *Let* (3.7.7) *and* (4.5.26) *and the assumptions of Lemma 4.5.23 hold with $p > 2$. Then the collection $\{\mathcal{E}^{\rm SG}_{\varepsilon h}\}_{\varepsilon,h>0}$ Γ-converges to $\mathcal{E}^{\rm SG}$ in the weak topology on bounded subsets of $(W^{1,p}(\Omega;\mathbb{R}^d)\cap H^2(\Omega;\mathbb{R}^d))$ in the sense* (A.3.33) *provided the following stability criterion is satisfied:*

$$\frac{h^{2(2^*-p)/(2^*p-2p)}}{\varepsilon} \to 0. \tag{4.5.35}$$

Proof. The liminf-condition $\liminf \mathcal{E}^{\rm SG}_{\varepsilon h}(y_{\varepsilon h}) \ge \mathcal{E}^{\rm SG}(y)$ for any weakly converging sequence $y_{\varepsilon h} \to y$ weakly holds unconditionally. This can be seen by showing this convergence for some lower estimate of $\mathcal{E}^{\rm SG}_{\varepsilon h}$, namely for $\mathcal{E}^{\rm SG}_{\varepsilon} \le \mathcal{E}^{\rm SG}_{\varepsilon h}$ with $\mathcal{E}^{\rm SG}_{\varepsilon}$ defined by (4.5.31). Actually, $\mathcal{E}^{\rm SG}_{\varepsilon}$ penalizes the constraints occurring in $\mathcal{E}$ defined by (3.7.3) together with the local-non-selfpenetration constraint involved in φ and this estimate then essentially follows from the weak lower-semicontinuity of $\mathcal{E}^{\rm SG}$.

The limsup-condition $\limsup \mathcal{E}^{\rm SG}_{\varepsilon h}(y_{\varepsilon h}) \le \mathcal{E}^{\rm SG}(y)$ for any y and some weakly converging sequence $y_{\varepsilon h} \to y$ needs an explicit construction of such a recovery sequence and, in our case, holds only conditionally under the stability criterion (4.5.35). For y given, the recovery sequence can be taken as

$$y_{\varepsilon h} = \mathcal{P}^{(2)}_h y \tag{4.5.36}$$

with $\mathcal{P}^{(2)}_h$ from (3.7.1) for $k = 2$; actually $y_{\varepsilon h}$ is independent of ε. We need to prove that

$$\begin{aligned}\limsup_{\varepsilon,h\to 0}\big(\mathcal{E}^{\rm SG}_{\varepsilon h}(y_{\varepsilon h}) - \mathcal{E}^{\rm SG}(y)\big) \le \limsup_{\varepsilon\to 0}\big(\mathcal{E}^{\rm SG}_{\varepsilon}(y) - \mathcal{E}^{\rm SG}(y)\big)\\ + \lim_{\varepsilon,h\to 0}\big(\mathcal{E}^{\rm SG}_{\varepsilon h}(y_{\varepsilon h}) - \mathcal{E}^{\rm SG}_{\varepsilon}(y)\big) \le 0.\end{aligned} \tag{4.5.37}$$

[12] Here we apply the formula

$$|(\alpha-\beta)^+ - (\gamma-\delta)^+| \le |\alpha-\gamma| + |\beta-\delta|$$

valid for every $\alpha,\beta,\gamma,\delta \in \mathbb{R}$. We also recall that if $a \in \mathbb{R}$ then $a^+ := \max(a,0) = (a+|a|)/2$.

If $\mathcal{E}^{\rm SG}(y)=+\infty$, then (4.5.37) is trivial because its left-hand side is $-\infty$. If $\mathcal{E}^{\rm SG}(y)<+\infty$, then $\limsup(\mathcal{E}^{\rm SG}_\varepsilon(y)-\mathcal{E}^{\rm SG}(y))\le 0$ follows from $\mathcal{E}^{\rm SG}_\varepsilon\le\mathcal{E}^{\rm SG}$. Therefore, we only need to prove that the last term in (4.5.37) is zero. More specifically, we can estimate

$$\begin{aligned}\big|\mathcal{E}^{\rm SG}_\varepsilon(y)-\mathcal{E}^{\rm SG}_\varepsilon(y_{\varepsilon h})\big| &\le L_\rho\Big(\frac{1}{\varepsilon}\big\|y-y_{\varepsilon h}\big\|_{W^{1,p}(\Omega;\mathbb{R}^d)}+\big\|y-y_{\varepsilon h}\big\|_{H^2(\Omega;\mathbb{R}^d)}\Big)\\ &\qquad +\big|{\rm meas}_d(y(\Omega))-{\rm meas}_d(y_{\varepsilon h}(\Omega))\big|\\ &\le\frac{L_\rho}{\varepsilon}\big\|y-\mathcal{P}^{(2)}_h y\big\|_{W^{1,p}(\Omega;\mathbb{R}^d)}+L_\rho\big\|y-\mathcal{P}^{(2)}_h y\big\|_{H^2(\Omega;\mathbb{R}^d)}\\ &\le L_\rho C\frac{h^{2(2^*-p)/(2^*p-2p)}}{\varepsilon}\big\|y\big\|_{H^2(\Omega;\mathbb{R}^d)}+|{\rm meas}_d(y(\Omega))-{\rm meas}_d(\mathcal{P}^{(2)}_h y(\Omega))|+o(1)\end{aligned}$$

where we used (3.7.1a) with k=2 and p=2, and also we used (3.7.1b) for k=2, p=2 and for l=1 and 2 to obtain $\|y-\mathcal{P}^{(2)}_h y\|_{W^{1,2}(\Omega;\mathbb{R}^d)}\le C_{2,1,2}h\|y\|_{H^2(\Omega;\mathbb{R}^d)}$ and $\|y-\mathcal{P}^{(2)}_h y\|_{W^{1,2^*}(\Omega;\mathbb{R}^d)}\le N\|y-\mathcal{P}^{(2)}_h y\|_{H^2(\Omega;\mathbb{R}^d)}\le NC_{2,2,2}\|y\|_{H^2(\Omega;\mathbb{R}^d)}$ with N denoting the norm of the embedding $W^{1,2}(\Omega)\subset L^{2^*}(\Omega)$, which can be further interpolated to obtain

$$\big\|y-\mathcal{P}^{(2)}_h y\big\|_{W^{1,p}(\Omega;\mathbb{R}^d)}\le Ch^{2(2^*-p)/(2^*p-2p)}\big\|y\big\|_{H^2(\Omega;\mathbb{R}^d)}$$

provided $2\le p\le 2^*$, cf. Theorem B.3.14. The term $|{\rm meas}_d(y(\Omega))-{\rm meas}_d(\mathcal{P}^{(2)}_h y(\Omega))|$ converges to zero for $h\to 0$ due to Lemma 3.7.2. □

Corollary 4.5.25. *Let the assumptions of Proposition 4.5.24 hold. Assume that* $\{\tilde y\}_{\varepsilon,h>0}\subset W^{1,p}(\Omega;\mathbb{R}^d)\cap H^2(\Omega;\mathbb{R}^d)$ *is a sequence of minimizers of* $\mathcal{E}^{\rm SG}_\varepsilon$. *Then every cluster point in the weak topology of* $W^{1,p}(\Omega;\mathbb{R}^d)\cap H^2(\Omega;\mathbb{R}^d)$ *is a minimizer of* $\mathcal{E}^{\rm SG}$.

Sketch of the proof. It follows from Proposition A.3.14. □

Remark 4.5.26. The recovery sequence (4.5.36) converges even strongly and therefore the weak Γ-convergence proved in Proposition 4.5.24 is, in fact, even the so-called Mosco convergence.

Remark 4.5.27. In [361], this Γ-convergence was combined with plastic strain under the Kröner-Lee-Liu decomposition (3.6.10) with the "incompressible" inelastic strain $F_{\rm IN}$ occurring typically in plasticity models, and use for evolution under time-dependent Dirichlet boundary conditions, possibly involving also a unilateral Signorini contact on the boundary.

4.5.4 Ferromagnetic and ferroelectric nonsimple materials

We now generalize model presented in Section 3.5 in a way that the incompressibility and the Heisenberg constraint (3.5.3) are avoided. Like [471] we use the concept of 2nd-grade nonsimple materials which also allow for avoiding the polyconvexity condition (3.5.5d) and thus to address very general magnetostrictive materials.

Avoiding the (in any case disputable) Heisenberg constraint, formally there is no difference between magnetization and polarization and the rest of this section can be interpreted as a model for ferroelectric materials with m, $h_{\rm dem}$, ϕ, and μ_0 standing for polarization, depolarizing field, electrostatic potential, and for vacuum permittivity, instead of magnetization, demagnetizing field, magnetostatic potential, and vacuum permeability, respectively.

In contrast to the mixed Lagrangian/Eulerian formulation used in Sect. 3.5, we now use a fully Lagrangian formulation (i.e. also the magnetization $m = m(x)$ is now considered in the reference configuration), cf. e.g. [494]. It is more compatible with the concept of internal variables as an attribute of the material itself and, if their evolution is considered (in Part II cf. Remark 9.7.5), it is more natural to formulate flow-rules in the reference configurations. Sometimes, the demagnetizing (resp. depolarizing) field necessarily related to actual spatial configuration can be neglected and then the fully Lagrangian model is especially simpler.

We will use here m for the magnetization in the reference configuration (in contrast to m in Section 3.5) and h for the outer magnetic field in the space (as in Section 3.5), while $m_{\rm s}^y$ will be now the magnetization in the actual deformed configuration and $h_{\rm ext,r}$ will be the magnetic field seen from the reference configuration, cf. also Figure 4.5. Recalling that $x \in \Omega$ and that $x^y := y(x)$), these quantities are

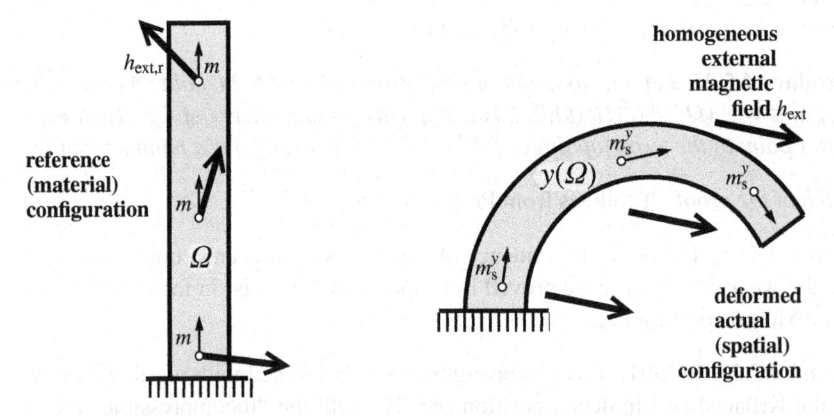

Fig. 4.5 An illustration how the magnetization and magnetic field transforms. The elastically soft but magnetically hard specimen with a homogeneous magnetization m fixed at the bottom by Dirichlet condition is deformed in a homogeneous external magnetic field $h_{\rm ext}$. Alternatively, the interpretation of polarization and electric field in case of elastic ferroelectric material can be considered, too.

transformed as

$$m_{\rm s}^y(x^y) = \frac{Fm(x)}{\det F} \quad \text{and} \quad h_{\rm ext,r}(x) = F^\top h_{\rm ext}(x^y) \quad \text{with } F = \nabla y(x) \tag{4.5.38}$$

for $x \in \Omega$; see e.g. [494, Formula (34)] for the latter transform. Let us note that the pulled-back field $h_{\rm ext,r}$ is defined only on Ω, in contrast to $h_{\rm ext}$ which is defined on the whole "universe" $\mathbb{R}^d$. Then, for example, for the Zeeman energy it holds

$$\int_\Omega h_{\rm ext,r}(x)\cdot m(x)\,{\rm d}x = \int_\Omega \big([\nabla y]^\top(x) h_{\rm ext}(y(x))\big)\cdot m(x)\,{\rm d}x$$
$$= \int_\Omega \big([\nabla y]^\top(x) h_{\rm ext}(y(x))\big)\cdot\big(\det(\nabla y(x))[\nabla y]^{-1}(x) m_{\rm s}^y(y(x))\big){\rm d}x$$
$$= \int_\Omega h_{\rm ext}(y(x))\cdot m_{\rm s}^y(y(x))\det(\nabla y(x))\,{\rm d}x = \int_{\Omega^y} h_{\rm ext}(x^y)\cdot m_{\rm s}^y(x^y)\,{\rm d}x^y, \quad (4.5.39)$$

where the last equation employs the substitution-in-integral calculus, cf. (B.3.18). Actually, the second form, i.e. $\int_\Omega([\nabla y]^\top(x)h_{\rm ext}(y(x)))\cdot m(x)\,{\rm d}x$ is most natural because it combines the given field $h_{\rm ext}$ with the magnetization in the reference configuration where all equations are formulated within the Lagrangian approach, as it is particularly desirable for evolution problems, cf. Remark 9.7.5 on p. 472. The Maxwell equation for the magnetostatic potential $\phi_{y,m}$ in the actual spatial configuration is now ${\rm div}(\mu_0\nabla\phi_{y,m} - \chi_{\Omega^y} m_{\rm s}) = 0$ with $m_{\rm s}$ from (4.5.38), i.e.

$${\rm div}\Big(\mu_0\nabla\phi_{y,m} - \chi_{\Omega^y}\Big[\frac{[\nabla y]m}{\det(\nabla y)}\Big](y^{-1})\Big) = 0 \quad \text{in } \mathbb{R}^d, \quad (4.5.40)$$

where χ_{Ω^y} is the characteristic function of the deformed configuration Ω^y, cf. (3.5.2). Assumptions of Propositions 4.5.28 below ensure that $y^{-1}:\Omega^y\to\Omega$ is continuous and therefore measurable. The energy of the demagnetizing field is

$$\frac{\mu_0}{2}\int_{\mathbb{R}^d}|\nabla\phi_{y,m}(z)|^2\,{\rm d}z = \frac12\int_{\Omega^y} h_{\rm dem}(z)\cdot m_{\rm s}(z)\,{\rm d}z \quad \text{with} \quad h_{\rm dem} = \nabla\phi_{y,m}; \quad (4.5.41)$$

this equality is obtained by testing (4.5.40) by $\phi_{y,m}$. Like (4.5.38), we can push back the demagnetizing field $h_{\rm dem}$ into the reference configuration, i.e. $h_{\rm dem,r}(x) = [\nabla y]^\top(x)h_{\rm dem}(x)$. Again, in contrast to $h_{\rm dem}$ which is defined on the whole "universe" $\mathbb{R}^d$, the pulled-back field $h_{\rm dem,r}$ is defined only on Ω. Like (4.5.39), the energy of the demagnetizing field written in the reference configuration is

$$\frac{\mu_0}{2}\int_{\mathbb{R}^d}\big|\nabla\phi_{y,m}(z)\big|^2\,{\rm d}z = \frac12\int_\Omega [\nabla y(x)]^\top\nabla\phi_{y,m}^y(x)\cdot m(x)\,{\rm d}x. \quad (4.5.42)$$

Noteworthy, the potential $\phi_{y,m}$ pulled back gives a function $\phi_{y,m,{\rm r}} := \phi_{y,m}^y : \Omega\to\mathbb{R}$ which serves as a potential of $h_{\rm dem,r}$, i.e. $h_{\rm dem,r} = \nabla\phi_{y,m,{\rm r}}$ because obviously $\nabla\phi_{y,m,{\rm r}} = \nabla\phi_{y,m}^y = [\nabla y]^\top[\nabla\phi_{y,m}]^y = [\nabla y]^\top h_{\rm dem}^y$. Also noteworthy, the equality (4.5.39) itself would hold if (4.5.38) would be simplified by taking $m_{\rm s}^y(x) = m(x)/\det F$ and $h_{\rm ext,r}(x) = h_{\rm ext}^y(x)$.

We consider the specific stored energy $\varphi:\Omega\times\mathbb{R}^{d\times d}\times\mathbb{R}^d\to\mathbb{R}$ in the reference configuration describing the material and supposed to be frame indifferent in the sense

$$\varphi(x,RF,m) = \varphi(x,F,m) \qquad \text{for all } R\in{\rm SO}(d), \quad (4.5.43)$$

which is different from (3.5.5b), cf. also Remark 4.5.29 below. This *frame indifference* is guaranteed (and, in fact, equivalent) if $\varphi(x,F,m) = \varphi_{\rm M}(x,C,m)$ for some "material" energy $\tilde\varphi_{\rm M} : \Omega\times\mathbb{R}^{d\times d}\times\mathbb{R}^d \to \mathbb{R}\cup\{+\infty\}$ with $C = F^\top F$. Of course, employing the decomposition[13] with $R \in {\rm SO}(d)$ and U symmetric positive definite (thus $U = \sqrt{C}$), the frame indifference means that $\varphi(x,F,m) = \tilde\varphi_{\rm M}(x,U,m)$ for some (another) "material" energy $\tilde\varphi_{\rm M}$.

Considering a specimen loaded by an external magnetic (or, in the ferroelectric interpretation, electric) field $h_{\rm ext}$ and mechanically fixed on $\Gamma_{\rm D}$ as in Figure 4.5, the overall stored energy then takes the form

$$\mathcal{E}(y,m) = \begin{cases} \displaystyle\int_\Omega\Big(\varphi(\nabla y,m) + (1+|\nabla^2 y|^{p-2})\mathbb{H}\nabla^2 y \vdots \nabla^2 y + \frac{\kappa}{2}|\nabla m|^2 \\ \displaystyle\qquad + [\nabla y]^\top\Big(\frac{1}{2}\nabla\phi^y - h_{\rm ext}^y\Big)\cdot m\Big)\,{\rm d}x \quad \text{with} \\ \displaystyle\phi^y = \frac{\Delta^{-1}}{\mu_0}\Big({\rm div}\Big(\chi_{\Omega^y}\Big[\frac{[\nabla y]m}{\det(\nabla y)}\Big](y^{-1})\Big)\Big) \quad \text{if } \int_\Omega \det(\nabla y)\,{\rm d}x \le {\rm meas}_d(y(\Omega)), \\ \qquad\qquad \text{and if } y = y_{\rm D} \text{ on } \Gamma_{\rm D}, \\ +\infty \qquad\qquad \text{otherwise.} \end{cases} \tag{4.5.44}$$

Actually, $\phi \in H^1(\mathbb{R}^d)$ is just the (unique) solution to (4.5.40). Note that we also used (4.5.39) and thus the only occurrence of y^{-1} remains in (4.5.40), which anyhow forces us to involve the global constraint (3.4.7) in (4.5.44) provided we do not want to ignore influence of the demagnetizing (or depolarizing) field. Let us also note that, in contrast to the model formulated in the mixed Lagrangian/Eulerian setting (2.3.15) in Section 3.5, the exchange-energy coefficient $\kappa = \kappa(x)$ is now a material property related to the gradient of m as an internal variable of the material itself.

Proposition 4.5.28 (Existence of minimizing configurations). *Let $p > d$, $\varphi : \Omega\times\mathbb{R}^{d\times d}\times\mathbb{R}^d \to \mathbb{R}\cup\{+\infty\}$ be continuous and satisfying the coercivity $\varphi(F,m) \ge \varepsilon|F|^p + \varepsilon|m|^2 + \varepsilon/(\det F)^q$ for some $\varepsilon > 0$ and with $q > pd/(p-d)$ for $\det F > 0$ while $\varphi(F,m) = +\infty$ otherwise, and let $h \in L^2(\mathbb{R}^d;\mathbb{R}^d)$, $\mathbb{H}$ be positive definite, $\kappa > 0$, ${\rm meas}_{d-1}(\Gamma_{\rm D}) > 0$ and $y_{\rm D}$ allow an extension to $W^{2,p}(\Omega;\mathbb{R}^d)$ making $\mathcal{E}(\cdot,0)$ finite. Then $\mathcal{E}$ from (4.5.44) possesses a minimizer (y,m) on $W^{2,p}(\Omega;\mathbb{R}^d)\times H^1(\Omega;\mathbb{R}^d)$.*

Proof. Similarly as in Theorem 3.5.2, we use the direct method and consider a minimizing sequence $\{(y_k,m_k)\}_{k\in\mathbb{N}}$ for the functional (4.5.44) with the corresponding magnetostatic potentials $\phi_k = \phi_{y_k,m_k}$ solving (4.5.40) with $(y,m) = (y_k,m_k)$. By the assumed coercivity, this sequence is bounded in $W^{2,p}(\Omega;\mathbb{R}^d)\times H^1(\Omega;\mathbb{R}^d)$. Therefore, by the Banach selection principle, we can consider a weakly converging subsequence, denoting by (y,u) its limit.

The weak lower semicontinuity of the functional $(y,m) \mapsto \int_\Omega\varphi(\nabla y,m) + (1+|\nabla^2 y|^{p-2})\mathbb{H}\nabla^2 y \vdots \nabla^2 y + \frac{\kappa}{2}|\nabla m|^2\,{\rm d}x$ follows by Theorem 3.3.1.

Furthermore, we need to show the weak continuity of the mapping $y \mapsto \chi_{\Omega^y}[[\nabla y]m/\det(\nabla y)](y^{-1})$ as a mapping $W^{2,p}(\Omega;\mathbb{R}^d) \to L^2(\mathbb{R}^d;\mathbb{R}^d)$. By Theo-

[13] Realize the calculus $C = F^\top F = U^\top R^\top RU = U^\top U = U^2$ for the polar decomposition of $F = RU$.

rem 2.5.3, we can rely on uniform local invertibility of y_k in the sense that $\det(\nabla y_k) \geq \varepsilon$ on $\bar{\Omega}$ for some $\varepsilon > 0$. Moreover, due to the fact that $\{y_k\} \subset W^{1,\infty}(\Omega;\mathbb{R}^d)$ is uniformly bounded we get the estimate $\sup_{k\in\mathbb{N}} |\nabla y_k|^d/\det \nabla y_k \leq \sup_{k\in\mathbb{N}} |\nabla y_k|^d/\varepsilon < \infty$. Hence, distortions K_{y_k} (see (3.4.8)) are uniformly bounded and, in view of Theorem 3.4.11, we get that y_k and consequently also y are continuous, injective everywhere, and open maps in Ω, hence homeomorphisms. Then we can use the continuity and the at most quadratic growth of $(F,m) \mapsto Fm/\det F$ on the set $\{(F,m) \in \mathbb{R}^{d\times d} \times \mathbb{R}^d;\ \det F \geq \varepsilon\}$. Thus we have proved

$$\chi_{\Omega^{y_k}}\Big[\frac{[\nabla y_k]m_k}{\det(\nabla y_k)}\Big](y_k^{-1}) \to \chi_{\Omega^y}\Big[\frac{[\nabla y]m}{\det(\nabla y)}\Big](y^{-1}) \qquad \text{weakly in } L^2(\mathbb{R}^d;\mathbb{R}^d). \tag{4.5.45}$$

We can then use the linearity of the operator $\Delta^{-1}\mathrm{div}$ considered in the weak (distributional) sense from $L^2(\mathbb{R}^d;\mathbb{R}^d) \to H^1(\mathbb{R}^d)$. Thus we know that $\phi_{y_k,m_k} \to \phi_{y,m}$ weakly in $H^1(\mathbb{R}^d)$. Eventually, realizing (4.5.42), we use the weak lower-semicontinuity of the functional $\phi \mapsto \int_{\mathbb{R}^d} |\nabla\phi(z)|^2 \mathrm{d}z$ on $H^1(\mathbb{R}^d)$. □

Results from [471] show that finer arguments based only on $\det(\nabla y) > 0$ e.a. on Ω allow for admitting $p = 2$ and, avoiding usage of [245], even for a simpler linear 2nd-grade model as far as the hyperstress concerns.

Remark 4.5.29 (Lagrangian versus mixed Eulerian/Lagrangian setting). It is noteworthy to realize the relation to the setting used in Section 3.5 which, following e.g. [260, 471], worked with the magnetization in the deformed configuration m_s. Let us now denote the energy φ used here by φ_{L} while the energy used in Section 3.5 by φ_{EL}. In contrast to (4.5.43) for φ_{L}, the frame indifference for φ_{EL} means that $\varphi_{\mathrm{EL}}(RF, R\tilde{m}) = \varphi_{\mathrm{EL}}(F,\tilde{m})$ for all $R \in \mathrm{SO}(d)$. Both approaches are mutually equivalent: taking $\varphi_{\mathrm{EL}}(F,\tilde{m}) := \tilde{\varphi}_{\mathrm{M}}(F^\top F, F^\top \tilde{m})$ with some "material" stored energy $\tilde{\varphi}_{\mathrm{M}} : \Omega \times \mathbb{R}^{d\times d} \times \mathbb{R}^d \to \mathbb{R} \cup \{+\infty\}$ and with $\tilde{m}$ a placeholder for m_s^y, cf. Remark 3.5.3, is the same as taking $\varphi_{\mathrm{L}}(F,m) := \varphi_{\mathrm{M}}(C,m) = \tilde{\varphi}_{\mathrm{M}}(C, Cm/\sqrt{\det C})$ with $m = (\det F)F^{-1}m_\mathrm{s}^y$ the pull-back magnetization according (4.5.38). Indeed,

$$\begin{aligned}\varphi_{\mathrm{L}} F,m) &:= \varphi_{\mathrm{M}}(C,m) = \tilde{\varphi}_{\mathrm{M}}\Big(C, \frac{Cm}{\sqrt{\det C}}\Big)\\ &= \tilde{\varphi}_{\mathrm{M}}\Big(F^\top F, (\det F)F^\top F\frac{F^{-1}m_\mathrm{s}^y}{\det F}\Big) = \tilde{\varphi}_{\mathrm{M}}(F^\top F, F^\top m_\mathrm{s}^y) = \varphi_{\mathrm{EL}} F,\tilde{m}).\end{aligned}$$

Alternatively, taking $\varphi_{\mathrm{EL}}(F,\tilde{m}) := \tilde{\varphi}_{\mathrm{M}}(F^\top F, F^{-1}\tilde{m})$ is the same as taking $\varphi_{\mathrm{L}}(F,m) := \varphi_{\mathrm{M}}(C,m) = \tilde{\varphi}_{\mathrm{M}}(C, m/\sqrt{\det C})$ because

$$\begin{aligned}\varphi_{\mathrm{L}}(F,m) &:= \varphi_{\mathrm{M}}(C,m) = \tilde{\varphi}_{\mathrm{M}}(C, \frac{m}{\sqrt{\det C}})\\ &= \tilde{\varphi}_{\mathrm{M}}\Big(F^\top F, (\det F)F^{-1}\frac{m_\mathrm{s}^y}{\det F}\Big) = \tilde{\varphi}_{\mathrm{M}}(F^\top F, F^{-1}m_\mathrm{s}^y) = \varphi_{\mathrm{EL}}(F,\tilde{m}).\end{aligned}$$

Anyhow, although both approaches are equivalent as far as the stored energy concerns, evolution formulated for the reference magnetization m is more amenable to mathematical analysis than for the magnetization in the deformed configuration m_s.

Remark 4.5.30 (Exchange energy in the actual configuration). There is no general agreement whether the exchange energy is to be related with the material itself as in (4.5.44) or should count with the *gradient in the space configuration* as already in (3.5.1). This may depend on particular applications, ranging from ferromagnetic crystals to ferromagnetic immersion in polymer matrix. In the latter case, $\frac{1}{2}\kappa|\nabla m|^2$ in (4.5.44) would replace by $\frac{1}{2}\kappa|\nabla((\det F)F^{-\top}m)|^2 = \frac{1}{2}\kappa|\nabla((\mathrm{Cof}\,F)m)|^2 = |(\mathrm{Cof}'F)\cdot\nabla Fm + (\mathrm{Cof}\,F)\nabla m|^2$ with $F = \nabla y$. The existence of global minimizers of (4.5.44) modified by this way is again by the direct method, exploiting the strong convergence of $\mathrm{Cof}\,F$, $\mathrm{Cof}'F$, and m, and thus the weak convergence of $(\mathrm{Cof}'\nabla y_k)\cdot\nabla^2 y_k m_k + (\mathrm{Cof}\,\nabla y_k)\nabla m_k$ for minimizing sequences $\{(y_k, m_k)\}$ provided φ is suitably coercive.[14]

Remark 4.5.31 (Weak formulation of the Poisson equation (4.5.40)). Using Lemma 9.7.1 later on p. 467 here even simplified as y^{-1} is single-valued, one can formulate (4.5.40) weakly in a lucid way as

$$\int_{\mathbb{R}^d} \mu_0 \nabla\phi_{y,m}(x^y)\cdot\nabla v(x^y)\,\mathrm{d}x^y = \int_{\Omega} m\cdot\nabla v(y(x))\,\mathrm{d}x. \tag{4.5.46}$$

Considering $v\in C^1(\mathbb{R}^d)$, one can make limit passage in this formulation even more explicitly than done in the proof of Proposition 4.5.28 in (4.5.45).

Exercise 4.5.32. Derive the Euler-Lagrange equation corresponding to the actual-gradient theory from Remark 4.5.30.

4.6 Modeling of microstructure in shape-memory materials

As we already know, in elasticity theory, it is assumed that experimentally observed patterns are minimizers or stable states of some energy. Single crystals of shape memory alloys[15] in particular have a preferred high-temperature lattice structure

[14] Actually, from this exchange-energy term, we can read the boundedness of $G := (\mathrm{Cof}'F)\cdot\nabla Fm + (\mathrm{Cof}\,F)\nabla m$ in $L^2(\Omega;\mathbb{R}^{d\times d})$, so that also $\nabla m = (\mathrm{Cof}\,F)^{-1}(G - (\mathrm{Cof}'F)\cdot\nabla Fm) = (\det F)F^\top(G - (\mathrm{Cof}'F)\cdot\nabla Fm)$ is bounded in $L^r(\Omega;\mathbb{R}^{d\times d})$ for $r < +\infty$ if φ is coercive in the sense $\varphi(F,m) \geq \varepsilon(|F|^p + |m|^q) - 1/\varepsilon$ with $p > 2$ and $q \geq 2$, so that the minimizing sequences $\{(y_k, m_k)\}$ have $\nabla((\mathrm{Cof}\,\nabla y_k)m_k) = (\mathrm{Cof}'\nabla y_k)\cdot\nabla^2 y_k m_k + (\mathrm{Cof}\,\nabla y_k)\nabla m_k$.

[15] In fact, these alloys are very special mixtures of rather precise stochiometric composition where particular atoms are connected with their neighbors by chemical bonds and thus each single crystal (in a possibly polycrystalline alloy) is a giant macromolecule. Such alloys are called intermetalics.

called austenite and a preferred low-temperature lattice structure called martensite.[16] The austenitic phase has a higher symmetry and only one phase variant while the martensitic phase exists in several symmetry related phases/variants; the mixing of these different phases can lead to the formation of complex microstructure and creation of it by cooling of austenite is called *martensitic phase transformation*. The stored energy density $\varphi : \mathbb{R}^{3\times3} \to \mathbb{R}$ is minimized on wells $\mathrm{SO}(3)U_i$, $i = 0,\dots,M$, defined by M positive definite and symmetric matrices $U_0,\dots,U_M$, each corresponding to austenite and M variants of martensite, respectively. By the choice of reference configuration, we may furthermore assume $U_0 := \mathbb{I}$ (the identity), i.e. the stress-free strain of austenite is described just by the special orthogonal group SO(3), while $\mathrm{SO}(3)U_i$, $1 \le i \le M$, denotes stress free strains of martensite. It generically holds that martensitic variants are rank-one connected, which means that there is $R \in \mathrm{SO}(3)$ (depending on i and j) such that $\mathrm{rank}\,(RU_i - U_j) = 1$ for $1 \le i < j \le M$. This indicates that φ is not rank-one convex and therefore it cannot be polyconvex. More seriously, minimizers of the functional of the elastic energy typically do not exist in Sobolev spaces.

To overcome this drawback, models have been considered where interfacial energy is taken into account. Such models have been e.g. used to estimate the scaling of the minimal energy and to derive typical length scales of patterns. The minimal scaling of the energy of an austenite-martensite interface has been studied by R.V. Kohn, S. Müller, and S. Conti in [125, 279, 280] for a two-dimensional model problem, the three-dimensional case and more realistic models have been investigated e.g. in [103, 104, 110, 275, 276, 545]. In these models, either a BV-penalization of the interfacial has been used or a penalization of some L^p–norm for the Hessian of the deformation function. In general, the specific form of the energy is, however, not clear from physical considerations. In the literature, necessary and sufficient conditions for the specific form of the interfacial energy have been investigated recently which allow for the existence of minimizers [189, 408]. Recently, M. Šilhavý has introduced a notion of interface polyconvexity and has proved that this notion is sufficient to ensure existence of minimizers for the corresponding static problem [484, 485].

If interfacial energy is not taken into account, then global minimizers of the energy in general do not exist. A way out is to use relaxation methods, searching for the so-called quasiconvex envelope of the specific stored energy [135, 381] or using Young measures [271, 272, 294, 359]; cf. Sect. D.1 on p. 549. These methods, however, require that the φ has polynomial growth and coercivity which prevents us of including orientation preservation and physically relevant growth of φ. Note that $F \mapsto F^\top F$ maps the whole group O(3) of orthogonal matrices with determinant ±1 onto the same point. Thus, for example, $F \mapsto |F^\top F - \mathbb{I}|$ is minimized on two energy wells, i.e., on SO(3) and also on $\mathrm{O}(3)\backslash\mathrm{SO}(3)$. However, the latter set is not acceptable in elasticity since corresponding deformations do not preserve the orientation. Additionally, notice that, for example, considering arbitrary $Q \in \mathrm{O}(3) \setminus \mathrm{SO}(3)$ and an arbitrary $R \in \mathrm{SO}(3)$ such that Q and R are rotations around the same axis of the

[16] Such shape memory alloys, as e.g. Ni-Ti, Cu-Al-Ni, Ni-Mn-Ga, or In-Th, have various technological applications, often in their polycrystalline forms; for an overview see e.g. [262].

Cartesian system then $\text{rank}(Q-R)=1$, i.e. Q and R are rank-one connected and determinant changes its sign on the line segment $[Q;R]$. Convex combinations of rank-one connected matrices play a key role in relaxation approaches to the variational calculus [32, 33, 135, 293]. This shows that is it important but also not straightforward to ensure that solutions are physically sound, in the sense that they preserve orientation.

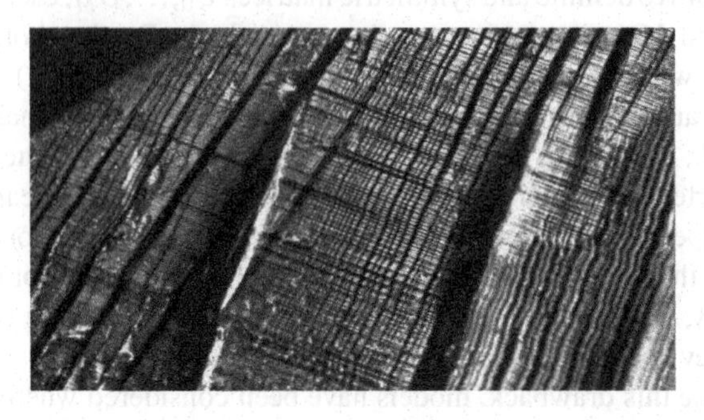

Fig. 4.6 Microstructure in orthorombic martensite in a single-crystal of Cu-Al-Ni as observed in an optic microscope.
Courtesy of Hanuš Seiner (Institute of Thermomechanics, Czech Academy of Sciences).

A stored energy density of a shape-memory alloy is modeled as a nonnegative function which is zero if and only if the deformation gradient $F \in \text{SO}(d)U_i$, $i = 0,\dots M$, where U_i are symmetric matrices representing the stress-free strains of the i-th variant of martensite (if $i > 0$) or the austenite (if $i = 0$). One possible choice is to consider

$$\varphi(F) := \min_i \varphi_i(F), \qquad (4.6.1)$$

where $\varphi_i \geq 0$ is a stored energy density of a martensite or the austenite. Consequently, φ is generically non-quasiconvex and the existence of minimizers to the corresponding elastic energy cannot be guaranteed. A natural option is to search for a relaxed problem in the sense of Section 4.4. or to modify φ resort to nonsimple materials as e.g., 2nd-grade materials or gradient-polyconvex ones. Here we review such basic approaches to mathematical problems of shape memory materials. The first one is about the extension of a notion of solutions from Sobolev maps to measures and it is discussed in Subsection 4.6.1.

4.6.1 Twinning equation

If we do not want to add any kind of surface energy to our problem and want to work exclusively with simple materials then existence of minimizers cannot be guaranteed in general for non-quasiconvex energy densities. On the other hand, to calculate $Q\varphi$, the quasiconvex envelope of φ from (4.6.1) is usually out of reach. Nevertheless, we can still get some information about it. In particular, we know that if $\varphi(A)=\varphi(B)=0$ for two matrices $\mathbb{R}^{d\times d}$ and $\mathrm{rank}(A-B)=1$ then $Q\varphi(\lambda A+(1-\lambda)B)=0$ for every $0\le\lambda\le1$ because $Q\varphi$ is rank-one convex as proved in Theorem 4.1.3. In order to motivate our next proposition, assume that we have a smooth deformation $y:\bar{\Omega}\to\mathbb{R}^d$ whose gradient, ∇y, takes precisely two values $F_1,F_2\in\mathbb{R}^{d\times d}$ and, moreover, let regions of different gradient values are separated by a planar interface with the unit normal $\vec{n}\in\mathbb{R}^d$. We get by continuity of y that planar gradient must be continuous, in other words, $F_1(\mathbb{I}-\vec{n}\otimes\vec{n}) = F_2(\mathbb{I}-\vec{n}\otimes\vec{n})$. This means $F_1-F_2=(F_1-F_2)\vec{n}\otimes\vec{n}$,i.e., $\mathrm{rank}(F_1-F_2)\le1$, i.e., the surface gradient is continuous on $\bar{\Omega}$. Denoting $a:=(F_1-F_2)\vec{n}\in\mathbb{R}^d$, we get that

$$F_1-F_2=a\otimes\vec{n}. \tag{4.6.2}$$

Proposition 4.6.1 (Double-valued gradient [32]). *Let $\Omega\subset\mathbb{R}^d$ be open and connected and let $y\in W^{1,\infty}(\Omega;\mathbb{R}^d)$ be such that, for some $F_1,F_2\in\mathbb{R}^{d\times d}$, it holds*

$$\nabla y(x)=\begin{cases}F_1 & \text{if } x\in\Omega_{F_1}\subset\Omega,\\ F_2 & \text{if } x\in\Omega\setminus\Omega_{F_1}.\end{cases} \tag{4.6.3}$$

Let $\mathrm{meas}_d(\Omega_{F_1})<\mathrm{meas}_d(\Omega)$. Then $F_1-F_2=a\otimes\vec{n}$ for some $a,\vec{n}\in\mathbb{R}^d$. Moreover, $y(x)=y_0+F_2x+\theta(x)\vec{n}$ where $y_0\in\mathbb{R}^d$, $y_0\cdot a=0$, and $\theta\in W^{1,\infty}(\Omega)$ is such that $\nabla\theta=\chi_{\Omega_{F_1}}\vec{n}$ where $\chi_{\Omega_{F_1}}$ is the characteristic function of Ω_{F_1} in Ω.

Proof. We follow [32, Proof of Prop. 1]. Denote $z(x):=y(x)-F_2x$ for all $x\in\Omega$. Hence, $\nabla z=\nabla y-F_2=\chi_{\Omega_{F_1}}(F_1-F_2)$. Let $\eta\in C_0^\infty(\Omega)$ be such that $\int_{\Omega_{F_1}}\nabla\eta(x)\,\mathrm{d}x\neq0$. As ∇z is curl-free in the sense of distributions we have

$$0=\int_\Omega\Big(\frac{\partial z_i}{\partial x_j}\frac{\partial\eta}{\partial x_k}-\frac{\partial z_i}{\partial x_k}\frac{\partial\eta}{\partial x_j}\Big)\mathrm{d}x=((F_1)_{ij}-(F_2)_{ij})n_k-((F_1)_{ik}-(F_2)_{ik})n_j\,\mathrm{d}x,$$

where

$$\vec{n}:=\frac{\int_{\Omega_{F_1}}\nabla\eta(x)\,\mathrm{d}x}{\big|\int_{\Omega_{F_1}}\nabla\eta(x)\,\mathrm{d}x\big|}.$$

Hence, F_1-F_2 is a rank-one matrix and we see that $F_1-F_2=a\otimes\vec{n}$ where $a:=(F_1-F_2)\vec{n}$. Moreover, $\nabla z=\chi_{\Omega_{F_1}}(F_1-F_2)=a\otimes\vec{n}$. If $b\in\mathbb{R}^d$ is such that $a\cdot b=0$ then also $\nabla(z\cdot b)=0$. Hence, $z(x)-z(x_0)$ is parallel to a where $x_0\in\Omega$ is fixed. As we can assume that $a\neq0$, we get the representation $y(x)=y_0+F_2x+\theta(x)\vec{n}$ for $\nabla\theta(x)=\chi_{\Omega_{F_1}}(x)\vec{n}$ for almost all $x\in\Omega$. □

We see that if the gradient of a Lipschitz map takes only two different values on Ω then the interface between them must be planar, the *surface gradient* is continuous across the interface and the difference of the two values is a matrix of rank one. As we already know, the *principle of frame indifference* requires that a stored energy density φ is minimized on energy wells and not at isolated points. Thus, a question arises whether there exist planar interfaces between various variants of martensite or between martensite and austenite. In other words, given invertible $F_1, F_2 \in \mathbb{R}^{d\times d}$, we search for two rotations $R_1, R_2 \in \mathrm{SO}(d)$ and for two vectors $a, \vec{n} \in \mathbb{R}^d$ such that the following *twinning equation* is satisfied:

$$R_1F_1 - R_2F_2 = a\otimes\vec{n}. \tag{4.6.4}$$

First of all, we can assume that F_1 and F_2 are symmetric for otherwise we apply polar decompositions which is unique as F_1 and F_2 are invertible. Premultiplying (4.6.4) by $R_2^\top$ and postmultiplying it by F_2^{-1} we get

$$R_2^\top R_1 F_1 F_2^{-1} = \mathbb{I} + b\otimes m, \tag{4.6.5}$$

where $b := R_2^\top a$ and $m := F_2^{-\top}\vec{n} = F_2^{-1}\vec{n}$. Putting $C := F_1^{-1}F_2^2F_1^{-1}$ the previous equation reads

$$C = (\mathbb{I} + m\otimes b)(\mathbb{I} + b\otimes m). \tag{4.6.6}$$

Proposition 4.6.2 (Solvability of the twinning equation [32]). *Let $d = 3$ and let $C \in \mathbb{R}^{d\times d} \neq \mathbb{I}$ be symmetric with eigenvalues $\lambda_1 \le \lambda_2 \le \lambda_3$. Then there are vectors $b, m \in \mathbb{R}^3$ such that* (4.6.6) *holds if and only if $0 \le \lambda_1$ and $\lambda_2 = 1$. Moreover, solutions satisfying* $\det(\mathbb{I} + b\otimes m) > 0$ *are given by the following formulas:*

$$b = \alpha\Big(\sqrt{\frac{\lambda_3(1-\lambda_1)}{\lambda_3-\lambda_1}}e_1 + \beta\sqrt{\frac{\lambda_1(\lambda_3-1)}{\lambda_3-\lambda_1}}e_3\Big) \quad \text{and}$$
$$m = \frac{1}{\alpha}\Big(\frac{\sqrt{\lambda_3}-\sqrt{\lambda_1}}{\sqrt{\lambda_3-\lambda_1}}\Big)\big(-\sqrt{1-\lambda_1}e_1 + \beta\sqrt{\lambda_3-1}e_3\big), \tag{4.6.7}$$

where $\alpha \neq 0$, e_1, e_3 are unit eigenvectors corresponding to λ_1 and λ_3, respectively, and $\beta \in \{-1, 1\}$.

Proof. We follow [32]. We first show necessity. If $v \in \mathbb{R}^3$ is perpendicular to b as well as to m then $Cv = v$, i.e., C has the unit eigenvalue. The quadratic form

$$v \mapsto (C-\mathbb{I}){:}(v\otimes v) = (v\cdot m)(2(v\cdot b) + |b|^2(v\cdot m))$$

is indefinite so that eigenvalues of $C-\mathbb{I}$ are both positive and negative. Consequently, $\lambda_1 < 1 < \lambda_3$. As $0 < \det C = \lambda_1\lambda_3$ we get that all eigenvalues of C must be positive.

Now we prove sufficiency. As $C \neq \mathbb{I}$ it follows that $b \neq 0$. If b and m satisfy (4.6.6) then $Cb = \pm\sqrt{\det C}(b + |b|^2 m)$. Then $m = (\pm C/\sqrt{\det C} - \mathbb{I})b/|b|^2$. Altogether,

$$C = \mathbb{I} + \frac{1}{|b|^2}\Big(\frac{Cb\otimes Cb}{\det C} - b\otimes b\Big). \tag{4.6.8}$$

Expressing all terms involved in (4.6.8) in the orthonormal basis of eigenvectors of C we have $b := (b_1, b_2, b_3)$ and $Cb = (\lambda_1 b_1, b_2, \lambda_3 b_3)$, and $C = \mathrm{diag}(\lambda_1, 1, \lambda_3)$. Therefore, necessarily $b_2 = 0$, and

$$\frac{b_1^2}{|b|^2} = \frac{\lambda_3(1-\lambda_1)}{\lambda_3-\lambda_1} \quad \text{and} \quad \frac{b_3^2}{|b|^2} = \frac{\lambda_1(\lambda_3-1)}{\lambda_3-\lambda_1},$$

which concludes the proof because (4.6.7) immediately follows. □

Exercise 4.6.3. Decide whether $\vec{n}$ in the proof of Proposition 4.6.1 depends on the choice of η.

Exercise 4.6.4 (Solvability of the twinning equation for $d = 2$). Show that if $d = 2$ then a necessary and sufficient condition on $F_1, F_2 \in \mathbb{R}^{2\times 2}$ so that (4.6.4) has a solution is that $|F_1 F_2^{-1}|^2 \geq 1 + \det^2 F_1 F_2^{-1}$. In particular, if $\det F_1 = \det F_2$ a solution always exists because $|F_1 F_2^{-1}|^2 \geq 2\det F_1 F_2^{-1} = 2$.

4.6.2 Rank-one convex envelope

Mathematical modeling of shape memory alloys is a vibrant research area where various relaxation procedures developed in Section 4.4 can be used. Obviously, even if stored energy densities of single variants of the martensite and of the austenite are polyconvex, formula (4.6.1) generically destroys polyconvexity properties of φ. Hence, if we formulate a corresponding minimization problem as in (4.4.6) then nonexistence of a solution is expected. If φ_i satisfies (4.4.3) and (4.4.4) for every $i = 0, \ldots, M$ and the functional $\mathcal{F}$ representing work of external forces fulfills (4.4.5) we can formulate the relaxed problem as in (4.4.14) and Theorem 4.4.3 can be readily applied. Equivalently, we can formulate the relaxed problem by means of the quasiconvex envelope of φ, $\mathrm{Q}\varphi$, as in (4.4.42) and use Theorem 4.4.7. Nevertheless, both approaches are purely theoretical and are of a very limited use in numerical approximations, for instance. Stress-free and zero energy (i.e. natural) states of martensitic variants typically satisfy the twinning equation (4.6.4) with U_i and U_j with $i \neq j$ in the place of F_1 and F_2. This provides us with some information about $\mathrm{Q}\varphi$, namely that $\mathrm{Q}\varphi = 0$ on the rank-one connected line segment joining energy wells $R_i U_i$ and $R_j U_j$ for some $R_i, R_j \in \mathrm{SO}(d)$, i.e., on the segment whose endpoints satisfy

$$\forall 1 \leq i < j \leq M: \qquad \mathrm{rank}(R_i U_i - R_j U_j) = 1. \tag{4.6.9}$$

On the other hand, if we define the *rank-one convex envelope*, $\mathrm{R}\varphi$, of $\varphi : \mathbb{R}^{d\times d} \to \mathbb{R}$ as the largest rank-one convex function not greater than φ, i.e.,

$$\mathrm{R}\varphi := \sup\{\psi : \mathbb{R}^{d\times d} \to \mathbb{R} \text{ rank-one convex}; \psi \leq \varphi\}. \tag{4.6.10}$$

we can define a variational problem for an energy functional with the stored energy density $\mathrm{R}\varphi$. As quasiconvexity implies rank-one convexity then $\mathrm{R}\varphi \ge \mathrm{Q}\varphi$ and this inequality can be strict for particular functions and points as there are rank-one convex functions which are not quasiconvex, cf. Theorem 4.1.7. Let (4.4.3), (4.4.4), (4.4.5), and (4.4.7) hold where $\mathcal{F} : W^{1,p}(\Omega;\mathbb{R}^d) \to \mathbb{R}$ is weakly continuous. We define

$$\mathcal{E}_{\mathrm{R}}(y) := \int_\Omega \mathrm{R}\varphi(\nabla y(x))\,\mathrm{d}x - \mathcal{F}(y) \tag{4.6.11}$$

and the problem

$$\left.\begin{array}{l}\text{minimize } \mathcal{E}_{\mathrm{R}}(y),\\ \text{subject to } y \in \mathcal{A}.\end{array}\right\} \tag{4.6.12}$$

Problem (4.6.12), in general, does not posses any solution because $\mathcal{E}_{\mathrm{R}}$ is not weakly lower semicontinuous on $W^{1,p}(\Omega;\mathbb{R}^d)$. On the other hand, in view of Theorem 4.4.3 we immediately see that

$$\inf_{y\in\mathcal{A}} \mathcal{E}(y) = \inf_{y\in\mathcal{A}} \mathcal{E}_{\mathrm{R}}(y) = \min_{y\in\mathcal{A}} \mathcal{E}_{\mathrm{Q}}(y). \tag{4.6.13}$$

Thus, (4.6.12) can correctly describe the energetic minimum of the relaxed problem. Moreover, the rank-one convex envelope is computationally easier to handle due to the following result.

Proposition 4.6.5 (Rank-one convex envelope). [17] *Let $\varphi : \mathbb{R}^{d\times d} \to [0,+\infty)$. Then, for every $A \in \mathbb{R}^{d\times d}$, it holds*

$$\mathrm{R}\varphi(A) = \lim_{k\to\infty} \mathrm{R}_k\varphi(A), \tag{4.6.14}$$

where $\mathrm{R}_0\varphi := \varphi$ and

$$\mathrm{R}_{k+1}\varphi(A) := \inf\Big\{\lambda \mathrm{R}_k\varphi(A_0) + (1-\lambda)\mathrm{R}_k\varphi(A_1);\ 0 \le \lambda \le 1,\\ A = \lambda A_0 + (1-\lambda)A_1,\ \mathrm{rank}(A_1 - A_0) \le 1\Big\},\quad k \in \mathbb{N}\cup\{0\}. \tag{4.6.15}$$

Proof. We essentially follow the proof of [135, Thm. 6.10]. Notice that $\mathrm{R}_k\varphi \ge \mathrm{R}_{k+1}\varphi \ge 0$ for all $k \in \mathbb{N}$. Let us denote $\mathrm{S}\varphi := \inf_{k\in\mathbb{N}} \mathrm{R}_k\varphi$. It follows fro the definition that $\mathrm{S}(\mathrm{R}\varphi) = \mathrm{R}\varphi$ and, consequently, $\mathrm{R}\varphi \le \mathrm{S}(\mathrm{R}\varphi) \le S\varphi \le \varphi$. Hence, it remains to show that $\mathrm{S}\varphi$ is rank-one convex. Take $A_0, A_1 \in \mathbb{R}^{d\times d}$ and $\lambda \ge 0$ as in the proposition, and $\epsilon > 0$ arbitrary. By the definition of the infimum, there is $i \in \mathbb{N}$ such that $\mathrm{S}\varphi(A_n) + \epsilon > \mathrm{R}_i\varphi(A_n)$ for $n = 0, 1$. Therefore,

$$\lambda \mathrm{S}\varphi(A_0) + (1-\lambda)\mathrm{S}\varphi(A_1) + \epsilon \ge \lambda \mathrm{R}_i\varphi(A_0) + (1-\lambda)\mathrm{R}_i\varphi(A_1)\\ \ge \mathrm{R}_{i+1}\varphi(\lambda A_0 + (1-\lambda)A_1) \ge \mathrm{S}\varphi(\lambda A_0 + (1-\lambda)A_1).$$

Arbitrariness of $\epsilon > 0$, of $A_0, A_1 \in \mathbb{R}^{d\times d}$, and of $\lambda \in [0,1]$ shows rank-one convexity of $S\varphi$. □

[17] See [277, Part II], or also [135, Sec. 6.4].

Although Proposition 4.6.5 provides us with an algorithm how to calculate $\mathrm{R}\varphi$, the main drawback is that, generally, $\mathrm{R}\varphi \neq \mathrm{R}_k\varphi$ for every $k \in \mathbb{N}$. We refer to [338] for further details related to this issue. Nevertheless, in view of Proposition 4.6.1 one can approximate $\mathrm{R}_k\varphi$ in terms of a laminated microstructure similar to the one on Figure 4.6. This motives us to consider the following minimization problem for $k \geq 0$:

$$\left.\begin{array}{l}\text{minimize } \mathcal{E}_{\mathrm{R}_k}(y),\\ \text{subject to } y \in \mathcal{A},\end{array}\right\} \tag{4.6.16}$$

where

$$\mathcal{E}_{\mathrm{R}_k}(y) := \int_\Omega \mathrm{R}_k\varphi(\nabla y(x))\,\mathrm{d}x - \mathcal{F}(y). \tag{4.6.17}$$

While (4.6.16) is not solvable in a generic situation, it still provides us with the correct value of the minimum attained in the relaxed problem (4.4.42). Indeed, we again have the following chain of equalities which holds for every $k \geq 0$:

$$\inf_{y\in\mathcal{A}} \mathcal{E}(y) = \inf_{y\in\mathcal{A}} \mathcal{E}_{\mathrm{R}_k}(y) = \inf_{y\in\mathcal{A}} \mathcal{E}_{\mathrm{R}}(y) = \min_{y\in\mathcal{A}} \mathcal{E}_{\mathrm{Q}}(y). \tag{4.6.18}$$

We will now explain how one can formulate (4.6.16) in a way more convenient for numerical solution based on approximation of microstructures by iterated laminates, having thus a specific interpretation and allowing for aposteriori reconstruction of (at least certain types of) microstructure. We will confine ourselves on the the 2nd-order laminate which means that we will stick to $k = 2$. Assume that we are given $y \in \mathcal{A}$ and we want to evaluate $R_2\varphi(\nabla y)$. To this end, we find two matrices $A_0, A_1 \in L^p(\Omega;\mathbb{R}^{d\times d}$ with $\mathrm{rank}(A_0 - A_1) \leq 1$ and $\lambda \in L^\infty(\Omega)$ such that $0 \leq \lambda \leq 1$ such that

$$\nabla y = \lambda A_0 + (1-\lambda)A_1\,. \tag{4.6.19}$$

We can equivalently write $A_1 - A_0 = q\otimes r$, where $q \in L^p(\Omega;\mathbb{R}^d)$ and $r \in L^\infty(\Omega;\mathbb{R}^d)$ with $|r| = 1$. Then A_0 and A_1 can be calculated from (4.6.19) as $A_0 = \nabla y - (1-\lambda)q\otimes r$ and $A_1 = \nabla y + \lambda q\otimes r$. We continue further with a decomposition of A_0 and A_1. For A_0, A_1 given as above, write convex combinations

$$A_0 = \lambda_0 A_{00} + (1-\lambda_0)A_{01}\,,\quad A_1 = \lambda_1 A_{10} + (1-\lambda_1)A_{11}\,, \tag{4.6.20}$$

where $A_{01} - A_{00} = q_0\otimes r_0$, and $A_{11} - A_{10} = q_1\otimes r_1$ with $\lambda_i \in L^\infty(\Omega)$ with $0 \leq \lambda_i \leq 1$, $q_i \in L^p(\Omega;\mathbb{R}^d)$ and $r_i \in L^\infty(\Omega;\mathbb{R}^d)$ with $|r_i| = 1$ for $i = 0, 1$. We then have

$$\begin{aligned}\mathrm{R}_2\varphi(\nabla y) = \inf\{&\lambda\lambda_0\varphi(A_{00}) + \lambda(1-\lambda_0)\varphi(,A_{01})\\ &+ (1-\lambda)\lambda_1\varphi(A_{10}) + (1-\lambda)(1-\lambda_1)\varphi(A_{11})\}\end{aligned} \tag{4.6.21}$$

where

$$A_{00} = \nabla y - (1-\lambda)q\otimes r - (1-\lambda_0)q_0\otimes r_0\,,\ A_{01} = \nabla y - (1-\lambda)q\otimes r + \lambda_0 q_0\otimes r_0$$

and

$$A_{10} = \nabla y + \lambda q\otimes r - (1-\lambda_1)q_1\otimes r_1\,,\ A_{11} = \nabla y + \lambda q\otimes r + \lambda_1 q_1\otimes r_1\,,$$

and the infimum is taken over all unit vectors $r, r_0, r_1 \in L^\infty(\Omega;\mathbb{R}^d)$, all vectors $q, q_0, q_1 \in L^p(\Omega;\mathbb{R}^d)$, and all $\lambda, \lambda_0, \lambda_1 \in L^\infty(\Omega)$ ranging over the interval $[0,1]$. Moreover, it holds for all test fields competing in (4.6.21) that

$$\nabla y = \lambda\lambda_0 A_{00} + \lambda(1-\lambda_0)A_{01} + (1-\lambda)\lambda_1 A_{10} + (1-\lambda)(1-\lambda_1)A_{11}, \qquad (4.6.22)$$

so that ∇y is the weighted average of A_{00}, A_{01}, A_{10}, and A_11. Here ∇y is a macroscopic gradient which is calculated by means of microscopic gradients taking values A_{00}, A_{01}, A_{10}, and A_11 with probabilities corresponding to the coefficients in (4.6.22). Vectors r_0 and r_1 are normal to the interface created by adjacent values A_{00}, A_{01} and A_{10}, A_{11}, respectively. Finally, r is the normal vector to the interface created by the two first-order laminates as depicted in Figure 4.7. Obviously, we can iteratively repeat this procedure to obtain laminates of an arbitrary order. On the other hand, problems (4.6.16) do not introduce any length scale to the problem and laminated patterns just reflect oscillations of $\{\nabla y_j^k\}_{j\in\mathbb{N}}$ where $\{y_j^k\}_{j\in\mathbb{N}} \subset \mathcal{A}$ is a minimizing sequence of (4.6.16). Values of λ, λ_0, and λ_1 provide us, however, with correct volume fractions of particular martensitic phases or of the austenite. Another drawback is that relaxation is known only if φ has a polynomial growth as in (4.4.4) in Section 4.4.1. This prevent us for using mechanically relevant energy densities satisfying constraints on the determinant of the deformation gradient.

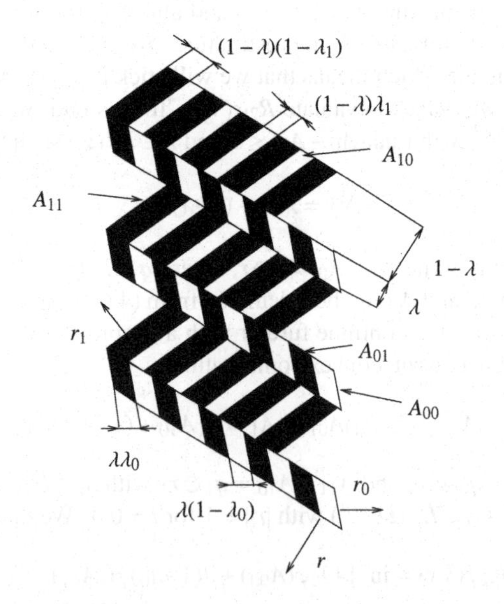

Fig. 4.7 2nd-order laminated microstructure and problem variables.

Computational studies with this approach for $k > 0$ can be found e.g. in [21, 57, 290, 293, 294, 396] using piecewise affine approximation of y and piecewise constant approximation of maps involved on the right-hand side of (4.6.21). See Figure 4.8 for a computational result on stress-induced transformation in NiTi.

Many computational and numerical studies were also done if $k = 0$ in (4.6.16), i.e., for the original minimization problem. First results appeared in [123, 124] and in [113, 274, 320, 321, 322].

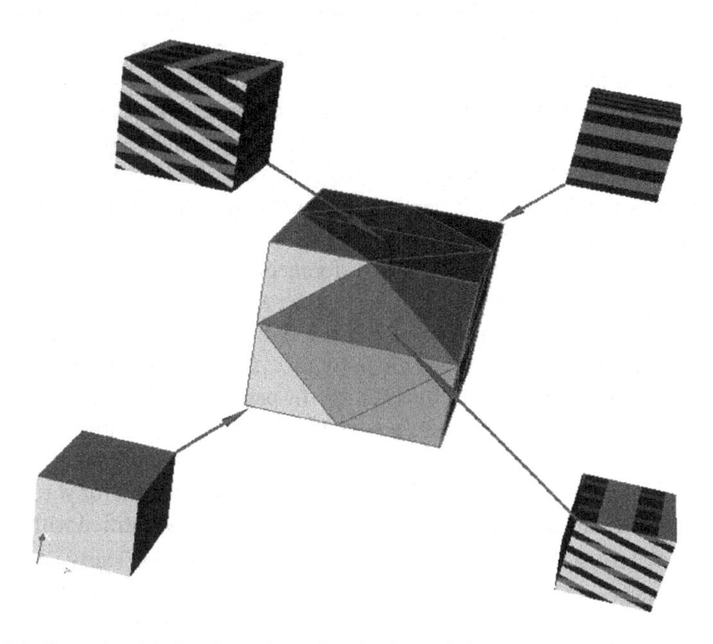

Fig. 4.8 A 3-dimensional 2nd-order x-dependent laminate (microstructure) schematically reconstructed at four different spots from the calculated elementwise constant Young measure as in Fig. 4.7 induced by mechanical loading.
Courtesy of Barbora Benešová (University of Würzburg).

Remark 4.6.6 (Relation to Young measures). If we define for almost every $x \in \Omega$ a probability measure

$$\begin{aligned}\nu_x := {} & \lambda(x)\lambda_0(x)\delta_{A_{00}(x)} + \lambda(x)(1-\lambda_0(x))\delta_{A_{01}(x)} \\ & + (1-\lambda(x))\lambda_1(x)\delta_{A_{10}(x)} + (1-\lambda(x))(1-\lambda_1(x))\delta_{A_{11}(x)},\end{aligned} \tag{4.6.23}$$

where $A_{00}, A_{01}, A_{10}, A_{11}, \lambda, \lambda_0$, and λ_1 are as in (4.6.22) then $\nu := \{\nu_x\}_{x \in \Omega}$ is a gradient Young measure called a laminate; cf. Section 4.4.1 for further details. Let us notice that its first moment $\bar{\nu} = \nabla y$ and that, for every *rank-one convex* function $\psi : \mathbb{R}^{d\times d} \to \mathbb{R}$ and almost every $x \in \Omega$, it holds

$$\int_{\mathbb{R}^{d\times d}} \psi(A)\nu_x(\mathrm{d}A) \geq \psi(\nabla y(x)). \tag{4.6.24}$$

Exercise 4.6.7. Verify formula (4.6.24).

4.6.3 Problems involving an interfacial energy

Recently a new static model of such materials appeared in [484, 485] which combines polyconvex stored energy densities of the austenite and single martensitic variants with a newly defined polyconvexity of interfacial energy between variants of martensite and austenite. Volume fractions of martensitic variants and of austenite play a role of additional design variables. Moreover, the resulting material microstructures have finite length scales. We will state the result for $d = 3$ and we invite the interested reader to reformulate the problem for $d = 2$ as an exercise. In contrast to the models in Section 4.6.2, here we can easily impose *orientation-preservation* of deformations as well as *injectivity*.

We assume that the specimen in its reference configuration is represented by a bounded Lipschitz domain $\Omega \subset \mathbb{R}^3$. We consider a shape memory alloy which again allows for M different variants of martensite. We assume that the region occupied by the i-th variant of martensite is given by $\Omega_i \subset \Omega$ for $1 \le i \le M$, while the region occupied by austenite is given by $\Omega_0 \subset \Omega$. In particular, the sets Ω_i are pairwise disjoint for $0 \le i \le M$. We also assume that the sets Ω_i are open, that $N := \Omega \bigcup_i \Omega_i$ is a zero (Lebesgue) measure set and that each Ω_i is of a *finite perimeter*, i.e., that its characteristic function has a bounded variation. The case $\Omega_i = \emptyset$ for some $0 \le i \le M$ is not excluded. The partition of Ω into $\{\Omega_i\}_{i=0}^M$ can be then identified with a mapping $z : \Omega \to \mathbb{R}^{M+1}$ such that $z_i(x) = 1$ if $x \in \Omega_i$ and $z_i(x) = 0$ else. Clearly, $\sum_{i=0}^M z_i(x) = 1$ for almost every $x \in \Omega$. We will call z the partition map corresponding to $\{\Omega_i\}_{i=0}^M$. Furthermore, we assume that the sets Ω_i have finite perimeter and equivalently the function z is of bounded variation. We hence consider $z \in \mathrm{BV}(\Omega, \mathbb{R}^{M+1})$ such that $z_i(x) \in \{0, 1\}$, $z_i(x)z_j(x) = 0$ for $i \neq j$, and $\sum_{i=0}^M z_i(x) = 1$ for a.a. $x \in \Omega$.

The total bulk energy of the specimen is then considered in the form

$$E_{\mathrm{b}}(y,z) := \int_\Omega \hat{\varphi}(\nabla y(x), z(x)) \, \mathrm{d}x \tag{4.6.25}$$

with the stored energy density $\hat{\varphi} : \mathbb{R}^{3\times 3} \times \mathbb{R}^{M+1} \to \mathbb{R} \cup \{+\infty\}$ of the specimen written as

$$\hat{\varphi}(F,z) := \sum_{i=0}^{M} z_i \varphi_i(F),$$

where φ_i, $0 \le i \le M$ is again the stored energy density of the i-th phase of the material. We assume that each φ_i is polyconvex hence in view of (3.1.1) we assume that

$$\varphi_i(F) := \begin{cases} \tilde{\varphi}_i(F, \mathrm{Cof}\, F, \det F) & \text{if } \det F > 0, \\ +\infty & \text{otherwise.} \end{cases} \tag{4.6.26}$$

for convex and lower semicontinuous functions $\tilde{\varphi}_i : \mathbb{R}^{19} \to \mathbb{R} \cup \{+\infty\}$. We use the following additional standard assumptions on the specific bulk energies φ_i[18]. For

[18] More generally, we could also assume (2.3.28), for instance.

$0 \le i \le M$, we assume that, for some $\varepsilon > 0$ and $p > 3$, it holds

$$\forall F \in \mathbb{R}^{3\times 3}: \quad \varphi_i(F) \ge \varepsilon |F|^p - 1/\varepsilon, \tag{4.6.27a}$$

$$\forall R \in \mathrm{SO}(3),\ F \in \mathbb{R}^{3\times 3}: \ \varphi_i(RF) = \varphi_i(F), \tag{4.6.27b}$$

$$\lim_{\det F \to 0_+} \varphi_i(F) = +\infty. \tag{4.6.27c}$$

We will consider the interfacial energy as the one introduced by M. Šilhavý in [485]. Let $\mathbb{F}$ be a placeholder for a surface gradient, $\nabla_{\mathrm{S}} y$, of the deformation $y : \Omega \to \mathbb{R}^3$. We assume that the specific interfacial energy f_{ij} between the two different phases $i, j \in \{0, \ldots, M\}$ with $n \in \mathbb{R}^3$ the unit normal to the mutual interface can be written in the form

$$f_{ij}(\mathbb{F}, n) = \hat{f}_i(\mathbb{F}, n) + \hat{f}_j(\mathbb{F}, n), \tag{4.6.28}$$

where $n \in \mathbb{R}^3$ is a unit vector such that $\mathbb{F}n = 0$. This is, indeed, possible because $\det \nabla_{\mathrm{S}} y = \det \mathbb{F} = 0$. We assume

$$\hat{f}_i(\mathbb{F}, n) := \psi_i^{\mathrm{R}}(n, \mathbb{F} \times n, \mathrm{Cof}\,\mathbb{F} n), \tag{4.6.29}$$

where the functions $\psi_i^{\mathrm{R}} : \mathbb{R}^{3\times 3\times 3} \to \mathbb{R}$ are nonnegative convex and positively one-homogeneous for $i = 1, \ldots, M$. Here, $F \times n : \mathbb{R}^3 \to \mathbb{R}^3$ is for any $F \in \mathbb{R}^{3\times 3}$ and any $n, a \in \mathbb{R}^3$ defined as $(F \times n)a := F(n \times a)$. As in [485], we assume for $0 \le i \le M$ that

$$\forall F \in \mathbb{R}^{3\times 3},\ n \in \mathbb{R}^3,\ |n|=1,\ R \in \mathrm{SO}(3): \ \hat{f}_i(RF, \vec{n}) = \hat{f}_i(F, \vec{n}) \ \text{ and } \ \hat{f}_i(F, n) = \hat{f}_i(F, -n). \tag{4.6.30}$$

We introduce a subspace of $W^{1,p}(\Omega;\mathbb{R}^3) \times \mathrm{BV}(\Omega;\mathbb{R}^{M+1})$ of functions with "finite interfacial energy", using a slightly modified version of [484, Def. 3.1]. It is given as follows:

Definition 4.6.8 (States with finite interfacial energy). For any pair $y, z \in W^{1,p}(\Omega;\mathbb{R}^3) \times \mathrm{BV}(\Omega;\mathbb{R}^{M+1})$ let $S_i = \partial^* \Omega_i \cap \Omega$ where $\partial^* \Omega_i$ is the measure-theoretic boundary of Ω_i with the outer measure-theoretic normal n_i. We denote by $\mathcal{U} \subset W^{1,p}(\Omega;\mathbb{R}^3) \times \mathrm{BV}(\Omega;\mathbb{R}^{M+1})$ the set of all pairs $(y, z) \in W^{1,p}(\Omega;\mathbb{R}^3) \times \mathrm{BV}(\Omega;\mathbb{R}^{M+1})$ such that for every $0 \le i \le M$ there exists a measure $J_i := (b_i, H_i, p_i) \in \mathfrak{M}(\Omega;\mathbb{R}^3) \times \mathfrak{M}(\Omega;\mathbb{R}^{3\times 3}) \times \mathfrak{M}(\Omega;\mathbb{R}^3)$ where for every $v \in C_0(\Omega;\mathbb{R}^3)$

$$\begin{aligned} b_i &:= n_i \mathscr{H}^2|_{S_i}, \\ \int_\Omega v H_i(\mathrm{d}x) &:= -\int_{\Omega_i} \nabla y(x) \,\mathrm{curl}\, v(x) \,\mathrm{d}x, \ \text{ and} \\ \int_\Omega v p_i(\mathrm{d}x) &:= \int_{\Omega_i} \mathrm{Cof}\,\nabla y(x) : \nabla v(x) \,\mathrm{d}x. \end{aligned} \tag{4.6.31}$$

With the notation of Definition 4.6.8, we define the interfacial energy as

$$\mathcal{E}_{\rm int}(y,z) := \begin{cases} \sum_{i=0}^{M} \int_{\Omega} \psi_i^{\rm R}\left(\frac{{\rm d}J_i}{{\rm d}|J_i|}\right) |J_i|({\rm d}x) & \text{for } (y,z) \in \mathcal{U}, \\ +\infty & \text{else.} \end{cases} \tag{4.6.32}$$

Here $|J_i|$ denotes the total variation of the measure J_i for $0 \le i \le M$. Finally, we will assume that there is $\varepsilon > 0$ such that for $0 \le i \varepsilon M$ it holds that

$$\forall A \in R^3 \times \mathbb{R}^{3\times3} \times \mathbb{R}^3 : \ \psi_i^{\rm R}(A) \ge \varepsilon |A| . \tag{4.6.33}$$

We can now define the total elastic energy of the specimen $E : \mathcal{U} \to \mathbb{R} \cup \{+\infty\}$ as

$$\mathcal{E}(y,z) := \mathcal{E}_{\rm b}(y,z) + \mathcal{E}_{\rm int}(y,z) - \mathcal{F}(y), \tag{4.6.34}$$

where $\mathcal{F} : W^{1,p}(\Omega) \to \mathbb{R}$ is weakly upper semicontinuous satisfying for some $K > 0$ and, all $y \in W^{1,p}(\Omega;\mathbb{R}^3)$, we have

$$\mathcal{F}(y) \le K\Big(\|y\|^{\tilde{p}}_{W^{1,p}(\Omega;\mathbb{R}^3)} + 1\Big). \tag{4.6.35}$$

Theorem 4.6.9 (Existence of minimizing configurations [484, 485]). *Let $p > 3$ and $y_{\rm D} \in W^{1,p}(\Omega;\mathbb{R}^3)$ be fixed. Let* (4.6.26)–(4.6.27c) *and* (4.6.30)–(4.6.33)*, and* (4.6.35) *hold, and furthermore*

$$\mathcal{A} := \{(y,z) \in \mathcal{U};\ y = y_{\rm D} \ \textit{on}\ \Gamma_{\rm D},\ \det \nabla y > 0 \ \textit{a.e. in}\ \Omega \ \textit{and}\ (3.4.7)\ \textit{holds for}\ d = 3\}$$

be nonempty, and there be $(y,z) \in \mathcal{A}$ such that $\mathcal{E}(y,z) < +\infty$. Then there exists a minimizer of $\mathcal{E}$ on $\mathcal{A}$.

Sketch of the proof. Let $\{(y_k,z^k)\} \subset \mathcal{A}$ be a minimizing sequence which exists as $\mathcal{A} \neq \emptyset$. We get by reflexivity of $W^{1,p}(\Omega;\mathbb{R}^3)$ and by Theorems 3.2.1 and 3.2.2 that it holds for a (not relabeled) subsequence

$$\begin{aligned} y_k &\to y && \text{weakly in } W^{1,p}(\Omega;\mathbb{R}^3), \\ \operatorname{Cof}\nabla y_k &\to \operatorname{Cof}\nabla y && \text{weakly in } L^{p/2}(\Omega;\mathbb{R}^{3\times3}), \text{ and} \\ \det\nabla y_k &\to \det\nabla y && \text{weakly in } L^{p/3}(\Omega). \end{aligned} \tag{4.6.36}$$

We also obtain that $\det\nabla y > 0$ almost everywhere in Ω and that y satisfies (3.4.7) by the same arguments as in the proof of Theorem 3.4.6. Theorem B.3.6 ensures that $y = y_{\rm D}$, too.

The growth condition (4.6.33) implies that $\{z^k\} \subset {\rm BV}(\Omega;\mathbb{R}^{M+1})$ is bounded, so that (for a subsequence) $z^k \to z$ in $L^1(\Omega;\mathbb{R}^{M+1})$. where $z \in {\rm BV}(\Omega;\mathbb{R}^{M+1})$ again defines a partition of Ω. Additionally, for every $) \le i \le M$ it holds that $\chi_{\Omega_i^k} \to \chi_{\Omega_i}$ in $L^1(\Omega)$ if $k \to \infty$ where χ_{Ω_i} is a characteristic function of $\Omega_i \subset \Omega$ where Ω_i is of a finite perimeter. Moreover, (4.6.33) yield that (for a subsequence) and all $0 \le i \le M$

$$(b_i^k, H_i^k, p_i^k) \to (\beta_i, \kappa_i, \pi_i) \text{ weakly* in } \mathfrak{M}(\Omega;\mathbb{R}^3) \times \mathfrak{M}(\Omega;\mathbb{R}^{3\times3}) \times \mathfrak{M}(\Omega;\mathbb{R}^3) \tag{4.6.37}$$

if $k \to \infty$. The strong L^1 convergence of $\{\chi_{\Omega_i^k}\}_{k\in\mathbb{N}}$ implies that $\beta_i := n_i \mathscr{H}^2_{|S_i}$ where n_i is a measure-theoretic normal to the measure-theoretic boundary $S_i := \partial^*\Omega_i \cap \Omega$. The weak convergence in (4.6.36) together with (4.6.31) imply that $\kappa_i = H_i$ and $\pi_i = p_i$ for every admissible i. This shows that $(y,z) \in \mathcal{U}$. The strong convergence of z^k to z in $L^1(\Omega;\mathbb{R}^{M+1})$ and polyconvexity of φ_i leads to the inequality $\mathcal{E}_{\rm b}(y,z) \le \liminf_{k\to\infty} \mathcal{E}_{\rm b}(y_k,z^k)$ in view of Theorem 3.3.1.

Finally, we have

$$\liminf_{j\to\infty} \mathcal{E}_{\rm int}(y_j,z^j) \ge \mathcal{E}_{\rm int}(y,z)$$

due to Theorem B.1.2. Thus, $(y,z) \in W^{1,p}(\Omega;\mathbb{R}^d) \times {\rm BV}(\Omega;\mathbb{R}^{M+1})$. Following the proof of Theorem 3.4.6 we see that $(y,z) \in \mathcal{U}$. Namely, y is injective almost everywhere in Ω and $\det\nabla y > 0$ almost everywhere in Ω. □

Remark 4.6.10 (Stored energy density with a minimum on SO(d)F_i*).* In literature, examples of stored energy density functions in nonlinear elasticity are usually minimized on SO(d). In the context of shape-memory alloys, the stored energy density is minimized on SO(d)F_i, $F_i \ne F_j$, $i,j = 0,\dots,M$. To construct such energy densities explicitly, we can now proceed as follows. Assume that $\varphi : \mathbb{R}^{d\times d} \to \mathbb{R}\cup\{+\infty\}$ is minimized on SO(d) and that $\varphi(F) = \hat\varphi(C)$ for some function $\hat\varphi : \mathbb{R}^{d\times d}_{sym} \to \mathbb{R}\cup\{+\infty\}$ and $C = F^\top F$ the right-Cauchy-Green tensor. Considering the polar decomposition of $F_i \in \mathbb{R}^{d\times d}$ with $\det F_i > 0$, we can write $F_i = R_i U_i$ where R_i is a rotation and U_i is symmetric and positive definite matrix. Note that $C_i = U_i^2$. Bearing this in mind, we define the energy of the i-th variant via a shift

$$\varphi_i(F) := \varphi(FU_i^{-1}) = \hat\varphi(U_i^{-1}CU_i^{-1}),$$

which is clearly minimized on SO(d)F_i. Notice also that if φ is polyconvex, so is φ_i. See also [544] for various constructions of stored energy densities for crystals undergoing phase transition.

Exercise 4.6.11. Modify φ in (4.6.1) by adding a term containing the second-order deformation gradient of y or a term containing $\nabla[{\rm Cof}\,\nabla y]$ and show existence of minimizers for the corresponding energy functional with suitable Dirichlet boundary conditions and loads by external forces.

4.7 Elastic bodies under monopole long-range interactions

Let us present another interesting situation where the reference and the space configurations come inevitably into considerations together like in the previous section 3.5 where dipolar long-range interactions have been addressed. This arises when mass density and the self-induced gravitational field is taken into consideration. This gravitational field must necessarily be counted in the space rather than in the reference configuration. On the other hand, the stored energy $\varphi : \Omega\times\mathbb{R}^{d\times d} \to \mathbb{R}$ and the monopolar density $m : \Omega \to \mathbb{R}$ are prescribed in the material configuration. The canonical choices are:

$$m = \begin{cases} \varrho & \text{mass density (attractive interaction),} \\ q & \text{electric charge (repulsive interaction).} \end{cases} \tag{4.7.1}$$

We will consider the physically relevant case $d = 3$ and an "external" monopolar density $m_{\rm ext} : \mathbb{R}^3 \to \mathbb{R}$ considered given in the space configuration. In the case of self-gravitating bodies, this external mass density gives rise to the so-called *tidal forces*. The *gravitational* or *electrostatic potential* ϕ considered in the space configuration satisfies[19]

$$-\kappa \Delta\phi = \chi_{\Omega^y} \frac{m(y^{-1})}{\det \nabla y(y^{-1})} + m_{\rm ext} \quad \text{with } \kappa = \begin{cases} -1/(4\pi \times \text{gravitational constant}), \\ \text{vacuum electric permittivity,} \end{cases} \tag{4.7.2}$$

to be solve on the whole space $\mathbb{R}^3$ with χ_S denoting the characteristic function of a set $S \subset \mathbb{R}^3$, referring respectively to the options (4.7.1). Note the convention that κ is negative in (4.7.2) in case of attractive (i.e. gravitational) interactions, which is also the usual geophysical convention leading to (4.7.2) in the form $\Delta\phi = 4\pi(m_{\rm s} + m_{\rm ext})/\times$gravitational constant and also note how the sign of κ changes the character of (4.7.3) below in ϕ from concave (for $\kappa > 0$) to convex (for $\kappa < 0$).

The energy of the monopoles in the potential field in the reference configuration Ω is $m(x)\phi(y(x))/\det(\nabla y(x))$ at a given reference point $x \in \Omega$.

We further involve the concept of local non-selfpenetration together with the Ciarlet-Nečas condition to guarantee injectivity of the deformation, which is needed because both the material and the space coordinates are involved in the problem. The energy of the self-induced potential field in naturally counted in the space coordinates over the whole Universe. The overall energy is thus

$$\mathcal{E}(y,\phi) = \begin{cases} \displaystyle\int_\Omega \varphi(\nabla y) + (1+|\nabla^2 y|^{p-2})\mathbb{H}\nabla^2 y \vdots \nabla^2 y \,{\rm d}x + \int_{\Omega^y} \frac{m(y^{-1})\phi}{[\det \nabla y](y^{-1})} {\rm d}x \\ \displaystyle\qquad + \int_{\mathbb{R}^3} m_{\rm ext} - \frac{\kappa}{2}|\nabla\phi|^2 \,{\rm d}x \quad \text{if } \int_\Omega \det(\nabla y)\,{\rm d}x \le {\rm meas}_3(y(\Omega)), \\ \qquad\qquad\qquad\qquad\qquad\qquad \text{and if } y = y_{\rm D} \text{ on } \Gamma_{\rm D}, \\ +\infty \qquad\qquad\qquad\qquad\qquad \text{otherwise.} \end{cases} \tag{4.7.3}$$

Note that we fixed the elastic body by the Dirichlet condition. Note also that the Ciarlet-Nečas condition ensures injectivity so that y^{-1} is single-valued almost everywhere in Ω^y.

The related variational problem seeks rather the critical point only. This is a particular case of repulsive interactions when the integrand is concave in $\nabla\phi$, so (y,ϕ) is sought such that $\mathcal{E}(\cdot,\phi)$ is minimized while $\mathcal{E}(y,\cdot)$ is maximized. In the attractive

[19] In the SI-units, the gravitational constant is approximately $6.674 \times 10^{-11}\,{\rm m^3 kg^{-1} s^{-2}}$ while the permittivity of vacuum is about $8.8542 \times 10^{-12}\,{\rm Fm^{-1}}$ (or ${\rm A^2 s^4 kg^{-1} m^{-3}}$).

interactions, thus global minimizer of $\mathcal{E}$ is to be sought. In any case, the Euler-Lagrange equation arising as a unique critical point of $\mathcal{E}(y,\cdot)$ is just the weak formulation of (4.7.2).

Existence of solutions is to be proved by the direct method. We exploit Healey-Krömer theorem 2.5.3 used to bound $1/\det\nabla y$ occurring in (4.7.3). First, the coercivity is automatically ensured only for repulsive interactions, where the last two integrals in (4.7.3) sums up to[20]

$$\int_{\Omega^y}\frac{m(y^{-1})\phi}{[\det\nabla y](y^{-1})}\mathrm{d}x+\int_{\mathbb{R}^3}m_{\rm ext}\phi-\frac{\kappa}{2}|\nabla\phi|^2\,\mathrm{d}x=\int_{\mathbb{R}^3}\frac{\kappa}{2}|\nabla\phi|^2\,\mathrm{d}x \tag{4.7.4}$$

when the constraint (4.7.2) is taken into account. This reveals that $\mathcal{E}$ is coercive on the manifold of pairs $(y,\phi)\in W^{2,p}(\Omega;\mathbb{R}^3)\times H^1(\mathbb{R}^3)$ satisfying (4.7.2). This non-affine manifold can equivalently be written in the form of the pairs (y,ϕ) satisfying the variational constraint

$$\int_{\mathbb{R}^3}\kappa\nabla\phi(x^y)\cdot\nabla v(x^y)\,\mathrm{d}x^y=\int_\Omega m(x)v(y(x))\,\mathrm{d}x \tag{4.7.5}$$

for any $v\in C^1(\mathbb{R}^3)$. This manifold is closed under the weak convergence, exploiting the important fact that $y\mapsto v\circ y$ is weakly continuous on $W^{2,p}(\Omega;\mathbb{R}^3)$ if v is continuous.

However, in the case of attractive interactions, these last two integrals give just opposite sign, showing that the coercivity (boundedness from below on the mentioned manifold) is not automatic. In general, minimizing sequences may shrink the mass towards a single point (a nonrelativistic "black hole") and the energy (4.7.4) blows up to $-\infty$; cf. also Remark 5.7.13 below on p. 191.[21]

[20] Testing (4.7.2) by ϕ gives, by Green's formula together with the "boundary" condition $\phi(\infty)=0$, that $\int_{\mathbb{R}^3}\kappa|\nabla\phi|^2\,\mathrm{d}x=\int_{\Omega^y}([m/\det\nabla y](y^{-1})\phi\,\mathrm{d}x+\int_{\mathbb{R}^3}m_{\rm ext})\phi\,\mathrm{d}x$.

[21] A possible gravitational instability has been recognized in planetary astrophysics already a long time ago. In general, such a collapse can be prevented by enough rigid elastic response in cooperation with rather small mass (or in a certain way perhaps also by dynamical extension of the problem), but it is not clear how to formulate such condition in an explicit way.

Chapter 5
Linearized elasticity

> The first mathematician to consider the nature of resistance of solids to rupture was Galileo ... Undoubtedly, the two great landmarks are the discovery of Hook's law in 1660 and the formulation of the general equations by Navier (1821).
>
> AUGUSTUS EDWARD HOUGH LOVE (1863–1940)

Rather special, but anyhow frequently occurred situations in many engineering applications exhibit deformations whose gradient is relatively very close to identity. In other words, displacement gradient is very small and often even the displacement itself is relatively small. It allows us, with reasonable accuracy, to neglect higher order terms and simplified a lot of aspects related to geometrical nonlinearities very substantially. It also facilitates computational algorithms substantially, and there is no wonder that most engineering computations in solid mechanics are just based on such hypotheses. As we will see, linearized elasticity provides us with unique solutions and strongly relies on convexity assumptions.

5.1 Concept of small strains

We know that the *Green-Lagrange strain* tensor is defined as $E = \frac{1}{2}(C - \mathbb{I})$ where $C = F^\top F$ is the right Cauchy-Green tensor, cf. (1.1.28). If we write $F = \nabla u + \mathbb{I}$, where u is a displacement we get for $|\nabla u|$ small that

$$E = \frac{1}{2}(\nabla u + (\nabla u)^\top) + o(|\nabla u|).$$

Then we define the linearized strain tensor, also called *small-strain tensor*, as $e(u) = \mathrm{sym}(\nabla u)$, i.e.

$$e(u) = \frac{1}{2}\nabla u + \frac{1}{2}(\nabla u)^\top. \tag{5.1.1}$$

Assume that $U \in \mathbb{R}^{d\times d}$ is such that $\det(\mathbb{I} + U) > 0$. We develop for smooth φ

M. Kružík and T. Roubíček, *Mathematical Methods in Continuum Mechanics of Solids*, Interaction of Mechanics and Mathematics,
https://doi.org/10.1007/978-3-030-02065-1_5

$$\varphi(x,\mathbb{I}+U)=\varphi(x,\mathbb{I})+\partial_F\varphi(x,\mathbb{I}):U+\frac{1}{2}[\partial^2_{FF}\varphi(x,\mathbb{I}):U]:U+o(|U|^2)$$
$$=\varphi(x,\mathbb{I})+S:U+\frac{1}{2}[\partial^2_{FF}\varphi(x,\mathbb{I}):U]:U+o(|U|^2). \quad (5.1.2)$$

Assume now that $\varphi(x,\mathbb{I})=0$ and that $y=\mathrm{id}$ is a stress-free state of the material, i.e. $S=0$, then we have that

$$\varphi(x,\mathbb{I}+U)=\frac{1}{2}[\partial^2_{FF}\varphi(x,\mathbb{I}):U]:U+o(|U|^2).$$

The leading term on the right-hand side depends only on the symmetric part of U, i.e.,

$$\varphi(x,\mathbb{I}+U)=\frac{1}{2}[\partial^2_{FF}\varphi(x,\mathbb{I}):(U+U^\top)]:(U+U^\top)+o(|U|^2).$$

Defining a *tensor of elastic constants* (*moduli*), denoted by $\mathbb{C}$, by

$$\mathbb{C}(x):=\partial^2_{FF}\varphi(x,\mathbb{I}) \quad (5.1.3)$$

we write a stress strain relation for a symmetric matrix $U:=e(u)$ as

$$\sigma=\mathbb{C}e(u). \quad (5.1.4)$$

This relationship is known as the *Hooke law*. In fact, it can be shown by various symmetries that there are only 21 independent constants in $\mathbb{C}$ if $d=3$ and 6 constants for $d=2$. Indeed, as $e(u)$ as well as σ are both symmetric, we immediately see that 36 independent entries of $\mathbb{C}$ will do if $d=3$. This means that we can swap i with j and also k with l without changing the entry $\mathbb{C}_{ijkl}$. These properties are called *minor symmetries*. Moreover, as $\mathbb{C}$ is defined by means of (5.1.3), the symmetry in ij and kl follows (being called *major symmetry*). Therefore,

$$\mathbb{C}_{ijkl}=\mathbb{C}_{klij}=\mathbb{C}_{jikl}. \quad (5.1.5)$$

The second equality in (5.1.5) ensures that the stress $\sigma=\mathbb{C}e$ is symmetric[1] while the first equality makes $e\mapsto\mathbb{C}e:\mathbb{R}^{d\times d}_{\mathrm{sym}}\to\mathbb{R}^{d\times d}_{\mathrm{sym}}$ symmetric which makes the operator $e\mapsto\mathbb{C}e:L^2(\Omega;\mathbb{R}^{d\times d}_{\mathrm{sym}})\to L^2(\Omega;\mathbb{R}^{d\times d}_{\mathrm{sym}})$ potential; cf. (A.3.6) of p. 495. This (quadratic) potential is $e\mapsto\int_\Omega\frac{1}{2}\mathbb{C}e{:}e\,\mathrm{d}x$.

Hence, finally we only need 21 entries in three dimensions, as mentioned. These can be seen as entries of a symmetric matrix 6×6. The reasoning for two dimensions is analogous.

In order to simplify the notation there is a convenient way how to rewrite the four-index tensor $\mathbb{C}$ as a standard matrix with two indices, see Table 5.1; this is the so-called Voigt notation [527]. We then write $\mathbb{C}_{\alpha\beta}:=\mathbb{C}_{ijkl}$ according to the following table: Analogously, we replace $2e(u)_{kl}$ by $e(u)_\beta$ and σ_{ij} by σ_α.

[1] Thus means $\sigma_{ij}=\sum_{k,l=1}^d\mathbb{C}_{ijkl}e_{kl}(u)=\sum_{k,l=1}^d\mathbb{C}_{jikl}e_{kl}(u)=\sigma_{ji}$.

ij or kl	11	22	33	23 or 32	13 or 31	12 or 21
α or β	1	2	3	4	5	6

Table 5.1 Voigt's notation for the replacement of two indices by one. (Here $d = 3$.)

If the material is homogeneous and isotropic $\mathbb{C}$ reduces to two positive quantities λ and G as used already in (2.4.3). In this case, $\mathbb{C}_{ijkl} = \lambda\delta_{ij}\delta_{kl} + G(\delta_{ik}\delta_{jl} + \delta_{il}\delta_{jk})$, i.e.,

$$\sigma = \lambda(\mathrm{div}\, u)\mathbb{I} + 2Ge(u) \tag{5.1.6}$$

and in this case we speak about *Lamé material*, cf. also its geometrically-nonlinear variant (2.4.1). The Lamé constants can then be expressed as (2.4.2).

If $\Omega \subset \mathbb{R}^d$ is a bounded Lipschitz domain and $\Gamma_{\mathrm{D}} \subset \Gamma$ is such that $\mathrm{meas}_{d-1}(\Gamma_{\mathrm{D}}) > 0$ we define

$$V := \{v \in H^1(\Omega;\mathbb{R}^d);\ v := 0 \text{ on } \Gamma_{\mathrm{D}}\} \tag{5.1.7}$$

and equip it with the strong topology of $H^1(\Omega;\mathbb{R}^d)$. Given referential body forces with the density $f : \Omega \to \mathbb{R}^d$ and referential surface forces with the density $g : \Gamma_{\mathrm{N}} \to \mathbb{R}^d$, we define $B : V \times V \to \mathbb{R}$ and $L : V \to \mathbb{R}$ as

$$B(u,v) := \int_\Omega \mathbb{C}e(u){:}e(v)\,\mathrm{d}x \quad \text{and} \quad L(v) := \int_\Omega f\cdot v\,\mathrm{d}x + \int_{\Gamma_{\mathrm{N}}} g\cdot v\,\mathrm{d}S\,. \tag{5.1.8}$$

Under the basic data qualification

$$\Gamma_{\mathrm{D}}, \Gamma_{\mathrm{N}} \subset \Gamma \text{ disjoint with}$$
$$\Gamma_{\mathrm{D}} \text{ (relatively) open and such that } \mathrm{meas}_{d-1}(\Gamma_{\mathrm{D}}) > 0, \tag{5.1.9a}$$
$$f \in L^{2^{*\prime}}(\Omega;\mathbb{R}^d) \ \text{ and } \ g \in L^{2^{\#\prime}}(\Gamma_{\mathrm{N}};\mathbb{R}^d), \tag{5.1.9b}$$

we have the following assertion which states that a minimizer of of the underlying potential is a weak solution to the equilibrium equations (5.1.10) and vice versa.

Theorem 5.1.1. *Let $\Omega \subset \mathbb{R}^d$ be a bounded Lipschitz domain and (5.1.9) hold. Let further B be coercive in the sense that $B(v,v) \geq \varepsilon\|v\|_V^2$ for all $v \in V$ and for some $\varepsilon > 0$ independent of v. A weak solution $u \in V$ of the linear boundary-value problem*

$$-\mathrm{div}\,\sigma = f \qquad \text{in } \Omega, \tag{5.1.10a}$$
$$u = 0 \qquad \text{on } \Gamma_{\mathrm{D}}, \tag{5.1.10b}$$
$$\sigma\vec{n} = g \qquad \text{on } \Gamma_{\mathrm{N}} \tag{5.1.10c}$$

is a unique minimizer of $\mathcal{E} : V \to \mathbb{R}$ defined as

$$\mathcal{E}(v) := \frac{1}{2}B(v,v) - L(v) \tag{5.1.11}$$

and satisfies the equation

$$B(u,v) = L(v) \quad \textit{for all } v \in V. \tag{5.1.12}$$

Proof. Notice that L is a continuous linear operator on V in view of Theorems B.3.2 and B.3.6 and that B is a continuous bilinear form on V in the sense that $|B(v,w)| \le K_B\|v\|_V\|w\|_V$ for some $K_B > 0$ and all $v, w \in V$ as can be easily seen from the Hölder inequality (B.2.5). We use the Green formula (B.3.14) specialized for a symmetric tensor σ and every $v \in V$:

$$\int_\Omega \operatorname{div} S \cdot v \, dx = -\int_\Omega S : \nabla v \, dx + \int_{\Gamma_N} S\vec{n} \cdot v \, dS = -\int_\Omega S : e(v) \, dx + \int_{\Gamma_N} S\vec{n} \cdot v \, dS.$$

Applying this identity to $S := \sigma = \mathbb{C}e(u)$, we immediately get formula (5.1.12) because of (5.1.10c).

Using coercivity and continuity of B we get

$$\varepsilon\|v\|^2 \le B(v,v) \le K_B\|v\|^2.$$

Hence, B is an inner product over V making it just a Hilbert space with the norm $\|v\|_H = \sqrt{B(v,v)}$. Moreover, $\|\cdot\|_H$ and $\|\cdot\|$ are equivalent. By the Lax-Milgram lemma A.1.6 there is only one $u \in V$ such that $L(v) = B(u,v)$ for all $v \in V$. Thus u is the unique solution. We show that it is the minimizer of $\mathcal{E}$ and that every minimizer of $\mathcal{E}$ satisfies (5.1.12).

Let us notice that, for every $v \in V$, it holds

$$\mathcal{E}(u+v) - \mathcal{E}(u) = B(u,v) - L(v) + \frac{1}{2}B(v,v).$$

Therefore, if $B(u,v) = L(v)$ then $\mathcal{E}(u+v) - \mathcal{E}(u) \ge 0$ and u is a minimizer.

Conversely, if u is a minimizer of $\mathcal{E}$ and $v \in V$ is such that $B(u,v) - L(v) \ne 0$ then without loss of generality we may suppose that $B(u,v) - L(v) < 0$ (replace v by $-v$ if necessary). Then, for $\vartheta > 0$ small enough, we would have

$$0 > \mathcal{E}(u+\vartheta v) - \mathcal{E}(u) = \vartheta(B(u,v) - L(v)) + \frac{\vartheta^2}{2}B(v,v),$$

a contradiction. □

We see that the key assumption of Theorem 5.1.1 is the coercivity of B on V. A necessary condition for it is the uniform positive definiteness of $\mathbb{C} \in L^\infty(\Omega; \mathrm{SLin}(\mathbb{R}^{d\times d}_{\mathrm{sym}}))$ with $\mathrm{SLin}(\mathbb{R}^{d\times d}_{\mathrm{sym}})$ from (2.5.25), i.e., we assume that, for almost all $x \in \Omega$ and all $e \in \mathbb{R}^{d\times d}_{\mathrm{sym}}$,

$$\exists \varepsilon_{\mathbb{C}} > 0 \ \forall_{\text{a.a.}} x \in \Omega \ \forall\, e \in \mathbb{R}^{d\times d}_{\mathrm{sym}} : \quad \mathbb{C}(x)e : e \ge \varepsilon_{\mathbb{C}}|e|^2. \tag{5.1.13}$$

The fact that this is also a sufficient condition follows from the nontrivial inequality, called Korn's inequality, discussed in the following section.

5.2 Korn's inequality

Let us start with a simple case showing main phenomenon behind one inequality which is essential in several problems of solid mechanics (as well as of fluid mechanics), first formulated by A. Korn [283] in 1906, although not rigorously proved. For any $u\in C_0^2(\Omega;\mathbb{R}^d)$ (i.e. zero traces on Γ), we have

$$\begin{aligned}\int_\Omega e(u):e(u)\,\mathrm{d}x &= \frac{1}{4}\int_\Omega\Big(\frac{\partial u_i}{\partial x_j}+\frac{\partial u_j}{\partial x_i}\Big)\Big(\frac{\partial u_i}{\partial x_j}+\frac{\partial u_j}{\partial x_i}\Big)\,\mathrm{d}x\\ &= \frac{1}{2}\int_\Omega\sum_{i,j=1}^d\Big(\frac{\partial u_i}{\partial x_j}\Big)^2\,\mathrm{d}x+\frac{1}{2}\int_\Omega\Big(\frac{\partial u_i}{\partial x_j}\Big)\Big(\frac{\partial u_j}{\partial x_i}\Big)\,\mathrm{d}x.\end{aligned}$$

Moreover, applying twice integration by parts to the second term on the right-hand side, we get (keep in mind that $u = 0$ on Γ)

$$\int_\Omega\Big(\frac{\partial u_i}{\partial x_j}\Big)\Big(\frac{\partial u_j}{\partial x_i}\Big)\,\mathrm{d}x = -\int_\Omega u_i\Big(\frac{\partial^2 u_j}{\partial x_i\partial x_j}\Big)\,\mathrm{d}x = \int_\Omega\Big(\frac{\partial u_i}{\partial x_i}\Big)\Big(\frac{\partial u_i}{\partial x_i}\Big)\,\mathrm{d}x \ge 0\,.$$

Hence,

$$\frac{1}{2}\int_\Omega\sum_{i,j=1}^d\Big(\frac{\partial u_i}{\partial x_j}\Big)^2\,\mathrm{d}x \le \int_\Omega e(u):e(u)\,\mathrm{d}x\,.$$

Due to the density of $C_0^2(\Omega;\mathbb{R}^d)$ in $H_0^1(\Omega;\mathbb{R}^d)$, we have proved the following result.

Theorem 5.2.1 (Korn's first inequality). *Let $\Omega\subset\mathbb{R}^d$ be a domain and $u\in H_0^1(\Omega;\mathbb{R}^d)$. Then*

$$\|\nabla u\|^2_{L^2(\Omega;\mathbb{R}^{d\times d})} \le 2\|e(u)\|^2_{L^2(\Omega;\mathbb{R}^{d\times d})}\,. \tag{5.2.1}$$

We thus see that the L^2-norm of the symmetric part of the gradient controls the L^2-norm of the whole gradient. This is surprising as the symmetric part has only $d(d-1)$ components whole the whole gradient has d^2 components. The calculation above naturally extends to the whole $H_0^1(\Omega;\mathbb{R}^d)$ by density of smooth mappings in this space. Formula (5.2.1) is called the first *Korn's inequality* [402], see also [117].

This result is a consequence of *the second Korn's inequality*:

Theorem 5.2.2 (Korn's second inequality I [283]). *Let $\Omega\subset\mathbb{R}^d$ be a bounded domain with a Lipschitz boundary. Then there is a constant $K>0$ such that, for each $v\in H^1(\Omega;\mathbb{R}^d)$, it holds*

$$\|v\|^2_{H^1(\Omega;\mathbb{R}^d)} \le K\big(\|v\|^2_{L^2(\Omega;\mathbb{R}^d)}+\|e(v)\|^2_{L^2(\Omega;\mathbb{R}^{d\times d})}\big)\,. \tag{5.2.2}$$

Hence, the norm $v \mapsto (\|v\|^2_{L^2(\Omega;\mathbb{R}^d)} + \|e(v)\|^2_{L^2(\Omega;\mathbb{R}^{d\times d})})^{1/2}$ *is equivalent to* $\|\cdot\|_{H^1(\Omega;\mathbb{R}^d)}$ *on* $H^1(\Omega;\mathbb{R}^d)$.

The following version of the second Korn's inequality is similar to Theorem 1.1.12.

Theorem 5.2.3 (Korn's second inequality II [282]). *Let* $\Omega \subset \mathbb{R}^d$ *be a bounded domain with a Lipschitz boundary. Then there is a constant* $K > 0$ *such that for each* $v \in H^1(\Omega;\mathbb{R}^d)$ *there is a skew symmetric matrix* $A \in \mathbb{R}^{d\times d}$ *so that*

$$\big\|\nabla v - A\big\|^2_{L^2(\Omega;\mathbb{R}^d)} \le K\big\|e(v)\big\|^2_{L^2(\Omega;\mathbb{R}^{d\times d})}. \tag{5.2.3}$$

Proof of Theorem 5.2.2. We show that $H^1(\Omega;\mathbb{R}^d)$ coincides with

$$K(\Omega;\mathbb{R}^d) = \{v \in L^2(\Omega;\mathbb{R}^d);\ e(v) \in L^2(\Omega;\mathbb{R}^{d\times d})\}.$$

Clearly, $K(\Omega;\mathbb{R}^d) \supset H^1(\Omega;\mathbb{R}^d)$, however, the opposite inclusion is far not obvious. The norm $v \mapsto (\|v\|^2_{L^2(\Omega;\mathbb{R}^d)} + \|e(v)\|^2_{L^2(\Omega;\mathbb{R}^{d\times d})})^{1/2}$ makes $K(\Omega;\mathbb{R}^d)$ a Hilbert space. We have

$$\partial^2_{x_j x_k} v_i = \partial_{x_j} e_{ik}(v) + \partial_{x_k} e_{ij}(v) - \partial_{x_i} e_{jk}(v). \tag{5.2.4}$$

So, if $v \in K(\Omega;\mathbb{R}^d)$ then $e_{ij}(v) \in L^2(\Omega)$ and $\partial_{x_k} e_{ij}(v) \in H^{-1}(\Omega)$. Hence $\partial^2_{x_j x_k} v_i \in H^{-1}(\Omega)$ and, by Lemma B.3.7, $\nabla v \in L^2(\Omega;\mathbb{R}^{d\times d})$. Thus, we see that $K(\Omega;\mathbb{R}^d) = H^1(\Omega;\mathbb{R}^d)$ (element-wise). Moreover, the embedding $H^1(\Omega;\mathbb{R}^d)$ into $K(\Omega;\mathbb{R}^d)$ is continuous and surjective. The proof is finished by an application of the closed-graph theorem to the identity map: $H^1(\Omega;\mathbb{R}^d) \to K(\Omega;\mathbb{R}^d)$. Notice that the identity is the bijection (i.e. injective and surjective) $H^1(\Omega;\mathbb{R}^d) \to K(\Omega;\mathbb{R}^d)$ which is continuous hence is inverse is also continuous, see Theorem A.1.5. □

It is to be noticed that (5.2.4) serves for a modification of (5.2.2) towards *Korn's inequality in 2nd-grade nonsimple materials* as

$$\big\|v\big\|^2_{H^2(\Omega;\mathbb{R}^d)} \le K\Big(\big\|v\big\|^2_{L^2(\Omega;\mathbb{R}^d)} + \big\|\nabla e(v)\big\|^2_{L^2(\Omega;\mathbb{R}^{d\times d\times d})}\Big). \tag{5.2.5}$$

and, using also the Poincaré inequality with $\mathrm{meas}_{d-1}(\Gamma_{\mathrm{N}}) > 0$, for

$$\big\|v\big\|^2_{H^2(\Omega;\mathbb{R}^d)} \le K\Big(\big\|v|_{\Gamma_{\mathrm{D}}}\big\|^2_{L^2(\Gamma_{\mathrm{D}};\mathbb{R}^d)} + \big\|\nabla e(v)\big\|^2_{L^2(\Omega;\mathbb{R}^{d\times d\times d})}\Big). \tag{5.2.6}$$

Actually, the generalization of Theorem 5.2.2 for a general $1 < p < +\infty$ holds, too; cf. [218, 386]. On the other hand, for $p = 1$ the Korn inequality does not hold, cf. a counterexample by D. Ornstein [404]. Various other generalizations have been designed for different purposes. Let us mention one generalization we will use later in Section 9.3.

Theorem 5.2.4 (Neff's and Pompe's generalization of Korn's inequality). [2] *Let* $1 < p < +\infty$, $\mathfrak{M}_{d-1}(\Gamma_{\rm D}) > 0$, *and* $F \in C(\bar{\Omega};\mathbb{R}^{d\times d})$ *with* $\det F \geq \varepsilon > 0$ *on* Ω. *Then there is* $K > 0$ *such that, for all* $u \in W^{1,p}(\Omega;\mathbb{R}^d)$ *with zero trace on* $\Gamma_{\rm D}$, *it holds*

$$\big\|\nabla u\big\|_{L^p(\Omega;\mathbb{R}^{d\times d})} \leq K\big\|{\rm sym}((\nabla u)F)\big\|_{L^p(\Omega;\mathbb{R}^{d\times d})}. \tag{5.2.7}$$

where sym$(\cdot)$ *denotes the symmetric part of a matrix. This estimate holds uniformly (in the sense that* $K > 0$ *is independent of* F*) for* F*'s ranging over a bounded set of* $W^{1,p}(\bar{\Omega};\mathbb{R}^{d\times d})$, $p > d$.[3]

5.3 Existence results in linearized elasticity

Having Korn's inequality at our disposal we can now show that there exists a unique solution to the equilibrium problem of linearized elasticity.

Lemma 5.3.1 (Closeness of V**).** *Let* $\Omega \subset \mathbb{R}^d$ *be a Lipschitz bounded domain, let* $\Gamma_{\rm D} \subset \Gamma$ *be measurable with* ${\rm meas}_{d-1}(\Gamma_{\rm D}) > 0$. *Then* V *is a closed subspace of* $H^1(\Omega;\mathbb{R}^d)$ *and* $v \mapsto \|e(v)\|_{L^2(\Omega;\mathbb{R}^{d\times d})}$ *is a norm equivalent to the* $H^1(\Omega;\mathbb{R}^d)$*-norm on* V.

Proof. Closeness of V follows from the continuity of the trace operator. Let us show that $v \mapsto \|e(v)\|_{L^2}$ is a norm on V. Let $e(v) = 0$. Then (5.2.4) implies that v is linear in x and we get that $v = (\nabla v)x + a$ for some $a \in \mathbb{R}^d$ and $\nabla v = -(\nabla v)^\top$, i.e., it is a skew symmetric matrix. Therefore, in view of Exercise 2.1.1 there is $b \in \mathbb{R}^d$, the axial vector of ∇v such that for almost all $x \in \Omega$

$$v(x) = a + b \times x$$

Now it is easy to see that if $v \in V$ and $e(v) = 0$ then $v = 0$. Namely, consider the set $\omega := \{x \in \Omega;\ v(x) = 0\}$. We have that

$$\omega = \begin{cases} \emptyset & \text{if } b = 0 \text{ and } a \neq 0 \\ & \text{or if } b \neq 0 \text{ and } a\cdot b \neq 0, \\ \{x \in \mathbb{R}^d;\ x := (b\times a)/|b|^2 + bt,\ t \in \mathbb{R}\} & \text{if } a\cdot b = 0 \text{ and } b \neq 0. \end{cases}$$

Hence, ${\rm meas}_d(\omega) = 0$ because $d \geq 2$. It is clear that $\|e(v)\|_{L^2(\Omega;\mathbb{R}^{d\times d})} \leq C\|v\|_{H^1(\Omega;\mathbb{R}^d)}$ for $v \in V$. Suppose now that there is $\{v_k\} \subset V$ such that $\|v_k\|_{H^1(\Omega;\mathbb{R}^d)} = 1$ and

[2] Actually, P. Neff [391] proved it only for $p = 2$ and $d = 3$ for $F \in C^1(\bar{\Omega};\mathbb{R}^{d\times d})$ with ${\rm curl}\, F \in C^2(\bar{\Omega};\mathbb{R}^{d\times d})$ while W. Pompe [419] proved even a more general form than (5.2.7).

[3] Actually, this last assertion can be seen by a compactness/contradiction argument applied to (5.2.7) and is contained in [362].

$\|e(v_k)\|_{L^2(\Omega;\mathbb{R}^{d\times d})} \to 0$. Thus by the compact embedding v_k converges in $L^2(\Omega;\mathbb{R}^d)$ (up to a subsequence) and because $e(v_k) \to 0$ in $L^2(\Omega;\mathbb{R}^{d\times d}_{\rm sym})$ we get that the sequence $\{v_k\}$ is Cauchy with respect to the norm $\|v\|_{L^2(\Omega;\mathbb{R}^d)} + \|e(v)\|_{L^2(\Omega;\mathbb{R}^{d\times d})}$. By Korn's inequality this norm is equivalent to the norm on $H^1(\Omega;\mathbb{R}^d)$ and therefore it converges to $v\in V$. Then $e(v) = 0$ and therefore, by the first part, $v = 0$. This is not possible because we supposed that $\|v_k\|_{H^1(\Omega;\mathbb{R}^d)} = 1$. □

Hence, we can finally state the existence result is linearized elasticity.

Theorem 5.3.2 (Existence of a solution). *If $\Gamma_{\rm D}$, $\Gamma_{\rm N}$, f, and g are as in (5.1.9) and if* (5.1.13) *holds then the bilinear form B defined in* (5.1.8) *is coercive and continuous. Assume that the other assumptions of Theorem 5.1.1 are also satisfied. Then there is a unique (weak) solution to the boundary value problem* (5.1.10).

Exercise 5.3.3 (Other boundary conditions – elastic support). Relying on a modified Korn's inequality $\|v\|^2_{H^1(\Omega;\mathbb{R}^d)} \le K(\|v|_\Gamma\|^2_{L^2(\Gamma_{\rm N};\mathbb{R}^d)} + \|e(v)\|^2_{L^2(\Omega;\mathbb{R}^{d\times d})})$ for $\text{meas}_{d-1}(\Gamma_{\rm N}) > 0$ instead of (5.2.2), consider $\sigma\vec{n} + \mathbb{B}u = g$ on Γ with $\mathbb{B}\in L^\infty(I;\mathbb{R}^{d\times d}_{\rm sym})$ instead of (5.1.10b,c) and realize the modifications necessary in this section.

5.4 Nonsimple materials at small strains

The concept of 2nd-grade nonsimple materials, introduced already in the context of large strains in Sect. 2.5, translates also at small strains. The physically relevant departing point can be the following quadratic functional

$$u \mapsto \int_\Omega \frac{1}{2}\mathbb{C}e(u){:}e(u) + \frac{1}{2}\mathbb{H}\nabla e(u)\vdots\nabla e(u) - f{\cdot}u\,{\rm d}x - \int_{\Gamma_{\rm N}} g{\cdot}u\,{\rm d}S \tag{5.4.1}$$

as the stored energy to be minimized on $H^2(\Omega;\mathbb{R}^d)$ subject, for example, to $u = w_{\rm D}$ on $\Gamma_{\rm D}$. Here $\mathbb{H}$ is a 4th-order tensor[4] as in (2.5.24), in particular $\mathbb{H}\in{\rm SLin}(\mathbb{R}^{d\times d}_{\rm sym})$ but now $[\mathbb{H}\nabla e]_{ijk} := \mathbb{H}_{jklm}\partial_{x_i} e_{lm}$, which slightly differs from usage of $\mathbb{H}$ in Sect. 2.5. Note that, for $\mathbb{H} = 0$, the functional (5.4.1) reduces to $u \mapsto \frac{1}{2}B(u,u) - L(u)$ from Theorem 5.1.1 describing linear elastic simple materials at small strains under load.

The Euler-Lagrange equation for (5.4.1) leads to the the integral identity

$$\int_\Omega \mathbb{C}e(u){:}e(v) + \mathbb{H}\nabla e(u)\vdots\nabla e(v)\,{\rm d}x = \int_\Omega f{\cdot}v\,{\rm d}x + \int_{\Gamma_{\rm N}} g{\cdot}v\,{\rm d}S \tag{5.4.2}$$

to hold for all $v\in H^2(\Omega;\mathbb{R}^d)$ with $v|_{\Gamma_{\rm D}} = 0$ and to be interpreted as a weak formulation of a boundary-value problem to be still identified. Using symmetry of $\mathbb{C}$, this integral identity can equally be written as

[4] In contrast to Sect. 2.5, the frame indifference is not any issue here, so that a general 6th-order tensor for $\mathbb{H}$ could be considered, too.

$$\int_\Omega \mathbb{C}e(u):\nabla v + \mathbb{H}\nabla e(u) \vdots \nabla e(v)\,\mathrm{d}x = \int_\Omega f\cdot v\,\mathrm{d}x + \int_{\Gamma_\mathrm{N}} g\cdot v\,\mathrm{d}S.$$

By using Green's formula (B.3.14) and by the symmetry of $\mathbb{H}$ which ensures $\mathrm{div}(\mathbb{H}\nabla e(u))$ valued in $\mathbb{R}^{d\times d}_{\mathrm{sym}}$, we further get

$$\int_\Omega -\mathrm{div}(\mathbb{C}e(u))\cdot v - \mathrm{div}(\mathbb{H}\nabla e(u)):\nabla v\,\mathrm{d}x = \int_\Omega f\cdot v\,\mathrm{d}x + \int_{\Gamma_\mathrm{N}} g\cdot v\,\mathrm{d}S$$
$$- \int_\Gamma (\mathbb{C}e(u)):(v\otimes\vec{n}) + (\mathbb{H}\nabla e(u)) \vdots (\nabla v\otimes\vec{n})\,\mathrm{d}S. \qquad (5.4.3)$$

Using Green's formula once more, we obtain

$$\int_\Omega -\mathrm{div}(\mathbb{C}e(u))\cdot v + \mathrm{div}^2(\mathbb{H}\nabla e(u))\cdot v\,\mathrm{d}x = \int_\Omega f\cdot v\,\mathrm{d}x + \int_{\Gamma_\mathrm{N}} g\cdot v\,\mathrm{d}S$$
$$- \int_\Gamma (\mathbb{C}e(u)):(v\otimes\vec{n}) + (\mathbb{H}\nabla e(u)) \vdots (\nabla v\otimes\vec{n}) - \mathrm{div}(\mathbb{H}\nabla e(u)):(v\otimes\vec{n})\,\mathrm{d}S\,, \qquad (5.4.4)$$

where "div^2" means "div Div". Now we need to re-write the term $\int_\Gamma (\mathbb{H}\nabla e(u)) \vdots (\nabla v\otimes \vec{n})\,\mathrm{d}S$. We use a general decomposition $\nabla v = \frac{\partial v}{\partial \vec{n}}\vec{n} + \nabla_\mathrm{S} v$ on Γ. Thus:

$$\int_\Gamma (\mathbb{H}\nabla e(u)) \vdots (\vec{n}\otimes\nabla v)\,\mathrm{d}S = \int_\Gamma \big((\mathbb{H}\nabla e(u)):(\vec{n}\otimes\vec{n})\big)\frac{\partial v}{\partial\vec{n}} + (\mathbb{H}\nabla e(u)) \vdots (\vec{n}\otimes\nabla_\mathrm{S} v)\,\mathrm{d}S$$
$$= \int_\Gamma \big((\mathbb{H}\nabla e(u)):(\vec{n}\otimes\vec{n})\big)\frac{\partial v}{\partial\vec{n}} - \mathrm{div}_\mathrm{S}\big((\mathbb{H}\nabla e(u))\cdot\vec{n}\big)v$$
$$+ (\mathrm{div}_\mathrm{S}\vec{n})\big((\mathbb{H}\nabla e(u)):(\vec{n}\otimes\vec{n})\big)v\,\mathrm{d}S.$$

Here we used a "surface" Green-type formula:

$$\int_\Gamma w:((\nabla_\mathrm{S} v)\otimes\vec{n})\,\mathrm{d}S = \int_\Gamma (\mathrm{div}_\mathrm{S}\vec{n})(w:(\vec{n}\otimes\vec{n}))v - \mathrm{div}_\mathrm{S}(w\vec{n})v\,\mathrm{d}S,$$

cf. also (B.3.25) on p. 520. Thus (5.4.4) can be re-written as:

$$\int_\Omega -\mathrm{div}(\mathbb{C}e(u))\cdot v + \mathrm{div}^2(\mathbb{H}\nabla e(u))\cdot v\,\mathrm{d}x = \int_\Omega f\cdot v\,\mathrm{d}x + \int_{\Gamma_\mathrm{N}} g\cdot v\,\mathrm{d}S$$
$$- \int_\Gamma (\mathbb{C}e(u)):(v\otimes\vec{n}) + \big((\mathbb{H}\nabla e(u)):(\vec{n}\otimes\vec{n})\big)\frac{\partial v}{\partial\vec{n}} - \mathrm{div}_\mathrm{S}\big((\mathbb{H}\nabla e(u))\cdot\vec{n}\big)v$$
$$+ (\mathrm{div}_\mathrm{S}\vec{n})\big((\mathbb{H}\nabla e(u)):(\vec{n}\otimes\vec{n})\big)v - \mathrm{div}(\mathbb{H}\nabla e(u)):(v\otimes\vec{n})\,\mathrm{d}S. \qquad (5.4.5)$$

From this, assuming smooth domain Ω, we can read the underlying boundary-value problem, corresponding to the above weak formulation (5.4.2), now in the classical formulation:

$$\mathrm{div}(\mathbb{C}e(u) - \mathrm{div}(\mathbb{H}\nabla e(u))) + f = 0 \qquad \text{on } \Omega, \qquad (5.4.6a)$$
$$(\mathbb{C}e(u) - \mathrm{div}(\mathbb{H}\nabla e(u)))\vec{n} - \mathrm{div}_\mathrm{S}\big((\mathbb{H}\nabla e(u))\cdot\vec{n}\big) = g \qquad \text{on } \Gamma_\mathrm{N}, \qquad (5.4.6b)$$

$$u = w_D \qquad \text{on } \Gamma_D, \tag{5.4.6c}$$

$$(\mathbb{H}\nabla e(u)){:}(\vec{n}\otimes\vec{n}) = 0 \qquad \text{on } \Gamma. \tag{5.4.6d}$$

The particular equations (5.4.6a,b,d) can be obtained from the integral identity (5.4.5) by choosing subsequently v with a compact support in Ω, then v with $\partial v/\partial\vec{n} = 0$ on Γ and $v|_{\Gamma_D} = 0$ on Γ_D, and eventually a general v with $v|_{\Gamma_D} = 0$ on Γ_D. Note that the term $(\mathrm{div}_S\vec{n})((\mathbb{H}\nabla e(u)){:}(\vec{n}\otimes\vec{n}))$ containing the curvature of Γ, i.e. $-\frac{1}{2}\mathrm{div}_S\vec{n}$, which occurs in (5.4.5) eventually vanish due to (5.4.6d) and thus does not occur in the left-hand side (5.4.6b). Actually, the left-hand side (5.4.6b) identifies the true *traction vector* which expands the standard traction vector in simple materials $(\mathbb{C}e(u))\vec{n}$ by two surface terms, resulting to

$$\text{traction vector} = \Big(\mathbb{C}e(u) - \mathrm{Div}(\mathbb{H}\nabla e(u))\Big)\vec{n} - \mathrm{div}_S\Big((\mathbb{H}\nabla e(u))\cdot\vec{n}\Big). \tag{5.4.7}$$

where we also used the calculus $\mathrm{div}(\mathbb{H}\nabla e(u)){:}(v\otimes\vec{n}) = (\mathrm{div}(\mathbb{H}\nabla e(u))\vec{n})\cdot v$. Let us summarize this section:

Definition 5.4.1 (Weak formulation of (5.4.6)). The function $u \in H^2(\Omega;\mathbb{R}^d)$ is called a weak solution to the boundary-value problem (5.4.6) if the integral identity (5.4.2) holds for any $v \in H^2(\Omega;\mathbb{R}^d)$ with $v|_{\Gamma_D} = 0$ and also (5.4.6c) holds.

By the direct method, relying on Korn's inequality and the strict convexity of the functional (5.4.1), we have the following assertion:

Proposition 5.4.2 (Existence/uniqueness of the weak solution to (5.4.6)). *Let $\mathbb{C},\mathbb{H} \in L^\infty(\Omega;\mathrm{SLin}(\mathbb{R}^{d\times d}_{\mathrm{sym}}))$ be positive definite, $d = 2,3$, $f \in L^1(\Omega;\mathbb{R}^d)$, $g \in L^1(\Gamma;\mathbb{R}^d)$, and $u_D \in H^{3/2}(\Gamma_D;\mathbb{R}^d)$ with $\mathrm{meas}_{d-1}\Gamma_D > 0$. Then there exists a unique weak solution according to Definition 5.4.1.*

Remark 5.4.3 (Nonlocal nonsimple materials). Recalling the concept of nonlocal materials from Sect. 2.5.2, one may think about replacing the local "nonsimple" quadratic $\mathbb{H}$-term in (5.4.1) by the quadratic functional

$$\mathscr{H}(\nabla e(u)) := \int_{\Omega\times\Omega} \mathfrak{K}(x,\tilde{x})\nabla e(u(x)-u(\tilde{x}))\,\vdots\,\nabla e(u(x)-u(\tilde{x}))\,\mathrm{d}\tilde{x}\mathrm{d}x \tag{5.4.8}$$

like in (2.5.34). Formally, one can repeat the above calculations with replacing the hyperstress $\mathbb{H}\nabla e(u)$ by the nonlocal hyperstress $\mathfrak{H}(\nabla e(u))$ with $\mathfrak{H}$ from (2.5.36) now applied on $\nabla e(u)$ instead of $\nabla^2 y$. The boundary-value problem (5.4.6) with $\mathbb{H}\nabla e(u)$ replaced by $\mathfrak{H}(\nabla e(u))$ is however to be understood really formally only, cf. the discussion in [1, Sect. 6] about some "hidden" additional boundary condition. The really essential departing point is thus the weak formulation rather than the classical boundary-value problem.

Exercise 5.4.4. Prove Proposition 5.4.2.

Exercise 5.4.5 (Other boundary conditions). Instead of a simply supported body, consider the clamped body, i.e. (5.4.6c) together with $\nabla u{\cdot}\vec{n}=0$ and realize the modifications necessary in this section.

5.5 Materials involving inelastic strains

The analog of the Kröner-Lee-Liu multiplicative decomposition (3.6.10) in the small-strain context is the Green-Naghdi [227] *additive decomposition*:

$$e(u) = e_{\rm EL} + e_{\rm IN}. \tag{5.5.1}$$

We saw in Section 5.1 that the small-strain tensor $e(u)$ arises, when introducing the displacement $u = y - \text{identity}$, from the Green-Lagrange strain tensor $E = \frac{1}{2}(F^\top F - \mathbb{I}) = \frac{1}{2}(\nabla u)^\top + \frac{1}{2}\nabla u + \frac{1}{2}(\nabla u)^\top\nabla u$ with $\mathbb{I}$ the identity matrix by neglecting the higher-order term $\frac{1}{2}(\nabla u)^\top\nabla u$, which is legitimate if ∇u is small. In combination with inelasticity, the arguments leading to (5.5.1) are

$$y = \text{identity} + u \quad\text{and}\quad F_{\rm IN} = \mathbb{I} + \Pi \quad\text{with}\quad u \text{ and } \Pi \text{ small.} \tag{5.5.2}$$

By using this ansatz in the Kröner-Lee-Liu multiplicative decomposition (3.6.10) we obtain $F_{\rm EL} = (\mathbb{I}{+}\nabla u)(\mathbb{I}{+}\Pi)^{-1}$ and then

$$\begin{aligned} E &= \frac{1}{2}(F_{\rm EL}{}^\top F_{\rm EL} - \mathbb{I}) = \frac{1}{2}(\mathbb{I}{+}\Pi)^{-\top}(\mathbb{I}{+}\nabla u)^\top(\mathbb{I}{+}\nabla u)(\mathbb{I}{+}\Pi)^{-1} - \frac{1}{2}\mathbb{I} \\ &= \frac{1}{2}(\mathbb{I} - \Pi^{-\top})(\mathbb{I}{+}(\nabla u)^\top)(\mathbb{I}{+}\nabla u)(\mathbb{I}{-}\Pi) - \frac{1}{2}\mathbb{I} + \text{ higher order terms} \\ &= e(u) - \frac{1}{2}(\Pi^\top + \Pi) + \text{ higher order terms.} \end{aligned} \tag{5.5.3}$$

In the quasistatic case with plasticity under small deformation, a rigorous passage from the multiplicative to the additive decompositions was performed in [365].

However, in some applications, the inelastic strain hardly can be considered small, although the elastic strain is still small. Thus also the displacement need not be small although. E.g. in geophysical applications, two elastic blocks can mutually move along flat fault region (say, in this direction of x_1-axis, while x_2-axis is the normal to this fault), cf. e.g. [323, 325]. Within thousands or millions of years, the displacement in the x_1-direction might indeed be large while the elastic strain can well be considered small. In such *stratified configuration*, only u_1 and $\frac{\partial u_1}{\partial x_2}$ are really large. Then also $[(\nabla u)^\top\nabla u]_{ij} = \sum_{k=1}^d \frac{\partial u_k}{\partial x_i}\frac{\partial u_k}{\partial x_j}$ may be large for $i = 2 = j$. This might (but may not!) be compensated by the combination with plastic slip. In the following assertion we consider a mere homogeneous stress-free shift in x_1-direction:

Proposition 5.5.1 (Green-Naghdi-type additive decomposition (5.5.1)**).** *Let us assume, like in (5.5.2),* $y = \text{identity} + u$ *and* $F_{\rm IN} = \mathbb{I}{+}\Pi$ *with*

$$F_{\rm EL} - \mathbb{I} \textit{ is small and } u_2 = u_3 = 0, \Pi_{ij} = 0 \textit{ if } i \neq 1 \textit{ and } j \neq 2. \tag{5.5.4}$$

Then the elastic Green-Lagrange tensor $E_{\rm EL} = \frac{1}{2}(F_{\rm EL}^\top F_{\rm EL} - \mathbb{I})$ *equals up to higher-order terms to the symmetric part of* $e(u) - e_{\rm IN}$ *provided* $F = \mathbb{I} + \nabla u$ *and* $e_{\rm IN} = \Pi$.

Proof. The assumption $\det F_{\rm pl} = 1$ in (5.5.4) causes that also Π_{11} vanishes, which further causes also that also $\frac{\partial u_1}{\partial x_1}$, together with $\frac{\partial u_1}{\partial x_2} - \Pi_{12}$ and $\frac{\partial u_1}{\partial x_3}$, are small because $F_{\rm EL} = (\mathbb{I}+\nabla u)(\mathbb{I}+\Pi)^{-1} \sim \mathbb{I}$ implies that

$$\nabla u \sim \Pi\,; \tag{5.5.5}$$

note that $\mathbb{I}+\Pi$ is indeed invertible because of the special form of Π. The additive splitting (5.5.1) can be seen when considering all small variables as zero and then to calculate

$$\begin{aligned}
E_{\rm EL} &= \frac{1}{2}(F_{\rm EL}^\top F_{\rm EL} - \mathbb{I}) = \frac{1}{2}((\mathbb{I}+\Pi)^{-\top}(\mathbb{I}+\nabla u)^\top(\mathbb{I}+\nabla u)(\mathbb{I}+\Pi)^{-1} - \mathbb{I}) \\
&= \frac{1}{2}\begin{pmatrix} 1 & \Pi_{12} & 0 \\ 0 & 1 & 0 \\ 0 & 0 & 1 \end{pmatrix}^{-\top}\begin{pmatrix} 1+\frac{\partial u_1}{\partial x_1} & \frac{\partial u_1}{\partial x_2} & \frac{\partial u_1}{\partial x_3} \\ 0 & 1 & 0 \\ 0 & 0 & 1 \end{pmatrix}^{\top}\begin{pmatrix} 1+\frac{\partial u_1}{\partial x_1} & \frac{\partial u_1}{\partial x_2} & \frac{\partial u_1}{\partial x_3} \\ 0 & 1 & 0 \\ 0 & 0 & 1 \end{pmatrix}\begin{pmatrix} 1 & \Pi_{12} & 0 \\ 0 & 1 & 0 \\ 0 & 0 & 1 \end{pmatrix}^{-1} - \frac{1}{2}\mathbb{I} \\
&= \frac{1}{2}\begin{pmatrix} 1 & 0 & 0 \\ -\Pi_{12} & 1 & 0 \\ 0 & 0 & 1 \end{pmatrix}\begin{pmatrix} 1+\frac{\partial u_1}{\partial x_1} & 0 & 0 \\ \frac{\partial u_1}{\partial x_2} & 1 & 0 \\ \frac{\partial u_1}{\partial x_3} & 0 & 1 \end{pmatrix}\begin{pmatrix} 1+\frac{\partial u_1}{\partial x_1} & \frac{\partial u_1}{\partial x_2} & \frac{\partial u_1}{\partial x_3} \\ 0 & 1 & 0 \\ 0 & 0 & 1 \end{pmatrix}\begin{pmatrix} 1 & -\Pi_{12} & 0 \\ 0 & 1 & 0 \\ 0 & 0 & 1 \end{pmatrix} - \frac{1}{2}\mathbb{I} \\
&= \frac{1}{2}\begin{pmatrix} (1+\frac{\partial u_1}{\partial x_1})^2 & (\frac{\partial u_1}{\partial x_2}-\Pi_{12})(1+\frac{\partial u_1}{\partial x_1}) & (1+\frac{\partial u_1}{\partial x_1})\frac{\partial u_1}{\partial x_3} \\ (\frac{\partial u_1}{\partial x_2}-\Pi_{12})(1+\frac{\partial u_1}{\partial x_1}) & 1+(\frac{\partial u_1}{\partial x_2}-\Pi_{12})^2 & (\frac{\partial u_1}{\partial x_2}-\Pi_{12})\frac{\partial u_1}{\partial x_3} \\ (1+\frac{\partial u_1}{\partial x_1})\frac{\partial u_1}{\partial x_3} & (\frac{\partial u_1}{\partial x_2}-\Pi_{12})\frac{\partial u_1}{\partial x_3} & 1+(\frac{\partial u_1}{\partial x_3})^2 \end{pmatrix} - \frac{1}{2}\mathbb{I} \\
&= \frac{1}{2}\begin{pmatrix} 2\frac{\partial u_1}{\partial x_1} & \frac{\partial u_1}{\partial x_2} & \frac{\partial u_1}{\partial x_3} \\ \frac{\partial u_1}{\partial x_2} & 0 & 0 \\ \frac{\partial u_1}{\partial x_3} & 0 & 0 \end{pmatrix} - \frac{1}{2}\begin{pmatrix} 0 & \Pi_{12} & 0 \\ \Pi_{12} & 0 & 0 \\ 0 & 0 & 0 \end{pmatrix} \\
&\quad + \frac{1}{2}\begin{pmatrix} (\frac{\partial u_1}{\partial x_1})^2 & (\frac{\partial u_1}{\partial x_2}-\Pi_{12})\frac{\partial u_1}{\partial x_1} & \frac{\partial u_1}{\partial x_1}\frac{\partial u_1}{\partial x_3} \\ (\frac{\partial u_1}{\partial x_2}-\Pi_{12})\frac{\partial u_1}{\partial x_1} & (\frac{\partial u_1}{\partial x_2}-\Pi_{12})^2 & (\frac{\partial u_1}{\partial x_2}-\Pi_{12})\frac{\partial u_1}{\partial x_3} \\ \frac{\partial u_1}{\partial x_1}\frac{\partial u_1}{\partial x_3} & (\frac{\partial u_1}{\partial x_2}-\Pi_{12})\frac{\partial u_1}{\partial x_3} & (\frac{\partial u_1}{\partial x_3})^2 \end{pmatrix}.
\end{aligned} \tag{5.5.6}$$

The last term is of a higher order since $\frac{\partial u_1}{\partial x_1}$, $\frac{\partial u_1}{\partial x_2} - \Pi_{12}$, and $\frac{\partial u_1}{\partial x_3}$ are small. Altogether, this yields $E_{\rm EL} \sim e(u) - \frac{1}{2}(\Pi^\top + \Pi)$, which justifies usage of the small strain and the additive splitting in (5.5.1) provided $\Pi = e_{\rm IN}$. The last equation relies on that, as already showed, $\nabla u - \Pi$ vanishes. Altogether, (5.5.6) yields $E = e(u) - \frac{1}{2}(\Pi^\top + \Pi)$, which justifies usage of the small strain and the additive splitting in (5.5.1) provided $\Pi = e_{\rm IN}$. □

Let us still comment that the assumption about incompressibility of the plastic strain, i.e. $\det F_{\rm pl} = 1$ in (5.5.4), is quite usual in plasticity, called also *isochoric plasticity*. Let us also note that the standard frame-indifference qualification of the stored energy implies that it depends on the right Cauchy-Green tensor $F^\top F$ and therefore on the symmetric Green-Lagrange tensor E, ignoring thus the antisymmetric part of $e(u){-}e_{\rm IN}$, which explains the assertion of Proposition 5.5.1.

5.6 Electro-elastic and magnetostrictive materials

Interesting coupled problems occur when electro/magnetic long-range interactions are considered. This may be due to monopolar or dipolar interactions, related to electric charges or polarization/magnetization, respectively.

Let us start with the monopolar interaction due to built-in electric charge $q = q(x)$; note that magnetic charges do not exist. It may be combined with the piezoelectric effects. Such materials are industrially exploited in various sensors or actuators. The so-called *direct piezoelectric effect* implies that some electric charge is induced in the piezoelectric material under mechanical load. Conversely, if some electric field is imposed, the material reacts with mechanical deformation which is referred to as a *converse piezoelectric effect*. In the linear setting, the constitutive relations read as [527]:

$$\sigma = \mathbb{C}e - \mathbb{P}^\top \vec{e}, \qquad \text{(converse piezoelectric effect)} \tag{5.6.1a}$$

$$\vec{d} = \mathbb{E}\vec{e} + \mathbb{P}e + qu, \qquad \text{(direct piezoelectric effect)} \tag{5.6.1b}$$

where $\mathbb{E}$ is the 2nd-order permittivity tensor, $\mathbb{P}$ is the 3rd-order piezoelectric tensor with its transposition $\mathbb{P}^\top$ understood as $[\mathbb{P}^\top]_{ijk} = \mathbb{P}_{kij}$, and $\vec{e}$ and $\vec{d}$ are intensity of electric field and the electric induction (also called electric displacement involving here also the displaced charge qu), respectively. Such (non-polarizable) electro-elastic materials are also called *piezoelectric materials* and, in contrast to Sect. 4.5.4, the spontaneous polarization and depolarizing electric field is not considered. The balance equations are then

$$\operatorname{div}\sigma + f = q\vec{e} \qquad \text{in } \Omega, \qquad \text{(force equilibrium)} \tag{5.6.2a}$$

$$\operatorname{div}\vec{d} = q \qquad \text{in } \Omega, \qquad \text{(Gauss law in electrostatics)} \tag{5.6.2b}$$

$$\text{with (5.6.1) using } e = \operatorname{sym}(\nabla u) \text{ and } \vec{e} = -\nabla\phi, \tag{5.6.2c}$$

where ϕ is the electrostatic potential. Thus (5.6.2) represents the system for u and ϕ, to be completed by some boundary conditions, e.g.

$$\sigma\cdot\vec{n} + ku = ku_\flat \quad \text{and} \quad \vec{d}\cdot\vec{n} + c\phi = c\phi_\flat \quad \text{on } \Gamma, \tag{5.6.2d}$$

where $u_\flat$ and $\phi_\flat$ are the (given) external displacement and electric potential, respectively, while k is the elastic constant and c the electric capacity of the boundary support. The right-hand side of (5.6.2a) is the so-called *Lorenz force* caused by influence of the electric field on the build-in electric charge.

The specific bulk free energy inducing the constitutive equations (5.6.1) is

$$\psi(u,e,\vec{e},\phi) = \frac{1}{2}\mathbb{C}e{:}e - \frac{1}{2}\mathbb{E}\vec{e}\cdot\vec{e} - \mathbb{P}e\cdot\vec{e} - qu\cdot\vec{e} + q\phi.$$

Then indeed $\sigma = \partial_e\psi(e,\vec{e})$ and $\vec{d} = -\partial_{\vec{e}}\psi(u,e,\vec{e})$. Interestingly, the underlying quadratic bulk potential is then $\int_\Omega \psi(u,e(u),-\nabla\phi,\phi)\,dx$ and it is convex in u but concave in ϕ. The overall potential governing the boundary-value problem (5.6.2) is then

$$(u,\phi) \mapsto \int_\Omega \frac{1}{2}\mathbb{C}e(u){:}e(u) + \mathbb{P}e(u)\cdot\nabla\phi - \frac{1}{2}\mathbb{E}\nabla\phi\cdot\nabla\phi - f\cdot u + q\phi - qu\cdot\nabla\phi\,dx + \int_\Gamma \frac{k}{2}|u|^2 - ku_\flat\cdot u - \frac{c}{2}\phi^2 + c\phi_\flat\phi\,dS. \quad (5.6.3)$$

The variational structure of (5.6.2) then relies on minimization of (5.6.3) with respect to u and its maximization with respect to ϕ.

Proposition 5.6.1 (Existence of a weak solution to (5.6.2)). *Let $\mathbb{C}$ and $\mathbb{E}$ be positive definite, $f \in L^{2^{*\prime}}(\Omega;\mathbb{R}^d)$, $q \in L^{2^{*\prime}}(\Omega)$, $u_\flat \in L^{2^{\sharp\prime}}(\Gamma;\mathbb{R}^d)$, $\phi_\flat \in L^{2^{\sharp\prime}}(\Gamma)$, and let $c > 0$ and $k > 0$. Then the boundary-value problem (5.6.2) possesses a unique weak solution $(u,\phi) \in H^1(\Omega;\mathbb{R}^d) \times H^1(\Omega)$. Even, the dependence of (u,ϕ) on the data $(f,q,u_\flat,\phi_\flat)$ is linear and continuous as a mapping $L^{2^{*\prime}}(\Omega;\mathbb{R}^d) \times L^{2^{*\prime}}(\Omega) \times L^{2^{\sharp\prime}}(\Gamma;\mathbb{R}^d) \times L^{2^{\sharp\prime}}(\Gamma) \to H^1(\Omega;\mathbb{R}^d) \times H^1(\Omega)$.*

Proof. The convex/concave potential (5.6.3) considered on the Cartesian product $H^1(\Omega;\mathbb{R}^d) \times H^1(\Omega)$ satisfies the conditions of Theorem A.3.7 on p. 496, which then ensures existence of at least one saddle point (u,ϕ). The mapping $(f,q,u_\flat,\phi_\flat) \mapsto (u,\phi)$ is obviously linear, so that the uniqueness and continuity can be proved just by apriori estimates. Here we can test (5.6.2a) by u and (5.6.2b) by ϕ. Using the boundary conditions (5.6.2d) and Young's inequality, these tests give

$$\begin{aligned}\int_\Omega \mathbb{C}e(u){:}e(u) + \mathbb{P}^\top\nabla\phi{:}e(u) + \mathbb{E}\nabla\phi\cdot\nabla\phi - \mathbb{P}e(u)\cdot\nabla\phi\,dx + \int_\Gamma k|u|^2 + c\phi^2\,dS \\ = \int_\Omega f\cdot u + q\phi\,dx + \int_\Gamma ku_\flat\cdot u + \phi_\flat\phi\,dS \\ \le C_\epsilon\|f\|^2_{L^{2^{*\prime}}(\Omega;\mathbb{R}^d)} + C_\epsilon\|q\|^2_{L^{2^{*\prime}}(\Omega)} + C_\epsilon\|u_\flat\|^2_{L^{2^{\sharp\prime}}(\Gamma;\mathbb{R}^d)} \\ + C_\epsilon\|\phi_\flat\|^2_{L^{2^{\sharp\prime}}(\Gamma)} + \epsilon\|u\|^2_{H^1(\Omega;\mathbb{R}^d)} + \epsilon\|\phi\|^2_{H^1(\Omega)}\end{aligned}$$

with any $\epsilon > 0$ to be chosen sufficiently small and with C_ϵ depending on ϵ. The desired estimate of (u,ϕ) in $H^1(\Omega;\mathbb{R}^d) \times H^1(\Omega)$ then uses Korn's inequality and Poincaré's inequality with $\Gamma_N = \Gamma$, and the identity $\mathbb{P}^\top\vec{e}{:}e = \mathbb{P}e\cdot\vec{e}$ so that $\mathbb{P}$ does not influence this estimate at all. □

The *magnetostrictive materials* just take $q = 0$ as there are no magnetic charges, otherwise the model copies the above electro-elastic model when $\vec{d}$ is the magnetic induction, $\vec{e}$ is the intensity of magnetic field, and $\mathbb{E}$ is the permeability tensor.

Let us now address still the dipolar long-range interactions that occur in ferroic materials as already mentioned in Sect. 4.5.4. The spontaneous polarization or magnetization and depolarizing or demagnetizing field are then considered. Instead of the scalar monopole (electric charge) $q = q(x)$, we now consider a given vector-valued dipole (polarization or magnetization) $\vec{q} = \vec{q}(x)$, and instead of (5.6.3), we consider the energy

$$(u,\phi) \mapsto \int_\Omega \frac{1}{2}\mathbb{C}e(u){:}e(u) - \frac{1}{2}\mathbb{E}\nabla\phi{\cdot}\nabla\phi - f{\cdot}u + \vec{q}{\cdot}\nabla\phi + (\vec{q}\otimes u){:}\nabla^2\phi\,\mathrm{d}x \\ + \int_\Gamma \frac{k}{2}|u|^2 - ku_\flat{\cdot}u - \frac{c}{2}\phi^2 + c\phi_\flat\phi\,\mathrm{d}S. \tag{5.6.4}$$

Again, it exhibits a convex/concave structure. Instead of (5.6.2) with (5.6.1) with $\mathbb{P} = 0$, we now have

$$\mathrm{div}(\mathbb{C}e(u)) + f = (\nabla\vec{q})^\top\vec{e} \qquad \text{in } \Omega, \tag{5.6.5a}$$

$$\mathrm{div}(\mathbb{E}\vec{e} + \vec{q} + \mathrm{div}(\vec{q}\otimes u)) = 0 \quad \text{with} \quad \vec{e} = -\nabla\phi \qquad \text{in } \Omega, \tag{5.6.5b}$$

$$\mathbb{C}e(u){\cdot}\vec{n} + ku = ku_\flat \;\text{ and }\; (\mathbb{E}\vec{e} + \vec{q} + \mathrm{div}(\vec{q}\otimes u)){\cdot}\vec{n} + c\phi = c\phi_\flat \qquad \text{on } \Gamma, \tag{5.6.5c}$$

where ϕ is the potential of a depolarizing or demagnetizing. Thus (5.6.5) represents the system for u and ϕ. The variational structure of (5.6.5) again relies on minimization of (5.6.4) with respect to u and its maximization with respect to ϕ.

Remark 5.6.2 (Elimination of saddle-point structure). The saddle convex/concave structure of the energies (5.6.3) and (5.6.4) is only "optical" and disappears when making maximization with respect to ϕ which is, in fact, the partial Legendre transform, cf. (A.3.16) on p. 498. In case of (5.6.3), this maximization gives, as its Euler-Lagrange equation, the boundary-value problem

$$\mathrm{div}(\mathbb{E}\nabla\phi - \mathbb{P}e(u) + qu) + q = 0 \qquad \text{in } \Omega, \tag{5.6.6a}$$

$$(\mathbb{E}\nabla\phi - \mathbb{P}e(u) - qu)\vec{n} + c\phi = c\phi_\flat \qquad \text{on } \Gamma, \tag{5.6.6b}$$

Testing (5.6.6a) by ϕ, we can (at least formally) write

$$-\int_\Omega q\phi\,\mathrm{d}x = \int_\Omega \phi\,\mathrm{div}(\mathbb{E}\nabla\phi - \mathbb{P}e(u) + qu)\,\mathrm{d}x \\ = \int_\Omega (\mathbb{P}e(u) - \mathbb{E}\nabla\phi - qu){\cdot}\nabla\phi\,\mathrm{d}x + \int_\Gamma \phi(\mathbb{E}\nabla\phi - \mathbb{P}e(u) + qu)\vec{n}\,\mathrm{d}S \\ = \int_\Omega (\mathbb{P}e(u) - \mathbb{E}\nabla\phi - qu){\cdot}\nabla\phi\,\mathrm{d}x + \int_\Gamma c(\phi_\flat - \phi)\phi\,.$$

We can then subtract it from (5.6.3) to obtain $-\frac{1}{2}\mathbb{E}\nabla\phi{\cdot}\nabla\phi - (-\mathbb{E}\nabla\phi{\cdot}\nabla\phi) = \frac{1}{2}\mathbb{E}\nabla\phi{\cdot}\nabla\phi$ as well as $-\frac{c}{2}\phi^2 - (-c\phi^2) = \frac{c}{2}\phi^2$ and to cancel the nonconvex terms $\pm qu{\cdot}\nabla\phi$, i.e. to obtain the convexity jointly in (u,ϕ) on the linear manifold determined by (5.6.6), namely

$$(u,\phi)\mapsto\int_\Omega\frac{1}{2}\mathbb{C}e(u){:}e(u)+\frac{1}{2}\mathbb{E}\nabla\phi\cdot\nabla\phi-f\cdot u+2q\phi\,\mathrm{d}x+\int_\Gamma\frac{k}{2}|u|^2-ku_\flat\cdot u+\frac{c}{2}\phi^2\,\mathrm{d}S.$$

Let us note that it is independent of $\phi_\flat$ but, in fact, $\phi_\flat$ enters the problem through the affine manifold (5.6.6).

Remark 5.6.3 (Maxwell stress). The Lorenz force $q\vec{e}$ in (5.6.2a) can be obtained as a divergence of a so-called Maxwell stress tensor

$$\sigma_M=\vec{e}\otimes\vec{d}-\frac{1}{2}(\vec{d}\cdot\vec{e})\mathbb{I}.$$

Indeed, using symmetry of $\mathbb{E}$ and of $\nabla\vec{e}$ (cf. (5.6.5b)), assuming that $\mathbb{E}$ is constant, and applying the Gauss law (5.6.2b), we can calculate

$$\begin{aligned}\operatorname{div}\sigma_M&=(\operatorname{div}\vec{d})\vec{e}+(\nabla\vec{e})\vec{d}-\frac{1}{2}((\nabla^\top\vec{d})\vec{e}+(\nabla^\top\vec{e})\vec{d})\\&=-q\vec{e}+(\nabla\vec{e})\mathbb{E}\vec{e}-\frac{1}{2}((\nabla\vec{e}\mathbb{E})^\top+(\mathbb{E}\nabla\vec{e})^\top)\vec{e}=-q\vec{e}.\end{aligned}$$

Exercise 5.6.4. Like already considered in Sections 3.5 and 4.5.4, consider the magnetization (and also polarization) not a-priori fixed and the potential (5.6.4) enhanced by terms like $\mathbb{P}e\cdot\vec{q}+\varphi(\vec{q})+\frac{1}{2}\kappa|\nabla\vec{q}|^2$ with some $\mathbb{P}\in\mathbb{R}^{d\times d\times d}$ as in (5.6.1), $\varphi:\mathbb{R}^d\to\mathbb{R}$ smooth and suitably coercive, and $\kappa>0$. Enhance the boundary-value problem (5.6.5) correspondingly, and prove existence of weak solutions.

5.7 Poroelastic materials

We will further revisit the problems in Sect. 3.6 but now, at small strains, we will have a convex structure of the underlying space at disposal. Again, we will distinguish the general *steady-state* situations (with the state together with all specific dissipation rates independent of time) and purely *static* situations (with all specific dissipation rates zero, as in Sect. 3.6). The small-strain concept will allow us to use the Schauder fixed-point method and to obtain a lot of existence results also for the steady-state non-static situation in Sect. 5.7.3 and various generalization of it in Sect. 5.7.4.[5]

[5] See also [462].

5.7.1 Examples of convex stored energies

Like in (3.6.3), we consider the flux of the diffusant governed by a *generalized Fick law*

$$j = -\mathbb{M}\nabla\mu \quad \text{with} \quad \mu = \partial_c\varphi \tag{5.7.1}$$

where $\mathbb{M} = \mathbb{M}(c)$ is a mobility matrix and $\varphi = \varphi(e, c)$ is the stored energy dependent, beside the small strain e, also on the concentration c of the diffusant. The momentum equilibrium equation coupled with the continuity equation $\operatorname{div} j = 0$ now results to the system:

$$\operatorname{div}\sigma + f = 0 \qquad \text{with} \quad \sigma = \partial_e\varphi(e(u), c), \tag{5.7.2a}$$

$$\operatorname{div}(\mathbb{M}(c)\nabla\mu) = 0 \qquad \text{with} \quad \mu = \partial_c\varphi(e(u), c). \tag{5.7.2b}$$

It is to be accompanied by suitable boundary conditions, e.g.

$$u = 0 \ \text{ on } \Gamma_{\rm D}, \qquad \sigma\vec{n} = g \ \text{ on } \Gamma_{\rm N}, \tag{5.7.3a}$$

$$\mathbb{M}(c)\nabla\mu\cdot\vec{n} + \alpha\mu = \alpha\mu_{\rm ext} \ \text{ on } \Gamma, \tag{5.7.3b}$$

where $\mu_{\rm ext} = \mu_{\rm ext}(x)$ is the chemical potential of the external environment and $\alpha = \alpha(x) \geq 0$ is the phenomenological coefficient of permeability of the boundary.

As in Sect. 3.6, two alternatives can be distinguished, namely that the strain is directly influenced by a concentration c of the diffusant which diffuses through the material or the diffusant has its own pressure which influences the overall stress and eventually indirectly the strain, too. See again Figure 3.4 on p. 74 for various microscopical explanation. These two options microscopically reflect that the diffusant does not directly contribute to the pressure but can change the reference configuration by influencing the atomic grid (as in the mentioned metal-hydrid transformation) or the the diffusant flows though pores under its pressure and, from the macroscopical viewpoint, the resulted porous solid is phenomenologically homogenized.

The simplest form of the free energy in the first option is

$$\varphi(e, c) = \frac{1}{2}\mathbb{C}e_{\rm EL}{:}e_{\rm EL} + k\Big(c\Big(\ln\frac{c}{c_{\rm eq}} - 1\Big) + \delta_{[0,+\infty)}(c)\Big) \quad \text{with} \quad e_{\rm EL} := e - Ec + Ec_{\rm eq}, \tag{5.7.4a}$$

with $k \geq 0$ a coefficient weighting chemical versus mechanical effects and $E \in \mathbb{R}^{d\times d}_{\rm sym}$ a matrix of the swelling coefficients (which, in isotropic materials, is an identity matrix up to a coefficient) and with $c_{\rm eq}$ an equilibrium concentration minimizing $\varphi(0, \cdot)$. This yields the stress

$$\sigma = \partial_e\varphi(e, c) = \mathbb{C}e_{\rm EL} = \mathbb{C}(e - Ec + Ec_{\rm eq}) \tag{5.7.4b}$$

and the chemical potential $\mu = \partial_c\varphi(e,c) = k\ln c - E{:}\sigma$ for $c > 0$, or, taking into account that $\varphi(e,\cdot) : \mathbb{R} \to \mathbb{R}\cup\{+\infty\}$ is a proper convex function which is nonsmooth, rather

$$\mu \in \partial_c\varphi(e,c) = \begin{cases} \{k\ln(c/c_{\rm eq}) - E{:}\sigma\} & \text{if } c > 0, \\ \emptyset & \text{if } c \le 0; \end{cases} \tag{5.7.4c}$$

note that the last term $E{:}\sigma$ is the pressure.

The latter option relies on the idea that the diffusant (fluid) fully occupies the pores (i.e. so-called saturated flow) whose volume is proportional to tr $e(u)$ through a coefficient $\beta > 0$ and its pressure is

$$p_{\rm FLUID} = M(\beta{\rm tr}\, e(u) - c + c_{\rm eq}) \tag{5.7.5}$$

with M and β the *Biot modulus* and the *Biot coefficient* as in (3.6.9). In addition, $c_{\rm eq}$ denotes the equilibrium concentration considered here fixed, and it has the meaning of the porosity of the material. This pressure is then summed up with the stress $\sigma_{\rm el}$ in the elastic solid. In isotropic materials, the total stress is then $\sigma = \sigma_{\rm el} + \beta p_{\rm FLUID}\mathbb{I}$. In such simplest variant, it leads to the potential

$$\varphi(e,c) = \frac{1}{2}\mathbb{C}e{:}e + \frac{1}{2}M(\beta{\rm tr}\, e{-}c{+}c_{\rm eq})^2 + k\Big(c\Big(\ln\frac{c}{c_{\rm eq}} - 1\Big) + \delta_{[0,+\infty)}(c)\Big), \tag{5.7.6a}$$

which yields the stress and the chemical potential

$$\sigma = \partial_e\varphi(e,c) = \mathbb{C}e + \beta M(\beta{\rm tr}\, e{-}c{+}c_{\rm eq})\mathbb{I} \quad \text{and} \tag{5.7.6b}$$

$$\mu \in \partial_c\varphi(e,c) = \begin{cases} \{M(c{-}\beta{\rm tr}\, e{-}c_{\rm eq}) + k\ln(c/c_{\rm eq})\} & \text{if } c > 0, \\ \emptyset & \text{if } c \le 0. \end{cases} \tag{5.7.6c}$$

Then, choosing still a standard ansatz $\mathbb{M}(c) = c\mathbb{M}_0$, (5.7.1) turns into

$$j = -\mathbb{M}\nabla\mu = \underbrace{-\,c\mathbb{M}_0\nabla p}_{\substack{\text{Darcy}\\ \text{law}}}\ \underbrace{-\mathbb{M}_0 k\nabla c}_{\substack{\text{Fick}\\ \text{law}}} \quad \text{with } p = \begin{cases} E{:}\mathbb{C}(e{-}Ec{+}Ec_{\rm eq}) & \text{in case (5.7.4)}, \\ M(\beta{\rm tr}\, e{-}c{+}c_{\rm eq}) & \text{in case (5.7.6)} \end{cases}$$

provided $c > 0$. Depending on k, either Darcy's mechanism or the Fick's one may dominate, as indicated on Figure 3.4. Note also that $|p| = \mathscr{O}(|E|)$ in the case (5.7.4). An interesting phenomenon is that the equilibrium concentration $c_{\rm eq}$ does not influence $\nabla\mu$ and can influence the solution only through the boundary conditions (5.7.3). The mass-conservation equation (5.7.8b) reveals that the pressure gradient ∇p is needed and it also reveals an "optical" difficulty because there is no obvious estimate on ∇p.

Yet, the standard definition of a weak solution to the boundary-value problem (5.7.2)–(5.7.3) avoids explicit occurrence of ∇p, and it indeed works if $\mathbb{M}$ is constant; note that the fixed-point argument used in the proofs of all "non-static" Propositions 5.7.5–5.7.12 becomes rather trivial because the distribution of the chemical potential μ is then fully determined by $\mu_{\rm ext}$ in the boundary condition (5.7.18b).

Quite surprisingly, the steady-state models allow for a lot of results without any regularization even when $\mathbb{M}$ depends on c, in contrast to the evolution situations, cf. Sections 7.6 and 8.6.

5.7.2 Static problems

In analog with Section 3.6 and here also as a preliminary step towards general steady-state non-static situations, we begin with special steady-state problems where the dissipation rate is not only constant in time but just zero. Then all transport processes vanish and, like in Section 3.6, these problems enjoy a full variational structure at least in the sense that some (if not all) of their solutions can be obtained by such way as critical points. As we already said, such problems are called *static*. We will see later in Sections 7.6 and 8.6 that the diffusion is related with the dissipation rate (and entropy and heat production rate), namely that the overall dissipation (or also the heat-production) rate is $\int_\Omega \mathbb{M}\nabla\mu{\cdot}\nabla\mu\,\mathrm{d}x$. This implies $\nabla\mu$ everywhere on Ω, and in particular μ constant if Ω is connected (let us denote this constant by $\bar\mu$) and (5.7.18b) reducing to $\alpha(\bar\mu-\mu_{\rm ext})=0$ on Γ.

As in Section 3.6, solvability of such problem requires either the system to be in equilibrium with the external environment or to be isolated. The former option is rather trivial as it fixes $\mu=\bar\mu$ and then one can eliminate c provided $\varphi(e,\cdot)$ is convex. As in (3.6.6) we can even write a bit more specifically

$$c=\partial_\mu\varphi^*(e,\bar\mu)\,. \tag{5.7.7}$$

Note that, even if $\varphi(e,\cdot)$ is not smooth as in the examples in Section 5.7.1, $\varphi^*(e,\cdot)$ is indeed single-valued if the natural requirement $\partial^2_{cc}\varphi>0$ holds.

Again, the latter option (i.e. the boundary permeability coefficient $\alpha=0$) is more interesting. When the profile $e=e(u(x))$ is known, the overall amount of diffusant $\int_\Omega c\,\mathrm{d}x=C_{\rm total}$ depends monotonically on $\bar\mu$ due to (5.7.7), which allows us to specify $\bar\mu$ if $C_{\rm total}$ is given. In this isolated situation, it is again natural to prescribe the total amount of diffusant (3.6.7), i.e. $\int_\Omega c\,\mathrm{d}x=C_{\rm total}$. Actually, in evolution variant of such system, $C_{\rm total}$ is determined by the initial conditions, cf. (7.6.1) on p. 316.

The general steady-state system (3.6.1) then modifies to

$$\mathrm{div}\,\partial_e\varphi(e(u),c)+f=0, \tag{5.7.8a}$$

$$\partial_c\varphi(e(u),c)\ni\bar\mu=\ \text{some constant}, \tag{5.7.8b}$$

to be coupled with (3.6.7) and with the boundary conditions (5.7.3a). Let us recall that $\varphi(e,\cdot)$ is allowed to be nonsmooth so that $\partial_c\varphi$ maybe set-valued so that (5.7.8b) is an inclusion rather than equation. The mentioned variational structure consists in the following constrained minimization problem is of a certain relevance:

$$\left.\begin{array}{l}\text{Minimize } (u,c) \mapsto \displaystyle\int_\Omega \varphi(e(u),c) - f\cdot u\,\mathrm{d}x - \int_{\Gamma_\mathrm{N}} g\cdot u\,\mathrm{d}S \\ \text{subject to } \displaystyle\int_\Omega c\,\mathrm{d}x = C_\mathrm{total}, \quad u\in H^1_\mathrm{D}(\Omega;\mathbb{R}^d), \quad c\in L^1(\Omega). \end{array}\right\} \tag{5.7.9}$$

In the following theorem, we again use the concept of variational solutions rather than weak solutions.

Proposition 5.7.1 (Existence of static solutions). *Let* $\varphi: \mathbb{R}^{d\times d}_\mathrm{sym}\times\mathbb{R}\to\mathbb{R}\cup\{+\infty\}$ *be convex, lower semicontinuous, and coercive in the sense that* $\varphi(e,c)\ge\epsilon|e|^2+\epsilon|c|^{1+\epsilon}$ *for some* $\epsilon>0$, $\mathrm{meas}_{d-1}(\Gamma_\mathrm{D})>0$, $f\in L^{2^{*\prime}}(\Omega;\mathbb{R}^d)$, *and* $g\in L^{2^{\#\prime}}(\Gamma_\mathrm{N};\mathbb{R}^d)$. *Moreover, let the body is isolated (i.e.* $\alpha=0$*) and the overall content* C_total *be given, assuming* $\varphi(0,C_\mathrm{total}/\mathrm{meas}_d(\Omega))<+\infty$. *Then the boundary-value problem (5.7.8) with boundary conditions (5.7.3a) and with the side condition (3.6.7) possesses at least one variational solution* $(u,c)\in H^1_\mathrm{D}(\Omega;\mathbb{R}^d)\times L^{1+\varepsilon}(\Omega)$ *whose chemical potential is constant over* Ω *in the sense that there exists a constant* $\bar\mu\in\partial_c\varphi(e(u),c)$ *a.e. on* Ω.

Proof. We use the direct method for minimization of the convex coercive functional subject to the affine constraint (5.7.9). Note that the assumption $\varphi(0,C_\mathrm{total}/\mathrm{meas}_d(\Omega))<+\infty$ guarantees that this problem is feasible, i.e. its admissible set contains at least one pair $(u,c):=(0,C_\mathrm{total}/\mathrm{meas}_d(\Omega))$. We thus obtain a minimizer $(u,c)\in H^1_\mathrm{D}(\Omega;\mathbb{R}^d)\times L^{1+\varepsilon}(\Omega)$.

As the functional in (5.7.9) is convex, lower semicontinuous, and the constraint is affine, introducing the Lagrange multiplier $\bar\mu$ to the scalar-valued constraint (3.6.7) which is involved in (5.7.9), this problem is equivalent to finding a critical point of the functional (a so-called Lagrangian):

$$(u,c,\bar\mu)\mapsto J(u,c,\bar\mu):=\int_\Omega \varphi(e(u),c)-f\cdot u-\bar\mu c\,\mathrm{d}x-\int_{\Gamma_\mathrm{N}} g\cdot u\,\mathrm{d}S+C_\mathrm{total}\bar\mu. \tag{5.7.10}$$

If this functional is smooth, from putting the Gâteaux derivative with respect to c zero, we can read that $\partial_c\varphi(e(u),c)=\bar\mu$ on Ω, i.e. the chemical potential is constant. In a general nonsmooth case, by disintegration of the condition $\partial_c J(u,c,\bar\mu)\ni 0$ we obtain the inclusion (5.7.8b) a.e. on Ω. From putting $\partial_{\bar\mu}J(u,c,\bar\mu)=0$, we can read that $\int_\Omega -c\,\mathrm{d}x+C_\mathrm{total}=0$, i.e. that the affine constraint (3.6.7) is satisfied. Eventually, if the Gâteaux derivative $\partial_u J(u,c,\bar\mu)$ exists, it must be zero, and we can read the equilibrium equation (5.7.8a) with the boundary conditions (5.7.3a) in the weak formulation. In a general case, (5.7.8a)–(5.7.3a) holds formally in a variational sense, i.e. the solution u is considered as a minimizer of $J(\cdot,c,\bar\mu)$. □

The above proof reveals the role of the scalar $\bar\mu$ as the Lagrange multiplier to the scalar constraint $\int_\Omega c\,\mathrm{d}x=C_\mathrm{total}$ and that it is a vital part of the solution.

If φ is strictly convex (as e.g. in the examples from Sect. 5.7.1) the minimization problem (5.7.9) has a unique solution, although the relation to solutions of the static problem in question may be more delicate, cf. Remark 5.7.3 below.

An interesting and useful generalization of the above basic scenario is towards a *multi-component fluid* with $m\ge 2$ components which can even be *electrically charged* with specific charges $z=(z_1,...,z_m)$. Also the elastic medium can be charged by some dopant with the specific charge $z_{\rm DOP}$. In the static problems there are no chemical reactions. An example for such situations is a poroelastic polymer negatively charged during the manufacturing process filled by water and ionized hydrogen, (i.e. protons) consisting a two-component diffusant.[6] The scalar-valued chemical potential is now to be replaced by an $\mathbb{R}^m$-valued *electro-chemical potential*

$$\mu=\partial_c\varphi(e,c)+z\phi$$

with ϕ the electrostatic potential. In static problems, again μ is constant. The system (5.7.8) then augments to

$$\begin{aligned}
&{\rm div}\,\partial_e\varphi(e(u),c)+f=z_{\rm DOP}\nabla\phi && \text{on } \Omega, && (5.7.11a)\\
&\partial_c\varphi(e,c)+z\phi\ni\mu=\text{ some constant} && \text{on } \Omega, && (5.7.11b)\\
&{\rm div}(\varepsilon\nabla\phi)+z\cdot c+z_{\rm DOP}={\rm div}(z_{\rm DOP}u) && \text{on } \mathbb{R}^d, && (5.7.11c)\\
&\int_\Omega c\,{\rm d}x=C_{\rm total}=\text{a given constant}\in\mathbb{R}^m. && && (5.7.11d)
\end{aligned}$$

The right-hand side $z_{\rm DOP}\nabla\phi$ in (5.7.11a) is the (negative) *Lorenz force* acting on a charged elastic solid in the electrostatic field. In (5.7.11c), $\varepsilon=\varepsilon(x)>0$ denotes the permittivity and the equation (5.7.11c) itself is the rest of the full Maxwell system if all evolution and magnetic effects are neglected. Note that (5.7.11c) is to be solved on the whole universe with the natural "boundary" condition $\phi(\infty)=0$, assuming naturally that z, c, and $z_{\rm DOP}$ are extended on $\mathbb{R}^d\backslash\Omega$ by zero. Actually, the physical units are fixed for notational simplicity in such a way that the Faraday constant (which should multiply the charges in (5.7.11c)) equals 1.

It is interesting that the underlying potential is not convex and, instead of a minimizer as in (5.7.9), we are now to seek a more general critical point, namely a saddle point solving the variational problem:

$$\left.\begin{aligned}
&\text{Min/max } (u,c,\phi)\mapsto\int_\Omega\Big(\varphi(e(u),c)+(z\cdot c+z_{\rm DOP})\phi+z_{\rm DOP}\nabla\phi\cdot u\\
&\qquad\qquad -f\cdot u\Big){\rm d}x-\int_{\mathbb{R}^d}\frac{\varepsilon}{2}|\nabla\phi|^2\,{\rm d}x-\int_{\Gamma_{\rm N}}g\cdot u\,{\rm d}S,\\
&\text{subject to}\int_\Omega c\,{\rm d}x=C_{\rm total},\ \ u\in H^1_{\rm D}(\Omega;\mathbb{R}^d),\ \ c\in L^1(\Omega;\mathbb{R}^m),\ \ \phi\in H^1(\mathbb{R}^d).
\end{aligned}\right\}\quad(5.7.12)$$

This convex/concave structure is sometimes referred under the name of *electrostatic Lagrangian* [422, Sect.3.2] and is consistent with a convex structure of the internal energy, cf. the argumentation in Remark 5.7.7 or (8.7.6)–(8.7.7) on p. 407.

Proposition 5.7.2. *Let* $\varphi:\mathbb{R}^{d\times d}_{\rm sym}\times\mathbb{R}^m\to\mathbb{R}\cup\{+\infty\}$ *and* $\Gamma_{\rm D}$, f, *and* g *be qualified as in Proposition 5.7.1, and moreover let* $z_{\rm DOP}\in L^\infty(\Omega)$, $z\in L^\infty(\Omega;\mathbb{R}^m)$, *and* $\varepsilon\in L^\infty(\mathbb{R}^d)$

[6] An important engineering application is hydrogen polymer-electrolyte fuel cells used in Apollo project where the negatively-charged poroelastic polymer is Nafion®, cf. e.g. [299, 422, 538].

have a positive infimum. Again, let the body be isolated (i.e. $\alpha = 0$) and the overall contents $C_{\rm total} \in \mathbb{R}^m$ be given, assuming again $\varphi(0, C_{\rm total}/{\rm meas}_d(\Omega)) < +\infty$. Then the boundary-value problem (5.7.11) with boundary conditions (5.7.3a) possesses at least one variational solution $(u,c) \in H^1_{\rm D}(\Omega;\mathbb{R}^d) \times L^{1+\varepsilon}(\Omega;\mathbb{R}^m)$ whose $\mathbb{R}^m$-valued electrochemical potential is constant over Ω in the sense that there exists a constant $\bar{\mu} \in \partial_c\varphi(e(u),c) + z\phi$ a.e. on Ω.

Proof. Existence of a saddle point in this problem is to be seen by the classical von Neumann theorem A.3.7 as well as that it yields some solution to the system (5.7.11) with μ being the (vector-valued) Lagrange multiplier to the constraint in (5.7.12). The assumption $\varphi(0, C_{\rm total}/{\rm meas}_d(\Omega)) < +\infty$ again guarantees the feasibility of (5.7.12). □

Remark 5.7.3 (Uniqueness). Interestingly, it is not clear whether the solution of the static problem in Proposition 5.7.1 is unique, even if φ is strictly convex so that the solution obtained in the proof of Proposition 5.7.1 by solving (5.7.9) is unique. Analogous comment is relevant to the problem in Proposition 5.7.2. As far as all steady-state but non-static problems, the nonuniqueness is rather to be expected.

Remark 5.7.4 (Nonconvex energy φ). Some applications uses nonconvex energies in terms of the concentration c. A double-well $\varphi(e,\cdot)$ may model two-phase systems. The capillarity regularization like (3.6.12) is then needed. This results here to the constraint minimization problem instead of (5.7.9):

$$\left.\begin{array}{l}\text{Minimize } (u,c) \mapsto \displaystyle\int_\Omega \varphi(e(u),c) + \frac{\kappa}{2}|\nabla c|^2 - f\cdot u\,{\rm d}x - \int_{\Gamma_{\rm N}} g\cdot u\,{\rm d}S \\ \text{subject to } \displaystyle\int_\Omega c\,{\rm d}x = C_{\rm total}, \quad u \in H^1_{\rm D}(\Omega;\mathbb{R}^d), \quad c \in H^1(\Omega)\end{array}\right\} \tag{5.7.13}$$

with $\kappa > 0$ presumably small. The existence of its minimizer is by a direct-method argument. Then one can again use the 1st-order optimality conditions to show that it solves the static variant of the system. Note that, for the steady-state nonstatic variant, the arguments based on the Schauder fixed point in Proposition 5.7.5 below fails in this case and thus the existence of a solution in the non-static steady-case is not granted, similarly as in the large strains in Section 3.6. Autonomous oscillations in evolutionary situations documented in literature suggest that there is a factual reason for possible non-existence of steady states.

5.7.3 Steady states of stress-assisted diffusion

The peculiarity behind the non-static steady-state problem (3.6.1) is that it mixes stored energy and the dissipation energy, cf. (8.7.6) below. Thus one should not expect a simple variational structure which is usual in problems governed merely by stored energy. To illustrate this peculiarity more, let us consider $\varphi(e,c) = \frac{1}{2}\mathbb{C}(e-Ec)$:

$(e-Ec)$ and write formally the underlying operator when ignoring the boundary condition, i.e.

$$\begin{pmatrix} u \\ c \end{pmatrix} \mapsto \begin{pmatrix} -\mathrm{div}(\mathbb{C}(e(u)-Ec)) \\ -\Delta(\mathbb{C}E:(Ec-e(u))) \end{pmatrix}.$$

This linear operator is obvously nonsymmetric (thus does not have any potential) and nonmonotone (and even not pseudomonotone[7]) due to the 3rd-order term $-\Delta\mathbb{C}E:e(u)$. Therefore, standard methods does not seem to be applicable. Yet, advantageously, the variational structure of the static problems in Section 5.7.2 can be combined with a carefully constructed fixed point.

Proposition 5.7.5 (Existence of steady states). *Let $\varphi : \mathbb{R}^{d\times d}_{\mathrm{sym}} \times \mathbb{R} \to \mathbb{R} \cup \{+\infty\}$ be lower semicontinuous, strictly convex and coercive in the sense that $\varphi(e,c) \geq \epsilon|e|^2 + \epsilon|c|^q$ for some $q > 2^{*\prime}$ and $\epsilon > 0$, $\mathbb{M} : \mathbb{R} \to \mathbb{R}^{d\times d}$ is continuous, bounded, and uniformly positive definite, $\alpha \geq 0$ with $\alpha > 0$ on a positive-measure part of Γ, $\mathrm{meas}_{d-1}(\Gamma_{\mathrm{D}}) > 0$, $f \in L^{2^{*\prime}}(\Omega;\mathbb{R}^d)$, $g \in L^{2^{\sharp\prime}}(\Gamma_{\mathrm{N}};\mathbb{R}^d)$, and $\mu_{\mathrm{ext}} \in L^{2^{\sharp\prime}}(\Gamma)$. Then the boundary-value problem (3.6.1) with boundary conditions (5.7.3) possesses at least one variational solution $(u,c) \in H^1_{\mathrm{D}}(\Omega;\mathbb{R}^d) \times L^q(\Omega)$ with the corresponding chemical potential $\mu \in H^1(\Omega) \cap L^\infty(\Omega)$.*

Proof. We construct the single-valued mapping $\tilde{\mu} \mapsto (u,c) \mapsto \mu$ for which the Schauder fixed-point theorem will be used. First, fixing $\tilde{\mu} \in H^1(\Omega)$, we solve

$$\left.\begin{array}{l} \text{Minimize } \ (u,c) \mapsto \displaystyle\int_\Omega \varphi(e(u),c) - \tilde{\mu}c - f\cdot u \,\mathrm{d}x - \int_{\Gamma_{\mathrm{N}}} g\cdot u \,\mathrm{d}S \\ \text{subject to } \ u \in H^1_{\mathrm{D}}(\Omega;\mathbb{R}^d) \ \text{ and } \ c \in L^q(\Omega). \end{array}\right\} \tag{5.7.14}$$

Note that the term $\tilde{\mu}c$ is integrable and $(\tilde{\mu},c) \mapsto \tilde{\mu}c : H^1(\Omega) \times L^q(\Omega) \to L^1(\Omega)$ is (weak,weak)-continuous due to the condition $q > 2^{*\prime}$ and the Rellich theorem B.3.3 on p. 515. Due to the assumed strict convexity of φ, this problem has a unique solution (u,c). It is also important that this solution depends continuously on $\tilde{\mu}$ in the sense that $\tilde{\mu} \mapsto (u,c) : H^1(\Omega) \to H^1_{\mathrm{D}}(\Omega;\mathbb{R}^d) \times L^q(\Omega)$ is (weak,strong)-continuous, which can be seen when exploiting the assumed strict convexity of φ, cf. [522].

Having $c \in L^1(\Omega)$, we then

$$\left.\begin{array}{l} \text{Minimize } \ \mu \mapsto \displaystyle\int_\Omega \frac{1}{2}\mathbb{M}(c)\nabla\mu\cdot\nabla\mu \,\mathrm{d}x + \int_{\Gamma_{\mathrm{N}}} \frac{\alpha}{2}\mu^2 - \alpha\mu_{\mathrm{ext}}\mu \,\mathrm{d}S \\ \text{subject to } \ \mu \in H^1(\Omega). \end{array}\right\} \tag{5.7.15}$$

Due to the assumed positive definiteness of $\mathbb{M}$ and the (partial) positivity of α, the problem (5.7.15) has a unique solution μ. It is important that the mapping

[7] A bounded (in general nonlinear) operator A between a Banach space and its dual is called pseudomonotone if, for any sequence $u_k \to u$ weakly, $\limsup_{k\to\infty}\langle A(u_k), u_k - u\rangle \leq 0$ implies $\langle A(u), u - v\rangle \leq \liminf_{k\to\infty}\langle A(u_k), u_k - v\rangle$. Pseudomonotonicity is a relevant generalization of both monotonicity and a compact-operator concept ensuring surjectivity of A, i.e. existence of solutions to the equation $A(u) = f$.

$c \mapsto \mu : L^1(\Omega) \to H^1(\Omega)$ is (strong,weak)-continuous. Actually, even (strong,strong)-continuity can easily be proved but it is not needed for our fixed-point argument.

It should be emphasized that μ from (5.7.15) does not need to be a chemical potential corresponding to (u,c). Yet, we will show that it is if $\mu = \tilde{\mu}$. Such pair $(\mu,\tilde{\mu})$ does exist due to the Schauder fixed point theorem. Here we also used that the solution μ ranges an a-priori bounded in $H^1(\Omega)$ because $\mathbb{M}(\cdot)$ is assumed uniformly positive definite.

The 1st-order optimality conditions for (5.7.14) compose from the partial Gâteaux derivatives with respect to u and to c to vanish. The former condition means the Euler-Lagrange equation representing the weak formulation of the boundary-value problem:

$$\operatorname{div} \partial_e \varphi(e(u),c) + f = 0 \qquad \text{on } \Omega, \tag{5.7.16a}$$

$$u = 0 \text{ on } \Gamma_{\mathrm{D}} \quad \text{and} \quad \sigma \vec{n} = g \text{ on } \Gamma_{\mathrm{N}}, \tag{5.7.16b}$$

while the latter conditions written for $\tilde{\mu} = \mu$ yields

$$\partial_c \varphi(e(u),c) - \mu \ni 0 \qquad \text{on } \Omega. \tag{5.7.17}$$

The 1st-order optimality conditions for (5.7.15) means the Euler-Lagrange equation representing the weak formulation of the boundary-value problem:

$$\operatorname{div}(\mathbb{M}(c)\nabla\mu) = 0 \qquad \text{on } \Omega, \tag{5.7.18a}$$

$$\mathbb{M}(c)\nabla\mu \cdot \vec{n} + \alpha\mu = \alpha\mu_{\mathrm{ext}} \qquad \text{on } \Gamma. \tag{5.7.18b}$$

Altogether, (5.7.16)–(5.7.18) reveal that (u,c) solves the boundary-value problem (3.6.1)–(5.7.3). More precisely, if the mentioned Gâteaux derivative do not exist, (5.7.16)–(5.7.17) holds not in the weak but the variational sense as solution to (5.7.14). □

Let us now investigate the steady-state variant of the static electrically-charged multi-component problem. As in (5.7.11), we consider both the elastic solid and (some of) the components of the diffusant to be electrically charged. This electrically-charged model results to

$$\operatorname{div} \partial_e \varphi(e(u),c) + f = z_{\mathrm{DOP}} \nabla\phi \qquad \text{on } \Omega, \tag{5.7.19a}$$

$$\operatorname{div}(\mathbb{M}(c)\nabla\mu) + r(c) = 0 \quad \text{with } \mu \in \partial_c \varphi(e(u),c) + z\phi \qquad \text{on } \Omega, \tag{5.7.19b}$$

$$\operatorname{div}(\varepsilon\nabla\phi) + z{\cdot}c + z_{\mathrm{DOP}} = \operatorname{div}(z_{\mathrm{DOP}} u) \qquad \text{on } \mathbb{R}^d, \tag{5.7.19c}$$

with $r = r(c)$ the rate of chemical reactions, to be completed by the boundary conditions (5.7.3) and $\phi(\infty) = 0$.

Proposition 5.7.6. *Let again $\varphi : \mathbb{R}^{d\times d}_{\mathrm{sym}} \times \mathbb{R}^m \to \mathbb{R} \cup \{+\infty\}$ be lower semicontinuous, strictly convex, and now even uniformly convex in c and coercive in the sense that, for some $\epsilon > 0$,*

$$
\begin{aligned}
&\forall (e_1,c_1),(e_2,c_2)\in\mathbb{R}^{d\times d}_{\rm sym}\times\mathbb{R}^m \ \ \forall m_1\in\partial_c\varphi(e_1,c_1),\ m_2\in\partial_c\varphi(e_2,c_2):\\
&\quad \epsilon(|c_1|^{q-1}-|c_2|^{q-1})(|c_1|-|c_2|)\le(\partial_e\varphi(e_1,c_1)-\partial_e\varphi(e_2,c_2)):(e_1-e_2)\\
&\qquad\qquad\qquad\qquad\qquad\qquad\qquad\qquad +(m_1-m_2)\cdot(c_1-c_2) \qquad (5.7.20a)\\
&\exists q>2^{*\prime}\ \ \forall (e,c)\in\mathbb{R}^{d\times d}_{\rm sym}\times\mathbb{R}^m: \qquad \varphi(e,c)\ge\epsilon|e|^2+\epsilon|c|^q. \qquad (5.7.20b)
\end{aligned}
$$

Let further $\mathbb{M}:\mathbb{R}^m\to\mathbb{R}^{d\times d\times m}$ *be continuous, bounded, and uniformly positive definite,* $r:\mathbb{R}^m\to\mathbb{R}^m$ *continuous and bounded,* $\varepsilon\in L^\infty(\mathbb{R}^d)$ *have a positive infimum,* $\alpha\ge 0$ *with* $\alpha>0$ *on a positive-measure part of* Γ, $z_{\rm DOP}\in L^\infty(\Omega)$, $z\in L^\infty(\Omega;\mathbb{R}^m)$, $\mathrm{meas}_{d-1}(\Gamma_{\rm D})>0$, $f\in L^{2^{*\prime}}(\Omega;\mathbb{R}^d)$, $g\in L^{2^{\sharp\prime}}(\Gamma_{\rm N};\mathbb{R}^d)$, *and* $\mu_{\rm ext}\in L^{2^{\sharp\prime}}(\Omega;\mathbb{R}^m)$. *Then the boundary-value problem (5.7.19) with boundary conditions (5.7.3) and* $\phi(\infty)=0$ *possesses at least one variational solution* $(u,c)\in H^1_{\rm D}(\Omega;\mathbb{R}^d)\times L^q(\Omega;\mathbb{R}^m)$ *with the corresponding electrochemical potential* $\mu\in H^1(\Omega;\mathbb{R}^m)$.

Proof. We construct the single-valued mapping $\tilde\mu\mapsto(u,c,\phi)\mapsto\mu$ for which the Schauder fixed-point theorem will be used. First, we fix $\tilde\mu\in H^1(\Omega;\mathbb{R}^m)$ and, being motivated by (5.7.12), we modify (5.7.14) as

$$
\left.
\begin{aligned}
&\text{Min/max}\quad (u,c,\phi)\mapsto\int_\Omega\Big(\varphi(e(u),c)+(z\cdot c+z_{\rm DOP})\phi+z_{\rm DOP}\nabla\phi\cdot u\\
&\qquad\qquad\qquad -\tilde\mu\cdot c-f\cdot u\Big)\,\mathrm{d}x-\int_{\mathbb{R}^d}\frac{\varepsilon}{2}|\nabla\phi|^2\,\mathrm{d}x-\int_{\Gamma_{\rm N}}g\cdot u\,\mathrm{d}S,\\
&\text{subject to}\ \ u\in H^1_{\rm D}(\Omega;\mathbb{R}^d),\ \ c\in L^q(\Omega;\mathbb{R}^m),\ \ \phi\in H^1(\mathbb{R}^d).
\end{aligned}
\right\}\qquad (5.7.21)
$$

Due to the assumed strict convexity of φ, this problem has a unique solution (u,c,ϕ) which depends continuously on $\tilde\mu$ in the sense that $\tilde\mu\mapsto(u,c,\phi):H^1(\Omega;\mathbb{R}^m)\to H^1_{\rm D}(\Omega;\mathbb{R}^d)\times L^q(\Omega;\mathbb{R}^m)\times H^1(\mathbb{R}^d)$ is (weak,strong)-continuous. More in detail, the existence of this saddle point is the classical von Neumann theorem A.3.7. The uniqueness can be proved by analyzing the optimality conditions (5.7.19a) and (5.7.19c) together with $\partial_c\varphi(e(u),c)+z\phi\ni\tilde\mu$ written as

$$
m+z\phi=\tilde\mu\qquad\text{for some }\ m\in\partial_c\varphi(e(u),c), \qquad (5.7.22)
$$

considered for two solutions (u_i,c_i,ϕ_i), $i=1,2$, and subtracted. Using the abbreviation $u_{12}=u_1-u_2$, $c_{12}=c_1-c_2$, $\phi_{12}=\phi_1-\phi_2$, and $m_{12}=m_1-m_2$, this results to the system

$$
\begin{aligned}
&\mathrm{div}(\partial_e\varphi(e(u_1),c_1)-\partial_c\varphi(e(u_2),c_2))=z_{\rm DOP}\nabla\phi_{12} &&\text{on }\Omega, &&(5.7.23a)\\
&\mathrm{div}(\varepsilon\nabla\phi_{12})+z\cdot c_{12}=\mathrm{div}(z_{\rm DOP}u_{12}) &&\text{on }\mathbb{R}^d, &&(5.7.23b)\\
&m_{12}+z\phi_{12}=0 &&\text{on }\Omega, &&(5.7.23c)
\end{aligned}
$$

with the homogeneous boundary conditions for (5.7.23a), i.e. $u_{12}=0$ on $\Gamma_{\rm D}$ and $(\partial_e\varphi(e(u_1),c_1)-\partial_e\varphi(e(u_2),c_2))\vec n=0$ on $\Gamma_{\rm N}$. Testing (5.7.23a) by u_{12} and also (5.7.23c) by c_{12} (integrated it over Ω) and (5.7.23b) by ϕ_{12} (integrated it over $\mathbb{R}^d$) gives

$$
\begin{aligned}
&\int_\Omega \big(\partial_e\varphi(e(u_1),c_1)-\partial_e\varphi(e(u_2),c_2)\big):e(u_{12})+m_{12}\cdot c_{12}\,\mathrm{d}x+\int_{\mathbb{R}^d}\varepsilon|\nabla\phi_{12}|^2\,\mathrm{d}x\\
&\quad=\int_\Omega z_{\mathrm{DOP}}\nabla\phi_{12}\cdot u_{12}-z\cdot c_{12}\,\phi_{12}\,\mathrm{d}x+\int_{\mathbb{R}^d} z\cdot c_{12}\phi_{12}+\mathrm{div}(z_{\mathrm{DOP}}u_{12})\phi_{12}\,\mathrm{d}x\\
&\quad=\int_{\mathbb{R}^d} z_{\mathrm{DOP}}\nabla\phi_{12}\cdot u_{12}-z\cdot c_{12}\,\phi_{12}+z\cdot c_{12}\phi_{12}-\mathrm{div}(z_{\mathrm{DOP}}u_{12})\phi_{12}\,\mathrm{d}x\\
&\quad=\int_{\mathbb{R}^d}\mathrm{div}(z_{\mathrm{DOP}}u_{12}\phi_{12})\,\mathrm{d}x=\big[z_{\mathrm{DOP}}u_{12}\phi_{12}\big](\infty)=0.
\end{aligned}
\tag{5.7.24}
$$

We used that $z=0$ on $\mathbb{R}^d\backslash\Omega$ and then cancellation of the terms $\pm z\cdot c_{12}\phi_{12}$ as well as that $\phi_{12}(\infty)=0$ and z_{DOP} compactly supported. From the strict monotonicity of $\partial\varphi$, we can easily see uniqueness of the saddle point of (5.7.21), needed for the Schauder fixed point. Moreover, the mentioned continuity is to be proved by taking two right-hand sides $\tilde\mu_i$, $i=1,2$, in (5.7.22). This gives rise the additional term $\int_\Omega\tilde\mu_{12}\cdot c_{12}\,\mathrm{d}x$ on the right-hand side of (5.7.24), which can be estimated by using the Hölder inequality and then, by the assumption (5.7.20a) and the uniform convexity of the $L^q(\Omega;\mathbb{R}^m)$-space, again obtain the desired (weak,strong)-continuity of $\tilde\mu\mapsto c: H^1(\Omega;\mathbb{R}^m)\to L^q(\Omega;\mathbb{R}^m)$; cf. e.g. [456, Chap. 2].

Having $c\in L^1(\Omega;\mathbb{R}^m)$, we again solve (5.7.15) now in addition with a term $-r(c)\cdot\mu$ and, as in the proof of Proposition 5.7.5, we prove that the mapping $c\mapsto\mu: L^1(\Omega;\mathbb{R}^m)\to H^1(\Omega;\mathbb{R}^m)$ is (strong,weak)-continuous. Actually, even (strong,weak)-continuity can easily be proved but it is not needed for our fixed-point argument.

It should be emphasized that μ from (5.7.15) does not need to be a chemical potential corresponding to (u,c). Yet, it is if $\mu=\tilde\mu$. Such pair $(\mu,\tilde\mu)$ does exists due to the Schauder fixed point theorem. Here we also used that the solution μ ranges an a-priori bounded subset of $H^1(\Omega)$ because of the assumed uniform positive definiteness of $\mathbb{M}(\cdot)$ and the boundedness of the reaction rates $r(\cdot)$. □

Remark 5.7.7 (Elimination of saddle-point structure). Making maximization with respect to ϕ in (5.7.12) or in (5.7.21) may eliminate the ϕ variable, cf. also e.g. [539, Sect.49.2] for a general viewpoint. Using (5.7.11c) together with the calculus

$$
\begin{aligned}
\int_\Omega z\phi\cdot c+z_{\mathrm{DOP}}\phi+z_{\mathrm{DOP}}\nabla\phi\cdot u\,\mathrm{d}x&=\int_{\mathbb{R}^d}\phi\Big(z\cdot c+z_{\mathrm{DOP}}-\mathrm{div}(z_{\mathrm{DOP}}u)\Big)\mathrm{d}x\\
&=-\int_{\mathbb{R}^d}\phi\,\mathrm{div}(\varepsilon\nabla\phi)=\int_{\mathbb{R}^d}\varepsilon|\nabla\phi|^2\,\mathrm{d}x,
\end{aligned}
\tag{5.7.25}
$$

the convex/concave problem (5.7.21) turns into the convex constrained problem:

$$
\left.
\begin{aligned}
&\text{Minimize}\ \ (u,c,\phi)\mapsto\int_\Omega\varphi(e(u),c)-\tilde\mu\cdot c-f\cdot u\,\mathrm{d}x\\
&\qquad\qquad\qquad\qquad+\int_{\mathbb{R}^d}\frac{\varepsilon}{2}|\nabla\phi|^2\,\mathrm{d}x-\int_{\Gamma_{\mathrm{N}}}g\cdot u\,\mathrm{d}S,\\
&\text{subject to}\ \ \mathrm{div}(\varepsilon\nabla\phi)+z\cdot c+z_{\mathrm{DOP}}=\mathrm{div}(z_{\mathrm{DOP}}u)\ \text{ on }\mathbb{R}^d,\\
&\qquad\qquad u\in H^1_{\mathrm{D}}(\Omega;\mathbb{R}^d),\ \ c\in L^1(\Omega;\mathbb{R}^m),\ \ \phi\in H^1(\mathbb{R}^d).
\end{aligned}
\right\}
\tag{5.7.26}
$$

Remark 5.7.8 (Towards electroneutrality). If $\varepsilon > 0$ is constant, we can further eliminate the electrostatic potential by introducing the electrical induction $\vec{d} = \varepsilon\nabla\phi$ and, defining the Banach space

$$L^2_{\mathrm{rot},\varepsilon}(\mathbb{R}^d;\mathbb{R}^d) := \Big\{ \vec{d} \in L^2(\mathbb{R}^d;\mathbb{R}^d);\ \exists \phi \in H^1(\mathbb{R}^d):\ \vec{d} = \varepsilon\nabla\phi \text{ in the sense of distributions} \Big\},$$

we can rewrite (5.7.26) as

$$\left.\begin{array}{ll}\text{Minimize} & (u,c,\vec{d}) \mapsto \displaystyle\int_\Omega \varphi(e(u),c) - \tilde{\mu}\cdot c - f\cdot u\,\mathrm{d}x \\ & \qquad\qquad + \displaystyle\int_{\mathbb{R}^d} \frac{1}{2\varepsilon}|\vec{d}|^2\,\mathrm{d}x - \int_{\Gamma_\mathrm{N}} g\cdot u\,\mathrm{d}S, \\ \text{subject to} & \operatorname{div}\vec{d} + z\cdot c + z_{\mathrm{DOP}} = \operatorname{div}(z_{\mathrm{DOP}}u) \ \text{ on } \mathbb{R}^d, \\ & u \in H^1_\mathrm{D}(\Omega;\mathbb{R}^d), \quad c \in L^1(\Omega;\mathbb{R}^m), \quad \vec{d} \in L^2_{\mathrm{rot},\varepsilon}(\mathbb{R}^d;\mathbb{R}^d). \end{array}\right\} \tag{5.7.27}$$

It reveals an asymptotics for $\varepsilon \to 0$: namely, assuming $\varepsilon(x) = \epsilon\varepsilon_0(x)$, the space $L^2_{\mathrm{rot},\varepsilon}(\mathbb{R}^d;\mathbb{R}^d)$ is independent of ϵ and then $\|\vec{d}\|_{L^2(\mathbb{R}^d;\mathbb{R}^d)} = \mathscr{O}(\epsilon^{1/2})$ for $\epsilon \to 0$ and thus

$$\|z\cdot c + z_{\mathrm{DOP}} - \operatorname{div}(z_{\mathrm{DOP}}u)\|_{H^{-1}(\mathbb{R}^d;\mathbb{R}^d)} = \mathscr{O}(\epsilon^{1/2}).$$

In particular, in the limit, one may expect the *electroneutrality*, i.e. $z\cdot c + z_{\mathrm{DOP}} - \operatorname{div}(z_{\mathrm{DOP}}u) = 0$. Without any rigorous justification, this ansatz is indeed often used in computational implementation if the specimen size is substantially bigger than the so-called Debye length to avoid spatially extremely stiff problems arising for small permittivity ε, cf. e.g. [209].

Remark 5.7.9 (Minimum entropy production principle). If the phenomenological coefficients are supposed to be constants, stationary non-equilibrium states can be characterized by a minimum of the entropy production, compatible with the external constraints imposed on the system, see [149, Sect. 3.1]. This interesting principle is manifested here in a bit trivial manner because, if $\mathbb{M}$ is constant, the system decouples and μ is uniquely determined and fixed, and the minimization of entropy production (in this isothermal case) is seen (up to the factor 1/2) just in (5.7.12).

Remark 5.7.10 (An alternative model: mechanically moved diffusant). A conceptually consistent (and perhaps sometimes more natural) modification replaces the term $z_{\mathrm{DOP}}\nabla\phi\cdot u$ in (5.7.21) by $(z\cdot c + z_{\mathrm{DOP}})\nabla\phi\cdot u$. Then the diffusant is closely bonded with the elastic medium so that it is involved also in the Lorenz force on the right-hand side of (5.7.19a), which now looks as $(z\cdot c + z_{\mathrm{DOP}})\nabla\phi$ while the electro-chemical potential ϕ in (5.7.19b) augments by the term $z(\nabla\phi\cdot u)$. The diffusion equation (5.7.19b) then looks as

$$\dot{c} - \operatorname{div}\Big(\underbrace{S(c)\nabla\phi}_{\substack{\text{Ohm's}\\\text{law}}} + \underbrace{S(c)([\nabla u]\nabla\phi + (\nabla^2\phi)\cdot u)}_{\substack{\text{an additional "mechanical"}\\\text{electric current}}} + \underbrace{M(c)\nabla\partial_c\varphi(e(u),c))}_{\substack{\text{swelling}\\\text{effects}}} \Big) = 0, \tag{5.7.28}$$

where $S(c) = zM(c)$ in the *Ohm law* is the electric conductivity.

5.7.4 Anisothermal steady-state problems

We already mentioned that the diffusion equation (5.7.8b) or (5.7.19b) is related to the dissipation rather than the stored energy. Thermodynamically, the dissipation rate (i.e. here $\mathbb{M}\nabla\mu\cdot\nabla\mu$) leads to the heat production which might substantially influence temperature if the specimen is large or/and the produced heat cannot be transferred away sufficiently fast. In turn, variation of temperature may influence the dissipation mechanism and the stored energy too, and thus gives rise to a thermo-mechanically coupled system.

The free energy φ as well as the mobility tensor $\mathbb{M}$ now may depend on temperature, let us denote it by θ. The original system (3.6.1) then augments as

$$\operatorname{div}\sigma + f = 0 \qquad \text{with} \quad \sigma = \partial_e\varphi(e(u),c,\theta), \tag{5.7.29a}$$
$$\operatorname{div}(\mathbb{M}(c,\theta)\nabla\mu) = 0 \qquad \text{with} \quad \mu \in \partial_c\varphi(e(u),c,\theta), \tag{5.7.29b}$$
$$\operatorname{div}(\mathbb{K}(c,\theta)\nabla\theta) + \mathbb{M}(c,\theta)\nabla\mu\cdot\nabla\mu = 0 \tag{5.7.29c}$$

to be solved on a bounded Lipschitz domain $\Omega \subset \mathbb{R}^d$. Note that (5.7.29c) involves the *Fourier law*, saying that the heat flux equals $-\mathbb{K}(c,\theta)\nabla\theta$. This system should be completed by suitable boundary conditions, e.g.

$$u = 0 \ \text{ on } \Gamma_{\mathrm{D}}, \qquad \sigma\vec{n} = g \ \text{ on } \Gamma_{\mathrm{N}}, \tag{5.7.30a}$$
$$\mathbb{M}(c,\theta)\nabla\mu\cdot\vec{n} + \alpha\mu = \alpha\mu_{\mathrm{ext}} \ \text{ on } \Gamma, \tag{5.7.30b}$$
$$\mathbb{K}(c,\theta)\nabla\theta\cdot\vec{n} + \gamma\theta = \gamma\theta_{\mathrm{ext}} \ \text{ on } \Gamma. \tag{5.7.30c}$$

In (5.7.29c) and (5.7.30c), $\mathbb{K} = \mathbb{K}(c,\theta)$ denotes a *heat-conductivity tensor*.

In this scalar case $m = 1$, an interesting transformation[8] is based on the formula $\operatorname{div}(av) = a\operatorname{div}v + \nabla a\cdot v$. One can indeed rely on

$$\operatorname{div}(\mu\mathbb{M}(c,\theta)\nabla\mu) = \mu\underbrace{\operatorname{div}(\mathbb{M}(c,\theta)\nabla\mu)}_{=0 \text{ by } (5.7.29b)} + \mathbb{M}(c,\theta)\nabla\mu\cdot\nabla\mu = \mathbb{M}(c,\theta)\nabla\mu\cdot\nabla\mu. \tag{5.7.31}$$

If $\mu \in L^\infty(\Omega)$, the left-hand side of (5.7.31) indeed can be tested by functions from $H^1(\Omega)$ and thus lives in $H^1(\Omega)^*$. In the scalar case, one has the information $\mu \in L^\infty(\Omega)$ at disposal due to the maximum principle if the external chemical potential μ_{ext} is in $L^\infty(\Gamma)$. Thus, instead of (5.7.29c), one can equivalently consider

$$\operatorname{div}(\mathbb{K}(c,\theta)\nabla\theta + \mu\mathbb{M}(c,\theta)\nabla\mu) = 0. \tag{5.7.32}$$

Proposition 5.7.11 (Existence of steady states). *Let $\varphi : \mathbb{R}^{d\times d}_{\mathrm{sym}}\times\mathbb{R}\times\mathbb{R} \to \mathbb{R}\cup\{+\infty\}$ be lower semicontinuous and coercive in the sense that $\varphi(e,c,\theta) \ge \epsilon|e|^2 + \epsilon|c|^q$ for some $q > 2^{*\prime}$ and $\epsilon > 0$ and $\varphi(\cdot,\cdot,\theta)$ be strictly convex for any $\theta \in \mathbb{R}$ and $\theta \mapsto \varphi(\cdot,\cdot,\theta)$ be continuous in the sense of Γ-convergence, and let $\mathbb{M},\mathbb{K} : \mathbb{R}^2 \to \mathbb{R}^{d\times d}$ be continuous, bounded, and uniformly positive definite, $\gamma \ge 0$, $\mu_{\mathrm{ext}} \in L^\infty(\Gamma)$, $\theta_{\mathrm{ext}} \in L^{2^{\sharp\prime}}(\Gamma)$, and α, f,*

[8] Such transformation is used also in a steady-state thermistor problem, cf. e.g. [456, Sect. 6.4].

and g be as in Proposition 5.7.5. Then the boundary-value problem (5.7.29)–(5.7.30) possesses at least one variational solution $(u,c,\theta)\in H^1_{\rm D}(\Omega;\mathbb{R}^d)\times L^q(\Omega)\times H^1(\Omega)$ *with the corresponding chemical potential* $\mu\in H^1(\Omega)\cap L^\infty(\Omega)$.

Proof. We construct the single-valued mapping $(\tilde\mu,\tilde\theta)\mapsto(u,c,\theta)\mapsto\mu$ for which the Schauder fixed-point theorem will be used. First, fixing $\tilde\mu\in H^1(\Omega)\cap L^\infty(\Omega)$ and $\tilde\theta\in H^1(\Omega)$, we solve

$$\left.\begin{array}{l}\text{Minimize }\ (u,c)\mapsto\displaystyle\int_\Omega\varphi(e(u),c,\tilde\theta)-\tilde\mu c-f\cdot u\,{\rm d}x-\int_{\Gamma_{\rm N}}g\cdot u\,{\rm d}S\\ \text{subject to }\ u\in H^1_{\rm D}(\Omega;\mathbb{R}^d)\ \text{ and }\ c\in L^q(\Omega).\end{array}\right\}\tag{5.7.33}$$

Due to the assumed strict convexity of φ, this problem has a unique solution (u,c). It is also important that this solution depends continuously on $(\tilde\mu,\tilde\theta)$ in the sense that $(\tilde\mu,\tilde\theta)\mapsto(u,c):H^1(\Omega)^2\to H^1_{\rm D}(\Omega;\mathbb{R}^d)\times L^q(\Omega)$ is (weak,strong)-continuous. In particular, we use the assumed Γ-convergence meaning that the set-valued mapping $\theta\mapsto{\rm epi}\,\varphi(\cdot,\cdot,\theta):=\{(e,c,a)\in\mathbb{R}^{d\times d}_{\rm sym}\times\mathbb{R}^m\times(\mathbb{R}\cup\{+\infty\});\ \varphi(e,c,\theta)\le a\}$ is continuous in the Hausdorff sense, and also the Rellich compactness theorem B.3.3 so that the functional

$$H^1_{\rm D}(\Omega;\mathbb{R}^d)\times L^q(\Omega)\to\mathbb{R}\cup\{+\infty\}\ :\ (u,c)\mapsto\int_\Omega\varphi(e(u),c,\tilde\theta)-\tilde\mu c\,{\rm d}x$$

Γ-converges if $(\tilde\mu,\tilde\theta)$ converges weakly in $H^1(\Omega)^2$. From this and the strict convexity, the desired continuity of $(\tilde\mu,\tilde\theta)\mapsto(u,c)$ is seen.

Further, we solve the boundary-value problem (5.7.18) now with $\mathbb{M}(c,\tilde\theta)$ instead of $\mathbb{M}(c)$. In addition, assuming $\mu_{\rm ext}\in L^\infty(\Gamma)$, we can use the maximum principle yielding the estimate ${\rm ess\,inf}\,\mu_{\rm ext}(\Gamma)\le\mu(x)\le{\rm ess\,sup}\,\mu_{\rm ext}(\Gamma)$ for a.a. $x\in\Omega$. This estimate is independent of $(\tilde\mu,\tilde\theta)$, as well as the estimates $\|\mu\|_{H^1(\Omega)}\le C$ and $\|\theta\|_{H^1(\Omega)}\le C$ provided C is large enough.

Eventually, having c, μ, and $\tilde\theta$ at disposal, we solve

$$\left.\begin{array}{l}\text{Minimize }\ \theta\mapsto\displaystyle\int_\Omega\Big(\frac12\mathbb{K}(c,\tilde\theta)\nabla\theta+\mu\mathbb{M}(c,\tilde\theta)\nabla\mu\Big)\cdot\nabla\theta\,{\rm d}x\\ \qquad\qquad\qquad\qquad\qquad\qquad+\displaystyle\int_\Gamma\frac{\gamma}{2}\theta^2-\gamma\theta_{\rm ext}\theta\,{\rm d}S\\ \text{subject to }\ \theta\in H^1(\Omega).\end{array}\right\}\tag{5.7.34}$$

For the (strong×strong×weak,weak)-continuity of the mapping $(c,\tilde\theta,\tilde\mu)\mapsto\theta$, it is important that the weak convergence of μ in $H^1(\Omega)$ implies that ${\rm div}(\mu\mathbb{M}(c,\tilde\theta)\nabla\mu)$ converges weakly in $H^1(\Omega)^*$ so that the weak convergence temperatures in $H^1(\Omega)$ easily follows.

This allows to execute the Schauder fixed point for the mapping $(\tilde\mu,\tilde\theta)\mapsto(\mu,\theta):H^1(\Omega)^2\to H^1(\Omega)^2$ in the weak topology.[9] □

[9] In fact, even strong convergence of μ's solving (5.7.18) now with $\mathbb{M}=\mathbb{M}(c,\tilde\theta)$ and thus also of θ's solving (5.7.32) with $\mathbb{K}=\mathbb{K}(c,\tilde\theta)$ and $\mathbb{M}=\mathbb{M}(c,\tilde\theta)$ can be proved but, in contrast with the proof of Proposition 5.7.12, we will not need it here.

The thermodynamical completion of the electrically-charged multicomponent system combines (5.7.19) with (5.7.29), resulting to

$$\operatorname{div}\partial_e\varphi(e(u),c,\theta)+f=z_{\rm DOP}\nabla\phi \qquad \text{on } \Omega, \tag{5.7.35a}$$
$$\operatorname{div}(\mathbb{M}(c,\theta)\nabla\mu)+r(c,\theta)=0 \quad \text{with} \quad \mu\in\partial_c\varphi(e(u),c,\theta)+z\phi \qquad \text{on } \Omega, \tag{5.7.35b}$$
$$\operatorname{div}(\mathbb{K}(c,\theta)\nabla\theta)+\mathbb{M}(c,\theta)\nabla\mu{:}\nabla\mu+h(c,\theta)=\mu{\cdot}r(c,\theta) \qquad \text{on } \Omega, \tag{5.7.35c}$$
$$\operatorname{div}(\varepsilon\nabla\phi)+z{\cdot}c+z_{\rm DOP}=\operatorname{div}(z_{\rm DOP}u) \qquad \text{on } \mathbb{R}^d. \tag{5.7.35d}$$

The right-hand side $\mu{\cdot}r(c,\theta)$ of (5.7.35c) represents the (negative) heat production where $h=h(c,\theta)$ denotes the heat-production rate due to chemical reactions. Exploiting (5.7.35b), the calculus (5.7.31) modifies to $\operatorname{div}(\mu{\cdot}\mathbb{M}(c,\theta)\nabla\mu)=\mathbb{M}(c,\theta)\nabla\mu{:}\nabla\mu-\mu{\cdot}r(c,\theta)$ so that the heat-transfer problem (5.7.35c) turns again to (5.7.32). The multi-component system (5.7.35) is more complicated than (5.7.29) because the maximum principle for μ and the variational structure for the heat-transfer equation (even if transformed into (5.7.32)) is not at disposal, however.

Proposition 5.7.12. *Let $\varphi:\mathbb{R}^{d\times d}_{\rm sym}\times\mathbb{R}^m\times\mathbb{R}\to\mathbb{R}\cup\{+\infty\}$ be lower semicontinuous and coercive in the sense that $\varphi(e,c,\theta)\ge\epsilon|e|^2+\epsilon|c|^q$ for some $q>2^{*\prime}$ and $\epsilon>0$ and $\varphi(\cdot,\cdot,\theta)$ be strictly convex satisfying (5.7.20a) uniformly for any $\theta\in\mathbb{R}$ and $\theta\mapsto\varphi(\cdot,\cdot,\theta)$ be continuous in the sense of Γ-convergence, $\varepsilon\in L^\infty(\mathbb{R}^d)$ have a positive infimum, and let $\mathbb{M}:\mathbb{R}^m\times\mathbb{R}\to\mathbb{R}^{d\times m}$ and $\mathbb{K}:\mathbb{R}^m\times\mathbb{R}\to\mathbb{R}^d$ be continuous, bounded, and uniformly positive definite, $h:\mathbb{R}^m\times\mathbb{R}\to\mathbb{R}$ be continuous and bounded, $\gamma\ge0$, and $\theta_{\rm ext}\in L^1(\Gamma)$. Moreover, let $\mu_{\rm ext}$, ε, α, $z_{\rm DOP}$, z, f, and g be as in Prop. 5.7.6. Then the boundary-value problem (5.7.35) with boundary conditions (5.7.3) and $\phi(\infty)=0$ possesses at least one variational solution $(u,c,\theta,\phi)\in H^1_{\rm D}(\Omega;\mathbb{R}^d)\times L^q(\Omega;\mathbb{R}^m)\times W^{1,p}(\Omega)\times H^1(\mathbb{R}^d)$ with any $1\le p<d'$ with the corresponding electrochemical potential $\mu\in H^1(\Omega;\mathbb{R}^m)$.*

Sketch of the proof. We organize the Schauder fixed point for a composed single-valued mapping

$$(\tilde\mu,\tilde\theta)\mapsto(u,c,\phi){:}H^1(\Omega;\mathbb{R}^m)\times W^{1,p}(\Omega)\to H^1_{\rm D}(\Omega;\mathbb{R}^d)\times L^q(\Omega;\mathbb{R}^m)\times H^1(\mathbb{R}^d), \tag{5.7.36a}$$
$$(c,\tilde\theta)\mapsto\mu:L^q(\Omega;\mathbb{R}^m)\times W^{1,p}(\Omega)\to H^1(\Omega;\mathbb{R}^m),\ \text{and eventually} \tag{5.7.36b}$$
$$(c,\tilde\theta,\mu)\mapsto\theta:L^q(\Omega;\mathbb{R}^m)\times W^{1,p}(\Omega)\times H^1(\Omega;\mathbb{R}^m)\to W^{1,p}(\Omega). \tag{5.7.36c}$$

For (5.7.36a), the minimization variational problem (5.7.33) in the previous proof is to be replaced by the saddle-point problem (5.7.21) but now with $\varphi=\varphi(\cdot,\cdot,\tilde\theta)$. The uniqueness of its solution is again due to (5.7.23)–(5.7.24) but now with $\varphi(e(u_i),c_i,\tilde\theta)$ instead of $\varphi(e(u_i),c_i)$, as well as the (weak,weak×strong×weak)-continuity of (5.7.36a).

Moreover, for (5.7.36b), we solve again the minimization problem (5.7.15) with $\mathbb{M}=\mathbb{M}(c,\tilde\mu)$ and additionally with the term $\mu{\cdot}r(c,\tilde\theta)$. In contrast to the proofs of Propositions 5.7.5–5.7.11, we now need the (strong×weak,strong)-continuity of (5.7.36b), which follows standardly by the uniform convexity of the functional in (5.7.15).

Eventually, for (5.7.36c), instead of the minimization problem (5.7.34) whose infimum might be $-\infty$ because now $\tilde{\mu}\cdot\mathbb{M}(c,\tilde{\theta})\nabla\tilde{\mu}\cdot\nabla\theta \notin L^1(\Omega)$ in general since the L^∞-estimate on $\tilde{\mu}$ (and thus on $\tilde{\mu}$ too) is not at disposal, we should solve (5.7.35c) with the boundary condition (5.7.30c) with $\mathbb{K}=\mathbb{K}(c,\tilde{\theta})$, $\mathbb{M}=\mathbb{M}(c,\tilde{\theta})$, and the heat sources $h(c,\tilde{\theta})+\mu\cdot r(c,\tilde{\theta})$, by the non-variational method. By the classical Stampacchia [493] transposition method, see also e.g. [456, Section 3.2.5], this linear boundary-value problem has a unique weak solution θ which belongs to $W^{1,p}(\Omega)$ with any $1\le p< d' = d/(d-1)$. The (strong×weak×strong,weak)-continuity of (5.7.36c) is obvious.

Altogether, the Schauder fixed-point relies on the weak continuity of the mapping $(\tilde{\mu},\tilde{\theta})\mapsto(\mu,\theta): H^1(\Omega;\mathbb{R}^m)\times W^{1,p}(\Omega)\to H^1(\Omega;\mathbb{R}^m)\times W^{1,p}(\Omega)$. □

Remark 5.7.13 (Nonvariational structure of the heat equation). Interestingly, the equation (5.7.35c) with the boundary condition (5.7.30b) with $\mathbb{K}=\mathbb{K}(c,\tilde{\theta})$, $\mathbb{M}=\mathbb{M}(c,\tilde{\theta})$, and $r=r(c,\tilde{\theta})$ fixed, is not a variational problem for θ, in contrast to (5.7.32). Formally, the underlying potential may exist[10] but, although it is convex, it cannot serve for any the direct method because its minimum is $-\infty$ for a general $\mu\in H^1(\Omega)$ and $d\ge 2$ for which $\mathbb{M}\nabla\mu\cdot\nabla\mu\in L^1(\Omega)$ may not belong to $H^1(\Omega)^*$.[11]

[10] One can see that $\theta\mapsto\int_\Omega\frac12\mathbb{K}(c,\tilde{\theta})\nabla\theta\cdot\nabla\theta-(\mathbb{M}(c,\tilde{\theta})\nabla\mu\cdot\nabla v)\theta\,\mathrm{d}x+\int_\Gamma\frac12\alpha_2\theta^2+\alpha_2\theta_{\rm ext}\theta\,\mathrm{d}S$ serves formally as such a potential. Even for temperature dependent $\mathbb{M}=\mathbb{M}(c,\theta)$, one should use a primitive function to $\theta\mapsto\mathbb{M}(c,\theta)\nabla\mu\cdot\nabla v$, while for temperature dependent $\mathbb{K}=\mathbb{K}(c,\theta)$, one can make a Kirchhoff transformation.

[11] As the right-hand side $h:=\mathbb{M}\nabla\mu\cdot\nabla\mu\in L^1(\Omega)$ in the heat-transfer problem, one can consider the situation that this heat source $h\notin H^1(\Omega)^*$ in general, which means $\|h\|_{H^1(\Omega)^*}=\sup_{\theta\in W^{1,\infty}(\Omega),\,\|\theta\|_{H^1(\Omega)}\le1}\int_\Omega h\theta\,\mathrm{d}x=+\infty$, which further means $\int_\Omega h\theta_k\,\mathrm{d}x\to+\infty$ for some $\theta_k\in W^{1,\infty}(\Omega)$ such that $\int_\Omega\frac12\mathbb{K}\nabla\theta_k\cdot\nabla\theta_k\,\mathrm{d}x+\int_\Gamma\frac12\alpha_2\theta^2\,\mathrm{d}S\le\frac12$, so that $\int_\Omega\frac12\mathbb{K}\nabla\theta_k\cdot\nabla\theta_k-h\theta_k\,\mathrm{d}x+\int_\Gamma\frac12\alpha_2\theta^2\,\mathrm{d}S\to-\infty$. Here used a suitable equivalent norm on $H^1(\Omega)$, namely $(\int_\Omega\mathbb{K}\nabla\theta_k\cdot\nabla\theta_k\,\mathrm{d}x+\int_\Gamma\alpha_2\theta^2\,\mathrm{d}S)^{1/2}$ involving $\mathbb{K}>0$ and $\alpha_1>0$.

Part II

EVOLUTION PROBLEMS

Now we turn our attention to evolution problems arising when solid bodies are allowed to evolve in time from an initial non-equilibrium state or/and are subjected to some time-dependent external loading. In contrast to Part I, a *mass density* $\varrho = \varrho(x) > 0$ is to be consider together with the related inertial effects. Moreover, in addition to describe properly the phenomenology how the material stores energy, it becomes equally important to describe the phenomenology how materials dissipate mechanical energy. Eventually, in some situations the dissipated energy (often contributed to the heat production) may influence substantially temperature which, in turn, may influence mechanical properties. This leads to a fully coupled thermodynamical systems which we will address in Chapter 8.

In contrast to most of engineering or physical studies in (computational) continuum mechanics which ignores energy balance or do not use it systematically to justify numerical stability or even convergence of computational simulations (often even openly revealed as nonconverging), we will focus on models possessing an explicit energy balance and exploit it to derive a-priori estimates which can be interpreted as stability of particular (usually not specified in detail here) numerical discretization schemes that can further be implemented on computers.

In qualified cases, these estimates may ensure convergence of such numerical schemes when discretization refines. Such limits are then (suitably defined) solutions to the continuum-mechanical problems. In particular, existence of such solutions can be thus proved even by a constructive way. Let us emphasize that existence of solutions is certainly a very basic attribute of the continuum-mechanical models and is not automatic even if the solutions are defined in a suitably weak sense. Nonexistence may be related with a wrong model itself, or with a particular physical phenomena leading to a non-existence like a blow up of solutions reflecting some catastrophic event. In most of engineering or physical studies, even this basic solution-existence attribute of the particular models is completely ignored.

The general philosophy applied throughout most of Part II is to suppress technical details related with particular approximation methods usually needed for evolution problems, cf. Section C.2. Thus, we focus on formal derivation of a-priori estimates (usually based on physical energetics underlying the particular problems) which yields corresponding data qualification and allows to specify function spaces where the solution is to be found.

Except the last Chapter 9, we will assume only small strains as we did in static situations in Chapter 5. Thorough the whole Part II, t will denote time, mostly considered as ranging over the interval $I := [0, T]$, and the dot $(\cdot)^{\boldsymbol{\cdot}}$ will denote the time derivative (partial or total).

Chapter 6
Linear rheological models at small strains

The mountains flowed before the Lord.
Prophetess Judge DEBORAH (~ 1200 BC)

Τὰ πάντα ῥεῖ (~ everything flows)
HERACLITUS OF EPHESUS (535-475 BC)

Rheology is a discipline studying *relaxation*[1] processes in materials and, related to them, the way how materials dissipate energy. Substantial dissipation may typically arise in sudden change of external load (in solids) or, conversely, in a long lasting constant load (like in fluids), or in combination of both. The distinction between solids and fluids is, from the purely mechanical viewpoint, not much lucid. For example, in geophysics, rocks are considered as fluids because they cannot permanently withstand a constant shear load. But they manifest its fluidic character only in observation time scale of millions of years, while in the man's observation time scale of years, they are well solid, as we all know from our everyday experience. Actually, this paradox is counted in a so-called *Deborah number* (sometimes denoted by the Hebrew letter 'daleth', ד) defined as a ratio between the relaxation time and the observation time. The difference between solids and fluids is thus reflected by this number: large $ד \gg 1$ means that the medium can be well understood as *solid* while small $0 < ד \ll 1$ indicates rather a *fluid*.

The word rheology itself arose from the Greek word "Rhei", referring to the Heraclitus' philosophical observation that everything flows (in Greek "Ta Panta Rhei") who generalized[2] the Jewish Prophetess Deborah's hypotheses about flowing mountains. In our context this means rather that everything flows and can be interpreted that the Deborah number is never infinite, although sometimes being very large.[3]

At small strains when the natural reference configuration and thus also its variations (and thus time derivatives) are well defined, we can well distinguish between linear rheological models and nonlinear ones. In this chapter, we will treat only the former ones, confining only on rather few basic variants. In contrast to the next

[1] It is to be emphasized that the word "relaxation" is used in many different meanings (also in medicine and psychology) and, in particular in this book, another meaning has already already occur in Definition 4.4.1 on p. 105.

[2] Likely Heraclitus did not know Deborah's statement as in an Old Testament scripture from the Book of Judges Chapter 5, verse 5.

[3] Actually, it was a chemist Eugene C. Bingham (along with a civil engineer Markus Reiner) who invented the name "rheology" in 1926. Cf. M. Reiner's historical reminiscence [433].

M. Kružík and T. Roubíček, *Mathematical Methods in Continuum Mechanics of Solids*, Interaction of Mechanics and Mathematics,
https://doi.org/10.1007/978-3-030-02065-1_6

Chapter 7, we will conceptually pronounce the relation between stress and strain in terms of differential relations without considering additional (internal) variables

We will consider mass of the medium, which gives rise to inertial forces during evolution and kinetic energy. Involving both stored and kinetic energy allows for a transfer between these two energies. This may lead to (periodic or chaotic) *vibrations* or *waves*, depending whether a substantial energy transport in space is suppressed or is manifested, respectively. Waves may exhibit effects as reflection, refraction, dispersion, attenuation, etc. Cf. Remark 6.3.5 for some (from many) types of waves in elastic continua. Such dynamical events may be very difficult to capture and specialized sophisticated computational techniques are needed,[4] which however is our of the scope of this book.

6.1 Stored energy versus kinetic and dissipation energies

Thorough this chapter, we will consider a (with exception only few remarks) linear (possibly anisotropic) material at small strains in isothermal situations. A departing point is the general *balance of forces* between *inertial force* $\varrho\ddot{u}$, the stress tensor in the material σ, and the external bulk force[5] f which read as:

$$\varrho\ddot{u} - \operatorname{div}\sigma = f. \tag{6.1.1}$$

A basic ideally nondissipative element is an "elastic *spring*" which can store mechanical energy with a stress-tensor response σ determined by a 4th-order tensor of elastic moduli $\mathbb{C} = [\mathbb{C}_{ijkl}]_{i,j,k,l=1}^{d}$ by the Hooke's law as follows:

$$\sigma_{ij} = \sum\nolimits_{k,l=1}^{d} \mathbb{C}_{ijkl} e_{kl}(u), \tag{6.1.2}$$

cf. (5.1.6) and Figure 6.1(left). We assume the major symmetry relations (5.1.5) for $\mathbb{C}$, i.e. $\mathbb{C} \in \mathrm{SLin}(\mathbb{R}^{d\times d}_{\mathrm{sym}})$.

In Figure 6.1, we use another ideally nondissipative element, namely a mass, which can "store" the kinetic energy. With (6.1.2), the general force balance (6.1.1) is a hyperbolic equation. It can be derived by the *Hamilton variational principle* [240] for *conservative systems* (i.e. those which conserves mechanical energy) which says that, among all admissible motions on a fixed time interval $[0,T]$, the actual motion is such that the integral

---

[4] Distinction between waves versus mere vibrations typically consists in higher frequencies to be captured numerically in wave computations, in contrast to usual vibration computations.

[5] In fact, f as well as $\varrho\ddot{u}$ are rather force density, with physical units N/m^d.

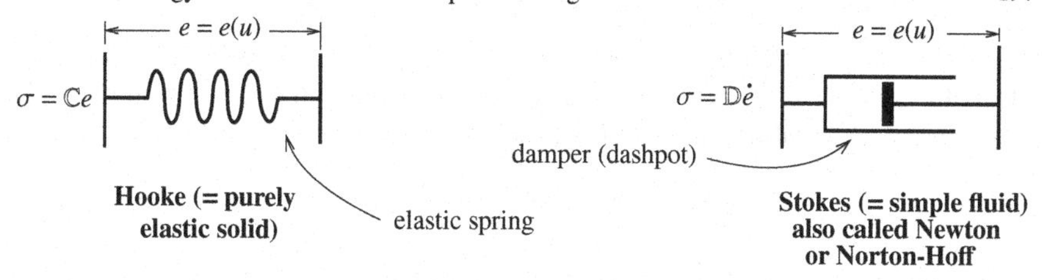

Fig. 6.1 Basic rheological elements: schematic 0-dimensional illustration (uniaxial representation) to be build into the force equilibrium (6.1.1).

$$\int_0^T \mathscr{L}(t,u,\dot{u})\mathrm{d}t \quad \text{is stationary (i.e. } u \text{ is its } \textit{critical point}\text{)}, \tag{6.1.3}$$

where $\dot{u} = \frac{\partial}{\partial t}u$ and $\mathscr{L}(t,u,\dot{u})$ is the *Lagrangian* defined by

$$\mathscr{L}(t,u,\dot{u}) := \mathscr{T}(\dot{u}) - \mathcal{E}(u) + \langle F(t), u\rangle \qquad \text{with} \tag{6.1.4a}$$

$$\mathscr{T}(\dot{u}) := \underbrace{\int_\Omega \frac{\varrho}{2}|\dot{u}|^2\,\mathrm{d}x}_{\text{kinetic energy}} \quad \text{and} \quad \mathcal{E}(u) := \underbrace{\int_\Omega \frac{1}{2}\sum\nolimits_{k,l=1}^d \mathbb{C}_{ijkl}e_{kl}(u)e_{ij}(u)\,\mathrm{d}x}_{\text{stored energy}}, \tag{6.1.4b}$$

where here $\langle F(t), u\rangle := \int_\Omega f(t)\cdot u\,\mathrm{d}x$. Then (6.1.3) leads after by-part integration in time to

$$\partial_u \mathscr{L}(t,u,\dot{u}) - \frac{\mathrm{d}}{\mathrm{d}t}\partial_{\dot{u}}\mathscr{L}(t,u,\dot{u}) = 0 \tag{6.1.5}$$

which is (6.1.1).

In reality, mechanical systems are not conservative in the sense that they dissipate mechanical energy. Such systems are called dissipative. Mechanism that leads to positive *dissipation energy* in the mechanical energy balance may have different microscopical origin, ranging from mere viscous-like processes to various activated processes. In reality, the mechanical energy is dissipated by this way to heat which is, except Chapter 8, considered here as transferred away so that it does not influence mechanical processes or possibly also to an irreversible structural changes[6] of the material as damage or hardening considered later in Chapter 7.

A basic dissipative element is a linear *damper* (called also a *dashpot*) which dissipates mechanical energy with a stress-tensor response σ determined through a tensor of viscous moduli $\mathbb{D} = [\mathbb{D}_{ijkl}]_{i,j,k,l=1}^d \in \mathrm{Lin}(\mathbb{R}^{d\times d}_{\mathrm{sym}})$ as follows:

$$\sigma_{ij} = \sum\nolimits_{k,l=1}^d \mathbb{D}_{ijkl}e_{kl}(\dot{u}), \tag{6.1.6}$$

[6] Sometimes, such irreversible changes can, however, be effectively considered as a contribution to the stored energy, too.

cf. Figure 6.1(right). Then (6.1.1) forms a parabolic equation in terms of velocities. In contrast to $\mathbb{C}$, we will sometimes need only the minor symmetry, i.e. $\mathbb{D} \in \mathrm{Lin}(\mathbb{R}^{d\times d}_{\mathrm{sym}})$; recalling the notation $\mathrm{Lin}(V)$ now with $V = \mathbb{R}^{d\times d}_{\mathrm{sym}}$, we have

$$\mathrm{Lin}(\mathbb{R}^{d\times d}_{\mathrm{sym}}) := \{\mathbb{C} \in \mathbb{R}^{(d\times d)\times(d\times d)} : \ \mathbb{C}_{ijkl} = \mathbb{C}_{jilk}\} \tag{6.1.7}$$

Such rheological models correspond to *Stokes' fluids* (or, in the context of solids, also called Newton's rheological models or a Norton-Hoff's or perfectly viscous solid) which indeed cannot store any energy. In particular, the term *viscosity*, borrowed from fluid mechanics, is used for solids, too.

Materials which combine elastic and viscous responses are called *viscoelastic* (or, in engineering literature sometime also *anelastic*) and we will present basic variants in the following Sections 6.4–6.7. Occasionally, we will use an intuitive "language" for describing various (also nonlinear) rheologies,[7] summarized in Table 6.1.

symbol	element or operator	meaning
$_{\mathbb{C}}$	Hooke (elastic spring) – Fig. 6.1-left	$\sigma = \mathbb{C}e$
$^{\alpha}_{\mathbb{C}}$	Hooke (elastic spring) damageable	$\sigma = \mathbb{C}(\alpha)e$
$_{\mathbb{D}}$	Stokes (dashpot) – Fig. 6.1-right	$\sigma = \mathbb{D}\dot{e}$
$^{\alpha}_{\mathbb{D}}$	Stokes (dashpot) depending on damage (or on temperature)	$\sigma = \mathbb{D}(\alpha)\dot{e}$ (or $\sigma = \mathbb{D}(\theta)\dot{e}$)
$_{K}$	plastic element used in Section 7.4	$N_S(\sigma) \ni \dot{e}$
$^{\alpha}_{K}$	plastic element depending on damage (or on temperature)	$N_{K(\alpha)}(\sigma) \ni \dot{e}$ (or $N_{K(\theta)}(\sigma) \ni \dot{e}$)
$^{c}_{E}$	swelling (or thermal) expansion element used in Sect. 7.6 and Chap. 8	$e = \mathbb{E}c$ (or $e = \mathbb{E}\theta$)
$_{\mathbb{H}}$	2nd-grade nonsimple spring	$\sigma = -\mathrm{div}(\mathbb{H}\nabla e)$
||	parallel arrangement	stresses to sum
—	serial arrangement	strains to sum
( )	parentheses	usual preference of operators

Table 6.1 A "language" for a brief description of various rheological models.

The operators "||" and "— " are commutative and associative. In this "language", for example the model (5.7.4) from Sect. 5.7 would read as $_{\mathbb{C}}$— $^{c}_{E}$.

From the mechanical and mathematical viewpoints, the basic classification of composed models distinguishes how the left-hand side in the diagrams like in Fig. 6.1 is connected with the right-hand side: (later being interpreted how the possible Dirichlet loading is acting on the inertial mass):

[7] Occasionally, this "language" is in parts used in literature, e.g. [524].

$$\text{mechanically}: \begin{cases} \text{a connected chain of springs exists} & \sim \text{ solid rheologies,} \\ \text{otherwise} & \sim \text{ fluid rheologies,} \end{cases}$$

$$\text{mathematically}: \begin{cases} \text{a connected chain of daspots exists} & \sim \text{ parabolic models,} \\ \text{otherwise} & \sim \text{ hyperbolic models,} \end{cases}$$

where the adjectives "parabolic" versus "hyperbolic" can also be translated as "well dissipative" versus conservative or only weakly dissipative. Summarizing most of this Chapter 6, Table 6.2 sorts the main models from this viewpoint.

\	parabolic	hyperbolic
solid	Kelvin-Voigt (Sect.6.4)	merely elastic (Sec. 6.3) or standard solid (Sec. 6.5)
fluid	Stokes or Jeffreys (Sect.6.6)	Maxwell (Sect.6.6)

Table 6.2 Basic rheological models classified from the mathematical and mechanical viewpoints.

Remark 6.1.1. Sometimes, it is useful to realize that the 2nd-order equation (6.1.1) can be equivalently written as the system of two 1st-order equations

$$\dot{u} = v, \qquad \varrho\dot{v} - \operatorname{div}\sigma = f. \tag{6.1.8}$$

6.2 Concepts of weak solutions and energy balance

As already said, we will consider a fixed time horizon $T > 0$. Thorough the whole Part II, we will use the abridged notation

$$Q := I{\times}\Omega, \quad \Sigma := I{\times}\Gamma, \quad \Sigma_{\rm D} := I{\times}\Gamma_{\rm D}, \quad \Sigma_{\rm N} := I{\times}\Gamma_{\rm N} \quad \text{with } I = [0,T]. \tag{6.2.1}$$

We use $\sigma^\top = \sigma$, cf. also the concept (1.2.12) at large strain on p. 19. Furthermore, we prescribe suitable boundary conditions. To this goal, we consider disjoint partition (up to zero $d{-}1$ dimensional measure) of the boundary Γ into two parts $\Gamma_{\rm D}$ and $\Gamma_{\rm N}$, i.e.

$$\Gamma_{\rm D} \cap \Gamma_{\rm N} = \emptyset \quad \text{and} \quad \text{meas}_{d-1}(\Gamma \setminus (\Gamma_{\rm D} \cup \Gamma_{\rm N})) = 0.$$

Then, at a current time $t \in I$, we assume the Dirichlet and the Neumann boundary conditions:

$$u|_{\Gamma_{\rm D}} = u_{\rm D}(t,\cdot) \qquad \text{on } \Gamma_{\rm D}, \tag{6.2.2a}$$

$$\sigma\vec{n} = g(t,\cdot) \qquad \text{on } \Gamma_{\rm N}, \tag{6.2.2b}$$

called also *hard-device* and *soft-device* loading, respectively. Even for a specific rheological relation between σ and $e(u)$, the evolution system governed by the boundary-value problem (6.1.1)–(6.2.2) would still allow for very many solutions and typically some other conditions are to be prescribed to select specific solutions. The basic scenario is to describe suitable conditions on the state of the system at one given time instance,[8] say $t = 0$. Then we spoke about an *initial-boundary-value problem.* In this paragraph, the basic option are the initial conditions

$$u|_{t=0} = u_0 \quad \text{and} \quad \dot{u}|_{t=0} = v_0, \tag{6.2.3}$$

although later we will see that materials with rather fluidic rheology or with internal variables will need still to prescribe further initial conditions (typically on the stress or on the mentioned internal variables).

Asuming (for a moment) smoothess of u) and using Green's formula (B.3.14) and the boundary conditions (6.2.2), we obtain

$$\begin{aligned}\int_\Omega ({\rm div}\,\sigma)v\,{\rm d}x &= \int_\Gamma \vec{n}^\top\sigma v\,{\rm d}S - \int_\Omega \sigma{:}\nabla v\,{\rm d}x\\ &= \int_{\Gamma_{\rm N}} \vec{n}^\top\sigma v\,{\rm d}S - \int_\Omega \sigma{:}e(v) + \underbrace{\sigma{:}\frac{\nabla v-(\nabla v)^\top}{2}}_{=0}{\rm d}x\\ &= \int_{\Gamma_{\rm N}} g{\cdot}v\,{\rm d}S - \int_\Omega \sigma{:}e(v)\,{\rm d}x,\end{aligned} \tag{6.2.4}$$

for any v such that $v|_{\Gamma_{\rm D}} = 0$. This identity leads to a general concept of a *weak solution* $(u,\sigma) \in W^{2,1}(I;H^1_{\rm D}(\Omega;\mathbb{R}^d)^*)\times L^2(Q;\mathbb{R}^{d\times d}_{\rm sym})$ with the Banach space

$$H^1_{\rm D}(\Omega;\mathbb{R}^d) := \big\{u\in H^1(\Omega;\mathbb{R}^d);\ u|_{\Gamma_{\rm D}} = 0\big\}, \tag{6.2.5}$$

satisfying the specified constitutive relation between $e(u)$ and σ and with appropriate initial conditions u_0 and v_0 in (6.2.3) to be specified case by case. More specifically, by testing (6.1.1) by $v = v(t,\cdot)$ with zero traces on $\Gamma_{\rm D}$ and by using (6.2.4), we obtain

$$\forall v\in H^1_{\rm D}(\Omega;\mathbb{R}^d)\ \ \forall_{\rm a.a.}t\in I:$$

$$\langle \sqrt{\varrho}\ddot{u}, \sqrt{\varrho}v\rangle + \int_\Omega \sigma{:}e(v)\,{\rm d}x = \int_\Omega f{\cdot}v\,{\rm d}x + \int_{\Gamma_{\rm N}} g{\cdot}v\,{\rm d}S \tag{6.2.6}$$

assuming ϱ smooth, where $\langle\cdot,\cdot\rangle$ denotes the duality between the Banach space $H^1_{\rm D}(\Omega;\mathbb{R}^d)$ defined in (6.2.5) and its dual where the acceleration $\ddot{u}$ is assumed to be

[8] Other option is to prescribe conditions on parts of the state at different time instants a-priori chosen or even to not a-priori chosen if some other quantities (typically the total energy) is prescribed; this occurs in various periodic problem with or without a prescribed period.

valued. The initial conditions must be involved separately in addition to this identity. Alternatively, we define a *very weak solution* $(u,\sigma)\in H^1(I;L^2(\Omega;\mathbb{R}^d))\times L^2(Q;\mathbb{R}^{d\times d}_{\rm sym})$ by integrating in time over $I=[0,T]$ and making one by-part integration in time

$$\forall v\in L^2(I;H^1_{\rm D}(\Omega;\mathbb{R}^d))\cap H^1(I;L^2(\Omega;\mathbb{R}^d)),\ \ v|_{t=T}=0:$$

$$\int_Q \sigma{:}e(v)-\varrho\dot{u}\cdot\dot{v}\,{\rm d}x{\rm d}t=\int_\Omega \varrho v_0\cdot v(0,\cdot)\,{\rm d}x+\int_Q f\cdot v\,{\rm d}x{\rm d}t+\int_{\Sigma_{\rm N}} g\cdot v\,{\rm d}S\,{\rm d}t \tag{6.2.7}$$

provided we count with the initial condition $\dot{u}|_{t=0}=v_0$, while the other possible initial conditions as here $u|_{t=0}=u_0$ must be involved separately in a definition of a weak solution. Let us note that we have used the abbreviated notation (6.2.1).

Another, even weaker concept emerges after making still one by-part integration in time more:

$$\forall v\in L^2(I;H^1_{\rm D}(\Omega;\mathbb{R}^d))\cap H^2(I;L^2(\Omega;\mathbb{R}^d)),\ \ v|_{t=T}=\dot{v}|_{t=T}=0:$$

$$\int_Q \sigma{:}e(v)+\varrho u\cdot\ddot{v}\,{\rm d}x{\rm d}t=\int_\Omega v_0\cdot v(0,\cdot)-u_0\cdot\dot{v}(0,\cdot)\,{\rm d}x+\int_Q f\cdot v\,{\rm d}x{\rm d}t+\int_{\Gamma_{\rm N}} g\cdot v\,{\rm d}S\,{\rm d}t \tag{6.2.8}$$

provided we count the initial conditions (6.2.3), while the other possible initial conditions (imposed on various internal variables) will have to be involved separately.

The formal energy balance, which also leads to a strategy yielding a-priori energy estimates, is to test (6.1.1) by *velocity* $\dot{u}$, which is indeed legitimate if $\dot{u}_{\rm D}=0$. One is then to use (6.2.4) with $v=\dot{u}$. However, if $u_{\rm D}$ is not constant in time, we cannot test (6.1.1) directly by $\dot{u}$ but we should use $v=(u-\bar{u}_{\rm D})^{\boldsymbol{\cdot}}$ where $\bar{u}_{\rm D}$ is the extension of $u_{\rm D}$ inside Ω. Therefore $\bar{u}_{\rm D}|_{\Gamma_{\rm D}}=u_{\rm D}$ and $v|_{\Gamma_{\rm D}}=(u-\bar{u}_{\rm D})^{\boldsymbol{\cdot}}|_{\Gamma_{\rm D}}=(u_{\rm D}-u_{\rm D})^{\boldsymbol{\cdot}}=0$ hence we can indeed use (6.2.4) for such v. Then one gets the *energy balance*[9]

$$\underbrace{\frac{\rm d}{{\rm d}t}\int_\Omega\frac{\varrho}{2}|\dot{u}|^2\,{\rm d}x}_{\text{kinetic energy}}+\underbrace{\int_\Omega\sigma{:}e(\dot{u})\,{\rm d}x}_{\substack{\text{rate of dissipation and}\\ \text{rate of stored energy}}}=\underbrace{\int_\Omega f\cdot\dot{u}\,{\rm d}x+\int_{\Gamma_{\rm N}}g\cdot\dot{u}\,{\rm d}S+\langle\vec{\tau}_{\rm D},\dot{u}_{\rm D}\rangle_{\Gamma_{\rm D}}}_{\text{power of external loading}} \tag{6.2.9}$$

where we have used the abbreviation

$$\langle\vec{\tau}_{\rm D},v\rangle_{\Gamma_{\rm D}}:=\int_\Omega\varrho\ddot{u}\cdot\bar{v}+\sigma{:}e(\bar{v})-f\cdot\bar{v}\,{\rm d}x-\int_{\Gamma_{\rm N}}g\cdot\bar{v}\,{\rm d}S \tag{6.2.10}$$

[9] If $\dot{u}$ is not enough regular to be put into (6.2.6), the energy balance (6.2.9) usually holds only as an inequality only.

where $\bar{v}$ denotes some extension of v (which is considered as defined on $\Sigma_{\rm D} = I \times \Gamma_{\rm D}$ only) on the whole Q. In fact, the right-hand side (and thus also the left-hand side) of (6.2.10) does not depend on a particular extension: indeed, considering two extensions $\bar{v}_1$ and $\bar{v}_2$, the difference of the right-hand sides vanishes by using (6.2.4) for $z = \bar{v} := \bar{v}_1 - \bar{v}_2$ and the balance equation (6.1.1):

$$\int_\Omega \varrho\ddot{u}\cdot\bar{v} + \sigma{:}e(\bar{v}) - f\cdot\bar{v}\,{\rm d}x - \int_{\Gamma_{\rm N}} g\cdot\bar{v}\,{\rm d}S \overset{\substack{(6.2.4)\\ \text{used}}}{=} \int_\Omega \varrho\ddot{u}\cdot\bar{v} - ({\rm div}\,\sigma)\cdot\bar{v} - f\cdot\bar{v}\,{\rm d}x \overset{\substack{(6.1.1)\\ \text{used}}}{=} 0.$$

Hence the vector field $\vec{\tau}_{\rm D} = \vec{\tau}_{\rm D}(\ddot{u},\sigma,f,g)$ on $\Gamma_{\rm D}$ from (6.2.10) defines a linear functional $v \mapsto \langle\vec{\tau}_{\rm D}, v\rangle_{\Gamma_{\rm D}}$ on traces v's, depending on $\ddot{u}$, σ, f, and g at a current time t. The meaning of this functional is the *traction vector* (or briefly just traction) on $\Gamma_{\rm D}$.

6.3 Purely elastic material

The very simplest solid model is the mentioned purely elastic nondissipative solid governed by the Hooke law (6.1.2), now considered under the general load as schematically shown in Figure 6.2. It is conservative, nondissipative and leads to

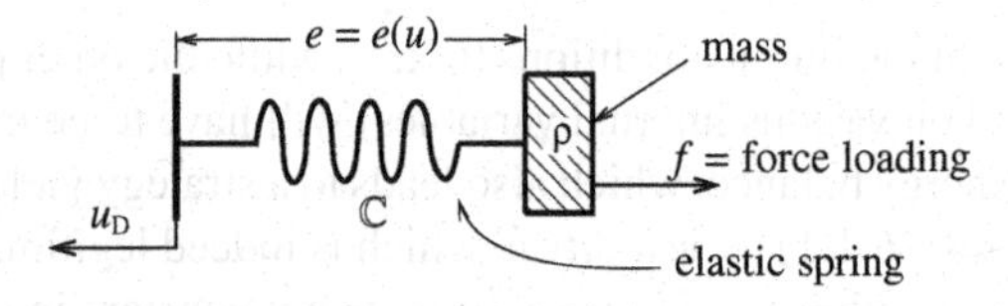

Fig. 6.2 Schematic 0-dimensional illustration of the Hook element (cf. Figure 6.1-left) to be build into the force equilibrium (6.1.1) with the inertia of the mass acting very schematically directly on e; in fact, in d-dimensional situations ($d \geq 1$) inertia (as well as the Dirichlet load $u_{\rm D}$ and the force f) acts rather on u and thus only indirectly on $e = e(u)$.

a hyperbolic problem. Here we consider it linear, otherwise serious mathematical difficulties related to very complex phenomena of wave propagation in nonlinear nondissipative media would occur. Anyhow, even the linear variant is illustrative and useful, although merely conservative systems are always a certain idealization.

Identifying $\sigma{:}e(\dot{u})$ in (6.2.9) for our simple model $\sigma = \mathbb{C}e$ now reveals the specific stored energy and the dissipation rate as:

$$\varphi(e) = \frac{1}{2}\mathbb{C}e{:}e \qquad \text{and} \qquad \xi = 0. \tag{6.3.1}$$

Proposition 6.3.1 (Elastodynamic problem: existence, uniqueness). *Let* $\mathbb{C} \in {\rm SLin}(\mathbb{R}^{d\times d}_{\rm sym})$ *be positive definite, the mass density satisfies*

$$\varrho \in L^\infty(\Omega), \quad \text{ess inf}\,\varrho > 0, \tag{6.3.2}$$

and let further the loading and the initial conditions satisfy

$$f \in L^1(I;L^2(\Omega;\mathbb{R}^d)), \quad g \in W^{1,1}(I;L^{2^{\sharp'}}(\Gamma_N;\mathbb{R}^d)), \tag{6.3.3a}$$

$$u_D \in W^{1,1}(I;H^{1/2}(\Gamma_D;\mathbb{R}^d)) \text{ possessing}$$
$$\text{an extension } \bar{u}_D \in W^{1,1}(I;H^1(\Omega;\mathbb{R}^d)) \cap W^{2,1}(I;L^2(\Omega;\mathbb{R}^d)), \tag{6.3.3b}$$

$$u_0 \in H^1(\Omega;\mathbb{R}^d), \text{ and } v_0 \in L^2(\Omega;\mathbb{R}^d). \tag{6.3.3c}$$

Then the initial-boundary-value problem (6.1.1)–(6.2.2)–(6.2.3) with $\sigma = \mathbb{C}e$ *has a unique weak solution* (u,σ) *such that*

$$u \in L^\infty(I;H^1(\Omega;\mathbb{R}^d)) \cap W^{1,\infty}(I;L^2(\Omega;\mathbb{R}^d)), \tag{6.3.4a}$$

$$\sigma \in L^\infty(I;L^2(\Omega;\mathbb{R}^{d\times d})). \tag{6.3.4b}$$

If (6.3.2) is strengthened to

$$\varrho \in L^\infty(\Omega) \cap W^{1,2^*2/(2^*-2)}(\Omega) \quad \text{and} \quad \operatorname{ess\,inf} \varrho > 0, \tag{6.3.5}$$

then also

$$\ddot{u} \in L^1(I;L^2(\Omega;\mathbb{R}^d)) + L^\infty(I;H^1_D(\Omega;\mathbb{R}^d)^*). \tag{6.3.6}$$

Proof. The test by $(u-\bar{u}_D)^{\boldsymbol{\cdot}}$ as in (6.2.9) with $\langle \vec{\tau}_D, \dot{u}_D\rangle_D$ substituted from (6.2.10) leads to an energy balance through the by-part integration in time, cf. Lemma C.2.2 on p. 535. Actually, it may be formal and some approximation is needed to make it legitimate. E.g. restriction of u on a finite-dimensional subspace of $H^1(\Omega;\mathbb{R}^d)$ makes it straightforwardly legal, which is the essence of the Faedo-Galerkin method (cf. Section C.2.4) and it leads even to a conceptual numerical strategy when employing the *finite-element method*, cf. Section 3.7.1. Actually, the resulted system, i.e. an initial-value problem for a system of ordinary-differential equations, possesses a solution and automatically conserves energy like (although only formally) the original system (6.2.9).

After integration over a time interval $[0,t]$ and using by-part integration for the terms $\ddot{u}\cdot\dot{\bar{u}}_D$ on $[0,t]\times\Omega$ and $g\cdot\dot{u}$ on $[0,t]\times\Gamma_N$, this strategy gives:

$$\begin{aligned}\int_\Omega \frac{\varrho}{2}|\dot{u}(t)|^2 + \frac{1}{2}\mathbb{C}e(u(t)):e(u(t))\,dx &= \int_\Omega \frac{\varrho}{2}|v_0|^2 + \frac{1}{2}\mathbb{C}e(u_0)):e(u_0))\,dx \\ &+ \int_0^t\!\int_\Omega f\cdot\dot{u} - \varrho\dot{u}\cdot\ddot{\bar{u}}_D + \mathbb{C}e(u):e(\dot{\bar{u}}_D) - f\cdot\dot{\bar{u}}_D\,dxdt - \int_0^t\!\int_{\Gamma_N} \dot{g}\cdot u\,dS\,dt \\ &+ \int_\Omega \varrho\dot{u}(t)\cdot\dot{\bar{u}}_D(t) - \varrho\dot{u}(0)\cdot\dot{\bar{u}}_D(0)\,dx + \int_{\Gamma_N} g(t)\cdot u(t) - g(0)\cdot u_0\,dS;\end{aligned} \tag{6.3.7}$$

for notational simplicity, we use (and will use) the same letter "t" in a two-fold position in $\int_0^t ....dt$. From this, we can read the a-priori estimates in (6.3.4a) by Young and Gronwall inequalities, combined with estimation of the trace operator $u \mapsto u|_{\Gamma_N}$. Then (6.3.4b) for $\sigma = \mathbb{C}e(u)$ follow.

The proof of uniqueness of the weak solution to the continuous (non-discretized) problem is rather delicate because $\sqrt{\varrho}\ddot{u}$ is not in duality with $\sqrt{\varrho}\dot{u}$ and thus test by $\dot{u}$ is not fully legitimate. Using linearity of the problem, we are to show that necessarily $u = 0$ if $f = 0$, $g = 0$, $u_{\rm D} = 0$, and $u_0 = v_0 = 0$. Considering $s \in I$, we take[10]

$$v(t) := \begin{cases} -\displaystyle\int_t^s u(r)\,{\rm d}r & \text{for } 0 \le t \le s, \\ 0 & \text{for } s < t \le T. \end{cases} \tag{6.3.8}$$

Let us note that $\dot{v} = u$ on $[0, s)$ and that $v \in L^\infty(I; H^1_{\rm D}(\varOmega;\mathbb{R}^d)) \cap W^{1,\infty}(I; L^2(\varOmega;\mathbb{R}^d))$ and also $v(T) = 0$ so that v is a legitimate test function for the weak formulation (6.2.7). We can use the calculus

$$\int_Q \varrho\dot{u}\cdot\dot{v}\,{\rm d}x{\rm d}t = \int_0^s\!\int_\varOmega \varrho\dot{u}\cdot u\,{\rm d}x{\rm d}t = \int_\varOmega \frac{\varrho}{2}|u(s)|^2\,{\rm d}x \quad \text{and} \tag{6.3.9a}$$

$$\int_Q \sigma{:}e(v)\,{\rm d}x{\rm d}t = \int_0^s\!\int_\varOmega \mathbb{C}e(\dot{v}){:}e(v)\,{\rm d}x{\rm d}t = -\int_\varOmega \frac{1}{2}\mathbb{C}e(v(0)){:}e(v(0))\,{\rm d}x; \tag{6.3.9b}$$

here we used that $u(0) = 0$ and $v(s) = 0$. Thus (6.2.7) with its right-hand side zero yields for all $s \in I$ that

$$\int_\varOmega \frac{\varrho}{2}|u(s)|^2 + \frac{1}{2}\mathbb{C}e(v(0)){:}e(v(0))\,{\rm d}x = 0, \tag{6.3.10}$$

from which $u = 0$ follows.

The estimate (6.3.6) is by comparison, realizing that $\ddot{u} = (f + {\rm div}\,\sigma)/\varrho$ and then

$$\begin{aligned}\Big\|\frac{{\rm div}\,\sigma}{\varrho}\Big\|_{L^\infty(I;H^1_{\rm D}(\varOmega;\mathbb{R}^d)^*)} &= \sup_{\|v\|_{L^1(I;H^1_{\rm D}(\varOmega;\mathbb{R}^d))}\le 1} \int_Q \frac{{\rm div}\,\sigma}{\varrho}\cdot v\,{\rm d}x{\rm d}t \\ &= \sup_{\|v\|_{L^1(I;H^1_{\rm D}(\varOmega;\mathbb{R}^d))}\le 1} \int_Q \frac{\sigma}{\varrho}{:}e(v) - \frac{\sigma}{\varrho^2}{:}(\nabla\varrho\otimes v)\,{\rm d}x{\rm d}t + \int_{\varSigma_{\rm N}} \frac{g}{\varrho}\cdot v\,{\rm d}S\,{\rm d}t \end{aligned} \tag{6.3.11}$$

and then to estimate it by the already obtained estimates (6.3.4); here the assumption $\nabla\varrho \in L^{2^*2/(2^*-2)}(\varOmega;\mathbb{R}^d)$ has been used. $\square$

More regular data imply more regular response, while relying only on (6.3.2). Here, in the inviscid material, the test by $\dddot{u}$ does not work because it would lead to the term $\mathbb{C}e(u){:}e(\dddot{u})$ which has a "bad" sign. Yet, after differentiation in time first, this test is doable, leading to:

Proposition 6.3.2 (Regularity and energy conservation). *Let again* $\mathbb{C} \in {\rm SLin}(\mathbb{R}^{d\times d}_{\rm sym})$ *be positive definite,* ϱ *satisfy (6.3.2), and now moreover*

[10] This test is usually credited to O.A. Ladyzhenskaya [300, Sect. IV.3], cf. also e.g. [437, Sect. 11.2.3].

$$f \in W^{1,1}(I;L^2(\Omega;\mathbb{R}^d)),\quad g \in W^{2,1}(I;L^{2^{\sharp'}}(\Gamma_{\rm N};\mathbb{R}^d)), \tag{6.3.12a}$$

$u_{\rm D} \in W^{2,1}(I;H^{1/2}(\Gamma_{\rm D};\mathbb{R}^d))$ *possessing*

$$\textit{an extension } \bar{u}_{\rm D} \in W^{2,1}(I;H^1(\Omega;\mathbb{R}^d)) \cap W^{3,1}(I;L^2(\Omega;\mathbb{R}^d)), \tag{6.3.12b}$$

$$u_0 \in H^2(\Omega;\mathbb{R}^d),\ \textit{and}\ v_0 \in H^1(\Omega;\mathbb{R}^d). \tag{6.3.12c}$$

Then the weak solution (u,σ) to the initial-boundary-value problem (6.1.1)–(6.2.2)–(6.2.3) with $\sigma = \mathbb{C}e(u)$ satisfies also

$$u \in W^{1,\infty}(I;H^1(\Omega;\mathbb{R}^d)) \cap W^{2,\infty}(I;L^2(\Omega;\mathbb{R}^d)), \tag{6.3.13a}$$

$$\sigma \in W^{1,\infty}(I;L^2(\Omega;\mathbb{R}^{d\times d})), \tag{6.3.13b}$$

$$\operatorname{div}(\mathbb{C}e(u)) \in L^\infty(I;L^2(\Omega;\mathbb{R}^d)). \tag{6.3.13c}$$

Moreover, this solution is unique and the energy conservation (in the sense (6.2.9) integrated in time) holds.

Sketch of the proof. As said, we differentiate the system (6.1.1) with (5.1.4) and also the boundary conditions (6.2.2) in time. As this system is linear, we can perform the strategy from the proof of the above Proposition 6.3.1 adapted for $\dot{u}$ and $\dot{f}$ and $\dot{g}$ in place of u and f and g, respectively. The only peculiarity is that we need to qualify initial conditions so that $\mathbb{C}e(\dot{u}(0)) \in L^2(\Omega;\mathbb{R}^{d\times d}_{\rm sym})$ and $\ddot{u}(0) = (\operatorname{div}(\mathbb{C}e(u(0))) + f(0))/\varrho \in L^2(\Omega;\mathbb{R}^d)$, for which we use the assumptions $u_0 \in H^2(\Omega;\mathbb{R}^d)$ and $v_0 \in H^1(\Omega;\mathbb{R}^d)$. Eventually, (6.3.13c) is simply by $\operatorname{div}(\mathbb{C}e(u)) = \varrho\ddot{u} - f$ holding now on Q a.e.

The test of the original system by $(u-\bar{u}_{\rm D})^{\boldsymbol{\cdot}} \in L^\infty(I;H^1(\Omega;\mathbb{R}^d))$ is now legitimate and this test function is now surely in duality with the inertial term $\varrho\ddot{u} \in L^\infty(I;L^2(\Omega;\mathbb{R}^d))$. Thus the energy conservation is proved. Performing the proof for differences of two solutions, also the uniqueness can be seen. □

Let us note that the smoothness (6.3.5) of ϱ used in Proposition 6.3.1 has not been now needed, being compensated by the smoothness of the loading. If Ω and $\mathbb{C} = \mathbb{C}(x)$ are qualified to provided H^2-regularity (which is, e.g., for smooth Ω and $\mathbb{C}$ homogeneous Lamé material), then (6.3.13c) gives also $u \in L^\infty(I;H^2(\Omega;\mathbb{R}^d))$.

Example 6.3.3 (Self-gravitational load). An interesting system arises when the force f is not a-priori given but is derived from the self-induced gravitational field which is to be considered if the total mass $\int_\Omega \varrho\,{\rm d}x$ is considerably large. Having in mind a (simplified purely mechanical) model of an "elastic planet" vibrating under the self-induced *gravitational force*, we consider the elasto-dynamic system

$$\varrho\ddot{u} - \operatorname{div}(\mathbb{C}e(u)) = f \quad \text{with} \quad f = -\varrho\nabla\phi \qquad \text{on } Q, \tag{6.3.14a}$$

$$\Delta\phi = \mathfrak{g}(\varrho - \operatorname{div}(\varrho u)) \qquad \text{on } I\times\mathbb{R}^d, \tag{6.3.14b}$$

together with the initial conditions (6.2.3) and the boundary condition (6.2.2b) with $g = 0$ on $\Gamma_{\rm N} := \Gamma$ and $\phi(\infty) = 0$, where ϕ stands here for the (self-induced) *gravitational potential* using the convention in geophysics that not $-\Delta$ but $+\Delta$ is at

(6.3.14b), the mass density $\varrho = \varrho(x)$ is considered 0 for $x \in \mathbb{R}^d \backslash \Omega$, and $\mathfrak{g}$ is the (normalized) gravitational constant[11]. The corresponding *Lagrangian* (6.1.4a) is now autonomous (i.e. time independent with $F(t) = 0$) in the form $\mathscr{L}(u,\phi,\dot{u}) := \mathscr{T}(\dot{u}) - \mathcal{E}(u,\phi)$ with $\mathcal{E}(u,\phi) = \int_\Omega \frac{1}{2}\mathbb{C}e(u):e(u) + \varrho\phi + \varrho u\cdot\nabla\phi \,\mathrm{d}x + \int_{\mathbb{R}^d} \frac{1}{2\mathfrak{g}}|\nabla\phi|^2 \,\mathrm{d}x$ and yields the system (6.3.14) via the Hamilton variational principle (6.1.3). This convex $\mathcal{E}$ actually yields a saddle structure on the affine manifold formed by the gravitation-potential equation (6.3.14b). Indeed, it is seen from (5.7.31) now, using (6.3.14b), in the variant

$$\int_\Omega \varrho\nabla\phi\cdot u + \varrho\phi \,\mathrm{d}x = \int_\Omega \phi(\varrho - \mathrm{div}(\varrho u))\,\mathrm{d}x = \int_\Omega \frac{1}{\mathfrak{g}}\phi\Delta\phi\,\mathrm{d}x = -\int_\Omega \frac{1}{\mathfrak{g}}|\nabla\phi|^2\,\mathrm{d}x.$$

Testing the particular equations in (6.3.14) by $\dot{u}$ and $\dot{\phi}$ and using

$$\int_\Omega \varrho\nabla\phi\cdot\dot{u} - \mathrm{div}(\varrho u)\dot{\phi}\,\mathrm{d}x = \int_\Omega \varrho\nabla\phi\cdot\dot{u} + \varrho u\cdot\nabla\dot{\phi}\,\mathrm{d}x = \frac{\mathrm{d}}{\mathrm{d}t}\int_\Omega \varrho\nabla\phi\cdot u\,\mathrm{d}x,$$

we obtain (at least formally) the energy balance for this conservative system:

$$\frac{\mathrm{d}}{\mathrm{d}t}\Big(\int_\Omega \frac{\varrho}{2}|\dot{u}|^2 + \frac{1}{2}\mathbb{C}e(u):e(u) + \varrho\phi + \varrho\nabla\phi\cdot u\,\mathrm{d}x + \int_{\mathbb{R}^d}\frac{1}{2\mathfrak{g}}|\nabla\phi|^2\,\mathrm{d}x\Big) = 0 \qquad (6.3.15)$$

from which the a-priori estimates follows providing $\varrho \in L^\infty(\Omega)$ with Ω bounded, cf. also Exercise 6.3.13 below. Interestingly, the system (6.3.14) is not much good approximation, which can be seen in homogenous incompressible materials for which $\mathrm{div}(\varrho u) = \varrho\,\mathrm{div}\,u = 0$ and the system (nonphysically) decouples. A more precise 2nd-order approximation would look as [12]

$$\varrho\ddot{u} - \mathrm{div}(\mathbb{C}e(u)) = f \quad \text{with} \quad f = -\varrho(\nabla\phi - [\nabla^2\phi]u) \qquad \text{on } Q, \qquad (6.3.16a)$$

$$\Delta\phi = \mathfrak{g}\big(\varrho - \mathrm{div}(\varrho u) + \tfrac{1}{2}\mathrm{div}^2(\varrho u\otimes u)\big) \qquad \text{on } I\times\mathbb{R}^d. \qquad (6.3.16b)$$

[11] In SI physical units in the dimension $d = 3$, $\mathfrak{g} \doteq 4\pi\cdot 6.674\times 10^{-11}\mathrm{m}^3\mathrm{kg}^{-1}\mathrm{s}^{-2}$.

[12] It is interesting to see realize that the term $\varrho - \mathrm{div}(\varrho u)$ in (6.3.14b) arises from (4.7.3) on p. 158 after a linearization: actually, the expansion us to 2nd-order terms looks as

$$\int_{\Omega^y} \frac{\varrho(y^{-1}(x^y))\phi(x^y)}{\det\nabla y[y^{-1}(x^y)]}\,\mathrm{d}x^y = \int_\Omega \frac{\varrho(y^{-1}(y(x)))\phi(y(x))}{\det\nabla y[y^{-1}(y(x))]}\det\nabla y(x)\,\mathrm{d}x$$
$$= \int_\Omega \varrho(x)\phi(y(x))\,\mathrm{d}x = \int_\Omega \varrho(x)\phi(x+u(x))\,\mathrm{d}x \sim$$
$$\sim \int_\Omega \varrho(x)(\phi(x) + \nabla\phi(x)\cdot u(x)) + \frac{1}{2}\nabla^2\phi(x):(u(x)\otimes u(x))\,\mathrm{d}x$$
$$= \int_\Omega \varrho(x)\phi(x)\,\mathrm{d}x - \int_{\mathbb{R}^d}\phi(x)\big(\mathrm{div}(\varrho(x)u(x)) - \frac{1}{2}\mathrm{div}^2(\varrho(x)u(x)\otimes u(x))\big)\,\mathrm{d}x$$

for small displacement u, from which (6.3.16) follows, while (6.3.14) follows when neglecting the 2nd-order term.

Example 6.3.4 (Centrifugal and Coriolis forces). One can also augment the gravitational force f in (6.3.14a) by the so-called *Coriolis*[13] and *centrifugal forces* i.e.

$$f = -\varrho(\nabla\phi + 2\omega\times\dot{u} + \omega\times(\omega\times(x+u))) \tag{6.3.17}$$

with a given angular velocity $\omega\in\mathbb{R}^3$ (cf. Figure 7.16 on p. 356) and the vectorial cross product "×" from (A.2.2). The enhanced gravitational force in (6.3.17) is sometimes called a *gravity force*.

Remark 6.3.5 (Elastodynamic problem: wave propagation). Typical mode of evolution of the elastic nondissipative medium are vibrations (like periodic oscillations) or waves. Basic modes of waves are *P- and S-waves*, respectively, referring respectively to the pressure and the shear waves, alternatively called also the primary (or also compressional waves) and the secondary waves because the former one arrives earlier that the latter one from a source to an observer. In the isotropic medium with the Lamé constants λ and G, the speed of the P-waves (=sound speed) is $\sqrt{M/\varrho}$ where $M = \lambda + 2G$ is the so-called *P-wave modulus*[14] while S-waves propagate always more slowly with the speed $\sqrt{G/\varrho}$. Let us still note that the system (6.1.1) for this purely elastic isotropic material is sometimes called *Navier* (or, particularly in the static case, Navier-Cauchy) *equation.*[15]

Remark 6.3.6 (Nonsimple materials: wave dispersion). Later, e.g. in Sect. 7.5.3, we will find useful to involve strain gradient as used already in the static case in Sec. 5.4. In the dynamical situations, this concept is related with *dispersion of waves*, which refers to the phenomenon that waves of different wavelengths travel at different speeds or, related to this, that impulses (which contain more than just one frequency in their Fourier transform in space) change their shape within propagation. This may be caused by various nonlinear effects or, here, in the linear situations by the higher gradients. In the *2nd-grade nonsimple materials* governed by the stored energy $\varphi_{\rm E}(e,\nabla e) = \frac{1}{2}\mathbb{C}e{:}e + \frac{1}{2}\mathbb{H}\nabla e \,\vdots\, \nabla e$, we can demonstrate this phenomenon in the one-dimensional homogeneous medium occupying the whole space $\Omega = \mathbb{R}$, which leads to the linear hyperbolic problem

$$\varrho\ddot{u} - \partial_x(C\partial_x u - \partial_x H\partial^2_{xx}u) = 0 \quad \text{on } [0,+\infty)\times\mathbb{R}, \tag{6.3.18}$$

with the initial conditions $u = u_0$ and $\dot{u} = v_0$ on $\mathbb{R}$. Considering the initial conditions $u_0 = \sin(x/\lambda)$ and $v_0 = \omega\cos(x/\lambda)$ yields the wave $u(t,x) = \sin(\omega t + x/\lambda)$ with the

---

[13] The Coriolis force, named after the French mathematician and mechanical engineer Gaspard-Gustave de Coriolis (1792-1843), is in a position of a "fictitious" force caused by the rotation of the body Ω in the inertial system with the fixed *angular velocity* ω. A similar fictitious character applies to the centrifugal force, except that it contribute directly to the energetics by the power $\varrho\omega\times(\omega\times(x+u))\cdot\dot{u}$.

[14] The P-wave modulus is related with the uniaxial-strain waves and is defined as $\sigma_{11} = Me_{11}$ if $e_{ij} = 0$ for $i \neq 1 \neq j$.

[15] Originally, Claude L.M.H. Navier (1785–1836) derived it in 1821 for a 2-dimensional plain-stress problem.

angular frequency ω and the wavelength λ. Substituting this sinusoidal-wave ansatz into (6.3.18), we can see that it solves this problem provided the relation

$$\varrho\omega^2 - \frac{C}{\lambda^2} - \frac{H}{\lambda^4} = 0 \tag{6.3.19}$$

is satisfied. Realizing that the speed of the wave is $v = \omega\lambda$, we thus obtain

$$v = \sqrt{\frac{C}{\varrho} + \frac{H}{\varrho\lambda^2}}. \tag{6.3.20}$$

For simple materials (i.e. for $H = 0$), we obtain the constant speed $v_{\text{SIMPLE}} = \sqrt{C/\varrho}$, i.e. no dispersion; cf. also Remark 6.3.5. For $H > 0$, we can see that the value v_{SIMPLE} is relevant only for very long waves (so-called long-wavelength limit) while high frequency waves propagate faster. Such media where high frequency components travel faster than the lower ones are called *anomalously dispersive*. Let us still mention the 3rd-grade nonsimple materials where (6.3.18) modifies as

$$\varrho\ddot{u} - \partial_x\Big(C\partial_x u - \partial_x(H_1\partial^2_{xx}u - H_2\partial^4_{xxxx})u\Big) = 0 \tag{6.3.21}$$

and (6.3.19) modifies as $\varrho\omega^2 - C/\lambda^2 - H_1/\lambda^4 - H_2/\lambda^6 = 0$. Such model allows for $H_1 < 0$ and then for the effect that wave speed can increase with wavelength at least in some wavelength range, which is referred to as a *normal dispersion*; cf. [64, Ch. 6] or [308] also for various other dispersive wave equations.

Remark 6.3.7 (Nonlocal materials: wave dispersion). In the case of (1+γ)-grade nonlocal simple materials or (2+γ)-grade nonlocal nonsimple materials, (6.3.18) modifies as

$$\varrho\ddot{u} - \partial_x\Big(C\partial_x u + \int_{\mathbb{R}} H(x-\tilde{x})\partial_x u(\tilde{x})\,\mathrm{d}\tilde{x}\Big) = 0 \quad \text{or}$$

$$\varrho\ddot{u} - \partial_x\Big(C\partial_x u - \partial_x \int_{\mathbb{R}} H(x-\tilde{x})\partial^2_{xx}u(\tilde{x})\,\mathrm{d}\tilde{x}\Big) = 0,$$

respectively. Confining ourselves on the (natural) symmetric case as used also for (2.5.36), i.e. $H(x) = H(-x)$, then (6.3.19) modifies for[16]

---

[16] Here we use the trigonometric identity $\int_{\mathbb{R}} H(\tilde{x})\sin(x+\tilde{x})\,\mathrm{d}\tilde{x} = H_0\sin(x+x_0)$ with H_0 satisfying

$$H_0^2 = \int_{\mathbb{R}\times\mathbb{R}} H(\tilde{x})H(x)\cos(x-\tilde{x})\,\mathrm{d}\tilde{x}\mathrm{d}x = \int_{\mathbb{R}\times\mathbb{R}} H(\tilde{x})H(x)\big(\cos x\cos\tilde{x} + \sin x\sin\tilde{x}\big)\mathrm{d}\tilde{x}\mathrm{d}x$$
$$= \int_{\mathbb{R}\times\mathbb{R}} H(\tilde{x})H(x)\cos x\cos\tilde{x}\,\mathrm{d}\tilde{x}\mathrm{d}x = \Big(\int_{\mathbb{R}} H(x)\cos x\,\mathrm{d}x\Big)^2$$

and $x_0 = \arctan(\int_{\mathbb{R}} H(x)\sin x\,\mathrm{d}x / \int_{\mathbb{R}} H(x)\cos x\,\mathrm{d}x)$, realizing that the symmetry of H which causes $x_0 = 0$. This yields (6.3.22) if H is supported on a neighborhood of 0 of the radius below $\pi/2$ so that the integral $\int_{\mathbb{R}} H(x)\cos x\,\mathrm{d}x$ is non-negative but, in fact, (6.3.22) can be proved generally by using Fourier's transform, cf. [219, 265].

$$\varrho\omega^2 - \frac{C}{\lambda^2} - \frac{\int_{\mathbb{R}} H(x)\cos x\,\mathrm{d}x}{\lambda^{2k}} = 0 \tag{6.3.22}$$

with k referring to the $(k+\gamma)$-grade materials. In contrast to (6.3.20), this may give *normal dispersion* provided $\int_{\mathbb{R}} H(x)\cos x\,\mathrm{d}x \le 0$, which may occur even if $H \ge 0$ is singular at $x = 0$ to comply with (2.5.28) or (2.5.37), cf. also [175, 265] for a thorough discussion and specific examples of the kernels H.

Remark 6.3.8 (Explicit time discretization of the elastodynamic problem). Considering the elastodynamic problem (6.3.24) in situations where high-frequency (i.e. small wave length) waves are to be computed , the implicit time-discretizations like in Exercise 6.3.11 are computationally cumbersome especially in 3-dimensional problems.[17] Hence *explicit time discretizations* combined with the above FEM-Faedo-Galerkin method has been proved efficient, avoiding solution of large algebraic systems which dramatically slow down computation by implicit time-discretisation schemes. The simplest explicit scheme is[18]

$$\varrho\frac{u_{\tau h}^{k+1} - 2u_{\tau h}^{k} + u_{\tau h}^{k-1}}{\tau^2} + C_h u_{\tau h}^k = f_{\tau h}^k \tag{6.3.23}$$

with C_h denoting the stiffness matrix arising by a FEM-approximation with the mesh size $h > 0$ of the boundary-value problem $\mathrm{div}(\mathbb{C}e(u)) = 0$ on Ω with $\vec{n}^\top\mathbb{C}e(u) = 0$ on Γ. Alternatively, expressing the elastodynamic in terms of velocity/stress as

$$\dot\sigma = \mathbb{C}e(v) \quad \text{and} \quad \varrho\dot v - \mathrm{div}\,\sigma = f,$$

one can apply a so-called “leap-frog” (*staggered*) *scheme*[19]

$$\frac{\sigma_{\tau h}^{k+1} - \sigma_{\tau h}^{k}}{\tau} = \mathbb{C}E_h(v_{\tau h}^k) \quad \text{and} \quad \varrho\frac{v_{\tau h}^{k+1} - v_{\tau h}^{k}}{\tau} - E_h^\top\sigma_{\tau h}^{k+1} = f_{\tau h}^{k+1}$$

with E_h a discretisation of $e(\cdot)$-operator. In both cases, the a-priori estimates and convergence for $\tau \to 0$ and $h \to 0$ needs a so-called *Courant-Friedrichs-Lewy (CFL) condition* [130] which typically bounds $\tau = \mathscr{O}(h_{\min})$ with $h_{\min}$ the smallest element size on a FEM discretization with the mesh size h, and energy conservation is satisfied only asymptotically for $\tau \to 0$.

Exercise 6.3.9 (Selectivity of the very weak formulations). Recover the initial-boundary-value problem (6.1.1)–(6.2.2)–(6.2.3) from the very weak formulation (6.2.7) or (6.2.8) for (u,σ) enough regular.

---

[17] In computational mechanics, so-called transient versus truly dynamical problems (i.e. low-frequency vibrations versus high-frequency waves, respectively) are distinguished and different numerical methods are used.

[18] A numerical integration leading to a diagonalization of the mass matrix, called lumping (see e.g. [543, App. I]), used also in (6.3.23) is an important ingredient for efficient explicit methods.

[19] Sometimes, this staggered scheme is also referred as a Newmark's one, and it can be combined with sophisticated FEM discretisation, cf. e.g. [53].

Exercise 6.3.10 (Energy conservation). Considering the homogeneous boundary-value problem

$$\varrho\ddot{u} - \mathrm{div}(\mathbb{C}e(u)) = 0 \quad \text{on } Q, \qquad \vec{n}^\top \mathbb{C}e(u) = 0 \quad \text{on } \Sigma_\mathrm{N} \tag{6.3.24}$$

with also $\Gamma_\mathrm{D} = \emptyset$ and the initial conditions (6.2.3), write the weak formulation (6.2.7) and show the energy conservation by choosing a suitable test function, assuming the regularity[20] $u \in C^1_\mathrm{weak}(I;L^2(\Omega;\mathbb{R}^d)) \cap C_\mathrm{w}(I;H^1(\Omega;\mathbb{R}^d))$ and $\varrho\ddot{u} \in L^2(I;H^1(\Omega;\mathbb{R}^d)^*)$.

Exercise 6.3.11 (Energy-conserving time discretization). Writing (6.3.11) as the system (6.1.8) and applying the *Crank-Nicolson time discretization* with the time step $\tau > 0$ as in Remark C.2.5 on p. 541 to obtain the recursive boundary-value problem for the couple (u_τ^k, v_τ^k):

$$\frac{u_\tau^k - u_\tau^{k-1}}{\tau} = \frac{v_\tau^k + v_\tau^{k-1}}{2}, \qquad \varrho\frac{v_\tau^k - v_\tau^{k-1}}{\tau} - \mathrm{div}\Big(\mathbb{C}e\Big(\frac{u_\tau^k + u_\tau^{k-1}}{2}\Big)\Big) = 0 \quad \text{on } \Omega, \tag{6.3.25a}$$

$$\vec{n}^\top \mathbb{C}e(u_\tau^k) = 0 \quad \text{on } \Gamma_\mathrm{N}, \tag{6.3.25b}$$

to be solved successively for $k = 1, ...,$ starting with $u_0^\tau = u_0$ and $v_\tau^0 = v_0$, show existence of the weak solution $(u_\tau^k, v_\tau^k) \in H^1(\Omega;\mathbb{R}^d) \times L^2(\Omega;\mathbb{R}^d)$ provided $u_0 \in H^1(\Omega;\mathbb{R}^d)$ and $v_0 \in L^2(\Omega;\mathbb{R}^d)$, and further show the *energy conservation* in this discrete scheme:

$$\int_\Omega \frac{\varrho}{2}|v_\tau^k|^2 + \frac{1}{2}\mathbb{C}e(u_\tau^k){:}e(u_\tau^k)\,\mathrm{d}x = \text{constant}. \tag{6.3.26}$$

Exercise 6.3.12 (A normal dispersion). Perform the energetic a-priori estimate for (6.3.21) with $H_1 < 0$. Think about a modification for $d \geq 2$.

Exercise 6.3.13 (Self-gravitational load). Complete the system (6.3.14) with the initial conditions. Realize the analog with the system (5.7.19a,c), formulate the energetics and perform the a-priori estimates, relying on $\|\phi\|_{L^2(\mathbb{R}^d)} \leq C\|\nabla\phi\|_{L^2(\mathbb{R}^d;\mathbb{R}^d)}$ for $\phi \in H^1(\mathbb{R}^d)$ such that $\lim_{|x|\to\infty}\phi(x) = 0$. Realize the corresponding modifications in the analysis when (6.3.17) is considered.

Exercise 6.3.14 (Continuous dependence on ϱ). Assume the data qualification as in Proposition 6.3.2 and take some constants $0 < \varrho_\mathrm{min} < \varrho_\mathrm{max} < +\infty$. Prove that the mapping $\varrho \mapsto (u,\sigma)$ where (u,σ) is the unique solution to the initial-boundary-value problem (6.1.1)–(6.2.2)–(6.2.3) with $\sigma = \mathbb{C}e(u)$ is weakly* continuous on the convex subset $\{\varrho_\mathrm{min} \leq \varrho(\cdot) \leq \varrho_\mathrm{max} \text{ on } \Omega\} \subset L^\infty(\Omega)$ to the topologies indicated in (6.3.13). Show existence of solutions to the minimization problem

[20] In concrete rheological models, such regularity can be ensured under sufficiently gentle loading, cf. Proposition 6.4.2 or 6.6.2.

$$
\left.\begin{array}{ll}
\text{minimize} & \displaystyle\int_\Sigma |u-u_{\rm d}|^2\,{\rm d}x{\rm d}t \\
\text{subject to} & u\in W^{1,\infty}(I;H^1(\Omega;\mathbb{R}^d))\cap W^{2,\infty}(I;L^2(\Omega;\mathbb{R}^d)) \\
& \qquad\qquad\qquad \text{solves (6.1.1)–(6.2.2)–(6.2.3)}, \\
& \text{with } \varrho\in L^\infty(\Omega),\quad \varrho_{\min}\le\varrho(\cdot)\le\varrho_{\max} \ \text{ for a.a. } x\in\Omega,
\end{array}\right\} \tag{6.3.27}
$$

where $u_{\rm d}\in L^2(\Sigma;\mathbb{R}^d)$. Modify the analysis for the variant that the velocity $v_{\rm d}$ rather than displacement is measured, and the cost functional in (6.3.27) by $\int_\Sigma|\dot u-v_{\rm d}|^2\,{\rm d}x{\rm d}t$ with $v_{\rm d}\in L^2(\Sigma;\mathbb{R}^d)$ is then considered. Try to modify the analysis still in order to replace the cost functional in (6.3.27) by $\int_\Sigma|\ddot u-a_{\rm d}|^2\,{\rm d}x{\rm d}t$ with $a_{\rm d}\in L^2(\Sigma;\mathbb{R}^d)$ a given acceleration.

Remark 6.3.15 (Seismic tomography). The optimization problem (6.3.27) can be interpreted as an inverse problem like actually used in the so-called *seismic tomography* to identify mass density in the Earth interior from the amplitude $u_{\rm d}$ or velocity $v_{\rm d}$ of seismic waves measured by seismometers on the Earth surface. Actually, some seismometers (called accelerometers) measure even rather acceleration $a_{\rm d}$ than velocity.

6.4 Kelvin-Voigt viscoelastic material

A basic example of rheological models of viscoelastic response of materials is a so-called *Kelvin-Voigt material* based on an arrangement of a spring and a damper parallelly so that the stresses sum up. In the "language" from Table 6.1, it reads as $\text{spring}_{\rm C}\,\|\,\text{damper}_{\rm D}$ and is governed by the constitutive relations:

$$
\sigma := \sigma_1+\sigma_2 := \mathbb{D}e(\dot u)+\mathbb{C}e(u), \tag{6.4.1}
$$

cf. Figure 6.3. Actually, this is the simplest rheological model for which the total

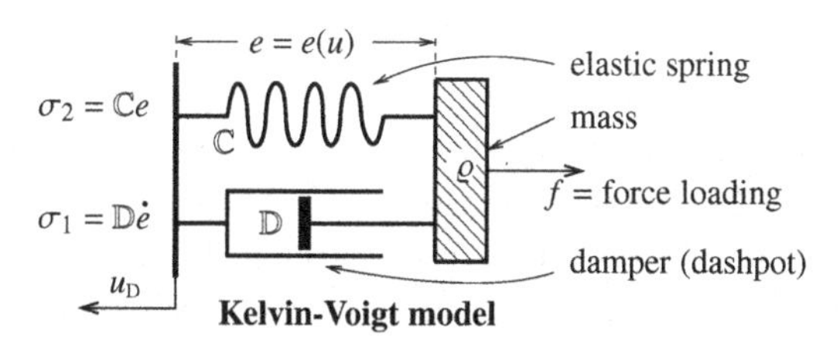

Fig. 6.3 The simplest visco-elastic solid-type rheological model combining only two elements from Fig. 6.1 in parallel (a 0-dimensional scheme).

strain e is covered by the chain of elastic springs (here only one) and simultaneously by the chain of dampers (again here only one). The former attribute characterizes models of solids while the latter attribute causes efficient attenuation which is advantageous from the mathematical viewpoint, cf. also Remark 6.4.5 below. This is why this model is the most often one which is used in mathematical literature, although there is an objection against such sort of materials because they exhibit unbounded (unphysical) stress response on jumping hard-device loading, cf. Figure 6.9 on p. 241 below.

To get the energy balance, we must substitute for $\sigma{:}e(\dot u)$ in (6.2.9). In this case of Kelvin-Voigt's material, just multiply (6.4.1) by $e(\dot u)$ to get formally

$$\sigma{:}e(\dot u) = \mathbb{D}e(\dot u){:}e(\dot u) + \mathbb{C}e(u){:}e(\dot u) = \xi(e(\dot u)) + \frac{\partial}{\partial t}\varphi(e(u)) \tag{6.4.2}$$

where the (specific) stored energy φ and the (specific) *dissipation rate* ξ are

$$\varphi(e) = \frac{1}{2}\mathbb{C}e{:}e \quad \text{and} \quad \xi(\dot e) = \mathbb{D}\dot e{:}\dot e. \tag{6.4.3}$$

Hence substituting it into (6.2.9) gives

$$\underbrace{\frac{\mathrm{d}}{\mathrm{d}t}\int_\Omega \frac{\varrho}{2}|\dot u|^2\,\mathrm{d}x}_{\text{kinetic energy}} + \underbrace{\int_\Omega \xi(e(\dot u))\,\mathrm{d}x}_{\text{rate of dissipation}} + \underbrace{\frac{\mathrm{d}}{\mathrm{d}t}\int_\Omega \varphi(e(u))\,\mathrm{d}x}_{\text{stored energy}}$$
$$= \int_\Omega f\cdot\dot u\,\mathrm{d}x + \int_{\Gamma_\mathrm{N}} g\cdot\dot u\,\mathrm{d}S + \big\langle \vec\tau_\mathrm{D}, \dot u_\mathrm{D}\big\rangle_{\Gamma_\mathrm{D}} = \begin{array}{l}\text{power}\\ \text{of external}\\ \text{loading}\,.\end{array} \tag{6.4.4}$$

Again, we impose the initial conditions (6.2.3) and assume the data qualification

$$\mathbb{D}\mathrm{Lin}(\mathbb{R}^{d\times d}_{\mathrm{sym}}),\ \mathbb{C}\mathrm{SLin}(\mathbb{R}^{d\times d}_{\mathrm{sym}})\ \text{ positive definite on } \mathbb{R}^{d\times d}_{\mathrm{sym}}, \tag{6.4.5a}$$
$$\varrho \in L^\infty(\Omega), \qquad \operatorname{ess\,inf}\varrho > 0, \tag{6.4.5b}$$
$$f \in L^2(I;L^{2^{*\prime}}(\Omega;\mathbb{R}^d)) + L^1(I;L^2(\Omega;\mathbb{R}^d)), \quad g \in L^2(I;L^{2^{\sharp\prime}}(\Gamma_\mathrm{N};\mathbb{R}^d)), \tag{6.4.5c}$$
$$u_\mathrm{D} \in H^1(I;H^{1/2}(\Gamma_\mathrm{D};\mathbb{R}^d))\ \text{ possessing}$$
$$\text{an extension } \bar u_\mathrm{D} \in H^1(I;H^1(\Omega;\mathbb{R}^d)) \cap W^{2,1}(I;L^2(\Omega;\mathbb{R}^d)), \tag{6.4.5d}$$
$$u_0 \in H^1(\Omega;\mathbb{R}^d), \quad \text{and} \quad v_0 \in L^2(\Omega;\mathbb{R}^d), \tag{6.4.5e}$$

where we used the notation $p' = p/(p-1)$, and also (B.3.7) and (B.3.12) with $p = 2$ for 2^* and $2^\sharp$ respectively. For the Sobolev-Slobodetskiĭ space $H^{1/2}(\Gamma_\mathrm{D})$ see Remark B.3.17 on p. 521. A-priori estimates can be obtained, at least formally,[21] by a test by $(u-\bar u_\mathrm{D})^{\boldsymbol\cdot}$ as in (6.2.9) with $\langle \vec\tau_\mathrm{D}, \dot u_\mathrm{D}\rangle_\mathrm{D}$ substituted from (6.2.10). After integration over a time interval $[0,t]$ and using by-part integration for the term $\ddot u\cdot\dot{\bar u}_\mathrm{D}$, this strategy gives:

$$\int_\Omega \frac{\varrho}{2}|\dot u(t)|^2 + \frac{1}{2}\mathbb{C}e(u(t)){:}e(u(t))\,\mathrm{d}x + \int_0^t\int_\Omega \mathbb{D}e(\dot u){:}e(\dot u)\,\mathrm{d}x\mathrm{d}t$$

[21] Cf. Remarks 6.4.9 and 6.4.10.

$$
\begin{aligned}
&= \int_\Omega \frac{\varrho}{2}|v_0|^2 + \frac{1}{2}\mathbb{C}e(u_0)):e(u_0))\,\mathrm{d}x \\
&+ \int_0^t\!\!\int_\Omega f\cdot\dot u - \varrho\dot u\cdot\ddot u_{\mathrm D} + \big(\mathbb{D}e(\dot u) + \mathbb{C}e(u)\big):e(\ddot u_{\mathrm D})\,\mathrm{d}x\mathrm{d}t + \int_0^t\!\!\int_{\Gamma_{\mathrm N}} g\cdot\dot u\,\mathrm{d}S\,\mathrm{d}t \\
&+ \int_\Omega \varrho\dot u(t)\cdot\dot u_{\mathrm D}(t) - \varrho\dot u(0)\cdot\dot u_{\mathrm D}(0)\,\mathrm{d}x - \int_0^t\!\!\int_\Omega f\cdot\dot u_{\mathrm D}\,\mathrm{d}x\mathrm{d}t - \int_0^t\!\!\int_{\Gamma_{\mathrm N}} g\cdot\dot u_{\mathrm D}\,\mathrm{d}S\,\mathrm{d}t; \qquad (6.4.6)
\end{aligned}
$$

again, for notational simplicity, we have used the same letter "t" in a two-fold position in $\int_0^t ....\mathrm{d}t$. Then, for $f = f_1 + f_2$ with $f_1 \in L^2(I;L^{2^{*\prime}}(\Omega;\mathbb{R}^d))$ and $f_2 \in L^1(I;L^2(\Omega;\mathbb{R}^d))$, we proceed by (6.4.5a) and through Korn's and Hölder's and Young's inequalities to

$$
\begin{aligned}
&\frac{\varrho_{\min}}{2}\big\|\dot u(t)\big\|^2_{L^2(\Omega;\mathbb{R}^d)} + \varepsilon\big\|\nabla u(t)\big\|^2_{L^2(\Omega;\mathbb{R}^{d\times d})} + \varepsilon\int_0^t \big\|\nabla\dot u\big\|^2_{L^2(\Omega;\mathbb{R}^d)}\,\mathrm{d}t \\
&\quad\le \frac{\varrho_{\max}}{2}\big\|v_0\big\|^2_{L^2(\Omega;\mathbb{R}^d)} + C\big\|\nabla u_0\big\|^2_{L^2(\Omega;\mathbb{R}^d)} \\
&\quad+ \int_0^t C\big\|f_1\big\|^2_{L^{2^{*\prime}}(\Omega;\mathbb{R}^d)} + \delta\big\|\nabla\dot u\big\|^2_{L^2(\Omega;\mathbb{R}^{d\times d})} + \big\|f_2\big\|_{L^2(\Omega;\mathbb{R}^d)}\Big(\frac14 + \big\|\dot u\big\|^2_{L^2(\Omega;\mathbb{R}^d)}\Big)\mathrm{d}t \\
&\quad+ \varrho_{\max}\int_0^t \big\|\ddot u_{\mathrm D}\big\|_{L^2(\Omega;\mathbb{R}^d)}\Big(\frac14 + \big\|\dot u\big\|^2_{L^2(\Omega;\mathbb{R}^d)}\Big)\mathrm{d}t \qquad (6.4.7a) \\
&\quad+ \int_0^t \delta\big\|\nabla\dot u\big\|^2_{L^2(\Omega;\mathbb{R}^{d\times d})} + \big\|\nabla u\big\|^2_{L^2(\Omega;\mathbb{R}^{d\times d})} + C\big\|\nabla\ddot u_{\mathrm D}\big\|^2_{L^2(\Omega;\mathbb{R}^{d\times d})}\,\mathrm{d}t \\
&\quad+ \int_0^t C\big\|g\big\|^2_{L^{2^{\sharp\prime}}(\Gamma_{\mathrm N};\mathbb{R}^d)} + \delta\big\|\nabla\dot u\big\|^2_{L^2(\Omega;\mathbb{R}^{d\times d})} + \delta\big\|\dot u\big\|^2_{L^2(\Omega;\mathbb{R}^d)}\,\mathrm{d}t + \delta\big\|\dot u(t)\big\|^2_{L^2(\Omega;\mathbb{R}^d)} \\
&\quad+ C\big\|\dot u_{\mathrm D}(t)\big\|^2_{L^2(\Omega;\mathbb{R}^d)} - \int_\Omega \varrho\dot u(0)\cdot\dot u_{\mathrm D}(0)\,\mathrm{d}x - \int_0^t\!\!\int_\Omega f\cdot\dot u_{\mathrm D}\,\mathrm{d}x\mathrm{d}t \\
&\quad- \int_0^t\!\!\int_{\Gamma_{\mathrm N}} g\cdot\dot u_{\mathrm D}\,\mathrm{d}S\,\mathrm{d}t + C\big\|u(t)\big\|^2_{L^2(\Omega;\mathbb{R}^{d\times d})} + C\int_0^t \big\|\dot u\big\|^2_{L^2(\Omega;\mathbb{R}^d)}\,\mathrm{d}t, \qquad (6.4.7b)
\end{aligned}
$$

where $\varrho_{\min} := \operatorname{ess\,inf}\varrho$, $\varepsilon, \delta > 0$ are small numbers, and C a generic constant, ε depending on $\mathbb{C}$, $\mathbb{D}$, $\varrho_{\max} := \|\varrho\|_{L^\infty(\Omega)}$, and Ω. The last two terms in (6.4.7b) come from the Korn's inequality, see Theorem 5.2.2 on p. 165. Then we choose δ so that the particular terms can be absorbed in the left-hand side, and then use Gronwall's inequality if, in (6.4.7b), one estimates

$$
\|u(t)\|^2_{L^2(\Omega;\mathbb{R}^{d\times d})} = \bigg\|\int_0^t \dot u\,\mathrm{d}t + u_0\bigg\|^2_{L^2(\Omega;\mathbb{R}^{d\times d})} \le 2T\int_0^t \|\dot u\|^2_{L^2(\Omega;\mathbb{R}^{d\times d})}\,\mathrm{d}t + 2\|u_0\|^2_{L^2(\Omega;\mathbb{R}^{d\times d})}.
$$

Moreover, assuming also a certain smoothness of the mass density profile, namely (6.3.5), one can prove the "dual" estimate for the acceleration by comparison by using a test by $v \in L^2(I;H^1_{\mathrm D}(\Omega;\mathbb{R}^d)) \cap L^\infty(I;L^2(\Omega;\mathbb{R}^d))$ with $H^1_{\mathrm D}(\Omega;\mathbb{R}^d)$ from (6.2.5) and estimating

$$
\begin{aligned}
\big\|\ddot{u}\big\|_{L^2(I;H^1_{\rm D}(\Omega;\mathbb{R}^d)^*)+L^1(I;L^2(\Omega;\mathbb{R}^d))} &:= \sup_{\|v\|_{L^2(I;H^1_{\rm D}(\Omega;\mathbb{R}^d))\cap L^\infty(I;L^2(\Omega;\mathbb{R}^d))}\le 1} \langle \ddot{u},v\rangle \\
&\le \max(N,\varrho_{\max}) \sup_{\|v\|_{L^2(I;H^1_{\rm D}(\Omega;\mathbb{R}^d))\cap L^\infty(I;L^2(\Omega;\mathbb{R}^d))}\le 1} \langle \ddot{u},\varrho v\rangle \\
&= \max(N,\varrho_{\max}) \sup_{\|v\|_{L^2(I;H^1_{\rm D}(\Omega;\mathbb{R}^d))\cap L^\infty(I;L^2(\Omega;\mathbb{R}^d))}\le 1} \int_Q \sigma{:}e(v) - f{\cdot}v\,{\rm d}x{\rm d}t - \int_{\Sigma_{\rm N}} g{\cdot}v\,{\rm d}S\,{\rm d}t,
\end{aligned}
\tag{6.4.8}
$$

where N is the norm[22] of the linear operator $v \mapsto \varrho v : H^1(\Omega;\mathbb{R}^d) \to H^1(\Omega;\mathbb{R}^d)$; here we used (6.4.5b) and the Hölder inequality and the already obtained estimates that bound σ in $L^2(Q;\mathbb{R}^{d\times d}_{\rm sym})$. Cf. also the estimation (6.3.11).

Similarly, we can prove also the estimate for $\sqrt{\varrho}\ddot{u} \in L^2(I;H^1_{\rm D}(\Omega;\mathbb{R}^d)^*) + L^1(I;L^2(\Omega;\mathbb{R}^d))$ by estimating

$$
\begin{aligned}
\big\|\sqrt{\varrho}\ddot{u}\big\|_{L^2(I;H^1_{\rm D}(\Omega;\mathbb{R}^d)^*)+L^1(I;L^2(\Omega;\mathbb{R}^d))} &:= \sup_{\|v\|_{L^2(I;H^1_{\rm D}(\Omega;\mathbb{R}^d))\cap L^\infty(I;L^2(\Omega;\mathbb{R}^d))}\le 1} \int_0^T \langle \sqrt{\varrho}\ddot{u},v\rangle\,{\rm d}t \\
&\le \max(N,\sqrt{\varrho_{\max}}) \sup_{\|v\|_{L^2(I;H^1_{\rm D}(\Omega;\mathbb{R}^d))\cap L^\infty(I;L^2(\Omega;\mathbb{R}^d))}\le 1} \int_0^T \langle \sqrt{\varrho}\ddot{u},\sqrt{\varrho}v\rangle\,{\rm d}t
\end{aligned}
\tag{6.4.9}
$$

and then the last equality in (6.4.8) again applies, where now N is the norm of the linear operator[23] $v \mapsto \sqrt{\varrho}v : H^1(\Omega;\mathbb{R}^d) \to H^1(\Omega;\mathbb{R}^d)$. In (6.4.9), $\langle\cdot,\cdot\rangle$ is the duality between $H^1_{\rm D}(\Omega;\mathbb{R}^d)^*$ and $H^1_{\rm D}(\Omega;\mathbb{R}^d)$. This is then used to put $\sqrt{\varrho}\ddot{u}$ into duality with $\sqrt{\varrho}\dot{u}$. Alternatively, like (6.3.11), we can see it when executing the estimate

$$
\begin{aligned}
\int_0^T \langle \sqrt{\varrho}\ddot{u},v\rangle\,{\rm d}t &= \int_0^T \Big\langle \frac{{\rm div}\,\sigma + f}{\sqrt{\varrho}},v\Big\rangle\,{\rm d}t \\
&= \int_{\Sigma_{\rm N}} g\cdot\frac{v}{\sqrt{\varrho}}\,{\rm d}S\,{\rm d}t + \int_Q \frac{f}{\sqrt{\varrho}}\cdot v - \sigma{:}\nabla\frac{v}{\sqrt{\varrho}}\,{\rm d}x{\rm d}t \\
&= \int_{\Sigma_{\rm N}} g\cdot\frac{v}{\sqrt{\varrho}}\,{\rm d}S\,{\rm d}t + \int_Q \frac{f}{\sqrt{\varrho}}\cdot v - \sigma{:}\Big(\frac{\nabla v}{\sqrt{\varrho}} - \frac{\sqrt{\varrho}\otimes v}{2\,\sqrt[3/2]{\varrho}}\Big)\,{\rm d}x{\rm d}t
\end{aligned}
$$

and then we use (6.3.5) provided $v \in L^2(I;H^1_{\rm D}(\Omega;\mathbb{R}^d)) \cap L^\infty(I;L^2(\Omega;\mathbb{R}^d))$ in the Hölder inequality.

Proposition 6.4.1 (Kelvin-Voigt model: existence of a solution). *Under the data qualification* (6.4.5)*, there is a weak solution* (u,σ) *to the initial-boundary-value problem* (6.1.1)–(6.2.2)–(6.2.3)–(6.4.1) *such that*

---

[22] Here we have to estimate $\nabla(\varrho v) = \varrho\nabla v + v\otimes\nabla\varrho$ in $L^2(\Omega;\mathbb{R}^{d\times d})$, using the bound of $v \in L^{2^*}(\Omega;\mathbb{R}^d)$ and that $\nabla\varrho \in L^{2^*2/(2^*-2)}(\Omega;\mathbb{R}^d)$, cf. the assumption (6.3.5).

[23] Here we use the calculus $\nabla(\sqrt{\varrho}v) = \sqrt{\varrho}\nabla v - v\otimes\nabla\varrho/(2\sqrt{\varrho})$ together with $v\otimes\nabla\varrho \in L^2(\Omega;\mathbb{R}^{d\times d})$ as already used in (6.4.8) and with $1/\sqrt{\varrho} \in L^\infty(\Omega)$, cf. the assumption (6.3.5).

$$u \in H^1(I;H^1(\Omega;\mathbb{R}^d)) \cap W^{1,\infty}(I;L^2(\Omega;\mathbb{R}^d)), \tag{6.4.10a}$$

$$\sigma \in L^2(Q;\mathbb{R}^{d\times d}), \tag{6.4.10b}$$

$$\varrho\ddot{u} \in L^2(I;H^1_{\mathrm{D}}(\Omega;\mathbb{R}^d)^*) + L^1(I;L^2(\Omega;\mathbb{R}^d)). \tag{6.4.10c}$$

If also (6.3.5) holds, then

$$\sqrt{\varrho}\ddot{u} \in L^2(I;H^1_{\mathrm{D}}(\Omega;\mathbb{R}^d)^*) + L^1(I;L^2(\Omega;\mathbb{R}^d)) \tag{6.4.10d}$$

and this solution is unique and satisfies the energy conservation (6.4.4) integrated in time.

Sketch of the proof. First, one can construct an approximate solution, typically by space or by time discretization, cf. Remarks 6.4.9 or 6.4.10 below. By the estimation strategy (6.4.7) adapted to the approximate solutions, one proves (6.4.10a,b) uniformly with respect to the discretization parameter.

Then choose weakly* convergent subsequences. Linearity of the problem then allows for the limit passage in suitable weak formulation. Then (6.4.10c) is proved by (6.4.8). Also, executing the estimation strategy (6.4.7) for the difference of two weak solutions, one obtain uniqueness.

Having proved that u solves (6.1.1) with (6.4.1) and (6.2.2) and (6.2.3) and using that $\sqrt{\varrho}\ddot{u}$ is in duality with $\sqrt{\varrho}\dot{u}$ due to (6.4.9) and that $\mathbb{C}^{1/2}e(\dot{u})$ is in duality with $\mathbb{C}^{1/2}e(u)$,[24] we can make rigorously the test of (6.1.1) by $\dot{u}$ to obtain the energy equality (6.4.4). □

Proposition 6.4.2 (Regularity). *Strengthening the previous data qualification (6.4.5) to $\mathbb{D} \in \mathrm{SLin}(\mathbb{R}^{d\times d}_{\mathrm{sym}})$ and to*

$$f \in L^2(Q;\mathbb{R}^d), \quad v_0 \in H^1(\Omega;\mathbb{R}^d), \quad g \in W^{1,1}(I;L^2(\Gamma_{\mathrm{N}};\mathbb{R}^d)) + H^1(I;L^{p^{\sharp\prime}}(\Gamma_{\mathrm{N}};\mathbb{R}^d)), \tag{6.4.11}$$

and assuming, for simplicity, the Dirichlet boundary conditions affine in time so that $\ddot{u}|_{\Gamma_{\mathrm{D}}} = 0$, it holds that

$$\ddot{u} \in L^2(Q;\mathbb{R}^d) \quad \textit{and} \quad \dot{u} \in L^\infty(I;H^1(\Omega;\mathbb{R}^d)). \tag{6.4.12}$$

Moreover, this solution is unique and satisfies the energy conservation (6.4.4) integrated in time. Assuming $\mathbb{D} = \tau_{\mathrm{R}}\mathbb{C}$ with some relaxation time $\tau_{\mathrm{R}} > 0$ and $\Omega \subset \mathbb{R}^d$ smooth, we have also

$$\dot{u} \in L^2(I;H^2(\Omega;\mathbb{R}^d)). \tag{6.4.13}$$

---

[24] Here both $\mathbb{C}^{1/2}e(\dot{u})$ and $\mathbb{C}^{1/2}e(u)$ belong to $L^2(Q;\mathbb{R}^{d\times d}_{\mathrm{sym}})$. Thus it holds

$$\int_\Omega \mathbb{C}e(u){:}e(\dot{u})\,\mathrm{d}x = \int_\Omega \mathbb{C}^{1/2}e(\dot{u}){:}\mathbb{C}^{1/2}e(u)\,\mathrm{d}x = \frac{\mathrm{d}}{\mathrm{d}t}\int_\Omega \frac{1}{2}\mathbb{C}^{1/2}e(u){:}\mathbb{C}^{1/2}e(u)\,\mathrm{d}x = \frac{\mathrm{d}}{\mathrm{d}t}\int_\Omega \frac{1}{2}\mathbb{C}e(u){:}e(u)\,\mathrm{d}x.$$

Sketch of the proof. One can launch a test by $\ddot{u}$ if $\mathbb{D}\in\mathrm{SLin}(\mathbb{R}^{d\times d}_{\mathrm{sym}})$ and thus the dissipative force is induced by a potential. This test gives

$$\begin{aligned}
&\int_\Omega \frac{1}{2}\mathbb{D}e(\dot{u}(t)):e(\dot{u}(t))\,\mathrm{d}x + \int_0^t\!\!\int_\Omega \varrho|\ddot{u}|^2\,\mathrm{d}x\mathrm{d}t\\
&= \int_\Omega \frac{1}{2}\mathbb{D}e(v_0):e(v_0)\,\mathrm{d}x + \int_0^t\!\!\int_\Omega f\cdot\ddot{u} - \mathbb{C}e(u):e(\ddot{u})\,\mathrm{d}x\mathrm{d}t + \int_0^t\!\!\int_{\Gamma_{\mathrm{N}}} g\cdot\ddot{u}\,\mathrm{d}S\,\mathrm{d}t\\
&= \int_\Omega \frac{1}{2}\mathbb{D}e(v_0):e(v_0) + \mathbb{C}e(u_0):e(v_0)\,\mathrm{d}x - \int_{\Gamma_{\mathrm{N}}} g(0)\cdot v_0\,\mathrm{d}S - \int_\Omega \frac{1}{2}\mathbb{C}e(u(t)):e(\dot{u}(t))\,\mathrm{d}x\\
&\qquad + \int_0^t\!\!\int_\Omega f\cdot\ddot{u} + \mathbb{C}e(\dot{u}):e(\dot{u})\,\mathrm{d}x\mathrm{d}t - \int_0^t\!\!\int_{\Gamma_{\mathrm{N}}} \dot{g}\cdot\dot{u}\,\mathrm{d}S\,\mathrm{d}t + \int_{\Gamma_{\mathrm{N}}} g(t)\cdot\dot{u}(t)\,\mathrm{d}S\\
&\le \varepsilon\big\|e(\dot{u}(t))\big\|^2_{L^2(\Omega;\mathbb{R}^{d\times d}_{\mathrm{sym}})} + C_\varepsilon\|e(u(t))\|^2_{L^2(\Omega;\mathbb{R}^{d\times d}_{\mathrm{sym}})} + \int_{\Gamma_{\mathrm{N}}} g(t)\cdot\dot{u}(t)\,\mathrm{d}S\\
&\qquad + \int_0^t\Big(C\big\|e(\dot{u})\big\|^2_{L^2(\Omega;\mathbb{R}^{d\times d}_{\mathrm{sym}})} + N\|\dot{g}\|_{L^2(\Gamma_{\mathrm{N}};\mathbb{R}^d)}\big(1+\|\dot{u}\|^2_{H^1(\Omega;\mathbb{R}^d)}\big) + C_\varepsilon\|f\|^2_{L^2(\Omega;\mathbb{R}^d)}\\
&\qquad + \varepsilon\big\|\ddot{u}\big\|^2_{L^2(\Omega;\mathbb{R}^d)}\Big)\mathrm{d}t + \int_\Omega \frac{1}{2}\mathbb{D}e(v_0):e(v_0) + \mathbb{C}e(u_0):e(v_0)\,\mathrm{d}x + \int_{\Gamma_{\mathrm{N}}} g\cdot v_0\,\mathrm{d}S
\end{aligned}$$

with N the norm of the trace operator $H^1(\Omega)\to L^2(\Gamma_{\mathrm{N}})$. Alternatively, we can estimate $\int_0^t\int_{\Gamma_{\mathrm{N}}}\dot{g}\cdot\dot{u}\,\mathrm{d}S\,\mathrm{d}t \le \varepsilon\|\dot{u}\|^2_{L^{p^\sharp}(\Sigma_{\mathrm{N}};\mathbb{R}^d)} + C_\varepsilon\|\dot{g}\|^2_{L^{p^{\sharp\prime}}(\Sigma_{\mathrm{N}};\mathbb{R}^d)}$. Thus (6.4.12) can be obtained when using (6.4.11).

Now, $\varrho\ddot{u}\in L^2(Q;\mathbb{R}^d)$ is in duality with $\dot{u}$ and thus the energy conservation and uniqueness hold even without assuming (6.3.5). Moreover, realizing that $\mathrm{div}(\mathbb{D}e(\dot{u}) + \mathbb{C}e(u)) = f - \varrho\ddot{u} \in L^2(Q;\mathbb{R}^d)$ and assuming $\mathbb{D} = \tau_{\mathrm{R}}\mathbb{C}$ with some relaxation time $\tau_{\mathrm{R}} > 0$, one can use the H^2-regularity for the generalized Lamé system on smooth domains to get $\tau_{\mathrm{R}}\dot{u} + u \in L^2(I;H^2_{\mathrm{loc}}(\Omega;\mathbb{R}^d))$ so that (6.4.13) follows. □

Proposition 6.4.3 (Further regularity). *Let (6.4.5a,b) hold while the loading be qualified as*

$$f\in W^{1,1}(I;L^2(\Omega;\mathbb{R}^d)) + H^1(I;L^{2^{*\prime}}(\Omega;\mathbb{R}^d)),\quad g\in H^1(I;L^{2^{\sharp\prime}}(\Gamma_{\mathrm{N}};\mathbb{R}^d)), \tag{6.4.14a}$$

$$u_{\mathrm{D}}\in H^2(I;H^{1/2}(\Gamma_{\mathrm{D}};\mathbb{R}^d)) \text{ possessing}$$
$$\text{an extension } \bar{u}_{\mathrm{D}}\in H^2(I;H^1(\Omega;\mathbb{R}^d))\cap W^{3,1}(I;L^2(\Omega;\mathbb{R}^d)), \tag{6.4.14b}$$

while the initial conditions be qualified as in (6.3.12c). Then the solution satisfies (6.3.13a) together with $\ddot{u}\in L^2(I;H^1(\Omega;\mathbb{R}^d))$.

Sketch of the proof. Like in the proof of Proposition 6.3.2, we first differentiate the boundary-value problem in time and then perform the test by $(u-\bar{u}_{\mathrm{D}})^{\cdot\cdot}$, executing essentially the estimation strategy (6.4.6)–(6.4.7) but for the time derivatives. □

Remark 6.4.4 (Higher regularity). As the system is linear, one can differentiate it in time k-times and perform tests by $\frac{\partial^{k+1}}{\partial t^{k+1}}u$ or $\frac{\partial^{k+2}}{\partial t^{k+2}}u$ to get higher regularity with any $k\ge 1$ under an appropriate qualification of the data.

Remark 6.4.5 (Parabolic character). Strong dissipation through "parallelly" arranged damping term makes the second-order equation (6.1.1) in fact a parabolic system in terms of velocity.

Remark 6.4.6 (Dirichlet condition). An alternative approach to Dirichlet conditions is a shift by $\bar{u}_{\rm D}$. The new variable $\tilde{u} := u - \bar{u}_{\rm D}$ then satisfied homogeneous Dirichlet condition on $\Gamma_{\rm D}$. The equation (6.1.1) with (6.4.1) in terms of $\tilde{u}$ results to

$$\varrho\ddot{\tilde{u}} - {\rm div}\Big(\mathbb{D}e(\dot{\tilde{u}}) + \mathbb{C}e(\tilde{u})\Big) = \tilde{f} := f - \varrho\ddot{\bar{u}}_{\rm D} + {\rm div}\,\bar{\sigma}_{\rm D}$$
$$\text{with}\quad \bar{\sigma}_{\rm D} := \mathbb{D}e(\dot{\bar{u}}_{\rm D}) + \mathbb{C}e(\bar{u}_{\rm D}). \tag{6.4.15}$$

Sometimes, $\bar{\sigma}_{\rm D}$ is referred to as a *loading stress*. The qualification of $\bar{u}_{\rm D}$ to get $\tilde{f} \in L^2(I;H^1_{\rm D}(\Omega;\mathbb{R}^d)^*) + L^1(I;L^2(\Omega;\mathbb{R}^d))$, cf. (6.4.5c), is again just (6.4.5d).

Remark 6.4.7 (Hamilton principle for dissipative systems). The Hamilton variation principle adapted for nonconservative systems with holonomic constraints $C(u) = 0$ (cf. also Bedford [54, Section 1.3]) says that the integral

$$\int_0^T \mathscr{T}(\dot{u}) - \mathcal{E}(u) + \langle F, u\rangle + \pi C(u)\,{\rm d}t \quad \text{is stationary} \tag{6.4.16}$$

with $\mathcal{E}$ being the stored energy and $F = -R'(\dot{u})$ being a nonconservative force with R denoting the (Rayleigh's *pseudo)potential* of the *dissipative force*, i.e. here $R(v) = \int_\Omega \frac{1}{2}\mathbb{D}e(v){:}e(v)\,{\rm d}x$. This gives the abstract *force-equilibrium equation*

$$\mathscr{T}'\ddot{u} + \mathcal{E}'(u) = F + [C']^*\pi \quad \text{with} \quad C(u) = 0, \tag{6.4.17}$$

where the apostrophe indicates the Gâteaux differential, cf. (A.3.2) on p. 494, and C forms the constraint[25] and $[C']^*$ is the adjoint operator to C', with π playing the role of the Lagrange multiplier to the constraint $Cu = 0$; i.e. here $[C']^* : W^{-1/2,2}(\Gamma;\mathbb{R}^d) \to W^{-1,2}(\Omega;\mathbb{R}^d)$ and π is the traction vector $\vec{t}_{\rm D}$. Cf. also Bedford [54, Section 3.1.1].

Remark 6.4.8 (Quasistatic problems). Often, inertial effects are neglected in many applications, i.e. $\varrho = 0$. This makes the problem usually easier, but certain cautious is advisable. It can be demonstrated on the elastically supported viscoelastic body loaded by a bulk force:

$$-{\rm div}(\mathbb{D}e(\dot{u}) + \mathbb{C}e(u)) = f \qquad \text{on } Q, \tag{6.4.18a}$$
$$(\mathbb{D}e(\dot{u}) + \mathbb{C}e(u))\vec{n} + Au = Au_\flat \qquad \text{on } \Sigma, \tag{6.4.18b}$$
$$u(0) = u_0 \qquad \text{on } \Omega \tag{6.4.18c}$$

with $A \in L^\infty(\Gamma;\mathbb{R}^{d\times d}_{\rm sym})$ a positive-definite matrix of moduli the elastic support on the boundary. Of course, also the tensors $\mathbb{D} \in {\rm Lin}(\mathbb{R}^{d\times d}_{\rm sym})$ and $\mathbb{C} \in {\rm SLin}(\mathbb{R}^{d\times d}_{\rm sym})$ are considered positive definite. The basic energetic estimates obtained by a test by $\dot{u}$ gives

[25] Here C is affine and involves the Dirichlet boundary conditions, considered now constant in time, i.e. $C(u) := u|_\Gamma - u_{\rm D}$.

$u \in L^\infty(I;H^1(\Omega;\mathbb{R}^d))$ via Korn's inequality but only $e(\dot u) \in L^2(Q;\mathbb{R}^{d\times d}_{\rm sym})$, while the estimate on $\dot u$ is now not obtained by this test because the boundary support is merely elastic, not viscoelastic. In particular, we must qualify $f \in W^{1,1}(I;L^{2^{*\prime}}(\Omega;\mathbb{R}^d))$ in order to be able to perform the by-part integration in time to estimate

$$\int_0^t\!\!\int_\Omega f\cdot\dot u\,{\rm d}x{\rm d}t = \int_\Omega f(T)\cdot u_0 - f(0)\cdot u_0\,{\rm d}x - \int_0^t\!\!\int_\Omega \dot f\cdot u\,{\rm d}x{\rm d}t$$
$$\le \varepsilon\|u(T)\|^2_{L^{2^*}(\Omega;\mathbb{R}^d)} + \int_0^t \|\dot f\|_{L^{2^{*\prime}}(\Omega;\mathbb{R}^d)}\Big(1+\|u\|^2_{L^{2^*}(\Omega;\mathbb{R}^d)}\Big){\rm d}t$$
$$+\frac{1}{4\varepsilon}\|f(T)\|^2_{L^{2^{*\prime}}(\Omega;\mathbb{R}^d)} + \|f(0)\|_{L^{2^{*\prime}}(\Omega;\mathbb{R}^d)} + \|u_0\|^2_{L^{2^*}(\Omega;\mathbb{R}^d)},$$

and then, for sufficiently small $\varepsilon > 0$, to absorb $\varepsilon\|u(T)\|^2_{L^{2^*}(\Omega;\mathbb{R}^d)}$ in the left-hand side after using the Korn inequality and to use the Gronwall inequality. Similarly, also $u_\flat \in W^{1,1}(I;L^{2^{\sharp\prime}}(\Gamma;\mathbb{R}^d))$ must be assumed. In addition, we need $u_0 \in H^1(\Omega;\mathbb{R}^d)$. Of course, here the system is linear so further options are open for higher-order regularity tests provided the data f, $u_\flat$, and u_0 are smoother.

Remark 6.4.9 (Faedo-Galerkin method). The test by $v = (u-\bar u_{\rm D})^{\boldsymbol\cdot}$ used for (6.4.6) is rather formal until one knows that such v is in duality with $\varrho\ddot u$ and the by-part integration in time can rigorously be employed, cf. Lemma C.2.2 on p. 535. To execute the estimation strategy (6.4.6)–(6.4.7) motivated by the physical energetics (6.4.4) is to use a suitable approximation for which the used calculus holds and then make a limit passage. The most straightforward option is an approximation by replacing the infinite-dimensional function spaces where the states of the system live in particular times, i.e. here $H^1(\Omega;\mathbb{R}^d)$ for $u(t)$, by (a sequence of) finite-dimensional subspaces, cf. Section C.2.4 for a general approach. Here the mentioned test requires still an approximation of $\bar u_{\rm D}$ into such finite-dimensional spaces. The convergence towards continuous problem is then facilitated by the density assumption (C.2.55) in $V = H^1(\Omega;\mathbb{R}^d)$. The "dual" estimate $\varrho\ddot u$ can be obtained by a comparison from the Galerkin identity, cf. Remark C.2.9 on p. 547.

Remark 6.4.10 (Rothe method). An alternative method of quite universal usage of constructing an approximation of the initial-boundary-value problem (6.1.1)–(6.2.2)–(6.2.3)–(6.4.1) is to make a discretization in time, considering $\tau > 0$ a time step. We put $u^0_\tau = u_0$, $u^{-1}_\tau = u_0 - \tau v_0$, and for $k = 1, ..., T/\tau$, and consider the recursive boundary-value problem for u^k_τ:

$$\varrho\frac{u^k_\tau - 2u^{k-1}_\tau + u^{k-2}_\tau}{\tau^2} - {\rm div}\,\sigma^k_\tau = f^k_\tau$$
$$\text{with}\quad \sigma^k_\tau = \mathbb{D}e\Big(\frac{u^k_\tau - u^{k-1}_\tau}{\tau}\Big) + \mathbb{C}e(u^k_\tau) \qquad \text{on } \Omega, \tag{6.4.19a}$$
$$u^k_\tau|_{\Gamma_{\rm D}} = u^k_{{\rm D},\tau} \qquad \text{on } \Gamma_{\rm D}, \tag{6.4.19b}$$
$$\vec n^\top\sigma^k_\tau = g^k_\tau \qquad \text{on } \Gamma_{\rm N}, \tag{6.4.19c}$$

where $f_\tau^k := f(k\tau)$, $g_\tau^k := g(k\tau)$, and $u_{\mathrm{D},\tau}^k := u_\mathrm{D}(k\tau)$. Its (unique) weak solution exists in $H^1(\Omega;\mathbb{R}^d)$ if assuming positive-definiteness of $\mathbb{C}$ and $\mathbb{D}$. Then test by $u_\tau^k - u_\tau^{k-1} - u_{\mathrm{D},\tau}^k + u_{\mathrm{D},\tau}^{k-1}$ gives the expected energy estimate by using the discrete "by-part" summations

$$(u_\tau^k - 2u_\tau^{k-1} + u_\tau^{k-2})\cdot(u_\tau^k - u_\tau^{k-1}) \geq \frac{1}{2}\big|u_\tau^k - u_\tau^{k-1}\big|^2 - \frac{1}{2}\big|u_\tau^{k-1} - u_\tau^{k-2}\big|^2, \quad \text{and} \tag{6.4.20a}$$

$$\mathbb{C}e(u_\tau^k):e(u_\tau^k - u_\tau^{k-1}) \geq \frac{1}{2}\mathbb{C}e(u_\tau^k):e(u_\tau^k) - \frac{1}{2}\mathbb{C}e(u_\tau^{k-1}):e(u_\tau^{k-1}), \tag{6.4.20b}$$

cf. (C.2.40) and (C.2.14) with $\|\cdot\|_H$ being replaced by the seminorm $\mathbb{C}^{1/2}e(\cdot)$ pointwise. Using the piecewise-constant and the piecewise affine interpolants $\bar{u}_\tau$ and u_τ, cf. (C.2.15) on p. 536, one thus obtains the discrete variant of the energy balance (6.4.4)), but here as an inequality only. For u_D constant in time, it looks as:

$$\begin{aligned}&\int_\Omega \frac{\varrho}{2}\big|\dot{u}_\tau(t)\big|^2 + \frac{1}{2}\mathbb{C}e(u_\tau(t)):e(u_\tau(t))\,\mathrm{d}x + \int_0^t\!\!\int_\Omega \xi(e(\dot{u}_\tau))\,\mathrm{d}x\mathrm{d}t \\ &\leq \int_0^t\!\!\int_\Omega \bar{f}_\tau\cdot\dot{u}_\tau\,\mathrm{d}x\mathrm{d}t + \int_0^t\!\!\int_{\Gamma_\mathrm{N}} \bar{g}_\tau\cdot\dot{u}_\tau\,\mathrm{d}S\,\mathrm{d}t + \int_\Omega \frac{\varrho}{2}|v_0|^2 + \frac{1}{2}\mathbb{C}e(u_0):e(u_0)\,\mathrm{d}x \end{aligned} \tag{6.4.21}$$

for any $t = k\tau \in I$ with $k\in\mathbb{N}$. For the weak formulation of the type (6.2.7), one then uses (C.2.37) on p. 542. Thus we obtain the discrete variant of (6.2.7), namely

$$\begin{aligned}\int_Q \mathbb{D}e(\dot{u}_\tau):e(\bar{v}_\tau) + \mathbb{C}e(\bar{u}_\tau):e(\bar{v}_\tau) - \bar{f}_\tau\cdot\bar{v}_\tau - \varrho\dot{u}_\tau(\cdot-\tau)\cdot\dot{v}_\tau\,\mathrm{d}x\mathrm{d}t& \\ + \int_\Omega \varrho\dot{u}_\tau(T)\cdot v_\tau(T)\,\mathrm{d}x = \int_\Omega \varrho\dot{u}_0\cdot v_\tau(0)\,\mathrm{d}x&\end{aligned} \tag{6.4.22}$$

where u_τ is assumed to be defined also on $[-\tau,0]$ by using the value $u_\tau(-\tau) = u_\tau^{-1} = u_0 - \tau v_0$ and where $\bar{v}_\tau$ and v_τ denote respectively the piecewise constant and the piecewise affine interpolants of the $\{v^k\}_{k=0}^{T/\tau}$ on the equidistant partition of $[0,T]$ with the time step τ, see again (C.2.15) on p. 536.

Remark 6.4.11 (Variational structure of (6.4.19)). Assume $\mathbb{D}\in\mathrm{SLin}(\mathbb{R}_{\mathrm{sym}}^{d\times d})$ as in (C.2.51a) on p. 545, the (unique) solution to (6.4.19) can be obtained through the following variational problem:

$$\left.\begin{aligned}&\text{minimize} \quad \int_\Omega \tau^2\varrho\Big|\frac{u-2u_\tau^{k-1}+u_\tau^{k-2}}{\tau^2}\Big|^2 + \tau\mathbb{D}e\Big(\frac{u-u_\tau^{k-1}}{\tau}\Big):e\Big(\frac{u-u_\tau^{k-1}}{\tau}\Big) \\ &\qquad\qquad\qquad + \mathbb{C}e(u):e(u) - 2f_\tau^k\cdot u\,\mathrm{d}x - \int_{\Gamma_\mathrm{N}} 2g_\tau^k\cdot u\,\mathrm{d}S \\ &\text{subject to} \quad u\in H^1(\Omega;\mathbb{R}^d), \qquad u|_{\Gamma_\mathrm{D}} = u_{\mathrm{D},\tau}^k\,. \end{aligned}\right\} \tag{6.4.23}$$

If $\varrho = 0$ and u_D constant in time, we could get an a-priori estimate by comparing the value at u_τ^k with value at u_τ^{k-1}. Indeed, relying that u_τ^k is a global minimizer of the convex problem (6.4.23), it yields

$$\int_\Omega \frac{1}{2}\tau\mathbb{D}e\Big(\frac{u_\tau^k-u_\tau^{k-1}}{\tau}\Big):e\Big(\frac{u_\tau^k-u_\tau^{k-1}}{\tau}\Big)+\frac{1}{2}\mathbb{C}e(u_\tau^k):e(u_\tau^k)\,\mathrm{d}x$$
$$\leq \int_\Omega \frac{1}{2}\mathbb{C}e(u_\tau^{k-1}):e(u_\tau^{k-1}) - f_\tau^k\cdot\frac{u_\tau^k-u_\tau^{k-1}}{\tau}\,\mathrm{d}x. \qquad (6.4.24)$$

This method must be modified if u_D varies in time, and does not work for $\varrho > 0$. Moreover, in contrast to (6.4.21), it gives a factor $\frac{1}{2}$ in front of the dissipated energy, which yields an a-priori estimate but only a sub-optimal energy estimate, cf. (6.4.4) and also Remark C.2.3 on p.540.

Remark 6.4.12 (Energy equality and strong convergence). Assume now $\mathbb{D} \in \mathrm{SLin}(\mathbb{R}^{d\times d}_{\mathrm{sym}})$ and, for simplicity, u_D constant in time. The weak convergence of an approximate solution, say u_τ, gives by weak lower/upper semicontinuity and by (6.4.21) that

$$\int_Q \xi(e(\dot u))\,\mathrm{d}x\mathrm{d}t \leq \liminf_{\tau\to 0}\int_Q \xi(e(\dot u_\tau))\,\mathrm{d}x\mathrm{d}t \leq \limsup_{\tau\to 0}\int_Q \xi(e(\dot u_\tau))\,\mathrm{d}x\mathrm{d}t$$
$$\leq \limsup_{\tau\to 0}\frac{1}{2}\int_\Omega \varrho|v_0|^2+\mathbb{C}e(u_0):(u_0)-\varrho|\dot u_\tau(T)|^2$$
$$-\,\mathbb{C}e(u_\tau(T)):e(u_\tau(T))\,\mathrm{d}x+\int_Q \bar f_\tau\cdot\dot u_\tau\,\mathrm{d}x\mathrm{d}t+\int_{\Sigma_\mathrm{N}}\bar g_\tau\cdot\dot u_\tau\,\mathrm{d}S\,\mathrm{d}t$$
$$\leq \frac{1}{2}\int_\Omega \varrho|v_0|^2+\mathbb{C}e(u_0):(u_0)-\varrho|\dot u(T)|^2-\mathbb{C}e(u(T)):e(u(T))\,\mathrm{d}x$$
$$+\int_Q f\cdot\dot u\,\mathrm{d}x\mathrm{d}t+\int_{\Sigma_\mathrm{N}} g\cdot\dot u\,\mathrm{d}S\,\mathrm{d}t=\int_Q \xi(e(\dot u))\,\mathrm{d}x\mathrm{d}t, \qquad (6.4.25)$$

where the last equality just uses (6.4.4) (for $\dot u_\mathrm{D}=0$) proved in Proposition 6.4.1. Thus $\int_Q \xi(e(\dot u_\tau))\,\mathrm{d}x\mathrm{d}t \to \int_Q \xi(e(\dot u))\,\mathrm{d}x\mathrm{d}t$ and, using the uniform convexity of the Banach space $L^2(Q;\mathbb{R}^{d\times d})$, also $e(\dot u_\tau)$ converges strongly in $L^2(Q;\mathbb{R}^{d\times d})$. For $\dot u_\mathrm{D}\neq 0$ we can proceed as in (6.4.15).

Remark 6.4.13 (Wave attenuation and dispersion). In contrast to wave propagation in conservative purely elastic materials as in Remarks 6.3.6 and 6.3.7, the computations are more complicated in dissipative viscoelastic materials where (possible) dispersion is combined with attenuation and complex-number calculus is advantageously used. Let us demonstrate the analysis in one-dimensional case, taking the ansatz $u=\mathrm{e}^{\mathrm{i}(wt+x/\lambda)}$ with the angular frequency $w=\omega+\mathrm{i}\gamma$ now considered complex with $\omega,\gamma\in\mathbb{R}$ and the real-valued wavelength λ; here $\mathrm{i}=\sqrt{-1}$ denotes the imaginary unit. Then $\varrho\ddot u=-\varrho w^2 u$, $C\partial^2_{xx}u=-Cu/\lambda^2$, and $D\partial^3_{xxt}u=-\mathrm{i}Dwu/\lambda^2$, so that the one-dimensional dispersive wave equation

$$\varrho\ddot u-\partial_x(D\partial^2_{xt}u+C\partial_x u)=0 \quad \text{on } [0,+\infty)\times\mathbb{R}$$

gives the algebraic condition $\varrho w^2-\mathrm{i}Dw/\lambda^2-C/\lambda^2=0$. When substituting $w=\omega+\mathrm{i}\gamma$ so that $w^2=\omega^2-\gamma^2+2\mathrm{i}\omega\gamma$, we obtain two algebraic equations for the real and the imaginary part each, called *Kramers-Kronig's relations*. More specifically, here

$$\varrho(\omega^2-\gamma^2)=\frac{C}{\lambda^2}-D\frac{\gamma}{\lambda^2}\quad\text{and}\quad 2\varrho\gamma=\frac{D}{\lambda^2}\,. \tag{6.4.26}$$

From the latter equation we can read that $\gamma = D/(2\varrho\lambda^2)$ and then the former equation yields $\omega^2 = C/(\varrho\lambda^2) - D^2/(4\varrho^2\lambda^4)$. Realizing that the speed of wave is $v = \omega\lambda$, we obtain

$$v=\sqrt{\frac{C}{\varrho}-\frac{D^2}{4\varrho^2\lambda^2}}\,,$$

which gives a *normal dispersion* for sufficiently long waves (such that $\lambda > \lambda_{\rm CRIT} := D/\sqrt{4\varrho C}$), in contrast to (6.3.20), i.e. waves with long lengths propagate faster than those with short lengths. Waves with the length $\lambda_{\rm CRIT}$ or shorter are so fast attenuated that they cannot propagate at all. Let us note that the coefficient γ determining the attenuation is here inversely proportional to the square power of the wave length. In other words, the so-called *Q-factor*, defined in our mechanical system as $\omega/(2\gamma) = \varrho\lambda v/D$, is low when $\lambda \to \lambda_{\rm CRIT}+$.

Remark 6.4.14 (Nonlinear materials: low-frequency large-amplitude loading). Generalization for nonlinear response in φ is possible provided $\mathbb{D}\in{\rm SLin}(\mathbb{R}^{d\times d}_{\rm sym})$ so that the linearly-responding dissipative forces have a potential and φ is convex (then Minty's trick comes into play) or uniformly convex. In the latter case, strong convergence of some approximate solutions can be obtained from the estimate (here performed formally):

$$\begin{aligned}
&\Big(a(\|e(u_\tau)\|_{L^p(Q;\mathbb{R}^{d\times d})}-a(\|e(u)\|_{L^p(Q;\mathbb{R}^{d\times d})})\Big)\Big(\|e(u_\tau)\|_{L^p(Q;\mathbb{R}^{d\times d})}-\|e(u)\|_{L^p(Q;\mathbb{R}^{d\times d})}\Big)\\
&\quad\le\int_Q\big(\varphi'(e(u_\tau))+\mathbb{D}e(\dot u_\tau)-\varphi'(e(u))-\mathbb{D}e(\dot u)\big){:}e(u_\tau-u)\,{\rm d}x{\rm d}t\\
&\quad=-\int_0^T\langle\varrho\ddot u_\tau,u_\tau-u\rangle\,{\rm d}t-\int_Q\big(\varphi'(e(u))+\mathbb{D}e(\dot u)\big){:}e(u_\tau-u)\,{\rm d}x{\rm d}t
\end{aligned}\tag{6.4.27}$$

with some increasing $a:(0,\infty)\to\mathbb{R}$ related to the uniformly convexity of φ. Eventually, the last expression in (6.4.27) can be pushed to 0 if one uses that $\ddot u_\tau$ is in duality with $\dot u_\tau$ so that

$$\begin{aligned}
\limsup_{\tau\to0}-\int_0^T\langle\varrho\ddot u_\tau,u_\tau-u\rangle\,{\rm d}t&=-\liminf_{\tau\to0}\frac12\langle\varrho\dot u_\tau(T),\dot u_\tau(T)\rangle+\lim_{\tau\to0}\int_Q\varrho\dot u_\tau\cdot\dot u_\tau\,{\rm d}x{\rm d}t\\
&\quad+\lim_{\tau\to0}\frac12\langle\varrho\dot u_\tau(0),\dot u_\tau(0)\rangle+\lim_{\tau\to0}\int_0^T\langle\varrho\ddot u_\tau,u\rangle\,{\rm d}t\le0
\end{aligned}$$

where we also used Aubin-Lions' theorem B.4.10 for $\dot u_\tau\to\dot u$ strongly in $L^2(Q;\mathbb{R}^d)$, weak convergence $\dot u_\tau(T)\to\dot u(T)$ in $L^2(\Omega;\mathbb{R}^d)$, as well as weak convergence $\varrho\ddot u_\tau\to\ddot u$ in a space which is dual to the space where u lives. Fine approximation technicalities have been omitted in (6.4.27), however.

Remark 6.4.15 (Nonlinear materials: high-frequency small-amplitude loading). Assume the dissipative force $D(\dot{e})$ with $D : \mathbb{R}^{d\times d}_{\rm sym} \to \mathbb{R}^{d\times d}_{\rm sym}$ uniformly monotone (even possibly nonpotential) instead of $\mathbb{D}\dot{e}$. Then strong convergence of some approximate solutions can be obtained from the estimate (here performed formally):

$$
\begin{aligned}
&\frac{1}{2}\min\varrho(\Omega)\big\|\dot{u}_\tau(T){-}\dot{u}(T)\big\|^2_{L^2(\Omega;\mathbb{R}^d)} + c\big\|e(\dot{u}_\tau{-}\dot{u})\big\|^2_{L^2(Q;\mathbb{R}^{d\times d})}\\
&\le \int_0^T \langle\varrho\ddot{u}_\tau - \varrho\ddot{u}, \dot{u}_\tau - \dot{u}\rangle + \langle D(e(\dot{u}_\tau)) - D(e(\dot{u})), e(\dot{u}_\tau{-}\dot{u})\rangle + \langle\mathbb{C}e(u_\tau{-}u), e(\dot{u}_\tau{-}\dot{u})\rangle\,\mathrm{d}t\\
&= \int_0^T \langle f - \varrho\ddot{u}, \dot{u}_\tau - \dot{u}\rangle - \langle D(e(\dot{u})) + \mathbb{C}e(u), e(\dot{u}_\tau{-}\dot{u})\rangle\,\mathrm{d}t,
\end{aligned}
$$

which can be limited to zero if $u_\tau \to u$ weakly, obtaining the strong convergence $e(\dot{u}_\tau) \to e(\dot{u})$ and thus $e(u_\tau) \to e(u)$, too. Having strong convergence, we can pass to the limit through the nonlinear term $D(\cdot)$ easily.

Remark 6.4.16 (Doubly nonlinear models). The Kelvin-Voigt materials with both φ and ξ non-quadratic has been treated in [204] and, under a weaker data qualification[26], in [98].

Exercise 6.4.17. Realize that (6.4.7a) used the qualification of $u_{\rm D}$ such that $\bar{u}_{\rm D} \in W^{2,1}(I;L^2(\Omega;\mathbb{R}^d))$. Modify it for $\bar{u}_{\rm D} \in W^{2,1}(I;L^2(\Omega;\mathbb{R}^d)) + H^2(I;L^{2^*\prime}(\Omega;\mathbb{R}^d))$ as assumed in (6.4.5d).

Exercise 6.4.18. Go through the arguments preceding Proposition 6.4.1 and modify them to a rigorous proof by the Faedo-Galerkin approximation.

Exercise 6.4.19. Consider the Faedo-Galerkin approximation and reorganize the argument for strong convergence in Remark 6.4.12 by estimating directly $\int_Q \xi(e(\dot{u}_k - \dot{u}))\,\mathrm{d}x\mathrm{d}t$.

Exercise 6.4.20. Show how comparing the value of (6.4.23) for u^k_τ versus u^{k-1}_τ expands (C.2.51) and why this estimation does not lead to desired a-priori estimates.

6.5 Standard solid models: Poynting-Thomson-Zener materials

Let us now present a model involving two springs and one damper, a so-called 3-parameter *standard linear solid*, also called a *Poynting-Thomson* [420] or a *Zener* [540] rheological model, sometimes also a standard *Boltzmann* model (or *material*). There are two option how to organize it, either $\text{spring}_{\mathbb{C}_1} - (\text{spring}_{\mathbb{C}_2} \,\|\, \text{damper}_{\mathbb{D}})$ or $(\text{spring}_{\tilde{\mathbb{C}}_1} - \text{damper}_{\tilde{\mathbb{D}}}) \,\|\, \text{spring}_{\tilde{\mathbb{C}}_2}$, cf. also Figure 6.4.

[26] The dissipation has been assumed of a polynomial coercivity ≥ 2 while the stored energy has been assumed of a polynomial coercivity ≤ 2 in [98].

A general feature of the rheological models of *solids* is that there exists a connected path in the rheological diagrams composed from elastic springs which overbridges the total strain, while its absence characterized models of *fluids*. The standard-solid model is thus the simplest viscoelastic model of solids except the Kelvin-Voigt model. The absence of any connected path in the rheological diagram composed from dashpots makes the problem rather hyperbolic than parabolic, cf. Remarks 6.4.5 vs. 6.6.4. Here, in contrast to the Kelvin-Voigt model, the problem is rather hyperbolic.

Let us first investigate the former option, i.e. the Poynting-Thomson model. The constitutive equations for this model are

$$\sigma = \mathbb{C}_1 e_1 = \sigma_1 + \sigma_2, \qquad e = e_1 + e_2, \qquad \sigma_1 = \mathbb{C}_2 e_2, \quad \text{and} \quad \sigma_2 = \mathbb{D}\dot{e}_2 \tag{6.5.1}$$

in the notation from Figure 6.4(left). We want to eliminate the "internal" quantities

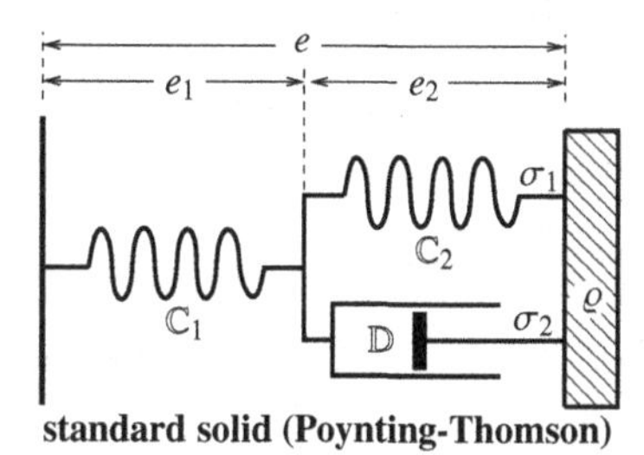

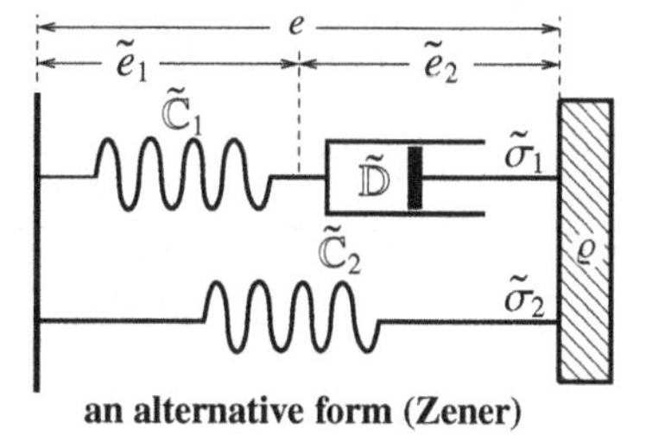

Fig. 6.4 Two alternatives for organizing two elastic springs and one damper (dashpot):
Left: Hooke's element from Fig. 6.1 in series with Kelvin-Voigt's model from Fig. 6.3.
Right: Hooke's element in parallel with Maxwell's model from Fig. 6.5 on p. 229 below.

e_1, e_2, σ_1, and σ_2, and write a constitutive relation between σ and ε only. After some algebraic manipulation, we obtain:

$$\begin{aligned}
\dot{e} = \dot{e}_1 + \dot{e}_2 = \mathbb{C}_1^{-1}\dot{\sigma} + \mathbb{D}^{-1}\sigma_2 &= \mathbb{C}_1^{-1}\dot{\sigma} + \mathbb{D}^{-1}(\sigma - \sigma_1)\\
&= \mathbb{C}_1^{-1}\dot{\sigma} + \mathbb{D}^{-1}(\sigma - \mathbb{C}_2 e_2)\\
&= \mathbb{C}_1^{-1}\dot{\sigma} + \mathbb{D}^{-1}(\sigma - \mathbb{C}_2(e - e_1))\\
&= \mathbb{C}_1^{-1}\dot{\sigma} + \mathbb{D}^{-1}(\sigma - \mathbb{C}_2(e - \mathbb{C}_1^{-1}\sigma)).
\end{aligned} \tag{6.5.2}$$

Altogether, after multiplication by $\mathbb{C}_1$, we obtain

$$\dot{\sigma} + \mathbb{C}_1\mathbb{D}^{-1}(\mathbb{I} + \mathbb{C}_2\mathbb{C}_1^{-1})\sigma = \mathbb{C}_1\dot{e} + \mathbb{C}_1\mathbb{D}^{-1}\mathbb{C}_2 e. \tag{6.5.3}$$

To specify energy balance (6.2.9), we must identify the rate of dissipation and the stored energy contained in $\sigma{:}e(\dot{u})$. To this goal, let us multiply the first row in (6.5.2) by σ and use (6.5.1):

$$\begin{aligned}\sigma{:}\dot{e} = \sigma{:}\mathbb{C}_1^{-1}\dot{\sigma} + \sigma{:}\mathbb{D}^{-1}\sigma_2 &= \sigma{:}\mathbb{C}_1^{-1}\dot{\sigma} + (\sigma_1+\sigma_2){:}\mathbb{D}^{-1}\sigma_2\\ &= \sigma{:}\mathbb{C}_1^{-1}\dot{\sigma} + \sigma_2{:}\mathbb{D}^{-1}\sigma_2 + \sigma_1{:}\dot{e}_2\\ &= \sigma{:}\mathbb{C}_1^{-1}\dot{\sigma} + \sigma_2{:}\mathbb{D}^{-1}\sigma_2 + e_2{:}\mathbb{C}_2\dot{e}_2. \end{aligned}\tag{6.5.4}$$

Comparing it with (6.2.9), we can now identify the specific stored energy φ and the specific dissipation rate ξ in the energy balance (6.6.4) as

$$\varphi(\sigma, e_2) = \frac{1}{2}\mathbb{C}_1^{-1}\sigma{:}\sigma + \frac{1}{2}\mathbb{C}_2 e_2{:}e_2, \tag{6.5.5a}$$

$$\xi(\sigma_2) = \mathbb{D}^{-1}\sigma_2{:}\sigma_2 \quad \text{or alternatively} \quad \xi(\dot{e}_2) = \mathbb{D}\dot{e}_2{:}\dot{e}_2; \tag{6.5.5b}$$

for the last formula, we used $\sigma_2 = \mathbb{D}\dot{e}_2$, cf. (6.5.1).

To obtain the energetics (which will serve also for the a-priori estimates), we must again substitute for $\sigma{:}e(\dot{u})$ from (6.5.4) to (6.2.9). Considering the Newton boundary conditions together with the Dirichlet loading $u_{\rm D}$, as in (6.2.9), this gives (at least formally):

$$\frac{\mathrm{d}}{\mathrm{d}t}\bigg(\int_\Omega \underbrace{\frac{\varrho}{2}|\dot{u}|^2 + \frac{1}{2}\mathbb{C}_1^{-1}\sigma{:}\sigma + \frac{1}{2}\mathbb{C}_2 e_2{:}e_2}_{\text{kinetic and stored energy}}\,\mathrm{d}x\bigg) + \int_\Omega \underbrace{\mathbb{D}^{-1}\sigma_2{:}\sigma_2}_{\text{dissipation rate}}\,\mathrm{d}x$$
$$= \int_\Omega f\cdot\dot{u}\,\mathrm{d}x + \int_{\Gamma_{\rm N}} g\cdot\dot{u}\,\mathrm{d}S + \langle \vec{\tau}_{\rm D}, \dot{u}_{\rm D}\rangle_{\Gamma_{\rm D}} = \begin{array}{c}\text{total power}\\ \text{of external}\\ \text{loading}\end{array}, \tag{6.5.6}$$

where $\vec{\tau}_{\rm D}$ is the traction vector defined by (6.2.10). The first integral (6.5.6) which is under time derivative indicates that we will need three initial conditions: on $\dot{u}$, σ, and on u to determine $e_2 = e(u) - \mathbb{C}_1^{-1}\sigma$. Hence, beside (6.2.3), we prescribe also $\sigma(0,\cdot) = \sigma_0$.

Proposition 6.5.1 (Poynting-Thomson model: existence of solutions). *Let ϱ satisfy (6.3.5), the loading and the initial conditions satisfy (6.3.3) together with $\sigma_0 \in L^2(\Omega;\mathbb{R}^{d\times d}_{\rm sym})$, and let further*

$$\mathbb{D}, \mathbb{C}_1, \mathbb{C}_2 \ \textit{positive definite on } \mathbb{R}^{d\times d}_{\rm sym}$$
$$\textit{with } \mathbb{C}_1, \mathbb{C}_2 \in \mathrm{SLin}(\mathbb{R}^{d\times d}_{\rm sym}). \tag{6.5.7}$$

Then the system (6.1.1) and (6.5.3) with $e = e(u)$ with the boundary conditions (6.2.2) and the initial conditions (6.2.3) and $\sigma(0,\cdot) = \sigma_0$ possesses a weak solution (u,σ) such that

$$\sigma, e_2 \in L^\infty(I; L^2(\Omega;\mathbb{R}^{d\times d})), \quad \dot{u} \in L^\infty(I; L^2(\Omega;\mathbb{R}^d)), \quad \sigma_2 \in L^2(Q;\mathbb{R}^{d\times d}), \tag{6.5.8a}$$

$$u \in L^\infty(I; H^1(\Omega;\mathbb{R}^d)), \quad \ddot{u} \in L^1(I; L^2\Omega;\mathbb{R}^d)) + L^\infty(I; H^1(\Omega;\mathbb{R}^d)^*), \tag{6.5.8b}$$

$$\dot{e}_2 \in L^2(Q;\mathbb{R}^{d\times d}), \tag{6.5.8c}$$

where $\sigma_2 = \mathbb{D}\dot{e} - \mathbb{D}\mathbb{C}_1^{-1}\dot{\sigma}$ and $e_2 = e - \mathbb{C}_1^{-1}\sigma$.

Sketch of the proof. First, one can construct an approximate solution, typically by space or by time discretization; we omit the corresponding technicalities in this proof.

To obtain the a-priori estimates independent of the (here unspecified) discretization parameter, we integrate (6.5.6) over $[0,s]$ written for the (here unspecified) approximate solution over $[0,s]$ typically as an inequality, and make by-part integration of the boundary terms $g\cdot\dot u$, and then estimate it as:

$$\begin{aligned}\int_{\Gamma_{\rm N}} g(s)\cdot u(s)\,{\rm d}S - \int_0^s\int_{\Gamma_{\rm N}} \dot g\cdot u\,{\rm d}S\,{\rm d}t\\ \le \|g(s)\|_{L^{2^{\sharp\prime}}(\Gamma_{\rm N};\mathbb{R}^d)}\|u(s)\|_{L^{2^\sharp}(\Gamma_{\rm N};\mathbb{R}^d)} + \int_0^s \|\dot g(t)\|_{L^{2^{\sharp\prime}}(\Gamma_{\rm N};\mathbb{R}^d)}\|u(t)\|_{L^{2^\sharp}(\Gamma_{\rm N};\mathbb{R}^d)}\,{\rm d}t\\ \le \frac{N}{2\varepsilon_0}\big\|g(s,\cdot)\big\|^2_{L^{2^{\sharp\prime}}(\Gamma_{\rm N};\mathbb{R}^d)} + \varepsilon_0\big\|u(s,\cdot)\big\|^2_{L^2(\Omega;\mathbb{R}^d)} + \varepsilon_0\big\|e(s,\cdot)\big\|^2_{L^2(\Omega;\mathbb{R}^{d\times d})}\\ + N\int_0^s \big\|\dot g\big\|_{L^{2^{\sharp\prime}}(\Gamma_{\rm N};\mathbb{R}^d)}\Big(\frac14 + \big\|e\big\|^2_{L^2(\Omega;\mathbb{R}^{d\times d})} + \|u\|^2_{L^2(\Omega;\mathbb{R}^d)}\Big)\,{\rm d}t\end{aligned} \tag{6.5.9}$$

with N the constant in the estimate

$$\|u\|^2_{L^{2^{\sharp\prime}}(\Gamma_{\rm N})} = \Big(\int_{\Gamma_{\rm N}} |u|^{2^{\sharp\prime}}{\rm d}S\Big)^{2/2^{\sharp\prime}} \le N\int_\Omega |e(u)|^2 + |u|^2\,{\rm d}x$$

for $u\in H^1(\Omega;\mathbb{R}^d)$. The needed estimate of $\|e(u(s,\cdot))\|^2_{L^2(\Omega;\mathbb{R}^{d\times d})}$ can be obtained when (6.5.6) is modified by replacing $\frac12\mathbb{C}_1^{-1}\sigma:\sigma = \frac12\mathbb{C}_1e_1:e_1$ by using the first equality in (6.5.1). Then we can control $\|e(s,\cdot)\|^2_{L^2(\Omega;\mathbb{R}^{d\times d})} \le 2\|e_1(s,\cdot)\|^2_{L^2(\Omega;\mathbb{R}^{d\times d})} + 2\|e_2(s,\cdot)\|^2_{L^2(\Omega;\mathbb{R}^{d\times d})}$ by the assumed positive definiteness of both $\mathbb{C}_1$ and $\mathbb{C}_2$. By the Gronwall inequality we can get the estimates[27] (6.5.8a). Here we used also the qualification of the loading and the initial conditions in (6.3.3) with $\sigma_0\in L^2(\Omega;\mathbb{R}^{d\times d}_{\rm sym})$. We need also $e_2(0,\cdot)\in L^2(\Omega;\mathbb{R}^{d\times d})$ but this is already determined by prescribing $\sigma(0,\cdot)$ and $u(0,\cdot)$.[28]

From the last estimate in (6.5.8a) and from the Korn inequality and via the comparison, we obtain $\varrho\ddot u = f+{\rm div}\,\sigma\in L^1(I;L^2\Omega;\mathbb{R}^d)) + L^\infty(I;H^1(\Omega;\mathbb{R}^d)^*)$, and then by using (6.3.5) and the estimate , we also obtain the estimate $\ddot u = (f+{\rm div}\,\sigma)/\varrho$ in (6.5.8b).

Further, as $\sigma_2 = \mathbb{D}\dot e_2$, cf. (6.5.1), we have $\xi(\sigma_2) = \mathbb{D}^{-1}\sigma_2:\sigma_2 = \mathbb{D}\dot e_2:\dot e_2$. Using the alternative form of $\xi = \xi(\dot e_2)$ in (6.5.5b), we also obtain (6.5.8c).

For $u_{\rm D}$ nonconstant in time, we can use the substitution $u-\bar u_{\rm D}$ instead of u. Testing by the time derivative to see the energy balance, we see (6.5.6). Then the resulted right-hand-side term $\langle\vec t_{\rm D},\dot u_{\rm D}\rangle_{\Gamma_{\rm D}}$ means, in view of (6.2.10), $\int_\Omega\varrho\ddot u\cdot\dot{\bar u}_{\rm D} + \sigma:e(\dot{\bar u}_{\rm D}) - f\cdot\dot{\bar u}_{\rm D}\,{\rm d}x - \int_{\Gamma_{\rm N}} g\cdot\dot{\bar u}_{\rm D}\,{\rm d}S$. This terms can then be estimated when using the qualification (6.3.3b) of $u_{\rm D}$.

---

[27] Here we used also $e = \mathbb{C}_1^{-1}\sigma + e_2 \in L^\infty(I;L^2(\Omega;\mathbb{R}^{d\times d}))$.

[28] Here we used that $e_1(0,\cdot) = \mathbb{C}_1^{-1}\sigma(0,\cdot)$ and thus $e_2(0,\cdot) = e(u(0,\cdot)) - \mathbb{C}_1^{-1}\sigma(0,\cdot)$.

Then, choosing a weakly* convergent subsequence and using linearity of the problem, we can show existence of a weak solution to the continuous problem. □

Proposition 6.5.2 (Regularity, uniqueness, energy conservation). *Let (6.5.7) hold together with (6.4.5b) and let (6.3.3) be strengthened (6.3.12), and let* ϱ *satisfy (6.3.2) and* $\sigma_0 \in L^2(\Omega;\mathbb{R}^{d\times d}_{\rm sym})$. *Then the weak solution* (u,σ) *from Proposition 6.5.1 is regular in the sense* $\dot{u} \in L^\infty(I;H^1(\Omega;\mathbb{R}^d)) \cap W^{1,\infty}(I;L^2(\Omega;\mathbb{R}^d))$ *and* $\dot\sigma, \dot e_2 \in L^\infty(I;L^2(\Omega;\mathbb{R}^{d\times d}))$, *is unique, and the energy is conserved in the sense that (6.5.6) holds integrated on any time interval* $[0,t] \subset I$:

$$
\begin{aligned}
&\int_\Omega \frac{\varrho}{2}|\dot u(t)|^2 + \frac12 \mathbb{C}_1^{-1}\sigma(t){:}\sigma(t) + \frac12 \mathbb{C}_2 e_2(t){:}e_2(t)\,\mathrm{d}x \\
&+\int_0^t\!\!\int_\Omega \mathbb{D}^{-1}\sigma_2{:}\sigma_2\,\mathrm{d}x\mathrm{d}t = \int_0^t\!\!\int_\Omega f\cdot\dot u\,\mathrm{d}x\mathrm{d}t + \int_0^t\!\!\int_{\Gamma_{\rm N}} g\cdot\dot u\,\mathrm{d}S\,\mathrm{d}t \\
&+\langle \vec\tau_{\rm D}, \dot u_{\rm D}\rangle_{\Gamma_{\rm D}} + \int_\Omega \frac{\varrho}{2}|v_0|^2 + \frac12 \mathbb{C}_1^{-1}\sigma_0{:}\sigma_0 + \frac12 \mathbb{C}_2 e_{2,0}{:}e_{2,0}\,\mathrm{d}x
\end{aligned}
\tag{6.5.10}
$$

with $e_{20} = e(u_0) - \mathbb{C}_1^{-1}\sigma_0$.

Sketch of the proof. We derive the energy balance for time derivatives, i.e. differentiate (6.5.1) in time and use time derivatives of test functions. As a result, we obtain (6.5.10) but with one time derivative more at each quantity. To read the mentioned a-priori estimates, we need to have also $\dot\sigma(0) \in L^2(\mathbb{R}^{d\times d}_{\rm sym})$, which can be obtained from (6.5.3) written at time $t = 0$.

Using again linearity of the problem and executing the above estimation strategy for the difference of two weak solutions, one obtain the uniqueness by testing by the difference of the velocities which are now in duality with accelerations, as well as with the divergence of stresses. These duality arguments also facilitate the rigorous prove of the energy conservation (6.5.10). □

Remark 6.5.3 (Elimination of σ_2 *and* e_2*).* Let us emphasize that we actually worked in terms of σ and e only. E.g., the variables σ_2 and e_2 occurring e.g. in the energy balance (6.5.6) could be eliminated because obviously $e_2 = e(u) - \mathbb{C}_1^{-1}\sigma$ and $\sigma_2 = \mathbb{D}\dot e_2 = \mathbb{D}e(\dot u) - \mathbb{D}\mathbb{C}_1^{-1}\dot\sigma$. This would result to the energy balance

$$
\begin{aligned}
&\frac{\mathrm{d}}{\mathrm{d}t}\Big(\int_\Omega \frac{\varrho}{2}|\dot u|^2 + \frac12\mathbb{C}_1^{-1}\sigma{:}\sigma + \frac12\mathbb{C}_2(\mathbb{D}e(\dot u) - \mathbb{D}\mathbb{C}_1^{-1}\dot\sigma){:}(\mathbb{D}e(\dot u) - \mathbb{D}\mathbb{C}_1^{-1}\dot\sigma)\mathrm{d}x\Big) \\
&+\int_\Omega \mathbb{D}(e(\dot u) - \mathbb{C}_1^{-1}\dot\sigma){:}(e(\dot u) - \mathbb{C}_1^{-1}\dot\sigma)\mathrm{d}x = \int_\Omega f\cdot\dot u\,\mathrm{d}x + \int_{\Gamma_{\rm N}} g\cdot\dot u\,\mathrm{d}S + \langle\vec\tau_{\rm D}, \dot u_{\rm D}\rangle_{\Gamma_{\rm D}}.
\end{aligned}
$$

Actually, one can understand the variables σ_2 and e_2 here as only abbreviations to make some formulas shorter. Also the initial condition of e_2 could be prescribed in terms of u and σ as $e(u_0) - \mathbb{C}_1^{-1}\sigma_0$, as done in (6.5.10).

Remark 6.5.4 (Alternative and equivalent Zener's model). The latter mentioned arrangement of 3-parameter standard solid ($\mathbb{C}_1$ — $\mathbb{D}$) $\|$ $\mathbb{C}_2$ leads to the relations:

$$e = \tilde{e}_1 + \tilde{e}_2, \quad \sigma = \tilde{\sigma}_1 + \tilde{\sigma}_2, \quad \tilde{\sigma}_2 = \tilde{\mathbb{C}}_2 e, \quad \tilde{\sigma}_1 = \tilde{\mathbb{C}}_1 \tilde{e}_1 = \tilde{\mathbb{D}} \dot{\tilde{e}}_2; \tag{6.5.11}$$

cf. Figure 6.4. Again we want to eliminate the "internal" quantities $\tilde{e}_1$, $\tilde{e}_2$, $\tilde{\sigma}_1$, and $\tilde{\sigma}_2$, and write a constitutive relation between σ and ε only. It holds

$$\sigma = \tilde{\sigma}_1 + \tilde{\sigma}_2 = \tilde{\sigma}_1 + \tilde{\mathbb{C}}_2 e \quad \text{and} \quad \dot{e} = \dot{\tilde{e}}_1 + \dot{\tilde{e}}_2 = \tilde{\mathbb{C}}_1^{-1} \dot{\tilde{\sigma}}_1 + \tilde{\mathbb{D}}^{-1} \tilde{\sigma}_1. \tag{6.5.12}$$

Multiplying the last equation by $\tilde{\mathbb{C}}_1$ from the left and adding it with $\tilde{\mathbb{C}}_2 \dot{e} = \dot{\tilde{\sigma}}_2 = \dot{\sigma} - \dot{\tilde{\sigma}}_1$, we will eliminate $\dot{\tilde{\sigma}}_1$:

$$(\tilde{\mathbb{C}}_1 + \tilde{\mathbb{C}}_2)\dot{e} = \dot{\sigma} + \tilde{\mathbb{C}}_1 \tilde{\mathbb{D}}^{-1} \tilde{\sigma}_1 = \dot{\sigma} + \tilde{\mathbb{C}}_1 \tilde{\mathbb{D}}^{-1}(\sigma - \tilde{\sigma}_2) = \dot{\sigma} + \tilde{\mathbb{C}}_1 \tilde{\mathbb{D}}^{-1}(\sigma - \tilde{\mathbb{C}}_2 e),$$

and therefore again we obtain an equation like (6.5.3) but now with

$$\dot{\sigma} + \tilde{\mathbb{C}}_1 \tilde{\mathbb{D}}^{-1} \sigma = (\tilde{\mathbb{C}}_1 + \tilde{\mathbb{C}}_2)\dot{e} + \tilde{\mathbb{C}}_1 \tilde{\mathbb{D}}^{-1} \tilde{\mathbb{C}}_2 e. \tag{6.5.13}$$

Remark 6.5.5 (Limit cases). Let us note that the standard solid covers both the Kelvin-Voigt material (for $\mathbb{C}_1 = +\infty$ and $\tilde{\mathbb{C}}_1 = +\infty$ on Figure 6.4 in the sense $\mathbb{C}_1^{-1} = 0 = \tilde{\mathbb{C}}_1^{-1}$) and the Maxwell material (for $\mathbb{C}_2 = 0 = \tilde{\mathbb{C}}_2$ on Figure 6.4) which will be addressed in the next section.

Remark 6.5.6 (Wave dispersion and attenuation in standard solids). Let us illustrate how the analysis like in Remark 6.4.13 becomes more complicated. To this goal, in the one-dimensional case, let us first formulate the system (6.1.1) with (6.5.3) or (6.5.13) a single equation, now in terms of σ rather than u. When differentiating the system $\varrho\ddot{u} = \partial_x\sigma$ and $\tau_1\dot{\sigma} + \sigma = C(\tau_2\partial_x\dot{u} + \partial_x u)$ with $\tau_1, \tau_2 > 0$ being some relaxation times, we obtain $\varrho\partial_x\ddot{u} = \partial^2_{xx}\sigma$ and $\tau_1\dddot{\sigma} + \ddot{\sigma} = C(\tau_2\partial_x\dddot{u} + \partial_x\ddot{u})$. Then we can eliminate u and eventually obtain the dispersive wave equation

$$\tau_1\varrho\dddot{\sigma} + \varrho\ddot{\sigma} = \tau_2 C\partial^2_{xx}\dot{\sigma} + C\partial^2_{xx}\sigma \quad \text{on } [0, +\infty) \times \mathbb{R}. \tag{6.5.14}$$

Like in Remark 6.4.13 we take the ansatz $\sigma = \mathrm{e}^{\mathrm{i}(wt + x/\lambda)}$ with the complex angular frequency $w = \omega + \mathrm{i}\gamma$ with $\omega, \gamma \in \mathbb{R}$ and the wavelength $\lambda \geq 0$. Then $\dddot{\sigma} = -\mathrm{i}w^2\sigma$, $\ddot{\sigma} = -w^2\sigma$, $\partial^2_{xx}\dot{\sigma} = -\mathrm{i}\sigma/\lambda^2$, and $\partial^2_{xx}\sigma = -\sigma/\lambda^2$, so that (6.5.14) leads to the algebraic condition $\mathrm{i}\tau_1\varrho w^2 + \varrho w^2 - \mathrm{i}\tau_2 C/\lambda^2 - \tau_2 C/\lambda^2 = 0$. When substituting $w = \omega + \mathrm{i}\gamma$ so that $w^2 = \omega^2 - \gamma^2 + 2\mathrm{i}\omega\gamma$ and $w^3 = \omega^3 - 3\omega\gamma^2 + \mathrm{i}(3\omega^2\gamma - \gamma^3)$, we obtain two algebraic equations, i.e. *Kramers-Kronig's relations*, here as

$$\tau_1(\omega^3 - 3\omega\gamma^2) - \omega^2 + \gamma^2 = \frac{C}{\varrho\lambda^2} \quad \text{and} \quad \tau_1(3\omega^2\gamma - \gamma^3) + 2\omega\gamma = \frac{\tau_2 C}{\varrho\lambda^2}. \tag{6.5.15}$$

Merging these two equations, we can eliminate λ and write

$$\tau_1(\omega^3 - 3\omega\gamma^2) - \omega^2 + \gamma^2 = \frac{\tau_1}{\tau_2}(3\omega^2\gamma - \gamma^3) + \frac{2}{\tau_2}\omega\gamma, \tag{6.5.16}$$

from which we can read γ as a function of ω, and eventually from (6.5.15) also λ as a function of ω. All the algebraic equations (6.5.15)–(6.5.16) are cubic so, in principle, solvable although the explicit formulas are cumbersome; cf. e.g. [106, 301] for discussion of dispersion in standard linear solids.

Exercise 6.5.7. Comparing (6.5.13) with (6.5.3), show that both models are actually equivalent to each other. Derive explicit transformation of parameters.

Exercise 6.5.8. Derive the stored energy φ and the dissipation rate ξ for the alternative model (6.5.13).

6.6 Fluid-type rheologies: Maxwell and Jeffreys materials

Even solids can sometimes exhibit a fluidic-type behavior, as already mentioned together with its relation to the Deborah number when relatively small. The combination of mere Stokes fluid as on Fig. 6.1-right, which itself cannot store energy and thus waves cannot propagate in such media, with some elastic elements allows for merging both fluidic character with capacity of exhibiting vibrations or wave propagation.

The simplest viscoelastic fluidic-type rheology is the Maxwell one [346]. The so-called *Maxwell material* is based on an arrangement of a spring and a damper in series, i.e. ⩚$_{\rm C}$—⊩$_{\rm D}$. In the notation of Figure 6.5, the strains sum up, i.e.:[29]

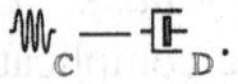

$$e(\dot{u}) := \dot{e}_1 + \dot{e}_2 = \mathbb{C}^{-1}\dot{\sigma} + \mathbb{D}^{-1}\sigma. \tag{6.6.1}$$

or, when $\mathbb{C}$ and $\mathbb{D}$ commute with each other, written without inversion of the tensors also as

$$\mathbb{D}\dot{\sigma} + \mathbb{C}\sigma = \mathbb{C}\mathbb{D}e(\dot{u}). \tag{6.6.2}$$

The situation that $\mathbb{C}$ and $\mathbb{D}$ commute occurs trivially if $\mathbb{D} = \tau_{\rm R}\mathbb{C}$ for a relaxation time $\tau_{\rm R} > 0$ called in this case sometimes the *Maxwell time*. To get the energy balance, we multiply (6.6.1) by σ. By this way, we obtain

$$\sigma{:}e(\dot{u}) = \frac{\partial}{\partial t}\frac{1}{2}\mathbb{C}^{-1}\sigma{:}\sigma + \mathbb{D}^{-1}\sigma{:}\sigma. \tag{6.6.3}$$

Hence substituting it into (6.2.9) gives

[29] In terms of Figure 6.1-left, we can write $e(u) = e_1 + e_2$ with $\mathbb{C}e_1 := \sigma$ and $\mathbb{D}\dot{e}_2 = \sigma$. Differentiating the first and the second equation allows for substituting which eventually yields (6.6.1).

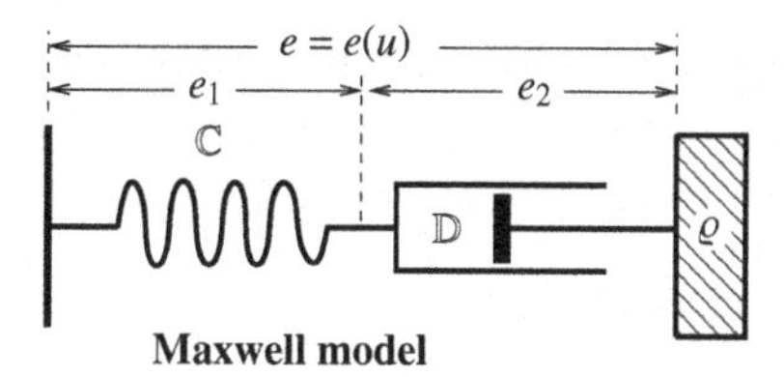

Fig. 6.5 A schematic 1-dimensional illustration of the second simplest rheological model combining two elements from Fig. 6.1 in series. Simultaneously it is the simplest hyperbolic fluid, cf. Table 6.2 on p. 199.

$$\underbrace{\frac{\mathrm{d}}{\mathrm{d}t}\int_\Omega \frac{\varrho}{2}|\dot{u}|^2\,\mathrm{d}x}_{\text{kinetic energy}} + \underbrace{\int_\Omega \xi(\sigma)\,\mathrm{d}x}_{\text{dissipation rate}} + \underbrace{\frac{\mathrm{d}}{\mathrm{d}t}\int_\Omega \varphi(\sigma)\,\mathrm{d}x}_{\text{stored energy}} = \underbrace{\int_\Omega f\cdot\dot{u}\,\mathrm{d}x + \int_{\Gamma_\mathrm{N}} g\cdot\dot{u}\,\mathrm{d}S + \langle \vec{\tau}_\mathrm{D}, \dot{u}_\mathrm{D}\rangle_{\Gamma_\mathrm{D}}}_{\text{power of external loading}} \tag{6.6.4}$$

with $\vec{\tau}_\mathrm{D}$ from (6.2.10), where now the specific stored energy φ and the specific dissipation-energy rate ξ are

$$\varphi(\sigma) := \frac{1}{2}\mathbb{C}^{-1}\sigma{:}\sigma, \qquad \xi(\sigma) := \mathbb{D}^{-1}\sigma{:}\sigma. \tag{6.6.5}$$

Following a concept that the dissipation rate is expressed in terms of state rates, we can alternatively write

$$\varphi(e, e_2) := \frac{1}{2}\mathbb{C}(e-e_2){:}(e-e_2), \qquad \xi(\dot{e}_2) := \mathbb{D}\dot{e}_2{:}\dot{e}_2. \tag{6.6.6}$$

In contrast to the Kelvin-Voigt material where two initial conditions were to prescribe, cf. (6.2.3), we will now need three initial conditions:

$$u(0,\cdot) = u_0, \qquad v(0,\cdot) = v_0, \qquad \sigma(0,\cdot) = \sigma_0. \tag{6.6.7}$$

An important difference from Kelvin-Voigt's material is that we cannot expect any estimate on $\nabla\dot{u}$ now under the basic loading qualification (6.3.3); the situation may be different for a more "gentle" loading, cf. Proposition 6.6.2 below. Therefore, a weak solution must be defined carefully. We say that $(u,\sigma) \in H^1(I; L^2(\Omega;\mathbb{R}^d) \times L^2(Q;\mathbb{R}^{d\times d}_{\mathrm{sym}})$ is a very weak solution if (6.2.7) holds and if the rheology relation (6.6.1) together with the initial condition $\sigma(0,\cdot) = \sigma_0$ is satisfied in the distributional sense, i.e.

$$\forall z \in H^1(Q;\mathbb{R}^{d\times d}_{\mathrm{sym}}),\ z(T,\cdot) = 0: \int_Q \dot{u}\cdot\mathrm{div}\, z\,\mathrm{d}x\mathrm{d}t = \int_Q \mathbb{C}^{-1}\sigma{:}\dot{z} - \mathbb{D}^{-1}\sigma{:}z\,\mathrm{d}x\mathrm{d}t + \int_\Omega \mathbb{C}^{-1}\sigma_0(x){:}z(0,x)\,\mathrm{d}x,$$

where $\mathbb{C}^{-1}$ and $\mathbb{D}^{-1}$ are sometimes called an elastic and a viscous *compliance tensors*.

Proposition 6.6.1 (Maxwell model: existence/uniqueness of weak solutions). *Let again the data qualification (6.3.3) hold. Then there is a unique weak solution (u,σ) to the initial-boundary-value problem (6.1.1)–(6.2.2)–(6.6.7) with (6.6.1) such that*

$$u \in L^\infty(I;H^1(\Omega;\mathbb{R}^d)), \tag{6.6.8a}$$

$$\dot{u} \in L^\infty(I;L^2(\Omega;\mathbb{R}^d)), \tag{6.6.8b}$$

$$\ddot{u} \in L^\infty(I;H^1(\Omega;\mathbb{R}^d)^*) + L^1(I;L^2(\Omega;\mathbb{R}^d)), \tag{6.6.8c}$$

$$\sigma \in L^\infty(I;L^2(\Omega;\mathbb{R}^{d\times d}_{\rm sym})) \cap W^{1,\infty}(I;H^{-1}(\Omega;\mathbb{R}^{d\times d}_{\rm sym})). \tag{6.6.8d}$$

Sketch of the proof. As $u_{\rm D}$ is assumed constant in time, we can read the a-priori estimates from (6.6.4) when estimating

$$\frac{\mathrm{d}}{\mathrm{d}t}\int_\Omega \frac{\varrho}{2}|\dot{u}|^2 + \varphi(\sigma)\,\mathrm{d}x + \int_\Omega \xi(\sigma)\,\mathrm{d}x = \int_\Omega f\cdot\dot{u}\,\mathrm{d}x + \int_{\Gamma_{\rm N}} g\cdot\dot{u}\,\mathrm{d}S$$
$$\leq \|f\|_{L^2(\Omega;\mathbb{R}^d)}\Big(\frac{1}{4} + \|\dot{u}\|^2_{L^2(\Omega;\mathbb{R}^d)}\Big) + \int_{\Gamma_{\rm N}} g\cdot\dot{u}\,\mathrm{d}S. \tag{6.6.9}$$

For simplicity, let us first consider the special case $g = 0$. Then we get immediately the a-priori estimates (6.6.8b) and the first part of (6.6.8d) and, by comparison, also (6.6.8c). Integrating (6.6.1) over $[0,s]$ one gets

$$e(s,\cdot) = \mathbb{C}^{-1}\sigma(s,\cdot) + \mathbb{D}^{-1}\int_0^s \sigma(t,\cdot)\,\mathrm{d}t + e(u_0) - \mathbb{C}^{-1}\sigma_0. \tag{6.6.10}$$

As $\sigma_0 \in L^2(\Omega;\mathbb{R}^{d\times d})$, we have $e(u) \in L^\infty(I;L^2(\Omega;\mathbb{R}^{d\times d}))$, hence through Korn's inequality (Theorem 5.2.2) also (6.6.8a). The second estimate in (6.6.8d) concerning $\dot\sigma = \mathbb{C}e(\dot u) - \mathbb{C}\mathbb{D}^{-1}\sigma \in L^\infty(I;H^1(\Omega;\mathbb{R}^d)^*)$ is by comparison from (6.6.1), namely we use the already obtained estimates for $\dot u$ and σ in

$$\|\dot\sigma\|_{L^\infty(I;H^{-1}(\Omega;\mathbb{R}^{d\times d}))} = \sup_{v\in L^1(I;H^1(\Omega;\mathbb{R}^{d\times d}))}\int_Q \dot{u}\cdot{\rm div}(\mathbb{C}v) + \mathbb{C}\mathbb{D}^{-1}\sigma : v\,\mathrm{d}x\mathrm{d}t.$$

If $g \neq 0$, then exploiting (6.6.9) is more complicated. As we do not have $\nabla\dot{u}$ estimated in Ω, we cannot estimate $\dot u$ on $\Gamma_{\rm N}$, and we must make a by-part integration in time for the last term in (6.6.9). Thus, after integration over $[0,s]$ like we did for (6.5.6), this gives:

$$\frac{\varrho}{2}\|\dot{u}(s,\cdot)\|^2_{L^2(\Omega;\mathbb{R}^d)} + \int_\Omega \varphi(\sigma(s,\cdot))\,\mathrm{d}x + \int_0^s\int_\Omega \xi(\sigma)\,\mathrm{d}x\mathrm{d}t$$
$$\leq \int_0^s \|f\|_{L^2(\Omega;\mathbb{R}^d)}\Big(\frac{1}{4} + \|\dot{u}\|^2_{L^2(\Omega;\mathbb{R}^d)}\Big)\mathrm{d}t + \int_{\Gamma_{\rm N}} g(s)\cdot u(s)\,\mathrm{d}S$$
$$- \int_0^s\int_{\Gamma_{\rm N}} \dot{g}\cdot u\,\mathrm{d}S\,\mathrm{d}t + \frac{\varrho}{2}\|v_0\|^2_{L^2(\Omega;\mathbb{R}^d)} + \int_\Omega \varphi(\sigma_0)\,\mathrm{d}x - \int_{\Gamma_{\rm N}} g(0)\cdot u_0\,\mathrm{d}S. \tag{6.6.11}$$

Now, through Poincaré's inequality and Korn's inequality (Theorems 5.2.2 and B.3.15), we estimate the boundary term by using by-part integration as in (6.5.9). In contrast to the standard-solid model, we now do not have a direct control on $e(s,\cdot)$ in L^2-norm but we can estimate it from (6.6.10):

$$
\begin{aligned}
&\big\|e(s,\cdot)\big\|^2_{L^2(\Omega;\mathbb{R}^{d\times d})} = \Big\|\mathbb{C}^{-1}\sigma(s,\cdot)+\mathbb{D}^{-1}\int_0^s\sigma(t,\cdot)\,\mathrm{d}t+e(u_0)-\mathbb{C}^{-1}\sigma_0\Big\|^2_{L^2(\Omega;\mathbb{R}^{d\times d})}\\
&\le C\Big(\big\|\sigma(s,\cdot)\big\|^2_{L^2(\Omega;\mathbb{R}^{d\times d})}+\Big\|\int_0^s\sigma(t,\cdot)\,\mathrm{d}t\Big\|^2_{L^2(\Omega;\mathbb{R}^{d\times d})}+\big\|u_0\big\|^2_{H^1(\Omega;\mathbb{R}^d)}+\big\|\sigma_0\big\|^2_{L^2(\Omega;\mathbb{R}^{d\times d})}\Big)\\
&\le C\Big(\big\|\sigma(s,\cdot)\big\|^2_{L^2(\Omega;\mathbb{R}^{d\times d})}+T\int_0^s\big\|\sigma(t,\cdot)\big\|^2_{L^2(\Omega;\mathbb{R}^{d\times d})}\,\mathrm{d}t+\big\|u_0\big\|^2_{H^1(\Omega;\mathbb{R}^d)}+\big\|\sigma_0\big\|^2_{L^2(\Omega;\mathbb{R}^{d\times d})}\Big).
\end{aligned}
\tag{6.6.12}
$$

Now we can add (6.6.12) multiplied by $\varepsilon>0$ to (6.6.11). Also we can add $\|u(s)\|^2_{L^2(\Omega;\mathbb{R}^d)} = \|\int_0^s\dot u\,\mathrm{d}t+u_0\|^2_{L^2(\Omega;\mathbb{R}^d)} \le 2T\int_0^s\|\dot u\|^2_{L^2(\Omega;\mathbb{R}^d)}\,\mathrm{d}t+2\|u_0\|^2_{L^2(\Omega;\mathbb{R}^d)}$ to (6.6.11). Moreover, we substitute (6.6.12) into (6.5.9) and then into (6.6.11). Using the coercivity of φ and ξ, i.e. $\sigma\mathbb{C}^{-1}\sigma\ge\alpha|\sigma|^2$ and $\sigma\mathbb{D}^{-1}\sigma\ge\beta|\sigma|^2$ with some $\alpha>0$ and $\beta>0$, we choose $\varepsilon<\min(\alpha/C,\beta(CT))$. Then we choose $\varepsilon_0<\varepsilon/2$. Then (6.6.8) follows by Gronwall's inequality.

The proof of the uniqueness is delicate due to lack of duality in between $\sqrt{\varrho}\ddot u$ and $\sqrt{\varrho}\dot u$, similarly as in purely elastic materials, cf. Proposition 6.3.1. To this goal, we formulate the problem in terms of the stress σ: applying $e(\cdot)$ on (6.1.1) gives

$$
e(\ddot u)-e\Big(\frac{\mathrm{div}\,\sigma}{\varrho}\Big)=e\Big(\frac{f}{\varrho}\Big),
\tag{6.6.13}
$$

and differentiating (6.6.1) in time allows to substitute for $e(\ddot u)=\mathbb{C}^{-1}\ddot\sigma+\varrho\mathbb{D}^{-1}\dot\sigma$, which gives

$$
\mathbb{C}^{-1}\ddot\sigma+\mathbb{D}^{-1}\dot\sigma-e\Big(\frac{\mathrm{div}\,\sigma}{\varrho}\Big)=e\Big(\frac{f}{\varrho}\Big).
\tag{6.6.14}
$$

Of course, now u and σ stands for the difference of two weak solutions and we are to prove that $u=0$ and $\sigma=0$. We use again the Ladyzhenskaya-type test [300, Sect. IV.3] similarly as (6.3.8) adapted here for non-conservative case like in [143, Sect. 5.1]. Namely, we test (6.6.14) by

$$
\hat\sigma(t):=\begin{cases}-\displaystyle\int_t^s\sigma(r)\,\mathrm{d}r & \text{for } 0\le t\le s,\\[1ex] 0 & \text{for } s<t\le T.\end{cases}
\tag{6.6.15}
$$

Let us note that $\dot{\hat\sigma}=\sigma$ on $[0,s)$ and that $\hat\sigma\in W^{1,\infty}(I;L^2(\Omega;\mathbb{R}^{d\times d}_{\mathrm{sym}}))$ so that it is a legitimate test function for a weak formulation of (6.6.14) which even allows for a by-part integration. This gives

$$\int_Q \mathbb{C}^{-1}\dot{\sigma}:\hat{\sigma}\,\mathrm{d}x\mathrm{d}t = \int_0^s\int_\Omega \mathbb{C}^{-1}\dot{\sigma}:\sigma\,\mathrm{d}x\mathrm{d}t = \int_\Omega \frac{1}{2}\mathbb{C}^{-1}\sigma(s):\sigma(s)\,\mathrm{d}x$$

$$\int_Q \mathbb{D}^{-1}\sigma:\hat{\sigma}\,\mathrm{d}x\mathrm{d}t = \int_0^s\int_\Omega \mathbb{D}^{-1}\ddot{\hat{\sigma}}:\hat{\sigma}\,\mathrm{d}x\mathrm{d}t$$

$$= -\int_0^s\int_\Omega \mathbb{D}^{-1}\dot{\hat{\sigma}}:\dot{\hat{\sigma}}\,\mathrm{d}x\mathrm{d}t = -\int_0^s\int_\Omega \mathbb{D}^{-1}\sigma:\sigma\,\mathrm{d}x\mathrm{d}t, \quad \text{and}$$

$$\int_Q \frac{\mathrm{div}\,\sigma}{\varrho}\cdot\mathrm{div}\,\hat{\sigma}\,\mathrm{d}x\mathrm{d}t = \int_0^s\int_\Omega \frac{\mathrm{div}\,\dot{\hat{\sigma}}}{\varrho}\cdot\mathrm{div}\,\hat{\sigma}\,\mathrm{d}x\mathrm{d}t = -\int_\Omega \frac{1}{2\varrho}|\mathrm{div}\,\hat{\sigma}(0)|^2\,\mathrm{d}x,$$

here we used that both $\sigma(0) = 0$ and $\hat{\sigma}(s) = 0$. Thus, using $\frac{1}{2\varrho}|\mathrm{div}\,\hat{\sigma}(0)|^2 \geq 0$, the definition (6.2.7) with its right-hand side zero yields for all $s\in\bar{I}$ that

$$\int_\Omega \frac{1}{2}\mathbb{C}^{-1}\sigma(s):\sigma(s) + \int_\Omega \frac{1}{2\varrho}|\mathrm{div}\,\hat{\sigma}(0)|^2\,\mathrm{d}x = \int_0^s\int_\Omega \mathbb{D}^{-1}\sigma:\sigma\,\mathrm{d}x\mathrm{d}t, \qquad (6.6.16)$$

from which $\sigma = 0$ follows by the Gronwall inequality. From (6.6.1), we then obtain also $e(\dot{u}) = 0$. In view of $u(0,\cdot) = 0$, we have also $e(u) = 0$ and, by Korn's inequality, eventually also $u = 0$.

For u_D non-constant in time, we must again test by $\dot{u} - \dot{u}_\mathrm{D}$ and, like we did in the proof of Proposition 6.5.1, we use the energy balance (6.6.4) with $\int_\Omega \varrho\ddot{u}\cdot\dot{\bar{u}}_\mathrm{D} + \sigma{:}\, e(\dot{\bar{u}}_\mathrm{D}) - f\cdot\dot{\bar{u}}_\mathrm{D}\,\mathrm{d}x - \int_{\Gamma_\mathrm{N}} g\cdot\dot{\bar{u}}_\mathrm{D}\,\mathrm{d}S$. We then use by-part integration in time for the term $\varrho\ddot{u}\cdot\dot{u}_\mathrm{D}$ as in (6.4.6) while the term $\sigma{:}e(\dot{u}_\mathrm{D})$ can be estimated as

$$\int_\Omega \sigma{:}e(\dot{u}_\mathrm{D})\,\mathrm{d}x \leq \big(1 + \|\sigma\|^2_{L^2(\Omega;\mathbb{R}^{d\times d})}\big)\big\|e(\dot{u}_\mathrm{D})\big\|_{L^2(\Omega;\mathbb{R}^{d\times d})}$$

and then treated by Gronwall inequality. Alternatively, by substitution $\tilde{u} = u - \bar{u}_\mathrm{D}$ into (6.1.1) and (6.6.10), one gets the transformed equation

$$\varrho\ddot{\tilde{u}} - \mathrm{div}\,\sigma = \tilde{f} := f - \varrho\ddot{\bar{u}}_\mathrm{D}, \qquad e(\tilde{u}) = \mathbb{C}^{-1}\sigma + \int_0^s \mathbb{D}^{-1}\sigma - e(\bar{u}_\mathrm{D})\,\mathrm{d}t + e(u_0) - \mathbb{C}^{-1}\sigma_0.$$

This reveals why the qualification (6.3.3b) is needed to have $\tilde{f} \in L^1(I;L^2(\Omega;\mathbb{R}^d))$ as in (6.3.3a) and $e(\bar{u}_\mathrm{D}) \in L^\infty(I;L^2(\Omega;\mathbb{R}^{d\times d}))$, i.e. $u_\mathrm{D} \in L^\infty(I;H^{1/2}(\Gamma_\mathrm{D};\mathbb{R}^d))$. □

Proposition 6.6.2 (Regularity, uniqueness, and energy conservation). *Let $\dot{f} \in L^1(I;L^2(\Omega;\mathbb{R}^d))$ while any boundary loading be ignored, and let the initial conditions satisfy*

$$\sigma_0 \in H^1(\Omega;\mathbb{R}^{d\times d}), \qquad v_0 \in H^1(\Omega;\mathbb{R}^d), \qquad f \in W^{1,1}(I;L^2(\Omega;\mathbb{R}^d)). \qquad (6.6.17)$$

Then the weak solution satisfy also

$$\|\sigma\|_{W^{1,\infty}(I;L^2(\Omega;\mathbb{R}^{d\times d}))} \leq C, \qquad \|u\|_{W^{2,\infty}(I;L^2(\Omega;\mathbb{R}^d))} \leq C, \qquad (6.6.18a)$$

$$u \in W^{1,\infty}(I;H^1(\Omega;\mathbb{R}^d)), \qquad (6.6.18b)$$

$$\dddot{u} \in L^1(I;L^2(\Omega;\mathbb{R}^d)) + L^\infty(I;H_0^1(\Omega;\mathbb{R}^d)^*). \tag{6.6.18c}$$

Moreover, this (unique) weak solution conserves energy.

Sketch of the proof. More regular data allows for stronger results based on the regularity. Namely, one can differentiate in time (6.1.1), test it by $\ddot{u}$, and use (6.6.1) also differentiated in time, i.e.

$$\frac{\mathrm{d}}{\mathrm{d}t}\int_\Omega \frac{\varrho}{2}|\ddot{u}|^2\,\mathrm{d}x + \int_\Omega \dot{\sigma}{:}e(\ddot{u})\,\mathrm{d}x = \int_\Omega \dot{f}\cdot\ddot{u}\,\mathrm{d}x \quad \& \quad e(\ddot{u}) = \mathbb{C}^{-1}\ddot{\sigma} + \mathbb{D}^{-1}\dot{\sigma}, \tag{6.6.19}$$

which altogether gives

$$\frac{\mathrm{d}}{\mathrm{d}t}\int_\Omega \frac{\varrho}{2}|\ddot{u}|^2 + \mathbb{C}^{-1}\dot{\sigma}{:}\dot{\sigma}\,\mathrm{d}x + \int_\Omega \mathbb{D}^{-1}\dot{\sigma}{:}\dot{\sigma}\,\mathrm{d}x = \int_\Omega \dot{f}\cdot\ddot{u}\,\mathrm{d}x, \tag{6.6.20}$$

which allows for estimate (6.6.18a) provided that $\dot{f} \in L^1(I;L^2(\Omega;\mathbb{R}^d))$ and $\ddot{u}(0) \in L^2(\Omega;\mathbb{R}^d)$ and $\dot{\sigma}(0) \in L^2(\Omega;\mathbb{R}^{d\times d})$. From $\ddot{u}(0) = \operatorname{div}(\sigma(0)) + f(0)$ and from $\dot{\sigma}(0) = \mathbb{C}e(\dot{u}(0)) - \mathbb{C}\mathbb{D}^{-1}\sigma(0)$, cf. (6.6.1), we can see that we need to qualify the initial conditions so that, altogether, we need (6.6.17). From (6.6.10), we still see $e(u) \in W^{1,\infty}(I;L^2(\Omega;\mathbb{R}^{d\times d}))$ so that we get also (6.6.18b). Moreover, we also have (6.6.18c) from comparison of $\dddot{u} = (\dot{f} + \operatorname{div}\dot{\sigma})/\varrho$. Here we did not deal with any boundary loading, however.

For the energy conservation, it is now important that $\varrho\ddot{u} \in L^\infty(I;L^2(\Omega;\mathbb{R}^d))$ is certainly in duality with $\dot{u} \in L^\infty(I;H^1(\Omega;\mathbb{R}^d))$, and $\dot{\sigma} \in L^\infty(I;L^2(\Omega;\mathbb{R}^{d\times d}))$ is certainly in duality with $\sigma \in L^\infty(I;L^2(\Omega;\mathbb{R}^{d\times d}))$. These arguments (when applied for differences of two solutions) also facilitate the proof of uniqueness, even in a simpler way than in Proposition 6.6.1 above. □

Remark 6.6.3 (Initial displacement). Let us also note that by prescribing σ_0 we have prescribed $e_1(0,\cdot) = \mathbb{C}^{-1}\sigma_0$ and prescribing also u_0 we have prescribed $e(u_0)$ and hence also the initial condition on the damper $e_2(0,\cdot) = e(u_0) - \mathbb{C}^{-1}\sigma_0$, as needed to some extent. In fact, the force response is not influenced by a specific choice of u_0, however.

Remark 6.6.4 (Wave attenuation and dispersion). Like in Remarks 6.4.13 and 6.5.6, let us illustrate the analysis of wave propagation in Maxwell materials in 1-dimensional situations when (6.6.14) can be written as

$$\frac{1}{C}\ddot{\sigma} + \frac{1}{D}\dot{\sigma} - \frac{1}{\varrho}\partial^2_{xx}\sigma = 0 \quad \text{on } [0,+\infty)\times\mathbb{R}. \tag{6.6.21}$$

Using the ansatz $\sigma = \mathrm{e}^{\mathrm{i}(wt+x/\lambda)}$ with the complex angular frequency $w = \omega + \mathrm{i}\gamma$ and the real-valued wavelength λ like in Remark 6.4.13, we obtain $w^2/C - \mathrm{i}w/D - 1/(\varrho\lambda^2) = 0$. In terms of the real-valued coefficients ω and γ, it gives $(\omega^2-\gamma^2)/C + 2\mathrm{i}\omega\gamma/C - \mathrm{i}\omega/D + \gamma/D - 1/(\varrho\lambda^2) = 0$, from which we obtain the *Kramers-Kronig relations* here as

$$\frac{1}{C}(\omega^2-\gamma^2)+\frac{1}{D}\gamma-\frac{1}{\varrho\lambda^2}=0 \quad \text{and} \quad \frac{2}{C}\gamma=\frac{1}{D}, \tag{6.6.22}$$

cf. (6.4.26). Now the latter equation tells us that the attenuation coefficient $\gamma = C/(2D)$ is independent of frequency, which is related to the mentioned low-attenuation character of Maxwell materials, in contrast to e.g. Kelvin-Voigt materials. The former equation in (6.6.22) yields $\omega^2 = C/(\varrho\lambda^2) - C^2/(4D^2)$. Realizing that the speed of wave is $v = \omega\lambda$, we obtain $v = \sqrt{C/\varrho - (C/4D)^2\lambda^2}$ for $\lambda \le \lambda_{\rm crit} := 2D/\sqrt{\varrho C}$; interestingly, waves longer than $\lambda_{\rm crit}$ cannot propagate through such 1-dimensional Maxwellien media. When expressed in terms of the frequency, after some algebra we obtain

$$v=\sqrt{\frac{C}{\varrho+C^2/(2D\omega)^2}},$$

which reveals an *anomal dispersion*, i.e. the i.e. the high-frequency waves propagate faster than low-frequency ones. Let us note that, in the ultra high frequency asymptotics $v = \sqrt{C/\varrho} - \mathscr{O}(1/\omega^2)$ for $\omega \to +\infty$ approaches the velocity in purely elastic solids, cf. Remark 6.3.5. In fact, (6.6.21) is a so-called telegraph equation which is known to exhibit a *hyperbolic character* with only the mentioned weak attenuation (in particular contrasting with e.g. in Kelvin-Voigt materials where high-frequency vibrations or waves which are highly attenuated), cf. also Table 6.2 on p. 199. For analysis of Maxwellian materials in multidimensional we refer to [137, Chap. 6].

Remark 6.6.5 (Nonlinear material: high-frequency small-amplitude loading). Like in Remark 6.4.15, assume the dissipative force $D(e)$ with $D:\mathbb{R}^{d\times d}_{\rm sym}\to\mathbb{R}^{d\times d}_{\rm sym}$ uniformly monotone (even possibly nonpotential) instead of $\mathbb{D}e$. Then (6.6.19) involves $e(\ddot u) = \mathbb{C}^{-1}\ddot\sigma + [D^{-1}]'(\sigma)\dot\sigma$ and (6.6.20) looks as

$$\frac{\rm d}{{\rm d}t}\int_\Omega \frac{\varrho}{2}|\ddot u|^2+\frac{1}{2}\mathbb{C}^{-1}\dot\sigma{:}\dot\sigma\,{\rm d}x+\int_\Omega [D^{-1}]'(\sigma)\dot\sigma{:}\dot\sigma\,{\rm d}x=\int_\Omega \dot f\cdot\ddot u\,{\rm d}x, \tag{6.6.23}$$

which again yields all the estimates of Proposition 6.6.2 provided $[D^{-1}]'$ is uniformly coercive. The limit passage in possible approximate solution (u_k,σ_k) (assumed to satisfy $e(\dot u_k) = \mathbb{C}^{-1}\dot\sigma_k + D^{-1}(\sigma_k)$ and $\varrho\ddot u_k - {\rm div}\,\sigma_k = f_k$ and $\sigma_k(0) = \sigma_0$) can be performed by using uniform monotonicity of D^{-1} and linearity and positive definiteness of the stored energy: indeed, for a subsequence converging weakly* to some (u,σ), we have

$$\begin{aligned}
&\varepsilon\|\dot\sigma_k(T)-\dot\sigma(T)\|^2_{L^2(\Omega;\mathbb{R}^{d\times d})}+\varepsilon\|\sigma_k-\sigma\|^2_{L^2(Q;\mathbb{R}^{d\times d})}\\
&\quad\le\int_Q \mathbb{C}^{-1}(\dot\sigma_k-\dot\sigma){:}(\sigma_k-\sigma)+(D^{-1}(\sigma_k)-D^{-1}(\sigma)){:}(\sigma_k-\sigma)\,{\rm d}x{\rm d}t\\
&\quad=\int_Q e(\dot u_k){:}(\sigma_k-\sigma)-(\mathbb{C}^{-1}\dot\sigma+D^{-1}(\sigma)){:}(\sigma_k-\sigma)\,{\rm d}x{\rm d}t
\end{aligned}$$

$$
\begin{aligned}
&= \int_Q -\dot{u}_k \cdot \mathrm{div}(\sigma_k - \sigma) - (\mathbb{C}^{-1}\dot{\sigma} + D^{-1}(\sigma)) : (\sigma_k - \sigma)\,\mathrm{d}x\mathrm{d}t \\
&= \int_Q \dot{u}_k \cdot (f_k - \varrho\ddot{u}_k - f + \varrho\ddot{u}) - (\mathbb{C}^{-1}\dot{\sigma} + D^{-1}(\sigma)) : (\sigma_k - \sigma)\,\mathrm{d}x\mathrm{d}t \qquad (6.6.24)
\end{aligned}
$$

where we used the linearity of the force balance equation $\varrho\ddot{u}_k - \mathrm{div}\,\sigma_k = f_k$ which allows for limit passage via weak* convergence in $L^\infty(I; L^2(\Omega; \mathbb{R}^d))$ to $\varrho\ddot{u} - \mathrm{div}\,\sigma = f$, cf. (6.6.18a), and then use the strong convergence $\dot{u}_k \to \dot{u}$ in $L^2(I; L^2(\Omega; \mathbb{R}^d))$ by using (6.6.18b) with (6.6.18a) and Aubin-Lions compact embedding theorem B.4.10. Thus $\sigma_k \to \sigma$ in $L^2(I; L^2(\Omega; \mathbb{R}^{d\times d}))$ and thus also $D^{-1}(\sigma_k) \to D^{-1}(\sigma)$.

Let us now come to the *Jeffreys rheologies*[30] [263]. They combine two dampers (dashpots) and one elastic spring. Like in Section 6.5, there are two options how to make it, either

$$
\big(\mathbb{C} - \mathbb{D}_2\big) \,\|\, \mathbb{D}_1 \qquad \text{or} \qquad \big(\tilde{\mathbb{C}} \,\|\, \tilde{\mathbb{D}}_1\big) - \tilde{\mathbb{D}}_2 \qquad (6.6.25)
$$

cf. also Figure 6.6. Sometimes, Jeffreys' rheologies are referred to also as an *anti-Zener rheological model (material)*, cf. [328].

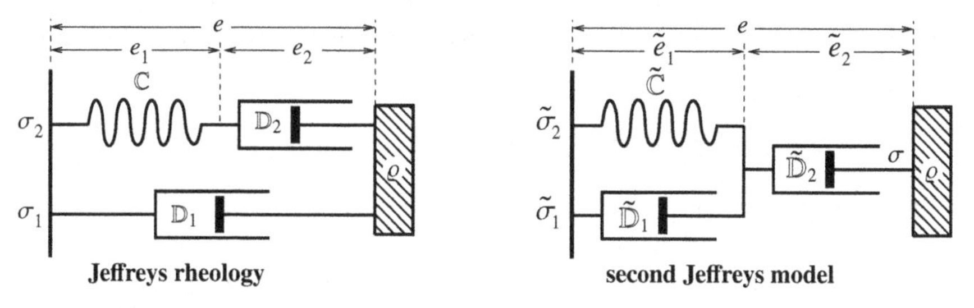

Fig. 6.6 A schematic 1-dimensional illustration of two more complicated rheological models using two dampers and one spring:
Left: Maxwell' model from Fig. 6.5 in parallel with the Stokes model from Fig. 6.1.
Right: a combination of Kelvin-Voigt's model from Fig. 6.3 in series with the Stokes model.

The constitutive equations for the former variant are

$$
\sigma = \sigma_1 + \sigma_2, \qquad e = e_1 + e_2, \qquad \sigma_1 = \mathbb{D}_1\dot{e}, \qquad \sigma_2 = \mathbb{D}_2\dot{e}_2 = \mathbb{C}e_1 \qquad (6.6.26)
$$

when using the notation from Figure 6.6-left. We want to eliminate the "internal" quantities e_1, e_2, σ_1, and σ_2, and write a constitutive relation between σ and ε only. It holds:

$$
\dot{e} = \dot{e}_1 + \dot{e}_2 = \mathbb{C}^{-1}\dot{\sigma}_2 + \mathbb{D}_2^{-1}\sigma_2
$$

[30] Sir Harold Jeffreys (1891-1989) was an English mathematician, statistician, geophysicist, and astronomer.

$$
\begin{aligned}
&= \mathbb{C}^{-1}(\dot\sigma-\dot\sigma_1)+\mathbb{D}_2^{-1}(\sigma-\sigma_1)\\
&= \mathbb{C}^{-1}(\dot\sigma-\mathbb{D}_1\ddot e)+\mathbb{D}_2^{-1}(\sigma-\mathbb{D}_1\dot e).
\end{aligned}
\tag{6.6.27}
$$

Altogether, after multiplication by $\mathbb{C}$, we obtain

$$
\dot\sigma+\mathbb{C}\mathbb{D}_2^{-1}\sigma=\mathbb{D}_1 e(\ddot u)+\mathbb{C}(\mathbb{I}+\mathbb{D}_2^{-1}\mathbb{D}_1)e(\dot u).
\tag{6.6.28}
$$

The energetics (6.2.9) needs to identify $\sigma : e(\dot u)$. To this goal, let us multiply (6.6.27) by $\sigma-\mathbb{D}_1\dot e$ from the right. This gives straightforwardly

$$
\sigma{:}\dot e=\frac{\partial}{\partial t}\frac{1}{2}\mathbb{C}^{-1}(\sigma-\mathbb{D}_1\dot e){:}(\sigma-\mathbb{D}_1\dot e)+\mathbb{D}_2^{-1}(\sigma-\mathbb{D}_1\dot e){:}(\sigma-\mathbb{D}_1\dot e)+\mathbb{D}_1\dot e{:}\dot e.
$$

Comparing it with (6.2.9), we can now identify the specific stored energy φ and the specific dissipation rate ξ in terms of the "observable" variables (i.e. the total strain e and the total stress σ) and their rates as

$$
\varphi(\sigma,\dot e)=\frac{1}{2}\mathbb{C}^{-1}(\sigma-\mathbb{D}_1\dot e){:}(\sigma-\mathbb{D}_1\dot e),\qquad \xi(\sigma,\dot e)=\mathbb{D}_1\dot e{:}\dot e+\mathbb{D}_2^{-1}(\sigma-\mathbb{D}_1\dot e){:}(\sigma-\mathbb{D}_1\dot e).
$$

Noteworthy, both these energies mix the rate $\dot e$ and the stress σ. Following the concept that dissipation rate depends indeed on rates like in (6.6.6), we can alternatively write

$$
\varphi(e,e_2)=\frac{1}{2}\mathbb{C}(e-e_2){:}(e-e_2)\quad\text{ and }\quad \xi(\dot e,\dot e_2)=\mathbb{D}_1\dot e{:}\dot e+\mathbb{D}_2\dot e_2{:}\dot e_2.
\tag{6.6.29}
$$

Let us now investigate the second variant in (6.6.25), i.e. the serial combination of the Kelvin-Voigt and the Stokes elements from Figure 6.6-right. Again, we want to derive the relation similar to (6.6.28). The Kelvin-Voigt element ($_{\tilde{\mathbb{C}}}\,\|\,_{\tilde{\mathbb{D}}_1}$) gives the total stress $\sigma=\sigma_1+\sigma_2=\tilde{\mathbb{D}}_1\dot{\tilde e}_1+\tilde{\mathbb{C}}\tilde e_1$ to be in equilibrium with the stress on the Maxwell element $_{\tilde{\mathbb{D}}_2}$, i.e. $\sigma=\tilde{\mathbb{D}}_2\dot{\tilde e}_2$. Moreover, $e(u)=\tilde e_1+\tilde e_2$. We then eliminate the internal strains $\tilde e_1$ and $\tilde e_2$. More specifically,

$$
\begin{aligned}
\dot e=\dot{\tilde e}_1+\dot{\tilde e}_2&=\tilde{\mathbb{D}}_1^{-1}(\sigma-\tilde{\mathbb{C}}\tilde e_1)+\tilde{\mathbb{D}}_2^{-1}\sigma\\
&=\tilde{\mathbb{D}}_1^{-1}(\sigma-\tilde{\mathbb{C}}(e-\tilde e_2))+\tilde{\mathbb{D}}_2^{-1}\sigma\\
&=(\tilde{\mathbb{D}}_1^{-1}+\tilde{\mathbb{D}}_2^{-1})\sigma-\tilde{\mathbb{D}}_1^{-1}\tilde{\mathbb{C}}e+\tilde{\mathbb{D}}_1^{-1}\tilde{\mathbb{C}}\tilde e_2
\end{aligned}
\tag{6.6.30}
$$

so that, differentiating in time allows for elimination of $\tilde e_2$ and thus, after multiplication by $\tilde{\mathbb{D}}_1$, we obtain

$$
(\mathbb{I}+\tilde{\mathbb{D}}_1\tilde{\mathbb{D}}_2^{-1})\dot\sigma+\tilde{\mathbb{C}}\tilde{\mathbb{D}}_2^{-1}\sigma=\tilde{\mathbb{D}}_1 e(\ddot u)+\tilde{\mathbb{C}}e(\dot u).
\tag{6.6.31}
$$

It has the analogous form as (6.6.28). The stored energy and the dissipation rate for this alternative Jeffreys model is

$$\varphi(e,\tilde{e}_2)=\frac{1}{2}\tilde{\mathbb{C}}(e-\tilde{e}_2):(e-\tilde{e}_2),\qquad \xi(\dot{e},\dot{\tilde{e}}_2)=\tilde{\mathbb{D}}_1(\dot{e}-\dot{\tilde{e}}_2):(\dot{e}-\dot{\tilde{e}}_2)+\tilde{\mathbb{D}}_2\dot{\tilde{e}}_2:\dot{\tilde{e}}_2. \quad (6.6.32)$$

Analogously to the former variant, we can express also these energies without the internal variables when realizing that $\tilde{\sigma}_2=\sigma-\tilde{\sigma}_1=\sigma-\tilde{\mathbb{D}}_1\dot{\tilde{e}}_1=\sigma-\tilde{\mathbb{D}}_1(\dot{e}-\dot{\tilde{e}}_2)=(\mathbb{I}+\tilde{\mathbb{D}}_1\tilde{\mathbb{D}}_2^{-1})\sigma-\tilde{\mathbb{D}}_1\dot{e}$, so that

$$\varphi(\sigma,\dot{e})=\frac{1}{2}\mathbb{C}^{-1}((\mathbb{I}+\tilde{\mathbb{D}}_1\tilde{\mathbb{D}}_2^{-1})\sigma-\tilde{\mathbb{D}}_1\dot{e}):((\mathbb{I}+\tilde{\mathbb{D}}_1\tilde{\mathbb{D}}_2^{-1})\sigma-\tilde{\mathbb{D}}_1\dot{e})\quad\text{and}\quad (6.6.33a)$$

$$\xi(\sigma,\dot{e})=\tilde{\mathbb{D}}_1^{-1}(\tilde{\mathbb{D}}_1\dot{e}-\tilde{\mathbb{D}}_1\tilde{\mathbb{D}}_2^{-1}\sigma):(\tilde{\mathbb{D}}_1\dot{e}-\tilde{\mathbb{D}}_1\tilde{\mathbb{D}}_2^{-1}\sigma)+\tilde{\mathbb{D}}_2^{-1}\sigma:\sigma. \quad (6.6.33b)$$

To see the relation between (6.6.28) and (6.6.31), let us multiply (6.6.28) by $\mathbb{D}_1^{-1}$ and (6.6.31) by $\tilde{\mathbb{D}}_1^{-1}$ to obtain $\mathbb{D}_1^{-1}\dot{\sigma}+\mathbb{D}_1^{-1}\mathbb{C}\mathbb{D}_2^{-1}\sigma=e(\ddot{u})+\mathbb{D}_1^{-1}\mathbb{C}(\mathbb{I}+\mathbb{D}_2^{-1}\mathbb{D}_1)e(\dot{u})$ and $(\tilde{\mathbb{D}}_1^{-1}+\tilde{\mathbb{D}}_2^{-1})\dot{\sigma}+\tilde{\mathbb{D}}_1^{-1}\tilde{\mathbb{C}}\tilde{\mathbb{D}}_2^{-1}\sigma=e(\ddot{u})+\tilde{\mathbb{D}}_1^{-1}\tilde{\mathbb{C}}e(\dot{u})$. Comparing them, we can see the relation between these alternatives as:

$$\mathbb{D}_1^{-1}=\tilde{\mathbb{D}}_1^{-1}+\tilde{\mathbb{D}}_2^{-1},\quad \mathbb{D}_1^{-1}\mathbb{C}\mathbb{D}_2^{-1}=\tilde{\mathbb{D}}_1^{-1}\tilde{\mathbb{C}}\tilde{\mathbb{D}}_2^{-1},\quad \mathbb{C}\mathbb{D}_1^{-1}+\mathbb{C}\mathbb{D}_2^{-1}=\tilde{\mathbb{D}}_1^{-1}\tilde{\mathbb{C}} \quad (6.6.34)$$

provided that the tensors $\mathbb{D}_1$, $\mathbb{D}_2$, and $\mathbb{C}$ commute with each other. Knowing the tensors $\tilde{\mathbb{D}}_1$, $\tilde{\mathbb{D}}_2$, and $\tilde{\mathbb{C}}$, we can determine $\mathbb{D}_1=(\tilde{\mathbb{D}}_1^{-1}+\tilde{\mathbb{D}}_2^{-1})^{-1}$, and then $\mathbb{C}\mathbb{D}_2^{-1}=\mathbb{D}_1\tilde{\mathbb{D}}_1^{-1}\tilde{\mathbb{C}}\tilde{\mathbb{D}}_2^{-1}$ and also $\mathbb{C}\mathbb{D}_1^{-1}=\tilde{\mathbb{D}}_1^{-1}\tilde{\mathbb{C}}-\mathbb{C}\mathbb{D}_2^{-1}$. Thus eventually also $\mathbb{D}_2$ is determined. Let us note that then $\mathbb{C}=\mathbb{D}_1\tilde{\mathbb{D}}_1^{-1}\tilde{\mathbb{C}}$ so that all tensors $\mathbb{D}_1$, $\mathbb{D}_2$, and $\mathbb{C}$ are positive definite if $\tilde{\mathbb{D}}_1$, $\tilde{\mathbb{D}}_2$, and $\tilde{\mathbb{C}}$ are so.

Remark 6.6.6. In the model from Figure 6.6-left, one can uniquely determine the partial strains e_1 and e_2 from the "observable" variables σ and e. Namely, using $\sigma_2=\sigma-\mathbb{D}_1\dot{e}$, we can easily see that $e_1=\mathbb{C}^{-1}\sigma_2=\mathbb{C}^{-1}\sigma-\mathbb{C}^{-1}\mathbb{D}_1\dot{e}$ and $e_2=e-e_1=e+\mathbb{D}_1\mathbb{C}^{-1}\dot{e}-\mathbb{C}^{-1}\sigma$.

Exercise 6.6.7. For the Maxwell model, derive the estimate of $\ddot{u}$ in (6.6.8). Perform a test by $\ddot{u}$. Realize the difficulties with the term $\mathbb{C}\sigma:\ddot{\sigma}$.

Exercise 6.6.8. Evaluate the energy rate $\sigma:e(\dot{u})$ for the second Jeffreys model (6.6.31) and, in particular, derive the formula (6.6.32). Derive also the a-priori estimates.

Exercise 6.6.9 (From Maxwell to Stokes). Perform a limit passage from the Maxwell model towards the Stokes fluid when $\mathbb{C}\to\infty$.

Exercise 6.6.10 (From Jeffreys to Maxwell). Show that the Jeffreys model for $\mathbb{D}_1=0$ turns to be the Maxwell one. Show the weak convergence of the respective solutions if $\mathbb{D}_1\to 0$. Moreover, show the strong convergence provided the limit solution u is regular in the sense $e(\dot{u})\in L^2(Q;\mathbb{R}^{d\times d})$, cf. (6.6.18b).

Exercise 6.6.11 (From Jeffreys to Kelvin-Voigt). Show that the Jeffreys model for $\mathbb{D}_2 \to \infty$ turns to be (in the limit) the Kelvin-Voigt one. Show the strong convergence of the respective solutions and that the energy dissipated on the $\mathbb{D}_2$ converges to 0.

Exercise 6.6.12 (From standard solid to Maxwell trace-free creep). Consider isotropic standard solid from Fig. 6.4-left with $\mathbb{C}_2$ determined by the bulk and shear moduli (K,G), cf. (6.7.9) below. Then make the limit passage towards the deviatoric (trace-free) creep by $K \to \infty$ (so that $\mathrm{tr}\, e_2 \to 0$) while $G \to 0$.

Exercise 6.6.13 (Energy-conserving time discretization). By using *Crank-Nicolson scheme* analogous to Exercise 6.3.11, devise a time discretization that would conserve energy, i.e. imitate (6.6.4) keeping it as an equality

$$\int_\Omega \frac{\varrho}{2}|v_\tau^k|^2 + \frac{1}{2}\mathbb{C}^{-1}\sigma_\tau^k{:}\sigma_\tau^k\,\mathrm{d}x + \sum_{l=1}^{k}\int_\Omega \mathbb{D}^{-1}\frac{\sigma_\tau^k{+}\sigma_\tau^{k-1}}{2}:\frac{\sigma_\tau^k{+}\sigma_\tau^{k-1}}{2}\,\mathrm{d}x$$
$$= \int_\Omega \frac{\varrho}{2}|v_0|^2 + \frac{1}{2}\mathbb{C}^{-1}\sigma_0{:}\sigma_0\,\mathrm{d}x + \sum_{l=1}^{k}\Big(\int_\Omega f_\tau^k{\cdot}v_\tau^k\,\mathrm{d}x + \int_{\Gamma_\mathrm{N}} g_\tau^k{\cdot}v_\tau^k\,\mathrm{d}S + \Big\langle \tau_\tau^k, \frac{u_{\mathrm{D},\tau}^k{-}u_{\mathrm{D},\tau}^{k-1}}{\tau}\Big\rangle\Big)$$

with $v_\tau^l = \frac{u_\tau^l - u_\tau^{l-1}}{\tau}$. Make analogous energy-conserving scheme also for the Jeffreys' models.

Exercise 6.6.14 (Generalized Maxwell material). Analyze the rheological model arising by merging two Maxwell materials in parallel, i.e.

$$(\mathbb{C}_1 - \mathbb{D}_1) \,\|\, (\mathbb{C}_2 - \mathbb{D}_2).$$

Modify it for more than two Maxwell materials in parallel; actually, such rheology is sometimes used in geophysics to fit better the attenuation of seismic waves propagating in viscoelastic rocks.

6.7 More complex 4-parameter rheologies

It is now not difficult to consider still more and more complicated rheologies. Combination of 2 elastic elements and 2 dampers makes possible several options. Let us select two:

$$\mathbb{D}_1 \,\|\, \big(\mathbb{C}_1 - (\mathbb{C}_2 \,\|\, \mathbb{D}_2)\big) \qquad \text{and} \qquad \mathbb{D}_1 - \mathbb{C}_1 - (\mathbb{C}_2 \,\|\, \mathbb{D}_2).$$

The former one is solid, while the latter one is fluid, sometimes called *Burgers* (or also *Andrade*) model [12, 100]. They are depicted on Figure 6.7.

Using the transformation from Remark 6.5.4, cf. Fig. 6.4, we can equivalently rewrite the former option by replacing $\mathbb{C}_1 - (\mathbb{C}_2 \,\|\, \mathbb{D}_2)$ by $\tilde{\mathbb{C}}_1 \,\|\, (\tilde{\mathbb{C}}_2 - \tilde{\mathbb{D}}_2)$ so that we obtain $(\text{⊣⊢}_{\mathbb{D}_1}\|\text{≋}_{\tilde{\mathbb{C}}_1})\|(\text{∿}_{\tilde{\mathbb{C}}_2}\!-\!\text{⊣⊢}_{\tilde{\mathbb{D}}_2}) = \text{⊣⊢}_{\mathbb{D}_1}\|\text{≋}_{\tilde{\mathbb{C}}_1}\|(\text{∿}_{\tilde{\mathbb{C}}_2}\!-\!\text{⊣⊢}_{\tilde{\mathbb{D}}_2})$, which is the parallelly merged Kelvin-Voigt and Maxwell models.

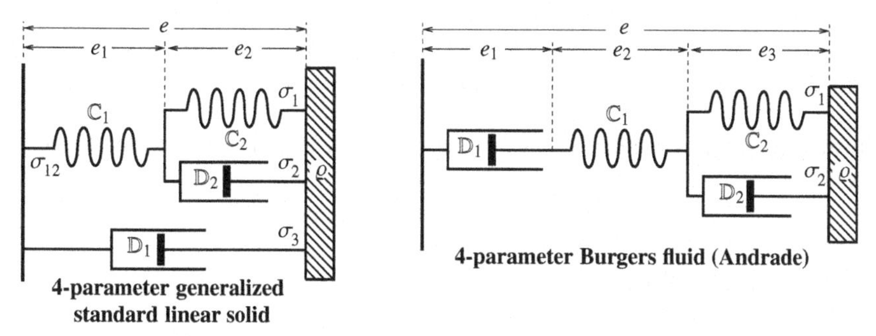

Fig. 6.7 Still a more complicated 4-parameter rheology.
Left: standard solid parallel with Stokes model.
Right: standard solid in series with Stokes model (or equivalently Kelvin-Voigt's model in series with Maxwell's model).

Let us focus on the solid model on Fig. 6.7-left. Denoting $\sigma_{12} = \sigma_1 + \sigma_2$, in view of (6.5.3) we obtain the constitutive equations

$$\sigma = \sigma_{12} + \sigma_3, \qquad \mathbb{D}_2\mathbb{C}_1^{-1}\dot{\sigma}_{12} + (\mathbb{I} + \mathbb{C}_2\mathbb{C}_1^{-1})\sigma_{12} = \mathbb{D}_2\dot{e} + \mathbb{C}_2 e, \qquad \sigma_3 = \mathbb{D}_1\dot{e}. \tag{6.7.1}$$

The first equation gives $\mathbb{D}_2\mathbb{C}_1^{-1}\dot{\sigma} = \mathbb{D}_2\mathbb{C}_1^{-1}\dot{\sigma}_{12} + \mathbb{D}_2\mathbb{C}_1^{-1}\dot{\sigma}_3$, and substituting $\dot{\sigma}_3 = \mathbb{D}_1\ddot{e}$ and $\mathbb{D}_2\mathbb{C}_1^{-1}\dot{\sigma}_{12}$ from (6.7.1), i.e.,

$$\begin{aligned}\mathbb{D}_2\mathbb{C}_1^{-1}\dot{\sigma}_{12} &= \mathbb{D}_2\dot{e} + \mathbb{C}_2 e - (\mathbb{I} + \mathbb{C}_2\mathbb{C}_1^{-1})\sigma_{12}\\ &= \mathbb{D}_2\dot{e} + \mathbb{C}_2 e - (\mathbb{I} + \mathbb{C}_2\mathbb{C}_1^{-1})(\sigma - \sigma_3)\\ &= \mathbb{D}_2\dot{e} + \mathbb{C}_2 e - (\mathbb{I} + \mathbb{C}_2\mathbb{C}_1^{-1})(\sigma - \mathbb{D}_1\dot{e}),\end{aligned}$$

we eventually obtain $\mathbb{D}_2\mathbb{C}_1^{-1}\dot{\sigma} = \mathbb{D}_2\dot{e} + \mathbb{C}_2 e - (\mathbb{I}{+}\mathbb{C}_2\mathbb{C}_1^{-1})(\sigma{-}\mathbb{D}_1\dot{e}) + \mathbb{D}_2\mathbb{C}_1^{-1}\mathbb{D}_1\ddot{e}$, i.e.

$$\mathbb{D}_2\mathbb{C}_1^{-1}\dot{\sigma} + (\mathbb{I}{+}\mathbb{C}_2\mathbb{C}_1^{-1})\sigma = \mathbb{D}_2\mathbb{C}_1^{-1}\mathbb{D}_1\ddot{e} + (\mathbb{D}_2{+}\mathbb{D}_1{+}\mathbb{C}_2\mathbb{C}_1^{-1}\mathbb{D}_1)\dot{e} + \mathbb{C}_2 e. \tag{6.7.2}$$

Like in Kelvin-Voigt's material, the additional damper brings more viscosity arranged advantageously to yield a "parabolic-regularization". Indeed, the estimation scenario based on the energy balance (6.5.6) works quite equally with $\mathbb{D} := \mathbb{D}_2$, with σ_{12} instead of σ, and with an additional term $\|\mathbb{D}_1\dot{e}:\dot{e}\|_{L^1(\Omega)}$ on the left-hand side of (6.5.6). This additional term yields and estimate of $\dot{e}$ in $L^2(I;L^2(\Omega;\mathbb{R}^{d\times d}))$ which eventually, through the Korn inequality, improves (6.5.8b) for

$$\dot{u} \in L^2(I;H^1(\Omega;\mathbb{R}^d)). \tag{6.7.3}$$

Also, it allows for the loading qualification (6.4.5) instead of (6.3.3).

In terms of e_2 and e, the material from Figure 6.7-left is described by the equations

$$\sigma = \mathbb{D}_1\dot{e} + \mathbb{C}_1(e - e_2), \tag{6.7.4a}$$

$$\mathbb{D}_2\dot{e}_2 = \mathbb{C}_1(e - e_2) - \mathbb{C}_2 e_2. \tag{6.7.4b}$$

Testing (6.7.4a) by $\dot{e}$ gives the rate of stored energy and dissipation used in (6.2.9) as

$$\sigma{:}\dot{e} = \mathbb{D}_1\dot{e}{:}\dot{e} + \mathbb{C}_1(e - e_2){:}\dot{e} \tag{6.7.5}$$

while testing (6.7.4b) by $\dot{e}_2$ gives

$$\mathbb{D}_2\dot{e}_2{:}\dot{e}_2 + \mathbb{C}_2 e_2{:}\dot{e}_2 = \mathbb{C}_1(e - e_2){:}\dot{e}_2 = \mathbb{C}_1(e - e_2){:}\dot{e} - \mathbb{C}_1(e - e_2){:}(\dot{e} - \dot{e}_2). \tag{6.7.6}$$

Subtracting (6.7.6) from (6.7.5), we can reveal its energetics as

$$\sigma{:}\dot{e} = \underbrace{\underbrace{\mathbb{D}_1\dot{e}{:}\dot{e}}_{=:\,\xi_1(\dot{e})} + \underbrace{\mathbb{D}_2\dot{e}_2{:}\dot{e}_2}_{=:\,\xi_2(\dot{e}_2)}}_{\text{dissipation rate } \xi(\dot{e},\dot{e}_2)} + \frac{\partial}{\partial t}\underbrace{\Big(\frac{1}{2}\mathbb{C}_1(e - e_2){:}(e - e_2) + \frac{1}{2}\mathbb{C}_2 e_2{:}e_2\Big)}_{=:\,\varphi(e,e_2) \,=\, \text{stored energy}}. \tag{6.7.7}$$

Remark 6.7.1 (Elimination of e_2). Similarly as in Remark 6.5.3, one can formulate everything in terms of σ and e (and their time derivatives) only. More specifically, $e_2 = e - \mathbb{C}_1\sigma_{12} = e - \mathbb{C}_1(\sigma - \sigma_3) = \mathbb{D}\mathbb{C}^{-1}\dot{e} + e - \mathbb{C}_1\sigma$. Yet, the initial condition for e_2 needed to fix the energy balance cannot be avoided because $\dot{e}$ is not well controlled in $L^\infty(I; L^2(\Omega; \mathbb{R}^{d\times d}))$. It is also seen from Fig. 6.7-left where prescription of the position of two pistons in the $\mathbb{D}_1$-$\mathbb{D}_2$-dampers cannot be guessed from the "outer" variables only, cf. also Remark 6.6.6.

It is illustrative to realize how particular rheological models respond on jump loading. This is schematically depicted on Figures 6.8 and 6.9.

Some media combine solid spherical (=volumetric) response with fluidic shear response. We can then speak about certain mixed viscoelastic solid fluids. Actually, in the context of solid mechanics, fluids (like rocks or soil) are sometimes defined as media which cannot permanently withstand shear stress. Yet, the resistance to the permanent spherical stress is still rather expected and then such mixed models are relevant. To distinguish these geometrical aspects, the total strain is decomposed to the *spherical* (also called hydrostatic) *strain* and the *deviatoric* (also called shear) *strain*:

$$e = \mathrm{sph}\,e + \mathrm{dev}\,e \quad \text{with} \quad \mathrm{sph}\,e = \frac{\mathrm{tr}\,e}{d}\mathbb{I} \quad \text{and} \quad \mathrm{dev}\,e = e - \frac{\mathrm{tr}\,e}{d}\mathbb{I}. \tag{6.7.8}$$

Let us note that the deviatoric and the spherical strains from (6.7.8) are orthogonal to each other and the deviatoric strain is trace-free, i.e. $\mathrm{tr}(\mathrm{dev}\,e) = 0$, cf. (A.2.11) on p. 490. In terms of this decomposition, the isotropic (Lamé) material at small strains

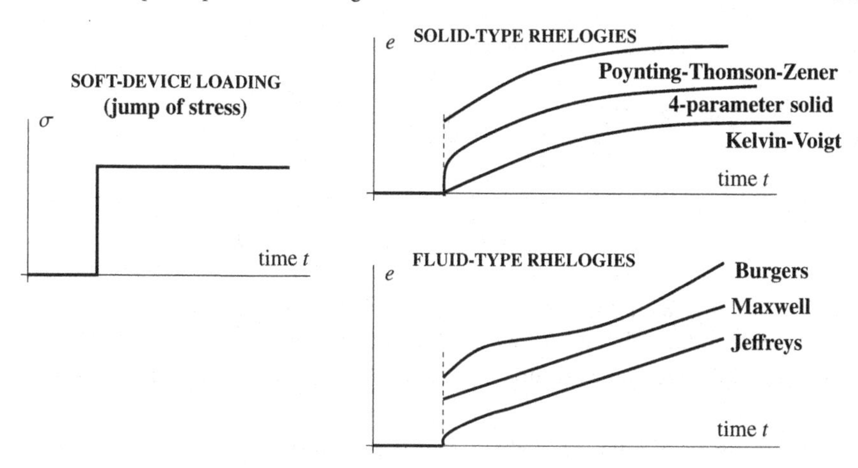

Fig. 6.8 A schematic comparison of strain response of various rheological models on the jump of stress (i.e. a "soft-device" loading). Solid-type models are characterized by a bounded response while fluid-type models keeps e growing.

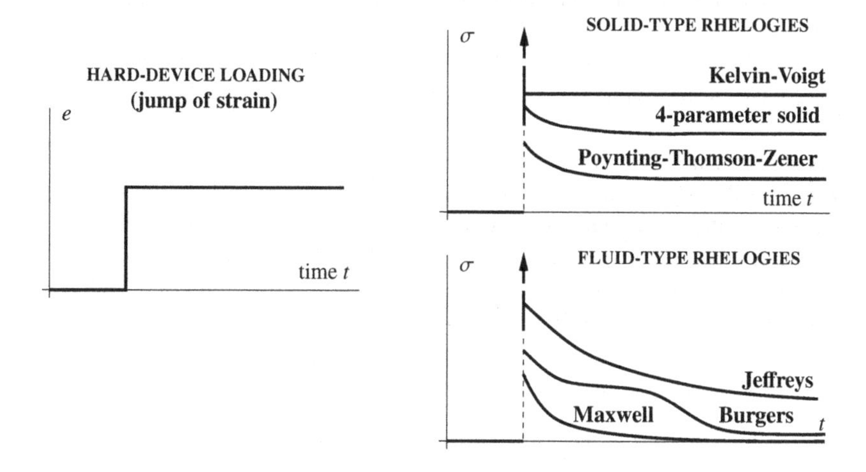

Fig. 6.9 A schematic comparison of stress response of various models on the jump of strain (i.e. a "hard-device" loading). Fluid-type models are characterized by vanishing response for $t \to \infty$ while solid-type models hold stress permanently. The Kelvin-Voigt and 4-parameter solids as well as the Jeffreys fluid model have unbounded stress response at the jump time, to be interpreted that they did not allow for jumping strains (or, in mathematical language, they are well dissipative and exhibit good regularizing effects on strains); note that exactly these 3 rheologies yielded a continuous response in Fig.6.8 and those for which the highest order of time derivatives of σ is less that the highest order of time derivatives of e in the constitutive relation, cf. (6.4.1), (6.6.28), and (6.7.2).

(5.1.6) has the quadratic stored energy[31]

$$\varphi(e) = \frac{1}{2}\lambda(\mathrm{tr}\,e)^2 + G|e|^2 = \frac{1}{2}\Big(\lambda + \frac{2}{d}G\Big)(\mathrm{tr}\,e)^2 + G\Big|e - \frac{\mathrm{tr}\,e}{d}\mathbb{I}\Big|^2$$
$$= \frac{1}{2}\Big(d\lambda + 2G\Big)|\mathrm{sph}\,e|^2 + G\big|\mathrm{dev}\,e\big|^2$$
$$= \frac{d}{2}K|\mathrm{sph}\,e|^2 + G|\mathrm{dev}\,e|^2 \quad \text{with } K = \lambda + \frac{2}{d}G. \tag{6.7.9}$$

where K is called the *bulk modulus* and λ is again the (first) Lamé coefficient and G is the shear modulus. In (6.7.9), we used the identity $|\mathrm{sph}\,e|^2 = (\frac{\mathrm{tr}\,e}{d})^2\mathbb{I}{:}\mathbb{I} = (\frac{\mathrm{tr}\,e}{d})^2 d = (\mathrm{tr}\,e)^2/d$ and the mentioned orthogonality of $\mathrm{sph}\,e$ and $\mathrm{dev}\,e$. Then the stress writes as:

$$\sigma = \varphi'(e) = \lambda(\mathrm{tr}\,e)\mathbb{I} + 2Ge =: \sigma_{\mathrm{dev}} + \sigma_{\mathrm{sph}} \quad \text{with}$$
$$\sigma_{\mathrm{dev}} = 2G\mathrm{dev}\,e \quad \text{and} \quad \sigma_{\mathrm{sph}} = dK\mathrm{sph}\,e. \tag{6.7.10}$$

For isotropic linear materials, it is useful to introduce the shorthand notation "ISO" which relates the general *moduli tensor* in this *isotropic* case with the bulk and shear moduli:

$$\mathbb{C} = \mathrm{ISO}(K,G) \quad \text{with} \quad \mathbb{C}_{ijkl} = K\delta_{ij}\delta_{kl} + G\Big(\delta_{ik}\delta_{jl} + \delta_{il}\delta_{jk} - \frac{2}{d}\delta_{ij}\delta_{kl}\Big). \tag{6.7.11}$$

As already said, we can combine various rheological models. For example, considering the Jeffreys model (6.6.28) in its isotropic variant in terms of the (K,G)-moduli, i.e. with $\mathbb{C} = \mathrm{ISO}(K_{\mathrm{E}},G_{\mathrm{E}})$ and also $\mathbb{D}_n = \mathrm{ISO}(K_{\mathrm{v,n}},G_{\mathrm{v,n}})$ for $n = 1,2$, we can easily design menagerie of combined materials. E.g. for $G_{\mathrm{E}} = 0$ and $G_{\mathrm{v},2} = 0$, we obtain the purely elastic isotropic solid the viscoelastic Jeffreys fluid in deviatoric response combined with a purely viscous Stokes fluid in volumetric response, namely

$$\sigma = \sigma_{\mathrm{dev}} + \sigma_{\mathrm{sph}} \quad \text{with} \quad \sigma_{\mathrm{dev}} = 2G_{\mathrm{v},1}\mathrm{dev}\,\dot{e} \quad \text{and}$$
$$\dot{\sigma}_{\mathrm{sph}} + \frac{K_{\mathrm{E}}}{K_{\mathrm{v},2}}\sigma_{\mathrm{sph}} = dK_{\mathrm{v},1}\,\mathrm{sph}\,\ddot{e} + dK_{\mathrm{E}}\frac{K_{\mathrm{v},1}+K_{\mathrm{v},2}}{K_{\mathrm{v},2}}\mathrm{sph}\,\dot{e}, \tag{6.7.12}$$

cf. Figure 6.10. Let us note that σ_{dev} is trace free if $\mathrm{tr}\sigma_{\mathrm{dev}}|_{t=0} = 0$ is ensured by the initial conditions because the coefficients in the differential equation for σ_{dev} in (6.7.12) are scalars (not tensors) so that obviously $\mathrm{tr}\sigma_{\mathrm{dev}}$ cannot evolve. Thus the decomposition $\sigma = \sigma_{\mathrm{dev}} + \sigma_{\mathrm{sph}}$ in (6.7.12) again splits the *stress* tensor to its *deviatoric* part and its *volumetric* part (=*pressure*), similarly as strains split in (6.7.8).

Also note that, equivalently, we can understand the model (6.7.12) as one the Stokes damper determined by $(K_{\mathrm{v},1}, G_{\mathrm{v},1})$ acting on the total strain e parallel to the Maxwell model determined by $(G_{\mathrm{E}}, G_{\mathrm{v},2})$ acting on σ_{sph} only.

Sending $K_{\mathrm{v},2} \to \infty$, we obtain a combination of the viscoelastic solid in the Kelvin-Voigt rheology (cf. Exercise 6.6.11) with the mentioned viscous Stokes fluid

[31] The second equality in (6.7.9) uses the calculus $|e - (\mathrm{tr}\,e)\mathbb{I}/d|^2 = |e|^2 - 2(e{:}\mathbb{I})(\mathrm{tr}\,e)/d + (\mathrm{tr}\,e)^2(\mathbb{I}{:}\mathbb{I})/d^2 = |e|^2 - 2(\mathrm{tr}\,e)^2/d + (\mathrm{tr}\,e)^2/d = |e|^2 - (\mathrm{tr}\,e)^2/d$.

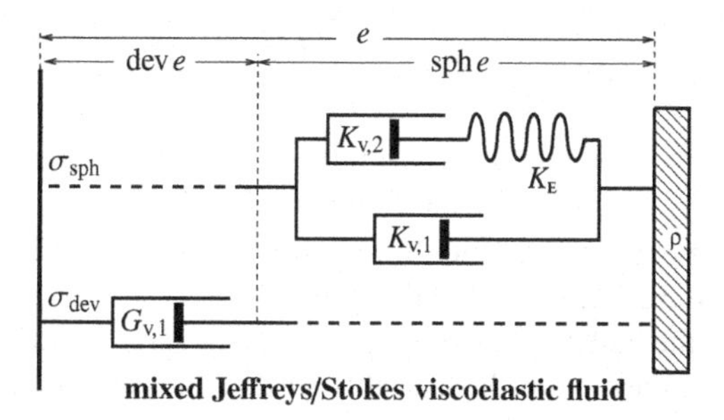

Fig. 6.10 An example of a mixed rheology acting differently on spherical (volumetric) and the deviatoric strains: here Jeffreys/Stokes viscoelastic fluid. Omitting the $K_{v,2}$-damper by sending $K_{v,2} \to \infty$ gives the Boger viscoelastic fluid which allows for the P-wave propagation facilitated by the ϱ-mass and $\mathbb{C}$-spring while avoiding any volumetric creep and any S-wave propagation.

where (6.7.12) turns to

$$\sigma = \sigma_{\rm dev} + \sigma_{\rm sph} \quad \text{with} \quad \sigma_{\rm dev} = 2G_{v,1}{\rm dev}\,\dot{e} \quad \text{and} \quad \sigma_{\rm sph} = dK_{v,1}{\rm sph}\,\dot{e} + dK_{\rm E}{\rm sph}\,e. \tag{6.7.13}$$

This concept is well motivated by long time scale response within which the remarkable creep effects are expected in shear while should be forbidden in volumetric part not to conflict with the concept of nonselfpenetration, although in infinitesimally small strains any selfpenetration is excluded by definition. Sometimes, such materials are called *Boger viscoelastic fluids* [75]. Such (idealized) material may model the liquid oceans as well as the fluidic outer core[32] of our planet Earth where only P-waves can propagate while S-waves are not penetrating this region, cf. Remark 6.3.5. In fact, the viscosity $(K_{v,1}, G_{v,1})$ is considered only very small (of the order 10^{-2}–10^{2}Pa s) in the real fluidic nickel-iron outer core (comparing to the Maxwell-type viscosity of the order 10^{14}–10^{22}Pa s or higher at the solid inner core or mantle). It motivates to consider the asymptotics when these remaining viscosities $(K_{v,1}, G_{v,1})$ approaches zero. In the limit, it yields an inviscid, purely elastic model where the Hooke elasticity counts only with the spherical response while the shear-stress-free response imitates the ideal inviscid fluid, cf. Exercise 6.7.7 below. Such materials are called *elastic fluids*, or more specifically just compressible inviscid fluids. These fluids allow still for propagation of *P-waves* (whose speed is $\varrho^{-1/2}K_{\rm E}^{1/2}$) while S-waves are completely excluded. In modeling of seismic waves, sometimes variation of self-induced gravitational field is relevant in particular in ultra-low-frequency waves [533], cf. Exercise 6.3.13 above.

Remark 6.7.2 (Capillarity in elastic fluids). The concept of *nonsimple materials* leads to capillarity when applied to elastic fluids. The (enhanced) stored energy

[32] The outer core in the Earth is a 2200 km thick layer between the (rather solid) inner core of the radius about 1300 km and the (rather solid) mantle, cf. also Figure 7.16 on p. 356.

$$\varphi_{\rm E}(e) = \frac{d}{2}K|{\rm sph}\,e|^2 + \frac{d}{2}K_{\rm H}\big|\nabla({\rm sph}\,e)\big|^2. \tag{6.7.14}$$

When the boundary $\varGamma_{\rm N}$ is a free surface under a pressure p, the loading functional contains a term $\int_{\varGamma_{\rm N}} pu\cdot\vec{n}\,{\rm d}S$ and the boundary conditions contains $p({\rm div}_{\rm S}\vec{n})\vec{n}$, cf. (C.1.7b). Thus we can see a traction force proportional to the mean curvature (i.e. to $-\frac{1}{2}{\rm div}_{\rm S}\vec{n}$) of the free surface, called *capillary force* (or *pressure*). Related to this surface effects, in the bulk the last term in (6.7.14) for $K_{\rm H} > 0$ also leads to *dispersion* of P-waves, cf. Remark 6.3.6. In fact, this surface pressure arises in solids in the same way, which is why the adjective capillary is sometimes borrowed for the strain gradient terms in nonsimple solids, too.

Remark 6.7.3 (More general rheology models). Sometimes, materials exhibit many relaxation times. Iterating the basic rheologies is sometimes referred to as a *fractional rheology* elements, see e.g. [19, 327, 449]. For example, the fractional variant of Jeffreys' rheology looks as

$$\mathbb{C}_0 - (\mathbb{D}_1 \,\|\, (\mathbb{C}_1 - (\mathbb{D}_2 \,\|\, (\mathbb{C}_2 - \cdots\, \mathbb{D}_{n-1} \,\|\, (\mathbb{C}_n - \mathbb{D}_n)\cdots))).$$

For $n \to \infty$, such models results to integral instead of differential constitutive laws and may lead to differential constitutive relations like (6.4.1), (6.5.3), (6.6.2), (6.6.28), or (6.7.2) but with fractional instead of classical integer time derivatives. Similar spirit is the *materials with memory* which, in our linear variant, is governed by the convolutory integral

$$\sigma(t) = \int_{-\infty}^{t} \mathscr{C}(s-t)e(s)\,{\rm d}s \tag{6.7.15}$$

with some kernel $\mathscr{C} : (-\infty,0] \to {\rm SLin}(\mathbb{R}^{d\times d}_{\rm sym})$ which, in general, can also be a distribution. For example, $\mathscr{C}(t) = \mathbb{C}\delta_0(t) + \mathbb{D}{\rm e}^{t/\tau_{\rm R}}$ with δ_0 the Dirac measure at $t = 0$ and some relaxation time $\tau_{\rm R} > 0$ would give the *standard solid* model; indeed, in this case, (6.7.15) becomes $\sigma(t) = \mathbb{C}e(t) + \int_{-\infty}^{t} \mathbb{D}{\rm e}^{(s-t)/\tau_{\rm R}} e(s)\,{\rm d}s$ so that, differentiating it in time, one gets

$$\dot\sigma(t) = \mathbb{C}\dot e(t) + \mathbb{D}e(t) - \int_{-\infty}^{t} \frac{1}{\tau_{\rm R}}\mathbb{D}{\rm e}^{(s-t)/\tau_{\rm R}} e(s)\,{\rm d}s = \mathbb{C}_0\dot e(t) + \mathbb{D}e(t) + \frac{\mathbb{C}e(t)-\sigma(t)}{\tau_{\rm R}}.$$

The resulted relation $\tau_{\rm R}\dot\sigma + \sigma = \tau_{\rm R}\mathbb{C}_0\dot e + (\mathbb{C}+\tau_{\rm R}\mathbb{D})e$ is then of the standard-solid type, cf. (6.5.3). Of course, (6.7.15) is much more general.

Exercise 6.7.4. Modify (6.2.2b) for the Robin-type boundary condition (=an elastic support) $\sigma(t,\cdot)\vec{n} = \alpha(u_{\rm D} - u)$ for $t \in I$ on $\varGamma_{\rm N}$, and derive (6.2.9).

Exercise 6.7.5. Derive the model of the type $\ddot\sigma + \dot\sigma + \sigma = \ddot e + \dot e$ for the *Burgers'* rheology (Fig. 6.7-right).

Exercise 6.7.6 (From Jeffreys to Boger viscoelastic fluids). In the notation (6.7.12), make the limit passage with $K_{\rm v,2} \to \infty$ to arrive to the model (6.7.13).

Exercise 6.7.7 (From Boger to elastic fluids). Consider the model from (6.7.13) and assume that $f \in L^1(I;L^2(\Omega;\mathbb{R}^d))$, $u_0 \in H^1(\Omega;\mathbb{R}^d)$, and $v_0 \in L^2(\Omega;\mathbb{R}^d)$, the load on the boundary vanishes, for simplicity. Send $(K_{\mathrm{V},1}, G_{\mathrm{V},1})$ to 0, and prove that the solution $(u_\varepsilon, \sigma_\varepsilon)$ corresponding to the Boger model with $(K_{\mathrm{V},1}, G_{\mathrm{V},1}) =: (\varepsilon K_{\mathrm{V}}, \varepsilon G_{\mathrm{V}})$ converges for $\varepsilon \to 0$ to the weak solution of the elastic-fluid model; here modify the weak formulation (6.2.7) for the limit u appropriately. Further, show that this limit solution is unique. Also prove that $\operatorname{dev}\sigma_\varepsilon \to 0$ strongly in $L^2(Q;\mathbb{R}^{d\times d}_{\mathrm{dev}})$.

Exercise 6.7.8 (From Jeffreys to Boger viscoelastic fluids II). If even the loading is qualified so that $f \in W^{1,1}(I;L^2(\Omega;\mathbb{R}^d))$, $u_0 \in H^2(\Omega;\mathbb{R}^d)$, and $v_0 \in H^1(\Omega;\mathbb{R}^d)$, prove also a regularity of the solution from Exercise 6.7.6 and that the energy conservation holds for this limit u. Furthermore, prove that

$$\operatorname{sph} e(u_\varepsilon(t)) \to \operatorname{sph} e(u(t)) \qquad \text{strongly in } L^2(\Omega;\mathbb{R}^{d\times d}_{\mathrm{sym}}) \text{ and} \tag{6.7.16a}$$

$$\dot{u}_\varepsilon(t) \to \dot{u}(t) \qquad \text{strongly in } L^2(\Omega;\mathbb{R}^d) \tag{6.7.16b}$$

for all $t \in I$, and that the energy dissipated by the viscous attenuation for u_ε is not only bounded but even converges to 0 when $\varepsilon \to 0$, i.e.

$$\lim_{\varepsilon\to 0} \int_Q \varepsilon d K_{\mathrm{V}} |\operatorname{sph}(e(\dot{u}_\varepsilon))|^2 + 2\varepsilon G_{\mathrm{V}} |\operatorname{dev}(e(\dot{u}_\varepsilon))|^2 \,\mathrm{d}x\mathrm{d}t = 0. \tag{6.7.17}$$

Exercise 6.7.9 (From Boger to incompressible fluids). Consider again the model from (6.7.13) and, assuming $\operatorname{div} v_0 = 0$, make a limit passage $K_{\mathrm{E}} \to \infty$ or $K_{\mathrm{V},1} \to \infty$. In addition, consider also the so-called convective term $\varrho(\dot{u}\cdot\nabla)\dot{u}$ in the force-equilibrium equation, being actually the *convective derivative* of the velocity instead of the mere partial derivative, cf. (7.7.24) below. Perform this limit passage towards so-called *Navier-Stokes fluid*:

$$\varrho\dot{v} + \varrho(v\cdot\nabla)v - \operatorname{div}(G_{\mathrm{V},1} e(v)) = f, \quad \operatorname{div} v = 0, \quad \text{with } v = \dot{u}.$$

Exercise 6.7.7 (From Burger to elastic fluids). Consider the model from (6.7.15) and assume that $f \in L^1(I;L^2(\Omega;\mathbb{R}^d))$, $u_0 \in H^1(\Omega;\mathbb{R}^d)$, and $v_0 \in L^2(\Omega;\mathbb{R}^d)$, the load on the boundary vanishes for simplicity. Send (K_M, G_M) to 0, and prove that the solution $(u_\varepsilon, [illegible])$ corresponding to the Burger model with $(K_M, G_M) = (\varepsilon K, \varepsilon G)$ converges for $\varepsilon \to 0$ to the weak solution of the elastic-fluid model; here modify the weak formulation (6.2.7) for the limit appropriately. Further, show that this limit solution is unique. Also prove that $\mathrm{dev}\,\sigma_\varepsilon \to 0$ strongly in $L^2(Q;\mathbb{R}^{d\times d}_{\mathrm{dev}})$.

Exercise 6.7.8 (From Jeffreys to Boger viscoelastic fluids [illegible]). If even the loading is qualified so that $f \in W^{1,1}(I;L^2(\Omega;\mathbb{R}^d))$, $u_0 \in H^2(\Omega;\mathbb{R}^d)$, and $v_0 \in H^1(\Omega;\mathbb{R}^d)$, prove the regularity of the solution from Exercise 6.7.6 and that the energy conservation holds for this limit u. Furthermore, prove that

$$\mathrm{sph}\, e(u_\varepsilon(t)) \to \mathrm{sph}\, e(u(t)) \quad \text{strongly in } L^2(\Omega;\mathbb{R}^{d\times d}_{\mathrm{sph}}) \text{ and} \tag{6.7.16a}$$

$$\dot{u}_\varepsilon(t) \to \dot{u}(t) \quad \text{strongly in } L^2(\Omega;\mathbb{R}^d) \tag{6.7.16b}$$

for all $t \in I$, and that the energy dissipated by the viscous attenuation term is not only bounded but even converges to 0 when $\varepsilon \to 0$, i.e.

$$\lim_{\varepsilon \to 0} \int_Q [illegible]\, \mathrm{dev}\, e(\dot{u}_\varepsilon)|^2 \, \mathrm{d}x\mathrm{d}t = 0. \tag{6.7.17}$$

Exercise 6.7.9 (From Burger to incompressible fluids). Consider again the model from (6.7.15) and, assuming $\mathrm{div}\, [illegible] = 0$, make a limit passage $K_E \to \infty$ or $K_M \to \infty$. In addition, consider also the so-called convective term $\varrho(v \cdot \nabla)v$ in the force-equilibrium equation, being actually the convective derivative of the velocity instead of the mere partial derivative, cf. (7.7.24) below. Perform this limit passage towards so-called Navier-Stokes fluid:

$$\varrho(\dot{v} + (v\cdot\nabla)v) - [illegible] = f, \quad \mathrm{div}\, v = 0, \quad \text{with } v = \dot{u}.$$

Chapter 7
Nonlinear materials with internal variables at small strains

> An element of viscoelastic material, for instance, may exhibit only its strain component as observed coordinates, while a great many hidden internal variables may contribute to its properties.[1]
>
> MAURICE ANTHONY BIOT (1905-1985)

> Internal variables ... are supposed to account for the complex internal microscopic processes that occur in the material and manifest themselves at a macroscopic scale in the form of dissipation. [2]
>
> GÉRARD A. MAUGIN (1944-2016)

The concept of *internal variables*[3] (sometimes called also *internal parameters* as in [239, 395]) dates back to P. Duhem [166], to the Nobel prize winner P.W. Bridgman [90], and to C. Eckart [171].[4] Although it may seem a bit artificial, this concept opens extremely fruitful possibilities for modeling in continuum mechanics and often has a direct motivation and mechanical or other physical interpretation.

A particular interpretation represents a so-called *phase-field* concept which originated in alloy solidification models but is now widely used in many other applications, cf. a topical review [495].

Even in cases when elimination of internal variables is possible as we actually focus in the preceding chapter, the option to hold them in the problem may be of an interest from both physical and computational reasons. E.g. time-integration of higher-order differential equations which we saw in the previous chapter may be difficult and replacing them by first-order system (although with additional degrees of freedom corresponding to the internal variables) can be an advantageous option, cf. e.g. Remark 7.5.21 below. Also prescribing the initial conditions is simpler, and the analysis of the problem more lucid in particular if nonlinearities are involved.

---

[1] See [70].

[2] See [343].

[3] The adjective "internal" refers to the attribute that such variables are rather "hidden", not directly subjected to outer loading and outer observation, although there are exceptions from this understanding. They do not exhibit any inertia and are thus governed by differential equations or inclusions of the 1st order in time.

[4] Cf. G.A. Maugin [344] for a thorough historical survey.

M. Kružík and T. Roubíček, *Mathematical Methods in Continuum Mechanics of Solids*, Interaction of Mechanics and Mathematics,
https://doi.org/10.1007/978-3-030-02065-1_7

Importantly, the concept of internal variables facilitates invention of *gradient theories* which can reflect various spatially nonlocal scales and allows for rigorous mathematical treatment of nonlinear situations.

7.1 Concept of generalized standard materials

We actually already used this concept in the previous chapter when using the strain e_2 in a position of an internal variable in some materials, cf. Figures 6.4, 6.5, 6.6, and 6.7-left. Yet, its position was sometimes rather only formal, cf. Remarks 6.5.3, 6.6.6, and 6.7.1. E.g., in the last case, in terms of $\varphi = \varphi(e,z)$ with $z = e_2$ introduced in (6.7.7) and the so-called *potential of dissipative forces* $\zeta = \zeta(\dot{e},\dot{z})$ given by $\zeta(\dot{e},\dot{z}) = \frac{1}{2}\mathbb{D}_1\dot{e}{:}\dot{e} + \frac{1}{2}\mathbb{D}_2\dot{z}{:}\dot{z}$, we can write (6.7.4) now in terms of a general internal variable z as:

$$\sigma = \partial_{\dot{e}}\zeta(\dot{e},\dot{z}) + \partial_e\varphi(e,z), \qquad \text{(generalized Kelvin-Voigt model)} \qquad (7.1.1a)$$

$$\partial_{\dot{z}}\zeta(\dot{e},\dot{z}) + \partial_z\varphi(e,z) \ni 0, \qquad \text{(flow rule for internal variables)} \qquad (7.1.1b)$$

to be considered together with the "geometrical" relation $e = e(u)$ and the equations for the displacement u:

$$\varrho\ddot{u} - \operatorname{div}\sigma = f \qquad \text{(force balance)} \qquad (7.1.1c)$$

as in (6.1.1). In (7.1.1b), $-\partial_z\varphi(e,z)$ stands for the so-called thermodynamical *driving force* for the internal-variable evolution. This ansatz reflects the philosophy that our loading f does not directly affect the internal variable dynamics (7.1.1b). We will consider here z as a general vector, say in $\mathbb{R}^m$ (without excluding $m = 1$) rather than a matrix to cover wider applications. Thus we will write e.g. $\partial_z\varphi{\cdot}\dot{z}$ instead of $\partial_z\varphi{:}\dot{z}$.

The potentials φ and ζ need not be quadratic, and the set of internal variables z may have a different meaning than only the matrix e_2 from the mentioned examples in Chapter 6. For $\partial_{\dot{e}}\zeta = 0$, it generalizes the standard (linear) solid from Figure 6.4 to, in particular, a nonlinear damper and nonlinear springs. Such *generalized standard solids*, called also rather *generalized standard materials*, has been introduced by B. Halphen and Q.S. Nguyen [239], see also e.g. [5, 235]. Even, $\zeta(\dot{e},\cdot)$ is standardly convex but need not be smooth and then its subdifferential $\partial_{\dot{z}}\zeta$ is set-valued, which is why we wrote an inclusion in (7.1.1b) rather than just an equation. Typically, the nondifferentiable point is zero rate, which can model *activated processes* where the driving force must reach certain threshold to start evolving the internal variable in question; more specifically, no evolution of z takes place if the driving force $-\partial_z\varphi$ ranges interior of $\partial_{\dot{z}}\zeta(\dot{e},0)$. Equivalently, the (in general) doubly-nonlinear inclusion (7.1.1b) is often written as a *flow rule*

$$\dot{z} \in [\partial_{\dot{z}}\zeta(\dot{e},\cdot)]^{-1}(-\partial_z\varphi(e,z)); \tag{7.1.2}$$

we will freely call "flow rule" also the form (7.1.1b). Quite typically,[5] the additive decoupling

$$\zeta(\dot{e},\dot{z}) = \zeta_{\rm VI}(\dot{e}) + \zeta_{\rm IN}(\dot{z}) \tag{7.1.3}$$

holds and then the *dissipative force* $\partial_{\dot{z}}\zeta$ is independent of $\dot{e}$ so that still equivalently, (7.1.1b) can be written as

$$\dot{z} \in \partial\zeta_{\rm IN}^*(-\partial_z\varphi(e,z)) \tag{7.1.4}$$

where $\zeta_{\rm IN}^*$ is the Legendre-Fenchel conjugate (cf. (A.3.16) on p. 498) to $\zeta_{\rm IN}$, called sometimes also an *inelastic potential.*

The energy balance for (7.1.1) can formally be obtained by testing (7.1.1c) by the velocity $\dot{u}$ to obtain the general energy balance (6.2.9) to which $\sigma:\dot{e}$ is to be substituted. To this goal, we multiply (7.1.1a) by $\dot{e}$ and (7.1.1b) by $\dot{z}$, which gives respectively $\sigma:\dot{e} = \partial_{\dot{e}}\zeta(\dot{e},\dot{z}):\dot{e} + \partial_e\varphi(e,z):\dot{e}$ and $\partial_{\dot{z}}\zeta(\dot{e},\dot{z})\cdot\dot{z} + \partial_z\varphi(e,z)\cdot\dot{z} = 0$, and by summing it, we eventually obtain

$$\sigma:\dot{e} = \xi(\dot{e},\dot{z}) + \partial_e\varphi(e,z):\dot{e} + \partial_z\varphi(e,z)\cdot\dot{z} = \xi(\dot{e},\dot{z}) + \frac{\partial}{\partial t}\varphi(e,z), \tag{7.1.5}$$

with the dissipation rate

$$\xi(\dot{e},\dot{z}) = \partial_{\dot{e}}\zeta(\dot{e},\dot{z}):\dot{e} + \partial_{\dot{z}}\zeta(\dot{e},\dot{z})\cdot\dot{z}. \tag{7.1.6}$$

The energy balance (6.2.9) now takes the form

$$\frac{\rm d}{{\rm d}t}\int_\Omega \frac{\varrho}{2}|\dot{u}|^2 + \varphi(e(u),z)\,{\rm d}x + \int_\Omega \xi(e(\dot{u}),\dot{z})\,{\rm d}x{\rm d}t = \int_\Omega f\cdot\dot{u}\,{\rm d}x + \int_{\Gamma_{\rm N}} g\cdot\dot{u}\,{\rm d}S + \langle\vec{\tau}_{\rm D},\dot{u}_{\rm D}\rangle_{\Gamma_{\rm D}} \tag{7.1.7}$$

with the traction vector $\vec{\tau}_{\rm D}$ and the duality $\langle\cdot,\cdot\rangle_{\Gamma_{\rm D}}$ from (6.2.10).

For example, the additively decoupled potential of dissipative forces (7.1.3) with $\zeta_{\rm VI}$ positively p-homogeneous (in the sense $\zeta_{\rm VI}(a\dot{e}) = p\zeta_{\rm VI}(\dot{e})$ for any $a \geq 0$) and $\zeta_{\rm IN}$ positively q-homogeneous leads to the dissipation rate $\xi(\dot{e},\dot{z}) = p\zeta_{\rm VI}(\dot{e}) + q\zeta_{\rm IN}(\dot{z})$ which is convex for $p \geq 1$ or $q \geq 1$ but different from $\zeta(\dot{e},\dot{z})$ if $p > 1$ or $q > 1$.

Thus, these dissipative potentials do not exactly occur in the energetics (in contrast to the stored-energy potential) and are sometimes called rather *pseudopotentials.*

Remark 7.1.1 (Minimum dissipation-potential principle). Formally, (7.1.2) can be written as the first-order optimality condition for the functional $\dot{z} \mapsto \zeta_{\rm IN}(\dot{z}) + \partial_z\varphi(e,z)\cdot$

[5] A natural exception from (7.1.3) in the spirit of (7.3.14) below is e.g. in Exercise 7.4.10 on p. 284.

process	internal variable	influenced objects	section
isochoric plasticity (+possibly hardening)	plastic strain (+ isotropic hardening)	$e_{\rm dev}$	7.4
damage or aging	volume fraction (density) of voids / microcracks	φ or/and ζ	7.5
swelling or diffusion in poroelastic media	concentration of diffusant	$e_{\rm sph}$	7.6
damage with plasticity	density of voids and plastic strain	$e_{\rm dev}$, φ, or/and ζ	7.7.1
swelling or diffusion in poro-elasto-plastic media	concentration and plastic strain	$e = e_{\rm sph} + e_{\rm dev}$	7.7.2

Table 7.1 Various processes treated in this Chapter 7 and the sort of objects (strains or energies) which they typically influence, referring to the decomposition $e = e_{\rm sph} + e_{\rm dev}$ in (6.7.8).

$\dot z$. Up to a constant $\partial_e\varphi(e,z){:}\dot e$, this functional is also $\frac{\rm d}{{\rm d}t}\varphi(e,z)+\zeta_{\rm IN}(\dot z) = \partial_e\varphi(e,z){:}\dot e + \partial_z\varphi(e,z){\cdot}\dot z+\zeta_{\rm IN}(\dot z)$. Denoting by $L(e,z,v) := \frac{\rm d}{{\rm d}t}\varphi(e,z)+\zeta_{\rm IN}(v)$, one can (a bit formally) say $\dot z$ minimizes $L(e,z,\cdot)$; cf. [236] etc. This principle expresses a certain "laziness" of the systems which by themselves do not like to lose unnecessarily their energy.

Exercise 7.1.2. Write Burger's rheology from Figure 6.7(right) in the form of (7.1.1).

7.2 Weak formulations of internal-variable flow rules

Confining ourselves to (7.1.1) with the additively split dissipation (7.1.3) without substantial restriction of applications, we are to deal with the system

$$\varrho\ddot u - {\rm div}(\zeta_{\rm VI}'(e(\dot u)) + \partial_e\varphi(e(u),z)) = f, \qquad \text{on } Q, \tag{7.2.1a}$$

$$\partial\zeta_{\rm IN}(\dot z) + \partial_z\varphi(e(u),z) \ni 0 \qquad \text{on } Q. \tag{7.2.1b}$$

We consider an initial/boundary value problem for it by prescribing

$$\sigma\vec n = g \qquad \text{on } \Sigma_{\rm N}, \tag{7.2.1c}$$

$$u = u_{\rm D} \qquad \text{on } \Sigma_{\rm D}, \tag{7.2.1d}$$

$$u(0) = u_0, \quad \dot u(0) = v_0, \quad z(0) = z_0 \qquad \text{in } \Omega. \tag{7.2.1e}$$

The flow rule written in the classical formulation as the inclusion (7.2.1b) to hold a.e. on Q deserves a weak formulation. There are various options that may serve differently in dependence of specific situations.

Using the definition of the subdifferential (A.3.14) on p. 497, we can write the flow rule (7.2.1b) as

$$\langle \partial_z \varphi(e,z), w - \dot{z} \rangle + \zeta_{\text{IN}}(w) \ge \zeta_{\text{IN}}(\dot{z}) \quad \forall w. \tag{7.2.2}$$

Here, instead of the scalar product on $\mathbb{R}^m$ as in (7.2.25), we consider more generally a duality between suitable function spaces on Q. In case of Lebesgue space, (7.2.2) then takes the form

$$\int_Q \partial_z \varphi(e,z) \cdot (w - \dot{z}) + \zeta_{\text{IN}}(w) \, \mathrm{d}x\mathrm{d}t \ge \int_Q \zeta_{\text{IN}}(\dot{z}) \, \mathrm{d}x\mathrm{d}t \quad \forall w \in C(\bar{Q}; \mathbb{R}^m). \tag{7.2.3}$$

The weak formulation of (7.2.1) then combines it with the force equilibrium (6.1.1) with σ from (7.1.1a) in the weak sense (6.2.6).

In some situations, $\partial_z \varphi$ and $\dot{z}$ are not in duality with each other in any reasonable function spaces and then (7.2.3) would not work. In these cases, still another weak formulation can be cast by by employing the *chain-rule formula* and substituting

$$\int_Q \partial_z \varphi(e(u),z) \cdot \dot{z} \, \mathrm{d}x\mathrm{d}t = \int_\Omega \varphi(e(u(T)),z(T)) - \varphi(e(u_0),z_0) \, \mathrm{d}x - \int_Q \partial_e \varphi(e(u),z) \cdot e(\dot{u}) \, \mathrm{d}x\mathrm{d}t. \tag{7.2.4}$$

Thus we obtain the following solution concept:

Definition 7.2.1 (Weak solution). The couple (u,z) with $u \in H^1(I; L^2(\Omega; \mathbb{R}^d))$ and $z \in C_{\mathrm{w}}(I; L^1(\Omega; \mathbb{R}^m))$ is a weak solution to the initial-boundary-value problem for the dynamical system with internal variable (7.2.1) if $\sigma = \zeta'_{\text{VI}}(e(\dot{u})) + \partial_e \varphi(e(u),z) \in L^1(Q; \mathbb{R}^{d \times d})$, $\partial_z \varphi(e(u),z) \in L^1(Q; \mathbb{R}^m)$, $\varphi(e(u(T)),z(T)) \in L^1(\Omega)$, $\zeta_{\text{IN}}(\dot{z}) \in L^1(Q)$, the force equilibrium (7.2.1a) is satisfied in the weak sense (6.2.7), i.e.

$$\int_Q \sigma : e(v) - \varrho \dot{u} \cdot \dot{v} \, \mathrm{d}x\mathrm{d}t + \int_\Omega \varrho \dot{u}(T) \cdot v(T) \, \mathrm{d}x = \int_\Omega \varrho v_0 \cdot v(0) \, \mathrm{d}x + \int_Q f \cdot v \, \mathrm{d}x\mathrm{d}t + \int_{\Sigma_{\mathrm{N}}} g \cdot v \, \mathrm{d}S \, \mathrm{d}t \tag{7.2.5a}$$

for all $v \in L^2(I; H^1_{\mathrm{D}}(\Omega; \mathbb{R}^d)) \cap H^1(I; L^2(\Omega; \mathbb{R}^d))$, and

$$\int_Q \partial_z \varphi(e(u),z) \cdot w + \zeta_{\text{IN}}(w) \, \mathrm{d}x\mathrm{d}t + \int_\Omega \varphi(e(u_0),z_0) \, \mathrm{d}x \ge \int_Q \zeta_{\text{IN}}(\dot{z}) + \partial_e \varphi(e(u),z) \cdot e(\dot{u}) \, \mathrm{d}x\mathrm{d}t + \int_\Omega \varphi(e(u(T)),z(T)) \, \mathrm{d}x \tag{7.2.5b}$$

holds for all smooth test functions w, and the remaining initial conditions $u(0) = u_0$ and $z(0) = z_0$ are satisfied (possibly in a suitable weak sense).

Let us note that (7.2.5a) slightly differs from (6.2.7) and allows for substituting $v = \dot{u}$ (up to a shift if time-varying Dirichlet conditions are involved), which makes the proof of possible energy conservation on $[0,T]$ easier, if needed.

Yet, in some situations like in Sect. 7.4.3 below, even $\partial_e \varphi$ and $e(\dot{u})$ are not in duality with each other. Then one can further still substitute

$$\int_Q \partial_e\varphi(e(u),z):e(\dot u)\,\mathrm{d}x\mathrm{d}t = \int_Q f\cdot\dot u - \xi_{\rm VI}(e(\dot u))\,\mathrm{d}x\mathrm{d}t + \int_{\Sigma_{\rm N}} g\cdot\dot u\,\mathrm{d}S\,\mathrm{d}t \\ + \int_0^T \langle\vec{\tau}_{\rm D},\dot u_{\rm D}\rangle_{\Gamma_{\rm D}}\mathrm{d}t + \int_\Omega \frac{\varrho}{2}|v_0|^2 - \frac{\varrho}{2}|\dot u(T)|^2\,\mathrm{d}x, \quad (7.2.6)$$

which arises by testing the momentum equation (force balance (7.2.1)) by $\dot u$; here $\vec{\tau}_{\rm D}$ is the traction vector defined by (6.2.10). Merging (7.2.4) with (7.2.6) turns (7.2.3) into

$$\int_Q \partial_z\varphi(e(u),z)\cdot\dot z\,\mathrm{d}x\mathrm{d}t = \int_\Omega \frac{\varrho}{2}\big|\dot u(T)\big|^2 + \varphi(e(u(T)),z(T)) - \frac{\varrho}{2}|v_0|^2 - \varphi(e(u_0),z_0)\,\mathrm{d}x \\ + \int_Q \xi_{\rm VI}(\dot e) - f\cdot\dot u\,\mathrm{d}x\mathrm{d}t - \int_{\Sigma_{\rm N}} g\cdot\dot u\,\mathrm{d}S\,\mathrm{d}t - \int_0^T \langle\vec{\tau}_{\rm D},\dot u_{\rm D}\rangle_{\Gamma_{\rm D}}\mathrm{d}t; \quad (7.2.7)$$

here $\vec{\tau}_{\rm D}$ is the traction vector defined by (6.2.10). Then, as devised in [454], the substitution of (7.2.7) into (7.2.2) and using also (7.1.6) eventually gives the variational inequality (7.2.8) in the following conceptual:

Definition 7.2.2 (Weak solution II). The couple (u,z) with $u\in H^1(I;L^2(\Omega;\mathbb{R}^d))$ and $z\in C_{\rm w}(I;L^1(\Omega;\mathbb{R}^m))$ is a weak solution to the initial-boundary-value problem for the dynamical system with internal variable (7.2.1) if $\sigma = \zeta'_{\rm VI}(e(\dot u)) + \partial_e\varphi(e(u),z) \in L^1(Q;\mathbb{R}^{d\times d})$, $\partial_z\varphi(e(u),z)\in L^1(Q;\mathbb{R}^m)$, $\varphi(e(u(T)),z(T))\in L^1(\Omega)$, $\xi(e(\dot u),\dot z) - f\cdot\dot u \in L^1(Q)$, $g\cdot\dot u\in L^1(\Sigma_{\rm N})$, the force equilibrium (7.2.1a) is satisfied in the weak sense (7.2.5a), and

$$\int_Q \partial_z\varphi(e(u),z)\cdot w + \zeta_{\rm IN}(w)\,\mathrm{d}x\mathrm{d}t \ge \int_Q \xi(e(\dot u),\dot z)\,\mathrm{d}x\mathrm{d}t \\ + \int_\Omega \frac{\varrho}{2}\big|\dot u(T)\big|^2 + \varphi(e(u(T)),z(T)) - \frac{\varrho}{2}|v_0|^2 - \varphi(e(u_0),z_0)\,\mathrm{d}x \\ - \int_Q f\cdot\dot u\,\mathrm{d}x\mathrm{d}t - \int_{\Sigma_{\rm N}} g\cdot\dot u\,\mathrm{d}S\,\mathrm{d}t - \int_0^T \langle\vec{\tau}_{\rm D},\dot u_{\rm D}\rangle_{\Gamma_{\rm D}}\mathrm{d}t \quad (7.2.8)$$

holds for all smooth test functions w, and the remaining initial conditions $u(0)=u_0$ and $z(0)=z_0$ are satisfied (possibly in a suitable weak sense).

By the Fenchel relation (A.3.17) on p. 498, the flow rule (7.2.1b) is equivalent to

$$\zeta_{\rm IN}(\dot z) + \zeta^*_{\rm IN}\big(-\partial_z\varphi(e(u),z)\big) + \partial_z\varphi(e(u),z)\cdot\dot z = 0. \quad (7.2.9)$$

We now use the substitution (7.2.4) similarly as we did in (7.2.5b). This leads to:

Definition 7.2.3 (DeGiorgi weak formulation). The couple (u,z) with $u\in H^1(I;L^2(\Omega;\mathbb{R}^d))$ and $z\in C_{\rm w}(I;L^1(\Omega;\mathbb{R}^m))$ is a weak solution to the initial-boundary-value problem for the dynamical system with internal variable (7.2.1) if $\sigma = \zeta'_{\rm VI}(e(\dot u)) + \partial_e\varphi(e(u),z)\in L^1(Q;\mathbb{R}^{d\times d})$, $\partial_z\varphi(e(u),z)\in L^1(Q;\mathbb{R}^m)$, $\varphi(e(u(T)),z(T))\in L^1(\Omega)$, $\zeta_{\rm IN}(\dot z)\in L^1(Q)$, the force equilibrium (7.2.1a) is satisfied in the weak sense (7.2.5a) together with the initial conditions for u and $\dot u$, and also $z(0)=z_0$ holds, and the following *energy-dissipation balance* holds:

$$\int_\Omega \varphi(e(u(T)),z(T))\,\mathrm{d}x + \int_Q \zeta_{\mathrm{IN}}(\dot z) + \zeta_{\mathrm{IN}}^*\big(-\partial_z\varphi(e(u),z)\big)\,\mathrm{d}x\mathrm{d}t$$
$$= \int_Q \partial_e\varphi(e(u),z){:}e(\dot u)\,\mathrm{d}x\mathrm{d}t + \int_\Omega \varphi(e(u_0),z_0)\,\mathrm{d}x. \quad (7.2.10)$$

It is notable that the only scalar identity (7.2.10) contains enough information to recover all the information behind the flow rule (7.2.1b). Sometimes it is advantageous still to make the substitution (7.2.6). This leads to the identity

$$\int_\Omega \frac{\varrho}{2}|\dot u(T)|^2 + \varphi(e(u(T)),z(T))\,\mathrm{d}x\mathrm{d}x + \int_Q \xi_{\mathrm{VI}}(e(\dot u)) + \zeta_{\mathrm{IN}}(\dot z) + \zeta_{\mathrm{IN}}^*\big(-\partial_z\varphi(e(u),z)\big)\,\mathrm{d}x\mathrm{d}t$$
$$= \int_\Omega \frac{\varrho}{2}|v_0|^2 + \varphi(e(u_0),z_0)\,\mathrm{d}x + \int_Q f\cdot\dot u\,\mathrm{d}x\mathrm{d}t + \int_{\Sigma_\mathrm{N}} g\cdot\dot u\,\mathrm{d}S\,\mathrm{d}t + \int_0^T \langle \vec\tau_\mathrm{D}, \dot u_\mathrm{D}\rangle_{\Gamma_\mathrm{D}}\mathrm{d}t \quad (7.2.11)$$

which is to replace (7.2.10) in Definition 7.2.3. Let us note that, by (7.2.9), at least formally $\zeta_{\mathrm{IN}}(\dot z) + \zeta_{\mathrm{IN}}^*(-\partial_z\varphi(e(u),z)) = \langle -\partial_z\varphi(e(u),z),\dot z\rangle = \langle \partial\zeta_{\mathrm{IN}}(\dot z),\dot z\rangle = \xi_{\mathrm{IN}}(\dot z)$ so that (7.2.11) represents an energy balance (7.1.7) but written in terms of ζ_{IN} and ζ_{IN}^* (involving explicitly the driving force $\partial_z\varphi(e,z)$) in place of the dissipation rate ξ_{IN} (which does not involve $\partial_z\varphi$).

In some cases, $\dot u$ may not have well defined trace on Σ. Then, one can modify (7.2.5b), (7.2.8), or (7.2.11) by substituting

$$\int_{\Sigma_\mathrm{N}} g\cdot\dot u\,\mathrm{d}S\,\mathrm{d}t = \int_{\Gamma_\mathrm{N}} g(0)\cdot u_0 - g(T)\cdot u(T)\,\mathrm{d}S - \int_{\Sigma_\mathrm{N}} \dot g\cdot u\,\mathrm{d}S\,\mathrm{d}t. \quad (7.2.12)$$

A particular situation of rate-independent flow rules deserves a special attention. When the time-scale of relaxation of internal variables (whose evolution might still be activated, however) is substantially smaller than the time-scale of outer loading variation, it is a reasonable approximation to consider the convex nonnegative dissipation potential ζ_{IN} as positively homogeneous of degree 1. This means that

$$\zeta_{\mathrm{IN}} = \delta_{S_{\mathrm{EL}}}^* \quad (7.2.13)$$

for some convex closed set S_{EL} containing zero, cf. Figure 7.1; this set is called an (abstract) *elasticity domain*, while its boundary describes a *yield*-like *stress*. Then

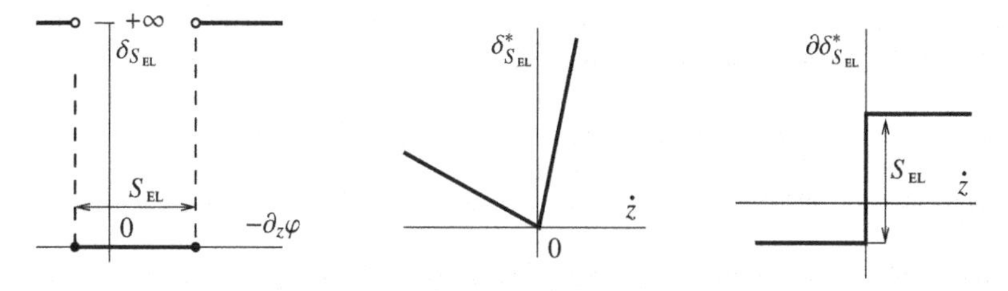

Fig. 7.1 Illustrative example of a set $S_{\mathrm{EL}} \ni 0$, its indicator function $\delta_{S_{\mathrm{EL}}}$, its conjugate $\delta^*_{S_{\mathrm{EL}}}$ which is a 1-homogeneous function, and $\partial\delta^*_{S_{\mathrm{EL}}}$ which is 0-homogeneous maximally monotone, its inverse being the set-valued normal-cone mapping $N_{S_{\mathrm{EL}}} = \partial\delta_{S_{\mathrm{EL}}}$.

$\zeta_{\rm IN}^* = \delta_{S_{\rm EL}}$ and (7.2.9) is the equivalent to

$$\zeta_{\rm IN}(\dot z) + \langle \partial_z\varphi(e(u),z), \dot z\rangle \le 0 \quad \text{and} \quad -\partial_z\varphi(e(u),z) \in S_{\rm EL}. \tag{7.2.14}$$

Relying on 0-homogeneity of $\partial\zeta_{\rm IN}$ and, in particular, on $\partial\zeta_{\rm IN}(\dot z) \subset \partial\zeta_{\rm IN}(0)$, the latter relation means[6] $\partial\zeta_{\rm IN}(0) + \partial_z\varphi(e(u),z) \ni 0$, so that $\zeta_{\rm IN}(\tilde z) + \langle\partial_z\varphi(e(u),z), \tilde z\rangle \ge \zeta_{\rm IN}(0) = 0$. When replacing $\tilde z$ by $\tilde z - z$, it reads as $\zeta_{\rm IN}(\tilde z - z) + \langle\partial_z\varphi(e(u),z), \tilde z - z\rangle \ge 0$ and, assuming $\varphi(e,\cdot)$ is convex, we use $\langle\partial_z\varphi(e(u),z), \tilde z - z\rangle + \varphi(e(u),z) \le \varphi(e(u),\tilde z)$ and we can eliminate the $\partial_z\varphi$-term and arrive to the so-called *semi-stability*

$$\zeta_{\rm IN}(\tilde z - z(t)) + \varphi(e(u(t)),\tilde z) \ge \varphi(e(u(t)), z(t)). \tag{7.2.15}$$

Let us note also that, in this 1-homogeneous case, $\zeta_{\rm IN}^*(-\partial_z\varphi(e(u),z)) = 0$ and (7.2.11) takes the form of the usual energy balance.

Since $\dot z$ is a measure in general in this 1-homogeneous case, its variation with respect to the functional $\zeta_{\rm IN}$, i.e. $\zeta_{\rm IN}(\dot z)$ is no longer in $L^1(Q)$, in contrast to what was assumed in Definition 7.2.1–7.2.3. Therefore, it is "analytically" desirable to avoid this variation by replacing $\int_Q \zeta_{\rm IN}(\dot z)\,{\rm d}x{\rm d}t$ by

$$\mathrm{Var}_{S_{\rm EL}}(z;I) := \sup \sum_{i=1}^{N} \int_\Omega \zeta_{\rm IN}(z(t_i,x) - z(t_{i-1},x))\,{\rm d}x \quad \text{with } \zeta_{\rm IN} = \delta_{S_{\rm EL}}^* \tag{7.2.16}$$

where the supremum is taken over all partitions of the type $0 = t_0 < t_1 < ... < t_N = T$, $N \in \mathbb{N}$. This applies directly to Definition 7.2.1. and similarly also to Definition 7.2.2 where the overall dissipation $\int_Q \zeta(e(\dot u),\dot z)\,{\rm d}x{\rm d}t$ is now $\int_Q \zeta_{\rm VI}(e(\dot u)) + \zeta_{\rm IN}(\dot z)\,{\rm d}x{\rm d}t$ in the inequality (7.2.8). Similar modification concerns also (7.2.10).

This last option can further be modified when implementing the above mentioned equality $\zeta_{\rm IN}^*(-\partial_z\varphi(e(u),z)) = 0$ and employing the semistability (7.2.15). The combination of (7.2.11) with (7.2.15) and with the momentum equation (7.2.1a,c,d) formulated weakly then forms a certain definition of weak solutions:[7]

Definition 7.2.4 (Energetic solutions). The couple (u,z) with $u \in H^1(I;L^2(\Omega;\mathbb{R}^d))$ and $z \in \mathrm{BV}(I;L^1(\Omega;\mathbb{R}^m))$ is a weak solution to the initial-boundary-value problem for the dynamical system with internal variable (7.2.1) if $\sigma = \zeta_{\rm VI}'(e(\dot u)) + \partial_e\varphi(e(u),z) \in L^1(Q;\mathbb{R}^{d\times d})$, $\varphi(e(u(T)),z(T)) \in L^1(\Omega)$, the force equilibrium (7.2.1a) with $\dot u(0) = v_0$ is satisfied in the weak sense (7.2.5a), the initial conditions $u(0) = u_0$ and $z(0) = z_0$ hold, and the following energy inequality holds

[6] Here we used the calculus $\partial\zeta_{\rm IN}(0) = \partial\delta_{S_{\rm EL}}^*(0) = [\partial\delta_{S_{\rm EL}}]^{-1}(0) = S_{\rm EL}$, cf. (A.3.22)–(A.3.23) on p. 499 and Figure 7.1.

[7] The energetic-solution concept was invented for purely rate-independent systems by A. Mielke et al. [366, 367], and later combined with inertia and viscosity as (7.2.1) in [454], both variants being developed in a lot works, cf. in particular [353, 360, 363] and [448]. The original concept considers (7.2.17b) for all $t \in I$ but our definition is slightly more applicable with essentially the same power.

$$\int_\Omega \frac{\varrho}{2}|\dot u(T)|^2 + \varphi(e(u(T)),z(T))\,\mathrm{d}x\mathrm{d}x + \int_Q \xi_{\rm VI}(e(\dot u))\,\mathrm{d}x\mathrm{d}t + \mathrm{Var}_{\zeta_{\rm EL}}(z;I)$$
$$\le \int_\Omega \frac{\varrho}{2}|v_0|^2 + \varphi(e(u_0),z_0)\,\mathrm{d}x + \int_Q f\cdot\dot u\,\mathrm{d}x\mathrm{d}t + \int_{\Sigma_{\rm N}} g\cdot\dot u\,\mathrm{d}S\,\mathrm{d}t + \int_0^T \langle \vec\tau_{\rm D}, \dot u_{\rm D}\rangle_{\Gamma_{\rm D}}\mathrm{d}t \tag{7.2.17a}$$

and the semistability (7.2.15) integrated over Ω holds for a.a. $t\in I$ and for all $\tilde z$:

$$\int_\Omega \zeta_{\rm IN}(\tilde z - z(t)) + \varphi(e(u(t)),\tilde z)\,\mathrm{d}x \ge \int_\Omega \varphi(e(u(t)),z(t))\,\mathrm{d}x. \tag{7.2.17b}$$

Remark 7.2.5 (Upper energy-dissipation inequality). Since it always holds $\zeta_{\rm IN}(\dot z) + \zeta_{\rm IN}^*(-\partial_z\varphi(e(u),z)) + \langle\partial_z\varphi(e(u),z),\dot z\rangle \ge 0$ due to the definition of the convex conjugate $\zeta_{\rm IN}^*$ to $\zeta_{\rm IN}$, cf. (A.3.18), it suffices to require only inequality in (7.2.9) and then also in (7.2.10) and (7.2.11), provided the chain rules we used for the substitutions which led to (7.2.10) and (7.2.11) rigorously hold. Otherwise, to maintain reasonable selectivity of this definition, (7.2.10) and (7.2.11) are to be modified as only inequalities but holding on a.a. subintervals of I including those starting at $t = 0$, cf. [356, 358]. Considering (7.2.10) and (7.2.11) as inequalities, one can see a straightforward relation between Definitions 7.2.2 and 7.2.3: the variational inequality (7.2.8) remains valid if one takes infimum over all w of its left-hand side which can be evaluated pointwise as

$$\inf_w \partial_z\varphi(e(u),z)\cdot w + \zeta_{\rm IN}(w) = -\sup_w\big(-\partial_z\varphi(e(u),z)\cdot w - \zeta_{\rm IN}(w)\big) = -\zeta_{\rm IN}^*(-\partial_z\varphi(e(u),z)).$$

This reveals that (7.2.11) as an inequality is essentially just (7.2.8).

Remark 7.2.6 (Gradient theories for internal variables). If a gradient theory for the internal variables is considered in the sense that an enhanced stored energy $\varphi_{\rm E}$ is considered typically in the form

$$\varphi_{\rm E}(e,z,\nabla z) = \varphi(e,z) + \varphi_1(\nabla z), \tag{7.2.18}$$

then the corresponding driving force is to be understood as a differential in a functional space and the flow rule (7.2.1b) should be augmented to the following evolutionary boundary-value problem

$$\partial\zeta_{\rm IN}(\dot z) + \partial_z\varphi(e(u),z) \ni \mathrm{div}\varphi_1'(\nabla z) \qquad \text{on } Q, \tag{7.2.19a}$$
$$\varphi_1'(\nabla z)\cdot\vec n = 0 \qquad \text{on } \Sigma. \tag{7.2.19b}$$

The weak formulations (7.2.2) and (7.2.9) are to be understood in functional spaces rather than locally pointwise on Q. Thus their integrated variants (7.2.3) or (7.2.5b) or (7.2.8) as well as (7.2.10) or (7.2.11) and also the semistability (7.2.17b) should involve the corresponding gradient terms. More specifically, (7.2.3) should be augmented by $\varphi_1'(\nabla z){:}\nabla(w-\dot z)$ on the left-hand side, while (7.2.5b) should be modified as

$$\int_Q \partial_z\varphi(e(u),z)\cdot w + \zeta_{\rm IN}(w) + \varphi_1'(\nabla z):\nabla w \,\mathrm{d}x\mathrm{d}t + \int_\Omega \varphi(e(u_0),z_0) + \varphi_1(\nabla z_0)\,\mathrm{d}x$$
$$\geq \int_Q \zeta_{\rm IN}(\dot z) + \partial_e\varphi(e(u),z)\cdot e(\dot u)\,\mathrm{d}x\mathrm{d}t + \int_\Omega \varphi(e(u(T)),z(T)) + \varphi_1(\nabla z(T))\,\mathrm{d}x \tag{7.2.20}$$

and analogously also for (7.2.8). The DeGiorgi formulation (7.2.10) is to be augmented as

$$\int_\Omega \varphi(e(u(T)),z(T)) + \varphi_1(\nabla z(T))\,\mathrm{d}x + \int_Q \zeta_{\rm IN}(\dot z) + \zeta_{\rm IN}^*(-\partial_z\varphi(e(u),z))\,\mathrm{d}x\mathrm{d}t$$
$$= \int_Q \partial_e\varphi(e(u),z):e(\dot u)\,\mathrm{d}x\mathrm{d}t + \int_\Omega \varphi(e(u_0),z_0) + \varphi_1(\nabla z_0)\,\mathrm{d}x \tag{7.2.21}$$

and analogously also (7.2.11) and (7.2.17a), while the semistability (7.2.17b) should augment for

$$\int_\Omega \zeta_{\rm IN}(\tilde z - z(t)) + \varphi(e(u(t)),\tilde z) + \varphi_1(\nabla\tilde z)\,\mathrm{d}x \geq \int_\Omega \varphi(e(u(t)),z(t)) + \varphi_1(\nabla z(t))\,\mathrm{d}x. \tag{7.2.22}$$

See Section 7.3.2 for more mathematical details.

Remark 7.2.7 (More general dissipation of internal variables). Sometimes, it is useful to consider $\zeta_{\rm IN} = \zeta_{\rm IN}(z,\dot z)$, and then the inclusion

$$\partial_{\dot z}\zeta_{\rm IN}(z,\dot z) + \partial_z\varphi(e(u),z) - \mathrm{div}(\varphi_1'(\nabla z)) \ni 0. \tag{7.2.23}$$

A-priori estimates based on the test by $\dot z$ are the same. Limit passage in the term $\int_Q \zeta_{\rm IN}(z,\dot z)\,\mathrm{d}x\mathrm{d}t$ in (7.2.8) is as before only with the Aubin-Lions compact embedding theorem B.4.10 for z in addition. Yet, for rate-independent cases when $\zeta_{\rm IN}(z,\cdot)$ is 1-homogeneous and $\dot z$ may be a general measure, this generalization is very nontrivial and cumbersome, cf. in particular [353, 360].

Remark 7.2.8 (Nonsmooth stored energy $\varphi(e,\cdot)$). Alternatively, especially if $\varphi(e,\cdot)$ is nonsmooth, we can use (7.2.2) or (7.2.8) with $\sigma_{\rm IN}$ in place of $\partial_z\varphi(e(u),z)$ and then, assuming convexity of $\varphi(e,\cdot)$, formulate the inclusion $\sigma_{\rm IN} \in \partial_z\varphi(e(u),z)$ weakly:

$$\int_Q \sigma_{\rm IN}\cdot(z-v) + \varphi(e(u),v)\,\mathrm{d}x\mathrm{d}t \geq \int_Q \varphi(e(u),z)\,\mathrm{d}x\mathrm{d}t. \tag{7.2.24}$$

Because of the term $\varphi(e(u),v)$ on the left-hand side of (7.2.24), we now have more benefit from a possible strong convergence of $e(u)$ than before in (7.2.8); cf. the technique in Section 7.3.2 below. Let us note that rate-independent flow rules may simplify the situation if formulated appropriately since $\partial_z\varphi(e(u),z)$ does not occur in Definition 7.2.4.

Remark 7.2.9 (Nonpotential dissipative forces). Relying on maximal monotonicity[8] of the convex subdifferential $\partial\zeta_{\rm IN} : \mathbb{R}^m \rightrightarrows \mathbb{R}^m$, the flow rule (7.2.1b) can be written as

$$(\omega + \partial_z\varphi(e,z))\cdot(w - \dot{z}) \geq 0 \quad \forall w, \omega \in \mathbb{R}^m \text{ such that } \omega \in \partial\zeta_{\rm IN}(w). \tag{7.2.25}$$

This can be treated similarly as (7.2.2) but allows for a generalization for maximal monotone mappings not necessarily having a potential in place of $\partial\zeta_{\rm IN}$, i.e. for dissipative forces not having a dissipation pseudopotential.

7.3 Basic strategies for analysis in brief

Before illustrating the abstract concept of generalized standard materials on selected examples, let us present some mathematical results of, perhaps, a general use.

The basic qualitative attribute is the existence of solutions in a suitable weak sense. The important aspects of the (here usually only sketched) proofs are also conceptual hints for a constructive approximate procedure that is stable (in the sense of specific apriori estimates) and convergent. In qualified cases, some other attribute is of importance, namely the energy conservation of the weak solution. This may not be entirely automatic and can be significant, not only for the noteworthy information itself, but occasionally it also results in stronger modes of convergence or a possibility to generalize the models into a full anisothermal thermodynamical framework, as done later in Chapter 8.

The simplest situation but still with interesting applications is when the stored energy is quadratic like in the previous Chapter 6. We will present some results in Section 7.3.1. Analysis of materials with non-quadratic $\varphi(e,z)$ requires either linear dissipation (i.e. a quadratic dissipation potential) or involving gradients in those variables at which φ is not quadratic or even nonconvex. We will treat these options in Section 7.3.2.

7.3.1 Quadratic stored energies

Let us first consider an extension of the purely hyperbolic linear elastodynamic problem (6.3.1) from Section 6.3 to materials with internal variables still acting linearly and still keeping the hyperbolic character of the problem, i.e. the stored energy remains quadratic and there is no attenuation acting directly on the total strain:

[8] A monotone (in general set-valued) mapping is said maximal if its graph is maximal in the class of graphs of monotone set-valued mappings. Convex subdifferentials of proper lower-semicontinuous functionals are examples of maximal-monotone mappings.

$$\varphi(e,z) = \frac{1}{2}\mathbb{A}\binom{e}{z}:\binom{e}{z} \quad \text{with } \mathbb{A} = \begin{pmatrix} \mathbb{C} & -\mathbb{X} \\ -\mathbb{X}^\top & \mathbb{B} \end{pmatrix} \quad \text{and} \quad \zeta(\dot{e},\dot{z}) = \zeta_{\text{IN}}(\dot{z}), \tag{7.3.1}$$

where $\mathbb{X}$ is a $(d\times m)$-matrix related to cross-effects between the strain e and the internal variable z. We have now $\partial_e\varphi(e,z) = \mathbb{C}e - \mathbb{X}z$ and $\partial_z\varphi(e,z) = \mathbb{B}z - \mathbb{X}^\top e$ and the system (7.2.1) reads as

$$\varrho\ddot{u} - \operatorname{div}(\mathbb{C}e(u)) = f - \operatorname{div}(\mathbb{X}z), \tag{7.3.2a}$$

$$\partial\zeta_{\text{IN}}(\dot{z}) + \mathbb{B}z \ni \mathbb{X}^\top e(u). \tag{7.3.2b}$$

The weak formulation (7.2.5a) of the force balance (momentum equation) now reads as:

$$\int_Q (\mathbb{C}e(u) - \mathbb{X}z):e(v) - \varrho\dot{u}\cdot\dot{v}\,\mathrm{d}x\mathrm{d}t + \int_\Omega \varrho\dot{u}(T)\cdot v(T)\,\mathrm{d}x = \int_\Omega \varrho v_0\cdot v(0)\,\mathrm{d}x + \int_Q f\cdot v\,\mathrm{d}x\mathrm{d}t + \int_{\Sigma_{\text{N}}} g\cdot v\,\mathrm{d}S\,\mathrm{d}t. \tag{7.3.3}$$

We consider the boundary and initial conditions (7.2.1c-e). The variational inequality (7.2.8) now looks as:

$$\int_Q (\mathbb{B}z - \mathbb{X}^\top e(u))\cdot w + \zeta_{\text{IN}}(w)\,\mathrm{d}x\mathrm{d}t \geq \int_Q \xi_{\text{IN}}(\dot{z}) - f\cdot\dot{u}\,\mathrm{d}x\mathrm{d}t + \frac{1}{2}\int_\Omega \varrho|\dot{u}(T)|^2 + \mathbb{A}\binom{e(u(T))}{z(T)}:\binom{e(u(T))}{z(T)} - \varrho|v_0|^2 - \mathbb{A}\binom{e(u_0)}{z_0}:\binom{e(u_0)}{z_0}\mathrm{d}x - \int_{\Sigma_{\text{N}}} g\cdot\dot{u}\,\mathrm{d}S\,\mathrm{d}t - \int_0^T \langle\vec{\tau}_{\text{D}}, \dot{u}_{\text{D}}\rangle_{\Gamma_{\text{D}}}\mathrm{d}t. \tag{7.3.4}$$

The conventional Galerkin method, being expected to yield as system of ordinary differential equations, needs a regularization (smoothing) of the dissipation potential ζ_{IN} which is, in general, nonsmooth. Rather for simplicity, assume that this regularization, let us denote is by $\zeta_{\text{IN},\varepsilon} \in C^1(\mathbb{R}^m)$, Γ-converges to ζ_{IN}, i.e.

$$\limsup_{\varepsilon\to 0} \zeta_{\text{IN},\varepsilon}(z) \leq \zeta_{\text{IN}}(z) \tag{7.3.5a}$$

$$\liminf_{\varepsilon\to 0,\ \tilde{z}\to z} \zeta_{\text{IN},\varepsilon}(\tilde{z}) \geq \zeta_{\text{IN}}(z) \tag{7.3.5b}$$

for all $z\in\mathbb{R}^m$; cf. Definition A.3.13 on p. 501. An example for $\zeta_{\text{IN},\varepsilon}$ is the Yosida approximation of ζ_{IN}; then (7.3.5) is satisfied provided $\zeta_{\text{IN}} : \mathbb{R}^m \to [0,+\infty]$ is convex lower semicontinuous. Using this approximation, (7.1.1) becomes a system of equations. Let us denote by $(u_{\varepsilon,k}, z_{\varepsilon,k})$ its Galerkin approximation, considering still the initial/boundary conditions (7.2.1c-e). In this simple case with only one nonlinearity $\partial\zeta_{\text{IN}}$, the joint convergence for $\varepsilon\to 0$ and $k\to\infty$ can be proved.

Proposition 7.3.1 (Existence of weak solutions: inviscid case). *Let ϱ satisfies (6.3.5), $\mathbb{A}$ from (7.3.1) be symmetric, positive definite, $\zeta_{\text{IN}}(z) \geq \epsilon|z|^p$ with $p > 1$, and the loading and the initial conditions satisfy (6.3.3) and $z_0 \in L^2(\Omega;\mathbb{R}^m)$.*

Then the Galerkin approximation of the ε-regularized system (7.2.1) with $\varphi(e,z) = (e,z)^\top\mathbb{A}(e,z)$ *and* $\zeta_{\rm VI} = 0$ *has a solution* $(u_{\varepsilon,k}, z_{\varepsilon,k})$ *satisfying the a-priori estimates*

$$\|u_{\varepsilon,k}\|_{W^{1,\infty}(I;L^2(\Omega;\mathbb{R}^d))\cap L^\infty(I;H^1(\Omega;\mathbb{R}^d))} \le C, \tag{7.3.6a}$$

$$\|z_{\varepsilon,k}\|_{W^{1,p}(I;L^p(\Omega;\mathbb{R}^m))\cap L^\infty(I;L^2(\Omega;\mathbb{R}^m))} \le C. \tag{7.3.6b}$$

For $\varepsilon \to 0$ *and* $k \to \infty$*, in terms of a selected subsequence,* $u_{\varepsilon,k} \to u$ *and* $z_{\varepsilon,k} \to z$ *weakly* in the topologies indicated in (7.3.6) and* $u \in W^{1,\infty}(I;L^2(\Omega;\mathbb{R}^d)) \cap L^\infty(I;H^1(\Omega;\mathbb{R}^d))$ *and* $z \in W^{1,p}(I;L^p(\Omega;\mathbb{R}^m)) \cap L^\infty(I;L^2(\Omega;\mathbb{R}^m))$ *forms a weak solution of (7.2.1).*

Sketch of the proof. Existence of the solution of the underlying ordinary-differential system behind the regularized Galerkin approximation is standard. From (7.1.7), one can read the a-priori estimates (7.3.6). Then we write the Galerkin identity in the form of Definition 7.2.2 exploiting the convexity of $\zeta_{\rm IN,\varepsilon}$, i.e. in particular we will see the inequality (7.3.4) with $\zeta_{\rm IN,\varepsilon}$ instead of ζ written for $(u_{\varepsilon,k}, z_{\varepsilon,k})$.

The limit passage with a approximate solutions is simple by the weak continuity in (7.3.3) and by weak semicontinuity in the variational inequality (7.3.4). The only delicate term is $\liminf_{\varepsilon\to 0,\, k\to\infty} \int_Q \zeta_{\rm IN,\varepsilon}(z_{\varepsilon,k})\,\mathrm{d}x\mathrm{d}t \ge \int_Q \zeta_{\rm IN}(z)\,\mathrm{d}x\mathrm{d}t$ for which (7.3.5b) together with Fatou lemma is used. □

A special when $\zeta_{\rm IN}$ is positively homogeneous of degree 1 is related with the rate-independent evolution of the internal variable z. Then $p = 1$ in (7.3.6b) and, in the limit, $\dot z$ is a measure on $\bar Q$ and the integral $\int_Q \xi_{\rm IN}(\dot z)\,\mathrm{d}x\mathrm{d}t$ is turned to be rather a variation of the measure $\dot z$ with respect to $\zeta_{\rm IN}(\cdot)$. Again, we rely on the weak* semicontinuity of the functional $\dot z \mapsto \int_Q \xi_{\rm IN}(\dot z)\,\mathrm{d}x\mathrm{d}t$ or rather of $z \mapsto \mathrm{Var}_{S_{\rm EL}}(z;I)$ in (7.2.17a).

Another generalization is weakening of the positive definiteness of $\mathbb{A}$. Some problems like perfect plasticity in Sect. 7.4.3 rather relay on a positive semi-definiteness of $\mathbb{A}$ only. More specifically, let us assume

$$\exists \epsilon > 0\ \forall (e,z) \in \mathbb{R}^{d\times d}_{\rm sym} \times \mathbb{R}^{d\times d}: \quad \mathbb{A}\binom{e}{z} \vdots \binom{e}{z} \ge \epsilon |e - Ez|^2 \tag{7.3.7}$$

with some linear mapping $E : \mathbb{R}^m \to \mathbb{R}^{d\times d}_{\rm sym}$. Then, from the coercivity $\inf \zeta_{\rm IN}(\cdot)/|\cdot|^p > 0$, one obtain $z \in L^\infty(I;L^p(\Omega;\mathbb{R}^m))$ if also $z_0 \in L^p(\Omega;\mathbb{R}^m)$, so that $e(u) - Ez \in L^\infty(I;L^2(\Omega;\mathbb{R}^{d\times d}))$ implies $e(u) \in L^\infty(I;L^{\min(2,p)}(\Omega;\mathbb{R}^{d\times d}))$. By Korn's inequality, $u \in L^\infty(I;W^{1,\min(2,p)}(\Omega;\mathbb{R}^d))$.

The additional dissipation $\zeta_{\rm VI}$ acting on the total strain yields a further estimate on $\dot e$ and may bring $\dot u$ into duality with $\ddot u$. If $\zeta_{\rm VI}$ is quadratic, say $\zeta_{\rm VI}(\dot e) = \frac{1}{2}\mathbb{D}_1\dot e{:}\dot e$, then (7.3.3) augments by $\mathbb{D}_1\dot e(u){:}e(v)$, which allows for a limit passage by the weak continuity. If $\zeta_{\rm VI}$ is not quadratic but uniformly convex (so that $\partial\zeta_{\rm VI}$ is uniformly monotone, then one can prove first strong convergence of (here Galerkin) approximate solutions like in Remark 6.4.15 but here executed simultaneously for the flow rule for z:

Proposition 7.3.2 (Existence of weak solutions: viscosity involved). *Let the assumptions of Proposition 7.3.1 hold except that the smooth uniformly convex potential ζ_{VI} now do not vanish, with $\zeta_{VI}(\cdot)/(1+|\cdot|^2)$ bounded and with $\inf_{e_1\neq e_2}(\zeta'_{VI}(e_1)-\zeta'_{VI}(e_2))/|e_1-e_2|^2 =: c_{DIS}$ positive. Then the Galerkin approximation of the ε-regularized system (7.2.1) with $\varphi(e,z)=(e,z)^\top\mathbb{A}(e,z)$ has a solution $(u_{\varepsilon,k},z_{\varepsilon,k})$ satisfying the a-priori estimates*

$$\|u_{\varepsilon,k}\|_{W^{1,\infty}(I;L^2(\Omega;\mathbb{R}^d))\cap H^1(I;H^1(\Omega;\mathbb{R}^d))} \le C, \tag{7.3.8a}$$

$$\|z_{\varepsilon,k}\|_{W^{1,p}(I;L^p(\Omega;\mathbb{R}^m))\cap L^\infty(I;L^2(\Omega;\mathbb{R}^m))} \le C. \tag{7.3.8b}$$

For $\varepsilon\to 0$ and $k\to\infty$, in terms of a selected subsequence, $u_{\varepsilon,k}\to u$ (resp. $z_{\varepsilon,k}\to z$) strongly (resp. weakly) in the topologies indicated in (7.3.8) and $u\in W^{1,\infty}(I;L^2(\Omega;\mathbb{R}^d))\cap H^1(I;H^1(\Omega;\mathbb{R}^d))$ and $z\in W^{1,p}(I;L^p(\Omega;\mathbb{R}^m))\cap L^\infty(I;L^2(\Omega;\mathbb{R}^m))$ forms a weak solution of (7.2.1).*

Proof. The existence of the Galerkin approximation of the regularized (smoothened) system is as in the proof of Proposition 7.3.1, and also (7.3.8) can be read from (7.1.7). Then one selects a subsequence $u_{\varepsilon,k}\to u$ and $z_{\varepsilon,k}\to z$ weakly* in the topologies indicated in (7.3.8).

Further, using the strong monotonicity of $\zeta'_{VI}:\mathbb{R}^{d\times d}_{sym}\to\mathbb{R}^{d\times d}_{sym}$, we prove the strong convergence[9] of $e(\dot u_{\varepsilon,k})$ in $L^2(Q;\mathbb{R}^{d\times d})$ and of $\dot u_{\varepsilon,k}$ in $L^\infty(I;L^2(\Omega;\mathbb{R}^d))$:

$$\begin{aligned}
&\frac12\inf\varrho(\cdot)\big\|\dot u_{\varepsilon,k}(t)-\dot u(t)\big\|^2_{L^2(\Omega;\mathbb{R}^d)}+c_{DIS}\big\|e(\dot u_{\varepsilon,k}-\dot u)\big\|^2_{L^2([0,t]\times\Omega;\mathbb{R}^{d\times d})}\\
&\le\int_0^t\varrho(\ddot u_{\varepsilon,k}-\varrho\ddot u)\cdot(\dot u_{\varepsilon,k}-\dot u)+(\zeta'_{VI}(e(\dot u_{\varepsilon,k}))-\zeta'_{VI}(e(\dot u))):e(\dot u_{\varepsilon,k}-\dot u)\\
&\qquad+(\partial\zeta_{IN}(\dot z_{\varepsilon,k})-\partial\zeta_{IN}(\dot z))\cdot(\dot z_{\varepsilon,k}-\dot z)+\frac12\mathbb{A}\binom{e(u_{\varepsilon,k}-u)}{z_{\varepsilon,k}-z}:\binom{e(u_{\varepsilon,k}-u)}{z_{\varepsilon,k}-z}\,\mathrm{d}t\\
&=\int_0^t\Big((f-\varrho\ddot u)\cdot(\dot u_{\varepsilon,k}-\dot u)-(\zeta'_{VI}(e(\dot u))+\mathbb{C}e(u)-\mathbb{X}z):e(\dot u_{\varepsilon,k}-\dot u)\\
&\qquad-(\partial\zeta_{IN}(\dot z)-\mathbb{X}^\top e(u)+\mathbb{B}z)\cdot(\dot z_{\varepsilon,k}-\dot z)\Big)\,\mathrm{d}t\to 0
\end{aligned}\tag{7.3.9}$$

for any $t\in I$. More in detail, if $\partial\zeta_{IN}$ is set-valued, then $\partial\zeta_{IN}(\dot z)$ means that a suitable selection is considered. And also, (7.3.9) is rather a conceptual strategy, being to make more precise by approximating u and z into the finite-dimensional subspaces used for the Galerkin approximation. □

For regularity, we can execute a higher-order test like in Proposition 6.3.2 or 6.6.2. Thus we obtain:

Proposition 7.3.3 (Regularity and energy conservation). *Let ϱ satisfy (6.3.2), the loading and the initial conditions satisfy (6.3.12) if $\zeta_{VI}=0$ or (6.4.14) if $\inf_{e\neq0}\zeta_{VI}(e)/|e|^2>0$ with $z_0\in H^1(\Omega;\mathbb{R}^m)$. Let also $\partial\zeta^*_{IN}$ has at most linear growth or*

[9] If ζ_{VI} were merely convex, the limit passage can be executed through mere weak convergence by Minty's trick exploiting strong monotonicity of ζ'_{VI}.

$\mathbb{B}z_0 - \mathbb{X}^\top e(u_0) \in -\mathrm{int}(\partial\zeta_{IN}(0))$. *Then the weak solution (u,z) from Proposition 7.3.1 or 7.3.2 satisfies also:*

$$u \in W^{2,\infty}(I;L^2(\Omega;\mathbb{R}^d)) \cap W^{1,\infty}(I;H^1(\Omega;\mathbb{R}^d)), \tag{7.3.10a}$$

$$z \in W^{1,\infty}(I;L^2(\Omega;\mathbb{R}^m)), \tag{7.3.10b}$$

and this solution is unique and conserves energy, i.e. (7.1.7) holds integrated over $[0,t]$ for any $t \in I$. Moreover, if ζ_{VI} is uniformly convex, then

$$\ddot{u} \in H^2(I;H^1(\Omega;\mathbb{R}^d)). \tag{7.3.10c}$$

Eventually, if ζ_{VI} is either uniformly convex with ζ_{VI}'' bounded or if $\zeta_{VI} = 0$, then also

$$\dddot{u} \in L^1(I;L^2(\Omega;\mathbb{R}^d)) + L^\infty(I;H_0^1(\Omega;\mathbb{R}^d)^*). \tag{7.3.10d}$$

Sketch of the proof. Consider either $\zeta_{\mathrm{VI}} = 0$ as in Proposition 7.3.1 or $\zeta_{\mathrm{VI}}(\dot{e}) = \frac{1}{2}\mathbb{D}\dot{e}:\dot{e} + \zeta_0(\dot{e})$ with some convex ζ_0 and with $\mathbb{D}$ positive definite. We can benefit from the quadratic structure of φ from (7.4.2). Differentiating the momentum equation and testing by $\ddot{u}$, like in (6.6.19) and using $\partial^2\zeta_0(\dot{e})(\ddot{e}):\ddot{e} \geq 0$, we get

$$\frac{\mathrm{d}}{\mathrm{d}t}\int_\Omega \frac{\varrho}{2}|\ddot{u}|^2\,\mathrm{d}x + \int_\Omega (\mathbb{D}e(\ddot{u}) + \mathbb{C}e(\dot{u}) - \mathbb{X}\dot{z}):e(\ddot{u})\,\mathrm{d}x \\ \leq \int_\Omega \dot{f}\cdot\ddot{u}\,\mathrm{d}x + \int_{\Gamma_\mathrm{N}} \dot{g}\cdot\ddot{u}\,\mathrm{d}S + \langle \dot{\vec{r}}_\mathrm{D}, \ddot{u}_\mathrm{D}\rangle_{\Gamma_\mathrm{D}}. \tag{7.3.11}$$

Differentiating the flow rule and testing by $\ddot{z}$, and using formally[10] $\partial^2\zeta_{\mathrm{IN}}(\dot{z})(\ddot{z})\cdot\ddot{z} \geq 0$, we obtain $(\mathbb{B}\dot{z} - \mathbb{X}^\top e(\dot{u}))\cdot\ddot{z} \leq 0$. Summing it, we obtain

$$\frac{\mathrm{d}}{\mathrm{d}t}\int_\Omega \frac{\varrho}{2}|\ddot{u}|^2 + \frac{1}{2}\mathbb{A}\begin{pmatrix} e(\dot{u}) \\ \dot{z}\end{pmatrix}:\begin{pmatrix} e(\dot{u}) \\ \dot{z}\end{pmatrix}\mathrm{d}x + \int_\Omega \mathbb{D}e(\ddot{u}):e(\ddot{u})\,\mathrm{d}x \\ = \int_\Omega \dot{f}\cdot\ddot{u}\,\mathrm{d}x + \int_{\Gamma_\mathrm{N}} \dot{g}\cdot\ddot{u}\,\mathrm{d}S + \langle \dot{\vec{r}}_\mathrm{D}, \ddot{u}_\mathrm{D}\rangle_{\Gamma_\mathrm{D}}. \tag{7.3.12}$$

Here, if $\mathbb{D} = 0$, one must use after an integration over $[0,t]$ also the by-part integration of the term $\dot{g}\cdot\ddot{u}$, similarly as in (6.3.7). It is important that, to facilitate usage of the Gronwall inequality, we must ensure

$$\ddot{u}(0) = \frac{\mathrm{div}\,\sigma(0) + f(0)}{\varrho} \\ = \frac{\mathrm{div}((\mathbb{C}e(u_0)) - \mathbb{X}z_0) + \zeta_{\mathrm{VI}}'(e(v_0)) + f(0)}{\varrho} \in L^2(\Omega;\mathbb{R}^d) \quad \text{and} \tag{7.3.13a}$$

$$\dot{z}(0) = \partial\zeta_{\mathrm{IN}}^*(\mathbb{X}^\top e(u_0) - \mathbb{B}z_0) \in L^2(\Omega;\mathbb{R}^m), \tag{7.3.13b}$$

[10] Very formally, we use that the (set-valued) Jacobian $\partial^2\zeta_{\mathrm{IN}}(\dot{z})$ is positive definite. This very formal argument can be made rigorous most easily by using the Rothe method, cf. e.g. [456, Sect.11.1.1].

cf. (7.3.2b), and $\dot{u}(0) = v_0 \in H^1(\Omega;\mathbb{R}^d)$ assumed in (6.3.12c). Here we simply qualify[11] $\dot{z}(0) = 0$. This eventually yields the estimates (7.3.10a-c).

The energy conservation now follows from that $\ddot{u} \in L^\infty(I;L^2(\Omega;\mathbb{R}^d))$ certainly makes $\sqrt{\varrho}\ddot{u}$ in duality with $\sqrt{\varrho}\dot{u}$ and that $(e(\dot{u}),\dot{z}) \in L^\infty(I;L^2(\Omega;\mathbb{R}^{d\times d}_{\rm sym}\times\mathbb{R}^m))$ is certainly in duality with $\varphi'(e(u),z)$, so that the by-part integration in time leading to (7.1.7) integrated over $[0,t]$.

Moreover, testing the difference of weak formulations for two weak solutions of the equilibrium equation by the difference of solutions is now legitimate and testing the weak formulations of the flow-rule for z by the opposite solution (cf., e.g., [456, Lemma 10.1]), we can show uniqueness of the weak solution.

Eventually, the last estimate (7.3.10d) is by comparison $\dddot{u} = (\mathrm{div}(\mathbb{C}e(\dot{u}) - \mathbb{X}\dot{z} + \zeta_{\rm VI}''(e(\dot{u}))e(\ddot{u})) + \dot{f})/\varrho$, using also (7.3.10c). □

Remark 7.3.4 (Rothe method). The alternative approximation method by time discretization is quite competitive because it does not need too urgently the regularization of $\zeta_{\rm IN}$. The particular boundary-value problems on each time level are then variational inequalities rather than equations if $\zeta_{\rm IN}$ is not smooth, but efficient numerical techniques (after further space discretization) exist.

Exercise 7.3.5 (Energy-conserving time discretization). Devise a time discretization that would conserve energy by the *Crank-Nicolson formula*, i.e. imitate the energy balance (7.2.17a) in its discrete variant as an equality

$$\int_\Omega \frac{\varrho}{2}|v_\tau^{k-1/2}|^2 + \frac{1}{2}\mathbb{A}\binom{e(u_\tau^k)}{z_\tau^k}:\binom{e(u_\tau^k)}{z_\tau^k}\,\mathrm{d}x + \sum_{l=1}^{k}\int_\Omega \zeta_{\rm IN}(v_\tau^{l-1/2})\,\mathrm{d}x$$
$$= \int_\Omega \frac{\varrho}{2}|v_0|^2 + \frac{1}{2}\mathbb{A}\binom{e(u_0)}{z_0}:\binom{e(u_0)}{z_0}\,\mathrm{d}x$$
$$+\sum_{l=1}^{k}\Big(\int_\Omega f_\tau^l\cdot v_\tau^{l-1/2}\,\mathrm{d}x + \int_\Gamma g_\tau^l\cdot v_\tau^{l-1/2}\,\mathrm{d}S\Big) \quad \text{with } v_\tau^{k-1/2} = \frac{u_\tau^k - u_\tau^{k-1}}{\tau}.$$

7.3.2 Nonquadratic, nonconvex stored energies, gradient theories

An alternative useful situation considers nonlinear static response governed by a nonquadratic, possibly nonconvex potential $\varphi = \varphi(e,z)$. Particular situations are:

[11] Alternatively, assuming $\zeta_{\rm IN}$ is coercive in the sense $\zeta_{\rm IN}(z) \geq \epsilon|z|^p$ with $p \geq 2$ so that $\partial\zeta_{\rm IN}^*$ has at most a linear growth, i.e. $\zeta_{\rm IN}^*$ has at most a quadratic growth, we can omit the requirement (6.3.12c) with $z_0 \in H^1(\Omega;\mathbb{R}^m)$.

(A) φ convex possibly nonquadratic and while the dissipative response is linear, i.e. ζ quadratic, cf. also Remark 6.4.14; this allows for using monotonicity methods.
(B) $\varphi(\cdot,z)$ quadratic while $\zeta_{\rm VI}$ nonquadratic and a gradient theory for z-variable is used; the method of Proposition 7.3.2 can be used and φ and even $\varphi(e,\cdot)$ can be nonconvex.
(C) $\varphi(\cdot,z)$ convex but nonquadratic while $\zeta_{\rm VI}$ quadratic, and again $\varphi(e,\cdot)$ can be even nonconvex.

Ad (A), we consider a quadratic dissipation potential allowing even cross-effects in the dissipative part, namely

$$\zeta(\dot e,\dot z) := \frac{1}{2}\mathbb{D}\begin{pmatrix} e(\dot u)\\ \dot z\end{pmatrix}\colon\begin{pmatrix} e(\dot u)\\ \dot z\end{pmatrix} \tag{7.3.14}$$

for some $\mathbb{D}\in\mathbb{R}^{(d\times d+m)^2}$ symmetric positive definite. Considering a Galerkin approximation (u_k,z_k), the essence of this situation is seen from the following estimate where the test by (u_k-u,z_k-z) is used:

$$\begin{aligned}
\int_\Omega \frac{1}{2}\mathbb{D}\begin{pmatrix} e(u_k(t){-}u(t))\\ z_k(t){-}z(t)\end{pmatrix}\colon\begin{pmatrix} e(u_k(t){-}u(t))\\ z_k(t){-}z(t)\end{pmatrix}\mathrm{d}x &= \int_0^t\!\!\int_\Omega \mathbb{D}\begin{pmatrix} e(\dot u_k{-}\dot u)\\ \dot z_k{-}\dot z\end{pmatrix}\colon\begin{pmatrix} e(u_k{-}u)\\ z_k{-}z\end{pmatrix}\mathrm{d}x\mathrm{d}t\\
\le \int_0^t\!\!\int_\Omega \mathbb{D}\begin{pmatrix} e(\dot u_k{-}\dot u)\\ \dot z_k{-}\dot z\end{pmatrix}\colon&\begin{pmatrix} e(u_k{-}u)\\ z_k{-}z\end{pmatrix} + (\partial_e\varphi(e(u_k),z_k)-\partial_e\varphi(e(u),z))\colon e(u_k{-}u)\\
&+(\partial_z\varphi(e(u_k),z_k)-\partial_z\varphi(e(u),z))\cdot(z_k{-}z)\,\mathrm{d}x\mathrm{d}t\\
= \int_0^t\!\!\int_\Omega f\cdot(u_k{-}u) - \mathbb{D}\begin{pmatrix} e(\dot u)\\ \dot z\end{pmatrix}\colon&\begin{pmatrix} e(u_k{-}u)\\ z_k{-}z\end{pmatrix} - \partial_e\varphi(e(u),z)\colon e(u_k{-}u)\\
-\partial_z\varphi(e(u),z)\cdot(z_k{-}z) + \varrho\dot u_k\cdot(\dot u_k-\dot u)\,\mathrm{d}x\mathrm{d}t &+ \int_0^t\!\!\int_{\Gamma_{\rm N}} g\cdot(u_k-u)\,\mathrm{d}S\,\mathrm{d}t\\
&- \int_\Omega \varrho\dot u_k(t)\cdot(u_k(t)-u(t))\,\mathrm{d}x \to 0.
\end{aligned} \tag{7.3.15}$$

When taken (u,z) an approximation of the limit of a selected converging subsequence $\{(u_k,z_k)\}_{k\in\mathbb{N}}$ living in the finite-dimensional subspaces, the mentioned test of the Galerkin solutions is legitimate and, in addition, the indicated convergence towards 0 in (7.3.15) can be achieved; without loss of generality, we can assume that the initial conditions live in all these finite-dimensional subspaces. Moreover, for this convergence we use the weak convergence of $\dot u_k(t)\to\dot u(t)$ in $L^2(\Omega;\mathbb{R}^d)$ and the strong convergence of $u_k(t)\to u(t)$ in $L^2(\Omega;\mathbb{R}^d)$; here we use the Rellich compact embedding $H^1(\Omega)\Subset L^2(\Omega)$. Thus, considering all $t\in I$, we can see the strong convergence

$$u_k\to u \quad\text{in } L^\infty(I;H^1(\Omega;\mathbb{R}^d)) \quad\text{and}\quad z_k\to z \quad\text{in } L^\infty(I;L^2(\Omega;\mathbb{R}^m)).$$

The limit passage towards weak solution to the system

$$\varrho\ddot u - {\rm div}(\mathbb{D}_{ee}e(\dot u)+\partial_e\varphi(e(u),z)) = f - {\rm div}(\mathbb{D}_{ze}\dot z) \quad\text{and} \tag{7.3.16a}$$

$$\mathbb{D}_{zz}\dot z + \partial_z\varphi(e(u),z) = \mathbb{D}_{ez}e(\dot u) \tag{7.3.16b}$$

together with the corresponding boundary conditions, where we used the notation $\mathbb{D} = \begin{pmatrix} \mathbb{D}_{ee} & -\mathbb{D}_{ze} \\ -\mathbb{D}_{ez} & \mathbb{D}_{zz} \end{pmatrix}$.

For the scenario (B) and (C), we consider now the option that we regularize the problem by involving gradients in φ. Options are to involve gradients in the internal variables as in (7.2.18) or in strains (using the concept of nonsimple materials from Sect. 5.4 or both.[12] The modeling goal is to involve some internal length-scale while the mathematical goal is to make a certain regularization so that the qualification of the data can be weakened. Typically, a certain nonconvexity of the stored energy can be allowed by this way. Here we will consider the former option, i.e. an enhanced stored energy $\varphi_{\rm E}$ from (7.2.18) with $\varphi_1(Z) := \frac{1}{2}\mathbb{Z}Z{:}Z$. The 4th-order symmetric tensor $\mathbb{Z} = \mathbb{Z}(x)$, namely $\mathbb{Z} \in L^\infty(\Omega; {\rm SLin}(\mathbb{R}^{d\times m}))$, positive definite uniformly in $x \in \Omega$, determines a certain length-scale in spatial profiles of the internal-variable field z. This yields an additional term ${\rm div}\,(\mathbb{Z}\nabla z)$ in (7.2.1b).

In case of (B) when $\varphi(\cdot, z)$ is quadratic, we consider $\varphi(e,z) = \frac{1}{2}\mathbb{C}(z)e{:}e$ with $\mathbb{C}: \mathbb{R}^m \to {\rm SLin}(\mathbb{R}^{d\times d}_{\rm sym})$, so that altogether

$$\varphi_{\rm E}(e, z, \nabla z) = \varphi(e,z) + \frac{1}{2}\mathbb{Z}\nabla z{:}\nabla z. \tag{7.3.17}$$

We thus arrive to the system[13]

$$\varrho\ddot{u} - {\rm div}\,(\zeta'_{\rm VI}(e(\dot{u})) + \mathbb{C}(z)e(u)) = f, \tag{7.3.18a}$$
$$\partial\zeta_{\rm IN}(\dot{z}) + \mathbb{C}'(z)e(u){:}e(u) - {\rm div}(\mathbb{Z}\nabla z) \ni 0. \tag{7.3.18b}$$

Now, beside the initial/boundary conditions (7.2.1c-e), we must prescribe boundary conditions for z. Usually, just homogeneous Neumann condition $(\mathbb{Z}\nabla z)\cdot\vec{n} = 0$ is considered on the whole boundary Σ, cf. (7.2.19b). The mathematical treatment relies on the linearity of (7.3.18a) in terms of u but, on the other hand, (7.3.18b) is nonlinear in terms of $e = e(u)$. A prototype of this structure is a damage-type problems which will be treated in Sect. 7.5.2. The latter option, i.e. involving the strain gradient, will be discussed in Sect. 7.5.3.

As to the scenario (C), we will consider an enhanced stored energy $\varphi_{\rm E}$ and linear Kelvin-Voigt attenuation by the viscous moduli $\mathbb{D}$, i.e.

$$\varphi_{\rm E}(e, z, \nabla z) = \varphi(e,z) + \frac{1}{2}\mathbb{Z}\nabla z{:}\nabla z \quad \text{and} \quad \zeta_{\rm VI}(\dot{e}) = \frac{1}{2}\mathbb{D}\dot{e}{:}\dot{e}. \tag{7.3.19}$$

We arrive to the system

$$\varrho\ddot{u} - {\rm div}\,(\mathbb{D}e(\dot{u}) + \partial_e\varphi(e(u), z)) = f, \tag{7.3.20a}$$
$$\partial\zeta_{\rm IN}(\dot{z}) + \partial_z\varphi(e(u), z) - {\rm div}(\mathbb{Z}\nabla z) \ni 0 \tag{7.3.20b}$$

[12] The so-called "gradient theories" in solids has been originated probably by E. Aifantis, see [4].

[13] The term $-{\rm div}\,(\mathbb{Z}\nabla z)$ in (7.3.18b) arises as the Gâteaux differential of $z \mapsto \int_\Omega \frac{1}{2}\mathbb{Z}\nabla z{:}\nabla z\,{\rm d}x$.

to be accompanied again with the initial/boundary conditions (7.2.1c-e) with $(\mathbb{Z}\nabla z)\cdot\vec{n} = 0$ on Σ. We assume the coercivity and growth of φ and "p-monotonicity" of $\partial_e\varphi(\cdot,z)$ uniformly with respect to z in the sense:[14]

$$\varphi : \mathbb{R}^{d\times d}_{\rm sym}\times\mathbb{R}\to\mathbb{R} \quad \text{continuously differentiable,} \tag{7.3.21a}$$

$$\forall(e,\tilde{e},z)\in\mathbb{R}^{d\times d}_{\rm sym}\times\mathbb{R}^{d\times d}_{\rm sym}\times\mathbb{R}^m: \quad \varphi(e,z)\geq c_0(|e|^p+|z|^q), \tag{7.3.21b}$$

$$|\varphi'(e,z)|\leq C(1+|e|^{p-1}+|z|^{q-1}), \quad \text{and} \tag{7.3.21c}$$

$$(\partial_e\varphi(e,z)-\partial_e\varphi(\tilde{e},z))\geq c(|e|^{p-2}e-|\tilde{e}|^{p-2}\tilde{e}){:}(e-\tilde{e}). \tag{7.3.21d}$$

We consider the smoothing of $\zeta_{\rm IN}$ as in (7.3.5), if needed, and then the Galerkin approximation $(u_{\varepsilon,k},z_{\varepsilon,k})$.

Proposition 7.3.6 (Existence of weak solutions to (7.3.20)). *Let (6.4.5) without considering $\mathbb{C}$ hold but with $u_0\in L^{\max(2,p)}(\Omega;\mathbb{R}^{d\times d}_{\rm sym})$, (7.3.21), and $z_0\in L^q(\Omega;\mathbb{R}^m)\cap H^1(\Omega;\mathbb{R}^m)$ hold, and let $\zeta_{\rm IN}:\mathbb{R}^m\to\mathbb{R}\cup\{+\infty\}$ be convex lower semicontinuous with $\zeta_{\rm IN}(\cdot)\geq\varepsilon|\cdot|^q$ for some $\varepsilon>0$. Then the Galerkin approximation $(u_{\varepsilon,k},z_{\varepsilon,k})$ satisfies the a-priori estimates*

$$\|\dot{u}_{\varepsilon,k}\|_{L^\infty(I;L^2(\Omega;\mathbb{R}^d))}\leq C, \quad \|e(u_{\varepsilon,k})\|_{L^\infty(I;L^p(\Omega;\mathbb{R}^{d\times d}))\cap H^1(I;L^2(\Omega;\mathbb{R}^{d\times d}))}\leq C,$$
$$\|z_{\varepsilon,k}\|_{L^\infty(I;L^q(\Omega;\mathbb{R}^m)\cap H^1(\Omega;\mathbb{R}^m))\cap H^1(I;L^2(\Omega))}\leq C. \tag{7.3.22a}$$

Any weakly converging subsequence in the topologies indicated in (7.3.22) yields as its limit (u,z) a weak solution to the initial-boundary-value problem (7.3.20) with (7.2.1c-e) and $(\mathbb{Z}\nabla z)\cdot\vec{n}=0$ on Σ, and moreover*

$$e(u_{\varepsilon,k})\to e(u) \qquad \textit{strongly in } L^2(Q;\mathbb{R}^{d\times d}_{\rm sym}). \tag{7.3.23}$$

Proof. Consider the Galerkin approximation $(u_{\varepsilon,k},z_{\varepsilon,k})$. Basic a-priori estimates of this model can be based on the basic energetic estimate as in (7.1.5), i.e. the test of the balance equation (6.1.1) by the velocity $\dot{u}$ and the flow rule by $\dot{z}$. We estimate the right-hand side by the Hölder, Young, and Gronwall inequalities. Thus we get bounds in (7.3.22).

In addition, by comparison, we can obtain the estimate for $\ddot{u}$ in $L^2(I;W^{1,p}(\Omega;\mathbb{R}^d)^*)+L^2(I;H^1(\Omega;\mathbb{R}^d)^*)$ in the sense of Remark C.2.9 on p.547.

Then, using (7.3.21d), we can prove strong convergence (7.3.23): Consider a subsequence of $(u_{\varepsilon,k},z_{\varepsilon,k})$ converging to (u,z) weakly*. By p-monotonicity (7.3.21d) of φ', by testing the equilibrium equation (6.1.1) with σ from (7.1.1a) by $u_{\varepsilon,k}-u$, we have

$$c\Big(\|e(u_{\varepsilon,k})\|^{p-1}_{L^p(Q;\mathbb{R}^{d\times d})}-\|e(u)\|^{p-1}_{L^p(Q;\mathbb{R}^{d\times d})}\Big)\Big(\|e(u_{\varepsilon,k})\|_{L^p(Q;\mathbb{R}^{d\times d})}-\|e(u)\|_{L^p(Q;\mathbb{R}^{d\times d})}\Big)$$
$$\leq c\int_Q\Big(|e(u_{\varepsilon,k})|^{p-2}e(u_{\varepsilon,k})-|e(u)|^{p-2}e(u)\Big){:}(e(u_{\varepsilon,k})-e(u))\,\mathrm{d}x\mathrm{d}t$$

---

[14] An example $\varphi(e,z)=\phi(z)|e|^p$ with $p>1$ satisfies (7.3.21d) with $c:=p\inf_{\mathbb{R}^m}\phi(\cdot)>0$.

$$\leq \int_Q (\partial_e\varphi(e(u_{\varepsilon,k}),z_{\varepsilon,k}) - \partial_e\varphi(e(u),z_{\varepsilon,k})):e(u_{\varepsilon,k}-u)\,\mathrm{d}x\mathrm{d}t$$
$$= \int_Q f\cdot(u_{\varepsilon,k}-u) - (\mathbb{D}e(\dot{u}_{\varepsilon,k}) + \partial_e\varphi(e(u),z_{\varepsilon,k})):e(u_{\varepsilon,k}-u) + \varrho\dot{u}_{\varepsilon,k}\cdot(\dot{u}_{\varepsilon,k}-\dot{u})\,\mathrm{d}x\mathrm{d}t$$
$$+ \int_{\Sigma_\mathrm{N}} g\cdot(u_{\varepsilon,k}-u)\,\mathrm{d}S\,\mathrm{d}t - \int_\Omega \varrho\dot{u}_{\varepsilon,k}(T)\cdot(u_{\varepsilon,k}(T)-u(T))\,\mathrm{d}x; \qquad (7.3.24)$$

here the first inequality is due to the Hölder inequality. Then we estimate limit superior from above, showing that it is nonpositive (so that, in fact, the limit exists and equals to 0). To this goal we use the linearity of the dissipation to have[15]

$$\liminf_{\varepsilon\to0,k\to\infty} \int_Q \mathbb{D}e(\dot{u}_{\varepsilon,k}):e(u_{\varepsilon,k}-u)\,\mathrm{d}x\mathrm{d}t$$
$$= \liminf_{\varepsilon\to0,k\to\infty} \frac{1}{2}\int_\Omega \mathbb{D}e(u_{\varepsilon,k}(T)):e(u_{\varepsilon,k}(T)) - \mathbb{D}e(u_0):e(u_0)\,\mathrm{d}x - \int_Q \mathbb{D}e(\dot{u}_{\varepsilon,k}):e(u)\,\mathrm{d}x\mathrm{d}t$$
$$\geq \int_\Omega \mathbb{D}e(u(T)):e(u(T)) - \mathbb{D}e(u_0):e(u_0)\,\mathrm{d}x - \int_Q \mathbb{D}e(\dot{u}):e(u)\,\mathrm{d}x\mathrm{d}t$$
$$= \int_Q \mathbb{D}e(\dot{u}):e(u-u)\,\mathrm{d}x\mathrm{d}t = 0; \qquad (7.3.25)$$

here we have used that the strain rate $e(\dot{u}) \in L^2(Q;\mathbb{R}^{d\times d}_{\mathrm{sym}})$ is certainly in duality with $e(u) \in L^\infty(I;L^{\max(2,p)}(\Omega;\mathbb{R}^{d\times d}_{\mathrm{sym}}))$. Also, we have used strong convergence of $\dot{u}_{\varepsilon,k}$ due to the Aubin-Lions' theorem B.4.10 as well as weak convergence of $\dot{u}_{\varepsilon,k}(T)$ in $L^2(\Omega;\mathbb{R}^d)$ and strong convergence of $u_{\varepsilon,k}(T)$ in $L^2(\Omega;\mathbb{R}^d)$; here we use the compact embedding $H^1(\Omega) \Subset L^2(\Omega)$ similarly as we did for (7.3.15). Also, we use the strong convergence in $z_{\varepsilon,k}$ by Aubin-Lions theorem B.4.10 so that $\partial_e\varphi(e(u),z_{\varepsilon,k}) \to \partial_e\varphi(e(u),z_{\varepsilon,k})$ in $L^{p'}(Q;\mathbb{R}^{d\times d}_{\mathrm{sym}})$ while $e(u_{\varepsilon,k}-u) \to 0$ weakly in $L^p(Q;\mathbb{R}^{d\times d}_{\mathrm{sym}})$ so that $\int_Q \partial_e\varphi(e(u),z_{\varepsilon,k}):e(u_{\varepsilon,k}-u)\,\mathrm{d}x\mathrm{d}t \to 0$.

Actually, both (7.3.24) and (7.3.25) represent rather a conceptual strategy and, in the Galerkin approximation, u as well as the Dirichlet boundary conditions should be first approximated to be valued in the finite-dimensional subspaces used for the Galerkin method.

From $\|e(u_{\varepsilon,k})\|_{L^p(Q;\mathbb{R}^{d\times d})} \to \|e(u)\|_{L^p(Q;\mathbb{R}^{d\times d})}$ and from the weak convergence $e(u_{\varepsilon,k}) \to e(u)$, we can see that $e(u_{\varepsilon,k}) \to e(u)$ strongly in $L^p(Q;\mathbb{R}^{d\times d})$ exploiting uniform convexity of the Banach space $L^p(Q;\mathbb{R}^{d\times d})$ when equipped with the standard norm, cf. Theorem A.3.8 on p. 497.

Limit passage of the approximate solution $(u_{\varepsilon,k},z_{\varepsilon,k})$ in the equilibrium equation (7.3.18a) is then easy. For the limit passage in (7.3.18b), we use the weak formulation (7.2.8) which here takes the form

$$\int_Q \partial_z\varphi(e(u_{\varepsilon,k}),z_{\varepsilon,k})\cdot w + \mathbb{Z}\nabla z_{\varepsilon,k}:\nabla w + \zeta_{\mathrm{IN},\varepsilon}(w)\,\mathrm{d}x\mathrm{d}t$$

[15] Alternatively, the $\mathbb{D}$-term can be treated by monotonicity as in Remark 6.4.14.

$$\geq \int_Q \mathbb{D}e(\dot{u}_{\varepsilon,k}):e(\dot{u}_{\varepsilon,k}) + \zeta_{\text{IN},\varepsilon}(\dot{z}_{\varepsilon,k}) - f\cdot\dot{u}_{\varepsilon,k}\,\mathrm{d}x\mathrm{d}t$$
$$+ \int_\Omega \Big(\frac{\varrho}{2}|\dot{u}_{\varepsilon,k}(T)|^2 + \varphi(e(u_{\varepsilon,k}(T)), z_{\varepsilon,k}(T)) + \frac{1}{2}\mathbb{Z}\nabla z_{\varepsilon,k}(T):\nabla z_{\varepsilon,k}(T) - \frac{\varrho}{2}|v_0|^2$$
$$- \varphi(e(u_0), z_0) - \frac{1}{2}\mathbb{Z}\nabla z_0:\nabla z_0\Big)\mathrm{d}x - \int_{\Sigma_\mathrm{N}} g\cdot\dot{u}_{\varepsilon,k}\,\mathrm{d}S\,\mathrm{d}t - \int_0^T \langle \vec{\tau}_{k,\mathrm{D}}, \dot{u}_\mathrm{D}\rangle_{\Gamma_\mathrm{D}}\mathrm{d}t. \quad (7.3.26)$$

The limit passage towards (7.2.8) is then easy through continuity or semi-continuity, using also (7.3.5). □

Examples of nonconvex energies at small strains in damage models are in Section 7.5 or in Section 7.6.2 below.

Exercise 7.3.7. Modify the proof of Proposition 7.3.6 for $\varphi(\cdot,z)$ merely convex, omitting (7.3.21d).

7.3.3 Brief excursion into rate-independent processes

The attribute of rate independence reflects that a time scale of the variables (or some of variables, here internal ones) is much faster than the time scale of outer loading or of other variables and, with a certain accuracy, it is reasonable to consider these "rate-independent variables" to evolve arbitrarily fast if the driving force is big enough. To explain this attribute, let us fix e-variable in the flow rule (7.3.20b), which then results to

$$\partial\zeta_{\text{IN}}(\dot{z}) + \partial_z\hat{\varphi}(t,z) \ni \operatorname{div}(\mathbb{Z}\nabla z) \qquad \text{with} \qquad \hat{\varphi}(t,z) := \varphi(e(t),z). \qquad (7.3.27)$$

A system governed by (7.3.27) is called *rate-independent* if it is invariant under monotone reparametrizations of time.[16] It is related (or rather equivalent) with the attribute that the potential of dissipative forces ζ_{IN} is 1-*homogeneous*, meaning that $\zeta_{\text{IN}}(av) = a\zeta_{\text{IN}}(v)$ for $a \geq 0$. In fact, it is equivalent with the ansatz (7.2.13). A consequence is that $\dot{z}$ is to be expected rather as a measure on $\bar{Q}$ and its variation with respect to ζ_{IN} which now enters the definition of the weak solution is to be written as an integral with respect to this measure, i.e. $\int_{\bar{Q}} \zeta_{\text{IN}}(\cdot)\dot{z}(\mathrm{d}x\mathrm{d}t)$.

A reliable and relatively simple theory exists for such (7.3.27) with $\hat{\varphi}(t,\cdot)$ convex, while a richer and multi-conceptual theory is naturally needed for $\hat{\varphi}(t,\cdot)$ nonconvex or for a more general $\zeta_{\text{IN}} = \zeta_{\text{IN}}(z,\dot{z})$ with $\zeta_{\text{IN}}(z,\cdot)$ 1-homogeneous for any z, cf. [353, 360] and Remark 7.3.12. We will confine ourselves to the convex case but consider (7.3.27) only as a subsystem combined with a rate-dependent part (6.1.1), i.e. $\ddot{u} - \operatorname{div}\sigma = f$ with $\sigma = \zeta'_{\text{VI}}(\dot{e}) + \partial_e\varphi(e,z)$ with $e = e(u)$.

Let us present the rather simplest variant of existence results in this case. To this goal, let us still introduce the notion of a so-called *mutual recovery sequence*

[16] This means that, for any monotone continuous $r: I \to \mathbb{R}$, each solution $z: I \to \mathbb{R}^m$ to (7.3.27) yield a solution $z_r := z\circ r : r(I) \to \mathbb{R}^m$ to $\partial\zeta_{\text{IN}}(\dot{z}_r) + \partial_z\hat{\varphi}(r(t), z_r) \ni 0$.

$\{\hat{z}_k\}_{k\in\mathbb{N}}$: for arbitrary $\hat{z}\in\mathbb{R}^m$ and also for $(u_k,z_k)\to(u,z)$ for $k\to\infty$ in the mode to be specified, we require

$$\limsup_{k\to\infty}\int_\Omega \zeta_{\rm IN}(\hat{z}_k-z_k)+\varphi(e(u_k),\hat{z}_k)+\frac{1}{2}\mathbb{Z}\nabla\hat{z}_k{:}\nabla\hat{z}_k-\varphi(e(u_k),z_k)-\frac{1}{2}\mathbb{Z}\nabla z_k{:}\nabla z_k\,{\rm d}x$$
$$\le\int_\Omega \zeta_{\rm IN}(\hat{z}-z)+\varphi(e(u),\hat{z})+\frac{1}{2}\mathbb{Z}\nabla\hat{z}{:}\nabla\hat{z}-\varphi(e(u),z)-\frac{1}{2}\mathbb{Z}\nabla z{:}\nabla z\,{\rm d}x, \quad (7.3.28)$$

cf. [363]. This requirement needs an explicit construction which is to be tailored case by case, and might be even very complicated and also depends on the mentioned mode of convergence at disposal.

Proposition 7.3.8 (Existence of weak or energetic solutions, regularity). *Let $\zeta_{\rm VI}(\dot{e})=\frac{1}{2}\mathbb{D}\dot{e}{:}\dot{e}$, $\zeta_{\rm VI}$ be 1-homogeneous with $\zeta_{\rm VI}(\dot{z})\ge\epsilon|\dot{z}|$ for $\epsilon>0$, φ satisfy (7.3.21a-c), $\mathbb{Z}\in{\rm SLin}(\mathbb{R}^{d\times m})$ be positive definite, (6.4.5) not considering $\mathbb{C}$ hold but with $u_0\in W^{1,\max(2,p)}(\Omega;\mathbb{R}^{d\times d}_{\rm sym})$, and $z_0\in H^1(\Omega;\mathbb{R}^m)$ hold. Then:*

(i) The initial-boundary-value problem (7.3.20) with (7.2.1c-e) and with $(\mathbb{Z}\nabla z)\cdot\vec{n}=0$ on Σ possesses weak solutions.

(ii) If there are mutual recovery sequences in the sense (7.3.28) with respect to the strong×weak convergence on $W^{1,p}(\Omega;\mathbb{R}^d)\times H^1(\Omega;\mathbb{R}^m)$, then there is an energetic solution.

(iii) If φ is quadratic and the loading is smooth in the sense (6.4.14) and also (6.4.5a,b) holds, and $\partial\zeta^*_{\rm IN}$ has at most linear growth (i.e. if $S_{\rm EL}$ is bounded, cf. Fig. 7.1-right) or $\partial_z\varphi(e(u_0),z_0)\in-{\rm int}(\partial\zeta_{\rm IN}(0))$ then the solution (u,z) is regular in the sense (7.3.10). Moreover, then this solution is unique and even $\mathbb{Z}=0$ and $z_0\in L^2(\Omega;\mathbb{R}^m)$ can be considered when $H^1(\Omega;\mathbb{R}^m)$ is replaced by $L^2(\Omega;\mathbb{R}^m)$.

Sketch of the proof. Let us now use the Rothe method, which is more natural for the rate-independent processes than the Galerkin method; we use an equidistant partition with the time step $\tau>0$ in the formula (C.2.35) written now for (u,z) instead of u although a more sophisticated schemes as in Remark C.2.8 could be used, too. By a direct method involving the (nonsmooth) functional (C.2.36), we thus obtain an approximate solution (u_τ,z_τ).

The limit passage in the force equilibrium (7.2.5a) written for the approximate solutions in terms of the interpolants (see (C.2.15) on p. 536) as

$$\int_Q\big(\partial_e\varphi(e(\bar{u}_\tau),\bar{z}_\tau)+\mathbb{D}e(\dot{u}_\tau)\big){:}e(\bar{v}_\tau)-\varrho\dot{u}_\tau\cdot\dot{v}_\tau\,{\rm d}x{\rm d}t$$
$$=\int_\Omega\varrho v_0\cdot v(0)\,{\rm d}x+\int_Q\bar{f}_\tau\cdot v_\tau\,{\rm d}x{\rm d}t+\int_{\Sigma_{\rm N}}\bar{g}_\tau\cdot v_\tau\,{\rm d}S\,{\rm d}t \qquad (7.3.29)$$

for $v\in C(I;W^{1,p}(\Omega;\mathbb{R}^d))$, $v|_{\Sigma_{\rm D}}=0$, $v|_{t=T}=0$, $\bar{v}_\tau$ and v_τ denote its piecewise-constant and piecewise-affine interpolants, respectively; cf. (C.2.39) on p. 542.

The limit passage in the inequality (7.2.8) written for the approximate solutions (realizing also that $\xi_{\rm IN}(\cdot)=\zeta_{\rm IN}(\cdot)$ due to the 1-homogeneity) as

$$\int_Q \partial_z\varphi(e(\overline{u}_\tau),\overline{z}_\tau)\cdot w+\mathbb{Z}\nabla\overline{u}_\tau:\nabla w+\zeta_{\rm IN}(w)\,{\rm d}x{\rm d}t \geq \int_Q \xi_{\rm VI}(e(\dot{u}_\tau))+\zeta_{\rm IN}(\dot{z}_\tau)\,{\rm d}x{\rm d}t$$
$$+\int_\Omega\Big(\frac{\varrho}{2}\big|\dot{u}_\tau(T)\big|^2+\varphi(e(u_\tau(T)),z_\tau(T))+\frac{1}{2}\mathbb{Z}\nabla z_\tau(T):\nabla z_\tau(T)-\frac{\varrho}{2}|v_0|^2-\varphi(e(u_0),z_0)$$
$$-\frac{1}{2}\mathbb{Z}\nabla z_0:\nabla z_0\Big){\rm d}x-\int_Q\overline{f}_\tau\cdot\dot{u}_\tau\,{\rm d}x{\rm d}t-\int_{\Sigma_{\rm N}}\overline{g}_\tau\cdot\dot{u}_\tau\,{\rm d}S\,{\rm d}t-\int_0^T\langle\vec{\tau}_{{\rm D},\tau},\dot{u}_{\rm D}\rangle_{\Gamma_{\rm D}}{\rm d}t \quad (7.3.30)$$

is by the weak* lower semicontinuity of the functional $\dot{z}\mapsto\int_{\bar{Q}}\zeta_{\rm IN}(\cdot)\dot{z}({\rm d}x{\rm d}t)$ on $\mathfrak{M}(\bar{Q})$, while the other terms are standard either by weak lower semicontinuity or by weak continuity $\partial_z\varphi(e(\overline{u}_\tau),\overline{z}_\tau)\to\partial_z\varphi(e(u),z)$ due to gradient theory for z-variable (hence strong convergence $\overline{z}_\tau\to z$) and also uniform convexity of $\zeta_{\rm VI}$ so that $e(\overline{u}_\tau)\to e(u)$ as in (7.3.15).

As for (ii), like (7.2.14)–(7.2.15) but here with the gradient term and integrated, we obtain semistability for the time-discrete solution, i.e.

$$\int_Q\varphi(e(\overline{u}_\tau),\overline{z}_\tau)+\frac{1}{2}\mathbb{Z}\nabla\overline{z}_\tau\cdot\nabla\overline{z}_\tau\,{\rm d}x{\rm d}t\leq\int_Q\varphi(e(\overline{u}_\tau),\tilde{z})+\zeta_{\rm IN}(\tilde{z}-\overline{z}_\tau)+\frac{1}{2}\mathbb{Z}\nabla\tilde{z}\cdot\nabla\tilde{z}\,{\rm d}x{\rm d}t \quad (7.3.31)$$

for any $\tilde{z}\in L^\infty(I;H^1(\Omega;\mathbb{R}^m))$. The limit passage in (7.3.31) is due to the mentioned strong convergence $e(\overline{u}_\tau)\to e(u)$ and the assumed existence of a mutual recovery sequence.

From (7.3.29) written for $w=0$ gives the energy inequality (7.2.17a) written for the approximate solution, when using also the identity $\int_Q\zeta_{\rm IN}(\dot{z}_\tau)\,{\rm d}x{\rm d}t={\rm Var}_{S_{\rm EL}}(\overline{z}_\tau;I)$. Then we can pass to the limit by weak lower semicontinuity.

By the Helly's selection-principle (Theorem B.4.13 on p. 526), we can also rely on that

$$\forall t\in I:\qquad z_\tau(t)\to z(t)\quad\text{weakly in }\;H^1(\Omega;\mathbb{R}^m). \quad (7.3.32)$$

In view of the definition (7.2.16), for any $\epsilon>0$ there is a partition $(t_i)_{i=1}^N$ of I such that ${\rm Var}_{S_{\rm EL}}(z;I)\leq\sum_{i=1}^N\int_\Omega\zeta_{\rm IN}(z(t_i,x)-z(t_{i-1},x))\,{\rm d}x-\epsilon$. Due to (7.3.32) and the weak lower semicontinuity of $\zeta_{\rm IN}$ on $H^1(\Omega;\mathbb{R}^m)$, we have $\limsup_{\tau\to0}\zeta_{\rm IN}(\overline{z}_\tau(t_i)-\overline{z}_\tau(t_{i-1}))\to\zeta_{\rm IN}(z(t_i)-z(t_{i-1}))$. Then we use that ${\rm Var}_{S_{\rm EL}}(\overline{z}_\tau;I)=\sum_{i=1}^N\zeta_{\rm IN}(\overline{z}_\tau(t_i)-\overline{z}_\tau(t_{i-1}))$ provided $\tau<\min_{1\leq i\leq N}t_i-t_{i-1}$.

As for (iii), the regularity is proved by differentiation of the equations in time and testing by the second time derivative of the solution. The only nonlinear term is the inelastic dissipation in z-variable, and it does not corrupt the estimate because of its good sign, i.e. formally $(\partial^2\zeta_{\rm IN}(\dot{z}_\tau))^{\boldsymbol\cdot}\cdot\dddot{z}_\tau=\partial^2\zeta_{\rm IN}(\dot{z}_\tau)\ddot{z}_\tau\cdot\dddot{z}_\tau\geq0$ because $\zeta_{\rm IN}$ is convex. More rigorously, we take the difference of the time-discrete scheme and this scheme shifted by $-\tau$, and test it be the second time difference, which give the term $(\partial\zeta_{\rm IN}(\dot{z}_\tau)-\partial\zeta_{\rm IN}(\dot{z}_\tau(\cdot-\tau)))\cdot(\dot{z}_\tau-\dot{z}_\tau(\cdot-\tau))/\tau\geq0$ due to the monotonicity of $\partial\zeta_{\rm IN}$. Here we also used the imposed qualification of the initial condition, cf. also the proof of Proposition 7.3.3. The uniqueness of the solution is as in the linear materials (see e.g. Proposition 6.5.2) not being corrupted by the nonlinearity $\partial\zeta_{\rm IN}$ which is monotone. □

Let us consider a convex closed set $S_{\rm EL} \subset \mathbb{R}^m$ containing origin 0 as used already in (7.2.13); in general, the set $S_{\rm EL}$ need not be bounded, neither 0 lies in its interior. Considering $\delta_{S_{\rm EL}}$ the indicator functional of $S_{\rm EL}$, its Legendre-Fenchel conjugate $\delta^*_{S_{\rm EL}}$ is

$$\delta^*_{S_{\rm EL}}(\dot{z}) = \sup_{\sigma\in\mathbb{R}^m} \sigma\cdot\dot{z} - \delta_{S_{\rm EL}}(\sigma) = \sup_{\sigma\in S_{\rm EL}} \sigma\cdot\dot{z}, \tag{7.3.33}$$

cf. (A.3.16) on p. 498. The important attribute is that $\delta^*_{S_{\rm EL}}$ is 1-homogeneous (so that its subdifferential $\partial\delta^*_{S_{\rm EL}}$ is 0-homogeneous (which corresponds to a rate-independent response of this particular rheological element) and nonsmooth at 0 (which corresponds to activation character of the evolution process of this internal variable), cf. again Fig. 7.1 on p. 253. As in (7.2.25), we can write the flow rule (7.1.1b) equivalently as

$$(\omega + \partial_z\varphi(e,z))\cdot(w - \dot{z}) \geq 0 \quad \forall w, \omega \in \mathbb{R}^m: \quad \omega \in \partial\delta^*_{S_{\rm EL}}(w). \tag{7.3.34}$$

In particular, (7.1.1b) for $w = 0$ yields $\omega\cdot\dot{z} \leq -\partial_z\varphi(e,z)\cdot\dot{z}$ for all $\omega \in \partial\delta^*_{S_{\rm EL}}(0)$ from which it follows that

$$\partial_z\varphi(e,z)\cdot\dot{z} = \max_{\omega\in S_{\rm EL}} \omega\cdot\dot{z} \qquad \text{a.e. on } Q, \tag{7.3.35}$$

where we also used the identity[17]

$$\begin{aligned}\partial\delta^*_{S_{\rm EL}}(0) &= \{\dot{z}\in\mathbb{R}^m;\ \forall v\in\mathbb{R}^m:\ \delta^*_{S_{\rm EL}}(v) \geq \dot{z}\cdot v\} \\ &= \Big\{\dot{z}\in\mathbb{R}^m;\ \forall v\in\mathbb{R}^m:\ \sup_{\sigma\in S_{\rm EL}} \sigma\cdot v \geq \dot{z}\cdot v\Big\} = S_{\rm EL}\end{aligned} \tag{7.3.36}$$

and the fact that always the "driving force" $\sigma_z := -\partial_z\varphi \in S_{\rm EL}$ because, by the flow-rule (7.1.1b) with (7.1.3) considered with $\zeta_{\rm IN} = \delta^*_{S_{\rm EL}}$, we have

$$\sigma_z := -\partial_z\varphi(e,z) \in \partial\delta^*_{S_{\rm EL}}(\dot{z}) \subset \partial\delta^*_{S_{\rm EL}}(0) = S_{\rm EL} \tag{7.3.37}$$

where also the degree-1 homogeneity of $\delta^*_{S_{\rm EL}}$ is used. The identity (7.3.35) says that the dissipation due to the "back stress" σ_z is maximal provided that the rate $\dot{z}\in\mathbb{R}^m$ is kept fixed while the vector of "driving stresses" ω (of physical dimension $\mathrm{Jm^{-3}}$=Pa) varies freely over all admissible driving stresses from $S_{\rm EL}$. This just resembles so-called Hill's *maximum-dissipation principle* [251].

In terms of (7.1.2), the flow rule (7.1.1b) can be expressed as

[17] The first equation in (7.3.36) uses also $\delta^*_{S_{\rm EL}}(0) = 0$. The middle equation is just (7.3.33). The last equation in (7.3.36) follows from that, on the one hand, for any $\dot{z} \in S_{\rm EL}$, it holds $\sup_{\sigma\in S_{\rm EL}} \sigma\cdot v \geq \dot{z}\cdot v$ for any $v\in\mathbb{R}^m$, and, on the other hand, for any $\dot{z}\in\mathbb{R}^m \setminus S_{\rm EL}$, by Hahn-Banach theorem A.1.4, there is some linear functional v which separates $\dot{z}$ from $S_{\rm EL}$, which means $\sup_{\sigma\in S_{\rm EL}} \sigma\cdot v < \dot{z}\cdot v$.

$$\dot{z} \in [\partial\delta^*_{S_{\rm EL}}]^{-1}(-\partial_z\varphi(e,z)) = \partial\delta_{S_{\rm EL}}(-\partial_z\varphi(e,z)) = N_{S_{\rm EL}}(-\partial_z\varphi(e,z)) \tag{7.3.38}$$

where $N_{S_{\rm EL}}(\sigma_z)$ denotes the normal cone to $S_{\rm EL}$ at the "driving stress" $\sigma_z = -\partial_z\varphi(e,z)$, i.e.

$$N_{S_{\rm EL}}(\sigma_z) := \begin{cases} \{\dot{z}\in\mathbb{R}^m;\ \forall w \in S_{\rm EL}:\ \dot{z}\cdot(w-\sigma_z) \le 0\} & \text{if } \sigma_z \in S_{\rm EL}, \\ \emptyset & \text{if } \sigma_z \notin S_{\rm EL}. \end{cases} \tag{7.3.39}$$

In particular, if $-\partial_z\varphi(e,z)$ is in the interior of $S_{\rm EL}$, then $N_{S_{\rm EL}}(-\partial_z\varphi(e,z)) = \{0\}$, so that the internal variable z does not evolve unless the driving force $\sigma_z = -\partial_z\varphi(e,z)$ reaches a certain threshold, i.e. here the boundary of $S_{\rm EL}$. We call such a *process* as *activated*. Then, alternatively, the maximum-dissipation principle can be expressed as normality in the sense that the plastic-strain rate $\dot{z}$ belongs to the cone of outward normals to the elasticity domain $S_{\rm EL}$, see also [318] or [241, Sect.3.2], [340, Sect.2.4.4] or [486, Sect.2.6]. On the other hand, the maximum-dissipation principle itself does not say much and yield the really complete information only when combined with a (local) stability saying that[18]

$$\forall v\in\mathbb{R}^m:\quad \zeta_{\rm IN}(v) \ge \partial_z\varphi\cdot v. \tag{7.3.40}$$

If combined with the maximum-dissipation principle, then indeed $-\partial_z\varphi \in \partial\zeta_{\rm IN}(\dot{z})$ is seen.

In view of Definition 7.2.4, the semistability can be understood as (and, for some advanced results which needs energy conservation, really is) a useful and vital property to be proved in particular cases. To this goal, one has usually to pass to the limit in some approximate solution, say (u_τ, z_τ) with a (here just abstract) parameter $\tau > 0$ ranging a countable set with the accumulation point 0, for which semistability holds and to construct a mutual recovery sequence in the sense (7.3.28). A typical approximation way that leads to semistability is a time-discretization and then global minimization in the discretized flow rule. Here, it leads to minimization of the functional

$$z \mapsto \int_\Omega \zeta_{\rm IN}(z - z_\tau^{k-1}) + \varphi(e(u_\tau^k), z)\,{\rm d}x, \tag{7.3.41}$$

assuming that $e(u_\tau^k)$ known from a *fractional-step method*, cf. Remark C.2.8 on p. 544. Assuming convexity and lower-semicontinuity of $\varphi(e,\cdot)$, a minimizer denoted by z_τ^k typically does exist by weak-compactness arguments, and satisfies

$$\int_\Omega \zeta_{\rm IN}(z_\tau^k - z_\tau^{k-1}) + \varphi(e(u_\tau^k), z_\tau^k)\,{\rm d}x \le \int_\Omega \zeta_{\rm IN}(\hat{z} - z_\tau^{k-1}) + \varphi(e(u_\tau^k), \hat{z})\,{\rm d}x \tag{7.3.42}$$

[18] By the definition of the subdifferential $\partial\zeta_{\rm IN}(\dot{z}) \ni -\xi \ \Leftrightarrow\ \forall v\in\mathbb{R}^m:\ \zeta_{\rm IN}(v) \ge \xi\cdot(v-\dot{z}) + \zeta_{\rm IN}(\dot{z})$ so that, substituting $v \to v + \dot{z}$, we can see that $\xi\cdot v \le \zeta_{\rm IN}(v+\dot{z}) - \zeta_{\rm IN}(\dot{z}) \le R(v)$ where the last "triangle" inequality is due to 1-homogeneity of $\zeta_{\rm IN}$.

from this, by using 1-homogeneity of $\zeta_{\rm IN}$, the approximate semistability follows[19]

$$\int_\Omega \zeta_{\rm IN}(\hat z - \bar z_\tau(t)) + \varphi(e(\bar u_\tau(t)),\hat z) - \varphi(e(\bar u_\tau(t)),\bar z_\tau(t))\,{\rm d}x \ge 0 \tag{7.3.43}$$

for any $\hat z$ when using the notation $(\bar u_\tau, \bar z_\tau)$ for the piecewise constant approximate interpolants, cf. (C.2.15) on p. 536.

Let us consider rather for notational simplicity (cf. Remark 7.3.11 below) $\mathbb{Z} = 0$ in the following assertion:

Proposition 7.3.9 (Energy equality). *Let the loading and the initial conditions satisfy (6.4.5) (assuming, rather for simplicity, the linear dissipation $\zeta'_{VI} = \mathbb{D}$), let $|\partial_e\varphi(e,z)| \le C(1+|e|+|z|^{q/2})$ and $\partial_e\varphi(\cdot,z)$ be Lipschitz continuous uniformly with respect to z, and let*

$$\sqrt{\varrho}\ddot u, f \in L^2(I;H^1(\Omega;\mathbb{R}^d)^*), \quad \sigma \in L^2(Q;\mathbb{R}^{d\times d}_{\rm sym}), \quad g \in L^2(I;L^{2^{\#\prime}}(\Gamma;\mathbb{R}^d)). \tag{7.3.44}$$

the initial conditions $z_0 \in L^q(\Omega;\mathbb{R}^m)$ be semi-stable with respect to u_0 in the sense

$$\int_\Omega \varphi(e(u_0),z_0)\,{\rm d}x \le \int_\Omega \varphi(e(u_0),v) + \zeta_{IN}(v-z_0)\,{\rm d}x \tag{7.3.45}$$

for all $v \in L^q(\Omega;\mathbb{R}^m)$. Let further (u,z) be an energetic solution in accord with Definition 7.2.4. Then the equality in (7.2.8) with $w = 0$ and with $\int_Q \zeta(e(\dot u),\dot z)\,{\rm d}x{\rm d}t = \int_Q \zeta_{VI}(e(\dot u))\,{\rm d}x{\rm d}t + {\rm Var}_{\mathfrak{S}_{\rm EL}}(z;I)$ holds. In other words, (7.2.17a) holds as an equality.

Proof. Testing (7.2.5a) with σ from (7.1.1a) with (7.1.3) by $v = \dot u$, we obtain

$$\int_Q (\zeta'_{\rm VI}(e(\dot u)) + \partial_e\varphi(e(u),z)){:}e(\dot u)\,{\rm d}x{\rm d}t - \int_0^T \langle \sqrt{\varrho}\dot u, \sqrt{\varrho}\ddot u\rangle\,{\rm d}t + \int_\Omega \varrho|u(T)|^2{\rm d}x$$
$$= \int_\Omega \varrho|u_0|^2{\rm d}x + \int_Q f\cdot\dot u\,{\rm d}x{\rm d}t + \int_{\Sigma_{\rm N}} g\cdot\dot u\,{\rm d}S\,{\rm d}t + \int_0^T \langle \vec\tau_{\rm D},\dot u_{\rm D}\rangle_{\Gamma_{\rm D}}{\rm d}t \tag{7.3.46}$$

with $\langle\cdot,\cdot\rangle$ denoting the duality between $H^1(\Omega;\mathbb{R}^d)$ and $H^1(\Omega;\mathbb{R}^d)^*$; here we used the qualification (7.3.44) and a continuous extension of the identity (7.2.5a). Using again (7.3.44) to make by-part integration $\int_0^T \langle \sqrt{\varrho}\dot u, \sqrt{\varrho}\ddot u\rangle\,{\rm d}t = \frac12\int_\Omega \varrho|u(T)|^2 - \varrho|u_0|^2{\rm d}x$ legitimate, we arrive at the energy balance for the viscoelastic part, i.e.

$$\int_\Omega \frac{\varrho}{2}\big|\dot u(T)\big|^2\,{\rm d}x + \int_Q \Big(\zeta'_{\rm VI}(e(\dot u)) + \partial_e\varphi(e(u),z)\Big){:}e(\dot u)\,{\rm d}x{\rm d}t$$
$$= \int_\Omega \frac{\varrho}{2}|v_0|^2\,{\rm d}x + \int_Q f\cdot\dot u\,{\rm d}x{\rm d}t + \int_{\Sigma_{\rm N}} g\cdot\dot u\,{\rm d}S\,{\rm d}t + \int_0^T \langle \vec\tau_{\rm D},\dot u_{\rm D}\rangle_{\Gamma_{\rm D}}{\rm d}t. \tag{7.3.47}$$

[19] From (7.3.42) we can see that $\int_\Omega \zeta_{\rm IN}(\hat z - z_\tau^k) + \varphi(e(u_\tau^k),\hat z) - \varphi(e(u_\tau^k),z_\tau^k)\,{\rm d}x \ge 0$ for any $\hat z$ when using also the 1-homogeneity of $\zeta_{\rm IN}$ for $\zeta_{\rm IN}(\hat z - z_\tau^{k-1}) - \zeta_{\rm IN}(z_\tau^k - z_\tau^{k-1}) \le \zeta_{\rm IN}(\hat z - z_\tau^k)$.

We consider now $\varepsilon > 0$, and a partition $0 = t_0^\varepsilon < t_1^\varepsilon < \ldots < t_{k_\varepsilon}^\varepsilon = T$ with $\max_{i=1,\ldots,k_\varepsilon}(t_i^\varepsilon - t_{i-1}^\varepsilon) \le \varepsilon$. Moreover, as (7.2.15) integrated over Ω holds a.e. $t \in I$ and by (7.3.45) also at $t = 0$, we can consider the above partition so that the semi-stability holds at t_i^ε for each $i = 0, \ldots, k_\varepsilon - 1$. Using this semi-stability of z at time t_{i-1}^ε gives, when tested by $v := z(t_i^\varepsilon)$, the estimate

$$\begin{aligned}\int_\Omega \varphi(e(u(t_{i-1}^\varepsilon)), z(t_{i-1}^\varepsilon))\,\mathrm{d}x &\le \int_\Omega \varphi(e(u(t_{i-1}^\varepsilon)), z(t_i^\varepsilon)) + \zeta_{\rm IN}(z(t_i^\varepsilon) - z(t_{i-1}^\varepsilon))\,\mathrm{d}x \\ &= \int_\Omega \Big(\varphi(e(u(t_i^\varepsilon)), z(t_i^\varepsilon)) + \zeta_{\rm IN}(z(t_i^\varepsilon) - z(t_{i-1}^\varepsilon)) \\ &\qquad - \int_{t_{i-1}^\varepsilon}^{t_i^\varepsilon} \partial_e\varphi(e(u(t)), z(t_i^\varepsilon)) : e(\dot u)\,\mathrm{d}t\Big)\mathrm{d}x. \end{aligned} \tag{7.3.48}$$

Summing (7.3.48) for $i = 1, \ldots, k_\varepsilon$ and assuming that $\{t_i^\varepsilon\}_{i=1}^{k_\varepsilon - 1}$ are chosen so that $\dot u(t_i^\varepsilon) \in H^1(\Omega;\mathbb{R}^d)$ are well defined, we obtain

$$\begin{aligned}&\int_\Omega \varphi(e(u(T)), z(T)) - \varphi(e(u_0), z_0)\,\mathrm{d}x + \mathrm{Var}_{S_{\rm EL}}(z;[0,T]) \\ &\quad\ge \sum_{i=1}^{k_\varepsilon} \int_{t_{i-1}^\varepsilon}^{t_i^\varepsilon} \int_\Omega \partial_e\varphi(e(u(t)), z(t_i^\varepsilon)) : e(\dot u)\,\mathrm{d}x\mathrm{d}t \\ &\quad\ge \sum_{i=1}^{k_\varepsilon - 1} \int_{t_{i-1}^\varepsilon}^{t_i^\varepsilon} \int_\Omega \partial_e\varphi(e(u(t)), z(t_i^\varepsilon)) : e(\dot u)\,\mathrm{d}x\mathrm{d}t - \delta_\varepsilon \\ &\quad= \sum_{i=1}^{k_\varepsilon - 1} (t_i^\varepsilon - t_{i-1}^\varepsilon) \int_\Omega \partial_e\varphi(e(u(t_i^\varepsilon)), z(t_i^\varepsilon)) : e(\dot u(t_i^\varepsilon))\,\mathrm{d}x \\ &\qquad + \sum_{i=1}^{k_\varepsilon - 1} \int_{t_{i-1}^\varepsilon}^{t_i^\varepsilon} \int_\Omega \Big(\partial_e\varphi(e(u(t)), z(t_i^\varepsilon)) - \partial_e\varphi(e(u(t_i^\varepsilon)), z(t_i^\varepsilon))\Big) : e(\dot u)\,\mathrm{d}x\mathrm{d}t \\ &\qquad + \sum_{i=1}^{k_\varepsilon - 1} \int_{t_{i-1}^\varepsilon}^{t_i^\varepsilon} \int_\Omega \partial_e\varphi(e(u(t_i^\varepsilon)), z(t_i^\varepsilon)) : e(\dot u - \dot u(t_i^\varepsilon))\,\mathrm{d}x\mathrm{d}t - \delta_\varepsilon \\ &\quad=: S_1^\varepsilon + S_2^\varepsilon + S_3^\varepsilon - \delta_\varepsilon \quad \text{with } \delta_\varepsilon := \bigg| \int_{[t_{k_\varepsilon - 1}^\varepsilon, T] \times \Omega} \partial_e\varphi(e(u(t)), z(T)) : e(\dot u(t))\,\mathrm{d}x\mathrm{d}t \bigg|. \end{aligned} \tag{7.3.49}$$

Now the goal is to evaluate the limit when refining the partitions of I and $\varepsilon \to 0$.

We have still a freedom to choose the partition $\{t_i^\varepsilon\}_{i=1}^{k_\varepsilon}$ in such a way that the Riemann sum S_1^ε approaches the corresponding Lebesgue integral,[20] namely

$$\lim_{\varepsilon \to 0} S_1^\varepsilon = \int_Q \partial_e\varphi(e(u(t)), z(t)) : e(\dot u)\,\mathrm{d}x\mathrm{d}t. \tag{7.3.50}$$

---

[20] For rather subtle arguments see [140, Lemma 4.12] or [200, Lemma 4.5], following the idea of Hahn [237] from 1915.

As to S_2^ε, using the Lipschitz continuity of $\partial_e\varphi(\cdot,z):\mathbb{R}^{d\times d}_{\rm sym}\to\mathbb{R}^{d\times d}_{\rm sym}$ with ℓ denoting the Lipschitz constant, we can estimate

$$
\begin{aligned}
|S_2^\varepsilon| &\le \sum_{i=1}^{k_\varepsilon}\int_{t_{i-1}^\varepsilon}^{t_i^\varepsilon} \ell\big\|e(u(t)-u(t_i^\varepsilon))\big\|_{L^2(\Omega;\mathbb{R}^{d\times d})}\big\|e(\dot u)\big\|_{L^2(\Omega;\mathbb{R}^{d\times d})}\,\mathrm{d}t\\
&\le \ell\max_{i=1,\dots,k_\varepsilon}\max_{t\in[t_{i-1}^\varepsilon,t_i^\varepsilon]}\big\|e(u(t)-u(t_i^\varepsilon))\big\|_{L^2(\Omega;\mathbb{R}^{d\times d})}\big\|e(u)\big\|_{W^{1,1}(I;L^2(\Omega;\mathbb{R}^{d\times d}))}.
\end{aligned}
\tag{7.3.51}
$$

Since certainly $e(u)\in W^{1,1}(I;L^2(\Omega;\mathbb{R}^{d\times d}))$, the "max max"-term tends to zero with $\varepsilon\to0$, hence $\lim_{\varepsilon\to0}S_2^\varepsilon=0$. As to S_3^ε, by Fubini's theorem, we can estimate

$$
\begin{aligned}
|S_3^\varepsilon| &= \bigg|\sum_{i=1}^{k_\varepsilon}\int_\Omega \partial_e\varphi(e(u(t_i^\varepsilon)),z(t_i^\varepsilon)):e\Big(u(t_i^\varepsilon)-u(t_{i-1}^\varepsilon)-(t_i^\varepsilon-t_{i-1}^\varepsilon)\dot u(t_i^\varepsilon)\Big)\,\mathrm{d}x\bigg|\\
&\le \big\|\partial_e\varphi(e(u),z)\big\|_{L^\infty(I;L^2(\Omega;\mathbb{R}^{d\times d}))}\sum_{i=1}^{k_\varepsilon}\big\|e(u(t_i^\varepsilon)-u(t_{i-1}^\varepsilon)-(t_i^\varepsilon-t_{i-1}^\varepsilon)\dot u(t_i^\varepsilon))\big\|_{L^2(\Omega;\mathbb{R}^{d\times d})}.
\end{aligned}
$$

Let us note that $u(t_i^\varepsilon)-u(t_{i-1}^\varepsilon)\in H^1(\Omega;\mathbb{R}^d)$ although particular terms are in $W^{1,p}(\Omega;\mathbb{R}^d)$ need not belong to $H^1(\Omega;\mathbb{R}^d)$ if $p<2$ and that the assumed growth of $\partial_e\varphi$ together with the assumed growth and Lipschitz continuity of $\partial_e\varphi$ indeed guarantees $\partial_e\varphi(e(u),z)\in L^\infty(I;L^2(\Omega;\mathbb{R}^{d\times d}))$.

We have still a freedom to choose the partition $\{t_i^\varepsilon\}_{i=1}^{k_\varepsilon}$ in such a way that also $\lim_{\varepsilon\to0}S_3^\varepsilon=0$.

Eventually, $\lim_{\varepsilon\to0}\delta_\varepsilon=0$ because the integrand $\int_\Omega\partial_e\varphi(e(u(\cdot)),z(T)):e(\dot u(\cdot))\,\mathrm{d}x$ in δ_ε in (7.3.49) is absolutely continuous in time and $t_{k_\varepsilon-1}^\varepsilon\to T$ for $\varepsilon\to0$. This allows for a limit passage in (7.3.49) for $\varepsilon\to0$, which gives

$$
\begin{aligned}
\int_\Omega &\varphi(e(u(T)),z(T))-\varphi(e(u_0),z_0)\,\mathrm{d}x\\
&+\mathrm{Var}_{S_{\rm EL}}(z;[0,T])\ge\int_Q\partial_e\varphi(e(u(t)),z(t)):e(\dot u)\,\mathrm{d}x\mathrm{d}t.
\end{aligned}
\tag{7.3.52}
$$

Summing it with (7.3.47), we obtain the opposite inequality in (7.2.17a). Therefore (7.2.17a) is shown to hold as an equality. □

Remark 7.3.10. An interesting observation is that if the dissipative force has a potential ζ and if dissipated energy is equal to this potential, then this potential must be 1-homogeneous, see e.g. [360, Proposition 3.2.4].

Remark 7.3.11 (Gradient theories in z). The proof of Proposition 7.3.9 can be straightforwardly modified if $\mathbb{Z}\ne0$, being positive definite and when assuming $|\partial_e\varphi(e,z,Z)|\le C(1+|e|+|z|^{q^*/2}+|Z|^{q/2})$ and $\partial_e\varphi(\cdot,z,Z)$ Lipschitz continuous uniformly with respect to (z,Z). Compactness in z is then to be exploited.

Remark 7.3.12 (More general dissipation). Yet, the generalization for $\zeta_{\rm IN}$ from Remark 7.2.7 is not right possible because $\int_Q\zeta_{\rm IN}(z,\dot z)\,\mathrm{d}x\mathrm{d}t$ may bring a measure $\dot z$ with a discontinuous function z.

The concept of the weak (or local) solutions in the general rate-independent context, i.e. for $\hat{\varphi}(t,\cdot)$ nonconvex or for a more general $\zeta_{\rm IN} = \zeta_{\rm IN}(z,\dot{z})$ with $\zeta_{\rm IN}(z,\cdot)$ in (7.3.27), has a drawback that it yields a rather wide class of solutions in general. In particular, the energy balance need not be guaranteed; cf. [360].

Exercise 7.3.13 (Variants of the convergence proof). Modify the arguments in the proof of Proposition 7.3.8 if (7.3.21d) is assumed and $\partial_e\varphi(e,\cdot)$ is affine while the (7.3.28) is considered with respect to the strong×weak convergence. Considering further a gradient theory for z, allow for $\partial_e\varphi(e,\cdot)$ not affine and (7.3.28) weakened. Hint: Use the arguments like (7.3.24) to show strong convergence of strains. For the gradient theory in z, use the compact embedding and strong convergence also of the internal variable z.

7.4 Elastoplasticity, viscoelastoplasticity, and creep

The phenomenon of *plasticity* usually refers both to a permanent deformation (and strain and stress) remaining in an unloaded specimen after its previous sufficiently large loading and to an activated character of the process leading to such permanent deformation in the sense that there is no such a remaining permanent effects is the loading intensity do not exceed some activation threshold. The internal speed of such plastification processes can be qualitatively much faster than the time scale of the external loading or other processes (like transport of some fluid or heat inside the medium) and then it can be modeled as *rate independent*. Sometimes, fast plastification brings additional dissipative effects and then rate-dependent models of plasticity are often considered, too. Plastification without any activation threshold and without hardening (or at least without hardening for the shear, i.e. in the deviatoric part) is referred to as *creep* and can be essentially identified with Maxwell rheology.

At small strains, those processes can still be well described by quadratic stored-energy potentials and thus fit with Sect. 7.3.1. Microscopical mechanisms behind plasticity are mutual shifts of atomic layers in metallic crystals or grains due to dislocations movement or mutual shift of layers in some noncrystalline materials like rocks or soils (where the word "plasticity" is replaced rather by inelastic deformation) or polymers (where rather creep than plasticity occur). In any case, these processes do not essentially influence volume but rather shear strain $\operatorname{dev} e$ only, and are thus accompanied by the adjective *isochoric*.

7.4.1 Viscoplasticity and hardening in plasticity

An illustrative but mathematically relatively simple situation is a model of a single-threshold isochoric linearized viscoplasticity with a linear *kinematic hardening* in isotropic materials. The elastic moduli tensor $\mathbb{C}$ is then determined by its bulk and shear moduli (K_E, G_E), i.e. $\mathbb{C} = \text{ISO}(K_E, G_E)$, acting on the orthogonal decomposition of the elastic strain $e_{EL} = \text{sph}\, e_{EL} + \text{dev}\, e_{EL}$ with $\text{sph}\, e_{EL} := (\text{tr}\, e_{EL})\mathbb{I}/d$ analogously as (6.7.8). Here, likewise in (5.5.1) on p. 171, the total strain $e(u)$ decomposes additively as $e(u) = e_{EL} + \pi$ where π denotes the trace-free *plastic strain* valued in $\mathbb{R}^{d\times d}_{\text{dev}}$ with

$$\mathbb{R}^{d\times d}_{\text{dev}} := \{A \in \mathbb{R}^{d\times d};\ A^\top = A,\ \text{tr}(A) = 0\}, \tag{7.4.1}$$

Then the number of internal variables is $m = d(d+1)/2 - 1$. As the trace of the plastic deformation π vanishes, i.e. the plastic strain preserves volume, which is related with the mentioned isochoric model.

In this simplest plasticity model, the internal variable z is in the position of this trace-free plastic strain and enters the model rather formally through the linear operator E being just the embedding $\mathbb{R}^{d\times d}_{\text{dev}} \to \mathbb{R}^{d\times d}_{\text{sym}}$ so that $\pi = Ez$.[21] Then, in terms of the notation from Sect. 7.1, we consider

$$\begin{aligned}\varphi(e,z) = \frac{1}{2}\mathbb{C}e_{EL}{:}e_{EL} + \frac{1}{2}\mathbb{H}z{\cdot}z &= \frac{1}{2}K_E(\text{tr}\, e_{EL})^2 + G_E|\text{dev}\, e_{EL}|^2 + \frac{1}{2}H|z|^2\\ &= \frac{d}{2}K_E|\text{sph}\, e|^2 + G_E|\text{dev}\, e - Ez|^2 + \frac{1}{2}H|z|^2,\end{aligned} \tag{7.4.2a}$$

$$\zeta_{VI}(\dot e) = 0, \qquad \zeta_{IN}(\dot z) = \frac{1}{2}D|\dot z|^2 + s_{YLD}|\dot z|. \tag{7.4.2b}$$

A special case $s_{YLD} = 0$, $D > 0$, and $H = 0$ reduces to the creep model, cf. Exercise 6.6.12, while $s_{YLD} > 0$ and $D = 0$ gives a rate-independent plasticity addressed in Sections 7.4.2 and 7.4.3 below. In general, here the stresses write as:

$$\begin{aligned}\sigma = \sigma_{\text{sph}} + \sigma_{\text{dev}} \quad \text{with} \quad &\sigma_{\text{sph}} = K_E(\text{tr}\, e_{EL})\mathbb{I} = dK_E\text{sph}\, e(u)\\ \text{and} \quad &\sigma_{\text{dev}} = 2G_E\text{dev}\, e_{EL} = 2G_E(\text{dev}\, e(u) - \pi).\end{aligned} \tag{7.4.3}$$

The general system (7.2.1a,b) now reads as $\varrho\ddot u - \text{div}\,\sigma = f$ with $\sigma = \mathbb{C}(e(u) - Ez)$ and $D\dot z + s_{YLD}\text{Dir}(\dot z) \ni E^*\sigma - Hz$. More specifically, in terms of π,

$$\varrho\ddot u - \text{div}\,\sigma = f \qquad \text{with } \sigma = dK_E\text{sph}\, e(u) + 2G_E(\text{dev}\, e(u) - \pi), \tag{7.4.4a}$$

$$D\dot\pi + s_{YLD}\text{Dir}(\dot\pi) + 2G_E\pi \ni 2G_E\text{dev}\, e(u) - H\pi, \tag{7.4.4b}$$

where the last term $H\pi$ is sometimes called a *back-stress* and where the "direction-mapping" $\text{Dir} : \mathbb{R}^{d\times d}_{\text{dev}} \rightrightarrows \mathbb{R}^{d\times d}_{\text{dev}}$ denotes the set-valued subdifferential of $|\cdot|$, i.e.

$$\text{Dir}(\dot\pi) = \begin{cases}\dot\pi/|\dot\pi| & \text{if } \dot\pi \neq 0,\\ \text{the unit ball} & \text{if } \dot\pi = 0.\end{cases} \tag{7.4.5}$$

We will use the notation $\text{⊪}_{s_{YLD}}$ for the dry-friction-like dissipating element described by the potential (7.3.33) with $S_{EL} = \{|\cdot| \le s_{YLD}\}$. The rheology of the material

[21] Alternatively, this can be included into our model by considering π valued in $\mathbb{R}^{d\times d}_{\text{sym}}$ and the initial condition π_0 trace-free, and by the elastic domain $S_{EL} \subset \mathbb{R}^{d\times d}_{\text{sym}}$ unbounded so that $\zeta_{IN}(\dot\pi) = \delta^*_{S_{EL}}(\dot\pi) = +\infty$ if $\text{tr}(\dot\pi) \neq 0$.

governed by (7.4.2) can be written in terms of the language from Table 6.1 on p. 198 as $_{G_E}$ — ($_{s_{\rm YLD}}$ || $_H$ || $_D$) in its deviatoric part and, as whole, can be schematically displayed by the following figure 7.2.

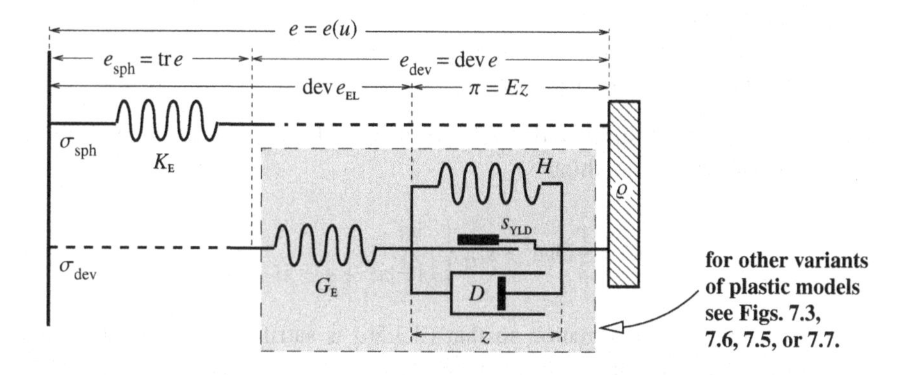

Fig. 7.2 Isochoric viscoplasticity with kinematic hardening. For the orthogonal decomposition $e = e_{\rm sph} + e_{\rm dev}$ see (6.7.8). The isotropic elastic-moduli tensor $\mathbb{C} = {\rm ISO}(K_{\rm E}, G_{\rm E})$ is determined by the bulk modulus $K_{\rm E}$ and the shear modulus $G_{\rm E}$ from (6.7.9).

We will consider the initial/boundary conditions (7.2.1c,d) with π in place of z. The weak formulation (7.2.5b) of the plastic flow rule results here to the following variational inequality:

$$\int_Q ((H+2G_{\rm E})\pi - 2G_{\rm E}{\rm dev}\, e(u)){:}w + \frac{1}{2}D|w|^2 + s_{\rm YLD}|w|\,{\rm d}x{\rm d}t + \int_\Omega \frac{1}{2}\mathbb{C}e_{{\rm EL},0}{:}e_{{\rm EL},0}\,{\rm d}x$$
$$\geq \int_Q \frac{1}{2}D|\dot{\pi}|^2 + s_{\rm YLD}|\dot{\pi}| + \mathbb{C}(e(u){-}\pi){:}e(\dot{u})\,{\rm d}x{\rm d}t + \int_\Omega \frac{1}{2}\mathbb{C}e_{\rm EL}(T){:}e_{\rm EL}(T)\,{\rm d}x$$
$$- \int_Q f\cdot\dot{u}\,{\rm d}x{\rm d}t - \int_{\Sigma_{\rm N}} g\cdot\dot{u}\,{\rm d}S\,{\rm d}t - \int_0^T \langle \vec{\tau}_{\rm D}, \dot{u}_{\rm D}\rangle_{\Gamma_{\rm D}}{\rm d}t, \tag{7.4.6}$$

where we have used the abbreviation $e_{\rm EL} = e(u){-}\pi$ and $e_{{\rm EL},0} = e(u_0){-}\pi_0$, and also $\mathbb{C} = {\rm ISO}(K_{\rm E}, G_{\rm E})$, and where $\vec{\tau}_{\rm D}$ is the traction vector from (6.2.10).

Alternatively and more universally,[22] we can therefore exploit the weak formulation (7.2.8) of the plastic flow rule, which here results to the variational inequality:

$$\int_Q ((H+2G_{\rm E})\pi - 2G_{\rm E}({\rm dev}\, e(u)){:}w + \frac{1}{2}D|w|^2 + s_{\rm YLD}|w|\,{\rm d}x{\rm d}t$$
$$+ \int_Q f\cdot\dot{u}\,{\rm d}x{\rm d}t + \int_{\Sigma_{\rm N}} g\cdot\dot{u}\,{\rm d}S\,{\rm d}t + \int_0^T \langle \vec{\tau}_{\rm D}, \dot{u}_{\rm D}\rangle_{\Gamma_{\rm D}}{\rm d}t + \int_\Omega \frac{\varrho}{2}|v_0|^2 + \frac{1}{2}\mathbb{C}e_{{\rm EL},0}{:}e_{{\rm EL},0}\,{\rm d}x$$
$$\geq \int_Q \frac{1}{2}D|\dot{\pi}|^2 + s_{\rm YLD}|\dot{\pi}|\,{\rm d}x{\rm d}t + \int_\Omega \frac{\varrho}{2}\big|\dot{u}(T)\big|^2 + \frac{1}{2}\mathbb{C}e_{\rm EL}(T){:}e_{\rm EL}(T). \tag{7.4.7}$$

[22] For perfect plasticity (which arises when $D = 0 = H$ and will be discussed later in Sect.,7.4.3) $e(\dot{u})$ is not well qualified and (7.4.6) is not analytically suitable.

Proposition 7.4.1 (Viscoelastoplasticity: existence, uniqueness, regularity). *Let ϱ satisfies (6.3.5), $K_{\rm E}$, $G_{\rm E}$, D, and H be positive, and the loading and the initial condition satisfy (6.3.3). Then:*

(i) The problem (7.4.4) with the initial/boundary conditions (7.2.1c,d) possesses a weak solution.

(ii) Moreover, if the loading is more regular in the sense (6.3.12), the weak solution is unique and conserves energy.

Sketch of the proof. It suffices to use Proposition 7.3.1. As $\zeta_{\rm IN}$ from (7.4.2b) is finite, we can use a simple smoothing

$$\zeta_{{\rm IN},\varepsilon}(\dot\pi) = \frac{1}{2}D|\dot\pi|^2 + s_{\rm YLD}\begin{cases} |\dot\pi| & \text{if } |\dot\pi| \ge \varepsilon, \\ \frac{1}{2}\Big(\frac{1}{\varepsilon}|\dot\pi|^2 + \varepsilon\Big) & \text{if } |\dot\pi| < \varepsilon. \end{cases} \tag{7.4.8}$$

This is an upper approximation so that (7.3.5b) is satisfied automatically while (7.3.5a) is easy, too; actually this fits with Example A.3.16 on p. 502. The existence is then by Proposition 7.3.1.

The regularity, uniqueness, and energy conservation is directly due to Proposition 7.3.3. □

Remark 7.4.2 (Nonlinear Maxwell rheology). It is worth realizing the connection to the linear Maxwell rheology as presented in Section 6.6 when the dissipation potential $\zeta = \frac{1}{2}\xi$ from (6.6.6) is generalized for a possibly nonquadratic and even nonsmooth $\zeta_{\rm IN} = \zeta_{\rm IN}(\dot\pi)$. Instead of $\dot\pi = \dot e_2 = \mathbb{D}^{-1}\sigma$ in (6.6.1), we have $\dot\pi = [\partial\zeta_{\rm IN}]^{-1}(\sigma) = \partial\zeta_{\rm IN}^*(\sigma)$ with $\zeta_{\rm IN}^*$ denoting the inelastic potential, cf. (A.3.17), so that (6.6.1) reads as $e(\dot u) \in \mathbb{C}^{-1}\dot\sigma + \partial\zeta_{\rm IN}^*(\sigma)$ to be coupled with the force equilibrium $\varrho\ddot u - {\rm div}\,\sigma = f$. Or, in terms of the original dissipation potential, $\partial\zeta_{\rm IN}(e(\dot u) - \mathbb{C}^{-1}\dot\sigma) \ni \sigma$.

Remark 7.4.3 (Wave attenuation and dispersion in gradient Maxwell materials). The internal-variable setting of the Maxwell materials suggested in the previous Remark 7.4.2 allows for considering the gradient-theory on this internal variable and to modify considerations in Remark 6.6.4. Confining to the 1-dimensional linear case, the system $\varrho\ddot u = \partial_x\sigma$ with (6.6.1) can be written as $\varrho\ddot u - \partial_x C(\partial_x u - \pi) = 0$ and $D\dot\pi = C(\partial_x u - \pi)$ and, considering the mentioned gradient theory, the flow rule for π is to be enhanced as $D\dot\pi = C(\partial_x u - \pi) + \kappa\partial_{xx}^2\pi$. We now eliminate[23] u to obtain the dispersive wave equation

$$\frac{1}{C}\dddot\pi + \frac{1}{D}\ddot\pi - \frac{1}{\varrho}\partial_{xx}^2\dot\pi - \frac{\kappa}{CD}\partial_{xx}^2\ddot\pi + \frac{\kappa}{\varrho D}\partial_{xxxx}^4\pi = 0 \quad \text{on } [0,+\infty)\times\mathbb{R}. \tag{7.4.9}$$

Using the ansatz $\pi = {\rm e}^{{\rm i}(wt+x/\lambda)}$ with the complex angular frequency $w = \omega + {\rm i}\gamma$ and the real-valued wavelength λ like in Remark 6.4.13, we obtain ${\rm i}w^3/C + w^2/D - {\rm i}w/(\varrho\lambda^2) + \kappa w^2/(CD\lambda^2) - \kappa/(\varrho D\lambda^4) = 0$. In terms of the real-valued coefficients ω and γ, it gives the *Kramers-Kronig relations* here as

[23] Indeed, by differentiating the first equation in x so that $\varrho\partial_x\ddot u = \partial_{xx}^2 C(\partial_x u - \pi)$ and by realizing that $\varrho\partial_x\ddot u = \varrho D\dddot\pi/C + \varrho\ddot\pi - \kappa\varrho\partial_{xx}^2\ddot\pi/C$ and $\partial_{xx}^2 C(\partial_x u - \pi) = D\partial_{xx}^2\dot\pi - \kappa\partial_{xxxx}^4\pi$, we obtain (6.6.21).

$$\gamma\frac{3\omega^2-\gamma^2}{C}-\frac{\omega^2-\gamma^2}{D}-\frac{\gamma}{\varrho\lambda^2}+\kappa\frac{\omega^2-\gamma^2}{CD\lambda^2}+\frac{\kappa}{\varrho D\lambda^4}=0 \quad \text{and} \tag{7.4.10a}$$

$$\frac{\omega^2-3\gamma^2}{C}+\frac{2\gamma}{D}-\frac{1}{\varrho\lambda^2}-\frac{2\gamma\kappa}{CD\lambda^2}=0\,. \tag{7.4.10b}$$

Let us note that for $\kappa=0$, (7.4.9) results (after being integrated in time) to a telegraph equation like (6.6.21). Yet, in contrast to (6.6.22) or (6.4.26), the algebraic system (7.4.10) is difficult to solve.

Exercise 7.4.4. Make a limit passage for vanishing hardening, i.e. for $H\to 0$, towards an elastic perfectly viscoplastic model spring_{C}—($\text{slider}_{\kappa}\|\text{dashpot}_{\text{D}}$), also called the *Bingham-Norton material.*

7.4.2 Rate-independent plasticity with hardening

A lot of modeling aspects related with linearized plasticity are better explained when the rate-dependent effects (except perhaps inertia) are suppressed. Then we speak about a rate-independent plasticity.

Let us first illustrate it on the model from Figure 7.2 where we just put $D=0$. A certain justification is by the limit passage with $D\to 0$.[24] The resulted non-viscous rheological scheme $D=0$ is depicted on Figure 7.3.[25] The suitable weak-solution

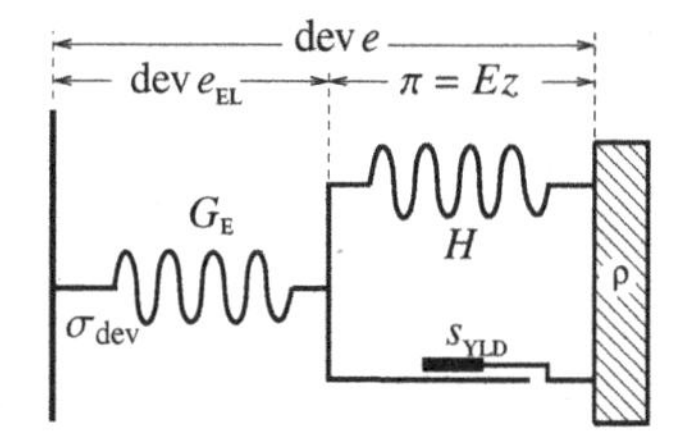

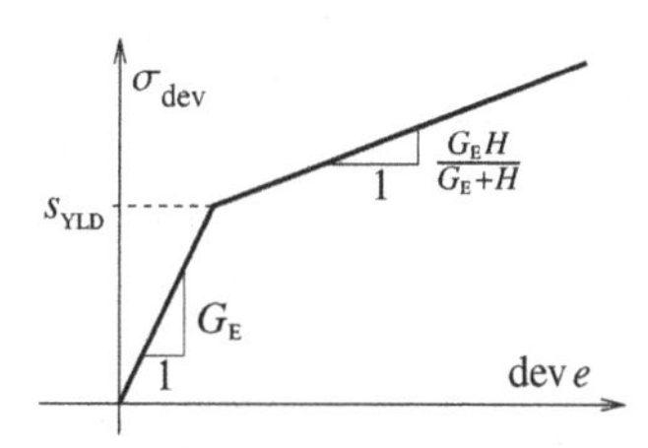

Fig. 7.3 Inviscid elasto-plastic rheology with a linear *kinematic hardening* (=a special rate-independent case of the rheology from Fig. 7.2 with $D=0$):
Left: a schematic diagram for the deviatoric part, the spherical one being as in Fig. 7.2.
Right: stress-strain response on "monotone" shear loading, starting from $e(0)=0=z(0)$.

---

[24] Another justification of the rate-independent model arises from re-scaling time, considering very slow loading. In the limit, one then can expect fully rate-independent model where not only viscous but also inertial effects disappears. This is rather nontrivial phenomenon rigorously justified for quadratic stored energies like here in the linearized plasticity, we refer to [360, Sect. 5.1.2.2].

[25] Cf. also [266, Fig.23.3]. If the plastic element is active, then the corresponding stress σ_{PL} has the magnitude s_{YLD}, and one can derive from $G_{\text{E}}(e-Ez)=\sigma_{\text{dev}}=Hz+\sigma_{\text{PL}}$ that $z=(G_{\text{E}}\text{dev}\,e-\sigma_{\text{PL}})/(G_{\text{E}}+H)$ so that $\sigma_{\text{dev}}=H(G_{\text{E}}+H)^{-1}(G_{\text{E}}\text{dev}\,e-\sigma_{\text{PL}})+\sigma_{\text{PL}}$, and thus we can see that the slope of the stress/strain response is $H(G_{\text{E}}+H)^{-1}G_{\text{E}}$ after the stress reaches the yield stress s_{YLD}, as schematically depicted on Figure 7.3(right).

concept can modify Definition 7.2.2 to involve the momentum equation in the weak sense (7.2.5a), i.e. now

$$\int_Q (dK_{\rm E}{\rm sph}\,e(u) + 2G_{\rm E}({\rm dev}\,e(u) - \pi)) : e(v) - \varrho\dot u\cdot\dot v\,{\rm d}x{\rm d}t + \int_\Omega \varrho\dot u(T)\cdot v(T)\,{\rm d}x = \int_\Omega \varrho v_0\cdot v(0)\,{\rm d}x + \int_Q f\cdot v + \int_{\Sigma_{\rm N}} g\cdot v\,{\rm d}S\,{\rm d}t \tag{7.4.11a}$$

to be satisfied for all $v \in C^1(\bar Q;\mathbb{R}^d)$ with $v(T) = 0$ and $v|_{\Sigma_{\rm D}} = 0$, and also $u(0) = u_0$ is to hold on Ω, while the inequality (7.2.8) now reads, for any $w \in L^2(Q;\mathbb{R}^{d\times d}_{\rm dev})$, as

$$\begin{aligned}\int_Q (2G_{\rm E}(\pi - {\rm dev}\,e(u)) + H\pi) : w + s_{\rm YLD}|w|\,{\rm d}x{\rm d}t \ge {\rm Var}_{S_{\rm EL}}(z;I) - \int_Q f\cdot\dot u_\varepsilon\,{\rm d}x{\rm d}t\\ + \int_\Omega \Big(\frac{\varrho}{2}|\dot u(T)|^2 + \frac{d}{2}K_{\rm E}|{\rm sph}\,e(u(T))|^2 + G_{\rm E}|{\rm dev}\,e(u(T)) - \pi(T)|^2 + \frac12 H|\pi(T)|^2\\ - \frac{\varrho}{2}|v_0|^2 - \frac{d}{2}K_{\rm E}|{\rm sph}\,e(u_0)|^2 + G_{\rm E}|{\rm dev}\,e(u_0) - \pi_0|^2 + \frac12 H|\pi_0|^2\Big){\rm d}x\\ - \int_{\Sigma_{\rm N}} g\cdot\dot u\,{\rm d}S\,{\rm d}t - \int_0^T \langle\vec\tau_{\rm D},\dot u_{\rm D}\rangle_{\Gamma_{\rm D}}\,{\rm d}t,\end{aligned} \tag{7.4.11b}$$

where the variation ${\rm Var}_{S_{\rm EL}}(\pi;I)$ of $\pi : I \to L^1(\Omega;\mathbb{R}^{d\times d}_{\rm dev})$ with respect to the potential $\delta^*_{S_{\rm EL}}$ is defined by (7.2.16) with $\zeta_{\rm IN} = \delta^*_{S_{\rm EL}}$ with $S_{\rm EL} = \{|\cdot| \le s_{\rm YLD}\}$.

Proposition 7.4.5 (Towards elastoplasticity via vanishing viscosity). *Let ϱ satisfy (6.3.2), $K_{\rm E} \ge 0$, $G_{\rm E}, H > 0$, and the loading and the initial conditions satisfy (6.3.3) and $\pi_0 \in L^2(\Omega;\mathbb{R}^{d\times d}_{\rm dev})$. Further, let $(u_\varepsilon,\pi_\varepsilon)$ denote the unique weak solution to the system (7.4.4) with $D_\varepsilon = \varepsilon D$. Then the following a-priori estimates hold*

$$\|u_\varepsilon\|_{W^{1,\infty}(I;L^2(\Omega;\mathbb{R}^d))\cap L^\infty(I;H^1(\Omega;\mathbb{R}^d))} \le C, \tag{7.4.12a}$$

$$\|\ddot u_\varepsilon\|_{L^1(I;L^2(\Omega;\mathbb{R}^d))+L^\infty(I;H^1_0(\Omega;\mathbb{R}^d)^*)} \le C, \tag{7.4.12b}$$

$$\|\pi_\varepsilon\|_{L^\infty(I;L^2(\Omega;\mathbb{R}^{d\times d}))\cap W^{1,1}(I;L^1(\Omega;\mathbb{R}^{d\times d}))} \le C, \tag{7.4.12c}$$

$$\|\dot\pi_\varepsilon\|_{L^2(Q;\mathbb{R}^{d\times d})} \le C/\sqrt{\varepsilon}, \tag{7.4.12d}$$

and there is a subsequence of $\{(u_\varepsilon,\pi_\varepsilon)\}_{\varepsilon>0}$ converging weakly in the topologies indicated in (7.4.12a–c) and any (u,π) obtained by such a limit is a weak solution to (7.4.4) with $D = 0$ in the sense (7.4.11).*

Proof. The a-priori estimates (7.4.12) follows by the standard energetic test by $(\dot u_\varepsilon,\dot\pi_\varepsilon)$, using the by-part integration as in (6.4.6).

Further, we make a selection of a subsequence converging weakly* in the topologies indicated in (7.4.12a–c). Here, an important ingredient is that, by the former estimate in (7.4.12d), we have $\{\pi_\varepsilon\}_{\varepsilon>0}$ bounded in ${\rm BV}(I;L^1(\Omega;\mathbb{R}^{d\times d}_{\rm dev}))$ and, by Helly's selection-principle Theorem B.4.13 on p. 526, we can consider a subsequence such that $\pi_\varepsilon(t) \to \pi(t)$ weakly* in $\mathfrak{M}(\bar\Omega;\mathbb{R}^{d\times d}_{\rm dev})$ for any $t \in I$, and, in particular, for $t = T$, too. By (7.4.12c), we can assume even $\pi_\varepsilon(T) \to \pi(T)$ weakly in $L^2(\Omega;\mathbb{R}^{d\times d}_{\rm dev})$ and also $\dot u_\varepsilon(T) \to \dot u(T)$ weakly in $L^2(\Omega;\mathbb{R}^d)$ and $e(u_\varepsilon(T)) \to e(u(T))$ weakly in $L^2(\Omega;\mathbb{R}^{d\times d}_{\rm sym})$.

The limit passage in the integral identity (7.4.11a) written for $(u_\varepsilon,\pi_\varepsilon)$ instead of (u,π) is easy just by the weak convergence, as well as the limit passage in the initial condition $u_\varepsilon(0) = u_0$ because $u_\varepsilon(t) \to u(t)$ weakly in $H^1(\Omega;\mathbb{R}^d)$ for any $t\in I$, and, in particular, for $t = 0$, too. The inequality (7.4.11b) written for the "viscous" solution $(u_\varepsilon,\pi_\varepsilon)$ now looks as

$$
\begin{aligned}
\int_Q &(2G_{\rm E}(\pi_\varepsilon - {\rm dev}\,e(u_\varepsilon)) + H\pi_\varepsilon):w + s_{\rm YLD}|w| + \varepsilon D|w|^2\,{\rm d}x{\rm d}t\\
&\ge \int_Q s_{\rm YLD}|\dot\pi_\varepsilon| + \varepsilon D|\dot\pi_\varepsilon|^2 - f\cdot\dot u_\varepsilon\,{\rm d}x{\rm d}t + \int_\Omega \frac{\varrho}{2}\big|\dot u_\varepsilon(T)\big|^2 + \frac12 K_{\rm E}|{\rm sph}\,e(u_\varepsilon(T))|^2\\
&\quad + G_{\rm E}|{\rm dev}\,e(u_\varepsilon(T)) - \pi_\varepsilon(T)|^2 + \frac12 H|\pi_\varepsilon(T)|^2\\
&\qquad - \frac{\varrho}{2}|v_0|^2 - \frac12 K_{\rm E}|{\rm sph}\,e(u_0)|^2 + G_{\rm E}|{\rm dev}\,e(u_0) - \pi_0|^2 + \frac12 H|\pi_0|^2\,{\rm d}x\\
&\qquad\qquad - \int_{\Sigma_{\rm N}} g\cdot\dot u_\varepsilon\,{\rm d}S\,{\rm d}t - \int_0^T \langle\vec\tau_{\rm D},\dot u_{\rm D}\rangle_{\Gamma_{\rm D}}{\rm d}t. \qquad (7.4.13)
\end{aligned}
$$

Omitting the term $\varepsilon D|\dot\pi_\varepsilon|^2 \ge 0$ does not corrupt the inequality (7.4.13). For a fixed w, we have clearly $\int_Q \varepsilon D|w|^2\,{\rm d}x{\rm d}t \to 0$. The limit of the other terms can be made by continuity or semicontinuity. Thus (7.4.11b) is obtained. In particular, careful arguments based on Helly's selection principle allow for a weak* convergence of π_ε to π in ${\rm BV}(I;L^1(\Omega;\mathbb{R}^{d\times d}_{\rm dev}))$, hence $\dot\pi$ is an $L^1(\Omega;\mathbb{R}^{d\times d}_{\rm dev})$-valued measure on I. By weak* lower semicontinuity, one has

$$
\liminf_{\varepsilon\to0}\int_Q s_{\rm YLD}|\dot\pi_\varepsilon|\,{\rm d}x{\rm d}t = {\rm Var}_{S_{\rm EL}}(z;I).
$$

□

Moreover, under gentle loading ensuring regularity from Proposition 7.3.3, one can even show that the viscous energy $\varepsilon D|\dot\pi_\varepsilon|^2$ converges to 0. More specifically, taking (7.4.7) with εD instead of D and $(u_\varepsilon,\pi_\varepsilon,e_{{\rm EL},\varepsilon})$ instead of $(u_\varepsilon,\pi_\varepsilon,e_{\rm EL})$, one can use it to estimate $\limsup_{\varepsilon\to0}\varepsilon D|\dot\pi_\varepsilon|^2$ from above and then make the limit passage with $\varepsilon\to0$, showing that the limit is zero due to energy conservation.

Proposition 7.4.6 (Regularity, uniqueness, energy conservation). *Let, in addition to the assumptions of Proposition 7.4.5, also* $f\in W^{1,1}(I;L^2(\Omega;\mathbb{R}^d))$, $u_0\in H^2(\Omega;\mathbb{R}^d)$, $v_0\in H^1(\Omega;\mathbb{R}^d)$, $\pi_0\in H^1(\Omega;\mathbb{R}^{d\times d}_{\rm dev})$, *and* $G_{\rm E}|{\rm dev}\,e(u_0)-\pi_0| < s_{\rm YLD}$. *Then, the weak solution* (u,π) *is unique, conserves energy and satisfies*

$$
\dot u\in W^{1,\infty}(I;L^2(\Omega;\mathbb{R}^d))\cap L^\infty(I;H^1(\Omega;\mathbb{R}^d)), \qquad (7.4.14a)
$$

$$
\dot e,\dot\pi\in L^\infty(I;L^2(\Omega;\mathbb{R}^{d\times d}_{\rm sym})),\quad {\rm div}\,\sigma\in L^\infty(I;L^2(\Omega;\mathbb{R}^d)), \qquad (7.4.14b)
$$

$$
\begin{aligned}
\forall_{a.a.}t\in I\ \forall\tilde\pi\in L^2(\Omega;\mathbb{R}^{d\times d}_{\rm dev}):\quad &\int_\Omega G_{\rm E}|{\rm dev}\,e(t)-\pi(t)|^2 + \frac12 H|\pi(t)|^2\\
&\le \int_\Omega s_{\rm YLD}|\pi(t)-\tilde\pi| + G_{\rm E}|{\rm dev}\,e(t)-\tilde\pi|^2 + \frac12 H|\tilde\pi|^2\,{\rm d}x. \qquad (7.4.14c)
\end{aligned}
$$

Sketch of the proof. We differentiate the system in time and test it by $(\dddot{u},\ddot{\pi})$. More in detail, exploiting the formal inequality $\partial^2|\cdot|\dot{\pi}:\ddot{\pi}\ge 0$, here

$$\int_\Omega \frac{\varrho}{2}|\ddot{u}(t)|^2+\frac{1}{2}\mathbb{C}\dot{e}_{\rm EL}(t):\dot{e}_{\rm EL}(t)+\frac{1}{2}H|\dot{\pi}(t)|^2\,{\rm d}x+\int_0^t\!\!\int_\Omega D|\ddot{\pi}|^2\,{\rm d}x$$
$$\le\int_0^t\!\!\int_\Omega \dot{f}\cdot\ddot{u}\,{\rm d}x{\rm d}t+\int_\Omega\frac{\varrho}{2}|\ddot{u}(0)|^2+\frac{1}{2}\mathbb{C}\dot{e}_{\rm EL}(0):\dot{e}_{\rm EL}(0)+\frac{1}{2}H|\dot{\pi}(0)|^2\,{\rm d}x \quad (7.4.15)$$

with $\mathbb{C}={\rm ISO}(K_{\rm E},G_{\rm E})$. The boundedness of the right-hand side, which needs to control $\ddot{u}(0)$, $e_{\rm EL}(0)$, and $\dot{\pi}(0)$, is proven like in the proof of Proposition 7.3.3. From this, (7.4.14a,b) can be seen.

In particular, $\int_Q|\dot{\pi}|\,{\rm d}x{\rm d}t$ is a conventional Lebesgue integral since $|\dot{\pi}|\in L^\infty(I;L^2(\Omega;\mathbb{R}^{d\times d}_{\rm sym}))\subset L^1(Q;\mathbb{R}^{d\times d}_{\rm sym})$, so that (u,π) is the weak solution according Definition 7.2.1 or 7.2.2.

We thus also have the flow-rule satisfied for a.a. $t\in I$, i.e. $s_{\rm YLD}{\rm Dir}(\dot{\pi}(t))\ni 2G_{\rm E}{\rm dev}\,e(u(t){-}\pi(t))-H\pi(t)$. Since $\partial\delta^*_{S_{\rm EL}}(\dot{\pi}(t))\subset\partial\delta^*_{S_{\rm EL}}(0)$ due to the 1-homogeneity and nonnegativity of $\delta^*_{S_{\rm EL}}$, we obtain $2G_{\rm E}{\rm dev}\,e(u(t){-}\pi(t))-H\pi(t)\in s_{\rm YLD}{\rm Dir}(0)$. By the manipulation used already for (7.2.15) but now integrated over Ω, we obtain the semistability (7.4.14c).

The energy conservation can be shown via Proposition 7.3.9. Here, we rely on that we have proved the semistability.

The uniqueness can be obtained by considering two solutions, subtracting the corresponding weak formulations, and test it by the difference of the time derivatives of these solutions. An important fact is that both $\varrho\ddot{u}$ and ${\rm div}\,\sigma$ with $\sigma=\mathbb{C}(e(\dot{u}){-}\dot{\pi})$ are now in duality with $\varrho\dot{u}$, so that the by-part-integration formula can be used. □

After suppressing viscous effects in the plastic flow rule, let us present some alternative or more complicated models. The first modification consists in different sort of hardening, so called *isotropic hardening*. The vector of the internal variables can be chosen as $z:=(\pi,\eta)\in L^2(\Omega;\mathbb{R}^{d\times d}_{\rm dev})\times L^2(\Omega)=:Z$ and is therefore composed from the plastic strain π and a hardening variable η. Then $k=d(d{+}1)/2$. Here we used the notation $\mathbb{R}^{d\times d}_{\rm dev}$ from (7.4.1). We postulate the stored energy and general dissipation potential as

$$\varphi(e,z)\equiv\varphi(e,\pi,\eta):=\frac{d}{2}K_{\rm E}|{\rm sph}\,e|^2+G_{\rm E}|{\rm dev}\,e-\pi|^2+\frac{b}{2}\eta^2, \quad (7.4.16a)$$

$$\zeta_{\rm IN}(\dot{z}):=\delta^*_{S_{\rm EL}}(\dot{\pi},\dot{\eta}) \quad (7.4.16b)$$

where $b>0$ a hardening variable which is now the only non-vanishing component of $\mathbb{C}_2$. The isotropic hardening is a unidirectional process; actually, $\dot{\eta}\ge 0$ is ensured by a suitable choice of $S_{\rm EL}$.

In reality, an activated response with rather smear-out thresholds (=yield stresses) than one single $s_{\rm YLD}$ in Figure 7.3 is often observed. An option to implement several thresholds is a "serial" arrangement of the elementary elasto-plastic elements, cf. also [340, Sect.5.1.3] or [93, 423], sometimes called the *Iwan plasticity model* [257]. It needs to multiplicate the internal variables $z=(\pi_1,\pi_2,...,\pi_n)$ valued in

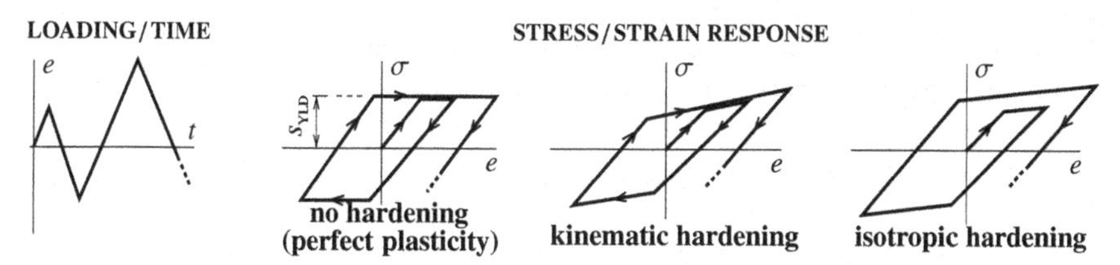

Fig. 7.4 Schematic response on strain loading cycling in time with gradually increasing amplitude (left) of the stress in plastic material without hardening and with kinematic or isotropic hardening.

$(\mathbb{R}^{d\times d}_{\rm dev})^n$. Considering again the kinematic hardening, the rheological scheme can be written as

$$\text{spring}_{\mathbb{C}} - \big(\text{slider}_{s_{\rm YLD,1}} \,\|\, \text{spring}_{H_1}\big) - \big(\text{slider}_{s_{\rm YLD,2}} \,\|\, \text{spring}_{H_2}\big) - \cdots - \big(\text{slider}_{s_{\rm YLD,\mathit{n}}} \,\|\, \text{spring}_{H_n}\big)$$

with $\mathbb{C} = \mathrm{ISO}(K_{\rm E}, G_{\rm E})$ and the plastic elements affecting only the deviatoric response. Thus, beside (7.4.29a,c), one obtains

$$\varphi(e,z) = \varphi(e,\pi_1,...,\pi_n) := \frac{d}{2}K_{\rm E}|\mathrm{sph}\,e|^2 + G_{\rm E}\Big|\mathrm{dev}\,e - \sum_{i=1}^{n}\pi_i\Big|^2 + \frac{1}{2}\sum_{i=1}^{n}H_i|\pi_i|^2 \tag{7.4.17a}$$

$$\zeta_{\rm IN}(\dot z) = \zeta_{\rm IN}(\dot\pi_1,...,\dot\pi_n) := \sum_{i=1}^{n} s_{{\rm YLD},i}|\dot\pi_i| \tag{7.4.17b}$$

Assuming $s_{\rm YLD,1} < s_{\rm YLD,2} \cdots < s_{{\rm YLD},n}$, the particular plastification processes are successively activated by increasing stress. The slope after the first plastification process is activated $G_1 = G_{\rm E}H_1/(G_{\rm E}+H_1)$ as in Fig. 7.3, while in general we have the recursive formula

$$G_i = G_{i-1}H_i/(G_{i-1}+H_i), \qquad G_0 = G_{\rm E}. \tag{7.4.18}$$

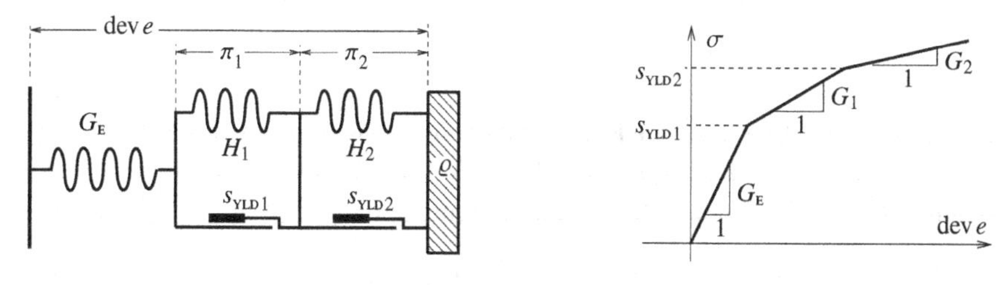

Fig. 7.5 Multiple-yield-stress elasto-plastic Prandtl-Reuss' rheology corresponding to (7.4.17a). Left: a schematic rheological diagram for the deviatoric part, the spherical one being as in Fig. 7.2. Right: stress-strain response on "monotone" loading, starting from $e(0) = \pi_1(0) = \pi_2(0) = 0$ with G_1 and G_2 from (7.4.18).

Example 7.4.7 (Combined hardening). One can combine kinematic with isotropic hardening η as in the model (7.4.16):

$$\varphi(e,z) \equiv \varphi(e,\pi,\eta) := \frac{d}{2}K_{\rm E}|{\rm sph}\,e|^2 + G_{\rm E}|{\rm dev}\,e - \pi|^2 + \frac{1}{2}H|\pi|^2 + \frac{1}{2}b\eta^2. \tag{7.4.19}$$

The flow rule now looks as:

$$\partial\zeta_{\rm IN}(\dot\pi,\dot\eta) + ((\mathbb{C}{+}H\mathbb{I})\pi,\, b\eta) \ni (\mathbb{C}e, 0). \tag{7.4.20}$$

The previous special cases are obtained when $H = 0$ and $b > 0$ (linear isotropic hardening), or $b = 0$ (kinematic hardening). Always, as $z \in L^\infty(I;Z)$, we obtain $\pi \in L^\infty(I;L^2(\Omega;\mathbb{R}^{3\times3}_{\rm dev}))$. This is an advantageous consequence of the hardening effects that plastic strain is bounded in a reflexive space (and not in a non-reflexive bounded-deformation space as it would be without hardening).

Remark 7.4.8 (Energy conservation without regularity). Adding a term $\mathbb{D}e(\dot u)$ into σ in (7.4.4), i.e. considering $\zeta_{\rm VI}(\dot e) = \frac{1}{2}\mathbb{D}\dot e{:}\dot e$ in (7.4.2b), would allow for proving uniqueness and energy conservation, and for usage the concept energetic solutions, see Definition 7.2.4. Without regularity in time like in Proposition 7.4.6, the semistability in the limit, i.e. (7.4.14c), can be proved by explicit construction of a *mutual recovery sequence*, namely

$$\hat\pi_k := \hat\pi + \pi_k - \pi \tag{7.4.21}$$

for $\pi_k \to \pi$ weakly in $L^2(\Omega;\mathbb{R}^{d\times d}_{\rm dev})$. This is to be used for the convergence of the time discretization (Rothe's method) rather than the Galerkin approximation, as we did in the proof of Proposition 7.3.8 but now without the gradient of the plastic strain.[26] We refer to [360, Sect. 5.2.1] for details. Although this rheological model, i.e. ($\mathbb{C}$—($_K\|_H$))$\|_{\mathbb{D}}$, is often used, the Kelvin-Voigt-type viscosity involved on the total strain may however be less conceptional than on the elastic strain, cf. Exercise 7.4.10. This may also be understand as the strain- versus stress-driven plasticity, as articulated in [360, Remark 5.2.2].

Exercise 7.4.9. Realize how Figure 7.3 modifies for limit cases $H = 0$ or $H = +\infty$. Similarly, think also about for $\mathbb{C} = +\infty$ and $H = 0$, which is called ideal *rigid-plastic material.*

Exercise 7.4.10 (Plasticity in Kelvin-Voigt viscoelastic materials). Modify the above plasticity-with-hardening model (i.e. (7.4.2) with $D = 0$) for the rheological scheme ($_{\mathbb{C}}\|_{\mathbb{D}}$)—($_K\|_H$). Realize that the corresponding dissipation potential

$$\zeta(\dot e,\dot\pi) = \frac{1}{2}\mathbb{D}\dot e_{\rm EL}{:}\dot e_{\rm EL} + \frac{1}{2}H|\dot\pi| = \frac{1}{2}\mathbb{D}(\dot e{-}\dot\pi){:}(\dot e{-}\dot\pi) + \frac{1}{2}H|\dot\pi|$$

[26] This so-called binominal trick makes $\zeta_{\rm IN}(\hat\pi_k - \pi_k) = \zeta(\hat\pi - \pi)$ simply constant, hence convergent, while the difference $\varphi(u_k,\pi_k) - \varphi(u_k,\hat\pi_k)$ converges due to the quadratic structure of φ.

does not comply with the splitting (7.1.3). Derive the energy-based a-priori estimates as well as the regularity estimates like (7.4.14a,b) and outline the related analysis.

7.4.3 Perfect plasticity

Further "simplification" after suppressing viscous effects in the flow rule is to omit also hardening. In fact, the hardening is a phenomenon related rather with metals where microscopically it is caused by "locking" of moving dislocations, while in some other materials like rocks there is no hardening. Hence, suppressing hardening is well motivated in some case, as well as the resulted models with no hardening, called *Prandtl elastic-perfect plasticity*, as schematically depicted on Figure 7.6. As

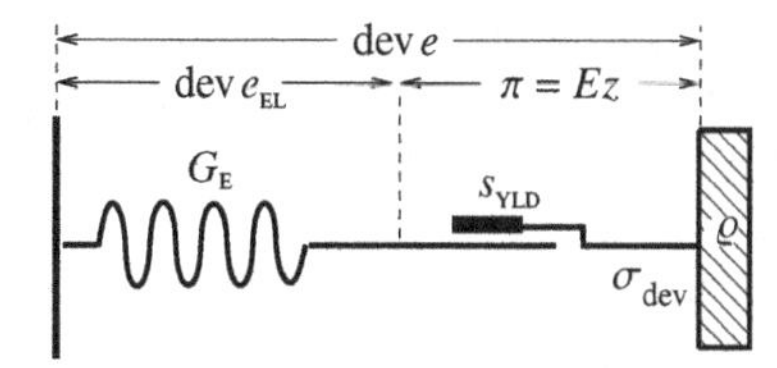

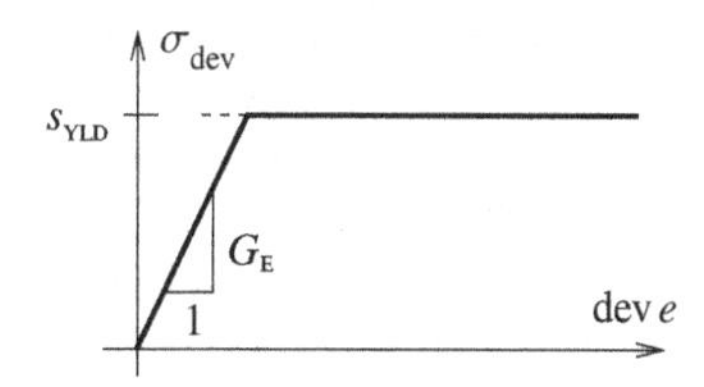

Fig. 7.6 Elasto-plastic Prandtl-Reuss' perfect plasticity.
Left: a rheological diagram for the deviatoric part, the spherical one being as in Fig. 7.2.
Right: stress-strain response on "monotone" shear loading, starting from $e(0)$=$z(0)$=0.

the hardening $H = 0$ here, we have coercivity of φ only in the directions $e(u)-\pi$, however, while the coercivity of π is controlled only through the L^1-type dissipation, and then also $e(u)$ has an L^1-character and, due to possible concentration into a measure supported on $(d-1)$-dimensional manifolds, slip bands can develop. This is an essential phenomenon which makes the analysis quite technical: a nonreflexive, so-called *bounded-deformation space* must replace the usual H^1-spaces, and loading must be qualified carefully, cf. Remark 7.4.13 below. More specifically, this space, defined by

$$\mathrm{BD}^p(\Omega;\mathbb{R}^d) := \{u \in L^p(\Omega;\mathbb{R}^d);\ \ e(u) \in \mathfrak{M}(\Omega;\mathbb{R}^{d\times d}_{\mathrm{sym}})\},$$

has been invented with $p = 1$ by P.-M. Suquet [497], cf. also [509], and its usage for perfect plasticity in the purely quasistatic variant is mainly due to G. Dal Maso et al [139]. A combination of the perfect plasticity with inertia (where the natural choice is $p = 2$) was then first studied in [16] and later in the spirit of Proposition 7.4.11 e.g. in [111].

The convergence in the linear force equilibrium (7.4.4a) is again simple, while for the plastic flow-rule we must either use some weaker concept of solutions or use a nontrivial *Kohn-Temam duality* pairing

$$[\mathrm{dev}\,\sigma{:}\pi](\Omega) = \int_\Omega \sigma{:}(\pi{-}e(u)) - \mathrm{div}\,\sigma\cdot u\,\mathrm{d}x \tag{7.4.22}$$

which needs a smooth boundary.[27] We define the distribution $\mathrm{dev}\,\sigma : \mathrm{dev}\,e(u)$ on Ω as

$$\int_\Omega z[\mathrm{dev}\,\sigma{:}\mathrm{dev}\,e(u)](\mathrm{d}x) := -\int_\Omega z\,\mathrm{div}\,\sigma\cdot u\,\mathrm{d}x - \int_\Omega z\frac{\mathrm{tr}\,\sigma}{d}\cdot\mathrm{div}\,u\,\mathrm{d}x - \int_\Omega \sigma{:}(u\odot\nabla z)\,\mathrm{d}x$$

for every $z\in C^\infty(\Omega)$ compactly supported, where $\odot$ stands for the symmetrized tensor product, namely $a\odot b := (a\otimes b+b\otimes a)/2$ for $a,b\in\mathbb{R}^d$. By [278, Theorem 3.2], it follows that $[\mathrm{dev}\,\sigma{:}\mathrm{dev}\,e(u)]$ is a bounded Radon measure on Ω.

To avoid these nontrivial technicalities and allow only Lipschitz (possibly nonsmooth) domains while accepting only basic existence results, we can rely on some weaker concept of solutions, e.g. (7.2.8), i.e. here:

$$\begin{aligned}\int_Q 2G_{\mathrm{E}}(\pi-\mathrm{dev}e(u))){:}w + s_{\mathrm{YLD}}|w|\,\mathrm{d}x\mathrm{d}t \geq \int_{\bar{Q}} s_{\mathrm{YLD}}|\cdot|\dot{\pi}(\mathrm{d}x\mathrm{d}t)\\ +\int_\Omega \frac{\varrho}{2}\big|\dot{u}(T)\big|^2 + G_{\mathrm{E}}|\mathrm{dev}e(u(T)) - \pi(T)|^2 - f(T)\cdot u(T)\,\mathrm{d}x\\ -\int_\Omega \frac{\varrho}{2}|\dot{u}_0|^2 + G_{\mathrm{E}}|\mathrm{dev}e(u_0)-\pi_0|^2 - f(0)\cdot u_0\,\mathrm{d}x - \int_Q f\cdot\dot{u}\,\mathrm{d}x\mathrm{d}t\end{aligned} \tag{7.4.23}$$

for all $w\in L^2(Q;\mathbb{R}^{d\times d}_{\mathrm{dev}})$. In this sense, an existence of a weak solution to this degenerate hyperbolic system can thus be obtained; cf. also [111] for a similar study but with vanishing kinematic (instead of isotropic) hardening.

An analog of Proposition 7.4.5 relies on our suitable Definition 7.2.2 of the weak solution. Considering kinematic hardening makes a relatively straightforward modification of Proposition 7.4.5 where the vanishing quantity occurs now not in the dissipation but in the stored energy[28]:

Proposition 7.4.11 (Towards perfect plasticity via vanishing hardening). *Let $f\in L^1(I;L^2(\Omega;\mathbb{R}^d))$, $u_0\in W^{1,1}(\Omega;R^d)$ and $\pi_0\in L^1(\Omega;\mathbb{R}^{d\times d}_{\mathrm{dev}})$ such that $e(u_0){-}\pi_0\in L^1(\Omega;\mathbb{R}^{d\times d}_{\mathrm{sym}})$, and $v_0\in L^2(\Omega;\mathbb{R}^d)$. Moreover, let us denote by $(u_\varepsilon,\pi_\varepsilon)$ a solution to the problem (7.4.4) with $D=0$ and $H=\varepsilon H_1$ with $H_1>0$ and with the initial conditions $u_{0,\varepsilon}\in H^1(\Omega;\mathbb{R}^d)$ and $\pi_{0,\varepsilon}\in L^2(\Omega;\mathbb{R}^{d\times d}_{\mathrm{dev}})$ such that $u_{0,\varepsilon}\to u_0$ in $\mathrm{BD}^1(\Omega;\mathbb{R}^d)$ and $\pi_{0,\varepsilon}\to\pi_0$ in $L^1(\Omega;\mathbb{R}^{d\times d}_{\mathrm{dev}})$. Then the following a-priori estimates hold*

$$\|u_\varepsilon\|_{W^{1,\infty}(I;L^2(\Omega;\mathbb{R}^d))} \leq C \quad\text{and}\quad \|e(u_\varepsilon)\|_{L^\infty(I;L^1(\Omega;\mathbb{R}^{d\times d}))} \leq C, \tag{7.4.24a}$$

$$\|\pi_\varepsilon\|_{L^\infty(I;L^1(\Omega;\mathbb{R}^{d\times d}))\cap W^{1,1}(I;L^1(\Omega;\mathbb{R}^{d\times d}))} \leq C, \tag{7.4.24b}$$

$$\|e(u_\varepsilon)-\pi_\varepsilon\|_{L^\infty(I;L^2(\Omega;\mathbb{R}^{d\times d}))} \leq C, \tag{7.4.24c}$$

[27] More specifically, (7.4.22) needs the C^2-regularity of Γ to apply [278, Prop. 2.5] in general, while for $d=2$, owing to the results in [199], Γ Lipschitz suffices.

[28] Vanishing isotropic hardening is a bit more technical because the structure of internal variables is changed in the limit. In the purely rate-independent case without inertia, we refer to [360, Sect. 4.3.1.2].

$$\|\pi_\varepsilon\|_{L^\infty(I;L^2(\Omega;\mathbb{R}^{d\times d}))} \leq C/\sqrt{\varepsilon}, \tag{7.4.24d}$$

$$\|\ddot{u}_\varepsilon\|_{L^1(I;L^2(\Omega;\mathbb{R}^d))+L^\infty(I;H_0^1(\Omega;\mathbb{R}^d)^*)} \leq C. \tag{7.4.24e}$$

For $\varepsilon\to 0$, the sequence $\{(u_\varepsilon,\pi_\varepsilon)\}_{\varepsilon>0}$ contains a subsequence converging weakly in the topologies indicated in (7.4.24a-c,e) and a limit (u,π) satisfies*

$$u \in W^{1,\infty}(I;L^2(\Omega;\mathbb{R}^d)) \cap L^\infty(I;\mathrm{BD}^2(\Omega;\mathbb{R}^d)), \tag{7.4.25a}$$

$$\pi \in L^\infty(I;\mathfrak{M}(\Omega;\mathbb{R}^{d\times d}_{\mathrm{dev}})) \cap \mathrm{BV}(I;L^1(\Omega;\mathbb{R}^{d\times d}_{\mathrm{dev}})), \tag{7.4.25b}$$

$$e(u)-\pi \in W^{1,\infty}(I;L^2(\Omega;\mathbb{R}^{d\times d}_{\mathrm{sym}})). \tag{7.4.25c}$$

Moreover, each (u,π) obtained by this way is a weak solution to the problem (7.4.4) with $D=0$ and $H=0$ in the sense that the force equilibrium is satisfied in the weak sense (7.2.5a), the variational inequality (7.4.23) holds for all $w\in L^2(Q;\mathbb{R}^{d\times d}_{\mathrm{dev}})$, and $u(0)=u_0$ and $\pi(0)=\pi_0$.

Sketch of the proof. By the usual energetic test by $\dot{u}_\varepsilon$ and $\dot{\pi}_\varepsilon$ one obtains the a-priori estimates (7.4.24a-d). More in detail, from the plastic-dissipation term, we obtain the $W^{1,1}$-estimate in (7.4.24b) from which we then obtain the second estimate in (7.4.24b) by the triangle and Hölder's inequalities

$$\begin{aligned}\|e(u_\varepsilon)\|_{L^1(\Omega;\mathbb{R}^{d\times d})} &\leq \|e(u_\varepsilon)-\pi_\varepsilon\|_{L^1(\Omega;\mathbb{R}^{d\times d})} + \|\pi_\varepsilon\|_{L^1(\Omega;\mathbb{R}^{d\times d})}\\ &\leq \sqrt{\mathrm{Meas}_d(\Omega)}\|e(u_\varepsilon)-\pi_\varepsilon\|_{L^2(\Omega;\mathbb{R}^{d\times d})} + \|\pi_\varepsilon\|_{L^1(\Omega;\mathbb{R}^{d\times d})}.\end{aligned}$$

Then, (7.4.24e) follows by comparison $\ddot{u}_\varepsilon = (\mathrm{div}\,\sigma_\varepsilon + f)/\varrho$ since $\sigma_\varepsilon = \mathbb{C}(e(u_\varepsilon)-\pi_\varepsilon)$ is controlled in $L^\infty(I;L^2(\Omega;\mathbb{R}^{d\times d}_{\mathrm{sym}}))$ due to (7.4.24c).

Let us modify suitably Definition 7.2.2. Adding $\varepsilon H_1\pi_\varepsilon$ to the left-hand side of (7.2.1b) and thus also $\int_Q \varepsilon H_1\pi_\varepsilon:(w-\dot{\pi}_\varepsilon)\,\mathrm{d}x\mathrm{d}t$ to the left-hand side of (7.2.3), we will see the same term also on the left-hand side of (7.2.8) written for the solution $(u_\varepsilon,\pi_\varepsilon)$. Forgetting $\int_Q \varepsilon H_1\pi_\varepsilon:\dot{\pi}_\varepsilon\,\mathrm{d}x\mathrm{d}t = \frac{\varepsilon}{2}\int_\Omega H|\pi_\varepsilon(T)|^2\,\mathrm{d}x \geq 0$, we can see that the the "viscous" solution satisfies

$$\begin{aligned}\int_Q (\varepsilon H_1\pi_\varepsilon + 2G_{\mathrm{E}}(\pi_\varepsilon - \mathrm{dev}\,e(u_\varepsilon))):w + s_{\mathrm{YLD}}|w|\,\mathrm{d}x\mathrm{d}t \geq \int_Q s_{\mathrm{YLD}}|\dot{\pi}_\varepsilon|\,\mathrm{d}x\mathrm{d}t\\ + \int_\Omega \frac{\varrho}{2}\big|\dot{u}_\varepsilon(T)\big|^2 + G_{\mathrm{E}}|\mathrm{dev}\,e(u_\varepsilon(T)) - \pi_\varepsilon(T)|^2 - \frac{\varrho}{2}|v_0|^2 - G_{\mathrm{E}}|\mathrm{dev}\,e(u_0)-\pi_0|^2\,\mathrm{d}x\\ - \int_Q f\cdot\dot{u}_\varepsilon\,\mathrm{d}x\mathrm{d}t - \int_0^T \langle\vec{\tau}_{\mathrm{D}},\dot{u}_{\mathrm{D}}\rangle_{\Gamma_{\mathrm{D}}}\mathrm{d}t.\end{aligned} \tag{7.4.26}$$

As usual, we made a selection of weakly* convergent subsequence respecting the topologies indicated in (7.4.24a-c,e). Here, an important ingredient is that, by (7.4.24d), we have $\{\pi_\varepsilon\}_{\varepsilon>0}$ bounded in $\mathrm{BV}(I;L^1(\Omega;\mathbb{R}^{d\times d}_{\mathrm{dev}}))$ and, by Helly's selection-principle Theorem B.4.13 on p. 526, we can consider a subsequence such that $\pi_\varepsilon(t)\to\pi(t)$ weakly* in $\mathfrak{M}(\bar{\Omega};\mathbb{R}^{d\times d}_{\mathrm{dev}})$ for any $t\in I$, and in particular for $t=T$, too.

Due to the inertial effects, we have also $\dot{u}_\varepsilon(t) \to \dot{u}(t)$ weakly in $L^2(\Omega;\mathbb{R}^d)$ for any $t\in I$, and in particular for $t = T$, too. Thus also $u_\varepsilon(T) \to u(T)$ weakly* in $\mathrm{BD}^2(\Omega;\mathbb{R}^d)$, too.

The estimate in (7.4.24d) ensures $\|\varepsilon H_1\pi_\varepsilon\|_{L^2(Q;\mathbb{R}^{d\times d})} = \varepsilon H_1\|\pi_\varepsilon\|_{L^2(Q;\mathbb{R}^{d\times d})} = \mathscr{O}(\sqrt{\varepsilon})$ so that, in particular, $\int_Q \varepsilon H_1\pi_\varepsilon{\cdot}w\,\mathrm{d}x\mathrm{d}t \to 0$. The limit of the other terms can be made by continuity or semicontinuity. In particular, $\liminf_{\varepsilon\to0}\int_Q s_{\text{YLD}}|\dot{\pi}_\varepsilon|\,\mathrm{d}x\mathrm{d}t \geq s_{\text{YLD}}|\dot{\pi}| = s_{\text{YLD}}\mathrm{Var}(\pi;I)$ where $\mathrm{Var}(\pi;I)$ is the total variation is defined by (7.2.16) with $\zeta_{\text{IN}} = s_{\text{YLD}}|\cdot|$. Thus (7.4.26) is obtained. □

Proposition 7.4.12 (Regularity, uniqueness, energy conservation). *Let, in addition, (6.4.14) with (6.3.12c) hold. Then the estimates (7.3.10a,b) now improves for*

$$u \in W^{2,\infty}(I;L^2(\Omega;\mathbb{R}^d))\cap W^{1,\infty}(I;\mathrm{BD}^2(\Omega;\mathbb{R}^d)), \tag{7.4.27a}$$

$$\pi \in W^{1,\infty}(I;\mathfrak{M}(\Omega;\mathbb{R}^{d\times d}_{\text{dev}})), \tag{7.4.27b}$$

$$e(u)-\pi \in W^{1,\infty}(I;L^2(\Omega;\mathbb{R}^{d\times d}_{\text{sym}})). \tag{7.4.27c}$$

Moreover, this weak solution is unique and, if also the initial condition π_0 is semi-stable in the sense that

$$\int_\Omega G_{\text{E}}|\mathrm{dev}e(u_0)-\pi_0|^2\,\mathrm{d}x \leq \int_\Omega G_{\text{E}}|\mathrm{dev}e(u_0)-\tilde{\pi}|^2\,\mathrm{d}x + \int_\Omega s_{\text{YLD}}|\cdot|[\pi_0{-}\tilde{\pi}](\mathrm{d}x)$$

for all $\tilde{\pi} \in \mathfrak{M}(\Omega;\mathbb{R}^{d\times d}_{\text{dev}})$, then this solution also conserves energy.

Sketch of the proof. The a-priori estimates (7.4.27) can be obtained when differentiating both equations/inclusions in (7.4.4) considered now with $D = 0$ and $H = 0$ in time and testing by $\ddot{u}$ and $\ddot{\pi}$, respectively. Here we used, written formally, that $[\mathrm{Dir}(\dot{\pi})]^{\boldsymbol{\cdot}}{:}\ddot{\pi} = [\partial^2|\cdot|](\dot{\pi})\ddot{\pi}{:}\ddot{\pi} \geq 0$.

As now $\varrho\dddot{u}\!\in\! L^\infty(I;L^2(\Omega;\mathbb{R}^d))$ is in duality with $\dot{u} \in L^\infty(I;\mathrm{BD}^2(\Omega;\mathbb{R}^d)) \subset L^1(I;L^2(\Omega;\mathbb{R}^d))$ and $\sigma \in W^{1,\infty}(I;L^2(\Omega;\mathbb{R}^{d\times d}_{\text{sym}}))$ is in duality with $e(u) - \pi \in W^{1,\infty}(I;L^2(\Omega;\mathbb{R}^{d\times d}_{\text{sym}})) \subset L^1(I;L^2(\Omega;\mathbb{R}^{d\times d}_{\text{sym}}))$, we can consider two weak solutions (u_1,π_1) and (u_2,π_2), subtract the respective weak formulations and test it by $\dot{u}_1{-}\dot{u}_2$ and $\dot{\pi}_1{-}\dot{\pi}_2$. We thus obtain

$$\begin{aligned}
\frac{\mathrm{d}}{\mathrm{d}t}\int_\Omega \frac{\varrho}{2}|\dot{u}_1{-}\dot{u}_2|^2 + G_{\text{E}}|\mathrm{dev}\,e(u_1{-}u_2)-\pi_1{+}\pi_2|^2 + \frac12 K_{\text{E}}|\mathrm{sph}\,e(u_1{-}u_2)|^2\,\mathrm{d}x&\\
\leq \frac{\mathrm{d}}{\mathrm{d}t}\int_\Omega \frac{\varrho}{2}|\dot{u}_1{-}\dot{u}_2|^2 + G_{\text{E}}|\mathrm{dev}\,e(u_1{-}u_2)-\pi_1{+}\pi_2|^2 + \frac12 K_{\text{E}}|\mathrm{sph}\,e(u_1{-}u_2)|^2\,\mathrm{d}x&\\
+\int_0^t\!\!\int_\Omega s_{\text{YLD}}(\mathrm{Dir}(\dot{\pi}_1){-}\mathrm{Dir}(\dot{\pi}_2)){:}(\dot{\pi}_1{-}\dot{\pi}_2)\,\mathrm{d}x\mathrm{d}t = 0,&
\end{aligned} \tag{7.4.28}$$

from which we can see $(u_1,\pi_1) = (u_2,\pi_2)$, i.e. the uniqueness. The last integral in (7.4.28) is non-negative and, in fact, rather formal.

The energy conservation is rather technical. One is to prove that the weak solution is also an energetic solution, so that in particular the semistability holds for all $t\in I$, i.e.

$$\int_\Omega G_{\rm E}|{\rm dev}\, e(u(t)) - \pi(t)|^2\,{\rm d}x \le \int_\Omega G_{\rm E}|{\rm dev}\, e(u(t)) - \tilde\pi|^2\,{\rm d}x + \int_\Omega s_{\rm YLD}|\cdot|\,[\pi(t){-}\tilde\pi]({\rm d}x),$$

for all $\tilde\pi \in \mathfrak{M}(\Omega;\mathbb{R}^{d\times d}_{\rm dev})$. Then one is to use Proposition 7.3.9. □

We already mentioned that in reality, a "more round" activated response with rather smear-out thresholds than a single one in Figure 7.6 often occurs. Here, in the context of perfect plasticity, we can consider the model like (7.4.29) but omitting the hardening. Thus, for Prandtl elastic-perfectly-plastic elements (i.e. $H_i = 0$), the equations governing the evolution of $(u,z) = (u,\pi_1,...,\pi_n)$ are

$$\varrho\ddot u - {\rm div}\,\sigma = f \qquad \text{with} \quad \sigma = dK_{\rm E}{\rm sph}\, e(u) + \textstyle\sum_{i=1}^n \sigma_{{\rm dev},i}, \tag{7.4.29a}$$

$$s_{{\rm YLD},i}{\rm Dir}(\dot\pi_i) \ni \sigma_{{\rm dev},i} := 2G_{{\rm E},i}({\rm dev}\, e(u){-}\pi_i). \tag{7.4.29b}$$

Assuming $s_{{\rm YLD},1}/G_{{\rm E},1} < s_{{\rm YLD},2}/G_{{\rm E},2} < \cdots < s_{{\rm YLD},n}/G_{{\rm E},n}$, the dry-friction elements are subsequently activated by gradually increasing loading and the monotone-loading curve exhibits multiple slope brakes, as schematically depicted on Figure 7.7 for $n = 3$. Let us note that, like in a single-threshold case, this material is not coercive if $s_{{\rm YLD},n} < +\infty$.

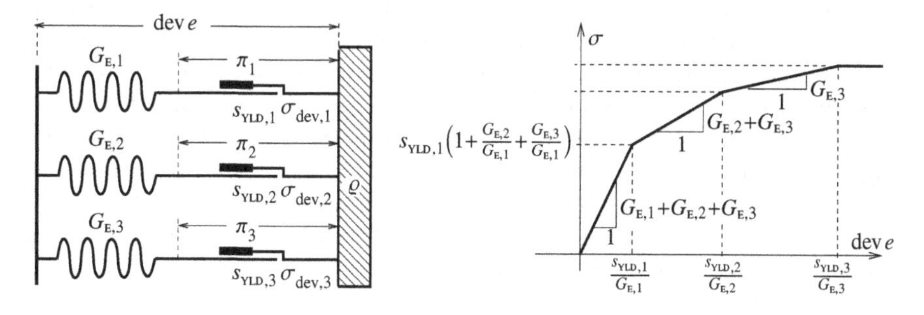

Fig. 7.7 Multiple-yield-stress elastic-perfectly-plastic Prandtl's rheology.
Left: a schematic rheological diagram ($n = 3$).
Right: stress-strain response on "monotone" loading, starting from $e(0) = \pi_1(0) = ... = \pi_n(0) = 0$.

Remark 7.4.13 (Safe load). From Fig. 7.6, one intuitively see that the perfectly-plastic material withstand only a limited stress, which causes that only a limited force load can be applied to guarantee bounded strain response. This is particularly relevant in the quasistatic problems considering $\varrho = 0$ where it is referred as a *safe-load* condition. In the dynamical problems, the bulk load can be possibly large but still the traction load in a boundary condition like $\sigma\vec n = g$ must be sufficiently small. This is seen from the energy estimate

$$\int_\Omega \frac{\varrho}{2}|\dot u(t)|^2 + G_{\rm E}|e(u(t)) - \pi(t)|^2{\rm d}x + \int_{[0,t]\times\Omega} s_{\rm YLD}|\cdot|\dot\pi({\rm d}x{\rm d}t)$$

$$\leq \int_0^t \Big(\int_\Omega f\cdot\dot u \,\mathrm{d}x + \int_\Gamma g\cdot\dot u \,\mathrm{d}S \Big)\mathrm{d}t + \int_\Omega \frac{\varrho}{2}|v_0|^2 + G_{\rm E}|e(u_0) - \pi_0|^2 \mathrm{d}x$$

$$\leq \frac{1}{2\min\varrho}\|f\|^2_{L^1(I;L^2(\Omega;\mathbb{R}^d))} + \frac{1}{2}\min\varrho \int_0^t \|u\|^2_{L^2(\Omega;\mathbb{R}^d)}\,\mathrm{d}t + \int_\Gamma g(t)\cdot u(t)\,\mathrm{d}S$$

$$- \int_0^t\int_\Gamma \dot g\cdot u \,\mathrm{d}S\,\mathrm{d}t - \int_\Gamma g(0)\cdot u_0 \,\mathrm{d}S + \int_\Omega \frac{\varrho}{2}|v_0|^2 + G_{\rm E}|e(u_0) - \pi_0|^2 \mathrm{d}x$$

when applying the fact that functions from $\mathrm{BD}^1(\Omega;\mathbb{R}^d)$ has traces in $L^1(\Gamma)$ and the estimate $\|u|_\Gamma\|_{L^1(\Gamma)} \leq C\|u\|_{\mathrm{BD}^1(\Omega;\mathbb{R}^d)}$ holds. Then one can use Gronwall's inequality provided g is enough small in the $W^{1,1}(I;L^\infty(\Gamma;\mathbb{R}^d))$ norm. This is a natural requirement since otherwise some shear bands can occur on the boundary in an uncontrolled way.

Exercise 7.4.14 (Perfect plasticity in Kelvin-Voigt viscoelastic materials). Analyze the model $({}_{\rm C}\|{}_{\rm D})$—${}_{S_{\rm YLD}}$.

7.5 Various damage models

The simplest examples of nonconvex stored energy $\varphi(e,z)$ are models of damage. The most typical models use z as just a scalar-valued (i.e. $k = 1$) non-negative internal variable, which we will denote by α, having the interpretation as a phenomenological volume fraction of microcracks or microvoids manifested macroscopically as a certain weakening of the elastic response. This concept was invented by L.M. Kachanov [267] and Yu.N. Rabotnov [425]. There are two conventions: damaging means α increasing and $\alpha = 1$ means maximal damage (which is used in engineering[29] or e.g. also in geophysics) or, conversely, damaging means α decreasing and $\alpha = 0$ means maximal damage (which is used in mathematical literature and also here).

The variety of damage models is quite large. Basic alternatives are unidirectional damage (i.e. no healing is allowed, relevant in most engineering materials) versus reversible damage (i.e. a certain reconstruction of the material is possible, relevant e.g. in rock mechanics in the time scales of thousands years or more). Likewise in plasticity models in Section 7.4, another alternative is rate dependent damage versus rate independent one. And, of course, damage models can be incorporated into various viscoelastic models, and damage can influence not only the stored energy but also the dissipation potential.

Most mathematical models at small strains consider the stored energy $\varphi(e,\alpha)$ quadratic in terms of e. On the other hand, a non-quadratic $\varphi(\cdot,\alpha)$ is a reasonable ansatz in particular that damage may act very differently on compression than on tension, and on shear load. Usually, in concrete- or masonry-, or rock-type materials,

[29] Often, d is used instead of α and elastic moduli subjected to damage are considered premultiplied with the factor $1-d$, and often in a certain jargon such models are refereed as $(1-d)$-models.

compression does not cause damage while tension may cause it relatively easily, as well as shear, cf. Example 7.5.6. In most models, the gradient of the damage variable is needed from mathematical reasons and, related to it, has also a desirable modeling interpretation allowing to control some length-scale in damage profiles.

We begin with a relatively simple model of damage in Kelvin-Voigt viscoelastic solid. The ansatz (7.3.19) with (7.1.3) now takes a more concrete form:

$$\varphi_{\rm E}(e,\alpha,\nabla\alpha) = \varphi(e,\alpha) - \varphi_{\rm DAM}(\alpha) + \delta_{[0,1]}(\alpha) + \frac{\kappa}{p}|\nabla\alpha|^p \quad \text{and} \tag{7.5.1a}$$

$$\zeta(\alpha;\dot e,\dot\alpha) = \zeta_{\rm VI}(\alpha;\dot e) + \zeta_{\rm IN}(\dot\alpha) \quad \text{with} \quad \zeta_{\rm VI}(\alpha;\dot e) = \frac{1}{2}\mathbb{D}(\alpha)\dot e{:}\dot e, \tag{7.5.1b}$$

with $\varphi_{\rm DAM}$ standing for the specific energy of damage which (if $\varphi_{\rm DAM}$ is increasing) gives rise to a driving force for healing and with $\delta_{[0,1]}(\cdot):\mathbb{R}\to\{0,\infty\}$ denoting the indicator function of the interval $[0,1]$ where the damage variable is assumed to take its values, and with $\zeta_{\rm IN}(\dot\alpha)$ to be specified later in particular models. It is natural to consider the stored energy $\varphi(\cdot,\alpha)$ convex with minimum at $e=0$, which is why we split the $\varphi_{\rm DAM}$-term out of φ.

The system (7.3.18) takes the form

$$\varrho\ddot u - {\rm div}\big(\mathbb{D}(\alpha)e(\dot u) + \partial_e\varphi(e(u),\alpha)\big) = f \qquad \text{in } Q, \tag{7.5.2a}$$

$$\partial\zeta_{\rm IN}(\dot\alpha) + \partial_\alpha\varphi(e(u),\alpha) + r_{\rm c} \ni \varphi_{\rm DAM}'(\alpha) + {\rm div}(\kappa|\nabla\alpha|^{p-2}\nabla\alpha) \qquad \text{with } r_{\rm c}\in N_{[0,1]}(\alpha) \qquad \text{in } Q, \tag{7.5.2b}$$

where $r_{\rm c}$ is a "reaction" force (pressure) to the constraint $0\le\alpha\le1$. Omitting Dirichlet boundary conditions for simplicity, we consider the initial-boundary value problem

$$(\mathbb{D}(\alpha)e(\dot u) + \partial_e\varphi(e(u),\alpha))\vec n = g \quad \text{and} \quad \nabla\alpha\cdot\vec n = 0 \qquad \text{on } \Sigma, \tag{7.5.2c}$$

$$u(0)=u_0, \quad \dot u(0)=v_0, \quad \alpha(0)=\alpha_0 \qquad \text{on } \Omega. \tag{7.5.2d}$$

The energetics can be obtained by testing (7.5.2a) by $\dot u$ and (7.5.2b) by $\dot\alpha$. This test yields, at least formally,

$$\int_\Omega \frac{\varrho}{2}|\dot u(t)|^2 + \varphi(e(u(t)),\alpha(t)) - \varphi_{\rm DAM}(\alpha(t)) + \frac{\kappa}{p}|\nabla\alpha(t)|^p\,{\rm d}x + \int_0^t\!\!\int_\Omega \mathbb{D}(\alpha)e(\dot u){:}e(\dot u) + \dot\alpha\zeta_{\rm IN}(\dot\alpha)\,{\rm d}x{\rm d}t$$
$$= \int_\Omega \frac{\varrho}{2}|v_0|^2 + \varphi(e(u_0),\alpha_0) - \varphi_{\rm DAM}(\alpha_0) + \frac{\kappa}{p}|\nabla\alpha_0|^p\,{\rm d}x + \int_0^t\!\!\int_\Omega f\cdot\dot u\,{\rm d}x{\rm d}t + \int_0^t\!\!\int_\Sigma g\cdot\dot u\,{\rm d}x{\rm d}t. \tag{7.5.3}$$

The weak solution to (7.5.2b) containing two inclusions is now to combine the inequalities (7.2.5b) with an inequality for $r_{\rm c}\in N_{[0,1]}(\alpha)$. Of course, it should be accompanied with the weak formulation of (7.5.2a) as usual, e.g. using (7.2.5a), which altogether here leads to:

Definition 7.5.1 (Weak solutions). The triple $(u,\alpha,r_{\rm c}) \in H^1(I;H^1(\Omega;\mathbb{R}^d)) \times C_{\rm w}(I;W^{1,p}(\Omega))\times L^2(Q)$ is a weak solution to the initial-boundary-value problem (7.5.2) if $\partial_\alpha\varphi(e(u),\alpha), \xi_{\rm IN}(\dot\alpha) \in L^1(Q)$, $0\le\alpha\le 1$ a.e. on Q, and

$$\forall v\in L^2(I;H^1_{\rm D}(\Omega;\mathbb{R}^d))\cap H^1(I;L^2(\Omega;\mathbb{R}^d)),\ \ v|_{t=T}=0:$$

$$\int_Q \sigma{:}e(v)-\varrho\dot u{\cdot}\dot v\,{\rm d}x{\rm d}t+\int_\Omega \varrho\dot u(T){\cdot}v(T,\cdot)\,{\rm d}x = \int_\Omega \varrho v_0{\cdot}v(0,\cdot)\,{\rm d}x+\int_Q f{\cdot}v\,{\rm d}x{\rm d}t+\int_{\Sigma_{\rm N}} g{\cdot}v\,{\rm d}S\,{\rm d}t, \tag{7.5.4a}$$

$$\forall w\in L^2(I;H^1(\Omega)):$$

$$\int_Q (\partial_\alpha\varphi(e(u),\alpha)+r_{\rm c}-\varphi_{\rm DAM}'(\alpha))w+\kappa|\nabla\alpha|^{p-2}\nabla\alpha{\cdot}\nabla w+\zeta_{\rm IN}(w)\,{\rm d}x{\rm d}t$$
$$\ge\int_Q \zeta_{\rm IN}(\dot\alpha)+\partial_e\varphi(e(u),\alpha){:}e(\dot u)\,{\rm d}x{\rm d}t+\int_\Omega\Big(\varphi(e(u(T)),\alpha(T))-\varphi_{\rm DAM}(\alpha(T))$$
$$+\frac{\kappa}{p}|\nabla\alpha(T)|^p-\varphi(e(u_0),\alpha_0)+\varphi_{\rm DAM}(\alpha_0)-\frac{\kappa}{p}|\nabla\alpha_0|^p\Big){\rm d}x, \tag{7.5.4b}$$

$$\forall v\in L^2(Q),\ \ 0\le v\le 1:\qquad \int_Q r_{\rm c}(v-\alpha)\,{\rm d}x{\rm d}t\ge 0, \tag{7.5.4c}$$

for $\sigma=\mathbb{D}(\alpha)e(\dot u)+\partial_e\varphi(e(u),\alpha)\in L^2(Q;\mathbb{R}^{d\times d}_{\rm sym})$ in (7.5.4a), and if also the resting initial conditions hold, i.e. $u|_{t=0}=u_0$ and $\alpha|_{t=0}=\alpha_0$.

Remark 7.5.2 (An alternative weak formulation). For (7.5.4b), we used the definition of the subdifferential $\partial\xi_{\rm IN}(\dot\alpha)$ so that the inclusion $\partial_e\varphi(e(u),\alpha)-\varphi_{\rm DAM}'(\alpha)+r_{\rm c}-{\rm div}(\kappa|\nabla\alpha|^{p-2}\nabla\alpha)\in\partial\xi_{\rm IN}(\dot\alpha)$ yields

$$\int_Q (\partial_\alpha\varphi(e(u),\alpha)-\varphi_{\rm DAM}'(\alpha))(w-\dot\alpha)+r_{\rm c}w+\kappa\nabla\alpha{\cdot}\nabla w+\zeta_{\rm IN}(w)\,{\rm d}x{\rm d}t$$
$$\ge\int_Q\zeta_{\rm IN}(\dot\alpha)\,{\rm d}x{\rm d}t+\int_\Omega\Big(\frac{\kappa}{p}|\nabla\alpha(T)|^p-\frac{\kappa}{p}|\nabla\alpha_0|^p\Big){\rm d}x, \tag{7.5.5}$$

when we have used Green's formula for the gradient term and substituted $\int_Q\kappa|\nabla\alpha|^{p-2}\nabla\alpha{\cdot}\nabla\dot\alpha\,{\rm d}x{\rm d}t=\int_\Omega\kappa|\nabla\alpha(T)|^p-\varphi_{\rm DAM}(\alpha(T))-\kappa|\nabla\alpha_0|^p\,{\rm d}x$, and $\int_Q r_{\rm c}\dot\alpha\,{\rm d}x{\rm d}t=\int_\Omega\delta_{[0,1]}(\alpha(T))-\delta_{[0,1]}(\alpha_0)\,{\rm d}x=0$ provided $0\le\alpha_0(\cdot)\le1$ on Ω. Also, we substituted $\int_Q\varphi_{\rm DAM}'(\alpha)\dot\alpha\,{\rm d}x{\rm d}t=\int_\Omega\varphi_{\rm DAM}(\alpha(T))-\varphi_{\rm DAM}(\alpha_0)\,{\rm d}x$ but this is not needed because, except rate-independent damage as in Remark 7.5.11, $\dot\alpha\in L^1(Q)$ is in duality with $\varphi_{\rm DAM}'(\alpha)\in L^\infty(Q)$. We arrive to (7.5.4b) when substitute further the (possibly not integrable) term $\int_Q\partial_\alpha\varphi(e(u),\alpha)\dot\alpha\,{\rm d}x{\rm d}t$ from the chain rule $\int_Q\partial_\alpha\varphi(e(u),\alpha)\dot\alpha\,{\rm d}x{\rm d}t=\int_\Omega\varphi(e(u(T)),\alpha(T))-\varphi(e(u_0),\alpha_0)\,{\rm d}x-\int_Q\partial_e\varphi(e(u),\alpha){:}e(\dot u)\,{\rm d}x{\rm d}t$. If both $\partial_\alpha\varphi(e(u),\alpha)$ and $\dot\alpha$ live in $L^2(Q)$, we do not need to make this substitution and, instead of (7.5.4b), use directly (7.5.5).

7.5.1 Damage in general nonlinear Kelvin-Voigt materials

We first analyze the model with the viscous moduli tensor $\mathbb{D}$ considered (a bit misconceptually) not subjected to damage, which will be generally needed for proving the strong convergence (7.5.9) except one particular situation in Proposition 7.5.5(ii), and for (7.5.14) and (7.5.15), too; for a damageable viscosity see Sections 7.5.2 and 7.5.3 below. Let us note that, as $e(\dot{u})$ is well controlled in the Kelvin-Voigt model, we do not need to use the more complicated inequality (7.2.8) instead of (7.5.4b). We impose relatively strong data qualifications which allows for a relatively simple analysis and still quite reasonable applications. More specifically, for some $\epsilon > 0$, we assume:

$$\varphi(\cdot,\alpha):\mathbb{R}^{d\times d}_{\rm sym}\to\mathbb{R}^+ \text{ convex, smooth, and coercive: } \varphi(e,\alpha)\ge\epsilon|e|^2, \tag{7.5.6a}$$

$$\varphi_{\rm DAM}:[0,1]\to\mathbb{R} \text{ continuously differentiable,} \tag{7.5.6b}$$

$$\big|\partial_e\varphi(e,\alpha)\big|\le C(1+|e|), \qquad \big|\partial_\alpha\varphi(e,\alpha)\big|\le C(1+|e|), \tag{7.5.6c}$$

$$\zeta_{\rm IN}:\mathbb{R}\to\mathbb{R}^+ \text{ convex, } \quad \epsilon|v|^2\le\zeta_{\rm IN}(v)\le(1+|v|^2)/\epsilon, \tag{7.5.6d}$$

$$\alpha_0\in H^1(\Omega), \quad 0\le\alpha_0\le1 \text{ a.e. on } \Omega. \tag{7.5.6e}$$

Without going into technical details, let us assume some approximate solution $(u_k,\alpha_k,r_{{\rm c},k})_{k\in\mathbb{N}}$ is at disposal, e.g. by a H^2-conformal[30] Galerkin's method; for simplicity, we do not make any regularization of the inclusion (7.5.2b) and admit that the resulted system of ordinary differential inclusions (rather than equations) can effectively be solved. Mostly, the simplest choice of a linear gradient-damage model works, so also here we consider $p = 2$.

Proposition 7.5.3 (Damage in Kelvin-Voigt materials: weak solutions). *Let (7.5.6) hold, ϱ, $\mathbb{D}$, and the loading satisfy (6.4.5a-c,e). Then the approximate solution $(u_k,\alpha_k,r_{{\rm c},k})_{k\in\mathbb{N}}$ satisfy the estimates*

$$\|u_k\|_{W^{1,\infty}(I;L^2(\Omega;\mathbb{R}^d))\cap H^1(I;H^1(\Omega;\mathbb{R}^d))}\le C, \tag{7.5.7a}$$

$$\|\alpha_k\|_{H^1(I;L^2(\Omega))\cap L^\infty(I;H^1(\Omega))\cap L^\infty(Q)}\le C, \tag{7.5.7b}$$

$$\|\Delta\alpha_k\|_{L^2(Q)}\le C \quad \text{and} \quad \|r_{{\rm c},k}\|_{L^2(Q)}\le C. \tag{7.5.7c}$$

Moreover, each limit $(u,\alpha,r_{\rm c})$ of a selected subsequence converging weakly in topologies specified in (7.5.7) is a weak solution to the initial-boundary-value problem (7.5.2) in the sense of Definition 7.5.1 with Remark 7.5.2, and even conserves energy in the sense that (7.5.3) holds for any $t\in I$.*

Proof. From (7.5.3) written for the Galerkin solution, the estimates (7.5.7a,b) follows by Young's and Gronwall's inequalities. The estimate (7.5.7b) follows by the

[30] Here, we assume to take a base for the Galerkin method in such a way that $\Delta\alpha_k$ is still valued in the respective finite-dimensional subspace; it needs special finite-dimensional spaces spanned on the eigen-functions of the Δ-operator on Ω.

test of (7.5.2b) by $\Delta\alpha_k$. Here it is important that $\partial\zeta_{\rm IN}(\dot\alpha_k)\in L^2(Q)$ due to the growth restriction (7.5.6d) and that $\partial_\alpha\varphi(e(u_k),\alpha_k)\in L^2(Q)$ due to the growth restriction (7.5.6c) and that, formally,

$$\int_\Omega r_{{\rm c},k}\Delta\alpha_k\,{\rm d}x = \int_\Omega \partial\delta_{[0,1]}(\alpha_k)\Delta\alpha_k\,{\rm d}x$$
$$= -\int_\Omega \nabla[\partial\delta_{[0,1]}(\alpha_k)]\cdot\nabla\alpha_k\,{\rm d}x = -\int_\Omega \partial^2\delta_{[0,1]}(\alpha_k)\nabla\alpha_k\cdot\nabla\alpha_k\,{\rm d}x \le 0 \qquad (7.5.8)$$

because the (generalized) Hessian $\partial^2\delta_{[0,1]}(\cdot)$ of the convex function $\delta_{[0,1]}(\cdot)$ is positive semi-definite.[31] By comparison, we then obtain also $r_{{\rm c},k}$ estimated in $L^2(Q)$.

After making apriori estimates, we perform the convergence proof. Here we need strong convergence of $e(u_k)$ to pass to the limit in the nonlinear term $\partial_\alpha\varphi(e(u_k),\alpha_k)$ towards (7.5.4b). To this goal, we proceed like in (7.3.15) but modified here by using strong convergence of α_k:[32]

$$\int_\Omega \frac12\mathbb{D}e(u_k(t){-}u(t)){:}e(u_k(t){-}u(t))\,{\rm d}x$$
$$+\int_0^t\!\!\int_\Omega (\partial_e\varphi(e(u_k),\alpha_k)-\partial_e\varphi(e(u),\alpha_k)){:}e(u_k{-}u)\,{\rm d}x{\rm d}t$$
$$= \int_0^t\!\!\int_\Omega (\mathbb{D}e(\dot u_k)+\partial_e\varphi(e(u_k),\alpha_k)-\mathbb{D}e(\dot u)-\partial_e\varphi(e(u),\alpha_k)){:}e(u_k{-}u)\,{\rm d}x{\rm d}t$$
$$= \int_0^t\!\!\int_\Omega (f-\varrho\ddot u_k)\cdot(u_k{-}u)-(\mathbb{D}e(\dot u)+\partial_e\varphi(e(u),\alpha_k)){:}e(u_k{-}u)\,{\rm d}x{\rm d}t$$
$$= \int_0^t\!\!\int_\Omega f\cdot(u_k{-}u)+\varrho\dot u_k\cdot(\dot u_k{-}\dot u)-(\mathbb{D}e(\dot u)+\partial_e\varphi(e(u),\alpha)){:}e(u_k{-}u)\,{\rm d}x{\rm d}t$$
$$+\int_0^t\!\!\int_\Omega (\partial_e\varphi(e(u),\alpha){-}\partial_e\varphi(e(u),\alpha_k)){:}e(u_k{-}u)\,{\rm d}x{\rm d}t$$
$$-\int_\Omega \varrho\dot u_k(t)\cdot(u_k(t){-}u(t))\,{\rm d}x \to 0. \qquad (7.5.9)$$

For the convergence to 0 in (7.5.9), we used that $\dot u_k\to\dot u$ strongly $L^2(Q;\mathbb{R}^d)$ due to the Aubin-Lions theorem, relying on the estimate (7.5.7a), and also $\dot u_k(t)$ is bounded in $L^2(\Omega;\mathbb{R}^d)$ while $u_k(t)\to u(t)$ strongly in $L^2(\Omega;\mathbb{R}^d)$ due to the Rellich compact embedding $H^1(\Omega)\subset L^2(\Omega)$, and also $\partial_e\varphi(e(u),\alpha_k)\to\partial_e\varphi(e(u),\alpha)$ strongly in $L^2(Q;\mathbb{R}^{d\times d})$ due to the continuity of the Nemytskiĭ mapping induced by $\partial_e\varphi(e(u,\cdot)$ and $\alpha_k\to\alpha$ in $L^2(Q)$ again just by the Rellich theorem since both $\dot\alpha_k$ and $\nabla\alpha_k$ is estimated in L^2-spaces. Thus, from (7.5.9), we obtain $e(u_k(t))\to e(u(t))$ strongly in $L^2(\Omega;\mathbb{R}^{d\times d})$ for all $t\in I$. This we used with $t=T$ for the limit passage towards (7.5.4b), although, in fact, weak convergence would suffice for it. Also, we use it

[31] See e.g. [360, Remark 4.3.20].

[32] Actually, (7.5.9) is to be understood rather as a conceptual strategy. Particular approximation method needs usually some further technicalities. E.g. the mentioned Galerkin method needs some strongly converging approximation of the limit u to make $u_k{-}u$ legitimate for the used test of the Galerkin approximation of the force balance.

for a general $t \in I$ to obtain $e(u_k) \to e(u)$ strongly in $L^2(Q;\mathbb{R}^{d\times d})$ by the Lebesgue theorem B.2.2 to be used for (7.5.4b), too.

The limit passage in (7.5.4c) is quite simple because $r_{c,k}$ converges weakly in $L^2(Q)$ while α_k converges strongly in $L^2(Q)$ by the Aubin-Lions theorem. Also the constrains $0 \le \alpha_k \le 1$ are preserved in the limit.

For the energy conservation, using (6.3.5), we exploit that $\sqrt{\varrho}\ddot{u} \in L^2(I;H^1(\Omega;\mathbb{R}^d)^*)$ is in duality with $\sqrt{\varrho}\dot{u} \in L^2(I;H^1(\Omega;\mathbb{R}^d))$ and also $\operatorname{div}(\mathbb{D}e(\dot{u}) + \partial_e\varphi(e(u),\alpha)) \in L^2(I;H^1(\Omega;\mathbb{R}^d)^*)$ is in duality with $\dot{u} \in L^2(I;H^1(\Omega;\mathbb{R}^d))$. Also, we exploit that $\Delta\alpha \in L^2(Q)$ is in duality with $\dot{\alpha} \in L^2(Q)$, as proved in (7.5.7b,c). Actually, this is used to justify the *chain-rule formula*:

$$\int_Q \dot{\alpha}\,\operatorname{div}(\kappa\nabla\alpha)\,\mathrm{d}x\mathrm{d}t = \int_\Omega \frac{\kappa}{2}|\nabla\alpha_0|^2 - \frac{\kappa}{2}|\nabla\alpha(T)|^2\mathrm{d}x. \tag{7.5.10}$$

In fact, this formula is quite nontrivial and can be proved e.g. by mollifying α with respect to the spatial variables.[33] In addition, we have used the chain-rule $\int_0^t\int_\Omega r_c\dot{\alpha}\,\mathrm{d}x\mathrm{d}t = \int_\Omega \delta_{[0,1]}(\alpha(t)) - \delta_{[0,1]}(\alpha(0))\,\mathrm{d}x = 0$ since $r_c \in N_{[0,1]}(\alpha)$ is in $L^2(Q)$ and thus in duality with $\dot{\alpha} \in L^2(Q)$, and since $0 \le \alpha_0 \le 1$ is assumed. Here we still have used that $r_c \in N_{[0,1]}(\alpha)$ has been already proved. □

Proposition 7.5.4 (Regularity). *Let (7.5.6) hold together with (6.4.11) and*

$$\big|\partial^2_{ee}\varphi(e,\alpha)\big| \le C, \qquad \big|\partial^2_{e\alpha}\varphi(e,\alpha)\big| \le C. \tag{7.5.11}$$

Then:

(i) the solutions (u,α,r_c) to (7.5.2) satisfy (7.5.7) together with

$$\dot{u} \in H^1(I;L^2(\Omega;\mathbb{R}^d)) \cap L^\infty(I;H^1(\Omega;\mathbb{R}^d)), \quad \textit{and} \tag{7.5.12a}$$

$$\operatorname{div}(\mathbb{D}e(\dot{u}) + \partial_e\varphi(e(u),\alpha)) \in L^2(Q;\mathbb{R}^d). \tag{7.5.12b}$$

(ii) If also (6.4.14a) together with $\operatorname{div}\partial_e\varphi(e(u_0),\alpha_0) + f(0) = 0$ and $v_0 \in H^2(\Omega;\mathbb{R}^d)$ hold, then even

$$\ddot{u} \in L^\infty(I;L^2(\Omega;\mathbb{R}^d)) \cap L^2(I;H^1(\Omega;\mathbb{R}^d)). \tag{7.5.13}$$

Sketch of the proof. We perform the regularity test for (7.5.2a) itself similarly as in Proposition 6.4.2 by the test by $\ddot{u}$. This test is legitimate in the Galerkin approximation. After integration over the time interval $[0,t]$, we obtain

---

[33] More in detail, denoting by M_η the mollification operator, for $\alpha_\eta := M_\eta\alpha$ we have $\dot{\alpha}_\eta = M_\eta\dot{\alpha} \in L^2(I;C^1(\Omega))$. By standard calculus, we obtain that (7.5.10) holds for the mollified function α_η, namely: $\int_Q \dot{\alpha}_\eta\operatorname{div}(\kappa\nabla\alpha_\eta)\,\mathrm{d}x\mathrm{d}t = \int_\Omega \frac{\kappa}{2}|\nabla(M_\eta\alpha_0)|^2 - \frac{\kappa}{2}|\nabla\alpha_\eta(T)|^2\,\mathrm{d}x$. We then obtain (7.5.10) by letting $\eta \to 0$, using the fact that $\operatorname{div}(\kappa\nabla\alpha_\eta) \to \operatorname{div}(\kappa\nabla\alpha)$ in $L^2(Q)$, $\dot{\alpha}_\eta \to \dot{\alpha}$ in $L^2(Q)$, and $\nabla\alpha_\eta(T) \to \nabla\alpha(T)$ in $L^2(\Omega;\mathbb{R}^d)$.

$$\int_0^t\int_\Omega \varrho|\ddot{u}_k|^2\,\mathrm{d}x\mathrm{d}t + \int_\Omega \frac{1}{2}\mathbb{D}e(\dot{u}_k(t)){:}e(\dot{u}_k(t))\,\mathrm{d}x$$
$$= \int_0^t\int_\Omega f\cdot\ddot{u}_k - \partial_e\varphi(e(u_k),\alpha_k){:}e(\ddot{u}_k)\,\mathrm{d}x\mathrm{d}t$$
$$+\frac{1}{2}\int_\Omega \mathbb{D}e(v_0){:}e(v_0)\,\mathrm{d}x + \int_0^t\int_\Gamma g\cdot\ddot{u}_k\,\mathrm{d}S\,\mathrm{d}t$$
$$= \int_0^t\int_\Omega f\cdot\ddot{u}_k + \big(\partial^2_{ee}\varphi(e(u_k),\alpha_k)e(\dot{u}_k) + \partial^2_{e\alpha}\varphi(e(u_k),\alpha_k)\dot{\alpha}_k\big){:}e(\dot{u}_k)\,\mathrm{d}x\mathrm{d}t$$
$$+\int_\Omega \frac{1}{2}\mathbb{D}e(v_0){:}e(v_0) + \partial_e\varphi(e(u_0),\alpha_0){:}e(v_0)$$
$$-\partial_e\varphi(e(u_k(t)),\alpha_k(t)){:}e(\dot{u}_k(t))\,\mathrm{d}x + \int_0^t\int_\Gamma g\cdot\ddot{u}_k\,\mathrm{d}S\,\mathrm{d}t. \qquad (7.5.14)$$

The last boundary term is to be treated by the by-part integration

$$\int_0^t\int_\Gamma g\cdot\ddot{u}_k\,\mathrm{d}S\,\mathrm{d}t = \int_\Gamma g(t)\cdot\dot{u}_k(t)\,\mathrm{d}S - \int_0^t\int_\Gamma \dot{g}\cdot\dot{u}_k\,\mathrm{d}S\,\mathrm{d}t - \int_\Gamma g(0)\cdot\dot{u}_0\,\mathrm{d}S$$

exploiting $g\in W^{1,1}(I;L^{2^{\sharp\prime}}(\Gamma;\mathbb{R}^d))$ as assumed in (6.4.11). Then, using (7.5.11) and the already obtained estimates of $e(\dot{u}_k)$ and $\dot{\alpha}_k$, by Young's and Gronwall's inequalities, we get the information (7.5.12a), and from the equation (7.5.2a) by comparison, i.e. $\mathrm{div}(\mathbb{D}e(\dot{u}_k)+\partial_e\varphi(e(u_k),\alpha_k)) = \varrho\ddot{u}_k - f$, we get also (7.5.12b).

The other regularity (ii) can be obtained by differentiating (7.5.2a) in time and by the test by $\dddot{u}$ as in Propositions 6.3.2 and (6.4.3). This gives

$$\int_\Omega \frac{\varrho}{2}|\ddot{u}_k(t)|^2\,\mathrm{d}x + \int_0^t\int_\Omega \mathbb{D}e(\ddot{u}_k){:}e(\ddot{u}_k)\,\mathrm{d}x\mathrm{d}t$$
$$= \int_0^t\int_\Omega \dot{f}\cdot\ddot{u}_k - \big(\partial^2_{ee}\varphi(e(u_k),\alpha_k)e(\dot{u}_k) + \partial^2_{e\alpha}\varphi(e(u_k),\alpha_k)\dot{\alpha}_k\big){:}e(\ddot{u}_k)\,\mathrm{d}x\mathrm{d}t$$
$$+\int_\Omega \frac{\varrho}{2}|\ddot{u}_k(0)|^2\,\mathrm{d}x + \int_0^t\int_\Gamma \dot{g}\cdot\ddot{u}_k\,\mathrm{d}S\,\mathrm{d}t$$
$$\le \int_0^t \|\dot{f}(t)\|_{L^2(\Omega;\mathbb{R}^d)}\big(1+\|\ddot{u}_k(t)\|^2_{L^2(\Omega;\mathbb{R}^d)}\big)\mathrm{d}t$$
$$+\frac{1}{\epsilon}\Big(\|\partial^2_{ee}\varphi(e(u_k),\alpha_k)\|^2_{L^\infty(Q;\mathbb{R}^{(d\times d)\times(d\times d)})}\|e(\dot{u}_k)\|^2_{L^2(Q;\mathbb{R}^{d\times d})}$$
$$+\|\partial^2_{e\alpha}\varphi(e(u_k),\alpha_k)\|^2_{L^\infty(Q)}\|\dot{\alpha}_k\|^2_{L^2(Q)} + N^2\|\dot{g}\|^2_{L^2(I;L^{2^{\sharp\prime}}(\Gamma;\mathbb{R}^d))}\Big)$$
$$+\frac{1}{2}\|\varrho\|_{L^\infty(\Omega)}\|\ddot{u}_k(0)\|^2_{L^2(\Omega)} + \epsilon\big(\|\ddot{u}_k\|^2_{L^2([0,t]\times\Omega;\mathbb{R}^d)} + \|e(\ddot{u}_k)\|^2_{L^2([0,t]\times\Omega;\mathbb{R}^{d\times d})}\big) \qquad (7.5.15)$$

with N the norm of the embedding $H^1(\Omega;\mathbb{R}^d)\to L^{2^\sharp}(\Gamma;\mathbb{R}^d)$ if $H^1(\Omega;\mathbb{R}^d)$ is equipped with the norm $u\mapsto(\int_\Omega|u|^2+|e(u)|^2\mathrm{d}x)^{1/2}$ through Korn's inequality, and with $\epsilon>0$ taken small so that the last two terms can be absorbed in (dominated by) the first terms in (7.5.15) by using $\inf\varrho$ and $\mathbb{D}$ positive (definite). Let us note that here we again needed the assumptions (7.5.6c). We further use our qualification of the initial conditions so that

$$\ddot{u}_k(0) = \frac{1}{\varrho}\Big(\mathrm{div}(\mathbb{D}e(v_0) + \partial_e\varphi(e(u_0),\alpha_0)) + f(0)\Big) = \frac{1}{\varrho}\Big(\mathrm{div}(\mathbb{D}e(v_0) + f(0)\Big) \in L^2(\Omega;\mathbb{R}^d)$$

This eventually gives (7.5.13) by Gronwall's inequality. □

Let us note an important modeling attribute of the presented model that, if $\varphi_{\mathrm{DAM}} = 0$, and $\varphi(e,\cdot)$ is nondecreasing, and $\partial\zeta_{\mathrm{IN}}(v)/v$ is large for $v>0$, then effectively there is nearly no driving force for healing. This might be useful for some applications. A more rigorous option to prevent healing, i.e. to ensure really $\dot\alpha \le 0$ and thus a *uni-directional* (sometimes called irreversible[34]) *damage*, is to consider $\zeta_{\mathrm{IN}}(v) = +\infty$ for $v>0$, as (7.5.16b) below.

A noteworthy variant of this model allows for putting simply $r_{\mathrm{c}} = 0$ and omit (7.5.4c). Two noteworthy options for this scenario arise when

$$\partial_\alpha\varphi(e,0) = 0, \quad \partial_\alpha\varphi(e,1) = 0, \quad \text{and} \quad \varphi'_{\mathrm{DAM}}(1) = 0, \qquad \text{(healing allowed)} \qquad (7.5.16a)$$

or

$$\partial_\alpha\varphi(e,0) = 0, \quad \zeta_{\mathrm{IN}}(\dot\alpha) = +\infty \quad \text{for } \dot\alpha > 0. \qquad \text{(unidirectional damage)} \qquad (7.5.16b)$$

Then, using the initial damage $0 \le \alpha_0(\cdot) \le 1$ a.e. on Ω, the constraints $0 \le \alpha \le 1$ are satisfied within the whole evolution even without considering r_{c} in (7.5.2b).

Weakening the estimate (7.5.7c), the condition (7.5.6c) can be weakened, cf. (7.5.18) below, so that we can admit, in particular, a natural strain energy $\frac{1}{2}\mathbb{C}(\alpha)e{:}e$ although this quadratic ansatz bears even a special treatment as in Sect. 7.5.2 below. A more specific example for the situation (7.5.16b) complying still with the convexity and coercivity in (7.5.6d) is

$$\zeta_{\mathrm{VI}}(\dot e) = \mathbb{D}\dot e{:}\dot e, \qquad \zeta_{\mathrm{IN}}(\dot\alpha) = -\mathfrak{g}_{\mathrm{c}}\dot\alpha + \frac{1}{2}\nu|\dot\alpha|^2 + \delta_{[0,+\infty)}(\dot\alpha) \qquad (7.5.17)$$

where $\mathfrak{g}_{\mathrm{c}} > 0$ is the energy (in J/m^d) needed for complete *damage* of the unit volume to the material, sometimes also called a *fracture toughness*, and ν is a viscosity-like coefficient, reflecting that fast damaging dissipates more energy than slow if $\nu > 0$. We use again a Galerkin approximation (generally, in contrast to Proposition 7.5.3, not necessarily H^2-conformal), denoting by $(u_k,\alpha_k)_{k\in\mathbb{N}}$ the approximate solutions.

Proposition 7.5.5 (Damage with $r_{\mathrm{c}} \equiv 0$, uniqueness, regularity). *Let (7.5.6a-c,e-g), (7.5.16) hold, ϱ satisfy (6.3.2), and let also*

$$|\partial_e\varphi(e,\alpha)| \le C(1+|e|) \quad \textit{and} \quad |\partial_\alpha\varphi(e,\alpha)| \le C(1+|e|^2). \qquad (7.5.18)$$

[34] The adjective "irreversible" often means that the process is dissipative, which would not distinguish the damage combined with healing or the uni-directional damage.

Then:

(i) *The mentioned Galerkin approximate solutions do exist and satisfy the a-priori estimates*

$$\|u_k\|_{W^{1,\infty}(I;L^2(\Omega;\mathbb{R}^d))\cap H^1(I;H^1(\Omega;\mathbb{R}^d))} \le C, \tag{7.5.19a}$$

$$\|\alpha_k\|_{H^1(I;L^2(\Omega))\cap L^\infty(I;H^1(\Omega))\cap L^\infty(Q)} \le C, \tag{7.5.19b}$$

The sequence $\{(u_k,\alpha_k)\}_{k\in\mathbb{N}}$ *possesses subsequences converging weakly* in the topologies indicated in (7.5.19) and also* $\{e(u_k)\}_{k\in\mathbb{N}}$ *converges (in terms of these subsequences) strongly in* $L^2(Q;\mathbb{R}^{d\times d})$*. The limit of each such a subsequence solves the initial-boundary-value problem (7.5.2) weakly in the sense of Definition 7.5.1 with Remark 7.5.2 with* $r_c = 0$*.*

(ii) *If (7.5.16b) holds and* $\mathbb{D} : [0,1] \to \mathrm{SLin}(\mathbb{R}^{d\times d}_{\mathrm{sym}})$ *be continuous, monotone (nondecreasing) with respect to the Löwner ordering, and with* $\mathbb{D}(0)$ *positive definite, then (i) holds, too.*

(iii) *In case of (7.5.16a) and in case* ϱ *satisfies also (6.3.5), this solution enjoys* $\mathrm{div}(\kappa\nabla\alpha) \in L^2(Q)$ *and* $\partial\zeta_{IN}(\dot\alpha) \in L^2(Q)$*, and conserves energy in the sense that the equality (7.5.3) holds for any* $t \in I$*.*

(iv) *Moreover, this weak solution is unique provided also*

$$\varphi' : \mathbb{R}^{d\times d}_{\mathrm{sym}} \times [0,1] \to \mathbb{R}^{d\times d}_{\mathrm{sym}} \times \mathbb{R} \quad \text{is Lipschitz continuous, and} \tag{7.5.20a}$$

$$\partial\zeta_{IN} \text{ uniformly monotone, i.e. } \exists \epsilon > 0\ \forall v_1, v_2 \in \mathbb{R} :$$
$$s_1 \in \partial\zeta_{IN}(v_1),\ s_2 \in \partial\zeta_{IN}(v_2) \ \Rightarrow\ (s_1 - s_2)(v_1 - v_2) \ge \epsilon(v_1 - v_2)^2. \tag{7.5.20b}$$

In particular, the whole sequence converges.

(v) *Assuming further, in addition to (6.4.11) and (7.5.11), that*

$$|\partial^2_{\alpha\alpha}\varphi(e,\alpha)| \le C(1+|e|^{1-2/2^*}), \tag{7.5.21a}$$

$$\partial^3_{ee\alpha}\varphi \text{ and } \partial^3_{e\alpha\alpha}\varphi \text{ bounded}, \quad \alpha_0 \in H^3(\Omega), \quad u_0 \in H^2(\Omega;\mathbb{R}^d), \tag{7.5.21b}$$

then also $\alpha \in W^{1,\infty}(I;H^1(\Omega)) \cap H^2(I;L^2(\Omega))$*.*

Let us note that in the case of (7.5.16b), as we will not prove the energy conservation for the damage flow rule, we do not claim the energy conservation for the whole system.

Let us also note that the Lipschitz continuity (7.5.20a) is compatible rather with (7.5.6c) than with (7.5.18). However, the attribute of uniqueness may not be even desired in some situations: when a symmetric specimen with two notches is stretched, it is experimentally well documented that always only the crack (damage) will start developing only from one of these two notches, giving the evidence that, by symmetry, at least two different solutions should be admitted by truly relevant models.

Proof of Proposition 7.5.5. As to (i), we perform the test of (7.5.2a,b) in the Galerkin approximation by $(\dot{u}_k, \dot{\alpha}_k)$. By using (7.5.6a-c,e-g), this gives the estimates (7.5.19). By comparison from (7.5.2a), we obtain also the bound for $\sqrt{\varrho}\ddot{u}_k$ in $L^2(I; H^1(\Omega;\mathbb{R}^d)^*)$; more precisely, this bound is valid only in Galerkin-induced seminorms or for the Hahn-Banach extension, cf. Remark C.2.9. After selection of weakly* convergent subsequences, we prove the strong convergence of $e(u_k)$ as in (7.5.9), and then we can pass to the limit by (semi)continuity towards the variational inequality (7.5.4b).

As to (ii), the a-priori estimates do not change by making $\mathbb{D}$ dependent on α, while (7.5.9) is to be modified by employing also the estimate

$$\int_\Omega \frac{1}{2}\mathbb{D}(\alpha_k(t))e(u_k(t)-u(t)):e(u_k(t)-u(t))\,\mathrm{d}x$$
$$= \int_0^t\!\!\int_\Omega \mathbb{D}(\alpha_k)e(\dot{u}_k-\dot{u}):e(u_k-u) + \frac{1}{2}\dot{\alpha}_k\mathbb{D}'(\alpha_k)e(u_k-u):e(u_k-u)\,\mathrm{d}x\mathrm{d}t$$
$$\le \int_0^t\!\!\int_\Omega \mathbb{D}(\alpha_k)e(\dot{u}_k-\dot{u}):e(u_k-u)\,\mathrm{d}x\mathrm{d}t \tag{7.5.22}$$

because $\dot{\alpha}_k\mathbb{D}'(\alpha_k)e(u_k-u):e(u_k-u) \le 0$ a.e. on Q since $\dot{\alpha}_k \le 0$ due to (7.5.16b) and $\mathbb{D}'(\cdot)$ is positive semidefinite due to the assumption of monotone dependence of $\mathbb{D}(\cdot)$.

Then the limit passage in the damage flow rule is easy when realizing that, due to the second condition in (7.5.18), $\partial_\alpha\varphi(e(u_k),\alpha_k) \to \partial_\alpha\varphi(e(u),\alpha)$ strongly in $L^1(Q)$.

As to (iii), by comparison we have $\mathrm{div}(\kappa\nabla\alpha) \in \partial\zeta_{\mathrm{IN}}(\dot{\alpha}) + \partial_\alpha\varphi(e(u),\alpha) - \varphi'_{\mathrm{DAM}}(\alpha) \in L^2(Q)$ because, due to (7.5.6d), the (in general set-valued) mapping $\partial\zeta_{\mathrm{IN}}$ maps bounded sets of $L^2(Q)$ into bounded sets and also that $\partial_\alpha\varphi(e(u),\alpha) \in L^2(Q)$ due to the growth assumption (7.5.6c). The energy conservation of the limit then relies on that $\sqrt{\varrho}\ddot{u}$ and $\mathrm{div}\,\sigma$ are in duality with $\sqrt{\varrho}\dot{u}$ and $\dot{u} \in L^2(I; H^1(\Omega;\mathbb{R}^d))$, respectively, and that $\mathrm{div}(\kappa\nabla\alpha)$ is in duality with $\dot{\alpha} \in L^2(Q)$.

As to (iv), take two weak solutions (u_1,α_1) and (u_2,α_2), subtract the weak formulations and test it successively by $\dot{u}_{12}$ and $\dot{\alpha}_{12}$, using the abbreviation $u_{12} := u_1 - u_2$ and $\alpha_{12} := \alpha_1 - \alpha_2$. Summing them together and using the uniform monotonicity of $\partial\zeta_{\mathrm{IN}}$, we obtain

$$\frac{\mathrm{d}}{\mathrm{d}t}\int_\Omega \frac{\varrho}{2}|\dot{u}_{12}|^2 + \frac{\kappa}{2}|\nabla\alpha_{12}|^2\,\mathrm{d}x + \int_\Omega \mathbb{D}e(\dot{u}_{12})\cdot e(\dot{u}_{12}) + \epsilon\dot{\alpha}_{12}^2\,\mathrm{d}x$$
$$\le \int_\Omega f\cdot\dot{u}_{12} - (\partial_e\varphi(e(u_1),\alpha_1) - \partial_e\varphi(e(u_2),\alpha_2)):e(\dot{u}_{12})$$
$$- (\partial_\alpha\varphi(e(u_1),\alpha_1) - \varphi'_{\mathrm{DAM}}(\alpha_1) - \partial_\alpha\varphi(e(u_2),\alpha_2) + \varphi'_{\mathrm{DAM}}(\alpha_1))\dot{\alpha}_{12}\,\mathrm{d}x + \int_\Gamma g\cdot\dot{u}_{12}\,\mathrm{d}S$$
$$\le \ell\Big(\|e(u_{12})\|_{L^2(\Omega;\mathbb{R}^{d\times d})} + \|\alpha_{12}\|_{L^2(\Omega)}\Big)\Big(\|e(\dot{u}_{12})\|_{L^2(\Omega;\mathbb{R}^{d\times d})} + \|\dot{\alpha}_{12}\|_{L^2(\Omega)}\Big)$$
$$+ \|f\|_{L^2(\Omega;\mathbb{R}^d)}\Big(1 + \|\dot{u}_{12}\|^2_{L^2(\Omega;\mathbb{R}^d)}\Big), \tag{7.5.23}$$

where ϵ is from (7.5.20b) and ℓ is the Lipschitz constant from (7.5.20a). Then the uniqueness follows by Gronwall's inequality, counting with $u_{12}|_{t=0} = 0$, $\dot{u}_{12}|_{t=0} = 0$, and $\alpha_{12}|_{t=0} = 0$.

As to (v), formally, or in the Galerkin approximation even rigorously, differentiating the damage flow rule in time and testing it by $\ddot{\alpha}$ yields

$$\int_\Omega \frac{\kappa}{2}|\nabla\dot{\alpha}(t)|^2\,\mathrm{d}x + \int_0^t\!\!\int_\Omega \partial^2\zeta_{\rm IN}(\dot{\alpha})|\ddot{\alpha}|^2\,\mathrm{d}x\mathrm{d}t \\ = \int_\Omega \frac{\kappa}{2}|\nabla\dot{\alpha}_0|^2\,\mathrm{d}x - \int_0^t\!\!\int_\Omega \partial^2_{e\alpha}\varphi(e(u),\alpha)e(\dot{u})\ddot{\alpha} + \partial^2_{\alpha\alpha}\varphi(e(u),\alpha)\dot{\alpha}\ddot{\alpha}. \tag{7.5.24}$$

Moreover, the assumption (7.5.21a) guarantees[35] that $\nabla\dot{\alpha}_0 \in L^2(\Omega;\mathbb{R}^d)$, we can use the already obtained basic energetic $L^2(Q)$-estimates of $e(\dot{u})$ and $\dot{\alpha}$ which in particular make $\varphi(e(u),\alpha)$ controlled in $L^{2^*2/(2^*-2)}(Q)$ due to the growth assumption (7.5.21a) on $\partial^2_{\alpha\alpha}\varphi$. Then, from (7.5.24), we obtain additionally the claimed information $\dot{\alpha} \in L^\infty(I;H^1(\Omega)) \cap H^1(I;L^2(\Omega))$. □

Example 7.5.6 (Damageable stored energies satisfying (7.5.6c) and (7.5.11)). The latter growth restriction in (7.5.6c) excludes energies of the type $\varphi(e,\alpha) = \frac{1}{2}\mathbb{C}(\alpha)e{:}e$ but facilitates the analysis without introducing strain gradients and without imposing (7.5.16). Examples of the stored energy satisfying (7.5.6c) might be

$$\varphi(e,\alpha) = \frac{1}{2}K|\mathrm{sph}\,e|^2 + \frac{G(\alpha)|\mathrm{dev}\,e|^2}{\sqrt{1+\epsilon|\mathrm{dev}\,e|^2}} \quad \text{or} \tag{7.5.25}$$

$$\varphi(e,\alpha) = \frac{d}{2}\Big(K(1)(\mathrm{tr}^-e)^2 + \frac{K(\alpha)(\mathrm{tr}^+e)^2}{\sqrt{1+\epsilon|\mathrm{tr}\,e|^2}}\Big) + \frac{G(\alpha)|\mathrm{dev}\,e|^2}{\sqrt{1+\epsilon|\mathrm{dev}\,e|^2}} \tag{7.5.26}$$

with $\mathrm{tr}^+e := \max(0,\mathrm{tr}\,e)$ and $\mathrm{tr}^-e := \min(0,\mathrm{tr}\,e)$. The first one (7.5.25) undergoes damage only under shear loading while (7.5.26) does it also under tension using the damageable bulk modulus $K = K(\alpha)$ but not in compression using the ever undamaged bulk modulus $K = K(1)$. The first option is well applicable in situations when the material is safely compressed and the volumetric tension effectively cannot occur in the model, like in geophysical models under big self-gravitational compression. The (small) regularizing parameter $\epsilon > 0$ makes the tension and shear stress bounded if $|e|$ is (very) large and makes the growth restriction on $\partial_\alpha\varphi$ in (7.5.6c) satisfied, while $\varepsilon = 0$ is admitted for the model (7.5.16).

Remark 7.5.7 (Fracture toughness). Let us illustrate heuristically how the flow-rule $\zeta_{\rm IN}(\dot{\alpha}) + \partial_\alpha\varphi(e,\alpha) \ni 0$ with the initial condition $0 < \alpha(0) = \alpha_0 \le 1$ operates when the loading gradually increases for example when $\varphi(e,\alpha) = \frac{1}{2}\mathbb{C}(\alpha)e{:}e$ with $\mathbb{C}(\cdot)$ nondecreasing, and $\zeta_{\rm IN}$ is from (7.5.17) with $\nu = 0$. The important question is when the damage in the material (with $\alpha = \alpha_0$) starts developing. This is a very nontrivial problem in fracture mechanics and basic dilemma is whether the stress or rather the energy is responsible for triggering damage, cf. also Remark 7.5.18. In our model, $\partial_\alpha\varphi(e,\alpha_0) = \frac{1}{2}\mathbb{C}'(\alpha_0)e{:}e$ increases and $\dot{\alpha} = 0$ until $\partial_\alpha\varphi$ reaches the threshold $\mathfrak{g}_\mathrm{c}$. Relying on the stress concept, we can express the equality $\frac{1}{2}\mathbb{C}'(\alpha_0)e{:}e = \mathfrak{g}_\mathrm{c}$ in term

[35] Here we rely on the equation (7.5.2b) with $r_\mathrm{c} = 0$ when the ∇-operator is applied, which gives $\nabla\dot{\alpha}_0 = \partial^2\zeta_{\rm IN}(\dot{\alpha})^{-1}[\nabla\kappa\Delta\alpha_0 - \partial^3_{ee\alpha}\varphi(e(u_0),\alpha_0)\nabla e(u_0) - \partial^3_{e\alpha\alpha}\varphi(e(u_0),\alpha_0)\nabla\alpha_0]$ when (7.5.21b) is taken into account.

of the stress $\sigma = \mathbb{C}(\alpha_0)e$. This gives the equality $\frac{1}{2}\mathbb{C}(\alpha_0)^{-\top}\mathbb{C}'(\alpha_0)\mathbb{C}(\alpha_0)^{-1}\sigma{:}\sigma = \mathfrak{g}_c$. For the example (7.5.25) with $\epsilon = 0$, the stress is $\sigma = \partial_e\varphi(e,\alpha) = \lambda(\mathrm{tr}\,e)\mathbb{I} + 2G(\alpha)e = \sigma_{\rm sph} + \sigma_{\rm dev}$ with $\sigma_{\rm sph} = dK\mathrm{sph}\,e$ and $\sigma_{\rm dev} = 2G(\alpha)\mathrm{dev}\,e$, cf. (6.7.10) on p. 242. The driving force for damage evolution expressed in therm of the actual stress is

$$\partial_\alpha\varphi(e,\alpha) = G'(\alpha)|\mathrm{dev}\,e|^2 = \frac{G'(\alpha)}{4G(\alpha)^2}|\sigma_{\rm dev}|^2.$$

In this simple case, the criterion $\partial_\alpha\varphi(e,\alpha_0) = \mathfrak{g}_c$ gives

$$|\sigma_{\rm dev}| = 2G(\alpha_0)\sqrt{\frac{\mathfrak{g}_c}{G'(\alpha_0)}} =: \sigma_{\rm DAM}(\alpha_0) = \text{"effective fracture stress"}. \tag{7.5.27}$$

If $G(\cdot)G'(\cdot)^{-1/2}$ is increasing (in particular if $G(\cdot)$ is concave), and the loading is via stress rather than displacement, damage then accelerates when started so that the rupture happens immediately (if any rate and spatial-gradient effects are neglected).

Remark 7.5.8 (Discontinuity in damage). We actually needed only mere $L^1(Q)$-compactness of α for which it would suffice to replace the term $\frac{1}{2}\kappa\|\nabla\alpha\|^2_{L^2(\Omega;\mathbb{R}^d)}$ in the stored energy $\int_\Omega \varphi_{\rm E}(e(u),\alpha,\nabla\alpha)$ by $\frac{1}{2}\kappa|\alpha|^2_{\gamma,2}$ with the Gagliardo semi-norm $|\cdot|_{\gamma,2}$ from (B.3.23). For $0<\gamma<1/2$, it would ensure even the $L^2(Q)$-compactness through the compact embedding $H^\gamma(\Omega)\subset L^2(Q)$ with the Aubin-Lions theorem and simultaneously allow for jump of damage profiles on sharp $(d-1)$-interfaces. Of course, a generalization using a singular scalar-valued kernel $\Omega\times\Omega\to\mathbb{R}$ satisfying the scalar analog of (2.5.28) would work, too.

Exercise 7.5.9 (Semiconvexity). Show that the potentials (7.5.25) and (7.5.26) are semiconvex.

7.5.2 Damage in linear Kelvin-Voigt materials

Let us now analyze the model with the (partly) damageable viscosity in the case that $\partial_e\varphi(\cdot,\alpha)$ is linear, i.e. $\varphi(e,\alpha) = \frac{1}{2}\mathbb{C}(\alpha)e{:}e$. In particular, the assumption (7.5.6c) is not satisfied. Let us investigate the special situation that $\mathbb{D}(\alpha) = \mathbb{D}_0 + \tau_{\rm R}\partial_e\varphi(\cdot,\alpha)$ with a relaxation time $\tau_{\rm R} > 0$ possibly dependent on $x\in\Omega$, cf. [304] or also [360, Sect.5.1.1 and 5.2.5] for the rate-independent unidirectional damage.

We again consider the ansatz (7.5.1) with $p = 2$ and, rather for simplicity, the scenarios (7.5.16), which means now $\mathbb{C}'(0) = 0$ and possibly also $\mathbb{C}'(1) = 0$. Again we consider an H^2-conformal Galerkin approximation. We allow for a *complete damage* in the elastic response, although a resting Stokes-type viscosity due to D_0 is needed for the following:

Proposition 7.5.10 (Damageable viscosity, existence, regularity). *Let the ansatz (7.5.1) with $p = 2$ be considered, and (6.3.2) and (7.5.6c,e-g) supposed, (7.5.16) hold, and let*

$$\mathbb{C}\in C([0,1];\mathrm{SLin}(\mathbb{R}^{d\times d}_{\rm sym}))\ \textit{be positive-semidefinite valued,} \tag{7.5.28a}$$

$$\mathbb{D}(\cdot) = \mathbb{D}_0 + \tau_{\rm R}\mathbb{C}(\cdot)\ \textit{with}\ \tau_{\rm R}\geq 0\ \textit{and}\ \mathbb{D}_0\ \textit{symmetric positive-definite.} \tag{7.5.28b}$$

Then the Galerkin approximation (u_k,α_k) *satisfies the a-priori estimates (7.5.19). For selected subsequences, we have*

$$u_k\to u \quad \textit{weakly* in}\ H^1(I;H^1(\Omega;\mathbb{R}^d))\cap W^{1,\infty}(I;L^2(\Omega;\mathbb{R}^d))\ \textit{and} \tag{7.5.29a}$$

$$\textit{strongly in}\ L^2(I;H^1(\Omega;\mathbb{R}^d))\cap H^1(I;L^2(\Omega;\mathbb{R}^d)),\ \textit{and} \tag{7.5.29b}$$

$$\alpha_k\to\alpha \quad \textit{weakly* in}\ H^1(I;L^2(\Omega))\cap L^\infty(I;H^1(\Omega))\cap L^\infty(Q), \tag{7.5.29c}$$

and every such a limit (u,α) *is a weak solution to the initial-boundary-value problem (7.5.2) with* $\partial_e\varphi(e,\alpha)=\mathbb{C}(\alpha)e$, $p=2$, *and* $r_{\rm c}\equiv 0$. *If, in addition* $f\in L^1(I;L^2(\Omega;\mathbb{R}^d))$ *and* $g\in W^{1,1}(I;L^{2^\sharp{}'}(\Gamma;\mathbb{R}^d))$, *and* $v_0\in H^1(\Omega;\mathbb{R}^d)$, *and the unidirectional damage evolution (7.5.16b) holds, and* $\mathbb{C}(\cdot)$ *be nondecreasing with respect to the Löwner ordering (cf. Example A.3.11 on p. 499), then*

$$\dot u\in L^\infty(I;H^1(\Omega;\mathbb{R}^d))\cap H^1(I;L^2(\Omega;\mathbb{R}^d)). \tag{7.5.30}$$

Proof. Again, the apriori estimates (7.5.19) can be obtained by standard energetic test by $\dot u_k$ and $\dot\alpha_k$. After selecting a subsequence weakly* converging in the sense (7.5.29a,c) and using the Aubin-Lions theorem for the damage and then continuity of the superposition operator induced by $\mathbb{C}(\cdot)$, we can pass to the limit first in the semilinear force-equilibrium equation. We put $w:=u+\tau_{\rm R}\dot u$ and write the limit equation as

$$\frac{\varrho}{\tau_{\rm R}}\dot w-{\rm div}(\mathbb{D}_0 e(\dot u)+\mathbb{C}(\alpha)e(w)) = f+\frac{\varrho}{\tau_{\rm R}}\dot u \tag{7.5.31}$$

accompanied with the initial/boundary conditions (7.5.2c,d).

For the damage flow rule, we need the strong convergence of $\{e(u_k)\}_{k\in\mathbb{N}}$, however. Furthermore, we denote $w_k:=u_k+\tau_{\rm R}\dot u_k$ and, using the linearity of $\partial_e\varphi(\cdot,\alpha)$, write the Galerkin approximation of the force equilibrium as

$$\frac{\varrho}{\tau_{\rm R}}\dot w_k-{\rm div}(\mathbb{D}_0 e(\dot u_k)+\mathbb{C}(\alpha_k)e(w_k)) = f+\frac{\varrho}{\tau_{\rm R}}\dot u_k \tag{7.5.32}$$

accompanied with the initial/boundary conditions (7.5.2c,d) written for the Galerkin approximation. Then we subtract (7.5.31) and (7.5.32), and test it by w_k-w. Assuming also (without loss of generality) that $u_0,v_0\in V_1\subset V_k\subset H^1(\Omega;\mathbb{R}^d)$ as used for the Galerkin approximation, we test (7.5.32) by w_k, and integrate over the time interval $[0,t]$. This gives[36]

[36] In fact, (7.5.33) is again rather conceptual strategy and still a strong approximation of (u,w) is needed to facilitate usage of the Galerkin identity and convergence-to-zero of the additional terms thus arising.

$$\int_\Omega \frac{\varrho}{2\tau_{\rm R}}|w_k(T){-}w(T)|^2 + \frac12 \mathbb{D}_0 e(u_k(T){-}u(T)){:}e(u_k(T){-}u(T))\,{\rm d}x \\ + \int_Q \mathbb{D}_0\tau_{\rm R} e(\dot u_k{-}\dot u){:}e(\dot u_k{-}\dot u) + \mathbb{C}(\alpha_k)e(w_k{-}w){:}e(w_k{-}w)\,{\rm d}x{\rm d}t \\ = \int_Q (\mathbb{C}(\alpha_k){-}\mathbb{C}(\alpha))e(w){:}e(w_k{-}w) + \frac{\varrho}{\tau_{\rm R}}(\dot u_k{-}\dot u){\cdot}(w_k{-}w)\,{\rm d}x{\rm d}t \to 0. \tag{7.5.33}$$

Here we used that $\dot u_k{-}\dot u \to 0$ strongly in $L^2(Q;\mathbb{R}^d)$ by the Aubin-Lions theorem and also that $\mathbb{C}(\alpha_k){-}\mathbb{C}(\alpha))e(w) \to 0$ strongly in $L^2(Q;\mathbb{R}^{d\times d}_{\rm sym})$. This gives (7.5.29b).

The limit passage in the damage variational inequality towards (7.5.4b) is then easy as before by (semi)continuity.

As to the regularity (7.5.30), instead of $\ddot u_k$ as in Proposition 6.4.2, we now can perform the test of (7.5.32) by $\dot w_k = \dot u_k + \tau_{\rm R}\ddot u_k$. Using the calculus $\mathbb{C}(\alpha)e(w){:}e(\dot w) = \frac12\frac{\rm d}{{\rm d}t}\mathbb{C}(\alpha)e(w){:}e(w) - \dot\alpha\mathbb{C}'(\alpha)e(w){:}e(w)$ and realizing that always $\dot\alpha\mathbb{C}'(\alpha)e(w){:}e(w) \le 0$ due to the unidirectional evolution assumption (7.5.16b) and the assumed monotonicity of $\mathbb{C}(\cdot)$, after integration[37] over $[0,t]$, it leads to a modification of (7.5.14):

$$\int_\Omega \frac12\tau_{\rm R}\mathbb{D}_0 e(\dot u_k(t)){:}e(\dot u_k(t)) + \frac12\mathbb{C}(\alpha_k(t))e(w_k(t)){:}e(w_k(t))\,{\rm d}x \\ + \int_0^t\!\!\int_\Omega \frac{\varrho}{\tau_{\rm R}}|\dot w_k|^2 + \mathbb{D}_0 e(\dot u_k){:}e(\dot u_k)\,{\rm d}x{\rm d}t \\ \le \int_0^t\!\!\int_\Omega \Big(f{+}\frac{\varrho}{\tau_{\rm R}}\dot u_k\Big){\cdot}\dot w_k\,{\rm d}x{\rm d}t + \int_0^t\!\!\int_\Sigma g{\cdot}\dot w_k\,{\rm d}S\,{\rm d}t \\ + \int_\Omega \frac12\tau_{\rm R}\mathbb{D}_0 e(\dot u_0){:}e(\dot u_0) + \frac12\mathbb{C}(\alpha_0)e(w_0){:}e(w_0)\,{\rm d}x. \tag{7.5.34}$$

The last term containing $\int_0^t\int_\Sigma \tau_{\rm R} g{\cdot}\ddot u_k\,{\rm d}S\,{\rm d}t$ is to be treated again by using a by-part integration in time. Then, by Gronwall's inequality, (7.5.30) for u_k independently of $k{\in}N$, and thus also for the limit u, follows. □

The concept of bulk damage can (asymptotically) imitate the philosophy of fracture along surfaces (*cracks*) provided the damage stored energy $\varphi_{\rm DAM}$ is big. A popular ansatz takes $\varphi_{\rm DAM}(\alpha) = \frac12\mathfrak{g}_{\rm c}(1{-}\alpha)^2/\varepsilon$ and $\kappa = \varepsilon\mathfrak{g}_{\rm c}$ in the basic model (7.5.1) with $p = 2$, with $\mathfrak{g}_{\rm c}$ denoting the energy of fracture and with ε controlling a "characteristic" width of the *phase-field fracture* zones. This width is supposed to be small with respect to the size of the whole body. Considering the quadratic ansatz, this model is governed (up to the forcing f and g) by the stored energy functional

$$\mathcal{E}(u,\alpha) := \int_\Omega \varphi_{\rm E}(e,\alpha,\nabla\alpha) := \int_\Omega \gamma(\alpha)\mathbb{C}e{:}e \\ + \underbrace{\mathfrak{g}_{\rm c}\Big(\frac{1}{2\varepsilon}(1{-}\alpha)^2 + \frac{\varepsilon}{2}|\nabla\alpha|^2\Big)}_{\text{crack surface density}}\,{\rm d}x \quad \text{with}\quad \underbrace{\gamma(\alpha) = \frac{(\varepsilon/\varepsilon_0)^2{+}\alpha^2}{2}}_{\text{degradation function}} \tag{7.5.35}$$

[37] Let us note that $\dot\alpha\mathbb{C}'(\alpha)e(w){:}e(w)$ is not a-priori integrable, so that the inequality (7.5.34) is to be rigorously proved by some regularization.

with $\varepsilon_0 > 0$.[38] This is known as the so-called *Ambrosio-Tortorelli functional*.[39] For a small length-scale parameter $\varepsilon > 0$, it allows for a nearly complete damage but with a tendency to be localized on very small volumes along evolving surfaces/cracks where the fracture propagates while elsewhere $\alpha \sim 1$ not to make the term $(1-\alpha)^2/\varepsilon$ too large, considering still the dissipation potential $\zeta_{\rm IN}(\dot\alpha) = \delta^*_{[0,+\infty)}(\dot\alpha) = \delta_{(-\infty,0]}(\dot\alpha)$. This model is useful for crack propagation but, like the Griffith model, the drawback for small $\varepsilon > 0$ is that the crack initiation is unrealistically difficult.[40] The term "crack surface density" in (7.5.35) is sometimes used in engineering literature while the damage variable α being called a *phase field*, cf. e.g. [351, 502] where also various modifications of the stored energy (7.5.35) and various dissipation potentials have been devised, although without any rigorous analytical justification of the resulted models. The important fact behind (7.5.35) is that $[\varphi_{\rm E}]'_\alpha(e,\alpha,\nabla\alpha) = 0$ for $\alpha = 0$ so that, like in the model (7.5.16), the constraint $\alpha \ge 0$ need not be explicitly involved.

Remark 7.5.11 (Rate-independent damage). In case of the activated unidirectional inviscid damage, the 1-homogeneous dissipation potential is:

$$\zeta_{\rm IN}(\dot\alpha) = \delta^*_{[-\mathfrak{g}_{\rm c},\infty)}(\dot\alpha) = -\mathfrak{g}_{\rm c}\dot\alpha + \delta_{[0,\infty)}(\dot\alpha). \tag{7.5.36}$$

Let us note that the qualification (7.5.6d) is not satisfied. Yet, we can benefit from the general advantage of the energetic formulation of the flow rule that $\partial_\alpha\varphi$, and thus here also $r_{\rm c}$, does not explicitly occur. The only essential step is an explicit construction of a *mutual recovery sequence*, cf. (7.3.28). For $p > d$, the following construction works:

$$\hat\alpha_k = \big(\hat\alpha - \|\alpha_k - \alpha\|_{C(\bar\Omega)}\big)^+. \tag{7.5.37}$$

Let us also note that always $0 \le \hat\alpha_k \le \alpha_k$. This follows from $\hat\alpha \le \alpha$ (while the opposite case is not interesting as $\mathcal{R}(\hat\alpha - \alpha)$ would be infinite and thus the mutual-recovery-sequence condition (7.3.28) is trivially satisfied) and from that obviously $\alpha \le \alpha_k + \|\alpha_k - \alpha\|_{C(\bar\Omega)}$ a.e. since $p > d$ was assumed so that $W^{1,p}(\Omega) \subset C(\bar\Omega)$, which implies $\hat\alpha - \|\alpha_k - \alpha\|_{C(\bar\Omega)} \le \alpha_k$, and then from $\alpha_k \ge 0$ we can see that $(\hat\alpha - \|\alpha_k - \alpha\|_{C(\bar\Omega)})^+ \le \alpha_k$. Simultaneously also $\hat\alpha_k \to \hat\alpha$ strongly in $H^1(\Omega)$ because

[38] The physical dimension of ε_0 as well as of ε is meters while the physical dimension of $\mathfrak{g}_{\rm c}$ is J/m^2.

[39] In the static case, this approximation was proposed by Ambrosio and Tortorelli [10, 11] in fact for the scalar Mumford-Shah functional [383] and the asymptotic analysis for $\varepsilon \to 0$ was rigorously executed, inspired by a now classical example in phase transition [372]. The generalization for the vectorial case is due to Focardi [188]. Later, it was extended for evolution situation, namely for a rate-independent cohesive damage, in [221], see also also [81, 82, 304, 360] where also inertial forces are sometimes considered.

[40] In fact, as $\varphi'_{\rm DAM}(1) = 0$, the initiation of damage has zero threshold and is happening even on very low stress but then, if $\varepsilon > 0$ is very small, stops and high stress is needed to continue damaging.

$$\nabla\hat{\alpha}_k(x) = \begin{cases} \nabla\hat{\alpha}(x) & \text{if } \hat{\alpha}(x) > \|\alpha_k - \alpha\|_{C(\bar{\Omega})}, \\ 0 & \text{if } \hat{\alpha}(x) \le \|\alpha_k - \alpha\|_{C(\bar{\Omega})}, \end{cases} \qquad \text{for a.a. } x \in \Omega, \tag{7.5.38}$$

and because $\|\alpha_k - \alpha\|_{C(\bar{\Omega})} \to 0$ since $\alpha_k \rightharpoonup \alpha$ in $H^1(\Omega) \Subset C(\bar{\Omega})$ if $d = 1$. Indeed, this yields $\nabla\hat{\alpha}_k(x) \to \nabla\hat{\alpha}(x)$ for a.a. $x \in \Omega$ and we obtain $|\nabla\hat{\alpha}_k - \nabla\hat{\alpha}|^2 \to 0$ in $L^1(\Omega)$ by Lebesgue's theorem B.2.2 when using the integrable majorant

$$\left|\nabla\hat{\alpha}_k - \nabla\hat{\alpha}\right|^2 \le 2(|\nabla\hat{\alpha}_k|^2 + |\nabla\hat{\alpha}|^2) \le 4|\nabla\hat{\alpha}|^2.$$

If $p \le d$, one can use a more sophisticated construction of the mutual recovery sequence, cf. [360] or [511, 512], respectively.[41]

Remark 7.5.12 (Fully rate-independent damage model). If there is no inertia and viscosity in the force equilibrium equation, the whole problem is then fully rate-independent and then the fully-implicit time discretization implemented by using (at least theoretically) a global minimization jointly in (u, α) leads to an approximation of the energetic solution. The mutual recovery sequence is then consider the pairs $(\hat{u}_k, \hat{\alpha}_k)$ with $\hat{u}_k = u_k$. Yet, in this situation, also other concepts of solutions may be considered which might be more physically relevant than the energetic-solution concept, cf. [360].

Remark 7.5.13 (Mode-sensitive fracture). Combining (7.5.26) with the crack surface density from (7.5.35), one can distinguish the fracture by tension while mere compression does not lead to fracture, cf. [351]. By this way, one can also distinguish the *Mode I (fracture* by opening) from *Mode II* (fracture by shear). In fact, this modifies the driving force in dependence of the fracture mode. An alternative option how to distinguish Mode I from Mode II is in the dissipation potential, reflecting the experimental observation that Mode II needs (dissipates) more energy than Mode I. Thus one can take a state-dependent $\zeta_{\text{IN}} = \zeta_{\text{IN}}(e; \dot{\alpha})$ e.g. as

$$\zeta_{\text{IN}}(e; \dot{\alpha}) = \begin{cases} -\mathfrak{g}_c(\text{sph}\,e, \text{dev}\,e)\dot{\alpha} + \nu\dot{\alpha}^2 & \text{if } \dot{\alpha} \le 0, \\ +\infty & \text{if } \dot{\alpha} > 0, \end{cases} \tag{7.5.39}$$

when putting $\mathfrak{g}_c = \mathfrak{g}_c(\text{sph}\,e, \text{dev}\,e) > 0$ to be rather small if $\text{tr}\,e \gg |\text{dev}\,e|$ (which indicates Mode I) and bigger if $|\text{tr}\,e| \ll |\text{dev}\,e|$ (i.e. Mode II), or very large if $\text{tr}\,e \ll -|\text{dev}\,e|$ (compression leading to no fracture). Of course, combination of both alternatives is possible, too. On top of it, one can also consider a combination with other dissipative processes triggered only in Mode II, a prominent example being isochoric plasticity with hardening, cf. Figure 7.10-left on p. 331.

Remark 7.5.14 (Fractional-step time discretization). The damage problem typically involves the stored energies $\varphi = \varphi(e, \alpha)$ which are separately convex (or even sepa-

[41] The other modeling option of the rate-independent damage might be to consider $\zeta_{\text{IN}}(\dot{\alpha}) = \delta^*_{[-\mathfrak{g}_c, D]}(\dot{\alpha})$ with $D > 0$ expectedly large. Then the construction of the mutual recovery sequence is much simpler, namely if suffices to take $\hat{\alpha}_k = \hat{\alpha}$ because we do not need to care about the constraint $\hat{\alpha}_k \le \alpha_k$ and $\hat{\alpha}_k - \alpha_k = \hat{\alpha} - \alpha_k \to \hat{\alpha} - \alpha$ strongly in $L^1(\Omega)$.

rately quadratic). This encourages for an illustration of the *fractional-step method*, also called *staggered scheme*, cf. Remark C.2.8 on p. 544. In addition, to suppress a unwanted numerical attenuation within vibration, the time discretization of the inertial term by the Crank-Nicholson scheme a'la (C.2.54) on p. 546 can also be considered, leading to an energy-conserving discrete scheme even with an easy possibility to vary time-step $\tau = \tau^k$ within $k \in \mathbb{N}$. Altogether, it leads to the recursive boundary-value decoupled problems:

$$\frac{u_\tau^k - u_\tau^{k-1}}{\tau} = v_\tau^{k-1/2} := \frac{v_\tau^k + v_\tau^{k-1}}{2}, \tag{7.5.40a}$$

$$\varrho \frac{v_\tau^k - v_\tau^{k-1}}{\tau} - \operatorname{div}\Big(\mathbb{D}(\alpha_\tau^{k-1}) e(v_\tau^{k-1/2}) + \partial_e \varphi(e(u_\tau^{k-1/2}), \alpha_\tau^{k-1})\Big) = f_\tau^k \tag{7.5.40b}$$

$$\partial \zeta_{\text{IN}}\Big(\frac{\alpha_\tau^k - \alpha_\tau^{k-1}}{\tau}\Big) + \partial_\alpha \varphi(e(u_\tau^k), \alpha_\tau^{k-1/2}) + N_{[0,1]}(\alpha_\tau^k) \ni \kappa \Delta \alpha_\tau^{k-1/2} \tag{7.5.40c}$$

with $u_\tau^{k-1/2} := \frac{1}{2} u_\tau^k + \frac{1}{2} u_\tau^{k-1}$ and $\alpha_\tau^{k-1/2} := \frac{1}{2} \alpha_\tau^k + \frac{1}{2} \alpha_\tau^{k-1}$, considered on Ω while completed with the corresponding boundary conditions. It is to be solved recursively for $k = 1, ..., T/\tau$, starting with

$$u_\tau^0 = u_0, \qquad v_\tau^0 = v_0, \qquad \alpha_\tau^0 = \alpha_0. \tag{7.5.41}$$

Remark 7.5.15 (Implicit time discretization). Some applications needs to reflect the coupled character of the problem in the truly coupled discrete fully-implicit scheme, i.e. (7.5.40b) with $\partial_e \varphi(e(u_\tau^k), \alpha_\tau^k)$ instead of $\partial_e \varphi(e(u_\tau^k), \alpha_\tau^{k-1})$. This is indeed often solved in engineering, but only an approximate solution can be expected by some iterative procedures.[42] In general, such schemes do not seem numerically stable, however. More specifically, except cases when φ is semiconvex like in Remark C.2.4 on p. 540, the a-priori estimates are not available. The semiconvexity here with respect to the $(H^1 \times L^2)$-norm can be exploited provided the Kelvin-Voigt viscosity is used, as it is indeed considered in the model (7.5.2) when $\mathbb{D}$ is qualified as (7.5.28). More precisely, the discretization in question is still not really fully implicit since not $\mathbb{D}(\alpha_\tau^k)$ but $\mathbb{D}(\alpha_\tau^{k-1})$ as in (7.5.40b) is used in order the resulted recursive problems have potentials. Actually, it needs still a modification of using $p > d$ in (7.5.1) instead of $p = 2$ to ensure convergence.[43] These underlying potential, cf. also (C.2.44), are strongly convex for the time-step $\tau > 0$ small enough and, assuming also a conformal space discretization, the iterative solvers have guaranteed convergence towards a unique (globally minimizing) solution of the implicit scheme (7.5.40). The men-

[42] This scheme is known under the label "monolithic" and solved iteratively e.g. by the Newton-Raphson (or here equivalently SQP = sequential quadratic programming) method with no guaranteed convergence, however.

[43] The presence of $\mathbb{D}(\alpha_\tau^{k-1})$ instead of $\mathbb{D}(\alpha_\tau^k)$ brings difficulties in proving the strong convergence of rates, because the "Rothe" analog of the argumentation (7.5.31) does not work, cf. Exercise 7.5.25. The mentioned modification which makes $\varphi_{\mathrm{E}}(e, \alpha)$ nonquadratic is algorithmically acceptable because $\varphi_{\mathrm{E}}(\cdot, \cdot, \nabla \alpha)$ is not quadratic anyhow.

tioned local semiconvexity is ensured[44] for φ from Example 7.5.6 with $K(\cdot)$ and $G(\cdot)$ uniformly convex or from (7.5.35) with γ uniformly convex, and $-\varphi_{\rm DAM}$ convex. In the cases (7.5.25) or (7.5.26), Proposition 7.5.5(iv) yields even uniqueness so that the whole sequence (not only selected subsequences) converges for the time-step $\tau \to 0$ to the weak solution. This scheme precisely conserves energy at least during regimes when damage does not evolve. Let us finally remark that, in the quasistatic case $\varrho = 0$ even with non-semiconvex φ's, one can attempt to seek such solutions which are simultaneously the global minimizers of the underlying potential, cf. (C.2.19), which then can yield a-priori estimates, Remark C.2.3 on p. 540, but finding global minimizers in nonconvex problems is always difficult and usually without guaranty.

Example 7.5.16 (Crack approximation). Let us illustrate the *phase-field fracture* model (7.5.35) on a computational experiment with a 2-dimensional elastic specimen with one circular void in the middle where the stress is concentrated during the stretching, cf. Figure 7.8-left. The concentrating stress eventually leads to a crack initiation and propagation, as in Fig. 7.8-right.[45]

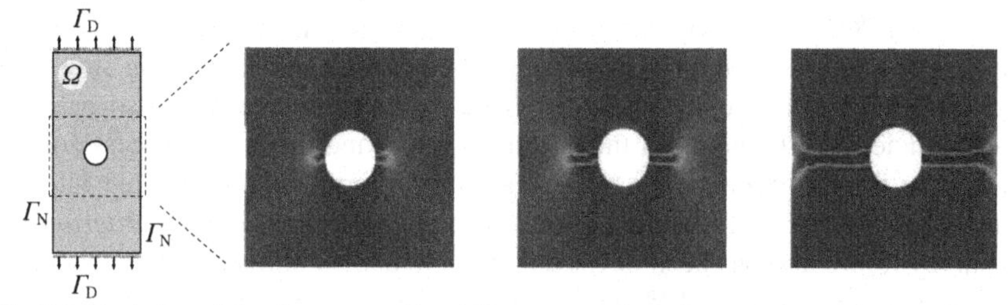

Fig. 7.8 Evolution of the damage profile within rupture of a 2-dimensional specimen (left) under gradually increasing stretching by the hard-device loading on the boundary $\Gamma_{\rm D}$, displayed at 3 snapshots of the selected area around the hole where the crack occurs due to the stress concentration.
Courtesy of Jose Reinoso (University of Seville)

Remark 7.5.17 (Crack approximation revisited). It should be emphasized that, in the "crack limit" for $\varepsilon \to 0$, the phase-field fracture model from Remark 7.5.16 approximates the true infinitesimally thin cracks rather in Griffith's variant (i.e. competition of energies), which works realistically for crack propagation but might have

[44] Note that φ's from (7.5.25) or (7.5.26) below have $\partial^2_{ee}\varphi$ positive definite (although not uniformly) and $\partial^2_{\alpha\alpha}\varphi$ nonnegative, while $\partial^2_{e\alpha}\varphi$ is bounded on $\{(e,\alpha) \in \mathbb{R}^{d\times d}_{\rm sym} \times [0,1]\}$. Hence $\partial^2\varphi + c{\rm diag}(\mathbb{I}, 1)$ is uniformly positive definite for c big enough.

[45] The fully implicit scheme has been used, as in Remark 7.5.15 below. Yet, several shortcuts in computational experiments have been adopted for these calculations, namely viscosity and inertial effects have not been implemented (so that it is, in fact, the quasistatic rate-independent model a'la Remark 7.5.11 but the inclusion has been solved iteratively so that only a critical point of the underlying nonconvex functional has been found, without guaranteed convergence of the iterative solver). No attempt for a global minimization that would have a capacity to converge towards the energetic solution has been performed, cf. (7.3.41) to be here made for (u, α) simultaneously, however. See [434, 435] for implementation details also about the used finite-element method.

unrealistic difficulties with crack initiation, while scaling of the fracture energy to 0 if $\varepsilon \to 0$ might lead to opposite effects, cf. also the discussion e.g. in [313]. This is partly reflected by that, in its rate-independent variant, the damage and phase-field fracture models admit many various solutions of very different characters, as presented in [360].[46] Various modifications have been devised. For example, Bourdin at al. [83] used $\varphi_{\rm DAM}(\alpha) = 3\mathfrak{g}_{\rm c}\alpha/(8\varepsilon)$ and $\kappa = 3\mathfrak{g}_{\rm c}\varepsilon/4$ in (7.5.1) again with $p = 2$ and $\mathfrak{g}_{\rm c}$ the energy of fracture in J/m^2 and with ε controlling a "characteristic" width of the phase-field fracture zone. This presumably exhibits a similar undesired behavior as (7.5.35) when $\varepsilon \to 0$ and leads to consider $\varepsilon > 0$ as another parameter (without intention to put it 0) in addition to $\mathfrak{g}_{\rm c}$ to tune the model. Various modifications of the degradation function from (7.5.35) therefore appeared in literature. E.g. a cubic degradation function γ has been used e.g. in [79, 521]. Inspired by (7.5.27), keeping still the original motivation $\varepsilon \to 0$, one can think about some γ convex increasing with $\gamma'(1) = \mathscr{O}(1/\varepsilon)$. The mentioned cubic ansatz is not compatible with these requirement. Some more sophisticated γ's have been devised in [476, 534]. We have thus more independent parameters than only $\mathfrak{g}_{\rm c}$ and ε in (7.5.35) to specify the lengh-scale of damage zone, fracture propagation and fracture initiation.

Remark 7.5.18 (Stress or energy?). In addition to FFM, in fracture (or in general damage) mechanics, there is a disputation whether only sufficiently big stress can lead to rupture or (in reminiscence to Grifith's concept) whether (also or only) some sufficiently big energy in the specimen or around the crack process zone is needed for it. A certain standpoint is that both criteria should be taken into account. This concept is nowadays referred as a *coupled stress-energy criterion*, cf. the survey [530], and has been devised and implemented in many variants in engineering literature, cf. e.g. [128, 213, 313, 331, 332], always without any analysis of numerical stability and convergence and thus computational simulations based on these models, whatever practical applications they have, stay in the position of rather speculative playing with computers. Here, this can be reflected by making ζ dependent on the strain energy. Having in mind FFM, one can think to let $\zeta_{\rm IN} = \zeta_{\rm IN}(\tilde{\varepsilon};\dot{\alpha})$ with $\tilde{\varepsilon}(x) = \int_\Omega k(x-\tilde{x})\varphi(e(u(\tilde{x})),\alpha(\tilde{x}))\,{\rm d}\tilde{x}$ for some kernel $k : \mathbb{R}^d \to \mathbb{R}^+$, making $-\partial_{\dot{\alpha}}\zeta_{\rm IN}(\tilde{\varepsilon};\dot{\alpha})$ larger is $\tilde{\varepsilon}$ is small.

Remark 7.5.19 (Nonlocal damage). In [49], an averaged damage χ is suggested to degrade (visco)elastic moduli rather than the local damage α. One specific option sometimes used in engineering literature is to put $\chi - \kappa\Delta\chi = \alpha$ for $\kappa > 0$ presumably small and with homogeneous Neumann boundary conditions. To put it into an energetically and mathematically justified context, we consider

$$\varphi_{\rm E}(e,\alpha,\chi,\nabla\chi) = \varphi(e,\chi) - \varphi_{\rm DAM}(\chi) + \delta_{[0,1]}(\chi) + \frac{\kappa}{2}|\nabla\chi|^2 + \frac{1}{2\varepsilon}|\alpha-\chi|^2 \quad \text{and} \tag{7.5.42a}$$

$$\zeta(\alpha;\dot{e},\dot{\alpha},\dot{\chi}) = \zeta_{\rm VI}(\chi;\dot{e},\dot{\chi}) + \zeta_{\rm IN}(\dot{\alpha}) \quad \text{with} \quad \zeta_{\rm VI}(\chi;\dot{e}) = \frac{1}{2}\mathbb{D}(\chi)\dot{e}:\dot{e} + \frac{\varepsilon}{2}\dot{\chi}^2. \tag{7.5.42b}$$

---

[46] In the dynamic variant, the influence of overall stored energy during fast rupture (which may be taken into account during quasistatic evolution) seems eliminated to some extent anyhow, because of finite speed of propagation of information about it.

The system (7.5.2) now modifies as

$$\varrho\ddot{u}-\mathrm{div}\big(\mathbb{D}(\chi)e(\dot{u})+\partial_e\varphi(e(u),\chi)\big)=f \qquad \text{in } Q, \qquad (7.5.43\text{a})$$

$$\partial\zeta_{\rm IN}(\dot{\alpha})\ni(\chi-\alpha)/\varepsilon \qquad \text{in } Q, \qquad (7.5.43\text{b})$$

$$\varepsilon\dot{\chi}-\kappa\Delta\chi+(\chi-\alpha)/\varepsilon+\varphi_{\rm DAM}'(\chi)+N_{[0,1]}(\chi)\ni\partial_\chi\varphi(e,\chi) \qquad \text{in } Q. \qquad (7.5.43\text{c})$$

For small $\varepsilon>0$, necessarily the difference $\alpha-\chi$ must be small in $L^\infty(I;L^2(\Omega))$-norm and, when summing (7.5.43b) with (7.5.43c), we obtain essentially the original model (7.5.2). When $\zeta_{\rm IN}$ is degree-1 homogeneous, we can rely on linearity of the static part and treat (7.5.43b) by the theory of rate-independent processes like we did for linearized plasticity in (7.4.21) even without any gradient of damage needed. On the other hand, (7.5.43c) involves (even set-valued) nonlinearity in the static part but is otherwise linear parabolic in the leading terms.

Remark 7.5.20 (Damage in Jeffreys materials). A model of damage in Jeffreys viscoelastic material with only a partial influence on the viscous part, namely $(\,{}_{\mathbb{C}}^{\alpha}\,\|\,{}_{\mathbb{D}_1}^{\alpha})\!-\!{}_{\mathbb{D}_2}$, can be cast analogously. Referring to Fig. 6.6-right on p. 235 with the internal variable $e_2=\pi$, the ansatz (7.3.19) now takes a more concrete form:

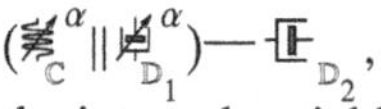

$$\varphi_{\rm E}(e,\pi,\alpha,\nabla\alpha)=\varphi(e-\pi,\alpha)-\varphi_{\rm DAM}(\alpha)+\delta_{[0,1]}(\alpha)+\frac{\kappa}{2}|\nabla\alpha|^2 \quad \text{and} \qquad (7.5.44\text{a})$$

$$\zeta_{\rm VI}(\dot{e},\dot{\pi})=\frac{1}{2}\mathbb{D}_1(\alpha)(\dot{e}-\dot{\pi}){:}(\dot{e}-\dot{\pi})+\frac{1}{2}\mathbb{D}_2\dot{\pi}{:}\dot{\pi}, \qquad (7.5.44\text{b})$$

with $\delta_{[0,1]}(\cdot):\mathbb{R}\to\{0,\infty\}$ the indicator function of the interval $[0,1]$ where the damage variable is assumed to take its values. The system (7.3.18) takes the form

$$\varrho\ddot{u}-\mathrm{div}\big(\mathbb{D}_1(\alpha)(e(\dot{u})-\dot{\pi})+\partial_e\varphi(e(u)-\pi,\alpha)\big)=f \qquad \text{in } Q, \quad (7.5.45\text{a})$$

$$(\mathbb{D}_1(\alpha)+\mathbb{D}_2)\dot{\pi}=\partial_e\varphi(e(u)-\pi,\alpha)+\mathbb{D}_1(\alpha)e(\dot{u}) \qquad \text{in } Q, \quad (7.5.45\text{b})$$

$$\partial\zeta_{\rm IN}(\dot{\alpha})+\partial_\alpha\varphi(e(u)-\pi,\alpha)+r_{\rm c}\ni\varphi_{\rm DAM}'(\alpha)+\kappa\Delta\alpha \ \text{ with } r_{\rm c}\in N_{[0,1]}(\alpha) \quad \text{in } Q, \quad (7.5.45\text{c})$$

where $r_{\rm c}$ is a "reaction" force (pressure) to the constraint $0\le\alpha\le1$. We consider the initial-boundary-value problem with the boundary and initial conditions

$$(\mathbb{D}_1(\alpha)(e(\dot{u})-\dot{\pi})+\partial_e\varphi(e(u)-\pi,\alpha))\vec{n}=g \quad \text{and} \quad \nabla\alpha\cdot\vec{n}=0 \qquad \text{on } \Sigma, \quad (7.5.45\text{d})$$

$$u(0)=u_0, \quad \dot{u}(0)=v_0, \quad \pi(0)=\pi_0, \quad \alpha(0)=\alpha_0 \qquad \text{on } \Omega. \quad (7.5.45\text{e})$$

The weak solution according Definitions 7.2.1 or 7.2.2 is to be now cast as if the splitting u vs z would be rather (u,π) vs $z=\alpha$. Therefore, (7.5.4b) reflecting Definition 7.2.1 is to be modified for

$$\int_Q(\partial_\alpha\varphi(e(u)-\pi,\alpha)+r_{\rm c}-\varphi_{\rm DAM}'(\alpha))w+\kappa\nabla\alpha\cdot\nabla w+\zeta_{\rm IN}(w)\,\mathrm{d}x\mathrm{d}t$$
$$\ge\int_Q\xi_{\rm IN}(\dot{\alpha})+\partial_e\varphi(e(u)-\pi,\alpha){:}(e(\dot{u})-\dot{\pi})\,\mathrm{d}x\mathrm{d}t+\int_\Omega\Big(\varphi(e(u(T))-\pi(T),\alpha(T))$$

$$+\frac{\kappa}{2}|\nabla\alpha(T)|^2 - \varphi(e(u_0){-}\pi_0,\alpha_0) - \frac{\kappa}{2}|\nabla\alpha_0|^2\Big)\mathrm{d}x \tag{7.5.46}$$

to be valid for all $w \in L^2(I;H^1(\Omega))$. If Definition 7.2.2 were used, then the resulting variational inequality would look as

$$\int_Q \big(\partial_\alpha\varphi(e(u){-}\pi,\alpha) + r_{\rm C} - \varphi'_{\rm DAM}(\alpha)\big)w + \kappa\nabla\alpha\cdot\nabla w + \zeta_{\rm IN}(w)\,\mathrm{d}x\mathrm{d}t$$
$$\geq \int_Q \mathbb{D}_1(\alpha)(e(\dot u){-}\dot\pi):(e(\dot u){-}\dot\pi) + \mathbb{D}_2\dot\pi:\dot\pi + \xi_{\rm IN}(\dot\alpha) - f\cdot\dot u\,\mathrm{d}x\mathrm{d}t - \int_\Sigma g\cdot\dot u\,\mathrm{d}S\,\mathrm{d}t$$
$$+\int_\Omega\Big(\frac{\varrho}{2}\big|\dot u(T)\big|^2 + \varphi(e(u(T)){-}\pi(T),\alpha(T)) + \frac{\kappa}{2}|\nabla\alpha(T)|^2$$
$$-\frac{\varrho}{2}|v_0|^2 - \varphi(e(u_0){-}\pi_0,\alpha_0) - \frac{\kappa}{2}|\nabla\alpha_0|^2\Big)\mathrm{d}x. \tag{7.5.47}$$

The analysis through the strong convergence (7.5.9) in combination with (7.5.22) works quite equally.[47]

Remark 7.5.21 (Damage in Jeffreys materials II). Let us now consider a special case of the model from Remark 7.5.20 such that the product $\mathbb{D}_1^{-1}(\alpha)\mathbb{C}(\alpha) = \mathbb{D}_1^{-1}(1)\mathbb{C}(1)$ is independent of α. One can then eliminate π like we did for (6.6.31), obtaining the system

$$\dot u = v, \qquad \varrho\dot v - \mathrm{div}\,\sigma = f, \tag{7.5.48a}$$
$$(\mathbb{I}{+}\mathbb{D}_1(\alpha)\mathbb{D}_2^{-1})\dot\sigma + \mathbb{C}(\alpha)\mathbb{D}_2^{-1}\sigma = \mathbb{D}_1(\alpha)e(\dot v)$$
$$+\mathbb{C}(\alpha)e(v) - \dot\alpha\mathbb{D}_1(\alpha)[\mathbb{D}_1^{-1}]'(\alpha)\sigma, \tag{7.5.48b}$$
$$\partial\zeta_{\rm IN}(\dot\alpha) + \partial_\alpha\varphi(\sigma,e(\dot u),\alpha) + r_{\rm C} \ni \varphi'_{\rm DAM}(\alpha) + \kappa\Delta\alpha$$
$$\text{with } r_{\rm C} \in N_{[0,1]}(\alpha). \tag{7.5.48c}$$

The last term in (7.5.48b) is controlled in $L^1(Q;\mathbb{R}^{d\times d})$ and resulted from time-differentiation used when deriving (6.6.31). The stored energy used in (7.5.48c), expressed when eliminating π-variable, is $\varphi(\sigma,\dot e,\alpha) = \frac{1}{2}\mathbb{C}^{-1}(\alpha)\sigma_{\rm EL}:\sigma_{\rm EL}$ with the elastic stress $\sigma_{\rm EL} = (\mathbb{I}{+}\mathbb{D}_1(\alpha)\mathbb{D}_2^{-2})\sigma - \mathbb{D}_1(\alpha)\dot e$, cf. (6.6.33a). Interestingly, if $\mathbb{D}_1^{-1}(\alpha)\mathbb{C}(\alpha)$ were dependent on α, the procedure leading to (6.6.31) by elimination of the creep-strain internal variable π would not work. Let us note that, for "$\mathbb{D}_2 \to \infty$" meaning that $\mathbb{D}_2^{-1} \to 0$, we obtain the Kelvin-Voigt rheology undergoing damage: indeed, (7.5.48b) when first multiplied by $\mathbb{D}_1^{-1}(\alpha)$ turns to be $\mathbb{D}_1^{-1}(\alpha)\dot\sigma = e(\dot v) + \mathbb{D}_1^{-1}\mathbb{C}e(v) - \dot\alpha[\mathbb{D}_1^{-1}]'(\alpha)\sigma$, which can be written as $[\mathbb{D}_1^{-1}(\alpha)\sigma]^{\boldsymbol\cdot} = e(\dot v) + \mathbb{D}_1^{-1}\mathbb{C}e(\dot u)$ and clearly gives the Kelvin-Voigt model when integrated in time; note that again we used that $\mathbb{D}_1^{-1}\mathbb{C}$ is independent of α. A particular case is $\mathbb{D}_1 = \tau_{\rm R}\mathbb{C}$ used already before, but

[47] Instead of the total strain $e(u)$, the strategy (7.5.9) is to be executed for the elastic strain $e_{\rm EL} = e(u){-}\pi$ and both viscosities $\mathbb{D}_1$ and $\mathbb{D}_2$ are to be involved, as well as an approximation of both (7.5.45a) and (7.5.45b) is to be used. Importantly, $\mathbb{D}_2$ is not damageable so that, when (7.5.45b) tested by π, the term $\mathbb{D}_2\dot\pi:\pi$ can be integrated out while $\mathbb{D}_1(\alpha)(\dot\pi{-}e(\dot u)):\pi$ is to be treated by (7.5.22) when summed up with the $\mathbb{D}_1$-term coming from (7.5.45a).

this α-independence of $\mathbb{D}_1^{-1}\mathbb{C}$ may be designed separately for the spherical and the deviatoric part. Also note, that under this condition, φ can be slightly simplified as

$$\varphi(\sigma,\dot{e},\alpha)=\frac{1}{2}(\mathbb{C}^{-1}(\alpha)\sigma+\mathbb{C}^{-1}\mathbb{D}_1\mathbb{D}_2^{-2}\sigma-\mathbb{C}^{-1}\mathbb{D}_1\dot{e}):((\mathbb{I}+\mathbb{D}_1(\alpha)\mathbb{D}_2^{-1})\sigma-\mathbb{D}_1(\alpha)\dot{e}).$$

The resulting driving force for damage in (7.5.48c) is then

$$\begin{aligned}\partial_\alpha\varphi(\sigma,\dot{e},\alpha)=&\ \frac{1}{2}[\mathbb{C}^{-1}]'(\alpha)\sigma:((\mathbb{I}+\mathbb{D}_1(\alpha)\mathbb{D}_2^{-1})\sigma-\mathbb{D}_1(\alpha)\dot{e})\\&+\frac{1}{2}(\mathbb{C}^{-1}(\alpha)\sigma+\mathbb{C}^{-1}\mathbb{D}_1\mathbb{D}_2^{-2})\sigma-\mathbb{C}^{-1}\mathbb{D}_1\dot{e}):\mathbb{D}_1'(\alpha)(\mathbb{D}_2^{-1}\sigma-\dot{e}).\quad(7.5.49)\end{aligned}$$

In contrast to (7.5.45) which straightforwardly leads to efficient decoupled time-discretization strategies even without the condition on α-independence of $\mathbb{D}_1^{-1}\mathbb{C}$, the system (7.5.48) is "more coupled" because $\dot{\alpha}$ occurs also in (7.5.48b). This shows other advantages of the internal-variable concept.

Remark 7.5.22 (Damage in Maxwell material). The Maxwell model is to use the weak formulation based on (7.2.8) rather than on (7.2.5b) because $e(\dot{u})$ is not well controlled. Yet, there are problems to execute the strategy (7.5.9) because the strong convergence of $\dot{u}$ is not directly at disposal. Also the regularity like Proposition 6.6.2 cannot be simply executed if $\mathbb{C}$ is not constant, being subjected to damage. This is not much surprising because damageable Maxwell materials with inertia leads to a nonlinear hyperbolic system.

Exercise 7.5.23 (Failure of semiconvexity). Show that $(e,\alpha)\mapsto\frac{1}{2}\alpha\mathbb{C}e{:}e:\mathbb{R}^{d\times d}_{\rm sym}\times[0,1]\to\mathbb{R}^+$ with $\mathbb{C}$ positive definite is not semiconvex.

Exercise 7.5.24 (A strengthened semiconvexity). Show that the potential $(e,\alpha)\mapsto\frac{1}{2}g(\alpha)\mathbb{C}e{:}e+\frac{1}{2}K|e|^2$ with $g(\cdot)$ smooth, positive, and strictly convex and with $\mathbb{C}$ positive semidefinite is convex for all K large enough; in particular, $(e,\alpha)\mapsto\frac{1}{2}g(\alpha)\mathbb{C}e{:}e$ is semiconvex.

Exercise 7.5.25 (Convergence of the implicit scheme). Consider the scheme (7.5.40)–(7.5.41) modified as in Remark 7.5.15 and prove the strong convergence of $e(\overline{u}_\tau)\to e(u)$ in $L^2(Q;\mathbb{R}^{d\times d}_{\rm sym})$.

Exercise 7.5.26. Write the underlying potential for the coupled implicit scheme from Remark 7.5.15. Do the same if $\partial_e\varphi(e(u_\tau^k),\alpha_\tau^{k-1/2})$ is used instead of $\partial_e\varphi(e(u_\tau^k),\alpha_\tau^{k-1})$ or $\partial_e\varphi(e(u_\tau^k),\alpha_\tau^k)$.

7.5.3 Damage in nonsimple viscoelastic or elastic materials

Exploiting the concept of nonsimple materials from Sect. 2.5, involving a strain-gradient term in the stored energy, allows for some interesting variants of damage models if the hyperstress term is not subjected to damage, otherwise a similar situation as in the previous Section 7.5.1 holds. This is obviously slightly misconceptual but sometimes acceptable regularization. More specifically, this allows us to address:

1) damage in Maxwell materials, cf. Remark 7.5.22 above, and
2) damage in purely elastic materials, possibly even without damage gradient.

And, of course, also a nonconvex stored energy $\varphi(\cdot,\alpha)$ is allowed. It should be mentioned that all these models focus to particular aspects while the highest-order "nonsimple" elastic-moduli term is not subjected to damage (like was $\mathbb{D}$ in (7.5.2)), which is in some sense a bit misconceptual and the limit towards simple material will not be under scrutiny.

The first model, dealing with the damageable Maxwelian rheology of the type $(\mathbb{C}^{\alpha}\|\mathbb{H})$—$\mathbb{D}^{\alpha}$, considers the stored energy dependent on the elastic strain $e_{\rm EL}$, namely

$$\varphi_{\rm EL}(e_{\rm EL},\alpha)=\frac{1}{2}\mathbb{C}(\alpha)e_{\rm EL}{:}e_{\rm EL}+\frac{1}{2}\mathbb{H}\nabla e_{\rm EL}\,\vdots\,\nabla e_{\rm EL}-\varphi_{\rm DAM}(\alpha)+\delta_{[0,1]}(\alpha)+\frac{\kappa}{2}|\nabla\alpha|^2.$$

Thus the actual stored energy as a function of e and π and the dissipation potential governing the model are

$$\varphi(e,\pi,\alpha)=\varphi_{\rm EL}(e{-}\pi,\alpha)=\frac{1}{2}\mathbb{C}(\alpha)(e{-}\pi){:}(e{-}\pi)+\frac{1}{2}\mathbb{H}\nabla(e{-}\pi)\,\vdots\,\nabla(e{-}\pi)$$
$$-\varphi_{\rm DAM}(\alpha)+\delta_{[0,1]}(\alpha)+\frac{\kappa}{2}|\nabla\alpha|^2, \qquad (7.5.50a)$$

$$\zeta(\alpha;\dot\pi,\dot\alpha)=\frac{1}{2}\mathbb{D}(\alpha)\dot\pi{:}\dot\pi+\zeta_{\rm IN}(\dot\alpha). \qquad (7.5.50b)$$

The motivation for introducing the $\mathbb{H}$-term is the difficulties mentioned in Remark 7.5.22. We assume $\mathbb{D}(\cdot)$ continuous, positive definite, so that incomplete damage in Stokes-type viscosity is allowed, and $\mathbb{C}(\cdot)$ continuous, positive semidefinite so that complete damage in elastic part is allowed, and $\varrho>0$ and $\mathbb{H}$ positive definite. Also (7.5.6b,d,e) is assumed, so we consider the variant allowing healing.

For the needed strong convergence of the elastic strain $e(u){-}\pi$, we now use $e(u){-}\pi\in L^\infty(I;H^2(\Omega;\mathbb{R}^{d\times d}_{\rm sym}))$ together with some information about its time derivative. Namely, we have $\dot\pi\in L^2(Q;\mathbb{R}^{d\times d}_{\rm sym})$ due to the positive definiteness of $\mathbb{D}$ and also $e(\dot u)\in L^\infty(I;H^{-1}(\Omega;\mathbb{R}^{d\times d}_{\rm sym}))$ due to the inertia and the estimate of $\dot u\in L^\infty(I;L^2(\Omega;\mathbb{R}^d))$; more in detail, we can estimate

$$\|e(\dot u)\|_{L^\infty(I;H^{-1}(\Omega;\mathbb{R}^{d\times d}_{\rm sym}))}=\sup_{\|v\|_{L^1(I;H^1_0(\Omega;\mathbb{R}^{d\times d}_{\rm sym}))}\le1}\int_Q e(\dot u){:}v\,{\rm d}x{\rm d}t$$
$$=\sup_{\|v\in\|_{L^1(I;H^1_0(\Omega;\mathbb{R}^{d\times d}_{\rm sym}))}\le1}\int_Q \dot u{:}{\rm div}\,v\,{\rm d}x{\rm d}t\le\|\dot u\|_{L^\infty(I;L^2(\Omega;\mathbb{R}^d))}\|v\|_{L^1(I;L^2(\Omega;\mathbb{R}^d))}.$$

Then also the elastic strain rate $(e(u){-}\pi)^{\boldsymbol\cdot}$ is controlled in $L^\infty(I;H^{-1}(\Omega;\mathbb{R}^{d\times d}_{\rm sym}))$. By (7.5.8), we get control over $\Delta\alpha$ and thus also on $r_{\rm c}\in N_{[0,1]}(\alpha)$ in $L^2(Q)$.

Other variants of this model could be possible, based on the scenarios (7.5.16) while leading to $r_c = 0$.

The second mentioned option, i.e. the gradient damage in purely elastic inviscid rheology $\mathbin{\not\leq}^{\alpha}_{\mathbb{C}} \,\|\, \mathbin{\leq}_{\mathbb{H}}$, is a special of the previous Maxwellien creep model for $\pi \equiv 0$, which can be formally obtained by putting "$\mathbb{D} = \infty$" and the initial condition $\pi_0 = 0$.

Eventually, let us now come to the last mentioned variant of this third model for *damage without gradient*. The important attribute is that $\mathbb{C}(\cdot)$ and $\zeta_{\text{IN}}(\cdot)$ are affine, except the constraints $\alpha \geq 0$ and $\dot{\alpha} \leq 0$ so that the weak convergence of α's will do a needed job if the energetic-solution concept is employed. More specifically, we consider

$$\varphi(e,\alpha) = \frac{1}{2}\alpha\mathbb{C}e{:}e + \frac{1}{2}\mathbb{H}\nabla e \,\vdots\, \nabla e + \delta_{[0,1]}(\alpha), \tag{7.5.51a}$$

$$\zeta(\alpha;\dot{e},\dot{\alpha}) = \frac{1}{2}(\alpha + \epsilon)\mathbb{D}\dot{e}{:}\dot{e} - \mathfrak{g}_c\dot{\alpha} + \delta_{(-\infty,0]}(\dot{\alpha}). \tag{7.5.51b}$$

Let us note that the $\mathbb{C}$-term itself can undergo a complete damage relying on the non-vanishing inertia, and also one consider purely *elastic material*, i.e. $\mathbb{D} = 0$. We deal with the rate-independent unidirectional damage which still involves the dissipation potential $\zeta(\dot{\alpha}) = \delta^*_{[-a,\infty)}(\dot{\alpha})$ as in (7.5.17) with $\nu = 0$ and exploit the definition of the energetic solution Definition 7.2.4. Then the system in the classical formulation takes the form

$$\varrho\ddot{u} - \text{Div}(\alpha\mathbb{C}e(u) - \text{Div}(\mathbb{H}\nabla e(u))) = f \qquad \text{in } Q, \tag{7.5.52a}$$

$$\partial\zeta(\dot{\alpha}) + \frac{1}{2}\mathbb{C}e(u){:}e(u) + r_c \ni 0 \quad \text{with} \quad r_c \in N_{[0,1]}(\alpha) \qquad \text{in } Q, \tag{7.5.52b}$$

$$(\alpha\mathbb{C}e(u) - \text{Div}(\mathbb{H}\nabla e(u)))\vec{n} - \text{div}_{\text{S}}((\mathbb{H}\nabla^2 u)\cdot\vec{n}) = g \quad \text{and} \quad (\mathbb{H}\nabla^2 u){:}(\vec{n}\otimes\vec{n}) = 0 \qquad \text{on } \Sigma, \tag{7.5.52c}$$

$$u|_{t=0} = u_0 \quad \dot{u}|_{t=0} = v_0, \quad \text{and} \quad \alpha|_{t=0} = \alpha_0 \qquad \text{in } \Omega. \tag{7.5.52d}$$

The weak formulation then uses the concept of energetic solution not involving explicitly the multiplier r_c from (7.5.52b), since ∂_z is not explicitly involved in Definition 7.2.4):

Definition 7.5.27. The pair $(u,\alpha) \in C_{\text{w}}(I;H^2(\Omega;\mathbb{R}^d)) \times \text{BV}(I;L^1(\Omega))$ with $\dot{u} \in C_{\text{w}}(I;L^2(\Omega;\mathbb{R}^d))$ will be called a weak/energetic solution to the initial-boundary-value problem (7.5.52) if the semistability (7.2.15), i.e. here

$$\int_\Omega \alpha(t)\mathbb{C}e(u(t)){:}e(u(t)) + \delta_{[0,1]}(\alpha(t))\,\text{d}x \leq \int_\Omega v\mathbb{C}e(u(t)){:}e(u(t)) + 2\zeta_{\text{IN}}(v{-}\alpha(t))\,\text{d}x, \tag{7.5.53a}$$

is valid for all $v \in L^\infty(\Omega)$ with $0 \leq v \leq 1$ at a.a. $t \in I$, and the weakly formulated momentum equation (7.5.52a) together with the initial and boundary conditions (7.5.52c,d), i.e.

$$\int_Q \alpha\mathbb{C}e(u):e(v) + \mathbb{H}\nabla e(u)\vdots\nabla e(v) - \varrho\dot{u}\cdot\dot{v}\,\mathrm{d}x\mathrm{d}t$$
$$= \int_Q f\cdot v\,\mathrm{d}x\mathrm{d}t + \int_\Sigma g\cdot v\,\mathrm{d}S\,\mathrm{d}t + \int_\Omega v_0\cdot v(0)\,\mathrm{d}x\mathrm{d}t, \qquad (7.5.53\mathrm{b})$$

holds for all v smooth with $v(T) = 0$, and eventually the energy balance is valid:

$$\int_\Omega \frac{\varrho}{2}|\dot{u}(T)|^2 + \frac{1}{2}\alpha(T)\mathbb{C}e(u(T)):e(u(T)) + \frac{1}{2}\mathbb{H}\nabla e(u(T))\vdots\nabla e(u(T)) + a\alpha(T)\,\mathrm{d}x\mathrm{d}t$$
$$= \int_\Omega \frac{\varrho}{2}|v_0|^2 + \frac{1}{2}\alpha(T)\mathbb{C}e(u_0):e(u_0) + \frac{1}{2}\mathbb{H}\nabla e(u_0)\vdots\nabla e(u_0) + a\alpha_0\,\mathrm{d}x\mathrm{d}t$$
$$+ \int_Q f\cdot\dot{u}\,\mathrm{d}x\mathrm{d}t + \int_\Sigma g\cdot\dot{u}\,\mathrm{d}S\,\mathrm{d}t \qquad (7.5.53\mathrm{c})$$

together with $u(0) = u_0$ holds.

For approximation, we can use e.g. the Rothe method combined with the fractional-step splitting because φ from (7.5.51) is separately convex. Assuming $\mathbb{H}$ positive definite and $\mathbb{C}$ positive semidefinite, $f \in L^1(I;L^2(\Omega;\mathbb{R}^d))$, $g \in W^{1,1}(I;L^{2^{*\sharp}}(\Gamma))$, $u_0 \in H^2(\Omega)$, $v_0 \in L^2(\Omega)$, $\alpha_0 \in L^\infty(\Omega)$ with $0 \le \alpha_0 \le 1$, and $\varrho > 0$, the approximate solution satisfy the following a-priori bounds:

$$\|\bar{u}_\tau\|_{L^\infty(I;H^2(\Omega;\mathbb{R}^d))} \le C, \qquad \|u_\tau\|_{W^{1,\infty}(I;L^2(\Omega))} \le C, \qquad (7.5.54\mathrm{a})$$
$$\|\bar{\alpha}_\tau\|_{L^\infty(Q)\cap\mathrm{BV}(I;L^1(\Omega))} \le C. \qquad (7.5.54\mathrm{b})$$

The limit passage in the approximate momentum equation is simple when using the Aubin-Lions theorem B.4.10 generalized for $\dot{\bar{u}}_\tau \in \mathfrak{M}(I;L^2(\Omega))$, cf. Lemma B.4.11 which gives strong convergence $e(\bar{u}_\tau) \to e(u)$ in $L^{1/\varepsilon}(I;L^{2^*-\varepsilon}(\Omega;\mathbb{R}^{d\times d}))$ for any $0 < \varepsilon \le$ ...., so that the stress $\underline{\alpha}_\tau\mathbb{C}e(\bar{u}_\tau):e(\bar{u}_\tau)$ converges weakly to $\alpha\mathbb{C}e(u):e(u)$ in $L^{1/\varepsilon}(I;L^{2^*/2-\varepsilon}(\Omega;\mathbb{R}^{d\times d}))$.

Now we prove (7.5.53a) integrated over time. For any $(\hat{u},\hat{\alpha})$ with $0 \le \hat{\alpha} \le \alpha$, we choose the so-called *mutual recovery sequence*:

$$\hat{u}_\tau := \bar{u}_\tau \quad \text{and} \quad \hat{\alpha}_\tau := \begin{cases} \underline{\alpha}_\tau\hat{\alpha}/\alpha & \text{where } \alpha > 0, \\ 0 & \text{where } \alpha = 0, \end{cases} \qquad (7.5.55)$$

cf. (7.3.28) integrated over time. Let us note that always $0 \le \hat{\alpha}_\tau \le \underline{\alpha}_\tau$ and, if one considers $\bar{u}_\tau \to u$ weakly in $H^2(\Omega;\mathbb{R}^d)$ and $\underline{\alpha}_\tau \to \alpha$ weakly* in $L^\infty(\Omega)$, then also $\hat{u}_\tau \to \hat{u}$ weakly in $H^2(\Omega;\mathbb{R}^d)$ and $\hat{\alpha}_\tau \to \hat{\alpha}$ weakly* in $L^\infty(\Omega)$. When disintegrating in time,[48] we get (7.5.53a) at a.e. $t \in I$.

[48] More in detail, assuming that (7.5.53a) does not hold for $t \in \hat{I} \subset I$ with some $\hat{v}(t)$ with $\mathrm{meas}_d(I_0) > 0$, we consider $v(t) = \hat{v}(t)$ if $t \in \hat{I}$ and $v(t) = \alpha(t)$ if $t \in I\setminus\hat{I}$ to get a contradiction with (7.5.53a) integrated over time. To execute this strategy rigorously, we should take a measurable selection from the set-valued mapping of the nonempty sets of all such $\hat{v}$'s at each given time t.

The limit passage in the energy equation towards (7.5.53c) is by lower weak semicontinuity as an inequality "$\le$" and then by Proposition 7.3.9 the opposite inequality is to be proved.

Noteworthy, a viscous alternative considering a rate-dependent damage (7.5.17) with $\nu > 0$ which does not seem to work without damage gradient in $\varphi_{\rm E}$. The weak solution should then be defined by the standard variational inequality integrated over time to allow for using the by-part integration to eliminate the term $\int_Q r_{\rm c}\dot\alpha\,{\rm d}x{\rm d}t$, cf. Definition 7.2.2 combined with the inequality $\int_Q r_{\rm c}(v-\alpha)\,{\rm d}x{\rm d}t \ge 0$ for any $0 \le v \le 1$, as in Remark 7.2.8. Here it would be a problem in passing to the limit in $\int_Q r_{\rm c}\alpha\,{\rm d}x{\rm d}t$ because of the only weak convergence of both $r_{\rm c}$ and α. This rather shows difficulty of the problem and the mathematical strength of the energetic-solution concept.

Remark 7.5.28 (Dynamic fracture in purely elastic materials by phase field). One can slightly modify the model (7.5.35) by considering the enhanced stored energy as

$$\varphi_{\rm E}(e,\alpha,\nabla e,\nabla\alpha) := \frac{(\varepsilon/\varepsilon_0)^2+\alpha^2}{2}\mathbb{C}e{:}e + \mathfrak{g}_{\rm c}\Big(\frac{1}{2\varepsilon}(1-\alpha)^2 + \frac{\varepsilon}{2}|\nabla\alpha|^2\Big) + \frac{\varepsilon}{2\varepsilon_0}\mathbb{H}\nabla e\,\vdots\,\nabla e. \tag{7.5.56}$$

One can then omit the Kelvin-Voigt viscosity, i.e. consider $\mathbb{D} = 0$. The fractional-step discretization as in Remark 7.5.14 applies again smoothly due to the separate convexity of $\varphi_{\rm E}$, i.e. convexity of both $\varphi_{\rm E}(\cdot,\alpha,\cdot,\nabla\alpha)$ and and $\varphi_{\rm E}(e,\cdot,\nabla e,\cdot)$. The fully-implicit discretization like in Remark 7.5.15 now must rely on the strict convexity of the kinetic energy instead of the viscosity, which however affect rather u than $e(u)$, so that it works only after the additional space discretization. and a robust convergence of the iterative scheme is guaranteed only for sufficiently small time step on dependence of the space discretization. For the convergence towards the weak solution of the continuous problem, the time step must approach zero sufficiently fast with respect how the space discretization is refined. Note also that Korn's inequality for 2nd-grade nonsimple materials (5.2.5) has to be used to obtain the H^2-estimate for the displacement.

Remark 7.5.29 (Nonlocal simple theory). Actually, instead of the full strain gradient, only a nonlocal theory with a "fractional" strain gradient like in Sect. 2.5.2 can be used for most of the results in this section, i.e. instead of $\frac{1}{2}\int_\Omega \mathbb{H}\nabla e\,\vdots\,\nabla e\,{\rm d}x$ we can use the quadratic functional $\mathscr{H}$ from (2.5.27) with the kernel $\mathbb{K}(x,\tilde x)$ satisfying (2.5.28) with any $0 < \gamma < 1$; even arbitrary small positive γ can serve for the desired compactness of e in $L^{1/\epsilon}(I;L^2(\Omega;\mathbb{R}^{d\times d}_{\rm sym}))$.

7.6 Darcy/Fick flows in swelling or poroelastic solids

The evolution variant of the steady-state models from Section 5.7 of diffusion governed by *chemical potential* μ interacting with mechanical properties is of an ultimate interest in many applications. There are various option both as far as the stored energy and as far as the viscoelastic rheology. We confine ourselves to the isotropic materials where the concentration of diffusant influences the spherical strains rather than the deviatoric ones.

The diffusion problem has, to some extent, a similar structure as the damage problem (in the spherical part) but there are essential differences: the dissipation potential is typically quadratic but nonlocal and state-dependent.

7.6.1 A convex-energy model

We first present a relatively simple model with the stored energy convex and having a controlled growth.

When avoiding the formulation using the nonlocal dissipation potential, the partial differential equations governing the problem, arising as an immediate evolution variant of (5.7.16)–(5.7.18), take the form

$$\varrho\ddot{u} - \mathrm{div}\,\sigma = f \qquad \text{with} \quad \sigma = \partial_e\varphi(e(u),c) + \partial_{\dot{e}}\zeta(e(\dot{u}),\dot{c}), \tag{7.6.1a}$$

$$\dot{c} - \mathrm{div}(\mathbb{M}(e(u),c)\nabla\mu) = 0 \qquad \text{with} \quad \mu = \partial_c\varphi(e(u),c) + \partial_{\dot{c}}\zeta(e(\dot{u}),\dot{c}), \tag{7.6.1b}$$

accompanied by boundary conditions, e.g.

$$u = u_\mathrm{D} \quad \text{on } \Gamma_\mathrm{D}, \qquad \sigma\vec{n} = g \quad \text{on } \Gamma_\mathrm{N}, \tag{7.6.2a}$$

$$\mathbb{M}(e(u),c)\nabla\mu\cdot\vec{n} = \alpha(\mu_\mathrm{ext}-\mu) \quad \text{on } \Gamma; \tag{7.6.2b}$$

as in (5.7.18b), with some phenomenological boundary-permeability coefficient $\alpha \geq 0$ and a prescribed external chemical potential μ_ext.

Let us note that here μ contains also rates – let us agree to call it therefore a *viscous chemical potential*, which is the original Gurtin's idea [233] to reflect transition kinetics in particular if φ describes a multiphase flow (and is nonconvex in c). The diffusive part $-\mathrm{div}(\mathbb{M}(e,c)\nabla\mu)$ in (7.6.1b) contributes to the dissipative part, cf. Remark 7.6.4 below. The energetics of this model is revealed when testing (7.6.1a) by $\dot{u}$ and the first equation in (7.6.1b) by μ from the second equation in (7.6.1b) tested by $\dot{c}$. This gives respectively

$$\int_\Omega \varrho\ddot{u}\cdot\dot{u} + (\partial_e\varphi(e(u),c) + \partial_{\dot{e}}\zeta(e(\dot{u}),\dot{c})):e(\dot{u})\,\mathrm{d}x = \int_\Omega f\cdot\dot{u}\,\mathrm{d}x + \int_{\Gamma_\mathrm{N}} g\cdot\dot{u}\,\mathrm{d}S + \langle\vec{\tau}_\mathrm{D},\dot{u}_\mathrm{D}\rangle_{\Gamma_\mathrm{D}}, \tag{7.6.3a}$$

$$\int_\Omega \dot{c}\mu + \mathbb{M}(e,c)\nabla\mu\cdot\nabla\mu\,\mathrm{d}x + \int_\Gamma \alpha\mu^2\,\mathrm{d}S = \int_\Gamma \alpha\mu_\mathrm{ext}\mu\,\mathrm{d}S \quad \text{and} \tag{7.6.3b}$$

$$\int_\Omega \dot{c}\mu\,\mathrm{d}x = \int_\Omega \partial_c\varphi(e(u),c)\dot{c} + \partial_{\dot{c}}\zeta(e(\dot{u}),\dot{c})\dot{c}\,\mathrm{d}x. \tag{7.6.3c}$$

Using the calculus $\partial_e\varphi(e(u),c):e(\dot{u}) + \partial_c\varphi(e(u),c)\dot{c} = \frac{\partial}{\partial t}\varphi(e(u),c)$ gives the *energy balance*

$$\underbrace{\frac{\mathrm{d}}{\mathrm{d}t}\int_\Omega \frac{\varrho}{2}|\dot{u}|^2 + \varphi(e(u),c)\,\mathrm{d}x}_{\text{kinetic and stored energy}} + \underbrace{\int_\Omega \xi(e(\dot{u}),\dot{c}) + \mathbb{M}(e,c)\nabla\mu\cdot\nabla\mu\,\mathrm{d}x + \int_\Gamma \alpha\mu^2\,\mathrm{d}S}_{\text{rate of dissipation energy}}$$

$$= \underbrace{\int_\Omega f\cdot\dot{u}\,\mathrm{d}x + \int_{\Gamma_\mathrm{N}} g\cdot\dot{u}\,\mathrm{d}S + \langle\vec{\tau}_\mathrm{D},\dot{u}_\mathrm{D}\rangle_{\Gamma_\mathrm{D}}}_{\text{power of external chemo--mechanical load}} + \underbrace{\int_\Gamma \alpha\mu_\mathrm{ext}\mu\,\mathrm{d}S,}_{\text{power of the boundary fluid flux}} \tag{7.6.4}$$

with the dissipation rate $\xi(\dot{e},\dot{c}) = \partial_{\dot{e}}\zeta(\dot{e},\dot{c}):\dot{e} + \partial_{\dot{c}}\zeta(\dot{e},\dot{c})\dot{c}$, cf. also (8.1.6) below.

We still consider the initial conditions

$$u(0) = u_0, \qquad \dot{u}(0) = v_0, \qquad c(0) = c_0 \qquad \text{on } \Omega, \tag{7.6.5}$$

and consider a special quadratic ansatz for the dissipation potential:

$$\zeta(\dot{e},\dot{c}) = \frac{1}{2}\mathbb{D}\dot{e}:\dot{e} + \frac{1}{2}\tau_\mathrm{R}\dot{c}^2. \tag{7.6.6}$$

Definition 7.6.1 (Flow/poroelastic problem: weak solutions). The triple $(u,c,\mu)\in H^1(I;H^1(\Omega;\mathbb{R}^d))\times L^2(Q)\times L^2(I;H^1(\Omega))$ is called a weak solution to the initial-boundary-value problem (7.6.1)–(7.6.2) with the initial conditions (7.6.5) with ζ from (7.6.6) if $u(0) = u_0$ and if:

(α) the following integral identity holds for all $\tilde{u}\in H^1(Q;\mathbb{R}^d)$:

$$\int_Q (\mathbb{D}e(\dot{u}) + \partial_e\varphi(e(u),c)):e(\tilde{u}) - \varrho\dot{u}\cdot\dot{\tilde{u}}\,\mathrm{d}x\mathrm{d}t$$
$$= \int_\Omega \varrho v_0\cdot\tilde{u}(0)\,\mathrm{d}x + \int_Q f\cdot\tilde{u}\,\mathrm{d}x\mathrm{d}t + \int_{\Sigma_\mathrm{N}} g\cdot\tilde{u}\,\mathrm{d}x\mathrm{d}t, \tag{7.6.7a}$$

(β) the following integral identity holds for all $\tilde{\mu}\in H^1(Q)$, $\tilde{\mu}(T) = 0$:

$$\int_Q \mathbb{M}(e,c)\nabla\mu\cdot\nabla\tilde{\mu} - c\dot{\tilde{\mu}}\,\mathrm{d}x\mathrm{d}t + \int_\Sigma \alpha\mu\tilde{\mu}\,\mathrm{d}S\,\mathrm{d}t = \int_\Omega c_0\tilde{\mu}(0)\,\mathrm{d}x + \int_\Sigma \alpha\mu_\mathrm{ext}\tilde{\mu}\,\mathrm{d}S\,\mathrm{d}t, \tag{7.6.7b}$$

(γ) and the following variational inequality holds for all $\tilde{c}\in L^2(Q)$:

$$\int_Q \varphi(e(u),\tilde{c})\,\mathrm{d}x\mathrm{d}t \ge \int_Q \varphi(e(u),c) + (\mu - \tau_\mathrm{R}\dot{c})(\tilde{c} - c)\,\mathrm{d}x\mathrm{d}t. \tag{7.6.7c}$$

We impose the following data qualification:

$$\varphi \text{ continuously differentiable, convex, and uniformly convex in } c, \tag{7.6.8a}$$

$$\varphi(e,c) \ge \epsilon c^2, \quad |\varphi(e,c)| \le C(1+|e|^2+c^2), \quad |\partial\varphi(e,c)| \le C(1+|e|+|c|), \tag{7.6.8b}$$

$$\mathbb{D} \text{ positive definite}, \quad \tau_\mathrm{R} \ge 0, \quad \mathbb{M}: \mathbb{R}^{d\times d}_\mathrm{sym}\times\mathbb{R} \to \mathbb{R}^{d\times d}_\mathrm{sym} \text{ continuous},$$

$$\text{bounded, and} \qquad \inf_{e\in\mathbb{R}^{d\times d}_{\rm sym},\ c\in\mathbb{R},\ j\in\mathbb{R}^d,\ |j|=1} \mathbb{M}(e,c)j{\cdot}j>0. \tag{7.6.8c}$$

We use the Faedo-Galerkin approximation by exploiting some finite-dimensional subspaces of $H^1(\Omega;\mathbb{R}^d)$ for (7.6.1a) and of $H^1(\Omega)$ for the first equation in (7.6.1b), and also the second equation in (7.6.1b) is to be discretized by using the same finite-dimensional space in order to facilitate the test leading to energy balance of the discrete scheme, i.e. the discrete analog of (7.6.4). We denote by $k\in\mathbb{N}$ the discretization parameter and the thus constructed approximate solution by (u_k,c_k,μ_k).

Proposition 7.6.2 (Existence of weak solutions). *Let (7.6.8) hold, ϱ, the loading and the initial conditions satisfy (6.4.5b-e) together with $\alpha\in L^\infty(\Gamma)$ positive and $\mu_{\rm ext}\in L^2(I;L^{2^{\sharp'}}(\Gamma_{\rm N}))$ and $c_0\in L^2(\Omega)$. Then (u_k,c_k,μ_k) does exist and satisfy the a-priori estimates*

$$\|u_k\|_{W^{1,\infty}(I;L^2(\Omega;\mathbb{R}^d))\cap H^1(I;H^1(\Omega;\mathbb{R}^d))\cap W^{2,1}(I;H^1(\Omega;\mathbb{R}^d)^*)}\le C, \tag{7.6.9a}$$

$$\|\mu_k\|_{L^2(I;H^1(\Omega))}\le C, \tag{7.6.9b}$$

$$\|\dot c_k\|_{L^2(I;H^1(\Omega)^*)}\le C \quad \textit{and, if } \tau_{\rm R}>0, \textit{ also } \quad \|\dot c_k\|_{L^2(Q)}\le C, \tag{7.6.9c}$$

and selected subsequences converge as follows:

$$u_k\to u \qquad \textit{weakly in } W^{1,\infty}(I;L^2(\Omega;\mathbb{R}^d))\cap H^1(I;H^1(\Omega;\mathbb{R}^d)), \tag{7.6.10a}$$

$$c_k\to c \qquad \textit{weakly* in } L^\infty(I;L^2(\Omega))\cap H^1(I;H^1(\Omega)^*), \tag{7.6.10b}$$

$$\mu_k\to \mu \qquad \textit{weakly in } L^2(I;H^1(\Omega)), \tag{7.6.10c}$$

$$e(u_k(t))\to e(u(t)) \qquad \textit{strongly in } L^2(\Omega;\mathbb{R}^{d\times d}_{\rm sym}) \textit{ and} \tag{7.6.10d}$$

$$c_k(t)\to c(t) \qquad \textit{strongly in } L^2(\Omega) \textit{ for any } t\in I. \tag{7.6.10e}$$

Moreover, any (u,c,μ) thus obtained is a weak solution to the initial-boundary-value problem (7.6.1)–(7.6.2)–(7.6.5) according Definition 7.6.1.

Proof. The existence of the Galerkin solution can be argued by the successive prolongation argument for the underlying system of ordinary-differential equations for (u_k,c_k), relying on the a-priori estimate obtained below.[49]

From (7.6.4), one can read the a-priori estimates (7.6.9) for the approximate solution (u_k,c_k,μ_k). More in detail, for the last term in (7.6.4), we use the inequality for estimation of

$$\begin{aligned}\int_\Gamma \alpha\mu_{\rm ext}\mu\,{\rm d}S &\le \frac{1}{\varepsilon}\|\alpha\mu_{\rm ext}\|^2_{L^2(\Gamma)}+\varepsilon\|\mu\|^2_{L^2(\Gamma)}\\ &\le \frac{1}{\varepsilon}\|\mu_{\rm ext}\|^2_{L^2(\Gamma)}+\varepsilon N_\Gamma\|\mu\|^2_{L^2(\Omega)}+\varepsilon N_\Gamma\|\nabla\mu\|^2_{L^2(\Omega;\mathbb{R}^d)}\end{aligned}$$

[49] Here we also use that the discretized equation $\mu_k=\partial_c\varphi(e(u_k),c_k)+\tau_{\rm R}\dot c_k$ allows for elimination of μ_k so that one has indeed a system of ordinary-differential equations for (u_k,c_k) at disposal.

with N_Γ the norm of the trace operator $H^1(\Omega) \to L^2(\Gamma)$; note that the last term can be absorbed by the dissipation potential when choosing $\varepsilon > 0$ small and then the estimates are obtained by a Gronwall inequality. Yet, it does not give any estimate on ∇c and thus no "compactness in c", which prevents convergence in an (unspecified) approximate solutions because $\partial_e\varphi$, $\partial_c\varphi$, and $\mathbb{M}$ depend nonlinearly on c, in general.

By Banach's selection principle, we then choose a subsequence so that (7.6.10a-c) holds. We further execute the strong-convergence strategy like (7.5.9) but, relying on joint convexity of φ, for both e and c simultaneously.[50] This facilitates our model without any concentration gradient. More specifically, we have:[51]

$$\begin{aligned}
&\int_\Omega \frac{1}{2}\mathbb{D}e(u_k(t){-}u(t)){:}e(u_k(t){-}u(t)) + \frac{1}{2}\tau_{\rm R}(c_k(t){-}c(t))^2\,\mathrm{d}x \\
&\qquad + \int_0^t\!\!\int_\Omega (\partial_e\varphi(e(u_k),c_k) - \partial_e\varphi(e(u),c)){:}e(u_k{-}u) \\
&\qquad\qquad + (\partial_c\varphi(e(u_k),c_k) - \partial_c\varphi(e(u),c))(c_k{-}c)\,\mathrm{d}x\mathrm{d}t \\
&= \int_0^t\!\!\int_\Omega \Big(\mathbb{D}e(\dot u_k) + \partial_e\varphi(e(u_k),c_k) - \mathbb{D}e(\dot u) - \partial_e\varphi(e(u),c)\Big){:}e(u_k{-}u) \\
&\qquad\qquad + \Big(\tau_{\rm R}\dot c_k + \partial_c\varphi(e(u_k),c_k) - \tau_{\rm R}\dot c - \partial_c\varphi(e(u),c)\Big)(c_k{-}c)\,\mathrm{d}x\mathrm{d}t \\
&= \int_0^t\!\!\int_\Omega f{\cdot}(u_k{-}u) + \varrho\ddot u_k{\cdot}(\dot u_k{-}\dot u) + \mu_k(c_k{-}c) \\
&\qquad - \Big(\mathbb{D}e(\dot u) + \partial_e\varphi(e(u),c)\Big){:}e(u_k{-}u) - (\tau_{\rm R}\dot c + \partial_c\varphi(e(u),c))(c_k{-}c)\,\mathrm{d}x\mathrm{d}t \\
&\qquad\qquad - \int_\Omega \varrho\dot u_k(t){\cdot}(u_k(t){-}u(t))\,\mathrm{d}x \;\to\; 0.
\end{aligned} \tag{7.6.11}$$

Like in (7.5.9), for the convergence to 0 in (7.6.11), we used that $\dot u_k \to \dot u$ strongly $L^2(Q;\mathbb{R}^d)$ due to the Aubin-Lions theorem and also $\dot u_k(t)$ is bounded in $L^2(\Omega;\mathbb{R}^d)$ while $u_k(t) \to u(t)$ strongly in $L^2(\Omega;\mathbb{R}^d)$ due to the Rellich compact embedding $H^1(\Omega) \subset L^2(\Omega)$. For $\mu_k(c_k{-}c) \to 0$ weakly in $L^1(Q)$, we use that due to the Aubin-Lions theorem, showing that

$$c_k \to c \quad \text{strongly in } L^2(I;H^1(\Omega)^*) \cong L^2(I;H^1(\Omega))^* \tag{7.6.12}$$

due to the estimate of $\{c_k\}_{k\in\mathbb{N}}$ in $L^\infty(I;L^2(\Omega)) \cap H^1(I;H^1(\Omega)^*)$ and the compact embedding $L^2(\Omega) \subset H^1(\Omega)^*$ obtained as an adjoint operator via the Rellich theorem, while $\{\mu_k\}_{k\in\mathbb{N}}$ is bounded in $L^2(I;H^1(\Omega))$. For $\int_Q -\tau_{\rm R}\dot c_k(c_k{-}c) = \frac{1}{2}\int_\Omega \tau_{\rm R}c_0^2 - \tau_{\rm R}c_k(T)^2\,\mathrm{d}x + \int_Q \tau_{\rm R}\dot c_k c\,\mathrm{d}x\mathrm{d}t$, we use the weak upper semicontinuity and the weak convergence $c_k(T) \to c(T)$ in $L^2(\Omega)$. Let us note also that we need the growth condition (7.6.8b) so that $\partial_e\varphi(e(u),c) \in L^2(Q;\mathbb{R}^{d\times d}_{\rm sym})$ and $\partial_c\varphi(e(u),c) \in L^2(Q)$. Altogether, (7.6.10d,e) has been proved.

---

[50] The strong-convergence strategy only for e would need ∇c to be controlled, likewise (7.5.9) used $\nabla\alpha$. This is used later in Sect. 7.6.2.

[51] Actually, (7.6.11) is rather conceptual strategy and a strongly-converging approximation of the limit (u,c) is needed to make usage of the Galerkin integral identities tested by $u_k - u$ and $c_k - c$ legal.

The convergence is now quite simple, using continuity of all involved Nemytskiĭ (or just composition) operators, i.e. $\mathbb{M}$, $\partial_e\varphi$, $\partial_c\varphi$, and $\varphi(\cdot,\tilde{c})$ used in (7.6.7c). Also, if $\tau_{\rm R} > 0$, we use the weak lower semicontinuity for the term $\int_Q \tau_{\rm R}\dot{c}c\,{\rm d}x{\rm d}t = \int_\Omega \frac{1}{2}\tau_{\rm R}c(T)^2 - \frac{1}{2}\tau_{\rm R}c_0\,{\rm d}x$ in (7.6.7c). $\square$

Example 7.6.3 (Biot poroelastic model). The linear Kelvin-Voigt viscoelastic material combined with the Biot poroelastic model as in Sect. 5.7.1, having the structure $_{\mathbb{C}}\|_{\mathbb{D}}\|(_{B_{\beta,M}} - {}^{c}_{\mathbb{E}_\beta})$, is an illustration of the particular choice of φ and ζ which are quadratic. In the isotropic case where the elastic-moduli tensor $\mathbb{C}$ is determined by the bulk and the shear moduli $(K_{\rm E},G_{\rm E})$ and similarly the viscous-moduli tensor $\mathbb{D}$ is determined by $(K_{\rm V},G_{\rm V})$, the quadratic stored energy and the dissipation potential looks as:

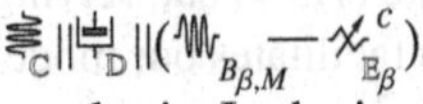

$$\varphi(e,c) = \frac{1}{2}K_{\rm E}|{\rm sph}\,e|^2 + \frac{1}{2}B_{\beta,M}|{\rm sph}\,e - \mathbb{E}_\beta c|^2 + G_{\rm E}|{\rm dev}\,e|^2 + \frac{1}{2}b_0c^2, \tag{7.6.13a}$$

$$\zeta(\dot{e},\dot{c}) = \frac{1}{2}K_{\rm V}|{\rm sph}\,\dot{e}|^2 + G_{\rm V}|{\rm dev}\,\dot{e}|^2 + \tau_{\rm R}\dot{c}^2. \tag{7.6.13b}$$

The relation with Sect. 5.7.1 is[52]

$$B_{\beta,M} = \beta^2 M/d \quad \text{and} \quad \mathbb{E}_\beta = \mathbb{I}/\beta \tag{7.6.14}$$

with M and β being the *Biot modulus* and the *Biot coefficient*, respectively, as used already in (5.7.6a); here for notational simplicity we put the equilibrium concentration $c_{\rm eq} = 0$. The total stress $\sigma = \mathbb{D}\dot{e} + \mathbb{C}e + \sigma_{\rm PORE}$ contains now a "pore stress" defined as $\sigma_{\rm PORE} = B_{\beta,M}({\rm sph}\,e - \mathbb{E}_\beta c)$. In terms of the fluid pressure $p_{\rm FLUID}$ from (5.7.5), we can also write $\sigma_{\rm PORE} = \beta p_{\rm FLUID}\mathbb{I}$. Multiplying it by $\mathbb{I}$, we can give the pore pressure

$$p_{\rm FLUID} = \frac{1}{\beta d}{\rm tr}\,\sigma_{\rm PORE}. \tag{7.6.15}$$

The overall model governed by (7.6.13) can schematically be depicted as in Figure 7.9.

Remark 7.6.4 (Gradient-flow structure of diffusion). The diffusion equation (7.6.1b) considered with the homogeneous Robin boundary conditions $\mathbb{M}\nabla\mu\cdot\vec{n} + \alpha\mu = 0$ on Γ can be rewritten in the form

$$\Delta^{-1}_{\mathbb{M}(e,c)}\dot{c} = \mu = \partial_c\varphi(e(u),c) + \partial_{\dot{c}}\zeta(e(\dot{u}),\dot{c}) \tag{7.6.16}$$

with using the notation $\Delta^{-1}_{\mathbb{M}} : f \mapsto \mu$ for the linear operator $H^1(\Omega)^* \to H^1(\Omega)$ defined by $\mu :=$ the weak solution to the equation ${\rm div}(\mathbb{M}\nabla\mu) = f$ with the (here homogeneous) Robin boundary condition $\mathbb{M}\nabla\mu\cdot\vec{n} + \alpha\mu = 0$. The *nonlocal dissipation potential* $R(e,c;\dot{u},\dot{c}) = \int_\Omega \zeta(e(\dot{u}),\dot{c}) + \frac{1}{2}|\mathbb{M}(e,c)^{1/2}\nabla\Delta^{-1}_{\mathbb{M}(e,c)}\dot{c}|^2\,{\rm d}x + \int_\Gamma \frac{1}{2}\alpha(\Delta^{-1}_{\mathbb{M}(e,c)}\dot{c})^2\,{\rm d}S$, cf. [360, Sect. 5.2.6].

---

[52] The coefficients (7.6.14) arise from the model (5.7.6a) by the simple algebra $\frac{1}{2}M(\beta{\rm tr}\,e - c)^2 = \frac{1}{2}M(\beta({\rm tr}\,e - c/\beta))^2 = \frac{1}{2}\frac{M}{d}(\beta({\rm tr}\,e - c/\beta)\mathbb{I})^2 = \frac{1}{2}\frac{\beta^2 M}{d}({\rm sph}\,e - c\mathbb{I}/\beta)^2 = \frac{1}{2}B_{\beta,M}|{\rm sph}\,e - \mathbb{E}_\beta c|^2$.

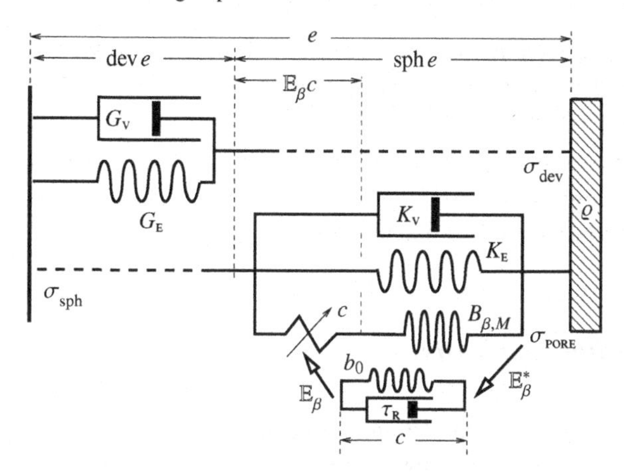

Fig. 7.9 The schematic Biot model from Sect. 5.7.1 with the decomposition $e(u) = e_{dev} + e_{sph}$ as in (6.7.8). The flow-rule for c of the abstract form $\partial_{\dot{c}}\zeta(\dot{e},\dot{c}) + \partial_c\varphi(e,c) = 0$ as in (7.2.1b), here reads as $\tau_R\dot{c} + b_0 c = \mathbb{E}^*_\beta\sigma_{PORE}$ with the pore stress $\sigma_{PORE} = B_{\beta,M}(\mathrm{sph}\, e - \mathbb{E}_\beta c)$ in the position of the driving force.

Remark 7.6.5 (Elimination of c). [53]In special cases, one can write the system (7.6.1) in terms of (u, p) instead of (u, c). Considering (5.7.6a) for $k = 0$ and $\mathbb{M}(\cdot)$ constant, from $p = M(\beta \mathrm{tr}\, e - c + c_E)$, cf. (5.7.5), we can eliminate $c = \beta \mathrm{tr}\, e - p/M + c_E$. For $\zeta = 0$, we thus obtain the system

$$\varrho\ddot{u} - \mathrm{div}(\mathbb{C}e(u)) - \beta\nabla p = f, \tag{7.6.17a}$$

$$\frac{1}{M}\dot{p} - \frac{\beta}{d}\mathrm{div}\,\dot{u} - \mathrm{div}(\mathbb{M}\nabla p) = 0. \tag{7.6.17b}$$

Remark 7.6.6 (Elimination of μ). For the inviscid model, i.e. $\mu = \partial_c\varphi$, one can write (7.6.1b) as $\dot{c} - \mathrm{div}(\tilde{\mathbb{M}}(e(u), c)\nabla c + \mathbb{M}(e(u), c)\partial^2_{ec}\varphi(e, c)\nabla e) = 0$ with $\tilde{\mathbb{M}}(e, c) = \partial^2_{cc}\varphi(e, c)\mathbb{M}(e, c)$.

Remark 7.6.7 (Softening by wetting). In some situations, elastic moduli are remarkably influenced by the concentration, i.e. $\mathbb{C} = \mathbb{C}(c)$. We know it well known from our everyday experience: various food ingredients we use at our kitchen (like rise, pasta, or legume) intentionally becomes much softer when wetted in water, beside inflating volume by swelling. Another common example might be so-called Egyptian cucumber[54] which is very rigid fiber composite when dry but it very easily absorbs water and then it becomes very soft, i.e. its elastic moduli drops many orders when wet. In this latter case, swelling effects are rather negligible comparing the dominant elastic-moduli dependence on c. A soft hydrogel contact lens is another highly

[53] For a system of such a form, we refer to [481].

[54] This plant is known also under the name Vietnamese luffa (or in Latin "Luffa aegyptiaca").

practical example. One then considers $\varphi(e,c)=\frac{1}{2}\mathbb{C}(c)e{:}e$. Then $\mu=\frac{1}{2}\mathbb{C}'(c)e{:}e$. This brings the problem closer to the damage problem which then differs only in terms of the dissipation potential.

Exercise 7.6.8. Modify the model by considering another viscous damping leading to the rheological model $((\wr_{\mathbb{C}_1}\|\sqcap_{\mathbb{D}_1})-\not\sim_{E}^{c})\|\wr_{\mathbb{C}_2}\|\sqcap_{\mathbb{D}_2}$ and formulate the stored energy φ as well as the dissipation potential ζ.

7.6.2 Involving capillarity

The non-negativity of concentration is a natural expectation, but model from the previous Sec. 7.6.1 does not allow to ensure $c\ge 0$. In principle, it might be ensured in the stored energy, growing to ∞ if $c\le 0$ but this would be in conflict with the growth requirement (7.6.8b), or by letting $\mathbb{M}(e,c)\to 0$ if $c\to 0+$, which is an usual engineering variant of the model but would be in conflict with (7.6.8c), Even, some applications ultimately need nonconvex stored energies in terms of c.

Standardly, the diffusion is governed by energies with even a non-polynomial growth, cf. the term $kc(\ln(c/c_{\rm E})-1)$ in φ in (5.7.4a) or (5.7.6a). This ensures $c>0$ because the derivative of this convex term blows to $-\infty$ if $c\searrow 0$ and then one can assume it equal to ∞ for $c<0$. In the latter case (5.7.6a), it yields the chemical potential

$$\mu=\partial_c\varphi(e,c)=\begin{cases}M(c-\beta\,{\rm tr}\,e-c_{\rm E})+k\ln(c/c_{\rm E}) & \text{for } c>0,\\ \emptyset & \text{for } c\le 0.\end{cases} \tag{7.6.18}$$

Sometimes nonconvex energies are used for example in a phase separation problems. Then rather the constraint $-1\le c\le 1$ is relevant. The logarithmic-type singularity can then be considered also

$$\varphi(c)=\begin{cases}k_0(c{+}1)\ln(c{+}1)+k_0(c{-}1)\ln(c{-}1)-k_1c^2 & \text{for } c\in[-1,1],\\ +\infty & \text{otherwise.}\end{cases} \tag{7.6.19}$$

Let us note that this function naturally respect the constraint $-1\le c\le 1$ and is nonconvex if k_1 is large enough.

One way to cope with this challenge is involving an elliptic regularization by adding $\frac{1}{2}\kappa|\nabla c|^2$ into the free energy, like we did in (3.6.12), leading to an additional term $-\kappa\Delta c$ augmenting the chemical potential. The coefficient $\kappa>0$ has a specific physical meaning related to capillarity effects. The enhanced free energy is then considered in the form

$$\varphi_{\rm E}(e,c,\nabla c)=\varphi(e,c)+\frac{\kappa}{2}|\nabla c|^2. \tag{7.6.20}$$

while again we use the dissipation potential (7.6.6). We modify (7.6.17) by augmenting μ in (7.6.1b) by $-\kappa\Delta c$. Thus, we consider the system:

$$\varrho\ddot{u} - \mathrm{div}\,\sigma = f \qquad \text{with} \quad \sigma = \partial_e\varphi(e(u),c) + \mathbb{D}e(\dot{u}), \tag{7.6.21a}$$

$$\dot{c} - \mathrm{div}(\mathbb{M}(e(u),c)\nabla\mu) = 0 \qquad \text{with} \quad \mu = \partial_c\varphi(e(u),c) + \tau_{\mathrm{R}}\dot{c} - \kappa\Delta c. \tag{7.6.21b}$$

For $\tau_{\mathrm{R}} = 0$, the diffusion part (7.6.21b) itself is called the *Cahn-Hilliard equation* [102] and, when combined here with elasticity, sometimes also the *Cahn-Larché system* [303].

Of course, the additional gradient requires one boundary condition more. Thus, (7.6.2b) is to be augmented. Considering $\kappa\nabla c\cdot\vec{n} = 0$ on Γ, the energetics (7.6.4) still holds when replacing φ with φ_{E} from (7.6.20).

Let us outline the analysis of (7.6.21). The Definition 7.6.1 of the weak solution now extends (7.6.7c) by the gradient term for

$$\int_Q \varphi(e(u),\tilde{c}) + \frac{\kappa}{2}|\nabla\tilde{c}|^2\,\mathrm{d}x\mathrm{d}t \geq \int_Q \varphi(e(u),c) + (\mu-\tau_{\mathrm{R}}\dot{c})(\tilde{c}-c) + \frac{\kappa}{2}|\nabla c|^2\,\mathrm{d}x\mathrm{d}t \tag{7.6.22}$$

to be valid for all $\tilde{c}\in L^2(I;H^1(\Omega))$ with the left-hand side taking possibly values $+\infty$ if $\varphi(e(u),\tilde{c})$ is not integrable; here we rely that naturally φ is bounded from below. Here we use this inequality for a possible non-polynomial growth of $\varphi(e,\cdot)$ like in (7.6.19), and then, instead of weak solution, one also speaks about a *variational solutions*.

We now modifies (7.6.8a,b)

$$\varphi \text{ continuously differentiable, } \varphi(\cdot,c) \text{ convex,} \tag{7.6.23a}$$

$$\varphi(e,c) \geq \epsilon c^2, \quad |\partial_e\varphi(e,c)| \leq C(1+|e|+|c|^{2^*/2}). \tag{7.6.23b}$$

Let us again consider the Faedo-Galerkin approximation of (7.6.21) like for Proposition 7.6.2, employing some finite-dimensional subspaces $U_k\times V_k\times W_k$ of $H^1_{\mathrm{D}}(\Omega;\mathbb{R}^d)\times H^1(\Omega)\times H^1(\Omega)$, indexed by k.

Proposition 7.6.9. *Let (7.6.23) and (7.6.8c) hold, ϱ, the loading and the initial conditions satisfy (6.4.5b-e) together with $j_{\mathrm{b}}\in L^2(I;L^{2^{\sharp'}}(\Gamma_{\mathrm{N}};\mathbb{R}^d))$ and $c_0\in H^1(\Omega)$, and $\kappa>0$. Then the Galerkin approximation (u_k,c_k,μ_k) does exist and satisfies the a-priori estimates*

$$\|u_k\|_{H^1(I;H^1(\Omega;\mathbb{R}^d))\cap W^{1,\infty}(I;L^2(\Omega;\mathbb{R}^d))} \leq C, \tag{7.6.24a}$$

$$\|c_k\|_{L^\infty(I;H^1(\Omega))} \leq C, \quad \textit{and} \quad \|\mu_k\|_{L^\infty(I;H^1(\Omega))} \leq C, \tag{7.6.24b}$$

$$\|\ddot{u}_k\|_{L^2(I;H^1(\Omega;\mathbb{R}^d)^*)} \leq C \qquad \textit{and} \qquad \|\dot{c}_k\|_{L^2(I;H^1(\Omega)^*)} \leq C, \tag{7.6.24c}$$

where $\ddot{u}_k$ and $\dot{c}_k$ mean suitable Hahn-Banach extensions of the time derivatives of the Galerkin solution, cf. Remark C.2.9 on p. 547. For selected subsequences, we have

$$u_k \to u \qquad \textit{strongly in } L^p(I;H^1(\Omega;\mathbb{R}^d)),\ 1\leq p<\infty \textit{ arbitrary, and}$$

$$\textit{weakly* in } W^{1,\infty}(I;L^2(\Omega;\mathbb{R}^d))\cap H^1(I;H^1(\Omega;\mathbb{R}^d)), \tag{7.6.25a}$$

$$c_k \to c \qquad \textit{weakly* in } L^\infty(I;H^1(\Omega)) \cap H^1(I;H^1(\Omega)^*), \tag{7.6.25b}$$

$$\mu_k \to \mu \qquad \textit{weakly in } L^2(I;H^1(\Omega)), \tag{7.6.25c}$$

and any (u,c,μ) thus obtained is a weak solution to the initial-boundary-value problem for the system (7.6.21) the boundary conditions (7.6.2) completed by $\kappa\nabla c\cdot\vec{n} = 0$ on Γ and the initial condition (7.6.5) has a weak/variational solution $(u,c,\mu) \in H^1(I;H^1(\Omega;H^1(\Omega;\mathbb{R}^d)) \times L^2(I;H^1(\Omega)) \times L^\infty(I;H^1(\Omega))$.

Sketch of the proof. The existence of a solution (u_k,c_k,μ_k) valued in $U_k \times V_k \times W_k$ of the initial-value problem for the underlying system of ordinary differential-algebraic equations follows directly by the a-priori estimates.[55] As for the mentioned a-priori estimates, they can be seen by testing the particular equations of (7.6.21) successively by $\dot{u}_k$, by μ_k and by $\dot{c}_k$. This gives (7.6.3) written for (u_k,c_k,μ_k) instead of (u,c,μ) with the second equality in (7.6.3b) augmented by the term $\int_\Omega \kappa\nabla c_k \cdot \nabla\dot{c}_k\,\mathrm{d}x$, and thus eventually the energy balance (7.6.4) written for (u_k,c_k,μ_k) with the first integral enhanced by $\frac{1}{2}\kappa|\nabla c_k|^2$; note that for this test in the Galerkin approximation, we need to have the finite-dimensional spaces for c and μ the same, i.e. $V_k = W_k$. By the Hölder and Young or Gronwall inequalities, we obtain the a-priori estimates (7.6.24a,b) and by comparison from the equations in (7.6.21) we obtain also the estimates (7.6.24c). If $\tau_\mathrm{R} > 0$, then even $\dot{c}_k$ is bounded even in $L^2(Q)$.

Having ∇c under control and again assuming $\mathbb{D}$ constant, we can now use the strong-convergence strategy (7.5.9) for $e(u_k)$. For this, the boundedness of $\partial_e\varphi(e(u),c_k)$ in $L^2(Q;\mathbb{R}^{d\times d}_\mathrm{sym})$ is used, being granted by the assumption (7.6.23b). Thus, for a subsequence, we thus have (7.6.25). For the limit passage in (the discrete variant of) (7.6.22), we use the strong lower semicontinuity of $(e,c) \mapsto \int_Q \varphi(e,c)\,\mathrm{d}x\mathrm{d}t$ and the weak lower semicontinuity of $c \mapsto \int_Q \frac{\kappa}{2}|\nabla c|^2\,\mathrm{d}x\mathrm{d}t$. The convergence in the term $\int_Q \mu c\,\mathrm{d}x\mathrm{d}t$ is now easy because we have strong convergence of c in $L^2(Q)$ due to Lions-Aubin's theorem based on the Rellich compact-embedding theorem B.3.3 as $H^1(\Omega) \subset L^2(\Omega)$. If $\tau_\mathrm{R} > 0$, then we can even use simply the Rellich theorem for $H^1(Q) \subset L^2(Q)$. □

Remark 7.6.10 (The constraint $c \geq 0$). An alternative variant to ensure non-negativity of the concentration without nonpolynomial growth like in ..... is just to use a hard constraint and augment φ by the indicator function $\delta_{[0,\infty)}(c)$, while φ may have a polynomial growth so that $\partial_c\varphi$ is controlled in $L^2(Q)$. Then the chemical potential contains the Lagrange multiplier to this constraint:

$$\mu = \partial_c\varphi(e,c) + \mu_\mathrm{r} - \kappa\Delta c \qquad \text{with} \quad \mu_\mathrm{r} \in \partial\delta_{[0,\infty)}(c) = N_{[0,\infty)}(c), \tag{7.6.26}$$

and allows for estimation of $\Delta c \in L^2(Q)$ by testing (7.6.26) by $-\Delta c$ and by using $\int_\Omega -\partial\delta_{[0,\infty)}(c)\Delta c\,\mathrm{d}x = \int_\Omega \nabla\partial\delta_{[0,\infty)}(c)\cdot\nabla c\,\mathrm{d}x = \int_\Omega \partial^2\delta_{[0,\infty)}(c)|\nabla c|^2\,\mathrm{d}x \geq 0$, written for-

[55] The Galerkin discretization of the elliptic problem in (7.6.21b) leads to the system of the type $A_k c_k + [\partial_c\varphi]_k(e(u_k),c_k) = \mu_k$ of nonlinear algebraic equations with a positive-definite matrix A_k and $[\partial_c\varphi]_k(e,\cdot)$ a monotone mapping, so that one can write "explicitly" write its unique solution as $c_k = [A_k + [\partial_c\varphi]_k(e,\cdot)]^{-1}(\mu_k)$ and thus eliminate c_k to obtain the system of ordinary-differential equations. This is related with the attribute of the differential-algebraic system has the index 1.

mally, which is the calculus we used already in (7.5.8). Then $\mu_{\rm r} = \mu - \partial_c\varphi(e,c) + \kappa\Delta c$ itself can be estimated in $L^2(Q)$. The weak formulation of the inclusion in (7.6.26) reads as

$$\forall \tilde c \in L^2(Q),\ \tilde c \geq 0: \qquad \int_Q \mu_{\rm r}(\tilde c - c)\,{\rm d}x{\rm d}t \geq 0. \tag{7.6.27}$$

The limit passage in the approximate inequality is simple because a subsequence of $\mu_{\rm r}$'s can be selected converging weakly in $L^2(Q)$ while c's converge weakly in $L^2(I;H^1(\Omega)) \cap H^1(I;H^1(\Omega)^*)$ and thus also strongly in $L^2(Q)$ by the Aubin-Lions theorem.

Remark 7.6.11 (Viscous Cahn-Hilliard equation). Sometimes, in analog to (7.6.20), one considers also the dissipation potential enhanced analogously, i.e.

$$\zeta_{\rm E}(\dot e, \dot c, \nabla\dot c) = \zeta(\dot e, \dot c) + \frac{1}{2}\hat\kappa|\nabla\dot c|^2, \tag{7.6.28}$$

so that the overall dissipation potential is $R(e,c;\dot u,\dot c) = \int_\Omega \zeta_{\rm E}(\dot e, \dot c, \nabla\dot c) + \frac12|\mathbb{M}(e,c)^{1/2}\nabla\Delta^{-1}_{\mathbb{M}(e,c)}\dot c|^2\,{\rm d}x$, cf. Remark 7.6.4. This model leads to the stress $\sigma = \partial_e\varphi(e(u),c) + \partial_{\dot e}\zeta(e(\dot u),\dot c)$ and the augmented viscous chemical potential $\mu = \partial_c\varphi(e(u),c) + \partial_{\dot c}\zeta(e(\dot u),\dot c) - \kappa\Delta c - \hat\kappa\Delta\dot c$. In case of (7.6.1b), such modification leads to the so-called viscous Cahn-Hilliard equation, following original Gurtin's ideas [233].

Remark 7.6.12 (Nonlocal Cahn-Hilliard equation). In the spirit of nonlocal simple materials as in Sect. 2.5.2, one can about replacing the gradient term $\frac12\int_\Omega \kappa(x)|\nabla c|^2\,{\rm d}x$ as in (7.6.20) by a nonlocal term $\frac14\int_\Omega \kappa(x,\tilde x)|c(x) - c(\tilde x)|^2\,{\rm d}x{\rm d}\tilde x$. This modification has been proposed and physically justified A. Giacomin and J.L. Lebowitz [220]. When the kernel κ has the singularity of the type (2.5.28) when $|x-\tilde x| \to 0$, this term has a similar compactifying character as the gradient term in (7.6.20). This model has been scrutinized in mathematical literature, cf. e.g. [1, 211, 286].

Exercise 7.6.13. Considering $\nu > 0$ a "viscosity" coefficient and a so-called *Brinkman's modification*[56] [91, 92] of (5.7.1), namely $\nu\Delta j - j = M\nabla\mu$ with $M > 0$ a constant, so that (in such an isotropic case) the boundary-value problem (7.6.21b)–(7.6.2b) turns into

$$\dot c = {\rm div}\, j \ \text{ and } \ \nu\Delta j - j = M\nabla\mu \quad \text{with } \ \mu = \partial_c\varphi(e(u),c) + \tau_{\rm R}\dot c - \kappa\Delta c \quad \text{on } \Omega, \tag{7.6.29a}$$

$$j\cdot\vec n + \alpha\mu = \alpha\mu_{\rm ext} \ \text{ and } \ \nabla j\cdot\vec n = 0 \quad \text{on } \Gamma. \tag{7.6.29b}$$

Write the energy balance and launch the analysis based on it.

[56] This correction term is used for a flow through granulated media where the grains of the media are porous themselves (a so-called double-porosity model).

7.6.3 Using nonsimple-material or phase-field concepts

There are other options to control gradient of c indirectly, in contrast to the capillarity model (7.6.20). In contrast to the capillarity regularization from the previous section 7.6.2, they require convexity of the stored energy in terms of c.

The first one is to exploit the nonsimple-material concept from Sect. 5.4, replacing (7.6.20) by

$$\varphi_{\text{E}}(e,c,G) = \varphi(e,c) + \frac{1}{2}\mathbb{H}G\vdots G \quad \text{with } G = \nabla e. \tag{7.6.30}$$

with $\mathbb{H}$ as in (5.4.1) on p. 168. The advantage of this regularization is that we can easily allow for $\varphi(\cdot,c)$ nonconvex and for the viscosity tensor $\mathbb{D}$ state dependent, i.e. $\mathbb{D} = \mathbb{M}(e,c)$. On the other hand, $\varphi(e,\cdot)$ must be uniformly convex. The resulted system in the classical formulation then reads as:

$$\varrho\ddot{u} - \text{div}(\sigma - \text{div}(\mathbb{H}\nabla e(u)) = f \quad \text{with } \sigma \text{ from (7.6.21a)}, \qquad \text{on } \Omega, \tag{7.6.31a}$$

$$u = 0 \qquad \text{on } \Gamma_{\text{D}}, \tag{7.6.31b}$$

$$(\sigma - \text{div}(\mathbb{H}\nabla e(u)))\vec{n} - \text{div}_{\text{S}}\big((\mathbb{H}\nabla e(u))\cdot\vec{n}\big) = g \qquad \text{on } \Gamma_{\text{N}}, \tag{7.6.31c}$$

$$(\mathbb{H}\nabla e(u)):(\vec{n}\otimes\vec{n}) = 0 \qquad \text{on } \Gamma, \tag{7.6.31d}$$

cf. (5.4.2), which is coupled with the continuity equation

$$\dot{c} - \text{div}(\mathbb{M}(e(u),c)\nabla\mu) = 0 \quad \text{with} \quad \mu \in \partial_c\varphi(e,c) \qquad \text{on } \Omega, \tag{7.6.31e}$$

$$\mathbb{M}(e(u),c)\nabla\mu\cdot\vec{n} + \alpha\mu = \alpha\mu_{\text{ext}} \qquad \text{on } \Gamma. \tag{7.6.31f}$$

We modify the assumptions (7.6.8a,b) as follows:

$$\varphi \text{ smooth}, \qquad \partial^2_{cc}\varphi \geq \varepsilon, \qquad |\partial^2_{ec}\varphi| \leq 1/\varepsilon, \tag{7.6.32a}$$

$$(u,c) \mapsto \int_\Omega \varphi_{\text{E}}(e(u),c,\nabla e(u))\,\mathrm{d}x \ \text{ coercive on } H^2_{\text{D}}(\Omega;\mathbb{R}^d)\times L^2(\Omega), \tag{7.6.32b}$$

where we used the notation

$$H^k_{\text{D}}(\Omega;\mathbb{R}^d) := \big\{u\in H^k(\Omega;\mathbb{R}^d);\ u|_{\Gamma_{\text{D}}} = 0\big\}. \tag{7.6.33}$$

here for $k = 2$. (For $k = 1$ see also (6.2.5).) Let us note that $\varphi : \mathbb{R}^{d\times d}_{\text{sym}} \times \mathbb{R} \to \mathbb{R} \cup \{+\infty\}$ from (7.6.37a) augmented by $\varphi(e,c) = +\infty$ for $c < 0$ satisfies (7.6.32) provided $\text{meas}_{d-1}(\Gamma_{\text{D}}) > 0$, where we used the Korn's inequality.

Proposition 7.6.14 (Existence of a solution to (7.6.31)). *Let (7.6.32) with (7.6.8c) hold and the loading and the initial conditions satisfy (6.4.5b-e) together with* $\alpha \in L^\infty(\Gamma)$ *positive and* $\mu_{\text{ext}} \in L^2(I;L^{2^{\sharp\prime}}(\Gamma_{\text{N}};\mathbb{R}^d))$ *and* $c_0 \in H^1(\Omega)$*, and* $\kappa > 0$*. hold, and let* $\mathbb{D} = \mathbb{D}(e,c)$ *with* $\mathbb{D} : \mathbb{R}^{d\times d}_{\text{sym}} \times \mathbb{R} \to \mathbb{R}^{d\times d}_{\text{sym}}$ *continuous, bounded, uniformly positive* *definite. Then there is a (weak) solution $(u,c)\in H^1(I;H^2_{\rm D}(\Omega;\mathbb{R}^d))\times L^2(I;H^1(\Omega))$ to the boundary-value problem (7.6.31) with the initial condition (7.6.5).*

Sketch of the proof. We can see that $\nabla\mu=[\partial^2_{cc}\varphi(e,c)](\nabla c)+[\partial^2_{ec}\varphi(e,c)](\nabla e)$. Therefore

$$\nabla c=[\partial^2_{cc}\varphi(e(u),c)]^{-1}\big(\nabla\mu-[\partial^2_{ec}\varphi(e(u),c)](\nabla e(u))\big). \tag{7.6.34}$$

From this, we can read an a-priori bound for ∇c in $L^2(\Omega;\mathbb{R}^d)$, while the Korn's inequality for 2nd-grade nonsimple materials (5.2.5) allows for the H^2-estimate of u uniformly in time. We now have strong convergence in both $e(u)$ and c just by Aubin-Lions' compactness theorem. The rest is then in lines of the proof of Proposition 7.6.9. □

It should be noted that, merging the concepts (7.6.20) and (7.6.30) for

$$\varphi_{\rm E}(e,c,G,\nabla c)=\varphi(e,c)+\frac12\mathbb{H}G\,\vdots\,G+\frac{\kappa}{2}|\nabla c|^2\quad\text{ with }\ G=\nabla e, \tag{7.6.35}$$

we would get the estimate on ∇c directly and (7.6.34) can be avoided. Therefore, a fully nonconvex φ can be admitted. This may be useful in some materials undergoing phase transitions.

An alternative option to coup with (or avoid) the concentration gradient is to introduce an auxiliary scalar field χ, called a *phase field*, which should be in most regimes of the model "nearly" the same as c but anyhow is more smooth in space and time. The idea of the phase field as an auxiliary variable to distinguish states that otherwise would be identical in other variables dates back to Ginzburg and Landau who called it an *order parameter* and used it to expand the thermodynamic state functions.[57]

Here we first choose a splitting

$$\varphi(e,c)=\varphi_{\rm ME}(e,c)+\varphi_{\rm CH}(c) \tag{7.6.36}$$

to the "mechanical" and "chemical" part. The goal of this split is to distinguish the part with is convex in c with possibly a nonpolynomial growth to ensure the constraint(s) on c from the resting part of φ which makes coupling of mechanical and chemical part but has a polynomial character. Then, instead of (7.6.20) or (7.6.30) with φ from (7.6.36), we use the modified stored energy

$$\varphi_{\rm E}(e,c,\chi,\nabla\chi):=\varphi_{\rm ME}(e,\chi)+\varphi_{\rm CH}(c)+\frac12K(c-\chi)^2+\frac{\kappa}{2}|\nabla\chi|^2 \tag{7.6.37a}$$

and then also the dissipation potential like (7.6.6), namely

$$\zeta(\chi;\dot e,\dot\chi)=\frac12\mathbb{D}(\chi)\dot e:\dot e+\frac12\delta\dot\chi^2. \tag{7.6.37b}$$

[57] Later, Cahn and Hilliard [102] used this idea of an order parameter for the alloy concentration.

Let us note that we can now admit dependence of the viscous moduli $\mathbb{D}$ on the state. Let us note that, in contrast to the other two already mentioned regularizations, this modification of φ depends on the particular choice of the splitting (7.6.36). The modeling expectation is that K is large so that the difference $|c{-}\chi|$ is rather small, or at least cannot be large at too much big regions in space/time domain.

The chemical potential is now

$$\mu \in \partial_c\varphi_{\rm CH}(c) + K(c{-}\chi) \tag{7.6.38}$$

and does not explicitly involve the strain e so that we do not need to prove strong convergence of e. The inclusion in (7.6.38) wants to emphasize that $\varphi_{\rm CH}$ may not be smooth and its convex subdifferential may be set-valued. The interpretation might be as a slightly compressible diffusant whose concentration is c and χ is a "geometrical" volume fraction. Let us note that, if $\kappa = 0$, then $\partial_\chi\varphi_{\rm E} = 0$ would mean $c - \chi = p/K$ with $p =$ being the pressure, e.g. $p = E{:}\mathbb{C}(e{-}Ec)$ or $M(\beta\,{\rm tr}\,e{-}c)$ in the models from Section 5.7.1. Substituting it into μ would give again $\mu = k\ln c - p$. For $\kappa > 0$, this does not hold exactly, and we should consider the functional derivative of the convex quadratic functional $\chi \mapsto \int_\Omega \varphi_{\rm E}(e(u), c, \chi, \nabla\chi)\,{\rm d}x$ whose unique critical point χ satisfies the elliptic boundary-value problem:

$$\partial_\chi\varphi_{\rm ME}(e(u),\chi) + K\chi - \kappa\Delta\chi = Kc \quad \text{on } \Omega, \qquad\qquad \frac{\partial\chi}{\partial\vec{n}} = 0 \quad \text{on } \Gamma, \tag{7.6.39}$$

The evolution system should now be completed with a flow rule for the new phase-field variable, considered naturally as $\partial_{\dot\chi}\zeta(\chi; e(\dot u), \dot\chi) + \partial_\chi\varphi_{\rm E}(e(u), c, \chi, \nabla\chi) - {\rm div}\partial_{\nabla\chi}\varphi_{\rm E}(e(u), c, \chi, \nabla\chi) = 0$ in order to be consistent with the general structure (7.1.1) in the gradient variant, cf. Remark 7.2.6. In view of (7.6.37), we thus arrive to the system:

$$\varrho\ddot u - {\rm div}(\mathbb{D}(\chi)e(\dot u) + \partial_e\varphi_{\rm ME}(e(u),\chi)) = f, \tag{7.6.40a}$$

$$\dot c - {\rm div}(\mathbb{M}(\chi)\nabla\mu) = 0 \qquad \text{with} \qquad \mu \in \partial_c\varphi_{\rm CH}(c) + K(c{-}\chi), \tag{7.6.40b}$$

$$\delta\dot\chi + \partial_\chi\varphi_{\rm ME}(e(u),\chi) + K\chi = Kc + \kappa\Delta\chi. \tag{7.6.40c}$$

The weak formulation of the inclusion in (7.6.40b) leads to the variational inequality

$$\int_Q \varphi_{\rm CH}(\tilde c)\,{\rm d}x{\rm d}t \ge \int_Q (\mu - K(c{-}\chi))(\tilde c{-}c) + \varphi_{\rm CH}(c)\,{\rm d}x{\rm d}t. \tag{7.6.41}$$

The existence of such a weak solution to the system (7.6.40) with (here nonspecified) boundary conditions and the initial conditions for u, $\dot u$, c and also χ can be shown e.g. by the Galerkin method. Under suitable (again here nonspecified) data qualification, we obtain the a-priori estimates:

$$\|u_k\|_{H^1(I;H^1(\Omega;\mathbb{R}^d))\cap W^{1,\infty}(I;L^2(\Omega;\mathbb{R}^d))\cap H^2(I;H^1(\Omega;\mathbb{R}^d)^*)} \le C, \tag{7.6.42a}$$

$$\|c_k\|_{L^\infty(I;L^2(\Omega))\cap H^1(I;H^1(\Omega)^*)} \le C \quad \text{and} \quad \|\mu_k\|_{L^2(I;H^1(\Omega))} \le C, \tag{7.6.42b}$$

$$\|\chi_k\|_{L^\infty(I;H^1(\Omega))\cap H^1(I;L^2(\Omega))} \le C. \tag{7.6.42c}$$

The variational inequality (7.6.41) for the Galerkin solution reads as

$$\int_Q \varphi_{\rm CH}(\widetilde{c})\,\mathrm{d}x\mathrm{d}t \ge \int_Q (\mu_k - K(c_k-\chi_k))(\widetilde{c}-c_k) + \varphi_{\rm CH}(c_k)\,\mathrm{d}x\mathrm{d}t. \tag{7.6.43}$$

and the convergence in it is like in the proof of Proposition 7.6.2, the only nontrivial issue is $\int_Q \mu_k c_k\,\mathrm{d}x\mathrm{d}t \to \int_Q \mu c\,\mathrm{d}x\mathrm{d}t$ for which again the Aubin-Lions theorem leading to (7.6.12) is to be used. Let us note also that the needed strong convergence of $e(u_k)$ needed now for the limit passage towards (7.6.40c) can be proved in a simpler way than (7.6.11), just

$$\begin{aligned}
&\int_\Omega \frac{1}{2}\mathbb{D}e(u_k(t)-u(t)){:}e(u_k(t)-u(t))\,\mathrm{d}x \\
&\qquad + \int_0^t\!\!\int_\Omega (\partial_e\varphi_{\rm ME}(e(u_k),\chi_k) - \partial_e\varphi_{\rm ME}(e(u),\chi_k)){:}e(u_k-u)\,\mathrm{d}x\mathrm{d}t \\
&= \int_0^t\!\!\int_\Omega \Big(\mathbb{D}e(\dot{u}_k) + \partial_e\varphi_{\rm ME}(e(u_k),\chi_k) - \mathbb{D}e(\dot{u}) - \partial_e\varphi_{\rm ME}(e(u),\chi_k)\Big){:}e(u_k-u)\,\mathrm{d}x\mathrm{d}t \\
&= \int_0^t\!\!\int_\Omega f\cdot(u_k-u) + \varrho\dot{u}_k\cdot(\dot{u}_k-\dot{u}) - \Big(\mathbb{D}e(\dot{u}) + \partial_e\varphi_{\rm ME}(e(u),\chi_k)\Big){:}e(u_k-u)\,\mathrm{d}x\mathrm{d}t \\
&\qquad\qquad - \int_\Omega \varrho\dot{u}_k(t)\cdot(u_k(t)-u(t))\,\mathrm{d}x \to 0,
\end{aligned}$$

when realizing the strong convergence of χ_k due to the Aubin-Lions theorem and thus also the strong convergence of $\partial_e\varphi_{\rm ME}(e(u),\chi_k)$ so that $\partial_e\varphi_{\rm ME}(e(u),\chi_k){:}e(u_k-u) \to 0$ weakly. Of course, continuity of $\varphi_{\rm ME}$ and convexity of $\varphi_{\rm ME}(\cdot,\chi)$ are need, together with suitable growth conditions, while $\varphi_{\rm ME}(e,\cdot$ may be nonconvex – we omitted these routine details here.

Exercise 7.6.15 (Incompressible fluid in poroelastic media). Use the Biot-type free energy

$$\varphi_M(e,c) = \frac{1}{2}\mathbb{C}e{:}e + \frac{1}{2}M(\beta\,\mathrm{tr}\,e-c)^2, \tag{7.6.44}$$

and make the limit passage for $M \to \infty$. Show that the limit solves the problem

$$\varrho\ddot{u} - \mathrm{div}\,\sigma = f - \beta\nabla p_{\rm FLUID} \qquad \text{with} \quad \sigma = \mathbb{C}e(u) + \partial_{\dot{e}}\zeta(e(\dot{u})), \tag{7.6.45a}$$

$$\mathrm{div}(\mathbb{M}(c)\nabla p_{\rm FLUID}) = \beta\,\mathrm{tr}\,e(\dot{u}). \tag{7.6.45b}$$

Let us note that the corresponding limit stored energy $\varphi_\infty(e,c)$ is $\frac{1}{2}\mathbb{C}e:e + \delta_{\{\beta\mathrm{tr}\,e=c\}}(e,c)$ so the chemical potential $\mu_\infty = \partial_c\varphi_\infty(e,c)$ is arbitrary provided the holonomic constraint $\beta\,\mathrm{tr}\,e = c$ holds and that $p_{\rm FLUID}$ is in the position of μ_∞; this constraint expresses the geometrical link between the rate of variation of the volume of the pores and the rate of variation of the fluid content. Let us note also that the steady-state variant of this problem is decoupled.

7.7 Examples of some combined processes

Let us still complete this chapter by illustration of some further combination of various models. Combination of various relatively simple processes often opens a surprisingly wide menagerie of models of very different character and usage.

7.7.1 Damage models combined with plasticity

Let us start with illustration of combination of plasticity and damage where many options can be identified. More specifically, damage can influence elasticity as before in Section 7.5 or here also plasticity through plastic yield stress, or both. Vice versa, plasticity can influence damage indirectly through influencing the stress and strain or directly through influencing activation threshold for damage, or both. Both these damage and plastic processes can be considered rate-independent[58] or with rate-dependent effects, damage in addition can be either unidirectional or with healing while plasticity can be either with or without hardening. Also small or large plastic strain can be distinguished, although still all these models can be formulated within the framework of small elastic strains. Eventually, gradients in plastic models can (or need not) be considered.

Therefore, the following options are just rather examples selected from this wide spool of models. Let us present a fairly general model $(\text{spring}^{\alpha}_{\mathbb{C}} \| \text{dashpot}^{\alpha}_{\mathbb{D}}) - (\text{slider}^{\pi,\alpha}_{s_{\rm YLD}} \| \text{spring}_{\mathbb{H}} \| \text{dashpot}_{\mathbb{K}})$, which basically modifies the damage in Jeffreys materials, i.e. the damageable creep model (7.5.45), and covers three scenarios from Figure 7.10-left and middle. Here, $\mathbb{C}$ and $\mathbb{D}$ are again the elasticity and viscosity tensor, while $\mathbb{H}$ stands for the hardening and $\mathbb{K}$ for viscoplastic tensor. Rather for notational simplicity, we do not consider isotropic hardening η so that the internal variables will be the plastic strain π and the damage α, i.e. $z = (\pi, \alpha)$. The main difference from (7.5.45) is that the evolution of the creep variable π is to be activated in the spirit which is standardly used in plasticity with the yield stress $s_{\rm YLD}$ dependent possibly on the plastification and damage, i.e. $s_{\rm YLD} = s_{\rm YLD}(z) = s_{\rm YLD}(\pi, \alpha)$. Then ζ should be augmented by a term like $s_{\rm YLD}|\dot\pi|$ and then (at least in some cases) it is desirable to augment φ by the term $|\nabla\pi|^2$.

Likewise in (5.5.1) on p. 171, we use the additive splitting of the strain, i.e. $e(u) = e_{\rm EL} + \pi$. Using the notation $z = (\pi, \alpha)$, we consider the stored energy $\varphi_{\rm E}$ enhanced

[58] If both damage and plasticity is considered rate-independent, the class of weak solutions becomes very wide and various concepts of weak solutions start playing an important role and selection of a specific concept becomes a vital part of the model itself. We will not consider such fully rate-independent model here, referring to [360, Sect.4.3].

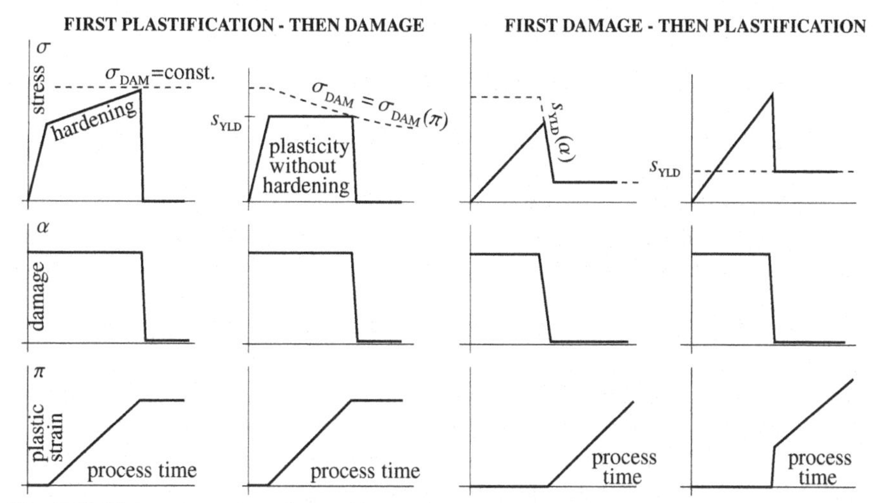

Fig. 7.10 Various scenarios of stress/damage/plastic-strain response under a hard-device load gradually increasing in time (i.e. loading by displacement):
Left: first plastification with hardening, then damage with constant toughness.
Middle-left: first plastification without hardening, then damage due to "plastified" toughness.
Middle-right: first damage, then plastification due to damageable yield stress.
Right: first damage of one (damageable) spring, then plastification due to increased stress on the other (plastifiable) spring in a model as (7.7.12a).

by the gradient terms and the dissipation potential $\zeta(z;\cdot)$ dependent on the internal variables as

$$\begin{aligned}\varphi_{\mathrm{E}}(e,z,\nabla z) = \frac{1}{2}\mathbb{C}(\alpha)(e{-}\pi):(e{-}\pi) - \varphi_{\mathrm{DAM}}(\alpha) + \delta_{[0,1]}(\alpha)\\ + \frac{1}{2}\mathbb{H}\pi:\pi + \frac{\kappa_1}{2}|\nabla\pi|^2 + \frac{\kappa_2}{p}|\nabla\alpha|^p,\end{aligned} \tag{7.7.1a}$$

$$\begin{aligned}\zeta(\alpha;\dot{e},\dot{z}) = \frac{1}{2}\mathbb{D}(\alpha)(\dot{e}{-}\dot{\pi}):(\dot{e}{-}\dot{\pi}) + \frac{1}{2}\mathbb{K}\dot{\pi}:\dot{\pi}\\ + s_{\mathrm{YLD}}(\pi,\alpha)|\dot{\pi}| + \delta^*_{[-\sigma_{\mathrm{DAM}}(\pi,\alpha),\sigma_{\mathrm{M}}]}(\dot{\alpha}) + \frac{\nu}{2}\dot{\alpha}^2,\end{aligned} \tag{7.7.1b}$$

with $\sigma_{\mathrm{M}} = +\infty$ for the unidirectional damage. Of course, when imposing (7.5.16), the explicit treatment of the constraint $0 \le \alpha \le 1$ can be suppressed. The resulted system looks as:

$$\varrho\ddot{u} - \mathrm{div}\,\sigma = f \quad \text{with} \quad \sigma = \mathbb{D}(\alpha)(e(\dot{u}){-}\dot{\pi}) + \mathbb{C}(\alpha)(e(u){-}\pi) \quad \text{on } Q, \tag{7.7.2a}$$

$$s_{\mathrm{YLD}}(\pi,\alpha)\mathrm{Dir}(\dot{\pi}) + \mathbb{K}\dot{\pi} + \mathbb{H}\pi \ni \mathrm{dev}\,\sigma + \kappa_1\Delta\pi \quad \text{on } Q, \tag{7.7.2b}$$

$$\begin{aligned}\partial\delta^*_{[-\sigma_{\mathrm{DAM}}(\pi,\alpha),\sigma_{\mathrm{M}}]}(\dot{\alpha}) + \chi\dot{\alpha} + \frac{1}{2}\mathbb{C}'(\alpha)(e(u){-}\pi):(e(u){-}\pi)\\ \ni \varphi'_{\mathrm{DAM}}(\alpha) + \mathrm{div}(\kappa_2|\nabla\alpha|^{p-2}\nabla\alpha) \quad \text{on } Q,\end{aligned} \tag{7.7.2c}$$

with "Dir" from (7.4.5), together with the boundary and initial conditions

$$(\mathbb{D}(\alpha)e(\dot{u}-\dot{\pi})+\mathbb{C}(\alpha)(e(u)-\pi))\vec{n}=g,\quad \kappa_1\nabla\pi\cdot\vec{n}=0,\quad \kappa_2\nabla\alpha\cdot\vec{n}=0\quad \text{on }\Sigma, \tag{7.7.3a}$$

$$u|_{t=0}=u_0,\quad \dot{u}|_{t=0}=v_0,\quad \pi|_{t=0}=\pi_0,\quad \alpha|_{t=0}=\alpha_0\qquad \text{on }\Omega. \tag{7.7.3b}$$

The weak formulation now involves two variational inequalities corresponding to the two set-valued nonsmooth terms in (7.7.1b).

For $s_{\text{YLD}}=0$ and $\kappa_1=0$, the system (7.7.2) describe damage in Jeffreys' viscoelastic materials like in Remark 7.5.20 or, if also $\mathbb{D}=0$, in Maxwell's viscoelastic materials as in Remark 7.5.22 where analytical problems resulted from $\kappa_1=0$ are mentioned.

Here, considering $\kappa_1>0$, existence of weak solutions to (7.7.2)–(7.7.3a) is granted by quite routine combination of arguments from Sections 7.4 and 7.5. In particular, a modification of the arguments in (7.5.9) leads to an additional term $s_{\text{YLD}}\dot{\pi}_k/|\dot{\pi}_k|:(\pi_k-\pi)$ which can be converged to 0 if $|\nabla\pi|^2$ is involved so that $\pi_k\to\pi$ strongly in $L^2(Q;\mathbb{R}^{d\times d}_{\text{dev}})$ by the Aubin-Lions theorem. Even $s_{\text{YLD}}=s_{\text{YLD}}(\alpha)$ can be well admitted. On the other hand, this compactness argument breaks in the combination of perfect plasticity with damage, i.e. when $s_{\text{YLD}}>0$ even independent of α but combined with $\mathbb{K}=0$, $\mathbb{H}=0$, and $\kappa_1=0$, which thus seems to make troubles in this basic setting linear in terms of the elastic strain $e_{\text{EL}}=e(u)-\pi$; later, suitable nonlinear modification will help, cf. (7.7.6) below.

Let us know discuss the scenarios outlined in Figure 7.10 in some particular aspects more in detail.

Scenario in Figure 7.10-left: To execute the feature that plastification starts first while leading to damage only later, one must set up the parameters of the model in such a way that the yield stress s_{YLD} is smaller than the fracture toughness as in (7.5.27).

The additional dissipation due to isochoric plastification is then achieved when damage is performed in a shear mode (i.e. Mode II) comparing to damage by opening (i.e. Mode I) where plastification is not triggered. When considering the isotropic stored energy (7.5.26) with adhesive damage (i.e. linearly-depending $K(\cdot)$ and $G(\cdot)$) and with $\varphi_{\text{DAM}}(\cdot)$ also linear and e_{EL} in place of e combined with the isotropic hardening $\mathbb{H}=H\mathbb{I}$ and with the dissipation potential (7.7.1b) with $s_{\text{YLD}}>0$ constant and $\sigma_{\text{DAM}}=0$, we altogether arrive at the model governed by

$$\varphi_{\text{E}}(e,\alpha,\pi,\nabla\alpha,\nabla\pi)=\frac{d}{2}\Big(K(\mathrm{tr}^-(e-\pi))^2+\alpha K(\mathrm{tr}^+(e-\pi))^2\Big)+\alpha G|\mathrm{dev}\,e-\pi|^2$$
$$-\mathfrak{g}_\text{c}\alpha+\delta_{[0,1]}(\alpha)+\frac{1}{2}H|\pi|^2+\frac{\kappa_1}{2}|\nabla\pi|^2+\frac{\kappa_2}{p}|\nabla\alpha|^p, \tag{7.7.4a}$$

$$\zeta(\alpha;\dot{e},\dot{\alpha},\dot{\pi})=\frac{1}{2}G_{\text{KV}}(\alpha)|\mathrm{dev}\,\dot{e}-\dot{\pi}|^2+\frac{1}{2}G_{\text{MX}}|\dot{\pi}|^2+s_{\text{YLD}}|\dot{\pi}|+\delta^*_{[0,+\infty)}(\dot{\alpha})+\frac{\nu}{2}\dot{\alpha}^2, \tag{7.7.4b}$$

Starting from undamaged material, the energy needed (dissipated) by damaging in opening without plastification is just the toughness $\mathfrak{g}_\text{c}$, while in shearing mode it is larger, namely $\mathfrak{g}_\text{c}+s_{\text{YLD}}(\sqrt{2G\mathfrak{g}_\text{c}}-s_{\text{YLD}})/H$ provided the parameters are tuned in a way to satisfy $\sqrt{G\mathfrak{g}_\text{c}/2}<s_{\text{YLD}}\le\sqrt{2G\mathfrak{g}_\text{c}}$. This was first devised for an interfacial delamination model [470], being inspired just by such bulk plasticity.

When build into the *phase-field fracture* model of the type (7.5.35), we obtain the *mode-sensitive fracture*, as already mentioned in Remark 7.5.13. Since the fracture toughness $\mathfrak{g}_c$ is scaled as $\mathscr{O}(1/\varepsilon)$ in (7.5.35), $s_{\rm YLD}$ in (7.7.4b) is to be scaled as $\mathscr{O}(1/\sqrt{\varepsilon})$.

Scenario in Figure 7.10-middle-left: An alternative to the previous scenario for materials where hardening is naturally avoided (as rocks of soils) is a plastifiable fracture toughness $\mathfrak{g}_c = \mathfrak{g}_c(|\pi|,\alpha)$. Starting from undamaged material with $\alpha_0 = 1$, the effective fracture stress (7.5.27) is now modified as

$$\sigma_{\rm DAM}(|\pi|,\alpha_0) := 2G(\alpha_0)\sqrt{\frac{\mathfrak{g}_c(|\pi|,\alpha_0)}{G'(\alpha_0)}} = \text{“effective fracture stress”}. \tag{7.7.5}$$

In the rock mechanics viewed in large-time scales, damage can undergo healing and it is a vital part of the model. In this case, a combination with plasticity is especially important (although sometimes misconceptually ignored or improperly modeled) because healing should go towards new, deformed configuration arising when counting evolved plastic strain and not to the original one. Simultaneously, hardening is naturally not any pronounced phenomenon in such materials.

Scenario in Figure 7.10-middle-right: Another important phenomenon advancing damage before plastification occurs when the yield-stress is damageable, i.e. $s_{\rm YLD} = s_{\rm YLD}(\alpha)$ as in (7.7.1b), and the original yield stress $s_{\rm YLD}(\alpha_0)$ is higher than than the fracture stress $\sigma_{\rm DAM}(0,\alpha_0)$ from (7.7.5) so that the fracture starts before plastification can be triggered.

A mathematically interesting case deserving also noteworthy applications when plastic shear-bands can become infinitesimally thin occurs when α lives "compactly" in $C(\bar{Q})$. Then one can think about dealing with the *perfect plasticity*, i.e. $\mathbb{H} = 0$, $\kappa_1 = 0$, and $\mathbb{K} = 0$ in (7.7.1) and, in view of Remark 7.4.13, also $g = 0$. To this goal, we take $p > d$. Altogether, we thus consider

$$\varphi_{\rm E}(e,z,\nabla z) = \frac{1}{2}\mathbb{C}(\alpha)(e-\pi):(e-\pi) - \varphi_{\rm DAM}(\alpha) + \frac{\kappa_2}{p}|\nabla\alpha|^p \quad\text{with}\quad z = (\pi,\alpha), \tag{7.7.6a}$$

$$\zeta(z;\dot{e},\dot{z}) = \frac{1}{2}\mathbb{D}(\alpha)(\dot{e}-\dot{\pi}):(\dot{e}-\dot{\pi}) + s_{\rm YLD}(\alpha)|\dot{\pi}| - \sigma_{\rm DAM}(\alpha)\dot{\alpha} + \chi\dot{\alpha}^2 + \delta_{(-\infty,0]}(\dot{\alpha}). \tag{7.7.6b}$$

The variational inequality like (7.4.7) for the damage-influenced perfect plasticity can be cast simultaneously for the plastic and the damage flow rules:

$$\begin{aligned}
&\int_Q s_{\rm YLD}(\alpha)|w| - \mathbb{C}(\alpha)e_{\rm EL}:w + \sigma_{\rm DAM}(\alpha)|v| + \chi v^2 - \frac{1}{2}v\mathbb{C}'(\alpha)e_{\rm EL}:e_{\rm EL}\\
&\quad + \kappa_2|\nabla\alpha|^{p-2}\nabla\alpha\cdot\nabla v\,\mathrm{d}x\mathrm{d}t \ge \int_Q \mathbb{D}\dot{e}_{\rm EL}:\dot{e}_{\rm EL} + \mathfrak{g}_c(\alpha)|\dot{\alpha}| + \chi\dot{\alpha}^2 - f\cdot\dot{u}\,\mathrm{d}x\mathrm{d}t\\
&\quad + \int_\Omega\Big(\frac{\varrho}{2}\big|\dot{u}(T)\big|^2 + \frac{1}{2}\mathbb{C}(\alpha)e_{\rm EL}(T):e_{\rm EL}(T) - \varphi_{\rm DAM}(\alpha(T)) + \frac{\kappa_2}{p}|\nabla\alpha(T)|^p - \frac{\varrho}{2}|v_0|^2\\
&\quad - \frac{1}{2}\mathbb{C}(\alpha_0)e_{\rm EL,0}:e_{\rm EL,0} + \varphi_{\rm DAM}(\alpha_0) - \frac{\kappa_2}{p}|\nabla\alpha_0|^p\Big)\mathrm{d}x + \int_{\bar{Q}} s_{\rm YLD}(\alpha)\,|\dot{\pi}|(\mathrm{d}x\mathrm{d}t)
\end{aligned} \tag{7.7.7}$$

for all $w \in C(Q;\mathbb{R}^{d\times d}_{\rm dev})$ and $v \in C^1(Q)$, where we have used the abbreviation $e_{\rm EL} = e(u)-\pi$ and now also $e_{\rm EL,0} = e(u_0)-\pi_0$. Analytically, the last integral in (7.7.7) uses the measure $|\dot\pi| \in \mathfrak{M}^+(\bar Q)$ as a total variation of $\dot\pi$ for the integration of $s_{\rm YLD}(\alpha) \in C(\bar Q)$. Moreover, if the damage evolution is unidirectional and $s_{\rm YLD}(\cdot)$ is nondecreasing, if $\mathbb{D}(\cdot) = \mathbb{D}_0 + \tau_{\rm R}\mathbb{C}(\cdot)$ with $\mathbb{C}(\cdot)$ nondecreasing with respect to the Löwner ordering, $s_{\rm YLD} = s_{\rm YLD}(\alpha)$, and the loading is enough regular, we perform an extension of the regularity estimate (7.5.34) by testing the plastic flow rule (7.7.2b) by $\dot\pi + \tau_{\rm R}\ddot\pi$. We use the (here formal) calculus

$$s_{\rm YLD}(\alpha){\rm Dir}(\dot\pi):\ddot\pi = \frac{\partial}{\partial t}\big(s_{\rm YLD}(\alpha)|\dot\pi|\big) - \dot\alpha s'_{\rm YLD}(\alpha)|\dot\pi| \geq \frac{\partial}{\partial t}\big(s_{\rm YLD}(\alpha)|\dot\pi|\big)$$

because $\dot\alpha s'_{\rm YLD}(\alpha)|\dot\pi| \leq 0$ when assuming $s_{\rm YLD}(\cdot)$ nondecreasing and using $\dot\alpha \leq 0$. Heuristically, considering $\mathbb{K}=0$, $\mathbb{H}=0$, and $\kappa_1 = 0$, this gives the estimate

$$\int_0^t\!\!\int_\Omega s_{\rm YLD}(\alpha)|\dot\pi| + \big(\mathbb{C}(\alpha)e_{\rm EL} + \mathbb{D}(\alpha)\dot e_{\rm EL}\big):(\dot\pi + \tau_{\rm R}\ddot\pi)\,{\rm d}x{\rm d}t \\ + \int_\Omega \tau_{\rm R} s_{\rm YLD}(\alpha(t))|\dot\pi(t)|\,{\rm d}x \leq \int_\Omega \tau_{\rm R} s_{\rm YLD}(\alpha_0)|\dot\pi_0|\,{\rm d}x. \tag{7.7.8}$$

Furthermore, we test the force equilibrium (7.7.2a) by $\dot u + \tau_{\rm R}\ddot u$, similarly as we did in (7.5.34). Summing it with (7.7.8) and denoting by $\varepsilon_{\rm el} := e_{\rm EL} + \tau_{\rm R}\dot e_{\rm EL}$, we obtain

$$\int_\Omega \frac{\tau_{\rm R}}{2}\mathbb{D}_0\dot e_{\rm EL}(t):\dot e_{\rm EL}(t) + \frac12\mathbb{C}(\alpha(t))\varepsilon_{\rm el}(t):\varepsilon_{\rm el}(t) + \tau_{\rm R}s_{\rm YLD}(\alpha(t))|\dot\pi(t)|\,{\rm d}x \\ + \int_0^t\!\!\int_\Omega \frac{\varrho}{\tau_{\rm R}}|\dot w|^2 + \mathbb{D}_0\dot e_{\rm EL}:\dot e_{\rm EL} + s_{\rm YLD}(\alpha)|\dot\pi|\,{\rm d}x{\rm d}t \leq \int_0^t\!\!\int_\Omega \Big(f + \frac{\varrho}{\tau_{\rm R}}\dot u\Big)\cdot\dot w\,{\rm d}x{\rm d}t \\ + \int_\Omega \frac{\tau_{\rm R}}{2}\mathbb{D}_0\dot e_{\rm EL,0}:\dot e_{\rm EL,0} + \frac12\mathbb{C}(\alpha_0)\varepsilon_{\rm el,0}:\varepsilon_{\rm el,0} + \tau_{\rm R}s_{\rm YLD}(\alpha_0)|\dot\pi_0|\,{\rm d}x, \tag{7.7.9}$$

where $\varepsilon_{\rm el,0} := e_{\rm EL,0} + \tau_{\rm R}\dot e_{\rm EL,0}$ with $e_{\rm EL,0} := e(u_0) - \pi_0$ and $\dot e_{\rm EL,0} := e(v_0) - \dot\pi_0$, and again $w := u + \tau_{\rm R}\dot u$. We have used also $\dot\alpha\mathbb{C}'(\alpha)\varepsilon_{\rm el}:\varepsilon_{\rm el} \leq 0$ similarly like we did for (7.5.34). We qualify the initial data additionally as $v_0 \in H^1(\Omega;\mathbb{R}^d)$ and consider the isotropic material, which allows us to see[59] that $\dot\pi(0) =: \dot\pi_0 \in L^2(\Omega;\mathbb{R}^{d\times d}_{\rm dev})$ and also $\dot e_{\rm EL,0} = e(v_0) - \dot\pi_0 \in L^2(\Omega;\mathbb{R}^{d\times d}_{\rm sym})$. From this, we obtain the a-priori estimates

$$\|u\|_{H^2(I;L^2(\Omega;\mathbb{R}^d))\cap W^{1,\infty}(I;{\rm BD}(\Omega;\mathbb{R}^d))} \leq C, \tag{7.7.10a}$$
$$\|e(u)-\pi\|_{H^2(I;L^2(\Omega;\mathbb{R}^{d\times d}_{\rm sym}))} \leq C, \tag{7.7.10b}$$
$$\|\pi\|_{W^{1,\infty}(I;\mathfrak{M}(\bar\Omega;\mathbb{R}^{d\times d}_{\rm dev})} \leq C, \tag{7.7.10c}$$
$$\|\alpha\|_{L^\infty(I;W^{1,p}(\Omega))\cap H^1(I;L^2(\Omega))} \leq C. \tag{7.7.10d}$$

Of course, the estimation strategy (7.7.8)–(7.7.9) is rather conceptual, while its

[59] Indeed, writing the plastic flow rule at $t = 0$, we have $s_{\rm YLD}(\pi_0,\alpha_0){\rm Dir}(\dot\pi_0) + \mathbb{D}(\alpha_0)\dot\pi_0 \ni \mathbb{D}(\alpha_0){\rm dev}\,e(v_0) + {\rm dev}(\mathbb{C}(\alpha_0)e_{\rm EL,0} \in L^2(\Omega;\mathbb{R}^{d\times d}_{\rm dev})$. From this, we obtain the desired estimate $\dot\pi_0 \in L^2(\Omega;\mathbb{R}^{d\times d}_{\rm dev})$.

rigorous execution is to be made rather on some approximation.[60] The problem is quite nontrivial due to the necessity of strong convergence of the elastic strains $e_{\rm EL}$ needed for the limit passage in the damage flow rule and simultaneously due to failure of energy conservation.[61] Although the usual energy-conservation based technique likewise e.g. (6.4.25), (7.5.33), or also (8.3.20) or (8.4.16) below does not seem to work here, one can estimate directly the difference between the (unspecified[62]) approximate solution (u_k,π_k) and its limit (u,π) conceptually as:[63]

$$\begin{aligned}\int_Q \mathbb{D}(\alpha_k)(\dot e_{{\rm EL},k}-\dot e_{\rm EL}):(\dot e_{{\rm EL},k}-\dot e_{\rm EL})\,{\rm d}x{\rm d}t&\\ +\int_\Omega \frac12\mathbb{C}(\alpha_k(T))(e_{{\rm EL},k}(T)-e_{\rm EL}(T)):(e_{{\rm EL},k}(T)-e_{\rm EL}(T))\,{\rm d}x&\\ \le \int_Q\int_\Omega \big(\mathbb{D}(\alpha_k)(\dot e_{{\rm EL},k}-\dot e_{\rm EL})+\mathbb{C}(\alpha_k)(e_{{\rm EL},k}-e_{\rm EL})\big):(\dot e_{{\rm EL},k}-\dot e_{\rm EL})\,{\rm d}x{\rm d}t&\\ =\int_Q\Big((f-\varrho\ddot u_k)\cdot(\dot u_k-\dot u)-(\mathbb{D}(\alpha_k)\dot e_{\rm EL}+\mathbb{C}(\alpha_k)e_{\rm EL}):(\dot e_{{\rm EL},k}-\dot e_{\rm EL})&\\ +s_{\rm YLD}(\alpha_k)(|\dot\pi|-|\dot\pi_k|)\Big)\,{\rm d}x{\rm d}t&\end{aligned}\tag{7.7.11}$$

where we used also the inequality[64] $\operatorname{dev}\sigma_k:(\dot\pi_k-\dot\pi)\ge s_{\rm YLD}(\alpha_k)(|\dot\pi_k|-|\dot\pi|)$ which follows from the plastic flow rule $s_{\rm YLD}(\alpha_k){\rm Dir}(\dot\pi_k)\ni\operatorname{dev}\sigma_k$ with σ_k from (7.7.2a) written for the approximate solution. The limit superior of the right-hand side in (7.7.11) can be made by weak* upper semicontinuity and estimated from above by zero (so that, in fact, the limit does exist and equals to zero), realizing also that $\alpha_k\to\alpha$ strongly in $C(\bar Q)$. From this, the desired strong convergence of $e_{{\rm EL},k}\to e_{\rm EL}$ follows. Then, in terms of subsequences, the limit passage towards a weak solution to (7.7.2)–(7.7.3) with $\mathbb{K}=\mathbb{H}=0$ and $\kappa_1=0$ is possible.

An illustration of this model is in Fig. 7.11, showing a computer experiment on a 2-dimensional rectangular domain with an initial partial damage on a narrow hor-

---

[60] We can consider a Galerkin approximation and then the estimates (7.7.10) holds even with ${\rm BD}(\Omega;\mathbb{R}^d)$ and $\mathfrak{M}(\bar\Omega;\mathbb{R}^{d\times d}_{\rm dev})$ replaced by $W^{1,1}(\Omega;\mathbb{R}^d)$ and $L^1(\Omega;\mathbb{R}^{d\times d}_{\rm dev})$, respectively, while the $H^2(I;L^2(\Omega;\mathbb{R}^d))$-estimate of $\ddot u$ holds for the Hahn-Banach extension. Alternatively, a time discretization with the fractional-step splitting of the variable (u,π) and α, which leads to a recursive convex minimization on ${\rm BD}(\Omega;\mathbb{R}^d)\times\mathfrak{M}(\Omega;\mathbb{R}^{d\times d}_{\rm dev})$ and on $W^{1,p}(\Omega)$.

[61] Assuming (7.5.28), one can prove the regularity for (u,π) as far as time derivatives, namely $u\in W^{1,\infty}(I;{\rm BD}(\Omega;\mathbb{R}^d))\cap H^2(I;L^2(\Omega;\mathbb{R}^d))$ and $\pi\in W^{1,\infty}(I;\mathfrak{M}(\bar\Omega;\mathbb{R}^{d\times d}_{\rm dev}))$, but the damage flow rule stays troublesome.

[62] Rather than a Galerkin approximation, one can think about some regularization e.g. by (vanishing) hardening or a damage gradient. Cf. also [144] for approximation by a fractional-step time discretization.

[63] The inequality in (7.7.11) is due to monotonicity of $\mathbb{C}(\cdot)$ with respect to the Löwner ordering so that $\frac{\partial}{\partial t}\frac12\mathbb{C}(\alpha_k)(e_{{\rm EL},k}-e_{\rm EL}):(e_{{\rm EL},k}-e_{\rm EL})\le\frac{\partial}{\partial t}\frac12\mathbb{C}(\alpha_k)(e_{{\rm EL},k}-e_{\rm EL}):(e_{{\rm EL},k}-e_{\rm EL})-\dot\alpha_k\mathbb{C}'(\alpha_k)(e_{{\rm EL},k}-e_{\rm EL}):(e_{{\rm EL},k}-e_{\rm EL})=\mathbb{C}(\alpha_k)(e_{{\rm EL},k}-e_{\rm EL}):(\dot e_{{\rm EL},k}-\dot e_{\rm EL})$.

[64] Note that $\operatorname{dev}\sigma_k\in L^2(Q;\mathbb{R}^{d\times d}_{\rm dev})$ here comes into a duality with $|\dot\pi_k|\in\mathfrak{M}(\bar Q;\mathbb{R}^{d\times d}_{\rm dev})$, which would not have a good sense in general while here it is legitimate but highly nontrivial, cf. [139, Prop. 2.2].

izontal layer in the center. The upper part is loaded to shift gradually to the right, while the lower part is forced to shift left, cf. Fig. 7.11.[65]

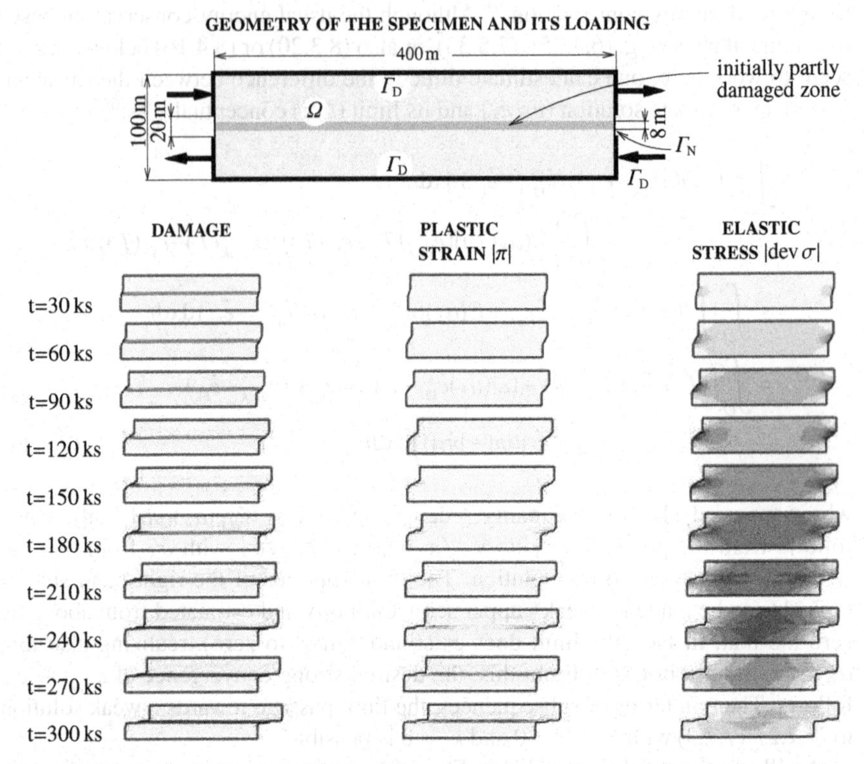

Fig. 7.11 Illustration to the scenario that, under gradually increasing load, first elastic stress rises, than damage start developing while decreasing the yield stress so that eventually also plastification starts and elastic stresses fall down, cf. Figure 7.10-middle-right. The displacement of the deformed domain is displayed magnified 12500×.

Courtesy of Jan Valdman (Czech Academy of Sciences, Prague)

Scenario in Figure 7.10-right: constant yield stress $s_{\rm YLD}$.

The last scenario can be realized when a more complicated rheology is considered. More specifically, the Zener-type construction ($\,_{s_{\rm YLD}}$ — $\,_{\mathbb{C}_1}$)||$\,^{\alpha}_{\mathbb{C}_2}$ like in Fig. 6.4 on p. 223 but with the plastic element instead of the viscous damper and with the second string $\mathbb{C}_2$ damageable may (for a sufficiently big yield stress $s_{\rm YLD}$) exhibit the effect that, under gradually increasing load, first the second spring breaks

[65] The undamaged isotropic material has Young modulus 25 GPA and Poisson ration 0.2 while the plastic yield stress falls from the undamaged state 2 MPa down nearly to 0, and the speed of loading was $\pm 30\,cm/year$, imitating a very fast shift of a tectonic fault; a more detailed description is in [468] where the above presented model is implemented with shortcuts $\varrho = 0$ (i.e. inertial effects neglected) and $\kappa_3 = 0$.

and then the stress on the plastic element jumps up and plastification (here without hardening) may either start immediately (and even the plastic strain may jump if rate-independent) or after some time. The stored energy is then considered as

$$\varphi(e,z) = \frac{1}{2}\mathbb{C}_1(e-\pi){:}(e-\pi) + \frac{1}{2}\mathbb{C}_2(\alpha)e{:}e - a(\alpha) \tag{7.7.12a}$$

and then $\varphi_{\rm E}(e,z) = \varphi(e,z) + \frac{1}{2}\kappa_1|\nabla\pi|^2 + \frac{1}{2}\kappa_2|\nabla\alpha|^2$, while the dissipation potential $\zeta = \zeta(\dot{e},\dot{z})$ can be taken similar (but simpler) as in (7.7.1b):

$$\zeta(\dot{e},\dot{z}) = \frac{1}{2}\mathbb{D}(\alpha)(\dot{e}-\dot{\pi}){:}(\dot{e}-\dot{\pi}) + \frac{1}{2}\mathbb{K}\dot{\pi}{:}\dot{\pi} + s_{\rm YLD}|\dot{\pi}| + \mathfrak{g}_{\rm c}|\dot{\alpha}| + \frac{\chi}{2}\dot{\alpha}^2, \tag{7.7.12b}$$

so that the resulted visco-elasto-plastic rheology is then

$$\Big(\big(\dashv\!\!\vdash_{s_{\rm YLD}} \,\|\, \dashv\!\vdash_{\mathbb{K}}\big) - \big(\wr_{\mathbb{C}_1} \,\|\, \dashv\!\vdash^{\alpha}_{\mathbb{D}}\big)\Big) \,\|\, \wr^{\alpha}_{\mathbb{C}_2}.$$

This type of model is used in masonry where the mortar between particular bricks may break while the subsequent sliding is subjected to some sort of friction, here controlled by $s_{\rm YLD}$ in the no-hardening plasticity model.

Remark 7.7.1 (Ductile crack approximation). A combination of damage in its *phase-field fracture* or crack approximation as in Remark (7.5.16) with plasticity is referred to (an approximation of) *ductile cracks*, in contrast to *brittle cracks* without possibility of plastification on the crack tips.

7.7.2 Poro-elasto-plastic materials

Another interesting combination of phenomena is plasticity or creep in poroelastic media, i.e. the combination of the plastic models from Section 7.4 with the poroelastic models from Section 7.6. Again, there are many variants that can be considered, depending on particular applications, ranging from swelling/creep in polymers to permanent inelastic deformation in poroelastic rocks during oil excavation, for example.

We present here (one particular class of such) models using rate-independent *isochoric plasticity* or *creep* with possibly kinematic hardening in poroelastic solids in Kelvin-Voigt-type rheology in the capillarity-variant as in (7.6.20), employing the *Biot poroelastic model* as in Example 7.6.3, influencing the volumetric part of the rheological model (while isochoric plasticity influences the deviatoric part). The internal variables are now the deviatoric plasticity strain π and the concentration c, i.e. $z = (\pi, c)$. Schematically, our model which combines the plastic rheology from Fig. 7.3 (in the Kelvin-Voigt variant) in the deviatoric part with the Biot model from Fig. 7.9 in the volumetric part is depicted on Figure 7.12. We use the gradient theory as far as the c-component of z concerns, i.e.

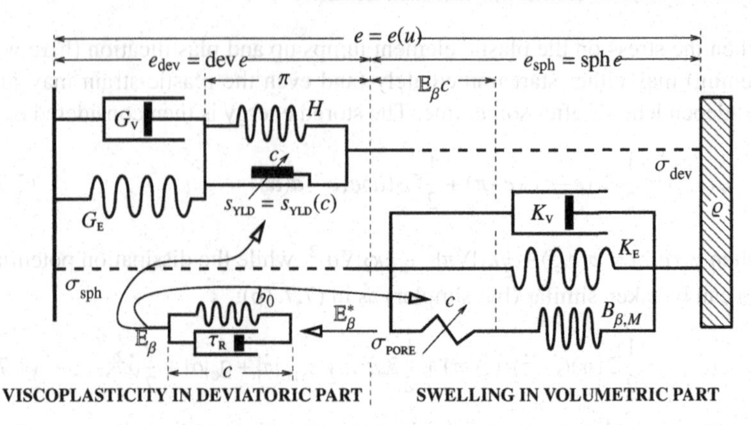

Fig. 7.12 The schematic poro-plastic model in the Kelvin-Voigt rheology, combining the plastic rheology from Fig. 7.3 (in the Kelvin-Voigt variant) in the deviatoric part with the Biot model from Fig. 7.9 in the volumetric part.

$$\varphi_{\rm E}(e,z,\nabla z) = \varphi(e,\pi,c) + \frac{\kappa}{2}|\nabla c|^2 \tag{7.7.13}$$

with the Biot model (7.6.13) combined with the plasticity (7.4.2) which, realizing that ${\rm sph}\,\pi = 0$, now results to

$$\varphi(e,\pi,c) = \frac{1}{2}K_{\rm E}|{\rm sph}\,e|^2 + G_{\rm E}|{\rm dev}\,e{-}\pi|^2 \\ + \frac{1}{2}B_{\beta,M}|{\rm sph}\,e - \mathbb{E}_\beta c|^2 + \frac{1}{2}H|\pi|^2 + \frac{1}{2}b_0 c^2, \tag{7.7.14a}$$

$$\zeta(c;\dot e,\dot\pi,\dot c) = \frac{1}{2}K_{\rm V}|{\rm sph}\,\dot e|^2 + G_{\rm V}|{\rm dev}\,\dot e - \dot\pi|^2 + s_{\rm YLD}(c)|\dot\pi| + \tau_{\rm R}\dot c^2, \tag{7.7.14b}$$

where $B_{\beta,M}$ and $\mathbb{E}_\beta$ are from (7.6.14).

The system of evolution equations/inclusions thus combines the viscous Cahn-Hilliard model (7.6.1) with the Biot ansatz (5.7.6) with the plasticity model (7.4.4). Written "compactly" without splitting the volumetric and deviatoric parts with $\mathbb{C}$ and $\mathbb{D}$ determined by the bulk and shear moduli $\mathbb{C} := {\rm ISO}(K_{\rm E}, G_{\rm E})$ and $\mathbb{D} := {\rm ISO}(K_{\rm V}, G_{\rm V})$ through "ISO" from (6.7.11), respectively, the system reads as

$$\varrho\ddot u - {\rm div}(\sigma_{\rm KV} + \sigma_{\rm PORE}) = f \quad \text{with } \sigma_{\rm PORE} = B_{\beta,M}({\rm sph}\,e - \mathbb{E}_\beta c) \\ \text{and } \sigma_{\rm KV} = \mathbb{C}(e(u){-}\pi) + \mathbb{D}(e(\dot u){-}\dot\pi), \tag{7.7.15a}$$

$$s_{\rm YLD}(c){\rm Dir}(\dot\pi) \ni {\rm dev}\,\sigma_{\rm KV} - H\pi, \tag{7.7.15b}$$

$$\dot c - {\rm div}(\mathbb{M}(c)\nabla\mu) = 0 \quad \text{with } \mu = \tau_{\rm R}\dot c + b_0 c - \mathbb{E}_\beta^*\sigma_{\rm PORE} - \kappa\Delta c, \tag{7.7.15c}$$

where ${\rm Dir} : \mathbb{R}^{d\times d}_{\rm dev} \rightrightarrows \mathbb{R}^{d\times d}_{\rm dev}$ as used in Sect. 7.4.1. We complete this system by the boundary and initial conditions

$$(\sigma_{\rm KV}{+}\sigma_{\rm PORE})\vec n = g, \quad \kappa\nabla c{\cdot}\vec n = 0, \quad \mathbb{M}(c)\nabla\mu = j_\flat \qquad \text{on } \Sigma, \tag{7.7.16a}$$

$$u(0)=u_0, \qquad \dot u(0)=v_0, \qquad \pi(0)=\pi_0, \qquad c(0)=c_0 \quad \text{in } \Omega. \tag{7.7.16b}$$

The energetics behind this model can be revealed standardly when testing (7.7.15a) by $\dot u$, (7.7.15b) by $\dot\pi$, and the equations in (7.7.15c) respectively by μ and $\dot c$. Like (7.6.4), and abbreviating the elastic strain $e_{\rm EL}=e(u)-\pi$ like (5.5.1) on p. 171 and now using $\mathrm{sph}\,e_{\rm EL}=\mathrm{sph}\,e(u)$, we have

$$\frac{\mathrm{d}}{\mathrm{d}t}\int_\Omega \underbrace{\frac{\varrho}{2}|\dot u|^2}_{\text{kinetic energy}} + \underbrace{\frac12\mathbb{C}e_{\rm EL}:e_{\rm EL}+\frac12 H|\pi|^2+\frac12 B_{\beta,M}\big|\mathrm{sph}\,e_{\rm EL}-\mathbb{E}_\beta c\big|^2+\frac12 b_0c^2+\frac{\kappa}{2}|\nabla c|^2}_{\text{mechano-chemical stored energy}}\,\mathrm{d}x$$
$$+\int_\Omega \underbrace{\mathbb{D}\dot e_{\rm EL}:\dot e_{\rm EL}+s_{\rm YLD}(c)|\dot\pi|}_{\text{dissipation via visco-plasticity}} + \underbrace{\mathbb{M}(c)\nabla\mu\cdot\nabla\mu+\tau_{\rm R}\dot c^2}_{\text{dissipation via diffusion}}\,\mathrm{d}x$$
$$=\underbrace{\int_\Omega f\cdot\dot u\,\mathrm{d}x+\int_\Gamma g\cdot\dot u+\mu j_\flat\,\mathrm{d}S}_{\text{total power of outer mechanical/chemical load}}\,. \tag{7.7.17}$$

The weak formulation combines Definition 7.6.1 with the the weak formulation (7.2.8) of the plastic flow rule (7.7.15b) can be cast like (7.4.7) except that the bulk force f is to be replaced by $f-\mathrm{div}(\mathbb{E}_\beta c)$.

Definition 7.7.2 (Weak solution to poro-elasto-plastic problem). The quadruple $(u,\pi,c,\mu)\in H^1(I;H^1(\Omega;\mathbb{R}^d))\times L^2(Q;\mathbb{R}^{d\times d}_{\rm dev})\times L^2(Q)\times L^2(I;H^1(\Omega))$ is called a weak solution to the initial-boundary-value problem (7.6.1)–(7.6.2) with the initial conditions (7.6.5) with ζ from (7.6.6) if $u(0)=u_0$, $u|_{\Sigma_{\rm D}}=u_{\rm D}$, and if:

(α) the following integral identity holds for all $\tilde u\in H^1(Q;\mathbb{R}^d)$ with $u|_{\Sigma_{\rm D}}=0$:

$$\int_Q(\mathbb{D}\dot e_{\rm EL}+\mathbb{C}e_{\rm EL}):e(\tilde u)-\varrho\dot u\cdot\dot{\tilde u}\,\mathrm{d}x\mathrm{d}t$$
$$=\int_\Omega\varrho v_0\cdot\tilde u(0)\,\mathrm{d}x+\int_Q f\cdot\tilde u\,\mathrm{d}x\mathrm{d}t+\int_{\Sigma_{\rm N}}g\cdot\tilde u\,\mathrm{d}x\mathrm{d}t, \tag{7.7.18a}$$

where we have used the abbreviation $e_{\rm EL}=e(u)-\pi$, and further

(β) the following variational inequality

$$\int_Q(H\pi-\mathbb{D}\dot e_{\rm EL}-\mathbb{C}e_{\rm EL}):\tilde\pi+s_{\rm YLD}(c)|\tilde\pi|\,\mathrm{d}x\mathrm{d}t$$
$$\ge\int_Q\mathbb{D}\dot e_{\rm EL}:\dot e_{\rm EL}+s_{\rm YLD}(c)|\dot\pi|-\mathrm{div}(\mathbb{E}_\beta c)\cdot\dot u\,\mathrm{d}x\mathrm{d}t$$
$$+\int_\Omega\Big(\frac12\mathbb{C}e_{\rm EL}(T):e_{\rm EL}(T)+\frac12 B_{\beta,M}|\mathrm{sph}\,e_{\rm EL}(T)|^2+\frac{\varrho}{2}\big|\dot u(T)\big|^2-\frac12\mathbb{C}e_{\rm EL,0}:e_{\rm EL,0}$$
$$-\frac12 B_{\beta,M}|\mathrm{sph}\,e(u_0)|^2-\frac{\varrho}{2}|v_0|^2\Big)\mathrm{d}x-\int_Q f\cdot\dot u-\int_\Sigma g\cdot\dot u\,\mathrm{d}S\,\mathrm{d}t \tag{7.7.18b}$$

holds for any $\tilde\pi\in L^2(Q;\mathbb{R}^{d\times d}_{\rm dev})$ and where $e_{\rm EL,0}=e(u_0)-\pi_0$, and also

(γ) the following integral identity holds for all $\tilde\mu,\tilde c\in H^1(Q)$ with $\tilde c(T)=\tilde\mu(T)=0$:

$$\int_Q \mathbb{M}(c)\nabla\mu\cdot\nabla\tilde{\mu} + (\mu - b_0 c - \mathbb{E}_\beta^*)\tilde{c} - c(\tau_R\dot{\tilde{c}} + \dot{\tilde{\mu}}) + \kappa\nabla c\cdot\nabla\tilde{c}\,\mathrm{d}x\mathrm{d}t$$
$$= \int_\Omega c_0(\tau_R\tilde{c}(0) + \tilde{\mu}(0))\,\mathrm{d}x + \int_\Sigma j_b\tilde{\mu}\,\mathrm{d}S\,\mathrm{d}t. \tag{7.7.18c}$$

Proposition 7.7.3 (Existence of weak solutions). *Let (6.3.3) with* $j_b \in L^2(I;L^{2^{\sharp'}}(\Gamma_N))$, $\pi_0 \in L^2(\Omega;\mathbb{R}^{d\times d}_{\mathrm{dev}})$, *and* $c_0 \in H^1(\Omega)$ *hold, and let*

$$K_E, G_E > 0,\quad K_V, G_V, B_{\beta,M} \ge 0,\quad H > 0,\quad \kappa, b_0 > 0,\quad \tau_R \ge 0, \tag{7.7.19a}$$
$$s_{\mathrm{YLD}} \in C(\mathbb{R}) \text{ bounded},\quad \inf_{c\in\mathbb{R}} s_{\mathrm{YLD}}(c) > 0, \tag{7.7.19b}$$
$$\mathbb{M} \in C(\mathbb{R};\mathbb{R}^{d\times d}_{\mathrm{sym}}) \text{ bounded},\qquad \inf_{c\in\mathbb{R},\ j\in\mathbb{R}^d,\ |j|=1} \mathbb{M}(c)j\cdot j > 0. \tag{7.7.19c}$$

Then there is a weak solution to the initial-boundary-value problem (7.7.15)–(7.7.16) such that

$$\|u\|_{W^{1,\infty}(I;L^2(\Omega;\mathbb{R}^d))\cap L^\infty(I;H^1(\Omega;\mathbb{R}^d))\cap H^2(I;H^1(\Omega;\mathbb{R}^d)^*)} \le C, \tag{7.7.20a}$$
$$\|\pi\|_{L^\infty(I;L^2(\Omega;\mathbb{R}^{d\times d}_{\mathrm{dev}}))} \le C, \tag{7.7.20b}$$
$$\|\mu\|_{L^2(I;H^1(\Omega))} \le C, \tag{7.7.20c}$$
$$\|c\|_{L^\infty(I;H^1(\Omega))\cap H^1(I;H^1(\Omega)^*)} \le C, \tag{7.7.20d}$$
$$\|\Delta c\|_{L^2(Q)} \le C. \tag{7.7.20e}$$

Moreover, if $\mathbb{D} = \mathrm{ISO}(K_V, G_V)$ *is positive definite or* $\tau_R > 0$, *then also respectively*

$$\|u\|_{H^1(I;H^1(\Omega;\mathbb{R}^d))} \le C \quad or \quad \|c\|_{H^1(I;L^2(\Omega))} \le C. \tag{7.7.20f}$$

In the case that both $\mathbb{D}$ *is positive definite and* $\tau_R > 0$, *then the energy is conserved in the sense that (7.7.17) integrated over* $[0,t]$ *holds as an equality.*

Proof. To be more specific but without going into details, we use the Faedo-Galerkin method applied to the system of equations obtained by smoothing (7.4.8) of the flow rule for π, taking also care about approximation of the diffusion by using the same finite-dimensional spaces for both equations in (7.7.15c) like in the proof of Proposition 7.6.2. When pushing the mentioned smoothing to the limit, the approximate solution satisfies analog of (7.7.18). In particular (7.7.18b) reads as the variational inequality

$$\int_Q (H\pi_k - \mathbb{D}\dot{e}_{\mathrm{EL},k} - \mathbb{C}e_{\mathrm{EL},k}):\tilde{\pi} + s_{\mathrm{YLD}}(c_k)|\tilde{\pi}|\,\mathrm{d}x\mathrm{d}t$$
$$\ge \int_Q \mathbb{D}\dot{e}_{\mathrm{EL},k}:\dot{e}_{\mathrm{EL},k} + s_{\mathrm{YLD}}(c_k)|\dot{\pi}_k| + \mathbb{E}_\beta c_k:\dot{e}_{\mathrm{EL},k} - \mathrm{div}\,\mathbb{E}_\beta c_k\cdot\dot{u}_k\,\mathrm{d}x\mathrm{d}t$$
$$+ \int_\Omega \Big(\frac{\varrho}{2}|\dot{u}_k(T)|^2 + \frac{1}{2}\mathbb{C}e_{\mathrm{EL},k}(T):e_{\mathrm{EL},k}(T) + \frac{1}{2}B_{\beta,M}|\mathrm{sph}\,e_{\mathrm{EL},k}(T)|^2 - \frac{\varrho}{2}|v_0|^2$$

$$-\frac{1}{2}\mathbb{C}e_{_{\rm EL,0}}\!:\!e_{_{\rm EL,0}}-\frac{1}{2}B_{\beta,M}|{\rm sph}\,e(u_0)|^2\Big){\rm d}x-\int_Q f\!\cdot\!\dot u_k-\int_\Sigma g\!\cdot\!\dot u_k\,{\rm d}S\,{\rm d}t \quad (7.7.21)$$

to hold for any $\tilde\pi\in L^2(Q;\mathbb{R}^{d\times d}_{\rm dev})$ with $\tilde\pi(t)$ from the particular finite-dimensional subspace of $L^2(\Omega;\mathbb{R}^{d\times d}_{\rm dev})$.

From (7.7.17) written for the Galerkin approximation, one can read the estimates $\|e_{_{\rm EL},k}\|_{L^\infty(I;L^2(\Omega;\mathbb{R}^{d\times d}))}\le C$ and $\|\mathbb{D}\dot e_{_{\rm EL},k}\|_{L^2(Q;\mathbb{R}^{d\times d})}\le C$ and also (7.7.20b,c) and the first estimates in (7.7.20a,d). Also the L^2-estimates (7.7.20f) on $e(\dot u_k)$ and $\dot c_k$ can be read from (7.7.17).

From $e(u_k)=e_{_{\rm EL},k}+E\pi_k$, we can then read also an estimate of $e(u_k)\in L^\infty(I;L^2(\Omega;\mathbb{R}^{d\times d}_{\rm sym}))$ and then the second estimate in (7.7.20a) by Korn's inequality. By comparison, using that also $\mathbb{D}\dot e_{_{\rm EL},k}+\mathbb{C}e_{_{\rm EL},k}\in L^2(Q;\mathbb{R}^{d\times d}_{\rm sym})$ is proved bounded, from (7.7.15c), one gets the last estimate in (7.7.20a). Also, we need ${\rm div}\,\mathbb{E}_\beta c_k$ to be estimated in $L^2(Q;\mathbb{R}^d)$, which follows from the estimate of ∇c_k in $L^2(Q;\mathbb{R}^d)$.

Moreover, by comparison from both two equations in (7.7.15c), one gets the last estimate in (7.7.20a) and the second estimates in (7.7.20d), and also the estimate (7.7.20e) for $\Delta c_k=(\tau_{\rm R}\dot c_k+b_0c_k-\mathbb{E}_\beta^*\sigma_{{\rm pore},k}-\mu_k)/\kappa$ is also the Galerkin approximation of (7.7.15c) uses subspaces of $H^2(\Omega)$. In the context for the Faedo-Galerkin method, all these estimates should be understood, e.g., in the sense of a suitable Hahn-Banach extension, cf. Remark C.2.9 on p. 547.

Then convergence of (7.7.21) for $k\to\infty$ towards (7.7.18b) is by weak continuity or lower semicontinuity, while the convergence towards the integral identities in (7.7.18a,c) is just by weak continuity combined with the Aubin-Lions compactness theorem.

Let us note also that $\mathbb{D}$ positive definite together with the hardening $H>0$ ensures both $\varrho\ddot u\in L^2(I;H^1(\Omega,\mathbb{R}^d)^*)$ and ${\rm div}(\sigma_{\rm KV}+\sigma_{_{\rm PORE}})\in L^2(I;H^1(\Omega,\mathbb{R}^d)^*)$ to be in duality with $\dot u\in L^2(I;H^1(\Omega,\mathbb{R}^d))$. Then we obtain energy conservation when taking into account that also $\dot c\in L^2(Q)$ is in duality with $\Delta c\in L^2(Q)$ so that we can use the chain-rule formula (7.5.10) □

Let us note that, having no viscosity in plastic flow rule, we do not have $\varrho\ddot u={\rm div}(\sigma_{\rm KV}+\sigma_{_{\rm PORE}})+f\in L^2(I;H^1(\Omega;\mathbb{R}^d)^*)$ in duality with $\dot u$ which is controlled only through $e(\dot u)=\dot e_{_{\rm EL}}+\dot\pi\in\mathfrak{M}(\bar Q;\mathbb{R}^{d\times d}_{\rm sym})$. Therefore, the energy conservation cannot be granted under the merely basic energetic estimates. Yet, the regularity assertion in Proposition 7.3.3 can be adapted to the visco-elasto-plastic part (7.7.15a,b) of our model (in the spirit of Proposition 7.4.6) because of the L^2-estimate of $\dot c$ due to the viscosity in the chemical potential (7.7.15c):

Proposition 7.7.4 (Regularity, energy conservation). *Let, in addition to (7.7.19), $\tau_{\rm R}>0$, and both $s_{_{\rm YLD}}$ and $\mathbb{M}$ be constant and also (6.4.14) with (6.3.12c) hold. Then there is a weak solution satisfying the additional a-priori estimates:*

$$\|u\|_{W^{2,\infty}(I;L^2(\Omega,\mathbb{R}^d))\cap W^{1,\infty}(I;H^1(\Omega,\mathbb{R}^d))}\le C, \qquad (7.7.22a)$$

$$\|\pi\|_{W^{1,\infty}(I;L^2(\Omega;\mathbb{R}^{d\times d}_{\rm dev}))}\le C. \qquad (7.7.22b)$$

Moreover, the energy conservation (7.7.17) rigorously holds for this solution.

Sketch of the proof. We use (7.3.11) but, like we used for (7.7.18b), with f now being replaced with $f - \mathrm{div}(\mathbb{E}_\beta c)$. In order to keep the possibility to make $\mathbb{M}$ dependent on c, we intentionally do not involve the diffusion equation although the quadratic nature of φ from (7.7.14a) would allow it. We have needed $f \in W^{1,1}(I; L^2(\Omega;\mathbb{R}^d))$ in (6.4.14) to estimate the term $\dot{f}\cdot\ddot{u}$, which now leads to estimation of the term $\mathrm{div}(\mathbb{E}_\beta\dot{c})\cdot\ddot{u}$.

Now $\varrho\ddot{u}$ is in duality with $\dot{u}$ and also $\dot{c}\in L^2(Q)$ is in duality with $\Delta c\in L^2(Q)$ so that we can use the chain-rule formula (7.5.10). This eventually ensures the energy conservation. □

Remark 7.7.5 (Perfectly plastic materials under diffusion). For vanishing hardening $H \to 0$, one expectedly approaching the perfect plasticity model. Here, like in the combination with damage in the previous Sect. 7.7.1, the mere existence of weak solution without caring for energy conservation is well doable provided the term $\frac{\kappa}{2}|\nabla c|^2$ in (7.7.13) is replaced (or augmented) by $\frac{\kappa}{p}|\nabla c|^p$ with $p > d$. Then the analytically important property of this model is that c is "compactly" in $C(\bar{Q})$ so that the term $s_{\rm YLD}(c)|\dot{\pi}|$ occurring in (7.7.18b) brings in duality a total variation of the measure $|\dot{\pi}|\in\mathfrak{M}(\bar{Q}) = C(\bar{Q})^*$ with $s_{\rm YLD}(c)\in C(\bar{Q})$.

7.7.3 Transport of internal variables by large displacement

Sometimes, the assumption about small plastic strain π is no longer realistic. Holding still the modeling ansatz of this Chapter 7 about small elastic strain $e_{\rm EL}$ may however allow for the total strain $e(u) = e_{\rm EL} + \pi$ possibly large if π is large, and thus also the displacement u may be possibly large. This additive ansatz is legitimate in stratified situations, cf. Proposition 5.5.1 on p. 171.

Internal variables are then transported to different spots than the original reference configuration. At this point, if the medium is not incompressible, it is important to distinguish between volume-dependent and volume-invariant variables. It means that they vary or are invariant provided the volume is compressed, respectively. It is slightly related with distinction between *intensive* or *extensive internal variables* (or physical quantities in general) according how the variables depend on addition of the volume (instead of multiplication).[66] The damage variable α meaning the ratio between damaged and undamaged material is thus and intensive property and is to be counted as volume-invariant. Similarly also the plastic strain π is an intensive internal variable and is to be counted as volume-invariant. On the other hand,

[66] The classification intensive vs extensive standardly relates to the independence or dependency of the properties upon the size or extent of the system. Typically e.g. temperature, pressure, porosity, concentration, or chemical potential are intensive properties of the system while e.g. energy, entropy, or heat content are extensive. Also the ratio of two extensive properties is scale-invariant, and is therefore an intensive property.

concentration (with the standard physical dimension mol/m^3) is an intensive variable but volume-dependent. Yet, if concentration is understood as a volume fraction (dimensionless), it is an intensive and volume-invariant variable. If the medium is compressed or expands, volume-invariant variables do not change, in contrast to volume-dependent ones. Thus, when transported in a moving medium, the transport equations must be designed differently for both mentioned class of variables unless the medium would be incompressible. A transport of a volume-dependent variable, let us denote it by z, is governed by

$$\dot{z} + \mathrm{div}(\dot{u}z) = \kappa\Delta z \tag{7.7.23}$$

while a transport of a volume-invariant variable would have the structure

$$\frac{\mathrm{D}z}{\mathrm{D}t} = \kappa\Delta z \qquad \text{with} \quad \frac{\mathrm{D}z}{\mathrm{D}t} = \dot{z} + \dot{u}\cdot\nabla z \tag{7.7.24}$$

standing for the *material derivative*, sometimes referred also as a *convective derivative*; cf. also (9.1.6) below. Here we adopted a gradient theory with a coefficient $\kappa > 0$. Let us note that, if $\mathrm{div}\,\dot{u} = 0$, then indeed (7.7.23) and (7.7.24) coincide with each other.

It is important that this gradient theory leads to a specific contribution σ_{K} to the stress tensor. Such stresses are motivated, in particular, primarily by balance energy and are known in incompressible-fluid mechanics under the name *Korteweg stresses* [284]. If a volume-dependent (i.e. extensive) variable is transported in a compressible medium, this stress takes the form

$$\sigma_{\mathrm{K,ext}} = \kappa\nabla z{\otimes}\nabla z - \kappa\Big(z\Delta z + \frac{1}{2}|\nabla z|^2\Big)\mathbb{I}. \tag{7.7.25}$$

Equivalently,[67] (7.7.25) can be written as $\sigma_{\mathrm{K,ext}} = \kappa\nabla z{\otimes}\nabla z - \frac{\kappa}{2}(\Delta z^2 - |\nabla z|^2)\mathbb{I}$. Such stress balances the "volume-dependent material time derivative" in the left-hand side (7.7.23) when tested by $\dot{z} + \mathrm{div}(\dot{u}z)$, which gives, when assuming an isolated system on the boundary $\partial\Omega$, that

$$\begin{aligned}\int_\Omega \big|\dot{z}+\mathrm{div}(\dot{u}z)\big|^2 + \frac{\kappa}{2}\frac{\partial}{\partial t}|\nabla z|^2\,\mathrm{d}x &= \int_\Omega \kappa\Delta z\mathrm{div}(\dot{u}z)\,\mathrm{d}x = \int_\Omega \kappa z\Delta z(\mathrm{div}\dot{u}) + \kappa\Delta z(\dot{u}\cdot\nabla z)\,\mathrm{d}x\\ &= \int_\Omega \kappa\Big(\frac{1}{2}|\nabla z|^2 + z\Delta z\Big)\mathrm{div}\,\dot{u} - \kappa(\nabla z{\otimes}\nabla z){:}e(\dot{u})\,\mathrm{d}x = \int_\Omega \sigma_{\mathrm{K,ext}}{:}e(\dot{u})\,\mathrm{d}x,\end{aligned} \tag{7.7.26}$$

where, for the last-but-one equality, we used the calculus

$$\int_\Omega \Delta z(\dot{u}\cdot\nabla z)\,\mathrm{d}x = \int_\Omega \frac{1}{2}|\nabla z|^2\mathrm{div}\,\dot{u} - (\nabla z{\otimes}\nabla z){:}e(\dot{u})\,\mathrm{d}x. \tag{7.7.27}$$

This last formula is important because it reveals that the "optically" high order term $\Delta z(\dot{u}\cdot\nabla z)$ is actually the desired structural-stress-type term $(\nabla z{\otimes}\nabla z - \frac{1}{2}|\nabla z|^2\mathbb{I}){:}e(\dot{u})$.

[67] Here we use the calculus $\Delta z^2 = \mathrm{div}\,(2z\nabla z) = 2z\Delta z + 2|\nabla z|^2$.

Then testing the momentum-equilibrium equation $\varrho\ddot{u} - \mathrm{div}(\sigma + \sigma_{\mathrm{K,ext}}) = f$ by $\dot{u}$, we enjoy cancellation of the terms $\int_\Omega \sigma_{\mathrm{K,ext}}{:}e(\dot{u})\,\mathrm{d}x$ and obtain the energy balance:

$$\int_\Omega \frac{\varrho}{2}|\dot{u}|^2 + \kappa^2|\Delta z|^2 + \sigma{:}e(\dot{u})\,\mathrm{d}x = \int_\Omega f\cdot\dot{u}\,\mathrm{d}x. \tag{7.7.28}$$

It reveals that $|\dot{z}+\mathrm{div}(\dot{u}z)|^2 = \kappa^2|\Delta z|^2$ is the specific dissipation rate due to convection (or possibly diffusion) of the volume-dependent internal variable z. Such a Korteweg stress (7.7.25) can be found relatively frequently in literature, cf. e.g. [88, 163, 246, 414].

On the other hand, when considering a transport of a volume-invariant (i.e. intensive) variable, we would have the structure (7.7.24) to be tested by $\frac{\mathrm{D}z}{\mathrm{D}t}$. It modifies the above considerations only by taking

$$\sigma_{\mathrm{K,int}} = \kappa\nabla z{\otimes}\nabla z - \frac{\kappa}{2}|\nabla z|^2\mathbb{I} \tag{7.7.29}$$

instead of (7.7.25) and, instead of (7.7.26), we would have

$$\begin{aligned}\int_\Omega \Big|\frac{\mathrm{D}z}{\mathrm{D}t}\Big|^2 + \frac{\kappa}{2}\frac{\partial}{\partial t}|\nabla z|^2\,\mathrm{d}x &= -\int_\Omega \kappa\Delta z(\dot{u}\cdot\nabla z)\,\mathrm{d}x\\ &= \int_\Omega \kappa(\nabla z{\otimes}\nabla z){:}e(\dot{u}) - \frac{\kappa}{2}|\nabla z|^2\mathrm{div}\,\dot{u}\,\mathrm{d}x = \int_\Omega \sigma_{\mathrm{K,int}}{:}e(\dot{u})\,\mathrm{d}x,\end{aligned} \tag{7.7.30}$$

see [162] for the Korteweg stress in the form (7.7.29). Again it reveals that $|\frac{\mathrm{D}z}{\mathrm{D}t}|^2 = \kappa^2|\Delta z|^2$ is the specific dissipation rate due to convection (or possibly diffusion) of the volume-invariant internal variable z and the energy balance (7.7.28) again holds. In particular, if the moving medium is incompressible, i.e. $\mathrm{tr}\,e(\dot{u}) = \mathrm{div}\,\dot{u} = 0$, both options of transport of volume-dependent or volume-invariant variable coincides with each other and, in particular, the term $\frac{\kappa}{2}|\nabla z|^2\mathrm{div}\,\dot{u}$ in (7.7.30) disappears and $\sigma_{\mathrm{K,int}}$ in (7.7.29) reduces to $\sigma_{\mathrm{K,int}} = \kappa\nabla z{\otimes}\nabla z$, which was used e.g. in [97, 316] for transport of mass or other internal variables. In geophysical models, such reduced Korteweg stress is used under the name a *structural stress*, cf. [324, 325].[68]

Example 7.7.6. Augmenting the transport equation (7.7.23) for a volume-dependent variable by a reaction term

$$\dot{z} + \mathrm{div}(\dot{u}z) = \kappa\Delta z - \varphi_{\mathrm{R}}'(z) \tag{7.7.31}$$

with some free-energy function φ_{R}, the energy balance results from the test of the transport equation by $\dot{z}+\mathrm{div}(\dot{u}z)$ and yields the term $\int_\Omega \varphi_{\mathrm{R}}'(z)(\dot{u}\cdot\nabla z + z\,\mathrm{div}\dot{u})\,\mathrm{d}x$ whose estimation essentially needs boundedness of φ_{R}', i.e. practically an at most linear growth of $\varphi_{\mathrm{R}} = \varphi_{\mathrm{R}}(z)$ if $|z| \to \infty$. A physically more natural model leading to energy conservation uses rather augmentation of the Korteweg stress (7.7.29) by a pressure-

[68] Actually, it is used even in compressible situations but without deriving any global energy balance analogous to (7.7.28).

like term to[69]

$$\sigma_{\rm K,ext} = \kappa\nabla z\otimes\nabla z - \Big(\kappa z\Delta z + \frac{\kappa}{2}|\nabla z|^2 + \varphi_{\rm R}(z) - z\varphi_{\rm R}'(z)\Big)\mathbb{I}. \qquad (7.7.32)$$

The energy balance again uses the test of the transport equation by $\dot z + {\rm div}(\dot u z)$ and expands (7.7.30) by the term $\int_\Omega \varphi_{\rm R}'(z){\rm div}(\dot u z)\,{\rm d}x = \int_\Omega \dot u\cdot\nabla\varphi_{\rm R}(z) + z\varphi_{\rm R}'(z){\rm div}\,\dot z\,{\rm d}x = \int_\Omega (z\varphi_{\rm R}'(z) - \varphi_{\rm R}(z))\,{\rm div}\,\dot u\,{\rm d}x$.[70] Analogous consideration holds for the volume-invariant-variable transport equation (7.7.24) augmented as $\frac{{\rm D}z}{{\rm D}t} = \kappa\Delta z - \varphi_{\rm R}'(z)$ to be tested by $\frac{{\rm D}z}{{\rm D}t}$, leading thus to (7.7.29) augmented by the pressure-like term $\varphi_{\rm R}(z)\mathbb{I}$, i.e.[71]

$$\sigma_{\rm K,int} = \kappa\nabla z\otimes\nabla z - \Big(\frac{\kappa}{2}|\nabla z|^2 + \varphi_{\rm R}(z)\Big)\mathbb{I}.$$

Example 7.7.7 (Large displacements). One can consider the elasto-plastic-damage model (7.7.2) again at small strains but under large displacements by replacing the partial time derivatives by the material time derivatives and involving a suitable Korteweg-type stress. It is important that both the plastic strain π and damage α are volume-invariant variables, and their transport is governed by the equation of the type (7.7.24). This gives the system:

$$\varrho\ddot u - {\rm div}(\partial_e\varphi(e_{\rm EL},\pi,\alpha) + \sigma_{\rm K,int}) = f \quad \text{with } e_{\rm EL} = e(u) - \pi, \quad \text{on } Q, \qquad (7.7.33a)$$

$$\partial_{\dot\pi}\zeta_1\Big(\pi,\alpha;\frac{{\rm D}\pi}{{\rm D}t}\Big) + \partial_\pi\varphi(e_{\rm EL},\pi,\alpha) \ni {\rm dev}\,\partial_e\varphi(e_{\rm EL},\pi,\alpha) + \kappa\Delta\pi \quad \text{on } Q, \qquad (7.7.33b)$$

$$\partial_{\dot\alpha}\zeta_2\Big(\pi,\alpha;\frac{{\rm D}\alpha}{{\rm D}t}\Big) + \partial_\alpha\varphi(e_{\rm EL},\pi,\alpha) \ni {\rm div}(\kappa_2|\nabla\alpha|^{p-2}\nabla\alpha) \quad \text{on } Q,$$

with the stored energy $\varphi = \varphi(e,\pi,\alpha)$ and with the material time derivatives $\frac{{\rm D}\pi}{{\rm D}t} = \dot\pi + \dot u\cdot\nabla\pi$ and and $\frac{{\rm D}\alpha}{{\rm D}t} = \dot\alpha + \dot u\cdot\nabla\alpha$. with the Korteweg-type stress as in (7.7.32) now

$$\sigma_{\rm K,int} = \kappa\nabla\pi\otimes\nabla\pi + \varkappa|\nabla\alpha|^{r-2}\nabla\alpha\otimes\nabla\alpha - \Big(\frac{\kappa}{2}|\nabla\pi|^2 + \frac{\varkappa}{2}|\nabla\alpha|^r + \varphi(e_{\rm EL},\pi,\alpha)\Big)\mathbb{I}, \quad (7.7.34)$$

cf. also Exercise 7.7.10 below. The energetics of this model can be revealed if testing (7.7.33) successively by $\dot u$, $\frac{{\rm D}\pi}{{\rm D}t}$, and $\frac{{\rm D}\alpha}{{\rm D}t}$. By this way, we obtain

$$\frac{\rm d}{{\rm d}t}\int_\Omega \frac{\varrho}{2}|\dot u|^2 + \varphi(e(u){-}\pi,\pi,\alpha) + \frac{\kappa}{2}|\nabla\pi|^2 + \frac{\varkappa}{2}|\nabla\alpha|^r\,{\rm d}x + \int_\Omega \xi_1\Big(\frac{{\rm D}\alpha}{{\rm D}t}\Big) + \xi_2\Big(\frac{{\rm D}\alpha}{{\rm D}t}\Big){\rm d}x$$
$$= \int_\Omega f\cdot\dot u + \dot u\cdot\sigma{:}\nabla e(u)\,{\rm d}x \quad \text{with } \sigma = \partial_e\varphi(e_{\rm EL},\pi,\alpha), \qquad (7.7.35)$$

with $\xi_1(\pi,\alpha;\dot\pi) = \dot\pi{:}\partial_{\dot\pi}\zeta_1(\pi,\alpha;\dot\pi)$ and $\xi_2(\pi,\alpha;\dot\alpha) = \dot\alpha\partial_{\dot\alpha}\zeta_2(\pi,\alpha;\dot\alpha)$ denoting the dissipation rates. The last term in (7.7.35) means componentwise $\sum_{i,j,k=1}^d \dot u_k\sigma_{ij}\frac{\partial e_{ij}}{\partial x_k}$ and its

[69] For the Korteweg-type stress as in (7.7.32) see [162].

[70] For simplicity, we assume here the boundary conditions $\varphi_{\rm R}'(z)\dot u\cdot\vec n = 0$ on Γ.

[71] Now we use the calculus $\int_\Omega \varphi_{\rm R}'(z)(\dot u\cdot\nabla z)\,{\rm d}x = \int_\Omega \dot u\cdot\nabla\varphi_{\rm R}(z)\,{\rm d}x = \int_\Omega -\varphi_{\rm R}(z)\,{\rm div}\,\dot u\,{\rm d}x$.

estimation needs σ bounded and also calls for a modification of the system (7.7.33) by using the 2nd-grade nonsimple material concept to control $\nabla e(u)$.

Exercise 7.7.8. Derive the formula (7.7.27).

Exercise 7.7.9. Derive the energy balance and basic a-priori estimate for the system of the force balance $\varrho\ddot{u} - \mathrm{div}\,(\sigma + \sigma_{\mathrm{K,int}}) = f$ coupled with the transport equation (7.7.31) with $\sigma_{\mathrm{K,int}}$ either from (7.7.29) or from (7.7.32), assuming φ_{R}' bounded for the former case.

Exercise 7.7.10. For the nonlinear transport equation $\frac{\mathrm{D}z}{\mathrm{D}t} = \kappa\mathrm{div}(|\nabla z|^{r-2}\nabla z)$ instead of (7.7.24), modify (7.7.29) so that the energy balance analogous to (7.7.30) hold and in particular prove that $\sigma_{\mathrm{K,int}} = \kappa|\nabla z|^{r-2}\nabla z\otimes\nabla z - \frac{\kappa}{2}|\nabla z|^r\mathbb{I}$, as actually already used in (7.7.34).

7.7.4 Solids interacting with viscoelastic fluids

There are certainly many possible applications of viscoelastic solid possibly undergoing some inelastic processes interacting with various viscoelastic fluids. Such problems are sometimes referred as a *fluid-structure interaction.*

Exploiting various limit passages we already made for linear rheological materials in Chapter 6, we can approximate the desired fluid by some solid-like rheology so that the approximating model can have the same form on both the solid and the fluid domains, being then called monolitic, cf. Proposition 7.7.11. We can thus approximate Boger, or Stokes, or elastic fluids from various types of rheologies possibly successively; cf. e.g. Exercises 6.6.9, 6.6.10, 6.6.12, 6.7.6, 6.7.7, or 6.7.9.

For example, we can start from the Kelvin-Voigt rheology with the elastic moduli $\mathbb{C}$ determined by $(K_{\mathrm{E}}, G_{\mathrm{E}})$ and viscous moduli $\mathbb{D}$ determined by $(K_{\mathrm{V}}, G_{\mathrm{V}})$ in the sense

$$\mathbb{C} = \mathrm{ISO}(K_{\mathrm{E}}, G_{\mathrm{E}}) \quad \text{and} \quad \mathbb{D} = \mathrm{ISO}(K_{\mathrm{V}}, G_{\mathrm{V}}) \tag{7.7.36}$$

with the notation “ISO” from (6.7.11), and being positive definite on the whole domain $\Omega = \Omega_{\mathrm{S}} \cup \Omega_{\mathrm{F}}$ but having different values on Ω_{S} and on Ω_{F}. Then sending $G_{\mathrm{E}} \to 0$ on Ω_{F}, we can approximate Boger viscoelastic fluid (6.7.13), or sending $K_{\mathrm{E}} \to \infty$ and $G_{\mathrm{E}} \to 0$ on Ω_{F}, we can attain the Stokes incompressible fluid, or sending $G_{\mathrm{E}} \to 0$ and $\mathbb{D} \to 0$ on Ω_{F}, we can approach the inviscid elastic fluid. A successive limit passage first $G_{\mathrm{E}} \to 0$ on Ω_{F} and then $G_{\mathrm{E}} \to 0$ on Ω_{F} is also possible, approximating the inviscid elastic fluid through the Boger viscoelastic fluid. Let us illustrate it for the mentioned Boger fluid on Ω_{F} interacting with the Kelvin-Voigt solid on Ω_{S}, which in the classical formulation reads as:

$$\varrho\ddot{u} - \mathrm{div}(\sigma_{\mathrm{sph}} + \sigma_{\mathrm{dev}}) = f \qquad \text{in } Q, \tag{7.7.37a}$$

$$\text{with } \sigma_{\mathrm{sph}} = dK_{\mathrm{V}}\mathrm{sph}\,e(\dot{u}) + dK_{\mathrm{E}}\mathrm{sph}\,e(u) \qquad \text{in } Q, \tag{7.7.37b}$$

$$\text{and}\quad \sigma_{\rm dev} = \begin{cases} 2G_{\rm v}{\rm dev}\,e(\dot u) + 2G_{\rm E}{\rm dev}\,e(u) & \text{in } Q_{\rm S},\\ 2G_{\rm v}{\rm sph}\,e(\dot u) & \text{in } Q_{\rm F}, \end{cases} \tag{7.7.37c}$$

$$u|_{\Sigma_{\rm D}} = u_{\rm D} \qquad \text{on } \Sigma_{\rm D}, \tag{7.7.37d}$$

$$(\sigma_{\rm sph}+\sigma_{\rm dev})\vec n = g \qquad \text{on } \Sigma_{\rm N}, \tag{7.7.37e}$$

$$u|_{t=0} = u_0, \quad \dot u|_{t=0} = v_0 \qquad \text{on } \Omega, \tag{7.7.37f}$$

where $Q_{\rm F} := I\times\Omega_{\rm F}$ and $Q_{\rm S} := I\times\Omega_{\rm S}$ and where $(K_{\rm v},G_{\rm v})$ and $K_{\rm E}$ may depend on x and, in particular, may be different on $\Omega_{\rm S}$ and on $\Omega_{\rm F}$.

As already said, the model (7.7.37) can be attained from the Kelvin-Voigt visco-elastic solid rheology, i.e. the initial-boundary-value problem (6.1.1)–(6.2.2)–(6.2.3) with $\sigma = \mathbb{D}e(\dot u) + \mathbb{C}e(u)$ as in (6.4.1) here with $\mathbb{D}$ and $\mathbb{C}$ from (7.7.36). This may be useful algorithmically when implementing the model on computers. Let us formulate it rigorously:

Proposition 7.7.11 (Monolithic approach to Kelvin-Voigt/Boger's model). *Let (6.4.5) be satisfied and let $u_\varepsilon \in H^1(I;H^1(\Omega;\mathbb{R}^d))\cap W^{1,\infty}(I;L^2(\Omega;\mathbb{R}^d))$ be the unique weak solution to the initial-boundary-value problem (6.1.1)–(6.2.2)–(6.2.3) with σ from (6.4.1) with (7.7.36) using the elastic moduli $\mathbb{C} = \mathbb{C}_\varepsilon$ with*

$$\mathbb{C}(x) = \mathbb{C}_\varepsilon(x) = \begin{cases} {\rm ISO}(K_{\rm E},G_{\rm E}) & \text{for } x\in\Omega_{\rm S},\\ {\rm ISO}(K_{\rm E},\varepsilon G_{\rm E}) & \text{for } x\in\Omega_{\rm F},\ \ \varepsilon>0, \end{cases} \tag{7.7.38}$$

while $\mathbb{D} = {\rm ISO}(K_{\rm v},G_{\rm v})$ is fixed. Then the whole sequence $\{u_\varepsilon\}_{\varepsilon>0}$ converges in the sense

$$u_\varepsilon \to u \quad \text{strongly in } H^1(I;H^1(\Omega;\mathbb{R}^d)) \text{ and weakly* in } W^{1,\infty}(I;L^2(\Omega;\mathbb{R}^d)),$$

and u is the unique weak solution to the initial-boundary-value problem (7.7.37).

Sketch of the proof. The basic energetic test yields $\|e(u_\varepsilon)\|_{L^\infty(I;L^2(\Omega;\mathbb{R}^{d\times d}))} = \mathscr{O}(1/\sqrt{\varepsilon})$ so that the term $\int_{Q_{\rm F}}\varepsilon G_{\rm E}{\rm dev}\,e(u_\varepsilon):{\rm dev}\,e(v)\,{\rm d}x{\rm d}t$ in a weak formulation of the problem is $\mathscr{O}(\sqrt{\varepsilon})$ and thus vanishes in the limit. Otherwise, the limit passage is simple by weak convergence, as the problem is linear. The claimed strong convergence in $H^1(I;H^1(\Omega;\mathbb{R}^d))$ is simply by testing the difference of weak formulations by $\dot u_\varepsilon - \dot u$, the resulting "elastic" term on the fluidic domain (which vanished in the limit) bears estimation "on the right-hand side" as

$$-\int_Q \varepsilon\mathbb{C}e(u_\varepsilon):e(\dot u_\varepsilon - \dot u)\,{\rm d}x{\rm d}t \le \int_Q \varepsilon\mathbb{C}e(u_\varepsilon):e(\dot u)\,{\rm d}x{\rm d}t + \int_\Omega \frac{\varepsilon}{2}\mathbb{C}e(u_0):e(u_0)\,{\rm d}x$$
$$\le \varepsilon|\mathbb{C}|\,\|e(u_\varepsilon)\|_{L^2(Q;\mathbb{R}^{d\times d})}\|e(\dot u)\|_{L^2(Q;\mathbb{R}^{d\times d})} + \frac{\varepsilon}{2}|\mathbb{C}|\,\|e(u_0)\|^2_{L^2(Q;\mathbb{R}^{d\times d})} = \mathscr{O}(\sqrt{\varepsilon}) \to 0.$$

The uniqueness of the limit is as in Proposition 6.4.1 as the viscosity is kept over all Ω and thus $\sqrt{\varrho}\ddot u$ is in duality with $\sqrt{\varrho}\dot u$ and ${\rm div}(\sigma_{\rm sph}+\sigma_{\rm dev})\in L^2(I;H^1(\Omega;\mathbb{R}^d)^*)$ is in duality with $\dot u\in L^2(I;H^1(\Omega;\mathbb{R}^d))$ so that testing the difference of two weak solutions by the difference of their velocities is legitimate and the corresponding chain-rule estimates work. □

The model (7.7.37) exhibits a notable feature that S-waves can propagate only in the solid part $\Omega_{\rm S}$ but not in the fluid part $\Omega_{\rm F}$ because there is not shear elasticity, and are strongly attenuated, depending on the viscosity shear modulus on $\Omega_{\rm F}$. This model can be subject to further limit process. E.g. one can suppress viscous effects both in the solid part and in the fluidic part, approaching the purely *elastic solid* and the inviscid *elastic fluid*, respectively. In particular, this suppresses S-waves in $\Omega_{\rm F}$ completely.

The weak formulation of (7.7.37) then becomes a bit delicate because only $\operatorname{sph} e(u)$ is well defined on Ω while $\operatorname{dev} e(u)$ is defined only on $\Omega_{\rm S}$. The appropriate weak formulation thus arises when testing (7.7.37a) by a smooth test function v with $v(T) = 0$ and $\operatorname{dev} e(v) = 0$ on $Q_{\rm F}$ and making by part integration in time and applying Green formula in space:

Definition 7.7.12 (Weak solution to inviscid solid-fluid problem). A weak solution to the initial-boundary-value problem (7.7.37) with $K_{\rm v} = G_{\rm v} = 0$ is understood as a triple $(u,\sigma_{\rm sph},\sigma_{\rm dev}) \in W^{1,\infty}(I;L^2(\Omega;\mathbb{R}^d)) \times L^\infty(I;L^2(\Omega;\mathbb{R}^{d\times d}_{\rm sym})) \times L^\infty(I;L^2(\Omega_{\rm S};\mathbb{R}^{d\times d}_{\rm dev}))$ such that the integral identity

$$\int_Q \sigma_{\rm sph}{:}\operatorname{sph} e(v)\,{\rm d}x{\rm d}t - \varrho\dot u{\cdot}\dot v\,{\rm d}x{\rm d}t + \int_{Q_{\rm S}} \sigma_{\rm dev}{:}\operatorname{dev} e(v)\,{\rm d}x{\rm d}t = \int_\Omega \varrho v_0{\cdot}v(0,\cdot)\,{\rm d}x + \int_Q f{\cdot}v\,{\rm d}x{\rm d}t + \int_\Sigma g{\cdot}v\,{\rm d}x{\rm d}t \tag{7.7.39}$$

holds for any $v \in H^1(Q;\mathbb{R}^d)$ with $v|_{\Sigma_{\rm D}} = 0$, $v(T) = 0$, and $\operatorname{dev} e(v) = 0$ on $Q_{\rm F}$, $\sigma_{\rm sph} = \operatorname{sph}(\mathbb{C}e(u))$ a.e. on Q and $\sigma_{\rm dev} = \operatorname{dev}(\mathbb{C}e(u))$ a.e. on $Q_{\rm S}$, the Dirichlet boundary condition (7.7.37d) holds, and also the initial condition $u(0,\cdot) = u_0$ holds a.e. on Ω.

Let us note that, controlling $\sigma_{\rm sph}$ and $\sigma_{\rm dev}$ in the above definition with $\mathbb{C} = \operatorname{ISO}(K,G)$, we have implicitly also included the information

$$\operatorname{div} u \in L^\infty(I;L^2(\Omega)) \quad \text{and} \quad u \in L^\infty(I;H^1(\Omega_{\rm S};\mathbb{R}^d)).$$

As the inviscid elastic fluid looses any shear resistance, we cannot loaded by the traction force g or the hard-device (Dirichlet-condition) load $u_{\rm D}$ on the part of the boundary Γ adjacent to $\Omega_{\rm F}$. Let us again formulate it rigorously:

Proposition 7.7.13 (Towards inviscid elastic fluid). *Let (6.3.3) with $g = 0$ hold, and let $u_\varepsilon \in H^1(I;H^1(\Omega;\mathbb{R}^d)) \cap W^{1,\infty}(I;L^2(\Omega;\mathbb{R}^d))$ be the unique weak solution to the initial-boundary-value (7.7.37) with $\mathbb{D} = \mathbb{D}_\varepsilon$ and with $\mathbb{C} = \mathbb{C}(x)$ fixed, namely*

$$\mathbb{D}_\varepsilon := \operatorname{ISO}(\varepsilon K_{\rm v},\varepsilon G_{\rm v}),\ \ \varepsilon > 0, \qquad \mathbb{C}(x) := \begin{cases} \operatorname{ISO}(K_{\rm E},G_{\rm E}) & \text{for } x \in \Omega_{\rm S}, \\ \operatorname{ISO}(K_{\rm E},0) & \text{for } x \in \Omega_{\rm F}. \end{cases} \tag{7.7.40}$$

Then, denoting $\sigma_\varepsilon := \mathbb{D}_\varepsilon e(\dot u_\varepsilon)+\mathbb{C}e(u_\varepsilon)$, *it holds:*
(i) *there is a subsequence such that:*

$$u_\varepsilon \to u \quad \text{weakly* in } W^{1,\infty}(I;L^2(\Omega;\mathbb{R}^d))\cap L^\infty(I;H^1(\Omega_{\rm S};\mathbb{R}^d)), \tag{7.7.41a}$$

$$\sigma_\varepsilon \to \sigma \quad \text{weakly* in } L^2(Q;\mathbb{R}^{d\times d}_{\rm sym}), \tag{7.7.41b}$$

$$\mathrm{dev}\,\sigma_\varepsilon \to 0 \quad \text{weakly in } L^2(Q_{\rm F};\mathbb{R}^{d\times d}_{\rm dev}), \text{ and} \tag{7.7.41c}$$

$$\mathbb{D}_\varepsilon e(\dot u_\varepsilon)\to 0 \quad \text{strongly in } L^2(Q;\mathbb{R}^{d\times d}_{\rm sym}) \tag{7.7.41d}$$

with $(\sigma_{\rm sph},\sigma_{\rm dev})$ *satisfying (7.7.37b,c) with* $K_{\rm v}=G_{\rm v}=0$. *Moreover, any* $(u,\sigma_{\rm sph},\sigma_{\rm dev})$ *obtained by this way is a weak solution to (7.7.37) with* $K_{\rm v}=G_{\rm v}=0$ *in the sense of Definition 7.7.12.*

(ii) *If, in addition, (6.3.12) holds, then the solution* $(u,\sigma_{\rm sph},\sigma_{\rm dev})$ *is unique and the whole sequence* $\{(u_\varepsilon,\sigma_{{\rm sph},\varepsilon},\sigma_{{\rm dev},\varepsilon})\}_{\varepsilon>0}$ *converges, and*

$$\mathrm{sph}\,e(u_\varepsilon(t))\to \mathrm{sph}\,e(u(t)) \quad \text{strongly in } L^2(\Omega;\mathbb{R}^{d\times d}_{\rm sym}) \text{ for all } t\in I, \tag{7.7.42a}$$

$$\mathrm{dev}\,e(u_\varepsilon(t))\to \mathrm{dev}\,e(u(t)) \quad \text{strongly in } L^2(\Omega_{\rm S};\mathbb{R}^{d\times d}_{\rm dev}) \text{ for all } t\in I, \tag{7.7.42b}$$

$$\dot u_\varepsilon(t)\to \dot u(t) \quad \text{strongly in } L^2(\Omega;\mathbb{R}^d) \text{ for all } t\in I, \tag{7.7.42c}$$

$$\sigma_{{\rm sph},\varepsilon} := \mathrm{sph}(\mathbb{D}_\varepsilon e(\dot u_\varepsilon)+\mathbb{C}_\varepsilon e(u_\varepsilon))\to\sigma_{\rm sph} \quad \text{in } L^2(Q;\mathbb{R}^{d\times d}_{\rm sym}), \tag{7.7.42d}$$

$$\sigma_{{\rm dev},\varepsilon} := \mathrm{dev}(\mathbb{D}_\varepsilon e(\dot u_\varepsilon)+\mathbb{C}_\varepsilon e(u_\varepsilon))\to\sigma_{\rm dev} \quad \text{in } L^2(Q;\mathbb{R}^{d\times d}_{\rm dev}), \text{ and} \tag{7.7.42e}$$

$$\int_Q \mathbb{D}_\varepsilon e(\dot u_\varepsilon){:}e(\dot u_\varepsilon)\,\mathrm{d}x\mathrm{d}t \to 0. \tag{7.7.42f}$$

Sketch of the proof. Testing by $\dot u_\varepsilon$ reveals the a-priori estimates

$$\|u_\varepsilon\|_{W^{1,\infty}(I;L^2(\Omega;\mathbb{R}^d))\cap L^\infty(I;H^1(\Omega_{\rm S};\mathbb{R}^d))}\le C \quad \text{and} \quad \|\mathrm{div}\,u_\varepsilon\|_{L^2(Q)}\le C, \tag{7.7.43a}$$

$$\|\mathbb{D}_\varepsilon e(\dot u_\varepsilon)\|_{L^2(Q;\mathbb{R}^{d\times d})}\le C\sqrt{\varepsilon} \quad \text{and} \quad \|e(\dot u_\varepsilon)\|_{L^2(Q;\mathbb{R}^{d\times d})}\le C/\sqrt{\varepsilon}. \tag{7.7.43b}$$

Here we used the data qualification and the estimation strategy from Proposition 6.3.1 not relying on the viscosity, so that C is independent of $\varepsilon>0$ in the estimates (7.7.43a). Let us note that we do not have estimated $e(u)$ on the whole Ω but only on $\Omega_{\rm S}$ while on $\Omega_{\rm F}$ only $\mathrm{tr}\,e(u)$ is estimated. Moreover, as $\|e(\dot u_\varepsilon)\|_{L^2(Q;\mathbb{R}^{d\times d})}=\mathscr{O}(1/\sqrt{\varepsilon})$, it holds $\|\mathbb{D}_\varepsilon e(\dot u_\varepsilon)\|_{L^2(Q;\mathbb{R}^{d\times d})}=\|\varepsilon\mathbb{D}_1 e(\dot u_\varepsilon)\|_{L^2(Q;\mathbb{R}^{d\times d})}=\mathscr{O}(\sqrt{\varepsilon})\to 0$, as expressed in (7.7.43b), so that the convergence (7.7.41d) is shown.

By Banach's selection principle, we take a subsequence converging weakly* in the spaces occurring in (7.7.43a) and prove (7.7.41).

The regularity is in the spirit of purely elastic materials, cf. Proposition 6.3.2 and also Exercise 6.7.7. Yet, we do not have estimated $\nabla\dot u$ on the whole Ω but only on $\Omega_{\rm S}$. More in detail, make differentiation of the system once in time and test it by $\ddot u$. It puts, in particular, $\sqrt{\varrho}\ddot u\in L^2([0,t];L^2(\Omega;\mathbb{R}^d))$ to duality with $\sqrt{\varrho}\dot u$, so that the by-part integration in time used within the "energetic" test by $\dot u$ is legitimate. This yields the energy conservation of the limit solution on any time interval $[0,t]$. For the strong convergence (6.7.16), still use

$$
\begin{aligned}
&\limsup_{\varepsilon\to 0}\int_\Omega \frac{\varrho}{2}|\dot u_\varepsilon(t)|^2+\frac{d}{2}K_{\rm E}|{\rm sph}\, e(u_\varepsilon(t))|^2\,{\rm d}x\\
&\le \limsup_{\varepsilon\to 0}\Big(\int_\Omega \frac{\varrho}{2}|\dot u_\varepsilon(t)|^2+\frac{d}{2}K_{\rm E}|{\rm sph}\, e(u_\varepsilon(t))|^2\,{\rm d}x\\
&\qquad\qquad +\varepsilon\int_0^t\!\!\int_\Omega dK_{\rm V}|{\rm sph}\, e(\dot u_\varepsilon)|^2+2G_{\rm V}|{\rm dev}\, e(\dot u_\varepsilon)|^2\,{\rm d}x{\rm d}t\Big)\\
&=\lim_{\varepsilon\to 0}\int_0^t\!\!\int_\Omega f\cdot\dot u_\varepsilon\,{\rm d}x{\rm d}t+\int_\Omega\frac{\varrho}{2}|v_0|^2+\frac{d}{2}K_{\rm E}|{\rm sph}\, e(u_0)|^2\,{\rm d}x\\
&=\int_0^t\!\!\int_\Omega f\cdot\dot u\,{\rm d}x{\rm d}t+\int_\Omega\frac{\varrho}{2}|v_0|^2+\frac{d}{2}K_{\rm E}|{\rm sph}\, e(u_0)|^2\,{\rm d}x\\
&\qquad\qquad\qquad =\int_\Omega\frac{\varrho}{2}|\dot u(t)|^2+\frac{d}{2}K_{\rm E}|{\rm sph}\, e(u(t))|^2\,{\rm d}x,
\end{aligned}
$$

where the last equality is the proved energy conservation on the time interval $[0,t]$. Since also $\dot u_\varepsilon(t)\to\dot u(t)$ and ${\rm sph}\, e(u_\varepsilon(t))\to{\rm sph}\, e(u(t))$ $L^2(\Omega)$-weakly and the opposite inequality holds always by the weak lower semicontinuity, the strong convergence (6.7.16) follows by the uniform convexity of the L^2-space, cf. Theorem A.3.8. Putting (6.7.16) again to the above calculations for $t=T$,

$$
\begin{aligned}
&\limsup_{\varepsilon\to 0}\varepsilon\int_Q dK_{\rm V}|{\rm sph}\, e(\dot u_\varepsilon)|^2+2G_{\rm V}|{\rm dev}\, e(\dot u_\varepsilon)|^2\,{\rm d}x{\rm d}t\\
&=\limsup_{\varepsilon\to 0}\Big(\int_\Omega\frac{\varrho}{2}|\dot u_\varepsilon(T)|^2+\frac{d}{2}K_{\rm E}|{\rm sph}\, e(u_\varepsilon(T))|^2\,{\rm d}x\\
&\qquad\qquad +\varepsilon\int_Q dK_{\rm V}|{\rm sph}\, e(\dot u_\varepsilon)|^2+2G_{\rm V}|{\rm dev}\, e(\dot u_\varepsilon)|^2\,{\rm d}x{\rm d}t\Big)\\
&\quad-\lim_{\varepsilon\to 0}\int_\Omega\frac{\varrho}{2}|\dot u_\varepsilon(T)|^2+\frac{d}{2}K_{\rm E}|{\rm sph}\, e(u_\varepsilon(T))|^2\,{\rm d}x=\lim_{\varepsilon\to 0}\int_Q f\cdot\dot u_\varepsilon\,{\rm d}x{\rm d}t\\
&\quad+\int_\Omega\frac{\varrho}{2}|v_0|^2+\frac{d}{2}K_{\rm E}|{\rm sph}\, e(u_0)|^2-\frac{\varrho}{2}|\dot u_\varepsilon(T)|^2-\frac{d}{2}K_{\rm E}|{\rm sph}\, e(u_\varepsilon(T))|^2\,{\rm d}x=0.
\end{aligned}
$$

Thus (6.7.17) follows. □

This approach can be combined with various inelastic processes, undergoing typically rather in the solid part. For example, one can consider damageable $G_{\rm E}=G_{\rm E}(\alpha)$ and $G_{\rm V}=G_{\rm V}(\alpha)$ like in Sect. 7.5.2 use Proposition 7.5.10. Considering $\Gamma_{\rm D}=\emptyset$ for simplicity, the initial-boundary-value problem (7.7.37) then augments as:

$$\varrho\ddot u-{\rm div}(\sigma_{\rm sph}+\sigma_{\rm dev})=f \qquad \text{in } Q, \tag{7.7.44a}$$

$$\text{with }\ \sigma_{\rm sph}=dK_{\rm V}{\rm sph}\, e(\dot u)+dK_{\rm E}{\rm sph}\, e(u) \qquad \text{in } Q, \tag{7.7.44b}$$

$$\text{and }\ \sigma_{\rm dev}=\begin{cases}2G_{\rm V}(\alpha){\rm dev}\, e(\dot u)+2G_{\rm E}(\alpha){\rm dev}\, e(u) & \text{in } Q_{\rm S},\\ 2G_{\rm V}{\rm sph}\, e(\dot u) & \text{in } Q_{\rm F},\end{cases} \tag{7.7.44c}$$

$$\partial\zeta_{\rm IN}(\dot\alpha)+G_{\rm E}'(\alpha)|{\rm dev}\, e(u)|^2\ni\varphi_{\rm DAM}'(\alpha)+\kappa\Delta\alpha \qquad \text{in } Q_{\rm S}, \tag{7.7.44d}$$

$$(\sigma_{\rm sph}+\sigma_{\rm dev})\vec n=0 \qquad \text{on } \Sigma, \tag{7.7.44e}$$

$$\nabla\alpha\cdot\vec{n}=0 \qquad \text{on } I\times\partial\Omega_{\rm S}, \tag{7.7.44f}$$

$$u|_{t=0}=u_0,\quad \dot{u}|_{t=0}=v_0,\quad \alpha|_{t=0}=\alpha_0, \qquad \text{on } \Omega, \tag{7.7.44g}$$

Let us note that damage becomes naturally irrelevant in the fluidic region $\Omega_{\rm F}$. The schematic rheology behind this model is depicted in Figure 7.13.

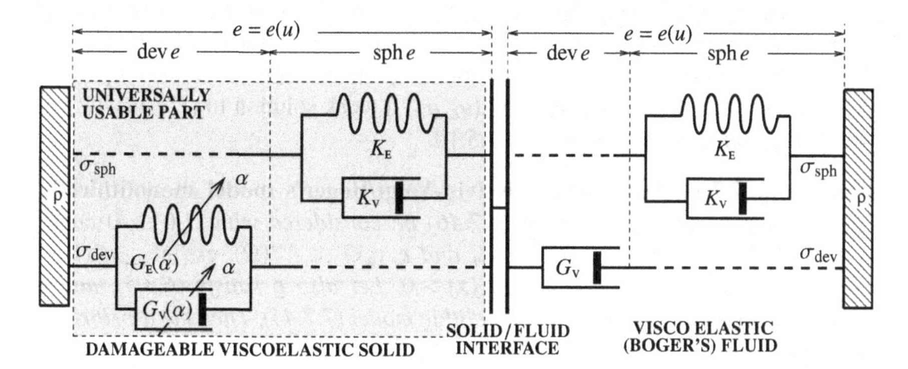

Fig. 7.13 *Schematic rheological diagram for solid-fluid interaction. The solid has a viscoelastic Kelvin-Voigt rheology subjected to damage α in the deviatoric part while the spherical (volumetric) part is not subjected to damage. The fluid is mere Stokes model in the deviatoric part while viscoelastic in spherical part, i.e. undamageable Boger's fluid. The left-hand part (in gray) can serve as an approximation for the fluid model under the scaling (7.7.46) in a monolithic model.*

We can again apply the monolithic approach, i.e. use the model as on $\Omega_{\rm S}$ but on the whole domain Ω and obtain (7.7.44) only as a limit by suitable scaling. More specifically,

$$\varrho\ddot{u}-{\rm div}(\mathbb{D}(\alpha)e(\dot{u})+\mathbb{C}_\varepsilon(\alpha)e(u))=f \qquad \text{in } Q, \tag{7.7.45a}$$

$$\partial\zeta_{{\rm in},\varepsilon}(\dot{\alpha})+\frac{1}{2}\mathbb{C}_\varepsilon'(\alpha)e(u){:}e(u)\ni\varphi_{\rm DAM}'(\alpha)+{\rm div}(\kappa_\varepsilon\nabla\alpha) \qquad \text{in } Q, \tag{7.7.45b}$$

$$(\mathbb{D}_\varepsilon(\alpha)e(\dot{u})+\mathbb{C}_\varepsilon(\alpha)e(u))\vec{n}=0, \quad \text{and} \quad \nabla\alpha\cdot\vec{n}=0 \qquad \text{on } \Sigma, \tag{7.7.45c}$$

$$u|_{t=0}=u_0,\quad \dot{u}|_{t=0}=v_0,\quad \alpha|_{t=0}=\alpha_0, \qquad \text{on } \Omega, \tag{7.7.45d}$$

with $\mathbb{D}(\cdot)$, $\mathbb{C}_\varepsilon(\cdot)$, κ_ε, and $\zeta_{{\rm in},\varepsilon}(\cdot)$ depending also on $x\in\Omega$, namely

$$\mathbb{D}(x,\alpha):=\begin{cases}{\rm ISO}(\tau_{\rm R}K_{\rm E},G_{\rm V}+\tau_{\rm R}G_{\rm E}(\alpha)) & \text{for } x\in\Omega_{\rm S},\\ {\rm ISO}(\tau_{\rm R}K_{\rm E},\tau_{\rm R}G_{\rm E}(1)) & \text{for } x\in\Omega_{\rm F}\end{cases} \tag{7.7.46a}$$

$$\mathbb{C}_\varepsilon(x,\alpha):=\begin{cases}{\rm ISO}(K_{\rm E},G_{\rm E}(\alpha)) & \text{for } x\in\Omega_{\rm S},\\ {\rm ISO}(K_{\rm E},\varepsilon G_{\rm E}(1)) & \text{for } x\in\Omega_{\rm F}\end{cases} \tag{7.7.46b}$$

$$\kappa_\varepsilon(x):=\begin{cases}\kappa,\\ \varepsilon\kappa,\end{cases} \qquad \zeta_{{\rm in},\varepsilon}(x,\dot{\alpha}):=\begin{cases}\zeta_{\rm IN}(\dot{\alpha}) & \text{for } x\in\Omega_{\rm S},\\ \varepsilon\zeta_{\rm IN}(\dot{\alpha}) & \text{for } x\in\Omega_{\rm F}\end{cases} \qquad \varepsilon>0. \tag{7.7.46c}$$

with some (presumably small) relaxation time $\tau_R > 0$. Let us note that the viscosity in not scaled by ε but is not damageable in the fluidic part so that, for $\varepsilon \to 0$, we expectedly come to the Boger viscoelastic fluid in Ω_F, as in (7.7.44a-c). The moduli K_E and G_E themselves can be x-dependent, so that they may differ on Ω_S and Ω_F, and the mass density ϱ may vary from Ω_S to Ω_F when (6.3.5) is considered only on each domain separately. Also κ and τ_R may be x-dependent.

Then we can use the technique of Proposition 7.5.10 modified here, counting with the nondamageable spherical part (which is simpler) and with vanishing but non-damageable elastic response on fluid domain (which modifies the technique from Proposition 7.5.10). Let us denote by $(u_\varepsilon, \alpha_\varepsilon)$ a weak solution to (7.7.44)–(7.7.46) which does exist just by Proposition 7.5.10.

Proposition 7.7.14 (Damageable Kelvin-Voigt/Boger's model monolithically). *Let (7.5.6a,e-g) hold, the scaling (7.7.46) be considered with $G_E(\cdot) \geq 0$ continuous and $G_V > 0$, let $\varphi'_{DAM} = 0$ on Ω_F, and $\kappa, \tau_R G_V, \in L^\infty(\Omega)$, $\operatorname{ess\,inf}_{x\in\Omega}\kappa(x) > 0$, $\operatorname{ess\,inf}_{x\in\Omega}\tau_R(x) > 0$, and $\operatorname{ess\,inf}_{x\in\Omega}G_V(x) > 0$. Let also ϱ satisfy (6.3.5), and let $(u_\varepsilon, \alpha_\varepsilon)$ be a weak solution to the monolithic model (7.7.44). Then the a-priori estimates hold:*

$$\|u_\varepsilon\|_{H^1(I;H^1(\Omega;\mathbb{R}^d))\cap W^{1,\infty}(I;L^2(\Omega;\mathbb{R}^d))} \leq C, \tag{7.7.47a}$$

$$\|\alpha_\varepsilon\|_{H^1(I;L^2(\Omega_S))\cap L^\infty(I;H^1(\Omega_S))\cap L^\infty(Q)} \leq C, \tag{7.7.47b}$$

$$\|\dot\alpha_\varepsilon\|_{L^2(Q_F)} \leq C/\sqrt{\varepsilon} \quad \text{and} \quad \|\nabla\alpha_\varepsilon\|_{L^\infty(I;L^2(\Omega_F;\mathbb{R}^3))} \leq C/\sqrt{\varepsilon}. \tag{7.7.47c}$$

Moreover, for selected subsequences with $\varepsilon \to 0$, we have

$$u_\varepsilon \to u \;\; \text{strongly in } H^1(I;H^1(\Omega;\mathbb{R}^d)) \cap W^{1,\infty}(I;L^2(\Omega;\mathbb{R}^d)), \tag{7.7.48a}$$

$$\alpha_\varepsilon|_{Q_S} \to \alpha \;\; \text{weakly* in } H^1(I;L^2(\Omega_S)) \cap L^\infty(I;H^1(\Omega_S)) \cap L^\infty(Q_S), \tag{7.7.48b}$$

and every such a limit (u,α) is a weak solution to the initial-boundary-value problem (7.7.45).

Proof. Similarly (but not exactly equally) as in (7.5.32), we can write the ε-model in the fluidic domain Ω_F as

$$\frac{\varrho}{\tau_R}\dot w_\varepsilon - \operatorname{div}(\mathbb{D}e(\dot u_\varepsilon) + \mathbb{C}_\varepsilon e(u_\varepsilon)) = f + \frac{\varrho}{\tau_R}\dot u_\varepsilon \qquad \text{on } Q_F \tag{7.7.49}$$

with $w_\varepsilon := u_\varepsilon + \tau_R \dot u_\varepsilon$ and with $\mathbb{D} := \operatorname{ISO}(\tau_R K_E, \tau_R G_E(1))$ and $\mathbb{C}_\varepsilon := \operatorname{ISO}(K_E, \varepsilon G_E(1))$ as in (7.7.46a,b) on Ω_F. It is important that this form complies with the test by w_ε for the a-priori estimates and by $w_\varepsilon - w$ for proving the strong convergence, similarly as we used for (7.5.33) and as we will need here for the damageable part of the model on Ω_S. Yet, the difference from (7.5.33) is that $\mathbb{D} \neq \tau_R\mathbb{C}$ here except an uninteresting case $\varepsilon = 1$.

More in detail, we first derive the a-priori estimates by testing (7.5.32) written with ε instead of k and (7.7.49) by w_ε. After analogous manipulation as in the proof of Proposition 7.5.10, we obtain the a-priori estimates (7.7.47).

Second, in view of (7.7.47a,b), we select a weakly* converging subsequence of $\{(u_\varepsilon,\alpha_\varepsilon)\}_{\varepsilon>0}$, denoting its weak* limit by (u,α).[72] Testing (7.7.49), by $w_\varepsilon - w$ we obtain an analog of (7.5.33):

$$\begin{aligned}&\int_{\Omega_{\rm F}} \frac{\varrho}{2\tau_{\rm R}}\big|w_\varepsilon(t)-w(t)\big|^2 + \frac12(\mathbb{D}+\tau_{\rm R}\mathbb{C}_\varepsilon)e(u_\varepsilon(t)-u(t)):e(u_\varepsilon(t)-u(t))\,{\rm d}x\\&\quad+\int_0^t\!\!\int_{\Omega_{\rm F}} \tau_{\rm R}\mathbb{D}e(\dot u_\varepsilon-\dot u):e(\dot u_\varepsilon-\dot u)+\mathbb{C}_\varepsilon e(u_\varepsilon-u):e(u_\varepsilon-u)\,{\rm d}x{\rm d}t\\&=\int_0^t\!\!\int_{\Omega_{\rm F}} \frac{\varrho}{\tau_{\rm R}}(\dot u_\varepsilon-\dot w)\cdot(w_\varepsilon-w)-(\mathbb{D}e(\dot u)+\mathbb{C}_\varepsilon e(u)):e(w_\varepsilon-w)\,{\rm d}x{\rm d}t\\&\quad+\int_0^t\!\!\int_{\Sigma_{\rm F}} g\cdot(w_\varepsilon-w)\,{\rm d}S\,{\rm d}t+\int_0^t\Big\langle(\mathbb{D}e(\dot u_\varepsilon)+\mathbb{C}_\varepsilon e(u_\varepsilon))\vec n_{\rm FS},w_\varepsilon-w\Big\rangle_{\rm SF}{\rm d}t,\end{aligned}\tag{7.7.50}$$

where $\langle\cdot,\cdot\rangle_{\rm SF}$ denotes the duality pairing between $H^{-1/2}(\bar\Omega_{\rm S}\cap\bar\Omega_{\rm L};\mathbb{R}^d)$ where tractions live and $H^{1/2}(\bar\Omega_{\rm S}\cap\bar\Omega_{\rm L};\mathbb{R}^d)$, and where $\vec n_{\rm FS}$ denotes the unit normal to the solid-fluid interface $\bar\Omega_{\rm S}\cap\bar\Omega_{\rm L}$ oriented from the fluid to the solid domain.

We now sum it with (7.5.33) written on $\Omega_{\rm S}$ with ε instead on Ω with k and with the term $\int_0^t\langle(\mathbb{D}_0e(\dot u_\varepsilon)+\mathbb{C}(\alpha_k)e(w_\varepsilon))\vec n_{\rm SF},w_\varepsilon-w\rangle_{\rm SF}{\rm d}t$ with $\vec n_{\rm SF}:=-\vec n_{\rm FS}$ and with $\mathbb{D}_0={\rm ISO}(0,G_{\rm v})$, realizing that this term cancels with the last term in (7.7.50); here we used that the tractions are equilibrated on the solid-fluid interface $\bar\Omega_{\rm S}\cap\bar\Omega_{\rm L}$. More rigorously, one should simply test (7.5.32) written on $\Omega_{\rm S}$ with ε instead of k and (7.7.49) together so that the terms on the interface $\bar\Omega_{\rm S}\cap\bar\Omega_{\rm L}$ which themselves do not converge to 0 simply do not appear. Then, using the arguments from the proof of Prop. 7.5.10, we eventually come to (7.7.48) on the whole domain Ω.

Now, the limit passage in the damage flow rule can be performed. We consider the weak formulation (7.5.5) for $r_{\rm c}=0$ and $p=2$ and realize, that $\varphi'_{\rm DAM}=0$ on $\Omega_{\rm F}$ and that $\partial_\alpha\varphi=0$ on $Q_{\rm F}$ so that we substitute for $\partial_\alpha\varphi(e(u),\alpha)\dot\alpha$ only on $Q_{\rm S}$, arriving thus to the variational inequality combining (7.5.5) on $\Omega_{\rm S}$ with (7.5.5) on $\Omega_{\rm F}$:

$$\begin{aligned}&\int_{Q_{\rm S}} G'_{\rm E}(\alpha_\varepsilon)|{\rm dev}\,e(u_\varepsilon)|^2w-\varphi'_{\rm DAM}(\alpha_\varepsilon)(w-\dot\alpha_\varepsilon)+\kappa\nabla\alpha_\varepsilon\cdot\nabla w+\zeta_{\rm IN}(w)\,{\rm d}x{\rm d}t\\&\quad+\varepsilon\int_{Q_{\rm F}}\kappa\nabla\alpha_\varepsilon\cdot\nabla w+\zeta_{\rm IN}(w)\,{\rm d}x{\rm d}t\ge\int_{Q_{\rm S}}\zeta_{\rm IN}(\dot\alpha_\varepsilon)+\mathbb{C}(\alpha_\varepsilon)e(u_\varepsilon):e(\dot u_\varepsilon)\,{\rm d}x{\rm d}t\\&\quad+\int_{\Omega_{\rm S}}\Big(\frac12\mathbb{C}(\alpha_\varepsilon(T))e(u_\varepsilon(T)):e(u_\varepsilon(T))+\frac\kappa2|\nabla\alpha_\varepsilon(T)|^2-\frac12\mathbb{C}(\alpha_0)e(u_0):e(u_0)\\&\qquad-\frac\kappa2|\nabla\alpha_0|^2\Big){\rm d}x+\varepsilon\int_{Q_{\rm F}}\zeta_{\rm IN}(\dot\alpha_\varepsilon)+\varepsilon\int_{\Omega_{\rm F}}\frac\kappa2|\nabla\alpha_\varepsilon(T)|^2-\frac\kappa2|\nabla\alpha_0|^2\,{\rm d}x\end{aligned}$$

The last two integrals are nonnegative and can be omitted without corrupting the inequality. Then we can pass to the limit with $\varepsilon\to0$, using also the estimates in (7.7.48b) so that

[72] We can make the weak limit passage in the semilinear equation (7.7.49) even by this weak* convergence, but more simply we can first improve it for the strong convergence.

$$\left|\varepsilon\int_{Q_F}\kappa\nabla\alpha_\varepsilon\cdot\nabla w+\zeta_{\rm IN}(w)\,{\rm d}x{\rm d}t\right|\le\varepsilon\|\kappa\|_{L^\infty(\Omega)}\|\nabla\alpha_\varepsilon\|_{L^2(Q;\mathbb{R}^d)}\|\nabla w\|_{L^2(Q;\mathbb{R}^d)}$$
$$+\varepsilon\|\zeta_{\rm IN}(w)\|_{L^1(Q)}=\mathscr{O}(\sqrt{\varepsilon})\to 0. \qquad (7.7.51)$$

Thus we obtain the variational inequality

$$\int_{Q_S}G'_{\rm E}(\alpha)|{\rm dev}\,e(u)|^2w-\varphi'_{\rm DAM}(\alpha)(w{-}\dot\alpha)+\kappa\nabla\alpha\cdot\nabla w+\zeta_{\rm IN}(w)\,{\rm d}x{\rm d}t$$
$$\ge\int_{\Omega_S}\frac12\mathbb{C}(\alpha(T))e(u(T)){:}e(u(T))+\frac{\kappa}{2}|\nabla\alpha(T)|^2-\frac12\mathbb{C}(\alpha_0)e(u_0){:}e(u_0)-\frac{\kappa}{2}|\nabla\alpha_0|^2\,{\rm d}x$$
$$+\int_{Q_S}\zeta_{\rm IN}(\dot\alpha)+\mathbb{C}(\alpha)e(u){:}e(\dot u)\,{\rm d}x{\rm d}t,$$

which is just the weak formulation of the initial-boundary value problem (7.7.44d,f,g) in the form (7.5.5). □

Let us illustrate a usage of the monolithic approach to a solid-fluid interaction problem motivated by geophysical applications, namely modeling of a rupture of a lithospheric fault, emitting seismic waves, propagation of these waves through a layered solid/fluid elastic continuum, and tectonic earthquakes. The geometry, boundary conditions, and partly also damage initial condition are depicted in Figure 7.14. The numerical implementation is made by the Crank-Nicholson formula

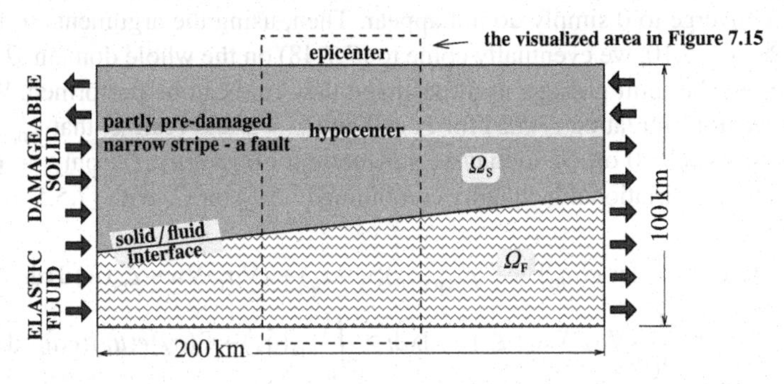

Fig. 7.14 A computational 2-dimensional domain for the experiment in Figure 7.15. Upper trapezoidal part is the solid from Figure 7.13-left while the lower part is the fluid from Figure 7.13-right. The upper solid part is shifted by the Dirichlet boundary conditions to the left while the lower solid/fluid part to the right. At time $t = 0$, the is "compact" only at the middle part (where the hypocenter of an earthquake will be) while the rest of the fault (i.e. both sides) is (partly) damaged.

for the 1st-order system (as in Exercise 6.3.11) but combined with the splitting in the sense of the fractional-step time discretization as in Remark 7.5.14, thus energy is conserved also in the discrete scheme. As for the space discretization, P1-finite elements have been used; cf. Fig. 3.6-upper/left on p. 82. In this way, an efficient numerical scheme leading to recursive linear-quadratic programming which conserves energy, is numerical stable, and convergent is obtained.[73] The results of an illustrative two-dimensional simulations are depicted in five selected snapshots in Figure 7.15. The loading by Robin boundary conditions is depicted on Figure 7.14, while the initial condition makes the fault partly damaged except its middle region where the rupture is to be simulated.[74]

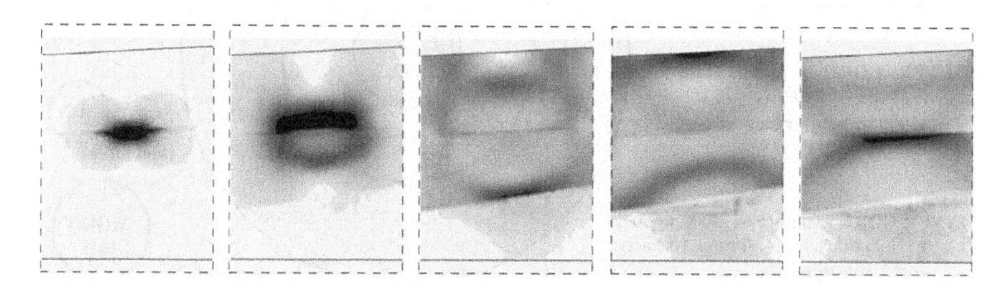

Fig. 7.15 *Simulations of a rupture of a fault in the hypocenter, cf. Figure 7.14, visualized via the space distribution of the kinetic energy (on a deformed domain with displacement magnified 100×) depicted at five selected snapshots: emission of a seismic (mainly S) wave in the hypocenter, its propagation and reflection on the solid/fluid interface, partly transforming into also into a weak P-wave, and eventually S-wave reaching also the surface (in an epicenter of the earthquake) and propagating to the sides.*

Courtesy of Roman Vodička (Technical University Košice, Slovakia)

Remark 7.7.15 (Global geophysical model). Prominent applications are in geophysics where it allows for coupling of global seismic-wave propagation models in a layered globe, cf. the schematic Figure 7.16, with seismic source models in the crust or the upper part of the Earth mantle where tectonic earthquakes on pre-existing lithospheric faults are the typical source generating seismic waves (besides volcanoes or various man-made sources). Thus, in addition, various inelastic processes can be undergone in the solid part, the above considered damage or phase-field fracture being only a minimal scenario. When ignoring a lot of internal variables[75] the above suggested model (7.7.44) combining damageable Kelvin-Voigt solid with Boger viscoelastic fluid is, of course, only a very simplified scenario but still relevant for short time-scales within undergoing rupture and subsequent seismicity. It

[73] Some shortcuts have been afforded for the computational simulations in Figure 7.15, in particular no viscosity has been implemented. Cf. [469] for more details.

[74] Other inelastic or viscous processes as plasticity/creep or swelling within a diffusant transport from Sections 7.4 and 7.6 can be combined similarly, remaining on effect only in the solid part $\Omega_{\rm S}$ after the limit towards the fluid on $\Omega_{\rm F}$.Even heat generation and transport as in Chapter 8 below on the whole domain Ω like in [461] can be relatively routinely considered because the problem still stays parabolic even in the Boger-fluid limit. A further limit passage with $K_{\rm v} \to 0$ and $G_{\rm v} \to 0$ on $\Omega_{\rm F}$ leads to the purely elastic fluid as in Proposition 7.7.13 interacting now with a nonlinear solid. However, this couples linear hyperbolic problem on $\Omega_{\rm F}$ with a parabolic but nonlinear problem on $\Omega_{\rm S}$, and therefore some difficulties are expectable and more involved techniques to be used; indeed, regularity estimates for a bit modified model with bounded elastic stress have used in [463].

[75] Often, e.g. porosity, water content, or breakage are considered in geophysical models of the crust, cf. e.g. [324, 325], together with the temperature in the fully thermodynamical context.

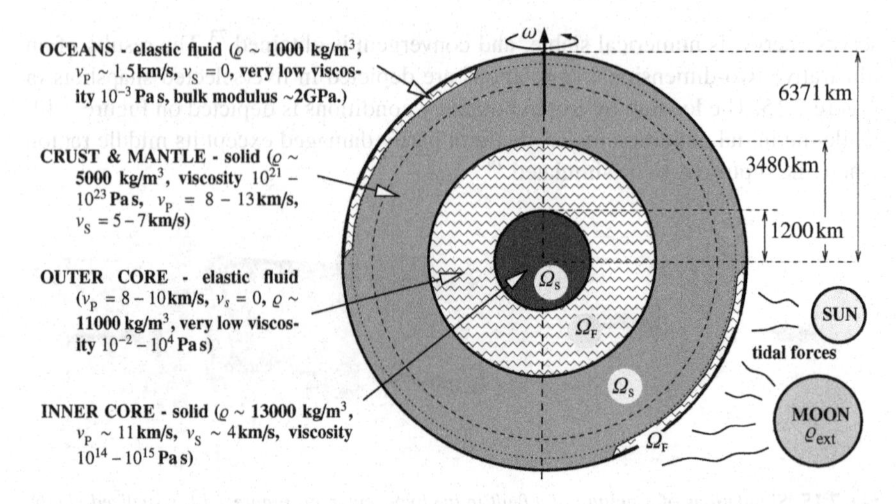

Fig. 7.16 The very basic layered structure of our planet Earth with some (only very round) geometrical and rheological data and the notation for the solid-type domains Ω_S and the fluid-type domains Ω_F. Here, v_P and v_S stand for the velocity of the P- and S-waves, respectively, $v_S = 0$ indicating that only the P- but not S-waves can propagate through the particular layer.

can be also coupled with self-gravitation (which is relevant effect in ultra-low frequency seismic waves), considering then (7.7.44) with f from (6.3.17) on p. 207 the *centrifugal* and the *Coriolis forces,* and with the equation for the gravitational field on the whole universe $\mathbb{R}^d$:

$$\operatorname{div}\Big(\frac{1}{\mathfrak{g}}\nabla\phi + \varrho u\Big) = \varrho + \varrho_{\rm ext}(t) \qquad \text{on } I\times\mathbb{R}^d, \tag{7.7.52}$$

with suitable "boundary" conditions for ϕ, typically $\phi(\infty) = 0$. The external time-varying mass density $\varrho_{\rm ext} = \varrho_{\rm ext}(t, x)$ allows for involvement of *tidal forces* arising from other bodies (stars, planets, moons) moving around.[76] The viscosity is very low.[77] It has therefore a good sense to investigate the asymptotics when this viscosity vanishes. For this, the technique from Proposition 7.7.13 is to be used. Let us note that the Coriolis force does not corrupt the a-priori estimates because it has no power while the centrifugal force can be handled by Gronwall's inequality, exploiting the inertia on the left hand side; here we can (although do not need to) use the fact that the *centrifugal force* has a quadratic *potential* $u \mapsto \frac{1}{2}\varrho(|\omega\cdot(x+u)|^2 - |\omega|^2|x+u|^2)$.

[76] The existence of weak solutions is then proved by merging Proposition 7.7.11 with the technique for damage in Section 7.5.1 with Exercise 6.3.13.

[77] Rather the Maxwellian viscosity is considered with moduli of the order $10^{21} - 10^{24}$ Pa s in the crust and mantle, and $10^{14} - 10^{15}$ Pa s, occasionally being combined with the Kelvin-Voigt rheology, leading thus to the Jeffreys model, cf. [325].

Chapter 8
Thermodynamics of selected materials and processes

> Heat ... may pass out of one body into another just as water may be poured from one vessel into another... We have therefore a right to speak of heat as of a measurable quantity, and to treat it mathematically like other measurable quantities so long as it continues to exist as heat.
>
> J. CLERK MAXWELL (1831-1879)

> It is often claimed that while mechanics is a mathematical theory, thermodynamics is a physical one. ... this distinction ... is historical only, in no way intrinsic to the regions of experience to which the two names refer.
>
> CLIFFORD A. TRUESDELL (1919-2000)

It is a basic principle (in fact, assumed rather by definition) that the total energy in a closed system conserves. Dissipation of mechanical energy with which we dealt in previous Chapters 6–7 then gives rise, beside possible irreversible changes of internal structure of materials contributing to the stored energy, also to a heat (and entropy) production. Except infinitesimally slow processes or very small bodies which are thermally stabilized with the environment, this dissipation of mechanical (or other, e.g. chemical) energy results to varying temperature and the heat transfer throughout the solid-body continuum. In turn, mechanical properties usually depend (sometimes even very substantially) on temperature both as well as stored and dissipation mechanisms concerns.

This gives rise to thermally coupled systems, requiring nontrivial mathematical techniques and a rigorous thermodynamical background. In particular, the fundamental *three Laws of Thermodynamics* should ultimately be respected by the non-isothermal models:

1st: total-*energy conservation*,

2nd: nondecreasing entropy in closed systems (the *Clausius-Duhem inequality*),

3rd: *positivity of temperature* (the *Nernst law*).

A lot of such coupled problems and models still wait for a suitable formulation

M. Kružík and T. Roubíček, *Mathematical Methods in Continuum Mechanics of Solids*, Interaction of Mechanics and Mathematics,
https://doi.org/10.1007/978-3-030-02065-1_8

amenable for rigorous mathematical analysis and solution. Here we confine ourselves rather to some examples.

The mathematical analysis of such coupled systems ultimately requires modern L^1-theory of the nonlinear heat equation, which was developed only relatively recently at the end of twentieth century [72, 73].

8.1 Thermodynamics of generalized standard materials

A general thermodynamic scheme applicable to generalized standard materials relies (instead of a stored energy) on the specific Helmholtz *free energy* and the *potential of dissipative forces*

$$\psi = \psi(e,z,\theta) \quad \text{and} \quad \zeta = \zeta(\dot{e},\dot{z}) \tag{8.1.1}$$

depending now also on temperature θ. From it, we define the elastic (=nondissipative) stress $\sigma_{\rm EL}$, driving force $\sigma_{\rm IN}$ for the evolution of z, and the so-called specific *entropy* respectively by

$$\sigma_{\rm EL} = \partial_e\psi, \qquad \sigma_{\rm IN} = \partial_z\psi, \qquad \mathfrak{s} = -\partial_\theta\psi. \tag{8.1.2}$$

The last equality is sometimes referred to as a *Gibbs' relation*. Further, we define the specific *internal energy* w by

$$w(e,z,\theta,\mathfrak{s}) := \psi(e,z,\theta) + \theta\mathfrak{s}. \tag{8.1.3}$$

Differentiating (8.1.3) in time and using the Gibb's relation in (8.1.2) gives

$$\dot{w} = \frac{\partial}{\partial t}\underbrace{\big(\psi(e,z,\theta)+\theta\mathfrak{s}\big)}_{\text{internal enery}} = \underbrace{\partial_e\psi{:}\dot{e} + \partial_z\psi{\cdot}\dot{z}}_{\substack{\text{mechanical}\\ \text{energy rate}}} + \underbrace{\partial_\theta\psi\dot{\theta} + \dot{\theta}\mathfrak{s}}_{\substack{=0 \text{ by}\\ (8.1.2)}} + \underbrace{\theta\dot{\mathfrak{s}}}_{\substack{\text{thermal ener-}\\ \text{gy rate}}}\,. \tag{8.1.4}$$

The further ingredient of the model is a *balance of the internal energy* which is postulated as[1]

$$\dot{w} = \sigma_{\rm EL}{:}\dot{e} + \sigma_{\rm IN}{\cdot}\dot{z} + \xi(\dot{e},\dot{z}) - {\rm div}\, j, \tag{8.1.5}$$

with the heat flux $j = j(\theta,e,z,\nabla\theta)$ and the rate of the heat production due to *dissipation rate* of the mechanical energy

[1] Cf. e.g. [201, Formula (3.11)].

$$\xi = \xi(\dot{e},\dot{z}) = \langle \zeta'(\dot{e},\dot{z}), (\dot{e},\dot{z})\rangle = \partial_{\dot{e}}\zeta(\dot{e},\dot{z}):\dot{e} + \partial_{\dot{z}}\zeta(\dot{e},\dot{z})\cdot\dot{z} \ge 0, \tag{8.1.6}$$

cf. also (7.1.6). Essentially, (8.1.5) is dictated by minimal entropy production and in particular say that the mere heat transfer does not contribute to the increase of internal energy, cf. Remark 8.1.5. Then the *balance of the thermal* part of the *internal energy* (i.e. the last term in (8.1.4), without counting the mechanical energy) is obtained by merging (8.1.4) with (8.1.5) and using (8.1.2), which gives the equation

$$\theta\dot{\mathfrak{s}} + \operatorname{div} j = \xi = \xi(\dot{e},\dot{z}). \tag{8.1.7}$$

Sometimes, (8.1.7) is called an *entropy equation*. The important fact is that the above procedure satisfies the 2nd thermodynamical law provided

$$j = j(\theta,e,z,\nabla\theta) := -\mathbb{K}(e,z,\theta)\nabla\theta \tag{8.1.8}$$

with the *heat-conductivity tensor* $\mathbb{K} = \mathbb{K}(e,z,\theta)$ positive definite, cf. also Remarks 8.1.3 and 8.1.5. This is the so-called *Fourier law* in the nonlinear anisotropic medium). Indeed, assuming $\theta > 0$ (cf. Remark 8.3.4 below) and dividing (8.1.7) by θ, we obtain:

$$\begin{aligned}\frac{\mathrm{d}}{\mathrm{d}t}\int_\Omega \mathfrak{s}(t,x)\,\mathrm{d}x &= \int_\Omega \frac{\xi + \operatorname{div}(\mathbb{K}\nabla\theta)}{\theta}\,\mathrm{d}x\\ &= \int_\Omega \frac{\xi}{\theta} - \mathbb{K}\nabla\theta\cdot\nabla\frac{1}{\theta}\,\mathrm{d}x + \int_\Gamma \frac{\mathbb{K}\nabla\theta}{\theta}\cdot\vec{n}\,\mathrm{d}S\\ &= \underbrace{\int_\Omega \frac{\xi}{\theta} + \frac{\mathbb{K}\nabla\theta\cdot\nabla\theta}{\theta^2}\,\mathrm{d}x}_{\text{entropy-production rate}} + \underbrace{\int_\Gamma \frac{-j}{\theta}\cdot\vec{n}\,\mathrm{d}S}_{\text{entropy flux through boundary}}.\end{aligned} \tag{8.1.9}$$

If the system is thermally isolated, i.e. $j = 0$ on Γ, the last integral vanishes and we obtain

$$\frac{\mathrm{d}}{\mathrm{d}t}\int_\Omega \mathfrak{s}(t,x)\,\mathrm{d}x \ge 0 \tag{8.1.10}$$

provided $\mathbb{K}$ is positive semidefinite and the dissipation rate ξ is non-negative, which are natural assumptions, cf. also Remark 8.1.5 below. This the so-called *Clausius-Duhem inequality*, saying that the total entropy $t \mapsto \int_\Omega \mathfrak{s}(t,x)\,\mathrm{d}x$ in isolated systems is nondecreasing, which articulates the *2nd law of thermodynamics*.

The Kelvin-Voigt type rheology (7.1.1) is to be modified as

$$\sigma = \sigma_{\mathrm{vi}} + \sigma_{\mathrm{EL}} = \partial_{\dot{e}}\zeta(\dot{e},\dot{z}) + \partial_e\phi(e,z,\theta). \tag{8.1.11}$$

Testing the equilibrium equation $\varrho\ddot{u} = \operatorname{div}\sigma + f$ (cf. (6.1.1)) by $\dot{u}$ and using the boundary conditions $\sigma\vec{n} = g$ gives

$$\int_\Omega \frac{\varrho}{2}|\dot{u}|^2 + \sigma{:}e(\dot{u})\,\mathrm{d}x = \int_\Omega f\cdot\dot{u}\,\mathrm{d}x + \int_\Gamma g\cdot\dot{u}\,\mathrm{d}S + \langle \vec{\tau}_\mathrm{D}, \dot{u}_\mathrm{D}\rangle_{\Gamma_\mathrm{D}}, \tag{8.1.12}$$

with the traction $\vec{\tau}_\mathrm{D}$ from (6.2.10), cf. the basic energy balance (6.2.9). Integrating (8.1.5) over Ω, using $\partial_z\psi(e,z,\theta) = -\partial_{\dot{z}}\zeta(\dot{e},\dot{z})$, cf. (7.1.1b), and (8.1.12), and also (8.1.6), we obtain

$$\underbrace{\frac{\mathrm{d}}{\mathrm{d}t}\int_\Omega \frac{\varrho}{2}|\dot{u}|^2 + w\,\mathrm{d}x}_{\text{total energy}} = \int_\Omega \frac{\varrho}{2}\frac{\partial}{\partial t}|\dot{u}|^2 + \underbrace{\sigma{:}\dot{e} - \partial_{\dot{e}}\zeta(\dot{e},\dot{z}){:}\dot{e} - \sigma_\mathrm{EL}{:}\dot{e}}_{=0\text{ due to }\sigma=\partial_{\dot{e}}\zeta(\dot{e},\dot{z})+\sigma_\mathrm{EL}}$$
$$+\underbrace{\sigma_\mathrm{EL}{:}\dot{e} - \partial_{\dot{z}}\zeta(\dot{e},\dot{z})\cdot\dot{z} + \xi(\dot{e},\dot{z}) - \operatorname{div} j}_{=\dot{w}\text{ due to (8.1.5)}}\,\mathrm{d}x$$
$$= \int_\Omega \frac{\varrho}{2}|\dot{u}|^2 + \sigma{:}e(\dot{u}) - \operatorname{div} j\,\mathrm{d}x = \underbrace{\int_\Omega f\cdot\dot{u}\,\mathrm{d}x + \int_\Gamma g\cdot\dot{u} + j\cdot\vec{n}\,\mathrm{d}S}_{\substack{\text{power of external mechanical}\\ \text{loading and heating}}}. \tag{8.1.13}$$

This reveals the *total energy balance* in terms of the sum of the kinetic and the internal energies.

Differentiating $\mathfrak{s} = -\partial_\theta\psi$ in (8.1.2) yields $\dot{\mathfrak{s}} = -\partial^2_{\theta\theta}\psi(e,z,\theta)\dot{\theta} - \partial^2_{e\theta}\psi(e,z,\theta){:}\dot{e} - \partial^2_{z\theta}\psi(e,z,\theta)\cdot\dot{z}$. Substituting it further into the equation (8.1.7), we obtain the *heat-transfer equation* for temperature:

$$c_\mathrm{v}\dot{\theta} + \operatorname{div} j = \xi(\dot{e},\dot{z}) + \theta\partial^2_{\theta e}\psi{:}\dot{e} + \theta\partial^2_{\theta z}\psi\cdot\dot{z}, \tag{8.1.14a}$$

with the *heat capacity*[2]

$$c_\mathrm{v} = c_\mathrm{v}(e,z,\theta) = -\theta\partial^2_{\theta\theta}\psi(e,z,\theta). \tag{8.1.14b}$$

The last two terms in (8.1.14a) represent *adiabatic heat* sources/sinks. Altogether, using also (8.1.8), we thus will treat the system

$$\varrho\ddot{u} - \operatorname{div}\big(\partial_{\dot{e}}\zeta(\dot{e},\dot{z}) + \partial_e\psi(e,z,\theta)\big) = f \qquad \text{with } e = e(u), \tag{8.1.15a}$$
$$\partial_{\dot{z}}\zeta(\dot{e},\dot{z}) + \partial_z\psi(e,z,\theta) = 0, \tag{8.1.15b}$$
$$c_\mathrm{v}(e,z,\theta)\dot{\theta} - \operatorname{div}(\mathbb{K}(e,z,\theta)\nabla\theta) = \xi(\dot{e},\dot{z}) + \theta\partial^2_{\theta e}\psi(e,z,\theta){:}\dot{e} + \theta\partial^2_{\theta z}\psi(e,z,\theta)\cdot\dot{z}. \tag{8.1.15c}$$

[2] A canonical form $\gamma_\mathrm{v}(\theta) = c_0\theta(\ln(\theta/\theta_0)+1)$ with c_0 a constant and θ_0 a reference temperature would lead to $\mathfrak{S} = c_0\ln(\theta/\theta_0)$ and $c_\mathrm{v}(e,z,\theta) = c_0$. General nonconstant heat capacity (8.1.14b) may, however, facilitate mathematical analysis in particular cases. More specifically, a growth of $c_\mathrm{v}(e,z,\cdot)$ is reflected by a decay of $\mathscr{C}_\mathrm{v}^{-1}(\cdot)$ which then allows for better estimates of the adiabatic terms.

The analysis of this system is based (disregarding technicalities with particular approximation methods) on the a-priori estimates obtained by testing (8.1.15) respectively by $\dot{u}$, $\dot{z}$, and 1. This "energetic" test is physically motivated and leads just to the energetics (8.1.13). This can be seen by using the calculus

$$\begin{aligned}\partial_e\psi(e,z,\theta){:}\dot{e}+\partial_z\psi(e,z,\theta){\cdot}\dot{z}+c_{\rm v}(e,z,\theta)\dot{\theta}-\theta\partial^2_{\theta e}\psi(e,z,\theta){:}\dot{e}-\theta\partial^2_{\theta z}\psi(e,z,\theta){\cdot}\dot{z}\\ =\partial_e\psi(e,z,\theta){:}\dot{e}+\partial_z\psi(e,z,\theta){\cdot}\dot{z}-\theta\frac{\partial}{\partial t}\partial_\theta\psi(e,z,\theta)\\ =\dot{\psi}-\partial_\theta\psi(e,z,\theta)\dot{\theta}-\theta\frac{\partial}{\partial t}\partial_\theta\psi(e,z,\theta)\\ =\frac{\partial}{\partial t}(\psi(e,z,\theta)-\theta\partial_\theta\psi(e,z,\theta))=\frac{\partial}{\partial t}(\psi+\theta\mathfrak{s})=\dot{w}.\end{aligned} \tag{8.1.16}$$

Then (8.1.13) is obtained when taking into account also the form (8.1.6) of the specific dissipation rate ξ and the boundary condition $\sigma\vec{n}=g$.

We again complete the system (8.1.15) by some boundary/initial conditions, e.g.

$$\big(\partial_{\dot{e}}\zeta(\dot{e},\dot{z})+\partial_e\psi(e,z,\theta)\big)\vec{n}=g \quad\text{and}\quad \mathbb{K}(e,z,\theta)\nabla\theta{\cdot}\vec{n}=h_\flat \qquad \text{on } \Sigma, \tag{8.1.17a}$$

$$u|_{t=0}=u_0, \qquad \dot{u}|_{t=0}=v_0, \qquad z|_{t=0}=z_0, \qquad \theta|_{t=0}=\theta_0 \qquad \text{in } \Omega. \tag{8.1.17b}$$

Using the concepts from Sect. 7.2 for the mechanical part (8.1.15a,b) and multiplying the thermal part (8.1.15c) by $1/c_{\rm v}(e,z,\theta)$, we can cast the following:

Definition 8.1.1 (Weak solutions to (8.1.15)-(8.1.17)). The triple (u,z,θ) is called a weak solution to the initial-boundary-value (8.1.15)-(8.1.17) if the mechanical part (8.1.15a,b) is satisfied in a weak form (7.2.5a) with σ from (8.1.11) and with $u(0)=u_0$ and, for all v smooth with $v(T)=0$, it holds[3]

$$\begin{aligned}\int_Q\frac{\mathbb{K}(e,z,\theta)\nabla\theta}{c_{\rm v}(e,z,\theta)}{\cdot}\nabla v-\theta\dot{v}-\frac{\mathbb{K}(e,z,\theta)\nabla\theta}{c^2_{\rm v}(e,z,\theta)}{\cdot}\Big(\partial_e c_{\rm v}(e,z,\theta){:}\nabla e(u)+\\ +\partial_z c_{\rm v}(e,z,\theta)\nabla z+\partial_\theta c_{\rm v}(e,z,\theta)\nabla\theta\Big)v\,{\rm d}x{\rm d}t\\ =\int_Q\frac{\xi(\dot{e},\dot{z})+\theta\partial^2_{\theta e}\psi(e,z,\theta){:}\dot{e}+\theta\partial^2_{\theta z}\psi(e,z,\theta){\cdot}\dot{z}}{c_{\rm v}(e,z,\theta)}v\,{\rm d}x{\rm d}t\\ +\int_\Sigma\frac{h_\flat v}{c_{\rm v}(e,z,\theta)}\,{\rm d}S\,{\rm d}t+\int_\Omega\theta_0 v(0)\,{\rm d}x\end{aligned} \tag{8.1.18}$$

with $e=e(u)$.

Let us note that ∇e and ∇z occur in (8.1.18) so that, in fact, this weak formulation works only if $\partial^3_{z\theta\theta}\psi=0$ and $\partial^3_{e\theta\theta}\psi=0$ so that $\partial_e c_{\rm v}=0$ and $\partial_z c_{\rm v}=0$, cf. Remark 8.1.7, or if the gradient theory would used for the strain (2nd-grade nonsimple material concept as in Sect. 5.4) and for the internal variable, respectively.

[3] Here for (8.1.18) we use the calculus $\int_\Omega{\rm div}(\mathbb{K}\nabla\theta)v/c_{\rm v}\,{\rm d}x=\int_\Gamma\mathbb{K}\nabla\theta{\cdot}\vec{n}v/c_{\rm v}\,{\rm d}S-\int_\Omega\mathbb{K}\nabla\theta{\cdot}\nabla(v/c_{\rm v})\,{\rm d}x$.

In situations when both $\partial^3_{z\theta\theta}\psi = 0$ and $\partial^3_{e\theta\theta}\psi = 0$, then c_v depends only on θ as in the mentioned Remark 8.1.7, one can easily replace the term $c_\mathrm{v}(\theta)\dot\theta$ in (8.1.15c) by $\mathscr{C}_\mathrm{v}(\theta)^\cdot$ with $\mathscr{C}_\mathrm{v}$ a primitive function to c_v and then introduce a new variable $w := \mathscr{C}_\mathrm{v}(\theta)$ and write (8.1.15) in terms of such "re-scaled" temperature. In mathematical literature, this is called an *enthalpy transformation*.

To implement the enthalpy transformation in our general situation, we would rather introduce the primitive function to $c_\mathrm{v}(e,z,\cdot)$ by

$$\mathscr{C}_\mathrm{v}(e,z,\theta) := \int_0^1 \theta c_\mathrm{v}(e,z,r\theta)\,\mathrm{d}r. \tag{8.1.19}$$

The heat-transfer equation (8.1.15c) then transforms to[4]

$$\dot\vartheta - \mathrm{div}(\mathbb{K}(e,z,\theta)\nabla\theta) = \xi(\dot e,\dot z) + \big(\theta\partial^2_{\theta e}\psi(e,z,\theta) + \partial_e\mathscr{C}_\mathrm{v}(e,z,\theta)\big){:}\dot e \\ + \big(\theta\partial^2_{\theta z}\psi(e,z,\theta) + \partial_z\mathscr{C}_\mathrm{v}(e,z,\theta)\big)\cdot\dot z \quad \text{with } \vartheta = \mathscr{C}_\mathrm{v}(e,z,\theta). \tag{8.1.20}$$

Using the definitions (8.1.3) and (8.1.19), one can easily see that

$$\partial_\theta(w-\vartheta) = \partial_\theta(\psi - \theta\partial_\theta\psi - \mathscr{C}_\mathrm{v}(e,z,\theta)) = -\theta\partial^2_{\theta\theta}\psi - c_\mathrm{v}(e,z,\theta) = 0 \tag{8.1.21}$$

by (8.1.14b). Thus ϑ differs from the internal energy w just by a function of (e,z), i.e. by the part of the internal energy which is purely mechanical. More specifically, counting with $\mathscr{C}_\mathrm{v}(e,z,0) = 0$ due to the definition (8.1.19), from (8.1.21) we can see that

$$w = \psi(e,z,0) + \vartheta. \tag{8.1.22}$$

The variable ϑ thus has an interpretation of the *thermal part of the internal energy*. More specifically, a general formula for $\mathscr{C}_\mathrm{v}(e,z,\theta) = \int_0^1 -\theta^2 r\partial^2_{\theta\theta}\psi(e,z,r\theta)\,\mathrm{d}r$ from (8.1.19) using lower derivatives is[5]

---

[4] Using a placeholder $q = (e,z)$, we can rely on the calculus:

$$\begin{aligned}[\mathscr{C}_\mathrm{v}(q,\theta)]^\cdot &= \int_0^1 c_\mathrm{v}(q,t\theta)\dot\theta + \theta\partial_\theta c_\mathrm{v}(q,t\theta)t\dot\theta + \theta\partial_q c_\mathrm{v}(q,t\theta)\dot q\,\mathrm{d}t \\ &= \Big(\int_0^1 c_\mathrm{v}(q,t\theta) + \partial_\theta c_\mathrm{v}(q,t\theta)t\theta\,\mathrm{d}t\Big)\dot\theta + \Big(\int_0^1 \theta\partial_q c_\mathrm{v}(q,t\theta)\,\mathrm{d}t\Big)\dot q \\ &= \Big(\int_0^1 \frac{\mathrm{d}}{\mathrm{d}t}(c_\mathrm{v}(q,t\theta)t)\,\mathrm{d}t\Big)\dot\theta + \partial_q\mathscr{C}_\mathrm{v}(q,\theta)\dot q = c_\mathrm{v}(q,\theta)\dot\theta + \partial_q\mathscr{C}_\mathrm{v}(q,\theta)\dot q\end{aligned}$$

with $\partial_q\mathscr{C}_\mathrm{v}(q,\theta) = \int_0^1 \theta\partial_q c_\mathrm{v}(q,t\theta)\,\mathrm{d}t$.

[5] Note the equality (8.1.23) clearly holds for $\theta = 0$ and also the partial derivative according θ of both sides is equal, namely $c_\mathrm{v} = -\theta\partial^2_{\theta\theta}$.

$$\mathscr{C}_{\rm v}(e,z,\theta) = \psi(e,z,\theta) - \theta\partial_\theta\psi(e,z,\theta) - \psi(e,z,0), \tag{8.1.23}$$

so that $w = \psi(e,z,\theta) - \theta\partial_\theta\psi(e,z,\theta)$, cf. (8.1.2)-(8.1.3). Thus (8.1.20) can also be written simply as

$$\begin{aligned}\dot\vartheta - \mathrm{div}(\mathbb{K}(e,z,\theta)\nabla\theta) = \xi(\dot e,\dot z) + (\partial_e\psi(e,z,\theta) - \partial_e\psi(e,z,0)){:}e(\dot u)\\ + (\partial_z\psi(e,z,\theta) - \partial_z\psi(e,z,0))\cdot\dot z \quad \text{with } \vartheta = \mathscr{C}_{\rm v}(e,z,\theta),\end{aligned} \tag{8.1.24}$$

cf. also [355].

We can now cast an alternative definition of a weak formulation which does not need ∇z and also less smoothness of ψ is needed comparing to Definition 8.1.1:

Definition 8.1.2 (Weak solutions to (8.1.15)-(8.1.17) based on (8.1.20)). The triple (u,z,θ) is called a weak solution to the initial-boundary-value (8.1.15)-(8.1.17) if the mechanical part (8.1.15a,b) is satisfied in a weak form as in Definition 8.1.1 and, for all v smooth with

$$\begin{aligned}\int_Q \mathbb{K}(e,z,\theta)\nabla\theta\cdot\nabla v - \mathscr{C}_{\rm v}(e,z,\theta)\dot v\,\mathrm{d}x\mathrm{d}t = \int_Q \Big(\xi(\dot e,\dot z)\\ + \big(\partial_e\psi(e,z,\theta) - \partial_e\psi(e,z,0)\big){:}\dot e + \big(\partial_z\psi(e,z,\theta) - \partial_z\psi(e,z,0)\big)\cdot\dot z\Big)v\,\mathrm{d}x\mathrm{d}t\\ + \int_\Sigma h_\flat v\,\mathrm{d}S\,\mathrm{d}t + \int_\Omega \mathscr{C}_{\rm v}(e_0,z_0,\theta_0)v(0)\,\mathrm{d}x\end{aligned} \tag{8.1.25}$$

with $e = e(u)$.

Remark 8.1.3 (Clausius-Duhem inequality). The ultimate departure point is the *entropy imbalance* together with the positivity of temperature θ, which is sometimes called a *local Clausius-Duhem inequality* and reads as:

$$\dot{\mathfrak{s}} + \mathrm{div}\frac{j}{\theta} \geq \frac{\xi}{\theta} \tag{8.1.26}$$

where j/θ is the entropy flux due to heat flux and ξ/θ is the rate of entropy production due to dissipation. If the equality holds (8.1.26), the process is called reversible, otherwise irreversible. Using the calculus $\theta\mathrm{div}(j/\theta) = \mathrm{div}j - (j\cdot\nabla\theta)/\theta$, this is equivalent to

$$\xi \leq \theta\dot{\mathfrak{s}} + \mathrm{div}j - \frac{j\cdot\nabla\theta}{\theta}, \tag{8.1.27}$$

which is sometimes called the *dissipation inequality*.Let us note that inequality (8.1.26) becomes the entropy equation (8.1.7) if the last term is omitted. This is related to that $-(j\cdot\nabla\theta)/\theta$ does not contribute to internal-energy increase and thus does not occur in the right-hand side of (8.1.5). Also, we can see that (8.1.26) is consistent with the local Clausius-Duhem inequality if the heat-conduction tensor $\mathbb{K}$ in (8.1.8) is positive semidefinite.

Remark 8.1.4 (One special ansatz). It is relatively typical that the termomechanical coupling, let us denote the corresponding potential by ψ_C, is separated from the purely mechanical stored energy φ as

$$\psi(e,z,\theta) := \varphi(e,z) + \psi_C(e,z,\theta) \quad \text{with} \quad \psi_C(e,z,0) = 0. \tag{8.1.28}$$

Then (8.1.23)–(8.1.24) simplifies as

$$\dot{\vartheta} - \operatorname{div}(\mathbb{K}(e,z,\theta)\nabla\theta) = \xi(\dot{e},\dot{z}) + \partial_e\psi_C(e,z,\theta):e(\dot{u}) + \partial_z\psi_C(e,z,\theta)\cdot\dot{z}$$
$$\text{with } \vartheta = \psi_C(e,z,\theta) - \theta\partial_\theta\psi_C(e,z,\theta).$$

In particular, the purely mechanical stored energy φ does not influence the heat production and transfer.

Remark 8.1.5 (Thermodynamical restrictions). Substituting (8.1.26) into the energy balance (8.1.5) written in a general form

$$\dot{w} = \tilde{\sigma}_{\text{EL}}:\dot{e} + \tilde{\sigma}_{\text{IN}}\cdot\dot{z} + \xi(\dot{e},\dot{z}) - \operatorname{div} j \tag{8.1.29}$$

with an unspecified $\tilde{\sigma}_{\text{EL}}$ and $\tilde{\sigma}_{\text{IN}}$ and using $\dot{w} = \frac{\partial}{\partial t}(\psi + \theta\mathfrak{s}) = \dot{\psi} + \theta\dot{\mathfrak{s}} + \dot{\theta}\mathfrak{s}$ with $\theta\dot{\mathfrak{s}}$ substituted from (8.1.27) considered as a basic departure point, one obtains

$$\tilde{\sigma}_{\text{EL}}:\dot{e} + \tilde{\sigma}_{\text{IN}}\cdot\dot{z} + \xi(\dot{e},\dot{z}) - \operatorname{div} j = \dot{w} = \dot{\psi} + \mathfrak{s}\dot{\theta} + \theta\dot{\mathfrak{s}}$$
$$\geq \dot{\psi} + \mathfrak{s}\dot{\theta} + \xi(\dot{e},\dot{z}) - \operatorname{div} j + \frac{j\cdot\nabla\theta}{\theta}. \tag{8.1.30}$$

Using also the calculus $\dot{\psi} = \partial_e\psi:\dot{e} + \partial_z\psi\cdot\dot{z} + \partial_\theta\psi\dot{\theta}$, this eventually yields the so-called *dissipation inequality*

$$0 \geq \dot{\psi} + \mathfrak{s}\dot{\theta} - \tilde{\sigma}_{\text{EL}}:\dot{e} - \tilde{\sigma}_{\text{IN}}\cdot\dot{z} + \frac{j\cdot\nabla\theta}{\theta}$$
$$= (\partial_e\psi - \tilde{\sigma}_{\text{EL}}):\dot{e} + (\partial_z\psi - \tilde{\sigma}_{\text{IN}})\cdot\dot{z} + (\partial_\theta\psi + \mathfrak{s})\dot{\theta} + \frac{j\cdot\nabla\theta}{\theta}. \tag{8.1.31}$$

The standard argument is that all the rates as well as the temperature gradient are completely independent of each other, so that satisfaction of (8.1.31) implies (and simultaneously requires) that

$$(\partial_e\psi - \tilde{\sigma}_{\text{EL}}):\dot{e} \leq 0, \quad (\partial_z\psi - \tilde{\sigma}_{\text{IN}})\cdot\dot{z} \leq 0, \quad \partial_\theta\psi + \mathfrak{s} = 0, \quad j\cdot\nabla\theta \leq 0. \tag{8.1.32}$$

In models from Chapters 6 and 7, we used $\tilde{\sigma}_{\text{EL}} = \sigma_{\text{EL}}(e,z,\theta) + \sigma_{\text{vi},1}(\dot{e})$ and $\tilde{\sigma}_{\text{IN}} = \sigma_{\text{IN}}(e,z,\theta) + \sigma_{\text{vi},2}(\dot{z})$. Then the first and the second relations in (8.1.32) imply

$$\partial_e\psi = \sigma_{\text{EL}}(e,z,\theta), \qquad \partial_z\psi = \sigma_{\text{IN}}(e,z,\theta), \tag{8.1.33a}$$

$$\sigma_{\mathrm{vi},1}(\dot{e}){:}\dot{e} \geq 0, \qquad \sigma_{\mathrm{vi},2}(\dot{z}){\cdot}\dot{z} \geq 0. \tag{8.1.33b}$$

In particular, (8.1.33a) and the third relation in (8.1.32) show that the 'definition' (8.1.2) is to be understood rather as a consequence of basic physical principles, namely of the 2nd law of thermodynamics. Also (8.1.33b) represents a thermodynamical restriction on the dissipation potential; it is ensured e.g. if $\sigma_{\mathrm{vi},i} = \zeta'_{\mathrm{vi},i}$ with a convex potential $\zeta_{\mathrm{vi},i} \geq 0$ with $\zeta_{\mathrm{vi},i}(0) = 0$ but it holds under more general conditions, too. Moreover, taking into account the Fourier law (8.1.8), the last inequality in (8.1.32) requires $\mathbb{K}$ positive semidefinite.

Remark 8.1.6 (Internal energy revisited). In view of (8.1.3) together with (8.1.2), it holds $\partial_\theta w(e,z,\theta,\mathfrak{s}) = \partial_\theta\psi(e,z,\theta) + \mathfrak{s} = 0$. Positivity of the heat capacity (8.1.14b) is a physically ultimate requirement and it is equivalent to the strict concavity of $\psi(e,z,\cdot)$. Then θ maximizes the internal energy $w(e,z,\cdot,\mathfrak{s})$. We can then express $\theta = \Theta(e,z,\mathfrak{s})$ and thus eliminate θ to express the "reduced" internal energy as $w_{\mathrm{R}}(e,z,\mathfrak{s}) = \psi(e,z,\Theta(e,z,\mathfrak{s})) + \mathfrak{s}\Theta(e,z,\mathfrak{s})$. In view of the definition (A.3.16) of the convex conjugate $[\cdot]^*$, the mentioned maximization of $\theta \mapsto w(e,z,\theta,\mathfrak{s}) + \theta\mathfrak{s} = -(-w(e,z,\theta,\mathfrak{s}) - \theta\mathfrak{s})$ gives the formula

$$w_{\mathrm{R}}(e,z,\mathfrak{s}) = -\big[-w(e,z,\cdot,\mathfrak{s})\big]^*(\mathfrak{s}). \tag{8.1.34}$$

Remark 8.1.7 (Partly linearized ansatz). Often, the free energy ψ takes the following partly linearized ansatz which often occurs in applications:

$$\psi(e,z,\theta) = \varphi(e,z) + \varphi_{\mathrm{C}}(e,z)\theta - \gamma_{\mathrm{v}}(\theta). \tag{8.1.35}$$

with some convex $\gamma_{\mathrm{v}} : \mathbb{R} \to \mathbb{R}$. It simplifies some formulas and arguments. Then $\partial^3_{e\theta\theta}\psi(e,z,\theta) = 0$ and $\partial^3_{z\theta\theta}\psi(e,z,\theta) = 0$ so that $\partial^2_{\theta\theta}\psi(e,z)$ and thus also the heat capacity from (8.1.14b) as well as its potential (8.1.19) depends only on temperature, i.e.

$$c_{\mathrm{v}}(\theta) = \theta\gamma''_{\mathrm{v}}(\theta) \quad \text{and} \quad \mathscr{C}_{\mathrm{v}}(\theta) = \theta\gamma'_{\mathrm{v}}(\theta) - \gamma_{\mathrm{v}}(\theta); \tag{8.1.36}$$

actually, it is a special form of $\mathscr{C}_{\mathrm{v}}$ from (8.1.19) up to a constant $\gamma_{\mathrm{v}}(0)$.[6] Both the entropy as well as the internal energy *separate the mechanical and the heat variables*, i.e.

$$\mathfrak{s} = \varphi_{\mathrm{C}}(e,z) + \gamma'_{\mathrm{v}}(\theta) \quad \text{and} \quad w = \psi + \theta\mathfrak{s} = \varphi(e,z) + \vartheta \quad \text{with} \quad \vartheta = \mathscr{C}_{\mathrm{v}}(\theta) \tag{8.1.37}$$

and obviously the energy balance (8.1.13) can be written

$$\underbrace{\frac{\mathrm{d}}{\mathrm{d}t}\int_\Omega \frac{\varrho}{2}|\dot{u}|^2 + \varphi(e(u),z) + \vartheta\,\mathrm{d}x}_{\text{total energy}} = \underbrace{\int_\Omega f\cdot\dot{u}\,\mathrm{d}x + \int_\Gamma g\cdot\dot{u} + h_\flat\cdot\vec{n}\,\mathrm{d}S}_{\text{power of external mechanical loading and heating}}. \tag{8.1.38}$$

[6] Note that, in view of (8.1.14b), we have $(\theta\partial\gamma_{\mathrm{v}}(\theta) - \gamma_{\mathrm{v}}(\theta))' = \theta\gamma''_{\mathrm{v}}(\theta) + \gamma'_{\mathrm{v}}(\theta) - \gamma'_{\mathrm{v}}(\theta) = \theta\gamma''_{\mathrm{v}}(\theta) = c_{\mathrm{v}}(\theta) = \mathscr{C}'_{\mathrm{v}}(\theta)$, hence $\mathscr{C}_{\mathrm{v}}$ differs from $\theta \mapsto \theta\gamma'_{\mathrm{v}}(\theta) - \gamma_{\mathrm{v}}(\theta)$ just by a constant.

Remark 8.1.8 (More general dissipation: nonsmooth or coupled). Making the potential ζ from (8.1.1) dependent on the state (e,z,θ) and possibly also coupling the rates $\dot{e}$ and $\dot{z}$, i.e. $\zeta=\zeta(e,z,\theta;\dot{e},\dot{z})$ is easily possible, too. Moreover, the convex function $\zeta(e,z,\theta;\cdot,\cdot)$ may be considered non-smooth at $(0,0)$. Then, e.g., the dissipation rate ξ in (8.1.6) is to be

$$\xi=\xi(e,z,\theta;\dot{e},\dot{z})=\big\langle\partial_{(\dot{e},\dot{z})}\zeta(e,z,\theta;\dot{e},\dot{z}),(\dot{e},\dot{z})\big\rangle \tag{8.1.39}$$

and, even in the mentioned nonsmoothness implying that $\partial_{(\dot{e},\dot{z})}\zeta(e,z,\theta;\cdot,\cdot)$ is set-valued, one is eligible to assume that $\xi(e,z,\theta;\cdot,\cdot)$ is single-valued. Also (some of) the equations in (8.1.15) might be inclusions. Typically it concerns the flow rule (8.1.15b) which then turns into the inclusion

$$\partial_{\dot{z}}\zeta(e(u),z,\theta;\dot{e},\dot{z})+\partial_z\psi(e(u),z,\theta)\ni 0. \tag{8.1.40}$$

Like in Remark 8.1.5, $\tilde{\sigma}_{\rm EL}=\sigma_{\rm EL}(e,z,\theta)+\sigma_{\rm vi,1}$ and $\tilde{\sigma}_{\rm IN}=\sigma_{\rm IN}(e,z,\theta)+\sigma_{\rm vi,2}$, and the first inequality in (8.1.32) gives the thermodynamical restriction (8.1.33a) while the two conditions (8.1.33b) are modified to yield one joint condition

$$\forall\sigma_{\rm vi,1}\in\partial_{\dot{e}}\zeta(e,z,\theta;\dot{e},\dot{z}),\ \sigma_{\rm vi,2}\in\partial_{\dot{z}}\zeta(e,z,\theta;\dot{e},\dot{z}):\quad \sigma_{\rm vi,1}:\dot{e}+\sigma_{\rm vi,2}\cdot\dot{z}\ge 0.$$

Remark 8.1.9 (More general dissipation: nonlocal). A weaker variant of the dissipation inequality (8.1.31) is obtained when integrated it over Ω. Then the first inequality in (8.1.32) is integrated, too. This allows for nonlocal dissipation. E.g. in terms of the internal variable, $\sigma_{\rm in,2}(\dot{z})$ can be nonlocal and the later condition in (8.1.33b) weakens to $\int_\Omega\sigma_{\rm in,2}(\dot{z})\cdot\dot{z}\,{\rm d}x\ge 0$.

Remark 8.1.10 (Holonomic constraints). Sometimes, the state of the system is subjected to constraints. The simplest case is a linear constraint involving (e,z), say

$$\Xi(e,z)\equiv\Xi_1 e+\Xi_2 z=0. \tag{8.1.41}$$

Then (8.1.15a,b) must be replaced by

$$\varrho\ddot{u}-{\rm div}\Big(\partial_{\dot{e}}\zeta(e(\dot{u}),\dot{z})+\partial_e\psi(e(u),z,\theta)\Big)=f-{\rm div}\,(\xi_{\rm VI}^*\lambda), \tag{8.1.42a}$$

$$\partial_{\dot{z}}\zeta(e((\dot{u}),\dot{z})+\partial_z\psi(e(u),z,\theta)=\xi_{\rm IN}^*\lambda, \tag{8.1.42b}$$

with a Lagrange multiplier λ. Then also the rates $\dot{e}$ and $\dot{z}$ are not independent, and (8.1.33a) must be modified. Namely, one should consider the constraint

$$\forall(\tilde{e},\tilde{z})\in\mathbb{R}^{d\times d}\times\mathbb{R}^m,\ \Xi(\tilde{e},\tilde{z})=0:\quad (\partial_e\psi-\sigma_{\rm el}(e,z)):\tilde{e}+(\partial_z\psi-\sigma_{\rm in}(e,z))\cdot\tilde{z}=0. \tag{8.1.43}$$

This implies

$$(\sigma_{\rm el}(e,z,\theta),\sigma_{\rm in}(e,z,\theta))=\Xi^*\partial_{(e,z)}\psi(e,z,\theta). \tag{8.1.44}$$

8.2 L^1-theory of the heat-transfer equation

We already pointed out the nonvariational structure of the heat-transfer problem in the steady-state situations when physically relevant heat sources with the mere L^1-integrability are considered, cf. Remark 5.7.13 on p. 191. This character is naturally exhibited in evolutionary situations, too. This urges for a rather sophisticated estimation technique which was developed only rather recently at the end of XX century and which is inevitably quite technical in particular when combined with mechanically coupled systems.

To demonstrate it, we first consider the initial-boundary-value problem for the heat-transfer equation itself, i.e.

$$c_{\rm v}(\theta)\dot{\theta} - {\rm div}(\mathbb{K}(t,x,\theta)\nabla\theta) = \xi_{\rm A}(t,x,\theta) \qquad \text{on } Q, \tag{8.2.1a}$$

$$\mathbb{K}(t,x,\theta)\nabla\theta = h_\flat(t,x) \qquad \text{on } \Sigma, \tag{8.2.1b}$$

$$\theta(x,0) = \theta_0(x) \qquad \text{on } \Omega, \tag{8.2.1c}$$

where $\xi_{\rm A} = \xi_{\rm A}(t,x,\theta)$ is in the position of the dissipation rate augmented possibly by adiabatic effects. Occasionally, we will use the enthalpy-like formulation, i.e. (8.2.1a) is replaced by Thus (8.2.1a) can be rewritten as

$$\dot{\vartheta} - {\rm div}(\mathbb{K}(t,x,\theta)\nabla\theta) = \xi_{\rm A}(t,x,\theta) \quad \text{with} \quad \vartheta = \mathscr{C}_{\rm v}(\theta) := \int_0^1 \theta c_{\rm v}(r\theta)\,dr. \tag{8.2.2}$$

Let us note that $\mathscr{C}_{\rm v}$ is from (8.1.19).

There are several possible scenarios how to qualify the data, depending on desired generality and on the chosen strategies of the proof, cf. also Remarks 8.2.2 and (8.2.4). The basic assumptions that allow for an existence of weak solutions are

$$\begin{aligned}
&c_{\rm v}:\mathbb{R}^+\to\mathbb{R}^+ \ \text{continuous},\\
&\exists\, 1\le\omega<d/(d{-}1),\ c_0,c_1>0,\ 0{\le}s{<}\omega/d,\\
&\qquad \forall(t,x,\theta){\in}Q{\times}\mathbb{R}^+ : \theta{\in}\mathbb{R}^+ : \quad c_0(1+\theta^{\omega-1})\le c_{\rm v}(\theta)\le c_1(1+\theta^s),
\end{aligned} \tag{8.2.3a}$$

$$\mathbb{K}: Q\times\mathbb{R}^+\to\mathbb{R}^{d\times d} \ \text{Carathéodory mapping, bounded, uniformly positive definite.} \tag{8.2.3b}$$

As already said, we will focus on the physically relevant L^1-theory, i.e.

$$|\xi_{\rm A}(t,x,\theta)|\le\xi_0(t,x)+C|\theta|^q \quad \text{with } \xi_0{\in}L^1(Q),\ 0\le q\le\omega, \tag{8.2.3c}$$

$$\vartheta_0 := \mathscr{C}_{\rm v}(\theta_0){\in}L^1(\Omega), \qquad h_\flat{\in}L^1(\Sigma). \tag{8.2.3d}$$

The restriction $q\le\omega$ in (8.2.3c) will be use to execute the Gronwall inequality for (8.2.10) while the upper bound $s<\omega/d$ in (8.2.3a) is needed for the $\nabla\vartheta$-estimation (8.2.21). This bound also restricts ω not to exceed $d' = d/(d-1)$ so that always $\omega-1\le s$ is needed in (8.2.3a).

Interpreting θ as the absolute temperature which should not be negative, another physically natural assumption important also for the analysis is the nonnegativity of all these sources. We however allow the right-hand side ξ_A to alternate sign, which is relevant in thermo-mechanically coupled systems where adiabatic-like effects may cause not only heating but also cooling, although still consistent with the non-negativity of temperature as a 3rd law of thermodynamics.

Proposition 8.2.1 (Temperature estimates). *Let (8.2.3) hold together with*

$$\xi_A(t,x,0) \geq 0, \quad h_\flat \geq 0, \text{ and } \vartheta_0 \geq 0. \tag{8.2.4}$$

Then the boundary-value problem (8.2.1) possesses a weak solution $\theta \geq 0$ such that

$$\theta \in L^\infty(I;L^\omega(\Omega)) \cap L^r(I;W^{1,r}(\Omega)) \cap L^p(Q) \text{ and} \tag{8.2.5a}$$

$$\vartheta = \mathscr{C}_V(\theta) \in L^\infty(I;L^1(\Omega)) \cap L^1(I,W^{1,1}(\Omega)) \cap W^{1,1}(I;H^{(d+3)/2}(\Omega)^*) \tag{8.2.5b}$$

for any $1 \leq r < \dfrac{2\omega+d}{\omega+d}$ *and* $1 \leq p < 1 + \dfrac{2\omega}{d}$. *More specifically,*

$$\|\nabla\theta\|_{L^r(Q;\mathbb{R}^d)} \leq C_r\big(1+\|\xi_A(\theta)\|_{L^1(Q)}\big)^{1/r} \quad \text{and} \tag{8.2.6a}$$

$$\|\theta\|_{L^p(Q)} \leq C_p\big(1+\|\xi_A(\theta)\|_{L^1(Q)}\big)^{1/p} \tag{8.2.6b}$$

with the constants C_r and C_p depending, apart of r and p, respectively, also on θ_0 and $h_\flat$, but not on $\xi_A = \xi_A(\theta)$.

Let us note that (8.2.6) cannot be expected to hold in the borderline cases $r = (2\omega+d)/(\omega+d)$ and $p = 1+2\omega/d$ and, from the proof below, one can see that the constants C_r and C_p in (8.2.6) blow us to ∞ if $r \to (2\omega+d)/(\omega+d)$ and $p \to 1+2\omega/d$, respectively.

Proof of Proposition 8.2.1. We divide the proof into five steps. For illustration we will use rather the time discretization (Rothe's method) which allows for only a one-step approximation, in contrast to the Galerkin method, cf. Remark 8.2.3 below.

Step 1 – approximate solution: We construct an approximate solution by the *Rothe method*, namely by the time discrete scheme

$$\frac{\vartheta_\tau^k - \vartheta_\tau^{k-1}}{\tau} - \mathrm{div}(\mathbb{K}_\tau^k(\theta_\tau^{k-1})\nabla\theta_\tau^k) = \xi_{A,\tau}^k(\theta_\tau^k) \quad \text{with } \vartheta_\tau^k = \mathscr{C}_V(\theta_\tau^k),$$

$$\mathbb{K}_\tau^k(x,\theta) := \frac{1}{\tau}\int_{\tau(k-1)}^{k\tau} \mathbb{K}(t,x,\theta)\,\mathrm{d}t \quad \text{and}$$

$$\xi_{A,\tau}^k(x,\theta) := \frac{1}{\tau}\int_{\tau(k-1)}^{k\tau} \frac{\xi_A(t,x,\theta)}{1+\tau|\xi_A(t,x,\theta)|}\,\mathrm{d}t, \tag{8.2.7a}$$

and with the boundary condition

$$\mathbb{K}_\tau^k(\theta_\tau^{k-1})\nabla\theta_\tau^k = h_{\flat,\tau}^k \quad \text{with } h_{\flat,\tau}^k(x) := \frac{1}{\tau}\int_{\tau(k-1)}^{k\tau} \frac{h_\flat(t,x)}{1+\tau|h_\flat(t,x)|}\,\mathrm{d}t. \tag{8.2.7b}$$

For notational simplicity, we omitted (and will omit) x while keeping t when important. The boundary-value problem (8.2.7) is to be solved recursively for $k = 1,2,...,T/\tau$, starting from

$$\vartheta_\tau^0 := \mathscr{C}_\mathrm{v}\Big(\frac{\theta_0}{1+\tau\theta_0}\Big). \tag{8.2.7c}$$

The existence of a conventional weak solution $\theta_\tau^k \in H^1(\Omega)$ to this boundary value problem is then easy to be shown; note that this semilinear problem has even a potential.

We can write (8.2.7)–(8.2.7c) in terms of our notation for the piecewise constant or affine interpolants (cf. (C.2.15) on p. 536) constructed here by the family $\{(\theta_\tau^k,\vartheta_\tau^k)\}_{k=0}^{T/\tau}$. The resulted initial-boundary-value problem then reads as:

$$\dot{\vartheta}_\tau - \mathrm{div}\big(\bar{\mathbb{K}}_\tau(\underline{\theta}_\tau)\nabla\bar{\theta}_\tau\big) = \bar{\xi}_{\mathrm{A},\tau}(\bar{\theta}_\tau) \qquad \text{in } Q, \tag{8.2.8a}$$

$$\bar{\mathbb{K}}_\tau(\underline{\theta}_\tau)\nabla\bar{\theta}_\tau\cdot\vec{n} = \bar{h}_{\flat,\tau} \qquad \text{on } \Sigma, \tag{8.2.8b}$$

$$\vartheta_\tau|_{t=0} = \vartheta_\tau^0 \qquad \text{on } \Omega. \tag{8.2.8c}$$

Step 2 – non-negativity: We now need to show that the solutions of (8.2.7) are non-negative. Let us extend $\xi_{\mathrm{A},\tau}^k(x,\theta) := \xi_{\mathrm{A},\tau}^k(x,0)$ for $\theta < 0$; note that such extension still keeps the Carathéodory property of $\xi_{\mathrm{A},\tau}^k$. Similarly, we extend $\mathscr{C}_\mathrm{v}(\theta) := \mathscr{C}_\mathrm{v}(0)$ for $\theta < 0$. Then we test (8.2.7) by the negative part $(\theta_\tau^k)^- := \min(\theta_\tau^k,0)$ of θ_τ^k. Realizing the definition (8.1.19), we have $\vartheta_\tau^k(\theta_\tau^k)^- = \mathscr{C}_\mathrm{v}(\theta_\tau^k)(\theta_\tau^k)^- = \mathscr{C}_\mathrm{v}(0)(\theta_\tau^k)^- \le 0$, and similarly $\xi_{\mathrm{A},\tau}^k(\theta_\tau^k)(\theta_\tau^k)^- = \xi_{\mathrm{A},\tau}^k(0)(\theta_\tau^k)^- \le 0$ and also $\vartheta_\tau^{k-1}(\theta_\tau^k)^- \le 0$ because $\theta_\tau^{k-1} \ge 0$. It gives the information $\theta_\tau^k \ge 0$ (since $h_\flat \ge 0$ and $\theta_0 \ge 0$). As this θ_τ^k does not take negative values, the mentioned extensions of ξ_A and $\mathscr{C}_\mathrm{v}$ are irrelevant, in fact.

Step 3 – basic a-priori estimates: Let us consider the non-negative solution of (8.2.7) and make a test by a constant (equal 1). Since $\theta \ge 0$, we obtain the estimates

$$\big\|\bar{\vartheta}_\tau\big\|_{L^\infty(I;L^1(\Omega))} \le C_1. \tag{8.2.9a}$$

$$\big\|\bar{\theta}_\tau\big\|_{L^\infty(I;L^\omega(\Omega))} \le C_1' \quad \text{with } \omega \ge 1 \text{ from (8.2.3a)}. \tag{8.2.9b}$$

More specifically, by the mentioned test we obtain

$$\begin{aligned}\frac{1}{\tau}\|\vartheta_\tau^k\|_{L^1(\Omega)} &= \frac{1}{\tau}\|\vartheta_\tau^{k-1}\|_{L^1(\Omega)} + \|\xi_{\mathrm{A},\tau}^k(\theta_\tau^k)\|_{L^1(\Omega)} + \|h_{\flat,\tau}^k\|_{L^1(\Gamma)}\\ &\le \frac{1}{\tau}\|\vartheta_\tau^{k-1}\|_{L^1(\Omega)} + \Big\|\int_{\tau(k-1)}^{k\tau}\xi_0(t,\cdot)\,\mathrm{d}t\Big\|_{L^1(\Omega)} + C\|\vartheta_\tau^k\|_{L^1(\Omega)}^{q/\omega} + \|h_{\flat,\tau}^k\|_{L^1(\Gamma)}\end{aligned} \tag{8.2.10}$$

where ξ_0, C, and q from (8.2.3c). From this, (8.2.9a) follows by the discrete Gronwall inequality while (8.2.9b) is then obtained by realizing that $\bar{\theta}_\tau = \mathscr{C}_{\rm V}^{-1}(\bar{\vartheta}_\tau)$ with $\mathscr{C}_{\rm V}^{-1}$ having at most $1/\omega$ growth.

Still (8.2.9) does not allow for a limit passage in the nonlinear the relation $\vartheta = \mathscr{C}_{\rm V}(\theta)$ in (8.3.8b) and in possibly nonlinear mappings $\theta \mapsto \mathbb{K}(t,x,\theta)$ and $\theta \mapsto \xi_{\rm A}(t,x,\theta)$, too.

Step 4 – further a-priori estimates: Now we prove an estimate of $\nabla\theta$ based on the test of the enthalpy equation (8.2.2) by $w_\epsilon(\theta_\tau)$ with an increasing nonlinear function $w_\epsilon : [0,+\infty) \to [0,1]$ defined by[7]

$$w_\epsilon(\theta) := 1 - \frac{1}{(1+\theta)^\epsilon}, \quad \epsilon > 0. \tag{8.2.11}$$

More specifically, we will execute this test on the discrete level, testing (8.2.7) by $w_\epsilon(\theta_\tau^k)$. We consider $1 \le r < 2$ and estimate the L^r-norm by Hölder's inequality as

$$\begin{aligned}\int_Q |\nabla\bar{\theta}_\tau|^r \,\mathrm{d}x\mathrm{d}t &= \int_Q (1+\bar{\theta}_\tau)^{(1+\epsilon)r/2} \frac{|\nabla\bar{\theta}_\tau|^r}{(1+\bar{\theta}_\tau)^{(1+\epsilon)r/2}} \,\mathrm{d}x\mathrm{d}t \\ &\le \Big(\int_Q (1+\bar{\theta}_\tau)^{(1+\epsilon)r/(2-r)} \,\mathrm{d}x\mathrm{d}t\Big)^{1-r/2} \Big(\int_Q \frac{|\nabla\bar{\theta}_\tau|^2}{(1+\bar{\theta}_\tau)^{1+\epsilon}} \,\mathrm{d}x\mathrm{d}t\Big)^{r/2} \\ &= C_{\epsilon,r,T} \underbrace{\Big(\int_0^T \big\|1+\bar{\theta}_\tau(t,\cdot)\big\|_{L^{(1+\epsilon)r/(2-r)}(\Omega)}^{(1+\epsilon)r/(2-r)} \,\mathrm{d}t\Big)^{1-r/2}}_{=: I^{(1)}_{r,\epsilon}(\bar{\theta}_\tau)} \underbrace{\Big(\int_Q \frac{|\nabla\bar{\theta}_\tau|^2}{(1+\bar{\theta}_\tau)^{1+\epsilon}} \,\mathrm{d}x\mathrm{d}t\Big)^{r/2}}_{=: I^{(2)}_\epsilon(\bar{\theta}_\tau)}\end{aligned} \tag{8.2.12}$$

with a constant $C_{\epsilon,r,T}$ dependent on ϵ, r, and T. Let us note that, using w_ϵ from (8.2.11), we can also write

$$I^{(2)}_\epsilon(\theta) = \int_Q \frac{1}{\epsilon} w'_\epsilon(\theta)|\nabla\theta|^2 \,\mathrm{d}x\mathrm{d}t. \tag{8.2.13}$$

Then we interpolate the Lebesgue space $L^{(1+\epsilon)r/(2-r)}(\Omega)$ between $W^{1,r}(\Omega)$ and $L^\omega(\Omega)$ in order to exploit the already obtained estimate in $L^\infty(I;L^\omega(\Omega))$. More specifically, by the Gagliardo-Nirenberg inequality (see (B.3.20) on p.519), we obtain

$$\begin{aligned}\big\|1+\bar{\theta}_\tau(t,\cdot)\big\|_{L^{(1+\epsilon)r/(2-r)}(\Omega)} &\le C_{\rm GN} \big\|1+\bar{\theta}_\tau(t,\cdot)\big\|_{L^\omega(\Omega)}^{1-\lambda} \Big(\big\|1+\bar{\theta}_\tau(t,\cdot)\big\|_{L^\omega(\Omega)} + \big\|\nabla\bar{\theta}_\tau(t,\cdot)\big\|_{L^r(\Omega;\mathbb{R}^d)}\Big)^\lambda \\ &\le C_{\rm GN} C_{1,\Omega}^{1-\lambda} \Big(C_{1,\Omega} + \big\|\nabla\bar{\theta}_\tau(t,\cdot)\big\|_{L^r(\Omega;\mathbb{R}^d)}\Big)^\lambda\end{aligned} \tag{8.2.14}$$

for

$$\frac{2-r}{(1+\epsilon)r} \ge \lambda\Big(\frac{1}{r} - \frac{1}{d}\Big) + \frac{1-\lambda}{\omega} \quad \text{with} \quad 0 < \lambda \le 1, \tag{8.2.15}$$

[7] This test was essentially proposed by E. Feireisl and J. Málek [186], simplifying the original idea of L. Boccardo and T. Gallouët [72, 73].

where $C_{1,\Omega} := \text{meas}_d(\Omega)^{1/\omega} + C_1'$ with C_1' from (8.2.9b). We rise (8.2.14) to the power $(1+\epsilon)r/(2-r)$ power, and choose $0 < \lambda \le 1$ in such a way to obtain the desired exponent $\lambda(1+\epsilon)r/(2-r) = r$, i.e. $\lambda := (2-r)/(1+\epsilon)$:

$$
\begin{aligned}
I_{r,\epsilon}^{(1)}(\bar\theta_\tau) &\le \int_0^T C_{\rm GN}^{(1+\epsilon)r/(2-r)} C_{1,\Omega}^{(1-\lambda)(1+\epsilon)r/(2-r)} \Big(C_{1,\Omega} + \big\|\nabla\bar\theta_\tau(t,\cdot)\big\|_{L^r(\Omega;\mathbb{R}^d)}\Big)^{\lambda(1+\epsilon)r/(2-r)} \mathrm{d}t \\
&\le \int_0^T C_{\rm GN}^{(1+\epsilon)r/(2-r)} C_{1,\Omega}^{(1-\lambda)(1+\epsilon)r/(2-r)} \Big(C_{1,\Omega} + \big\|\nabla\bar\theta_\tau(t,\cdot)\big\|_{L^r(\Omega;\mathbb{R}^d)}\Big)^{r} \mathrm{d}t \\
&\le C_2\Big(1 + \int_Q \big|\nabla\bar\theta_\tau\big|^r \mathrm{d}x\mathrm{d}t\Big).
\end{aligned}
\tag{8.2.16}
$$

Furthermore, we estimate $I_\epsilon^{(2)}(\bar\theta_\tau)$ in (8.2.12). Let us denote by $\mathcal{X}$ the primitive function of $w_\epsilon \circ \mathscr{C}_{\rm V}^{-1}$ with w_ϵ from (8.2.11) such that $\mathcal{X}(0) = 0$; note that $\mathcal{X}$ is convex because both w_ϵ and $\mathscr{C}_{\rm V}^{-1}$ are increasing. It is important here to have $w_\epsilon(\bar\theta_\tau(t,\cdot)) \in H^1(\Omega)$, hence it is a legal test function, because $0 \le \bar\theta_\tau(t,\cdot) \in H^1(\Omega)$ has already been proved and because w_ϵ is Lipschitz continuous on $[0,+\infty)$. We now use the discrete heat equation (8.2.8). Realizing that $w_\epsilon'(\theta) = \epsilon/(1+\theta)^{1+\epsilon}$ as used already in (8.2.13), we further estimate:

$$
\begin{aligned}
I_\epsilon^{(2)}(\bar\theta_\tau) &= \frac{1}{\epsilon}\int_Q w_\epsilon'(\bar\theta_\tau)|\nabla\bar\theta_\tau|^2 \mathrm{d}x\mathrm{d}t \\
&\le \frac{1}{\kappa_0\epsilon}\int_Q w_\epsilon'(\bar\theta_\tau)\bar{\mathbb{K}}_\tau(\underline{\theta}_\tau)\nabla\bar\theta_\tau\cdot\nabla\bar\theta_\tau \,\mathrm{d}x\mathrm{d}t = \frac{1}{\kappa_0\epsilon}\int_Q \bar{\mathbb{K}}_\tau(\underline{\theta}_\tau)\nabla\bar\theta_\tau\cdot\nabla w_\epsilon(\bar\theta_\tau)\,\mathrm{d}x\mathrm{d}t \\
&\le \frac{1}{\kappa_0\epsilon}\Big(\int_Q \bar{\mathbb{K}}_\tau(\underline{\theta}_\tau)\nabla\bar\theta_\tau\cdot\nabla w_\epsilon(\bar\theta_\tau)\,\mathrm{d}x\mathrm{d}t + \int_\Omega \mathcal{X}(\bar\theta_\tau(T,\cdot))\,\mathrm{d}x\Big) \\
&\le \frac{1}{\kappa_0\epsilon}\Big(\int_\Omega \mathcal{X}(\theta_0)\,\mathrm{d}x + \int_\Sigma \bar h_{\mathrm{b},\tau} w_\epsilon(\bar\theta_\tau)\,\mathrm{d}S\,\mathrm{d}t + \int_Q \bar\xi_{\mathrm{A},\tau}(\bar\theta_\tau) w_\epsilon(\bar\theta_\tau)\,\mathrm{d}x\mathrm{d}t\Big) \\
&\le \frac{1}{\kappa_0\epsilon}\Big(\big\|\vartheta_\tau^0\big\|_{L^1(\Omega)} + \big\|\bar h_{\mathrm{b},\tau}\big\|_{L^1(\Sigma)} + \big\|\bar\xi_{\mathrm{A},\tau}(\bar\theta_\tau)\big\|_{L^1(Q)}\Big) \\
&\le \frac{1}{\kappa_0\epsilon}\Big(\big\|\mathscr{C}_{\rm V}(\theta_0)\big\|_{L^1(\Omega)} + \big\|h_{\mathrm{b}}\big\|_{L^1(\Sigma)} + \big\|\xi_{\mathrm{A}}(\bar\theta_\tau)\big\|_{L^1(Q)}\Big) =: C_3\Big(1 + \big\|\xi_{\mathrm{A}}(\bar\theta_\tau)\big\|_{L^1(Q)}\Big).
\end{aligned}
\tag{8.2.17}
$$

The inequality on the 2nd line uses $\kappa_0 = \inf_{(t,x)\in Q,\ \theta\in\mathbb{R}^+,\ j\in\mathbb{R}^d,\ |j|=1} \mathbb{K}(t,x,\theta)j\cdot j$ which is assumed positive, cf. from (8.2.3b), while the inequality on the 4th line arises due to testing of (8.2.7) by using monotonicity of w_ϵ and of $\mathscr{C}_{\rm V}$, and hence the mentioned convexity of $\mathcal{X}$ so that the "discrete chain rule" holds:

$$
\frac{\mathcal{X}(\theta_\tau^k) - \mathcal{X}(\theta_\tau^{k-1})}{\tau} \le \frac{\vartheta_\tau^k - \vartheta_\tau^{k-1}}{\tau} w_\epsilon(\theta_\tau^k).
$$

The inequality on the 5th line of (8.2.17) uses $\vartheta_\tau^0 \le \mathscr{C}_{\rm V}(\theta_0)$ due to (8.2.7c) with the monotonicity of $\mathscr{C}_{\rm V}$, and $\int_0^T \bar h_{\mathrm{b},\tau}(t,x)\mathrm{d}t \le \int_0^T h_{\mathrm{b}}(t,x)\,\mathrm{d}t$ and similarly $\int_0^T |\bar\xi_{\mathrm{A},\tau}(t,x,\theta)|\mathrm{d}t \le \int_0^T |\xi_{\mathrm{A}}(t,x,\tau)|\,\mathrm{d}t$.

Using (8.2.16) risen to the power $1-r/2$ and (8.2.17) risen to the power $r/2$ for estimation of the right-hand side of (8.2.12) gives the estimate

$$\|\nabla\bar{\theta}_\tau\|^r_{L^r(Q;\mathbb{R}^d)} \le C_{\epsilon,r,T}C_2C_3\Big(1+\|\nabla\bar{\theta}_\tau\|^r_{L^r(Q;\mathbb{R}^d)}\Big)^{1-r/2}\Big(1+\|\xi_{\rm A}(\bar{\theta}_\tau)\|_{L^1(Q)}\Big)^{r/2}.$$

It is important that $\|\nabla\bar{\theta}_\tau\|_{L^r(Q;\mathbb{R}^d)}$ occurs on the left-hand side in the power r which is always higher than its power on the right-hand side $r(1-r/2)$ because $r>0$, and therefore the estimate of $\nabla\bar{\theta}_\tau$ is indeed obtained. More specifically, we obtain (8.2.6a) here for the approximate solution, and later inherited when passing to the limit, too.

Substituting this choice of $\lambda := (2-r)/(1+\epsilon)$ into (8.2.15),[8] one gets after some algebra the conditions

$$r \le \frac{2\omega+d-\epsilon d}{\omega+d}. \tag{8.2.18}$$

To obtain an "isotropic" estimate of θ, we still use the Sobolev embedding $W^{1,r}(\Omega)\subset L^{r^*}(\Omega)$ and interpolate the respective "anisotropic" spaces to see the embedding

$$L^r(I;L^{r^*}(\Omega))\cap L^\infty(I;L^\omega(\Omega)) \subset L^p(I;L^p(\Omega)) \cong L^p(Q) \tag{8.2.19}$$

for some $p\ge 1$ sufficiently small. This can be seen by using Proposition B.4.8 on p. 524. The condition (B.4.6) here reads as

$$\frac{1}{p} = \lambda\frac{d-r}{dr}+(1-\lambda)\frac{1}{\omega} \quad\text{ and }\quad \frac{1}{p} = \lambda\frac{1}{r}+(1-\lambda)\frac{1}{+\infty}.$$

The second relation gives $\lambda = r/p$ and then the first one yields

$$p = r\Big(1+\frac{\omega}{d}\Big) \tag{8.2.20}$$

and, in view of (8.2.18), one identifies the bound $p < 1+2\omega/d$. The exponent $1/p$ in (8.2.6b) results from $1/r$ used in (8.2.6a) and thus also in the estimate of $\|\theta\|_{L^r(I;L^{r^*}(\Omega))}$, and then in (8.2.19) written as (B.2.7) more specifically here

$$\|\theta\|_{L^p(Q)} \le C\|\theta\|^\lambda_{L^r(I;L^{r^*}(\Omega))}\|\theta\|^{1-\lambda}_{L^\infty(I;L^\omega(\Omega))} \le C'\Big(1+\|\xi_{\rm A}(\theta)\|_{L^1(Q)}\Big)^{\lambda/r}\|\theta\|^{1-\lambda}_{L^\infty(I;L^\omega(\Omega))}$$

used with our choice $\lambda = r/p$.

We can further read also the information about $\nabla\bar{\vartheta}_\tau$ from the chain rule:

$$\nabla\bar{\vartheta}_\tau = \nabla\mathscr{C}_{\rm V}(\bar{\theta}_\tau) = \mathscr{C}'_{\rm V}(\bar{\theta}_\tau)\nabla\bar{\theta}_\tau = c_{\rm V}(\bar{\theta}_\tau)\nabla\bar{\theta}_\tau. \tag{8.2.21}$$

[8] Note that $0<\lambda<1$ needed in (8.2.15) is automatically ensured by $1\le r<2$ and $\epsilon>0$.

From the already obtained estimates of $\nabla\bar{\theta}_\tau$ in $L^r(Q;\mathbb{R}^d)$ and $\bar{\theta}_\tau$ in $L^p(Q)$. we can see that $c_V(\bar{\theta}_\tau)$ is bounded in $L^{r'}(Q)$,[9] so that altogether the bound of $\nabla\bar{\vartheta}_\tau$ in $L^1(Q;\mathbb{R}^d)$ is obtained. Only here we used the upper bound on c_V in (8.2.3a) which needs the restriction $\omega < d/(d-1)$ not to be in conflict with the lower bound in (8.2.3a). Actually, this bound would hold if the growth of c_V in (8.2.3a) were maximal (i.e. $s = \omega/d$) but, as we put some "reserve" in (8.2.3a), we have actually the sequence $\{\nabla\bar{\vartheta}_\tau\}_{\tau>0}$ bounded in $L^{1+\epsilon}(Q;\mathbb{R}^d)$ for some $\epsilon > 0$.

Step 5 – convergence: By the Banach selection principle, we can consider a subsequence and some θ and ϑ such that, for some $\epsilon > 0$,

$$\bar{\theta}_\tau \to \theta \text{ weakly in } L^r(I;W^{1,r}(\Omega)) \cap L^p(Q) \quad \text{for any } 1 \le r < \frac{2\omega+d}{\omega+d},\ 1 \le p < \frac{2\omega+d}{d}, \tag{8.2.22a}$$

$$\bar{\vartheta}_\tau \to \vartheta \text{ weakly in } L^{1+\epsilon}(I;W^{1,1+\epsilon}(\Omega)) \cap L^p(Q) \text{ for any } 1 \le p < 1 + \frac{1}{d}. \tag{8.2.22b}$$

The bound for the exponent p in (8.2.22a) has been derived in Step 4 while the bound in (8.2.22b) due to the embedding $W^{1,1}(\Omega) \subset L^{d'}(\Omega)$ and then the interpolation $L^\infty(I;L^1(\Omega)) \cap L^1(I;L^{d'}(\Omega)) \subset L^{1+1/d}(Q)$.

We need however a strong convergence to pass to the limit through the nonlinearities $\mathbb{K}$, ξ_A, and $\mathscr{C}_V$. For this, we need some information about a time derivative of ϑ or of θ. Here, as we already estimated $\nabla\bar{\vartheta}_\tau$, we can exploit an estimate for $\dot{\vartheta}_\tau$ directly from the equation by comparison. (For the later option cf. Remark 8.2.4 below.) In view of (8.2.8), after using Green's formula and taking into account also the boundary conditions, we have the identity

$$\int_Q \dot{\vartheta}_\tau v \,\mathrm{d}x\mathrm{d}t = \int_Q \xi_{A,\tau}(\bar{\theta}_\tau)v - \bar{\mathbb{K}}_\tau(\underline{\theta}_\tau)\nabla\bar{\theta}_\tau\cdot\nabla v \,\mathrm{d}x\mathrm{d}t + \int_\Sigma \bar{h}_{b,\tau} v \,\mathrm{d}S\,\mathrm{d}t. \tag{8.2.23}$$

Using the test-functions v's from a set bounded in $L^\infty(Q)$ for which also ∇v's is bounded in $L^\infty(Q;\mathbb{R}^d)$, we can estimate the right-hand side of (8.2.23). In this way, we can estimate $\dot{\vartheta}_\tau$ in $L^1(I;H^{(d+3)/2}(\Omega)^*)$, relying on the embedding $H^{(d+3)/2}(\Omega) \subset W^{1,\infty}(\Omega)$.

This gives also the same estimate for the time derivative of $\bar{\vartheta}_\tau$ which, in fact, is in $\mathfrak{M}(I;H^{(d+3)/2}(\Omega)^*)$ rather than in $L^1(I;H^{(d+3)/2}(\Omega)^*)$. Anyhow, one can use the generalization of the Aubin-Lions theorem (cf. Lemma B.4.11) to see

$$\bar{\vartheta}_\tau \to \vartheta \quad \text{strongly in } L^p(Q) \quad \text{for any } 1 \le p < 1 + \frac{1}{d}. \tag{8.2.24}$$

Then, $\bar{\theta}_\tau = \mathscr{C}_V^{-1}(\bar{\vartheta}_\tau)$. Due to the lower bound in (8.2.3a), the growth of $\mathscr{C}_V^{-1}(\cdot)$ is at most $1/\omega$. By the continuity of $\mathscr{C}_V^{-1}$, from (8.2.24) we can conclude that

---

[9] More in detail, the growth restriction (8.2.3a) gives $c_V(\bar{\theta}_\tau) \in L^{p/s}(Q)$ here needed for $p/s > r'$, which implies $s < \omega/d$ used in (8.2.3a) when taken $r < (2\omega+d)/(\omega+d)$ into account.

$$\bar{\theta}_\tau = \mathscr{C}_{\mathrm{V}}^{-1}(\bar{\vartheta}_\tau) \to \mathscr{C}_{\mathrm{V}}^{-1}(\vartheta) = \theta \quad \text{strongly in } L^p(Q) \quad \text{for any } 1 \le p < \omega + \frac{\omega}{d}. \tag{8.2.25}$$

Actually, always $\omega+\omega/d \le 1+2\omega/d$ due to our restriction $\omega < d/(d{-}1)$ so that, in view of (8.2.22a), we have the strong convergence in (8.2.25) even for $p < 1+2\omega/d$ used in (8.2.22a).

For the interpolant $\underline{\theta}_\tau$, we obtain the same strong convergence as (8.2.25).[10] Also ϑ_τ converges to the same limit ϑ. The proof of convergence in (8.2.8) written in a weak formulation towards the weak formulation of the original problem (8.2.1) is then easy. □

Remark 8.2.2 (Alternative technique for $\nabla\vartheta$). We can alternatively estimate $\nabla\vartheta$ and use Aubin-Lions' theorem for ϑ in the relation $\theta = \mathscr{C}_{\mathrm{V}}^{-1}(\vartheta)$ in the Step 5. An easiest modification is to strengthen the assumption on $\mathbb{K}$ by requiring $\mathbb{K}(t,x,\theta)j{\cdot}j \ge \kappa_0 c_{\mathrm{V}}(\theta)|j|^2$ instead of mere uniform positive definiteness in (8.2.3b), and realizing $\nabla\vartheta = c(\theta)\nabla\theta$ so we can eliminate θ from (8.2.1) to obtain

$$\dot{\vartheta} - \mathrm{div}\Big(\frac{\mathbb{K}(t,x,\mathscr{C}_{\mathrm{V}}^{-1}(\vartheta))}{c_{\mathrm{V}}(\mathscr{C}_{\mathrm{V}}^{-1}(\vartheta))}\nabla\vartheta\Big) = \xi_{\mathrm{A}}(t,x,\mathscr{C}_{\mathrm{V}}^{-1}(\vartheta)).$$

We thus get directly the estimate on $\nabla\vartheta$ in $L^r(Q;\mathbb{R}^d)$ with $1 \le r < (2\omega{+}d)/(\omega{+}d)$ without using (8.2.21).

Remark 8.2.3 (Galerkin method[11]). The "nonlinear" tests in Steps 2 and 4 in the above proof cannot be executed for Faedo-Galerkin approximation. Anyhow, a successive two-step approximation works if the proof is suitably reorganized. Namely, taking $V_k \subset H^1(\Omega) =: V$ satisfying (C.2.55), one can use

$$\dot{\vartheta}_{\varepsilon k} - \mathfrak{L}_{\mathbb{K}(t,x,\theta_{\varepsilon k}),k}\theta_{\varepsilon k} = \mathfrak{h}_k(\xi_\varepsilon(\theta_{\varepsilon k}), h_{\flat,\varepsilon}) \quad \text{with } \vartheta_{\varepsilon k} = \mathscr{C}_{\mathrm{V}}(\theta_{\varepsilon k}) \tag{8.2.26a}$$

$$\text{and with } \xi_\varepsilon(t,x,\theta) := \frac{\xi_{\mathrm{A}}(t,x,\theta)}{1{+}\varepsilon|\xi_{\mathrm{A}}(t,x,\theta)|} \quad \text{and } h_{\flat,\varepsilon} := \frac{h_\flat}{1{+}\varepsilon|h_\flat|} \tag{8.2.26b}$$

with the initial condition $\theta_{\varepsilon k}(0) = \theta_{\varepsilon k,0}$ with $\theta_{\varepsilon k,0} \in V_k$ being an approximation of θ_0 such that $\theta_{\varepsilon k,0} \to \theta_{\varepsilon,0} := \theta_0/(1{+}\theta_0)$ in $L^{\omega+1}(\Omega)$ for $k \to \infty$. In (8.2.26a), the linear operator $\mathfrak{L}_{\mathbb{K},k} : V_k \to V_k^*$ and $\mathfrak{h}_k(\xi_{\mathrm{A}}, h_\flat) \in V_k^*$ are defined by

$$\langle \mathfrak{L}_{\mathbb{K},k}\theta, v\rangle = \int_\Omega \mathbb{K}\nabla\theta{\cdot}\nabla v\,\mathrm{d}x \quad \text{and} \quad \langle \mathfrak{h}_k(\xi_{\mathrm{A}}, h_\flat), v\rangle = \int_\Omega \xi_{\mathrm{A}} v\,\mathrm{d}x + \int_\Gamma h_\flat v\,\mathrm{d}S$$

[10] Note that $\underline{\theta}_\tau$ enjoys the same a-priori estimates as $\bar{\vartheta}_\tau$, and therefore must converges strongly (no further selection of a subsequence is needed as its limit must be again ϑ used in (8.2.22b) and again by the continuity of the Nemytskiĭ mapping we have $\underline{\theta}_\tau = \mathscr{C}_{\mathrm{V}}^{-1}(\bar{\mathfrak{I}}_\tau(\cdot{-}\tau), \underline{\vartheta}_\tau) \to \mathscr{C}_{\mathrm{V}}^{-1}(\vartheta) = \theta$.

[11] See e.g. [96, 455] for usage of a two-step Galerkin method in particular evolution problems.

for $v \in V_k$. This system of nonlinear ordinary differential equations (8.2.26) has a solution by standard arguments based on the test[12] by $\theta_{\varepsilon k}$ and these solutions converge (in terms of subsequences) for $k \to \infty$, cf. Sect. C.2.4. Any such limit $\theta_\varepsilon \in L^\infty(I;L^{\omega+1}(\Omega)) \cap L^2(I;H^1(\Omega))$ with $C_v(\theta_\varepsilon) \in L^\infty(I;L^{1+1/\omega}(\Omega)) \cap H^1(I;H^1(\Omega)^*)$ solves in the weak sense the parabolic initial-boundary-value problem

$$\dot{\vartheta}_\varepsilon - \operatorname{div}(\mathbb{K}(\theta_\varepsilon)\nabla\theta_\varepsilon) = \xi_\varepsilon(\theta_\varepsilon) \quad \text{with} \quad \vartheta_\varepsilon = \mathscr{C}_v(\theta_\varepsilon) \tag{8.2.27a}$$

$$\text{and with the boundary condition} \quad \mathbb{K}(\theta_\varepsilon)\nabla\theta_\varepsilon = h_{\flat,\varepsilon} \tag{8.2.27b}$$

with the initial condition $\theta_\varepsilon(0) = \theta_{\varepsilon,0}$ and with ξ_ε and $h_{\flat,\varepsilon}$ from (8.2.26b). Only after this, we can execute the mentioned nonlinear tests and obtain the a-priori estimates as in Steps 2 and 4, and eventually make the limit passage for $\varepsilon \to 0$ analogously as in Step 5.

Remark 8.2.4 (Estimation of $\dot\theta$). Actually, for the autonomous homogeneous heat-conductivity tensor $\mathbb{K} = \mathbb{K}(\theta)$, Step 5 in the above proof for the continuous problem (8.2.27) bears an alternative approach by estimating $\dot\theta_\varepsilon$ instead of $\nabla\vartheta_\varepsilon$ and then use the Aubin-Lions theorem on θ_ε instead of ϑ_ε. To this goal, we can write (8.2.27a) in terms of θ in the form

$$\dot\theta_\varepsilon = \frac{\operatorname{div}(\mathbb{K}(\theta_\varepsilon)\nabla\theta_\varepsilon) + \xi_\varepsilon(\theta_\varepsilon)}{c_v(\theta)}$$

so that

$$\int_Q \dot\theta_\varepsilon v \, dxdt = \int_Q \frac{\operatorname{div}(\mathbb{K}(\theta_\varepsilon)\nabla\theta_\varepsilon) + \xi_\varepsilon(\theta_\varepsilon)}{c_v(\theta)} v \, dxdt = \int_Q \bigg(\int_\Sigma \frac{h_{\flat,\varepsilon}}{c_v(\theta)} v \, dS \, dt$$
$$+ \bigg(\frac{c_v'(\theta_\varepsilon)\mathbb{K}(\theta_\varepsilon)\nabla\theta_\varepsilon \cdot \nabla\theta_\varepsilon}{c_v(\theta)^2} + \frac{\xi_\varepsilon(\theta_\varepsilon)}{c_v(\theta)} \bigg) v - \frac{\mathbb{K}(\theta_\varepsilon)\nabla\theta_\varepsilon}{c_v(\theta)} \cdot \nabla v \bigg) dxdt. \tag{8.2.28}$$

To estimate the particular terms in (8.2.28), we may assume e.g. $|c_v'(\theta)| \le Cc_v(\theta)^2/(1+\theta)^{1+\epsilon}$ for some $\epsilon > 0$, and then we can estimate

$$\int_Q \bigg| \frac{c_v'(\theta_\varepsilon)\mathbb{K}(\theta_\varepsilon)\nabla\theta_\varepsilon \cdot \nabla\theta_\varepsilon}{c_v(\theta)^2} \bigg| dxdt \le \int_Q \frac{|c_v'(\theta_\varepsilon)|(\sup|\mathbb{K}|)\,|\nabla\theta_\varepsilon|^2}{c_v(\theta)^2} dxdt$$
$$\le \int_Q C(\sup|\mathbb{K}|) \frac{|\nabla\theta_\varepsilon|^2}{(1+\theta)^{1+\varepsilon}} dxdt = C(\sup|\mathbb{K}|) I_\epsilon^{(2)}(\theta_\epsilon)$$

with $I_\epsilon^{(2)}$ used (and estimated) in (8.2.17). The resting terms in (8.2.28) are even simpler to estimate. Altogether, we thus obtain the estimate of $\dot\theta_\varepsilon$ in $L^1(I;H^{(d+3)/2}(\Omega)^*)$. In particular, we do not need the estimate on $\nabla\vartheta_\varepsilon$ like in (8.2.21), and thus the restriction $\omega < d/(d-1)$ and the upper bound in (8.2.3a) are not needed.

Exercise 8.2.5 (Sobolev with Hölder lead to Gagliardo and Nirenberg). Merge the Sobolev embedding $W^{1,r}(\Omega) \subset L^{r^*}(\Omega)$ with the Lebesgue-space interpolation

[12] Here we use $\dot\vartheta_{\varepsilon k}\theta_{\varepsilon k} = c_v(\theta_{\varepsilon k})\dot\theta_{\varepsilon k}\theta_{\varepsilon k} = C_2(\theta_{\varepsilon k})^\cdot$ with $C_2(\theta) := \int_0^1 r\theta^2 c_v(r\theta)\,dr$ having at least the $\omega+1$-polynomial growth.

(based essentially on Hölder's inequality) between $L^{r^*}(\Omega)$ and $L^{\omega}(\Omega)$ to obtain

$$W^{1,r}(\Omega) \subset L^{r^*}(\Omega) \subset L^p(\Omega) \subset L^{\omega}(\Omega) \tag{8.2.29}$$

as used in (8.2.14)–(8.2.15) with $p = (1+\epsilon)r/(2-r)$, referring to the *Gagliardo-Nirenberg inequality*. Do the same for the embedding

$$W^{2,r}(\Omega) \subset L^{r^{**}}(\Omega) \subset L^p(\Omega) \subset L^{\omega}(\Omega); \tag{8.2.30}$$

cf. also Theorem B.3.14 on p. 518 for $\beta = 0$ and $k = 1$ and 2, respectively.

8.3 Kelvin-Voigt materials with linear thermal expansion

The thermal dilation (expansion) of materials with increasing temperature is the typical phenomenon in most materials which may create (sometimes) very big mechanical stresses. The Kelvin-Voigt viscoelastic rheology is the simplest case on which the estimation technique can be well illustrated.[13] This rheology is a very special case of the general ansatz introduced in Section 8.1 because there is no internal variable z.

We assume the body occupying the domain Ω is made from isotropic viscoelastic and heat conductive linearly-responding material, again described in terms of the small strains here. The departure point is the specific Helmholtz free energy in the partly linearized form (8.1.35) considered here as:

$$\psi(e,\theta) := \frac{1}{2}\mathbb{C}e{:}e - (\theta-\theta_{\rm R})\mathbb{B}{:}e - \gamma_{\rm V}(\theta), \tag{8.3.1}$$

where $\mathbb{B} \in \mathbb{R}^{d\times d}_{\rm sym}$ and $\theta_{\rm R}$ is a reference temperature assumed constant for simplicity. The particular terms in (8.3.1) are related respectively to the elastic stored energy, temperature dilatation, and purely thermal contribution to the free energy.

Alternatively (but equivalently), (8.3.1) can be written in the form

$$\psi(e,\theta) := \frac{1}{2}\mathbb{C}(e-(\theta-\theta_{\rm R})\mathbb{E}){:}(e-(\theta-\theta_{\rm R})\mathbb{E}) - \frac{(\theta-\theta_{\rm R})^2}{2}\mathbb{C}\mathbb{E}{:}\mathbb{E} - \gamma_{\rm V}(\theta), \tag{8.3.2}$$

where $\mathbb{E}$ is the *thermal expansion* (dilatability) matrix. Obviously, we have simply $\mathbb{B} = \mathbb{C}\mathbb{E}$ in (8.3.1). Then the specific entropy defined by (8.1.2) and the heat capacity by (8.1.14b) are

$$\mathfrak{s} = \mathfrak{s}(e,\theta) = \mathbb{B}{:}e + \gamma_{\rm V}'(\theta) \quad \text{and} \quad c_{\rm V}(\theta) = \theta\gamma_{\rm V}''(\theta), \tag{8.3.3}$$

[13] Thermoviscoelasticity in this rheology was historically quite a difficult problem even at small strains mainly because of nonlinear coupling with the heat-transfer equation which has not a variational structure and special techniques had to be developed. After the pioneering work by C.M. Dafermos [136] in one dimension and only much later various three-dimensional studies occurred (cf. e.g. [71, 455]) when the L^1-theory for the nonlinear heat equation has been developed.

while the specific internal energy by (8.1.3) is

$$w = w(e,\theta) := \psi + \theta\mathfrak{s} = \frac{1}{2}\mathbb{C}e\!:\!e + \theta_{\mathrm{R}}\mathbb{B}\!:\!e - \gamma_{\mathrm{V}}(\theta) + \theta\gamma_{\mathrm{V}}'(\theta), \tag{8.3.4}$$

cf. Remark 8.1.7. Besides, we pose the standard kinetic energy $\frac{1}{2}\varrho|\dot{u}|^2$, we consider (in general possibly nonquadratic) convex potential of dissipative force $\zeta:\mathbb{R}^{d\times d}_{\mathrm{sym}}\to\mathbb{R}$ which satisfies

$$\zeta:\mathbb{R}^{d\times d}_{\mathrm{sym}}\to\mathbb{R} \ \text{ strictly convex,} \ \ \zeta(\dot{e}) \ge \varepsilon|\dot{e}|^q, \quad \big|\zeta'(\dot{e})\big| \le C(1+|\dot{e}|^{q-1}), \tag{8.3.5}$$

for some $1<q<+\infty$. This corresponds to the schematic rheology $({}_{\mathbb{C}}\!\!-\!\!{}_{\mathbb{B}}^{\,\theta})\,\|\,{}_{\mathbb{D}}^{\,\dot{e}}$ with possibly a nonlinear damper. We assume also $\mathbb{C}$ positive definite. The equilibrium equation balances the total stress $\sigma = \zeta(\dot{e}) + \partial_e\psi(e,\theta)$ with the inertial forces and outer mechanical loading f:

$$\varrho\ddot{u} - \mathrm{div}\,\sigma = f \quad \text{ with } \quad \sigma = \zeta'(e(\dot{u})) + \mathbb{C}e(u) - (\theta-\theta_{\mathrm{R}})\mathbb{B}\,. \tag{8.3.6}$$

Defining still the heat flux subjected to Fourier's law $j = -\mathbb{K}(e,\theta)\nabla\theta$, cf. (8.1.8), we complete (8.3.6) by the heat equation which results from (8.1.14) after taking into account $\partial^2_{\theta e}\psi = 0$ here:

$$c_{\mathrm{V}}(\theta)\dot{\theta} - \theta\mathbb{B}\!:\!e(\dot{u}) - \mathrm{div}\big(\mathbb{K}(e(u),\theta)\nabla\theta\big) = \xi(e(\dot{u})) := \zeta'(e(\dot{u}))\!:\!e(\dot{u})\,. \tag{8.3.7}$$

As in Section 8.1, we first write the original system (8.1.15) in terms of a suitably "rescaled" temperature (enthalpy), namely $\vartheta = \mathscr{C}_{\mathrm{V}}(\theta) := \int_0^\theta c_{\mathrm{V}}(r)\,\mathrm{d}r$, cf. (8.1.19). As (8.2.2), we transform the system (8.3.6)-(8.3.7) into:

$$\varrho\ddot{u} - \mathrm{div}\Big(\zeta'(e(\dot{u})) + \mathbb{C}e(u) - \theta\mathbb{B}\Big) = f, \tag{8.3.8a}$$

$$\dot{\vartheta} - \mathrm{div}\big(\mathbb{K}(e(u),\theta)\nabla\theta\big) = \zeta'(e(\dot{u}))\!:\!e(\dot{u}) + \theta\mathbb{B}\!:\!e(\dot{u}) \quad \text{with } \vartheta = \mathscr{C}_{\mathrm{V}}(\theta), \tag{8.3.8b}$$

where, for simplicity, we rely on θ_{R} constant so the $\mathrm{div}(\theta_{\mathrm{R}}\mathbb{B}) = 0$. We complete this system by some boundary conditions, e.g. an unsupported body exposed to traction $f_\flat$ and heated by an external boundary flux $h_\flat$, and initial conditions:

$$\vec{n}\cdot\sigma = f_\flat, \qquad \mathbb{K}(e(u),\theta)\nabla\theta\cdot\vec{n} = h_\flat \qquad \text{on } \Sigma, \tag{8.3.9a}$$

$$u(0,\cdot) = u_0, \qquad \dot{u}(0,\cdot) = v_0, \qquad \theta(0,\cdot) = \theta_0 \qquad \text{on } \Omega, \tag{8.3.9b}$$

where σ is from (8.3.6) so that, in particular, (8.3.9a) involves the reference temperature θ_{R}.

The energy balance can be obtained by multiplication of (8.3.8a) by $\dot{u}$ and (8.3.8b) by 1, and by using Green's formula both for (8.3.8a) and once for (8.3.8b): In other words, substituting w from (8.3.4) into (8.1.13) yields the *total-energy balance* as:

$$\frac{\mathrm{d}}{\mathrm{d}t}\int_\Omega \underbrace{\frac{\varrho}{2}|\dot{u}|^2}_{\substack{\text{kinetic}\\ \text{energy}}} + \underbrace{\vartheta}_{\substack{\text{thermal part}\\ \text{of internal}\\ \text{energy}}} + \underbrace{\frac{1}{2}\mathbb{C}e(u){:}e(u)}_{\substack{\text{stored}\\ \text{energy}}}\,\mathrm{d}x$$

$$= \underbrace{\int_\Omega f{\cdot}\dot{u}\,\mathrm{d}x + \int_\Gamma f_\flat{\cdot}\dot{u}\,\mathrm{d}S}_{\substack{\text{power of external}\\ \text{mechanical load}}} + \underbrace{\int_\Gamma h_\flat\,\mathrm{d}S}_{\substack{\text{total power of external}\\ \text{boundary heating}}} . \tag{8.3.10}$$

In what follows, we focus to three particular cases:

(A) nonlinear viscosity $\zeta'(\cdot)$ which enough attenuates fast processes so that $\omega = 1$, which in particular allows for c_v simply constant, and $\mathbb{K}$ uniformly positive definite, or

(B) the dissipation-force potential ζ is quadratic so that $\zeta'(\cdot) = \mathbb{D}\cdot$ is constant but the heat capacity c_v growing with temperature sufficiently fast, and $\mathbb{K}$ again uniformly positive definite, or

(C) a linear or general nonlinear viscosity $\zeta'(\cdot)$ as in (A) or (B) but $\mathbb{K}/c_\mathrm{v}$ uniformly positive definite as in Remark 8.2.2.

Notably, for the viscosity potential $\zeta(\cdot)$ with enough growth as in (A), the option (C) coincides with (A). To some extent natural basic option is constant heat capacity c_v and constant viscosity ζ', i.e. $\omega = 1$ and $q = 2$, which can easily be covered if $d = 1$, and nearly attained also if $d = 2$. In the physically relevant case $d = 3$, the lowest values of $\omega > 1$ and $q \geq 2$ are obtained in the variant (C) on the price that also $\mathbb{K}$ must depend on θ; e.g. $q = 2$ and $\omega \sim 6/5$ is allowed, cf. (8.3.24b).

In any case, we will first regularize the initial-boundary-value problem (8.3.8)–(8.3.9) similarly as we did in (8.2.7), using here a regularization parameter $\varepsilon > 0$:

$$\varrho\ddot{u} - \mathrm{div}\Big(\zeta'(e(\dot{u})) + \mathbb{C}e(u) - \theta\mathbb{B}\Big) = f \qquad \text{on } Q, \tag{8.3.11a}$$

$$\dot{\vartheta} - \mathrm{div}(\mathbb{K}(e(u),\theta)\nabla\theta) = \frac{\zeta'(e(\dot{u})){:}e(\dot{u}) + \theta\mathbb{B}{:}e(\dot{u})}{1+\varepsilon\zeta(e(\dot{u}))} \quad \text{with } \vartheta = \mathscr{C}_\mathrm{v}(\theta) \quad \text{on } Q, \tag{8.3.11b}$$

$$\vec{n}{\cdot}\sigma = f_\flat, \qquad \mathbb{K}(e(u),\theta)\nabla\theta{\cdot}\vec{n} = \frac{h_\flat}{1+\varepsilon h_\flat} \qquad \text{on } \Sigma, \tag{8.3.11c}$$

$$u(0,\cdot) = u_0, \qquad \dot{u}(0,\cdot) = v_0, \qquad \theta(0,\cdot) = \frac{\theta_0}{1+\varepsilon\theta_0} \qquad \text{on } \Omega. \tag{8.3.11d}$$

As to the first option (A), admitting nonlinear viscosity $\zeta'(\cdot)$ allows for a simple scenario which allows for choosing $\omega = 1$ in (8.2.3a) and thus includes, in particular, the possibility of the heat capacity c_v being constant provided the exponent q in (8.3.5) is sufficiently large, namely $q > 1 + d/2$. Let us note that this scenario allows for the linear viscosity (i.e. $q = 2$) only in 1-dimensional cases.[14] On the other hand, (8.2.3) now should be modified for $\mathbb{K} = \mathbb{K}(e,\theta)$ depending also on e and the qualification of the mechanical data is to be added. Let us summarize the modified/enhanced set of assumptions:

[14] Actually, this resembles the original result of C. Dafermos [136] from 1982.

$$c_v,\, 1/c_v : \mathbb{R} \to \mathbb{R}^+,\ \mathbb{K} : \mathbb{R}^{d\times d}_{\rm sym} \times \mathbb{R} \to \mathbb{R}^{d\times d} \ \text{continuous and bounded,} \tag{8.3.12a}$$

$$\mathbb{C} \in \mathrm{SLin}(\mathbb{R}^{d\times d}_{\rm sym}) \ \text{positive definite,}\ \ \mathbb{K}\ \text{uniformly positive definite,} \tag{8.3.12b}$$

$$\zeta : \mathbb{R}^{d\times d}_{\rm sym} \to \mathbb{R}\ \text{convex, smooth, and}$$

$$\exists \varepsilon > 0\ \forall e \in \mathbb{R}^{d\times d}_{\rm sym} :\ \ \varepsilon|e|^q \le \zeta'(e) \le C(1+|e|^q)/\varepsilon, \tag{8.3.12c}$$

$$f \in L^1(I; L^2(\Omega;\mathbb{R}^d)),\ \ f_\flat \in W^{1,1}(I; L^{2^{*\prime}}(\Gamma;\mathbb{R}^d)),\ \ h_\flat \in L^1(\Sigma),\ \ h_\flat \ge 0, \tag{8.3.12d}$$

$$u_0 \in W^{1,q}(\Omega;\mathbb{R}^d),\qquad v_0 \in L^2(\Omega;\mathbb{R}^d),\qquad \theta_0 \in L^1(\Omega),\qquad \theta_0 \ge 0. \tag{8.3.12e}$$

Proposition 8.3.1 (Thermo-visco-elasticity: existence of weak solutions). *Let (8.3.5) and (8.3.12) hold with $q > 1 + d/2$, and let ϱ satisfies (6.3.5). Then the initial-boundary-value problem (8.3.8)–(8.3.9) has a weak solution (u,θ,ϑ) with $u \in W^{1,q}(I; W^{1,q}(\Omega;\mathbb{R}^d)) \cap W^{1,\infty}(I; L^2(\Omega)) \cap W^{2,q'}(I; W^{1,q}(\Omega;\mathbb{R}^d)^*)$ and $\theta,\vartheta \in L^r(I; W^{1,r}(\Omega)) \cap L^\infty(I; L^1(\Omega))$ with $\vartheta \in W^{1,1}(I; H^{(d+3)/2}(\Omega)^*)$ which also conserve the total and the mechanical energy in the sense that (8.3.10) integrated over I and (8.3.14) hold.*

Sketch of the proof. We divide the (sketched) proof on the five steps analogously as in the proof of Proposition 8.2.1.

Step 1 – Approximate solution: Let us take a solution to the regularized problem (8.3.11) by $(u_\varepsilon,\theta_\varepsilon)$. Existence of such solutions can be proved by using the Rothe or the Galerkin or possibly Galerkin/Rothe methods; cf. Remark 8.3.5 below. Let us note that, for each $\varepsilon > 0$ fixed, the standard L^2-theory for the parabolic equation with all bulk-boundary and initial data bounded can be used.

Step 2 – Non-negativity of temperature: Test the enthalpy equation (8.3.11b) by ϑ_ε^-. It gives the information $\vartheta_\varepsilon \ge 0$ (since $h_\flat \ge 0$ and $\vartheta_0 \ge 0$).

Step 3 – Energetic estimates: From (8.3.10), obtain basic a-priori estimates. Since $\theta \ge 0$, (8.3.10) yields boundedness of (some approximate) solutions

$$\big\|u_\varepsilon\big\|_{L^\infty(I;H^1(\Omega;\mathbb{R}^d))} \le C,\quad \big\|\dot u_\varepsilon\big\|_{L^\infty(I;L^2(\Omega;\mathbb{R}^d))} \le C,\quad \big\|\theta_\varepsilon\big\|_{L^\infty(I;L^1(\Omega))} \le C. \tag{8.3.13}$$

Still it does not allow for a limit passage in the nonlinear terms $\theta_\varepsilon\mathbb{B}:e(\dot u_\varepsilon)/(1+\varepsilon\zeta(\dot u_\varepsilon))$ and $\zeta'(e(\dot u_\varepsilon)):e(\dot u)/(1+\varepsilon\zeta(\dot u_\varepsilon))$. And also a limit passage in the relation $\vartheta_\varepsilon = \mathscr{C}_v(\theta_\varepsilon)$ in (8.3.8b) is still not possible unless a very special case c_v constant.

Step 4 – Estimation of temperature gradient: We use the the mechanical energy balance integrated over I

$$\begin{aligned}\int_\Omega \frac{\varrho}{2}\big|\dot u_\varepsilon(T)\big|^2 + \frac12 \mathbb{C}e(u_\varepsilon(T)):e(u_\varepsilon(T))\,\mathrm{d}x + \int_Q \zeta'(e(\dot u_\varepsilon)):e(\dot u_\varepsilon)\,\mathrm{d}x\mathrm{d}t\\ = \int_Q \theta_\varepsilon\mathbb{B}:e(\dot u_\varepsilon) + f\cdot\dot u_\varepsilon\,\mathrm{d}x\mathrm{d}t + \int_\Sigma f_\flat\cdot\dot u_\varepsilon\,\mathrm{d}S\,\mathrm{d}t + \int_\Omega \frac{\varrho}{2}|v_0|^2 + \frac12\mathbb{C}e(u_0):e(u_0)\,\mathrm{d}x.\end{aligned} \tag{8.3.14}$$

Further, we use the "prefabricated" estimate (8.2.6b) here with the heat production/consumption rate $\xi = \xi(\dot e,\theta) = (\zeta'(\dot e) - \theta\mathbb{B}):\dot e/(1+\varepsilon\zeta'(\dot e)$, which gives

$$\begin{aligned}\|\theta_\varepsilon\|^p_{L^p(Q;\mathbb{R}^d)} &\le C_p^p\Big(1+\Big\|\frac{(\zeta'(e(\dot u_\varepsilon))-\theta_\varepsilon\mathbb{B}):e(\dot u_\varepsilon)}{1+\varepsilon\zeta(e(\dot u_\varepsilon))}\Big\|_{L^1(Q)}\Big)\\ &\le C_p^p\Big(1+\big\|(\zeta'(e(\dot u_\varepsilon))-\theta_\varepsilon\mathbb{B}):e(\dot u_\varepsilon)\big\|_{L^1(Q)}\\ &\le C_p^p\Big(1+\int_Q \zeta'(e(\dot u_\varepsilon)):e(\dot u_\varepsilon)+\big|\theta_\varepsilon\mathbb{B}:e(\dot u_\varepsilon)\big|\,\mathrm{d}x\mathrm{d}t\Big)\end{aligned}\tag{8.3.15}$$

and add it to (8.3.14) with a sufficiently small weight $0<\varkappa<1/C_p^p$. Forgetting the first integral in (8.3.14), this gives the estimate

$$\begin{aligned}(1-\varkappa C_p^p)\int_Q \zeta'(e(\dot u_\varepsilon)):e(\dot u_\varepsilon)\,\mathrm{d}x\mathrm{d}t+\varkappa\|\theta_\varepsilon\|^p_{L^p(Q;\mathbb{R}^d)} \le \int_Q f\cdot\dot u_\varepsilon\,\mathrm{d}x\mathrm{d}t+\int_\Sigma f_\flat\cdot\dot u_\varepsilon\,\mathrm{d}S\,\mathrm{d}t\\ +\int_\Omega \frac{\varrho}{2}|v_0|^2+\frac12\mathbb{C}e(u_0):e(u_0)\,\mathrm{d}x+\varkappa C_p^p+(1+\varkappa C_p^p)\int_Q\big|\theta_\varepsilon\mathbb{B}:e(\dot u_\varepsilon)\big|\,\mathrm{d}x\mathrm{d}t.\end{aligned}\tag{8.3.16}$$

The last term is to be estimated while the other right-hand side terms are already controlled due to (8.3.13). To this goal, assuming $p^{-1}+q^{-1}<1$, we can estimate

$$\int_Q\big|\theta_\varepsilon\mathbb{B}:e(\dot u_\varepsilon)\big|\,\mathrm{d}x\mathrm{d}t\le C_{\delta,p,q}+\delta\|\theta_\varepsilon\|^p_{L^p(Q)}+\delta\|e(\dot u_\varepsilon)\|^q_{L^q(Q;\mathbb{R}^{d\times d})}\tag{8.3.17}$$

with some $C_{\delta,p,q}$, and absorb the last two terms in the left-hand side of (8.3.16) by choosing $\delta>0$ sufficiently small, counting also the first assumption in (8.3.5). We thus eventually obtain the estimates

$$\|e(\dot u_\varepsilon)\|_{L^q(Q;\mathbb{R}^{d\times d})}\le C_2 \quad\text{with } q \text{ from (8.3.5)},\tag{8.3.18a}$$

$$\|\theta_\varepsilon\|_{L^p(Q)}\le C_p \quad\text{with } 1\le p<1+2/d\tag{8.3.18b}$$

and, using still the other "prefabricated" estimate (8.2.6a), also

$$\|\nabla\theta_\varepsilon\|_{L^r(Q;\mathbb{R}^d)}\le C_r \quad\text{with } 1\le r<(2+d)/(1+d)\tag{8.3.18c}$$

and eventually, by comparison from the force equilibrium, also

$$\|\sqrt{\varrho}\ddot u_\varepsilon\|_{L^{q'}(I;W^{1,q}(\Omega;\mathbb{R}^d)^*)}\le C;\tag{8.3.18d}$$

note p and r are from Proposition 8.2.1 with $\omega=1$. As for (8.3.18d), relying on (6.3.5), we estimate

$$\begin{aligned}\|\sqrt{\varrho}\ddot u_\varepsilon\|_{L^{q'}(I;W^{1,q}(\Omega;\mathbb{R}^d)^*)} &= \sup_{\|v\|_{L^q(I;W^{1,q}(\Omega;\mathbb{R}^d))}\le 1}\int_Q \sqrt{\varrho}\ddot u_\varepsilon v\,\mathrm{d}x\mathrm{d}t\\ = \max(N,\|\varrho\|^{1/2}_{L^\infty(\Omega)}) &\sup_{\|v\|_{L^q(I;W^{1,q}(\Omega;\mathbb{R}^d))}\le 1}\int_Q\Big((\zeta'(e(\dot u_\varepsilon))+\mathbb{C}e(u_\varepsilon)-\theta_\varepsilon\mathbb{B}):e(v)\\ &\qquad -f\cdot v\Big)\mathrm{d}x\mathrm{d}t+\int_\Sigma f_\flat\cdot v\,\mathrm{d}S\,\mathrm{d}t\end{aligned}\tag{8.3.19}$$

where N is the norm of the linear operator $v \mapsto \sqrt{\varrho}v : H^1(\Omega;\mathbb{R}^d) \to H^1(\Omega;\mathbb{R}^d)$ as in (6.4.9); here we used (8.3.12c) to guarantee $\zeta'(e(\dot{u})) \in L^{q'}(Q;\mathbb{R}^{d\times d}_{\rm sym})$ and (8.2.6b) for estimation $\int_Q \theta_\varepsilon \mathbb{B}:e(v)\,\mathrm{d}x\mathrm{d}t \le C + \|\theta_\varepsilon\|^p_{L^p(Q)} + \|e(v)\|^q_{L^q(Q;\mathbb{R}^{d\times d})}$ like we did in (8.3.17).

Step 5 – Convergence: First we chose the weakly* convergent subsequence respecting the a-priori estimates (8.3.13) and (8.3.18). As $\omega = 1$, we have also ϑ_ε estimated in $L^r(I;W^{1,r}(\Omega)) \cap L^p(Q)$ and also its time derivative is estimated, cf. (8.2.5b).

By Aubin-Lions' theorem, the limit passage in the relation $\vartheta = \mathscr{C}_{\rm v}(\theta)$ in (8.3.8b) is in the lines of (8.2.24)–(8.2.25). In particular, we can rely on that $\theta_\varepsilon = \mathscr{C}_{\rm v}^{-1}(\vartheta_\varepsilon)$ converges strongly in $L^p(Q)$.

To pass to the limit through the dissipative heat $\xi(e(\dot{u}_\varepsilon))$, we need to prove strong convergence of $e(\dot{u}_\varepsilon)$ in $L^q(Q;\mathbb{R}^{d\times d})$. Like in (6.4.25), for the dissipation rate $\xi(\dot{e}) := \zeta'(\dot{e}):\dot{e}$, we have

$$
\begin{aligned}
\int_Q \xi(e(\dot{u}))\,\mathrm{d}x\mathrm{d}t &\le \liminf_{\varepsilon\to 0} \int_Q \xi(e(\dot{u}_\varepsilon))\,\mathrm{d}x\mathrm{d}t \le \limsup_{\varepsilon\to 0} \int_Q \xi(e(\dot{u}_\varepsilon))\,\mathrm{d}x\mathrm{d}t \\
&\le \limsup_{\varepsilon\to 0} \frac{1}{2}\int_\Omega \varrho|v_0|^2 + \mathbb{C}e(u_0):(u_0) - \varrho\big|\dot{u}_\varepsilon(T)\big|^2 - \mathbb{C}e(u_\varepsilon(T)):e(u_\varepsilon(T))\,\mathrm{d}x \\
&\qquad\qquad - \int_Q \theta_\varepsilon \mathbb{B}:e(\dot{u}_\varepsilon)\,\mathrm{d}x\mathrm{d}t + \int_Q f\cdot\dot{u}_\varepsilon\,\mathrm{d}x\mathrm{d}t + \int_\Sigma g\cdot\dot{u}_\varepsilon\,\mathrm{d}S\,\mathrm{d}t \\
&\le \frac{1}{2}\int_\Omega \varrho|v_0|^2 + \mathbb{C}e(u_0):(u_0) - \varrho\big|\dot{u}(T)\big|^2 - \mathbb{C}e(u(T)):e(u(T))\,\mathrm{d}x \\
&\quad - \int_Q \theta\mathbb{B}:e(\dot{u})\,\mathrm{d}x\mathrm{d}t + \int_Q f\cdot\dot{u}\,\mathrm{d}x\mathrm{d}t + \int_\Sigma g\cdot\dot{u}\,\mathrm{d}S\,\mathrm{d}t = \int_Q \xi(e(\dot{u}))\,\mathrm{d}x\mathrm{d}t, \quad (8.3.20)
\end{aligned}
$$

from which we get $\lim_{\varepsilon\to 0}\int_Q \xi(e(\dot{u}_\varepsilon))\,\mathrm{d}x\mathrm{d}t = \int_Q \xi(e(\dot{u}))\,\mathrm{d}x\mathrm{d}t$ when using the strong convergence of $\theta_\varepsilon \to \theta$ in $L^p(Q)$. Hence, counting again the weak convergence $e(\dot{u}_\varepsilon) \to e(\dot{u})$ in $L^q(Q;\mathbb{R}^{d\times d})$, we obtain the strong convergence; here the strict convexity of ζ assumed in (8.3.5) is also important – see Theorem B.2.9. For the last equality in (8.3.20), the important fact is that $\varrho\ddot{u} \in L^2(I;W^{1,q}(\Omega;\mathbb{R}^d)^*)$ is in duality with $\dot{u} \in L^2(I;W^{1,q}(\Omega;\mathbb{R}^d))$ due to (8.3.18d). so that the test of the force-balance equation (8.3.8b) by $\dot{u}$ is legal and gives indeed the mechanical-energy equality, cf. Lemma C.2.2 on p. 535. □

The second mentioned interesting particular situation (B) is the linear viscosity, i.e. $q = 2$ and $\zeta'(\dot{e}) = \mathbb{D}\dot{e}$ as considered in (most of the) Section 6.4. This means $\zeta(\dot{e}) = \frac{1}{2}\mathbb{D}\dot{e}:\dot{e}$ and $\xi(\dot{e}) = \mathbb{D}\dot{e}:\dot{e}$. The exponent ω is subjected to little more restriction than only $\omega \ge 1$ to facilitate estimation of the adiabatic effects. More specifically, the assumptions (8.3.12) now should be modified

$$\mathbb{C} \in \mathrm{SLin}(\mathbb{R}^{d\times d}_{\rm sym}),\ \mathbb{D} \in \mathrm{Lin}(\mathbb{R}^{d\times d}_{\rm sym}) \text{ positive definite,} \tag{8.3.21a}$$

$$c_{\rm v} : \mathbb{R} \to \mathbb{R}^+,\ \ \mathbb{K} : \mathbb{R}^{d\times d}_{\rm sym} \times \mathbb{R} \to \mathrm{Lin}(\mathbb{R}^{d\times d}_{\rm sym}) \text{ continuous,} \tag{8.3.21b}$$

$$\mathbb{K} \text{ bounded, uniformly positive definite,} \tag{8.3.21c}$$

$$\exists \omega \geq 1,\ \omega > \frac{d}{2},\ c_0, c_1 > 0,\ \ \omega-1 \leq s < \frac{\omega}{d}\ \ \forall \theta \in \mathbb{R}^+ :$$
$$c_0(1+\theta^{\omega-1}) \leq c_v(\theta) \leq c_1(1+\theta)^s, \tag{8.3.21d}$$

$$f \in L^1(I;L^2(\Omega;\mathbb{R}^d)),\ \ f_\flat \in W^{1,1}(I;L^{2^{*\prime}}(\Gamma;\mathbb{R}^d)),\ \ h_\flat \in L^1(\Sigma),\ \ h_\flat \geq 0, \tag{8.3.21e}$$

$$u_0 \in H^1(\Omega;\mathbb{R}^d)),\ \ v_0 \in L^2(\Omega;\mathbb{R}^d),\ \ \vartheta_0 := \mathscr{C}_v(\theta_0) \in L^1(\Omega),\ \ \theta_0 \geq 0. \tag{8.3.21f}$$

Proposition 8.3.2 (Thermo-visco-elasticity - variant B). *Let (8.3.21) hold and ϱ satisfy (6.3.5). Then the initial-boundary-value problem (8.3.8)–(8.3.9) with $\zeta(\dot{e}) := \frac{1}{2}\mathbb{D}\dot{e} : \dot{e}$ has a weak solution (u,θ,ϑ) with $u \in H^1(I;H^1(\Omega;\mathbb{R}^d)) \cap W^{1,\infty}(I;L^2(\Omega)) \cap H^2(I;H^1(\Omega;\mathbb{R}^d)^*)$ and $\theta,\vartheta \in L^r(I;W^{1,r}(\Omega)) \cap L^\infty(I;L^1(\Omega))$ with $\vartheta \in W^{1,1}(I;H^{(d+3)/2}(\Omega)^*)$ which also conserve the total and the mechanical energy in the sense that (8.3.10) integrated over I and (8.3.14) hold.*

Sketch of the proof. We just outlined modifications in the proof of Proposition 8.3.1.

Steps 1 and 2 – Existence of approximate solutions, nonnegativity of θ: the same as in the proof of Proposition 8.3.1.

Step 3 – Energetic estimates: From (8.3.10), obtain basic a-priori estimates. Since $\theta \geq 0$, (8.3.10) yields boundedness of (some approximate) solutions again in the sense (8.3.13a,b) together with

$$\big\|\vartheta_\varepsilon\big\|_{L^\infty(I;L^1(\Omega))} \leq C \quad \text{and} \quad \big\|\theta_\varepsilon\big\|_{L^\infty(I;L^\omega(\Omega))} \leq C \tag{8.3.22}$$

Step 4 – Estimation of temperature gradient: One just uses (8.3.16) with $\zeta'(e(\dot{u})) = \mathbb{D}e(\dot{u})$ and $p > 2$ while $q = 2$ in (8.3.17). The estimates (8.3.18) now hold for $q = 2$ and p and r again as in Proposition 8.2.1 now for a general ω, i.e.

$$\big\|e(\dot{u}_\varepsilon)\big\|_{L^q(Q;\mathbb{R}^{d\times d})} \leq C \ \text{ with } q \text{ from (8.3.5)}, \tag{8.3.23a}$$
$$\|\theta_\varepsilon\|_{L^p(Q)} \leq C_p \ \text{ with } 1 \leq p < 1+2\omega/d \tag{8.3.23b}$$

and, using still the other "prefabricated" estimate (8.2.6a), also

$$\|\nabla\theta_\varepsilon\|_{L^r(Q;\mathbb{R}^d)} \leq C_r \ \text{ with } 1 \leq r < (2\omega{+}d)/(\omega{+}d) \tag{8.3.23c}$$

and eventually, by comparison from the force equilibrium, also

$$\|\ddot{u}_\varepsilon\|_{L^{q'}(I;W^{1,q}(\Omega;\mathbb{R}^d)^*)} \leq C. \tag{8.3.23d}$$

Let us note that the condition $p < 1+2\omega/d$ in (8.3.23b) is compatible with $p > 2$ needed for (8.3.17) with $q = 2$ only if $1+2\omega/d > 2$, i.e. $\omega > d/2$, as indeed assumed in (8.3.21d).

Step 5 - Convergence: As in (8.2.21), we estimate also $\nabla\vartheta_\varepsilon$ in $L^1(Q;\mathbb{R}^d)$ provided the growth restriction in (8.3.21d) with $s < \omega/d$ holds.

By (8.3.20) with $\xi(e(\dot u)) = \mathbb{D}e(\dot u){:}e(\dot u)$ we again see that $e(\dot u_\varepsilon) \to e(\dot u)$ strongly in $L^2(Q;\mathbb{R}^{d\times d}_{\rm sym})$. Then the limit passage through the nonlinearities in the heat equation is easy. $\square$

The last mentioned interesting particular situation (C) admits again the linear viscosity, i.e. $\zeta(\dot e) = \frac12\mathbb{D}\dot e:\dot e$ and $\xi(\dot e) = \mathbb{D}\dot e:\dot e$ as a particular case, but $\mathbb{K}$ qualified differently than in (B), causing the exponent ω subjected to little less restriction than $\omega > d/2$ needed for (B), and even no growth restriction to the heat capacity. More specifically, the assumptions (8.3.21c,d) are now modified as

$$(e,\theta) \mapsto \frac{\mathbb{K}(e,\theta)}{c_{\rm v}(\theta)} : \mathbb{R}^{d\times d}_{\rm sym} \times \mathbb{R} \to \mathbb{R}^{d\times d} \text{ bounded, continuous, and uniformly positive definite,} \tag{8.3.24a}$$

$$\exists \omega > \frac{dq}{(d{+}2)(q{-}1)},\ c_0 > 0,\ \forall \theta{\in}\mathbb{R}^+: \qquad c_{\rm v}(\theta) \ge c_0(1+\theta^{\omega-1}) \tag{8.3.24b}$$

with q from (8.3.5).

Proposition 8.3.3 (Thermo-visco-elasticity - variant C). *Let (8.3.5), (8.3.12d,e), and (8.3.24) hold, and ϱ satisfy (6.3.5). Then the initial-boundary-value problem (8.3.8)–(8.3.9) has a weak solution (u,θ,ϑ) with $u \in W^{1,q}(I;W^{1,q}(\Omega;\mathbb{R}^d)) \cap W^{1,\infty}(I;L^2(\Omega)) \cap W^{2,q'}(I;W^{1,q}(\Omega;\mathbb{R}^d)^*)$ and $\theta,\vartheta \in L^r(I;W^{1,r}(\Omega)) \cap L^\infty(I;L^1(\Omega))$ with $1 \le r < (2\max(1,\omega){+}d)/(\max(1,\omega){+}d)$ and $\vartheta \in W^{1,1}(I;H^{(d+3)/2}(\Omega)^*)$ which also conserve the total and the mechanical energy in the sense that (8.3.10) integrated over I and (8.3.14) hold.*

Sketch of the proof. We transform the initial-boundary-value problem (8.3.8)–(8.3.9) by using Remark 8.2.2. This results to to a problem formulated exclusively in terms of u and ϑ:

$$\varrho\ddot u - {\rm div}\Big(\zeta'(e(\dot u)) + \mathbb{C}e(u) + \mathscr{C}_{\rm v}^{-1}(\vartheta)\mathbb{B}\Big) = f \qquad \text{on } Q, \tag{8.3.25a}$$

$$\dot\vartheta - {\rm div}\Big(\frac{\mathbb{K}(e(u),\mathscr{C}_{\rm v}^{-1}(\vartheta))}{c_{\rm v}(\mathscr{C}_{\rm v}^{-1}(\vartheta))}\nabla\vartheta\Big) = \Big(\zeta'(e(\dot u)) - \mathscr{C}_{\rm v}^{-1}(\vartheta)\mathbb{B}\Big){:}e(\dot u) \qquad \text{on } Q, \tag{8.3.25b}$$

$$\vec n{\cdot}\sigma = f_\flat, \qquad \mathbb{K}(e(u),\mathscr{C}_{\rm v}^{-1}(\vartheta))\nabla\vartheta{\cdot}\vec n = h_\flat c_{\rm v}(\mathscr{C}_{\rm v}^{-1}(\vartheta)) \qquad \text{on } \Sigma, \tag{8.3.25c}$$

$$u(0,\cdot) = u_0, \qquad \dot u(0,\cdot) = v_0, \qquad \vartheta(0,\cdot) = \mathscr{C}_{\rm v}(\theta_0) \qquad \text{on } \Omega. \tag{8.3.25d}$$

We just outlined modification in the proof of Propositions 8.3.1 and 8.3.2.

We again first regularize (8.3.25) analogously as we did in (8.3.11) Relying on the coercivity of $c_{\rm v}(\theta) \ge c_0(1{+}\theta^{\omega-1})$ assumed in (8.3.24b), we have $\mathscr{C}_{\rm v}(\theta) \ge c_0(\theta{+}\omega\theta^\omega)$ and thus we control the growth of the inverse

$$\mathscr{C}_{\rm v}^{-1}(\vartheta) \le C_\omega(1+\vartheta)^{1/\omega} \tag{8.3.26}$$

with some C_ω. The estimate (8.3.17) with $p^{-1}+q^{-1} < 1$ now modifies as

$$\begin{aligned}\Big|\int_Q \mathscr{C}_{\rm v}^{-1}(\vartheta)\mathbb{B}{:}e(\dot u)\,{\rm d}x{\rm d}t\Big| &\le C_{\delta,p,q} + \delta\big\|\mathscr{C}_{\rm v}^{-1}(\vartheta)\big\|^p_{L^p(Q)} + \delta\big\|e(\dot u)\big\|^q_{L^q(Q;\mathbb{R}^{d\times d})}\\ &\le C'_{\delta,p,q} + \delta C_\omega^p\big\|1{+}\vartheta\big\|^{p/\max(1,\omega)}_{L^{p/\max(1,\omega)}(Q)} + \delta\big\|e(\dot u)\big\|^q_{L^q(Q;\mathbb{R}^{d\times d})}\end{aligned} \tag{8.3.27}$$

with some $C_{\delta,p,q}$ and $C'_{\delta,p,q}$. Then we apply Proposition 8.2.1 to the heat-transfer problem (8.3.25b) for ϑ, which gives here the estimate $L^{p/\max(1,\omega)}$-estimate for ϑ provided $p/\max(1,\omega) < 1+2/d$; cf. (8.2.6b). Realizing that (8.3.27) works for $p > q' = q/(q-1)$, we obtain the restriction $\omega(1+2/d) > q/(q-1)$ used in (8.3.24b). □

Remark 8.3.4 (Positivity of temperature). If $\vartheta_0 \geq \varepsilon > 0$, then one can prove positivity of ϑ during the whole evolution. One can estimate $\dot{\vartheta} - \mathrm{div}(\mathcal{K}(e(u),\vartheta)\nabla\vartheta) \geq -C|\mathscr{C}_{\mathrm{V}}^{-1}(\vartheta)|^2 \geq -C|\vartheta|^2$ for a sufficiently large C and C', and then compare it with the solution to the ordinary differential equation $\dot{\chi} = -C|\chi|^2$ with $\chi(0) = \varepsilon$. Obviously, $\chi(t) = 1/(C't + 1/\varepsilon)$. Subtracting the equations for ϑ and χ and testing it by $-(\vartheta-\chi)^-$, one gets $\vartheta \geq \chi > 0$.

Remark 8.3.5 (Approximation strategies). The rigorous approximation is a bit delicate. The Galerkin method applied directly to the nonregularized problem does not work smoothly due to various "nonlinear" tests needed, in particular the test by the negative part to prove non-negativity of temperature and the test by $\chi(\theta)$ to estimate the temperature gradient, as pointed out already in Remark 8.2.3. One should use a successive limit passage so that the heat-transfer equation is continuous in space and can be tested by nonlinear functions of ϑ.[15] The Rothe method works more straightforwardly, it allows here even a decoupling by fractional-step method but some regularization is needed to compensate the growth of the heat sources/sinks on the discrete level, cf. also (8.2.7). There are in principle two options: to increase the growth of the left-hand side or to decrease the growth of the right-hand side; here in (8.3.11) the latter option was employed.

Remark 8.3.6 (Modifications for non-simple materials). We saw that the linear heat equation (in particular $\omega = 1$) in case $d = 3$ would need more dissipation. Instead of nonlinear dissipation as in option (A) above, an option is to involve a higher-order term, using the concept of 2nd-grade nonsimple viscoelastic materials. For the 4-th order term thus arising, we would get (counting also respective Korn's inequality) $\|\nabla^2\dot{u}\|^2_{L^2(\Omega;\mathbb{R}^{d\times d\times d})}$ on the left-hand side of (8.3.17) and the last term in (8.3.17) can still be interpolated

$$\big\|e(\dot{u})\big\|_{L^p(\Omega;\mathbb{R}^{d\times d})} \leq C\big\|\dot{u}\big\|^{1-\nu}_{L^2(\Omega;\mathbb{R}^d)}\big\|\nabla^2\dot{u}\big\|^{\nu}_{L^2(\Omega;\mathbb{R}^{d\times d\times d})} \leq CC_2^{1-\nu}\big\|\nabla^2\dot{u}\big\|^{\nu}_{L^2(\Omega;\mathbb{R}^{d\times d\times d})}$$

for

$$\frac{1}{p} \geq \frac{1}{d} + \nu\Big(\frac{1}{2} - \frac{2}{d}\Big) + (1-\nu)\frac{1}{2} \quad \text{with} \quad \frac{1}{2} \leq \nu \leq 1. \tag{8.3.28}$$

For $d = 3$ and $p > 5/2$, it allows us to take $\nu = 13/20$ and then $\nu p < 2$ holds (at least if p is considered as close to 5/2, as we always can here, so that νp is close to 13/8),

[15] In fact, such limit passage needs a lot of technicalities, e.g. in case of the Galerkin scheme, a regularization to compensate a sub-linear growth of the dissipative and the adiabatic heat terms, and then a careful successive limit passage first with space discretization of the heat equation, then by the mentioned regularization, and finally in the space discretization of the mechanical part, cf. e.g. [96, 455] for details.

and we can absorb this term. Alternatively, another option is to modify the linear dissipation to some nonlinear one like in Remark 6.4.15 with a polynomial growth $> 5/2$, and then use Minty's trick in limit passage in the equilibrium equation before eventually proving strong convergence of $e(\dot{u})$ in $L^{5/2-\varepsilon}(Q;\mathbb{R}^{d\times d})$.

Remark 8.3.7 (Alternative model). Consistently with the free energy depending on the elastic strain $e_{\rm EL} = e(u) - \mathbb{B}\theta$, one can consider the dissipation potential $\frac{1}{2}\mathbb{D}\dot{e}_{\rm EL}:\dot{e}_{\rm EL}$ instead of $\frac{1}{2}\mathbb{D}e(\dot{u}):e(\dot{u})$. In other words, instead of the rheology $(\mathbb{C}$—$\mathbb{B}^{\theta})\,||\,\mathbb{D}$, one can consider the model $(\mathbb{C}\,||\,\mathbb{D})$—$\mathbb{B}^{\theta}$. Then $e_{\rm EL} + \mathbb{B}\theta - e(u) = 0$ is the holonomic constraint while $e_{\rm EL}$ is in the position of an internal variable which will be addressed in the next section.

Remark 8.3.8 (Melting/solidification). Making the elastic-moduli tensor $\mathbb{C}_\varepsilon$ in (7.7.38) dependent continuously on temperature instead of $x\in\Omega$, one can model the *solid/liquid phase transformation* during melting or solidification processes, with the liquid as the Boger fluid as in Proposition 7.7.11. Cf. also [389] for the static problem.

Exercise 8.3.9. In addition to strict convexity of ζ in (8.3.5), assume the derivative ζ' is not only strictly but uniformly monotone in the sense that $(\zeta'(\dot{e}_1)-\zeta'(\dot{e}_2)):(\dot{e}_1-\dot{e}_2) \geq \varepsilon|\dot{e}_1-\dot{e}_2|^q$. Prove the strong convergence of $e(\dot{u}_k)$ directly without using Theorem B.2.9 and modifying the arguments in (8.3.20).

8.4 Materials with internal variables

Let us now consider the general system (8.1.15). The peculiarity is that $c_{\rm v}$ depends also on the mechanical variables. This needs essentially usage of gradients of theses variables, cf. e.g. Definition 8.1.1 or Step 5 in the proof below.

For simplicity of the explanation and to avoid usage of the concept of nonsimple materials, we admit only dependence on the (unspecified) internal variables z, i.e. $c_{\rm v} = c_{\rm v}(z,\theta)$, cf. Remark 8.4.3 for the dependence on the strain e. In the smooth situations, this is related with the structural assumption $\partial^3_{e\theta\theta}\psi = 0$. For Step 5 in the proof below, we will need still linearity of the viscous stress, i.e. ζ quadratic in terms of $\dot{e}$. Together with the mentioned gradient theory, it leads to the ansatz

$$\psi_{\rm E}(e,z,\nabla z,\theta) = \psi(e,z,\theta) + \frac{\kappa}{2}|\nabla z|^2 \quad \text{with}$$
$$\psi(e,z,\theta) = \varphi(e,z) + \varphi_{\rm c}(e,z)\theta - \gamma_{\rm v}(z,\theta), \tag{8.4.1a}$$
$$\zeta(\dot{e},\dot{z}) = \frac{1}{2}\mathbb{D}\dot{e}:\dot{e} + \zeta_{\rm IN}(\dot{z}), \tag{8.4.1b}$$

cf. also (8.1.35). The heat capacity and its primitive is then

$$c_{\rm v}(z,\theta) = \theta\partial^2_{\theta\theta}\gamma_{\rm v}(z,\theta) \quad \text{and} \quad \mathscr{C}_{\rm v}(z,\theta) = \theta\partial_\theta\gamma_{\rm v}(z,\theta) - \gamma_{\rm v}(z,\theta); \tag{8.4.2}$$

cf. also (8.1.36). We will exploit the form (8.1.15a,b)-(8.1.20) and use the calculus[16]

$$c_V(z,\theta)\dot\theta = \frac{\partial}{\partial t}\mathscr{C}_V(z,\theta) - \partial_z\mathscr{C}_V(z,\theta)\cdot\dot z \quad \text{with} \quad \mathscr{C}_V(z,\theta) = \int_0^1 \theta c_V(z,r\theta)\,\mathrm{d}r, \tag{8.4.3}$$

the heat equation (8.1.24) then takes the simplified form

$$\begin{aligned}\dot\vartheta - \mathrm{div}(\mathbb{K}(z,\theta)\nabla\theta) &= \xi(e(\dot u),\dot z) + \theta\partial_e\varphi_C(e(u),z){:}e(\dot u)\\ &\quad + (\theta\partial_z\varphi_C(e(u),z) - \partial_z\gamma_V(z,\theta) + \partial_z\gamma_V(z,0))\cdot\dot z \quad \text{with } \vartheta = \mathscr{C}_V(z,\theta).\end{aligned} \tag{8.4.4}$$

Let us note that, possibly up to a constant, both definitions of $\mathscr{C}_V$ in (8.4.2) and (8.4.3) coincide with each other, the latter one normalizing $\mathscr{C}_V(z,0) = 0$.

Altogether, introducing still the abbreviation for a *thermal stress* and a *thermal driving force* as

$$\sigma_{\mathrm{th}}(e,z,\theta) := \theta\partial_e\varphi_C(e,z) \quad \text{and} \quad \sigma_{\mathrm{in}}(e,z,\theta) := \theta\partial_z\varphi_C(e,z) - \partial_z\gamma_V(z,\theta),$$

the system (8.1.15) now takes the form

$$\varrho\ddot u - \mathrm{div}\big(\mathbb{D}e(\dot u) + \partial_e\varphi(e(u),z) + \sigma_{\mathrm{th}}(e(u),z,\theta)\big) = f, \tag{8.4.5a}$$

$$\partial\zeta_{\mathrm{IN}}(\dot z) + \partial_z\varphi(e(u),z) + \sigma_{\mathrm{in}}(e(u),z,\theta) = \kappa\Delta z, \tag{8.4.5b}$$

$$\begin{aligned}\dot\vartheta - \mathrm{div}(\mathbb{K}(z,\theta)\nabla\theta) &= \xi(e(\dot u),\dot z) + \sigma_{\mathrm{th}}(e(u),z,\theta){:}e(\dot u)\\ &\quad + (\sigma_{\mathrm{in}}(e(u),z,\theta) + \partial_z\gamma_V(z,0))\cdot\dot z \qquad \text{with } \vartheta = \mathscr{C}_V(z,\theta).\end{aligned} \tag{8.4.5c}$$

We complete this system with the initial and boundary conditions (8.1.17) now together with the condition $\nabla z\cdot\vec n = 0$ on Σ.

Beside the total-energy balance (8.1.13), we here need also the mechanical-energy balance resulted by testing separately (8.4.5a,b) like in (7.1.7). Here it gives some additional terms, namely

$$\begin{aligned}\frac{\mathrm{d}}{\mathrm{d}t}\int_\Omega \frac{\varrho}{2}|\dot u|^2 + \varphi(e(u),z)\,\mathrm{d}x + \int_\Omega \xi(e(\dot u),\dot z)\,\mathrm{d}x\mathrm{d}t = \int_\Omega f\cdot\dot u\,\mathrm{d}x\\ + \int_\Gamma g\cdot\dot u\,\mathrm{d}S + \int_\Omega \sigma_{\mathrm{th}}(e(u),z,\theta){:}e(\dot u) + \sigma_{\mathrm{in}}(e(u),z,\theta)\cdot\dot z.\end{aligned} \tag{8.4.6}$$

There is a certain freedom in qualifying the data, as we also saw in the preceding section. Let us take a relatively simple set of assumptions, being also parallel to Variant (B) from the previous Section 8.3 due to the linear viscosity used here:

$$\varrho > 0, \ \ \kappa > 0, \ \ \mathbb{D} \in \mathrm{Lin}(\mathbb{R}^{d\times d}_{\mathrm{sym}}) \text{ positive definite}, \tag{8.4.7a}$$

$$\varphi(e,z) \ge c(|e|^2 + |z|), \quad (\partial_e\varphi(e_1,z) - \partial_e\varphi(e_2,z)){:}(e_1-e_2) \ge c|e_1-e_2|^2,$$

---

[16] Similar family of enthalpy transformations (parametrized by (t,x) instead of e) has been used in [397].

$$\exists C < +\infty: \quad |\partial_e \varphi(e,z)| \le C(1+|e|+|z|^{q^*/2}), \tag{8.4.7b}$$

$$\zeta_{\rm IN}: \mathbb{R}^m \to \mathbb{R}\cup\{+\infty\} \text{ strictly convex}, \quad \zeta_{\rm IN}(\dot z) \ge c|\dot z|^q,$$

$$\exists C < +\infty: \quad \zeta_{\rm IN}(\dot z) < +\infty \ \Rightarrow\ \xi_{\rm IN}(\dot z) = \partial\zeta_{\rm IN}(\dot z)\cdot\dot z \le C(1+|\dot z|^q), \tag{8.4.7c}$$

$$\varphi_{\rm C}(\cdot,z): \mathbb{R}^{d\times d}_{\rm sym} \to \mathbb{R} \text{ convex}, \quad |\varphi_{\rm C}'(e,z)| \le C, \tag{8.4.7d}$$

$$|\partial_z\gamma_{\rm V}(z,\theta)| \le C(1+\theta), \qquad |\partial^2_{z\theta}\gamma_{\rm V}(z,\theta)| \le C, \tag{8.4.7e}$$

$$\exists \omega \ge 1,\ \omega > \frac{d}{2},\ c_0, c_1 > 0,\ \ \omega-1 \le s < \frac{\omega}{d} \ \ \forall \theta\in\mathbb{R}^+:$$

$$c_0(1+\theta^{\omega-1}) \le c_{\rm V}(z,\theta) \le c_1(1+\theta)^s, \tag{8.4.7f}$$

$$\mathbb{K}: \mathbb{R}^m\times\mathbb{R}^+ \to \mathbb{R}^{d\times d} \text{ continuous, bounded, uniformly positive definite}, \tag{8.4.7g}$$

$$f\in L^1(I;L^2(\Omega;\mathbb{R}^d)),\quad f_{\rm b}\in W^{1,1}(I;L^{2^{*\prime}}(\Gamma;\mathbb{R}^d)),\quad h\in L^1(\Sigma), \qquad h\ge 0, \tag{8.4.7h}$$

$$u_0 \in H^1(\Omega;\mathbb{R}^d), \qquad v_0 \in L^2(\Omega;\mathbb{R}^d), \qquad z_0 \in H^1(\Omega;\mathbb{R}^m), \tag{8.4.7i}$$

$$\theta_0 > 0 \text{ such that } \vartheta_0 := \mathscr{C}_{\rm V}(z_0,\theta_0) \in L^1(\Omega). \tag{8.4.7j}$$

We now modify Proposition 8.3.2 appropriately.

Proposition 8.4.1 (Existence of weak solutions). *Let (8.4.7) and (8.3.12) hold with $q > 1 + d/2$. Then the initial-boundary-value problem for (8.4.5) with the initial and boundary conditions (8.1.17) together with $\nabla z\cdot\vec n = 0$ on Σ has a weak solution (u,z,θ,ϑ) with $u \in H^1(I;H^1(\Omega;\mathbb{R}^d)) \cap W^{1,\infty}(I;L^2(\Omega)) \cap H^2(I;H^1(\Omega;\mathbb{R}^d)^*)$, $z\in C_{\rm w}(I;H^1(\Omega;\mathbb{R}^m))$ with $\dot z\in L^q(Q;\mathbb{R}^m)$, $\theta \in L^r(I;W^{1,r}(\Omega)) \cap L^\infty(I;L^\omega(\Omega))$ and $\vartheta \in L^\infty(I;L^1(\Omega)) \cap W^{1,1}(I;H^{(d+3)/2}(\Omega)^*)$ which also conserves the total and the mechanical energy in the sense that (8.3.10) integrated over I and (8.3.14) hold.*

Sketch of the proof. We divide the (sketched) proof into five steps analogously as in the proof of Proposition 8.2.1.

Step 1 – Approximate solution: Assume some approximate solution, e.g. Rothe or Galerkin or possibly Galerkin/Rothe; cf. Remark 8.3.5.

Step 2 – Non-negativity of temperature: Test the enthalpy equation modified by ϑ^-. It gives the information $\vartheta \ge 0$ (since $h_{\rm b} \ge 0$ and $\vartheta_0 \ge 0$).

Step 3 – Energetic estimates: By an "energetic" test, i.e. by testing the equations in (8.4.5) respectively by $\dot u$, $\dot z$, and by 1, and exploiting the proved nonnegativity $\theta \ge 0$, one gets the a-priori estimates for (some approximate, e.g., Galerkin) solutions

$$\|u\|_{W^{1,\infty}(I;L^2(\Omega;\mathbb{R}^d))\cap L^\infty(I;H^1(\Omega;\mathbb{R}^d))} \le C, \tag{8.4.8a}$$

$$\|z\|_{L^\infty(I;H^1(\Omega;\mathbb{R}^m))} \le C, \tag{8.4.8b}$$

$$\|\vartheta\|_{L^\infty(I;L^1(\Omega))} \le C, \quad \|\theta\|_{L^\infty(I;L^\omega(\Omega))} \le C. \tag{8.4.8c}$$

Step 4 – Estimation of temperature gradient: We again use the "prefabricated" estimate (8.2.6b) here with the heat production/consumption rate as in the right-hand side of (8.4.5c), i.e. $\xi_{\rm A} = \xi_{\rm A}(e,z,\theta,\dot e,\dot z) = \xi(\dot e,\dot z) + \sigma_{\rm th}(e(u),z,\theta){:}\dot e + (\sigma_{\rm in}(e,z,\theta) + \partial_z\gamma_{\rm v}(z,0))\cdot\dot z$, which gives

$$\|\theta\|^p_{L^p(Q;\mathbb{R}^d)} \le C_p^p\Big(1 + \big\|\xi(e(\dot u),\dot z) + \sigma_{\rm th}(e(u),z,\theta){:}e(\dot u) + (\sigma_{\rm in}(e,z,\theta) + \partial_z\gamma_{\rm v}(z,0))\cdot\dot z\big\|_{L^1(Q)}\Big); \tag{8.4.9}$$

note that $\xi(e(\dot u),\dot z) = \mathbb{D}e(\dot u){:}e(\dot u) + \xi_{\rm IN}(\dot z)$ is controlled in $L^1(Q)$ due to the assumption (8.4.7c). We then add it with a sufficiently small weight $0 < \varkappa < 1/C_p^p$ to the mechanical energy balance (8.4.6) integrated over I. This gives the estimate

$$(1-\varkappa C_p^p)\int_Q \xi(e(\dot u),\dot z)\,{\rm d}x{\rm d}t + \varkappa\|\theta\|^p_{L^p(Q;\mathbb{R}^d)} \le \int_Q f\cdot\dot u\,{\rm d}x{\rm d}t + \int_\Sigma f_{\rm b}\cdot\dot u\,{\rm d}x{\rm d}t + \int_\Omega \frac{\varrho}{2}|v_0|^2 + \varphi(e(u_0),z_0) + \frac{\kappa}{2}|\nabla z_0|^2\,{\rm d}x + \varkappa C_p^p + (1+\varkappa C_p^p)\int_Q\Big|\sigma_{\rm th}(e(u),z,\theta){:}e(\dot u) + (\sigma_{\rm in}(e,z,\theta) + \partial_z\gamma_{\rm v}(z,0))\cdot\dot z\Big|\,{\rm d}x{\rm d}t. \tag{8.4.10}$$

The last term is to be estimated while the other right-hand-side terms are already controlled due to (8.3.13). To this goal, assuming $p^{-1} + \min(2,q)^{-1} < 1$, we note that the assumption (8.4.7d) ensures that

$$|\sigma_{\rm th}(e,z,\theta)| \le C(1+\theta) \quad\text{and}\quad |\sigma_{\rm in}(e,z,\theta)| \le C(1+\theta), \tag{8.4.11}$$

we then can estimate

$$\int_Q\Big|\sigma_{\rm th}(e(u),z,\theta){:}e(\dot u) + (\sigma_{\rm in}(e,z,\theta) + \partial_z\gamma_{\rm v}(z,0))\cdot\dot z\Big|\,{\rm d}x{\rm d}t \le C_{\delta,p,q} + \delta\|\theta\|^p_{L^p(Q)} + \delta\|e(\dot u)\|^2_{L^2(Q;\mathbb{R}^{d\times d})} + \delta\|\dot z\|^q_{L^q(Q;\mathbb{R}^m)} \tag{8.4.12}$$

with some $C_{\delta,p,q}$, where q is from (8.4.7c), and absorb the last two terms in the left-hand side of (8.3.16) by choosing $\delta > 0$ sufficiently small, counting also the first assumption in (8.4.7c).

We thus eventually obtain the estimates

$$\|e(\dot u)\|_{L^2(Q;\mathbb{R}^{d\times d})} + \|\dot z\|_{L^q(Q;\mathbb{R}^m)} \le C_2 \quad\text{with } q \text{ from (8.4.7c)} \tag{8.4.13a}$$

and, using still the other "prefabricated" estimates (8.2.6), also

$$\|\theta\|_{L^p(Q)} \le C_p \quad\text{with } 1 \le p < 1 + 2\omega/d, \tag{8.4.13b}$$

$$\|\nabla\theta\|_{L^r(Q;\mathbb{R}^d)} \le C_r \quad\text{with } 1 \le r < (2\omega{+}d)/(\omega{+}d)\text{, and also} \tag{8.4.13c}$$

$$\|\nabla\vartheta\|_{L^1(Q;\mathbb{R}^d)} \le C. \tag{8.4.13d}$$

Eventually, by comparison from the force equilibrium and from the heat equations, we obtain also

$$\|\ddot{u}\|_{L^{q'}(I;W^{1,q}(\Omega;\mathbb{R}^d)^*)} \le C \quad \text{and} \quad \|\dot{\vartheta}\|_{W^{1,1}(I;H^{(d+3)/2}(\Omega)^*)} \le C; \tag{8.4.13e}$$

note p and r are from Proposition 8.2.1 with ω from (8.4.7f). For (8.4.13d), we have used the calculus

$$\nabla\vartheta = \nabla\mathscr{C}_\text{V}(z,\theta) = \partial_z\mathscr{C}_\text{V}(z,\theta)\nabla z + \partial_\theta\mathscr{C}_\text{V}(z,\theta)\nabla\theta = \partial_z\mathscr{C}_\text{V}(z,\theta)\nabla z + c_\text{V}(z,\theta)\nabla\theta. \tag{8.4.14}$$

Then we use that $\partial_z\mathscr{C}_\text{V}(z,\theta)$ is bounded in $L^2(Q)$ due to the assumption (8.4.7e) and that $c_\text{V}(z,\theta)$ is bounded in $L^{r/s}(Q)$ due to the assumption (8.4.7f) and the already obtained estimate (8.4.13b), so that $c_\text{V}(z,\theta)\nabla\theta \in L^1(Q)$ due to (8.4.13c); cf. also the argumentation we already used for (8.2.21).

As for (8.4.13e), relying on (6.3.5), we estimate

$$\begin{aligned}\|\sqrt{\varrho}\ddot{u}\|_{L^2(I;H^1(\Omega;\mathbb{R}^d)^*)} &= \sup_{\|v\|_{L^2(I;H^1(\Omega;\mathbb{R}^d))}\le 1/\varrho} \int_Q \varrho\ddot{u}\,\mathrm{d}x\mathrm{d}t \\ &= \sup_{\|v\|_{L^2(I;H^1(\Omega;\mathbb{R}^d))}\le 1/\varrho} \int_Q (\mathbb{D}e(\dot{u}) + \partial_e\varphi(e(u),z) + \sigma_{\text{th}}(e(u),z,\theta)):e(v) - f\cdot v\,\mathrm{d}x\mathrm{d}t\end{aligned}$$

by using the growth of $\partial_e\varphi$ assumed in (8.4.7b) to have $\partial_e\varphi(e(u),z)$ controlled in $L^2(Q;\mathbb{R}^{d\times d}_{\text{sym}})$ and the boundedness of $\partial_e\varphi_\text{C}$ assumed in (8.4.7d) for estimating $|\int_Q \sigma_{\text{th}}(e(u),z,\theta):e(v)\,\mathrm{d}x\mathrm{d}t| = |\int_Q \theta\partial_e\varphi_\text{C}(e(u),z):e(v)\,\mathrm{d}x\mathrm{d}t| \le C + \|\theta\|^p_{L^p(Q)} + \|e(v)\|^2_{L^2(Q;\mathbb{R}^{d\times d})}$ like we did in (8.3.17).

Step 5 – Convergence: First we chose the weakly* convergent subsequence respecting the a-priori estimates (8.4.8a,b) and (8.4.13). As $\omega = 1$, we have also ϑ estimated in $L^r(I;W^{1.r}(\Omega)) \cap L^p(Q)$ and also its time derivative is estimated, cf. (8.2.5b).

By Aubin-Lions' theorem, the limit passage in the relation $\vartheta = \mathscr{C}_\text{V}(z,\theta)$ in (8.3.8b) is in the lines of (8.2.24)–(8.2.25). In particular, we can rely on that $\theta = [\mathscr{C}_\text{V}(z,\cdot)]^{-1}(\vartheta)$ converges strongly in $L^p(Q)$.

To pass to the limit through the dissipative heat $\xi(e(\dot{u}),\dot{z})$, we need to prove strong convergence of $e(\dot{u})$ in $L^2(Q;\mathbb{R}^{d\times d}_{\text{sym}})$ and of $\dot{z}$ in $L^q(Q;\mathbb{R}^m)$. As we avoided usage of ∇e but allowed still some coupling between e and (z,θ), we need first prove the strong convergence of $e(u)$. To this goal, we modify the arguments from Remark 6.4.14 on p. 221 relying uniform monotonicity of $\partial_e\varphi(\cdot,z)$ in the sense (8.4.7b) and on the convexity of $\varphi_\text{C}(\cdot,z)\theta$ and also on the linear dissipation in terms of $\dot{e}$:

$$\begin{aligned}c\|e(u_k{-}u)\|^2_{L^2(Q;\mathbb{R}^{d\times d})} &\le \int_Q (\partial_e\varphi(e(u_k),z_k) - \partial_e\varphi(e(u),z_k)):e(u_k-u)\,\mathrm{d}x\mathrm{d}t \\ &\le \int_Q (\partial_e\varphi(e(u_k),z_k) + \mathbb{D}e(\dot{u}_k) - \partial_e\varphi(e(u),z_k)) - \mathbb{D}e(\dot{u})):e(u_k-u)\,\mathrm{d}x\mathrm{d}t \\ &\qquad + (\sigma_{\text{th}}(e(u_k),z_k,\theta_k) - \sigma_{\text{th}}(e(u),z_k,\theta_k)):e(u_k-u)\,\mathrm{d}x\mathrm{d}t \\ &= -\int_0^T \langle\ddot{u}_k, u_k-u\rangle\,\mathrm{d}t - \int_Q \Big(\partial_e\varphi(e(u),z_k) + \mathbb{D}e(\dot{u}) \\ &\qquad - \sigma_{\text{th}}(e(u),z_k,\theta_k)\Big):e(u_k-u)\,\mathrm{d}x\mathrm{d}t\end{aligned}$$

$$= \int_Q \dot{u}_k \cdot (\dot{u}_k - \dot{u}) - \big(\partial_e \varphi(e(u), z_k) + \mathbb{D}e(\dot{u}) - \sigma_{\rm th}(e(u), z_k, \theta_k)\big) : e(u_k - u)\,{\rm d}x{\rm d}t$$
$$- \int_\Omega \dot{u}_k(T) \cdot (u_k(T) - u(T))\,{\rm d}x \to 0, \tag{8.4.15}$$

where $\langle\cdot,\cdot\rangle$ denotes the duality between $H^1(\Omega;\mathbb{R}^d)^*$ and $H^1(\Omega;\mathbb{R}^d)$. Like in (7.3.24), for the dissipation rate $\xi(\dot{e},\dot{z}) = \mathbb{D}\dot{e}:\dot{e} + \zeta'_{\rm VI}(\dot{z})\cdot\dot{z}$, we have

$$\int_Q \xi(e(\dot{u}),\dot{z})\,{\rm d}x{\rm d}t \le \liminf_{k\to 0} \int_Q \xi(e(\dot{u}_k),\dot{z}_k)\,{\rm d}x{\rm d}t \le \limsup_{k\to 0} \int_Q \xi(e(\dot{u}_k),\dot{z}_k)\,{\rm d}x{\rm d}t$$
$$\le \limsup_{k\to 0} \int_\Omega \frac{\varrho}{2}|v_0|^2 + \varphi(e(u_0),z_0) - \frac{\varrho}{2}\big|\dot{u}_k(T)\big|^2 - \varphi(e(u_k(T)),z_k(T))\,{\rm d}x$$
$$- \int_Q \sigma_{\rm th}(e(u_k),z_k,\theta_k):e(\dot{u}_k) + \sigma_{\rm in}(e(u_k),z_k,\theta_k)\cdot\dot{z}\,{\rm d}x{\rm d}t$$
$$+ \int_Q f\cdot\dot{u}_k\,{\rm d}x{\rm d}t + \int_\Sigma g\cdot\dot{u}_k\,{\rm d}S\,{\rm d}t$$
$$\le \int_\Omega \frac{\varrho}{2}|v_0|^2 + \varphi(e(u_0),z_0) - \frac{\varrho}{2}\big|\dot{u}(T)\big|^2 - \varphi(e(u(T)),z(T))\,{\rm d}x$$
$$- \int_Q \sigma_{\rm th}(e(u),z,\theta):e(\dot{u}_k) + \sigma_{\rm in}(e(u),z,\theta)\cdot\dot{z}\,{\rm d}x{\rm d}t$$
$$+ \int_Q f\cdot\dot{u}\,{\rm d}x{\rm d}t + \int_\Sigma g\cdot\dot{u}\,{\rm d}S\,{\rm d}t = \int_Q \xi(e(\dot{u}),\dot{z})\,{\rm d}x{\rm d}t, \tag{8.4.16}$$

from which we get $\lim_{k\to 0}\int_Q \xi(e(\dot{u}_k),\dot{z}_k)\,{\rm d}x{\rm d}t = \int_Q \xi(e(\dot{u}),\dot{z})\,{\rm d}x{\rm d}t$ when using the strong convergence of $\theta_k \to \theta$ in $L^p(Q)$. Hence, counting again the weak convergence $e(\dot{u}_k) \to e(\dot{u})$ in $L^2(Q;\mathbb{R}^{d\times d})$ and of $\dot{z}_k \to \dot{z}$ in $L^q(Q;\mathbb{R}^m)$, we obtain the strong convergence; here the strict convexity of ζ assumed in (8.4.7c) is also important, cf. Theorem B.2.9. For the last inequality in (8.4.16), it is important that we use a gradient theory for z-variable so that we have strong convergence in z by Aubin-Lions' theorem, while the strong convergence in $e(u)$ has already been proved in (8.4.15). For the last equality in (8.4.16), the important fact is that $\varrho\ddot{u} \in L^2(I;H^1(\Omega;\mathbb{R}^d)^*)$ is in duality with $\dot{u} \in L^2(I;H^1(\Omega;\mathbb{R}^d))$ due to (8.4.13e). so that the test of the force-balance equation (8.3.8b) by $\dot{u}$ is legal and gives indeed the mechanical-energy equality, cf. Lemma C.2.2 on p.535. □

Remark 8.4.2. Alternatively we may modify the option (C) used in Proposition 8.3.3, i.e. we eliminate θ by using exclusively $\vartheta = \mathscr{C}_{\rm V}(z,\theta)$. To this goal, let us define $\mathscr{C}_{\rm V}^{\rm inv}(z,\vartheta) := [\mathscr{C}_{\rm V}(z,\cdot)]^{-1}(\vartheta)$; note that $\theta = \mathscr{C}_{\rm V}^{\rm inv}(z,\vartheta)$. As in (8.3.25b), we rewrite (8.4.4) into the form

$$\dot{\vartheta} - {\rm div}(\mathscr{K}(z,\vartheta)\nabla\vartheta) = \xi(e(\dot{u}),\dot{z}) + \mathscr{A}_1(e(u),z,\vartheta):e(\dot{u}) + \mathscr{A}_2(e(u),z,\vartheta)\cdot\dot{z}$$
$$\text{with } \mathscr{K}(z,\vartheta) = \frac{\mathbb{K}(z,\theta)}{c_{\rm V}(z,\theta)} = \frac{\mathbb{K}(z,\mathscr{C}_{\rm V}^{\rm inv}(z,\vartheta))}{c_{\rm V}(z,\mathscr{C}_{\rm V}^{\rm inv}(z,\vartheta))},$$
$$\mathscr{A}_1(e,z,\vartheta) = \theta\partial_e\varphi_{\rm C}(e,z) = \mathscr{C}_{\rm V}^{\rm inv}(z,\vartheta)\partial_e\varphi_{\rm C}(e,z), \text{ and}$$
$$\mathscr{A}_2(e,z,\vartheta) = \theta\partial_z\varphi_{\rm C}(e,z) - \partial_z\gamma_{\rm V}(z,\theta) - \partial_z\gamma_{\rm V}(z,0)$$
$$= \mathscr{C}_{\rm V}^{\rm inv}(z,\vartheta)\partial_z\varphi_{\rm C}(e,z) - \partial_z\gamma_{\rm V}(z,\mathscr{C}_{\rm V}^{\rm inv}(z,\vartheta)) - \partial_z\gamma_{\rm V}(z,0).$$

Remark 8.4.3 (Dependence of $c_{\rm V}$ on the strain e). In general, the heat capacity may directly depend on the strain in some ferroic materials as in Section 4.5. Here we need the gradient model both for e and z, i.e. in particular the concept of 2nd-grade nonsimple materials; cf. the calculus (8.4.14) for $\nabla\vartheta$ or (8.2.28) for $\dot{\theta}$ where ∇e occurs in this case. On the other hand, it opens the possibility for a fully general coupling as well as avoiding convexity of $\varphi(\cdot,z,\theta)$ previously assumed in (8.4.7b,d). The (enhanced) free energy and dissipation potential can then be considered as:

$$\psi_{\rm E}(e,z,\theta,\nabla e,\nabla z) := \psi(e,z,\theta) + \frac{1}{2}\mathbb{H}\nabla e \,\vdots\, \nabla e + \frac{\kappa}{2}|\nabla z|^2 \quad \text{and}$$

$$\zeta_{\rm E}(\dot{e},\dot{z},\nabla e,\nabla z) := \zeta(\dot{e},\dot{z}) + \frac{1}{2}\mathbb{H}_{\rm V}\nabla\dot{e} \,\vdots\, \nabla\dot{e}.$$

The system (8.1.15a,b) with (8.1.24) now takes the form

$$\varrho\ddot{u} - {\rm div}\Big(\partial_{\dot{e}}\zeta(e(\dot{u}),\dot{z}) + \sigma_{\rm th}(e(u),z,\theta) - {\rm div}(\mathbb{H}\nabla e(u) + \mathbb{H}_{\rm V}\nabla e(\dot{u}))\Big) = f, \tag{8.4.17a}$$

$$\partial_{\dot{z}}\zeta(e(\dot{u}),\dot{z}) + \sigma_{\rm in}(e(u),z,\theta) = \kappa\Delta z, \tag{8.4.17b}$$

$$\begin{aligned}\dot{\vartheta} - {\rm div}(\mathbb{K}(e(u),z,\theta)\nabla\theta) &= \xi(e(\dot{u}),\dot{z}) + \mathbb{H}_{\rm V}\nabla e(\dot{u}) \,\vdots\, \nabla e(\dot{u}) \\ &\quad + (\sigma_{\rm th}(e(u),z,\theta) - \sigma_{\rm th}(e(u),z,0)):e(\dot{u}) \\ &\quad + (\sigma_{\rm in}(e(u),z,\theta) - \sigma_{\rm in}(e(u),z,0))\cdot\dot{z} \quad \text{with} \quad \vartheta = \mathscr{C}_{\rm V}(e(u),z,\theta),\end{aligned} \tag{8.4.17c}$$

where $\sigma_{\rm th}(e,z,\theta) := \partial_e\psi(e,z,\theta)$ and $\sigma_{\rm in}(e,z,\theta) := \partial_z\psi(e,z,\theta)$. Also the boundary condition (8.1.17a) expands correspondingly.

Remark 8.4.4 (Difficulties in thermal expansion in Maxwellien materials). The rheological model $\mathbb{C}$—$\overset{\theta}{\mathbb{E}}$—$\mathbb{D}$ is an example for difficulties in weakly dissipating materials, cf. also Remark 6.6.4. Then $e_{\rm EL} = e(u) - \mathbb{E}\theta - \pi$ and $\sigma = \mathbb{C}e_{\rm EL} = \mathbb{C}(e(u) - \mathbb{E}\theta - \pi)$ with π the creep strain. Then $\dot{\sigma} = \mathbb{C}e(\dot{u}) - \mathbb{C}\mathbb{E}\dot{\theta} - \mathbb{E}\dot{\pi}$ and then, taking $\sigma = \mathbb{D}\dot{\pi}$ into account, instead of (6.6.1), we obtain

$$e(\dot{u}) = \mathbb{C}^{-1}\dot{\sigma} + \mathbb{E}\dot{\theta} + \mathbb{D}^{-1}\sigma \tag{8.4.18}$$

while the energy balance (6.6.4) expands by $\sigma:\mathbb{E}\dot{\theta} = \frac{\partial}{\partial t}(\sigma:\mathbb{E}\theta) - \dot{\sigma}:\mathbb{E}\theta$. The free energy is

$$\psi(\sigma,\theta) = \frac{1}{2}\mathbb{C}^{-1}(\sigma + \mathbb{C}\mathbb{E}\theta):(\sigma + \mathbb{C}\mathbb{E}\theta) - \frac{1}{2}\theta^2\mathbb{C}\mathbb{E}:\mathbb{E} - \gamma_{\rm V}(\theta). \tag{8.4.19}$$

The entropy $\mathfrak{s} = -\partial_\theta\psi(\sigma,\theta) = \gamma_{\rm V}'(\theta) - \sigma:\mathbb{E}$ gives the entropy equation (8.1.7)–(8.1.8) as $\theta\dot{\mathfrak{s}} - {\rm div}(\mathbb{K}\nabla\theta) = \mathbb{D}^{-1}\sigma:\sigma$ which yields the heat equation:

$$c_{\rm V}(\theta)\dot{\theta} - {\rm div}(\mathbb{K}\nabla\theta) = \mathbb{D}^{-1}\sigma:\sigma + \dot{\sigma}:\mathbb{E}\theta. \tag{8.4.20}$$

The estimate of the last term in (8.4.20), i.e. the adiabatic heat due to thermal expansion, however cannot be executed since we do not have any estimate of $\dot{\sigma}$, cf. (6.6.8), while the regularity estimates analogous to Proposition 6.3.2 also cannot be executed.

Remark 8.4.5 (Difficulties in temperature-dependent Maxwellian creep). Similar difficulties as in the previous Remark 8.4.4 occurs if the temperature influences dissipative instead of conservative part. In view of the constitutional rheological relation (6.6.2) and the related dissipation rate (6.6.5) which is here simultaneously the heat-production rate, we formulate the system:

$$\varrho\ddot{u} - \mathrm{div}\,\sigma = f, \tag{8.4.21a}$$

$$\dot{\sigma} = \mathbb{C}e(\dot{u}) - \mathbb{D}^{-1}(\theta)\mathbb{C}\sigma, \tag{8.4.21b}$$

$$c_{\mathrm{v}}(\theta)\dot{\theta} - \mathrm{div}(\mathbb{K}(\theta)\nabla\theta) = \mathbb{D}^{-1}(\theta)\sigma{:}\sigma. \tag{8.4.21c}$$

This is to be completed by suitable boundary and initial conditions. The analysis requires strong convergence (e.g. of the Galerkin approximation) of σ. For this, one needs (8.4.21a) tested by u and (8.4.21b) by σ and $\varrho\ddot{u}$ in duality with $\varrho\dot{u}$ and with $\mathrm{div}\,\sigma$. Yet, this is not implied by basic energy-based estimates, while the regularity Proposition 6.6.2 would now give rise a term $\dot{\theta}[\mathbb{D}^{-1}](\theta)\mathbb{C}\sigma{\cdot}\dot{\sigma}$ which hardly could be estimated since $\dot{\theta}$ is not well controlled.

Remark 8.4.6 (Thermodynamics of transport under large displacements). Another interesting example exhibiting difficulties is when large displacements (still under small strains as in Section 7.7.3 are considered in Maxwell materials even without any thermal expansion. This allows for an interesting thermodynamical calculations. Temperature is then transported as a volume-invariant variable, thus, like in Example 7.7.7, one deals with a system

$$\varrho\ddot{u} - \mathrm{div}(\varphi'(e_{\mathrm{EL}}) + \sigma_{\mathrm{K,int}}) + g = 0 \quad \text{with} \quad e_{\mathrm{EL}} = e(u) - \pi, \tag{8.4.22a}$$

$$\mathbb{D}(\theta)\frac{\mathrm{D}\pi}{\mathrm{D}t} = \mathrm{dev}\,\varphi'(e_{\mathrm{EL}}) + \kappa\Delta\pi, \tag{8.4.22b}$$

$$c_{\mathrm{v}}(\theta)\frac{\mathrm{D}\theta}{\mathrm{D}t} = \mathrm{div}(\mathbb{K}(\theta)\nabla\theta) + \xi\Big(\theta;\frac{\mathrm{D}\pi}{\mathrm{D}t}\Big) + \theta\mathfrak{s}\,\mathrm{div}\dot{u} \tag{8.4.22c}$$

where $\frac{\mathrm{D}\pi}{\mathrm{D}t} = \dot{\pi} + \dot{u}{\cdot}\nabla\pi$ denotes the material derivative, and similarly also $\frac{\mathrm{D}\theta}{\mathrm{D}t}$, and $\psi(e,\theta) = \varphi(e) - \gamma_{\mathrm{v}}(\theta)$, $c_{\mathrm{v}}(\theta) = \theta\gamma_{\mathrm{v}}''(\theta)$, $\mathfrak{s} = \gamma_{\mathrm{v}}'(\theta)$ is the entropy, and where the heat-production rate and the Korteweg-type stress as in (7.7.32) are now

$$\xi(\theta;\dot{\pi}) = \mathbb{D}(\theta)\dot{\pi}{:}\dot{\pi} \quad \text{and} \quad \sigma_{\mathrm{K,int}} = \kappa\nabla\pi\otimes\nabla\pi - \Big(\frac{\kappa}{2}|\nabla\pi|^2 + \psi(e(u){-}\pi,\theta)\Big)\mathbb{I}, \tag{8.4.22d}$$

The last term in (8.4.22c) is a certain adiabatic heat. Eventually, we again rescale temperature $\vartheta = \mathscr{C}_{\mathrm{v}}(\theta) := \theta\gamma_{\mathrm{v}}'(\theta) - \gamma_{\mathrm{v}}(\theta)$, cf. (8.1.36), i.e. we again make an *enthalpy*-type *transformation*, here under advection in compressible continuum. Then (8.4.22c) transforms to[17]

[17] To see that (8.4.23) is indeed equivalent to (8.4.22c), it suffices to substitute ϑ and use the calculus

$$\dot{\vartheta} + \mathrm{div}(\dot{u}\vartheta) = \mathrm{div}(\mathbb{K}(\theta)\nabla\theta) + \xi\Big(\theta; \frac{\mathrm{D}\pi}{\mathrm{D}t}\Big) + (\vartheta - \theta\eta)\mathrm{div}\dot{u}, \tag{8.4.23}$$

let us note that ϑ is an extensive variable in contrast to θ or π which are intensive variables, cf. also (7.7.23) versus (7.7.24). Now we can test the transformed heat equation (8.4.23) by 1 and we obtain the balance of the overall heat energy

$$\frac{\mathrm{d}}{\mathrm{d}t}\int_\Omega \vartheta\,\mathrm{d}x = \int_\Omega \xi + \mathrm{div}(j - \dot{u}\vartheta) + (\vartheta - \theta\mathfrak{s})\mathrm{div}\dot{u}\,\mathrm{d}x$$
$$= \int_\Omega \xi + \gamma_{\mathrm{v}}(\theta)\mathrm{div}\dot{u}\,\mathrm{d}x + \int_\Gamma j_{\mathrm{ext}} - \vartheta\dot{u}\cdot\vec{n}\,\mathrm{d}S, \tag{8.4.24}$$

where $j = \mathbb{K}(\theta)\nabla\theta$ is the heat flux. Adding it to the mechanical-energy balance like (7.7.35), we can see cancellation of heat source terms, in particular also $\pm(r+\gamma_{\mathrm{v}}(\theta)\mathrm{div}\dot{u})$. Considering homogeneous boundary conditions for simplicity, we obtain at least formally the total energy balance:

$$\frac{\mathrm{d}}{\mathrm{d}t}\underbrace{\int_\Omega \frac{1}{2}\varrho|\dot{u}|^2 + \varphi(e_{\mathrm{EL}}) + \vartheta\,\mathrm{d}x}_{\text{kinetic, mechanical and heat energies in the bulk}} + \underbrace{\int_\Gamma (\varphi(e_{\mathrm{EL}}) + \vartheta)\dot{u}\cdot\vec{n}\,\mathrm{d}S}_{\text{thermomechanical energy flux thru }\Gamma} = \int_\Omega (\dot{u}\otimes\varphi'(e_{\mathrm{EL}}))\,\vdots\,\nabla e(u)\,\mathrm{d}x. \tag{8.4.25}$$

Realizing the Gibbs relation $\vartheta = \theta\mathfrak{s} - \gamma_{\mathrm{v}}(\theta) = \theta\gamma_{\mathrm{v}}'(\theta) - \gamma_{\mathrm{v}}(\theta)$ the heat equation in the entropy formulation (8.4.23) can be written in the form of an entropy equation as:[18]

$$\theta(\dot{\mathfrak{s}} + \mathrm{div}(\dot{u}\mathfrak{s})) = \xi - \mathrm{div}\,j \tag{8.4.26}$$

with the heat flux $j = -\mathbb{K}\nabla\theta$. Hence, we obtain the overall *entropy balance*

$$\frac{\mathrm{d}}{\mathrm{d}t}\underbrace{\int_\Omega \mathfrak{s}\,\mathrm{d}x}_{\text{total entropy}} = \int_\Omega \dot{\mathfrak{s}}\,\mathrm{d}x = \int_\Omega \frac{\xi - \mathrm{div}\,j}{\theta} - \mathrm{div}(\dot{u}\mathfrak{s})\,\mathrm{d}x$$
$$= \underbrace{\int_\Omega \frac{\xi}{\theta} + \frac{\mathbb{K}\nabla\theta\cdot\nabla\theta}{\theta^2}\,\mathrm{d}x}_{\text{entropy production in the bulk }\Omega} - \int_\Gamma \underbrace{\mathfrak{s}\dot{u}\cdot\vec{n}}_{\text{entropy flux through the boundary }\Gamma}\,\mathrm{d}S\,. \tag{8.4.27}$$

Counting that the heat-production rate r and the heat-conductivity tensor $\mathbb{K}$ are non-negative, we obtain the Clausius-Duhem inequality $\frac{\mathrm{d}}{\mathrm{d}t}\int_\Omega \mathfrak{s}\,\mathrm{d}x \geq -\int_\Gamma \mathfrak{s}\dot{u}\cdot\vec{n}\,\mathrm{d}S$ repre-

---

$$\frac{\mathrm{D}\vartheta}{\mathrm{D}t} + (\theta\eta - \vartheta)\mathrm{div}\dot{u} = \dot{\vartheta} + \dot{u}\cdot\nabla\vartheta + (\mathrm{div}\dot{u})\vartheta + (\theta\eta - \vartheta)\mathrm{div}\dot{u} = c_{\mathrm{v}}(\theta)\frac{\mathrm{D}\theta}{\mathrm{D}t} - \theta\mathfrak{s}\,\mathrm{div}\dot{u}.$$

[18] Here we use also the calculus

$$\dot{\vartheta} + \mathrm{div}(\dot{u}\vartheta)(\vartheta - \theta\mathfrak{s})\mathrm{div}\dot{u} = (\gamma_{\mathrm{v}}(\theta) - \theta\gamma_{\mathrm{v}}'(\theta))^{\boldsymbol{\cdot}} + \mathrm{div}(\dot{u}\gamma_{\mathrm{v}}(\theta) - \dot{u}\theta\gamma_{\mathrm{v}}'(\theta)) - \gamma_{\mathrm{v}}(\theta)\mathrm{div}\dot{u}$$
$$= \theta\dot{\mathfrak{s}} + \mathrm{div}(\dot{u}\gamma_{\mathrm{v}}(\theta)) - \theta\mathrm{div}(\dot{u}\gamma_{\mathrm{v}}'(\theta)) - \gamma_{\mathrm{v}}'(\theta)\dot{u}\cdot\nabla\theta - \gamma_{\mathrm{v}}(\theta)\mathrm{div}\dot{u}$$
$$= \theta(\dot{\mathfrak{s}} + \mathrm{div}(\dot{u}\mathfrak{s})) + \mathrm{div}(\dot{u}\gamma_{\mathrm{v}}(\theta)) - \dot{u}\cdot\nabla\gamma_{\mathrm{v}}(\theta) - \gamma_{\mathrm{v}}(\theta)\mathrm{div}\dot{u} = \theta(\dot{\mathfrak{s}} + \mathrm{div}(\dot{u}\mathfrak{s})),$$

senting the 2nd law of thermodynamics; of course, we rely on non-negativity (or rather positivity) of temperature to be proved. Like in Remark 8.4.4, an adiabatic-heat term arisen in (8.4.22c) does not allow for estimation because $\mathrm{div}\,\dot{u}$ is not controlled and, moreover, like in Example 7.7.7, the last term in (8.4.25) would need to enhance (8.4.22a) by the strain-gradient term.

Example 8.4.7 (Special partly linear case). Let us consider the material from Figure 6.7(left) on p. 239. Its thermally expansive variant uses φ from (6.7.7) and a linear $\varphi_{\rm c}$ so that, with $z = e_2$, (8.4.1) results into

$$\psi(e,z,\theta) := \frac{1}{2}\mathbb{C}_1(e-e_2):(e-e_2)+\frac{1}{2}\mathbb{C}_1 e_2:e_2 \\ -(\theta{-}\theta_{\rm R})\mathbb{B}_1(e-e_2)-(\theta{-}\theta_{\rm R})\mathbb{B}_2 e_2-\gamma_{\rm V}(\theta). \tag{8.4.28a}$$

$$\zeta(\dot{e},\dot{z}) := \frac{1}{2}\mathbb{D}_1\dot{e}:\dot{e}+\frac{1}{2}\mathbb{D}_2\dot{e}_2:\dot{e}_2\,. \tag{8.4.28b}$$

Then the heat equation (8.1.14) looks as

$$c_{\rm v}(\theta)\dot{\theta}+\mathrm{div}(j) = \mathbb{D}_1 e(\dot{u}):e(\dot{u})+\mathbb{D}_2\dot{e}_2:\dot{e}_2+\theta\mathbb{B}_1\dot{e}+\theta(\mathbb{B}_2{+}\mathbb{B}_1)\dot{e}_2\,.$$

All estimates can be performed similarly as in Section 8.3. In particular, note that we do not need to involve the gradient of z.

Exercise 8.4.8 (Temperature dependent viscosity). Consider $\mathbb{D} = \mathbb{D}(\theta)$ and realize the needed modifications the proof of Proposition 8.4.1, relying on a suitable generalization of Theorem B.2.9 for a situation $\int_\Omega \psi(\theta_k,q_k)\,\mathrm{d}x \to \int_\Omega \psi(\theta,q)\,\mathrm{d}x$ with $\psi(\cdot,q)$ continuous and $\psi(\theta,\cdot)$ strictly convex and $\theta_k \to \theta$ strongly.

Exercise 8.4.9 (Strain dependent heat capacity). Make a revision of the proof of Proposition 8.4.1 for the enhanced system (8.4.17) and realize the needed modifications.

8.5 Jeffreys-type rheology and thermoplasticity

The modification of the Maxwell rheology to a "better dissipative" Jeffreys rheology modifies the hyperbolic Maxwellian rheology (shown to be troublesome in Remarks 8.4.5 and 8.4.4) to the parabolic one, still keeping its character as a fluid rather than solid. This parabolic "regularization" smears out various difficulties from the previous section and opens thus possibilities for modeling various other phenomena.

We will first consider the model for *thermoviscoplasticity* as a thermodynamical expansion of the model (7.4.4), for simplicity without hardening, i.e. $H = 0$. Putting $s_{\rm YLD} = 0$ results to purely linear Jeffreys rheology, albeit in the deviatoric component only. We will consider the thermal coupling through the thermal expansion (acting on the spherical part of the model) and temperature dependence of the yield stress. The free energy and the dissipation potential governing this problem are then

$$\psi(e,\pi,\theta) = \frac{1}{2}\mathbb{C}e_{\rm EL}:e_{\rm EL} - \gamma_{\rm V}(\theta) \quad \text{with} \quad e = e_{\rm EL} + \pi + \mathbb{E}\theta, \tag{8.5.1a}$$

$$\zeta(\theta;\dot e,\dot\pi) = \frac{1}{2}\mathbb{D}(e(\dot u)-\dot\pi):(e(\dot u)-\dot\pi) + \frac{1}{2}D|\dot\pi|^2 + s_{\rm YLD}(\theta)|\dot\pi|. \tag{8.5.1b}$$

The resulted system of equations/inclusion of the abstract form (8.1.15a,b) with (8.1.24) now reads as:

$$\varrho\ddot u - {\rm div}\,\sigma_{\rm KV} = f, \quad \text{with} \quad \sigma_{\rm KV} = \mathbb{D}(e(\dot u)-\dot\pi) + \mathbb{C}(e(u)-\pi-\mathbb{E}\theta) \tag{8.5.2a}$$

$$D\dot\pi + s_{\rm YLD}(\theta){\rm Dir}(\dot\pi) \ni {\rm dev}\,\sigma_{\rm KV}, \tag{8.5.2b}$$

$$\dot\vartheta - {\rm div}(\mathbb{K}(\theta)\nabla\theta) = \mathbb{D}(e(\dot u)-\dot\pi):(e(\dot u)-\dot\pi) + D|\dot\pi|^2 + s_{\rm YLD}(\theta)|\dot\pi| + \theta\mathbb{E}:\mathbb{C}(e(\dot u)-\dot\pi) \quad \text{with } \vartheta = \mathscr{C}_{\rm V}(\theta) \tag{8.5.2c}$$

considered in Q, where, as before, $\mathscr{C}_{\rm V}(\theta) := \theta\partial_\theta\gamma_{\rm V}(\theta) - \gamma_{\rm V}(\theta)$. The schematic rheology is depicted in Figure 8.1. We complete the system (8.5.2) with boundary and initial conditions:

$$\sigma_{\rm KV}\vec n = f_\flat, \quad \mathbb{K}(\theta)\nabla\theta\cdot\vec n = h_\flat \qquad \text{on } \Sigma, \tag{8.5.3a}$$

$$u|_{t=0} = u_0, \quad \dot u|_{t=0} = v_0, \quad \pi|_{t=0} = \pi_0, \quad \theta|_{t=0} = \theta_0 \qquad \text{on } \Omega. \tag{8.5.3b}$$

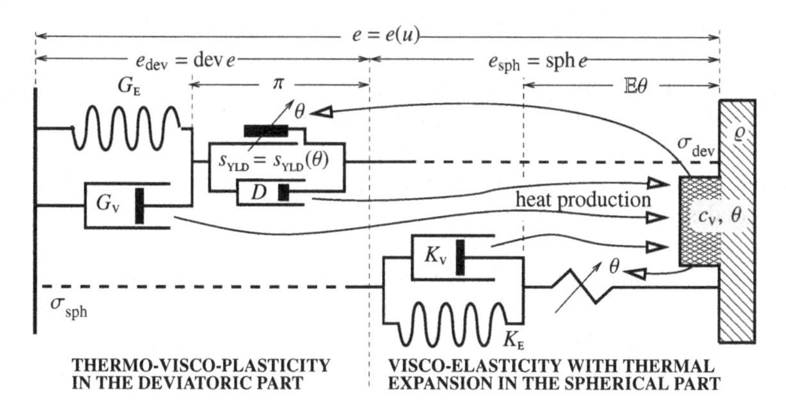

Fig. 8.1 The schematic rheological scheme for thermoviscoplasticity (8.5.2) with the decomposition $e(u) = e_{\rm dev} + e_{\rm sph}$ as in (6.7.8). Isochoric model of plastification concerns only the deviatoric part while the isotropic thermal expansion concerns only the spherical part.

The weak formulation of force equilibrium (8.5.2a) can rely on (7.2.5a) while the flow-rule inclusion (8.5.2b) can use Definition 7.2.1 and the heat-transfer equation (8.5.2c) can use Definition 8.1.2.

The energetics can be revealed by testing the equations in the system (8.5.2) in its weak formulation successively by $\dot u$, $\dot\pi$, and 1. Let us note that $\theta\mathbb{E}:e(\dot u) = \theta\mathbb{E}:(e(\dot u)-\dot\pi)$ due to the the orthogonality $\mathbb{E}\theta:\dot\pi = 0$.

We assume, together with (6.3.5) and (8.3.21), that $D>0$, $s_{\rm YLD}\ge 0$, and $\pi_0\in L^2(\Omega;\mathbb{R}^{d\times d}_{\rm dev})$. We first consider a regularization of the right-hand side of (8.5.2c) and $h_\flat$ and θ_0 in (8.5.3) like we did in (8.3.11), denoting the approximate solutions by $(u_\varepsilon,\pi_\varepsilon,\theta_\varepsilon)$ whose existence can be proved when relying on the conventional L^2-theory of the parabolic heat-transfer equation.

Testing (8.5.2) successively by $\dot u_\varepsilon$, $\dot\pi_\varepsilon$, and 1/2, and using the interpolation as in Proposition 8.3.2, we can read the estimates:

$$\|u_\varepsilon\|_{H^1(I;H^1(\Omega;\mathbb{R}^d))\cap W^{1,\infty}(I;L^2(\Omega;\mathbb{R}^d))}\le C, \tag{8.5.4a}$$

$$\|\pi_\varepsilon\|_{H^1(I;L^2(\Omega;\mathbb{R}^{d\times d}))}\le C, \tag{8.5.4b}$$

$$\|\theta_\varepsilon\|_{L^\infty(I;L^1(\Omega))}\le C\ \text{ and }\ \|\theta_\varepsilon\|_{L^\infty(I;L^\omega(\Omega))}\le C \tag{8.5.4c}$$

provided $\mathscr{C}_{\rm v}$ has at least an ω-polynomial growth, cf. (8.2.3a). The further estimates can be obtained by using (8.2.6a) and by comparison from (8.5.2a,c):

$$\|\nabla\theta_\varepsilon\|_{L^r(Q;\mathbb{R}^d)}\le C_r\quad\text{with }\ 1\le r<(2\omega+d)/(\omega+d), \tag{8.5.4d}$$

$$\big\|\sqrt{\varrho}\,\ddot u_\varepsilon\big\|_{L^2(I;H^1(\Omega;\mathbb{R}^d)^*)}\le C, \tag{8.5.4e}$$

$$\big\|\dot\vartheta_\varepsilon\big\|_{L^1(I;H^{(d+3)/2}(\Omega)^*)}\le C, \tag{8.5.4f}$$

Here we use $\omega>\max(1,d/2)$ so that, by (8.2.6b), $\{\theta_\varepsilon\}_{\varepsilon>0}$ is bounded in $L^2(Q)$ in order to have the adiabatic heat source/sink in (8.5.2c) in $L^1(Q)$. We then choose a weakly* convergent subsequence (in the topologies indicated in (8.5.4)) of approximate solutions $\{(u_\varepsilon,\pi_\varepsilon,\theta_\varepsilon)\}_{\varepsilon>0}$, denoting by (u,π,θ) its limit.

The important attribute of the model is that, due to viscosity $D>0$ in the plastic flow rule, we have $|\dot\pi_\varepsilon|\in L^2(Q)$ and thus the heat production rate $s_{\rm YLD}(\theta_\varepsilon)|\dot\pi_\varepsilon|\in L^2(Q)$ has a good sense. The energy conservation holds because, by comparison from (8.5.2a), we can see for a weak limit (u,π,θ) that $\sqrt{\varrho}\ddot u_\varepsilon=({\rm div}\,\sigma_{\rm KV}{+}f)/\sqrt{\varrho}\in L^2(H^1(\Omega;\mathbb{R}^d))$ is in duality with $\sqrt{\varrho}\dot u$ provided ϱ is qualified as (6.3.5) and that $\mathbb{C}^{1/2}e(\dot u)$ is in duality with $\mathbb{C}^{1/2}e(u)$. The mere mechanical energy conservation obtained by testing (8.5.2a,b) by $\dot u$ and $\dot\pi$ also facilitates the strong convergence of $\dot\pi_\varepsilon$ and of $e(\dot u_\varepsilon){-}\dot\pi_\varepsilon$ needed to pass to the limit in the right-hand side of (8.5.2c): indeed, we can estimate

$$\begin{aligned}
&\int_Q\mathbb{D}(e(\dot u){-}\dot\pi){:}(e(\dot u){-}\dot\pi)+D|\dot\pi|^2\,{\rm d}x{\rm d}t\\
&\le\limsup_{\varepsilon\to0}\int_Q\mathbb{D}(e(\dot u_\varepsilon){-}\dot\pi_\varepsilon){:}(e(\dot u_\varepsilon){-}\dot\pi_\varepsilon)+D|\dot\pi_\varepsilon|^2\,{\rm d}x{\rm d}t\\
&=\int_\Omega\frac{\varrho}{2}|v_0|^2+\frac12\mathbb{C}(e(u_0){-}\pi_0){:}(e(u_0){-}\pi_0)\,{\rm d}x+\lim_{\varepsilon\to0}\int_Q\theta_\varepsilon\mathbb{E}{:}\mathbb{C}(e(\dot u_\varepsilon){-}\dot\pi_\varepsilon)\,{\rm d}x{\rm d}t\\
&\quad-\liminf_{\varepsilon\to0}\int_\Omega\frac{\varrho}{2}|\dot u_\varepsilon(T)|^2+\frac12\mathbb{C}(e(u_\varepsilon(T)){-}\pi_\varepsilon(T)){:}(e(u_\varepsilon(T)){-}\pi_\varepsilon(T))\,{\rm d}x\\
&\quad+\frac12\mathbb{C}(e(u_0){-}\pi_0){:}(e(u_0){-}\pi_0)\,{\rm d}x+\int_Q s_{\rm YLD}(\theta_\varepsilon)|\dot\pi_\varepsilon|\,{\rm d}x{\rm d}t
\end{aligned}$$

$$
\begin{aligned}
&\le \int_Q \theta\mathbb{E}:\mathbb{C}(e(\dot{u})-\dot{\pi}) - s_{\text{YLD}}(\theta)|\dot{\pi}|\,\mathrm{d}x\mathrm{d}t + \int_\Omega\Big(\frac{\varrho}{2}|v_0|^2 - \frac{\varrho}{2}|\dot{u}(T)|^2 \\
&\quad + \frac{1}{2}\mathbb{C}(e(u_0)-\pi_0):(e(u_0)-\pi_0) - \frac{1}{2}\mathbb{C}(e(u(T))-\pi(T)):(e(u(T))-\pi(T))\Big)\mathrm{d}x \\
&\qquad = \int_Q \mathbb{D}(e(\dot{u})-\dot{\pi}):(e(\dot{u})-\dot{\pi}) + D|\dot{\pi}|^2\,\mathrm{d}x\mathrm{d}t,
\end{aligned}
$$

where the last equality is due to the mentioned mechanical energy conservation, cf. also (8.3.20). From this we can see that $\lim_{\varepsilon\to 0}\int_Q \mathbb{D}(e(\dot{u}_\varepsilon)-\dot{\pi}_\varepsilon):(e(\dot{u}_\varepsilon)-\dot{\pi}_\varepsilon) + D|\dot{\pi}_\varepsilon|^2\,\mathrm{d}x\mathrm{d}t = \int_Q \mathbb{D}(e(\dot{u})-\dot{\pi}):(e(\dot{u})-\dot{\pi}) + D|\dot{\pi}|^2\,\mathrm{d}x\mathrm{d}t$ and, since $e(\dot{u}_\varepsilon)\to e(\dot{u})$ and $\dot{\pi}_\varepsilon\to\dot{\pi}$ weakly in $L^2(Q;\mathbb{R}^{d\times d}_{\text{sym}})$, we obtain the claimed strong convergence. The resting limit passages are simple by weak convergence and compactness arguments. Thus, to summarize, we have obtained:

Proposition 8.5.1 (Weak solutions to thermoviscoplasticity). *Let the assumptions (6.3.5), (8.3.21), $D>0$, $s_{\text{YLD}}(\cdot)\ge 0$ continuous bounded, and $\pi_0\in L^2(\Omega;\mathbb{R}^{d\times d}_{\text{dev}})$ hold. Then the initial-boundary-value problem (8.5.2)–(8.5.3) has a weak solution (u,π,θ) satisfying (8.5.4) with ε omitted.*

Alternatively, we could use also the qualification (8.3.24) instead of (8.3.21) and the arguments from the proof of Proposition 8.3.3.

Another interesting model arises when the viscosity coefficient vanishes, i.e. $D\to 0$. As there is nor hardening neither plastic-strain gradient in the model (8.5.2), we obtain the *perfect plasticity*. The analytical difficulty is that the plastic dissipation rate $\dot{\pi}$ is (intentionally allowed to be) valued in measures on $\bar{\Omega}$ while temperature θ hardly can be expected continuous on $\bar{\Omega}$ so that the plastification heat production rate $s_{\text{YLD}}(\theta)|\dot{\pi}|$ with $|\dot{\pi}|$ meaning the total variation of the measure $\dot{\pi}$ occurring in (8.5.2c) does not have a well defined sense. We therefore have to regularize the model (8.5.2) with $D=0$ by using *phase-field* approach to overcome any direct coupling of temperature θ with the measure π. The original model (8.5.1) now modifies, using the free energy and the dissipation potential

$$
\begin{aligned}
&\psi_{\text{E}}(e,\pi,\chi,\theta,\nabla\chi) = \varphi(e,\pi,\chi) - \gamma_{\text{V}}(\theta) + \frac{1}{2\varkappa}|\chi-\theta|^2 + \frac{\kappa}{p}|\nabla\chi|^p \\
&\qquad\text{with }\ \varphi(e,\pi,\chi) = \frac{1}{2}\mathbb{C}e_{\text{EL}}:e_{\text{EL}}\ \text{ with }\ e = e_{\text{EL}}+\pi-\mathbb{E}\chi,
\end{aligned}
\tag{8.5.5a}
$$

$$
\zeta(\chi;\dot{e},\dot{\pi},\dot{\chi}) = \frac{1}{2}\mathbb{D}(\dot{e}-\dot{\pi}):(\dot{e}-\dot{\pi}) + s_{\text{YLD}}(\chi)|\dot{\pi}| + \frac{1}{2}\tau_{\text{R}}\dot{\chi}^2.
\tag{8.5.5b}
$$

The resulted system thus reads as:

$$
\varrho\ddot{u} - \mathrm{div}\,\sigma_{\text{KV}} = f, \quad \text{with} \quad \sigma_{\text{KV}} = \mathbb{D}(e(\dot{u})-\dot{\pi}) + \mathbb{C}(e(u)-\pi-\mathbb{E}\chi),
\tag{8.5.6a}
$$

$$
s_{\text{YLD}}(\chi)\mathrm{Dir}(\dot{\pi}) \ni \mathrm{dev}\,\sigma_{\text{KV}},
\tag{8.5.6b}
$$

$$
\tau_{\text{R}}\dot{\chi} + \mathbb{E}^\top : e_{\text{EL}} - \mathrm{div}(\kappa|\nabla\chi|^{p-2}\nabla\chi) = (\theta-\chi)/\varkappa,
\tag{8.5.6c}
$$

$$
\begin{aligned}
&\dot{\vartheta} - \mathrm{div}(\mathbb{K}(\theta)\nabla\theta) = \mathbb{D}(e(\dot{u})-\dot{\pi}):(e(\dot{u})-\dot{\pi}) \\
&\qquad\qquad + s_{\text{YLD}}(\chi)|\dot{\pi}| + \tau_{\text{R}}\dot{\chi}^2 - \theta\dot{\chi}/\varkappa \quad \text{with } \vartheta = \mathscr{C}_{\text{V}}(\theta)
\end{aligned}
\tag{8.5.6d}
$$

where, as before, $\mathscr{C}_{\rm V}(\theta) := \theta\partial_\theta\gamma_{\rm V}(\theta)-\gamma_{\rm V}(\theta)$. We again consider the boundary and initial conditions (8.5.3). The energetics can be revealed by testing (8.5.6) successively by $\dot u$, $\dot\pi$, $\dot\chi$, and 1.

Again we use a regularization of the right-hand side of (8.5.6d) and $h_\flat$ and θ_0 in (8.5.3) like we did in (8.3.11) Then testing (8.5.6c) by $\dot\chi$ is well possible, and gives in particular (8.5.7c). Altogether, this test yields the estimates:

$$\|u_\varepsilon\|_{L^\infty(I;{\rm BD}(\bar\Omega;\mathbb{R}^d))} \le C, \tag{8.5.7a}$$
$$\|\pi_\varepsilon\|_{L^\infty(I;\mathfrak{M}(\bar\Omega;\mathbb{R}^d))} \le C, \tag{8.5.7b}$$
$$\|\chi_\varepsilon\|_{H^1(I;L^2(\Omega))\cap L^\infty(I;W^{1,p}(\Omega))} \le C, \tag{8.5.7c}$$
$$\|e(u_\varepsilon)-\pi_\varepsilon\|_{H^1(I;L^2(\Omega;\mathbb{R}^{d\times d}))} \le C, \tag{8.5.7d}$$

together with (8.5.7c,d,f). In particular, (8.5.7c) yields χ "compactly" in $C(\bar Q)$, which makes legitimate the expression $s_{\rm YLD}(\chi)|\dot\pi|$ with $|\dot\pi|$ meaning the total variation of the measure $|\dot\pi|$. However, the peculiarity is that $e(\dot u)$ is not in $L^2(Q;\mathbb{R}^{d\times d}_{\rm sym})$ due to the perfect-plasticity model, and thus $\dot u$ is not in duality with $\ddot u$ and thus the energy conservation is not expected under the basic data qualification. This prevents limit passage for (unspecified) approximate solutions in the right-hand side of the heat equation (8.5.6d). Therefore, the higher-order test of the mechanical part (8.5.6a,b) like we did in Proposition 7.4.12 is needed provided $s_{\rm YLD}$ = is constant. More specifically, we differentiate (8.5.6a,b) in time and test it by $\ddot u$ and $\ddot\pi$. We have $\chi\mathbb{E}{:}e(\ddot u) = \chi\mathbb{E}{:}(e(\ddot u){-}\ddot\pi)$. This gives

$$\|\dot u_\varepsilon\|_{L^\infty(I;{\rm BD}(\bar\Omega;\mathbb{R}^d))} \le C, \tag{8.5.8a}$$
$$\|\dot\pi_\varepsilon\|_{L^\infty(I;\mathfrak{M}(\bar\Omega;\mathbb{R}^d))} \le C, \tag{8.5.8b}$$
$$\|e(\dot u_\varepsilon)-\dot\pi_\varepsilon\|_{H^1(I;L^2(\Omega;\mathbb{R}^{d\times d}))} \le C, \tag{8.5.8c}$$

It brings $\sqrt{\varrho}\ddot u$ in duality with $\sqrt{\varrho}\dot u$ and also $\sigma_{\rm KV}$ in duality with $e(\dot u)-\dot\pi$. This facilitates energy conservation, now using also on Proposition 7.3.9.

Proposition 8.5.2 (Energetic solutions to perfect thermoplasticity). *Let the assumptions (6.3.5), (8.3.21), $s_{\rm YLD}(\cdot)$ constant, and $\pi_0 \in \mathfrak{M}(\Omega;\mathbb{R}^{d\times d}_{\rm dev})$ hold. Then the initial-boundary-value problem (8.5.6) with (8.5.3) has a weak solution (u,π,θ) satisfying (8.5.7c,d,f), (8.5.7c), and (8.5.8) with ε omitted.*

Remark 8.5.3 (Quasistatic perfect plasticity). Based on energetic formulation, the quasistatic (i.e. $\varrho = 0$) variant of the model with thermal expansion and temperature independent yield stress can be treated directly without the phase-field regularization even in the inviscid variant (i.e. (8.5.2) with $D = 0$, resulting to the perfect plasticity), cf. [458].

8.6 Mass and heat transport in poroelastic solids

An interesting illustrative example is a thermodynamical expansion of the models from Section 7.6.2. For simplicity, we will not consider effects of thermal expansion and, except Remark 8.6.2, concentration dependence of the heat capacity. This corresponds to the structural assumptions $\partial^2_{e\theta}\psi = 0$ and $\partial^2_{c\theta}\psi = 0$. In fact, considering capillarity with the quadratic gradient term as in (7.6.20), it essentially dictates the form

$$\psi_{\mathrm{E}}(e,c,\nabla c,\theta) = \psi(e,c,\theta) + \frac{\kappa}{2}|\nabla c|^2 \quad \text{with} \quad \psi(e,c,\theta) = \varphi(e,c) - \gamma_{\mathrm{V}}(\theta). \tag{8.6.1a}$$

Another ingredient of the model, the dissipation potential, is taken from (7.6.6) modified here appropriately, again making the viscous-moduli tensor state-dependent and again advocating the viscous variant of the Cahn-Hilliard model:

$$\xi(c,\theta;\dot{e},\dot{c}) = \frac{1}{2}\mathbb{D}(c,\theta)\dot{e}:\dot{e} + \frac{1}{2}\tau_{\mathrm{R}}|\dot{c}|^2. \tag{8.6.1b}$$

Thus, allowing also for the mobility $\mathbb{M}$ as well as the heat-conductivity tensor $\mathbb{K}$ depending both on c and on θ, the *anisothermal* variant of the *viscous Cahn-Hilliard model* (7.6.21b) coupled with the mechanical force equilibrium results to

$$\varrho\ddot{u} - \mathrm{div}\,\sigma = f \qquad \text{with} \quad \sigma = \mathbb{D}(c,\theta)e(\dot{u}) + \partial_e\varphi(e(u),c), \tag{8.6.2a}$$

$$\dot{c} - \mathrm{div}(\mathbb{M}(c,\theta)\nabla\mu) = 0 \qquad \text{with} \quad \mu = \partial_c\varphi(e(u),c) + \tau_{\mathrm{R}}\dot{c} - \kappa\Delta c, \tag{8.6.2b}$$

$$\begin{aligned}\dot{\vartheta} - \mathrm{div}(\mathbb{K}(c,\theta)\nabla\theta) &= \mathbb{D}(c,\theta)e(\dot{u}):e(\dot{u}) \\ &\quad + \tau_{\mathrm{R}}\dot{c}^2 + \mathbb{M}(c,\theta)\nabla\mu\cdot\nabla\mu \qquad \text{with } \vartheta = \mathscr{C}_{\mathrm{V}}(\theta)\end{aligned} \tag{8.6.2c}$$

where, as before, $\mathscr{C}_{\mathrm{V}}(\theta) := \theta\partial_\theta\gamma_{\mathrm{V}}(\theta) - \gamma_{\mathrm{V}}(\theta)$. It should emphasized the viscosity (i.e. the relaxation time τ_{R} positive) is important in this anisothermal model to facilitate the chain-rule formula (7.5.10). Temperature dependence of the mobility tensor $\mathbb{M}$ is well known from our everyday experience: the process of swelling and softening of various food ingredients as mentioned in Remark 7.6.7 by wetting in water proceed much faster if warmed (boiled) than under room temperature.

This system is to be accompanied by suitable boundary conditions, e.g. (7.6.2) on p. 316 together with $\mathbb{K}(c,\theta)\nabla\theta\cdot\vec{n} = h_{\flat}$, and the natural initial conditions. Altogether,

$$\sigma\vec{n} = g, \quad \kappa\nabla c\cdot\vec{n} = 0, \quad \mathbb{M}(c)\nabla\mu = j_{\flat}, \quad \mathbb{K}(c,\theta)\nabla\theta\cdot\vec{n} = h_{\flat} \qquad \text{on } \Sigma, \tag{8.6.3a}$$

$$u(0) = u_0, \qquad \dot{u}(0) = v_0, \qquad c(0) = c_0, \qquad \theta(0) = \theta_0 \qquad \text{in } \Omega. \tag{8.6.3b}$$

The energetics of this system can be revealed by testing the particular equations in (8.6.2) respectively by $\dot{u}$, μ, and 1, and after applying Green's formula using also the boundary conditions (7.6.2). The *balance of chemo-mechanical energy* results from (8.6.2a,b):

$$\frac{\mathrm{d}}{\mathrm{d}t}\int_\Omega \underbrace{\frac{\varrho}{2}|\dot u|^2}_{\substack{\text{kinetic}\\ \text{energy}}} + \underbrace{\varphi(e(u),c) + \frac{\kappa}{2}|\nabla c|^2}_{\substack{\text{stored energy=}\\ \text{chemo-mechanical}\\ \text{energy}}}\,\mathrm{d}x + \int_\Omega \underbrace{\mathbb{D}(c,\theta)e(\dot u){:}e(\dot u) + \tau_{\mathrm{R}}\dot c^2 + \mathbb{M}(c,\theta)\nabla\mu\cdot\nabla\mu}_{\text{rate of dissipation}}\,\mathrm{d}x$$

$$= \underbrace{\int_\Omega f\cdot\dot u\,\mathrm{d}x + \int_{\Gamma_{\mathrm{N}}} g\cdot\dot u\,\mathrm{d}S + \langle\vec\tau_{\mathrm{D}},\dot u_{\mathrm{D}}\rangle_{\Gamma_{\mathrm{D}}}}_{\substack{\text{power of external}\\ \text{mechanical forcing}}} + \underbrace{\int_\Gamma j_{\flat}\mu\,\mathrm{d}S}_{\substack{\text{power of the boun-}\\ \text{dary fluid flux}}}\,. \tag{8.6.4}$$

Summing it with (8.6.2c) tested by 1 when respecting also the boundary condition $\mathbb{K}(c,\theta)\nabla\theta\cdot\vec n = h_\flat$ gives the total-energy balance

$$\underbrace{\frac{\mathrm{d}}{\mathrm{d}t}\int_\Omega \frac{\varrho}{2}|\dot u|^2 + \varphi(e(u),c) + \vartheta + \frac{\kappa}{2}|\nabla c|^2\,\mathrm{d}x}_{\text{total kinetic and internal energy}}$$

$$= \underbrace{\int_\Omega f\cdot\dot u\,\mathrm{d}x + \int_\Gamma g\cdot\dot u\,\mathrm{d}S + \langle\vec\tau_{\mathrm{D}},\dot u_{\mathrm{D}}\rangle_{\Gamma_{\mathrm{D}}}}_{\substack{\text{power of external}\\ \text{mechanical forcing}}} + \underbrace{\int_\Gamma j_\flat\mu + h_\flat\,\mathrm{d}S}_{\substack{\text{power of the boundary}\\ \text{fluid and heat fluxes}}}\,. \tag{8.6.5}$$

Let us illustrate the analysis on the simplest case that $\varphi(\cdot,c)$ is quadratic (as in the *Biot poroelastic model* in Example 7.6.3) so that the system (8.6.2) is semilinear, cf. Exercise 8.6.5 for a more general case. As there is no coupling through adiabatic terms, we can use Proposition 8.2.1 simply with $\omega = 1$. Altogether, beside standard assumptions for the mechanical part (6.4.5), we assume, for some $\epsilon > 0$, that:

$$\varphi:\mathbb{R}^{d\times d}_{\mathrm{sym}}\times\mathbb{R}\to\mathbb{R}\ \text{ continuous, }\ \varphi(\cdot,c):\mathbb{R}^{d\times d}_{\mathrm{sym}}\to\mathbb{R}\ \text{ quadratic,}$$
$$\forall e\in\mathbb{R}^{d\times d}_{\mathrm{sym}},\ c\in\mathbb{R}:\ \ \epsilon(|e|^2+c^2)\le|\varphi(e,c)\le(1+|e|^2+c^2)/\epsilon, \tag{8.6.6a}$$

$\mathbb{D},\mathbb{M}$ continuous, bounded, and uniformly positive definite in the sense

$$c_{\mathbb{D},\mathbb{M}} := \inf_{e\in\mathbb{R}^{d\times d}_{\mathrm{sym}},\ g\in\mathbb{R}^d,\ |e|=|g|=1,\ c,\theta\in\mathbb{R}} \mathbb{D}(c,\theta)e{:}e + \mathbb{M}(c,\theta)g\cdot g > 0, \tag{8.6.6b}$$

$$\epsilon/\theta \le \gamma_{\mathrm{v}}''(\theta) \le 1/\epsilon\theta,\quad \kappa>0,\quad \tau_{\mathrm{R}}>0, \tag{8.6.6c}$$

$$j_\flat\in L^2(I;L^{2^{\sharp\prime}}(\Gamma;\mathbb{R}^d)),\quad h_\flat\in L^1(\Sigma),\quad h_\flat\ge0, \tag{8.6.6d}$$

$$c_0\in H^1(\Omega),\quad \theta_0\in L^1(\Omega),\quad \theta_0\ge0. \tag{8.6.6e}$$

To be specific, let us again use Galerkin approximation of a regularization like in Remark 8.2.3. This means, instead of (8.6.2c), we consider

$$\dot\vartheta - \mathrm{div}(\mathbb{K}(c,\theta)\nabla\theta) = \mathbb{D}(c,\theta)e(\dot u){:}\frac{e(\dot u)}{1+\varepsilon|e(\dot u)|^2} + \frac{\tau_{\mathrm{R}}\dot c^2}{1+\varepsilon\dot c^2} + \frac{\mathbb{M}(c,\theta)\nabla\mu\cdot\nabla\mu}{1+\varepsilon|\nabla\mu|^2}\quad \text{with } \vartheta = \mathscr{C}_{\mathrm{v}}(\theta) \tag{8.6.7}$$

together with $h_\flat$ and θ_0 in (8.6.3) replaced by $h_\flat/(1+\varepsilon h_\flat)$ and $\theta_0/(1+\varepsilon\theta_0)$, respectively. Without loss of generality, we can assume the Galerkin method using the same finite-dimensional subspaces of $H^2(\Omega)$ for both equations in (8.6.2b) so that, in particular, Δc_k has a good sense. Before starting physically relevant estimations, one should make the limit passage in the heat-transfer equation based on L^2-theory, cf. Remark 8.3.5, while the estimates (8.6.8e) are to be considered carefully as in Sect. C.2.4. To avoid technicalities which copies previous sections, we therefore dare assume that, via the mentioned Galerkin method, we already obtained a weak solution to (8.6.2a,b) and (8.6.7) with (8.6.3) using $h_\flat/(1+\varepsilon h_\flat)$ and $\theta_0/(1+\varepsilon\theta_0)$ instead of $h_\flat$ and θ_0. Let us denote the solution of such problem by $(u_\varepsilon, c_\varepsilon, \mu_\varepsilon, \theta_\varepsilon)$.

Proposition 8.6.1 (Weak solutions to heat/mass transfer problem). *Let ϱ, f, g, u_0, v_0 satisfy (6.4.5b,c,e), and let (8.6.6) hold. Then the above devised approximate solution satisfies*

$$\|u_\varepsilon\|_{W^{1,\infty}(I;L^2(\Omega;\mathbb{R}^d))\cap H^1(I;H^1(\Omega;\mathbb{R}^d))} \le C, \tag{8.6.8a}$$

$$\|c_\varepsilon\|_{L^\infty(I;H^1(\Omega))\cap H^1(I;L^2(\Omega))} \le C \quad \textit{and} \quad \|\mu_\varepsilon\|_{L^2(I;H^1(\Omega))} \le C, \tag{8.6.8b}$$

$$\|\vartheta_\varepsilon\|_{L^\infty(I;L^1(\Omega))} \le C \quad \textit{and} \quad \|\theta_\varepsilon\|_{L^\infty(I;L^1(\Omega))} \le C. \tag{8.6.8c}$$

$$\|\nabla\theta_\varepsilon\|_{L^r(Q;\mathbb{R}^d)} \le C \quad \textit{and} \quad \|\theta_\varepsilon\|_{L^p(Q)} \le C, \tag{8.6.8d}$$

$$\|\varrho\ddot{u}_\varepsilon\|_{L^2(I;H^1(\Omega;\mathbb{R}^d)^*)} \le C, \quad \|\Delta c_\varepsilon\|_{L^2(Q)} \le C, \quad \|\dot{\vartheta}_\varepsilon\|_{L^1(I;H^{(d+3)/2}(\Omega)^*)} \le C. \tag{8.6.8e}$$

There is a subsequence converging weakly in the topologies indicated in (8.6.8) to some (u, c, μ, θ). Moreover, every such subsequence also exhibits*

$$e(\dot{u}_\varepsilon) \to e(\dot{u}) \qquad \textit{strongly in } L^2(Q;\mathbb{R}^{d\times d}_{\rm sym}), \tag{8.6.9a}$$

$$\dot{c}_\varepsilon \to \dot{c} \qquad \textit{strongly in } L^2(Q), \textit{ and} \tag{8.6.9b}$$

$$\nabla\mu_\varepsilon \to \nabla\mu \qquad \textit{strongly in } L^2(Q;\mathbb{R}^d). \tag{8.6.9c}$$

and every quadruple (u, c, μ, θ) obtained by such way is a weak solution to the initial-boundary-value problem (8.6.2)–(8.6.3) has a weak solution. Eventually, the chemo-mechanical as well as the total energies are conserved in the sense that (8.6.4) and (8.6.5) hold when integrated over I.

Proof. Consider an above specified approximate solution, i.e. a quadruple $(u_\varepsilon, c_\varepsilon, \mu_\varepsilon, \theta_\varepsilon)$. From (8.6.4) written for such approximate solution, one can read the a-priori estimates (8.6.8a,b) and, proving also $\vartheta_k \ge 0$, from (8.6.5) also (8.6.8c). The estimate of the temperature gradient is here particularly simple because there is no direct coupling between (e, c) and θ in the free energy and thus no adiabatic-like term in the heat-transfer equation, and one can use directly Proposition 8.2.1. This gives still the estimate (8.6.8d). Moreover, by comparison from (8.6.2) considered in an approximate form, we obtain also (8.6.8e).

Further step is the selection of a weak* convergent subsequence in the topology indicated in (8.6.8a-d), cf. Theorem A.1.3. By the Rellich theorem B.3.3, we have also strong convergence of $c_\varepsilon \to c$ in $L^2(Q)$ and, by the Aubin-Lions theorem B.4.10,

$\theta_\varepsilon \to \theta$ in $L^p(Q)$ with $1 \le p < 1+2/d$ from Proposition 8.2.1. The limit passage in the approximate chemo-mechanical part towards weakly formulated semilinear system (8.6.2a,b) is then easy.

For the limit passage towards (8.6.2c), we need to improve the weak convergence of $e(\dot{u}_\varepsilon)$, $\dot{c}_\varepsilon$, and of $\nabla\mu_\varepsilon$ to the strong convergence so that the right-hand side of the heat equation will converge in $L^1(Q)$. To this goal, we use the technique as in (8.3.20) or (8.4.16):

$$
\begin{aligned}
&\limsup_{\varepsilon\to 0}\int_Q \mathbb{D}(c_\varepsilon,\theta_\varepsilon)e(\dot{u}_\varepsilon):e(\dot{u}_\varepsilon)+\tau_{\rm R}\dot{c}_\varepsilon^2+\mathbb{M}(c_\varepsilon,\theta_\varepsilon)\nabla\mu_\varepsilon\cdot\nabla\mu_\varepsilon\,{\rm d}x{\rm d}t\\
&=\limsup_{\varepsilon\to 0}\int_\Sigma g\cdot\dot{u}_\varepsilon+j_{\rm b}\mu_\varepsilon\,{\rm d}S\,{\rm d}t+\int_Q(f-\varrho\ddot{u}_\varepsilon)\cdot\dot{u}_\varepsilon-\sigma_\varepsilon:e(\dot{u}_\varepsilon)+\tau_{\rm R}\dot{c}_\varepsilon^2-\dot{c}_\varepsilon\mu_\varepsilon\,{\rm d}x{\rm d}t\\
&=\limsup_{\varepsilon\to 0}\int_\Sigma g\cdot\dot{u}_\varepsilon+j_{\rm b}\mu_\varepsilon\,{\rm d}S\,{\rm d}t\\
&\qquad+\int_Q(f-\varrho\ddot{u}_\varepsilon)\cdot\dot{u}_\varepsilon-\partial_e\varphi(e_\varepsilon,c_\varepsilon):e(\dot{u}_\varepsilon)-(\partial_c\varphi(e_\varepsilon,c_\varepsilon)-\kappa\Delta c_\varepsilon)\dot{c}_\varepsilon\,{\rm d}x{\rm d}t\\
&=\limsup_{\varepsilon\to 0}\int_\Sigma g\cdot\dot{u}_\varepsilon+j_{\rm b}\mu_\varepsilon\,{\rm d}S\,{\rm d}t+\int_Q f\cdot\dot{u}_\varepsilon\,{\rm d}x{\rm d}t+\int_\Omega\Big(\varphi(e(u_0),c_0)+\frac{\kappa}{2}|\nabla c_0|^2\\
&\qquad+\frac{\varrho}{2}|v_0|^2-\varphi(e(u_\varepsilon(T),c_\varepsilon(T))-\frac{\kappa}{2}|\nabla c_\varepsilon(T)|^2-\frac{\varrho}{2}|\dot{u}_\varepsilon(T)|^2\Big){\rm d}x\\
&\le\int_\Sigma g\cdot\dot{u}+j_{\rm b}\mu\,{\rm d}S\,{\rm d}t+\int_Q f\cdot\dot{u}\,{\rm d}x{\rm d}t+\int_\Omega\Big(\varphi(e(u_0),c_0)\\
&\qquad+\frac{\kappa}{2}|\nabla c_0|^2+\frac{\varrho}{2}|v_0|^2-\varphi(e(u(T),c(T))-\frac{\kappa}{2}|\nabla c(T)|^2-\frac{\varrho}{2}|\dot{u}(T)|^2\Big){\rm d}x\\
&=\int_Q\mathbb{D}(c,\theta)e(\dot{u}):e(\dot{u})+\tau_{\rm R}\dot{c}^2+\mathbb{M}(c,\theta)\nabla\mu\cdot\nabla\mu\,{\rm d}x{\rm d}t. \qquad (8.6.10)
\end{aligned}
$$

For the last equality in (8.6.10), we have used that we already showed that (u,c,μ,θ) solves weakly (8.6.2a,b) and that we have rigorously the chemo-mechanical energy equality at disposal, i.e. (8.6.4) integrated over $[0,T]$. For this, the essential information is that $\sqrt{\varrho}\ddot{u}$ is in duality with $\sqrt{\varrho}\dot{u}$ as well as $\dot{c}$ is in duality with μ, and also that $\Delta c\in L^2(Q)$ is in duality with $\dot{c}\in L^2(Q)$, cf. the calculus (7.5.10).

From (8.6.10), we can see the desired strong convergence (8.6.9). More in detail, it holds

$$
\begin{aligned}
&\|e(\dot{u}_\varepsilon-\dot{u})\|^2_{L^2(Q;\mathbb{R}^{d\times d})}+\|\dot{c}_\varepsilon-\dot{c}\|^2_{L^2(Q)}+\|\nabla(\mu_\varepsilon-\mu)\|^2_{L^2(Q;\mathbb{R}^d)}\\
&\le\frac{1}{\min(c_{\rm D,M},\tau_{\rm R})}\int_Q\mathbb{D}(c_\varepsilon,\theta_\varepsilon)e(\dot{u}_\varepsilon-\dot{u}):e(\dot{u}_\varepsilon-\dot{u})\\
&\qquad+\tau_{\rm R}|\dot{c}_\varepsilon-\dot{c}|^2+\mathbb{M}(c_\varepsilon,\theta_\varepsilon)\nabla(\mu_\varepsilon-\mu)\cdot\nabla(\mu_\varepsilon-\mu)\,{\rm d}x{\rm d}t\\
&=\frac{1}{\min(c_{\rm D,M},\tau_{\rm R})}\int_Q\mathbb{D}(c_\varepsilon,\theta_\varepsilon)e(\dot{u}_\varepsilon):e(\dot{u}_\varepsilon)+\tau_{\rm R}\dot{c}_\varepsilon^2+\mathbb{M}(c_\varepsilon,\theta_\varepsilon)\nabla\mu_\varepsilon\cdot\nabla\mu_\varepsilon\\
&\qquad+\mathbb{D}(c_\varepsilon,\theta_\varepsilon)e(\dot{u}):e(\dot{u}-2\dot{u}_\varepsilon)+\tau_{\rm R}\dot{c}(\dot{c}-2\dot{c}_\varepsilon)\\
&\qquad+\mathbb{M}(c_\varepsilon,\theta_\varepsilon)\nabla\mu\cdot\nabla(\mu-2\mu_\varepsilon)\,{\rm d}x{\rm d}t\to 0 \qquad (8.6.11)
\end{aligned}
$$

with $c_{\mathbb{D},\mathbb{M}} > 0$ from (8.6.6b). The convergence towards 0 in (8.6.11) uses (8.6.10) together with the fact that $\mathbb{D}(c_\varepsilon,\theta_\varepsilon)e(\dot u) \to \mathbb{D}(c,\theta)e(\dot u)$ strongly in $L^2(Q;\mathbb{R}^{d\times d}_{\rm sym})$ while $e(\dot u{-}2\dot u_\varepsilon) \to -e(\dot u)$ weakly in $L^2(Q;\mathbb{R}^{d\times d}_{\rm sym})$ so that

$$\int_Q \mathbb{D}(c_\varepsilon,\theta_\varepsilon)e(\dot u){:}e(\dot u{-}2\dot u_\varepsilon)\,{\rm d}x{\rm d}t \to -\int_Q \mathbb{D}(c,\theta)e(\dot u){:}e(\dot u)\,{\rm d}x{\rm d}t \quad\text{and}$$
$$\int_Q \mathbb{M}(c_\varepsilon,\theta_\varepsilon)\nabla\mu{\cdot}\nabla(\mu{-}2\mu_\varepsilon)\,{\rm d}x{\rm d}t \to -\int_Q \mathbb{M}(c,\theta)\nabla\mu{\cdot}\nabla\mu\,{\rm d}x{\rm d}t, \tag{8.6.12}$$

and similarly, even more simply, also $\int_Q \tau_{\rm R}\dot c(\dot c{-}2\dot c_\varepsilon)\,{\rm d}x{\rm d}t \to -\int_Q \tau_{\rm R}\dot c^2\,{\rm d}x{\rm d}t$. Thus, (8.6.11) and therefore also (8.6.9) is proved. Having (8.6.9) at our disposal, we can pass to the limit in the heat-transfer equation. □

A combination with some other processes like plasticity, creep, or damage is straightforward, although especially damage makes the problem much more complicated in particular as far as convergence concerns, cf. also [460] for a combination with a rate-independent cohesive damage.

Remark 8.6.2 (Concentration-dependent heat capacity $c_{\rm v} = c_{\rm v}(c,\theta)$). The heat capacity can naturally be affected by the content of the fluid c. This can be modeled by modifying the free energy (8.6.1a) as

$$\psi(e,c,\nabla c,\theta) = \varphi(e,c) - \gamma_{\rm v}(c,\theta) + \frac{\kappa}{2}|\nabla c|^2. \tag{8.6.13}$$

The entropy $\mathfrak{s} = \partial_\theta\gamma_{\rm v}(c,\theta)$ now becomes c-dependent and substituting it into the so-called entropy equation (8.1.7) gives

$$c_{\rm v}(c,\theta)\dot\theta + {\rm div}\,j = \xi - (\partial_c\gamma_{\rm v}(c,\theta) - \partial_c\gamma_{\rm v}(c,0))\dot c \quad\text{with}\quad c_{\rm v}(c,\theta) = \theta\partial^2_{\theta\theta}\gamma_{\rm v}(c,\theta). \tag{8.6.14}$$

This reveals that, beside making the heat capacity $c_{\rm v}$ temperature dependent, the right-hand side of the heat-transfer equation augments by the adiabatic-like term $\theta\partial^2_{c\theta}\gamma_{\rm v}(c,\theta)\dot c = (\partial_c\gamma_{\rm v}(c,\theta) - \partial_c\gamma_{\rm v}(c,0))$. The internal energy $w = \psi + \mathfrak{s}\theta$ equals to $\varphi(e,c) + \frac{\kappa}{2}|\nabla c|^2 + \theta\partial_\theta\gamma_{\rm v}(c,\theta) - \gamma_{\rm v}(c,\theta)$. The chemical potential then become temperature dependent, namely it augments as $\mu = \partial_c\varphi(e,c,\nabla c) - \partial_c\gamma_{\rm v}(c,\theta) - \kappa\Delta c$. Testing the diffusion equation $\dot c = {\rm div}(\mathbb{M}(c,\theta)\nabla\mu)$ in (8.6.2b) by μ gives, in addition, the term $\partial_c\gamma_{\rm v}(c,\theta)\dot c$. In the overall energy balance, this terms balance the rate of thermal part of internal energy due to the obvious calculus, canceling the terms $\pm\partial_\theta\gamma_{\rm v}(c,\theta)\dot\theta$, yields

$$\frac{\partial}{\partial t}(\gamma_{\rm v}(c,\theta) - \theta\partial_\theta\gamma_{\rm v}(c,\theta)) = \partial_c\gamma_{\rm v}(c,\theta)\dot c - \theta\partial^2_{\theta\theta}\gamma_{\rm v}(c,\theta)\dot\theta - \theta\partial^2_{c\theta}\gamma_{\rm v}(c,\theta)\dot c.$$

Estimation of the mentioned adiabatic term $\theta\partial^2_{c\theta}\gamma_{\rm v}(c,\theta)\dot c$ needs a "viscosity" that would control $\dot c$, i.e. ζ should be coercive also in terms of $\dot c$ as in (8.6.1b) with $\tau_{\rm R} > 0$, leading to viscosity in μ in (8.6.2b).

Remark 8.6.3 (The heat capacity $c_{\rm v} = c_{\rm v}(c,\theta)$ and inviscid μ). Another example without viscosity in the chemical potential μ is $\varphi(e,e_1,c) = \frac{1}{2}\mathbb{C}_1 e_1 : e_1 + \frac{1}{2}\mathbb{C}_2 e : e$ with $e = e_1 + \mathbb{E}c$ and $\zeta(\dot e,\dot e_1) = \frac{1}{2}\mathbb{D}_1\dot e_1 : \dot e_1 + \frac{1}{2}\mathbb{D}_2\dot e : \dot e$ arising from a rheological model $((\lessgtr_{\mathbb{C}_1}\|\dashv\vdash_{\mathbb{D}_1})\text{—}\not\prec^{\,c}_{\mathbb{E}})\|\lessgtr_{\mathbb{C}_2}\|\dashv\vdash_{\mathbb{D}_2}$, cf. also Exercise 7.6.8. In particular, to control $\dot c$, the potential ζ does not necessarily contain a term like $|\dot c|^2$, relying on that $\dot c = \mathbb{E}^{-1}\cdot(\dot e - \dot e_1)$.

Remark 8.6.4 (Thermodynamical restrictions). The arguments in Remark 8.1.5 can be modified to justify the concept of chemical potential as a driving force. Considering the gradient theory, the vector of the internal variables is now $z = (c,g)$ with the holonomic linear constraint $\nabla c = g$. The internal energy balance (8.1.29) then modifies as

$$\dot w = \tilde\sigma_{\rm el}:\dot e + \tilde\sigma_{\rm dif}\dot c + \tilde\sigma_{\rm hyp}\cdot\dot g + \xi(\dot e,\dot c,\dot g) + \frac{1}{2}\big|\mathbb{M}(c,\theta)^{1/2}\nabla\Delta^{-1}_{\mathbb{M}(c,\theta)}\dot c\big|^2 - {\rm div}\, j \qquad \text{with } \dot g = \nabla\dot c.$$

with an unspecified driving force (resp. hyperforce) for diffusion $\tilde\sigma_{\rm dif}$ (resp. $\tilde\sigma_{\rm hyp}$) and with the nonlocal dissipation term $\frac{1}{2}|\mathbb{M}(c,\theta)^{1/2}\nabla\Delta^{-1}_{\mathbb{M}(c,\theta)}\dot c|^2$ as in Remark 7.6.4 from where we also postulate the general diffusion dynamics

$$\Delta^{-1}_{\mathbb{M}(c,\theta)}\dot c = \tilde\mu. \tag{8.6.15}$$

Realizing that $\tilde\mu = \partial_c\varphi$, it reveals the structure of diffusion as a gradient flow with a special nonlocal metric, cf. e.g. [214]. The mentioned nonlocal dissipation term then takes the form $\frac{1}{2}|\mathbb{M}\nabla\tilde\mu\cdot\nabla\tilde\mu|^2 = \frac{1}{2}|\mathbb{M}^{-1/2}\nabla\Delta^{-1}_{\mathbb{M}(c,\theta)}|^2$. The *dissipation inequality* (8.1.31) now reads as

$$\begin{aligned}0 \ge\ & (\tilde\sigma_{\rm el}-\partial_e\psi):\dot e + (\partial_c\psi-\tilde\sigma_{\rm dif})\dot c\\ & + (\partial_g\psi-\tilde\sigma_{\rm hyp})\cdot\dot g + (\partial_\theta\psi+\mathfrak{s})\dot\theta - \xi(\dot e,\dot c,\dot g) - \frac{1}{2}|\nabla\tilde\mu|^2 + \frac{j\cdot\nabla\theta}{\theta}.\end{aligned} \tag{8.6.16}$$

Like in (8.1.32), argumenting of independence of rates $\dot e$ and $\nabla\theta$ from $(\dot c,\dot g)$, we can read the condition $(\tilde\sigma_{\rm dif}-\partial_c\psi)\dot c + (\tilde\sigma_{\rm hyp}-\partial_{\nabla c}\psi)\cdot\dot g \le 0$. Like in Remark 8.1.10, the rates $(\dot c,\dot g)$ are not independent and one has to count the holonomic constraint $\dot g = \nabla\dot c$. An analogy of (8.1.43) must be formulated in a nonlocal way as

$$\forall(\tilde c,\tilde g)\in L^2(\Omega;\mathbb{R}\times\mathbb{R}^d),\ \ \nabla\tilde c = \tilde g:\quad \int_\Omega(\tilde\sigma_{\rm dif}-\partial_c\psi)\tilde c + (\tilde\sigma_{\rm hyp}-\partial_g\psi)\cdot\tilde g\,{\rm d}x = 0.$$

Eliminating g-variable and applying Green's formula yield

$$\int_\Omega(\tilde\sigma_{\rm dif}-\partial_c\psi)\tilde c - {\rm div}(\tilde\sigma_{\rm hyp}-\partial_g\psi)\tilde c\,{\rm d}x = \int_\Gamma(\partial_g\psi-\tilde\sigma_{\rm hyp})\cdot\vec n\tilde c\,{\rm d}S.$$

Taking a general test function $\tilde c$ with zero traces on Γ and taking into account our previous ansatz $\tilde\sigma_{\rm dif} = \sigma_{\rm dif}(e,c,\theta) + \partial_{\dot c}\zeta(\dot e,\dot c)$ and $\tilde\sigma_{\rm hyp} = \sigma_{\rm hyp}(e,c,g,\theta) +$

$\partial_{\nabla\dot{c}}\zeta(\dot{e},\dot{c},\nabla\dot{c})$, like in (8.1.33), we obtain the condition

$$\partial_c\psi - \operatorname{div}\partial_{\nabla c}\psi = \tilde{\sigma}_{\rm dif} - \operatorname{div}\tilde{\sigma}_{\rm hyp}, \tag{8.6.17a}$$

$$\partial_{\dot{c}}\zeta(\dot{e},\dot{c})\dot{c} \ge 0, \qquad \partial_{\nabla\dot{c}}\zeta(\dot{e},\dot{c},\nabla\dot{c})\cdot\nabla\dot{c} \ge 0. \tag{8.6.17b}$$

The relation (8.6.17a) justifies the form of the chemical potential $\mu = \partial_c\psi - \operatorname{div}\partial_{\nabla c}\psi$ as a driving force $\tilde{\sigma}_{\rm dif} - \operatorname{div}\tilde{\sigma}_{\rm hyp}$.

Exercise 8.6.5 (General convex φ). Consider a general convex φ in (8.6.2) and modify the proof of Proposition 8.6.1.

8.7 Electric charge transport in poroelastic solids

Other problems from Section 5.7.4 yield interesting examples in their dynamical variant. Let us illustrate it on the evolution variant of the steady-state system (5.7.35) on p. 190 describing electrically charged (with a given charged dopant profile $z_{\rm DOP} = z_{\rm DOP}(x)$ independent of time) deformable solid containing a multi-component *electrically charged diffusant*. Let us recall the convex/concave-type energy from (5.7.12) on p. 181:

$$(u,e,c,\phi,\vec{e}) \mapsto \varphi(e(u),c) + (z{\cdot}c{+}z_{\rm DOP})\phi - z_{\rm DOP}\vec{e}{\cdot}u - f{\cdot}u - \frac{\varepsilon}{2}|\vec{e}|^2,$$

where ϕ is the electrostatic potential and $\vec{e}$ the intensity of the electrostatic field.

For simplicity, let us assume here the partly linearized ansatz (8.1.35) and even suppress thermo-mechanical interactions by separating θ from the other variables. Therefore, the free energy is considered as

$$\psi_{\rm E}(u,e,c,\phi,\vec{e},\theta,\nabla c) = \psi(u,e,c,\phi,\vec{e},\theta) + \frac{\kappa}{2}|\nabla c|^2 \quad \text{with}$$
$$\psi(u,e,c,\phi,\vec{e},\theta) = \varphi(e,c) + (z{\cdot}c{+}z_{\rm DOP})\phi - z_{\rm DOP}\vec{e}{\cdot}u - \gamma_{\rm V}(\theta) - \frac{\varepsilon}{2}|\vec{e}|^2. \tag{8.7.1}$$

Realizing the holonomic-type constraints

$$e = \operatorname{sym}\nabla u \quad \text{and} \quad \vec{e} = -\nabla\phi, \tag{8.7.2}$$

and the *nonlocal dissipation potential* behind the diffusion problem as identified in Remark 7.6.4 on p. 320, the overall integrated functionals can be written in terms of (u,c,ϕ,θ) as

$$\begin{aligned}\Psi(u,c,\phi,\theta) &= \int_{\mathbb{R}^d} \psi_{\rm E}(u,e(u),c,\phi,\nabla\phi,\theta,\nabla c)\,{\rm d}x \\ &= \int_\Omega \Big(\varphi(e(u),c) + (z{\cdot}c{+}z_{\rm DOP})\phi + z_{\rm DOP}\nabla\phi{\cdot}u - \gamma_{\rm V}(\theta) \\ &\qquad + \frac{\kappa}{2}|\nabla c|^2\Big){\rm d}x - \int_{\mathbb{R}^d}\frac{\varepsilon}{2}|\nabla\phi|^2\,{\rm d}x,\end{aligned} \tag{8.7.3a}$$

$$\mathcal{R}(c,\theta;\dot{u},\dot{c}) = \mathcal{R}_1(c,\theta;\dot{u}) + \mathcal{R}_2(c,\theta;\dot{c})$$
$$\text{with}\quad \mathcal{R}_1(c,\theta;\dot{u}) = \int_\Omega \mathbb{D}(c,\theta)e(\dot{u}){:}e(\dot{u})\,\mathrm{d}x$$
$$\text{and}\quad \mathcal{R}_2(c,\theta;\dot{c}) = \int_\Omega \frac{1}{2}\tau_{\mathrm{R}}\dot{c}^2 + \frac{1}{2}|\mathbb{M}(c,\theta)^{1/2}\nabla\Delta^{-1}_{\mathbb{M}(c,\theta)}\dot{c}|^2\,\mathrm{d}x, \tag{8.7.3b}$$
$$\mathscr{T}(\dot{u}) = \int_\Omega \frac{\varrho}{2}|\dot{u}|^2\,\mathrm{d}x \tag{8.7.3c}$$

with $\Delta^{-1}_{\mathbb{M}}$ as in Remark 7.6.4.

Denoting the mechano-chemical-electrical state by $q = (u,c,\phi)$, the abstract momentum equation (6.4.17) on p. 217 considered here with q in place of u reads as

$$\mathscr{T}'\ddot{q} + \partial_{\dot{q}}\mathcal{R}(q,\theta;\dot{q}) + \partial_q\Psi(q,\theta) = \partial_q F(t)$$

with $F(t,\cdot)$ a linear functional defined by $F(t,q) = \int_\Omega f(t)\cdot u\,\mathrm{d}x + \int_\Gamma g\cdot u\,\mathrm{d}S$. Realizing that $\mathscr{T}(\dot{q}) = \mathscr{T}(\dot{u})$ and $\mathcal{R}(q,\theta;\dot{q}) = \mathcal{R}_1(q,\theta;\dot{u}) + \mathcal{R}_2(q,\theta;\dot{c})$, while $\Psi(q,\theta) = \Psi(u,c,\phi,\theta)$, it takes a more specific form when written component-wise as

$$\mathscr{T}'\ddot{u} + \partial_{\dot{u}}\mathcal{R}_1(c,\theta;\dot{u}) + \partial_u\Psi(u,c,\phi,\theta) = \partial_u F(t), \tag{8.7.4a}$$
$$\partial_{\dot{c}}\mathcal{R}_2(c,\theta;\dot{c}) + \partial_c\Psi(u,c,\phi,\theta) = 0, \tag{8.7.4b}$$
$$\partial_\phi\Psi(u,c,\phi,\theta) = 0. \tag{8.7.4c}$$

When taken into account the specific functionals (8.7.3), the equations (8.7.4) reads respectively as:

$$\varrho\ddot{u} - \mathrm{div}(\mathbb{D}(c,\theta)e(\dot{u}) + \partial_e\varphi(e(u),c)) = f - z_{\mathrm{DOP}}\nabla\phi, \tag{8.7.5a}$$
$$\dot{c} - \mathrm{div}(\mathbb{M}(c,\theta)\nabla\mu) = r(c,\theta) \quad \text{with } \mu = \partial_c\varphi(e(u),c) + \tau_{\mathrm{R}}\dot{c} + z\phi - \kappa\Delta c, \tag{8.7.5b}$$
$$\mathrm{div}(\varepsilon\nabla\phi) + z\cdot c + z_{\mathrm{DOP}} = \mathrm{div}(z_{\mathrm{DOP}}u). \tag{8.7.5c}$$

The usual thermodynamical procedure (cf. Remark 8.1.7 with $\varphi_{\mathrm{C}} = 0$) completes the mechano-chemo-electrical system (8.7.5a-c) still by the heat-transfer equation:

$$\dot{\vartheta} - \mathrm{div}(\mathbb{K}(c,\theta)\nabla\theta) = \mathbb{D}(c,\theta)e(\dot{u}){:}e(\dot{u}) + \tau_{\mathrm{R}}|\dot{c}|^2 + \mathbb{M}(c,\theta)\nabla\mu{:}\nabla\mu$$
$$+ h(c,\theta) - \mu\cdot r(c,\theta) \quad \text{with } \vartheta = \mathscr{C}_{\mathrm{V}}(\theta), \tag{8.7.5d}$$

where again $\mathscr{C}_{\mathrm{V}}(\theta) = \gamma_{\mathrm{V}}(\theta) - \theta\partial_\theta\gamma_{\mathrm{V}}(\theta)$. Let us note also that the partial derivatives of the free energy give the stress, entropy, chemical potential, and the *electric* induction (or *displacement*):

$$\sigma = \partial_e\psi(e,c), \quad \mathfrak{s} = -\partial_\theta\psi(\theta), \quad \mu = \partial_c\psi(e,c) - \mathrm{div}(\partial_{\nabla c}\psi_{\mathrm{E}}(\nabla c)), \quad \vec{d} = -\partial_{\vec{e}}\psi(\vec{e}).$$

Here $\vec{d}$ is equal to $\varepsilon\vec{e}-z_{\rm DOP}u$, which contains the displaced charge $z_{\rm DOP}u$ and justifies the name "electric displacement". Let us note that (8.7.5c) reads also as ${\rm div}\vec{d}+z{\cdot}c+z_{\rm DOP}=0$.

This is to be completed by suitable boundary conditions and by the natural initial conditions (8.6.3). The right-hand side $\mu{\cdot}r(c,\theta)$ of (8.7.5d) represents the (negative) heat production due to chemical reactions. The heat production and chemical-reaction rates can be naturally assumed to vanish at $\theta=0$, i.e. $h(c,0)=0$ and $r(c,0)=0$, and then one can prove $\theta\ge 0$. Let us note that, instead of a full Maxwell electromagnetic system, we kept the electrostatic equation (8.7.5c), which reflects the well acceptable modeling assumption that the thermo-chemical-mechanical processes are much slower than the electromagnetical processes and that the electric currents are rather small so that the magnetic effects can be neglected. The balance of the electro-chemical-mechanical energy can be revealed by testing the particular equations (8.7.5a,b,c) successively by $\dot{u}$, μ, $\dot{c}$, and $\dot{\phi}$:

$$\frac{\rm d}{{\rm d}t}\Big(\int_\Omega\frac{\varrho}{2}|\dot{u}|^2+\varphi(e(u),c)+\frac{\kappa}{2}|\nabla c|^2+z\phi{\cdot}c+z_{\rm DOP}\phi+z_{\rm DOP}\nabla\phi{\cdot}u\,{\rm d}x$$
$$-\int_{\mathbb{R}^d}\frac{\varepsilon}{2}|\nabla\phi|^2\,{\rm d}x\Big)+\int_\Omega\mathbb{D}(c,\theta)e(\dot{u}){:}e(\dot{u})+\tau_{\rm R}|\dot{c}|^2+\mathbb{M}(c,\theta)\nabla\mu{:}\nabla\mu\,{\rm d}x$$
$$=\int_\Omega f{\cdot}\dot{u}+r(c,\theta){\cdot}\mu\,{\rm d}x+\int_{\Gamma_{\rm N}}g{\cdot}\dot{u}\,{\rm d}S+\int_\Gamma j_{\rm b}{\cdot}\mu\,{\rm d}S. \tag{8.7.6}$$

Using again (8.7.5c) together with the calculus (5.7.31), and adding also (8.7.5d) tested by 1, we obtain the total energy balance

$$\frac{\rm d}{{\rm d}t}\Big(\int_\Omega\underbrace{\frac{\varrho}{2}|\dot{u}|^2}_{\substack{\text{kinetic}\\ \text{energy}}}+\underbrace{\varphi(e(u),c)+\mathscr{C}_{\rm v}(\theta)+\frac{\kappa}{2}|\nabla c|^2}_{\substack{\text{internal chemo-thermo-}\\ \text{-mechanical energy}}}\,{\rm d}x+\int_{\mathbb{R}^d}\underbrace{\frac{\varepsilon}{2}|\nabla\phi|^2}_{\substack{\text{electrostatic}\\ \text{energy}}}\,{\rm d}x\Big)$$
$$=\underbrace{\int_\Omega f{\cdot}\dot{u}+h(c,\theta)\,{\rm d}x+\int_{\Gamma_{\rm N}}g{\cdot}\dot{u}\,{\rm d}S}_{\substack{\text{power of mechanical load and of}\\ \text{the heat from chemical reactions}}}+\underbrace{\int_\Gamma j_{\rm b}{\cdot}\mu+h_{\rm b}\,{\rm d}S}_{\substack{\text{power of the boundary}\\ \text{fluid and heat fluxes}}}. \tag{8.7.7}$$

As always, the energetics (8.7.7) is a departure point for making the analysis of this system. An interesting effect here is that the dopant concentration $z_{\rm DOP}$ has been "optically" eliminated from (8.7.7) but, in fact, it is still involved through (5.7.31) acting as a constraint.

From the Maxwell theory of electromagnetic field, it is well known that electric current induces magnetic field. Interesting coupled thermodynamical problems arise when this magnetic field cannot be neglected. In general, the hyperbolic Maxwell equations are thus coupled with nonlinear models of solids, which represents mathematically extremely difficult problem. In some situations, when the material is highly electrically conductive (like metals), it is legitimate to neglect displacement current, which gives so-called eddy-current approximation of the full Maxwell system. This approximation has already a parabolic character and is thus mathematically well amenable.

Chapter 9
Evolution at finite strains

> ... advantage is obtained by treating the problems of mechanics by direct means and by considering from the beginning, the bodies of ... material points or particles, each moved by given forces. And nothing is more easy to modify and simplify by this consideration...
>
> JOSEPH-LOUIS LAGRANGE (1736-1813)

> ...the existence of a physical space, apart from any frame of reference, is an illusion. ... should be replaced by what is now called the principle of frame-indifference.
>
> WALTER NOLL (1925-2017)

This book has begun with problems at finite or large strains in Chapters 2–4; recall the Convention 1.1.1 on p. 6. Thus it is expectable to close it with such problems. In dynamical problems at large deformation and finite/large strains, many serious difficulties are pronounced and only very few results are at disposal, in contrast to the static situations. Essentially, the absence of a clearly identifiable reference configuration and of a linear geometry that would allow for defining a time derivative in a unique way is the source of these problems.

9.1 Time-dependent large deformations

Let us first focus on pure kinematics, using again the notation from Chapter 1. A time-dependent deformation (also called a *motion*) is a mapping $y : Q \to \mathbb{R}^d$. We denote $\Omega^{y(t,\cdot)} := y(t,\Omega)$ and it is the place occupied by the specimen at the time instant $t \in [0,T]$. The mapping $y(t,\cdot)$ is invertible, so that $y^{-1}(t,\cdot) : \Omega^{y(t,\cdot)} \to \Omega$. Considering the fixed time horizon T, the set

$$\mathcal{T} = \mathcal{T}(y,\Omega,[0,T]) := \{(t,x^y);\ x^y \in \Omega^{y(t,\cdot)}\ \&\ t \in [0,T]\} \tag{9.1.1}$$

is called the *trajectory* of the motion of the domain Ω through the deformation y. Let

M. Kružík and T. Roubíček, *Mathematical Methods in Continuum Mechanics of Solids*, Interaction of Mechanics and Mathematics,
https://doi.org/10.1007/978-3-030-02065-1_9

$$p : \mathcal{T} \to \Omega : (t, x^y) \mapsto x \tag{9.1.2}$$

be called the *reference mapping* of the motion . If there is a time-dependent field defined in the reference configuration, i.e. in Q, then we might be interested in how this field evolves in the deformed configuration $\Omega^{y(t,\cdot)}$. Conversely, a time-dependent field defined in the deformed configuration has its image in the reference configuration. Here we explain how to calculate corresponding time derivatives. We call

$$\dot{y}(t,x) = \partial_t y(t,x) \tag{9.1.3}$$

the *material velocity*. On the other hand, we can also define

$$v(t,x^y) := \dot{y}(t, p(t,x^y)) = \partial_t y(t, p(t,x^y)) \tag{9.1.4}$$

which is the *spatial* description of the *velocity*. Generally speaking, we distinguish a *material field*, which is a field defined on the cylinder $Q := [0,T] \times \Omega$ and a *spatial field* whose domain is $\mathcal{T}$. Hence, any field associated with the motion can be described either as a material or a spatial field. Injectivity of deformations allows us to change freely between these two descriptions. Indeed, if ψ is defined on Q, we write $\psi_{\rm S}(t,x^y) := \psi(t, p(t,x^y))$ and call $\psi_{\rm S}$ a *spatial description* of a material field, i.e. the Eulerian description. On the other hand, if a spatial field z is defined on $\mathcal{T}$ then we define $z_{\rm M}(t,x) := z(t, y(t,x))$ and call it a *material description* of a spatial field, i.e. the Lagrangian description.

Next, we define the *material time derivative* of the spatial field z and denote it $\frac{{\rm D}z}{{\rm D}t}$. It is defined in the following way: we first transform z to $z_{\rm M}$, i.e., to its material description, then take the time derivative, and put the resulting quantity to the spatial description. In other words, we calculate

$$\frac{{\rm D}z}{{\rm D}t}(t,x^y) := \partial_t z(t, y(t,x)). \tag{9.1.5}$$

As we will see in the following proposition, there is a relationship between the material time derivatives and the spatial time derivatives of spatial fields, showing in other words that the time derivative in the Lagrangian setting corresponds to the material derivative in the Euler coordinates:

Proposition 9.1.1 (Material vs. spatial time derivative). *Let $z \in C^1(\mathcal{T})$ and $y(\cdot;x) \in C^1([0,T];\mathbb{R}^d)$. Then*

$$\frac{{\rm D}z}{{\rm D}t}(t,x^y) = \partial_t z(t,x^y) + v(t,x^y)\cdot\partial_{x^y} z(t,x^y). \tag{9.1.6}$$

Proof. We evaluate the right-hand side of (9.1.5) using a chain rule for the partial derivative with respect to time and then we must express the result in terms of spatial variables (i.e. variables in the deformed configuration). Hence, we get, keeping in mind that $x^y = y(t,x)$, that

$$\begin{aligned}\partial_t z(t,y(t,x)) &= \partial_t z(t,x^y) + \partial_t y(t,x)\cdot\partial_{x^y} z(t,x^y)\\ &= \partial_t z(t,x^y) + \partial_t y(t,p(t,x^y))\cdot\partial_{x^y} z(t,x^y)\\ &= \partial_t z(t,x^y) + v(t,x^y)\cdot\partial_{x^y} z(t,x^y).\end{aligned}$$

The last equality is due to (9.1.4). □

We have the following Corollary which calculates the material derivative of a vector-valued spatial field. For its proof, we just apply the previous proposition to each component.

Corollary 9.1.2. *Let $z \in C^1(\mathcal{T};\mathbb{R}^d)$ and $y(\cdot;x) \in C^1([0,T];\mathbb{R}^d)$. Then*

$$\frac{\mathrm{D}z}{\mathrm{D}t}(t,x^y) = \partial_t z(t,x^y) + [\partial_{x^y} z(t,x^y)]v(t,x^y). \tag{9.1.7}$$

The term $\partial_t z(t,x^y)$ is the spatial time derivative of the spatial field z. Thus, the difference of material and spatial derivatives is the convective term $v(t,x^y)\cdot\partial_{x^y} z(t,x^y)$ for scalar fields or $[\partial_{x^y} z(t,x^y)]v(t,x^y)$ for vectorial fields.

Example 9.1.3. Consider $z(t,x^y) := x^y$, i.e., the spatial field expressing the position of a point in a deformed configuration. Applying Corollary 9.1.2 we can see that $\frac{\mathrm{D}z}{\mathrm{D}t}(t,x^y) = v(t,x^y)$. If z is independent of the spatial variable, i.e, it is spatially constant, then the convective term vanishes and the spatial time derivative coincides with the material time derivative.

Let us now discuss how transport-coefficient tensors transform. Let again z be the field whose gradient governs the transport, being in particular situations typically temperature, chemical potential, or electrostatic potential if energy, mass, or charge is to be transported, respectively. Let further $\omega^y \subset \Omega^y$ be a measurable set with a smooth boundary. Let $j^y : \partial\omega \to \mathbb{R}^d$ be a flux through $\partial\omega^y$. Let us define, for almost every $x \in \Omega$, the flux in the reference configuration as:

$$j(x) := (\det\nabla y(x))(\nabla y(x))^{-1} j^y(x^y) = (\mathrm{Cof}\,\nabla y(x))^\top j^y(x^y). \tag{9.1.8}$$

Then we have, in view of (1.1.20),

$$\begin{aligned}\int_{\partial\omega^y} j^y(x^y)\cdot\vec{n}^y\,\mathrm{d}S^y &= \int_{\partial\omega} (\mathrm{Cof}\,\nabla y(x))^{-\top} j(x)\cdot(\mathrm{Cof}\,\nabla y(x))\vec{n}(x)\,\mathrm{d}S\\ &= \int_{\partial\omega} j(x)\cdot\vec{n}(x)\,\mathrm{d}S.\end{aligned} \tag{9.1.9}$$

Formula (9.1.8) allows us to find how some material properties transform from the spatial (actual/deformed) configuration to the reference/undeformed one. Let us consider that $j^y(x^y) := -\mathbb{K}^y(x^y)\nabla^y z^y(x^y)$ for almost every $x^y \in \Omega^y$. Here $\mathbb{K}^y : \Omega^y \to \mathbb{R}^{d\times d}$ and $\theta^y : \Omega^y \to \mathbb{R}$ is a scalar quantity. We can think about θ^y as temperature whose transfer In concrete cases, this may represent Fourier's law (with $\mathbb{K}^y$ the heat-conductivity tensor in the spatial configuration) or Darcy/Fick law (with $\mathbb{K}^y$

the mobility tensor) or Ohm's law (with $\mathbb{K}^y$ the electric-conductivity tensor). Then, (9.1.8) yields (see also Figure 9.1)

$$j^y(x) = -\mathbb{K}^y(x^y)\nabla^y z^y(x^y) = (\mathrm{Cof}\,\nabla y(x))^{-\top} j(x). \tag{9.1.10}$$

Hence, for every $x \in \Omega$ we have

$$j(x) = -(\mathrm{Cof}\,\nabla y(x))^\top \mathbb{K}^y(x^y)\nabla^y z^y(x^y). \tag{9.1.11}$$

Given a spatial quantity $z^y : \Omega^y \to \mathbb{R}^d$ we define the associate reference quantity $z : \Omega \to \mathbb{R}^d$ (here again we implicitly presume injectivity of y) by

$$z(x) := z^y(x^y) = z^y(y(x)). \tag{9.1.12}$$

Then (recalling that ∇ denotes differentiation with respect to $x \in \Omega$ while ∇^y is the gradient in the deformed configuration with respect to x^y) we have in view of (9.1.12) that $\nabla z(x) = \nabla^y z^y(x^y)\nabla y(x) = (\nabla y(x))^\top \nabla^y z^y(x^y)$. Consequently,

$$(\nabla y(x))^{-\top}\nabla z(x) = \nabla^y z^y(x^y). \tag{9.1.13}$$

Plugging (9.1.13) into (9.1.11), we get

$$\begin{aligned} j(x) &= -[(\mathrm{Cof}\,\nabla y(x))^\top \mathbb{K}^y(y(x))(\nabla y(x))^{-\top}]\nabla z(x) \\ &= -\frac{(\mathrm{Cof}\,\nabla y(x))^\top \mathbb{K}^y(x^y)\mathrm{Cof}\,\nabla y(x)}{\det \nabla y(x)}\nabla z(x), \end{aligned} \tag{9.1.14}$$

which suggests the referential form of $\mathbb{K}^y$, denoted by $\mathcal{K}$ in order to express explicitly its dependence on ∇y, as

$$\mathcal{K}(x, \nabla y(x)) := \frac{(\mathrm{Cof}\,\nabla y(x))^\top \mathbb{K}^y(x^y)\mathrm{Cof}\,\nabla y(x)}{\det \nabla y(x)}, \tag{9.1.15}$$

which means (with x partly omitted) that $\mathcal{K}(x, \nabla y) = (\det \nabla y)(\nabla y)^{-1}\mathbb{K}^y(x^y)(\nabla y)^{-\top}$, where we recalled the formula $(\mathrm{Cof}\,F)^\top := \mathrm{Cof}\,F^\top = (\det F)F^{-1}$. One can see that, up to the $\det F$ factor, $\mathcal{K}$ is a tensorial pullback (1.1.11b) of $\mathbb{K}^y$. See Exercise 1.1.6. In this way, we obtain for almost all $x \in \Omega$ that

$$j(x) = -\mathcal{K}(x, \nabla y(x))\nabla z(x) \tag{9.1.16}$$

as the formula for flux in the reference configuration.

Let us suppose without loss of generality[1] that $0 \in \Omega^y$ and that a deformation y is followed by a rigid body rotation $R \in \mathrm{SO}(d)$ such that $\tilde{y} := Ry$. In view of formula (9.1.10), we see that

[1] It allows us to rotate Ω^y around the origin which simplifies our notation.

$$j^{\tilde{y}}(x^{\tilde{y}}) := -\mathbb{K}^{\tilde{y}}(x^{\tilde{y}})\nabla^{\tilde{y}} z^{\tilde{y}}(x^{\tilde{y}}). \tag{9.1.17}$$

However, assuming further that

$$j^{\tilde{y}}(x^{\tilde{y}}) = Rj^{y}(x^{y}) \text{ and } z^{\tilde{y}}(x^{\tilde{y}}) = z^{y}(x^{y}), \tag{9.1.18}$$

which implies

$$\nabla^{\tilde{y}} z^{\tilde{y}}(x^{\tilde{y}}) = R^{-\top}\nabla^{y} z^{y}(x^{y}), \tag{9.1.19}$$

we get, by plugging (9.1.19) into (9.1.17) and into (9.1.18), that

$$\mathbb{K}^{\tilde{y}}(x^{\tilde{y}})\nabla^{\tilde{y}} z^{\tilde{y}}(x^{\tilde{y}}) = \mathbb{K}^{\tilde{y}}(x^{\tilde{y}})R^{-\top}\nabla^{y} z^{y}(x^{y}) = R\mathbb{K}^{y}(x^{y})\nabla^{y} z^{y}(x^{y}).$$

Hence, finally

$$\forall R \in \mathrm{SO}(d): \mathbb{K}^{\tilde{y}}(x^{\tilde{y}}) = R\mathbb{K}^{y}(x^{y})R^{\top}. \tag{9.1.20}$$

This formula together with (9.1.15) shows that, for every $R \in \mathrm{SO}(d)$ and every $F \in \mathrm{GL}^{+}(d)$, it holds that $\mathcal{K}(x, RF) = \mathcal{K}(x, F)$, and, consequently,

$$\mathcal{K}(x, F) = \hat{\mathcal{K}}(x, F^{\top}F) \tag{9.1.21}$$

for some $\hat{\mathcal{K}} : \Omega \times \mathbb{R}^{d\times d}_{\mathrm{sym}} \to \mathbb{R}^{d\times d}_{\mathrm{sym}}$ by the polar decomposition formula (A.2.27).

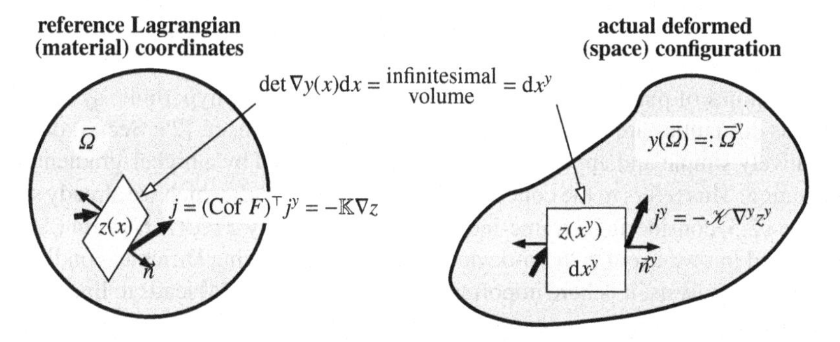

Fig. 9.1 The balance of the flux on an (infinitesimally small) domain D in the actual (spatial) configuration (right) pulled back into the reference configuration (left).

Like in the static situations in solids where the Lagrange description was simpler than Euler's description which uses the not a-priori known domain Ω^{y} as already pointed out in Sect 1.1.2, also in the evolution problems the Lagrange description is more natural.

Remark 9.1.4 (Material transport coefficients). In real situations, even when considering the *transport in the actual deforming configurations*, one must feed the model with transport coefficients, let us denote them by $\mathbb{K}_M = \mathbb{K}_M(x)$, that are known for particular materials at the point $x \in \Omega$. Then $\mathbb{K}^y = \mathbb{K}^y(x^y)$ occurring in (9.1.15) can be chosen as $\mathbb{K}^y(x^y) = \mathbb{K}_M(y^{-1}(x^y)) = \mathbb{K}_M(x)$ because $y^{-1}(x^y) = y^{-1}(y(x)) = x$ in Ω. It should be emphasized that this choice relies on the assumption that $\mathbb{K}_M$ is measured on material which is undeformed in the reference configuration and does not explicitly depend on F. Sometimes, $\mathbb{K}_M$ may depend also on some internal variable, say w (which may but may not be just z), i.e. $\mathbb{K}_M = \mathbb{K}_M(x,w)$. Then $\mathbb{K}^y = \mathbb{K}^y(x^y,w^y)$ in (9.1.15), and using (9.1.12) for w instead of z, we obtain

$$\mathbb{K}^y(x^y,w^y) = \mathbb{K}_M(x,w). \tag{9.1.22}$$

Of course, under the rotation by R as used in (9.1.20), the material tensor $\mathbb{K}_M$ rotates, as well.

9.2 Dynamical problems for nonsimple purely elastic materials

Let us first point out the fundamental difficulty. Namely, one may be tempted to qualify the stored energy in such a way that the stress would be quasimonotone in the sense $\int_\Omega (S(\nabla y_1) - S(\nabla y_2)):\nabla(y_1 - y_2)\,dx \geq 0$ for any $y_1, y_2 \in W_0^{1,\infty}(\Omega;\mathbb{R}^d)$. This is obviously a natural property that would allow for a limit passage via classical Minty's trick in the term $-\mathrm{div}\,S(\nabla y)$. Such a concept is unfortunately not much relevant in elasticity at large strains because such quasimonotone condition is not implied by polyconvexity.

Dynamics of purely elastic materials leads to nonlinear hyperbolic systems and serious difficulties are related to it and very little is known, cf. [29, Sect.3] or [30].[2] Relatively simple and applicable results can be obtained by a higher-gradient regularization. This refers to the concept of *nonsimple materials* which we already saw in Section 2.5, considering a frame-indifferent stored energy $\varphi = \varphi(F,G)$ as in (2.5.24). In accord to , we avoid a "hard-device" load by time-varying Dirichlet conditions.

For the analysis, it is here important that the ansatz (2.5.24) leads to linear hyperstresses. In this evolutionary case, the *inertial forces* $\varrho\ddot{y}$ play important role in particular when purely inviscid, conservative rheology is considered. Thus $\inf_{x\in\Omega}\varrho(x) > 0$ becomes a particularly important assumption in this context.

[2] E.g. in [153] one can read that "that only measure-valued solutions are so far within reach of present analytical techniques is a shortcoming of understanding of compactness properties for multi-dimensional conservation laws". We saw, however, in Section 4.4.4 that the concept of measure-valued solutions is extremely non-selective in generally and thus very questionable.

9.2.1 Basic isothermal model

In the Lagrangian coordinates, the evolution variant of the static problem from Section 2.5.1 leads to the initial-boundary-value problem, i.e. *force-equilibrium equations in the reference configuration*:

$$\varrho\ddot{y} - \operatorname{div} S = f \quad \text{with } S = \varphi'(\nabla y) - \operatorname{Div}(\mathbb{H}\nabla^2 y) \qquad \text{in } Q, \tag{9.2.1a}$$

$$S\vec{n} - \operatorname{div}_{\mathrm{S}}((\mathbb{H}\nabla^2 y)\cdot\vec{n}) = g \quad \text{and} \quad (\mathbb{H}\nabla^2 y):(\vec{n}\otimes\vec{n}) = 0 \qquad \text{on } \Sigma, \tag{9.2.1b}$$

$$y|_{t=0} = y_0 \quad \text{and} \quad \dot{y}|_{t=0} = v_0 \qquad \text{on } \Omega. \tag{9.2.1c}$$

where "Div" is from (2.5.3) on p. 40. The typical example of the right-hand sides is

$$f(t,x,y) = f_0(t,x) + f_1(y) \tag{9.2.2}$$

where $f_0 : Q \to \mathbb{R}^d$ is the dead body force while $f_1 : \mathbb{R}^d \to \mathbb{R}^d$ is the prescribed conservative body force, cf. (1.2.17a). When $\dot{y}$ has not well controlled trace on Γ, and in particular in this section, we must consider only a "dead" traction force $g = g(t,x)$ independent of $(y,\nabla y)$, in contrast to a general form (1.2.17b).

The weak formulation combines e.g. (6.2.7) with the concept of 2nd-grade nonsimple materials, cf. Sect. 2.5. The analog of the small-strain definition (6.2.7) leads to:

Definition 9.2.1 (Elastodynamics: weak solutions). We call $y \in C_{\mathrm{w}}(I;H^2(\Omega;\mathbb{R}^d)) \cap C^1_{\mathrm{w}}(I;L^2(\Omega;\mathbb{R}^d))$ a weak solution to the initial-boundary-value problem (9.2.1) if the integral identity

$$\int_Q \varphi'(\nabla y):\nabla v - \varrho\dot{y}\cdot\dot{v} + \sum_{i=1}^d \nabla^2 y_i:\mathbb{H}:\nabla^2 v_i \,\mathrm{d}x\mathrm{d}t \\ = \int_Q f\cdot v\,\mathrm{d}x\mathrm{d}t + \int_\Sigma g\cdot v\,\mathrm{d}S\,\mathrm{d}t + \int_\Omega \varrho v_0\cdot v(0,\cdot)\,\mathrm{d}x \tag{9.2.3}$$

holds for all $v \in L^2(I;H^2(\Omega;\mathbb{R}^d)) \cap H^1(I;L^2(\Omega;\mathbb{R}^d))$ with $v|_{t=T} = 0$, and also the initial condition $y(0,\cdot) = y_0$ is satisfied.

Proposition 9.2.2 (Existence of weak solutions). *Let $\varphi : \mathbb{R}^{d\times d} \to \mathbb{R}\cup\{+\infty\}$ be continuously differentiable with φ' satisfying the growth condition $|\varphi'(\cdot)| \le C(1+|\cdot|^{2^*-\epsilon})$ for some $\epsilon > 0$, ϱ satisfy (6.3.2), $\mathbb{H}$ be positive definite, and for the loading and initial conditions satisfy*

$$\text{(9.2.2) with } f_0 \in L^1(I;L^2(\Omega;\mathbb{R}^d)),\ f_1 \in C(\mathbb{R}^d,\mathbb{R}^d) \text{ bounded,} \tag{9.2.4a}$$

$$g \in W^{1,1}(I;L^{2^{*\sharp'}}(\Gamma;\mathbb{R}^d)),\quad y_0 \in H^2(\Omega;\mathbb{R}^d),\ \text{and } v_0 \in L^2(\Omega;\mathbb{R}^d). \tag{9.2.4b}$$

Then the problem (9.2.1) has a solution according Definition 9.2.1.

Sketch of the proof. Considering e.g. the Faedo-Galerkin discretization indexed by $k \in \mathbb{N}$ (referring also to Remark 9.2.7 below for an alternative technique) and using the test by $\dot{y}_k$, we obtain the a-priori estimates (as well as the existence of y_k by the successive prolongation):

$$\|y_k\|_{L^\infty(I;H^2(\Omega;\mathbb{R}^d))} \le K \quad \text{and} \quad \|\dot{y}_k\|_{L^\infty(I;L^2(\Omega;\mathbb{R}^d))} \le K. \tag{9.2.5}$$

For the boundary term, we needed to make the by-part integration because $\dot{y}_k$ does not have well defined trace on Γ, namely

$$\int_0^t\!\!\int_\Gamma g\cdot\dot{y}_k\,\mathrm{d}S\,\mathrm{d}t = \int_\Gamma g(t)\cdot y_k(t)\,\mathrm{d}S - \int_0^t\!\!\int_\Gamma \dot{g}\cdot y_k\,\mathrm{d}S\,\mathrm{d}t - \int_\Gamma g(0)\cdot y_0\,\mathrm{d}S,$$

and then also use the estimate $\|y_k\|_{L^{2^{*\sharp}}(\Gamma;\mathbb{R}^d)} \le K(\|y_k\|_{L^2(\Gamma;\mathbb{R}^d)} + \|\nabla^2 y\|_{L^2(\Gamma;\mathbb{R}^{d\times d\times d})})$.

By the Banach selection principle, in view of (9.2.5), we select a subsequence such that $y_k \to y$ weakly* in $L^\infty(I;H^2(\Omega;\mathbb{R}^d))$ and $\dot{y}_k \to \dot{y}$ weakly* in $L^\infty(I;L^2(\Omega;\mathbb{R}^d))$. Also $y_k \to y$ converges weakly* in $L^\infty(I;H^2(\Omega;\mathbb{R}^d))$. By Aubin-Lions' theorem B.4.10, we have compactness of ∇y's in $L^{1/\epsilon}(I;W^{1,2^*-\epsilon}(\Omega;\mathbb{R}^d))$. This means that $\nabla y_k \to \nabla y$ strongly in $L^{1/\epsilon}(I;W^{1,2^*-\epsilon}(\Omega;\mathbb{R}^d))$. This then allows for the limit passage in the nonlinear and (presumably) nonmonotone φ'-term. Here, one needs to employ the growth assumption on φ' so that, by the continuity of the Nemytskiĭ mapping induced by φ', we have $\varphi'(\nabla y_k) \to \varphi'(\nabla y)$ strongly in $L^1(Q;\mathbb{R}^{d\times d})$.

The other terms are linear and the weak convergence is easy to be used, so that altogether the limit towards (9.2.3) is accomplished. □

The growth assumption on φ' used in Proposition 9.2.2 allow in 3-dimensional problems potentials with polynomial growth up to $7-\epsilon$. In particular the St.Venant-Kirchhoff material (2.4.1) which has the 4th-order polynomial growth in F is covered.

Sometimes, we need even higher than H^2-regularization. This may be important for having a control over $1/\det\nabla y$, as used here in Sections 9.2.2–9.2.3, and later in Sections 9.6 or 9.7 below. Without using 3rd-grade nonsimple materials (which would lead to very complicated boundary conditions) and without using nonlinear 2nd-grade nonsimple materials leading to $W^{2,p}$-regularization with $p > d$ as used essentially in Healey-Krömer Theorem 2.5.3, we can employ nonlocal 2nd-grade nonsimple materials as in Sect. 2.5.2. This keeps linearity of the highest "nonsimple" term, which is important in our nonlinear hyperbolic problem, and simultaneously allows for control over $1/\det\nabla y$ in $L^\infty(Q)$ by using Corollary 2.5.11 on p. 50. This leads, instead of (9.2.1), to the boundary-value problem[3]

$$\varrho\ddot{y} - \operatorname{div} S = f(y) \quad \text{with} \quad S = \varphi'(\nabla y) - \operatorname{Div}(\mathfrak{H}\nabla^2 y) \qquad \text{in } Q, \tag{9.2.6a}$$

$$S\vec{n} - \operatorname{div}_S((\mathfrak{H}\nabla^2 y)\cdot\vec{n}) = g \quad \text{and} \quad (\mathfrak{H}\nabla^2 y){:}(\vec{n}\otimes\vec{n}) = 0 \qquad \text{on } \Sigma, \tag{9.2.6b}$$

[3] In fact, there is some additional fractional boundary condition omitted in (9.2.6b), as pointed out in [1], so the classical formulation is truly formal, cf. also Remark 5.4.3. The really important and, under respective assumptions, rigorous formulation is (9.2.8).

again with the initial condition (9.2.1c). Here, $\mathfrak{H}$ is from (2.5.36) on p. 49. The Definition 9.2.1 of the weak solution then modifies by using the bi-linear term $\int_0^T \mathscr{H}''(\nabla^2 y, \nabla^2 v)\,\mathrm{d}t = \int_Q \mathfrak{H}\nabla^2 y \vdots \nabla^2 v\,\mathrm{d}x\mathrm{d}t$ where $\mathscr{H}$ is the quadratic form inducing $\mathfrak{H} := \mathscr{H}'$ instead of $\int_Q \mathbb{H}\nabla^2 y \vdots \nabla^2 v\,\mathrm{d}x\mathrm{d}t$ in (9.2.3). More specifically, considering the kernel $\mathfrak{K} : \Omega\times\Omega \to \mathrm{SLin}(\mathbb{R}^{d\times d})$ and the quadratic form

$$\mathscr{H} : G \mapsto \frac{1}{4}\sum_{i=1}^{d}\int_{\Omega\times\Omega}(G_i(x)-G_i(\tilde{x})):\mathfrak{K}(x,\tilde{x}):(G_i(x)-G_i(\tilde{x}))\,\mathrm{d}x\mathrm{d}\tilde{x}$$
$$= \frac{1}{4}\int_{\Omega\times\Omega}\mathfrak{K}(x,\tilde{x})(G(x)-G(\tilde{x}))\vdots(G(x)-G(\tilde{x}))\,\mathrm{d}x\mathrm{d}\tilde{x} \qquad (9.2.7)$$

with G_i a placeholder for the Hessian $\nabla^2 y_i$ and thus G_{ijk} being a placeholder for $\partial^2_{x_jx_k} y_i$, we use $H^{2+\gamma}(\Omega;\mathbb{R}^d)$ in Definition 9.2.1 instead of $H^2(\Omega;\mathbb{R}^d)$ and, assuming the symmetry $\mathfrak{K}(x,\tilde{x}) = \mathfrak{K}(\tilde{x},x)$, instead of (9.2.3) we use

$$\int_Q \varphi'(\nabla y):\nabla v - \varrho\dot{y}\cdot\dot{v} + \mathfrak{H}\nabla^2 y \vdots \nabla^2 v\,\mathrm{d}x\mathrm{d}t$$
$$= \int_Q f(y)\cdot v\,\mathrm{d}x\mathrm{d}t + \int_\Sigma g\cdot v\,\mathrm{d}S\,\mathrm{d}t + \int_\Omega \varrho v_0\cdot v(0,\cdot)\,\mathrm{d}x \qquad (9.2.8a)$$
$$\text{with } [\mathfrak{H}\nabla^2 y](t,x) = \int_\Omega \mathfrak{K}(x,\tilde{x})(\nabla^2 y(t,x) - \nabla^2 y(t,\tilde{x}))\,\mathrm{d}\tilde{x}. \qquad (9.2.8b)$$

Let us demonstrate usage of Corollary 2.5.11 on this dynamical problems, assuming

$$\varphi : \mathrm{GL}^+(d) \to \mathbb{R}^+ \text{ continuously differentiable, and for some } \epsilon > 0:$$
$$\varphi(F) \geq \frac{\epsilon}{(\det F)^q}, \quad q > \frac{2d}{2\gamma+2-d} \quad \text{for some } \gamma > \frac{d}{2} - 1, \qquad (9.2.9a)$$
$$\mathscr{H} : H^\gamma(\Omega;\mathbb{R}^{d\times d\times d}) \to \mathbb{R} \ \text{ from (9.2.7)}$$
$$\text{with } \mathfrak{K} \text{ symmetric, satisfying (2.5.37)}, \qquad (9.2.9b)$$
$$f(t,x,y) = f_0(t,x) + f_1(y) \ \text{ with}$$
$$f_0 \in L^1(I;L^2(\Omega;\mathbb{R}^d)), \ \ f_1 : \mathbb{R}^d \to \mathbb{R}^d \text{ continuous bounded}, \qquad (9.2.9c)$$
$$g \in W^{1,1}(I;L^1(\Gamma;\mathbb{R}^d)), \qquad (9.2.9d)$$
$$y_0 \in H^{2+\gamma}(\Omega;\mathbb{R}^d), \ \min_{x\in\bar{\Omega}} \det\nabla y_0(x) > 0, \ \text{ and } \ \ v_0 \in L^2(\Omega;\mathbb{R}^d). \qquad (9.2.9e)$$

Proposition 9.2.3 (Evolution respecting local non-interpenetration). *Let (9.2.9) hold and ϱ satisfy (6.3.2). Then (9.2.6) with the initial conditions (9.2.1c) has a weak solution $y \in C_\mathrm{w}(I;H^{2+\gamma}(\Omega;\mathbb{R}^d)) \cap C^1_\mathrm{w}(I;L^2(\Omega;\mathbb{R}^d))$ such that*

$$\nabla y \in C(\bar{Q};\mathbb{R}^{d\times d}) \quad \textit{and} \quad \min_{(t,x)\in\bar{Q}} \det\nabla y > 0. \qquad (9.2.10)$$

Sketch of the proof. It is important that by the Healey-Krömer theorem in its modification as Corollary 2.5.11 on p. 50, for some positive $\varepsilon \leq \min_{x\in\bar{\Omega}} \det\nabla y_0(x)$, we have

$$\forall(t,x)\in\bar{Q}: \qquad \det\nabla y_k(t,x)\ge\varepsilon. \tag{9.2.11}$$

This holds by successive-continuation argument on the Galerkin level, and thus ∇y_k is valued in the definition domain of φ and the singularity of φ is not seen, and therefore the Lavrentiev phenomenon is excluded. The rest is then identical with the proof of Proposition 9.2.2 with $H^{2+\gamma}(\Omega;\mathbb{R}^d)$ in (9.2.5) instead of $H^2(\Omega;\mathbb{R}^d)$ and with $\varphi'(\nabla y_k)\to\varphi'(\nabla y)$ strongly even in any $L^p(Q;\mathbb{R}^{d\times d})$, $1\le p<+\infty$.[4] $\square$

We can also prove a regularity like we did at small strains in Proposition 6.3.2 on p. 204 by differentiating the equation (9.2.6a) in time and test it by $\dddot{y}$. We can present it for the nonlocal nonsimple model as in Proposition 9.2.3:

Proposition 9.2.4 (Regularity at large strains). *In addition to (9.2.9) hold and* $\inf_{x\in\Omega}\varrho(x)>0$*, let also* $\varphi:\mathrm{GL}^+(d)\to\mathbb{R}^+$ *have* φ''' *continuous,* $y_0\in H^{4+2\gamma}(\Omega;\mathbb{R}^d)$*,* $v_0\in H^{2+\gamma}(\Omega;\mathbb{R}^d)$*,* $f_0\in W^{1,1}(I;L^2(\Omega;\mathbb{R}^d))$*,* $f_1\in C^1(\mathbb{R}^d;\mathbb{R}^d)$*, and* $g\in W^{2,1}(I;L^1(\Gamma;\mathbb{R}^d))$*. Then the weak solution whose existence has been stated in Proposition 9.2.3 satisfies also*

$$y\in C^2_{\mathrm{w}}(I;L^2(\Omega;\mathbb{R}^d))\cap C^1_{\mathrm{w}}(I;H^{2+\gamma}(\Omega;\mathbb{R}^d))\cap W^{3,\infty}(I;H^{2+\gamma}(\Omega;\mathbb{R}^d)^*). \tag{9.2.12}$$

In addition, this solution conserves energy in the sense that, for ant $t\in I$*, it holds*

$$\int_\Omega\frac{\varrho}{2}|\dot{y}(t)|^2+\varphi(\nabla y(t))-f_1(y(t))+\mathscr{H}(\nabla^2y(t))\,\mathrm{d}x=\int_\Omega\Big(\frac{\varrho}{2}|v_0|^2+\varphi(\nabla y_0)$$
$$-f_1(y_0)+\mathscr{H}(\nabla^2y_0)\Big)\mathrm{d}x+\int_0^t\Big(\int_\Omega f_0\cdot\dot{y}\,\mathrm{d}x+\int_\Gamma g\cdot\dot{y}\,\mathrm{d}S\Big)\mathrm{d}t. \tag{9.2.13}$$

Sketch of the proof. Again, we use rather heuristic presentation. The tests used in this proof are legal in the Faedo-Galerkin approximation and then the obtained a-priori estimates are to be inherited in the limit.[5]

Differentiating (9.2.6a) with f from (9.2.2) in time and realizing that $\frac{\partial}{\partial t}(\mathrm{div}\,\varphi'(\nabla y))=\mathrm{div}(\varphi''(\nabla y){:}\nabla\dot{y})$ gives

$$\varrho\dddot{y}-\mathrm{div}\,\dot{S}=\dot{f}_0+f_1'(y)\dot{y}\quad\text{with }\dot{S}=\varphi''(\nabla y){:}\nabla\dot{y}-\mathrm{Div}(\mathfrak{H}\nabla^2\dot{y})\quad\text{in }Q, \tag{9.2.14a}$$
$$\dot{S}\vec{n}-\mathrm{div}_{\mathrm{S}}((\mathfrak{H}\nabla^2\dot{y})\cdot\vec{n})=\dot{g}\qquad\text{and}\qquad(\mathfrak{H}\nabla^2\dot{y}){:}(\vec{n}\otimes\vec{n})=0\quad\text{on }\Sigma. \tag{9.2.14b}$$

Testing by $\ddot{y}$, we obtain

$$\frac{\mathrm{d}}{\mathrm{d}t}\Big(\frac{\varrho}{2}\|\ddot{y}\|^2_{L^2(\Omega;\mathbb{R}^d)}+\mathscr{H}(\nabla^2\dot{y})\Big)$$
$$=\int_\Omega(\dot{f}_0+f_1'(y)\dot{y})\cdot\ddot{y}+\mathrm{div}(\varphi''(\nabla y){:}\nabla\dot{y})\cdot\ddot{y}\,\mathrm{d}x+\int_\Gamma\dot{g}\cdot\ddot{y}\,\mathrm{d}S$$
$$=\int_\Omega(\dot{f}_0+f_1'(y)\dot{y})\cdot\ddot{y}+\varphi''(\nabla y){:}\nabla^2\dot{y}\cdot\ddot{y}+\varphi'''(\nabla y){:}\nabla\dot{y}{:}\nabla^2y\cdot\ddot{y}\,\mathrm{d}x+\int_\Gamma\dot{g}\cdot\ddot{y}\,\mathrm{d}S$$

[4] In fact, it holds even in $C(\bar{Q};\mathbb{R}^{d\times d})$ if a modification of the Aubin-Lions theorem is used, cf. the Arzelà-Ascoli-type arguments in [456, Lemma 7.10].

[5] For a variant relying on a 3rd-grade nonsimple material, cf. also [429, Prop. 2].

$$\leq \Big(\|\dot{f}_0\|_{L^2(\Omega;\mathbb{R}^d)} + \|f_1'(y)\|_{L^\infty(\Omega;\mathbb{R}^{d\times d})}\|\dot{y}\|_{L^2(\Omega;\mathbb{R}^d)}$$
$$+ \|\varphi''(\nabla y)\|_{L^\infty(Q;\mathbb{R}^{(d\times d)\times(d\times d)})}\|\nabla^2\dot{y}\|_{L^2(\Omega;\mathbb{R}^{d\times d\times d})} + \|\varphi'''(\nabla y)\|_{L^\infty(Q;\mathbb{R}^{d^6})}\times$$
$$\times\|\nabla\dot{y}\|_{L^\infty(\Omega;\mathbb{R}^{d\times d})}\|\nabla^2 y\|_{L^2(\Omega;\mathbb{R}^{d\times d\times d})}\Big)\|\dddot{y}\|_{L^2(\Omega;\mathbb{R}^d)} + \int_\Gamma \dot{g}\cdot\dddot{y}\,\mathrm{d}S. \tag{9.2.15}$$

Assuming $\gamma > d/2-1$, we have used that $\nabla y \in L^\infty(Q;\mathbb{R}^{d\times d})$ and no growth restriction is needed on $\varphi \geq 0$, only the boundedness of φ''' on any set $\{F \in \mathrm{GL}^+(d);\ \det F \geq \varepsilon\}$, which is ensured if φ''' is continuous. Moreover, we use

$$\varrho\ddot{y}(0) = \mathrm{div}((\nabla y_0) - \mathrm{Div}(\mathfrak{H}\nabla^2 y_0)) + f(y_0) \in L^2(\Omega;\mathbb{R}^d)$$

due to the assumed smoothness of y_0. Also, we use the estimate $\|\dot{y}(t)\|_{L^2(\Omega;\mathbb{R}^d)} = \|v_0 + \int_0^t \ddot{u}\,\mathrm{d}t\|_{L^2(\Omega;\mathbb{R}^d)} \leq \|v_0\|_{L^2(\Omega;\mathbb{R}^d)} + \int_0^t \|\ddot{u}\|_{L^2(\Omega;\mathbb{R}^d)}\mathrm{d}t$. For the boundary term integrated over a time interval $[0,t]$, we use the by-part integration

$$\int_0^t\!\!\int_\Gamma \dot{g}\cdot\dddot{y}\,\mathrm{d}S\,\mathrm{d}t = \int_\Gamma \dot{g}(t)\cdot\dot{y}(t)\,\mathrm{d}S - \int_0^t\!\!\int_\Gamma \ddot{g}\cdot\dot{y}\,\mathrm{d}S\,\mathrm{d}t - \int_\Gamma \dot{g}(0)\cdot v_0\,\mathrm{d}S,$$

which can be easily controlled through the left hand-side in (9.2.15) which bounds the trace of $\dot{y}$ in $L^\infty(\Gamma;\mathbb{R}^d)$. Then, by the Gronwall inequality, we obtain $\ddot{y} \in C_{\mathrm{w}}(I;L^2(\Omega;\mathbb{R}^d))$ and $\dot{y} \in C_{\mathrm{w}}(I;H^{2+\gamma}(\Omega;\mathbb{R}^d))$, proving thus (9.2.12) is obtained.

From (9.2.14a), by comparison we obtain also the information about $\dddot{y}$. From this one can see also the weak continuity $\ddot{y}: I \to L^2(\Omega;\mathbb{R}^d)$ stated in (9.2.12).

Eventually, as now $\sqrt{\varrho}\ddot{y} \in L^\infty(I;L^2(\Omega;\mathbb{R}^d))$ is certainly in duality with $\sqrt{\varrho}\dot{y} \in L^\infty(Q;\mathbb{R}^d)$, similarly $\varphi'(\nabla y) \in L^\infty(Q;\mathbb{R}^{d\times d})$ is certainly in duality with $\nabla\dot{y} \in L^\infty(Q;\mathbb{R}^{d\times d})$, and also $\nabla^2\dot{y} \in L^\infty(I;H^\gamma(\Omega;\mathbb{R}^{d\times d\times d}))$ is in duality with $\mathfrak{H}\nabla^2 y \in L^\infty(I;H^\gamma(\Omega;\mathbb{R}^{d\times d\times d})^*)$, the by-part integration under the test by $\dot{y}$ is legitimate and (9.2.13) holds. □

Remark 9.2.5 (Regularity ignoring nonpenetration). Refining the estimation strategy (9.2.15) by using the Hölder inequality as

$$\|\nabla\dot{y}\otimes\nabla^2 y\|_{L^2(\Omega;\mathbb{R}^{(d\times d)\times(d\times d)})} \leq \|\nabla\dot{y}\|_{L^{6d/(d-2)}(\Omega;\mathbb{R}^{d\times d})}\|\nabla^2 y\|_{L^{3d/(d+1)}(\Omega;\mathbb{R}^{d\times d\times d})}$$

for some $p \geq 2$, we can alternatively consider $\gamma > (d-2)/3$, which is a bit smaller than in (9.2.9a) if $d \geq 3$ and which still ensures the embedding $H^{1+\gamma}(\Omega) \subset L^p(\Omega)$ and $H^\gamma(\Omega) \subset L^{2p/(p-2)}(\Omega)$, and $\varphi \in C^3(\mathbb{R}^{d\times d})$ without dealing with the local nonpenetration in Proposition 9.2.4, modifying also the arguments for the energy conservation.

Remark 9.2.6 (Simple elastic materials with nonlocal theories). In most situations, alternatively to the 2nd-grade nonsimple material concept, one can consider the nonlocal modification concept from Sect. 2.5.2 involving only the deformation gradient itself and thus simplifying the boundary conditions. Relying on (2.5.27), we again consider the boundary-value problem[6]

$$\varrho\ddot{y} - \operatorname{div} S = f(y) \quad \text{with} \quad S = \varphi'(\nabla y) + \mathfrak{H}\nabla y \qquad \text{in } Q, \tag{9.2.16a}$$

$$S\vec{n} = g \qquad \text{on } \Sigma \tag{9.2.16b}$$

with the initial condition (9.2.1c) and here with the nonlocal contribution to the stress corresponding to the potential (2.5.27) as in (2.5.31), i.e.

$$[\mathfrak{H}\nabla y](t,x) = \int_\Omega \mathbb{K}(x,\tilde{x})(\nabla y(t,x) - \nabla y(t,\tilde{x}))\,\mathrm{d}\tilde{x}, \quad (t,x)\in Q, \tag{9.2.17}$$

with some kernel $\mathbb{K} : \Omega\times\Omega \to \mathrm{SLin}(\mathbb{R}^{d\times d}_{\mathrm{sym}})$ as in (2.5.27) singular in the sense of (2.5.28). Testing (9.2.16a) by $\dot{y}$ and assuming the coercivity $\varphi(F) \geq \varepsilon|F|^p$ for some $\varepsilon > 0$ and $p > 1$, we obtain the estimates

$$\nabla y \in L^\infty(I; H^\gamma(\Omega;\mathbb{R}^{d\times d}) \cap L^p(\Omega;\mathbb{R}^{d\times d})), \quad \dot{y} \in L^\infty(I; L^2(\Omega;\mathbb{R}^d)). \tag{9.2.18}$$

Under a mathematically acceptable growth condition $|\sigma(F)| \leq C(1+|F|^{p-1})$, and under the assumption $p < 2d/(d-2\gamma)$, we can use the compact embedding $H^\gamma(\Omega) \subset L^p(\Omega)$ and Aubin-Lions' theorem B.4.10 to have "compactness" of ∇y in $L^p(Q;\mathbb{R}^{d\times d})$. This allows for a limit passage in the non-quasimonotone σ. The definition of the weak solution (9.2.8) then uses the bi-linear term $\int_0^T \mathscr{H}''(\nabla y, \nabla v)\,\mathrm{d}t$ where $\mathscr{H}$ is the quadratic form inducing $\mathfrak{H} := \mathscr{H}'$ instead of $\int_Q \mathbb{H}\nabla^2 y \vdots \nabla^2 y\,\mathrm{d}x\mathrm{d}t$. More specifically, assuming the symmetry $\mathbb{K}(x,\tilde{x}) = \mathbb{K}(\tilde{x},x)$, in view of (2.5.27), the weak solution is to satisfy the integral identity

$$\begin{aligned}\int_Q \Big(&\varphi'(\nabla y(t,x)) : \nabla v(t,x) - \varrho\dot{y}(t,x)\cdot\dot{v}(t,x)\\ &+ \Big(\int_\Omega \mathbb{K}(x,\tilde{x})\nabla(y(t,x) - y(t,\tilde{x}))\,\mathrm{d}\tilde{x}\Big) : \nabla v(t,x)\Big)\mathrm{d}x\mathrm{d}t\\ &= \int_Q f(y)\cdot v\,\mathrm{d}x\mathrm{d}t + \int_\Sigma g\cdot v\,\mathrm{d}S\,\mathrm{d}t + \int_\Omega \varrho v_0\cdot v(0,\cdot)\,\mathrm{d}x.\end{aligned} \tag{9.2.19}$$

Let us note that traces of ∇y may still have a good sense provided $\gamma < 1$ is sufficiently large.

Remark 9.2.7 (Rothe method). An approximation by the Rothe method is natural at least for the reason that it leads to minimization problems similar as treated in Sect. 4.5. Here it is important to allow for φ nonconvex and then introduction of an artificial viscosity which expectedly vanishes in the limit may compensate this nonconvexity provided φ is assumed *semiconvex* in the sense that

$$\exists K \in \mathbb{R}^+ : \qquad F \mapsto \varphi(F) + K|F|^2 : \mathbb{R}^{d\times d} \to \mathbb{R}\cup\{+\infty\} \text{ is convex}, \tag{9.2.20}$$

[6] Again, (9.2.16) is very formal, cf. in particular Remark 5.4.3 about certain "hidden" fractional boundary condition omitted in (9.2.16b).

cf. also (C.2.30). One can then use the backward-Euler formula with a viscosity-like regularization:

$$\varrho\frac{y_\tau^k-2y_\tau^{k-1}+y_\tau^{k-2}}{\tau^2}-\operatorname{div}S_\tau^k=f_\tau^k(y_\tau^k,\nabla y_\tau^k)$$
$$\text{with}\ \ S_\tau^k=\varphi'(\nabla y_\tau^k)-\operatorname{Div}\mathfrak{h}_\tau^k+\frac{\nabla y_\tau^k-\nabla y_\tau^{k-1}}{\sqrt{\tau}}\qquad\text{on }\Omega,\qquad(9.2.21\text{a})$$

$$S_\tau^k\vec{n}-\operatorname{div}_{\rm S}(\mathfrak{h}_\tau^k\cdot\vec{n})=g_\tau^k,\qquad \mathfrak{h}_\tau^k:(\vec{n}\otimes\vec{n})=0,\qquad\text{on }\Gamma,\qquad(9.2.21\text{b})$$

where here $\mathfrak{h}_\tau^k=\mathbb{H}\nabla^2y_\tau^k$. to be solved recursively for $k=1,2,...$, starting with $y_\tau^0=y_0$ and $y_\tau^{-1}=y_0-\tau v_0$ for $k=1$. Let us note that the last term in S_τ^k represents a linear viscosity with the coefficient $\sqrt{\tau}$. Such a viscosity would not be frame indifferent (cf. Sect. 9.3 below) but it vanishes when $\tau\to 0$ so it does not corrupt the model itself. Then, e.g., using the discrete by-part integration (C.2.16) analogously to (9.2.3), we have

$$\int_Q\Big(\varphi'(\nabla\bar{y}_\tau)+\sqrt{\tau}\nabla\dot{y}_\tau\Big):\nabla\bar{v}_\tau+\mathbb{H}\nabla^2\bar{y}_\tau\,\vdots\,\nabla^2\bar{v}_\tau-\varrho\dot{u}_\tau(\cdot-\tau)\cdot\dot{v}_\tau\,\mathrm{d}x\mathrm{d}t$$
$$=\int_Q\bar{f}_\tau\cdot\bar{v}_\tau\,\mathrm{d}x\mathrm{d}t+\int_\Sigma\bar{g}_\tau\cdot\bar{v}_\tau\,\mathrm{d}S\,\mathrm{d}t+\int_\Omega v_0\cdot v(0,\cdot)\,\mathrm{d}x\qquad(9.2.22)$$

and also the initial condition $y_\tau(0,\cdot)=y_0$, where $\dot{v}_\tau$ and $\bar{v}_\tau$ denote the piecewise affine and the piecewise constant interpolants of a smooth test function $v\in C^1(I;H^2(\Omega;\mathbb{R}^d))$ with $v|_{t=T}=0$. By the second estimate in (9.2.23b), we have $\|\sqrt{\tau}\nabla\dot{y}_\tau\|_{L^2(Q;\mathbb{R}^{d\times d})}=\mathscr{O}(\sqrt[4]{\tau})\to 0$ so that $\int_Q\sqrt{\tau}\nabla\dot{y}_\tau:\nabla\bar{v}_\tau\,\mathrm{d}x\mathrm{d}t\to 0$. Therefore the viscous regularizing term in (9.2.22) vanishes in the limit for $\tau\to 0$. Actually, one can prove that, if $\varphi:\mathbb{R}^{d\times d}\to\mathbb{R}\cup\{+\infty\}$ is semiconvex lower semicontinuous, $\inf_{x\in\Omega}\varrho(x)>0$, $\mathfrak{h}=\mathbb{H}\nabla^2y$ with $\mathbb{H}$ positive definite, $f\in L^1(I;L^2(\Omega;\mathbb{R}^d))$, $g\in W^{1,1}(I;L^{2^{\sharp'}}(\Gamma;\mathbb{R}^d))$, $y_0\in H^2(\Omega;\mathbb{R}^d)$ and $v_0\in L^2(\Omega;\mathbb{R}^d)$, then the weak (or, more precisely, rather variational) solution to the recursive boundary-value problem (9.2.21) does exist and satisfy the following a-priori estimates with C independent of τ:

$$\|y_\tau\|_{L^\infty(I;H^2(\Omega;\mathbb{R}^d))}\le C,\qquad(9.2.23\text{a})$$
$$\|\dot{y}_\tau\|_{L^\infty(I;L^2(\Omega;\mathbb{R}^d))}\le C\quad\text{and}\quad\|\dot{y}_\tau\|_{L^\infty(I;H^1(\Omega;\mathbb{R}^d))}\le C/\sqrt[4]{\tau},\qquad(9.2.23\text{b})$$

where y_τ denotes the piecewise affine continuous interpolant of $(y_\tau^0)_{k=0,...,T/\tau}$. It should however be pointed out that the semiconvexity is not much natural in finite-strain elasticity, cf. also Exercise 9.2.15. A certain implicit-explicit scheme that works for polyconvex materials (although with only polynomial growth) was devised in [153, 371].

Exercise 9.2.8. Prove existence and the apriori estimates (9.2.23) for the Rothe's approximation.

Exercise 9.2.9. Let us note that, without the inertial term, i.e. $\varrho = 0$, one can avoid the semiconvexity requirement by choosing such weak solution(s) to (9.2.21) which globally minimizes the underlying potential, cf. Remark C.2.3 on p. 540. Show that it does not work if $\varrho = 0$, however.

9.2.2 Heat transfer in the actual deformed configuration

The heat flux is governed again by the Fourier law but it should be formulated in the actual deformed configuration rather than in the reference one, taking thus the form $-\mathbb{K}^y(x^y,\theta^y)\nabla^y(\theta^y)$ with

$$\mathbb{K}^y = \mathbb{K}^y(x^y,\theta^y) = \mathbb{K}_\mathrm{M}(x,\theta) \tag{9.2.24}$$

with $\mathbb{K}_\mathrm{M}$ the *heat-conductivity tensor* as a given *material* property, cf. (9.1.22). This tensor should be pulled back into the reference configuration where all the equations are formulated and analyzed. The analogous transformation may (but need not) apply for the other transportation processes, in particular a mass transport governed by the Fick or the Darcy laws or a charge transport governed by *Ohm's law* if the diffusant or charge is not forced to follow some material-related conductive "channels" but can flow straight in the actual deformed configuration.

For a general purpose in *transport* processes *in the actual deforming configurations* or gradient theories, let us define the "diffusant-conserving" *tensorial pullback* $\mathscr{P}_F\mathbb{K}_\mathrm{M}$ of a material tensor $\mathbb{K}_\mathrm{M}\in\mathbb{R}^{d\times d}$ by means of a suitable linear pullback *operator* $\mathscr{P}_F : \mathbb{R}^{d\times d} \to \mathbb{R}^{d\times d}$ after being pushed forward by (9.1.22), namely

$$\mathscr{P}_F\mathbb{K}^y = \mathscr{P}_F\mathbb{K}_\mathrm{M} = (\det F)F^{-1}\mathbb{K}^y F^{-\top} = (\det F)F^{-1}\mathbb{K}_\mathrm{M}F^{-\top}, \tag{9.2.25}$$

which is (up to the determinant factor) the pullback (1.1.11b). Here, we use the *Fourier law transformed* (i.e. *pulled-back*) into the reference configuration, which involves the *heat-conductivity tensor* $\mathbb{K}_\mathrm{M}$, so that[7]

$$\mathscr{K}(x,F,\theta)\begin{cases} = \mathscr{P}_F\mathbb{K}_\mathrm{M}(x,\theta) = \dfrac{(\mathrm{Cof}\,F^\top)\mathbb{K}_\mathrm{M}(x,\theta)\,\mathrm{Cof}\,F}{\det F} & \text{if } \det F > 0, \\ \quad\text{not defined} & \text{if } \det F \le 0, \end{cases} \tag{9.2.26}$$

which we actually derived in detail already in (9.1.15), while the case $\det F \le 0$ being considered unphysical. This is the usual transformation of 2nd-order covariant tensors,[8] cf. also Figure 9.1 for an illustration.

---

[7] Indeed, if $v : \mathbb{R}^d \to \mathbb{R}^d$ is a material vector, then $F^{-\top}v$ is a spatial vector by (1.1.10). Since $\mathbb{K}^y(x^y) := \mathbb{K}_\mathrm{M}(x)$ is a spatial tensor, $\mathbb{K}^y(F^{-\top}v)$ is again a spatial vector. Finally, again due to (1.1.10), $F^{-1}(\mathbb{K}_\mathrm{M}(F^{-\top}v))$ is a material vector. Hence, $F^{-1}\mathbb{K}_\mathrm{M}F^{-\top}$ maps material vectors to material vectors, i.e., it is a material tensor being the pullback of the spatial tensor $\mathbb{K}^y$.

[8] In literature, this formula is often used in the isotropic case $\mathbb{K}_\mathrm{M}(\theta) = k(\theta)\mathbb{I}$ where (9.2.26) can easily be written by using the right Cauchy-Green tensor $C = F^\top F$ as $\det(C^{1/2})k(\theta)C^{-1} = (\det(F^\top)\det(F))^{1/2}k(\theta)(F^\top F)^{-1} = \det(F)F^{-1}(k(\theta)\mathbb{I})F^{-\top}$, cf. e.g. [164, Formula (67)] or [224, Formula (3.19)]. In fact, the C-dependence holds for the anisotropic case, too; cf. (9.1.21).

Considering the heat-conductivity tensor $\mathscr{K} = \mathscr{K}(x,F,\theta)$ from (9.2.26), the heat-transfer problem in the enthalpy formulation (8.2.2) with the initial/boundary conditions (8.2.1b,c) now reads in the reference configuration as

$$\dot{\vartheta} - \mathrm{div}(\mathscr{K}(x,F,\theta)\nabla\theta) = \xi(t,x,\theta) \quad \text{with} \quad \vartheta = \mathscr{C}_{\mathrm{v}}(\theta) \qquad \text{in } Q, \qquad (9.2.27a)$$

$$\mathscr{K}(x,F,\theta)\nabla\theta = h_\flat(t,x) \qquad \text{on } \Sigma, \qquad (9.2.27b)$$

$$\theta(x,0) = \theta_0(x) \qquad \text{on } \Omega, \qquad (9.2.27c)$$

where again $\mathscr{C}_{\mathrm{v}}(\theta) := \int_0^1 \theta c_{\mathrm{v}}(r\theta)\,\mathrm{d}r$ and where here $F = F(t,x)$ is a fixed field (later in the position of a deformation gradient, of course) assume to fulfill $\det F > 0$.

It will be useful to state an analog of Proposition 8.2.1 on p. 368, benefiting from having admitted the nonautonomous inhomogeneous heat-conductivity tensor $\mathbb{K}_{\mathrm{M}} = \mathbb{K}_{\mathrm{M}}(t,x,\theta)$ there. Here we use Proposition 8.2.1 for $\mathbb{K}_{\mathrm{M}}(t,x,\theta) := \mathscr{K}(x,F(t,x),\theta)$. Intuitively, if the material is very stretched and the heat conduction in the actual configuration is unchanged, the heat conduction viewed from the reference configuration drops down, and therefore to ensure the uniform positive definiteness of the pulled-back conductivity, we must control the deformation gradient. Simultaneously, also $1/\det F$ is to be controlled to make the formula (9.2.26) working because of the factor $1/\det F$ occurring in (9.2.25).

Corollary 9.2.10 (Heat transfer revisited). *Let (8.2.3a,b,d-e) and (8.2.4) hold, and let*

$$\mathbb{K}_{\mathrm{M}} : \Omega\times\mathbb{R}^+ \to \mathbb{R}^{d\times d} \ \ \textit{Carathéodory mapping, bounded, uniformly positive definite,} \qquad (9.2.28a)$$

$$F \in L^\infty(Q;\mathbb{R}^{d\times d}) \ \ \textit{given with} \ \ \operatorname*{ess\,inf}_{(t,x)\in Q} \det F(t,x) =: \delta > 0. \qquad (9.2.28b)$$

Then the boundary-value problem (9.2.27) with $\mathscr{K}$ from (9.2.26) possesses a weak solution $\theta \geq 0$ such that (8.2.5) and (8.2.6) hold again for any $1 \leq r < (2\omega{+}d)/(\omega{+}d)$ and $1 \leq p < 1+2\omega/d$ with $\omega \geq 1$ from (8.2.3a).

Sketch of the proof. One should only verify (8.2.3) for $\mathbb{K}(t,x,\theta) := \mathscr{K}(x,F(t,x),\theta)$ with $\mathscr{K}$ from (9.2.26) with $\mathbb{K}_{\mathrm{M}}$ and F satisfying (9.2.28). Denoting by κ_{M} the positive-definiteness constant for $\mathbb{K}_{\mathrm{M}}$ which is assumed positive by (9.2.28a), this is simply seen from the estimate

$$\inf_{|j|=1} j^\top\mathbb{K}(t,x,\theta)j \geq \inf_{|(\mathrm{Cof}\,F)j/\sqrt{\det F}|\geq 1/\|F/\sqrt{\det F}\|_{L^\infty(Q;\mathbb{R}^{d\times d})}} j^\top\mathbb{K}(t,x,\theta)j \qquad (9.2.29a)$$

$$= \bigg\|\frac{F}{\sqrt{\det F}}\bigg\|_{L^\infty(Q;\mathbb{R}^{d\times d})}^{-2} \inf_{|(\mathrm{Cof}\,F)j/\sqrt{\det F}|\geq 1} \frac{j^\top(\mathrm{Cof}\,F^\top)\mathbb{K}_{\mathrm{M}}(x,\theta)(\mathrm{Cof}\,F)j}{\det F}$$

$$= \bigg\|\frac{F}{\sqrt{\det F}}\bigg\|_{L^\infty(Q;\mathbb{R}^{d\times d})}^{-2} \inf_{|j|\geq 1} j^\top\mathbb{K}_{\mathrm{M}}(x,\theta)j \geq \frac{\kappa_{\mathrm{M}}}{\|F/\sqrt{\det F}\|_{L^\infty(Q;\mathbb{R}^{d\times d})}^2} \geq \frac{\kappa_{\mathrm{M}}\delta^2}{\|F\|_{L^\infty(Q;\mathbb{R}^{d\times d})}^2}, \qquad (9.2.29b)$$

where $\delta > 0$ is from (9.2.28b). For the first inequality, i.e. (9.2.29a), we used the identity

$$\frac{F}{\sqrt{\det F}}\frac{\operatorname{Cof} F^\top}{\sqrt{\det F}} = \mathbb{I},$$

cf. (A.2.17), so that

$$|j| = 1 \quad\Rightarrow\quad \Big\|\frac{F}{\sqrt{\det F}}\Big\|_{L^\infty(Q;\mathbb{R}^{d\times d})}\Big|\frac{(\operatorname{Cof} F(t,x))j}{\sqrt{\det F(t,x)}}\Big| \geq 1.$$

Altogether, (9.2.29) verifies the uniform positive definiteness assumption (8.2.3b) so that the assertion follows by Proposition 8.2.1. □

Remark 9.2.11 (Boundary heat flux y-dependent). Consistently with $f(t,x,y) = f_0(t,x) + f_1(y)$ or even with f_1 depending also on time, we might consider also the boundary heat flux with both "dead" referential and spatial contributions in the form $h_\flat(t,x,y) = h_{\flat,0}(t,x) + h_{\flat,0}(t,y)$. The generalization of the proof here and in what follows is straightforward.

9.2.3 Thermoelastic nonsimple materials by phase-field approach

In some situations like in thermally expanding materials or in ferroelastic materials (undergoing some "displacive" phase transformations), the temperature-dependence of stored energy $\psi = \psi(F,\theta)$ and the thermal coupling and the heat-transfer cannot be ignored. Yet, this standardly leads to adiabatic terms involving dilatation rate whose control, in turn, needs a parabolic-type viscoelastic model (e.g. of Kelvin-Voigt-type) as we saw already in Chapter 8. In the purely elastic materials (or also in hyperbolic-type viscoelastic models (e.g. of Maxwell type), we can use a *phase-field regularization* to overcome any direct coupling of temperature θ and the deformation gradient F in the stored energy and, instead, to couple F with a phase field χ which can be regularized in the reference configuration. Likewise (7.6.36), the modeling ingredient is the choice of a splitting $\psi(F,\theta) = \varphi(F,\theta) - \gamma_\mathrm{v}(\theta)$ and then, like like we did in (8.5.5a) or in other context also in (7.6.37a), we replace $\varphi(F,\theta)$ by $\varphi(F,\chi) + \frac{1}{2\varkappa}|\chi-\theta|^2$ with some fixed coefficient $\varkappa > 0$ presumable "small". This leads to the free energy

$$\psi_\mathrm{PF}(F,\chi,\theta) := \psi(F,\chi) + \frac{1}{2\varkappa}|\chi-\theta|^2 - \gamma_\mathrm{v}(\theta). \tag{9.2.30}$$

Again, we also employ the nonlocal concept as we did in (9.2.6), which will allow for a control of the determinant of the deformation gradient and, apart of the local non-selfpenetration, also the mechanically relevant treatment of the heat transport in the Lagrangian setting, cf. (9.2.26) above. This leads to the ansatz of the total free energy

$$(y,\chi,\theta) \mapsto \int_\Omega \psi_{\rm PF}(\nabla y,\chi,\theta) + \frac{\kappa}{2}|\nabla\chi|^2\,{\rm d}x + \mathscr{H}(\nabla^2 y) \tag{9.2.31}$$

with $\kappa > 0$ a fixed, presumably small coefficient and $\mathscr{H}$ from (2.5.34) also presumably small. A modeling intention is to achieve χ not much different from θ in most spots in particular solutions, which can intuitively be executed when taking also $\varkappa > 0$ small, although we will not have ambitions to study the asymptotics when $\varkappa \to 0$.

Considering still a dissipation potential $\int_\Omega \frac{1}{2}\tau_{\rm R}\dot\chi^2\,{\rm d}x$ with $\tau_{\rm R} > 0$ a (presumably small) relaxation time being a part of the model, the system (9.2.1) is then to be augmented by

$$\varrho\ddot y - {\rm div}\,S = f(y) \quad \text{with } S = \partial_F\varphi(\nabla y,\chi) - {\rm Div}(\mathfrak{H}\nabla^2 y) \qquad \text{in } Q, \tag{9.2.32a}$$

$$\tau_{\rm R}\dot\chi - \kappa\Delta\chi + \partial_\chi\varphi(\nabla y,\chi) = (\theta-\chi)/\varkappa \qquad \text{in } Q, \tag{9.2.32b}$$

$$\dot\vartheta - {\rm div}(\mathscr{K}(\nabla y,\theta)\nabla\theta) = \tau_{\rm R}\dot\chi^2 + \theta\dot\chi/\varkappa \quad \text{with} \quad \vartheta = \mathscr{C}_{\rm V}(\theta) \qquad \text{in } Q, \tag{9.2.32c}$$

$$S\vec n - {\rm div}_{\rm S}((\mathfrak{H}\nabla^2 y)\cdot\vec n) = g \quad \text{and} \quad (\mathfrak{H}\nabla^2 y):(\vec n\otimes\vec n) = 0 \qquad \text{on } \Sigma, \tag{9.2.32d}$$

$$\nabla\chi\cdot\vec n = 0 \quad \text{and} \quad \mathscr{K}(\nabla y,\theta) = h_\flat \qquad \text{on } \Sigma, \tag{9.2.32e}$$

$$y|_{t=0} = y_0, \qquad \dot y|_{t=0} = v_0, \qquad \chi|_{t=0} = \chi_0, \qquad \theta|_{t=0} = \theta_0 \qquad \text{on } \Omega, \tag{9.2.32f}$$

where again $\mathscr{C}_{\rm V}(\theta) := \int_0^1 \theta c_{\rm V}(r\theta)\,dr$ and $\mathscr{K}$ is determined by $\mathbb{K}_{\rm M}$ through (9.2.26).

For the analysis, we again qualify $\mathscr{H}$, f, g, y_0, and v_0 as in (9.2.9b-e) and moreover we assume:

$$\varphi : {\rm GL}^+(d)\times\mathbb{R} \to \mathbb{R}^+ \ \text{ continuously differentiable, and for some } \epsilon > 0:$$

$$\varphi(F,\chi) \ge \frac{\epsilon}{(\det F)^q}, \quad q > \frac{2d}{2\gamma+2-d} \quad \text{for some } \gamma > \frac{d}{2} - 1, \tag{9.2.33a}$$

$$|\partial_\chi\varphi(F,\chi)| \le K(F)(1+|\chi|^{2^*/2}) \quad \text{with } K : {\rm GL}^+ \to \mathbb{R} \text{ continuous}, \tag{9.2.33b}$$

$$c_{\rm V} : \mathbb{R} \to \mathbb{R}^+ \ \text{ continuous, satisfying } (8.4.7f), \tag{9.2.33c}$$

$$\mathbb{K}_{\rm M} : \Omega\times\mathbb{R} \to \mathbb{R}^{d\times d} \ \text{ Carathéodory mapping, uniformly positive definite}, \tag{9.2.33d}$$

$$\varrho > 0, \quad \tau_{\rm R} > 0, \quad \kappa > 0, \quad \varkappa > 0, \quad h_\flat \in L^1(\Sigma), \quad h_\flat \ge 0 \ \text{ on } \Sigma, \tag{9.2.33e}$$

$$\chi_0 \in H^1(\Omega), \quad \theta_0 \in L^1(\Omega), \quad \theta_0 \ge 0 \ \text{ on } \Omega. \tag{9.2.33f}$$

note that the again exponents q and γ are fitted with the assumptions in Corollary 2.5.11 on p. 50.

Proposition 9.2.12 (Existence of weak solutions). *Let (9.2.33) and (9.2.9b-e) hold. Then the initial-boundary value problem (9.2.32) possesses a weak solution* $(y,\chi,\theta,\vartheta)$ *with* $\theta \ge 0$.

Sketch of the proof. We will use the Faedo-Galerkin method[9] in a careful way combined with a regularization of the heat sources like we did in (8.2.7). We divide the proof into five steps.

Step 1 – approximate solutions: First, the mentioned regularization leads to the initial-boundary-value problem

$$\varrho\ddot{y} - \operatorname{div} S = f(y) \quad \text{with} \quad S = \partial_F\varphi(\nabla y,\chi) - \operatorname{Div}(\mathfrak{H}\nabla^2 y) \qquad \text{in } Q, \tag{9.2.34a}$$

$$\tau_{\rm R}\dot{\chi} - \kappa\Delta\chi + \partial_\chi\varphi(\nabla y,\chi) = (\theta-\chi)/\varkappa \qquad \text{in } Q, \tag{9.2.34b}$$

$$\dot{\vartheta} - \operatorname{div}(\mathscr{K}(\nabla y,\theta)\nabla\theta) = \frac{\tau_{\rm R}\dot{\chi}^2 + \theta\dot{\chi}}{\varkappa(1+\varepsilon\dot{\chi}^2)} \quad \text{with} \quad \vartheta = \mathscr{C}_{\rm v}(\theta) \qquad \text{in } Q, \tag{9.2.34c}$$

$$S\vec{n} - \operatorname{div}_{\rm S}((\mathfrak{H}\nabla^2 y)\cdot\vec{n}) = g \quad \text{and} \quad (\mathfrak{H}\nabla^2 y):(\vec{n}\otimes\vec{n}) = 0 \qquad \text{on } \Sigma, \tag{9.2.34d}$$

$$\kappa\nabla\chi\cdot\vec{n} = 0 \quad \text{and} \quad \mathscr{K}(\nabla y,\theta) = \frac{h_\flat}{1+\varepsilon h_\flat} \qquad \text{on } \Sigma \tag{9.2.34e}$$

$$y|_{t=0} = y_0, \quad \dot{y}|_{t=0} = v_0, \quad \chi|_{t=0} = \chi_0, \quad \theta|_{t=0} = \frac{\theta_0}{1+\varepsilon\theta_0} \qquad \text{on } \Omega. \tag{9.2.34f}$$

Then we make the Faedo-Galerkin approximation by exploiting some finite-dimensional subspaces of $H^{2+\gamma}(\Omega;\mathbb{R}^d)\cap W^{1,p}(\Omega)$ for (9.2.34a) and of $H^1(\Omega)$ for both equations (9.2.34b,c). We denote by $k\in\mathbb{N}$ the discretization parameter. We may assume that (y_0,v_0,χ_0) belongs to the finite-dimensional spaces used for the approximation for all $k\in\mathbb{N}$ while $\theta_0\in L^1(\Omega)$ is approximated by some $\theta_{0,k}\in L^2(\Omega)$ to belong to the respective finite-dimensional space and assume, without any loss of generality if the qualification (9.2.33f) and density of the finite-dimensional subspaces is assumed, that

$$\|\theta_{0,k}\|_{L^1(\Omega;\mathbb{R}^d)} \le K. \tag{9.2.35a}$$

The existence of the Galerkin solution, let us denote it by $(y_{\varepsilon k},\chi_{\varepsilon k},\theta_{\varepsilon k})$, can be argued by the successive prolongation argument, relying on the a-priori estimate obtained by means of testing the particular equations (9.2.32a,b,c) successively by $\dot{y}_{\varepsilon k}$, by $\dot{\chi}_{\varepsilon k}$, and by $\theta_{\varepsilon k}$. More specifically, we thus obtain the estimates

$$\big\|y_{\varepsilon k}\big\|_{L^\infty(I;H^{2+\gamma}(\Omega;\mathbb{R}^d)\cap W^{1,p}(\Omega;\mathbb{R}^d))\cap W^{1,\infty}(I;L^2(\Omega;\mathbb{R}^d))} \le K, \tag{9.2.36a}$$

$$\big\|1/\det\nabla y_{\varepsilon k}\big\|_{L^\infty(Q)} \le K, \tag{9.2.36b}$$

$$\big\|\chi_{\varepsilon k}\big\|_{L^\infty(I;H^1(\Omega))\cap H^1(I;L^2(\Omega))} \le K_\varepsilon, \tag{9.2.36c}$$

$$\big\|\theta_{\varepsilon k}\big\|_{L^\infty(I;L^2(\Omega))\cap L^2(I;H^1(\Omega))\cap H^1(I;H^1(\Omega)^*)} \le K_\varepsilon\,. \tag{9.2.36d}$$

It is important that, due to the assumption (9.2.33a) with (9.2.9b), we have at disposal the Healey-Krömer theorem in the variant of Corollary 2.5.11 so that, by

[9] As we need to control the determinant of the deformation gradient, we cannot impose semi-convexity assumption and cannot rely on a time discretization, cf. Remark 9.2.7.

the successive-prolongation argument, $\det\nabla y_{\varepsilon k} > 0$ satisfies with (9.2.36b) and the Lavrentiev phenomenon is excluded for the Galerkin procedure.

Step 2 – Limit passage in the Galerkin discretization: We now can pass to the limit $k \to \infty$ in terms of a selected subsequence. Here it is important that we need the strong convergence of $\dot{\chi}$ only, not of $\dot{y}$. This strong convergence is however easy since (9.2.32b) is the semilinear parabolic equation; cf. Exercise 9.2.16. Thus we obtain a weak solution $(y_\varepsilon, \chi_\varepsilon, \theta_\varepsilon, \vartheta_\varepsilon)$ to the regularized initial-boundary-value problem (9.2.34). The apriori estimates (9.2.36) are inherited for this solution, too.

Step 3 – Energetic estimates: Then, we can show that $\theta_\varepsilon \geq 0$ and perform the physically motivated test of the heat equation by 1 instead of by θ. Thus we obtain the estimates independent of ε, namely

$$\big\|y_\varepsilon\big\|_{L^\infty(I;H^{2+\gamma}(\Omega;\mathbb{R}^d))\cap W^{1,p}(\Omega;\mathbb{R}^d))\cap W^{1,\infty}(I;L^2(\Omega;\mathbb{R}^d))} \leq K, \tag{9.2.37a}$$

$$\big\|\chi_\varepsilon\big\|_{L^\infty(I;H^1(\Omega))\cap H^1(I;L^2(\Omega))} \leq K, \tag{9.2.37b}$$

$$\big\|\theta_\varepsilon\big\|_{L^\infty(I;L^\omega(\Omega))} \leq K, \quad \text{and} \quad \big\|\vartheta_\varepsilon\big\|_{L^\infty(I;L^1(\Omega))} \leq K. \tag{9.2.37c}$$

Step 4 – Further estimation of $\nabla\theta$: We again use the "prefabricated" estimate (8.2.6b) on p.368 here with the heat production/consumption rate as in the right-hand side of (8.4.17c), i.e. $\xi_{\rm A} = \xi_{\rm A}(\theta,\dot{\chi},\dot{z}) = \tau_{\rm R}\dot{\chi}^2 + \theta\dot{\chi}/\varkappa$. From (9.2.34b) tested by $\dot{\chi}$ and integrated over I, we obtain:

$$\begin{aligned}\int_Q \tau_{\rm R}\dot{\chi}_\varepsilon^2\,{\rm d}x{\rm d}t &\leq \int_Q \frac{1}{\varkappa}\theta_\varepsilon\dot{\chi}_\varepsilon - \partial_\chi\varphi(\nabla y_\varepsilon,\chi_\varepsilon)\dot{\chi}_\varepsilon\,{\rm d}x{\rm d}t + \int_\Omega \frac{\kappa}{2}|\nabla\chi_0|^2\,{\rm d}x\\ &\leq \epsilon\|\theta_\varepsilon\|_{L^p(Q)}^p + \frac{\tau_{\rm R}}{2}\|\dot{\chi}_\varepsilon\|_{L^2(Q)}^2 + C_\epsilon\end{aligned} \tag{9.2.38}$$

with $\epsilon > 0$ arbitrarily small and C_ϵ depending also on $\sup_{\varepsilon>0}\|\partial_\chi\varphi(\nabla y_\varepsilon,\chi_\varepsilon)\|_{L^2(Q)}$ which is finite due to the assumption (9.2.33b) and the a-priori bounds (9.2.37a,b). In particular, (9.2.38) relies on $p > 2$. We can further use Corollary 9.2.10 and the mentioned estimate (8.2.6a), namely

$$\begin{aligned}\|\nabla\theta_\varepsilon\|_{L^r(Q;\mathbb{R}^d)}^r + \|\theta_\varepsilon\|_{L^p(Q)}^p &\leq K_1\Big(1 + \Big\|\tau_{\rm R}\dot{\chi}_\varepsilon^2 + \frac{1}{\varkappa}\theta_\varepsilon\dot{\chi}_\varepsilon\Big\|_{L^1(Q)}\Big)\\ &\leq K_1\Big(1 + \int_Q \tau_{\rm R}\dot{\chi}_\varepsilon^2 + \Big|\frac{1}{\varkappa}\theta_\varepsilon\dot{\chi}_\varepsilon\Big|\,{\rm d}x{\rm d}t\Big)\\ &\leq K_1\Big(1 + \tau_{\rm R}\|\dot{\chi}_\varepsilon\|_{L^2(Q)}^2 + \frac{1}{\varkappa}\|\theta_\varepsilon\|_{L^2(Q)}^2 + \frac{1}{\varkappa}\|\dot{\chi}_\varepsilon\|_{L^2(Q)}^2\Big)\\ &\leq K_2\Big(1 + \tau_{\rm R}\|\dot{\chi}_\varepsilon\|_{L^2(Q)}^2 + \frac{1}{\varkappa}\|\dot{\chi}_\varepsilon\|_{L^2(Q)}^2\Big) + \frac{1}{2}\|\theta_\varepsilon\|_{L^p(Q)}^p,\end{aligned} \tag{9.2.39}$$

where the last inequality again uses $p > 2$. Then we add (9.2.39) to (9.2.38) with a sufficiently small positive weight, namely $1/(2K_2)$. Altogether, we obtain

$$\|\nabla\theta_\varepsilon\|_{L^r(Q;\mathbb{R}^d)}^r + \|\theta_\varepsilon\|_{L^p(Q)}^p \leq K$$

with K depending on p, τ_R, $\|\chi_0\|_{H^1(\Omega)}$, and $\sup_{\varepsilon>0}\|\partial_\chi\varphi(\nabla y_\varepsilon,\chi_\varepsilon)\|_{L^2(Q)}$. Using also (8.2.6a), we can see the estimate of $\nabla\theta_\varepsilon\in L^r(Q;\mathbb{R}^d)$.

Step 5 – limit passage in the regularization of the heat equation with $\varepsilon\to 0$: We need again the $L^2(Q)$-strong convergence of $\dot\chi_\varepsilon$ only, not of $\dot y_\varepsilon$. Then $\dot\chi_\varepsilon^2/(1+\dot\chi_\varepsilon^2)\to\dot\chi^2$ and also $\theta_\varepsilon\dot\chi_\varepsilon/(1+\dot\chi_\varepsilon^2)\to\theta\dot\chi$ strongly in $L^1(Q)$. The resting terms in the semi-linear system (9.2.34) are easy, using also the Aubin-Lions compactness theorem. □

Example 9.2.13 (Thermally expanding material). Advantageously, the $\varkappa$-term in (9.2.31) leaves the heat capacity independent of χ, namely $c_V(\theta)=-\theta\partial^2_{\theta\theta}\psi=\theta\gamma_V''(\theta)-\theta/\varkappa$. E.g., taking the convex function $\gamma_V(\cdot)$ as $\gamma_V(\theta)=c_{V,0}\theta(\ln\theta-1)+\frac{1}{2}\theta^2/\varkappa$ yields $c_V(\theta)=c_{V,0}$. For example, this ansatz allows for modeling the *thermal expansion* by taking $\varphi(F,\chi)=\hat\varphi(F_{\rm EL})$ for some "elastic" energy $\hat\varphi:\mathbb{R}^{d\times d}\to\mathbb{R}\cup\{+\infty\}$ with the multiplicative decomposition

$$F=F_{\rm EL}F_{\rm IN}\quad\text{with}\quad F_{\rm IN}=F_{\rm IN}(\chi)=\mathbb{I}/b(\chi). \tag{9.2.40}$$

cf. also (3.6.11) on p. 79. Then $\varphi(F,\chi)=\hat\varphi(b(\chi)F)$ the contribution to the stress S is $\partial_F\varphi(\nabla y,\chi)=b(\chi)\hat\varphi'(b(\chi)F)$ while $\partial_\chi\varphi(\nabla y,\chi)=b'(\chi)\hat\varphi'(b(\chi)\nabla y){:}\nabla y$.

Remark 9.2.14 (Macroscopical response and the role of gradient theories). Introducing gradients (or even "nonlocal gradients") into the stored energy may give rise to a certain controversy in particular because the respective coefficients are usually not reliably known or even the form of these gradient terms is disputable. In macroscopical bodies, these gradient terms are very small and represent rather a singular perturbation only, and in some situation they may play and important analytical without influencing essentially the macroscopical response, as articulated already in [457, Remark 5.6]. This phenomenon is well known from variational calculus while in evolution problems, it is more difficult to prove it rigorously (cf. some references in [457]) and generally it does not hold and even small singular perturbation may have a big influence in particular in systems which may loose their stability (as typical in stressed faults before rupture). Anyhow, at least some global energetics (i.e. some a-priori estimates survive even if the coefficients in the gradient terms tend to zero.

Exercise 9.2.15 (Semiconvexity at large-strain problems). Modify the model for St.Venant-Kirchhoff material (2.4.1) to have a bounded stress and a potential with a linear (instead of quadratic) growth and verify the semiconvexity property (9.2.20). Realize the difficulties with the semiconvexity for $\varphi(F)=v(\det F)$ with v convex.

Exercise 9.2.16. Prove the strong convergence of $\dot\chi_\varepsilon\to\dot\chi$ in the problem (9.2.32), using the test by $\dot\chi_\varepsilon-\dot\chi$ or alternatively using the energy conservation.

9.3 Kelvin-Voigt viscoelastic solids

We know that the Kelvin-Voigt rheological model sums up the viscous and the elastic stresses, cf. Sect. 6.4. At finite strains, the Kelvin-Voigt 2nd-grade nonsimple material gives the model

$$\varrho\ddot{y} - \operatorname{div} S = f(y) \quad \text{with} \quad S = S_{\rm VI}(\nabla y, \nabla\dot{y}) + S_{\rm EL}(\nabla y) - \operatorname{Div}(\mathfrak{H}\nabla^2 y), \tag{9.3.1}$$

where the involved *hyperstress* $\mathfrak{H}\nabla^2 y = \mathscr{H}'(\nabla^2 y)$ can be local, i.e. $\mathfrak{H}\nabla^2 y = \mathbb{H}\nabla^2 y$ as in Example 2.5.6 on p. 45, or nonlocal as in (2.5.36). In this Chapter 9, we thus use the notation $\mathfrak{H}$ as a linear operator acting on $\nabla^2 y$. A physically ultimate requirement is the *frame indifference* which for the stresses means that

$$S_{\rm VI}(RF, \dot{R}F{+}R\dot{F}) = RS_{\rm VI}(F,\dot{F}) \quad \text{and} \quad S_{\rm EL}(RF) = RS_{\rm EL}(F) \tag{9.3.2}$$

for any smoothly time-varying $R : t \mapsto R(t) \in {\rm SO}(d)$. It should be emphasized that classical linear viscosity we used for Kelvin-Voigt's type materials, i.e. $S_{\rm VI}(F,\dot{F}) = \mathbb{D}\dot{F}$, is not frame indifferent, which is Antman's objection [13].[10] For $S_{\rm VI}$ having a potential in the sense

$$S_{\rm VI}(F,\dot{F}) = \partial_{\dot{F}}\zeta(F,\dot{F}), \tag{9.3.3}$$

the *frame indifference* means that $\zeta(F,\dot{F}) = \zeta(RF,(RF)^{\boldsymbol{\cdot}})$, i.e.

$$\forall R \in C^1(I;\mathbb{R}^{d\times d}),\ R(t){\in}{\rm SO}(d): \quad \zeta(F,\dot{F}) = \zeta(RF,\dot{R}F{+}R\dot{F})$$

This essentially means that there is another potential $\hat{\zeta} : \mathbb{R}^{d\times d}\times\mathbb{R}^{d\times d} \to \mathbb{R}$ inducing a second stress tensor $\hat{S}_{\rm VI} : \mathbb{R}^{d\times d}\times\mathbb{R}^{d\times d} \to \mathbb{R}^{d\times d}$ (up to the factor 2, analogous to the 2nd Piola-Kirchhoff tensor) such that[11]

$$S_{\rm VI}(F,\dot{F}) = 2F\hat{S}_{\rm VI}(C,\dot{C}) \quad \text{with} \quad \hat{S}_{\rm VI}(C,\dot{C}) = \partial_{\dot{C}}\hat{\zeta}(C,\dot{C}).$$

Altogether, realizing that $\dot{C} = \dot{F}^\top F + F^\top\dot{F} = 2{\rm sym}(F^\top\dot{F})$,

$$S_{\rm VI}(F,\dot{F}) = 2F\partial_{\dot{C}}\hat{\zeta}(F^\top F, 2{\rm sym}(F^\top\dot{F})), \tag{9.3.4}$$

which is analog to (2.3.9) on p. 30. The relation with potential used in (9.3.3) is $\zeta(F,\dot{F}) = \hat{\zeta}(C,\dot{C})$. The simplest choice is that the viscosity is considered linear, which is the relevant ansatz for non-activated dissipative processes undergoing at

[10] Cf. also [122, 152] or [14, Chap.12].

[11] By Noll's theorem, $S_{\rm VI}(F,\dot{F}) = F\hat{S}_{\rm VI}(U,\dot{U})$ with $\hat{S}_{\rm VI}$ a symmetric tensor and $U^2 = F^\top F$, i.e. $F = RU$ with $R \in {\rm SO}(d)$. Clearly, as $C = F^\top F = U^2$, it holds $\dot{C} = 2U\dot{U}$ so that $\hat{S}(C,\dot{C}) = \hat{S}(U^2, 2U\dot{U})$, which reveals the form of S if $\hat{S}$ is given.

rather small rates (in contrast to activated processes like plasticity which leads to nonsmooth potentials homogeneous degree 1 in such a small rate approximation). This linear viscosity leads to a potential which is quadratic in terms of rates, which here means the rate of the right Cauchy-Green tensor $\dot{C} = (F^\top F)^{\boldsymbol{\cdot}} = 2\mathrm{sym}(F^\top \dot{F})$ rather than the rate of deformation gradient $\dot{F}$ itself, i.e. $\hat{\zeta}(C,\dot{C}) = \frac{1}{2}\mathbb{D}(C)\dot{C}:\dot{C}$. For $\mathbb{D} = \nu\mathbb{I}$ we obtain $S_{\rm VI}(F,\dot{F}) = 2\nu F(F^\top\dot{F} + \dot{F}^\top F)$, while for $\mathbb{D}(C) = \nu C^{-1}$ we obtain $S_{\rm VI}(F,\dot{F}) = 2\nu(\dot{F} + F^{-\top}\dot{F}^\top F)$, cf. [357]. In simple materials, such physical viscosity does not seem mathematically tractable unless restrictions on short-time analysis or special rate-independent situations, as pointed out also in [29, Sec. 3.2] and discussed in detail also in [357]. In view of (9.3.4), realizing that $\mathrm{sym}(F^\top\dot{F}) = \frac{1}{2}(F^\top\dot{F}+\dot{F}^\top F) = \frac{1}{2}\dot{C}$ and assuming symmetry of $\partial_{\dot{C}}\hat{\zeta}$ as in (9.3.6) below, we have the dissipation rate of this model[12]

$$\begin{aligned}\xi(F,\dot{F}) &= 2F\partial_{\dot{C}}\hat{\zeta}(C,\dot{C}):\dot{F} = 2\partial_{\dot{C}}\hat{\zeta}(C,\dot{C}):(F^\top\dot{F})\\ &= 2\partial_{\dot{C}}\hat{\zeta}(C,\dot{C}):\mathrm{sym}(F^\top\dot{F}) = \partial_{\dot{C}}\hat{\zeta}(C,\dot{C}):\dot{C} = \hat{\zeta}(C,\dot{C}).\end{aligned}\tag{9.3.5}$$

Again, things are substantially simpler in nonsimple materials. We expose basic estimates and convergence arguments for materials linear in rates of C, although a lot of geometrical nonlinearity arising from large strains is still a vital part of the model. More specifically, we consider $\hat{\zeta}(C,\dot{C})$ independent of C for simplicity, i.e.

$$\hat{\zeta}(C,\dot{C}) = \frac{1}{8}\mathbb{D}(C)\dot{C}:\dot{C} \quad \text{with} \quad \hat{\zeta}(C,\dot{C}) \geq \varepsilon|\dot{C}|^2 \quad \text{and} \tag{9.3.6a}$$

$$\mathbb{D} : \mathbb{R}^{d\times d}_{\rm sym} \to \mathrm{SLin}(\mathbb{R}^{d\times d}_{\rm sym}) \ \text{continuous and bounded.} \tag{9.3.6b}$$

Then $S_{\rm VI}(F,\dot{F}) = \hat{\mathbb{D}}(F)\mathrm{sym}(F^\top\dot{F})$ with $\hat{D}(F) := 2F\mathbb{D}(F^\top F)$, in analog to the small-strain model in Section 6.4 where $S_{\rm VI}(\dot{e}) = \mathbb{D}\dot{e} = \mathbb{D}\mathrm{sym}(\nabla\dot{u})$.

Altogether, in terms of $\hat{\zeta}$ and φ and $\mathscr{H}$ in the local variant from Sect. 2.5.1, the thus augmented force balance (9.3.1) accompanied with corresponding mixed Dirichlet/Neumann-type boundary conditions leads to the initial-boundary-value problem

$$\varrho\ddot{y} - \mathrm{div}\big(S_{\rm VI}(\nabla y,\nabla\dot{y}) + S_{\rm EL}(\nabla y) - \mathrm{Div}(\mathscr{H}'\nabla^2 y)\big) = f(t,x,y) \qquad \text{with}$$

$$S_{\rm VI}(F,\dot{F}) = \hat{\mathbb{D}}(F)\mathrm{sym}(F^\top\dot{F}) \ \text{ and } \ S_{\rm EL}(F) = \varphi'(F) \qquad \text{on } Q, \tag{9.3.7a}$$

$$(S_{\rm VI}(\nabla y,\nabla\dot{y}) + \varphi'(\nabla y) - \mathrm{Div}(\mathscr{H}'\nabla^2 y))\vec{n} - \mathrm{div}_{\rm S}((\mathscr{H}'\nabla^2 y)\cdot\vec{n}) = g \quad \text{on } S_{\rm N}, \tag{9.3.7b}$$

$$y = y_{\rm D} \qquad \text{on } S_{\rm D}, \tag{9.3.7c}$$

$$(\mathscr{H}'\nabla^2 y):(\vec{n}\otimes\vec{n}) = 0 \qquad \text{on } \Sigma, \tag{9.3.7d}$$

$$y\big|_{t=0} = y_0 \quad \text{and} \quad \dot{y}\big|_{t=0} = v_0 \qquad \text{on } \Omega \tag{9.3.7e}$$

with $f(t,x,y) = f_0(t,x) + f_1(y)$ as in (9.2.2).

[12] Here we used the algebra (A.2.14) together with the symmetry $\partial_{\dot{C}}\hat{\zeta}$.

The energetics can be revealed by testing (9.3.7a) by $\dot{y}$, assuming $\dot{y}_{\rm D} = 0$ for simplicity. We identify naturally the specific dissipation rate

$$\xi(F,\dot{F}) = S_{\rm vi}(F,\dot{F}):\dot{F} = 2F\partial_{\dot{C}}\hat{\zeta}(C,\dot{C}):\dot{F} = \hat{\mathbb{D}}(F)\,{\rm sym}(F^\top\dot{F}):{\rm sym}(F^\top\dot{F}). \quad (9.3.8)$$

By using Green's formula (twice in Ω and once on Γ) and by the boundary conditions (9.3.7d), we thus arrive at the *energy balance*

$$\underbrace{\frac{\rm d}{{\rm d}t}\Big(\int_\Omega \frac{\varrho}{2}|\dot{y}|^2 + \varphi(\nabla y)\,{\rm d}x + \mathscr{H}(\nabla^2 y)\Big)}_{\text{kinetic and stored energy}} + \underbrace{\int_\Omega \hat{\mathbb{D}}(\nabla y)\,{\rm sym}(\nabla y^\top\nabla\dot{y}):{\rm sym}(\nabla y^\top\nabla\dot{y})\,{\rm d}x}_{\text{dissipation rate}} = \underbrace{\int_\Omega f(y)\cdot\dot{y}\,{\rm d}x + \int_\Gamma g\cdot\dot{y}\,{\rm d}S}_{\text{power of external loading}}. \quad (9.3.9)$$

Let us start with the simple case where existence results can be based on the weak convergence without any knowledge of the rate of deformations gradient. The possible (even local) self-interpenetration is ignored (so that φ can have a polynomial growth) and confine ourselves to the physically relevant case $d \le 3$ where simply $H^2(\Omega) \subset L^\infty(\Omega)$. We thus impose the basic data qualification:

$$0 \le \varphi(F) \le c(1+|F|^{2^*-\epsilon}), \qquad |\varphi'(F)| \le c(1+|F|^{2^*-\epsilon}) \quad (9.3.10a)$$

$$\mathscr{H}(\nabla^2 y) = \int_\Omega \frac{1}{2}\mathbb{H}\nabla^2 y \,\vdots\, \nabla^2 y\,{\rm d}x \quad \text{with } \mathbb{H} \in {\rm SLin}(\mathbb{R}^{d\times d}_{\rm sym}) \text{ positive definite,} \quad (9.3.10b)$$

$$y_0 \in H^2(\Omega;\mathbb{R}^d), \quad v_0 \in L^2(\Omega;\mathbb{R}^d), \quad (9.3.10c)$$

while f and g are qualified as in (9.2.9c,d), assuming $d \le 3$.

Without going into details, assume we have at disposal an approximate solution y_k to (9.3.7), employing the Faedo-Galerkin method with the nested sequence of finite-dimensional subspaces $V_k \subset H^2(\Omega;\mathbb{R}^d)$, $k \in \mathbb{N}$. Without loss of generality, we may assume $y_0, v_0 \in V_k$.

Proposition 9.3.1 (Existence based on weak convergence of $\dot{C}$). *Under the assumptions (9.3.6), (9.3.10), (9.2.9c,d), and ϱ satisfying and ϱ satisfies (6.3.2), the Galerkin approximation $\{y_k\}_{k\in\mathbb{N}}$ satisfies the a-priori estimates:*

$$\|y_k\|_{L^\infty(I;H^2(\Omega;\mathbb{R}^d))\cap W^{1,\infty}(I;L^2(\Omega;\mathbb{R}^d))} \le K, \quad (9.3.11a)$$

$$\|{\rm sym}(\nabla y_k^\top\nabla\dot{y}_k)\|_{L^2(Q;\mathbb{R}^{d\times d})} \le K. \quad (9.3.11b)$$

The weak limit y in the topology indicated in (9.3.11a) of a selected subsequence satisfies also* ${\rm sym}(\nabla y_k^\top\nabla\dot{y}_k) \to {\rm sym}(\nabla y^\top\nabla\dot{y})$ *weakly in $L^2(Q;\mathbb{R}^{d\times d})$ and solves the initial-boundary-value problem (9.3.7). In particular, this problem possesses weak solutions $y \in L^\infty(I;H^2(\Omega;\mathbb{R}^d)) \cap W^{1,\infty}(I;L^2(\Omega;\mathbb{R}^d))$ such that also* ${\rm sym}(\nabla y^\top\nabla\dot{y}) \in L^2(Q;\mathbb{R}^{d\times d}_{\rm sym})$.

Proof. We first perform the test of (9.3.7a) in its Galerkin approximation by $\dot{y}_k$, we obtain (9.3.9) written for y_k instead of y. Integrating (9.3.9) over $[0,t]$ for $t\in I$ and making the by-part integration in time of the loading terms, we can read the basic a-priori estimates (9.3.11). This is already sufficient to prove existence of a weak solution to the initial-boundary-value problem (9.3.7) by applying the Galerkin method. The important fact is that, in view of (9.3.6), the Galerkin approximation of the viscous stress is $S_{\text{vi},k} = \hat{\mathbb{D}}(\nabla y_k)\text{sym}(\nabla y_k^\top \nabla \dot{y}_k)$ and that, due to (9.3.11b) in terms of subsequences, $\text{sym}(\nabla y_k^\top \nabla \dot{y}_k)$ converges weakly in $L^2(Q;\mathbb{R}^{d\times d}_{\text{sym}})$ to some limit, let us denote it by C°, and that also $\nabla y_k \to \nabla y$ strongly in $L^{1/\epsilon}(I;L^{2^*-\epsilon}(\Omega;\mathbb{R}^{d\times d}))$ for any $0<\epsilon\le 2^*-1$ due to (9.3.11a) and the Aubin-Lions theorem. Therefore $S_{\text{vi},k}$ converges weakly to $\hat{\mathbb{D}}(\nabla y)C^\circ$ in $L^{2-\epsilon}(I;L^{2^*2/(2^*+2)-\epsilon}(\Omega;\mathbb{R}^{d\times d}))$. We are to identify the limit C°. We have

$$\begin{aligned}\int_0^t \text{sym}(\nabla y_k^\top \nabla \dot{y}_k)\,\mathrm{d}t &= \int_0^t 2(\nabla y_k^\top \nabla y_k)^{\boldsymbol{\cdot}}\,\mathrm{d}t = 2\nabla y_k^\top(t)\nabla y_k(t) - 2\nabla y_0^\top \nabla y_0\\ &\to 2\nabla y^\top(t)\nabla y(t) - 2\nabla y_0^\top \nabla y_0 = \int_0^t \text{sym}(\nabla y^\top \nabla y)^{\boldsymbol{\cdot}}\,\mathrm{d}t \qquad (9.3.12)\end{aligned}$$

strongly in $L^{2^*/2-\epsilon}(\Omega;\mathbb{R}^{d\times d})$ for a.a. $t\in I$ because $\nabla y_k(t)\to\nabla y(t)$ weakly in $H^1(\Omega;\mathbb{R}^{d\times d})$ and hence strongly in $L^{2^*-\epsilon}(\Omega;\mathbb{R}^{d\times d})$; in fact, here a subsequence is to be chosen, cf. Proposition B.2.1(ii)-(iii) on p.510. Simultaneously,

$$\int_0^t \text{sym}(\nabla y_k^\top \nabla \dot{y}_k)\,\mathrm{d}t \to \int_0^t C^\circ\,\mathrm{d}t \quad \text{weakly in } L^2(\Omega;\mathbb{R}^{d\times d}) \qquad (9.3.13)$$

for each $t\in I$. Therefore, comparing (9.3.12) and (9.3.13), we can identify $C^\circ = (\nabla y^\top \nabla y)^{\boldsymbol{\cdot}} = \text{sym}(\nabla y^\top \nabla \dot{y})$. The resting limit passage in the weak formulation of (9.3.7) is easy as before for the inviscid material in (9.2.1). □

The above results can be used for the viscous expansion of the Lamé-type (St.Venant-Kirchhoff) material, see (2.4.3) on p. 36, which involves φ with the 4th-order polynomial growth and thus satisfies (9.3.10a) for $d\le 3$.

The next challenge is to prevent local self-interpenetration. As we cannot rely on the variational structure of the static part but must use the (evolution extension) of the corresponding Euler-Lagrange differential equations, we need to control the determinant of ∇y away from 0 to allow for the limit passage through Nemytskiĭ operators not having a polynomial growth. For this, Healey-Krömer's Theorem 2.5.3 in the form of Corollary 2.5.11 is to be used together with a nonlocal linear hyperstress as in (9.2.8b), i.e. $[\mathfrak{H}G](t,x) = \int_\Omega \mathfrak{K}(x,\tilde{x}) \vdots (G(t,x)-G(t,\tilde{x}))$. Now we can handle more general traction force as in (1.2.17b), namely

$$\begin{aligned}g(t,x,y,\nabla y) = g_0(t,x) + g_1(y,\nabla y) \quad &\text{with } g_0\in L^2(I;L^{2^{\sharp'}}(\Gamma;\mathbb{R}^d)) \text{ and}\\ &g_1:\mathbb{R}^d\times\mathbb{R}^{d\times d}\to\mathbb{R}^d \text{ continuous, bounded.} \qquad (9.3.14)\end{aligned}$$

One can (and even must) avoid the polynomial growth of φ and can obtain some additional result comparing Proposition 9.3.1, while also the boundedness[13] of $\mathbb{D}$ assumed in (9.3.6b) could be avoided:

Proposition 9.3.2 (Existence preventing local self-interpenetration). *Let the assumptions of Proposition 9.3.1 hold, together with (9.2.9b) and (9.3.14) hold while f, y_0, and v_0 satisfy (9.2.9c,e), and $\mathfrak{M}_{d-1}(\Gamma_\mathrm{D})>0$. Then the Galerkin approximation $\{y_k\}_{k\in\mathbb{N}}$ satisfies the a-priori estimates (9.3.11) with H^2 replaced by $H^{2+\gamma}$. Beside the weak* convergence of $y_k\to y$ in $L^\infty(I;H^{2+\gamma}(\Omega;\mathbb{R}^d))\cap W^{1,\infty}(I;L^2(\Omega;\mathbb{R}^d))$ as in Proposition 9.3.1 with y a solution to the initial-boundary-value problem (9.3.7), we have now also*

$$\nabla\dot{y}_k\to\nabla\dot{y}\quad\textit{weakly in }\ L^2(Q;\mathbb{R}^{d\times d}).\tag{9.3.15}$$

Sketch of the proof. In addition to the proof of Proposition 9.3.1, we now take nested finite-dimensional subspaces $V_k\subset V_{k+1}\subset H^{2+\gamma}(\Omega;\mathbb{R}^d)$ whose union is dense in $H^{2+\gamma}(\Omega;\mathbb{R}^d)$. We also approximate the initial data (y_0,v_0) by $(y_{0k},v_{0k})\in V_k^2$ so that $y_{0k}\to y_0$ in $H^{2+\gamma}(\Omega;\mathbb{R}^d)$ and $v_{0k}\to v_0$ in $L^2(\Omega;\mathbb{R}^d)$ strongly.

Further, we use Healey-Krömer's Theorem 2.5.3 in the form of Corollary 2.5.11, so that $\min\det\nabla y_k\ge\varepsilon>0$ is granted for some $\varepsilon>0$. In particular, it excludes the Lavrentiev phenomenon similarly as we did in Section 4.5.3.

This allows us to use Neff and Pompe's generalization of Korn's inequality in Theorem 5.2.4 on p. 167, relying also on the Dirichlet boundary condition. Thus we obtain the estimate $\|\nabla\dot{y}_k\|_{L^2(Q;\mathbb{R}^{d\times d})}\le K\|\mathrm{sym}(F^\top\nabla\dot{y}_k)\|_{L^2(Q;\mathbb{R}^{d\times d})}$ for any field $F\in C(\bar{Q};\mathbb{R}^{d\times d})$ with $\min\det F\ge\varepsilon>0$ with K depending on $\varepsilon>0$. Here we use it for $F=\nabla y_k$ to get control over $\{\nabla\dot{y}_k\}_{k\in\mathbb{N}}$ in $L^2(Q;\mathbb{R}^{d\times d})$. Then we can also estimate the boundary term without by-part integration as $\int_\Gamma g\cdot\dot{y}\,\mathrm{d}S\le N\|g\|_{L^{2^{\sharp\prime}}(\Gamma;\mathbb{R}^d)}\|\dot{y}\|_{L^{2^\sharp}(\Gamma;\mathbb{R}^d)}$ and then to use (9.3.14).

From this, (9.3.15) follows for the already selected subsequence. The limit passage towards the weak solution is then standardly using compactness by Aubin-Lions like in Sect. 9.2. $\square$

Remark 9.3.3 (Quasistatic evolution). When inertial effects are neglected (i.e. $\varrho=0$), one can prevent local self-interpenetration by using the local 2nd-grade nonsimple material concept, provided the nonlinear concept and Theorem 2.5.3 is employed. The convergence is then by Minty's trick: taking $\tilde{y}_k$ valued in the finite-dimensional space $V_k\subset W^{2,\infty}(\Omega;\mathbb{R}^d)$, we can estimate

$$\begin{aligned}0\le\int_Q(\mathscr{H}'(\nabla^2y_k)-\mathscr{H}'(\nabla^2\tilde{y}_k))\,\vdots\,\nabla^2(y_k-\tilde{y}_k)\,\mathrm{d}x\mathrm{d}t=\int_Q f(y_k)\cdot(y_k-\tilde{y}_k)\\-\partial_{\dot{C}}\hat{\zeta}(\nabla y_k^\top\nabla y_k,\nabla\dot{y}_k^\top\nabla y_k+\nabla y_k^\top\nabla\dot{y}_k):(\nabla y^\top\nabla(y_k-\tilde{y}_k))\\+\partial_F\varphi(\nabla y_k):\nabla(y_k-\tilde{y}_k)-\varepsilon\nabla\dot{y}_k:\nabla(y_k-\tilde{y}_k)-\mathscr{H}'(\nabla^2\tilde{y}_k)\,\vdots\,\nabla^2(y_k-\tilde{y}_k)\,\mathrm{d}x\mathrm{d}t\end{aligned}$$

[13] In fact, a certain polynomial growth of $\mathbb{D}(\cdot)$ can be allowed even for Proposition 9.3.1 provided the boundedness of $\nabla y\mathbb{D}(\nabla y^\top\nabla y)$ in $L^2(Q;\mathbb{R}^{(d\times d)^2})$ would be granted.

$$\to \int_Q f(y)\cdot(y-\tilde{y}_k) - \partial_{\dot{C}}\hat{\zeta}(\nabla y^\top\nabla y, \nabla\dot{y}^\top\nabla y+\nabla y^\top\nabla\dot{y}):(\nabla y^\top\nabla(y-\tilde{y}_k))$$
$$+\partial_F\varphi(\nabla y):\nabla(y-\tilde{y}_k) - \varepsilon\nabla\dot{y}:\nabla(y-\tilde{y}_k) - \mathscr{H}'(\nabla^2\tilde{y}_k)\vdots\nabla^2(y-\tilde{y}_k)\,\mathrm{d}x\mathrm{d}t, \qquad (9.3.16)$$

where we used monotonicity of $\mathscr{H}'$ due to the convexity assumed in (3.4.5) and the discrete force balance (9.3.7a) in its Galerkin approximation tested by $y_k-\tilde{y}_k$. The rest is then by limiting $\tilde{y}_k$ to $\tilde{y}$ considered arbitrary, by putting $\tilde{y} = y+\varepsilon\hat{y}$ with $\hat{y}$ considered arbitrary and putting ε to 0.

Remark 9.3.4 (Nonlocal simple materials). For Proposition 9.3.1, instead of the nonsimple material, we may assume the nonlocal simple hyperelastic material in the sense that the stress is enhanced by the contribution $\mathfrak{H}\nabla y$ from (9.2.17). When assuming the symmetric kernel $\mathbb{K}$ in (9.2.17) to be singular in the sense of (2.5.28), we have a similar compactness as needed in Proposition 9.3.1, based only on $y\in L^\infty(I;H^{1+\gamma}(\Omega;\mathbb{R}^d))$ instead of $L^\infty(I;H^2(\Omega;\mathbb{R}^d))$ like in Remarks 9.2.6 and (7.5.29).

Remark 9.3.5 (From large to small strains in Kelvin-Voigt material). It is an important question if a model of a viscoelastic material at small strain (see Section 6.4) can be derived from the finite-strain variant. This was achieved in [205] where the authors consider a second-grade (visco)elastic material with $\varepsilon|G|^p/K \le \mathscr{H}_\varepsilon(G) \le K\varepsilon|G|^p$ and $|\mathscr{H}'(G)| \le K|G|^{p-1}$ for some $K>1$, $\varepsilon>0$, and $p>d$, and all $G\in\mathbb{R}^{d\times d\times d}$. Here we consider the stress strain relation (9.3.1) with $\mathscr{H}_\varepsilon$ in the place of $\mathscr{H}$. Under suitable assumptions which are too involved to be stated here, it is shown in [205] that there exists a solution to a nonlinear viscoelastic model. Searching for the deformation in the form $y_\delta(t,x) := x+\delta u(t,x)$ where $\delta>0$ and taking into account suitably rescaled stored and dissipated energy in the parameter regime $\varepsilon\to 0$ if $\delta\to 0$ it is proved that u solves a small strain Kelvin-Voigt viscoelastic model. Moreover, solutions to the large strain model converge for $\delta\to 0$ in a suitable sense to the unique solution of the small strain problem. The main tools are the theory of minimizing movements and gradient flows in metric spaces [9].

Exercise 9.3.6. Prove existence of a weak solution to (9.3.1) by, say, the Galerkin method and the limit passage by Minty's trick if S would be assumed (physically not realistically) quasimonotone.

9.4 Maxwell-type viscoelastic materials, creep, and plasticity

Some models with internal parameters can straightforwardly be formulated and analyzed in the Lagrangian description. Let us demonstrate it first on Maxwellian-type rheological models.

An analog of the additive decomposition $e(u) = e_{\rm EL} + e_{\rm EL}$ (mentioned in Section 7.4), which would lead to the Maxwell rheology at small strains as considered in Section 6.6 if $\varphi(e,e_{\rm EL}) = \frac{1}{2}\mathbb{C}e_{\rm EL}{:}e_{\rm EL} = \frac{1}{2}\mathbb{C}(e-e_{\rm EL}){:}(e-e_{\rm EL})$ and $\zeta(\dot{e}_{\rm EL}) = \frac{1}{2}\mathbb{D}\dot{e}_{\rm EL}{:}\dot{e}_{\rm EL}$ is chosen as in (6.6.6), is here the Kröner-Lee-Liu *multiplicative decomposition* $F = F_{\rm EL}F_{\rm VI}$. The specific stored energy is now $\varphi = \hat{\varphi}(C, C_{\rm VI})$ with $C = F^\top F$ and $C_{\rm VI} = F_{\rm VI}^\top F_{\rm VI}$.

The material from Figure 7.2 with $\mathbb{D}_1 = 0$ (cf. also Figure 7.3) would use the stored energy $\varphi_{\rm EL}(F_{\rm EL}) + \varphi_{\rm HD}(\Pi)$ having the elastic and the hardening part, cf. also (7.4.2). Together with the Kröner-Lee-Liu *multiplicative decomposition* $\nabla y = F = F_{\rm EL}\Pi$, cf. (3.6.10), it results to

$$\varphi(F,\Pi) = \varphi_{\rm EL}(F\Pi^{-1}) + \varphi_{\rm HD}(\Pi); \tag{9.4.1}$$

here we denoted the viscous (or inelastic) strain $F_{\rm VI} = \Pi$. For analytical reasons specified below, we need the concept of nonsimple materials both as far as the elastic and the hardening energies concern, considering an enhanced stored energy

$$\varphi_{\rm E}(F,\Pi,G,\nabla\Pi) = \varphi(F,\Pi) + \frac{1}{2}\mathbb{H}G\vdots G + \frac{\kappa}{p}|\nabla\Pi|^p \quad \text{with} \quad G = \nabla F. \tag{9.4.2}$$

Even, if transport processes are involved (or if local selfpenetration would be avoided), we need the nonlocal generalization as in Sect. 2.5.2, which will concern the heat transfer in Sect. 9.4.2. Here, the ansatz (9.4.1)–(9.4.2) leads us to consider the stored energy

$$\begin{aligned}\mathcal{E}(y,\Pi) &= \int_\Omega \varphi_{\rm E}(\nabla y, \Pi, \nabla^2 y, \nabla\Pi)\,{\rm d}x \\ &= \int_\Omega \varphi_{\rm EL}(\nabla y\Pi^{-1}) + \varphi_{\rm HD}(\Pi) + \frac{1}{2}\mathbb{H}\nabla^2 y\vdots\nabla^2 y + \frac{\kappa}{p}|\nabla\Pi|^p\,{\rm d}x.\end{aligned} \tag{9.4.3}$$

The general plastic flow rule uses the dissipation potential $\zeta = \zeta(\Pi,\dot{\Pi})$ which is convex but possibly nonsmooth in terms of the plastic strain rate $\dot{\Pi}$, and reads as

$$\partial_{\dot{\Pi}}\zeta(\Pi,\dot{\Pi}) + S_{\rm IN} \ni 0 \quad \text{with} \quad S_{\rm IN} = \partial_\Pi \mathcal{E}(y,\Pi). \tag{9.4.4}$$

Frame-indifference of the above stored energy $\varphi(F,\Pi)$ from (9.4.1) is again required in terms of F, which that $\varphi_{\rm EL}$ is frame-indifferent in the usual sense (2.3.4). Here we also apply the requirement of an indifference of the above stored energy $\varphi(F,\Pi)$ from (9.4.1) and of dissipation potential $\zeta = \zeta(\Pi,\dot{\Pi})$ under previous plastic strain (also called *plastic indifference*[14]) in the sense that

$$\forall R \in {\rm SO}(d): \quad \varphi(RF,\Pi) = \varphi(RF,\Pi), \tag{9.4.5a}$$

$$\forall P \in {\rm GL}^+(d): \quad \varphi(FP,\Pi P) = \varphi(F,\Pi), \tag{9.4.5b}$$

$$\zeta(\Pi P,\dot{\Pi}P) = \zeta(\Pi,\dot{\Pi}). \tag{9.4.5c}$$

[14] The plastic indifference was articulated by J. Mandel [330] and later discussed e.g. in [352].

Let us note that (9.4.5a) needs the plastic indifference of the hardening,[15] i.e. $\varphi_{\rm HD}(P\cdot) = \varphi_{\rm HD}(\cdot)$ for all $P\in{\rm GL}^+(d)$, which is ensured if $\varphi_{\rm HD}(\Pi) = \mathfrak{f}_{\rm hd}(\Pi^{-1})$ for some $\mathfrak{f}_{\rm hd} : {\rm GL}^+(d) \to \mathbb{R}$. The condition (9.4.5c) implies (and at the same time is ensured by) the existence of a potential $\hat\zeta : \mathbb{R}^{d\times d} \to \mathbb{R}$ such that $\zeta(\Pi,\dot\Pi) = \hat\zeta(\dot\Pi\Pi^{-1})$; indeed, note that taking $P = \Pi^{-1}$, we obtain $\zeta(\Pi,\dot\Pi) = \zeta(\mathbb{I},\dot\Pi\Pi^{-1})$, which reveals that $\hat\zeta(\cdot) = \zeta(\mathbb{I},\cdot)$ and that always $\zeta(\Pi P,\dot\Pi P) = \hat\zeta(\dot\Pi P(\Pi P)^{-1}) = \hat\zeta(\dot\Pi\Pi^{-1}) = \zeta(\Pi,\dot\Pi)$ even for any $P\in{\rm GL}(d)$, cf. also [226]. In terms of the potential $\hat\zeta$, the flow rule (9.4.4) reads as

$$\partial\hat\zeta(\dot\Pi\Pi^{-1})\Pi^{-\top} + S_{\rm IN} \ni 0. \tag{9.4.6}$$

Let us note that the flow rule (9.4.6) multiplied by $\Pi^\top$ can equivalently be written also as

$$\partial\hat\zeta(\dot\Pi\Pi^{-1}) + S_{\rm IN}\Pi^\top \ni 0.$$

Testing it by $\dot\Pi\Pi^{-1}$ is then equivalent to testing the flow rule (9.4.6) by $\dot\Pi$. This then gives, using also the matrix algebra (A.2.14), that

$$\partial\hat\zeta(\dot\Pi\Pi^{-1}):(\dot\Pi\Pi^{-1}) = -S_{\rm IN}\Pi^\top:(\dot\Pi\Pi^{-1}) = -(S_{\rm IN}\Pi^\top\Pi^{-\top}):\dot\Pi = -S_{\rm IN}:\dot\Pi. \tag{9.4.7}$$

Yet, (9.4.6) and its test by $\dot\Pi$ is better fitted with usage of the Faedo-Galerkin method because the product $\dot\Pi\Pi^{-1}$ does not take values in general in the corresponding finite-dimensional subspaces used for the Galerkin method and is thus not a legitimate test function.

9.4.1 Isothermal creep model

The "materially" linear (but geometrically nonlinear) *Maxwell rheology* corresponds to $\hat\zeta$ quadratic and convex. Then, in particular, $\partial\hat\zeta = \hat\zeta'$ is single-valued linear operator, which simplifies a lot of notation and arguments. To model *creep*, it is natural to avoid hardening in the shear. This here means that we admit (and even will rely on) hardening only for $\det\Pi$ rather than for Π itself. Hardening in $\det\Pi$ in collaboration with the gradient of Π will serve for controlling invertibility of Π through the Healey-Krömer theorem 2.5.3 on p. 43.

The inelastic driving stress for the evolution of Π is

$$\begin{aligned} S_{\rm IN} &= \partial_\Pi\mathcal{E}(y,\Pi) \\ &= (\nabla y)^\top\varphi_{\rm EL}'((\nabla y)\Pi^{-1}):\Big(\frac{{\rm Cof}'\Pi^\top}{\det\Pi} - \frac{{\rm Cof}\,\Pi^\top\otimes{\rm Cof}\,\Pi}{(\det\Pi)^2}\Big) \\ &\qquad + \varphi_{\rm HD}'(\Pi) - {\rm div}(\kappa|\nabla\Pi|^{p-2}\nabla\Pi). \end{aligned} \tag{9.4.8}$$

[15] Note that $\varphi(FP,\Pi P) = \varphi_{\rm EL}(FP(\Pi P)^{-1}) + \mathfrak{f}_{\rm hd}(P(\Pi P)^{-1}) = \varphi_{\rm EL}(F\Pi^{-1}) + \mathfrak{f}_{\rm hd}(\Pi^{-1}) = \varphi(F,\Pi)$ even for any $P\in{\rm GL}(d)$.

Let us note that (9.4.10b) used the formulas $\Pi^{-1} = \mathrm{Cof}\,\Pi^\top/\det\Pi$ and $\det'\Pi = \mathrm{Cof}\,\Pi$, cf. Sect. A.2, so that

$$(\Pi^{-1})' = \frac{\mathrm{Cof}'\Pi^\top}{\det\Pi} - \frac{\mathrm{Cof}\,\Pi^\top \otimes \mathrm{Cof}\,\Pi}{(\det\Pi)^2}. \tag{9.4.9}$$

The overall considered system is then

$$\begin{aligned}
&\varrho\ddot{y} - \mathrm{div}\, S_{\rm EL} = f(y) \quad \text{with the Piola-Kirchhoff stress } S_{\rm EL}\\
&\qquad \text{so that } \mathrm{div}\, S_{\rm EL} = \partial_y \mathcal{E}(y,\Pi),\\
&\qquad \text{i.e. } S_{\rm EL} = \varphi_{\rm EL}'(\nabla y\Pi^{-1})\Pi^{-\top} - \mathrm{Div}(\mathbb{H}\nabla^2 y) && \text{in } Q, && (9.4.10a)\\
&(\hat{\zeta}'\dot{\Pi}\Pi^{-1})\Pi^{-\top} + S_{\rm IN} = 0 \quad \text{with } S_{\rm IN} \text{ in (9.4.8)} && \text{in } Q, && (9.4.10b)\\
&S_{\rm EL}\vec{n} - \mathrm{div}_{\rm S}((\mathbb{H}\nabla^2 y)\cdot\vec{n}) = g \quad \text{and} \quad \mathbb{H}\nabla^2 y:(\vec{n}\otimes\vec{n}) = 0 && \text{on } \Sigma, && (9.4.10c)\\
&(\nabla\Pi)\vec{n} = 0 \quad \text{on } \Sigma_{\rm N} \quad \text{and} \quad \Pi = \mathbb{I} && \text{on } \Sigma_{\rm D}, && (9.4.10d)\\
&y|_{t=0} = y_0, \quad \dot{y}|_{t=0} = v_0, \quad \Pi|_{t=0} = \Pi_0 && \text{on } \Omega. && (9.4.10e)
\end{aligned}$$

Let us note that $\mathrm{Cof}'\Pi$ is the 4th-order tensor so that $S_{\rm IN}$ in (9.4.10b) is indeed the 2th-order tensor.

Definition 9.4.1 (Weak formulation). The couple $(y,\Pi) \in C_{\rm w}(I;H^2(\Omega;\mathbb{R}^d)) \times C_{\rm w}(I;W^{1,p}(\Omega;\mathbb{R}^{d\times d}))$ will be called a weak solution to the initial-boundary-value problem (9.4.10) if $\dot{\Pi}\Pi^{-1} \in L^2(Q;\mathbb{R}^{d\times d})$ and if

$$\begin{aligned}
&\int_Q \varphi_{\rm EL}'((\nabla y)\Pi^{-1}):((\nabla v)\Pi^{-1}) + \mathbb{H}\nabla^2 y\,\vdots\,\nabla^2 v + \varrho y\cdot\ddot{v}\,\mathrm{d}x\mathrm{d}t\\
&\qquad = \int_\Omega \varrho v_0\cdot v(0) - \varrho y_0\cdot\dot{v}(0)\,\mathrm{d}x + \int_Q f(y)\cdot v\,\mathrm{d}x\mathrm{d}t + \int_\Sigma g\cdot v\,\mathrm{d}S\,\mathrm{d}t
\end{aligned} \tag{9.4.11a}$$

is satisfied for all $v \in C^2(Q;\mathbb{R}^d)$ with $v(T) = \dot{v}(T) = 0$ and if[16]

$$\begin{aligned}
&\int_Q (\hat{\zeta}'\dot{\Pi}\Pi^{-1}):(P\Pi^{-1}) + (\nabla y)^\top\varphi_{\rm EL}'((\nabla y)\Pi^{-1}):((\Pi^{-1})':P)\\
&\qquad\qquad + \varphi_{\rm HD}'(\Pi):P + \kappa|\nabla\Pi|^{p-2}\nabla\Pi\,\vdots\,\nabla P\,\mathrm{d}x\mathrm{d}t = 0
\end{aligned} \tag{9.4.11b}$$

is satisfied for any $P \in L^\infty(I;W^{1,p}(\Omega;\mathbb{R}^{d\times d}))$ with $P|_{\Sigma_{\rm D}} = 0$, and also the initial condition $\Pi(0) = \Pi_0$ as well as the Dirichlet boundary condition $\Pi = \mathbb{I}$ on $\Sigma_{\rm D}$ hold.

Let us summarize our assumptions on the data of the initial-boundary-value problem (9.4.10):

$$\varphi_{\rm EL}, \varphi_{\rm HD} : \mathbb{R}^{d\times d} \to \mathbb{R} \text{ continuously differentiable (when finite)}, \tag{9.4.12a}$$

$$\varphi_{\rm EL}(F) \geq 0 \quad \text{and} \quad |\varphi_{\rm EL}'(F)| \leq K(1+|F|^{2^*-1}), \tag{9.4.12b}$$

[16] Note that (9.4.11b) arises by multiplying (9.4.6) by P, and integrating it over Q and using the Green formula on Ω together with the matrix algebra (A.2.14).

$$\varphi_{\rm HD}(\Pi) \ge \begin{cases} \epsilon/(\det\Pi)^q & \text{if } \det\Pi > 0, \\ +\infty & \text{if } \det\Pi \le 0, \end{cases} \qquad q \ge \frac{pd}{p{-}d}, \qquad p > d, \tag{9.4.12c}$$

$$\hat\zeta : \mathbb{R}^{d\times d} \to \mathbb{R} \text{ quadratic, coercive,} \tag{9.4.12d}$$

$$\mathbb{H} \in \mathrm{SLin}(\mathbb{R}^{d\times d}_{\rm sym}) \text{ positive definite,} \tag{9.4.12e}$$

$$y_0 \in H^2(\Omega;\mathbb{R}^d), \quad \Pi_0 \in W^{1,p}(\Omega;\mathbb{R}^{d\times d}), \quad \varphi_{\rm HD}(\Pi_0) \in L^1(\Omega), \tag{9.4.12f}$$

while f, g, y_0, and v_0 are qualified as in (9.2.9c-e). Let us note that (9.4.12a) is compatible with the concept of creep which does not exhibit any hardening for shear flow. Let us note also that the last condition in (9.4.12f) together with (9.4.12a) and $p > d$ ensures that $\varphi_{\rm HD}(\Pi_0)$ is bounded, hence certainly integrable.

We again use the Galerkin mehod which allows optimal data qualification in comparison with time discretization, cf. Remark 9.2.7. We use collections of (nested) finite-dimensional spaces in $W^{2,\infty}(\Omega;\mathbb{R}^d)$ and in $W^{1,\infty}(\Omega;\mathbb{R}^{d\times d})$ for the discretization of y- and Π-variables, assuming density of their union in $H^2(\Omega;\mathbb{R}^d)$ and in $W^{1,p}(\Omega;\mathbb{R}^{d\times d})$; respectively. Without any loss of generality if the qualification (9.4.12f) is assumed, we can also assume that all the (nested) finite-dimensional spaces used for the Galerkin approximation of (9.4.10) contain both y_0 and v_0, and also Π_0. Let us denote the approximate solution resulted by this scheme by (y_k, Π_k).

Proposition 9.4.2 (A-priori estimates). *Let (9.4.12) hold and* $\mathrm{Meas}_{d-1}(\Sigma_{\rm D}) > 0$. *Then the Galerkin approximate solution of (9.4.10) exists and satisfies*

$$\|y_k\|_{L^\infty(I;H^2(\Omega;\mathbb{R}^d))\cap W^{1,\infty}(I;L^2(\Omega;\mathbb{R}^d))} \le K, \tag{9.4.13a}$$

$$\|\Pi_k\|_{L^\infty(I;W^{1,p}(\Omega;\mathbb{R}^{d\times d}))\cap H^1(I;L^2(\Omega;\mathbb{R}^{d\times d}))} \le K, \tag{9.4.13b}$$

$$\big\|1/\det\Pi_k\big\|_{L^\infty(Q)} \le K \quad \text{and} \quad \big\|\Pi_k^{-1}\big\|_{L^\infty(Q;\mathbb{R}^{d\times d})} \le K. \tag{9.4.13c}$$

Sketch of the proof. We test the momentum equation (in its Galerkin approximation) by $\dot y_k$ and the plastic flow rule (in its Galerkin approximation) by $\dot\Pi_k$. Using (9.4.7), we obtain the energy balance for the Galerkin approximation. From this, we can see the L^∞-bounds in when counting also with the assumed Dirichlet condition (9.4.10d) on Π, and also the bound $\|\dot\Pi_k\Pi_k^{-1}\|_{L^2(Q;\mathbb{R}^{d\times d}))} \le K$.

Then, exploiting (9.4.12c), we can use the Healey-Krömer theorem 2.5.3 on p. 43 now for the plastic strain instead of the deformation gradient. This gives (9.4.13c).

Then, using $L^\infty(I;W^{1,p}(\Omega;\mathbb{R}^{d\times d})) \subset L^\infty(Q;\mathbb{R}^{d\times d})$, we get also

$$\begin{aligned} \big\|\dot\Pi_k\big\|_{L^2(Q;\mathbb{R}^{d\times d})} &= \big\|(\dot\Pi_k\Pi_k^{-1})\Pi_k\big\|_{L^2(Q;\mathbb{R}^{d\times d})} \\ &\le \big\|\dot\Pi_k\Pi_k^{-1}\big\|_{L^2(Q;\mathbb{R}^{d\times d})}\big\|\Pi_k\big\|_{L^\infty(Q;\mathbb{R}^{d\times d}))}, \end{aligned} \tag{9.4.14}$$

which completes the proof of (9.4.13b). □

The convergence of the Galerkin approximation is easy for $\hat\zeta$ quadratic (i.e. for a linear creep) considered in this section.

Proposition 9.4.3 (Convergence of Galerkin approximation). *Let again (9.4.12) hold with* $\mathrm{Meas}_{d-1}(\Sigma_{\rm D})>0$. *Then, for a subsequence, we have*

$$y_k\to y \qquad \textit{weakly* in } L^\infty(I;H^2(\Omega;\mathbb{R}^d))\cap W^{1,\infty}(I;L^2(\Omega;\mathbb{R}^d)), \tag{9.4.15a}$$

$$\Pi_k\to\Pi \qquad \textit{weakly* in } L^\infty(I;W^{1,p}(\Omega;\mathbb{R}^{d\times d}))\cap H^1(I;L^2(\Omega;\mathbb{R}^{d\times d})) \tag{9.4.15b}$$

$$\textit{and strongly in } L^p(I;W^{1,p}(\Omega;\mathbb{R}^{d\times d})). \tag{9.4.15c}$$

Moreover, any (y,Π) obtained as such a limit is a weak solution to (9.4.10) due to Definition 9.4.1.

Sketch of the proof. Based on the a-priori estimates (9.4.13a,b), we select a subsequence of $\{(y_k,\Pi_k)\}_{k\in\mathbb{N}}$ weakly* converging to some (y,Π) in the topologies indicated in (9.4.13a,b). In particular, we have (9.4.15a,b). Taking $p>d$ into account, due to our estimates (9.4.13b),[17] we have also

$$\Pi_k\to\Pi \ \text{ and } \ \Pi_k^{-1}\to\Pi^{-1} \quad \text{strongly in } L^\infty(Q;\mathbb{R}^{d\times d}). \tag{9.4.16}$$

To prove (9.4.15c), we take some $\tilde\Pi_k$ valued in the finite-dimensional subspace used for the Galerkin approximation so that $\Pi_k-\tilde\Pi_k$ is a legitimate test function for this approximation. Without loss of generality, we may consider that $\Pi_k-\tilde\Pi_k\to0$ strongly in $L^\infty(I;W^{1,p}(\Omega;\mathbb{R}^{d\times d}))\cap H^1(I;L^2(\Omega;\mathbb{R}^{d\times d}))$ by the assumed density of of the collection of finite-dimensional subspaces used for the Galerkin approximation, so that also strongly in $L^\infty(Q;\mathbb{R}^{d\times d})$. Then we can estimate

$$\begin{aligned}
&\limsup_{k\to\infty} c_{d,p}\|\nabla\Pi_k-\nabla\Pi\|^p_{L^p(Q;\mathbb{R}^{d\times d\times d})}\\
&\le \limsup_{k\to\infty} c'_{d,p}\|\nabla\Pi_k-\nabla\tilde\Pi_k\|^p_{L^p(Q;\mathbb{R}^{d\times d\times d})}+\limsup_{k\to\infty} c'_{d,p}\|\nabla\tilde\Pi_k-\nabla\Pi\|^p_{L^p(Q;\mathbb{R}^{d\times d\times d})}\\
&\le \lim_{k\to\infty}\int_Q\big(|\nabla\Pi_k|^{p-2}\nabla\Pi_k-|\nabla\tilde\Pi_k|^{p-2}\nabla\tilde\Pi_k\big)\vdots\nabla(\Pi_k-\tilde\Pi_k)\,\mathrm{d}x\mathrm{d}t\\
&=\lim_{k\to\infty}\frac{1}{\kappa}\int_Q \nabla y_k^\top\varphi'_{\rm EL}(F_{{\rm EL},k}){:}(\Pi_k^{-1})'{:}(\Pi_k-\tilde\Pi_k)+(\hat\zeta'\dot\Pi_k\Pi_k^{-1}){:}((\Pi_k-\tilde\Pi_h)\Pi_k^{-1})\,\mathrm{d}x\mathrm{d}t\\
&\qquad\qquad -\int_Q|\nabla\tilde\Pi_k|^{p-2}\nabla\tilde\Pi_k)\vdots\nabla(\Pi_k-\tilde\Pi_k)\,\mathrm{d}x\mathrm{d}t=0,
\end{aligned} \tag{9.4.17}$$

where we abbreviated $F_{{\rm EL},k}=\nabla y_k\Pi_k^{-1}$ and $(\Pi_k^{-1})'=\mathrm{Cof}'\Pi_k^\top/\det\Pi_k-(\mathrm{Cof}\,\Pi_k^\top\otimes\mathrm{Cof}\,\Pi_k)/(\det\Pi_k)^2$. Let us note also that $\nabla y_k\in L^\infty(I;L^{2^*}(\Omega;\mathbb{R}^{d\times d}))$ and $\varphi'_{\rm EL}(F_{{\rm EL},k})\in L^\infty(I;L^{2^{*\prime}}(\Omega;\mathbb{R}^{d\times d}))$ due to (9.4.12b), so that $\nabla y_k^\top\varphi'_{\rm EL}(F_{{\rm EL},k})\in L^\infty(I;L^1(\Omega;\mathbb{R}^{d\times d}))$ and the integrals in (9.4.17) have a good sense.

Having (9.4.15)–(9.4.16) at disposal, the limit passage towards the integral identities (9.4.11) is then easy. □

[17] Here a modification of the Aubin-Lions theorem is used, cf. the Arzelà-Ascoli-type arguments in [456, Lemma 7.10].

Remark 9.4.4 (Jeffreys' material). The system of equations (9.4.10a,b) can be combined with the equation (9.3.7a) for an internal variable Π which, for the easier analysis, is again assumed isochoric. The Kelvin-Voigt model is thus combined with creep, obtaining thus the Jeffreys rheology. This can be done by two ways, corresponding to two variants in Figure 6.6 on p. 235. One option would yield the system

$$\varrho\ddot{y} - \mathrm{div}\,(S_{\mathrm{VI}}(\nabla y,\nabla\dot{y}) + S_{\mathrm{EL}}(\nabla y,\Pi) - \mathrm{div}\,(\mathbb{H}\nabla^2 y)) = f(y)$$
$$\text{with } S_{\mathrm{VI}}(F,\dot{F}) = F\mathbb{D}_1\mathrm{sym}(F^\top\dot{F})$$
$$\text{and } S_{\mathrm{EL}}(F,\Pi) = \Pi^{-\top}\mathbb{C}(\Pi^{-\top}F^\top F\Pi^{-1}-\mathbb{I})F\Pi^{-1}, \tag{9.4.18a}$$

$$\mathbb{D}_2(\dot{\Pi}\Pi^{-1})\Pi^{-\top} + S_{\mathrm{IN}}(\nabla y,\Pi) = 0$$
$$\text{with } S_{\mathrm{IN}}(F,\Pi) = F^\top\mathbb{C}(\Pi^{-\top}F^\top F\Pi^{-1}-\mathbb{I})F{:}(\Pi^{-1})'$$
$$+\varphi_{\mathrm{HD}}'(\Pi) - \mathrm{div}(\kappa|\nabla\Pi|^{p-2}\nabla\Pi). \tag{9.4.18b}$$

This corresponds to Figure 6.6-left with the stored energy and the dissipation potential energy (again with $F = \nabla y$ and $G = \nabla F = \nabla^2 y$):

$$\mathcal{E}(y,\Pi) = \int_\Omega \frac{1}{8}\mathbb{C}(\Pi^{-\top}F^\top F\Pi^{-1}-\mathbb{I}){:}(\Pi^{-\top}F^\top F\Pi^{-1}-\mathbb{I}) + \varphi_{\mathrm{HD}}(\Pi)$$
$$+\frac{1}{2}\mathbb{H}G\,\vdots\,G + \frac{\kappa}{p}|\nabla\Pi|^p\,\mathrm{d}x, \quad\text{and} \tag{9.4.19a}$$

$$\mathcal{R}(y,\Pi;\dot{y},\dot{\Pi}) = \int_\Omega \frac{1}{8}\mathbb{D}_1\mathrm{sym}(F^\top\dot{F}){:}\mathrm{sym}(F^\top\dot{F}) + \frac{1}{2}\mathbb{D}_2(\dot{\Pi}\Pi^{-1}){:}(\dot{\Pi}\Pi^{-1})\,\mathrm{d}x, \tag{9.4.19b}$$

cf. (4.5.20) and realize that $\dot{F}^\top F + F^\top\dot{F} = 2\mathrm{sym}(F^\top\dot{F})$.

Remark 9.4.5 (Limiting $\mathbb{C}\to\infty$). For large elastic moduli $\mathbb{C}$, Jeffreys' rheology is expected to approach Stokes' one. This can easily be proved at small strains where π replaces $e(u)$. Passing $\mathbb{C}$ "to infinity" in (9.4.19a) would result to $\Pi^{-\top}F^\top F\Pi^{-1} = \mathbb{I}$, so that $F^\top F = \Pi^\top\Pi$. This means that $\Pi = RF$ for some $R\in\mathrm{SO}(d)$, see Exercise 1.1.14, and also it implies that

$$\mathrm{sym}(F^\top\dot{F}) = \frac{1}{2}(F^\top F)^{\boldsymbol{\cdot}} = \frac{1}{2}(\Pi^\top\Pi)^{\boldsymbol{\cdot}} = \mathrm{sym}(\Pi^\top\dot{\Pi}).$$

Yet, it does not imply that $F = \Pi$. Here, Π indeed cannot be considered as in the position of F under such a limit, because the dissipation mechanisms, counting with $\mathrm{sym}(F^\top\dot{F})$ as in (9.3.4) and with $\dot{\Pi}\Pi^{-1}$, have a different form. This is related with that frame indifference act differently on the strain and on internal variables.

Remark 9.4.6 (Nonsimple-material concept). In fact, in (9.4.3) we used a rather misconceptual simplification that, although φ_{EL} in (9.4.1) depends on F_{EL}, we applied the gradient on F rather than on F_{EL}, i.e. we used $\frac{1}{2}\mathbb{H}\nabla F\,\vdots\,\nabla F$ in (9.4.2)–(9.4.3) rather than $\frac{1}{2}\mathbb{H}\nabla F_{\mathrm{EL}}\,\vdots\,\nabla F_{\mathrm{EL}}$ with $F_{\mathrm{EL}} = F\Pi^{-1}$. This alternative (conceptually more consistent) option would yield S_{EL} in (9.4.10a) using $\Pi^{-\top}\mathrm{Div}(\mathbb{H}\nabla(F\Pi^{-1}))$ instead of $\mathrm{Div}(\mathbb{H}\nabla F)$, i.e.

$$S_{\rm EL} = \varphi'_{\rm EL}(\nabla y\Pi^{-1})\Pi^{-\top} - \Pi^{-\top}{\rm Div}(\mathbb{H}\nabla(\nabla y\Pi^{-1})) \tag{9.4.20a}$$

while $S_{\rm IN}$ in (9.4.10b) expanding by $((\Pi^{-1})'F)^\top{\rm Div}(\mathbb{H}\nabla(F\Pi^{-1}))$ so that

$$S_{\rm IN} = ((\Pi^{-1})'\nabla y)^\top\big(\varphi'_{\rm EL}((\nabla y)\Pi^{-1}) - {\rm div}(\mathbb{H}\nabla(\nabla y\Pi^{-1}))\big) + \varphi'_{\rm HD}(\Pi) - {\rm div}(\kappa|\nabla\Pi|^{p-2}\nabla\Pi) \tag{9.4.20b}$$

with $(\Pi^{-1})'$ from (9.4.9). This would still allow for a-priori estimates[18] of $\nabla F = \nabla F_{\rm EL}\Pi + F_{\rm EL}\nabla\Pi$ but would make the limit passage in the plastic flow rule difficult and actually the analysis for such model seems still open.[19]

Exercise 9.4.7. Write the weak formulation of the problem suggested in Remark 9.4.6 and derive the formulas (9.4.20).

Exercise 9.4.8 (Nonlinear nonpotential creep). Instead of the linear $\hat\zeta'$ referring to (9.4.12d), consider a general continuous monotone mapping $Z:\mathbb{R}^{d\times d}\to\mathbb{R}^{d\times d}$ coercive and bounded in the sense $Z'(P){:}P\ge\epsilon|P|^2$ and $|Z(P)|\le(1+|P|)/\epsilon$ with some $\epsilon>0$. Modify the proof of Proposition 9.4.3 by using Minty's trick.

9.4.2 Visco-plasticity

Plasticity is typically an activated process, which is realized by the dissipation potentials $\hat\zeta$ which are convex but nonsmooth at 0. Also, in contrast to creep (where $\varphi_{\rm HD}$ was assumed coercive only weakly to control $\det\Pi$), hardening may be a relevant phenomenon which, when indeed considered, allows to consider purely Neumann boundary condition for Π.

As $\partial\hat\zeta$ is now set-valued, (9.4.10b) with $\hat\zeta'\dot\Pi\Pi^{-1}$ replaced by $\partial\hat\zeta(\dot\Pi\Pi^{-1})$ becomes an inclusion rather than equation, and the variational equation (9.4.11b) is to be formulated rather as a variational inequality. To this goal, it is convenient to use the Green formula together with by-part-integration formula like (9.5.17):

$$\int_Q({\rm div}(\kappa|\nabla\Pi|^{p-2}\nabla\Pi)\Pi^\top){:}(\dot\Pi\Pi^{-1})\,{\rm d}x{\rm d}t = \int_Q{\rm div}(\kappa|\nabla\Pi|^{p-2}\nabla\Pi){:}\dot\Pi\,{\rm d}x{\rm d}t = \int_\Omega\frac{\kappa}{p}|\nabla\Pi_0|^p - \frac{\kappa}{p}|\nabla\Pi(T)|^p\,{\rm d}x, \tag{9.4.21}$$

cf. also (7.5.10). Thus the set-valued variant of the flow rule (9.4.10b) admits a weak formulation as:[20]

---

[18] Denoting $G=\nabla(F\Pi^{-1}) = \nabla F\Pi^{-1} + F(\Pi^{-1})'\nabla\Pi$, from the energy bound we have $G\in L^\infty(I;L^2(\Omega;\mathbb{R}^{d\times d\times d}))$ and therefore also $\nabla F = (G - F(\Pi^{-1})'\nabla\Pi)\Pi\in L^\infty(I;L^2(\Omega;\mathbb{R}^{d\times d\times d}))$.

[19] One problem is the strong convergence simultaneously in ∇F and of $\nabla\Pi$ needed for the limit passage in the weak formulation of the force equilibrium and the plastic flow rule due to the newly arising terms.

[20] To obtain (9.4.22), we use the definition of the convex subdifferential (A.3.14) on p. 497 applied to the flow rule (9.4.10b). This gives formally the inequality $\hat\zeta(P)+S_{\rm IN}\Pi^\top{:}(P-\dot\Pi\Pi^{-1})\ge\hat\zeta(\dot\Pi\Pi^{-1})$

$$\int_Q \hat\zeta(P\Pi^{-1}) + \big(\varphi'_{\rm EL}((\nabla y)\Pi^{-1})\big)\Pi^\top :((\nabla y)(\Pi^{-1})') + \varphi'_{\rm HD}(\Pi)\big):(P-\dot\Pi)$$
$$+\kappa|\nabla\Pi|^{p-2}\nabla\Pi\vdots\nabla P\,{\rm d}x{\rm d}t + \int_\Omega \frac{\kappa}{p}|\nabla\Pi_0|^p\,{\rm d}x$$
$$\geq \int_Q \hat\zeta(\dot\Pi\Pi^{-1})\,{\rm d}x{\rm d}t + \int_\Omega \frac{\kappa}{p}|\nabla\Pi(T)|^p\,{\rm d}x \qquad (9.4.22)$$

which is to hold for any $P\in L^2(I;W^{1,p}(\Omega;\mathbb{R}^{d\times d}))$. Let us note that we do not use the Green formula on Ω in a simple way to obtain $\kappa|\nabla\Pi|^{p-2}\nabla\Pi:\nabla(P-\dot\Pi)$ because $\nabla\dot\Pi$ is not well defined. Thus we come to:

Definition 9.4.9 (Weak formulation). The couple $(y,\Pi)\in C_{\rm w}(I;H^2(\Omega;\mathbb{R}^d))\times C_{\rm w}(I;W^{1,p}(\Omega;\mathbb{R}^{d\times d}))$ will be called a weak solution to the initial-boundary-value problem (9.4.10) if $\dot\Pi\Pi^{-1}\in L^2(Q;\mathbb{R}^{d\times d})$ and if (9.4.11a) together with (9.4.22) hold for all $v\in C^2(Q;\mathbb{R}^d)$ with $v(T)=\dot v(T)=0$ and $P\in L^2(I;W^{2,\infty}(\Omega;\mathbb{R}^{d\times d}))$, respectively, as well as the initial condition $\Pi(0)=\Pi_0$ as well as the Dirichlet boundary condition $\Pi=\mathbb{I}$ on $\Sigma_{\rm D}$ hold (with $\Sigma_{\rm D}$ possibly empty).

We will again approximate the problem by the Galerkin method combined here with a regularization (smoothing) $\hat\zeta_\varepsilon$ of the convex nonsmooth potential $\hat\zeta$. We assume that this approximation satisfies

$$\forall\varepsilon>0:\ \ \hat\zeta_\varepsilon:\mathbb{R}^{d\times d}\to\mathbb{R}\ \text{ convex, continuously differentiable,} \qquad (9.4.23a)$$
$$\exists a_\zeta>0\ \ \forall P\in\mathbb{R}^{d\times d}:\ \ a_\zeta|P|^2\leq\hat\zeta_\varepsilon(P)\leq(1+|P|^2)/a_\zeta, \qquad (9.4.23b)$$
$$\lim_{\varepsilon\to0}\partial_P\hat\zeta_\varepsilon(0)=0\in\partial_P\hat\zeta(0)\ \text{ and }\ P\neq0\Rightarrow\lim_{\varepsilon\to0}\partial_P\hat\zeta_\varepsilon(P)=\partial_P\hat\zeta(P). \qquad (9.4.23c)$$

We then consider (9.4.10) but with $\hat\zeta'\dot\Pi\Pi^{-1}$ replaced by $\hat\zeta'_\varepsilon(\dot\Pi\Pi^{-1})$.

Proposition 9.4.10 (Weak solution to viscoplasticity). *Let $\hat\zeta:\mathbb{R}^{d\times d}\to\mathbb{R}^+$ be convex (possibly nonquadratic) with $a_\zeta|P|^2\leq\hat\zeta_\varepsilon(P)\leq(1+|P|^2)/a_\zeta$ for some $a_\zeta>0$, admitting an approximation $\hat\zeta_\varepsilon$ satisfying (9.4.23), and let again (9.4.12a-c,e-f) hold now with* ${\rm Meas}_{d-1}(\Sigma_{\rm D})>0$ *or* $\lim_{|\Pi|\to\infty}\varphi_{\rm HD}(\Pi)=+\infty$. *Then, the regularized problem, i.e. (9.4.10) with $\hat\zeta'\dot\Pi\Pi^{-1}$ replaced by $\hat\zeta'_\varepsilon(\dot\Pi\Pi^{-1})$, possesses a solution, let us denote it by $(y_\varepsilon,\Pi_\varepsilon)$. For $\varepsilon\to0$, these solutions converge in terms of subsequences in the sense*

$$y_\varepsilon\to y \quad \textit{weakly* in } L^\infty(I;H^2(\Omega;\mathbb{R}^d))\cap W^{1,\infty}(I;L^2(\Omega;\mathbb{R}^d)), \qquad (9.4.24a)$$
$$\Pi_\varepsilon\to\Pi \quad \textit{weakly* in } L^\infty(I;W^{1,p}(\Omega;\mathbb{R}^{d\times d}))\cap H^1(I;L^2(\Omega;\mathbb{R}^{d\times d})) \qquad (9.4.24b)$$
$$\textit{and strongly in } L^p(I;W^{1,p}(\Omega;\mathbb{R}^{d\times d})). \qquad (9.4.24c)$$

---

for all P. As Π is invertible, we can substitute $P\Pi^{-1}$ in place of P without destroying selectivity of this definition, so that we arrive at $\hat\zeta(P\Pi^{-1})+S_{\rm IN}\Pi^\top:(P-\dot\Pi)\Pi^{-1}\geq\hat\zeta(\dot\Pi\Pi^{-1})$, and using also the matrix algebra (A.2.14), also $\hat\zeta(P\Pi^{-1})+S_{\rm IN}:(P-\dot\Pi)\geq\hat\zeta(\dot\Pi\Pi^{-1})$. Then we apply Green's formula for the term ${\rm div}(\kappa|\nabla\Pi|^{p-2}\nabla\Pi):P$ while for ${\rm div}(\kappa|\nabla\Pi|^{p-2}\nabla\Pi):\dot\Pi$ we apply (9.4.21).

and any (y,Π) obtained as such a limit is weak solutions to (9.4.10) due to Definition 9.4.1 exist.

Sketch of the proof. We divide the proof into three steps.

Step 1 – Galerkin approximation: We apply the Galerkin approximation to the regularized problem, and denote its solution by $(y_{\varepsilon k},\Pi_{\varepsilon k})$. Without loss of generality, we now assume that the finite-dimensional subspaces used for the discretization of Π are subspaces of $W^{2,\infty}(\Omega;\mathbb{R}^{d\times d})$ so that the p-Laplacian has a good sense even on the discrete level. More specifically, the Galerkin identity for the regularized flow rule takes the form of (9.4.11b) here as:

$$\int_Q \hat{\zeta}_\varepsilon'(\dot{\Pi}_{\varepsilon k}\Pi_{\varepsilon k}^{-1}):(P\Pi_{\varepsilon k}^{-1})+\Big((\nabla y_{\varepsilon k})^\top\varphi_{\rm EL}'((\nabla y_{\varepsilon k})\Pi_{\varepsilon k}^{-1}):(\Pi_{\varepsilon k}^{-1})'+\varphi_{\rm HD}'(\Pi_{\varepsilon k})\Big):P \\ +\kappa|\nabla\Pi_{\varepsilon k}|^{p-2}\nabla\Pi_{\varepsilon k}\vdots\nabla P\,\mathrm{d}x\mathrm{d}t=0 \qquad (9.4.25)$$

is satisfied for any $P\in L^\infty(I;W^{1,p}(\Omega;\mathbb{R}^{d\times d}))$ valued in the finite-dimensional space used for the Galerkin approximation with $k\in\mathbb{N}$. Existence of a solution of the Galerkin approximation is as above for (9.4.10).

The a-priori estimates (9.4.13) now hold for $(y_{\varepsilon k},\Pi_{\varepsilon k})$ independently of $\varepsilon>0$. We further observe that

$$\int_Q \mathrm{div}(\kappa|\nabla\Pi_{\varepsilon k}|^{p-2}\nabla\Pi_{\varepsilon k}):P\,\mathrm{d}x\mathrm{d}t \\ =\int_Q\Big((\nabla y_\varepsilon)^\top\varphi_{\rm EL}'(\nabla y_\varepsilon\Pi_\varepsilon^{-1}):(\Pi_\varepsilon^{-1})'+\varphi_{\rm HD}'(\Pi_{\varepsilon k})+\hat{\zeta}_\varepsilon'(\dot{\Pi}_{\varepsilon k}\Pi_{\varepsilon k}^{-1})\Big):P\,\mathrm{d}x\mathrm{d}t \qquad (9.4.26)$$

for P valued in the finite-dimensional subspaces used for the Galerkin approximation $V_k^{d\times d}$. From this, we get in addition the estimate

$$\big|\mathrm{div}(\kappa|\nabla\Pi_{\varepsilon k}|^{p-2}\nabla\Pi_{\varepsilon k})\big|_{L^2(Q;\mathbb{R}^{d\times d})/\cong_k}\le K \qquad (9.4.27)$$

where we define the norm on the factor-space or equivalently the seminorm $|\cdot|_{L^2(Q;\mathbb{R}^{d\times d})/\cong_k}$ on $L^2(Q;\mathbb{R}^{d\times d})$ by

$$\big|\Xi\big|_{L^2(Q;\mathbb{R}^{d\times d})/\cong_k}=\sup_{\|v\|_{L^2(Q;\mathbb{R}^{d\times d})}\le1,\ v(t)\in V_k^{d\times d}\text{ for a.a. }t\in I}\int_Q\Xi:v\,\mathrm{d}x\mathrm{d}t.$$

Step 2 –Convergence of Galerkin approximation for $k\to\infty$: Fixing $\varepsilon>0$, we choose a subsequence of $\{(y_{\varepsilon k},\Pi_{\varepsilon k})\}_{k\in\mathbb{N}}$ weakly* converging for $k\to\infty$ in the topologies indicated in (9.4.13) and denote its limit by $(y_\varepsilon,\Pi_\varepsilon)$. Then, for a subsequence, we have

$$y_{\varepsilon k}\to y_\varepsilon \quad \text{weakly* in } L^\infty(I;H^2(\Omega;\mathbb{R}^d))\cap W^{1,\infty}(I;L^2(\Omega;\mathbb{R}^d)), \qquad (9.4.28a)$$

$$\Pi_{\varepsilon k}\to\Pi_\varepsilon \quad \text{weakly* in } L^\infty(I;W^{1,p}(\Omega;\mathbb{R}^{d\times d}))\cap H^1(I;L^2(\Omega;\mathbb{R}^{d\times d})) \qquad (9.4.28b)$$

$$\text{and strongly in } L^p(I;W^{1,p}(\Omega;\mathbb{R}^{d\times d})). \qquad (9.4.28c)$$

The strong convergence (9.4.28c) follows again by the argumentation (9.4.17). Let us note that we have also $\nabla\Pi_{\varepsilon k}(T) \to \nabla\Pi_\varepsilon(T)$ weakly in $L^p(\Omega;\mathbb{R}^{d\times d})$ due to the estimates in (9.4.13b). The a-priori estimates (9.4.13) are inherited for $(y_\varepsilon,\Pi_\varepsilon)$, too.

The limit passage in (9.4.25) can be done by mere weak convergence of $\dot\Pi_{\varepsilon k}\Pi_{\varepsilon k}^{-1}$ when exploiting potentiality of the monotone nonlinearity $\hat\zeta_\varepsilon'$ on the Galerkin approximation. Yet, one should realize that $\dot\Pi_{\varepsilon k}\Pi_{\varepsilon k}^{-1}$ is not valued in the respective finite-dimensional subspace. For this reason, we consider an approximation P_k valued in this space such that $P_k - \dot\Pi_{\varepsilon k}\Pi_{\varepsilon k}^{-1} \to 0$ strongly in $L^2(Q;\mathbb{R}^{d\times d})$. In particular, each P_k belongs also to $L^2(I;W^{1,p}(\Omega))$. Then (9.4.25) can be written in the form

$$\int_Q \hat\zeta_\varepsilon'(P_k){:}P + \Big((\nabla y_{\varepsilon k})^\top\varphi_{\rm EL}'((\nabla y_{\varepsilon k})\Pi_{\varepsilon k}^{-1}){:}(\Pi_{\varepsilon k}^{-1})' + \varphi_{\rm HD}'(\Pi_{\varepsilon k}) \\ - {\rm div}(\kappa|\nabla\Pi_{\varepsilon k}|^{p-2}\nabla\Pi_{\varepsilon k}) + E_k\Big){:}(P\Pi_{\varepsilon k})\,{\rm d}x{\rm d}t = 0 \qquad (9.4.29)$$

with the error $E_k := \hat\zeta_\varepsilon'(P_k) - \hat\zeta_\varepsilon'(\dot\Pi_{\varepsilon k}\Pi_{\varepsilon k}^{-1}) \to 0$ strongly in $L^2(Q;\mathbb{R}^{d\times d})$ since $\hat\zeta_\varepsilon'$ is assumed continuous with at most linear growth. Using convexity of $\hat\zeta_\varepsilon$ and the definition of the subdifferential $\partial\hat\zeta_\varepsilon$, we can further write (9.4.29) in the form of an inequality

$$\int_Q \hat\zeta_\varepsilon(P) + \Big((\nabla y_{\varepsilon k})^\top\varphi_{\rm EL}'((\nabla y_{\varepsilon k})\Pi_{\varepsilon k}^{-1}){:}(\Pi_{\varepsilon k}^{-1})' + \varphi_{\rm HD}'(\Pi_{\varepsilon k}) \\ - {\rm div}(\kappa|\nabla\Pi_{\varepsilon k}|^{p-2}\nabla\Pi_{\varepsilon k}) + E_k\Big){:}((P{-}P_k)\Pi_{\varepsilon k})\,{\rm d}x{\rm d}t \ge \int_Q \hat\zeta_\varepsilon(P_k)\,{\rm d}x{\rm d}t. \quad (9.4.30)$$

From (9.4.27) and from $P_k{-}\dot\Pi_{\varepsilon k}\Pi_{\varepsilon k} \to 0$ in $L^2(Q;\mathbb{R}^{d\times d})$, we can then obtain

$$\bigg|\int_Q {\rm div}(\kappa|\nabla\Pi_{\varepsilon k}|^{p-2}\nabla\Pi_{\varepsilon k}){:}(P_k - \dot\Pi_{\varepsilon k}\Pi_{\varepsilon k}^{-1})\Pi_{\varepsilon k}\,{\rm d}x{\rm d}t\bigg| \le \|\Pi_{\varepsilon k}\|_{L^\infty(Q;\mathbb{R}^{d\times d})}\times \\ \times\big\|{\rm div}(\kappa|\nabla\Pi_{\varepsilon k}|^{p-2}\nabla\Pi_{\varepsilon k})\big\|_{L^2(Q;\mathbb{R}^{d\times d})/\cong_k}\big\|P_k - \dot\Pi_{\varepsilon k}\Pi_{\varepsilon k}^{-1}\big\|_{L^2(Q;\mathbb{R}^{d\times d})} \to 0$$

and thus we can further estimate

$$\limsup_{k\to\infty}\int_Q \kappa|\nabla\Pi_{\varepsilon k}|^{p-2}\nabla\Pi_{\varepsilon k}\,\vdots\,\nabla(P_k\Pi_{\varepsilon k})\,{\rm d}x{\rm d}t \\ = \limsup_{k\to\infty}\int_Q {\rm div}(\kappa|\nabla\Pi_{\varepsilon k}|^{p-2}\nabla\Pi_{\varepsilon k}){:}\dot\Pi_{\varepsilon k}\,{\rm d}x{\rm d}t \\ + \lim_{k\to\infty}\int_Q {\rm div}(\kappa|\nabla\Pi_{\varepsilon k}|^{p-2}\nabla\Pi_{\varepsilon k}){:}(P_k - \dot\Pi_{\varepsilon k}\Pi_{\varepsilon k}^{-1})\Pi_{\varepsilon k}\,{\rm d}x{\rm d}t \\ = \limsup_{k\to\infty}\int_\Omega \frac{\kappa}{p}|\nabla\Pi_0|^p - \frac{\kappa}{p}|\nabla\Pi_{\varepsilon k}(T)|^p\,{\rm d}x \le \int_\Omega \frac{\kappa}{p}|\nabla\Pi_0|^p - \frac{\kappa}{p}|\nabla\Pi_\varepsilon(T)|^p\,{\rm d}x.$$

Using the weak lower semicontinuity of $\int_Q\hat\zeta_\varepsilon(\cdot)\,{\rm d}x{\rm d}t$ and that $P_k \to \dot\Pi_{\varepsilon k}\Pi_{\varepsilon k}^{-1}$ weakly in $L^2(Q;\mathbb{R}^{d\times d})$, we can pass to the limit in (9.4.30), obtaining

$$\int_Q \hat\zeta_\varepsilon(P) + \big(\varphi_{\rm EL}'((\nabla y_\varepsilon)\Pi_\varepsilon^{-1}))\Pi_\varepsilon^\top : ((\nabla y_\varepsilon)(\Pi_\varepsilon^{-1})') + \varphi_{\rm HD}'(\Pi_\varepsilon)\big){:}((P{-}\dot\Pi_\varepsilon)\Pi_\varepsilon) \\ + \kappa|\nabla\Pi_\varepsilon|^{p-2}\nabla\Pi_\varepsilon\,\vdots\,\nabla P\,{\rm d}x{\rm d}t + \int_\Omega \frac{\kappa}{p}|\nabla\Pi_0|^p\,{\rm d}x \\ \ge \int_Q \hat\zeta_\varepsilon(\dot\Pi_\varepsilon\Pi_\varepsilon^{-1})\,{\rm d}x{\rm d}t + \int_\Omega \frac{\kappa}{p}|\nabla\Pi_\varepsilon(T)|^p\,{\rm d}x. \qquad (9.4.31)$$

Step 3 –Convergence of the regularization for $\varepsilon \to 0$: Now, by substituting $P\Pi_\varepsilon^{-1}$ for P, we can rewrite (9.4.31) more in the form of the inequality (9.4.22), i.e.

$$\int_Q \hat{\zeta}_\varepsilon(P\Pi_\varepsilon^{-1}) + \big(\varphi'_{\rm EL}((\nabla y_\varepsilon)\Pi_\varepsilon^{-1}))\Pi_\varepsilon^\top : ((\nabla y_\varepsilon)(\Pi_\varepsilon^{-1})') + \varphi'_{\rm HD}(\Pi_\varepsilon)\big) : (P - \dot{\Pi}_\varepsilon)$$
$$+ \kappa|\nabla\Pi_\varepsilon|^{p-2}\nabla\Pi_\varepsilon \vdots \nabla P\,{\rm d}x{\rm d}t + \int_\Omega \frac{\kappa}{p}|\nabla\Pi_0|^p\,{\rm d}x$$
$$\geq \int_Q \hat{\zeta}_\varepsilon(\dot{\Pi}_\varepsilon\Pi_\varepsilon^{-1})\,{\rm d}x{\rm d}t + \int_\Omega \frac{\kappa}{p}|\nabla\Pi_\varepsilon(T)|^p\,{\rm d}x \tag{9.4.32}$$

not only for P's used for the Galerkin method but even for all $P \in L^2(I;W^{1,p}(\Omega;\mathbb{R}^{d\times d}))$. We take a weakly* converging subsequence of $\{(y_\varepsilon,\Pi_\varepsilon)\}_{\varepsilon>0}$. Again we can prove $\nabla\Pi_\varepsilon \to \nabla\Pi$ strongly in $L^p(Q;\mathbb{R}^{d\times d})$ by the argumentation like (9.4.17) used now for $\varepsilon \to 0$ instead of $k \to \infty$; no finite-dimensional-valued approximation of the limit is needed and the boundedness of the nonlinear $\hat{\zeta}'_\varepsilon(\dot{\Pi}_\varepsilon\Pi_\varepsilon^{-1})$ in $L^2(Q;\mathbb{R}^{d\times d})$ is sufficient to converge the term $\int_Q \hat{\zeta}'_\varepsilon(\dot{\Pi}_\varepsilon\Pi_\varepsilon^{-1}) : ((\Pi_\varepsilon - \Pi)\Pi_\varepsilon^{-1}))\,{\rm d}x{\rm d}t$ to zero. Altogether, (9.4.24) is at our disposal. The limit passage to (9.4.22) is then easy, exploiting $\liminf_{\varepsilon\to 0}\int_Q \hat{\zeta}_\varepsilon(\dot{\Pi}_\varepsilon\Pi_\varepsilon^{-1})\,{\rm d}x{\rm d}t \geq \int_Q \hat{\zeta}(\dot{\Pi}\Pi^{-1})\,{\rm d}x{\rm d}t$ and $\lim_{\varepsilon\to 0}\int_Q \hat{\zeta}_\varepsilon(P\Pi_\varepsilon^{-1})\,{\rm d}x{\rm d}t = \int_Q \hat{\zeta}(P\Pi^{-1})\,{\rm d}x{\rm d}t$ which is due to (9.4.23). The limit passage in the weak force equilibrium equation towards (9.4.11) is obvious. □

Example 9.4.11 (Gradient theory in the reference configuration). Considering the gradient $\nabla\Pi$ in the reference configuration in (9.4.3) may bring an unwanted effect of a tendency of widening the plastified zone within large plastic shift. This can be seen when assuming $p = 2$ and the stratified situation (=a plastified flat band homogeneous along x_1-axis of the width 2ϵ around the plane $\{x_1 = 0\}$ in a 2-dimensional rectangular domain $\Omega = [-1,1]^2$) and intentionally neglecting hardening, i.e. $\varphi_{\rm HD} = 0$ in (9.4.3). Assuming stress-free configuration (i.e. $F_{\rm EL} = \mathbb{I}$ and thus $\nabla y = F = \Pi$), the simplest profile of Π compatible with (9.4.3) is continuous piece-linear in x_2-coordinate, i.e. the simplest profile of y is continuously differentiable piece-quadratic in x_2-coordinate. Assuming a time-dependent shift (caused e.g. by boundary conditions) with a constant (for notational simplicity unit) velocity in the $\pm x_2$-directions in the upper/lower sides of Ω, the mentioned piece-quadratic deformation is

$$y(x) = \begin{pmatrix} y_1(x_1,x_2) \\ y_2(x_1,x_2) \end{pmatrix} \quad \text{with} \quad y_1(x_1,x_2) = \begin{cases} x_1 + t & \text{if } x_2 > \epsilon, \\ x_1 + 2t - 2t\dfrac{x_2}{\epsilon} + t\dfrac{x_2^2}{\epsilon^2} & \text{if } 0 \leq x_2 \leq \epsilon, \\ x_1 - 2t - 2t\dfrac{x_2}{\epsilon} - t\dfrac{x_2^2}{\epsilon^2} & \text{if } 0 \geq x_2 \geq -\epsilon, \\ x_1 - t & \text{if } x_2 < -\epsilon_1, \end{cases}$$
$$\text{and} \quad y_2(x_1,x_2) = x_2,$$

so that the piece-wise linear profile of the plastic strain is

$$\Pi \sim F = \nabla y = \begin{pmatrix} 1 & \partial y_1/\partial x_2 \\ 0 & 1 \end{pmatrix} \quad \text{with} \quad \frac{\partial y_1}{\partial x_2} = \begin{cases} 2t\dfrac{x_2-\epsilon}{\epsilon^2} & \text{if } 0 \le x_2 \le \epsilon, \\ 2t\dfrac{x_2+\epsilon}{\epsilon^2} & \text{if } 0 \ge x_2 \ge -\epsilon, \\ 0 & \text{if } |x_2| > \epsilon. \end{cases} \tag{9.4.33}$$

Therefore, in view of (9.4.33), $\nabla\Pi$ has only one nonvanishing entry, namely

$$\frac{\partial \Pi_{12}}{\partial x_2} \sim \begin{cases} 2t/\epsilon^2 & \text{if } |x_2| \le \epsilon, \\ 0 & \text{if } |x_2| > \epsilon. \end{cases} \tag{9.4.34}$$

Then the plastic-strain gradient term $\frac{1}{2}\kappa|\nabla\Pi|^2$ in (9.4.3) contributes by $4\kappa t^2/\epsilon^3$ when again assuming Ω to have a unit size. These calculations reveal roughly the scaling of the width of the plastified band as $\epsilon \sim \kappa^{1/3}t^{2/3}$. It is interesting (and not exactly desired) feature of the model that the first scaling is time dependent. If the coefficient κ is constant, then the plastified band would have tendency to widen within accommodation of large slip, which can be interpreted as some sort of kinematic hardening. Making κ decaying in time as t^{-2} (and thus the material properties varying in time) would suppress this hardening-like effect but making the model non-autonomous is certainly misconceptual.

Remark 9.4.12 (Gradient theory in actual configurations). To avoid the effect in the above Example 9.4.11, one can consider the *gradient in the space configuration* rather than in the reference configuration. Thus, instead of (9.4.3), one is to consider

$$\mathcal{E}(y,\Pi) = \int_\Omega \varphi_{\rm EL}((\nabla y)\Pi^{-1}) + \varphi_{\rm HD}(\Pi) + \frac{\kappa}{p}\big|(\nabla y)^{-\top}\nabla\Pi\big|^p\,\mathrm{d}x + \mathscr{H}(\nabla^2 y) \tag{9.4.35}$$

with $\mathscr{H}$ from (9.2.7). Here we used the pushforward (1.1.10). The system (9.4.10) then uses $\operatorname{div} S_{\rm EL} = \partial_y\mathcal{E}(y,\Pi)$ and $S_{\rm IN} = \partial_\Pi\mathcal{E}(\nabla y,\Pi)$, i.e.

$$S_{\rm EL} = \varphi'_{\rm EL}(\nabla y\Pi^{-1})\Pi^{-\top} - \operatorname{Div}(\mathfrak{H}\nabla^2 y) + \kappa|(\nabla y)^{-\top}\nabla\Pi|^{p-2}\nabla\Pi^\top(\nabla y)^{-1}(\nabla y)^{-\top}\nabla\Pi \quad \text{and}$$
$$S_{\rm IN} = (\nabla y)^\top\varphi'_{\rm EL}(\nabla y\Pi^{-1}){:}(\Pi^{-1})' + \varphi'_{\rm HD}(\Pi) - \operatorname{div}(\kappa|(\nabla y)^{-\top}\nabla\Pi|^{p-2}(\nabla y)^{-1}(\nabla y)^{-\top}\nabla\Pi).$$

The last term in $S_{\rm EL}$ is a *Korteweg*-like *stress* and, because of it, now the strong convergence in $\nabla\Pi$ is needed for the analysis. The nonlocal hyperstress $\mathfrak{H}\nabla^2 y$ arising from the $\mathcal{H}$-term in (9.4.35) allows for keeping control over $(\nabla y)^{-\top}$. Based on the uniform (with respect to ∇y) strong monotonicity of the mapping $\Pi \mapsto -\operatorname{div}(\kappa|(\nabla y)^{-\top}\nabla\Pi|^{p-2}(\nabla y)^{-1}(\nabla y)^{-\top}\nabla\Pi)$, we refer to Sect. 9.5.1 below for a similar analysis in the case of damage gradient.

Remark 9.4.13 (Isochoric plasticity/creep). The isochoric plasticity or creep from Sect. 7.4 at large strains would mean that $\det\Pi = 1$ in a position of the nonaffine holonomic constraint. The condition (9.4.5b,c) would then be weakened, replacing $\mathrm{GL}^+(d)$ by $\mathrm{SL}(d)$. Yet, the flow rule (9.4.6) would involve a "reaction force" to this constraint, resulting to

$$\partial\hat{\zeta}(\dot{\Pi}\Pi^{-1})\Pi^{-\top} + S_{\rm IN} \ni L{\rm Cof}\,\Pi \quad \text{and} \quad \det\Pi = 1,$$

where L valued in $\mathbb{R}^{d\times d}$ is a Lagrange multiplier; note that we used also the formula (A.2.20).[21] The mathematical analysis of such system seems open. A relevant model, which would possess the isochoric attribute at least approximately for small $\delta > 0$, is covered by the above analysis when considering the hardening term of the type

$$\varphi_{\rm HD}(\Pi) := \begin{cases} \dfrac{\delta}{\max(1,\det\Pi)^q} + \dfrac{(\det\Pi - 1)^2}{2\delta} & \text{if } \det\Pi > 0, \\ +\infty & \text{if } \det\Pi \le 0; \end{cases} \tag{9.4.36}$$

note that the minimum of this potential is attained just at the set SL(d) of the isochoric plastic strains, and that it complies with (9.4.12c) for $q \ge pd/(p-d)$ and also with the plastic-indifference condition (9.4.5c) because such $\varphi_{\rm HD}$ can be viewed as a function of Π^{-1} since $\det\Pi^{-1} = 1/\det\Pi$. The isochoric model falls into the general philosophy that materials may exhibit different rheological response in volumetric and shear evolution, which we used at small strains by involving the additive strain decomposition $e = {\rm sph}\,e + {\rm dev}\,e$ with ${\rm sph}\,e = {\rm tr}\,e\mathbb{I}$, which at large strain takes the analog of the multiplicative decomposition $F = F_{\rm vol}F_{\rm iso}$ with the volumetric part $F_{\rm vol} = (\det F)^{1/d}\mathbb{I}$ so that $F_{\rm iso} = F_{\rm vol}^{-1}F = (\det F)^{-1}F$ is isochoric.

9.4.3 Thermodynamics of visco-plasticity and creep

A combination of plasticity with the *heat transfer* is not so easy because in most applications the Fourier law is to be applied rather in the actual deformed configuration than in the reference one. Thus a pull-back procedure as in (9.2.26) is to be applied and $\det\nabla y$ is to be controlled away from zero. We already handled such a situation by the concept of nonlocal nonsimple materials. Here it is important that, in addition to the techniques from the previous Sections 9.4.1–9.4.2, we need (and will have) a strong convergence of the plastic dissipation rate (=heat production) due to (9.4.28c). For simplicity, we do not consider the coupling of temperature and the mechanical variables (y,Π) through the free energy but only through the dissipative processes, cf. Remark 9.4.17. This leads to the free-energy ansatz

$$\Psi(y,\Pi,\theta) = \int_\Omega \psi(\nabla y,\Pi,\theta) + \frac{\kappa}{p}|\nabla\Pi|^p\,{\rm d}x + \mathscr{H}(\nabla^2 y) \quad \text{with } \mathscr{H} \text{ from (9.2.7) and}$$
$$\text{with} \quad \psi(F,\Pi,\theta) = \varphi_{\rm EL}(F\Pi^{-1}) + \varphi_{\rm HD}(\Pi) - \gamma_{\rm V}(\theta). \tag{9.4.37}$$

For the dissipation, one can consider a natural phenomenon of a temperature dependent yield stress or of a temperature dependent creep modulus, i.e. $\hat{\zeta} = \hat{\zeta}(\theta;P)$ with P a placeholder for $\dot{\Pi}\Pi^{-1}$. Fixing $\hat{\zeta}(\cdot;0) = 0$, the dependence $\hat{\zeta}(\cdot;P)$ is typically de-

[21] For an alternative way of treating the nonlinear isochoric constraint (but again without mathematical analysis) see [78].

creasing, expressing the typical phenomenon that performing creep or plastification is easier at higher temperatures.

The standard thermodynamical procedure as presented in Section 8.1 will yield to an expansion of the isothermal system (9.4.10) into the form

$$\varrho\ddot{y} - \operatorname{div} S_{\rm EL} = f(y) \quad \text{with} \quad S_{\rm EL} = \varphi'_{\rm EL}(F\Pi^{-1})\Pi^{-\top} - \operatorname{Div}(\mathfrak{H}\nabla^2 y) \quad \text{in } Q, \tag{9.4.38a}$$

$$\begin{aligned}\partial_P\hat{\zeta}(\theta;\dot{\Pi}\Pi^{-1}) + S_{\rm IN}\Pi^\top \ni 0& \\ \text{with } S_{\rm IN} = (\nabla y)^\top\varphi'_{\rm EL}(F\Pi^{-1}):(\Pi^{-1})'& \\ +\varphi'_{\rm HD}(\Pi) - \operatorname{div}(\kappa|\nabla\Pi|^{p-2}\nabla\Pi)& \quad \text{in } Q,\end{aligned} \tag{9.4.38b}$$

$$c_{\rm v}(\theta)\dot{\theta} - \operatorname{div}(\mathscr{K}(\nabla y,\theta)\nabla\theta) = \partial_P\hat{\zeta}(\theta;\dot{\Pi}\Pi^{-1}):\dot{\Pi}\Pi^{-1} \quad \text{in } Q, \tag{9.4.38c}$$

$$\sigma\vec{n} - \operatorname{div}_{\rm S}((\mathfrak{H}\nabla^2 y)\cdot\vec{n}) = g \quad \text{and} \quad (\mathfrak{H}\nabla^2 y):(\vec{n}\otimes\vec{n}) = 0 \quad \text{on } \Sigma, \tag{9.4.38d}$$

$$\mathscr{K}(\nabla y,\theta)\nabla\theta\cdot\vec{n} = h_\flat \quad \text{on } \Sigma, \tag{9.4.38e}$$

$$y|_{t=0} = y_0, \qquad \dot{y}|_{t=0} = v_0, \qquad \Pi|_{t=0} = \Pi_0, \qquad \theta|_{t=0} = \theta_0 \quad \text{on } \Omega. \tag{9.4.38f}$$

with the boundary conditions (9.4.10d) for Π and with $\mathscr{K}(F,\theta) = \mathscr{P}_F\mathbb{K}_{\rm M}(\theta) = (\operatorname{Cof} F^\top)\mathbb{K}_{\rm M}(\theta)\operatorname{Cof} F/\det F$ from (9.2.26), as already considered in (9.2.32). Considering the nonlocal hyper-stress potential (9.2.7) and modifying (9.4.11a) in the spirit of (9.2.8), we arrive at:

Definition 9.4.14 (Weak solution to thermo-viscoplasticity/creep). The triple $(y,\Pi,\theta)\in C_{\rm w}(I;H^{2+\gamma}(\Omega;\mathbb{R}^d))\times C_{\rm w}(I;W^{1,p}(\Omega;\mathbb{R}^{d\times d}))\times C_{\rm w}(I;W^{1,1}(\Omega))$ will be called a weak solution to the initial-boundary-value problem (9.4.10) with (9.4.10d) and with $\mathfrak{H}$ from (2.5.36) on p. 49 if $\dot{\Pi}\Pi^{-1}\in L^2(Q;\mathbb{R}^{d\times d})$ and if

$$\begin{aligned}&\int_Q\Big(\varphi'_{\rm EL}((\nabla y)\Pi^{-1}):((\nabla v)\Pi^{-1}) + \varrho y\cdot\ddot{v} \\ &\quad + \sum_{i=1}^d\int_\Omega\nabla^2(y_i(t,x)-y_i(t,\tilde{x})):\mathfrak{K}(x,\tilde{x}):\nabla^2(v_i(t,x)-v_i(t,\tilde{x}))\,\mathrm{d}\tilde{x}\Big)\,\mathrm{d}x\mathrm{d}t \\ &\quad = \int_\Omega\varrho v_0\cdot v(0) - \varrho y_0\cdot\dot{v}(0)\,\mathrm{d}x + \int_Q f(y)\cdot v\,\mathrm{d}x\mathrm{d}t + \int_\Sigma g\cdot v\,\mathrm{d}S\,\mathrm{d}t\end{aligned} \tag{9.4.39a}$$

is satisfied for all $v\in C^2(Q;\mathbb{R}^d)$ with $v(T) = \dot{v}(T) = 0$ and if (9.4.22) holds with $\hat{\zeta}(\theta,P\Pi^{-1})$ and $\hat{\zeta}(\theta,\dot{\Pi}\Pi^{-1})$ in place of $\hat{\zeta}(P\Pi^{-1})$ and $\hat{\zeta}(\dot{\Pi}\Pi^{-1})$, respectively, and eventually

$$\begin{aligned}\int_Q\mathscr{K}(\nabla y,\theta)\nabla\theta\cdot\nabla v - C_{\rm v}(\theta)\dot{v}\,\mathrm{d}x\mathrm{d}t &= \int_Q(\partial_P\hat{\zeta}(\theta;\dot{\Pi}\Pi^{-1}):\dot{\Pi}\Pi^{-1})v\,\mathrm{d}x\mathrm{d}t \\ &\quad + \int_\Sigma h_\flat v\,\mathrm{d}S\,\mathrm{d}t + \int_\Omega C_{\rm v}(\theta_0)v(0)\,\mathrm{d}x\end{aligned} \tag{9.4.39b}$$

with $C_{\rm v}(\theta) = \int_0^1 r\theta c_{\rm v}(r\theta)\,\mathrm{d}r$ is satisfied for any $v\in C^1(\bar{Q})$ with $v(T) = 0$, and also the initial condition $\Pi(0) = \Pi_0$ as well as the Dirichlet boundary condition $\Pi = \mathbb{I}$ on $\Sigma_{\rm D}$ hold.

As we now deal with the pullback in the heat-transfer coefficient, we need to have controlled also $\det\nabla y$ in addition to $\det\Pi$. For this reason, we have consider the nonlocal nonsimple model and we will modify the qualification of (9.4.12b) in the spirit of (9.2.33a), namely

$$\begin{aligned}&\varphi_{\rm EL}:{\rm GL}^+(d)\to\mathbb{R}\ \ \text{continuously differentiable and, for some }\epsilon>0:\\&\varphi_{\rm EL}(F)\ge\frac{\epsilon}{(\det F)^q},\quad q>\max\Big(\frac{pd}{p-d},\frac{2d}{2\gamma+2-d}\Big)\ \text{ for some }\gamma>\frac{d}{2}-1\end{aligned}\qquad(9.4.40)$$

with p from (9.4.12c) and γ related to the qualification (9.2.9b) of $\mathscr{H}$.

For modeling the plasticity phenomenon, we again consider a regularization (smoothing) of $\hat\zeta$ as in (9.4.23), now counting the θ-dependence and the uniform convexity in P (i.e. uniform monotonicity of $\partial_P\hat\zeta$) needed for the mentioned strong convergence:

$$\hat\zeta_\varepsilon:\mathbb{R}\times\mathbb{R}^{d\times d}\to\mathbb{R}\ \text{ and }\ \partial_P\hat\zeta_\varepsilon:\mathbb{R}\times\mathbb{R}^{d\times d}\to\mathbb{R}^{d\times d}\ \text{ continuous,}\qquad(9.4.41a)$$

$$\begin{aligned}&\exists a_\zeta>0\ \ \forall\theta\in\mathbb{R},\ P,\tilde P\in\mathbb{R}^{d\times d}:\ \ a_\zeta|P|^2\le\hat\zeta_\varepsilon(\theta;P)\le(1+|P|^2)/a_\zeta,\\&\qquad(\hat\zeta_\varepsilon'(\theta;P)-\hat\zeta_\varepsilon'(\theta;\tilde P)):(P-\tilde P)\ge a_\zeta|P-\tilde P|^2,\end{aligned}\qquad(9.4.41b)$$

$$\begin{aligned}&\lim_{\varepsilon\to0}\partial_P\hat\zeta_\varepsilon(\theta;0)=0\in\partial_P\hat\zeta(\theta;0)\ \text{ and}\\&P\ne0\ \Rightarrow\ \lim_{\varepsilon\to0,\tilde\theta\to\theta}\partial_P\hat\zeta_\varepsilon(\tilde\theta;P)=\partial_P\hat\zeta(\theta;P).\end{aligned}\qquad(9.4.41c)$$

For the creep problem with $\hat\zeta(\theta;\cdot)$ smooth, such regularization could be avoided.

We now consider a regularized problem, arising from (9.4.38) also when the heat sources are regularized as in (9.2.34), facilitating usage of L^2-theory for the regularized heat problem and using the same ε as used for $\hat\zeta_\varepsilon$. Besides, we apply the enthalpy-transformation by substituting $\vartheta=C_{\rm v}(\theta)$ with $C_{\rm v}$ as in Definition 9.4.14, cf. also Remark 8.2.2. More specifically,

$$\varrho\ddot y-{\rm div}\,S_{\rm EL}=f(y)\quad\text{with }\ S_{\rm EL}\ \text{from (9.4.38}a)\qquad\text{in }Q,\qquad(9.4.42a)$$

$$\partial_P\hat\zeta_\varepsilon(C^{-1}(\vartheta);\dot\Pi\Pi^{-1})\Pi^{-\top}+S_{\rm IN}\ni0\ \ \text{with }S_{\rm IN}\text{ from (9.4.38}b)\qquad\text{in }Q,\qquad(9.4.42b)$$

$$\dot\vartheta-{\rm div}(\mathfrak{K}(\nabla y,\vartheta)\nabla\vartheta)=\frac{\partial_P\hat\zeta(C^{-1}(\vartheta);\dot\Pi\Pi^{-1}):\dot\Pi\Pi^{-1}}{1+\varepsilon|\dot\Pi\Pi^{-1}|^2}\qquad\text{in }Q,\qquad(9.4.42c)$$

$$\sigma\vec n-{\rm div}_{\rm S}((\mathfrak{H}\nabla^2y)\cdot\vec n)=g\qquad\text{on }\Sigma,\qquad(9.4.42d)$$

$$(\mathfrak{H}\nabla^2y):(\vec n\otimes\vec n)=0\ \text{ and }\ \mathfrak{K}(\nabla y,\theta)\nabla\theta\cdot\vec n=\frac{h_\flat}{1+\varepsilon h_\flat},\qquad\text{on }\Sigma,\qquad(9.4.42e)$$

$$y|_{t=0}=y_0,\quad\dot y|_{t=0}=v_0,\quad\Pi|_{t=0}=\Pi_0,\quad\vartheta|_{t=0}=C_{\rm v}\Big(\frac{\theta_0}{1+\varepsilon\theta_0}\Big)\qquad\text{on }\Omega,\qquad(9.4.42f)$$

again with the boundary conditions (9.4.10d) for Π. Here, we used $\mathfrak{K}(F,\vartheta)=\mathscr{K}(F,C^{-1}(\vartheta))/c_{\rm v}((\vartheta))=\mathscr{P}_F\mathbb{K}_{\rm M}(C^{-1}(\vartheta))/c_{\rm v}((\vartheta))$.

Proposition 9.4.15 (Convergence and existence of weak solutions). *Let, in addition to (9.4.40), (9.4.41) (9.2.9b), (9.4.12), also $\partial\hat\zeta$ be uniformly monotone in the sense*

$$\forall P,\tilde P\in\mathbb{R}^{d\times d},\ S\in\partial\hat\zeta(P),\ \tilde S\in\partial\hat\zeta(\tilde P):\quad (S-\tilde S):(P-\tilde P)\ge a_{\hat\zeta}|P-\tilde P|^2, \tag{9.4.43}$$

admitting the smoothing[22] *satisfying (9.4.41), and let $c_{\rm V}:\mathbb{R}\to\mathbb{R}$ and $\mathbb{K}_{\rm M}:\mathbb{R}\to\mathbb{R}^{d\times d}$ be continuous and bounded, and uniformly positive and positive definite, respectively. Then (9.4.42) with (9.4.10d) has a weak solution, let us denote it by $(y_\varepsilon,\Pi_\varepsilon,\vartheta_\varepsilon)$ and, in terms of subsequences, these solutions converge for $\varepsilon\to0$ in the sense (9.4.24) together with*

$$\dot\Pi_\varepsilon\Pi_\varepsilon^{-1}\to\dot\Pi\Pi^{-1}\qquad\text{strongly in }\ L^2(Q;\mathbb{R}^{d\times d}),\ \text{ and} \tag{9.4.44a}$$

$$\vartheta_\varepsilon\to\vartheta\qquad\text{weakly in }\ L^r(I;W^{1,r}(\Omega))\ \text{ with any }\ 1\le r<\frac{d+2}{d+1}, \tag{9.4.44b}$$

and every such a limit gives a weak solution (y,Π,θ) with $\theta=C_{\rm V}^{-1}(\vartheta)$ to the initial-boundary-value problem (9.4.10) with (9.4.10d) in the sense of Definition 9.4.14. In particular, these solutions do exist.

Sketch of the proof. We divide the proof into four steps.

Step 1 – Galerkin's approximation of the regularized problem: Existence for (9.4.42) and a-priori estimates can be seen by testing the particular equations by $\dot y_{\varepsilon k}$, $\dot\Pi_{\varepsilon k}$, and $\theta_{\varepsilon k}$. The peculiarity is that $\varphi_{\rm EL}$ controls $\det F_{\rm EL}=\det(F\Pi^{-1})$ while the hyperstress controls ∇F but not $\nabla F_{\rm EL}$, cf. Remark 9.4.6. Yet, we can estimate $\nabla F_{\rm EL}$ from ∇F and $\nabla\Pi$, realizing that

$$\nabla F_{\rm EL}=\nabla F\Pi^{-1}+F(\Pi^{-1})'\nabla\Pi.$$

Since $\nabla F\in L^\infty(I;H^\gamma(\Omega;\mathbb{R}^{d\times d\times d}))\subset L^\infty(I;L^{2d/(d-2\gamma)-\epsilon}(\Omega;\mathbb{R}^{d\times d\times d}))$ and $\nabla\Pi\in L^\infty(I;L^p(\Omega;\mathbb{R}^{d\times d\times d}))$, we have $\nabla F_{\rm EL}$ controlled in $L^\infty(I;L^r(\Omega;\mathbb{R}^{d\times d\times d}))$ with $r>\min(p,2d/(d-2\gamma)$.

Then, besides applying the Healey-Krömer theorem 2.5.3 on p. 43 for $\Pi_{\varepsilon k}$ while exploiting (9.4.12c) as before, we can use it for the second time also for $F_{{\rm EL},\varepsilon k}=F_{\varepsilon k}\Pi_{\varepsilon k}^{-1}$ when exploiting (9.4.40), obtaining $\det F_{{\rm EL},\varepsilon k}\ge\delta_1>0$. Having also $\det\Pi_{\varepsilon k}\ge\delta_2>0$ again by Theorem 2.5.3, exploiting now (9.4.12c), we get also

$$\det\nabla y_{\varepsilon k}=\det(F_{{\rm EL},\varepsilon k}\Pi_{\varepsilon k})=\det F_{{\rm EL},\varepsilon k}\det\Pi_{\varepsilon k}\ge\delta_1\delta_2>0.$$

by using the Cauchy-Binet formula (A.2.16). Altogether, we obtain a-priori estimates

$$\|y_{\varepsilon k}\|_{L^\infty(I;H^2(\Omega;\mathbb{R}^d))\cap W^{1,\infty}(I;L^2(\Omega;\mathbb{R}^d))}\le K, \tag{9.4.45a}$$

$$\|\Pi_{\varepsilon k}\|_{L^\infty(I;W^{1,p}(\Omega;\mathbb{R}^{d\times d}))\cap H^1(I;L^2(\Omega;\mathbb{R}^{d\times d}))}\le K, \tag{9.4.45b}$$

$$\big\|1/\det\Pi_{\varepsilon k}\big\|_{L^\infty(Q)}\le K\quad\text{and}\quad\big\|\Pi_{\varepsilon k}^{-1}\big\|_{L^\infty(Q;\mathbb{R}^{d\times d})}\le K, \tag{9.4.45c}$$

[22] An example for such $\hat\zeta$ can be $\hat\zeta(\theta,P)=\hat\zeta_1(\theta,P)+s_{\rm YLD}(\theta)|P|$ when $\hat\zeta_1(\theta,\cdot)$ is smooth and $\hat\zeta_\varepsilon$ uses the Yosida approximation of $|\cdot|$, cf. [465].

$$\|\vartheta_{\varepsilon k}\|_{L^\infty(I;L^2(\Omega))\cap L^2(I;H^1(\Omega))\cap H^1(I;H^1(\Omega)^*)} \leq K_\varepsilon; \tag{9.4.45d}$$

here the positive definiteness of $\mathscr{K}$ exploiting also (9.4.45c) was used, cf. the proof of Corollary 9.2.10. Also, (9.4.27) is at our disposal, based on the identity (9.4.26) now with $\hat{\zeta}'_\varepsilon(C_V^{-1}(\vartheta_{\varepsilon k}),\dot{\Pi}_{\varepsilon k}\Pi_{\varepsilon k}^{-1})$ in place of $\hat{\zeta}'_\varepsilon(\dot{\Pi}_{\varepsilon k}\Pi_{\varepsilon k}^{-1})$.

Step 2 – Convergence for $k\to\infty$: We take a subsequence so that (9.4.28) holds together with $\vartheta_{\varepsilon k}\to\vartheta_\varepsilon$ weakly* in the topologies indicating in (9.4.45d). Again, also the strong convergence (9.4.17) holds, giving here $\nabla\Pi_{\varepsilon k}\to\nabla\Pi_\varepsilon$ in $L^p(Q;\mathbb{R}^{d\times d})$. This allows us to pass to the limit in the force-equilibrium equation and the flow rule towards (9.4.42a,b) formulated weakly.[23]

In particular, certainly $\mathrm{div}(\kappa|\nabla\Pi_\varepsilon|^{p-2}\nabla\Pi_\varepsilon)$ exists in $L^{p'}(I;W^{1,p}(\Omega;\mathbb{R}^{d\times d})^*)$. To pass to the limit in the heat-transfer equation, we will now also need another estimate which can be obtained by comparison from the flow rule $\mathrm{div}(\kappa|\nabla\Pi_\varepsilon|^{p-2}\nabla\Pi_\varepsilon)=\Sigma_{\mathrm{pl},\varepsilon}+(\nabla y_\varepsilon)^\top\varphi'_{\mathrm{EL}}(\nabla y_\varepsilon\Pi_\varepsilon^{-1}):(\Pi_\varepsilon^{-1})'+\varphi'_{\mathrm{HD}}(\Pi_\varepsilon)$ which we will get as a limit from the Galerkin approximation. More specifically, we now have:

$$\big\|\mathrm{div}(\kappa|\nabla\Pi_\varepsilon|^{p-2}\nabla\Pi_\varepsilon)\big\|_{L^2(Q;\mathbb{R}^{d\times d})}\leq K \tag{9.4.46}$$

actually with the same K as in (9.4.27).

We already indicated that that the strong convergence of the plastification rate is now needed for solving the heat equation. To this goal, using (9.4.43), we can estimate

$$\begin{aligned}
&\limsup_{k\to\infty} a_\zeta\big\|\dot{\Pi}_{\varepsilon k}\Pi_{\varepsilon k}^{-1}-\dot{\Pi}_\varepsilon\Pi_\varepsilon^{-1}\big\|^2_{L^2(Q;\mathbb{R}^{d\times d})}\\
&\leq\limsup_{k\to\infty}\int_Q\big(\partial_P\hat{\zeta}_\varepsilon(\theta_{\varepsilon k},\dot{\Pi}_{\varepsilon k}\Pi_{\varepsilon k}^{-1})-\partial_P\hat{\zeta}_\varepsilon(\theta_{\varepsilon k},\dot{\Pi}_\varepsilon\Pi_\varepsilon^{-1})\big):(\dot{\Pi}_{\varepsilon k}\Pi_{\varepsilon k}^{-1}-\dot{\Pi}_\varepsilon\Pi_\varepsilon^{-1})\,\mathrm{d}x\mathrm{d}t\\
&\leq\limsup_{k\to\infty}\int_Q\partial_P\hat{\zeta}_\varepsilon(\theta_{\varepsilon k},\dot{\Pi}_{\varepsilon k}\Pi_{\varepsilon k}^{-1}):((\dot{\Pi}_{\varepsilon k}-\dot{\tilde{\Pi}}_k)\Pi_{\varepsilon k}^{-1})\,\mathrm{d}x\mathrm{d}t\\
&\qquad+\lim_{k\to\infty}\int_Q\partial_P\hat{\zeta}_\varepsilon(\theta_{\varepsilon k},\dot{\Pi}_{\varepsilon k}\Pi_{\varepsilon k}^{-1}):(\dot{\tilde{\Pi}}_k\Pi_{\varepsilon k}^{-1}-\dot{\Pi}_\varepsilon\Pi_\varepsilon^{-1})\,\mathrm{d}x\mathrm{d}t\\
&\qquad\qquad-\lim_{k\to\infty}\int_Q\partial_P\hat{\zeta}_\varepsilon(\theta_{\varepsilon k},\dot{\Pi}_\varepsilon\Pi_\varepsilon^{-1}):(\dot{\Pi}_{\varepsilon k}\Pi_{\varepsilon k}^{-1}-\dot{\Pi}_\varepsilon\Pi_\varepsilon^{-1}))\,\mathrm{d}x\mathrm{d}t\\
&=\lim_{k\to\infty}\int_Q\varphi'_{\mathrm{EL}}((\nabla y_{\varepsilon k})\Pi_{\varepsilon k}^{-1}):(\nabla y_{\varepsilon k}(\Pi_{\varepsilon k}^{-1})'(\dot{\Pi}_{\varepsilon k}-\dot{\tilde{\Pi}}_k))+\varphi'_{\mathrm{HD}}(\Pi_{\varepsilon k}):(\dot{\Pi}_{\varepsilon k}-\dot{\tilde{\Pi}}_k))\,\mathrm{d}x\mathrm{d}t\\
&\qquad\qquad-\liminf_{k\to\infty}\int_Q\kappa|\nabla\Pi_{\varepsilon k}|^{p-2}\nabla\Pi_{\varepsilon k}\,\vdots\,\nabla(\dot{\Pi}_{\varepsilon k}-\dot{\tilde{\Pi}}_k)\,\mathrm{d}x\mathrm{d}t\\
&=\lim_{k\to\infty}\int_Q\varphi'_{\mathrm{EL}}((\nabla y_{\varepsilon k})\Pi_{\varepsilon k}^{-1}):(\nabla y_{\varepsilon k}(\Pi_{\varepsilon k}^{-1})'(\dot{\Pi}_{\varepsilon k}-\dot{\tilde{\Pi}}_k))+\varphi'_{\mathrm{HD}}(\Pi_{\varepsilon k}):(\dot{\Pi}_{\varepsilon k}-\dot{\tilde{\Pi}}_k))\\
&\qquad-\mathrm{div}(\kappa|\nabla\Pi_{\varepsilon k}|^{p-2}\nabla\Pi_{\varepsilon k}):\dot{\tilde{\Pi}}_k\,\mathrm{d}x\mathrm{d}t+\limsup_{k\to\infty}\int_\Omega\frac{\kappa}{2}|\nabla\Pi_0|^2-\frac{\kappa}{2}|\nabla\Pi_{\varepsilon k}(T)|^2\,\mathrm{d}x
\end{aligned}$$

[23] Actually, not having (9.4.30)–(9.4.31) at our disposal, we can prove the strong convergence (9.4.47) first, and only after this step we can complete the limit passage in the flow rule more simply by using the strong convergence $\dot{\Pi}_{\varepsilon k}\Pi_{\varepsilon k}^{-1}-\dot{\Pi}_\varepsilon\Pi_\varepsilon^{-1}$.

$$\leq \int_\Omega \frac{\kappa}{p}|\nabla\Pi_0|^p - \frac{\kappa}{p}|\nabla\Pi_\varepsilon(T)|^p\,\mathrm{d}x - \int_Q \mathrm{div}(\kappa|\nabla\Pi_\varepsilon|^{p-2}\nabla\Pi):\dot{\Pi}_\varepsilon\,\mathrm{d}x\mathrm{d}t = 0, \tag{9.4.47}$$

where we used the estimate

$$\begin{aligned}
&\partial_P\hat{\zeta}_\varepsilon(\theta_{\varepsilon k},\dot{\Pi}_{\varepsilon k}\Pi_{\varepsilon k}^{-1}):(\dot{\tilde{\Pi}}_k\Pi_{\varepsilon k}^{-1}-\dot{\Pi}_\varepsilon\Pi_\varepsilon^{-1})\\
&= \partial_P\hat{\zeta}_\varepsilon(\theta_{\varepsilon k},\dot{\Pi}_{\varepsilon k}\Pi_{\varepsilon k}^{-1}):(\dot{\Pi}_\varepsilon(\Pi_{\varepsilon k}^{-1}-\Pi_\varepsilon^{-1})) + \partial_P\hat{\zeta}_\varepsilon(\theta_{\varepsilon k},\dot{\Pi}_{\varepsilon k}\Pi_{\varepsilon k}^{-1}):((\dot{\tilde{\Pi}}_k-\dot{\Pi}_\varepsilon)\Pi_{\varepsilon k}^{-1})\\
&\leq \big\|\partial_P\hat{\zeta}_\varepsilon(\theta_{\varepsilon k},\dot{\Pi}_{\varepsilon k}\Pi_{\varepsilon k}^{-1})\big\|_{L^2(Q;\mathbb{R}^{d\times d})}\Big(\big\|\dot{\Pi}_\varepsilon\big\|_{L^2(Q;\mathbb{R}^{d\times d})}\big\|\Pi_{\varepsilon k}^{-1}-\Pi_\varepsilon^{-1}\big\|_{L^\infty(Q;\mathbb{R}^{d\times d})}\\
&\qquad + \big\|\dot{\tilde{\Pi}}_k-\dot{\Pi}_\varepsilon\big\|_{L^2(Q;\mathbb{R}^{d\times d})}\big\|\Pi_{\varepsilon k}^{-1}\big\|_{L^\infty(Q;\mathbb{R}^{d\times d})}\Big) \to 0.
\end{aligned}$$

Moreover, we used also the calculus

$$\lim_{k\to\infty}\int_Q \mathrm{div}(\kappa|\nabla\Pi_{\varepsilon k}|^{p-2}\nabla\Pi_{\varepsilon k}):\dot{\tilde{\Pi}}_k\,\mathrm{d}x\mathrm{d}t = \int_Q \mathrm{div}(\kappa|\nabla\Pi_\varepsilon|^{p-2}\nabla\Pi_\varepsilon):\dot{\Pi}_\varepsilon\,\mathrm{d}x\mathrm{d}t$$

because of the bounds (9.4.27) which allows to consider $\mathrm{div}(\kappa|\nabla\Pi_{\varepsilon k}|^{p-2}\nabla\Pi_{\varepsilon k}) \to \mathrm{div}(\kappa|\Pi_\varepsilon|^{p-2}\nabla\Pi_\varepsilon)$ not only strongly in $L^{p'}(I;W^{1,p}(\Omega;\mathbb{R}^{d\times d})^*)$ but even weakly[24] while $\dot{\tilde{\Pi}}_k \to \dot{\Pi}_\varepsilon$ strongly in $L^2(Q;\mathbb{R}^{d\times d})$, and further, for the last equality in (9.4.47), we used (9.4.21) now for Π_ε. It is important that we have the estimates (9.4.45b-d) at disposal so that $\Pi_{\varepsilon k}^{-1} \to \Pi_\varepsilon^{-1}$ strongly in $L^\infty(Q;\mathbb{R}^{d\times d})$ and $\vartheta_{\varepsilon k} \to \vartheta_\varepsilon$ (and thus also $\theta_{\varepsilon k} = C_\mathrm{V}^{-1}(\vartheta_{\varepsilon k}) \to C_\mathrm{V}^{-1}(\vartheta_\varepsilon) = \theta_\varepsilon$ strongly in $L^{2^*-\epsilon}(Q)$, which has been used many times in (9.4.47). Also, $\nabla\Pi_{\varepsilon k}(T) \to \nabla\Pi_\varepsilon(T)$ weakly in $L^p(Q;\mathbb{R}^{d\times d})$ due to the estimate (9.4.45b) together with weak semicontinuity of the functional $\Pi \mapsto \int_\Omega \kappa|\nabla\Pi|^p\,\mathrm{d}x$ has been used for the last inequality in (9.4.47).

Having the strong convergence $\dot{\Pi}_{\varepsilon k}\Pi_{\varepsilon k}^{-1} \to \dot{\Pi}_\varepsilon\Pi_\varepsilon^{-1}$ proved, the limit passage in the heat equation towards (9.4.42c) formulated weakly is then easy.

Step 3 – Further estimates: Using Corollary 9.2.10, we have at disposal the prefabricated estimates on temperature based on the physical L^1-theory:

$$\big\|\nabla\vartheta\big\|_{L^r(Q;\mathbb{R}^d)} \leq C_r \qquad\text{and}\qquad \big\|\vartheta\big\|_{L^q(Q)} \leq C_q \tag{9.4.48}$$

with $1 \leq r < (d+2)/(d+1)$ and $1 \leq q < 1+2/d$ as in Proposition 8.2.1 when $\omega = 1$.

Step 4 – Limit passage with $\varepsilon \to 0$: One has just to modify also the argument in (9.4.47), for the term $\nabla\Pi_\varepsilon$ is not well defined but, on the other hand, omitting the approximation of the limit needed for the Galerkin approximation. Relying on the fact that $\mathrm{div}(\kappa|\nabla\Pi_\varepsilon|^{q-2}\nabla\Pi_\varepsilon) \in L^2(Q;\mathbb{R}^{d\times d})$, we have

$$\limsup_{\varepsilon\to 0} a_\zeta\big\|\dot{\Pi}_\varepsilon\Pi_\varepsilon^{-1} - \dot{\Pi}\Pi^{-1}\big\|^2_{L^2(Q;\mathbb{R}^{d\times d})}$$

[24] Here, we may consider a Hahn-Banach extension on $L^2(Q;\mathbb{R}^{d\times d})$ to keep the bounds (9.4.27) even in the full norm on $L^2(Q;\mathbb{R}^{d\times d})$.

$$
\begin{aligned}
&\le \limsup_{\varepsilon\to0}\int_Q\big(\partial_P\hat\zeta_\varepsilon(\theta_\varepsilon;\dot\Pi_\varepsilon\Pi_\varepsilon^{-1})-\partial_P\hat\zeta_\varepsilon(\theta;\dot\Pi\Pi^{-1})\big):(\dot\Pi_\varepsilon\Pi_\varepsilon^{-1}-\dot\Pi\Pi^{-1})\,\mathrm{d}x\mathrm{d}t\\
&=\lim_{\varepsilon\to0}\int_Q\nabla y_\varepsilon^\top\varphi_{\rm EL}'(\nabla y_\varepsilon\Pi_\varepsilon^{-1}):(\Pi_\varepsilon^{-1})':(\dot\Pi_\varepsilon-\dot\Pi)+\varphi_{\rm HD}'(\Pi_\varepsilon):(\dot\Pi_\varepsilon-\dot\Pi)\,\mathrm{d}x\mathrm{d}t\\
&\qquad-\liminf_{\varepsilon\to0}\int_Q\mathrm{div}(\kappa|\nabla\Pi_\varepsilon|^{q-2}\nabla\Pi_\varepsilon):(\dot\Pi_\varepsilon-\dot\Pi)\,\mathrm{d}x\mathrm{d}t\\
&\qquad\quad-\lim_{\varepsilon\to0}\int_Q\partial_P\hat\zeta_\varepsilon(\theta_\varepsilon;\dot\Pi\Pi^{-1}):(\dot\Pi_\varepsilon\Pi_\varepsilon^{-1}-\dot\Pi\Pi^{-1}))\,\mathrm{d}x\mathrm{d}t\\
&=\lim_{\varepsilon\to0}\int_Q\nabla y_\varepsilon^\top\varphi_{\rm EL}'(\nabla y_\varepsilon\Pi_\varepsilon^{-1}):(\Pi_\varepsilon^{-1})':(\dot\Pi_\varepsilon-\dot\Pi)+\varphi_{\rm HD}'(\Pi_\varepsilon):(\dot\Pi_\varepsilon-\dot\Pi)\\
&\qquad+\mathrm{div}(\kappa|\nabla\Pi_\varepsilon|^{q-2}\nabla\Pi_\varepsilon):\dot\Pi\,\mathrm{d}x\mathrm{d}t+\limsup_{\varepsilon\to0}\int_\Omega\frac{\kappa}{q}|\nabla\Pi_0|^q-\frac{\kappa}{q}|\nabla\Pi_\varepsilon(T)|^q\,\mathrm{d}x\\
&\le\int_\Omega\frac{\kappa}{q}|\nabla\Pi_0|^q-\frac{\kappa}{q}|\nabla\Pi(T)|^q\,\mathrm{d}x-\int_Q\mathrm{div}(\kappa|\nabla\Pi|^{q-2}\nabla\Pi):\dot\Pi\,\mathrm{d}x\mathrm{d}t=0.
\end{aligned}
$$

The other arguments copy Step 2, with the only slight modification that $\theta_{\varepsilon k}\to\theta_\varepsilon$ strongly in $L^q(Q)$ with q as in (9.4.48). □

Remark 9.4.16 (Concept of small elastic strains). If the Green-Lagrange elastic strain $E_{\rm EL}=\frac12F_{\rm EL}^\top F_{\rm EL}=\frac12(\Pi^{-\top}F^\top F\Pi^{-1}-\mathbb{I})$ is small, then $C=F^\top F\sim\Pi^\top\Pi$, cf. Remark 9.4.5. As the effective heat-conductivity tensor $\mathscr{K}$ depends rather on C, cf. (9.1.21), considering $\mathscr{K}(\Pi,\theta)=\mathscr{P}_\Pi\mathbb{K}_{\rm M}(\theta)$ in (9.4.38) instead of $\mathscr{K}(F,\theta)$ is then a reasonable modeling approximation. Some arguments can thus be simplified because we adopt the scenario from Sections 9.4.1 and 9.4.2 not controlling $\det\nabla y$. E.g. (9.4.40) can be replaced by (9.4.12b) only slightly strengthened as $|\varphi_{\rm EL}'(F)|\le K(1+|F|^{2^*/2-1})$ in order to have $\nabla y\varphi_{\rm EL}'(\nabla y\Pi^{-1})$ bounded in $L^2(Q;\mathbb{R}^{d\times d})$ instead of only $L^1(Q;\mathbb{R}^{d\times d})$ needed before, cf. [465].

Remark 9.4.17 (Thermal coupling more general). Considering a general free energy $\psi=\psi(F,\Pi,\theta)$ in (9.4.37) would lead to the heat capacity $c_{\rm v}(F,\Pi,\theta)=-\theta\partial^2_{\theta\theta}\psi(F,\Pi,\theta)$ dependent also on (F,Π) and to the adiabatic heat source/sink term $(\partial_F\psi(F,\Pi,\theta)-\partial_F\psi(F,\Pi,0)):\dot F+(\partial_\Pi\psi(F,\Pi,\theta)-\partial_\Pi\psi(F,\Pi,0)):\dot\Pi$. While $\dot\Pi$ is controlled through the visco-plastification/creep, occurrence of $\dot F$ would need to involve also the Kelvin-Voigt viscosity like in Section 9.3. In the case of creep, the thermodynamical model for the Jeffreys viscoelastic material from Remark 9.4.4 would thus be obtained.

9.5 Damage and its thermodynamics

Also other internal-variable models can be analyzed under large strains when formulated in the Lagrangian description. Let us demonstrate it on the damage model.

9.5.1 Isothermal damage model

Let us illustrate the techniques on the model with the stored energy considered like (7.5.50a) but without π-variable, i.e. enhanced by involving gradients of both deformation gradient (i.e. deformation Hessian) and of damage:

$$\varphi_{\rm E}(F,G,\alpha,\nabla\alpha) := \varphi(F,\alpha) + \frac{1}{2}\sum_{i,k,l,m,n=1}^{d} \mathbb{H}_{klmn}G_{ikl}G_{imn} + \delta_{[0,1]}(\alpha) + \frac{\kappa}{2}|\nabla\alpha|^2, \quad (9.5.1)$$

with G again a placeholder for the Hessian $\nabla^2 y$. Then we consider the initial-boundary-value problem for the system of an equation and an inclusion:

$$\varrho\ddot{y} - \operatorname{div} S = f(y) \quad \text{with} \quad S = \partial_F\varphi(\nabla y,\alpha) - \operatorname{Div}(\mathbb{H}\nabla^2 y) \qquad \text{in } Q, \quad (9.5.2a)$$

$$\partial\zeta(\dot{\alpha}) + \partial_\alpha\varphi(\nabla y,\alpha) + r_{\rm c} \ni \kappa\Delta\alpha \quad \text{with} \quad r_{\rm c} \in N_{[0,1]}(\alpha), \qquad \text{in } Q, \quad (9.5.2b)$$

$$S\vec{n} - \operatorname{div}_{\rm S}((\mathbb{H}\nabla^2 y)\cdot\vec{n}) = g \quad \text{and} \quad (\mathbb{H}\nabla^2 y):(\vec{n}\otimes\vec{n}) = 0 \qquad \text{on } \Sigma, \quad (9.5.2c)$$

$$\kappa\nabla\alpha\cdot\vec{n} = 0 \qquad \text{on } \Sigma, \quad (9.5.2d)$$

$$y|_{t=0} = y_0, \qquad \dot{y}|_{t=0} = v_0, \qquad \alpha|_{t=0} = \alpha_0 \qquad \text{on } \Omega. \quad (9.5.2e)$$

Let us expose some results for the viscous damage with a possible healing, assuming, as in (7.5.6), that for some $0 < \epsilon \le 2^*$, it holds

$$|\partial_F\varphi(F,\alpha)| \le K(1+|F|^{2^*-\epsilon}) \quad \text{and} \quad |\partial_\alpha\varphi(F,\alpha)| \le K(1+|F|^{2^*/2}). \quad (9.5.3a)$$

$$\zeta:\mathbb{R}\to\mathbb{R} \text{ convex with } \epsilon|v|^2 \le \zeta(v) \le (1+|v|^2)/\epsilon. \quad (9.5.3b)$$

Similarly as in Sect. 7.5.1, the approximation can be made by the H^2-conformal[25] Galerkin method. As in the mentioned section, we do not make any regularization of the inclusion (9.5.2b).

Proposition 9.5.1 (Weak solution to reference-gradient damage). *Let the assumptions (7.5.6b-e), and $\varphi \in C^1(\mathbb{R}^{d\times d}\times[0,1])$ and ζ satisfy (9.5.3), let ϱ satisfy (6.3.2), $\mathbb{H} \in \mathrm{SLin}(\mathbb{R}^{d\times d})$ be positive definite, and the loading and initial conditions satisfy (9.2.4) with also $\alpha_0 \in H^1(\Omega)$ so that $\varphi(\nabla y_0,\alpha_0) \in L^1(\Omega)$. Then the above mentioned Galerkin approximation (y_k,α_k) exists and satisfies the a-priori estimates*

$$\|y_k\|_{L^\infty(I;H^2(\Omega;\mathbb{R}^d))\cap W^{1,\infty}(I;L^2(\Omega;\mathbb{R}^d))} \le K, \quad (9.5.4a)$$

$$\|\alpha_k\|_{L^\infty(Q)\cap L^\infty(I;H^1(\Omega))\cap H^1(I;L^2(\Omega))} \le K, \quad (9.5.4b)$$

$$\|\Delta\alpha_k\|_{L^2(Q)} \le K \quad \textit{and} \quad \|r_{{\rm c},k}\|_{L^2(Q)} \le K. \quad (9.5.4c)$$

[25] Recall that this means that a base for the Galerkin method is such that $\Delta\alpha_k$ is still valued in the respective finite-dimensional subspaces.

Moreover, every weak converging subsequence of* $\{(y_k, \alpha_k)\}_{k\in\mathbb{N}}$ *for* $k \to \infty$ *yields in its limit a weak solution to the initial-boundary-value problem (9.5.2) with the inclusion (9.5.2b) holding even a.e. on* Q.

Sketch of the proof. Testing (9.5.2a,b) in its Galerkin approximation by $(\dot{y}_k, \dot{\alpha}_k)$, we obtain the energy-based a-priori estimates for the approximate solution (9.5.4a,b) together, by a comparison from (9.5.2b), with the estimates in (9.5.4c). More in detail, by (9.5.3a), $\partial_\alpha\varphi(\nabla y_k, \alpha_k) \in L^\infty(I; L^2(\Omega))$ while, by (9.5.3b), the set $\partial\zeta(\dot{\alpha}_k)$ is bounded in $L^2(Q)$ so that, by using the calculus (7.5.8), the test of the damage flow rule (9.5.2b) by $\Delta\alpha_k$ makes both $\Delta\alpha_k$ and $r_{c,k}$ controlled in $L^2(Q)$.

The selection of converging subsequences and the limit passage in the Galerkin integral identities is then analogous to methods in Sect. 7.5.3. □

Various modifications of this model could be possible, e.g. based on the scenarios (7.5.16) while leading to $r_c = 0$ or considering rate-independent damage while exploiting the concept of energetic solutions with semistability and the mutual recovery sequence as in (7.5.37).

A physically more relevant variant preventing local penetration would use the nonlocal nonsimple model, replacing the local quadratic form $G \mapsto \frac{1}{2}\mathbb{H}G\vdots G$ by $\mathscr{H}$ from (9.2.7) satisfying (9.2.9b). Then, like (9.4.40), the potential φ is to be qualified as

$$\varphi : \mathrm{GL}^+(d) \times [0,1] \to \mathbb{R} \quad \text{continuously differentiable and} \tag{9.5.5a}$$

$$\exists \epsilon > 0 \ \ \forall F \in \mathrm{GL}^+(d),\ \alpha \in [0,1]:$$

$$\varphi(F,\alpha) \geq \frac{\epsilon}{(\det F)^q} \quad \text{with} \quad q > \frac{2d}{2\gamma{+}2{-}d} \quad \text{for some} \ \ \gamma > \frac{d}{2} - 1 \tag{9.5.5b}$$

where γ related to the qualification (9.2.9b) of $\mathscr{H}$. As in these cases before, the local hyperstress $\mathbb{H}\nabla^2 y$ in (9.5.2) is to be replaced by the nonlocal hyperstress $\mathfrak{H}\nabla^2 y = \mathscr{H}'(\nabla^2 y)$ from (2.5.36) in order to facilitate usage of the Healey-Krömer theorem in the form of Corollary 2.5.11.

This allows for involving also various transport processes as the heat transfer (in the following Section 9.5.2) or mass transport (as in Section 9.6). Here, this also allows for an interesting (and often physically more relevant than the model (9.5.1)) modification of the damage model itself, when considering the *gradient of the damage in the actual configuration* rather than in the reference configuration in the spirit of Remark 9.4.12. To be more general, let us consider anisotropic gradient damage by replacing $\frac{\kappa}{2}|\nabla\alpha|^2$ by a general quadratic form $\frac{1}{2}\mathbb{K}\nabla\alpha\cdot\nabla\alpha$ with $\mathbb{K} = \mathbb{K}(x) \in \mathbb{R}^{d\times d}_{\mathrm{sym}}$ positive definite. Then, instead of (9.5.1), we consider

$$\varphi_{\mathrm{E}}(F,G,\alpha,\nabla\alpha) := \varphi(F,\alpha) + \mathscr{H}(G) + \frac{1}{2}\mathbb{K}(P(F)\nabla\alpha)\cdot(P(F)\nabla\alpha) \tag{9.5.6}$$

with $P : \mathrm{GL}^+(d) \to \mathbb{R}^{d\times d}$ a vectorial pushforward to be specified, depending on a model and satisfying

$$P, P^{-1}: \mathrm{GL}^+(d) \to \mathbb{R}^{d\times d} \text{ continuously differentiable.} \tag{9.5.7}$$

To avoid the Laplacian-like testing of the damage flow rule which would be problematic[26] in the Galerkin approximation (while direct usage of Rothe method in the dynamical case is also troublesome as outlined in Exercises 9.2.9 and 9.2.15), we adopt the scenario (7.5.16) which have allowed us to write $\varphi_{\rm E}$ in (9.5.6) without a term $\delta_{[0,1]}(\alpha)$ Then the system (9.5.2a-d) is modified as:

$$\varrho\ddot{y} - \mathrm{div}\,(S + S_{\rm K}(\nabla y, \nabla\alpha)) = f(y) \quad \text{with } S = \partial_F\varphi(\nabla y, \alpha) - \mathrm{Div}(\mathfrak{H}\nabla^2 y)$$
$$\text{and with } S_{\rm K}(F, \nabla\alpha) = P(F)^\top\mathbb{K}{:}P'(F){:}(\nabla\alpha\otimes\nabla\alpha) \quad \text{in } Q, \tag{9.5.8a}$$
$$\partial\zeta(\dot\alpha) + \partial_\alpha\varphi(\nabla y, \alpha) \ni \mathrm{div}(P^\top(\nabla y)\mathbb{K}P(\nabla y)\nabla\alpha) \quad \text{in } Q, \tag{9.5.8b}$$
$$S\vec{n} - \mathrm{div}_{\rm S}((\mathfrak{H}\nabla^2 y){\cdot}\vec{n}) = g \text{ and } (\mathfrak{H}\nabla^2 y){:}(\vec{n}\otimes\vec{n}) = 0 \quad \text{on } \Sigma, \tag{9.5.8c}$$
$$P^\top(\nabla y)\mathbb{K}P(\nabla y)\nabla\alpha{\cdot}\vec{n} = 0 \quad \text{on } \Sigma, \tag{9.5.8d}$$

together with the initial conditions (9.5.2e). Let us note that $P'(F)$ in (9.5.8a) is the 4th-order tensor so that the respective term is indeed the 2nd-order tensor, as expected in the stress. A natural option[27] to be considered is

$$P(F) = F^{-\top} = \mathrm{Cof}\,F/\det F, \tag{9.5.9}$$

which is just the push-forward (1.1.10) or (1.2.18). The resulted *Korteweg*-like *stress* $S_{\rm K}$ in (9.5.8a) then contains $P'(F) = \mathrm{Cof}'F/\det F - \mathrm{Cof}\,F\otimes\mathrm{Cof}\,F/(\det F)^2$. Let us note that the material tensor $\mathbb{K} = \mathbb{K}(x)$ is, in fact, pushed forward to $\mathbb{K}^y = \mathbb{K}^y(x^y)$ as in (9.1.22) before being pulled back $\mathscr{P}_F$ from (1.1.11b) to obtain $P^\top(F)\mathbb{K}P(F) = F^{-1}\mathbb{K}^y F^{-\top}$ occurring in (9.5.8b). Let us also note that such pullback differs from the "diffusant-conserving" pullback operator $\mathscr{P}_F$ used in (9.2.25) by the factor $\det F$.

A simplified model with likely similar effects in most cases where volume compression/expansion is not a pronounced effect may consider just $P(F) = \mathrm{Cof}\,F$ and then the Korteweg-like stress containing $P'(F) = \mathrm{Cof}'F$. In both cases, we need the nonlocal 2nd-grade nonsimple model which controls also $\det\nabla y$ to have $P(\nabla y)^{-1}$ at

---

[26] More specifically, testing $r_{{\rm c},k} \in N_{[0,1]}(\alpha)$ by $\mathrm{div}(P^\top(\nabla y_k)\mathbb{K}P(\nabla y_k)\nabla\alpha_k)$ instead of $\Delta\alpha_k$ would formally modify the calculus (7.5.8) as

$$\int_\Omega r_{{\rm c},k}\mathrm{div}\big(P^\top(\nabla y_k)\mathbb{K}P(\nabla y_k)\nabla\alpha_k\big)\mathrm{d}x = \int_\Omega \partial\delta_{[0,1]}(\alpha_k)\mathrm{div}\big(P^\top(\nabla y_k)\mathbb{K}P(\nabla y_k)\nabla\alpha_k\big)\mathrm{d}x$$
$$= -\int_\Omega \partial^2\delta_{[0,1]}(\alpha_k)(P(\nabla y_k)\nabla\alpha_k)^\top\mathbb{K}(P(\nabla y_k)\nabla\alpha_k)\,\mathrm{d}x \le 0.$$

Yet this test hardly can be legitimate in Galerkin approximation even if very special bases would be considered.

[27] More in details, the damage profile in the actual configuration $\alpha^y := \alpha\circ y^{-1}$ is subjected to the gradient in the actual configuration, i.e. $\nabla^y\alpha^y$. As we have the chain rule $\nabla\alpha = \nabla^y(\alpha^y\circ y) = (\nabla\alpha^y)\nabla y$, we can see that $\nabla^y\alpha^y = (\nabla y)^{-\top}\nabla\alpha = ((\mathrm{Cof}\,\nabla y)\nabla\alpha)/\det\nabla y$, which gives rise to the choice (9.5.9). For isotropic materials, i.e. $\mathbb{K} = \kappa\mathbb{I}$, the last term in (9.5.6) takes the form $\frac{1}{2}\kappa\nabla\alpha^\top C^{-1}\nabla\alpha$ with $C = F^\top F$ the right Cauchy-Green tensor, which is used in literature as e.g. [529, Eq. (49)].

disposal so that, from the estimate of $P(\nabla y)\nabla\alpha$, one can infer also an estimate of $\nabla\alpha$ needed for a limit passage in the Korteweg-like stress.

We can again analyze this system by the Galerkin method. Again, rather for simplicity, we do not make any regularization of the inclusion (9.5.8b).

Proposition 9.5.2 (Weak solution to spatial-gradient damage model). *Let $\varphi \in C^1(\mathrm{GL}^+(d)\times[0,1])$ satisfy the assumptions (9.5.3b) and (9.5.5), ϱ satisfy (6.3.2), $\mathbb{H}\in\mathrm{SLin}(\mathbb{R}^{d\times d}_{\mathrm{sym}})$ and $\mathbb{K}\in\mathbb{R}^{d\times d}_{\mathrm{sym}}$ be positive definite, P satisfy (9.5.7), and the loading and initial conditions satisfy (9.2.4) with also $\alpha_0\in H^1(\Omega)$ so that $\varphi(\nabla y_0,\alpha_0)\in L^1(\Omega)$. Then the above mentioned Galerkin approximation (y_k,α_k) exists and satisfies the a-priori estimates*

$$\|y_k\|_{L^\infty(I;H^2(\Omega;\mathbb{R}^d))\cap W^{1,\infty}(I;L^2(\Omega;\mathbb{R}^d))} \le K, \tag{9.5.10a}$$

$$\|\alpha_k\|_{L^\infty(Q)\cap L^\infty(I;H^1(\Omega))\cap H^1(I;L^2(\Omega))} \le K. \tag{9.5.10b}$$

Moreover, every weak converging subsequence of $\{(y_k,\alpha_k)\}_{k\in\mathbb{N}}$ for $k\to\infty$ yields in its limit a weak solution to the initial-boundary-value problem (9.5.8). In addition,* $\mathrm{div}(P^\top(\nabla y)\mathbb{K}P(\nabla y)\nabla\alpha)\in L^2(Q)$ *and the inclusion (9.5.2b) holds even a.e. on Q.*

Sketch of the proof. Again we use Galerkin approximation and denote the approximate solution by (y_k,α_k). Testing (9.5.8a,b) in its Galerkin approximation by $(\dot y_k,\dot\alpha_k)$, in addition to (9.5.4a), we obtain also the estimates

$$\|\alpha_k\|_{L^\infty(Q)} \le K \quad\text{and}\quad \|P(\nabla y_k)\nabla\alpha_k\|_{L^\infty(I;L^2(\Omega;\mathbb{R}^d))} \le K.$$

Based on the weak convergence $y_k\to y$ and $\alpha_k\to\alpha$ and the Aubin-Lions compactness arguments, we prove the convergence in the damage flow rule.

In contrast to the model (9.5.2), we need to prove the strong convergence of $\nabla\alpha_k$ in $L^2(Q;\mathbb{R}^d)$ because of the nonlinear Korteweg-like stress $S_{\mathrm{K}}(F,\nabla\alpha)=P(F)^\top\mathbb{K}{:}P'(F){:}(\nabla\alpha\otimes\nabla\alpha)$ in (9.5.8a). To this goal, we use the uniform (with respect to y) strong monotonicity of the mapping $\alpha\mapsto -\mathrm{div}(P^\top(\nabla y)\mathbb{K}P(\nabla y)\nabla\alpha)$. Taking $\widetilde\alpha_k$ an approximation of α valued in the respective finite-dimensional spaces used for the Galerkin approximation and converging to α strongly, we can test (9.5.8b) in its Galerkin approximation by $\alpha_k-\widetilde\alpha_k$ and use it in the estimate

$$\limsup_{k\to\infty}\int_Q P^\top(\nabla y_k)\mathbb{K}P(\nabla y_k)\nabla(\alpha_k-\widetilde\alpha_k)\cdot\nabla(\alpha_k-\widetilde\alpha_k)\,\mathrm{d}x\mathrm{d}t = \lim_{k\to\infty}\int_Q\Big(\partial_\alpha\varphi(\nabla y_k,\alpha_k)$$
$$+\partial\zeta(\dot\alpha_k)\Big)(\widetilde\alpha_k-\alpha_k) - P^\top(\nabla y_k)\mathbb{K}P(\nabla y_k)\nabla\widetilde\alpha_k\cdot\nabla(\alpha_k-\widetilde\alpha_k)\,\mathrm{d}x\mathrm{d}t = 0$$

because $\partial_\alpha\varphi(\nabla y_k,\alpha_k)+\partial\zeta(\dot\alpha_k)$ is bounded in $L^2(Q)$ while $\widetilde\alpha_k-\alpha_k\to 0$ strongly in $L^2(Q)$ by the Aubin-Lions compactness theorem and because $P^\top(\nabla y_k)\mathbb{K}P(\nabla y_k)\nabla\widetilde\alpha_k$ converges strongly in $L^2(Q;\mathbb{R}^d)$ while $\nabla(\alpha_k-\widetilde\alpha_k)\to 0$ weakly in $L^2(Q;\mathbb{R}^d)$. As $P^\top(\nabla y_k)\mathbb{K}P(\nabla y_k)$ is uniformly positive definite, we thus obtain that $\nabla(\alpha_k-\widetilde\alpha_k)\to 0$ strongly in $L^2(Q;\mathbb{R}^d)$, and thus $\nabla\alpha_k\to\nabla\alpha$ strongly in $L^2(Q;\mathbb{R}^d)$. Then we have the convergence in the Korteweg-like stress

$$S_{\mathbb{K}}(\nabla y_k,\nabla\alpha_k)=P(\nabla y_k)^\top\mathbb{K}{:}P'(\nabla y_k){:}(\nabla\alpha_k\otimes\nabla\alpha_k)\\ \to P(\nabla y)^\top\mathbb{K}{:}P'(\nabla y){:}(\nabla\alpha\otimes\nabla\alpha)=S_{\mathbb{K}}(\nabla y,\nabla\alpha)$$

even strongly in $L^p(I;L^1(\Omega;\mathbb{R}^{d\times d}))$ for any $1\le p<+\infty$. The limit passage in the force equilibrium towards (9.5.8a) formulated weakly is then straightforward. $\Box$

Remark 9.5.3 (Fracture approximation). One can combine the model (9.5.1) with the crack surface density from (7.5.35) or (7.5.56) to obtain a *phase-field* model of (approximation of) fracture at large strains, as used in engineering literature, cf. e.g. [7]. Also the alternative model (9.5.6) can be adopted by this way.

Remark 9.5.4 (Time discretization). Inspired from the small-strain model, cf. Remark 7.5.14, we can devise a decoupled time discretization of (9.5.2a,b) now again with a vanishing-viscosity regularization as used in (9.2.21):

$$\varrho\frac{y_\tau^k-2y_\tau^{k-1}+y_\tau^{k-2}}{\tau^2}-\mathrm{div}\,S_\tau^k=f_\tau^k(y_\tau^k,\nabla y_\tau^k)\\ \text{with }\ S_\tau^k=\partial_F\varphi(\nabla y_\tau^k,\alpha_\tau^{k-1})-\mathrm{Div}(\mathbb{H}\nabla^2y^k)+\frac{\nabla y_\tau^k-\nabla y_\tau^{k-1}}{\sqrt{\tau}}\qquad\text{on }\Omega,\quad(9.5.11a)$$

$$\partial\zeta\Big(\frac{\alpha_\tau^k-\alpha_\tau^{k-1}}{\tau}\Big)+\partial_\alpha\varphi(\nabla y_\tau^k,\alpha_\tau^k)\ni\kappa\Delta\alpha_\tau^k,\qquad\text{on }\Omega\quad(9.5.11b)$$

to be completed with the corresponding boundary conditions and to be solved recursively for $k=1,...,T/\tau$ with

$$y_\tau^0=y_0,\qquad y_\tau^{-1}=y_0-\tau v_0,\qquad \alpha_\tau^0=\alpha_0.\qquad(9.5.12)$$

Assuming semiconvexity of $F\mapsto\varphi(F,\alpha)$ uniformly with respect to α (in the sense that the constant K in (9.2.20) is independent of α) and convexity of $\alpha\mapsto\varphi(F,\alpha)$, we can perform the test of (9.5.11a) by $y_\tau^k-y_\tau^{k-1}$ and of (9.5.11b) by $\alpha_\tau^k-\alpha_\tau^{k-1}$. One thus obtains the estimates

$$\|y_\tau\|_{L^\infty(I;H^2(\Omega;\mathbb{R}^d))\cap W^{1,\infty}(I;L^2(\Omega;\mathbb{R}^d))}\le K,\qquad(9.5.13a)$$

$$\|\alpha_\tau\|_{L^\infty(I;H^1(\Omega))\cap H^1(I;L^2(\Omega))}\le K\ \text{ and }\ \|\Delta\alpha_\tau\|_{L^2(Q)}\le K.\qquad(9.5.13b)$$

The convergence in terms of subsequences then can rely on the Aubin-Lions theorem (possibly generalized as Lemma B.4.11) so that $\nabla\bar y_\tau$ and $\bar\alpha_\tau$ converge strongly in $L^{1/\epsilon}(I;L^{2^*-\epsilon}(\Omega;\mathbb{R}^d))$ and $L^{1/\epsilon}(I;L^{2^*-\epsilon}(\Omega))$, respectively. Yet, due to the inertial term together with stress non-monotonically dependent on the deformation gradient, the assumption of semiconvexity makes the Rothe method not of real usage in particular if also $\det\nabla y$ is to be controlled positive, cf. Exercise 9.2.15.

Remark 9.5.5 (Damage combined with plasticity). A combination of such *plasticity/creep* model with the *damage* model (9.5.2) is a relatively routine task as far as the formulation concerns, based now on the elastic stored energy $\varphi_{\rm EL}=\varphi_{\rm EL}(F_{\rm EL},\alpha)$ and the dissipation potential $\mathfrak{R}=\mathfrak{R}(\alpha;P,\dot\alpha)=\mathfrak{R}_{\rm PL}(\alpha;P)+\delta^*_{(-\infty,a]}(\dot\alpha)$:

$$\varrho\ddot{y} - \mathrm{div}\big(\partial_F\varphi_{\rm EL}(\nabla y\Pi^{-1},\alpha)\Pi^{-\top} - \mathrm{Div}(\mathbb{H}\nabla^2 y)\big) = f(y) \qquad \text{in } Q, \tag{9.5.14a}$$

$$\partial_P\mathfrak{R}_{\rm PL}(\alpha;\dot{\Pi}\Pi^{-1})\Pi^{-\top} + (\nabla y)^\top\partial_\Pi\varphi_{\rm EL}(\nabla y\Pi^{-1},\alpha){:}(\Pi^{-1})' + \varphi_{\rm HD}'(\Pi) \ni \mathrm{div}(\kappa_1|\nabla\Pi|^{p-2}\nabla\Pi) \qquad \text{in } Q, \tag{9.5.14b}$$

$$\partial\delta^*_{(-\infty,a]}(\dot{\alpha}) + \partial_\alpha\varphi_{\rm EL}(\nabla y\Pi^{-1},\alpha) = \kappa_2\Delta\alpha \qquad \text{in } Q, \tag{9.5.14c}$$

$$S\vec{n} - \mathrm{div}_{\rm S}((\mathbb{H}\nabla^2 y)\cdot\vec{n}) = g \qquad \text{on } \Sigma, \tag{9.5.14d}$$

$$(\mathbb{H}\nabla^2 y){:}(\vec{n}\otimes\vec{n}) = 0, \ \ (\nabla\Pi)\vec{n} = 0, \ \text{ and } \ \nabla\alpha\cdot\vec{n} = 0 \qquad \text{on } \Sigma, \tag{9.5.14e}$$

$$y|_{t=0} = y_0, \qquad \dot{y}|_{t=0} = v_0, \qquad \Pi|_{t=0} = \Pi_0, \qquad \alpha|_{t=0} = \alpha_0 \qquad \text{on } \Omega. \tag{9.5.14f}$$

For chosing concrete data, such models can then consider all scenarios discussed already at small strains in Sect. 7.7.1.

9.5.2 Anisothermal model

An anisothermal variant may consider natural coupling through a temperature-dependent healing, while suppressing other mechanisms like thermal expansion causing adiabatic effects. This leads to the initial-boundary-value problem:

$$\varrho\ddot{y} - \mathrm{div}\, S = f(y) \quad \text{with } \ S = \partial_F\varphi(\nabla y,\alpha) - \mathrm{Div}(\mathfrak{H}\nabla^2 y) \qquad \text{in } Q, \tag{9.5.15a}$$

$$\partial_{\dot{\alpha}}\zeta(\theta;\dot{\alpha}) - \kappa\Delta\alpha + \partial_\alpha\varphi(\nabla y,\alpha) = 0 \qquad \text{in } Q, \tag{9.5.15b}$$

$$c_{\rm v}(\theta)\dot{\theta} - \mathrm{div}(\mathscr{K}(\nabla y,\alpha,\theta)\nabla\theta) = \xi(\theta,\dot{\alpha}) \qquad \text{in } Q, \tag{9.5.15c}$$

$$S\vec{n} - \mathrm{div}_{\rm S}((\mathfrak{H}\nabla^2 y)\cdot\vec{n}) = g \quad \text{and} \quad \mathscr{K}(\nabla y,\theta)\nabla\theta\cdot\vec{n} = h_\flat \qquad \text{on } \Sigma, \tag{9.5.15d}$$

$$(\mathfrak{H}\nabla^2 y){:}(\vec{n}\otimes\vec{n}) = 0 \quad \text{and} \quad \kappa\nabla\alpha\cdot\vec{n} = 0 \qquad \text{on } \Sigma, \tag{9.5.15e}$$

$$y|_{t=0} = y_0, \qquad \dot{y}|_{t=0} = v_0, \qquad \alpha|_{t=0} = \alpha_0, \qquad \theta|_{t=0} = \theta_0 \qquad \text{on } \Omega. \tag{9.5.15f}$$

where the damage-dissipation rate is $\xi(\theta,\dot{\alpha}) = \partial_{\dot{\alpha}}\zeta(\theta;\dot{\alpha})\dot{\alpha}$. The dependence of the heat-conductivity tensor $\mathscr{K}$ on the deformation gradient comes naturally from considering the heat transfer in the actual deformed rather than the reference configuration, cf. (9.2.26).

In addition to the analysis of the isothermal system (9.5.2), we now need to prove strong convergence of $\dot{\alpha}$ to pass to the limit in the heat source $\xi(\theta,\dot{\alpha})$. We use the strong monotonicity of $\partial_{\dot{\alpha}}\zeta(\theta;\cdot)$. By (9.5.20b) tested by $\dot{\alpha}_k - \dot{\alpha}$, we obtain

$$\begin{aligned}
&\limsup_{\tau\to 0}\int_Q \big(\partial_{\dot{\alpha}}\zeta(\theta_k;\dot{\alpha}_k) - \partial_{\dot{\alpha}}\zeta(\theta_k;\dot{\alpha})\big)(\dot{\alpha}_k - \dot{\alpha})\,\mathrm{d}x\mathrm{d}t\\
&\quad = \limsup_{\tau\to 0}\int_Q \kappa\nabla\alpha_k\cdot\nabla(\dot{\alpha} - \dot{\alpha}_k) + \big(\partial_\alpha\varphi(\nabla y_k,\alpha_k) + \partial_{\dot{\alpha}}\zeta(\theta_k;\dot{\alpha})\big)(\dot{\alpha} - \dot{\alpha}_k)\,\mathrm{d}x\mathrm{d}t\\
&\quad \le \lim_{\tau\to 0}\int_Q -\kappa\dot{\alpha}_k\Delta\alpha_k + \big(\partial_\alpha\varphi(\nabla y_k,\alpha_k) + \partial_{\dot{\alpha}}\zeta(\theta_k;\dot{\alpha})\big)(\dot{\alpha} - \dot{\alpha}_k)\,\mathrm{d}x\mathrm{d}t
\end{aligned}$$

$$+\limsup_{\tau\to 0}\int_\Omega \frac{\kappa}{2}|\nabla\alpha_0|^2-\frac{\kappa}{2}|\nabla\alpha_k(T)|^2\,\mathrm{d}x$$
$$\le -\int_Q \kappa\dot\alpha\Delta\alpha\,\mathrm{d}x\mathrm{d}t+\int_\Omega \frac{\kappa}{2}|\nabla\alpha_0|^2-\frac{\kappa}{2}|\nabla\alpha(T)|^2\,\mathrm{d}x=0, \tag{9.5.16}$$

where we used the inequality of the type (C.2.17) together with the facts that both $\partial_\alpha\varphi(\nabla y_k,\alpha_k)$ and $\partial_{\dot\alpha}\zeta(\theta_k;\dot\alpha)$ converge strongly in $L^2(Q)$. The last equality in (9.5.16) relies on the regularity property $\Delta\alpha=\kappa^{-1}(\partial_{\dot\alpha}\zeta(\theta;\dot\alpha)+\partial_\alpha\varphi(\nabla y,\alpha))\in L^2(Q)$, which makes it in duality with $\dot\alpha\in L^2(Q)$ and which makes the following formula legitimate:

$$\int_Q \dot\alpha\Delta\alpha\,\mathrm{d}x\mathrm{d}t=\int_\Omega \frac{1}{2}|\nabla\alpha_0|^2-\frac{1}{2}|\nabla\alpha(T)|^2\,\mathrm{d}x; \tag{9.5.17}$$

cf. the arguments we already used for (7.5.10) or (9.4.21).

Remark 9.5.6 (Time discretization). We again apply the *fractional-step method* to the system (9.5.15a-c) in such a way that the system decouples, and use the vanishing-viscosity regularization as we did in (9.2.21). The resulted time-incremental boundary problems in the classical formulation read as:

$$\varrho\frac{y_\tau^k-2y_\tau^{k-1}+y_\tau^{k-2}}{\tau^2}-\operatorname{div}S_\tau^k=f_\tau^k(y_\tau^k,\nabla y_\tau^k)$$
$$\text{with}\ \ S_\tau^k=\partial_F\varphi(\nabla y_\tau^k,\alpha_\tau^{k-1})-\operatorname{Div}(\mathfrak{H}\nabla^2y^k)+\frac{\nabla y_\tau^k-\nabla y_\tau^{k-1}}{\sqrt\tau}\quad \text{in }Q, \tag{9.5.18a}$$
$$\partial_{\dot\alpha}\zeta\Big(\theta_\tau^{k-1};\frac{\alpha_\tau^k-\alpha_\tau^{k-1}}{\tau}\Big)-\kappa\Delta\alpha_\tau^k+\partial_\alpha\varphi(\nabla y_\tau^k,\alpha_\tau^k)=0,\quad \text{in }Q, \tag{9.5.18b}$$
$$\frac{\vartheta_\tau^k-\vartheta_\tau^{k-1}}{\tau}-\operatorname{div}(\mathscr{K}(\nabla y_\tau^k,\alpha_\tau^k,\theta_\tau^{k-1})\nabla\theta_\tau^k)$$
$$=\xi_\tau\Big(\theta_\tau^{k-1},\frac{\alpha_\tau^k-\alpha_\tau^{k-1}}{\tau}\Big)\ \ \text{with}\ \vartheta_\tau^k=\mathscr{C}_{\rm v}(\theta_\tau^k)\quad \text{in }Q, \tag{9.5.18c}$$
$$S_\tau^k\vec n-\operatorname{div}_{\rm S}((\mathfrak{H}\nabla^2y_\tau^k)\cdot\vec n)=g_\tau^k(y_\tau^k,\nabla y_\tau^k),\quad (\mathfrak{H}\nabla^2y_\tau^k):(\vec n\otimes\vec n)=0,\quad \text{on }\Sigma. \tag{9.5.18d}$$
$$\kappa\nabla\alpha_\tau^k\cdot\vec n=0,\quad\text{and}\quad \mathscr{K}(\nabla y_\tau^k,\alpha_\tau^k,\theta_\tau^{k-1})\nabla\theta_\tau^k\cdot\vec n=h_{\flat,\tau}^k\quad \text{on }\Sigma. \tag{9.5.18e}$$

where we used the regularization ξ_τ of the dissipation rate defined as

$$\xi_\tau(\theta,\alpha):=\frac{\xi(\theta,\alpha)}{1+\tau\xi(\theta,\alpha)}. \tag{9.5.19}$$

Let us note that this system is decoupled: first one can solve the boundary-value problem (9.5.18a,b,d) to obtain some (y_τ^k,α_τ^k) and then the boundary-value problem (9.5.18c,e) to obtain θ_τ^k. Also note that $0\le\xi_\tau(\cdot,\cdot)\le 1/\tau$, which facilitates simple arguments to show existence of a conventional weak solutions to these semilinear boundary-value problems. Using the notation from Sect. C.2.3 for the interpolants, the system (9.5.15a-c) will then look as

$$\varrho\ddot{\bar y}_\tau-\operatorname{div}\overline S_\tau=\bar f_\tau\quad\text{with}\ \ \overline S_\tau=\partial_F\varphi(\nabla\bar y_\tau,\bar\alpha_\tau)-\operatorname{Div}(\mathfrak{H}\nabla^2\bar y_\tau)+\sqrt\tau\nabla\dot y_\tau, \tag{9.5.20a}$$

$$\partial_{\dot\alpha}\zeta(\underline{\theta}_\tau;\dot\alpha_\tau) - \kappa\Delta\overline{\alpha}_\tau + \partial_\alpha\varphi(\nabla\overline{y}_\tau,\overline{\alpha}_\tau) = 0, \tag{9.5.20b}$$

$$\dot\vartheta_\tau - \operatorname{div}(\mathscr{K}(\nabla\overline{y}_\tau,\overline{\alpha}_\tau,\underline{\theta}_\tau)\nabla\overline{\theta}_\tau) = \xi_\tau(\underline{\theta}_\tau,\dot\alpha_\tau) \quad\text{with}\quad \overline{\vartheta}_\tau = \mathscr{C}_{\rm v}(\overline{\theta}_\tau) \tag{9.5.20c}$$

on Q with the boundary conditions (9.5.15d,e) discretized appropriately and with the initial conditions (9.5.15f) written for $(y_\tau,\alpha_\tau,\theta_\tau)$; for the notation $\ddot{\tilde u}_\tau$ see (C.2.16) on p. 536.

Exercise 9.5.7. Show solvability of the boundary-value problems (9.5.18a,b,d) and (9.5.18c,e) by the direct method.

9.6 Flow in poroelastic materials and its thermodynamics

Another illustrative model at large strains is the diffusion in poroelastic materials analogously as it was formulated already at small strains in Section 7.6.3. Due to a transformed transportation process, cf. (9.6.3) below, we need to have a control over the determinant of the deformation gradient and therefore we will need to use the concept of nonlocal 2nd-grade nonsimple materials as we did in (9.2.6). Thus, in the isothermal situation, we consider the ansatz for the overall stored energy

$$\mathcal{E}(y,c) := \int_\Omega \varphi(\nabla y,c)\,{\rm d}x + \mathscr{H}(\nabla^2 y) \tag{9.6.1}$$

with $\mathscr{H}$ from (2.5.36). This leads to the initial-boundary-value problem for the system of two equations expressing balance of momentum and of concentration of the diffusant. Here we consider

$$\varrho\ddot y - \operatorname{div} S = f(y) \quad\text{with}\quad S = \partial_F\varphi(\nabla y,c) - \operatorname{Div}(\mathfrak{H}\nabla^2 y) \qquad\text{in } Q, \tag{9.6.2a}$$

$$\dot c - \operatorname{div}(\mathscr{M}(\nabla y,c)\nabla\mu) = 0 \quad\text{with}\quad \mu = \partial_c\varphi(\nabla y,c) \qquad\text{in } Q, \tag{9.6.2b}$$

$$S\vec n - \operatorname{div}_{\rm S}((\mathfrak{H}\nabla^2 y)\cdot\vec n) = g \quad\text{and}\quad (\mathfrak{H}\nabla^2 y):(\vec n\otimes\vec n) = 0 \qquad\text{on } \Sigma, \tag{9.6.2c}$$

$$\mathscr{M}(\nabla y,c)\nabla\mu\cdot\vec n = j_\flat \qquad\text{on } \Sigma, \tag{9.6.2d}$$

$$y|_{t=0} = y_0, \qquad \dot y|_{t=0} = v_0, \qquad c|_{t=0} = c_0 \qquad\text{on } \Omega. \tag{9.6.2e}$$

Like (9.2.26), a reasonable (although not ultimate) modeling assumption is that the *mass transport* is in the *actual deforming configuration* rather than in the reference one, so that the *transformed* (generalized) *Fick law* (in particular covering also *Darcy law*) uses the matrix of mobility coefficients *pulled back*, i.e. is

$$\mathscr{M}(x,F,c) = \mathscr{P}_F\mathbb{M}(c) := \frac{(\operatorname{Cof}F^\top)\mathbb{M}(x,c)\operatorname{Cof}F}{\det F} \quad\text{if } \det F > 0, \tag{9.6.3}$$

where $\mathbb{M}$ is the diffusant mobility (depending possibly also on $x\in\Omega$) as a material property while for $\mathscr{P}_F$ see again (9.2.25).

The *energy balance* behind this model is revealed when testing (9.6.2a) and (9.6.2b) respectively by $\dot{y}$ and μ when substituting $\dot{c}\mu = \partial_c\varphi(\nabla y,c)\dot{c}$:

$$\frac{\mathrm{d}}{\mathrm{d}t}\bigg(\int_\Omega \frac{\varrho}{2}|\dot{y}|^2 + \varphi(\nabla y,c)\,\mathrm{d}x + \mathscr{H}(\nabla^2 y)\bigg) + \int_\Omega \mathscr{M}(\nabla y,c)\nabla\mu\cdot\nabla\mu\,\mathrm{d}x = \int_\Omega f(y)\cdot\dot{y}\,\mathrm{d}x + \int_\Gamma g\cdot\dot{y} + \mu j_\flat\cdot\vec{n}\,\mathrm{d}S. \tag{9.6.4}$$

Definition 9.6.1 (Weak solutions). The triple (y,c,μ) with $y\in L^\infty(I;H^{2+\gamma}(\Omega;\mathbb{R}^d))\cap W^{1,\infty}(I;L^2(\Omega;\mathbb{R}^d))$, $c\in C_\mathrm{w}(I;H^1(\Omega))$, and $\mu\in L^2(I;H^1(\Omega))$ is called a weak solution to the initial-boundary-value problem (9.6.2) if (9.2.8) holds with $\partial_F\varphi(\nabla y,c)$ instead of $\varphi'(\nabla y)$, and, for any $v\in C^1(\bar{Q})$ with $v(T)=0$, it holds

$$\int_Q \frac{\mathbb{M}(c)(\mathrm{Cof}\,\nabla y)\nabla\mu}{\det\nabla y}\cdot((\mathrm{Cof}\,\nabla y)\nabla v) - c\dot{v}\,\mathrm{d}x\mathrm{d}t = \int_\Omega c_0 v(0)\,\mathrm{d}x \tag{9.6.5}$$

with $\mu=\partial_c\varphi(\nabla y,c)$ a.e. on Q.

Proposition 9.6.2 (Existence of weak solutions). *Let us consider (9.6.1) with $\varphi:\mathbb{R}^{d\times d}\times\mathbb{R}\to\mathbb{R}$ a normal integrand satisfying the coercivity $\varphi(F,c)\ge\varepsilon|F|^{1/\varepsilon}+\varepsilon|c|^{1/\varepsilon}+\varepsilon/(\det F)^q$ for some $\varepsilon>0$ and with $q>2d/(2\gamma+2-d)$ for $\det F>0$ while $\varphi(F,c)=+\infty$ otherwise, and with growth restriction*

$$\forall K_1\ \exists K\ \forall(F,c),\ |F|\le K_1,\ \det F\ge 1/K_1:\quad |\varphi'(F,c)|\le K(1+|c|^{2^*-\epsilon}),$$

and $\varphi(F,\cdot)$ with being uniformly convex. Let us further assume that $\mathscr{H}$ satisfies (2.5.37) with $d/2-1<\gamma<1$, the effective mobility tensor (9.6.3) uses $\mathbb{M}:\Omega\times\mathbb{R}\to\mathbb{R}^{d\times d}_{\mathrm{sym}}$ Carathéodory, bounded, and uniformly positive definite holds, ϱ satisfy (6.3.2), and and the loading and initial conditions satisfy (9.2.4) with also $j_\flat\in L^2(I;L^{2^{\sharp\prime}}(\Gamma;\mathbb{R}^d))$ and $c_0\in H^1(\Omega)$ so that $\varphi(\nabla y_0,c_0)\in L^1(\Omega)$. Then (9.6.2) possesses a weak solution (y,c,μ) in the sense of Definition 9.6.1.

Proof. As we need to control the determinant of the deformation gradient occurring in (9.6.3), we cannot impose semi-convexity assumption and cannot rely on a time discretization. Therefore, we will again use the Faedo-Galerkin in a careful way like in the proof of Proposition 9.2.12. Here we use some finite-dimensional subspaces of $H^{2+\gamma}(\Omega;\mathbb{R}^d)\cap W^{1,p}(\Omega;\mathbb{R}^d)$ for (9.6.2a) and of $H^1(\Omega)$ for both equations in (9.6.2b). Again, $k\in\mathbb{N}$ will be the discretization parameter. The existence of the Galerkin solution of thus regularized problem, let us denote it by (y_k,c_k,μ_k), can be argued by the successive prolongation argument, relying on the a-priori estimate obtained by means of (9.6.4) below.[28] As before, we use Healey-Krömer's Theorem 2.5.3 in the form of Corollary 2.5.11 is to be used together with a nonlocal

[28] The structure of the discrete problem is an initial-value problem for a system of so-called differential-algebraic equations of the 1st degree, which means that no time differentiation of the algebraic constraint, i.e. $\mu-\partial_c\varphi(\nabla y,c)=0$ discretized by the Galerkin method, is needed to elimination of the fast variable, i.e. here μ_k, to obtain the underlying system of ordinary differential equations. Here it is particularly simple because this constraint is even explicit with respect to μ_k.

linear hyperstress as in (9.2.8b) to have $\det\nabla y_k$ uniformly positive to control the effective mobility (9.6.3).

Then, written (9.6.2) for (y_k, c_k, μ_k) instead of (y, c, μ), we make the test of (9.6.2a) by $\dot{y}_k$. Moreover, the test of the former equation in (9.6.2b) by μ_k is legitimate as well as of the latter equation in (9.6.2b) by $\dot{c}_k$. This leads successively to

$$\frac{\mathrm{d}}{\mathrm{d}t}\int_\Omega \frac{\varrho}{2}|\dot{y}_k|^2 + \frac{1}{2}\mathfrak{H}\nabla^2 y_k \vdots \nabla^2 y_k \,\mathrm{d}x + \int_\Omega \partial_F \varphi(\nabla y_k, c_k) : \nabla \dot{y}_k \,\mathrm{d}x = \int_\Omega f \cdot \dot{y}_k \,\mathrm{d}x + \int_\Gamma g \cdot \dot{y}_k \,\mathrm{d}S, \tag{9.6.6a}$$

$$\int_\Omega \dot{c}_k \mu_k + \mathscr{M}(\nabla y_k, c_k)\nabla \mu_k \cdot \nabla \mu_k \,\mathrm{d}x = \int_\Gamma \mu_k j_\flat \cdot \vec{n} \,\mathrm{d}S, \tag{9.6.6b}$$

$$\int_\Omega \partial_c \varphi(\nabla y_k, c_k)\dot{c}_k \,\mathrm{d}x = \int_\Omega \mu_k \dot{c}_k \,\mathrm{d}x, \tag{9.6.6c}$$

cf. also (7.6.3). Summing it up, we enjoy cancellation of the terms $\pm\mu_k\dot{c}_k$. Using further the calculus $\partial_F \varphi(\nabla y_k, c_k) : \nabla \dot{y}_k + \partial_c \varphi(\nabla y_k, c_k)\dot{c}_k = \frac{\partial}{\partial t}\varphi(\nabla y_k, c_k)$, we eventually obtain the energy identity (9.6.4) written for the Galerkin approximation. From this, we obtain the a-priori estimates

$$\|y_k\|_{L^\infty(I;H^{2+\gamma}(\Omega;\mathbb{R}^d))\cap W^{1,\infty}(I;L^2(\Omega;\mathbb{R}^d))} \le K \quad\text{and}\quad \Big\|\frac{1}{\det\nabla y_k}\Big\|_{L^\infty(Q)} \le K, \tag{9.6.7a}$$

$$\Big\|\frac{(\mathrm{Cof}\,\nabla y_k)\nabla\mu_k}{\sqrt{\det\nabla y_k}}\Big\|_{L^2(Q;\mathbb{R}^d)} \le K. \tag{9.6.7b}$$

From (9.6.7a), we have the bound $\nabla y_k \in L^\infty(Q;\mathbb{R}^{d\times d})$ so that, realizing that $(\mathrm{Cof}\,\nabla y_k)^{-1} = (\nabla y_k)^\top/\det\nabla y_k$, we have

$$\|\nabla\mu_k\|_{L^2(Q;\mathbb{R}^d))} = \Big\|(\nabla y_k)^\top \frac{(\mathrm{Cof}\,\nabla y_k)}{\det\nabla y_k}\nabla\mu_k\Big\|_{L^2(Q;\mathbb{R}^d)} \le \|\nabla y_k\|_{L^\infty(Q;\mathbb{R}^{d\times d})}\Big\|\frac{1}{\sqrt{\det\nabla y_k}}\Big\|_{L^\infty(Q)}\Big\|\frac{(\mathrm{Cof}\,\nabla y_k)\nabla\mu_k}{\sqrt{\det\nabla y_k}}\Big\|_{L^2(Q;\mathbb{R}^{d\times d})}. \tag{9.6.8}$$

Then we use (9.6.7b) to obtain the bound of $\nabla\mu_k$ in $L^2(Q;\mathbb{R}^d)$. Then also the estimate

$$\|c_k\|_{L^\infty(I;H^1(\Omega))} \le K \tag{9.6.9}$$

is due to (9.6.7a,b) through (7.6.34).

Then we select a weakly* convergent subsequence in the topologies indicated in (9.6.7) and (9.6.9). Moreover, by comparison we can obtain also the estimates for $\varrho\ddot{y}_k$, and from the first equation (9.6.2b) and (9.6.7b), we can also see that (a Hahn-Banach extension of) $\dot{c}_k$ is bounded in $L^2(I;H^1(\Omega)^*)$. Then one can use the Aubin-Lions lemma to get strong convergence both for

$$\nabla y_k \to \nabla y \quad\text{in } L^p(Q;\mathbb{R}^{d\times d}) \quad\text{and}\quad c_k \to c \text{ in } L^p(I;L^{2^*-\varepsilon}(\Omega))$$

for any $1 \le p < +\infty$ and $0 < \varepsilon < 2^* - 1$. The convergence towards the weak solution of (9.6.2) is then easy. The only peculiar term is the diffusion flux when considering the ansatz (9.6.3) and thus the weak formulation (9.6.5), for which we need to show that

$$\int_Q \Big(\mathbb{M}(c_k)\frac{\mathrm{Cof}\,\nabla y_k}{\sqrt{\det\nabla y_k}}\nabla\mu_k\Big)\cdot\Big(\frac{\mathrm{Cof}\,\nabla y_k}{\sqrt{\det\nabla y_k}}\nabla v\Big)\,\mathrm{d}x\mathrm{d}t \to \int_Q \Big(\mathbb{M}(c)\frac{\mathrm{Cof}\,\nabla y}{\sqrt{\det\nabla y}}\nabla\mu\Big)\cdot\Big(\frac{\mathrm{Cof}\,\nabla y}{\sqrt{\det\nabla y}}\nabla v\Big)\,\mathrm{d}x\mathrm{d}t \tag{9.6.10}$$

for any $v \in C^1(\bar{Q})$. Here we used that

$$\frac{\mathrm{Cof}\,\nabla y_k}{\sqrt{\det\nabla y_k}}\nabla\mu_k \to \frac{\mathrm{Cof}\,\nabla y}{\sqrt{\det\nabla y}}\nabla\mu \qquad \text{weakly in } L^2(Q;\mathbb{R}^d), \text{ and} \tag{9.6.11a}$$

$$\frac{\mathrm{Cof}\,\nabla y_k}{\sqrt{\det\nabla y_k}} \to \frac{\mathrm{Cof}\,\nabla y}{\sqrt{\det\nabla y}} \qquad \text{strongly in } L^2(Q;\mathbb{R}^{d\times d}) \tag{9.6.11b}$$

because $\mathrm{Cof}\,\nabla y_k \to \mathrm{Cof}\,\nabla y$ strongly in $L^p(Q;\mathbb{R}^{d\times d})$ and $1/\det\nabla y_k \to 1/\det\nabla y$ strongly in $L^p(Q)$ for any $1 \le p < +\infty$ due to the Aubin-Lions theorem together with the latter estimate in (9.6.7a), and that $\mathbb{M}(c_k)(\nabla y_k)^{-1} \to \mathbb{M}(c)(\nabla y)^{-1}$ weakly in $L^2(Q)$ thanks to the estimate (9.6.7b). $\square$

The energy balance (9.6.4) shows that diffusion is related to the specific dissipation rate $\mathscr{M}(\nabla y, c)\nabla\mu\cdot\nabla\mu$ which, of course, in the full thermodynamical context is the heat-production rate. If the produced heat cannot be neglected (or at least considered as transferred away fast so that its influence on temperature can be neglected), it may cause a substantial variation of temperature and it may give rise to a coupled problem e.g. by considering the transport coefficients dependent also on temperature, cf. also Sect. 5.7.4.

In contrast to the isothermal system (9.6.2), the analysis now needs strong convergence of $\nabla\mu$ due to the nonlinear heat-production term $\mathscr{M}\nabla\mu\cdot\nabla\mu$. Yet, unlike we did in (8.6.10), we do not want to rely on the energy conservation because we do not have any Kelvin-Voigt-type viscosity in the model and therefore we do not have $\ddot{y}$ in duality with $\dot{y}$ granted. In particular, the argumentation like (8.6.10) does not work. For this reason, we now will still need compactness in $\nabla^2 y$, cf. the argumentation in (9.6.14) below. Therefore this is further reason for which we augment the stored energy by the nonlocal term as in Sect. 2.5.2.

Here it leads to expansion of the system (9.6.2) by the heat-transfer equation together with the mentioned modification of the hyperstress:

$$\varrho\ddot{y} - \mathrm{div}(S - \mathrm{Div}(\mathfrak{H}\nabla^2 y)) = f(y) \quad \text{with } S = \partial_F\varphi(\nabla y, c) \qquad \text{in } Q, \tag{9.6.12a}$$

$$\dot{c} - \mathrm{div}(\mathscr{M}(\nabla y, c, \theta)\nabla\mu) = 0 \quad \text{with } \mu = \partial_c\varphi(\nabla y, c) \qquad \text{in } Q, \tag{9.6.12b}$$

$$c_\mathrm{v}(\theta)\dot{\theta} - \mathrm{div}(\mathscr{K}(\nabla y, c, \theta)\nabla\theta) = \mathscr{M}(\nabla y, c, \theta)\nabla\mu\cdot\nabla\mu \qquad \text{in } Q, \tag{9.6.12c}$$

$$S\vec{n} - \mathrm{div}_\mathrm{S}((\mathfrak{H}\nabla^2 y)\cdot\vec{n}) = g \quad \text{and} \quad (\mathfrak{H}\nabla^2 y):(\vec{n}\otimes\vec{n}) = 0 \qquad \text{on } \Sigma, \tag{9.6.12d}$$

$$\mathscr{M}(\nabla y,c,\theta)\nabla\mu\cdot\vec{n}=j_\flat \quad\text{and}\quad \mathscr{K}(\nabla y,c,\theta)\nabla\theta\cdot\vec{n}=h_\flat \qquad \text{on } \Sigma, \tag{9.6.12e}$$

$$y|_{t=0}=y_0,\quad \dot{y}|_{t=0}=v_0,\quad c|_{t=0}=c_0,\quad \theta|_{t=0}=\theta_0 \qquad \text{on } \Omega. \tag{9.6.12f}$$

where we assumed φ independent of temperature and $\mathscr{M}(x,F,c,\theta)=\mathscr{P}_F^{-1}\mathbb{M}_{\rm M}(x,c,\theta)$ and $\mathscr{K}(x,F,c,\theta)=\mathscr{P}_F^{-1}\mathbb{K}_{\rm M}(x,c,\theta)$ are the pull-backed diffusivity and heat-conductivity tensors, cf. (9.2.26), with $\mathbb{M}_{\rm M}=\mathbb{M}_{\rm M}(x,c,\theta)$ and $\mathbb{K}_{\rm M}=\mathbb{K}_{\rm M}(x,c,\theta)$ as a material property, and where $\mathfrak{H}\nabla^2 y$ is from (2.5.36). Of course, both $\mathbb{M}_{\rm M}$, $\mathbb{K}_{\rm M}:\Omega\times\mathbb{R}^2\to\mathbb{R}^{d\times d}_{\rm sym}$ are to be Carathéodory mappings, bounded, and uniformly positive definite.

Let us only outline the peculiar points in the analysis. We again assume the blow-up growth of $\varphi(F,c)$ if $\det F\to 0+$ is in Proposition 9.6.2 to make the effective transport coefficients $\mathscr{M}$ and $\mathscr{K}$ under control. The energy-based a-priori estimation of an approximate solution is as before. As to the ∇c-estimate, it is as we already made in (9.6.9).

In contrast to the isothermal problem (9.6.2), we now need to prove the strong convergence of ∇c_k. We use (9.6.12b) written for the Galerkin approximation in the form

$$\begin{aligned}\dot{c}_k &= {\rm div}\big(\mathscr{M}(\nabla y_k,c_k,\theta_k)\nabla\partial_c\varphi(\nabla y_k,c_k)\big)\\ &= {\rm div}\big(\mathcal{M}(\nabla y_k,c_k,\theta_k)\nabla c_k+\mathscr{M}(\nabla y_k,c_k,\theta_k)\partial^2_{Fc}\varphi(\nabla y_k,c_k)\nabla^2 y_k\big)\end{aligned} \tag{9.6.13}$$

where we used the abbreviation $\mathcal{M}(F,c,\theta):=\mathscr{M}(F,c,\theta)\partial^2_{cc}\varphi(F,c)$. Testing (9.6.13) by $c_k-\tilde{c}_k$ with $\tilde{c}_k$ an approximation of c valued in V_k and converging to c strongly in $L^2(I;H^1(\Omega))$, one can estimate:

$$\begin{aligned}&\limsup_{k\to\infty}\int_Q \mathcal{M}(\nabla y_k,c_k,\theta_k)\nabla(c_k-\tilde{c}_k)\cdot\nabla(c_k-\tilde{c}_k)\,{\rm d}x{\rm d}t\\ &=\lim_{k\to\infty}\int_Q \mathscr{M}(\nabla y_k,c_k,\theta_k)\partial^2_{Fc}\varphi(\nabla y_k,c_k)\nabla^2 y_k\cdot\nabla(\tilde{c}_k-c_k)-\dot{c}_k\tilde{c}_k\\ &\qquad -\mathcal{M}(\nabla y_k,c_k,\theta_k)\nabla\tilde{c}_k\cdot\nabla(\tilde{c}_k-c_k)\,{\rm d}x{\rm d}t+\limsup_{k\to\infty}\frac12\int_\Omega |c_0|^2-|c_k(T)|^2\,{\rm d}x\\ &\le -\int_0^T\langle\dot{c},c\rangle\,{\rm d}x{\rm d}t+\frac12\int_\Omega|c_0|^2-|c(T)|^2\,{\rm d}x=0,\end{aligned} \tag{9.6.14}$$

where $\langle\cdot,\cdot\rangle$ here denotes the duality between $H^1(\Omega)^*$ and $H^1(\Omega)$. We used that $\nabla^2 y_k\to\nabla^2 y$ strongly due to the nonlocal $H^{2+\gamma}$-regularization via Aubin-Lions theorem. Also it is important that $\dot{c}_k\to\dot{c}$ weakly in $L^2(I;H^1(\Omega)^*)$ when considering $\dot{c}_k$ as a Hahn-Banach extension and $c\in L^2(I;H^1(\Omega))$ is in duality with $\dot{c}$. Due to the assumed uniform positive definiteness of $\mathcal{M}$, (9.6.14) yields the strong convergence of $\nabla c_k\to\nabla c$ in $L^2(Q;\mathbb{R}^d)$.

For the heat equation, we need the strong convergence of $\nabla\mu_k$, which can be seen from $\nabla\mu_k=\nabla\partial_c\varphi(\nabla y_k,c_k)=\partial^2_{cc}\varphi(\nabla y_k,c_k)\nabla c_k+\partial^2_{Fc}\varphi(\nabla y_k,c_k)\nabla^2 y_k$, as used already in (9.6.13).

Then the limit passage towards a weak solution to the initial-boundary-value problem (9.6.12) is easy.

In case of substantial variations of concentration of diffusant, the heat capacity may substantially vary, too. One should then consider $c_V = c_V(c,\theta)$. This would augment the right-hand side of (9.6.12c) by the adiabatic-type term, cf. (8.6.14). The control of $\dot{c}$ would then be needed, which could be made through a modification of the Cahn-Hilliard equation (9.6.12b) as in Remark 7.6.11.

Remark 9.6.3 (Involving capillarity). One can consider ∇c-term in the free energy, leading to the Cahn-Hilliard capillarity model as in Section 7.6.2. At large strains, instead of the gradient in the reference configuration like in Section 7.6.2, in some applications it is more relevant to consider the gradient in the *actual deformed configurations*, cf. e.g. [254, 350]. This gives rise to the term $\frac{\kappa}{2}|F^{-\top}\nabla c|^2$ enhancing the stored energy and new contributions into μ and S in (9.6.2) which then enhance as

$$\mu = \partial_c\varphi(F,c) - \mathrm{div}(\kappa F^{-1}F^{-\top}\nabla c) \quad \text{and} \tag{9.6.15a}$$

$$S = \partial_F\varphi(F,c) - \mathrm{Div}(\mathfrak{H}G) + \kappa F^{-1}\Big(\frac{\mathrm{Cof}' F}{\det F} - \frac{\mathrm{Cof}\, F \otimes \mathrm{Cof}\, F}{(\det F)^2}\Big){:}(\nabla c\otimes\nabla c) \tag{9.6.15b}$$

with $F = \nabla y$ and $G = \nabla^2 y$, cf. (9.5.8a,b). The last term in (9.6.15b) is called a *Korteweg stress* and, for this stress, we need the strong convergence of ∇c like we proved by (9.6.14).

Exercise 9.6.4. Assuming semiconvexity[29] of $F \mapsto \varphi(F,c)$ with K in (9.2.20) independent of c and convexity of $c \mapsto \varphi(F,c)$, devise a numerically stable time-discrete approximation for (9.6.2).

9.7 Towards selfpenetration problem under long-range interactions

So far, the models presented in this chapter either ignored the selfpenetration issue at all or prevented only local non-selfpenetration while possible selfpenetration of distant parts (like on Fig. 3.2-right on p. 66) was allowed, without caring about possible lost of relevancy of the model during the evolution. Actually, considering so-called Signorini (self)contact preventing selfpenetration in dynamical problems at finite or large strains seems extremely difficult and is (to date of publishing this book) still an open problem.

In case of quasistatic evolution of problems with internal variables, a certain option often considered in literature is a purely elastic inviscid model for deformation (respecting global non-interpenetration by means of the Ciarlet-Nečas condition (3.4.7)) while some dissipation in internal variables. In particular, the rate-independent evolution has been scrutinized for plasticity [361] or [360, Remark 4.2.12].

[29] Semiconvexity is rather academical assumption for large-strain elasticity and not needed for the Galerkin method, cf. also Exercise 9.2.9.

In dynamical cases, engineering computational mechanics at finite (or even at "only" large) strains routinely ignores possible selfpenetration. This is obviously the simplest approach if only reference configuration can be considered. Yet, sometimes, also the actual time-evolving deformedconfiguration is to be taken into account and then possible overlapping brings difficulties. This occurs typically when some quasistatic long-range interactions are involved, leading to gravitational, electrostatic, demagnetizing, or depolarizing field to be ultimately considered in the actual space configuration.

It is thus worth to develop a relevant mathematics. One must cope with situations that the deformation y is not injective and its inverse y^{-1} may be set-valued in some points. It relies on a nontrivial results as Federer's area formula (B.3.16) again in cooperation with the Healey-Krömer theorem 2.5.3.

Let us demonstrate it on the monopolar repulsive interaction, represented by the spatial time-evolving electric charge density q_S related with the time-independent referential material monopolar density q by the formula

$$q_\mathrm{S}(t,x^y) := \sum\nolimits_{x\in[y(t,\cdot)]^{-1}(x^y)} \frac{q(x)}{\det\nabla y(t,x)} \qquad \text{for } x^y \in \mathbb{R}^d. \tag{9.7.1}$$

Actually, (9.7.1) is our (natural) modeling choice, saying that overlapping means just summing the referential densities when correspondingly compressed or inflating. Let us note that the number of summation may vary in time (and may be zero if $x^y \neq \Omega^y(t)$ so that $[y(t,\cdot)]^{-1}(x^y) = \emptyset$), which makes handling of the spatial density cumbersome. Yet, the following assertion facilitates it quite easily, being thus of an essential importance:

Lemma 9.7.1 (G. Tomassetti et al. [466]). *If $q \in L^1(\Omega)$ and $y : \Omega \to \mathbb{R}^d$ is Lipschitz continuous with* $\operatorname{ess\,inf}_{x\in\Omega}\det\nabla y(x) > 0$, *and $q_\mathrm{S}(z) = \sum_{x\in y^{-1}(z)} q(x)/\det\nabla y(x)$ holds for a.a. $z \in \mathbb{R}^d$, then, for any $v \in C(\mathbb{R}^d)$, it holds*

$$\int_{\mathbb{R}^d} q_\mathrm{S}(x^y)v(x^y)\,\mathrm{d}x^y = \int_\Omega q(x)v(y(x))\,\mathrm{d}x. \tag{9.7.2}$$

Proof. Since $v\circ y : \Omega \to \mathbb{R}$ is always constant on the set $y^{-1}(z)$, we can calculate

$$\begin{aligned}\int_{\mathbb{R}^d} q_\mathrm{S}(x^y)v(x^y)\,\mathrm{d}x^y &= \int_{\mathbb{R}^d}\Big(\sum_{x\in y^{-1}(z)} \frac{q(x)}{\det(\nabla y(x))}\Big)v(z)\,\mathrm{d}z\\ &= \int_{\mathbb{R}^d}\sum_{x\in y^{-1}(z)} \frac{q(x)v(y(x))}{\det\nabla y(x)}\,\mathrm{d}z = \int_\Omega q(x)v(y(x))\,\mathrm{d}x.\end{aligned} \tag{9.7.3}$$

The last equation is due to change-of-variables formula (B.3.16) used here for $g(x) = q(x)v(y(x))/\det\nabla y(x)$, relying also with the assumption that $1/\det\nabla y$ is bounded, so that g is integrable. □

Together with the kinetic energy $\mathscr{T} = \mathscr{T}(\dot{u})$ from (6.1.4b), like in (4.7.3) on p. 158, we now consider the stored (or, due to negative electrostatic energy, rather free) energy

$$\begin{aligned}\mathcal{E}(y,\phi) &:= \int_\Omega \varphi(\nabla y)\,\mathrm{d}x + \int_{\mathbb{R}^d} \Big(q_{\rm ext}(x^y) + \sum_{x\in y^{-1}(x^y)} \frac{q(x)}{\det(\nabla y(x))}\phi(x^y) \\ &\qquad - \frac{\varepsilon_0}{2}\big|\nabla\phi(x^y)\big|^2 - \frac{\varepsilon_1}{p}\big|\nabla\phi(x^y)\big|^p \Big)\,\mathrm{d}x^y + \mathscr{H}(\nabla^2 y) \qquad (9.7.4a)\\ &= \int_\Omega \varphi(\nabla y) + q(x)\phi(y(x))\,\mathrm{d}x \\ &\quad + \int_{\mathbb{R}^d} q_{\rm ext}(x^y)\phi(x^y) - \frac{\varepsilon_0}{2}\big|\nabla\phi(x^y)\big|^2 - \frac{\varepsilon_1}{p}\big|\nabla\phi(x^y)\big|^p\,\mathrm{d}x^y + \mathscr{H}(\nabla^2 y); \qquad (9.7.4b)\end{aligned}$$

the equality between (9.7.4a) and (9.7.4b) being due to Lemma 9.7.1. Moreover, the loading $\mathcal{F}(t)$ is considered as a functional $\langle\mathcal{F}(t),y\rangle = \int_\Omega f(t)\cdot y\,\mathrm{d}x + \int_\Sigma g(t)\cdot y\,\mathrm{d}S$. The constant $\varepsilon_1 > 0$ in (9.7.4b) plays a regularizing role and $\varepsilon_0 + |\vec{e}|^{p-2}\varepsilon_1$ bears an interpretation of a permittivity depending on the intensity of electric field $|\vec{e}|$; one then speaks about nonlinear[30] (electrically charged) dielectrics. Actually, this is a bit speculative model and here it serves rather to illustrate the technique and to reveal mathematical difficulties.

Considering this energetics, the Hamilton variation principle (6.1.3), i.e. stationarity of $(y,\phi) \mapsto \int_0^T \mathscr{T}(\dot{y}) - \mathcal{E}(t,y,\phi) + \langle\mathcal{F}(t),y\rangle\,\mathrm{d}x$, yields the system

$$\varrho\ddot{y} - {\rm div}\,S = f - q(\nabla\phi\circ y) \quad \text{ with } \quad S = \varphi'(\nabla y) - {\rm Div}(\mathfrak{H}\nabla^2 y) \quad \text{ on } Q, \qquad (9.7.5a)$$

$$\begin{aligned}&{\rm div}((\varepsilon_0 + \varepsilon_1|\nabla\phi(t,\cdot)|^{p-2})\nabla\phi(t,\cdot)) \\ &\qquad + \sum_{x\in y^{-1}(t,\cdot)} \frac{q(x)}{\det(\nabla y(t,x))} + q_{\rm ext} = 0 \qquad \text{for a.a. } t\in I \qquad \text{on } \mathbb{R}^d \qquad (9.7.5b)\end{aligned}$$

with $\nabla\phi$ understood as space derivatives in the actual space configuration while other time/space derivatives are in the reference configuration, and the nonlocal hyperstress $\mathfrak{H}\nabla^2 y$ is from (2.5.36). Cf. also the static small-strain problem (5.6.1)-(5.6.2). Of course, (9.7.5) is to be completed by the boundary conditions (9.2.6b) and $\phi(\infty) = 0$. Let us note that the force in (9.7.5a) can be seen from (9.7.4b) while the right-hand side of (9.7.5b) can be seen from the equivalent form (9.7.4a).

The particular peculiarity consists in the bulk force $[q\nabla\phi]\circ y$ in (9.7.5a) which makes difficulty in convergence of approximate solutions. For this reason, we better rewrite it as $[q\nabla\phi]\circ y = [\nabla y]^{-\top}\nabla\phi_{\rm r}$ with $\phi_{\rm r} := \phi^y := \phi\circ y : Q \to \mathbb{R}$ being the pulled-back potential in the reference configuration as used in Sect. 4.5.4. Then, in the weak formulation, we use the Green formula and the calculus

[30] In fact, (9.7.7a) holds independently of the nonlinearity, i.e. with $\varepsilon_1 = 0$ too, so that the actual configuration $\Omega^{y(t)}$ is moving in some a-priori known area $B \subset \mathbb{R}^d$ where the coefficient ε_1 should be positive, otherwise it can vanish and just a linear vacuum with the constant permittivity ε_0 can be considered on the rest of the Universe. We will need $\varepsilon_1 > 0$ on B only for analytical reasons to control ϕ in $W^{1,p}(B) \subset C(B)$. This allows for coping with the composition $\phi\circ y$ which would otherwise be problematic if ϕ would not be continuous.

$$\int_Q q(\nabla\phi\circ y)\cdot v\,\mathrm{d}x\mathrm{d}t = \int_Q qv\cdot[\nabla y]^{-\top}\nabla\phi_\mathrm{r}\,\mathrm{d}x\mathrm{d}t$$
$$= \int_\Sigma \phi_\mathrm{r}[\nabla y]^{-\top}:(qv\otimes\vec{n})\,\mathrm{d}S\,\mathrm{d}t - \int_Q \phi_\mathrm{r}\nabla(qv):[\nabla y]^{-\top}$$
$$+ q\phi_\mathrm{r}v\cdot\Big[\frac{\mathrm{Cof}}{\det}\Big]'(\nabla y)\vdots\nabla^2 y\,\mathrm{d}x\mathrm{d}t.$$

with $[\mathrm{Cof}/\det]'(F) = [(\cdot)^{-\top}]'(F) = \mathrm{Cof}'\,F/\det F - \mathrm{Cof}\,F\otimes\mathrm{Cof}\,F^\top/\det F^2$ like in (9.4.9). In this way, the weak formulation of (9.7.5a) reads as

$$\int_Q \varphi'(\nabla y):\nabla v + \mathfrak{H}\nabla^2 y\vdots\nabla^2 v - \varrho\dot{y}\cdot\dot{v} + \phi^y\nabla(qv):[\nabla y]^{-\top} + q\phi^y v\cdot\Big[\frac{\mathrm{Cof}}{\det}\Big]'(\nabla y)\vdots\nabla^2 y\,\mathrm{d}x\mathrm{d}t$$
$$= \int_Q f\cdot v\,\mathrm{d}x\mathrm{d}t + \int_\Sigma g\cdot y + \phi^y[\nabla y]^{-\top}:(qv\otimes\vec{n})\,\mathrm{d}S\,\mathrm{d}t + \int_\Omega v_0\cdot v(0)\,\mathrm{d}x \qquad (9.7.6a)$$

for all $v: Q\to\mathbb{R}^d$ smooth with $v(T)=0$ together with the initial condition $y(0)=y_0$, while $\partial_\phi\mathcal{E}(y,\phi)=0$ gives the weak formulation of (9.7.5b) as

$$\int_{I\times\mathbb{R}^d} (\varepsilon_0+\varepsilon_1|\nabla\phi(t,x^y)|^{p-2})\nabla\phi(t,x^y)\cdot\nabla v(t,x^y) - q_\mathrm{ext}(x^y)v(t,x^y)\,\mathrm{d}x^y\mathrm{d}t$$
$$= \int_Q q(x)v(t,y(t,x))\,\mathrm{d}x\mathrm{d}t \qquad (9.7.6b)$$

for any $v: I\times\mathbb{R}^d\to\mathbb{R}$ smooth. The particular peculiarity is that $q_\mathrm{s}(t,\cdot)$ from (9.7.1) is controlled only in $L^1(\mathbb{R}^d)$ but not in $L^2(\mathbb{R}^d)$ even if $q\in L^\infty(\Omega)$. In addition, we need ϕ to be continuous to facilitate the convergence in the term $\phi\circ y$ which would otherwise be problematic if ϕ would be only in some Sobolev space. Hence, instead of the L^1-theory from Sect. 8.2, we have to use a nonlinear Laplacian and a conventional $W^{1,p}$-weak-solution theory with $p>d$. Again, we will use the Galerkin approximation, denoting by (y_k,ϕ_k) the approximate solution.

Proposition 9.7.2 (Existence and approximability of solutions to (9.7.5)). *Let (9.2.9) be satisfied, $q\in W^{1,1}(\Omega)$ be fixed, and $q_\mathrm{ext}\in L^1(\mathbb{R}^d)$ independent of time. Let also $p>d$. Then the Galerkin approximate solutions (y_k,ϕ_k) do exist and satisfy the a-priori estimates*

$$\|y_k\|_{L^\infty(I;H^{2+\gamma}(\Omega;\mathbb{R}^d))\cap W^{1,\infty}(I;L^2(\Omega;\mathbb{R}^d))} \le K \quad \textit{and} \quad \Big\|\frac{1}{\det\nabla y_k}\Big\|_{L^\infty(Q)} \le K, \qquad (9.7.7a)$$
$$\|\phi_k\|_{L^\infty(I;W^{1,p}(\mathbb{R}^d))} \le K. \qquad (9.7.7b)$$

Moreover, for $k\to\infty$, (y_k,ϕ_k) converges (in terms of subsequences) to weak solutions to (9.7.5) with the initial/boundary condition (9.2.1c) and (9.2.6b) with $\phi(\infty)=0$ in the sense (9.7.6).

Proof. The a-priori estimates (9.7.7) can be seen from the boundedness of the kinetic and the stored energy when transformed by exploiting (9.7.5b) similarly as in (4.7.4). More specifically, we use (9.7.6b) in its Galerkin approximation tested by ϕ_k, which gives

$$\int_{\mathbb{R}^d} \varepsilon_0 |\nabla\phi_k(t,x^y)|^2 + \varepsilon_1 |\nabla\phi_k(t,x^y)|^p \,\mathrm{d}x^y
= \int_\Omega q(x)\phi_k(t,y_k(t,x))\,\mathrm{d}x + \int_{\mathbb{R}^d} q_{\rm ext}(x^y)\phi_k(t,x^y)\,\mathrm{d}x^y \tag{9.7.8}$$

for a.a. $t\in I$. Then we perform the energetic test of (9.7.5) in its Galerkin approximation by $\dot y_k$ and $\dot\phi_k$, which altogether gives

$$\begin{aligned}
&\frac{\mathrm{d}}{\mathrm{d}t}\Big(\int_\Omega \frac{\varrho}{2}|\dot y_k|^2 + \varphi(\nabla y_k)\,\mathrm{d}x + \int_{\mathbb{R}^d} \frac{\varepsilon_0}{2}|\nabla\phi_k|^2 + \frac{\varepsilon_1}{p'}|\nabla\phi_k|^p\,\mathrm{d}x^y + \mathscr{H}(\nabla^2 y_k)\Big)\\
&\overset{\text{due to (9.7.8)}}{=} \frac{\mathrm{d}}{\mathrm{d}t}\Big(\int_\Omega \frac{\varrho}{2}|\dot y_k|^2 + \varphi_k(\nabla y_k) + q(\phi_k\circ y_k)\,\mathrm{d}x\\
&\qquad\qquad - \int_{\mathbb{R}^d} \frac{\varepsilon_0}{2}|\nabla\phi_k|^2 + \frac{\varepsilon_1}{p}|\nabla\phi_k|^p - q_{\rm ext}\phi_k\,\mathrm{d}x^y + \mathscr{H}(\nabla^2 y_k)\Big)\\
&= \frac{\mathrm{d}}{\mathrm{d}t}\Big(\int_\Omega \frac{\varrho}{2}|\dot y_k|^2 + \varphi(\nabla y_k)\,\mathrm{d}x - \int_{\mathbb{R}^d} \frac{\varepsilon_0}{2}|\nabla\phi_k|^2 + \frac{\varepsilon_1}{p}|\nabla\phi_k|^p\,\mathrm{d}x^y + \mathscr{H}(\nabla^2 y_k)\Big)\\
&\qquad\qquad + \int_{\mathbb{R}^d} q_{\rm ext}\dot\phi_k + \dot q_{\rm ext}\phi_k\,\mathrm{d}x^y + \int_\Omega q(\nabla\phi_k\circ y_k)\dot y_k + q\dot\phi\circ y_k\,\mathrm{d}x\\
&\overset{\text{(9.7.5) tested by } \dot y_k \text{ and } \dot\phi_k}{=} \int_\Omega f\cdot\dot y_k\,\mathrm{d}x + \int_\Gamma g\cdot\dot y_k\,\mathrm{d}S + \int_{\mathbb{R}^d} \dot q_{\rm ext}\phi_k\,\mathrm{d}x^y,
\end{aligned} \tag{9.7.9}$$

where we used also the calculus $q(\phi\circ y)^\cdot = q(\phi\circ y)^\cdot = q(\nabla\phi\circ y)\dot y + q\dot\phi\circ y$ and also $\int_{\mathbb{R}^d}(q_s+q_{\rm ext})(\dot\phi\circ y)\,\mathrm{d}x^y = \frac{\mathrm{d}}{\mathrm{d}t}\int_{\mathbb{R}^d}\frac{1}{2}\varepsilon_0|\nabla\phi(x^y)|^2 + \frac{1}{p}\varepsilon_1|\nabla\phi(x^y)|^p\,\mathrm{d}x^y$. For simplicity, we assumed $q_{\rm ext}$ constant in time so that $\dot q_{\rm ext} = 0$ and the last term in (9.7.9) simply vanish. The term $\int_\Gamma g\cdot\dot y_k\,\mathrm{d}S$ in (9.7.9) is to be treated by the by-part integration when (9.7.9) is integrated in time over the interval $[0,t]$. By Gronwall inequality and by the Healey-Krömer theorem 2.5.3 in the form of Corollary 2.5.11, we obtain the a-priori estimates (9.7.7).

By Banach's selection principle, we can take a subsequence of $\{y_k\}_{k\in\mathbb{N}}$ converging weakly* in $L^\infty(I;H^{2+\gamma}(\Omega;\mathbb{R}^d)) \cap W^{1,\infty}(I;L^2(\Omega;\mathbb{R}^d))$ to some y. In particular, $\nabla y_k \to \nabla y$ in $L^{1/\varepsilon}(I;C(\bar\Omega;\mathbb{R}^{d\times d}))$ and $\nabla^2 y_k \to \nabla^2 y$ in $L^{1/\varepsilon}(I;L^2(\Omega;\mathbb{R}^{d\times d\times d}))$ with any $\varepsilon>0$ by the Aubin-Lions theorem[31] together with the compact embedding $H^{2+\gamma}(\Omega)\subset C^1(\bar\Omega)$. Also $(\nabla y_k)^{-\top} \to (\nabla y)^{-\top}$ in $L^{1/\varepsilon}(I;C(\bar\Omega;\mathbb{R}^{d\times d}))$, and therefore also its traces on Σ converges in $L^{1/\varepsilon}(I;C(\Gamma;\mathbb{R}^{d\times d}))$. By Proposition B.2.1(ii)–(iii), for a subsequence we have also the convergence $\nabla^2 y_k(t) \to \nabla^2 y(t)$ strongly in $L^2(\Omega;\mathbb{R}^{d\times d\times d})$ and $\nabla y_k(t)\to\nabla y(t)$ and $(\nabla y_k(t))^{-\top} \to (\nabla y(t))^{-\top}$ in $C(\bar\Omega;\mathbb{R}^{d\times d})$ and of the traces in $C(\Gamma;\mathbb{R}^{d\times d})$ for a.a. $t\in I$.[32]

[31] In fact, $\nabla^2 y_k \to \nabla^2 y$ in even a smaller space, namely in $L^{1/\varepsilon}(I;L^{2d/(d-2\gamma)-\varepsilon}(\Omega;\mathbb{R}^{d\times d\times d}))$ with any $0<\varepsilon\le 2d/(d-2\gamma)-1$.

[32] Actually, to see the analog of Proposition B.2.1 for Banach-space valued functions, it should be applied to the scalar-valued functions $t\mapsto \|\nabla^2 y_k(t) - \nabla^2 y(t)\|_{L^2(\Omega;\mathbb{R}^{d\times d\times d})}$ and $t\mapsto \|(\nabla y_k(t))^{-\top} - (\nabla y(t))^{-\top}\|_{C(\bar\Omega;\mathbb{R}^{d\times d})}$ etc.

A peculiarity is that any information about $\dot{\phi}_k$ seems problematic, and we thus avoid usage of it. The limit passage in the weakly formulated approximate Poisson equation, i.e.

$$\int_{\mathbb{R}^d} (\varepsilon_0+\varepsilon_1|\nabla\phi_k(t,x^y)|^{p-2})\nabla\phi_k(t,x^y)\cdot\nabla v(t,x^y) - q_{\rm ext}(t,x^y)v(t,x^y)\,{\rm d}x^y \\ = \int_\Omega q(x)v(t,y_k(t,x))\,{\rm d}x \qquad (9.7.10)$$

for any $t\in I$ towards (9.7.6b) not integrated in time is then easy since $v(t,\cdot)\circ y_k(t)\to v(t,\cdot)\circ y(t)$ for $v(t,\cdot)$ continuous. As the limit $\phi(t)\in W^{1,p}(\mathbb{R}^d)$ is determined uniquely, we do not need to select any other subsequences after already having selected the mentioned subsequence of $\{y_k\}_{k\in\mathbb{N}}$. Taking into account the compact embedding $W^{1,p}_{\rm loc}(\mathbb{R}^d)\subset C_{\rm loc}(\mathbb{R}^d)$ since $p>d$, and having thus proved the strong (i.e. uniform) convergence of $\phi_k(t)\to\phi(t)$ on $C(B)$ with $B\subset\mathbb{R}^d$ a sufficiently large ball such that $y_k(t,\Omega)\subset B$ for any $k\in\mathbb{N}$ and any $t\in I$, we have also $\phi_k(t,y_k(t))\to\phi(t,y(t))$ in $C(\bar{\Omega})$, cf. also Proposition B.2.11. Then we have also

$$\int_\Omega \phi_k(t,y_k(t))\nabla(qv(t)){:}[\nabla y_k(t)]^{-\top}{\rm d}x \to \int_\Omega \phi(t,y(t))\nabla(qv(t)){:}[\nabla y(t)]^{-\top}{\rm d}x, \quad (9.7.11a)$$

$$\int_\Omega q\phi_k(t,y_k(t))v(t)\cdot\Big[\frac{\rm Cof}{\rm det}\Big]'(\nabla y_k(t))\,\vdots\,\nabla^2 y_k(t)\,{\rm d}x \\ \to \int_\Omega q\phi(t,y(t))v(t)\cdot\Big[\frac{\rm Cof}{\rm det}\Big]'(\nabla y_k(t))\,\vdots\,\nabla^2 y_k(t)\,{\rm d}x, \quad \text{and also} \qquad (9.7.11b)$$

$$\int_\Gamma \phi_k(t,y_k(t))[\nabla y_k(t)]^{-\top}{:}((qv\otimes\vec{n})\,{\rm d}S \to \int_\Gamma \phi(t,y(t))[\nabla y(t)]^{-\top}{:}((qv\otimes\vec{n})\,{\rm d}S \quad (9.7.11c)$$

for a.a. $t\in I$. Note that, in (9.7.11a), we used that $\nabla q\in L^1(\Omega;\mathbb{R}^d)$ while, in (9.7.11c), we used that $q|_\Gamma\in L^1(\Sigma)$ due to (B.3.12) if $q\in W^{1,1}(\Omega)$, as assumed. Then we can integrate (9.7.11) over I, and prove the convergence by using Lebesgue theorem, relying on a common majorant which is even constant due to the $L^\infty(I)$-estimates at disposal. The convergence in the approximate form of (9.7.6a) towards its limit is then proved. □

Remark 9.7.3 (Evolving charge). One can consider the charge q not fixed but undergoing a possible evolution. Then $\mathcal{E}(y,\phi)$ in (9.7.4) is to be considered also as a function of q, i.e. $\mathcal{E}=\mathcal{E}(y,\phi,q)$. The partial derivative $\partial_q\mathcal{E}$ is then in the position of a "chemical", or here rather electrochemical potential μ, and equals here to just the (negative) electrostatic potential $\mu=-\phi^y$; this can be seen particularly from (9.7.4b). Considering the electric conductivity tensor $\mathbb{S}=\mathbb{S}(x)$ in the material configuration, the flux of q governed by the Fick law $q=\mathscr{S}(x,F)\nabla\mu$ turns here to be the *Ohm law*

$$\text{electric current} = -\mathscr{S}(x,\nabla y(x))\nabla\phi(y(x)) \\ \text{with}\ \ \mathscr{S}(x,F)=\mathscr{P}_F\mathbb{S}(x)=\frac{({\rm Cof}\,F)^\top\mathbb{S}(x){\rm Cof}\,F}{\det F} \qquad (9.7.12)$$

where the pull-back operator $\mathscr{P}_F : \mathbb{R}^{d\times d} \to \mathbb{R}^{d\times d}$ from (9.2.25). The nondiffusive Allen-Cahn type equation represents here electric-charge conservation:

$$\dot{q} + \operatorname{div}(\mathscr{S}(\nabla y)\nabla\phi^y) = 0, \qquad (9.7.13)$$

which couples with the system (9.7.5). Yet, the qualification of q valued in $W^{1,1}(\Omega)$ as used in Proposition 9.7.2 seems not realistic in this simple model.

Remark 9.7.4 (Attractive monopolar interactions). Another prominent example of a monopolar interaction is gravitation. Then q is mass density and ϕ is the gravitational field. The essential difference is that the gravitational constant occurs in (9.7.4) and (9.7.5b) with a negative sign, cf. also (4.7.2) on p. 158. The coercivity of the static stored energy is not automatic and, roughly speaking needs sufficiently small total mass of a medium sufficiently elastically tough, cf. also Sect. 4.7. It is needed also in the dynamical case where its weaker coercivity when $\|\cdot\|^2_{L^2(\Omega;\mathbb{R}^d)}$ is added with a sufficiently large weights suffices. On top of it, $\varepsilon_1 = 0$ is the only reasonable choice in such gravitation interaction, which is not covered by the proof of Proposition 9.7.2.

Remark 9.7.5 (Dipolar long-range interactions). Considering a vector-valued density $\vec{q} : \Omega \to \mathbb{R}^d$ and $\vec{q}_{\rm ext}(\cdot,t) : \mathbb{R}^d \to \mathbb{R}^d$ instead of just scalar valued does not change (9.7.1) and Lemma 9.7.1, but the stored energy (9.7.4b) is to be modified as:

$$\mathcal{E}(y,\phi) = \int_\Omega \varphi(\nabla y) - \vec{q}\cdot\nabla\phi(y)\,\mathrm{d}x + \int_{\mathbb{R}^d} \frac{\kappa}{2}\big|\nabla\phi(x^y)\big|^2 - \vec{q}_{\rm ext}\cdot\nabla\phi\,\mathrm{d}x^y + \mathscr{H}(\nabla^2 y);$$

cf. also (4.5.44) with (4.5.42). Instead of (9.7.5), the Hamiltonian variational principle then yields the system

$$\varrho\ddot{y} - \operatorname{div} S = f + [(\nabla^2\phi)\circ y]\vec{q} \quad \text{with} \quad S = \varphi'(\nabla y) - \operatorname{Div}(\mathfrak{H}\nabla^2 y) \quad \text{on } Q, \qquad (9.7.14a)$$

$$\kappa\Delta\phi = \operatorname{div}\Big(\sum\nolimits_{x\in y^{-1}(\cdot,t)} \frac{\vec{q}(x)}{\det(\nabla y(x,t))} + \vec{q}_{\rm ext}(\cdot,t)\Big) \quad \text{for a.a. } t\in I \qquad \text{on } \mathbb{R}^d \quad (9.7.14b)$$

The interpretation is of ϕ and $\vec{q}$ is the potential of magnetic field and magnetization in elastic ferromagnets, respectively, or alternatively electrostatic field and polarization in elastic piezoelectric materials with spontaneous polarization. The analysis is, however, even more problematic comparing to the repulsive monopolar case due to less regularity of the Poisson equation (9.7.14b) which has the divergence of an L^1-function in the right-hand side.[33]

[33] Actually, this difficulty is not seen in the a-priori estimation strategy (9.7.9) as well as in the limit passage in the Poisson equation (9.7.14b) written in the form (4.5.46) on p. 144. Yet, the difficulty occurs in the term $(\nabla^2\phi)\circ y$ in the right-hand side of (9.7.14a) because $\nabla^2\phi$ hardly can be continuous and the composition with y even does not need to be measurable.

Suggestions for further reading and some open problems

> Remember to look up at the stars and not down at your feet. Be curious. And however difficult life may seem, there is always something you can do and succeed at. It matters that you just don't give up.
>
> STEPHEN WILLIAM HAWKING (1942–2018)

It is worth putting this book into a broader context of state of art, both as far as the already existing literature, surrounding fields not addressed in this book, and further open challenges.

Other monographical expositions

There is a wast amount of more specialized textbooks and monographs exposing particular mechanical or mathematical concepts more in details or advancing them, or focusing only on some particular mechanical or mathematical aspects.

There are many thorough and general continuum-mechanical expositions without strong emphasis to mathematical aspects. Let us mention monographs by Y. Basar and D. Weichert [48], A. Bertram and R. Glüge [66], Y.I. Dimitrienko [157], M.E. Gurtin [231, 232], M.E. Gurtin, E. Fried, and L. Anand [234], P. Haupt [243], J.N. Reddy [432], or D.R. Smith [489], or C. Truesdell [518]. Mathematical treatment of nonlinear elasticity can be found in S. Antman [14] including problems for lower-dimensional structures, like rods or strings or in P.G. Ciarlet [114, 115, 116] where the the author treats problems ranging from three-dimensional elasticity over plates to shells. Mathematical modeling in continuum mechanics including a comprehensive introduction to the topic can be found in R. Temam and A. Miranville [510]. Computational approaches and application of elasticity are reviewed in M.H. Sadd [473]. Part I of this book is directly related to multidimensional and vectorial variational calculus. Mathematical aspects of such variational calculus (not always in the direct context with mechanics of deformable solids) is in many monographs, in particular B. Dacorogna [134, 135], G. Dal Maso [138], I. Ekeland and R. Temam [173], L.C. Evans [179], E. Giusti [222], C.B. Morrey [375], F. Rindler [440], M. Struwe [496], M. Šilhavý [483], or E. Zeidler [539]. Additionally, new mathematical results coming from other areas of mathematical analysis as e.g. theory of conformal maps have found applications in elasticity as far as injectivity and regularity of deformations concern, see S. Hencl and P. Koskela [248].

The notion of Γ-convergence is an important tool in many situations. In particular, the static phase-field crack approximation (i.e. the Ambrosio-Tortorelli functional and various generalizations) is in the monograph by A. Braides [84]. Another

M. Kružík and T. Roubíček, *Mathematical Methods in Continuum Mechanics of Solids*, Interaction of Mechanics and Mathematics,
https://doi.org/10.1007/978-3-030-02065-1

usage of the Γ-convergence is for relaxation in the sense of Definition 4.4.1. Here, a well-known monograph by B. Dacorogna's [135] provides a comprehensive exposition of the existence and relaxation results in the vectorial variational calculus in the context of nonlinear elasticity theory. Young measures and their applications in relaxation and the calculus of variation are the main topic of P. Pedregal's book [410], and G. Dozmann's and S. Müller's lecture notes [160] and [381], respectively. We also refer to the second author's monograph [452]. Nevertheless, techniques described there do not allow for physically relevant treatment of variational problems in elasticity. The issue is that the orientation-preservation of the admissible class of deformations may be lost.

Thermomechanics (in this book Chap. 8 and partly also Chap. 9) is usually addressed "only" as far as formulation of the problems, without truly mathematical analysis, although even physical justification of the formulation itself (in this book rather suppressed) might be highly nontrivial and sophisticated. Actually, the development started in particular in 19th century when R. Clausius realizes the entropy production in closed systems (2nd Law of Thermodynamics) and later augmentation to open systems have been a great cultural achievement of mankind, related e.g. with the names of J. van der Waals, E. Schrödinger, A. Einstein, W. Nernst, L. Onsager, or I. Prigoggine, to name at least some from more that 20 winners of Nobel prizes related to thermodynamics. Sometimes, the adjective "irreversible" and "nonequilibrium" are used in this context for dissipative processes in distributed-parameter settings (leading to partial rather than ordinary differential equations). Focusing to these irreversible nonequilibrium situations, the reader can find deep arguments in the monographs by S.R de Groot & P. Mazur [149], H.C. Öttinger [405], M. Šilhavý [483], or K. Wilmanski [532]. Particular focus on plasticity and fracture (thermo)mechanics is in G.A. Maugin [340].

Nonlocal (gradient) theories, leading to nonsimple materials and heavily used in some parts of this book, can be found e.g. in A. Berezovski, J. Engelbrecht, and G.A. Maugin [63], or A.C. Eringen [176], cf. also works of D.G.B. Edelen and C.B. Kafadar in [177] or of M. Šilhavý [483]. It may lead to desired physical phenomena dispersion of elastic waves (as briefly shown in this book) while yielding important mathematical benefits.

Mathematical methods for evolution problems use calculus of variations rather in a limited extent, exploiting the concept of critical points and Hamilton variational principle rather than mere minimization, cf. e.g., M. Struwe [496] or A. Bedford [54]. More universally and efficiently, it basically applies theory of linear or quasilinear parabolic or hyperbolic partial differential equations, or possibly variational inequalities if the governing potentials are nonsmooth, which is in cases that the evolution of internal variables is activated or unidirectional or these variables are explicitly constrained. There is a vast number of monographs and textbooks in this subject, in particular L.C. Evans [179], J.-L. Lions [317], or the second author's monograph [456]. Specialized monographs are for quasistatic rate-independent problems, exploiting calculus of variations quite significantly, see, e.g., A. Mielke and T. Roubíček [360].

Even relatively narrow fields may enjoy in fact a wide application and intensive research initiative worldwide. An example is certainly poroelasticity and poromechanics in general, see e.g. the monographs by J. Bear [52], S.C. Cowin [132], S. de Boer [145], A.H.-D. Cheng [112], or O. Coussy [131]. The damage mechanics is a topic of L.M. Kachanov's book [268] and it also can be found in M. Frémond's monograph [202, Chap. 6] or recent book by Z.P. Bažant and M. Jirásek [51].

Plasticity is another lively area of mathematical research. Here we would like to list books by A. Bertram [65], W. Han and B.D. Reddy [241], G.T. Houlsby and A.M. Puzrin [255], or R. Temam [509]. See also H.-D. Alber [5] where several other fairly general models of material constitutive laws are considered. The monograph Q.S. Nguyen [395] addresses continuum mechanics of solids addressed from many viewpoints and directions. Phenomena like plasticity, damage combined with viscoelasticity are thoroughly reviewed by G. Maugin in [340, 342]. An easy-to-read introduction to analysis of elasticity and elastoplasticity is [388] by J. Nečas and I. Hlaváček. An engineering point of view including experimental methods in viscoelasticity is taken in R.S. Lakes [301].

Other monographs concern phase transitions in solids, cf. K. Bhattacharya [68], M. Pitteri and G. Zanzotto [415], or M. Frémond [201, Chap. 13] and [202, Chap. 5], or I. Müller and W. Weiss [377, Chap. 13]. Mechanics coupled with (electro)magnetic effects are the topic of W.F. Brown [95], L. Dorfmann and R.W. Ogden [161].

Some other related areas.

Although it might seem that this book addresses a lot of aspects in continuum mechanics of solids, several important topics remained intentionally untouched, partly also to limit the size of this book.

In particular, we did not present unilateral problems in mechanics, related with variational inequalities and *nonsmooth mechanics* (except nonsmoothness in dissipation mechanism related to activated processes in solids). These unilateral problems arises typically as various contact problems on the boundary or inside on a 2-dimensional interfaces. A special monographs in this direction are in particular G. Duvaut and J.-L. Lions [170], M. Frémond [201, 202], or I. Hlaváček et al. [252].

A rich and active area of research are various *dimension-reduction* problems. The aim is to predict asymptotic mechanical behavior of thin structures like membranes, plates, shells, or thin films, if their thickness approaches zero and thus the original 3-dimensional geometry is reduced to only 2-dimensional one. To go even further, one obtains various 1-dimensional models of beams, rods, strings, trusses, or struts. This is obviously convenient from the computational point of view. In static situations, such models are (desired to be) justified by convergence of minimizers and/or *Γ-convergence*, which represents here the main underlying mathematical tools. Alternatively, some other asymptotic methods are used, giving less justification. Various models distinguish according whether they are linear or not and which sort of geometrical deformation they capture (bending or torsion etc.). To outline 3D-2D dimensional reductions, let us consider for $h > 0$ a cylindrical domain $\Omega_h := \omega \times (-h/2, h/2)$, where $\omega \subset \mathbb{R}^2$ is a bounded Lipschitz domain, and the stored energy

$$\mathcal{E}^h(\tilde{y}) := \int_{\Omega_h} \varphi(\nabla\tilde{y}(z))\,\mathrm{d}z\,.$$

A standard approach is to rescale the domain and to work with $\Omega := \Omega_1$ by introducing $x := (z_1, z_2, z_3/h)$ and $y(x) := \tilde{y}(z(x)) = \tilde{y}(x_1, x_2, hx_3)$ and to consider the elastic energy

$$y \mapsto \int_{\Omega} \varphi(\nabla_h y(x))\,\mathrm{d}x\,,$$

where $\nabla_h(y) := (\partial_{x_1} y, \partial_{x_2} y, \partial_{x_3}/h)$. Various scaling regimes of external forces $f^h \approx h^\alpha$

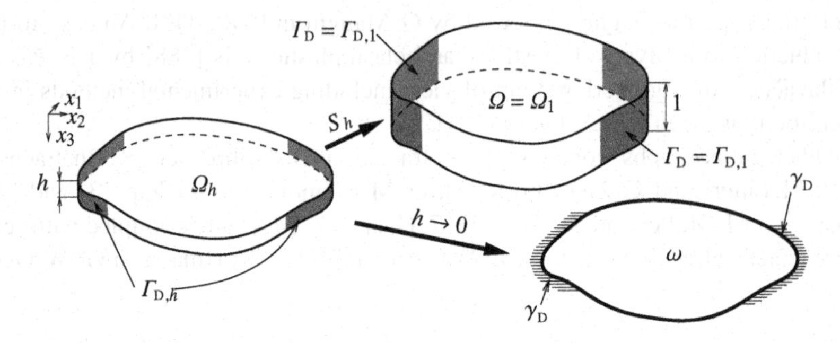

Schematic geometry of a 3-dimensional thin plate (with the thickness h) and the 2-dimensional plate arising for the limit $h \to 0$ clamped on (two parts of) the boundary γ_D, together also with the auxiliary domain arising by $h = 1$ used for derivation of the limit plate.

can be considered for $f^h : \omega \to \mathbb{R}^3$ leading to a minimization problem for

$$\mathcal{E}_h(y) := \int_{\Omega} \varphi(\nabla_h y(x)) - f^h \cdot y\,\mathrm{d}x\,,$$

where $\nabla_h := (\partial_{x_1}, \partial_{x_2}, h^{-1}\partial_{x_3})$. In [207], G. Friesecke, S. Müller, and R.D. James studied deformations with the scaling $\mathcal{E}_h/h \approx h^\beta$ for $\beta > 2$. They rigorously showed that if $\beta = 4$ the corresponding theory leads to von Kármán plate theory, if $\beta > 4$ they obtain the classical linear theory leading to the biharmonic equation for the out-of-plane component for isotropic energies. If $\beta \in (2,4)$ various constraints on deformations are present. Moreover, if $\beta = 1$ one gets usual membrane theory and, if $\beta = 2$, one obtains the Kirchhoff-Love plate theory. Another related approach due to Mindlin-Reissner, is an extension of Kirchhoff-Love plate theory that takes into account shear deformations through-the-thickness of a plate. Plates are original considered flat when unloaded. If the original geometry is substantially curved, a theory of shells (where curvature enters the model) becomes relevant. Cf. P.G. Ciarlet [115, 116] or J.N. Reddy [431] for a rigorous mathematical treatment.

A very particular area are waves in (visco)elastic or inelastic continua and their surfaces or interfaces. Many types of waves can be recognized as well as many related phenomena as reflection, refraction, polarization, dispersion, interference, attenuation, etc. Important applications as nondestructive testing of various materials in engineering, ultrasonic measurements of elastic moduli of anisotropic materials (specifically crystals), or detailed exploitation of planets, in particular our Earth of course, and their moons in seismology makes it a very particular area with many specialized monographs existing, cf. e.g. [63, 64, 301, 341]. In contrast to spatially nonlocal models in space used in this book, also models which are nonlocal in time are studied, cf. e.g. F. Mainardi [327].

The idea of drawing macroscopic properties of composite materials from their microscopic properties lies at the core of *homogenization theory*. Composite materials represent wide and highly applicable area of continuum mechanics (or micromechanics) of solids which is also not addressed in this book. This represents a typical two-scale problem. On the micro-scale, the such materials are (typically) isotropic but inhomogeneous, while on the macro-scale it is observed as homogeneous but anisotropic with possibly completely new properties. Typical examples are laminated materials or fiber-reinforced composites (e.g. carbon fibers in a polymer matrix or metal bars in concrete) or masonry walls etc. Yet e.g. composite materials with negative Poisson ratio or negative thermal expansion can be cast from materials with positive Poisson ratios or thermal-expansion coefficients. The mathematical method to pass rigorously from the micro- to macro-scale is called *homogenization*. Various specialized monographs deal with this mathematically. We cite the classical books by V.V. Jikov, S.M. Kozlov, and O.A. Olejnik [264], A. Braides and A. Defranceschi [86], D. Cioranescu and P. Donato [120], L. Tartar [507], G.W. Milton [368], and A. Bensoussan, J.L. Lions, and G. Papanicolaou [62], or E. Sanchez-Palencia [475] for thorough overviews. The Γ-convergence of integral functionals (see also monographs by A. Braides [85] or G. Dal Maso [138]) plays a key role in this research. The main objective is to study properties for $\varepsilon \to 0$ of

$$\mathcal{E}_\varepsilon(y_\varepsilon) := \int_\Omega \varphi\Big(x, \frac{x}{\varepsilon}, \nabla y_\varepsilon(x)\Big)\,\mathrm{d}x$$

where φ is $(0,1)^d$-periodic in the middle variable. Applications to composite materials and polycrystals can be found e.g. in [109] where also stochastic homogenization approaches appear. This modern area of analysis and probability aims at inclusion of randomness (e,g. impurities, defects, non-periodicity) into the problem which brings mathematical models closer to real materials. Variational methods and their applications to nonlinear partial differential equations are also discussed by L.C. Evans in [178] and in [179].

One particular sub-area is e.g. homogenization of polycrystals, having each grain anisotropic homogeneous, but macroscopically the material might be homogeneous if the shape and orientation of the grain is randomly distributed. Also, in fact, the models of flows in porous media, presented in this book on the macro-scale, might be a result of some homogenizations of pores in viscoelastic solid saturated by a diffusant (fluid) on micro-scale, cf. Figure 3.4-right on p. 74. It should be emphasized that homogenization is a very difficult topic and many homogenization studies in engineering are however only very heuristic and not rigorously justified.

Another topic which remained (with an exception of Sections 3.7 and 4.5.3 and some particular remarks in other sections) basically untouched is numerical methods and *computational mechanics*, although the proofs in this book are mostly based on constructive approximation methods and can thus serve as conceptual algorithms or, after some modification or completion, as a justification of existing engineering schemes usually carried out without ambitions to guarantee numerical stability and convergence. Specialized monographs are focused to mathematical aspects of numerical approximations, e.g. R. de Borst, M.A. Crisfield, J.J.C. Remmers, and C.V. Verhoosel [146] or O.C. Zienkiewicz and R.L. Taylor [543]. The other monographs are focused on algorithmical, computational, or modelling aspects in solid continuum mechanics, e.g. R.D. Cook, D.S. Malkus, M.E. Plesha, and R.J. Witt [127], M. Jirásek and Z.P. Bažant [266], M.H. Sadd [473], or J.C. Simo and T.J.R. Hughes [486].

When the continuum models are build and enough "robustly" implemented, in particular if they allow fast numerical solution preferably with unique solution, one may get further ambitions. One may think about *optimization of shape or material*, i.e. optimization of the domain Ω or of the coefficients in the model, and optimization of external loading or of initial conditions with respect to some specific criteria. These are known also under the name of *optimal (shape or material) design* and optimal control, respectively. Basic desired attribute of the model is continuity of its response to the varied to-be-optimized inputs. Yet, for an efficient implementation, also some sensitivity analysis is desirable, leading to an information about differential of this response mapping. Alternatively, knowing a response of the model on particular loading regimes e.g. from experiments, this optimization technique can also be exploited to *identification* of some *parameters* of the model which are not reliably known e.g. because they are not directly accessible to measurements. Such a task is also known under the name of *inverse problems*. In the context of solid mechanics, cf. e.g. [56, 242] for introduction into the technique, together with numerical approximation.

Open questions and problems

The above paragraphs and this book itself would not like to create a false impression that everything in continuum mechanics is solved and can be found in literature. The reality is rather opposite. In spite of long-lasting effort and many monographs and research articles published, similarly as in the continuum mechanics of fluids, also in the continuum mechanics of solids there are still many (sometimes quite fundamental) problems and questions which are not satisfactorily solved and answered.

In general, the approach to open problems in continuum mechanics exploits usually one of the following options:

(i) Development of adequate mathematical methods. Needless to say, this option is the most difficult and often even not completely realistic from factual reasons. Then, one can think about:
(ii) Suitable regularization together with corresponding asymptotical analysis and together with an attempt to devise an appropriate concept of "solution" that would be enough general to comply with the asymptotics at disposal – in fact, the well-accepted concepts of weak solutions or variational solutions (in contrast to classical solutions) are examples fitted with basic a-priori estimates. The concept of measure-valued solutions in Sect. 4.6.1 is a further example, although not so reliable working.[34] If even this does not work, then:
(iii) Without any asymptotical analysis, one may attempt to give a physical interpretation right to the regularized problem so that it may be consider as an "original" solvable problem – in fact, we saw this when inventing the concept of nonsimple materials or capillarity or various phase fields.

Apart of the options (ii)–(iii), at the time of publishing this book, many problems falling under the category (i) are still waiting for satisfactory solution.

We can say, with only a little exaggeration, that problems at small strains are already relatively well understood, and after some recent advances which were partly reported in this book in Chapters 5–8. Most fundamental open problems are therefore rather at large/finite strains. Such list may include (without any ambitions for completeness) the following problems:

(A) An energy $\varphi = \varphi(F)$ of simple materials that would be frame-indifferent, comply with the condition local non-selfinterpenetration $\varphi(F) \to \infty$ if $\det F \to 0+$, lead to a weakly lower-semicontinuous functional, and would be stable with respect to the Γ-convergence is not known, if it exists at all.[35] The quasiconvex energies that yield weakly lower-semicontinuous functionals must be, according to available results, of a polynomial growth but it is not compatible with the local non-selfpenetration; see Section 4.4.3. This fundamental challenge lasting many decades unanswered can be interpreted as a certain justification of the concept of nonsimple materials presented in Section 2.5 and used in Sections 4.5, 7.5, and 7.6, and in Chapter 9.
(B) Coincidence or a difference of quasi- and rank-one convexity in dimension $d = 2$, extending the famous Šverák's example [500] which proved the difference if $d \geq 3$.
(C) Perfect plasticity at large strains even in static case is completely open – here we conjecture that probably a small hardening or a small plastic-strain gradient will always be inevitable and reflects a real phenomena, in contrast to the small-strain case (Sect. 7.4.3) which then remains in a position of a mere mathematical idealization and an interesting asymptotic analysis for vanishing hardening or vanishing plastic-strain gradient.

[35] Let us remind that we saw that polyconvex energies may comply with local non-self-interpenetration but are not stable under Γ-convergence because the Γ-limit of the corresponding integral functionals yield only quasiconvex energies; cf. e.g. [85, Sec.12.2.1].

(D) Evolution at large strains in simple materials respecting local non-interpenetration, i.e. governed by potentials (e.g. polyconvex) blowing up if $\det F \to 0+$, having necessarily a nonpolynomial growth is not known (except quasistatic evolution and except particular results for small data or short times as e.g. [153, 314] or using questionable concepts of measure-valued solutions [152]) is not known. In particular, holding the attribute of deformation to be homeomorphic also during evolution is not known. Again, this may be interpreted as a certain justification of using the concept of nonsimple materials as we did in Chapter 9.

(E) Derivation of small-strain models from the large strains ones possesses already particular results. Linearized elasticity was shown to be a Γ-limit of a suitably rescaled functional of a nonlinear elastic energy with Dirichlet boundary conditions and bulk forces if the displacement gradient is vanishing in [141]. A similar program was undertaken in [205] where equations of linearized viscoelasticity are derived from the ones for nonlinear second-grade viscoelastic materials. In many other examples, however, a rigorous justification is missing, as, for instance, in linearized nonsimple materials.

(F) The local injectivity in linearized elasticity/elastoplasticity is usually only assumed implicitly without formulation appropriate constraints. See [195] for discussions on this topic.

(G) Energy conservation and thermodynamics of viscoelastic materials at large strains when strain-rate dissipative processes (as in Kelvin-Voigt's or Jeffreys' models) together with inertia are involved.

(H) Interaction of solids with fluids (known as *fluid-structure interaction*) is largely an open problem even at small strains, although some models rigorously justified have been presented in Sect. 7.7.4. Here, in general, the difficulty consists in merging the Eulerian and the Lagrangian descriptions suitable for fluids and solids, respectively.

(I) Characterization of the set of traces of orientation-preserving and possibly also injective Sobolev maps: while the the trace space is known for general Sobolev maps, it is not known what y_0 on $\Gamma_\mathrm{D} \subset \partial\Omega$ must satisfy to ensure that there is $y \in W^{1,p}(\Omega;\mathbb{R}^d)$ with $y = y_0$ on Γ_D and $\det \nabla y > 0$ a.e. in Ω.

(J) Relaxation methods for problems with non-polynomial growth of the energy density, in particular for problems respecting local non-interpenetration are still not satisfactorily developed. Some partial results have become available only recently. Here [58, 59, 126] is to be mentioned. The first reference, however, works only for bi-Lipschitz deformations[36] in the plane, while the latter one provides relaxation results only if the resulting quasiconvex envelope is already polyconvex, which cannot be expected, in general. This problem is also related to the fact that we do not have efficient tools to check quasiconvexity of a particular function. Nevertheless, see [67] for some conditions in special cases.

[36] A map $y : \Omega \to \mathbb{R}^d$ is bi-Lipschitz if both y and y^{-1} are Lipschitz continuous.

(K) Regularity of minimizers in static problems of large strain elastostatics. Can one show, for example, that minimizers are Lipschitz maps? See, e.g.,[498] for some results related to polyconvex energies.

Beside these truly fundamental general problems of a ultimate challenge and importance, there are a lot of particular "smaller" problems related to some concrete models. In particular, reviewing this book, the reader may identify the following problems:

(L) Phase-field models of fracture are used in engineering but basically analytically unjustified, in particular in the limit towards the infinitesimally thin cracks. Here, the limit of the Ambrosio-Tortorelli approximation as in Remark 7.5.16 on p. 307, known only at static situations, is open in the dynamical viscoelastic cases even at small strains, and it is even worse at large strains.
(M) The mentioned Γ-limit of the Ambrosio-Tortorelli approximation in the static situation leads to the Griffith-type fracture, but it is well recognized that this model suffers the drawback that infinitely large stresses may be needed to nucleate cracks when there are no pre-existing cracks. Scaling fracture toughness towards zero would allow for realistic initiation of fracture but leads to unrealistically easy crack propagation. To summarize, although many engineering models have been devised and successfully used, fracture mechanics still faces basic modeling dilemmas, there is no models that can truly satisfactorily capture both crack initiation and propagation.[37]
(N) Gradient damage even at small strains (possibly combined with gradient plasticity etc.) where also the gradient terms are subject to (at least partial) damage seems waiting for a rigorous analysis.
(O) Damage in purely elastic (i.e. inviscid) simple materials seems unjustified, too; note that this is not covered in Section 7.5. In particular, the phase-field approximation of fracture in these cases is open.
(P) The *complete damage* which admits a complete disintegration of the material leads mathematically to losses of some a-priori estimates and fatal mathematical difficulties. In quasistatic problems, it was treated in [80, 354] under Dirichlet loading, while only very partial results are obtained in dynamic situations where intuitively inertia should help even if visco-elastic resistance completely disappears, see [364]. In particular, complete damage at large strains seems completely open.
(Q) Analysis of the coupled problem (7.6.1) with a guaranty of non-negativity of concentration thorough an unlimited growth of the stored energy which would conflict with (7.6.8b), or without the regularizations made in Section 7.6.3.
(R) Perfect plasticity with damage in Sect. 7.7.1 heavily relied on unidirectionality of the damaging process, but allowing for healing is of an interest in some application and open.

[37] Cf. also Remark 7.5.16 on p. 307 or also the analysis of stress-intensity factors on singular domains by V. Maz'ya et al. [348].

(S) Perfect plasticity with temperature-dependent yield stress without the phase-field regularization (i.e. a limit in the model (8.5.6) for $\varkappa \to 0$) is another open problem.
(T) Gradient theories at large strains for the elastic rather than total strain seems to bring serious difficulties, cf. also Remark 9.4.6 on p. 440.
(U) Gradient theories at large strains in the actual rather than in a reference configuration (like for damage in (9.5.8) or for the Cahn-Hilliard model in Remark 9.6.3 or for the plastic strain in Remark 9.4.12) seem to be open for analysis for the gradient of the elastic or total strain itself.[38]
(V) Thermodynamics of the Kelvin-Voigt viscoelastic materials under large strains seems also open if the inertial effects are considered while the quasistatic situations, when these effects are neglected, is scrutinized in [362].
(W) Coping with non-selfinterpenetration in evolution problems at large strains is also waiting for some concepts/solution in dynamical situations when inertial forces are considered.[39]
(X) Allowing for possible selfinterpenetration in long-range interactions, as presented in Sect. 9.7 for repulsive monopolar interactions, seems difficult to be performed for (physically relevant) linear Poisson equation (9.7.5b) for ϕ, i.e. $p = 2$, because of the L^1-right-hand side in (9.7.5b) and because the composition $\phi \circ y$ is not easily amenable unless the potential ϕ is continuous. In particular, it concerns the attractive monopolar interactions due to gravitation where nonlinear Poisson equation does not have a reasonable sense at all. For dipolar interactions, the difficulties are even more pronounced.
(Y) The electrostatic or the dipolar interactions through depolarizing or demagnetizing field is, in fact, to be considered in the context of full electro-magnetic Maxwell system, possibly in electrically conductive continuum, including the electromagnetic field outside the body. Even if the displacive current inside the body is neglected (which is a justified approximation in highly conductive media) leading to a so-called eddy-current approximation, the Maxwell system outside the body is hyperbolic, although linear. Coupling of such linear hyperbolic system with typically nonlinear response of deformable elastic continuum with various internal variables seems not satisfactorily solved even in particular cases.
(Z) The numerical treatment of the Ciarlet-Nečas integral constraint (3.4.7) has not been satisfactorily devised. Namely, it not much clear how to calculate efficiently $\mathrm{meas}_d(y(\Omega))$, in particular, for higher-order finite elements.

[38] In the stored energy, we thus would see terms like $|(\nabla y)^{-\top}\nabla((\nabla y)^{-\top}\nabla y)|^2$ or $|\nabla((\nabla y)\Pi^{-1})|^2$, respectively.

[39] Techniques based on Euler-Lagrange equations need explicit handling of the reaction force arising from a possible self-contact, like it was, in the static case, scrutinised in [406, 479]. This seems possible at most in quasistatic evolutionar situations only, however.

Appendix A
Elements of abstract functional analysis

The goal of this appendix (and the other appendices, too) is to collect only very basic definitions and briefly expose concepts used in this book, without going into details. We intentionally try to avoid too abstract concepts (as topologies or even non-metrizable topologies) and stay in the conventional framework of sequences.

A.1 Ordering, metric and normed linear spaces, Banach spaces

A binary relation, denoted by $\leq$, on a set X is called an *ordering* if it is reflexive, transitive, and antisymmetric[1]. The ordering $\leq$ is called *linear* if $x_1 \leq x_2$ or $x_2 \leq x_1$ always hold for any $x_1, x_2 \in X$. Having two ordered sets X_1 and X_2 and a mapping $f : X_1 \to X_2$, we say that f is *non-decreasing* (resp. *non-increasing*) if $x_1 \leq x_2$ implies $f(x_1) \leq f(x_2)$ (resp. $f(x_1) \geq f(x_2)$).

We say that $x_1 \in X$ is the greatest element of the ordered set X if $x_2 \leq x_1$ for any $x_2 \in X$. Similarly, $x_1 \in X$ is the least element of X if $x_1 \leq x_2$ for any $x_2 \in X$. Further, we say that $x_1 \in X$ is *maximal* in the ordered set X if there is no $x_2 \in X$ such that $x_1 < x_2$. Note that the greatest element, if it exists, is always maximal but not conversely. Similarly, $x_1 \in X$ is *minimal* in X if there is no $x_2 \in X$ such that $x_1 > x_2$.

The ordering $\leq$ on X induces also the ordering on a subset A of X, given just by the restriction of the relation $\leq$. We say that $x_1 \in X$ is an upper bound of $A \subset X$ if $x_2 \leq x_1$ for any $x_2 \in A$. Analogously, $x_1 \in X$ is called a lower bound of A if $x_1 \leq x_2$ for any $x_2 \in A$. If every two elements $x_1, x_2 \in X$ possess both the least upper bound and the greatest lower bound, denoted respectively by $\sup(x_1, x_2)$ and $\inf(x_1, x_2)$ and called the *supremum* and the *infimum* of $\{x_1, x_2\}$, then the ordered set $(X, \leq)$ is called a *lattice*. Then the supremum and the infimum exist for any finite subset and is determined uniquely because the ordering is antisymmetric.

[1] This means respectively $x \leq x$ for any $x \in X$, $x_1 \leq x_2$ & $x_2 \leq x_3$ imply $x_1 \leq x_3$ for any $x_1, x_2, x_3 \in X$, and $x_1 \leq x_2$ & $x_2 \leq x_1$ imply $x_1 = x_2$.

M. Kružík and T. Roubíček, *Mathematical Methods in Continuum Mechanics of Solids*, Interaction of Mechanics and Mathematics,
https://doi.org/10.1007/978-3-030-02065-1

A function $d : X\times X \to \mathbb{R}$ is called a *distance* on X if, for all $x_1, x_2, x_3 \in X$, $d(x_1,x_2) \geq 0$, $d(x_1,x_2) = 0$ is equivalent to $x_1 = x_2$, and $d(x_1,x_2) \leq d(x_1,x_3) + d(x_3,x_2)$. A distance $d : X\times X \to \mathbb{R}$ is called a *metric* on X if $d(x_1,x_2) = d(x_2,x_1)$ for all $x_1, x_2 \in X$. The set X equipped with a metric d is called a *metric space*.

A mapping $x : \mathbb{N} \to X$ is called a *sequence*, and we write x_k instead of $x(k)$ and $\{x_k\}_{k\in\mathbb{N}} \subset X$ instead of $x(\mathbb{N}) \subset X$. A *sequence* is called *Cauchy*[2] if

$$\forall\varepsilon > 0\ \exists k \in \mathbb{N}\ \forall m,n \geq k: \quad d(x_m,x_n) \leq \varepsilon.$$

and *converging* if

$$\exists x \in X\ \forall\varepsilon > 0\ \exists k \in \mathbb{N}\ \forall m \geq k: \quad d(x_m,x) \leq \varepsilon;$$

then x is called the *limit* of the sequence $\{x_k\}_{k\in\mathbb{N}}$ and we write $x = \lim_{k\to\infty} x_k$ or shortly $x_k \to x$ for $k \to \infty$. The metric space X is called *complete* if every Cauchy sequence converges. A subset $A \subset X$ is called *closed* if every converging sequence $\{x_k\}_{k\in\mathbb{N}} \subset A$ has a limit belonging also to A. A subset $A \subset X$ is called *open* if $X\setminus A$ is closed. If $x \in A \subset X$, we say that A is a *neighborhood* of x if $A \supset \{\tilde{x} \in X;\ d(\tilde{x},x) \leq \varepsilon\}$ for some $\varepsilon > 0$.

We define the *interior*, the *closure*, and the *boundary* of a set A respectively by

$$\text{int}(A) := \{x \in X;\ \exists N \text{ a neighborhood of } x : \ N \subset A\}, \tag{A.1.1a}$$

$$\text{cl}(A) := \{x \in X;\ \exists\{x_k\}_{k\in\mathbb{N}} :\ x = \lim_{k\to\infty} x_k\} \tag{A.1.1b}$$

$$\partial A := \text{cl}(A) \setminus \text{int}(A). \tag{A.1.1c}$$

Having $A \subset B \subset X$, we say that A is *dense* in B if $\text{cl}(A) \supset B$.

A metric space X is *compact* if every sequence in X admits a subsequence[3] that converges in X. A set A is called *relatively compact* if the closure of A is compact in X.

A mapping $f : X \to Y$ between two metric spaces is called *continuous* if $f(x) = \lim_{k\to\infty} f(x_k)$ whenever $x = \lim_{k\to\infty} x_k$.

The following theorem is an important statement about extensions of continuous functions.

Theorem A.1.1 (Tietze's theorem [514]). *Assume that A is a closed subset of a metric space X and $f : A \to \mathbb{R}$ is continuous. Then there is a continuous map $g : X \to \mathbb{R}$ such that $f(x) = g(x)$ for all $x \in A$, i.e., g is an extension of the function f.*

The image of a compact set via a continuous mapping is compact. The metric space X is called *connected* if, for any $x_0, x_1 \in X$, there is a continuous mapping $x : [0,1] \to X$ with $x(0) = x_0$ and $x(1) = x_1$.

Theorem A.1.2 (Invariance of Domain Theorem [401]). *Let $d \in \mathbb{N}$. The image of a continuous injective mapping of an open set in $\mathbb{R}^d$ into $\mathbb{R}^d$ is open.*

---

[2] This concept was introduced by B. Bolzano a few years before A. Cauchy, cf. [506].

[3] A subsequence of $\{x_k\}_{k\in\mathbb{N}}$ is a sequence $\{\tilde{x}_k\}_{k\in\mathbb{N}}$ defined by $\tilde{x}_k := x_{n_k}$, where $n_1 < n_2 < \ldots$ is an increasing sequence of indices.

We say that a functional (or briefly a function) $f : X \to \mathbb{R} \cup \{+\infty\}$ is *lower semicontinuous* (resp. *upper semicontinuous*) with respect to the metric on X in question if

$$x_k \to x \quad \Rightarrow \quad f(x) \le \liminf_{k\to\infty} f(x_k) \quad \Big(\text{resp. } f(x) \ge \limsup_{k\to\infty} f(x_k)\Big) \tag{A.1.2}$$

Obviously, f is upper semicontinuous if and only if $-f$ is lower semicontinuous. This is also intimately related with canonical ordering of $\mathbb{R}$ and exploits the *limit inferior* of a sequence $(y_k)_{k\in\mathbb{N}} \subset \mathbb{R}$ (here used for $y_k = f(x_k)$) defined as $\sup_{k\in\mathbb{N}} \inf\{y_l;\ l \ge k\}$; if admitted to attain values in $\mathbb{R}\cup\{\infty\}$, this supremum always exists because $\mathbb{R}\cup\{\pm\infty\}$ is a complete lattice. Analogously, one defines *limit superior* of a sequence $(y_k)_{k\in\mathbb{N}} \subset \mathbb{R}$ as $\inf_{k\in\mathbb{N}} \sup\{y_l;\ l \ge k\}$.

We call V a (real) *linear space* if it is endowed by a binary operation $(v_1, v_2) \mapsto v_1 + v_2 : V \times V \to V$ which makes it a group[4] and furthermore it is equipped with a multiplication by scalars $(a, x) \mapsto ax : \mathbb{R} \times V \to V$ satisfying $(a_1 + a_2)v = a_1v + a_2v$, $a(v_1 + v_2) = av_1 + av_2$, $(a_1a_2)v = a_1(a_2v)$, and $1v = v$.

A non-negative, degree-1 homogeneous, sub-additive functional $\|\cdot\|_V : V \to \mathbb{R}$ on the linear space V is called a *semi-norm*. This functional is called a *norm* if it vanishes only at 0; often, we write briefly $\|\cdot\|$ instead of $\|\cdot\|_V$ if V is clear from the context.[5] The norm induces a metric $(v_1, v_2) \mapsto \|v_1 - v_2\|$ which further induces a convergence, called *strong*. A subset $A \subset V$ is called *bounded* if $\sup_{x\in A} \|x\| < \infty$. A linear space equipped with a norm is called a *normed linear space*. If there is a countable dense subset of V, we say that V is *separable*.[6] Complete normed linear spaces are also called *Banach spaces* [41].

An example of a separable Banach space is $\mathbb{R}^d$ endowed by the norm, denoted usually by $|\cdot|$ instead of $\|\cdot\|$, defined by $|s| = (\sum_{i=1}^d s_i^2)^{1/2}$; such a Banach space is called an d-dimensional *Euclidean space*.

If V is a normed linear space we define for $0 < \lambda < 1$ (resp. $\lambda = 1$), the mapping $f : X \to V$ is called *Hölder* (resp. *Lipschitz*) *continuous* if

$$\exists \ell\in\mathbb{R}\ \forall x, \tilde{x}\in X: \quad \|f(x) - f(\tilde{x})\|_V \le \ell\|x - \tilde{x}\|_X^\lambda. \tag{A.1.3}$$

The set of all Hölder/Lipschitz continuous functions is denoted by $C^{0,\lambda}(X;Y)$ and it is a Banach space if normed by

$$\|f\|_{C^{0,\lambda}(X;Y)} := \sup_{x\in X} \|f(x)\|_Y + \inf\{\ell\in\mathbb{R};\ \text{(A.1.3) holds}\}\,.$$

---

[4] This means $v_1 + v_2 = v_2 + v_1$, $v_1 + (v_2 + v_3) = (v_1 + v_2) + v_3$, $\exists 0 \in V$: $v + 0 = v$, and $\forall v_1 \in V\ \exists v_2$: $v_1 + v_2 = 0$.

[5] The mentioned properties means respectively: $\|v\| \ge 0$, $\|av\| = |a|\,\|v\|$, $\|u+v\| \le \|u\| + \|v\|$ for any $u, v \in V$ and $a \in \mathbb{R}$, and $\|v\| = 0 \Rightarrow v = 0$.

[6] If V' is separable, then the weak* convergence can be induced by a metric when restricted to bounded sets. An "explicit" choice of such a metric, denoted by $d_{\rm w}$, is $d_{\rm w}(u, \tilde{u}) := \sum_{k=1}^\infty 2^{-k} |\langle u - \tilde{u}, z_k\rangle| / (1 + |\langle u - \tilde{u}, z_k\rangle|)$ when taking some dense subset $\{z_k\}_{k\in\mathbb{N}}$ in a unit ball of V'.

A function $f : X \mapsto Y$ is called *locally Lipschitz* continuous if for every $x \in X$ there is a neighborhood such that f is Lipschitz continuous when restricted to this neighborhood.

Having two normed linear spaces V_1 and V_2 and a mapping $A : V_1 \to V_2$, we say that A is *continuous* if $v_k \to v$ in V_1 implies that $Av_k \to Av$ in V_2 as $k \to \infty$. and is a *linear operator* if it satisfies $A(a_1v_1 + a_2v_2) = a_1A(v_1) + a_2A(v_2)$ for any $a_1, a_2 \in \mathbb{R}$ and $v_1, v_2 \in V_1$. Often we write briefly Av instead of $A(v)$. If $V_1 = V_2$, a linear continuous operator $A : V_1 \to V_2$ is called a *projector* if $A \circ A = A$. The set of all linear continuous operators $V_1 \to V_2$ is denoted by $\mathrm{Lin}(V_1, V_2)$, being itself a normed linear space when equipped with the addition and the multiplication by scalars defined respectively by $(A_1 + A_2)v = A_1v + A_2v$ and $(aA)v = a(Av)$, and with the norm

$$\|A\|_{\mathrm{Lin}(V_1,V_2)} := \sup_{\|v\|_{V_1} \le 1} \|Av\|_{V_2} = \sup_{v \ne 0} \frac{\|Av\|_{V_2}}{\|v\|_{V_1}}. \tag{A.1.4}$$

If $V := V_1 = V_2$, we write shortly $\mathrm{Lin}(V) := \mathrm{Lin}(V_1, V_2)$.

As $\mathbb{R}$ itself is a normed linear space[7], we can consider the linear space $\mathrm{Lin}(V, \mathbb{R})$, being also denoted by V^* and called the *dual space* to V. The original space V is then called *pre-dual* to V^*. For an operator (=now a functional) $f \in V^*$, we write $\langle f, v\rangle$ instead of fv. The bilinear form $\langle \cdot, \cdot\rangle_{V^* \times V} : V^* \times V \to \mathbb{R}$ is called a canonical *duality pairing*. Instead of $\langle \cdot, \cdot\rangle_{V^* \times V}$, we often write briefly $\langle \cdot, \cdot\rangle$. Always, V^* is a Banach space if endowed by the norm (A.1.4), denoted often briefly $\|\cdot\|_*$ instead of $\|\cdot\|_{V^*}$, i.e. $\|f\|_* = \sup_{\|v\| \le 1}\langle f, v\rangle$. Obviously,

$$\langle f, u\rangle = \|u\| \Big\langle f, \frac{u}{\|u\|}\Big\rangle \le \|u\| \sup_{\|v\| \le 1} \langle f, v\rangle = \|f\|_* \|u\|. \tag{A.1.5}$$

For $A \in \mathrm{Lin}(V_1, V_2)$, we define the so-called *adjoint operator* $A^* \in \mathrm{Lin}(V_2^*, V_1^*)$ by $\langle A^* f_2, v_1\rangle := \langle f_2, Av_1\rangle$ where the former duality is $V_1^* \times V_1 \to \mathbb{R}$ while the later duality is $V_2^* \times V_2 \to \mathbb{R}$. For two Banach spaces V_1 and V_2 such that $V_1 = V_2^*$ and $V_2 = V_1^*$, we may say that an operator $A \in \mathrm{Lin}(V_1, V_2)$ is *symmetric* if $A = A^*$; let us denote the Banach space of such operators as $\mathrm{SLin}(V_1, V_2)$.

A sequence $\{u_k\}_{k \in \mathbb{N}} \subset V$ is called *weakly convergent* to some $u \in V$ if $\langle f, u_k\rangle \to \langle f, u\rangle$ for any $f \in V^*$. A sequence $\{f_k\}_{k \in \mathbb{N}} \subset V^*$ is called *weakly* convergent* to some $f \in V^*$ if $\langle f_k, u\rangle \to \langle f, u\rangle$ for any $u \in V$. We will then write $u_k \rightharpoonup u$ and $f_k \overset{*}{\rightharpoonup} f$. Any strongly convergent sequence in V is weakly convergent while the converse statement holds only if V is finite dimensional. The *weakly lower* (resp. upper) *semicontinuous* functionals on V are defined just as in (A.1.2) but with strong convergence replaced by the weak one.

If V possesses a predual V', we say that a sequence $\{u_k\}_{k \in \mathbb{N}}$ converges weakly* to u if $\lim_{k \to \infty}\langle u_k, z\rangle = \langle u, z\rangle$ for any $z \in V'$. Often, $V' = V^*$ (such spaces are called *reflexive*) and then the weak* and the weak convergences coincide.

---

[7] The conventional norm on $\mathbb{R}$ is the absolute value $|\cdot|$.

If V is a Banach space such that, for any $v \in V$, the functional $V \to \mathbb{R} : u \mapsto \|u+v\|^2 - \|u-v\|^2$ is linear, then V is called a *Hilbert space*. In this case, we define the *inner product* (also called *scalar product*) by

$$(u,v) := \frac{1}{4}\|u+v\|^2 - \frac{1}{4}\|u-v\|^2. \tag{A.1.6}$$

By the assumption, $(\cdot,\cdot) : V \times V \to \mathbb{R}$ is a bilinear form which is obviously symmetric[8] and satisfies $(u,u) = \|u\|^2$. E.g., the Euclidean space $\mathbb{R}^n$ is a Hilbert space. If V is a Hilbert space, then $(u \mapsto (f,u)) \in V^*$ for any $f \in V$, and the mapping $f \mapsto (u \mapsto (f,u))$ identified V with V^*. Then (A.1.5) turns into a so-called *Cauchy-Bunyakovskiĭ* inequality $(f,u) \le \|f\|\,\|u\|$.

The dual space V^* to any Hilbert space V is again a Hilbert space, which can (but need not) be identified with the original space V by considering $v \in V$ as a linear functional $u \mapsto (v,u)$. For a Hilbert spaces V, symmetric operators $V \to V$ are called *self-adjoint*, i.e. $A : V \to V$ is selfadjoint if $A = A^*$. The space of such operators is briefly denoted by $\mathrm{SLin}(V) := \mathrm{SL}(V,V)$.

For $A \in \mathrm{Lin}(V_1,V_2)$, we define the so-called *adjoint operator* $A^* \in \mathrm{Lin}(V_2^*,V_1^*)$ by $\langle A^* f_2, v_1\rangle := \langle f_2, Av_1\rangle$ where the former duality is $V_1^* \times V_1 \to \mathbb{R}$ while the later duality is $V_2^* \times V_2 \to \mathbb{R}$.

An important and, in our sequential variant, constructive tool needed for limit passages in various approximation methods is the following:

Theorem A.1.3 (Banach selection principle [43]).[9] *In a Banach space with a separable predual, any bounded sequence contains a weakly* convergent subsequence (or, in other words, it is relatively compact).*

The celebrated Hahn-Banach theorem is a fundamental tool of functional analysis. We recall that a linear manifold of a normed linear space is a subset which is closed under addition of elements and scalar multiplication.

Theorem A.1.4 (Hahn and Banach [42, 238]). *Let K be an open convex nonempty subset of normed linear space V and L be a linear manifold that does not intersect K. Then there is a closed hyperplane $\bar{L}$ such that $L \subset \bar{L}$ and $K \cap \bar{L} = \emptyset$. In other words, there is $f \in V^*$ such that $\langle f,u\rangle > \langle f,v\rangle$ whenever $u \in K$ and $v \in L$.*

---

[8] This means both $u \mapsto (u,v)$ and $v \mapsto (u,v)$ are linear functionals on V and $(u,v) = (v,u)$.

[9] This important theorem [43, Chap.VIII, Thm.3] enjoys various generalization and is sometimes referred under different names, typically as Alaoglu-Bourbaki's theorem. The proof of its basic sequential variant is quite constructive and simple when relying on compactness of closed bounded intervals in $\mathbb{R}$: Consider a sequence $\{f_k\}_{k\in\mathbb{N}}$ bounded in V^* and a countable dense subset $\{v_k\}_{k\in\mathbb{N}}$ in V, take v_1 and select an infinite subset $A_1 \subset \mathbb{N}$ such that the sequence of real numbers $\{\langle f_k, v_1\rangle\}_{k\in A_1}$ converges in $\mathbb{R}$ to some $f(v_1)$. Then take v_2 and select an infinite subset $A_2 \subset A_1$ such that $\{\langle f_k, v_2\rangle\}_{k\in A_2}$ converges to some $f(v_2)$, etc. for $v_3, v_4, \ldots$. A diagonalization procedure by taking l_k the first number in A_k which is greater than k shows that $\{\langle f_{l_k}, v_i\rangle\}_{k\in\mathbb{N}}$ converge to $f(v_i)$ for all $i \in \mathbb{N}$. The functional f is linear on $\mathrm{span}(\{v_i\}_{i\in\mathbb{N}})$ and bounded because $|f(v_i)| \le \lim_{k\to\infty} |\langle f_{l_k}, v_i\rangle| \le \limsup_{k\to\infty} \|f_k\|_* \|v_i\|$, and can be extended f on the whole V^* just by continuity.

Theorem A.1.5 (Bounded inverse theorem [168]). *If $A : X \to Y$ is a bijective continuous linear operator between Banach spaces X and Y, then the inverse operator $A^{-1} : Y \to X$ is continuous.*

The following *Lax-Milgram lemma* is a fundamental result with applications to linear PDEs, for instance.

Lemma A.1.6 (Lax-Milgram [306]). *Let V be a Hilbert space and $B : V\times V \to \mathbb{R}$ a continuous and bilinear[10] form, i.e., for some $K_B > 0$*

$$|B(v,w)| \le K_B\|v\|_V\|w\|_V, \text{ for all } v,w\in V \tag{A.1.7}$$

which is also coercive, i.e., there is $K > 0$

$$B(v,v) \ge K\|v\|_V^2, \text{ for all } v\in V. \tag{A.1.8}$$

Then for every $f \in V^*$ there exists a unique $u \in V$ satisfying

$$B(u,v) = \langle f,v\rangle, \text{ for all } v\in V. \tag{A.1.9}$$

A useful consequence of the Hahn-Banach theorem is usually called Mazur's lemma .

Lemma A.1.7 (Mazur's lemma [347]). *Let V be a Banach space and $\{u_k\} \subset V$ weakly converges to $u\in V$, i.e., $u_k \rightharpoonup u$ for $k \to \infty$. Then there is a function $N : \mathbb{N} \to \mathbb{N}$ and nonnegative reals $\lambda_j^k \ge 0$, $\sum_{j=k}^{N(k)} \lambda_j^k = 1$ such that $v_k \to u$ strongly in V as $k \to \infty$ where, for all $k \in \mathbb{N}$,*

$$v_k := \sum_{j=k}^{N(k)} \lambda_j^k u_j\,.$$

A.2 Finite-dimensional linear spaces and operators

Finite-dimensional linear spaces have finite base and can thus be understood as Euclidean spaces containing vectors and linear operators between them play special role. The linear operators which map $\mathbb{R}^d$ to $\mathbb{R}^d$ can be represented by matrices in $\mathbb{R}^{d\times d}$.

Here we summarize basic definitions and operations with vectors and matrices. If $a,b\in\mathbb{R}^d$ we define the dot product $a\cdot b\in\mathbb{R}$ as

$$a\cdot b := \sum_{i=1}^d a_i b_i, \tag{A.2.1}$$

[10] Sometimes also called "sesquilinear".

and, if $d = 3$, the cross product $a \times b \in \mathbb{R}^3$ with components (the summation convention applies)

$$(a \times b)_i = \varepsilon_{ijk} a_j b_k \,, \tag{A.2.2}$$

for $1 \le i \le 3$ where ε_{ijk} is the *Levi-Civita permutation symbol*

$$\varepsilon_{ijk} := \begin{cases} 1 & \text{if } (i,j,k) = (1,2,3),\ (2,3,1),\ (3,1,2)\,, \\ -1 & \text{if } (i,j,k) = (3,2,1),\ (1,3,2),\ (2,1,3)\,, \\ 0 & \text{otherwise.} \end{cases} \tag{A.2.3}$$

It is easy to see that $a \times b = 0$ if and only a and b are collinear and if $a, b \neq 0$ are not collinear then $a \times b$ is perpendicular to both a and b. Two vectors $a, b \in \mathbb{R}^d$ are perpendicular to each other if and only if $a \cdot b = 0$. If $a, b, c \in \mathbb{R}^3$ the following identity holds

$$a \times (b \times c) = b(a \cdot c) - c(a \cdot b)\,. \tag{A.2.4}$$

For a matrix $A \in \mathbb{R}^{d \times d}$, we denote by A_{ij} its entries with $i = 1, \dots, d$ numbering the rows and $j = 1, \dots, d$ the columns. The entries A_{ii} for $1 \le i \le d$ (no summation) are called *diagonal entries* . Their sum is the *trace* of A

$$\operatorname{tr} A := \sum_{i=1}^{d} A_{ii}\,.$$

The *identity matrix*, $\mathbb{I} \in \mathbb{R}^{d \times d}$ has entries which coincide with the *Kronecker's delta* , i.e.,

$$\mathbb{I}_{ij} := \delta_{ij}\,, \qquad \text{where} \quad \delta_{ij} := \begin{cases} 1 & \text{if } i = j, \\ 0 & \text{otherwise.} \end{cases} \tag{A.2.5}$$

The *dot-product* of two matrices $A \in \mathbb{R}^{d \times d}$ and $B \in \mathbb{R}^{d \times d}$, denoted by $A : B \in \mathbb{R}$, is defined as

$$A : B := \sum_{i,j=1}^{d} A_{ij} B_{ij}\,. \tag{A.2.6}$$

The *Frobenius norm* of A is just the norm defined by this dot-product, i.e., $|A|^2 := A : A$. For two matrices $A \in \mathbb{R}^{d \times d}$ and $B \in \mathbb{R}^{d \times d}$, the matrix multiplication $AB \in \mathbb{R}^{d \times d}$ is defined as

$$(AB)_{ij} := \sum_{k=1}^{d} A_{ik} B_{kj} \quad \text{with } i = 1, \dots, d \text{ and } j = 1, \dots, d. \tag{A.2.7}$$

We can identify every matrix $A \in \mathbb{R}^{d \times d}$ with a linear continuous map from $\mathbb{R}^d$ to $\mathbb{R}^d$ such that, for all $b \in \mathbb{R}^d$ and $1 \le i \le d$, it holds

$$(Ab)_i := \sum_{k=1}^{n} A_{ik} b_k \,. \tag{A.2.8}$$

Note also that the definition (A.2.7) corresponds to the composition of the linear operators $\mathbb{R}^d \to \mathbb{R}^d$ and $\mathbb{R}^d \to \mathbb{R}^d$ induced respectively by the matrices A and B.

The entries of the *transposition* of A, denoted $A^\top$, are defined by $A^\top_{ij} = A_{ji}$ for all $i, j = 1,\dots,d$. If $A = A^\top$, we say that A is *symmetric* while, if $A = -A^\top$, we call A *skew symmetric*. Every matrix $A \in \mathbb{R}^{d\times d}$ can be decomposed as a sum of its symmetric and skew symmetric parts, namely

$$A = \mathrm{sym}A + \mathrm{skew}A \quad \text{with } \mathrm{sym}A := \frac{A+A^\top}{2} \text{ and } \mathrm{skew}A := \frac{A-A^\top}{2}. \tag{A.2.9}$$

The subspace of symmetric matrices in $\mathbb{R}^{d\times d}$ is denoted $\mathbb{R}^{d\times d}_{\mathrm{sym}}$. Another decomposition of $(d\times d)$-matrices is on the deviatoric and the spherical parts

$$A = \mathrm{sph}A + \mathrm{dev}A \quad \text{with } \mathrm{sph}A := \frac{\mathrm{tr}A}{d}\mathbb{I} \text{ and } \mathrm{dev}A := A - \frac{\mathrm{tr}A}{d}\mathbb{I}. \tag{A.2.10}$$

Note that the deviatoric part from (6.7.8) is trace free and orthogonal to the spherical part; indeed, realizing that $\mathrm{tr}\,\mathbb{I} = \mathbb{I}{:}\mathbb{I} = d$, we have

$$\mathrm{tr}(\mathrm{dev}A) = \mathrm{tr}\Big(A - \frac{\mathrm{tr}A}{d}\mathbb{I}\Big) = \mathrm{tr}A - \frac{\mathrm{tr}A}{d}d = 0 \quad \text{and} \tag{A.2.11a}$$

$$\mathrm{sph}A{:}\mathrm{dev}A = \Big(\frac{\mathrm{tr}A}{d}\mathbb{I}\Big){:}\Big(A - \frac{\mathrm{tr}A}{d}\mathbb{I}\Big) = \frac{(\mathrm{tr}A)^2}{d} - \frac{(\mathrm{tr}A)^2}{d} = 0. \tag{A.2.11b}$$

The Frobenius norm of A defined above as $|A| = (A{:}A)^{1/2}$ can be then expressed as

$$|A| = \sqrt{\mathrm{tr}(A^\top A)}\,. \tag{A.2.12}$$

We have a useful identity valid for all matrices $A, B, C \in \mathbb{R}^{d\times d}$, namely,

$$A{:}(BC) = (B^\top A){:}C \quad \text{and also} \tag{A.2.13}$$

$$A{:}(BC) = \mathrm{tr}(AC^\top B^\top) = (AC^\top){:}B\,. \tag{A.2.14}$$

The *determinant* of $A \in \mathbb{R}^{d\times d}$ is denoted by $\det A \in \mathbb{R}$ and defined as

$$\det A := \begin{cases} (A\mathbf{e}_1 \times A\mathbf{e}_2)\cdot A\mathbf{e}_3 & \text{if } d = 3, \\ A_{11}A_{22} - A_{12}A_{21} & \text{if } d = 2\,, \end{cases} \tag{A.2.15}$$

where $\mathbf{e}_1 := (1,0,0)$, $\mathbf{e}_2 := (0,1,0)$, and $\mathbf{e}_3 := (0,0,1)$. The matrix A is *invertible* if $\det A \neq 0$. This is equivalent to the statement that the inverse of A, denoted A^{-1}, exists and $A^{-1}A = AA^{-1} = \mathbb{I}$. In fact, although not used in this book which is focused rather on problems with $d \leq 3$, the general definition uses the Levi-Civita permutation symbol $\varepsilon_{i_1,i_2,...,i_d}$ [11] to define $\det A = \sum_{i_1,i_2,...,i_d=1}^{d} \varepsilon_{i_1,i_2,...,i_d} A_{1i_1} \cdot A_{2i_2} \cdots A_{di_d}$.

For any $A, B \in \mathbb{R}^{d\times d}$, the following *Cauchy-Binet formula* holds:

$$\det(AB) = (\det A)(\det B)\,. \tag{A.2.16}$$

Further, for $d \geq 2$, the *cofactor matrix* Cof A of a matrix $A \in \mathbb{R}^{d\times d}$ is defined by $[\mathrm{Cof}\,A]_{ij} := (-1)^{i+j}\det C_{ij}$ where $C_{ij} \in \mathbb{R}^{(d-1)\times(d-1)}$ is the submatrix of A obtained from A by removing the i-th row and the j-th column for $i, j = 1, \ldots, d$. Therefore, $\mathbb{I}\det A = A(\mathrm{Cof}\,A)^\top = (\mathrm{Cof}\,A)^\top A$. Hence, if $A \in \mathbb{R}^{d\times d}$ is invertible then

$$\mathrm{Cof}\,A = (\det A)A^{-\top}\,. \tag{A.2.17}$$

Consequently, we get a simple formula for the matrix inverse called *Cramer's rule* for every $A \in \mathrm{GL}(d)$

$$A^{-1} = \frac{(\mathrm{Cof}\,A)^\top}{\det A} = \frac{\mathrm{Cof}\,A^\top}{\det A} \tag{A.2.18}$$

and also

$$\det \mathrm{Cof}\,A = (\det A)^{d-1} = \det A^{d-1}\,. \tag{A.2.19}$$

If $d = 3$ then written component-wise for $A := (A_{ij})_{i,j=1}^d$ it reads (no summation, counting the indices modulo 3, i.e., $4 \mapsto 1$, $5 \mapsto 2$)

$$(\mathrm{Cof}\,A)_{ij} := A_{i+1,j+1}A_{i+2,j+2} - A_{i+1,j+2}A_{i+2,j+1} \tag{A.2.20}$$

while if $d = 2$

$$\mathrm{Cof}\,A := \begin{pmatrix} A_{22} & -A_{21} \\ -A_{12} & A_{11} \end{pmatrix}. \tag{A.2.21}$$

The cofactor matrix can also be expressed as the
derivative of the determinant understood as a function $\det : \mathbb{R}^{d\times d} \to \mathbb{R}$, i.e.,

$$\mathrm{Cof}\,A = (\det A)'. \tag{A.2.22}$$

---

[11] In general dimensions $\varepsilon_{i_1,i_2,...,i_d}$ equals 1 or -1 if $(i_1, i_2, ..., i_d)$ is an even or an odd permutation of $(1, \ldots, d)$ and it equals 0 if two indices coincide.

The *Hadamard's inequality* gives a bound on the matrix determinant in terms of the Frobenius matrix norm. More specifically, if $A \in \mathbb{R}^{d\times d}$, then

$$|\mathrm{det}A| \le d^{d/2}|A|^d\,. \tag{A.2.23}$$

This formula together with (A.2.19) immediately yields that for every $A \in \mathbb{R}^{d\times d}$ with $d \ge 2$

$$|\mathrm{det}A| \le d^{d/2}|\mathrm{Cof}\,A|^{d/(d-1)}\,. \tag{A.2.24}$$

The *rank* of $A \in \mathbb{R}^{d\times d}$ is the dimension of the space $\{Ax;\ x\in\mathbb{R}^d\} \subset \mathbb{R}^d$. Having two vectors $a, b\in\mathbb{R}^d$ we define the *tensorial product* $a\otimes b \in \mathbb{R}^{d\times d}$ as

$$(a\otimes b)_{ij} = a_i b_j \quad \text{for all } i, j \in \{1,\dots,d\}. \tag{A.2.25}$$

If $a \ne 0$ and $b \ne 0$ then the rank of the matrix $a\otimes b$ is one. Conversely, any matrix of rank one (also called a *rank-one matrix*) can be written in the form $a\otimes b$ for some nonzero vectors $a, b \in \mathbb{R}^d$.

Moreover, GL(d) stands for the subset of invertible matrices in $\mathbb{R}^{d\times d}$, forming a so-called *general linear group*, and $\mathrm{GL}^+(d) \subset \mathrm{GL}(d)$ contains matrices with positive determinants. Matrices with the determinant equal to 1 form the set SL(d), forming a so-called *special linear group*. Further, the set of orthogonal matrices is denoted

$$\mathrm{O}(d) := \{A\in\mathrm{GL}(d):\ A^\top A = AA^\top = \mathbb{I}\,,\ |\mathrm{det}A| = 1\}.$$

Orthogonal matrices preserve the dot product, i.e., $Au\cdot Av = u\cdot v$ for every $u, v \in \mathbb{R}^d$, and every $A \in \mathrm{O}(d)$.

Finally SO(d) is the so-called *special orthogonal group* of rotation matrices with the determinant 1, i.e.,

$$\mathrm{SO}(d) := \{A\in\mathrm{SL}(d):\ A^\top A = AA^\top = \mathbb{I}\}. \tag{A.2.26}$$

If $A \in \mathbb{R}^{d\times d}$ is symmetric and $Av = \lambda v$ for some $\lambda \in \mathbb{R}$ and $v \in \mathbb{R}^d$ we say that λ is an *eigenvalue* of A and v is an *eigenvector* of A[12]

Every matrix $A \in \mathrm{GL}^+(d)$ can be decomposed as

$$A = RU = VR\,, \tag{A.2.27}$$

where $R \in \mathrm{O}(d)$ and $U = U^\top$ and $V = V^\top$ are *positive definite* matrices, i.e., for every $0 \ne x \in \mathbb{R}^d$ we have $x\cdot Ux > 0$ and $x\cdot Vx > 0$. Notice that

$$U = \sqrt{A^\top A} \text{ and } V = \sqrt{AA^\top} \tag{A.2.28}$$

[12] Eigenvalues and eigenvectors can be also defined for nonsymmetric matrices but then the eigenvalues are complex numbers, in general.

In this case, the eigenvalues $\{\lambda_i\}_{i=1}^d$ of $A^\top A$ and eigenvalues of $\{\bar{\lambda}_i\}_{i=1}^d$ of $AA^\top$ are positive and

$$U = \sum_{i=1}^{d} \sqrt{\lambda_i} v_i \otimes v_i \text{ and } V = \sum_{i=1}^{d} \sqrt{\bar{\lambda}_i} \bar{v}_i \otimes \bar{v}_i \,, \tag{A.2.29}$$

where $\{v_i\}$ is the unit eigenvector[13] corresponding to λ_i and $\bar{v}_i$ is the unit eigenvector corresponding to $\bar{\lambda}_i$. Again, here $1 \le i \le d$. Numbers $\sqrt{\lambda_i}$, for $1 \le i \le d$ are called singular values of A. The largest singular value is the so-called *spectral norm* of A. Symmetric matrices are positive (semi)definite if and only if all their *principal minors*[14] have positive (resp. non-negative) determinants, which is called *Sylvester's criterion.*

If $\mathbb{C} \in \mathbb{R}^{d\times d\times d\times d}$, i.e., $\mathbb{C} = (\mathbb{C}_{ijkl})_{i,j,k,l=1}^d$ and $A \in \mathbb{R}^{d\times d}$ we define $\mathbb{C}A \in \mathbb{R}^{d\times d}$ as

$$(\mathbb{C}A)_{ij} := \sum\nolimits_{k,l=1}^{d} \mathbb{C}_{ijkl} A_{kl} \,. \tag{A.2.30}$$

In view of (A.2.6) then

$$\mathbb{C}A{:}B := \sum\nolimits_{i,j,k,l=1}^{d} \mathbb{C}_{ijkl} A_{kl} B_{ij} \tag{A.2.31}$$

for every $B \in \mathbb{R}^{d\times d}$. We can also consider $\mathbb{H} \in \mathbb{R}^{d\times d\times d\times d}$ as a linear operator from $\mathbb{R}^{d\times d\times d} \to \mathbb{R}^{d\times d\times d}$ defined as

$$(\mathbb{H}G)_{ikl} := \sum\nolimits_{m,n=1}^{d} \mathbb{H}_{klmn} G_{imn} \tag{A.2.32}$$

and, consequently,

$$\mathbb{H}G \,\vdots\, G := \sum\nolimits_{i,k,l,m,n=1}^{d} \mathbb{H}_{klmn} G_{imn} G_{ikl} \,. \tag{A.2.33}$$

If $d = 3$ we also have (see [417, Formula (1.34)]), for all $A, B \in \mathbb{R}^{3\times 3}$, that

$$\mathrm{Cof}(A+B) = \mathrm{Cof}\,A + \mathrm{Cof}\,B + (\mathrm{Cof}\,A)'{:}B = \mathrm{Cof}\,A + \mathrm{Cof}\,B + (\mathrm{Cof}\,B)'{:}A \,. \tag{A.2.34}$$

A.3 Selected notions and results from nonlinear analysis

Nonlinear analysis plays a fundamental role in our book. Here we collect some basis tools and results which are frequently used.

---

[13] in the Frobenius norm

[14] A minor is principal if its diagonal is a part of the diagonal of the original matrix.

A.3.1 Fixed points, differentials, and monotone operators

An extremely powerful, although very nonconstructive theoretical tool for proving existence of solutions to various nonlinear equations (in particular those which are not related to any potential) is the following theorem, which tremendously generalizes a finite-dimensional variant, known as a Brouwer fixed-point theorem [94]:

Theorem A.3.1 (Schauder fixed point [477]). *A continuous mapping $T : S \to S$ with a closed, bounded, convex set S in a Banach space and with $T(S)$ relatively compact has a fixed point, i.e.*

$$\exists x \in S : \quad x = T(x).$$

In particular, it holds if S itself is compact; then $T(S)$ is automatically relatively compact.

The Banach-space structure allows further to say that $A : V_1 \to V_2$ is *directionally differentiable* if the directional derivative at u in the direction of (variation) v, defined as

$$\mathrm{D}A(u,v) = \lim_{\varepsilon \searrow 0} \frac{A(u+\varepsilon v) - A(u)}{\varepsilon}, \tag{A.3.1}$$

exists for any $u, v \in V_1$, and is *smooth* if it is directionally differentiable and $\mathrm{D}A(u,\cdot) : V_1 \to V_2$ is a linear continuous functional; then the *Gâteaux differential* (sometimes also called the *Gâteaux derivative*) $A'(u) \in \mathrm{Lin}(V_1,V_2)$ is defined by

$$\mathrm{D}A(u,v) := [A'(u)](v). \tag{A.3.2}$$

In the special case $V_2 = \mathbb{R}$, the mapping A is a functional, let us denote it by $\varPhi$ and write briefly $V = V_1$. If smooth, $\varPhi' \in \mathrm{Lin}(V,\mathbb{R}) \cong V^*$ is defined via (A.3.2) by

$$\langle \varPhi'(u), v \rangle := \mathrm{D}\varPhi(u,v). \tag{A.3.3}$$

Furthermore, $u \in V$ is called a *critical point* if

$$\varPhi'(u) = 0, \tag{A.3.4}$$

which is an abstract version of the *Euler-Lagrange equation*. In fact, (A.3.4) is a special case of the abstract operator equation

$$A(u) = f \quad \text{with } A : V \to V^*,\ f \in V^*, \tag{A.3.5}$$

provided $A = \varPhi' + f$ for some *potential* $\varPhi$ whose existence requires some symmetry of A: if A itself is Gâteaux differentiable and *hemicontinuous*[15], it has a potential if and only if it is *symmetric*, i.e.

[15] This is a very weak mode of continuity, requiring that $t \mapsto \langle A(u+tv), w \rangle$ is continuous.

$$\langle [A'(u)](v), w\rangle = \langle [A'(u)](w), v\rangle \tag{A.3.6}$$

for any $u, v, w \in V$; up to an additive constant; this potential is given by the formula

$$\Phi(u) = \int_0^1 \langle A(\lambda u), u\rangle \mathrm{d}\lambda. \tag{A.3.7}$$

The equation (A.3.4) is satisfied, e.g. if Φ attains its minimum[16] or maximum at u. Such u is called a *minimizer* or a maximizer, respectively. The former case is often connected with a *minimum-energy principle* that is assumed to govern many steady-state physical problems. The existence of solutions to (A.3.4) can thus often be based on the existence of a minimizer of Φ, which can rely on the following:

Theorem A.3.2 (Bolzano-Weierstrass principle).[17] *Every lower (resp. upper) semicontinuous function $X \to \mathbb{R}$ on a compact set X attains its minimum (resp. maximum) on this set.*

The Bolzano-Weierstrass theorem underlies the *direct method,*[18] invented essentially in [515], for proving existence of a solution to (A.3.4). We say that Φ is *coercive* if $\lim_{\|u\|\to\infty} \Phi(u)/\|u\| = +\infty$.

Theorem A.3.3 (Direct method).[19] *Let V have a pre-dual and $\Phi : V \to \mathbb{R}$ be weakly* lower semicontinuous, smooth, and coercive. Then* (A.3.4) *has a solution.*

Since continuous convex functionals are also weakly* lower semicontinuous, one gets a useful modification:

Theorem A.3.4 (Direct method II). *Let V have a pre-dual and let $\Phi : V \to \mathbb{R}$ be continuous, smooth, coercive, and convex. Then* (A.3.4) *has a solution.*

If Φ is furthermore *strictly convex* in the sense that $\Phi(\lambda u+(1-\lambda)v) < \lambda\Phi(u) + (1-\lambda)\Phi(v)$ for any $u \neq v$ and $0 < \lambda < 1$, then (A.3.4) has at most one solution.

We say that a nonlinear operator $A : V \to V^*$ is *monotone* if $\langle A(u)-A(v), u-v\rangle \geq 0$ for any $u, v \in V$. Monotonicity of a potential nonlinear operator implies convexity of its potential, and then Theorem A.3.4 implies:

---

[16] The proof is simple: suppose $\Phi(u) = \min \Phi(\cdot)$ but $\Phi'(u) \neq 0$, then for some $v \in V$ we would have $\langle \Phi'(u), v\rangle = \mathrm{D}\Phi(u, v) < 0$ so that, for a sufficiently small $\varepsilon > 0$, $\Phi(u+\varepsilon v) = \Phi(u)+\varepsilon\langle \Phi'(u), v\rangle + o(\varepsilon) < \Phi(u)$, a contradiction.

[17] In fact, this is rather a tremendous generalization of the original assertion by Bolzano [77] who showed, rather intuitively in times when only irrational but not transcendental numbers was known (because completeness of $\mathbb{R}$ and its local compactness was not rigorously known) that any real continuous function of a bounded closed interval is bounded. An essence, called the Bolzano-Weierstrass principle, is that every sequence in a compact set has a cluster point, i.e. a point whose each neighborhood contains infinitely many members of this sequence.

[18] This means that no approximation and subsequent convergence is needed.

[19] The proof relies on coercivity of Φ which allows for a localization on bounded sets and then, due to weak* compactness of convex closed bounded sets in V, on the Bolzano-Weierstrass' theorem.

Theorem A.3.5. *Let V be reflexive and $A : V \to V^*$ be monotone, hemicontinuous, coercive in the sense that* $\lim_{\|u\|\to\infty}\langle A(u),u\rangle = \infty$, *and possess a potential. Then, for any $f \in V^*$, the equation* (A.3.5) *has a solution.*

In fact, Theorem A.3.5 holds even for mappings not having a potential but its proof, due to Brézis [89], then relies on an approximation and on implicit, non-constructive fixed-point arguments.

The solutions to (A.3.4) do not need to represent global minimizers which we considered so far. Local minimizers, being consistent with physical principles of minimization of energy, would serve well, too. The same holds for maximizers. Critical points may, however, have a more complicated saddle-like character: For $\Phi : A\times B \to \mathbb{R}$ with A and B nonempty sets, we say that $(u,v)\in A\times B$ is a *saddle point* of Φ if

$$\forall \tilde{u}\in A \ \ \forall \tilde{v}\in B: \qquad \Phi(u,\tilde{v}) \le \Phi(u,v) \le \Phi(\tilde{u},v)\,. \tag{A.3.8}$$

Always, $\min_A(\sup\Phi(\cdot,B)) \ge \max_B(\inf\Phi(A,\cdot))$.

Proposition A.3.6 (Characterization of saddle points).[20] *Let A and B be nonempty sets and $\Phi : A\times B\to\mathbb{R}$. Then $(u,v)\in A\times B$ is a saddle point of Φ on $A\times B$ if and only if*

$$\begin{aligned}
&u\in \mathrm{Argmin}(\sup\Phi(\cdot,B)),\\
&v\in \mathrm{Argmax}(\inf\Phi(A,\cdot)),\\
&\min_A(\sup\Phi(\cdot,B)) = \max_B(\inf\Phi(A,\cdot)).
\end{aligned}$$

Theorem A.3.7 (Existence of saddle points; von Neumann [393], Ky Fan [182]).[21] *If the Banach spaces U and V are separable and reflexive, $A\subset U$ and $B\subset V$ are convex closed, and $\Phi : U\times V\to\mathbb{R}$ satisfy:*

$$\forall v\in V: \quad \Phi(\cdot,v) : U\to\mathbb{R} \text{ is convex, lower semicontinuous,} \tag{A.3.10a}$$

$$\forall u\in U: \quad \Phi(u,\cdot) : V\to\mathbb{R} \text{ is concave, upper semicontinuous,} \tag{A.3.10b}$$

$$\exists v\in B: \quad \Phi(\cdot,v) \text{ is coercive on } A, \text{ i.e. } \lim_{\|u\|\to\infty,\ u\in A} \Phi(u,v) = \infty, \tag{A.3.10c}$$

$$\exists u\in A: \quad -\Phi(u,\cdot) \text{ is coercive on } B, \text{ i.e. } \lim_{\|v\|\to\infty,\ v\in B} \Phi(u,v) = -\infty, \tag{A.3.10d}$$

then Φ has a saddle point on $A\times B$.

A.3.2 Elements of convex analysis

A set K in a linear space is called *convex* if $\lambda u + (1-\lambda)v \in K$ whenever $u,v \in K$ and $\lambda\in[0,1]$, and it is called a *cone* (with the vertex at the origin 0) if $\lambda v\in K$ whenever $v\in K$ and $\lambda\ge 0$. A functional $f : V\to\mathbb{R}$ is called *convex* if $f(\lambda u+(1-\lambda)v) \le \lambda f(u)+$

[20] See also [539, Sect. 49.2].

[21] See also [539, Ch. 49]. The original version [393] dealt with the finite-dimensional case only.

$(1-\lambda)f(v)$. If $u_1 \neq u_2$ implies (A.3.11) with strict inequality, then f is called *strictly convex*. A functional $f : V \to \mathbb{R}$ is convex (resp. lower semicontinuous) if and only if its *epigraph* $\mathrm{epi}(f) := \{(x,a) \in V\times\mathbb{R};\ a \geq f(x)\}$ is convex (resp. closed) subset of $V\times\mathbb{R}$. This is related to that a lower semicontinuous functional f is convex if and only if

$$\frac{1}{2}f(u_1)+\frac{1}{2}f(u_2) \geq f\Big(\frac{u_1+u_2}{2}\Big), \tag{A.3.11}$$

We say that a convex functional is *proper* if it is not identially ∞ and never takes value $-\infty$.

Let us call a Banach space V *strictly convex* if the sphere[22] in V does not contain any line segment. The space V is *uniformly convex* if

$$\forall\varepsilon>0\ \exists\delta>0\ \forall u,v\in V:\quad \left.\begin{array}{l}\|u\|=\|v\|=1\\ \|u-v\|\geq\varepsilon\end{array}\right\} \ \Rightarrow\ \Big\|\frac{1}{2}u+\frac{1}{2}v\Big\| \leq 1-\delta. \tag{A.3.12}$$

Any uniformly convex Banach space is strictly convex but not vice versa.

Theorem A.3.8 (K. Fan, I.L. Glicksberg).[23] *If V is uniformly convex, $u_k \to u$ weakly, and $\|u_k\| \to \|u\|$, then $u_k \to u$ strongly.*

Having a convex subset K of a Banach space V and $u \in K$, we define the *tangent cone* $T_K(u) \subset V$ by

$$T_K(u) := \mathrm{cl}\Big(\bigcup_{a>0} a(K-u)\Big).$$

Obviously, $T_K(u)$ is a closed convex cone.[24] Besides, we define the *normal cone* $N_K(v)$ as

$$N_K(u) := \{f\in V^*;\ \forall v\in T_K(u):\ \langle f,v\rangle \leq 0\}. \tag{A.3.13}$$

Again, the normal cone is always a closed convex cone in V^*.

A *domain* of a functional $F : V \to \mathbb{R}$ defined as $\mathrm{Dom}\,F := \{v\in V;\ F(v) < +\infty\}$ is convex or closed if F is convex or lower semicontinuous, respectively. The *subdifferential* of a convex functional $F : V \to \mathbb{R}$ is defined as a convex closed subset of V^*:

$$\partial F(v) := \{\xi\in V^*;\ \forall\tilde v\in V:\ F(v+\tilde v) \geq F(v)+\langle\xi,\tilde v\rangle\}. \tag{A.3.14}$$

Elements of the subdifferential are called *subgradients*. For two convex functionals $F_1, F_2 : V \to \mathbb{R}$, the *sum rule*

[22] A "sphere" $\{v\in V;\ \|v\| = \varrho\}$ (of the radius $\varrho > 0$) is a surface of a "ball" $\{v\in V;\ \|v\| \leq \varrho\}$.

[23] See Fan and Glicksberg [183] for thorough investigation and various modifications.

[24] Note that $v \in T_K(u)$ means precisely that $u + a_k v_k \in K$ for suitable sequences $\{a_k\}_{k\in\mathbb{N}} \subset \mathbb{R}$ and $\{v_k\}_{k\in\mathbb{N}} \subset V$ such that $\lim_{k\to\infty} v_k = v$.

$$\partial(F_1+F_2) = \partial F_1 + \partial F_2 \tag{A.3.15}$$

holds if some additional conditions are satisfied by F_1 and F_2. For example, if $V := \mathbb{R}^d$ a sufficient condition reads $\mathrm{relint}(\mathrm{Dom}\,F_1) \cap \mathrm{relint}(\mathrm{Dom}\,F_2) \neq \emptyset$; cf. [443][25]. In general, we only have $\partial(F_1+F_2) \supset \partial F_1 + \partial F_2$ simply by the definition (A.3.14); cf. also Example A.3.10.

Let V be a Banach space and $F : V \to \mathbb{R}\cup\{\infty\}$ a proper, lower semicontinuous, and convex functional. Its *Legendre-Fenchel transform* $F^* : V^* \to \mathbb{R}\cup\{\infty\}$ is defined via

$$F^*(\xi) := \sup\{\langle \xi, v\rangle - F(v);\ v\in V\}. \tag{A.3.16}$$

Also, F^* is called a *convex conjugate* functional to F; note that F^* is always convex, being the supremum of convex functions. The Fenchel equivalences for subdifferentials read[26]

$$\xi\in\partial F(v) \quad\Leftrightarrow\quad v\in\partial F^*(\xi) \quad\Leftrightarrow\quad F(v)+F^*(\xi) = \langle \xi, v\rangle. \tag{A.3.17}$$

Moreover, by the definition (A.3.16) of F^* we always have the Fenchel-Young inequality

$$F(v)+F^*(\xi) \geq \langle \xi, v\rangle \qquad \text{for all } v\in V \text{ and } \xi\in V^*. \tag{A.3.18}$$

For a convex proper lower semicontinuous F, we have

$$F^{**} := (F^*)^* = F. \tag{A.3.19}$$

A Banach linear space V is called *ordered* by a relation $\geq$ if this relation is an ordering and, in addition, it is compatible with the linear and topological structure.[27] It is easy to see that $D := \{u\in V;\ u\geq 0\}$ is a closed convex cone which does not contain a line. Conversely, having a closed convex cone $D\subset V$ which does not contain a line, the relation "$\geq$" defined by $u\geq v$ provided $u-v\in D$ makes V an ordered linear topological space.

The so-called negative polar cone $\{f\in V;\ \forall v\in V:\ \langle f, v\rangle \geq 0\}$ defines an ordering on V^* if it does not contain a line; this ordering is then called a *dual ordering*.

A general noteworthy regularization construction is a so-called *Yosida approximation*, defined for a proper convex function $F : V \to \mathbb{R}\cup\{+\infty\}$ by

---

[25] The abbreviation "relint" stands for the relative interior. It is the interior of the set within its affine hull.

[26] The inclusion $\xi \in \partial F(v)$ is equivalent to $\langle \xi, u\rangle - F(u) \leq \langle \xi, v\rangle - F(v)$ holding for any u, which is equivalent to $F^*(\xi) = \sup_u \langle \xi, u\rangle - F(u) = \langle \xi, v\rangle - F(v)$. Thus we obtain the Fenchel equality $F(v) + F^*(\xi) = \langle \xi, v\rangle$. By symmetry, we can derive backwards $v \in \partial F^*(\xi)$. Hence $[\partial F]^{-1} = \partial F^*$ is shown.

[27] This just means the following four properties: $u\geq 0$ and $a\geq 0$ imply $au\geq 0$, $u\geq 0$ and $v\geq 0$ imply $u{+}v\geq 0$, $u\geq v$ implies $u{+}w\geq v{+}w$ for any w, and $u_\xi\geq 0$ and $u_\xi\to x$ implies $x\geq 0$.

$$F_\varepsilon(v) := \inf_{\tilde{v}\in V} F(\tilde{v}) + \frac{1}{2\varepsilon}\|\tilde{v}-v\|^2. \tag{A.3.20}$$

Always, $F_\varepsilon \le F$ is convex, finite, of at most a quadratic growth, and $F_\varepsilon \to F$ pointwise for $\varepsilon \to 0$ provided F is lower semicontinuous; this holds generally on Banach spaces. If V and V^* are strictly convex, F_ε is Gâteaux differentiable with $F'_\varepsilon : V \to V^*$ (strong,weak)-continuous and bounded, cf. [456, Lemma 5.17]. If V is a Hilbert space (and in particular if it is an Euclidean space normed standardly), then F'_ε is even Lipschitz continuous.

Example A.3.9. Using a notation δ_K for the so-called *indicator function* defined as

$$\delta_K(v) := \begin{cases} 0 & \text{if } v \in K, \\ \infty & \text{otherwise,} \end{cases} \tag{A.3.21}$$

we illustrate the above definitions by the following formulas holding for every convex closed K:

$$\partial\delta_K = N_K \quad \text{and} \quad \partial\delta_K^* = T_K \quad \text{and} \quad [\partial\delta_K^*]^{-1} = \partial\delta_K^{**} = \partial\delta_K = N_K. \tag{A.3.22}$$

The last relation also implies $\partial\delta_K^* = [N_K]^{-1}$ so that, in particular,

$$\partial\delta_K^*(0) = [N_K]^{-1}(0) = K. \tag{A.3.23}$$

*Example A.3.10 (**Inequality in the sum rule**).* The equality (A.3.15) indeed needs a qualification of F_1 and F_2. In general, even a very extreme situation

$$V = [\partial(F_1{+}F_2)](v) \supsetneqq \partial F_1(v) + \partial F_2(v) = \emptyset + \emptyset$$

is easily possible. It occurs e.g. for $v = 0 \in \mathbb{R} = V$ for

$$F_1(v) = F_2(-v) = \begin{cases} -\sqrt{v} & \text{if } v \ge 0, \\ \infty & \text{if } v < 0; \end{cases}$$

note that $F_1 + F_2 = \delta_0$ while the derivative of F_1 tends to $-\infty$ for $v \to 0+$ and the derivative of F_2 tends to ∞ for $v \to 0-$ so that indeed both $\partial F_1(0) = \emptyset$ and $\partial F_2(0) = \emptyset$. And obviously $\mathrm{Dom}\, F_1 = [0,\infty)$ and $\mathrm{Dom}\, F_2 = (-\infty, 0]$ so that $(\mathrm{relintDom}\, F_1) \cap (\mathrm{relintDom}\, F_2) = \emptyset$ and the qualification for (A.3.15) we mentioned above indeed is not satisfied.

*Example A.3.11 (**Löwner ordering**).* The set of positive semidefinite $(d{\times}d)$-matrices forms a closed convex cone. The ordering of all $(d{\times}d)$-matrices by this cone is called Löwner's ordering.

Example A.3.12 ($\mathrm{Dom}(\partial F) \ne \mathrm{Dom}\, F$*).* Denoting by $\mathrm{Dom}(\partial F) := \{v \in V;\ \partial F(v) \ne \emptyset\}$ the *domain of the subdifferential*, one can have situations when $\mathrm{Dom}(\partial F) \ne \mathrm{Dom}\, F$.

An example is the function $F:\mathbb{R}\to\mathbb{R}\cup\{+\infty\}$

$$F(x) := \begin{cases} x(\ln x - 1) & \text{if } x>0, \\ 0 & \text{if } x=0, \\ +\infty & \text{if } x<0, \end{cases} \tag{A.3.24}$$

for which $F(0) = 0 < \infty$ but $\partial F(0) = \emptyset$. Notice that $\lim_{x\to 0_+} F'(x) = -\infty$.

A.3.3 Differential operators on vector and matrix fields

Here we define three main differential operators, namely the *gradient*, *curl*, and *divergence* of a vector and a matrix fields. If $\Omega \subset \mathbb{R}^d$ is a bounded domain and $v:\Omega\to\mathbb{R}^d$ is smooth, we define its gradient $\nabla v:\Omega\to\mathbb{R}^{d\times d}$ as the gradient of every component of v, i.e., for $i,j\in\{1,\ldots,d\}$

$$(\nabla v)_{ij} := \partial_{x_j} v_i\,. \tag{A.3.25}$$

The divergence of v, $\operatorname{div} v:\Omega\to\mathbb{R}$, reads

$$\operatorname{div} v := \sum\nolimits_{i=1}^{d} \partial_{x_i} v_i = \operatorname{tr}\nabla v \tag{A.3.26}$$

and for $d=3$ $\operatorname{curl} v:\Omega\to\mathbb{R}^3$ is a vector whose components satisfy, for $1\le i\le 3$

$$(\operatorname{curl} v)_i := \sum\nolimits_{j,k=1}^{3} \varepsilon_{ijk}\partial_{x_j} v_k\,, \tag{A.3.27}$$

where ε_{jjk} is the Levi-Civita symbol; cf. (A.2.3). If $d=2$, $\operatorname{curl} v:\Omega\to\mathbb{R}$

$$\operatorname{curl} v := \partial_{x_1} v_2 - \partial_{x_2} v_1\,. \tag{A.3.28}$$

If $A:\Omega\to\mathbb{R}^{d\times d}$ is a smooth matrix-valued map, its divergence, $\operatorname{div} A:\Omega\to\mathbb{R}^d$ is a vector defined, for $i\in\{1,\ldots,d\}$, as

$$(\operatorname{div} A)_i := \sum\nolimits_{j=1}^{d} \partial_{x_j} A_{ij}\,. \tag{A.3.29}$$

Similarly, if $d=3$, we calculate the curl of A, $\operatorname{curl} A:\Omega\to\mathbb{R}^{3\times 3}$, as

$$(\mathrm{curl}\,A)_{ij} := \sum_{k,m=1}^{3} \varepsilon_{jkm} \partial_{x_k} A_{im} \tag{A.3.30}$$

for $i, j \in \{1,2,3\}$ and again with the help the Levi-Civita symbol (A.2.3). In the case $d = 2$ $\mathrm{curl}\,A : \Omega \to \mathbb{R}^2$ and we have, for $1 \le i \le 2$, that

$$(\mathrm{curl}\,A)_i := \partial_{x_1} A_{i2} - \partial_{x_2} A_{i1}\,. \tag{A.3.31}$$

It is well-known that if $y : \Omega \to \mathbb{R}^d$ is smooth then

$$\mathrm{curl}\,\nabla y = 0 \quad \text{and} \quad \mathrm{div}\,\mathrm{curl}\,y = 0\,. \tag{A.3.32}$$

A.3.4 Γ-convergence

Γ-convergence is a notion of convergence for functionals which was introduced by E. De Giorgi [147, 148]. To motivate the definition, suppose that X is a metric space, and consider a family of functionals

$$F_\varepsilon : X \longrightarrow [-\infty, +\infty]$$

parametrized by a real parameter $\varepsilon > 0$. For a sequence $\{u_\varepsilon\} \subset X$ suppose that

$$F_\varepsilon(u_\varepsilon) = \min_{u \in X} F_\varepsilon(u),$$

that is, for each $\varepsilon > 0$, u_ε is a minimizer of the functional F_ε. It is quite a natural question to ask for the asymptotic behavior of

$$\lim_{\varepsilon \to 0} F_\varepsilon(u_\varepsilon),$$

i.e. whether there is an appropriate notion of a limit F of the sequence F_ε of functionals. It turns out that this notion is given by the concept of Γ-convergence.

Definition A.3.13 (Γ-convergence). Let X be a metric space and let $F_\varepsilon : X \to [-\infty, +\infty]$, $\varepsilon > 0$, $F : X \to [-\infty, \infty]$. Then we say that the sequence $\{F_\varepsilon\}_{\varepsilon>0}$ Γ-converges to F in X and write $F_\varepsilon \xrightarrow{\Gamma} F$ in X if the following two conditions hold:

$$\forall u \in X \ \forall \{u_\varepsilon\}_{\varepsilon>0} \subset X,\ u_\varepsilon \xrightarrow{X} u : \qquad \liminf_{\varepsilon \to 0} F_\varepsilon(u_\varepsilon) \ge F(u), \tag{A.3.33a}$$

$$\forall u \in X \ \exists \{u_\varepsilon\}_{\varepsilon>0} \subset X,\ u_\varepsilon \xrightarrow{X} u : \qquad \limsup_{\varepsilon \to 0} F_\varepsilon(u_\varepsilon) \le F(u), \tag{A.3.33b}$$

the sequence occurring in (A.3.33b) being called the *recovery sequence.*

Proposition A.3.14. *The notion of Γ-convergence enjoys the following properties:*
(i) The Γ-limit is lower semicontinuous on X.
(ii) The Γ-limit is stable under continuous perturbation, i.e. for G continuous,

$$F_\varepsilon \xrightarrow{\Gamma} F \qquad \Rightarrow \qquad F_\varepsilon + G \xrightarrow{\Gamma} F + G.$$

(iii) Stability of minimizing sequences: If $F_\varepsilon \xrightarrow{\Gamma} F$ and v_ε minimizes F_ε over X, then any cluster point of the sequence $\{v_\varepsilon\}_\varepsilon$ minimizes F.

Example A.3.15 (Γ-limit vs. pointwise limit). Consider $F_\varepsilon(x) = \cos(x/\varepsilon)$ for $x \in \mathbb{R}$. Assume that $\mathbb{R}$ is equipped with the Euclidean topology. We have $F_\varepsilon \xrightarrow{\Gamma} -1$. Notice, however, that pointwise limit of $\{F_\varepsilon\}$ does not exist.

Example A.3.16 (Monotone pointwise convergence from above). Assume that $F_\varepsilon \to F$ pointwise for F lower semicontinuous on X and assume that $\{F_\varepsilon\}$ is a monotone sequence, i.e., $F_{\varepsilon_1} \geq F_{\varepsilon_2} \geq F$ for every $\varepsilon_1 > \varepsilon_2 > 0$. Then $F_\varepsilon \xrightarrow{\Gamma} F$. Indeed, if $u_\varepsilon \to u$ in X, then

$$\liminf_{\varepsilon\to 0} F_\varepsilon(u_\varepsilon) \geq \liminf_{\varepsilon\to 0} F(u_\varepsilon) \geq F(u)$$

because of lower semicontinuity of F. On the other hand, the constant sequence $u_\varepsilon := u$ for all $\varepsilon > 0$ is a recovery sequence due to the pointwise convergence.

A little more complicated is the following result about the pointwise convergence from below.

Theorem A.3.17 (Monotone pointwise convergence from below [138]). *Assume that $F_\varepsilon \to F$ pointwise, $F_{\varepsilon_2} \leq F_{\varepsilon_1} \leq F$ for every $\varepsilon_2 > \varepsilon_1 > 0$. Let, moreover F_ε be lower semicontinuous for every $\varepsilon > 0$. Then $F_\varepsilon \xrightarrow{\Gamma} F$.*

Example A.3.18 (Γ-limit of convex functions). Assume that X is a normed linear space. Let $F_\varepsilon : X \to [-\infty, \infty]$ be convex for every $\varepsilon > 0$ and such that $F_\varepsilon \xrightarrow{\Gamma} F$. Then $F : X \to [-\infty, \infty]$ is convex and lower semicontinuous. Indeed, take $u, v \in X$ and $0 \leq \lambda \leq 1$. Let $\{u_\varepsilon\}_{\varepsilon>0} \subset X$ and $\{v_\varepsilon\}_{\varepsilon>0} \subset X$ be recovery sequences for u and v, respectively; cf. Definition A.3.13. Convexity of F_ε implies that $F_\varepsilon(\lambda u_\varepsilon + (1-\lambda)v_\varepsilon) \leq \lambda F_\varepsilon(u_\varepsilon) + (1-\lambda)F_\varepsilon(v_\varepsilon)$. Passing to the liminf for $\varepsilon \to 0$ yields, in view of (A.3.33a), that $F(\lambda u + (1-\lambda)v) \leq \lambda F(u) + (1-\lambda)F(v)$, where we exploited properties of the recovery sequences (A.3.33b) and that $u_\varepsilon + v_\varepsilon \to u + v$ in X. Lower semicontinuity of F follows from Proposition A.3.14.

Example A.3.19 (Dirichlet integral/Poisson equation). Consider

$$F_\varepsilon : H_0^1(\Omega) \to \mathbb{R} : u \mapsto \int_\Omega \frac{1}{2}|\nabla u|^2 - f_\varepsilon u \, \mathrm{d}x$$

with $f_\varepsilon \in L^{2^{*\prime}}(\Omega)$. If $f_\varepsilon \to f$ in $L^{2^{*\prime}}(\Omega)$ then, obviously, $F_\varepsilon \xrightarrow{\Gamma} F$ where

$$F : H_0^1(\Omega) \to \mathbb{R} : u \mapsto \int_\Omega \frac{1}{2}|\nabla u|^2 - fu\,\mathrm{d}x\,.$$

Indeed, if $u_\varepsilon \rightharpoonup u$ in $H_0^1(\Omega)$ then $u_\varepsilon \rightharpoonup u$ in $L^{2^*}(\Omega)$ which implies (A.3.33a). The constant sequence $u_\varepsilon := u$ serves as a recovery one to show (A.3.33b). However, if $f_\varepsilon \to f$ in $L^1(\Omega)$ with $\{f_\varepsilon\}_{\varepsilon>0} \subset L^{2^{*\prime}}(\Omega)$ and with $f \in L^1(\Omega) \setminus L^{2^{*\prime}}(\Omega)$ then $F = -\infty$. Realize that in this case $\|f_\varepsilon\|_{L^{2^*\prime}(\Omega)} \to +\infty$ for $\varepsilon \to 0$. It is enough to find a recovery sequence. First notice that if it holds for some $c > 0$ that

$$\inf_{\{u\in H_0^1(\Omega);\, \|u\|_{H_0^1(\Omega)}\le 1\}} F_\varepsilon(u) > -c$$

then inevitably (cf. [456, Exercise 3.4.1])

$$\|f_\varepsilon\|_{H^{-1}(\Omega)} = \sup_{\|u\|_{H_0^1(\Omega)}\le 1} \int_\Omega f_\varepsilon u\,\mathrm{d}x \le \sup_{\|u\|_{H_0^1(\Omega)}\le 1} \int_\Omega \frac{1}{2}|\nabla u|^2 + c \le \frac{1}{2} + c\,, \tag{A.3.34}$$

i.e., also $\|f_\varepsilon\|_{L^{2^*\prime}(\Omega)} < +\infty$ and, consequently, $\lim_{\varepsilon\to 0} \inf_{u\in H_0^1(\Omega);\|u\|_{H_0^1(\Omega)}\le 1} F_\varepsilon(u) = -\infty$ if $\|f_\varepsilon\|_{L^{2^*\prime}(\Omega)} \to +\infty$ for $\varepsilon \to 0$. Let $\tilde u_\varepsilon \in H_0^1$ with $\|\tilde u_\varepsilon\|_{H_0^1(\Omega)} \le 1$ be such that $\lim_{\varepsilon\to 0} F_\varepsilon(\tilde u_\varepsilon) = -\infty$. Then also $\int_\Omega f_\varepsilon u_\varepsilon\,\mathrm{d}x \to +\infty$. Assume without loss of generality that $g_\varepsilon^{-2} := \int_\Omega f_\varepsilon u_\varepsilon\,\mathrm{d}x > 0$ for every $\varepsilon > 0$. Take $u \in W_0^{1,\infty}(\Omega)$ and define $u_\varepsilon := u + g_\varepsilon \tilde u_\varepsilon$. It holds at least for a subsequence that $u_\varepsilon \rightharpoonup u$ because $\{\tilde u_\varepsilon\}_{\varepsilon>0}$ is uniformly bounded and $g_\varepsilon \to 0$ for $\varepsilon \to 0$. Obviously, $F_\varepsilon(u_\varepsilon) \to -\infty$ for $\varepsilon \to 0$. If $u \in H_0^1(\Omega) \setminus W_0^{1,\infty}(\Omega)$ we find a sequence of maps in $W_0^{1,\infty}(\Omega)$ which converges strongly to u in the topology of $H_0^1(\Omega)$ and apply a suitable diagonalization argument. Details arc left to the interested reader.

Example A.3.20 (Abstract "numerical" approximation). Let us consider the case

$$X_k \subset X_{k+1} \subset X \;\text{ and }\; F_k := F + \delta_{X_k}\,, \tag{A.3.35}$$

where δ_{X_k} is the indicator function of X_k, cf. (A.3.21). Further suppose that F is continuous and $\bigcup_{k\in\mathbb{N}} X_k$ is dense in X. Typically, it occurs in various numerical approximation where X_k are finite-dimensional manifolds. Then $\Gamma\text{–}\lim_{k\to\infty} F_k = F$. Indeed, (A.3.33a) holds because again $F_k \ge F_{k+1} \ge F$ as in Example A.3.16. For any $\hat u \in X$, there is $\hat u_k \in X_k$ such that $\hat u_k \to \hat u$. Then $\lim_{k\to\infty} F_k(\hat u_k) = \lim_{k\to\infty} F(\hat u_k) = F(\hat u)$ and also $\lim_{k\to\infty} \hat u_k = u$ in X so that $\{\hat u_k\}_{k\in\mathbb{N}}$ is a recovery sequence for (A.3.33a). Note that lower semicontinuity of F would not be sufficient. A simple counterexample is $F := \delta_{\{u\}}$ (the indicator function of $\{u\}$) with some $u \in X \setminus \bigcup_{k\in\mathbb{N}} X_k$ where $F_k \equiv +\infty$ obviously does not Γ-converge to F.

Example A.3.21 (Monotone approximation from below: penalty function). The situation $F_k \le F_{k+1} \le F_\infty$ can be illustrated by considering $A \subset X$ closed, X equipped with a metric d inducing a topology finer than (but not necessarily identical with) the topology of X, F lower semi-continuous, $F_\infty := F + \delta_A$, and

$$F_k(u) := F(u) + k\,\mathrm{dist}(u,A)^\alpha := F(u) + k \inf_{\tilde{u}\in A} d(u,\tilde{u})^\alpha \tag{A.3.36}$$

with $\alpha > 0$, where the last term is called a *penalty function* for the constraint "$u \in A$". Then Γ–$\lim_{k\to\infty} F_k = F_\infty$. The ($\Gamma$inf)-condition is trivial for $u \in A$ because then $F_\infty(u) = F(u) \le \liminf_{k\to\infty} F(u_k) \le \liminf_{k\to\infty} F_k(u_k)$ if using successively lower semicontinuity of F and that $F \le F_k$. If $u \notin A$, then $\mathrm{dist}(u,A) > 0$ because A is closed and there is k_0 such that $\mathrm{dist}(u_k,A) \ge \frac{1}{2}\mathrm{dist}(u,A) > 0$ for all $k \ge k_0$, and then

$$\begin{aligned}\liminf_{k\to\infty} F_k(u_k) &= \liminf_{k\to\infty} F(u_k) + \lim_{k\to\infty} k\,\mathrm{dist}(u,A)^\alpha \\ &\ge F(u) + \lim_{k\to\infty} k\frac{\mathrm{dist}(u,A)^\alpha}{2^\alpha} = +\infty = F_\infty(u).\end{aligned}$$

The condition (A.3.33b) is satisfied for the constant recovery sequence $\hat{u}_k = \hat{u}$: if $\hat{u} \in A$, then $F_k(\hat{u}_k) = F(\hat{u}) = F_\infty(\hat{u})$, and if $\hat{u} \notin A$, then $F_\infty(\hat{u}) = \infty$ and (A.3.33b) holds trivially.

Remark A.3.22 (Importance of correct rescaling). Instead of looking at $\{F_\varepsilon\}_{\varepsilon>0}$, we may look at $\{f(\varepsilon)F_\varepsilon\}_{\varepsilon>0}$, $f(\varepsilon) > 0$, and study Γ-convergence of this modified sequence. If v_ε minimizes F_ε, it will also minimizes $f(\varepsilon)F_\varepsilon$. However, this sequence might have a different Γ-limit. The problem thus arises of rescaling the functionals F_ε so as to obtain as much information as possible about the minimizers. For example, it may happen that $\lim_\varepsilon F_\varepsilon$ Γ-converges to a constant, which would give no information at all. Hence, to find a suitable scaling factor heavily influences obtained results.

Appendix B
Lebesgue and Sobolev function spaces

Mathematical continuum mechanics ultimately works with concrete suitable function space in the position of the abstract Banach spaces or Hilbert spaces. This book, being devoted to rather basic problems and related mathematical apparatus, confines in this Appendix on presentation of quite basic constructions and results only. Cf. e.g. Adams, Fournier [3], Giusti [222], Kufner, Fučík, John [298], Tartar [506], or Ziemer [542] for detailed expositions.

B.1 Basic notions from measure theory

For a set S and for a *σ-algebra*[1] $\mathfrak{S}$ of its subsets, a σ-additive[2] set function $\mu : \mathfrak{S} \to \mathbb{R} \cup \{\pm\infty\}$ is called a *measure*. The *variation* $|\mu|$ of μ is a function $\mathfrak{S} \to \mathbb{R} \cup \{\infty\}$ defined by

$$|\mu|(A) := \sup\Big\{ \sum_{i=1}^{n} |\mu(A_i)|;\ n\in\mathbb{N},\ \bigcup_{i=1}^{n} A_i \subset A,\ \forall 1{\le}i{\le}n,\ 1{\le}j{\le}n,\ i{\ne}j:\ A_i{\in}\mathfrak{S},\ A_i{\cap}A_j = \emptyset \Big\}.$$

We say that μ has a finite variation if the *total variation* $|\mu|(S)$ is finite. The measures on $(S, \mathfrak{S})$ that have finite total variations take only values from $\mathbb{R}$ and can naturally be added and multiplied by real numbers, which makes the set of all such measures a linear space. It can be normed by the variation $|\cdot|(S)$, which makes it a Banach space. We denote it by $\mathcal{M}(S, \mathfrak{S})$ (or simply $\mathcal{M}(S)$ if $\mathfrak{S}$ is to be self-understood).

If S is also a topological space, a set function μ is called *regular* if $\forall A{\in}\Sigma\ \forall\varepsilon{>}0$ $\exists A_1, A_2{\in}\Sigma$: $\mathrm{cl}(A_1){\subset}A{\subset}\mathrm{int}(A_2)$ and $|\mu|(A_2{\setminus}A_1){\le}\varepsilon$. The subspace of regular measures from $\mathcal{M}(S)$, called *Radon measures*, is the denoted by $\mathfrak{M}(S)$; it is a Banach space if

[1] We call $\mathfrak{S}$ an algebra if $\emptyset \in \mathfrak{S}$, $A \in \mathfrak{S} \Rightarrow S \setminus A \in \mathfrak{S}$, and $A_1, A_2 \in \mathfrak{S} \Rightarrow A_1 \cup A_2 \in \mathfrak{S}$. If even $\bigcup_{i\in\mathbb{N}} A_i \in \mathfrak{S}$ for any mutually disjoint $A_i \in \mathfrak{S}$, it is called a σ-algebra.

[2] This means $\mu(\bigcup_{i\in\mathbb{N}} A_i) = \sum_{i\in\mathbb{N}} \mu(A_i)$ for any mutually disjoint $A_i \in \mathfrak{S}$.

M. Kružík and T. Roubíček, *Mathematical Methods in Continuum Mechanics of Solids*, Interaction of Mechanics and Mathematics,
https://doi.org/10.1007/978-3-030-02065-1

normed by the variation $|\cdot|(S)$. An example of a σ-algebra is the *Borel σ-algebra*[3], denoted as $\mathfrak{B}(S)$. In the case $\mathfrak{S} = \mathfrak{B}(S)$, element of $\mathcal{M}(S,\mathfrak{S})$ are called *Borel measures*. A mapping $f : S \to \hat{S}$ between two sets S and $\hat{S}$ is called a *Borel mapping* if $f^{-1}(B) := \{a \in S; f(a) \in B\}$ is a Borel set for every open set $B \subset \hat{S}$. If S is compact, there is an isometrical isomorphism $f \mapsto \mu : C(S)^* \to \mathfrak{M}(S)$ is by setting $\langle f,v\rangle = \int_S v\mu(\mathrm{d}x)$ for all $v \in C(S)$; this is known as the *Riesz theorem*. Then $\|f\|_{C(S)^*} = |\mu|(S)$.

For a measure μ, we define its *positive variation* $\mu^+ := \frac{1}{2}|\mu| + \frac{1}{2}\mu$. If $\mu^+ = \mu$, we say that μ is *positive*. We denote by $\mathfrak{M}_1^+(S) = \{\mu \in \mathfrak{M}(S);\ \mu(S) = 1,\ \mu \text{ positive}\}$ the convex set of *probability measures*. For $s \in S$, a measure $\delta_s \in \mathfrak{M}_1^+(S)$ defined by $\langle \delta_s, v\rangle = v(s)$ is called *Dirac's measure* supported at $s \in S$.

One way of generating a σ-algebra on S is by an *outer measure* which is, by definition, a σ-subadditive monotone function $2^S \to \mathbb{R}\cup\{\infty\}$ vanishing on the empty set.[4] A subset E of S is called *μ-measurable* if $\mu(A) = \mu(A\cap E) + \mu(A\setminus E)$ for any $A \subset S$. The μ-measurable sets form a σ-algebra and μ restricted to the measurable sets is a *complete*[5] measure.

If $(S,\mathfrak{S})$ and $(X,\mathfrak{X})$ are measurable spaces, i.e., $\mathfrak{S}$ and $\mathfrak{X}$ are σ-algebras on S and X, respectively, then a mapping $f : S \to X$ is called *measurable* (more precisely $(\mathfrak{S},\mathfrak{X})$-measurable), if $f^{-1}(B) := \{s{\in}S;\ f(s){\in}B\} \in \mathfrak{S}$ for all $B \in \mathfrak{X}$.

For mapping $F : S \to 2^X$, where 2^X denotes the power set, we write $F : S \rightrightarrows X$ to indicate that F is a *set-valued map*, i.e., for all $s \in S$ we have $F(s) \subset X$. The set $\mathrm{Gr}F := \{(s,x);\ x{\in}F(s)\}$ is referred to as a *graph* of F. A set-valued map $F : S \rightrightarrows X$ is called *measurable*, (more precisely $(\mathfrak{S},\mathfrak{X})$-measurable) if for all $B \in \mathfrak{X}$ the preimage $F^{-1}(B) := \{s \in S;\ F(s)\cap B \neq \emptyset\}$ lies in $\mathfrak{S}$. A mapping $f : S \to X$ is a *selection* of F, if for all $s \in S$ we have $f(s) \in F(s)$.

For a set S, a σ-algebra $\mathfrak{S}$ of its subsets, and a measure $\mu : \mathfrak{S}\cup\{\pm\infty\}$, we say that $(S,\mathfrak{S},\mu)$ is a *σ-finite complete measure space* if there exists $\{S_k\}_{k\in\mathbb{N}} \subset \mathfrak{S}$ with $\mu(S_k) < \infty$ and $S = \bigcup_{k\in\mathbb{N}} S_k$ and such that $S_1 \subset S_2 \in \mathfrak{S}$ and $\mu(S_2) = 0$ implies $S_1 \in \mathfrak{S}$. Throughout the rest of this subsection, we assume $(S,\mathfrak{S},\mu)$ to be qualified in this way. On the product space $S{\times}X$ we use the product σ-algebra $\mathfrak{S}\otimes\mathfrak{B}(X)$, which is the smallest σ-algebra containing all cylinders $A{\times}B$ with $A \in \mathfrak{S}$ and $B \in \mathfrak{B}(X)$.

For a metric space (X,ρ) and for $U \subset X$ and any $d \geq 0$, we define[6]

$$\mathscr{H}^d(S) := \frac{\pi^{d/2}}{\Gamma(1{+}d/2)} \lim_{\delta\to 0} \inf_{\substack{S\subset\bigcup_{i=1}^\infty U_i \\ \mathrm{diam}(U_i)<\delta}} \sum_{i=1}^{\infty} \Big(\frac{\mathrm{diam}(U_i)}{2}\Big)^d \tag{B.1.1}$$

with $\mathrm{diam}(U)$ denoting the *diameter*, i.e. $\mathrm{diam}(U) := \sup\{\rho(x,y); x,y \in U\}$, and with Γ referring to the Gamma-function $\Gamma(z) := \int_0^\infty t^{z-1}e^{-t}\,\mathrm{d}t$. It can be seen that $\mathscr{H}^d(\cdot)$ is an outer measure and, by general theory, its restriction to the corresponding σ-

[3] This is defined as the smallest σ-algebra containing all open subsets of S.

[4] This means that $\mu : 2^S \to \mathbb{R}\cup\{\infty\}$ is to satisfy $\mu(\bigcup_{j=1}^\infty A_j) \leq \sum_{j=1}^\infty \mu(A_j)$ for any collection $(A_j)_{j\in\mathbb{N}}$, and $A \subset B \Rightarrow \mu(A) \leq \mu(B)$, and also $\mu(\emptyset) = 0$.

[5] This means that any subset of any null set is measurable: $S \subseteq N \in \Sigma$ and $\mu(N) = 0 \implies S \in \Sigma$.

[6] Note that the infimum in (B.1.1) depends monotonically on δ hence the limit does exist.

algebra of measurable sets is a measure. It is called the d-dimensional *Hausdorff measure* [244]. All Borel subsets of X are $\mathscr{H}^d$-measurable. A special case $\mathscr{H}^0(\cdot)$ is a particular example, called the *counting measure*, i.e. the measure which counts the number of elements of a given set:

$$\#(A) = \begin{cases} |A| & \text{if } A \text{ is finite,} \\ +\infty & \text{otherwise,} \end{cases} \tag{B.1.2}$$

where here $|A|$ denotes the number of elements (cardinality) of A.

Definition B.1.1 (Weak* convergence of measures). Let $\Omega \subset \mathbb{R}^d$ be a bounded domain. We say that a sequence $\{\mu_k\}_{k\in\mathbb{N}}$ of $\mathbb{R}^M$-valued Radon measures on Ω weakly* converges to a Radon measure μ on Ω. If for every $v \in C_0(\Omega;\mathbb{R}^M)$

$$\lim_{k\to\infty} \int_\Omega v(x)\mu_k(\mathrm{d}x) = \int_\Omega v(x)\mu(\mathrm{d}x)\,. \tag{B.1.3}$$

The following result will be important in Section 4.6.3. If $x \in \Omega$ and $\mu \in \mathfrak{M}$ then $\frac{\mathrm{d}\mu}{\mathrm{d}|\mu|}(x) := \lim_{r\to 0} \frac{\mu(B(x,r))}{|\mu|(B(x,r))}$ is the *density* or the *Radon-Nikodým derivative* of μ with respect to $|\mu|$.

Theorem B.1.2 (Lower semicontinuity, Y.G. Reshetnyak [438]).[7] *Let $\Omega \subset \mathbb{R}^d$ be a bounded domain. Let a sequence $\{\mu_k\}_{k\in\mathbb{N}}$ of $\mathbb{R}^M$-valued Radon measures on Ω weakly* converges to a Radon measure μ on Ω. and let $f : \Omega\times\mathbb{R}^M \to [0,+\infty]$ be a lower semicontinuous function and such that $f(x,\cdot)$ is convex and positively one-homogeneous for every $x \in \Omega$. Then*

$$\liminf_{k\to\infty} \int_\Omega f\Big(x, \frac{\mathrm{d}\mu_k}{\mathrm{d}|\mu_k|}(x)\Big)\,|\mu_k|(\mathrm{d}x) \geq \int_\Omega f\Big(x, \frac{\mathrm{d}\mu}{\mathrm{d}|\mu|}(x)\Big)\,|\mu|(\mathrm{d}x)\,.$$

B.2 Lebesgue spaces, Nemytskiĭ mappings

The d-dimensional *Lebesgue outer measure* $\mathrm{meas}_d(\cdot)$ on the Euclidean space $\mathbb{R}^d$, $d \geq 1$, is defined as

$$\mathrm{meas}_d(A) := \inf\Big\{\sum_{k=1}^{\infty}\prod_{i=1}^{d} b_i^k - a_i^k;\ A \subset \bigcup_{k=1}^{\infty}[a_1^k,b_1^k]\times\ldots\times[a_d^k,b_d^k],\ a_i^k \leq b_i^k\Big\}. \tag{B.2.1}$$

Measurable sets with respect to this outer measure are called Lebesgue measurable.[8] As $\mathrm{meas}_d(\cdot)$ is an outer measure, the collection Σ of Lebesgue measurable

[7] See also [8].

[8] This just means that $A \subset \mathbb{R}^d$ is Lebesgue measurable if $\mathrm{meas}_d(A) = \mathrm{meas}_d(A \cap S) + \mathrm{meas}_d(A \setminus S)$ for any subset $S \subset \mathbb{R}^d$. For example, all closed sets are Lebesgue measurable, hence every open set too, as well as their countable union or intersection, etc.

subsets of Ω forms a σ-algebra.[9] The function $\mathrm{meas}_d : \Sigma \to \mathbb{R} \cup \{+\infty\}$ is called the *Lebesgue measure*. In fact, *Lebesgue measurable* functions (as defined above by means of the outer Lebesgue measure) are exactly functions which are measurable with resepct to the Lebesgue measure.

Having a set $\Omega \in \Sigma$, we say that a property holds *almost everywhere* on Ω (in abbreviation a.e. on Ω) if this property holds everywhere on Ω with the possible exception of a set of Lebesgue-measure zero; referring to those x where this property holds, we also say that it holds at *almost all* $x \in \Omega$ (in abbreviation a.a. $x \in \Omega$).

A function $u : \mathbb{R}^d \to \mathbb{R}^m$ is called (Lebesgue)[10] *measurable* if $u^{-1}(A) := \{x \in \mathbb{R}^d;\ u(x) \in A\}$ is Lebesgue measurable for any $A \in \mathbb{R}^m$ open.

We call $u : \mathbb{R}^d \to \mathbb{R}^m$ *simple* if it takes only finite number of values $v_i \in \mathbb{R}^m$ and $u^{-1}(v_i) = \{x;\ u(x) = v_i\} \in \Sigma$; then we define the integral $\int_{\mathbb{R}^d} u(x)\,\mathrm{d}x$ naturally as $\sum_{\text{finite}} \mathrm{meas}_d(A_i)v_i$. Furthermore, a measurable u is called *integrable* if there is a sequence of simple functions $\{u_k\}_{k\in\mathbb{N}}$ such that $\lim_{k\to\infty} u_k(x) = u(x)$ for a.a. $x \in \Omega$ and $\lim_{k\to\infty} \int_{\mathbb{R}^d} u_k(x)\,\mathrm{d}x$ does exists in $\mathbb{R}$. Then, this limit is denoted by $\int_{\mathbb{R}^d} u(x)\,\mathrm{d}x$ and call it the (Lebesgue) *integral* of u. It is then independent of the particular choice of the sequence $\{u_k\}_{k\in\mathbb{N}}$. If $\Omega \subset \mathbb{R}^d$ is measurable, by $\int_\Omega u\,\mathrm{d}x$ we mean naturally $\int_{\mathbb{R}}^d \tilde{u}\,\mathrm{d}x$ where $\tilde{u} := \chi_\Omega u$. By χ_A, we mean the *characteristic function* defined by $\chi_A(x) := 1$ for $x \in A$ and $\chi_A(x) := 0$ for $x \notin A$.

We now consider $\Omega \subset \mathbb{R}^d$ measurable with $\mathrm{meas}_d(\Omega) < \infty$. The notions of measurability and the integral of functions $\Omega \to \mathbb{R}^m$ can be understood as before provided all these functions are extended on $\mathbb{R}^d \setminus \Omega$ by 0. By $L^p(\Omega;\mathbb{R}^m)$ we denote the set of all measurable functions[11] $u : \Omega \to \mathbb{R}^m$ such that $\|u\|_{L^p(\Omega;\mathbb{R}^m)} < \infty$, where

$$\|u\|_{L^p(\Omega;\mathbb{R}^m)} := \begin{cases} \sqrt[p]{\int_\Omega |u(x)|^p\,\mathrm{d}x} & \text{for } 1 \le p < \infty\,, \\ \operatorname{ess\,sup}_{x\in\Omega} |u(x)| & \text{for } p = \infty\,, \end{cases} \tag{B.2.2}$$

and $|\cdot|$ is the Euclidean norm on $\mathbb{R}^m$. The *essential supremum* "ess sup" in (B.2.2) is defined as

$$\operatorname*{ess\,sup}_{x\in\Omega} f := \inf_{\mathrm{meas}_d(N)=0}\ \sup_{x\in\Omega\setminus N} f(x). \tag{B.2.3}$$

Replacing mutually "sup" with "inf" in (B.2.3), we would obtain the *essential infimum*, denoted by "ess inf".

---

[9] In fact, Σ is the so-called Lebesgue extension of the Borel σ-algebra, created by adding all subsets of sets having the measure zero. It has, together with the function $\mathrm{meas}_d : \Sigma \to \mathbb{R} \cup \{\infty\}$, have (and are characterized by) the following four properties: (i) A open implies $A \in \Sigma$; (ii) $A = [a_1, b_1] \times ... \times [a_d, b_d]$ with $a_i \le b_i$ implies $\mathrm{meas}_d(A) = \prod_{i=1}^d (b_i - a_i)$; (iii) meas_d is countably additive, i.e. $\mathrm{meas}_d(\bigcup_{k\in\mathbb{N}} A_k) = \sum_{k\in\mathbb{N}} \mathrm{meas}_d(A_k)$ for any countable collection $\{A_k\}_{k\in\mathbb{N}}$ of pairwise disjoint sets $A_i \in \Sigma$; (iv) $A \subset B \in \Sigma$ and $\mathrm{meas}_d(B) = 0$ implies $A \in \Sigma$ and $\mathrm{meas}_d(A) = 0$.

[10] If not said otherwise, "measurable" means Lebesgue measurable.

[11] Here we however adopt the usual convention not to distinguish between functions that equal to each other a.e., so that, strictly speaking, $L^p(\Omega;\mathbb{R}^m)$ contains classes of equivalence of such functions.

The set $L^p(\Omega;\mathbb{R}^m)$, endowed by a pointwise addition and scalar multiplication, is a linear space. Besides, $\|\cdot\|_{L^p(\Omega;\mathbb{R}^m)}$ is a norm on $L^p(\Omega;\mathbb{R}^m)$ which makes it a Banach space, called a *Lebesgue space.*

If a measure $\mu \in \mathfrak{M}(\bar{\Omega})$ possesses a density $d_\mu \in L^1(\Omega)$, which means $\mu(A) = \int_A d_\mu(x)\mathrm{d}x$ for any measurable $A \subset \Omega$, then μ has a certain special property, namely it is *absolutely continuous* with respect to the Lebesgue measure[12] and also the converse assertion is true: every absolutely continuous measure possesses a density belonging to $L^1(\Omega)$. This fact is known as the *Radon-Nikodým theorem* [398, 426].

An important question is how to characterize concretely the dual spaces. The "natural" duality pairing comes from the inner product in L^2-spaces, which means $\langle u,v\rangle := \int_\Omega u(x)\cdot v(x)\,\mathrm{d}x$, where $u\cdot v := \sum_{i=1}^m u_i v_i$ is the inner product in $\mathbb{R}^m$. If $1<p<\infty$, then $L^p(\Omega;\mathbb{R}^m)$ is reflexive. From the algebraic *Young inequality*

$$ab \le \frac{1}{p}a^p + \frac{1}{p'}b^{p'} \tag{B.2.4}$$

one gets[13] the *Hölder inequality* [253][14]

$$\int_\Omega \big|u(x)\cdot v(x)\big|\,\mathrm{d}x \le \big\|u\big\|_{L^p(\Omega;\mathbb{R}^m)}\big\|v\big\|_{L^{p'}(\Omega;\mathbb{R}^m)} \tag{B.2.5}$$

where p' is the so-called *conjugate exponent* defined as

$$p' := \begin{cases} 1 & \text{for } p=\infty, \\ p/(p-1) & \text{for } 1<p<+\infty, \\ +\infty & \text{for } p=1. \end{cases} \tag{B.2.6}$$

In fact, the modification of (B.2.5) for $p=1$ or $p=\infty$ looks trivially as $\int_\Omega |u\cdot v|\mathrm{d}x \le \|u\|_{L^\infty(\Omega;\mathbb{R}^d)}\|u\|_{L^1(\Omega;\mathbb{R}^d)}$. From (B.2.5), it can be shown that the dual space is isometrically isomorphic with $L^{p'}(\Omega;\mathbb{R}^m)$ if $1\le p<\infty$. On the other hand, the dual space to $L^\infty(\Omega;\mathbb{R}^m)$ is substantially larger than $L^1(\Omega;\mathbb{R}^m)$ and rather "exotic" object with a nonmetrizable weak* topology certainly out of the scope of this book.[15] Hölder's inequality also allows for an *interpolation* between $L^{p_1}(\Omega)$ and $L^{p_2}(\Omega)$: for $p_1, p_2, p \in [1,\infty]$, $\lambda \in [0,1]$, it holds

$$\frac{1}{p} = \frac{\lambda}{p_1} + \frac{1-\lambda}{p_2} \quad \Rightarrow \quad \forall v \in L^{\max(p_1,p_2)}(\Omega):\ \ \|v\|_{L^p(\Omega)} \le \|v\|_{L^{p_1}(\Omega)}^{\lambda}\|v\|_{L^{p_2}(\Omega)}^{1-\lambda}. \tag{B.2.7}$$

---

[12] This means that $\forall\varepsilon>0\ \exists\delta>0\ \forall A\subset\Omega$ measurable: $\mathrm{meas}_d(A)\le\delta \Longrightarrow |\mu(A)|\le\varepsilon$.

[13] Just simply $\|u\|_{L^p(\Omega)}^{-1}\|v\|_{L^{p'}(\Omega)}^{-1}\int_\Omega |u\cdot v|\mathrm{d}x \le \frac{1}{p}\|u\|_{L^p(\Omega)}^{-p}\int_\Omega |u|^p\mathrm{d}x + \frac{1}{p'}\|v\|_{L^{p'}(\Omega)}^{-p'}\int_\Omega |v|^{p'}\mathrm{d}x = 1$.

[14] Originally, Hölder [253] states this in a less symmetrical form for sums in place of integrals.

[15] The elements of $L^\infty(\Omega;\mathbb{R}^m)^*$ are indeed very abstract objects and can be identified with finitely-additive measures vanishing on zero-measure sets, see Yosida and Hewitt [536].

Applying the algebraic Young inequality (B.2.4) to the right-hand side of (B.2.5), we obtain another important inequality, the integral *Young inequality*

$$\int_\Omega \big|u(x)\cdot v(x)\big|\,\mathrm{d}x \le \frac{1}{p}\int_\Omega |u(x)|^p\,\mathrm{d}x + \frac{1}{p'}\int_\Omega |v(x)|^{p'}\,\mathrm{d}x.$$

Important geometric property of the norm (B.2.2) is that, for $1 < p < \infty$, the space $L^p(\Omega;\mathbb{R}^m)$ is *uniformly convex*, cf. [121]. If $1 \le q \le p \le \infty$, the embeddings $C(\bar{\Omega};\mathbb{R}^m) \subset L^p(\Omega;\mathbb{R}^m) \subset L^q(\Omega;\mathbb{R}^m)$ are continuous; recall that $\mathrm{meas}_d(\Omega) < \infty$ is always assumed here. Moreover, for $p < \infty$ these embeddings are dense and then, as $C(\bar{\Omega};\mathbb{R}^m)$ is separable, $L^p(\Omega;\mathbb{R}^m)$ is separable, too. On the other hand, $L^\infty(\Omega)$ is not separable.

Further, we say that a sequence $u_k : \Omega \to \mathbb{R}^m$ *converges in measure* to u if

$$\forall \varepsilon > 0: \quad \lim_{k\to\infty} \mathrm{meas}_d\Big(\big\{x\in\Omega;\ \big|u_k(x)-u(x)\big| \ge \varepsilon\big\}\Big) = 0. \tag{B.2.8}$$

Naturally, the convergence a.e. means that $u_k(x) \to u(x)$ for a.a. $x\in\Omega$.

Proposition B.2.1 (Various modes of convergences).
(i) Any sequence converging a.e. converges also in measure.
(ii) Any sequence converging in measure admits a subsequence converging a.e.
(iii) Any sequence converging in $L^1(\Omega)$ converges in measure.

Theorem B.2.2 (Lebesgue [311]). *Let $\{u_k\}_{k\in\mathbb{N}} \subset L^1(\Omega)$ be a sequence converging a.e. to some u and $|u_k(x)| \le v(x)$ for some $v \in L^1(\Omega)$. Then u lives in $L^1(\Omega)$ and $\lim_{k\to\infty}\int_A u_k(x)\,\mathrm{d}x = \int_A u(x)\,\mathrm{d}x$ for any $A \subset \Omega$ measurable.*

Lemma B.2.3 (Fatou [184]). *Let $\{u_k\}_{k\in\mathbb{N}} \subset L^1(\Omega)$ be a sequence of non-negative functions*[16] *such that $\liminf_{k\to\infty}\int_\Omega u_k(x)\,\mathrm{d}x < \infty$. Then the function $x \mapsto \liminf_{k\to\infty} u_k(x)$ is integrable and*

$$\liminf_{k\to\infty}\int_\Omega u_k(x)\,\mathrm{d}x \ge \int_\Omega \Big(\liminf_{k\to\infty} u_k(x)\Big)\mathrm{d}x. \tag{B.2.9}$$

Useful generalizations of the Lebesgue dominated-convergence Theorem B.2.2 and the Fatou Lemma B.2.3 use sequences of upper or lower bounds that are uniformly integrable: a set $M \subset L^1(\Omega)$ (or, more general, of $L^1(\Omega;\mathbb{R}^m)$) is *uniformly integrable* if

$$\forall \varepsilon > 0\ \exists K \in \mathbb{R}^+: \qquad \sup_{u\in M}\int_{\{x\in\Omega;|u(x)|\ge K\}} |u(x)|\,\mathrm{d}x \le \varepsilon. \tag{B.2.10}$$

[16] Obviously, existence of a common integrable minorant can simplify (but weaken) this assertion.

Theorem B.2.4 (Vitali [525][17]). *Let $\{u_k\}_{k\in\mathbb{N}} \subset L^1(\Omega)$ be a sequence converging a.e. to some u. Then $u \in L^p(\Omega)$ and $u_k \to u$ in $L^p(\Omega)$ if and only if $\{|u_k|^p\}_{k\in\mathbb{N}}$ is uniformly integrable.*

Theorem B.2.5 (Fatou, generalized[18]). *The conclusion of Theorem B.2.3 holds if $u_k \geq 0$ is replaced by $u_k \geq v_k$ with $\{v_k\}_{k\in\mathbb{N}}$ being uniformly integrable.*

From Theorem A.1.3, it immediately follows that bounded sets in $L^p(\Omega;\mathbb{R}^m)$ are weakly or weakly* relatively sequentially compact provided $1 < p < \infty$ or $p = \infty$, respectively. For $p = 1$ the situation is far more delicate. In particular, Theorem B.2.2 says that the set $\{u_k;\ k \in \mathbb{N}\}$ is relatively weakly compact[19] in $L^1(\Omega)$.

Also, the following assertions are useful:

Lemma B.2.6 (Monotone convergence, G. Levi). *Let X be a measure space, and let $0 \leq f_1 \leq f_2 ...$ be a monotone increasing sequence of nonnegative measurable functions. Let $f : X \to \mathbb{R} \cup \{\infty\}$ be the function defined by $f(x) = \lim_{k\to\infty} f_k(x)$. Then f is measurable, and $\lim_{k\to\infty} \int_X f_k \,\mathrm{d}x = \int_X f \,\mathrm{d}x$.*

Theorem B.2.7 (Lusin). *Every measurable function is a continuous function on nearly all its domain. More specifically: let $\Omega \subset \mathbb{R}^d$ be a bounded domain and let $f : \Omega \to \mathbb{R}$ be a measurable function. Then given $\varepsilon > 0$, there exists a compact $E \subset \Omega$ such that f restricted to E is continuous and $\mu(\Omega \setminus E) < \varepsilon$. Note that E inherits the subspace topology from Ω; continuity of f restricted to E is defined using this topology.*

Theorem B.2.8 (Dunford and Pettis [167]). *Let $M \subset L^1(\Omega;\mathbb{R}^m)$ be bounded. Then the following statements are equivalent to each other:*
(i) M is relatively weakly compact in $L^1(\Omega;\mathbb{R}^m)$,
(ii) the set M is uniformly integrable.

The following result is a variant of Theorem A.3.8 for Lebesgue spaces.

Theorem B.2.9 (Visintin [523]). *Let $\Omega \subset \mathbb{R}^d$ be a bounded domain. Let $u_k \to u$ weakly in $L^1(\Omega;\mathbb{R}^d)$ as $k \to \infty$. Let further $\int_\Omega \psi(u_k(x))\,\mathrm{d}x \to \int_\Omega \psi(u(x))\,\mathrm{d}x \neq +\infty$ for a strictly convex and lower semicontinuous proper[20] function $\psi : \mathbb{R}^d \to \mathbb{R} \cup \{+\infty\}$. Then $u_k \to u$ and $\psi(u_k) \to \psi(u)$ strongly in $L^1(\Omega;\mathbb{R}^d)$ and $L^1(\Omega)$, respectively.*

---

[17] More precisely, in [525], the integration of summable series is investigated rather than mere sequences.

[18] See Ash [18, Thm.7.5.2], or also Klei and Miyara [273] or Saadoune and Valadier [472].

[19] Note that the linear hull of all characteristic functions χ_A with $A \subset \Omega$ measurable is dense in $L^\infty(\Omega) \cong L^1(\Omega)^*$, so that the sequence $\{u_k\}$, being bounded in $L^1(\Omega)$, converges weakly in $L^1(\Omega)$ and, as such, it is relatively sequentially weakly compact, hence by the Eberlein-Šmuljan theorem relatively weakly compact, too.

[20] It means that ψ is not identically $+\infty$.

A fundamental phenomenon that composition of two measurable mappings need not be measurable[21] can be handled in the context of integral functionals of the type $\int_\Omega a(x,u(x))\mathrm{d}x$ by qualifying $a:\Omega\times\mathbb{R}^m\to\mathbb{R}\cup\{\infty\}$ as a *normal integrand* if $a(x,\cdot):\mathbb{R}^m\to\mathbb{R}\cup\{\infty\}$ is lower semicontinuous for a.a. $x\in\Omega$ and a is measurable.[22] A special case of the normal integrand $a:\Omega\times\mathbb{R}^m\to\mathbb{R}\cup\{\infty\}$ with $a(x,\cdot):\mathbb{R}^m\to\mathbb{R}\cup\{\infty\}$ continuous for a.a. $x\in\Omega$ is called a Carathéodory integrand. Having guarandeed measurability of $a(x,u(x))$, we can ask about integrability. Scalar character of values plays no longer any role, so we can consider $a(x,\cdot):\mathbb{R}^m\to\mathbb{R}^{m_0}$, $m_0\geq 1$. Now, however, growth of $|a(x,\cdot)|$ play role and it is thus worth allowing some "anisotropy" by splitting $\mathbb{R}^m=\mathbb{R}^{m_1}\times...\times\mathbb{R}^{m_j}$ with some integers $j, m_1, ..., m_j$. We say that $a:\Omega\times\mathbb{R}^{m_1}\times...\times\mathbb{R}^{m_j}\to\mathbb{R}^{m_0}$ is a *Carathéodory mapping* if $a(\cdot,r_1,...,r_j):\Omega\to\mathbb{R}^{m_0}$ is measurable for all $(r_1,...,r_j)\in\mathbb{R}^{m_1}\times...\times\mathbb{R}^{m_j}$ and $a(x,\cdot):\mathbb{R}^{m_1}\times...\times\mathbb{R}^{m_j}\to\mathbb{R}^{m_0}$ is continuous for a.a. $x\in\Omega$. Then the so-called *Nemytskiĭ mappings* $\mathcal{N}_a$ maps functions $u_i:\Omega\to\mathbb{R}^{m_i}$, $i=1,...,j$, to a function $\mathcal{N}_a(u_1,...,u_j):\Omega\to\mathbb{R}^{m_0}$ defined by

$$[\mathcal{N}_a(u_1,...,u_j)](x)=a(x,u_1(x),...,u_j(x)).\tag{B.2.11}$$

If $m_0=1$, a is sometimes called a *Carathéodory integrand*, being a special case of a normal integrand.

Theorem B.2.10 (Nemytskiĭ mappings on Lebesgue spaces). *Let $a:\Omega\times\mathbb{R}^{m_1}\times...\times\mathbb{R}^{m_j}\to\mathbb{R}^{m_0}$ be a Carathéodory mapping and the functions $u_i:\Omega\to\mathbb{R}^{m_i}$, $i=1,...,j$, be measurable. Then $\mathcal{N}_a(u_1,...,u_j)$ is measurable. Moreover, if a satisfies also the growth condition*

$$\big|a(x,r_1,...,r_j)\big|\leq\gamma(x)+C\sum_{i=1}^{j}\big|r_i\big|^{p_i/p_0}\quad\textit{for some }\gamma\in L^{p_0}(\Omega),\tag{B.2.12}$$

with $1\leq p_i<\infty$, $1\leq p_0<\infty$, then $\mathcal{N}_a$ is a bounded continuous mapping $L^{p_1}(\Omega;\mathbb{R}^{m_1})\times...\times L^{p_j}(\Omega;\mathbb{R}^{m_j})\to L^{p_0}(\Omega;\mathbb{R}^{m_0})$. If some $p_i=\infty$, $i=1,...,j$, the same holds if the respective term $|\cdot|^{p_i/p_0}$ are replaced by any continuous function.

Sometimes, it is useful to let the Nemytskiĭ mapping itself varied. Let us only confine to one-argument mappings.

Proposition B.2.11 (Stability of Nemytskiĭ mappings[23]). *Let $\Omega\subset\mathbb{R}^d$ be a bounded domain and $\{u_k:\Omega\to\mathbb{R}^m\}_{k\in\mathbb{N}}$ converges strongly to u in $L^{p_1}(\Omega;\mathbb{R}^m)$, and $a_k:\Omega\times$*

---

[21] Indeed, consider a continuous function $a:[0,1]\to[0,2]$ with the continuous inverse which maps a Lebesgue measurable set E on a (Lebesgue) non-measurable set D, i.e., $a(E)=D$. If χ_E is the characteristic function of E in $[0,1]$ then it is measurable. However, $\chi_E(a^{-1})=\chi_D$ is the characteristic function of a non-measurable set D in $[0,2]$ and therefore is non-measurable. See [216, Ex. 39 on p. 109] for an example of a, E, and D. This shows that even the composition of the Lebesgue measurable function χ_E with the continuous (and therefore measurable) function a^{-1} is not measurable.

[22] More precisely, it means that, for each A open in $\mathbb{R}^m$, $\{x\in\Omega;\ \mathrm{epi}f(x,\cdot)\cap A\neq\emptyset\}$ is Lebesgue measurable, i.e. the multivalued mapping $t\mapsto\mathrm{epi}f(x,\cdot)$ is Lebesgue measurable.

[23] To prove the proposition, we realize (omitting x-variable for brevity) that

$\mathbb{R}^m \to \mathbb{R}$ *be a sequence of Carathéodory integrands satisfying (B.2.12) for* $j = 1$ *uniformly for* k *and converging to* $a : \Omega\times\mathbb{R}^m \to \mathbb{R}$ *in the sense*

$$\exists\gamma_k \to 0 \ \ in \ \ L^{p_0}(\Omega),\ C_k \to 0\ \forall (x,r)\in\Omega\times\mathbb{R}^m : \qquad\qquad\qquad\qquad \big|a_k(x,r) - a(x,r)\big| \le \gamma_k(x) + C_k|r|^{p_1/p_0}. \tag{B.2.13}$$

Then $\mathcal{N}_{a_k}(u_k) \to \mathcal{N}_a(u)$ *strongly in* $L^{p_0}(\Omega;\mathbb{R}^m)$.

Eventually, the following integral-calculus assertion is often used:

Theorem B.2.12 (Fubini [208]). *Considering two Lebesgue measurable sets* $\Omega_1 \subset \mathbb{R}^{n_1}$ *and* $\Omega_2 \subset \mathbb{R}^{n_2}$*, the following identity holds provided* $g\in L^1(\Omega_1\times\Omega_2)$ *(in particular, each of the following double-integrals does exist and is finite):*

$$\int_{\Omega_1\times\Omega_2} g(x_1,x_2)\,\mathrm{d}(x_1,x_2) = \int_{\Omega_1}\Big(\int_{\Omega_2} g(x_1,x_2)\,\mathrm{d}x_2\Big)\mathrm{d}x_1 = \int_{\Omega_2}\Big(\int_{\Omega_1} g(x_1,x_2)\,\mathrm{d}x_1\Big)\mathrm{d}x_2.$$

B.3 Sobolev spaces, embeddings, traces

We now consider $\Omega\subset\mathbb{R}^d$ open connected[24] with $\mathrm{meas}_d(\Omega) < \infty$; such a set is called a *domain*. Having a function $u \in L^p(\Omega)$, we define its *distributional derivative* $\partial^k u/\partial x_1^{k_1}...\partial x_d^{k_d}$ with $k_1+...+k_d = k$ and $k_i \ge 0$ for any $i = 1,...,d$ as a distribution such that

$$\forall g\in\mathscr{D}(\Omega): \qquad \Big\langle \frac{\partial^k u}{\partial x_1^{k_1}...\partial x_d^{k_d}}, g\Big\rangle = (-1)^k\int_\Omega u\,\frac{\partial^k g}{\partial x_1^{k_1}...\partial x_d^{k_d}}\,\mathrm{d}x, \tag{B.3.1}$$

where $\mathscr{D}(\Omega)$ stands for *infinitely differentiable functions with a compact support.* The d-tuple of the first-order distributional derivatives $(\frac{\partial}{\partial x_1}u, ..., \frac{\partial}{\partial x_d}u)$ is denoted by ∇u and called the *gradient* of u. For $p < \infty$, we define a *Sobolev space*

$$W^{1,p}(\Omega) := \{u\in L^p(\Omega);\ \nabla u\in L^p(\Omega;\mathbb{R}^d)\},\ \text{ equipped with the norm} \tag{B.3.2a}$$

$$\big\|u\big\|_{W^{1,p}(\Omega)} := \begin{cases}\Big(\big\|u\big\|^p_{L^p(\Omega)} + \big\|\nabla u\big\|^p_{L^p(\Omega;\mathbb{R}^d)}\Big)^{1/p} & \text{if } p<\infty\\ \max\big(\|u\|_{L^\infty(\Omega)}, \|\nabla u\|_{L^\infty(\Omega;\mathbb{R}^d)}\big) & \text{if } p=\infty.\end{cases} \tag{B.3.2b}$$

$$\int_\Omega \big|a_k(u_k) - a(u)\big|^{p_0}\,\mathrm{d}x \le p\int_\Omega \big|a_k(u_k) - a(u_k)\big|^{p_0}\,\mathrm{d}x + p\int_\Omega \big|a(u_k) - a(u)\big|^{p_0}\,\mathrm{d}x,$$

and then, due to (B.2.13), the first integral on the right-hand side bears estimation by $\int_\Omega \gamma_k^{p_0} + C_k|u_k|^{p_1}\,\mathrm{d}x \le \|\gamma_k\|^{p_0}_{L^{p_0}(\Omega)} + C_k\|u_k\|^{p_1}_{L^{p_1}(\Omega;\mathbb{R}^m)} \to 0$ while the second integral converges to zero due to Theorem B.2.10.

[24] The adjective "connected" means that that Ω cannot be represented as the union of two or more disjoint nonempty open subsets. In particular, any two points can be connected by a continuous curve and all functions on Ω with zero gradient must be constant.

As, by Rademacher's theorem, Lipschitz functions are a.e. differentiable, it holds $W^{1,\infty}(\Omega) = C^{0,1}(\Omega)$. on a Lipschitz domain Ω. An obvious example of a situation when $C^{0,1}(\Omega;\mathbb{R}^m) \neq W^{1,\infty}(\Omega;\mathbb{R}^m)$ is for the domain Ω from Figure 3.2(middle).

Analogously, for $k > 1$, we define

$$W^{k,p}(\Omega) := \{u \in L^p(\Omega;\mathbb{R}^m);\ \nabla^k u \in L^p(\Omega;\mathbb{R}^{d^k})\}, \tag{B.3.3}$$

where $\nabla^k u$ denotes the set of all k-th order partial derivatives of u understood in the distributional sense. The standard norm on $W^{k,p}(\Omega)$ is

$$\|u\|_{W^{k,p}(\Omega)} := \begin{cases} \Big(\|u\|^p_{L^p(\Omega)} + \|\nabla^k u\|^p_{L^p(\Omega;\mathbb{R}^{d^k})}\Big)^{1/p} & \text{if } p < \infty, \\ \max\big(\|u\|_{L^\infty(\Omega)}, \|\nabla^k u\|_{L^\infty(\Omega;\mathbb{R}^{d^k})}\big) & \text{if } p = \infty \end{cases} \tag{B.3.4}$$

which makes it a Banach space. Likewise for Lebesgue spaces, for $1 \le p < \infty$ the Sobolev spaces $W^{k,p}(\Omega)$ are separable and, if $1 < p < \infty$, they are uniformly convex, hence by Milman-Pettis' theorem also reflexive. The spaces of $\mathbb{R}^m$-valued functions, $W^{k,p}(\Omega;\mathbb{R}^d)$, are defined analogously[25]. For $p = 2$, the Sobolev spaces are Hilbert spaces and, for brevity, we write

$$H^k(\Omega;\mathbb{R}^d) := W^{k,2}(\Omega;\mathbb{R}^d). \tag{B.3.5}$$

To give a good sense to traces on the *boundary* $\Gamma := \partial\Omega := \bar{\Omega}\setminus\Omega$ with $\bar{\Omega} := \mathrm{cl}(\Omega)$, we must qualify Ω suitably. We say that Ω is a *domain of* C^k*-class* if there is a finite number of overlapping parts Γ_i of the boundary of Γ that are graphs of C^k-functions in local coordinate systems and Ω lies on one side of Γ.[26] Replacing C^k with $\mathrm{C}^{0,1}$, we say that Ω is of the $\mathrm{C}^{0,1}$-class or rather a *Lipschitz domain*.

The following property of a domain is called cone property (or sometimes cone condition). It is satisfied by bounded Lipschitz domains in $\mathbb{R}^d$:

Definition B.3.1 (Cone property). Let $B(x,1) \subset \mathbb{R}^d$ be a unit ball and let $B(\tilde{x};1/4)$ be such that $x \notin B(\tilde{x};1/4)$. Define $\mathfrak{C}_x := \{x+t(z-x):\ t>0\ \&\ z \in B(\tilde{x};1/4)\}$ be a cone with the vertex x. We say that $\Omega \subset \mathbb{R}^d$ has the cone property if there is a finite cone $\mathfrak{C}$ such that each $x \in \Omega$ is a vertex of a finite cone $\mathfrak{C}_x$ which is obtained from $\mathfrak{C}$ by a translation and a rotation.

Theorem B.3.2 (Sobolev embedding [490]). *The continuous embedding*

$$W^{1,p}(\Omega) \subset L^{p^*}(\Omega) \tag{B.3.6}$$

[25] This means $W^{k,p}(\Omega;\mathbb{R}^d) := \{(u_1,...,u_d);\ u_i \in W^{k,p}(\Omega),\ i = 1,...,d\}$.

[26] Written formally, it requires existence of transformation unitary matrices A_i and open sets $G_i \subset \mathbb{R}^{d-1}$ and $g_i \in C^k(\mathbb{R}^{d-1})$ such that each Γ_i can be expressed as $\Gamma_i = \{A_i\xi;\ \xi \in \mathbb{R}^d,\ (\xi_1,...,\xi_{d-1}) \in G_i,\ \xi_d = g_i(\xi_1,...,\xi_{d-1})\}$ and $\{A_i\xi;\ \xi \in \mathbb{R}^d,\ (\xi_1,...,\xi_{d-1}) \in G_i,\ g_i(\xi_1,...,\xi_{d-1})-\varepsilon < \xi_d < g_i(\xi_1,...,\xi_{d-1})\} \subset \Omega$ and simultaneously $\{A_i\xi;\ \xi \in \mathbb{R}^d,\ (\xi_1,...,\xi_{d-1}) \in G_i,\ g_i(\xi_1,...,\xi_{d-1}) < \xi_d < g_i(\xi_1,...,\xi_{d-1})+\varepsilon\} \subset \mathbb{R}^d \setminus \bar{\Omega}$ for some $\varepsilon > 0$.

holds provided the so-called Sobolev exponent p^* *is defined as*

$$p^* := \begin{cases} \dfrac{dp}{d-p} & \text{for } p<d\,, \\ \text{an arbitrarily large real} & \text{for } p=d\,, \\ \infty & \text{for } p>d\,. \end{cases} \tag{B.3.7}$$

Moreover, for $p>d$, even $W^{1,p}(\Omega)\subset C(\bar\Omega)$.[27] *Here $C(\bar\Omega)$ is the space of functions which are continuous on $\bar\Omega$.*

Theorem B.3.3 (Rellich, Kondrachov [281, 436][28]). *The compact embedding*

$$W^{1,p}(\Omega)\Subset L^{p^*-\epsilon}(\Omega)\,,\qquad \epsilon\in(0,p^*-1], \tag{B.3.8}$$

holds for p^* from (B.3.7). Moreover, for $p>d$, even $W^{1,p}(\Omega)\Subset C^{0,1-d/p}(\bar\Omega)$.

Re-iterating[29] Theorems B.3.2 or B.3.3, one gets the assertion (which, in fact, holds even for $k=\alpha\geq 0$ noninteger, i.e. for Sobolev-Slobodetskiĭ spaces with fractional derivatives as defined in (B.3.23) below):

Corollary B.3.4 (Higher-order Sobolev embedding).
(i) If $kp<d$, the following compact embedding holds:

$$W^{k,p}(\Omega)\Subset L^{dp/(d-kp)-\epsilon}(\Omega)\,,\qquad \epsilon\in(0,dp/(d-kp)], \tag{B.3.9}$$

while for $\epsilon=0$ this embedding is only continuous.
(ii) For $kp=d$, it holds $W^{k,p}(\Omega)\Subset L^q(\Omega)$ for any $q<\infty$.
(iii) For $kp>d$, it holds $W^{k,p}(\Omega)\Subset C^{k-[d/p]-1,\gamma}(\bar\Omega)$, where

$$\gamma := \begin{cases} [\frac{d}{p}]+1-\frac{d}{p} & \text{if } \frac{d}{p}\notin\mathbb{N}, \\ \text{any positive number} <1 & \text{otherwise.} \end{cases} \tag{B.3.10}$$

with $[a]$ denoting the integer part of $a\in\mathbb{R}$.

Theorem B.3.5 (Superposition in Sobolev spaces [335]). *Suppose that $d>1$ and that $\Omega\subset\mathbb{R}^d$ is bounded. Let $h:\mathbb{R}^d\to\mathbb{R}$ be a Borel function. Assume that $1\leq r\leq p<d$. If $y\in W^{1,p}(\Omega;\mathbb{R}^d)$ then $h(y)\in W^{1,r}(\Omega)$ if and only if h is locally Lipschitz and simultaneously there is $K>0$ such that*

[27] This means that each $u\in W^{1,p}(\Omega)$ admits a continuous extension on the closure $\bar\Omega$ of Ω.

[28] The pioneering Rellich's work has dealt with $p=2$ only.

[29] E.g., $W^{2,p}(\Omega)\subset W^{1,dp/(d-p)}(\Omega)$ by applying Theorem B.3.2 on first derivatives, and applying Theorem B.3.2 once again for $dp/(d-p)$ instead of p one comes to $W^{1,dp/(d-p)}(\Omega)\subset L^{dp/(d-2p)}(\Omega)$ provided $2p<d$. Repeating once again yields $W^{3,p}(\Omega)\Subset L^{dp/(d-3p)}(\Omega)$ provided $3p<d$, etc.

$$|h'| \le K(1+|\cdot|^q)\,, \tag{B.3.11}$$

where $q := d(p-r)/(r(d-p))$. If $d < p$ then $h(y) \in W^{1,p}(\Omega)$ if and only if h is locally Lipschitz.

Theorem B.3.6 (Trace operator). *There is exactly one linear continuous operator $T : W^{1,p}(\Omega) \to L^1(\Gamma)$ such that, for any $u \in C(\bar{\Omega})$, it holds $Tu = u|_\Gamma$ (=the restriction of u on Γ). Moreover,*

$$u \mapsto u|_\Gamma : W^{1,p}(\Omega) \to L^q(\Gamma) \text{ is } \begin{cases} \text{continuous} & \text{if } 1 \le q \le p^\sharp, \\ \text{compact} & \text{if } 1 \le q < p^\sharp, \end{cases}$$

provided the so-called Sobolev trace exponent *$p^\sharp$ is defined as*

$$p^\sharp := \begin{cases} \dfrac{dp-p}{d-p} & \text{for } p < d\,, \\ \text{an arbitrarily large real} & \text{for } p = d\,, \\ \infty & \text{for } p > d\,. \end{cases} \tag{B.3.12}$$

We call the operator T from Theorem B.3.6 the *trace operator*, and write simply $u|_\Gamma$ instead of Tu even if $u \in W^{1,p}(\Omega) \setminus C(\bar{\Omega})$. Then we define $W_0^{1,p}(\Omega) := \{v \in W^{1,p}(\Omega);\ v|_\Gamma = 0\}$. For $k > 1$, we define similarly $W_0^{k,p}(\Omega) := \{v \in W^{k,p}(\Omega);\ \nabla^i v \in W_0^{k-i,p}(\Omega;\mathbb{R}^{d_i}),\ i = 0,...,k-1\}$. And, of course, $H_0^k(\Omega) := W_0^{k,2}(\Omega)$. The dual space to $W_0^{k,p}(\Omega)$ is standardly denoted $W^{-k,p'}(\Omega)$ instead of $W_0^{k,p}(\Omega)^*$ and, of course, $H^{-k}(\Omega) := H_0^k(\Omega)^*$.

To prove Theorem 5.2.2 in Section 5, we will need the following lemma, cf. [170].[30]

Lemma B.3.7 (Lions' lemma). *Let $\Omega \subset \mathbb{R}^d$ be a bounded domain with a Lipschitz boundary. Let $v \in H^{-1}(\Omega) \cong H_0^1(\Omega)^*$ and $\partial_{x_i} v \in H^{-1}(\Omega)$ for all i. Then $v \in L^2(\Omega)$.*

Assuming Ω a Lipschitz domain, we denote by $\vec{n} = \vec{n}(x) \in \mathbb{R}^d$ the *unit outward normal* to the boundary Γ at a point $x \in \Gamma$; this is well defined $\mathscr{H}^{d-1}$-almost everywhere on Γ.[31] The multidimensional by-part integration

$$\int_\Omega \Big(v\frac{\partial z}{\partial x_i} + \frac{\partial v}{\partial x_i} z\Big)\mathrm{d}x = \int_\Gamma vz\vec{n}_i\,\mathrm{d}S \qquad \text{with } \mathrm{d}S := \mathscr{H}^{d-1}|_\Gamma \tag{B.3.13}$$

holds for any $v \in W^{1,p}(\Omega)$ and $z \in W^{1,p'}(\Omega)$ and for all $i = 1,..,d$. Considering $z = (z_1,...,z_d)$, writing (B.3.13) for z_i instead of z, and eventually summing it over $i = 1,...,d$ with abbreviating $\mathrm{div}\, z := \sum_{i=1}^d \frac{\partial}{\partial x_i} z_i$ the *divergence* of the vector field z, we arrive at the formula which we will often use:

---

[30] It has first appeared in [326, Footnote (22)] referring to J.L. Lions. The first proof of this result appeared, however, in [170].

[31] This normal can be defined by means of gradients of Lipschitz functions describing locally Γ as their graphs. By so-called Rademacher's theorem, these derivatives exist $\mathscr{H}^{d-1}$-almost everywhere on Γ.

Theorem B.3.8 (Green formula [229]).[32] *The following formula holds for any* $v \in W^{1,p}(\Omega)$ *and* $z \in W^{1,p'}(\Omega;\mathbb{R}^d)$*:*

$$\int_\Omega \big(v(\mathrm{div}\, z) + z\cdot\nabla v\big)\,\mathrm{d}x = \int_\Gamma v(z\cdot\vec{n})\,\mathrm{d}S. \tag{B.3.14}$$

Further, we mention two change-of-variables formulas. First we define the so-called *Lusin's condition.*

Definition B.3.9 (Lusin's condition N). Let $\Omega\subset\mathbb{R}^d$ be a bounded domain. Then $y:\Omega\to\mathbb{R}^d$ is said to satisfy Lusin's condition N if for every $\omega\subset\Omega$ such that $\mathrm{meas}_d(\omega)=0$ it holds that $\mathrm{meas}_d(y(\omega))=0$, as well.

We recall that if y is continuous then Lusin's condition N is necessary and sufficient to ensure that every measurable set is transformed by y to a measurable set.

To formulate change of variables results, let us still define, for any $z\in\mathbb{R}^d$ and $A\subset\mathbb{R}^d$, the *Banach indicatrix* $N(z,y,\omega)$ by

$$N(z,y,A) := \#\{x\in A;\ y(x)=z\}, \tag{B.3.15}$$

where the right-hand-side is the counting measure (B.1.2), i.e., the number of elements of the set $y^{-1}(z)=\{x\in A;\, y(x)=z\}$.

For $y:\mathbb{R}^d\to\mathbb{R}^d$ Lipschitz continuous and $g:\mathbb{R}^d\to\mathbb{R}$ integrable, let us formulate Federer's *area formula* [185, Thm. 3.2.3] (see also e.g. [180, Thm. 3.9]):

$$\int_{\mathbb{R}^d} g(x)\big|\det\nabla y(x)\big|\,\mathrm{d}x = \int_{\mathbb{R}^d}\sum_{x\in y^{-1}(z)} g(x)\,\mathrm{d}z. \tag{B.3.16}$$

As a special case for g the characteristic function of a domain Ω, we may little weaken the qualification of y and formulate:

Theorem B.3.10 (First change-of-variables formula [76][33]). *Let* $\Omega\subset\mathbb{R}^d$ *be a bounded domain. Let* $y:\Omega\to\mathbb{R}^d$ *be a continuous mapping satisfying Lusin's condition N. Assume that* y *is weakly differentiablea.e. in* Ω *and that* $\det\nabla y$ *is integrable in* $\mathbb{R}^d$*. Then the Banach indicatrix* $N(\cdot,y,\Omega)$ *is integrable in* $\mathbb{R}^d$ *and*

$$\int_\Omega |\det\nabla y(x)|\,\mathrm{d}x = \int_{\mathbb{R}^d} N(z,y,\Omega)\,\mathrm{d}z = \int_{y(\Omega)} N(z,y,\Omega)\,\mathrm{d}z\,. \tag{B.3.17}$$

[32] Putting $z=\nabla u$ into (B.3.14), we get $\int_\Omega v\Delta u+\nabla v\cdot\nabla u\,\mathrm{d}x = \int_\Gamma v\frac{\mathrm{d}u}{\mathrm{d}\vec{n}}\,\mathrm{d}S$ derived in [229]. In fact, (B.3.14) holds, by continuous extension, even under weaker assumptions, cf. [387].

[33] Also, without Luzin's condition, (B.3.17) is known as Federer's area formula, cf. [185, Thm. 3.2.3] where this formula is stated in a more general case when the dimension of the target space may be bigger than d, the dimension of the domain $\mathbb{R}^d$ and with $\mathrm{d}z$ referring to the Hausdorff measure.

In particular, the above assumptions of the Theorem are fulfilled and (B.3.17) *holds if* $y \in W^{1,p}(\Omega;\mathbb{R}^d)$ *for some* $p > d$ *is a continuous representative of the equivalence class.*

Another special case of (B.3.16) can be obtained for $g(x) = f(y(x))$ with some $f: y(\Omega) \to \mathbb{R}$:

Theorem B.3.11 (Second change-of-variables formula [76][34]). *Let* $\Omega \subset \mathbb{R}^d$ *be a bounded domain and let* $y \in W^{1,d}(\Omega;\mathbb{R}^d)$ *be continuous, satisfying Lusin's condition N, and such that* $\det \nabla y > 0$ *a.e. in* Ω. *Then for every* $f \in L^\infty(y(\Omega))$ *it holds that*

$$\int_\Omega f(y(x)) \det \nabla y(x)\, \mathrm{d}x = \int_{y(\Omega)} f(z) N(z,y,\Omega)\, \mathrm{d}z. \tag{B.3.18}$$

It is sometimes important to know that the pre-image of a null set is also a set of measure zero. Let us specify this property in:

Definition B.3.12 (Lusin's condition N^{-1}). Let $\Omega \subset \mathbb{R}^d$ be a bounded domain. Then $y: \Omega \to \mathbb{R}^d$ is said to satisfy Lusin's condition N^{-1} if for every $\omega \subset y(\Omega)$ such that $\mathrm{meas}_d(\omega) = 0$ it holds that $\mathrm{meas}_d(y^{-1}(\omega)) = 0$, as well.

The following result gives sufficient conditions on a map $y: \Omega \to \mathbb{R}^d$ ensuring validity of Lusin's condition N^{-1}.

Theorem B.3.13 (Validity of Lusin's condition N^{-1} [76]). *Let* $\Omega \subset \mathbb{R}^d$ *be a bounded domain. Let* $y: \Omega \to \mathbb{R}^d$ *be a continuous mapping satisfying Lusin's condition N. Assume that y is weakly differentiable a.e. in* Ω *and that* $\det \nabla y$ *is integrable in* $\mathbb{R}^d$ *and positive a.e. in* Ω. *Then y satisfies Lusin's condition* N^{-1}. *In particular, if* $y \in W^{1,p}(\Omega;\mathbb{R}^d)$ *for some* $p > d$ *is a continuous representative of the equivalence class and* $\det \nabla y > 0$ *a.e. in* Ω *then y satisfies Lusin's condition* N^{-1}.

Another result important especially for various coupled systems concerns a certain *interpolation* between the Sobolev space $W^{k,p}(\Omega)$ and the Lebesgue space $L^q(\Omega)$. Actually, it merges the Lebesgue-space interpolation (B.2.7) with the Sobolev embedding (B.3.8). Cf. Exercise 8.2.5 on p. 375. In general, it leads to:

Theorem B.3.14 (Gagliardo-Nirenberg inequality [210, 399, 400]). *Let* $\beta = \beta_1 + \ldots + \beta_d$, $\beta_1, \ldots, \beta_d \in \mathbb{N} \cup \{0\}$, $k \in \mathbb{N}$, *r, q, and p satisfy*

$$\frac{1}{r} = \frac{\beta}{d} + \lambda\Big(\frac{1}{p} - \frac{k}{d}\Big) + \frac{1-\lambda}{q}, \qquad \frac{\beta}{k} \le \lambda \le 1, \quad 0 \le \beta \le k-1, \tag{B.3.19}$$

then it holds

[34] In [185, Thm. 3.2.3], the formula (B.3.18) was stated for Lipschitz functions and with dz referring to the Hausdorff measure.

$$\left\| \frac{\partial^\beta v}{\partial x_1^{\beta_1} \ldots \partial x_d^{\beta_d}} \right\|_{L^r(\Omega)} \le C_{\rm GN} \|v\|_{W^{k,p}(\Omega)}^{\lambda} \|v\|_{L^q(\Omega)}^{1-\lambda}, \tag{B.3.20}$$

provided $k-\beta-d/p$ is not a negative integer (otherwise it holds for $\lambda = |\beta|/k$).

Theorem B.3.15 (Generalized Poincaré inequality). *Let $\Omega \in \mathbb{R}^d$ be an open bounded Lipschitz domain and let $1 \le p < +\infty$. Let further $\Gamma_{\rm D} \subset \partial\Omega$ be such that* $\text{meas}_{d-1}(\Gamma_{\rm D}) > 0$. *Then there is a constant $K > 0$ such that for every $v \in W^{1,p}(\Omega;\mathbb{R}^d)$ the following inequalities hold*

$$\int_\Omega |v(x)|^p \,{\rm d}x \le K\Big(\int_\Omega |\nabla v(x)|^p \,{\rm d}x + \Big| \int_\Omega v(x) \,{\rm d}x \Big| \Big) \tag{B.3.21}$$

and if $\Gamma \subset \partial\Omega$ is measurable and such that $\text{meas}_{d-1}(\Gamma) > 0$ *then*

$$\int_\Omega |v(x)|^p \,{\rm d}x \le K\Big(\int_\Omega |\nabla v(x)|^p \,{\rm d}x + \Big| \int_{\Gamma_{\rm D}} v(x) \,{\rm d}S \Big| \Big). \tag{B.3.22}$$

A generalization of Sobolev spaces for "fractional derivatives", i.e. a non-integer order $\alpha \ge 0$, is often useful for various finer investigations: We define the *Sobolev-Slobodeckiĭ space* [17, 210, 488], sometimes also called the Aronszajn or Gagliardo space,[35] as

$$W^{\alpha,p}(\Omega) := \Big\{ u \in W^{[\alpha],p}(\Omega);\ \big|\nabla^{[\alpha]} u\big|_{\alpha-[\alpha],p} < +\infty \Big\}$$

$$\text{where} \quad |G|_{\gamma,p} := \Big(\int_{\Omega\times\Omega} \frac{|G(x)-G(\xi)|^p}{|x-\xi|^{d+\gamma p}} \,{\rm d}x{\rm d}\xi \Big)^{1/p} \tag{B.3.23}$$

and where $[\alpha]$ denotes the integer part of α, so that $0 < \gamma < 1$. The functional $|\cdot|_{\gamma,p}$ is called *Gagliardo's semi-norm*. These space are Banach spaces and, for $p = 2$, even Hilbert spaces; in the latter case, we use the notation $H^\alpha(\Omega) := W^{\alpha,2}(\Omega)$. Note in particular that $|G|_{\gamma,2} < +\infty$ means that $(x,\xi) \mapsto |G(x)-G(\xi)|/|x-\xi|^{d/2+\gamma} \in L^2(\Omega \times \Omega)$. Using the integral representation of the scalar product in this Hilbert space and considering the norm $(\|\cdot\|_{L^2(\Omega)}^2 + |\cdot|_{\gamma,2}^2)^{1/2}$, we obtain the integral representation of the scalar product (A.1.6) in $H^\alpha(\Omega)$ as

$$(u,v) = \int_\Omega u(x)v(x)\,{\rm d}x + \int_{\Omega\times\Omega} \frac{(\nabla^{[\alpha]}u(x) - \nabla^{[\alpha]}u(\xi)) \vdots (\nabla^{[\alpha]}v(x) - \nabla^{[\alpha]}v(\xi))}{|x-\xi|^{d+2\gamma}} \,{\rm d}x{\rm d}\xi \tag{B.3.24}$$

[35] Cf. [156] for a survey, also historical.

for any $u,v\in H^\alpha(\Omega)$; note that the last integrand is then indeed from $L^1(\Omega\times\Omega)$, begin a product of two functions from $L^2(\Omega\times\Omega;\mathbb{R}^{d[\alpha]})$.

Theorem B.3.16 (Embedding of Sobolev-Slobodeckiĭ spaces). *Let $d\ge 1$, $\Omega\subset\mathbb{R}^d$ be a bounded Lipschitz domain. Let $1\le p<+\infty$ and $0<\alpha<1$ such that $\alpha p>d$. Then $W^{\alpha,p}(\Omega;\mathbb{R}^d)\Subset C^{0,\gamma}(\bar\Omega;\mathbb{R}^d)$ for $\gamma:=\alpha-d/p$.*

Let us still present basics from the differential calculus on the domain boundary.[36] Let $\Gamma:=\partial\Omega$ or its part. Let $\vec{n}$ be the outer unit normal to Γ. We define a projector $\mathbb{P}:=\mathbb{I}-\vec{n}\otimes\vec{n}$. Let v be a smooth vector field defined in a neighborhood of Γ. We define its the *surface gradient* of v as $\nabla_{\rm S}v:=\nabla v\mathbb{P}$. Similarly as in the standard situation, the trace of the surface gradient is called the *surface divergence* of v, i.e., $\mathrm{div}_{\rm S}v:=\mathrm{tr}\nabla_{\rm S}v$. It is easy to see that

$$\mathrm{div}_{\rm S}v=\mathbb{P}{:}\nabla v=\mathrm{div}\,v-\vec{n}\cdot(\nabla v)\vec{n}\,.$$

Further, $\partial_{\vec{n}}v:=(\nabla v)\vec{n}$ is the *normal derivative* of v, i.e., the derivative of components of v in the direction of $\vec{n}$. Thus, we have the following identities

$$\nabla v=\nabla_{\rm S}v+\partial_{\vec{n}}v\otimes\vec{n}$$

and applying the trace operator on both sides yields

$$\mathrm{div}\,v=\mathrm{div}_{\rm S}v+\partial_{\vec{n}}v\cdot\vec{n}\,.$$

The *curvature* $\mathbb{K}$ of Γ is defined as $\mathbb{K}:=-\nabla_{\rm S}\vec{n}$ and it is often called the *Weingarten tensor*. The half of the trace of $\mathbb{K}$ is the *mean curvature* κ, i.e., $\kappa:=\mathrm{tr}\,\mathbb{K}/2=-\mathrm{div}_{\rm S}\vec{n}/2$. Given a tensor field $\mathbb{A}:\Gamma\to\mathbb{R}^{3\times3}$, we set $\mathrm{div}_{\rm S}\mathbb{A}:\Gamma\to\mathbb{R}^3$ be the unique vector field such that $\mathrm{div}_{\rm S}(\mathbb{A}^\top a)=a{\cdot}\mathrm{div}_{\rm S}\mathbb{A}$ for all constant vector fields $a:\Gamma\to\mathbb{R}^3$. Written component-wise, we get for $i=1,\dots,d$ that $(\mathrm{div}_{\rm S}\mathbb{A})_i:=\partial_{x_k}\mathbb{A}_{ij}P_{kj}$.

This allows us to define the *Sobolev spaces $W^{k,p}(\Gamma)$ on the boundary Γ* very analogously as we did on Ω only with $\nabla_{\rm S}$ replacing ∇. The embeddings in Theorem B.3.2 and Corollary B.3.4 hold for these spaces except that d must be replaced by $d-1$.

A variant of the Green formula (B.3.14) on a curved smooth surface Γ is sometimes useful: With the above definitions, we have $\mathrm{div}_{\rm S}(gv)=g\mathrm{div}_{\rm S}v+v{\cdot}\nabla_{\rm S}g$, cf. e.g. [203, Formula (21)] for a vectorial case. If $\Omega\subset\mathbb{R}^d$ has a smooth boundary and $v\in C^1(\Gamma;\mathbb{R}^d)$ is tangential to $\Gamma\subset\partial\Omega$, i.e., $v\cdot\vec{n}=0$ and $g\in C^1(\Gamma)$ then the following version of the *Green formula on the surface* holds Γ:

$$\int_\Gamma g\,\mathrm{div}_{\rm S}v\,\mathrm{d}S+\int_\Gamma v{\cdot}\nabla_{\rm S}g\,\mathrm{d}S=\int_\Gamma \mathrm{div}_{\rm S}(gv)\,\mathrm{d}S=\int_{\partial\Gamma}(gv){\cdot}\vec{\nu}\,\mathrm{d}a \tag{B.3.25}$$

[36] Here we follow the exposition given in [203, 417]. Alternatively, pursuing the concept of fields defined exclusively on Γ, we can consider $g:\Gamma\to\mathbb{R}$ and extend it to a neighborhood of Γ and then again define $\nabla_{\rm S}g:=(\nabla g)P$ which, in fact, does not depend on the particular extension.

where $\vec{\nu}$ here is the unit outward normal to the $(d-2)$-dimensional boundary $\partial\Gamma$ of the $(d-1)$-dimensional surface Γ. Cf. also [418, 516]. The operator $\mathrm{div}_\mathrm{S}\nabla_\mathrm{S}$ is called the *Laplace-Beltrami operator*. If $v=0$ on $\partial\Gamma$ or if $\partial\Gamma=\emptyset$, then (B.3.25) implies for a smooth field $A\in C^1(\Gamma;\mathbb{R}^{d\times d})$ and $v\in C^1(\Gamma;\mathbb{R}^d)$ that (see [203, Formula (34)])

$$\int_\Gamma A{:}\nabla_\mathrm{S} v\,\mathrm{d}S = -\int_\Gamma(\mathrm{div}_\mathrm{S}A+2\kappa A\vec{n})\cdot v\,\mathrm{d}S\,, \tag{B.3.26}$$

where κ is the mean curvature of Γ. The further important concept expands the one-dimensional bounded variation, cf. (B.4.1) below, to d dimensions. If $\Omega\subset\mathbb{R}^d$ is a bounded domain we say that $z\in L^1(\Omega;\mathbb{R}^M)$, for some $M\ge1$, has a *bounded variation* and write that $z\in\mathrm{BV}(\Omega;\mathbb{R}^M)$ if the distributional derivative of z is representable by a finite Radon measure, i.e., if for every $1\le i\le M$ and $1\le j\le d$ it holds

$$\int_\Omega z_i\partial_{x_j}v(x)\,\mathrm{d}x = -\int_\Omega v(x)\,\mathrm{D}_jz_i(\mathrm{d}x) \tag{B.3.27}$$

for every $v\in C_c^1(\Omega)$, i.e., in the space of continuously differentiable maps on Ω. We then call $\mathfrak{M}(\Omega;\mathbb{R}^{M\times d})\ni\mathrm{D}z:=(\mathrm{D}_jz_i)_{i,j}$ the measure-valued gradient of z.

Let $\tilde\Omega\subset\Omega\subset\mathbb{R}^d$ be Lebesgue measurable sets and let $B(x,r):=\{a\in\mathbb{R}^d:|x-a|<r\}$. For $x\in\Omega$ we denote the *density* of $\tilde\Omega$ at x by $\theta(\tilde\Omega,x):=\lim_{r\to0}\mathcal{L}^3(\tilde\Omega\cap B(x,r))/\mathcal{L}^3(B(x,r))$ whenever this limit exists. A point $x\in\Omega$ is called a *point of density* of $\tilde\Omega$ if $\theta(\tilde\Omega,x)=1$. If $\theta(\tilde\Omega,x)=0$ for some $x\in\Omega$, then x is called a *point of rarefaction* of $\tilde\Omega$. The *measure-theoretic boundary* $\partial^*\tilde\Omega$ of $\tilde\Omega$ is the set of all points $x\in\Omega$ such that either $\theta(\tilde\Omega,x)$ does not exist or $\theta(\tilde\Omega,x)\notin\{0,1\}$. We call $\tilde\Omega$ a set of finite perimeter if $\mathcal{H}^2(\partial^*\tilde\Omega)<+\infty$. Let $\vec{n}\in\mathbb{R}^d$ be a unit vector and let $H(x,\vec{n}):=\{\tilde x\in\Omega:\ (\tilde x-x)\cdot\vec{n}<0\}$. We say that $\vec{n}$ is the (outer) *measure-theoretic normal* to $\tilde\Omega$ at x if $\theta(\tilde\Omega\cap H(x,-\vec{n}),x)=0$ and $\theta((\Omega\setminus\tilde\Omega)\cap H(x,\vec{n}),x)=0$. The measure-theoretic normal exists for $\mathscr{H}^{d-1}$ almost every point in $\partial^*\tilde\Omega$, see e.g. [180, 483].

Remark B.3.17 (Sobolev-Slobodeckiĭ space on the boundary). In fact, the trace operator $u\mapsto u|_\Gamma$ maps $W^{1,p}(\Omega)\to W^{1-1/p,p}(\Gamma)$, where the Sobolev-Slobodeckiĭ space $W^{1-1/p,p}(\Gamma)$ is defined as in (B.3.23) but now on an $(d-1)$-dimensional manifold Γ instead of d-dimensional domain Ω. Then, similarly as in Theorem B.3.2, we have the embedding $W^{1-1/p,p}(\Gamma)\subset L^{p^\sharp}(\Gamma)$, resp. $W^{1-1/p,p}(\Gamma)\Subset L^{p^\sharp-\epsilon}(\Gamma)$.

Remark B.3.18 (Convention). We use the shorthand but intuitive notation about exponents of the type $p^{**}:=(p^*)^*$ or $p^{**\prime}:=((p^*)^*)'$ or $p^{*\sharp}:=(p^*)^\sharp$ etc., referring respectively to the Lebesgue spaces occurring in the embedding $W^{2,p}(\Omega)\subset L^{p^{**}}(\Omega)$, in its adjoint operator, or in the trace operator $W^{2,p}(\Omega)\to L^{p^{*\sharp}}(\Gamma)$. Thus, e.g., we have $H^2(\Omega)\subset L^{2^{**}-\varepsilon}(\Omega)$ compactly for any $0<\varepsilon\le2^{**}-1$.

B.4 Banach-space-valued functions of time, Bochner spaces

We now define spaces of abstract functions on a bounded interval $I := [0,T] \subset \mathbb{R}$ valued in a Banach space V, invented by Bochner [74]. We say that $z : I \to V$ is *simple* if it takes only finite number of values $v_i \in V$ and $A_i := z^{-1}(v_i)$ is Lebesgue measurable; then $\int_0^T z(t)\,\mathrm{d}t := \sum_{\rm finite}\mathcal{L}^1(A_i)v_i$. We say that $z : I \to V$ is *Bochner measurable* if it is a point-wise limit (in the strong topology) of V of a sequence $\{z_k\}_{k\in\mathbb{N}}$ of simple functions; i.e. $z_k(t) \to z(t)$ for a.a. $t\in I$. The space of all bounded (everywhere defined) Bochner measurable mappings $z : I \to V$ is denoted by $\mathrm{B}(I;V)$. It is a linear space under pointwise multiplication/addition and, if equipped with the norm $\|z\|_{\mathrm{B}(I;V)} = \sup_{0\le t\le T}\|z(t)\|_V$, also a Banach space.[37]

Moreover, we say that $z : I \to V$ is *weakly measurable* if $\langle v^*, z(\cdot)\rangle$ is Lebesgue measurable for any $v^* \in V^*$. If $V = (V')^*$ for some Banach space V', then $z : I \to V$ is *weakly* measurable* if $\langle v^*, z(\cdot)\rangle$ is Lebesgue measurable for any $v^* \in V'$.

The *variation* of $z : I \to V$ with respect to the norm of V is defined as

$$\mathrm{Var}_V(z,I) := \sup_{0\le t_0<t_1<\ldots<t_n\le T,\ n\in\mathbb{N}} \sum\nolimits_{i=1}^n \big\|z(t_i)-z(t_{i-1})\big\|_V. \tag{B.4.1}$$

The subspace of mappings $z \in \mathrm{B}(I;V)$ with a *bounded variation* $\mathrm{Var}_V(z,I) < \infty$ is a Banach space if normed by $\|\cdot\|_{\mathrm{B}(I;V)} + \mathrm{Var}_V(\cdot,I)$, denoted by $\mathrm{BV}(I;V)$. It should be pointed out that, for I a closed interval, $\mathrm{BV}(\Omega)$ for $\Omega = \mathrm{int}\,I$ as defined in (B.3.27) on p. 521, ignores values on zero-measure sets and is thus obviously different from $\mathrm{BV}(I)$ as defined here.

We say that $z : I \to V$ is *absolutely continuous*[38] if, for each $\varepsilon > 0$, there is $\delta > 0$ such that $\sum_{k=1}^K \|z(t_k) - z(s_k)\|_V \le \varepsilon$ whenever $t_{k-1} \le s_k \le t_k \le T$ for $k = 1,\ldots,K \in \mathbb{N}$, $t_0 = 0$, and $\sum_{k=1}^K t_k - s_k \le \delta$. The space of absolutely continuous mappings $z : I \to V$ is denoted by $\mathrm{AC}(I;V)$. Always $\mathrm{AC}(I;V) \subset \mathrm{BV}(I;V)$.

A point $t\in I$ is called a *Lebesgue point* of $z : I \to V$ if

$$\lim_{h\to 0^+} \frac{1}{h}\int_{-h/2}^{h/2} \big\|z(t+\vartheta) - z(t)\big\|_V \mathrm{d}\vartheta = 0.$$

Analogously, a right Lebesgue point $t \in [0,T)$ means that $\lim_{h\to 0^+} \frac{1}{h}\int_0^h \|z(t+\vartheta) - z(t)\|_V \mathrm{d}\vartheta = 0$.

Theorem B.4.1 (Pettis [412].[39]). *If V is separable, then $z : I \to V$ is Bochner measurable if and only if it is weakly measurable.*

---

[37] Note that metrizability allows to work with sequences and that any Cauchy sequence $(z_k)_{k\in\mathbb{N}}$ in $\mathrm{B}(I;V)$ induces sequences $(z_k(t))_{k\in\mathbb{N}}$ that are Cauchy in V, and their limit $z(t)$ form also the limit u in $\mathrm{B}(I;V)$ which is attainable by a sequence of simple functions because each z_k is so; here a diagonalization procedure applies.

[38] If $V = \mathbb{R}^1$, this definition naturally coincides with the absolute-continuity with respect to the Lebesgue measure on I as defined on p. 509.

[39] In fact, [412] works with a general Banach space, showing equivalence of the Bochner measurable mappings with a.e. separably valued weakly measurable ones.

Considering simple functions $\{z_k\}_{k\in\mathbb{N}}$ as above, we call $z : I \to V$ *Bochner integrable* if $\lim_{k\to\infty} \int_0^T \|z(t) - z_k(t)\|_V \mathrm{d}t = 0$. Then $\int_0^T z(t)\,\mathrm{d}t := \lim_{k\to\infty} \int_0^T z_k(t)\,\mathrm{d}t$; this limit exists and is independent of the particular choice of the sequence $(z_k)_{k\in\mathbb{N}}$. Moreover, if V is separable, then a Bochner measurable function z is Bochner integrable if and only if $\|z(\cdot)\|_V$ is Lebesgue integrable. Then also $\big\|\int_0^T z(t)\,\mathrm{d}t\big\|_V \le \int_0^T \|z(t)\|_V \mathrm{d}t$. From this, we can see that $\lim_{h\to 0^+} \frac{1}{h}\int_{-h/2}^{h/2} z(t{+}\vartheta)\mathrm{d}\vartheta \to z(t)$ at each Lebesgue point $t\in I$; note that $\|z(t){-}\frac{1}{h}\int_{-h/2}^{h/2} z(t{+}\vartheta)\mathrm{d}\vartheta\|_V = \|\frac{1}{h}\int_{-h/2}^{h/2} z(t{+}\vartheta){-}z(t)\mathrm{d}\vartheta\|_V \le \frac{1}{h}\int_{-h/2}^{h/2} \|z(t{+}\vartheta){-}z(t)\|_V \mathrm{d}\vartheta$.

The following theorem states that the set of Lebesgue points has the full measure. An analogous assertion holds for right Lebesgue points, too.

Theorem B.4.2 (Lebesgue points in Bochner spaces). *If $z\in L^1(I;V)$. Then a.e. $t\in I$ is a Lebesgue point for z.*

Definition B.4.3 (Weakly continuous functions). A function $z : I \to V$ taking values in a Banach space V is weakly continuous if $\langle v^*, z(\cdot)\rangle : I \to \mathbb{R}$ is continuous for every $v^* \in V^*$.

Any $u\in C_{\mathrm{w}}(I;V) := \{I \to V$ weakly continuous$\}$ is an example of a Bochner integrable function. For such u, the Lebesgue integral $\int_0^T u(t)\,\mathrm{d}t$ can alternatively be defined by Riemann's manner, i.e. $\int_0^T u(t)\,\mathrm{d}t = \lim_{\mathrm{fineness}(\Pi)\to 0} \mathrm{Riem}(u,\Pi)$ over all partitions Π of I the form $0 = t_0 < t_1 < \ldots < t_{N-1} < t_N = T$, $N \in \mathbb{N}$, with $\mathrm{fineness}(\Pi) := \max_{j=1,\ldots,N}(t_j - t_{j-1})$, and where

$$\mathrm{Riem}(u,\Pi) := \sum_{j=1}^{N} u(t_j)(t_j{-}t_{j-1}) \tag{B.4.2}$$

is a *Riemann sum* for the integral $\int_0^T u(t)\,\mathrm{d}t$ with respect to the partition Π of I. This cannot hold for a general $u \in L^1(I;V)$ which is defined only a.e. on I. Anyhow, the following result still holds:

Theorem B.4.4 (Approximation by Riemann sums).[40] *For any $u \in L^1(I;V)$ with V a Banach space, there exists a sequence of partitions Π^m with* $\mathrm{fineness}(\Pi^m) \to 0$ *such that* $\mathrm{Riem}(u,\Pi^m) \to \int_0^T u(t)\,\mathrm{d}t$ *in V for $m \to \infty$.*

Theorem B.4.5 (Approximation of Lebesgue integrals [410]). *Let $\Omega \subset \mathbb{R}^d$ be an open domain with* $\mathrm{meas}_d(\partial\Omega) = 0$ *and let $\omega \subset \Omega$ be of the zero Lebesgue measure. Then, for $r_k : \Omega\setminus\omega \to (0,+\infty)$ and $\{f_k\}_{k\in\mathbb{N}} \subset L^1(\Omega)$, there exist a set of points $\{a_{ik}\} \subset \Omega\setminus\omega$ and positive numbers $\{\epsilon_{ik}\}$, $\epsilon_{ik} \le r_k(a_{ik})$ such that $\{a_{ik} + \epsilon_{ik}\bar{\Omega}\}$ are pairwise*

[40] See [140, Sect.4.4]. In the scalar-valued variant, this sort of results dates back to Hahn [237]. Moreover, the assertion holds in fact for a.a. sequences of partitions Π^m with $\mathrm{fineness}(\Pi^m) \to 0$. Note that we rely that u is defined everywhere on I and in particular $u(T)$ is defined, although its particular value is not important because the term $u(t_N)(t_N{-}t_{N-1})$ can always be made arbitrarily small by sending $t_{N-1} \to t_N = T$.

disjoint for each $k \in \mathbb{N}$, $\bar{\Omega} = \cup_i \{a_{ik} + \epsilon_{ik}\bar{\Omega}\} \cup \omega_k$ *with* $\mathrm{meas}_d(\omega_k) = 0$ *and, for any* $j \in \mathbb{N}$ *and any* $g \in L^\infty(\Omega)$*, it holds*

$$\lim_{k\to\infty} \sum_i f_j(a_{ik}) \int_{a_{ik}+\epsilon_{ik}\Omega} g(x)\,\mathrm{d}x = \int_\Omega f_j(x)g(x)\,\mathrm{d}x\,.$$

For $1 \le p < \infty$, a *Bochner space* $L^p(I;V)$ is the linear space (of classes with respect to equivalence a.e.) of Bochner integrable functions $z : I \to V$ satisfying $\int_0^T \|z(t)\|_V^p \mathrm{d}t < \infty$. This space is a Banach space if endowed with the norm

$$\big\|z\big\|_{L^p(I;V)} := \begin{cases} \Big(\int_0^T \big\|z(t)\big\|_V^p \,\mathrm{d}t\Big)^{1/p} & \text{if } 1 \le p < \infty, \\ \operatorname*{ess\,sup}_{t\in I} \big\|z(t)\big\|_V & \text{if } p = \infty. \end{cases} \tag{B.4.3}$$

If V has a predual, i.e. $V = (V')^*$ for some Banach space V', the notation $L^p_{\mathrm{w}*}(I;V)$ stands for space of weakly* measurable p-integrable (or, if $p = \infty$, essentially bounded) functions $I \to V$.

Proposition B.4.6 (Uniform convexity). *If* V *is uniformly convex and* $1 < p < \infty$*, then* $L^p(I;V)$ *is uniformly convex, too.*

Proposition B.4.7 (Dual space).
(i) If $p \in [1,\infty)$*, the dual space to* $L^p(I;V)$ *always contains* $L^{p'}(I;V^*)$ *and the equality holds if* V^* *is separable, the duality pairing being given by the formula*

$$\langle f, z\rangle_{L^{p'}(I;V^*)\times L^p(I;V)} := \int_0^T \langle f(t), z(t)\rangle_{V^*\times V}\,\mathrm{d}t\,. \tag{B.4.4}$$

Thus, if $p \in (1,\infty)$ *and* V *is reflexive and separable, then* $L^p(I;V)$ *is reflexive.*
(ii) Moreover, $L^\infty_{\mathrm{w}*}(I;V)$ *is dual to the space* $L^1(I;V')$*.*

Let us emphasize that, in general, $L^\infty_{\mathrm{w}*}(I;V)$ is not equal to $L^\infty(I;V)$. If V is separable reflexive, then $L^\infty(I;V) = L^\infty_{\mathrm{w}*}(I;V)$ by Pettis' theorem B.4.1, however.

Often, V itself is a Banach space of functions, say on Ω. Then the natural isomorphism $\tilde{z} \mapsto z$: $[z(t)](x) := \tilde{z}(t,x)$ identifies $L^p(I;V)$ with a space of functions on $I{\times}\Omega$. E.g., if $V = L^q(\Omega;\mathbb{R}^d)$, then we identify $L^p(I;L^q(\Omega;\mathbb{R}^d))$ with

$$\Big\{\tilde{z} : I{\times}\Omega \to \mathbb{R}^d;\ \int_0^T \Big(\int_\Omega \big|\tilde{z}(t,x)\big|^q \,\mathrm{d}x\Big)^{p/q} \mathrm{d}t < \infty\Big\}. \tag{B.4.5}$$

Likewise, $L^p(I;W^{1,q}(\Omega))$ is identified via this isomorphism with functions $\tilde{u} : I{\times}\Omega \to \mathbb{R}$ for which $\int_0^T \big(\int_\Omega |\tilde{u}(t,x)|^q + |\nabla\tilde{u}(t,x)|^q \mathrm{d}x\big)^{p/q} \mathrm{d}t < \infty$.

The following assertion iterates the interpolation (B.2.7) separately in space and in time:

Proposition B.4.8 (Interpolation between $L^{p_1}(I;L^{q_1}(\Omega))$ and $L^{p_2}(I;L^{q_2}(\Omega))$). *Let* $p_1, p_2, q_1, q_2 \in [1,\infty]$*,* $\lambda \in [0,1]$*, and* $z \in L^{p_1}(I;L^{q_1}(\Omega)) \cap L^{p_2}(I;L^{q_2}(\Omega))$*. Then*

$$\frac{1}{p}=\frac{\lambda}{p_1}+\frac{1-\lambda}{p_2} \quad \& \quad \frac{1}{q}=\frac{\lambda}{q_1}+\frac{1-\lambda}{q_2}$$
$$\Rightarrow \quad \|z\|_{L^p(I;L^q(\Omega))} \le \|z\|^{\lambda}_{L^{p_1}(I;L^{q_1}(\Omega))}\|z\|^{1-\lambda}_{L^{p_2}(I;L^{q_2}(\Omega))}. \qquad \text{(B.4.6)}$$

Considering Banach spaces V_0, V_1, ... V_k and $a: I{\times}V_1{\times}...{\times}V_k \to V_0$, let us still define the *Nemytskiĭ mappings* $\mathcal{N}_a$ again by the formula (B.2.11). The following generalization of Theorem B.2.10 holds:

Theorem B.4.9 (Nemytskiĭ mappings on Bochner spaces [319]). *Let V_0, V_1, ... V_k be separable Banach spaces, $a: I{\times}V_1{\times}...{\times}V_k \to V_0$ a Carathéodory mapping*[41] *and the growth condition*

$$\big\|a(x,r_1...,r_k)\big\|_{V_0} \le \gamma(x) + C\sum_{i=1}^{k}\|r_i\|_{V_i}^{p_i/p_0} \quad \textit{for some } \gamma\in L^{p_0}(I), \qquad \text{(B.4.7)}$$

hold with $p_0, p_1, ..., p_k$ as for (B.2.12). Then $\mathcal{N}_a$ maps $L^{p_1}(I;V_1){\times}...{\times}L^{p_k}(I;V_k)$ continuously into $L^{p_0}(I;V_0)$.

We denote by $(\cdot)^{\boldsymbol{\cdot}} = \frac{\mathrm{d}}{\mathrm{d}t}$ the *distributional derivative* of u understood as the abstract linear operator $\dot{u} \in \mathrm{Lin}(\mathscr{D}(I),(V,\mathrm{weak}))$ defined by

$$\forall \varphi \in \mathscr{D}(I): \qquad \dot{u}(\varphi) := -\int_0^T u\dot{\varphi}\,\mathrm{d}t, \qquad \text{(B.4.8)}$$

where again $\mathscr{D}(I)$ stands for smooth compactly supported functions on I. Then we define the *Sobolev-Bochner space* $W^{1,p}(I;V)$ by

$$W^{1,p}(I;V) := \{z\in L^1(I;V);\ \dot{z}\in L^p(I;V)\}. \qquad \text{(B.4.9)}$$

It is a Banach space if normed by $\|z\|_{W^{1,p}(I;V)} := \|z\|_{L^1(I;V)} + \|\dot{z}\|_{L^p(I;V)}$.

An important ingredient often used in evolution problems deals with compactness:

Theorem B.4.10 (Aubin and Lions [20, 317]). *Let V_1, V_2, V_3 be Banach spaces, V_1 be separable and reflexive, $V_1 \Subset V_2$ (a compact embedding), $V_2 \subset V_3$ (a continuous embedding), $1<p<\infty$, $1\le q\le\infty$. Then $L^p(I;V_1)\cap W^{1,q}(I;V_3) \Subset L^p(I;V_2)$, and the embedding is sequentially compact*[42].

In the context of rate-independent processes, it is important to have a generalization which allows for the time-derivative to be controlled only as a measure:

Lemma B.4.11 (Generalization for $\dot{u}$ a measure [456]).[43] *Assuming $V_1 \Subset V_2 \subset V_3$ (the compact and the continuous embeddings between Banach spaces, respectively),*

[41] Generalizing the finite-dimensional case, it means that, for a.a. $t\in I$, $a(t,\cdot): V_1{\times}...V_k \to V_0$ is to be (norm,norm)-continuous and $a(\cdot,r_1...,r_k): I\to V_0$ is to be measurable.

[42] This means that every bounded sequence is mapped into a relatively compact one. In fact, these original references cope with slightly different compactness concept and slightly stronger assumptions, e.g. reflexivity of V_3 or $1<q<\infty$.

[43] For $L^\infty_{\mathrm{w*}}(I;V_1)\cap \mathrm{BV}(I;V_3) \Subset L^p(I;V_2)$ see also [390].

V_1 reflexive, the Banach space V_3 having a predual space V_3', i.e. $V_3 = (V_3')^*$, and $1 < p < \infty$, it holds

$$L^p(I;V_1) \cap \mathrm{BV}(I;V_3) = \{u \in L^p(I;V_1);\ \dot{u} \in \mathfrak{M}(I;V_3)\} \Subset L^p(I;V_2) \qquad \text{(B.4.10)}$$

in the sense that bounded sets in $L^p(I;V_1) \cap \mathrm{BV}(I;V_3)$ are sequentially relatively compact in $L^p(I;V_2)$.

Lemma B.4.12 (Generalization for $\dot{u}$ valued in a locally convex space [456]). *The assertion of Theorem B.4.10 holds even if V_3 is only a metrizable Hausdorff locally convex space.*[44]

In particular, the previous Lemma B.4.11 combined with Proposition B.2.1(ii)–(iii) yields that every sequence bounded in $L^p(I;V_1) \cap \mathrm{BV}(I;V_3)$ possesses a subsequence converging a.e. in I strongly in V_2.[45] There is another selection principle yielding a subsequence converging only weakly* in V_3 but everywhere on I. In fact, this principle needs much less, namely only the BV-boundedness:

Theorem B.4.13 (Helly selection principle for Banach spaces[46]). *Let $V = V_0^*$ with a separable Banach space V_0. Then any bounded sequence in $\mathrm{BV}(I;V)$ contains a subsequence converging weakly* in V everywhere on I and the limit lives again in $\mathrm{BV}(I;V)$.*

This is, in fact, a generalizations of *Helly's selection principle* [247] which, in its classical version used often in probability theory for distribution functions, states that each bounded sequence of nondecreasing bounded functions on an interval possesses a subsequence that converges pointwise to a nondecreasing limit function.

Corollary B.4.14.[47] *Let V be as in Theorem B.4.13 and let V_1 be a reflexive Banach space continuously embedded into V and having a separable predual. Any bounded sequence in $C_{\mathrm{w}}(I;V_0) \cap \mathrm{BV}(I;V)$ contains a subsequence converging weakly in V_1 everywhere on I and the limit lives in $\mathrm{B}(I;V_0) \cap \mathrm{BV}(I;V)$.*

The following definitions lead to the statement of the *Vitali's covering theorem* which is an important tool in analysis.

---

[44] Without going into details, it means a linear space V_3 equipped with the collection of seminorms $|\cdot|_k$, $k \in \mathbb{N}$, such that $|v|_k = 0$ for all $k \in \mathbb{N}$ implies $v = 0$. Boundedness in $L^1(I;V_3)$ means that all seminorms $v \mapsto \int_0^t |v(t)|_k \,\mathrm{d}t$ are bounded.

[45] Indeed, by Lemma B.4.11 the sequence in question, say $(z_k)_{k\in\mathbb{N}}$, is relatively compact in $L^p(I;V_2)$ and thus, up to a subsequence, $z_k \to z$ in $L^p(I;V_2)$, i.e. $\|z_k(\cdot) - z(\cdot)\|_{V_2}^p \to 0$ in $L^1(I)$, and then, by Proposition B.2.1(ii)–(iii), by selecting further a subsequence it holds $\|z_k(t) - z(t)\|_{V_2}^p \to 0$ for a.a. $t \in I$.

[46] See [44, 373] for separable reflexive V, or [139, Lemma 7.2] for a general case.

[47] By $C_{\mathrm{w}}(I;V_0) \subset L^\infty(I;V_0)$, there is a subsequence converging weakly* in $L^\infty(I;V_0)$ and, in particular, the limit is Bochner measurable. By Theorem B.4.13, selecting further subsequence, we get the pointwise convergence weakly in V. At each particular time instance, its subsequences must converge in V_0 but their limits must be again the same as the limit in V.

Definition B.4.15 (Sets tending to x). We say that a sequence of measurable sets $\{\omega_i\}_i \subset \mathbb{R}^d$ tends to $x \in \mathbb{R}^d$ if $\omega_i \subset B(x, r_i)$ for all i and some $r_i > 0$, and there is $C > 0$ such that for all i $\text{meas}_d(\omega_i) \geq C r_i^d$ where $\lim_{i\to\infty} r_i = 0$.

Definition B.4.16 (Vitali's Cover [474]). We say that a family of sets $\{\Theta_j\}_j \subset \mathbb{R}^d$ covers $\Omega \subset \mathbb{R}^d$ in the sense of Vitali if for every $x \in \Omega$ there is a sequence of sets $\{A_i(x)\}_{i\in\mathbb{N}} \subset \{\Theta_j\}_j$ tending to $x \in \Omega$ in the sense of Definition B.4.15.

Theorem B.4.17 (Vitali's Covering Theorem [474, 526]). *Assume that a family of sets $\{\Theta_j\}_j \subset \mathbb{R}^d$ covers $\Omega \subset \mathbb{R}^d$ in the sense of Vitali. Then there is at most countable sequence of sets $\{A_i\}_{i\in\mathbb{N}} \subset \{\Theta_j\}_j$ such that* $\text{meas}_d(\Omega \setminus \cup_{i\in\mathbb{N}} A_i) = 0$.

Appendix C
Elements of modern theory of differential equations

Differential equations have served as a language for description of problems in mechanics many centuries. Their solutions historically meant rather finding a concrete analytical solution and, in spite of intensive activities during centuries, was limited to very particular cases only.

Modern theory based on suitable integral identities (making weak formulations) was developed during 20th century, philosophically based on the initiative of D. Hilbert [250] to describe physically motivated problems rather in terms of integral identities than the differential equations. Modern methods rely on efficient approximation methods and computer implementation of approximate solutions of broad classes of problems rather than finding an exact analytical solutions of a very specific problems only.

For a more detailed exposition the reader is referred to the monographs [179, 317, 456, 539]. For a brief survey of basic notions, concepts, and results in variational calculus see also [459].

C.1 Boundary-value problems in a weak formulation

Here, we present only very basic concepts of weak solutions to boundary-value problems, focusing on problems having a potential. Equipped with the theory of $W^{k,p}$-Sobolev spaces,we confine ourselves to problems exhibiting polynomial growth/coercivity.

C.1.1 *Second-order problems*

Always, $\Omega \subset \mathbb{R}^d$ denotes a bounded, Lipschitz domain with the boundary $\Gamma := \partial\Omega$ divided (up to $(d-1)$-zero measure) into two disjoint open subsets, Γ_{N} and Γ_{D}, one of them possibly being empty.

M. Kružík and T. Roubíček, *Mathematical Methods in Continuum Mechanics of Solids*, Interaction of Mechanics and Mathematics,
https://doi.org/10.1007/978-3-030-02065-1

Considering $u : \Omega \to \mathbb{R}^n$ as an unknown "solution", for Carathéodory integrands $\varphi : \Omega{\times}\mathbb{R}^n{\times}\mathbb{R}^{n\times d} \to \mathbb{R}$ and $\varphi_{\rm N} : \Gamma_{\rm N}{\times}\mathbb{R}^n \to \mathbb{R}$, we consider the integral functional

$$\Phi(u) = \int_\Omega \varphi(x,u,\nabla u)\,{\rm d}x + \int_{\Gamma_{\rm N}} \varphi_{\rm N}(x,u)\,{\rm d}S \tag{C.1.1a}$$

$$\text{on an affine closed manifold } \{u \in W^{1,p}(\Omega;\mathbb{R}^n);\ \ u|_{\Gamma_{\rm D}} = u_{\rm D}\} \tag{C.1.1b}$$

for a suitable given $u_{\rm D}$; in fact, existence of $\bar u_{\rm D} \in W^{1,p}(\Omega;\mathbb{R}^n)$ such $u_{\rm D} = \bar u_{\rm D}|_{\Gamma_{\rm D}}$ is to be required. Having in mind the underlying Sobolev space $W^{1,p}(\Omega;\mathbb{R}^n)$, we confine ourselves to a p-polynomial-type coercivity of the highest-order term and the corresponding growth restrictions on the partial derivatives $\partial_F\varphi$, $\partial_u\varphi$, and $\partial_u\varphi_{\rm N}$ with some $1 < p < +\infty$ and $1 \le q < +\infty$, i.e.

$$\partial_F\varphi(x,u,F){:}F + \partial_u\varphi(x,u,F){\cdot}u \ge \epsilon|F|^p + \epsilon|u|^q - C, \tag{C.1.2a}$$

$$\exists \gamma \in L^{p'}(\Omega):\ \ |\partial_F\varphi(x,u,F)| \le \gamma(x) + C|u|^{(p^*-\epsilon)/p'} + C|F|^{p-1}, \tag{C.1.2b}$$

$$\exists \gamma \in L^{p^{*\prime}}(\Omega):\ \ |\partial_u\varphi(x,u,F)| \le \gamma(x) + C|u|^{p^*-1-\epsilon} + C|F|^{p/p^{*\prime}}, \tag{C.1.2c}$$

$$\exists \gamma \in L^{p^{\sharp\prime}}(\Gamma):\ \ |\partial_u\varphi_{\rm N}(x,u)| \le \gamma(x) + C|u|^{p^\sharp-1-\epsilon} \tag{C.1.2d}$$

for some $\epsilon > 0$ and $C < +\infty$; we used F as a placeholder for ∇u. By Theorem B.2.10, we can see that (C.1.2b) ensures just continuity of $\mathscr{N}_{\partial_F\varphi} : L^{p^*-\epsilon}(\Omega;\mathbb{R}^n)\times L^p(\Omega;\mathbb{R}^{n\times d}) \to L^{p'}(\Omega;\mathbb{R}^{n\times d})$, and analogously also (C.1.2c) works for $\mathscr{N}_{\partial_u\varphi}$, while (C.1.2d) gives continuity of $\mathscr{N}_{\partial_u\varphi_{\rm N}} : L^{p^\sharp-\epsilon}(\Gamma;\mathbb{R}^n) \to L^{p^{\sharp\prime}}(\Gamma;\mathbb{R}^n)$. This, together with the embedding/trace Theorems B.3.2–B.3.6, reveals the motivation for the growth conditions (C.1.2b-d).

For $\epsilon \ge 0$, (C.1.2b-d) also ensures that the functional Φ from (C.1.1) is *Gâteaux differentiable* on $W^{1,p}(\Omega;\mathbb{R}^n)$. The abstract *Euler-Lagrange equation* (A.3.4) then leads to the integral identity

$$\int_\Omega \partial_F\varphi(x,u,\nabla u){:}\nabla v + \partial_u\varphi(x,u,\nabla u){\cdot}v\,{\rm d}x + \int_{\Gamma_{\rm N}} \partial_u\varphi_{\rm N}(x,u){\cdot}v\,{\rm d}S = 0 \tag{C.1.3}$$

for any $v \in W^{1,p}(\Omega;\mathbb{R}^n)$ such that $v|_{\Gamma_{\rm D}} = 0$; the notation ":" or "$\cdot$" means summation over two indices or one index, respectively. Completed by the Dirichlet condition on $\Gamma_{\rm D}$, this represents a *weak formulation* of the *boundary value problem* for a system of 2nd-order elliptic quasilinear equations:[1]

$$\operatorname{div}\partial_F\varphi(u,\nabla u) = \partial_u\varphi(u,\nabla u) \qquad \text{in } \Omega, \tag{C.1.4a}$$

$$\partial_F\varphi(u,\nabla u){\cdot}\vec n + \partial_u\varphi_{\rm N}(u) = 0 \qquad \text{on } \Gamma_{\rm N}, \tag{C.1.4b}$$

$$u|_\Gamma = u_{\rm D} \qquad \text{on } \Gamma_{\rm D}, \tag{C.1.4c}$$

[1] Assuming sufficiently smooth data as well as u, this can be seen by multiplying (C.1.4a) by v, using the Green formula $\int_\Omega(\operatorname{div}\Sigma){\cdot}v + \Sigma{:}v\,{\rm d}x = \int_\Gamma(\Sigma{\cdot}\vec n)v\,{\rm d}S$ with $\Sigma = \partial_F\varphi$, and using $v = 0$ on $\Gamma_{\rm D}$ and the boundary conditions (C.1.4b) on $\Gamma_{\rm N}$.

where x-dependence has been omitted for notational simplicity. The conditions (C.1.4b) and (C.1.4c) are called the *Robin* and the *Dirichlet boundary conditions*, respectively, and (C.1.4) is called the *classical formulation* of this boundary value problem. Any $u \in C^2(\bar{\Omega};\mathbb{R}^n)$ satisfying (C.1.4) is called a *classical solution* while $u \in W^{1,p}(\Omega;\mathbb{R}^n)$ satisfying (C.1.3) for any $v \in W^{1,p}(\Omega;\mathbb{R}^n)$ such that $v|_{\Gamma_\mathrm{D}} = 0$ is called a *weak solution*; note that much less smoothness is required for weak solutions.

C.1.2 Fourth-order problems

Higher-order problems can be considered analogously but the complexity of the problem grows with the order. Let us therefore use for illustration 4th order problems only, governed by an integral functional

$$\Phi(u) = \int_\Omega \varphi(x,u,\nabla u,\nabla^2 u)\,\mathrm{d}x + \int_{\Gamma_\mathrm{N}} \varphi_\mathrm{N}(x,u,\nabla u)\,\mathrm{d}S \tag{C.1.5a}$$

$$\text{on an affine closed manifold } \{u \in W^{2,p}(\Omega;\mathbb{R}^n);\ u|_{\Gamma_\mathrm{D}} = u_\mathrm{D}\} \tag{C.1.5b}$$

involving Carathéodory integrands $\varphi : \Omega\times\mathbb{R}^n\times\mathbb{R}^{n\times d}\times\mathbb{R}^{n\times d\times d} \to \mathbb{R}$ and $\varphi_\mathrm{N} : \Gamma_\mathrm{N}\times\mathbb{R}^n\times\mathbb{R}^{n\times d} \to \mathbb{R}$ and a given boundary data u_D Instead of (C.1.3), the abstract Euler-Lagrange equation (A.3.4) now leads to the integral identity:

$$\int_\Omega \partial_G\varphi(x,u,\nabla u,\nabla^2 u)\,\vdots\,\nabla^2 v + \partial_F\varphi(x,u,\nabla u,\nabla^2 u){:}\nabla v + \partial_u\varphi(x,u,\nabla u,\nabla^2 u)\cdot v\,\mathrm{d}x$$
$$+ \int_{\Gamma_\mathrm{N}} \partial_F\varphi_\mathrm{N}(x,u,\nabla u)\cdot\frac{\partial v}{\partial\vec{n}} + \partial_u\varphi_\mathrm{N}(x,u,\nabla u)\cdot v\,\mathrm{d}S = 0 \tag{C.1.6}$$

for any $v \in W^{2,p}(\Omega;\mathbb{R}^n)$ such that $v|_{\Gamma_\mathrm{D}} = 0$; the notation "$\vdots$" stands for summation over three indices and G is now a placeholder for $\nabla^2 u$ while F is a placeholder for ∇u. Completed by the Dirichlet condition on Γ_D, this represents a *weak formulation* of the *boundary value problem* for a system of 4th order elliptic quasilinear equations

$$\mathrm{div}^2\,\partial_G\varphi(u,\nabla u,\nabla^2 u) - \mathrm{div}\,\partial_F\varphi(u,\nabla u,\nabla^2 u) + \partial_u\varphi(u,\nabla u,\nabla^2 u) = 0 \quad \text{in } \Omega, \tag{C.1.7a}$$

with two natural (although quite complicated) *boundary conditions* prescribed on each part of the boundary, namely:[2]

$$\big(\mathrm{div}\,\partial_G\varphi(u,\nabla u,\nabla^2 u) - \partial_F\varphi(u,\nabla u,\nabla^2 u)\big)\cdot\vec{n} + \mathrm{div}_\mathrm{S}\big(\partial_G\varphi(u,\nabla u,\nabla^2 u)\vec{n}\big)$$
$$+ (\mathrm{div}_\mathrm{S}\vec{n})\partial_F\varphi_\mathrm{N}(u,\nabla u) + \partial_u\varphi_\mathrm{N}(u,\nabla u) = 0 \quad \text{on } \Gamma_\mathrm{N}, \tag{C.1.7b}$$

$$\partial_G\varphi(u,\nabla u,\nabla^2 u){:}(\vec{n}\otimes\vec{n}) + \partial_F\varphi_\mathrm{N}(u,\nabla u) = 0 \quad \text{on } \Gamma, \tag{C.1.7c}$$

[2] Actually, rather the term $-(\mathrm{div}_\mathrm{S}\vec{n})(\vec{n}^\top\partial_G\varphi(u,\nabla u,\nabla^2 u)\vec{n})$ is seen in (C.1.7b) but it equals to $(\mathrm{div}_\mathrm{S}\vec{n})\partial_u\varphi_\mathrm{N}(u,\nabla u)$ when (C.1.7c) is taken into account.

$$u\big|_{\Gamma} = u_{\rm D} \qquad \text{on } \Gamma_{\rm D}. \tag{C.1.7d}$$

Again, (C.1.7) is called the *classical formulation* of the boundary value problem in question, and its derivation from (C.1.6) is more involved than in Section C.1.1. One must use a general decomposition $\nabla v = \frac{\partial v}{\partial \vec{n}}\vec{n} + \nabla_{\rm S} v$ on Γ with $\nabla_{\rm S} v = \nabla v - \frac{\partial v}{\partial \vec{n}}\vec{n}$ being the surface gradient of v, cf. Sect. B.3. On a smooth boundary Γ, one can use another (now $(d-1)$-dimensional) Green-type formula on tangent spaces:

$$\int_{\Gamma} \mathfrak{h} \vdots (\vec{n} \otimes \nabla v)\, \mathrm{d}S = \int_{\Gamma} \big(\vec{n}^\top \mathfrak{h} \vec{n}\big) \frac{\partial v}{\partial \vec{n}} + \mathfrak{h} \vdots (\vec{n} \otimes \nabla_{\rm S} v)\, \mathrm{d}S$$
$$= \int_{\Gamma} \big(\vec{n}^\top \mathfrak{h} \vec{n}\big) \frac{\partial v}{\partial \vec{n}} - \mathrm{div}_{\rm S}(\mathfrak{h}\vec{n})\cdot v + (\mathrm{div}_{\rm S}\vec{n})\big(\vec{n}^\top \mathfrak{h} \vec{n}\big) v\, \mathrm{d}S \tag{C.1.8}$$

where $\mathrm{div}_{\rm S} = \mathrm{tr}(\nabla_{\rm S})$ with $\mathrm{tr}(\cdot)$ being the trace of a $(d-1)\times(d-1)$-matrix, denotes the $(d-1)$-dimensional surface divergence, cf. (B.3.25), so that $\mathrm{div}_{\rm S}\vec{n}$ is (up to a factor $-\frac{1}{2}$) the mean curvature of the surface Γ. We use (C.1.8) with the "hyperstress" $\mathfrak{h} = \partial_G \varphi(x, u, \nabla u, \nabla^2 u)$. Specifically, to derive (C.1.7), we apply the Green formula on Ω twice to (C.1.6) and then the surface Green formula (C.1.8), and then one can see (C.1.7a) when taken the test functions v with compact support in Ω, and then one can see (C.1.7b) when taking the test functions v with arbitrary traces but with $\partial v/\partial \vec{n} = 0$.[3] Eventually, taken v arbitrarily with nonvanishing $\partial v/\partial \vec{n}$ but with zero traces on $\Gamma_{\rm D}$, one obtains also (C.1.7c).

Comparing the variational formulation as critical points of (C.1.5) with the weak formulation (C.1.6) and with the classical formulation (C.1.7), one can see that although formally all formulations are equivalent to each other, the advantage of the variational formulations like (C.1.5) in its simplicity being obvious even when $\varphi_{\rm N} = \varphi_{\rm N}(x, u)$ does not depend on $F = \nabla u$ which makes (C.1.7b,c) simpler because $\partial_F \varphi_{\rm N} = 0$.

In (C.1.7), we considered mixed boundary conditions on both $\Gamma_{\rm D}$ and $\Gamma_{\rm N}$ dealing with zeroth or third derivatives combined with the second derivative. One can, however, think about prescribing zeroth and first derivatives (i.e. clamping the body on $\Gamma_{\rm D}$) or prescribing first- and and third-order derivatives.[4]

C.2 Abstract evolution problems

In Part II, there are various evolution problems formulated. In contrast to Section C.1, we confine ourselves exclusively to the initial-value problems on the ab-

[3] This is (up to density arguments) possible if Γ is smooth because on can locally rectify Γ and then reflect the arbitrary test function v outside Ω so that, when smoothened, $\partial v/\partial \vec{n} = 0$ because of the symmetry in the normal direction across Γ.

[4] The other two combinations, namely the zeroth and the third order derivatives or the first and the second order derivatives, are not natural from the variational viewpoint because they overdetermine some of the two boundary terms arising in the weak formulation (C.1.6).

stract level, using abstract Banach spaces. A combination with Section C.1 will routinely lead to initial-boundary-value problems for partial differential equations of parabolic or hyperbolic type.

Disregarding *direct methods*[5] whose application is rather limited and theoretical only, the usual "indirect" approach to their solutions consists in the four steps:

a) an approximation by discretization in space, or in time, or both in time and space,
b) existence of solutions of approximate problems (by results for static problems, for (abstract) ordinary differential equations, or by algebraic systems),
c) a-priori estimates (possibly interpreted as numerical stability if implemented on computers),
d) convergence towards suitably defined (weak) solutions within refining discretization.

We confine ourselves only on techniques with direct mechanical interpretation. For example, the abstract parabolic equation $\dot{u} + A(u) = f$ which is a special case of (C.2.1) below allows for a theory based on the test by u not needed even A to have a potential, yielding the identity $\frac{\mathrm{d}}{\mathrm{d}t}\frac{1}{2}\|u\|^2 + \langle A(u), u\rangle = \langle f, u\rangle$ whose particular terms however do not have a clear interpretation, in contrast to (C.2.2), and we will thus not consider it here.

The mentioned approximation method by discretization in in space is also called the *Galerkin* [212] (or in the evolution context *Faedo-Galerkin*) *method* while discretization in time is also called the *Rothe method* [450]. Both methods can serve as an efficient theoretical method for analyzing evolution problems, and simultaneously for designing efficient conceptual algorithms for numerical solution of such problems.

We illustrate the above outlined strategy first on an abstract initial-value problem for a doubly-nonlinear abstract evolution equation

$$\Psi'(\dot{u}) + \Phi'(u) = f(t) \quad \text{for } t \in I, \qquad u\big|_{t=0} = u_0 \tag{C.2.1}$$

and we will based the analysis on the "energetics" which can be revealed by testing (C.2.1) by $\dot{u}$, i.e. formally

$$\underbrace{\langle \Psi'(\dot{u}), \dot{u}\rangle}_{\substack{\text{rate of}\\ \text{dissipation}}} + \underbrace{\langle \Phi'(u), \dot{u}\rangle}_{\substack{=\frac{\mathrm{d}}{\mathrm{d}t}\Phi(u) = \text{rate of}\\ \text{energy stored}}} = \underbrace{\langle f(t), \dot{u}\rangle}_{\substack{\text{power of}\\ \text{loading}}}. \tag{C.2.2}$$

Moreover, considering inertial effects in mechanical systems leads to the second-order (possibly doubly-nonlinear) problems, augmenting the initial-value problem (C.2.25) in the following way:

---

[5] An example for the direct approach to the abstract parabolic problem $\dot{u} + \Phi'(u) = f$ with the initial condition $u|_{t=0} = u_0$ being a special case of (C.2.1) consists in minimization of the functional $u \mapsto \int_0^T \Phi(u) + \Phi^*(f - \dot{u}) - \langle f, u\rangle\,\mathrm{d}t + \frac{1}{2}\|u(T)\|_H^2$ subject to the mentioned boundary condition, cf. [453], which is referred in its non-variational form requiring still the value to be zero as a Brezis-Ekeland-Nayroles principle, cf. [459] also for various generalizations.

$$\mathscr{T}'\ddot{u} + \partial\Psi(\dot{u}) + \Phi'(u) \ni f(t) \quad \text{for } t\in I, \quad u\big|_{t=0} = u_0, \quad \dot{u}\big|_{t=0} = v_0, \tag{C.2.3}$$

with $\mathscr{T}$ denoting an abstract quadratic functional corresponding to the kinetic energy. Like in (C.2.25), the "energetics" can be revealed by testing (C.2.3) by $\dot{u}$, i.e. formally

$$\underbrace{\langle \mathscr{T}'\ddot{u},\dot{u}\rangle}_{=\frac{\mathrm{d}}{\mathrm{d}t}\mathscr{T}(\dot{u}) = \text{rate of kinetic energy}} + \underbrace{\langle \Psi'(\dot{u}),\dot{u}\rangle}_{\text{rate of dissipation}} + \underbrace{\langle \Phi'(u),\dot{u}\rangle}_{=\frac{\mathrm{d}}{\mathrm{d}t}\Phi(u) = \text{rate of energy stored}} = \underbrace{\langle f(t),\dot{u}\rangle}_{\text{power of loading}}. \tag{C.2.4}$$

C.2.1 Auxiliary tools: Gronwall inequality and by-part integration

The main ingredient for estimation of evolution systems in general is the following estimate:

Lemma C.2.1 (Gronwall inequality). [6] *Let $y(t) \le C + \int_0^t (a(\vartheta)y(\vartheta) + b(\vartheta))\mathrm{d}\vartheta$ for some $a,b \ge 0$ integrable, then:*

$$y(t) \le \Big(C + \int_0^t b(\theta)\mathrm{e}^{-\int_0^\theta a(\vartheta)\mathrm{d}\vartheta}\mathrm{d}\theta\Big)\mathrm{e}^{\int_0^t a(\theta)\mathrm{d}\theta} \tag{C.2.5}$$

For $a \ge 0$ constant, (C.2.5) simplifies to $y(t) \le \mathrm{e}^{at}(C + \int_0^t b(\vartheta)\mathrm{e}^{-a\vartheta}\mathrm{d}\vartheta) \le (C + \int_0^t b(\vartheta)\mathrm{d}\vartheta)\mathrm{e}^{aT}$ for $t \in I$.

The following *discrete* version of the *Gronwall inequality* for non-negative sequences $\{y_k\}_{k\ge0}$ and $\{(a_k,b_k)\}_{k\ge0}$ instead of functions of time will be often used:[7]

$$\Big(\forall l\ge0:\ y_l \le C + \tau\sum_{k=1}^{l-1}(a_k y_k + b_k)\Big) \ \Rightarrow\ \forall l\ge0:\ y_l \le \Big(C + \tau\sum_{k=1}^{l-1} b_k\Big)\mathrm{e}^{\tau\sum_{k=1}^{l-1} a_k}; \tag{C.2.6}$$

of course, for $l = 0$ it means $y_l \le C$. We used a factor $\tau > 0$ which can be omitted (by replacing simply τa_k and τb_k by a_k and b_k, respectively) but which plays a role of a time step for an equidistant dicretisation of the integrals in (C.2.5); note that the factor $\mathrm{e}^{-\int_0^\theta a(\vartheta)\mathrm{d}\vartheta}$ has been simply replaced by 1.

We will often use $a_k \equiv a$ constant, and then[8]

---

[6] It is also called the Bellman-Gronwall inequality according to the original works [230] (for $C = 0$, a,b constant) and [55] (for $C \ge 0$, $b = 0$, $a \in L^1(0,T)$).

[7] See e.g. Quarteroni and Valli [424, Lemma 1.4.2] or Thomée [513].

[8] Note that the premise in (C.2.7) implies $y_l \le (1 - a\tau)^{-1}(C + \tau b_1 + \tau\sum_{k=1}^{l-1}(ay_k + b_{k+1}))$ so that (C.2.6) can be used if $\tau < 1/a$.

$$\left.\begin{array}{l}\forall l\ge 0:\ \ y_l\le C+\tau\displaystyle\sum_{k=1}^{l}(ay_k+b_k)\\ \tau<1/a\end{array}\right\}\ \Rightarrow\ \forall l\ge 0:\ \ y_l\le\frac{e^{\tau la/(1-a\tau)}}{1-a\tau}\Big(C+\tau\sum_{k=1}^{l}b_k\Big). \tag{C.2.7}$$

Abstract setting for evolution problem often relies on the construction of a so-called evolution (also called Gelfand's) triple assuming V embedded continuously and densely into a Hilbert space H identified with its own dual, $H\equiv H^*$, we have also $H\subset V^*$ continuously. Indeed, denoting $i:V\to H$ the mentioned embedding, the adjoint mapping i^* (which is continuous) maps H^* into V^* and is injective because i just makes restriction of linear continuous functionals $H\to\mathbb{R}$ on the subset V and different continuous functional must remain different when restricted on such a dense subset. The mentioned identification of H with its own dual H^* yields altogether

$$V\subset H\equiv H^*\subset V^*; \tag{C.2.8}$$

the triple (V,H,V^*) is called a *Gelfand triple*. The duality pairing between V^* and V is then a continuous extension of the inner product on H, denoted by $(\cdot,\cdot)_H$, i.e. for $u\in H$ and $v\in V$ we have[9]

$$(u,v)_H=\langle u,v\rangle_{H^*\times H}=\langle u,iv\rangle_{H^*\times H}=\langle i^*u,v\rangle_{V^*\times V}=\langle u,v\rangle_{V^*\times V}\,. \tag{C.2.9}$$

Moreover, the embedding $H\subset V^*$ is dense.

An important tool is a generalization of the classical calculus of the by-part integration for abstract Banach-space-valued functions in the case of the Gelfand triple (C.2.8):

Lemma C.2.2 (By-part integration formula). *Let $V\subset H\cong H^*\subset V^*$. Then $L^p(I;V)\cap W^{1,p'}(I;V^*)\subset C(I;H)$ continuously and the following by-parts integration formula holds for any $u,v\in L^p(I;V)\cap W^{1,p'}(I;V^*)$ and any $0\le t_1\le t_2\le T$:*

$$(u(t_2),v(t_2))_H-(u(t_1),v(t_1))_H=\int_{t_1}^{t_2}\langle\dot u,v\rangle_{V^*\times V}+\langle u,\dot v\rangle_{V\times V^*}\,\mathrm{d}t. \tag{C.2.10}$$

In particular, the formula (C.2.10) for $u=v$ gives

$$\frac{1}{2}\big\|u(t_2)\big\|_H^2-\frac{1}{2}\big\|u(t_1)\big\|_H^2=\int_{t_1}^{t_2}\langle\dot u,u\rangle_{V^*\times V}\,\mathrm{d}t\,, \tag{C.2.11}$$

and shows that the function $t\mapsto\frac{1}{2}\|u(t)\|_H^2$ is absolutely continuous and its derivative exists a.e. on I and

[9] The equalities in (C.2.9) follow successively from the identification H with H^*, the embedding $V\subset H$, the definition of the adjoint operator i^*, and the identification of i^*u with u.

$$\frac{1}{2}\frac{\mathrm{d}}{\mathrm{d}t}\|u(t)\|_H^2 = \langle \dot{u}, u\rangle_{V^*\times V} \qquad \textit{for a.a. } t\in I. \tag{C.2.12}$$

A discrete variant of the by-part integration formula (C.2.10) for sequences $\{u^k\}_{k=0,...,K}$ and $\{v^k\}_{k=0,...,K}$ reads as a *by-part summation* formula[10]

$$(u^j, v_\tau^j) - (u^i, v_\tau^i) = \sum_{k=i+1}^{j} \left(\langle u^k - u^{k-1}, v_\tau^k\rangle + \langle u^{k-1}, v_\tau^k - v_\tau^{k-1}\rangle\right) \tag{C.2.13}$$

with the integers $0 \le i < j \le K$ while the discrete variant of (C.2.11) reads as

$$\frac{1}{2}\|u^j\|_H^2 - \frac{1}{2}\|u^i\|_H^2 \le \frac{1}{2}\|u^j\|_H^2 - \frac{1}{2}\|u^i\|_H^2 + \sum_{k=i+1}^{j} \|u^k - u^{k-1}\|_H^2 = \sum_{k=i+1}^{j} \langle u^k - u^{k-1}, u^k\rangle; \tag{C.2.14}$$

note that by ignoring a term $\sum_{k=i+1}^{j} \|u^k - u^{k-1}\|_H^2 \ge 0$ arising from the time discretization,[11] we obtain only the inequality in (C.2.14), in contrast to the equality in (C.2.11).

Considering a fixed time step $\tau > 0$ and $\{u_\tau^k\}_{k=0,...,K}$ with $K = T/\tau$, we define the piecewise-constant and the piecewise affine *interpolants* respectively by

$$\bar{u}_\tau(t) = u_\tau^k, \qquad \underline{u}_\tau(t) = u_\tau^{k-1}, \qquad \bar{\underline{u}}_\tau(t) = \frac{1}{2}u_\tau^k + \frac{1}{2}u_\tau^{k-1}, \quad \text{and} \tag{C.2.15a}$$

$$u_\tau(t) = \frac{t-(k-1)\tau}{\tau}u_\tau^k + \frac{k\tau - t}{\tau}u_\tau^{k-1} \qquad \text{for } (k-1)\tau < t \le k\tau. \tag{C.2.15b}$$

Similar meaning has also v_τ, $\bar{v}_\tau$, etc. In terms of these interpolants, one can write (C.2.13) as

$$(u_\tau(t_2), v_\tau(t_2)) - (u_\tau(t_1), v_\tau(t_1)) = \int_{t_1}^{t_2} \langle \dot{u}_\tau, \bar{v}_\tau\rangle + \langle \underline{u}_\tau, \dot{v}_\tau\rangle \,\mathrm{d}t, \tag{C.2.16}$$

for $t_1 = i\tau$ and $t_2 = j\tau$ with the integers $0 \le i < j \le T/\tau$, while (C.2.14) can be written as

$$\frac{1}{2}\|u_\tau(t_2)\|_H^2 - \frac{1}{2}\|u_\tau(t_1)\|_H^2 \le \int_{t_1}^{t_2} \langle \dot{u}_\tau, \bar{u}_\tau\rangle \,\mathrm{d}t. \tag{C.2.17}$$

---

[10] It can easily be seen from the simple algebra $(a_k - a_{k-1})b_k + a_{k-1}(b_k - b_{k-1}) = a_k b_k - a_{k-1}b_{k-1}$ when summing it for $k = i+1, ..., j$.

[11] Indeed, using $u := u^{k-1}$ and $v := u^k$ in (A.1.6) yields $(u^{k-1}, u^k) = \frac{1}{4}\|u^k + u^{k-1}\|_H^2 - \frac{1}{4}\|u^k - u^{k-1}\|_H^2 = \frac{1}{4}\|u^k\|_H^2 + \frac{1}{2}(u^{k-1}, u^k) + \frac{1}{4}\|u^{k-1}\|_H^2 - \frac{1}{4}\|u^k - u^{k-1}\|_H^2$ which further yields the used identity $(u^k - u^{k-1}, u^k) = \|u^k\|_H^2 - (u^{k-1}, u^k) = \frac{1}{2}\|u^k\|_H^2 - \frac{1}{2}\|u^{k-1}\|_H^2 + \frac{1}{2}\|u^k - u^{k-1}\|_H^2$.

Typically, all these functions are valued in V and so are also their time differences, hence the dualities in (C.2.13)–(C.2.17) are in fact scalar product in H, in contrast to (C.2.10)–(C.2.12).

C.2.2 First-order evolution problems: Rothe method

For the*Rothe method*, we consider a uniform partition of the time interval with the time step $\tau > 0$ with T/τ integer.

Let us first now discretize the doubly-nonlinear abstract equation (C.2.1) as

$$\Psi'\Big(\frac{u_\tau^k - u_\tau^{k-1}}{\tau}\Big) + \Phi'(u_\tau^k) = f_\tau^k := \int_{(k-1)\tau}^{k\tau} f(t)\,\mathrm{d}t, \quad k = 1,...,T/\tau, \quad u_\tau^0 = u_0. \tag{C.2.18}$$

This is also known as the *implicit* (or *backward*) *Euler formula* and u_τ^k for $k = 1,...,T/\tau$ approximate respectively the values $u(k\tau)$. One can apply the direct method by employing the recursive variational problem for the functional

$$u \mapsto \frac{\Phi(u)}{\tau} + \Psi\Big(\frac{u-u_\tau^{k-1}}{\tau}\Big) - \Big\langle f_\tau^k, \frac{u}{\tau}\Big\rangle. \tag{C.2.19}$$

Obviously, any critical point u (and, in particular, a minimizer) of this functional solves (C.2.18) and we put $u = u_\tau^k$. Coercivity and weak lower semicontinuity (as used in Chapters 3–5) are typical arguments behind existence of u_τ^k applied recursively for $k = 1,...,T/\tau$. Then a-priori estimates have to be derived. For some Banach spaces $V \subset H$, assuming

$$\exists \epsilon > 0\ \forall v \in V_1: \quad \Phi(u) \geq \epsilon\|u\|_V^p - 1/\epsilon \quad \& \quad \langle \Psi'(v), v\rangle \geq \epsilon\|v\|_{V_1}^q, \tag{C.2.20a}$$

$$f \in L^{q'}(I; V_1^*), \qquad u_0 \in V, \qquad \Phi(u_0) < \infty, \tag{C.2.20b}$$

from (C.2.2) we get the inequality

$$\epsilon\|\dot{u}\|_{V_1}^q + \frac{\mathrm{d}}{\mathrm{d}t}\Phi(u) \leq \langle f, \dot{u}\rangle \leq \frac{\epsilon}{2}\|\dot{u}\|_{V_1}^q + C_{q,\epsilon}\|f(t)\|_{V_1^*}^{q'} \tag{C.2.21}$$

with $C_{q,\epsilon}$ sufficiently large[12] so that $\frac{\epsilon}{2}\|\dot{u}\|_{V_1}^q + \frac{\mathrm{d}}{\mathrm{d}t}\Phi(u) \leq C_{q,\epsilon}\|f(t)\|_{V_1}^{q'}$. Integrating it over time, we obtain $\int_0^t \|\dot{u}\|_{V_1}^q\,\mathrm{d}t + \Phi(u(t)) \leq C_{q,\epsilon}\|f\|_{L^{q'}(I;V_1)}^{q'} + \Phi(u_0)$, from which we obtain the a-priori estimate:

$$\|u\|_{L^\infty(I;V)\cap W^{1,q}(I;V_1)} \leq C \tag{C.2.22}$$

[12] Using Hölder's inequality, one can identify more specifically $C_{q,\epsilon} \geq 2^{q-1}(q-1)q^{-q}/\epsilon^{q-1}$.

for some $C<\infty$. Alternative estimation may rely on a by-part integration of (C.2.21) before making the estimation of the right-hand side, i.e.

$$\begin{aligned}\epsilon\int_0^t \|\dot u\|_{V_1}^q\,\mathrm{d}t + \Phi(u(t)) &\le \langle f(t),u(t)\rangle + \int_0^t \langle \dot f,u\rangle\,\mathrm{d}t + \Phi(u_0) - \langle f(0),u_0\rangle\\ &\le \frac{\epsilon}{2}\|u(t)\|_V^p + \int_0^t \|\dot f(t)\|_{V^*}(1+\|u(t)\|_V^p)\,\mathrm{d}t\\ &\quad + C_{q,\epsilon}\|f(t)\|_{V^*}^{p'} + \Phi(u_0) - \langle f(0),u_0\rangle. \end{aligned}\tag{C.2.23}$$

Then (C.2.22) can be obtained when $f\in W^{1,1}(I;V^*)$.

Although (C.2.2) together with (C.2.21) is an important and lucid heuristic, we should emphasize that the test (C.2.2) is unfortunately only formal because, counting with that $\Phi'(u(\cdot))$ is valued in V^*, one can hardly expect $\dot u$ valued in V, and similarly for Ψ too. On the other hand, u itself is valued in V and the discrete analog of (C.2.2) is legal: instead of testing (C.2.1) by $\dot u$, we test (C.2.18) by the time difference $(u_\tau^k-u_\tau^{k-1})/\tau$ and thus, instead of (C.2.2), we obtain

$$\begin{aligned}\underbrace{\Big\langle \Phi'(u_\tau^k),\frac{u_\tau^k-u_\tau^{k-1}}{\tau}\Big\rangle}_{\ge \frac{\Phi(u_\tau^k)-\Phi(u_\tau^{k-1})}{\tau}\ \text{if }\Phi\text{ is convex}} + \underbrace{\Big\langle \Psi'\Big(\frac{u_\tau^k-u_\tau^{k-1}}{\tau}\Big),\frac{u_\tau^k-u_\tau^{k-1}}{\tau}\Big\rangle}_{\ge \epsilon\|\frac{u_\tau^k-u_\tau^{k-1}}{\tau}\|_{V_1}^q\ \text{due to (C.2.20a)}} &= \Big\langle f_\tau^k,\frac{u_\tau^k-u_\tau^{k-1}}{\tau}\Big\rangle\\ &\le C_{q,\epsilon}\|f_\tau^k\|_{V_1^*}^{q'} + \frac{\epsilon}{2}\Big\|\frac{u_\tau^k-u_\tau^{k-1}}{\tau}\Big\|_{V_1}^q. \end{aligned}\tag{C.2.24}$$

The convexity is exploited for $\langle\Phi'(u_\tau^k),u_\tau^k-u_\tau^{k-1}\rangle\ge\Phi(u_\tau^k)-\Phi(u_\tau^{k-1})$, cf. the inequality in (A.3.14) for $\xi=\Phi'(u_\tau^k)$, $v=u_\tau^k$, and $\tilde v=u_\tau^{k-1}-u_\tau^k$. Estimating the right-hand side as in (C.2.21), one obtains a-priori estimate of the type (C.2.21) for the discrete solution u_τ. In specific situations, some more sophisticated estimation techniques are to be considered.

Then convergence as $\tau\to0$ is to be proved by various methods, typically, a combination of the arguments based on weak lower semicontinuity or compactness is used, depending whether Ψ or Φ is quadratic or not; cf. e.g. [456, Sect. 11.1]. In addition, Ψ does not need to be smooth and, assuming Ψ convex and referring to the definition of the convex subdifferential (A.3.14), we can easily investigate the set-valued variational inclusion

$$\partial\Psi(\dot u)+\Phi'(u)\ni f(t)\quad\text{for }t\in I,\quad u\big|_{t=0}=u_0.\tag{C.2.25}$$

In (C.2.18), Ψ' replaces by $\partial\Psi$ while (C.2.19) remains the same.

In specific situations, the backward-Euler scheme (C.2.18) can be advantageously modified in various ways. E.g., in case $\Psi=\Psi_1+\Psi_2$, $\Phi=\Phi_1+\Phi_2$ and $f=f_1+f_2$, one can apply the *fractional-step method*, called in engineering literature alternatively as *staggered scheme*, alternatively to be understood as a Lie-Trotter (or *sequential*) *splitting* combined with the backward Euler formula:

$$\Psi_1\Big(\frac{u_\tau^{k-1/2}-u_\tau^{k-1}}{\tau}\Big)+\Phi_1'(u_\tau^{k-1/2}) = f_{1,\tau}^k := \int_{(k-1)\tau}^{k\tau} f_1(t)\,\mathrm{d}t, \tag{C.2.26a}$$

$$\Psi_2\Big(\frac{u_\tau^{k}-u_\tau^{k-1/2}}{\tau}\Big) + \Phi_2'(u_\tau^k) \ni f_{2,\tau}^k := \int_{(k-1)\tau}^{k\tau} f_2(t)\,\mathrm{d}t, \tag{C.2.26b}$$

with $k = 1,...,T/\tau$. Clearly, (C.2.26) leads to two variational problems that are to be solved in alternation.

In case $\Phi : V = U\times Z \to \mathbb{R}$, $\Psi = \Psi_1 + \Psi_2 : V = U\times Z \to \mathbb{R}$ with $\Psi_1 : U \to \mathbb{R}$, $\Psi_2 : Z \to \mathbb{R}$, $f = (g,h)$ and the couple (u,z) in place of u in (C.2.1), we are to deal with a system of two equations

$$\partial\Psi_1(\dot u) + \partial_u\Phi(u,z) = g, \qquad u(0) = u_0, \tag{C.2.27a}$$

$$\partial\Psi_2(\dot z) + \partial_z\Phi(u,z) = h, \qquad z(0) = z_0, \tag{C.2.27b}$$

with $\partial_u\Phi$ and $\partial_z\Phi$ denoting partial differentials, one can thus think also about the splitting $\Phi'-f = (\partial_u\Phi-g, \partial_z\Phi-h) = (\partial_u\Phi-g,0) + (0,\partial_z\Phi-h)$. Then the fractional-step method such as (C.2.26) yields a *semi-implicit scheme*[13]

$$\partial\Psi_1\Big(\frac{u_\tau^k-u_\tau^{k-1}}{\tau}\Big)+\partial_u\Phi(u_\tau^k,z_\tau^{k-1}) = g_\tau^k, \qquad u_\tau^0 = u_0, \tag{C.2.28a}$$

$$\partial\Psi_2\Big(\frac{z_\tau^k-z_\tau^{k-1}}{\tau}\Big)+\partial_z\Phi(u_\tau^k,z_\tau^k) = h_\tau^k, \qquad z_\tau^0 = z_0, \tag{C.2.28b}$$

again for $k = 1,...,T/\tau$. Note that the use of z_τ^{k-1} in (C.2.28a) decouples the system (C.2.28), in contrast to the backward-Euler formula which would use z_τ^k in (C.2.28a) and would not decouple the original system (C.2.27). The underlying variational problems for the functionals

$$u \mapsto \frac{\Phi(u,z_\tau^{k-1})}{\tau} + \Psi_1\Big(\frac{u-u_\tau^{k-1}}{\tau}\Big) - \Big\langle g_\tau^k, \frac{u}{\tau}\Big\rangle \quad\text{and}\quad z \mapsto \frac{\Phi(u_\tau^k,z)}{\tau} + \Psi_2\Big(\frac{z-z_\tau^{k-1}}{\tau}\Big) - \Big\langle h_\tau^k, \frac{z}{\tau}\Big\rangle$$

represent recursive alternating variational problems; these particular problems can be convex even if Φ itself is only separately convex[14]. Besides, under certain relatively weak conditions, this semi-implicit discretization is "numerically" stable.[15]

Of course, this decoupling method can be advantageously applied to nonsmooth situations and for u with more than 2 components, i.e. for systems of more than two equations or inclusions. Even more, the splitting like (C.2.27) may yield a "bi-

[13] Indeed, in (C.2.26), one is to take $(u_\tau^{k-1},z_\tau^{k-1})$, (u_τ^k,z_τ^{k-1}), and eventually (u_τ^k,z_τ^k) in place of u_τ^{k-1}, $u_\tau^{k-1/2}$, and u_τ^k, respectively.

[14] This means that only $\Phi(u,\cdot)$ and $\Phi(\cdot,z)$ are convex but not necessarily $\Phi(\cdot,\cdot)$.

[15] Cf. also e.g. [456, Rem. 8.25] for more details.

variational" structure of the decoupled incremental problems even if the original problem of the type $\dot{u}+A(u)\ni 0$ itself does not have potential.

Let us just remark that, if both Ψ and Φ are nonsmooth, one can again use (C.2.19) which then leads, similarly as (C.2.18), to the formula

$$\partial\Psi\Big(\frac{u_\tau^k-u_\tau^{k-1}}{\tau}\Big)+\partial\Phi(u_\tau^k)\ni f_\tau^k. \tag{C.2.29}$$

Remark C.2.3 (Global-minimization strategy). The energetically motivated estimate (C.2.22) is based on the attribute of u_τ^k being a critical point of the functional (C.2.19) and needs convexity of Φ. For global minimizers of (C.2.19), one can alternatively also compare the value of this functional with its value at u_τ^{k-1}, which gives the inequality

$$\frac{\Phi(u_\tau^k)}{\tau}+\Psi\Big(\frac{u_\tau^k-u_\tau^{k-1}}{\tau}\Big)\le\frac{\Phi(u_\tau^{k-1})}{\tau}+\underbrace{\Psi\Big(\frac{u_\tau^{k-1}-u_\tau^{k-1}}{\tau}\Big)}_{=0}+\Big\langle f_\tau^k,\frac{u_\tau^k-u_\tau^{k-1}}{\tau}\Big\rangle$$

even without assuming convexity of Φ. One thus obtains again (C.2.22) but with the dissipation rate replaced by its potential, which allows for the same usage as far as a-priori estimates concern. In particular, if Ψ is p-homogeneous, the factor p occurs in the modified estimate (C.2.22) as the dissipation rate multiplied by the factor p as $\langle\Psi'(\dot{u}),\dot{u}\rangle=p\Psi(\dot{u})$. In particular, both estimates coincides if Ψ is 1-homogeneous.

Remark C.2.4 (Semiconvex potential Φ).[16] The convexity of Φ assumed in (C.2.22) can be relaxed by assuming the functional Φ to be only *semiconvex* with respect to the norm of a Hilbert space H on which Ψ is coercive in the sense

$$\exists\varepsilon_0,\varepsilon_1>0:\ \ v\mapsto\Phi(v)+\|v\|_H^2/\varepsilon_0:V\to\mathbb{R}\ \text{ is convex and }\ \langle\Psi'(v),v\rangle\ge\varepsilon_1\|v\|^2. \tag{C.2.30}$$

Then, for τ sufficiently small, namely $0<\tau\le(2/\varepsilon_0)^{-2}$, the estimation (C.2.22) can be modified by using

$$\begin{aligned}
\Big\langle\Psi'\Big(\frac{u_\tau^k-u_\tau^{k-1}}{\tau}\Big)+\Phi'(u_\tau^k),u_\tau^k-u_\tau^{k-1}\Big\rangle&\ge\Big\langle\varepsilon_0\frac{u_\tau^k-u_\tau^{k-1}}{\tau}+\Phi'(u_\tau^k),u_\tau^k-u_\tau^{k-1}\Big\rangle\\
&=\Big\langle\varepsilon_0\frac{u_\tau^k}{\sqrt{\tau}}+\Phi'(u_\tau^k),u_\tau^k-u_\tau^{k-1}\Big\rangle-\Big\langle\varepsilon_0\frac{u_\tau^{k-1}}{\sqrt{\tau}},u_\tau^k-u_\tau^{k-1}\Big\rangle+\varepsilon_0\frac{1-\sqrt{\tau}}{\tau}\|u_\tau^k-u_\tau^{k-1}\|_H^2\\
&\ge\frac{\varepsilon_0}{2\sqrt{\tau}}\|u_\tau^k\|_H^2+\Phi(u_\tau^k)-\frac{\varepsilon_0}{2\sqrt{\tau}}\|u_\tau^{k-1}\|_H^2-\Phi(u_\tau^{k-1})\\
&\qquad-\Big\langle\varepsilon_0\frac{u_\tau^{k-1}}{\sqrt{\tau}},u_\tau^k-u_\tau^{k-1}\Big\rangle+\varepsilon_0\frac{1-\sqrt{\tau}}{\tau}\|u_\tau^k-u_\tau^{k-1}\|_H^2\\
&=\Phi(u_\tau^k)-\Phi(u_\tau^{k-1})+\varepsilon_0\tau\Big(1-\frac{\sqrt{\tau}}{2}\Big)\Big\|\frac{u_\tau^k-u_\tau^{k-1}}{\tau}\Big\|_H^2.
\end{aligned} \tag{C.2.31}$$

[16] See [456, Remark 8.24].

This refines Rothe's method and, in this respect, brings it closer to the Galerkin method which, when executing the estimation strategy (C.2.23), does not need any convexity of Φ at all; cf. Sect. C.2.4 below.

Remark C.2.5 (Higher-order formulas). The backward-Euler formula (C.2.18) or (C.2.29) serves well for theoretical purposes but computationally may not be optimal. A "cheap" modification is the so-called *Crank-Nicolson formula* [133] ((also called a *mid-point formula* or a *trapezoidal rule*):

$$\Psi'\Big(\frac{u_\tau^k - u_\tau^{k-1}}{\tau}\Big) + \Phi'\Big(\frac{u_\tau^k + u_\tau^{k-1}}{2}\Big) = \frac{f_\tau^k + f_\tau^{k-1}}{2} \tag{C.2.32}$$

or alternatively also $\Psi'\big(\frac{u_\tau^k - u_\tau^{k-1}}{\tau}\big) + \frac{1}{2}\Phi'(u_\tau^k) + \frac{1}{2}\Phi'(u_\tau^{k-1}) \ni \frac{1}{2}f_\tau^k + \frac{1}{2}f_\tau^{k-1}$, which is the same if Φ is quadratic. Note that, if Φ is quadratic and Ψ is p-homogeneous, (C.2.32) conserves energy even for the discrete solution in contrast to (C.2.18), namely the equality

$$p\Psi\Big(\frac{u_\tau^k - u_\tau^{k-1}}{\tau}\Big) + \frac{\Phi(u_\tau^k) - \Phi(u_\tau^{k-1})}{\tau} = \Big\langle \frac{f_\tau^k + f_\tau^{k-1}}{2}, \frac{u_\tau^k - u_\tau^{k-1}}{\tau}\Big\rangle. \tag{C.2.33}$$

holds just by testing (C.2.32) by $u_\tau^k - u_\tau^{k-1}$. Another computationally "cheap" option is to use a multi-level formula. The simplest one is the 3-level *Gear's formula* [215]:

$$\Psi'\Big(\frac{3u_\tau^k - 4u_\tau^{k-1} + u_\tau^{k-2}}{2\tau}\Big) + \Phi'(u_\tau^k) = f_\tau^k, \quad k \geq 2, \tag{C.2.34}$$

while for $k = 1$ one is to use (C.2.18). This formula approximates the time derivative with a higher order, may yield a better error estimate if a solution is enough regular, and may simultaneously have good stability properties.[17]

Remark C.2.6 (Linear dissipation Ψ'). If Ψ is quadratic and coercive on H, one can weaken (C.2.20a) to require $\langle \Phi'(u), u\rangle \geq \epsilon\|u\|_V^p - \|u\|_H^2/\epsilon$ and $\Psi'(v) \geq \epsilon\|v\|_H^2$. Then we can perform also the test by u itself, obtaining $\frac{\mathrm{d}}{\mathrm{d}t}\Psi(u) + \langle \Phi'(u), u\rangle = \langle f, u\rangle \leq \frac{\varepsilon}{2}\|u\|_V^p + C_{p,\varepsilon}\|f\|_{V^*}$, which gives in particular the estimate $\|u\|_{L^\infty(I;H)} \leq C$, which can be used in (C.2.21) so that again (C.2.22) holds now with $q = 2$.

[17] For a linear case, see [451].

C.2.3 Second-order evolution problems: time discretization

The Rothe method (i.e. fully implicit time-discretization by the backward-Euler formula) for (C.2.3) now uses more ultimately equidistant[18] partition with a time step $\tau > 0$ (again assuming $T/\tau \in \mathbb{N}$) and results in the formula

$$\mathscr{T}'\frac{u_\tau^k - 2u_\tau^{k-1} + u_\tau^{k-2}}{\tau^2} + \partial\Psi\Big(\frac{u_\tau^k - u_\tau^{k-1}}{\tau}\Big) + \Phi'(u_\tau^k) \ni f_\tau^k \tag{C.2.35a}$$

to be executed recursively for $k = 1, ..., T/\tau$, starting for $k = 1$ with

$$u_\tau^0 = u_0 \quad \text{and} \quad u_\tau^{-1} = u_0 - \tau v_0. \tag{C.2.35b}$$

The analog of (C.2.19) is now the recursive variational problem for the functional

$$u \mapsto \frac{\Phi(u)}{\tau} + \Psi\Big(\frac{u - u_\tau^{k-1}}{\tau}\Big) + \tau\mathscr{T}\Big(\frac{u - 2u_\tau^{k-1} + u_\tau^{k-2}}{\tau^2}\Big) - \Big\langle f_\tau^k, \frac{u}{\tau}\Big\rangle. \tag{C.2.36}$$

For the weak formulation arising by one by-part integration in time of the type (6.2.7) or (7.2.5a), one then uses (C.2.13) written for the time-differences

$$\begin{aligned}\tau\sum_{k=1}^{T/\tau}\Big\langle\frac{u_\tau^k - 2u_\tau^{k-1} + u_\tau^{k-2}}{\tau^2}, v_\tau^k\Big\rangle &= \Big\langle\frac{u_\tau^j - u_\tau^{j-1}}{\tau}, v_\tau^j\Big\rangle \\ &- \Big\langle\frac{u_\tau^i - u_\tau^{i-1}}{\tau}, v_\tau^i\Big\rangle - \tau\sum_{k=1}^{T/\tau}\Big\langle\frac{u_\tau^{k-1} - u_\tau^{k-2}}{\tau}, \frac{v_\tau^k - v_\tau^{k-1}}{\tau}\Big\rangle,\end{aligned} \tag{C.2.37}$$

so that (C.2.16) written for time-differences reads as

$$\int_0^T \langle \ddot{\tilde{u}}_\tau, \bar{v}_\tau\rangle\,\mathrm{d}t = \langle \dot{u}_\tau(T), v_\tau(T)\rangle - \langle \dot{u}_\tau(0), v_\tau(0)\rangle - \int_0^T \langle \dot{u}_\tau(\cdot - \tau), \dot{v}_\tau\rangle\,\mathrm{d}t,$$

where $\tilde{u}_\tau$ denotes the piece-wise quadratic interpolant of $\{u_\tau^k\}_{k=-1}^{T/\tau}$ such that

$$\tilde{u}_\tau(k\tau) = u_\tau^k \quad \text{and} \quad \ddot{\tilde{u}}_\tau(t) = \frac{u_\tau^k - 2u_\tau^{k-1} + u_\tau^{k-2})}{\tau^2} \quad \text{for } ((k-1)\tau, k\tau]. \tag{C.2.38}$$

When considering the continuous piecewise-affine interpolant u_τ extended on $(-\tau, 0]$ so that $u_\tau(-\tau) = u_\tau^{-1}$ from (C.2.35b) and Ψ is (for simplicity) smooth, the weak formulation of (C.2.35) reads as

$$\int_0^T \langle \Psi'(\dot{u}_\tau) + \Phi'(\bar{u}_\tau), \bar{v}_\tau\rangle - \langle \mathscr{T}'\dot{u}_\tau(\cdot - \tau), \dot{v}_\tau\rangle\,\mathrm{d}t = \langle \mathscr{T}'v_0, v_\tau(0)\rangle \tag{C.2.39}$$

[18] Actually, the discretization of the term $\mathscr{T}'\ddot{u}$ would be much more involved on non-equidistant partitions. Cf. also Remark C.2.5.

for any $v \in C(I;V)$ with $v(T)=0$, where $\bar{v}_\tau$ and v_τ denote its piecewise-constant and piecewise-affine interpolants, respectively. For Ψ nonsmooth, one should combine the equation (C.2.39) with an inequality arising from the subdifferential formulation, cf. (7.2.5b) or (7.2.8).

The analog of (C.2.14) in terms of time-differences to be used for H considered equipped by the (semi) norm $\mathscr{T}^{1/2}$ looks as

$$\frac{1}{2}\Big\|\frac{u_\tau^j - u_\tau^{j-1}}{\tau}\Big\|_H^2 - \frac{1}{2}\Big\|\frac{u_\tau^i - u_\tau^{i-1}}{\tau}\Big\|_H^2 \le \tau \sum_{k=i+1}^{j} \Big\langle \frac{u_\tau^k - 2u_\tau^{k-1} + u_\tau^{k-2}}{\tau^2}, \frac{u_\tau^k - u_\tau^{k-1}}{\tau}\Big\rangle. \quad \text{(C.2.40)}$$

Assuming again (C.2.20) with $v_0 \in H$ and the coercivity of the quadratic form $\mathscr{T}: H \to \mathbb{R}$ and testing (C.2.35) by the time-difference $u_\tau^k - u_\tau^{k-1}$, in terms of the interpolants (C.2.15) and $\ddot{\bar{y}}_\tau$, we obtain the a-priori estimates

$$\|u_\tau\|_{L^\infty(I;V)\cap W^{1,q}(I;V_1)\cap W^{1,\infty}(I;H)} \le C, \quad \text{(C.2.41a)}$$

$$\|\mathscr{T}'\ddot{\bar{u}}_\tau\|_{L^\infty(I;V^*)+L^2(I;V_1^*)} \le C \quad \text{and} \quad \|\mathscr{T}'\ddot{u}_\tau\|_{\mathfrak{M}(I;V^*)} \le C. \quad \text{(C.2.41b)}$$

Then convergence (in terms of subsequences) as $\tau \to 0$ towards weak solutions is to be proved by various methods, typically, a combination of the arguments based on weak lower semicontinuity or compactness is used, depending whether Ψ or Φ is quadratic or not; cf. e.g. [456, Sect. 11.3].

Remark C.2.7 (Energy-conserving schemes). The backward-Euler formula (C.2.35) serves well for theoretical purposes but, like said in Remark C.2.5, may not be optimal numerical. Even worse, due to the artificial numerical attenuation by the term $\tau^2\sum_{k=i+1}^{j}\|(u_\tau^k - 2u_\tau^{k-1} + u_\tau^{k-2})/\tau^2\|_H^2 \ge 0$ (= the difference of the right-hand and the left-hand parts) in (C.2.40), the formula (C.2.35) practically cannot be used for realistic calculation of wave propagation unless $\tau > 0$ is made extremely small. Many other methods have been devised for wave propagation. A simple method consists in application of the *Crank-Nicolson formula* to (C.2.3) transformed to the system of two 1st-order equations[19]

$$\dot{u} = v, \quad \mathscr{T}'\dot{v} + \Psi'(v) + \Phi'(u) = f(t) \quad \text{for } t \in I, \qquad u|_{t=0} = u_0, \quad v|_{t=0} = v_0. \quad \text{(C.2.42)}$$

This mid-point strategy results to the system for the couple (u_τ^k, v_τ^k):

$$\frac{u_\tau^k - u_\tau^{k-1}}{\tau} = \frac{v_\tau^k + v_\tau^{k-1}}{2}, \qquad u_\tau^0 = u_0, \quad \text{(C.2.43a)}$$

$$\mathscr{T}'\frac{v_\tau^k - v_\tau^{k-1}}{\tau} + \Psi'\Big(\frac{v_\tau^k + v_\tau^{k-1}}{2}\Big) + \Phi'\Big(\frac{u_\tau^k + u_\tau^{k-1}}{2}\Big) = f_\tau^k, \qquad v_\tau^0 = v_0, \quad \text{(C.2.43b)}$$

[19] This approximation was suggested in engineering literature e.g. in [407, 531], possibly even for nonlinear problems. Actually, it falls into a broader class of so-called Hilbert-Hughes-Taylor formulas widening the popular Newmark method [394] as a special choice of parameter, namely $\alpha = \beta = 1/2$ and $\gamma = 1$, cf. [249].

to be solved for $k = 1,...,T/\tau$. In contrast to (C.2.35), this system does not satisfy the symmetry condition (A.3.6) and thus does not have any potential, but eliminating v_τ^k by substituting $v_\tau^k = \frac{2}{\tau}(u_\tau^k + u_\tau^{k-1}) - v_\tau^{k-1}$ into (C.2.43b), one again obtains a potential problem for u_τ^k. To be more specific, instead of (C.2.36), u_τ^k is now a minimizer of the functional

$$u \mapsto \frac{2}{\tau}\Phi\Big(\frac{u+u_\tau^{k-1}}{2}\Big) + \Psi\Big(\frac{u-u_\tau^{k-1}}{\tau}\Big) + 2\tau\mathscr{T}\Big(\frac{u-\tau v_\tau^{k-1}-u_\tau^{k-1}}{\tau^2}\Big) - \Big\langle f_\tau^k, \frac{u}{\tau}\Big\rangle. \tag{C.2.44}$$

If not only $\mathscr{T}$ but also Φ is quadratic and if Ψ is p-homogeneous, by testing (C.2.43b) by $v_\tau^k + v_\tau^{k-1}$ and substituting also $v_\tau^k + v_\tau^{k-1} = \frac{2}{\tau}(u_\tau^k - u_\tau^{k-1})$ due to (C.2.43a), we obtain the equality

$$\frac{\mathscr{T}(v_\tau^k) - \mathscr{T}(v_\tau^{k-1})}{\tau} + p\Psi\Big(\frac{v_\tau^k + v_\tau^{k-1}}{2}\Big) + \frac{\Phi(u_\tau^k) - \Phi(u_\tau^{k-1})}{\tau} = \Big\langle f_\tau^k, \frac{v_\tau^k + v_\tau^{k-1}}{2}\Big\rangle, \tag{C.2.45}$$

which shows that the discrete scheme (C.2.43) conserves energy, cf. (C.2.4); in particular if $\Psi = 0$ and $f = 0$ then $\mathscr{T}(v_\tau^k) + \Phi(u_\tau^k) =$constant. Another advantage of (C.2.43) in comparison to (C.2.35) is a possibility to vary the time step during evolution adaptively with respect to suitable strategies, which is often desirable to capture different time scales occurring during evolution. Using the interpolants from (C.2.15), the scheme (C.2.43a) can be written analogously to (C.2.42) as

$$\dot u_\tau = \underline{\overline{v}}_\tau, \quad \& \quad \mathscr{T}'\dot v_\tau + \Psi'(\underline{\overline{v}}_\tau) + \Phi'(\underline{\overline{u}}_\tau) = \overline{f}_\tau \ \text{ for } t \in I, \qquad u_\tau\big|_{t=0} = u_0, \ \ v_\tau\big|_{t=0} = v_0, \tag{C.2.46}$$

while (C.2.45) gives, after summation for $k = 1,...,l$, the equality

$$\mathscr{T}(v_\tau(t)) + \Phi(u_\tau(t)) + p\int_0^t \Psi(\underline{\overline{v}}_\tau)\,\mathrm{d}t = \mathscr{T}(v_0) + \Phi(u_0) + \int_0^t \langle \overline{f}_\tau, \underline{\overline{v}}_\tau\rangle\,\mathrm{d}t \tag{C.2.47}$$

with $t = l\tau$. Realizing that $\underline{\overline{v}}_\tau$ is piecewise constant, let us note that $\int_0^t \langle \overline{f}_\tau, \underline{\overline{v}}_\tau\rangle \mathrm{d}t = \int_0^t \langle f, \underline{\overline{v}}_\tau\rangle \mathrm{d}t$ due to the definition of f_τ^k and thus of $\overline{f}_\tau$. Assuming coercivity and quadratic growth of Φ on V and of Ψ on V_1 and $\mathscr{T}$ on H, and furthermore $f \in L^2(I;V_1)$, $u_0 \in V$, and $v_0 \in H$, from we obtain the a-priori estimates

$$\|v_\tau\|_{L^\infty(I;H)\cap L^2(I;V_1)} \le C, \qquad \|u_\tau\|_{L^\infty(I;V)} \le C, \tag{C.2.48a}$$

$$\|\dot u_\tau\|_{L^\infty(I;H)\cap L^2(I;V_1)} \le C, \qquad \|\dot v_\tau\|_{L^\infty(I;V^*)+L^2(I;V_1^*)} \le C, \tag{C.2.48b}$$

the last one being by comparison from (C.2.46). Note however that v_τ is not the velocity corresponding to u_τ, i.e. $\overline{v}_\tau \neq \dot u_\tau$ in general, although in the limit $\overline{v}_\tau - \dot u_\tau \to 0$ for $\tau \to 0$ and the relation $\dot u_\tau = \underline{\overline{v}}_\tau$ must be taken into account as a vital part of (C.2.47).

Remark C.2.8 (Fractional-step strategies). The fractional-step method and in particular various semi-implicit variants of (C.2.29) and (C.2.35) are widely applicable for the 2nd-order problems, too. Let us demonstrate it on the system (C.2.27) modified by admitting inertia only in one variable, say u:

$$\mathscr{T}'\ddot{u}+\partial\Psi_1(\dot{u})+\partial_u\Phi(u,z)=g, \qquad u|_{t=0}=u_0, \quad \dot{u}|_{t=0}=v_0, \tag{C.2.49a}$$

$$\partial\Psi_2(\dot{z})+\partial_z\Phi(u,z)=h, \qquad z|_{t=0}=z_0, \tag{C.2.49b}$$

with $\Phi: V{\times}Z\to\mathbb{R}$, $\Psi_1: V_1\to\mathbb{R}$, and $\Psi_2: Z_1\to\mathbb{R}$ for some Banach spaces V, Z, V_1, and Z_1. Combining (C.2.28) with (C.2.43), one obtains the *semi-implicit scheme* for the decoupled system for the pair (u_τ^k, v_τ^k) and for z_τ^k:[20]

$$\frac{u_\tau^k-u_\tau^{k-1}}{\tau}=\frac{v_\tau^k+v_\tau^{k-1}}{2}, \qquad u_\tau^0=u_0, \tag{C.2.50a}$$

$$\mathscr{T}'\frac{v_\tau^k-v_\tau^{k-1}}{\tau}+\Psi_1'\Big(\frac{v_\tau^k+v_\tau^{k-1}}{2}\Big)+\partial_u\Phi\Big(\frac{u_\tau^k+u_\tau^{k-1}}{2},z_\tau^{k-1}\Big)=g_\tau^k, \qquad v_\tau^0=v_0, \tag{C.2.50b}$$

$$\Psi_2'\Big(\frac{z_\tau^k-z_\tau^{k-1}}{\tau}\Big)+\partial_z\Phi(u_\tau^k,z_\tau^k)=h_\tau^k, \qquad z_\tau^0=z_0. \tag{C.2.50c}$$

Eliminating v_τ^k by substituting $v_\tau^k=\frac{2}{\tau}(u_\tau^k+u_\tau^{k-1})-v_\tau^{k-1}$, both (C.2.50b) and (C.2.50c) get the variational structure with underlying potentials combining (C.2.19) and (C.2.44), namely

$$u\mapsto\frac{2}{\tau}\Phi\Big(\frac{u+u_\tau^{k-1}}{2},z_\tau^{k-1}\Big)+\Psi_1\Big(\frac{u-u_\tau^{k-1}}{\tau}\Big)+2\tau\mathscr{T}\Big(\frac{u-\tau v_\tau^{k-1}-u_\tau^{k-1}}{\tau^2}\Big)-\Big\langle g_\tau^k,\frac{u}{\tau}\Big\rangle, \tag{C.2.51a}$$

$$z\mapsto\frac{\Phi(u_\tau^k,z)}{\tau}+\Psi_2\Big(\frac{z-z_\tau^{k-1}}{\tau}\Big)-\Big\langle h_\tau^k,\frac{z}{\tau}\Big\rangle. \tag{C.2.51b}$$

Moreover, testing the particular equations (C.2.50b,c) successively by $(v_\tau^k+v_\tau^{k-1})/2$ and $z_\tau^k-z_\tau^{k-1}$, we obtain the estimates

$$\frac{\mathscr{T}(v_\tau^k)-\mathscr{T}(v_\tau^{k-1})}{\tau}+p_1\Psi_1\Big(\frac{v_\tau^k+v_\tau^{k-1}}{2}\Big)+\frac{\Phi(u_\tau^k,z_\tau^{k-1})-\Phi(u_\tau^{k-1},z_\tau^{k-1})}{\tau}=\Big\langle g_\tau^k,\frac{v_\tau^k+v_\tau^{k-1}}{2}\Big\rangle, \tag{C.2.52a}$$

$$p_2\Psi_2\Big(\frac{z_\tau^k-z_\tau^{k-1}}{\tau}\Big)+\frac{\Phi(u_\tau^k,z_\tau^k)-\Phi(u_\tau^k,z_\tau^{k-1})}{\tau}\le\Big\langle h_\tau^k,\frac{z_\tau^k-z_\tau^{k-1}}{\tau}\Big\rangle, \tag{C.2.52b}$$

provided we assume $\Phi(\cdot,z)$ quadratic and $\Phi(u,\cdot)$ convex and Ψ_i p_i-homogeneous. Summing it up, we benefit from the cancellation of the terms $\pm\Phi(u_\tau^k,z_\tau^{k-1})$ and obtain the energy-type estimate

$$\frac{\mathscr{T}(v_\tau^k)-\mathscr{T}(v_\tau^{k-1})}{\tau}+p_1\Psi_1\Big(\frac{v_\tau^k+v_\tau^{k-1}}{2}\Big)+p_2\Psi_2\Big(\frac{z_\tau^k-z_\tau^{k-1}}{\tau}\Big)+\frac{\Phi(u_\tau^k,z_\tau^k)-\Phi(u_\tau^{k-1},z_\tau^{k-1})}{\tau}\le\Big\langle g_\tau^k,\frac{v_\tau^k+v_\tau^{k-1}}{2}\Big\rangle+\Big\langle h_\tau^k,\frac{z_\tau^k-z_\tau^{k-1}}{\tau}\Big\rangle. \tag{C.2.53}$$

[20] Such componentwise-split Crank-Nicolson method is sometimes referred to as a 2nd-order Yanenko scheme [535].

Like in Remark C.2.5, we can write (C.2.50) more "compactly" as

$$\dot{u}_\tau = \underline{v}_\tau, \quad \mathscr{T}'\dot{v}_\tau + \Psi_1'(\underline{v}_\tau) + \partial_u\Phi(\underline{\overline{u}}_\tau, \underline{z}_\tau) = \overline{g}_\tau, \quad \Psi_2'(\dot{z}_\tau) + \partial_z\Phi(\overline{u}_\tau, \overline{z}_\tau) = \overline{h}_\tau, \quad \text{(C.2.54)}$$

to hold on the time interval I together with the initial conditions $u|_{t=0} = u_0$, $v|_{t=0} = v_0$, and $z|_{t=0} = z_0$, while (C.2.53) can be written after summation $k = 1,...,l$ as

$$\mathscr{T}(v_\tau(t)) + \Phi(u_\tau(t), z_\tau(t)) + \int_0^t p_1\Psi_1(\underline{v}_\tau) + p_2\Psi_2(\dot{z}_\tau)\,dt \\ \leq \mathscr{T}(v_0) + \Phi(u_0, z_0) + \int_0^t \langle\overline{g}_\tau, \underline{v}_\tau\rangle + \langle\overline{h}_\tau, \dot{z}_\tau\rangle\,dt$$

with $t = l\tau$. The a-priori estimates (C.2.48) hold now together with the estimate $\|z_\tau\|_{L^\infty(I;Z)\cap H^1(I;Z_1)} \leq C$ provided Φ is coercive on $V{\times}Z$ and also Φ_2 has a quadratic growth.

C.2.4 Faedo-Galerkin method

As already advertise, the alternative method to analyze evolution problems consist in discretization of V, referred to as a *Faedo-Galerkin method* [181, 212, 513], or mostly briefly as *Galerkin method*. In contrast to the Rothe method, it approximates the evolutionary problems by approximation of the Banach space V while keeping time continuous. In concrete problems, instead of boundary-value problems for partial differential equations or inequalities arising by Rothe method, Galerkin's method leads to initial-value problems for ordinary-differential equations.

Let us first present it on the first-order doubly-nonlinear problem (C.2.1). The method uses the countable number of subspaces $(V_k)_{k\in\mathbb{N}}$ satisfying

$$V_k \subset V_{k+1} \subset V \qquad \& \qquad \mathrm{cl}\Big(\bigcup_{k\in\mathbb{N}} V_k\Big) = V. \tag{C.2.55}$$

Typically, all V_k are finite-dimensional Euclidean spaces. The approximate variant of (C.2.1) is determined by

$$\forall t\in I\ \forall v\in V_k: \qquad \langle\Psi'(\dot{u}_k) + \Phi'(u_k), v\rangle = \langle f(t), v\rangle, \qquad u_k\big|_{t=0} = u_{0k}, \tag{C.2.56}$$

where $u_{0k} \in V_k$ is a suitable approximation of the original initial condition $u_0 \in V$. Using a special choice $v = u_k(t)$, we obtain the discrete analog of the energy balance (C.2.2).

Taking a base of V_k, say $\{v_{ki}\}_{i=1,...,n_k}$ with $n_k := \dim(V_k)$, and considering the ansatz $u_k(t) = \sum_{i=1}^{n_k} c_{ki}(t)v_{ki}$, (C.2.56) represents an initial-value problem for a system of n_k ordinary differential equations for the coefficients $(c_i)_{i=1,...,n_k}$.

Under the assumptions (C.2.20b), we obtain again the estimate (C.2.21) and then (C.2.22) for u_k. In contrast to the Rothe method (C.2.22), we do not need convexity of Φ. On the other hand, the Rothe method (C.2.29) or its various modifications as (C.2.26) or (C.2.28) can more immediately handle nonsmooth potentials while the Galerkin method in its standard variant needs some additional smoothing leading to ordinary differential equations instead of inclusions.

The second-order problem (C.2.3) results to

$$\forall t\in I\ \forall v\in V_k: \quad \langle \mathcal{T}'\ddot{u}_k + \Psi'(\dot{u}_k) + \Phi'(u_k), v\rangle = \langle f(t), v\rangle, \ \ u_k\big|_{t=0} = u_{0k}, \ \ \dot{u}_k\big|_{t=0} = v_{0k}, \tag{C.2.57}$$

which is, in fact, a system of $2n_k$ ordinary differential equations if we introduce also the rate variable $\dot{u}_k$ as in (C.2.42). Using a special choice $v = \dot{u}_k(t)$, we obtain the discrete analog of the energy balance (C.2.4).

Avoiding the rather technical manipulation of time differences in Rothe's method in Sect. C.2.3 brings a cost to pay: the 1st and 2nd time derivative of the Galerkin solution of (C.2.56) and of (C.2.57), respectively, do not live in the desired dual space $L^2(I;V)^*$ in general.

In particular, when the Aubin-Lions theorem B.4.10 is to be used, then rather its generalization (see Lemma B.4.12) should be employed with the seminorms naturally arising by using test-functions valued in the finite-dimensional subspaces used for the Galerkin method.

Remark C.2.9 (Handling of $\ddot{u}_k$ by the Hahn-Banach extension). In fact, in the latter case (C.2.57), $\mathcal{T}'\ddot{u}_k$ are linear functionals defined only on functions valued in the finite-dimensional subspaces V_k. Yet, on some conditions, they enjoy an extension complying with the expected a-priori estimates. More in detail, as $\mathcal{T}'\ddot{u}_k \in L^2(I;V_k) \subset L^2(I;V^*) \cong L^2(I;V)^*$ and, in qualified situations, there is a uniform bound $\|\mathcal{T}'\ddot{u}_k\|_{L^2(I;V_k)} \le C$ independent of k. One can consider an extension $\mathfrak{v}_k \in L^2(I;V)^*$ of $\ddot{u}_k$ according the Hahn-Banach theorem A.1.4 with $\|\mathfrak{v}_k\|_{L^2(I;V)^*} = \|\mathcal{T}'\ddot{u}_k\|_{L^2(I;V_k)} \le C$. Then, up to a subsequence, $\mathfrak{v}_k \rightharpoonup \mathfrak{v}$ in $L^2(I;V^*)$. Although in general $\mathfrak{v}_k \ne \mathcal{T}'\ddot{u}_k$, in the limit one has $\mathfrak{v} = \mathcal{T}'\ddot{u}$. This can be seen from

$$\begin{aligned}\langle \mathfrak{v}, v\rangle_{L^2(I;V)^*\times L^2(I;V)} \leftarrow \langle \mathfrak{v}_k, v\rangle_{L^2(I;V)^*\times L^2(I;V)} &= \int_0^T (\mathcal{T}'\ddot{u}_k, v)\,\mathrm{d}t \\ &= -\int_0^T (\mathcal{T}'\dot{u}_k, \dot{v})\,\mathrm{d}t \to -\int_0^T \langle \mathcal{T}'\dot{u}, \dot{v}\rangle_{V^*\times V}\,\mathrm{d}t = -\langle \mathcal{T}'\dot{u}, \dot{v}\rangle,\end{aligned}$$

which holds for any $v \in C_0^1(I;V_l)$ for a sufficiently large k, namely $k \ge l$. Eventually, one uses density of $\bigcup_{l\in\mathbb{N}} C_0^1(I;V_l)$ in $C_0^1(I;V)$ to see that $\mathfrak{v}$ is indeed the distributional derivative of $\mathcal{T}'\dot{u}$.

Appendix D
Parametrized measures

It is a well-known fact that a composition of a weakly converging sequence with a nonlinear function generically does not converge to the composition of a weak limit with this nonlinearity. We can easily demonstrate it on a simple example.

Example D.0.1. For an oscillating sequence $\{\xi_k\}_{k\in\mathbb{N}}$ with $\xi_k : (0,1) \to \{-1,1\}$ such that

$$\xi_k(x) := \begin{cases} 1 & \text{if } \sin(2\pi kx) > 0, \\ -1 & \text{otherwise,} \end{cases} \tag{D.0.1}$$

it is obvious that $\{\xi_k\}_{k\in\mathbb{N}}$ is bounded in $L^\infty(0,1)$ and, it is an easy exercise to show that it weakly* converges to 0. On the other hand, even a very simple nonlinearity as the modulus $|\cdot|$, for instance, does not commute with weak* convergence. Namely, notice that $|\xi_k(x)| = 1$ for $x \in (0,1)$ and, consequently, $\{|\xi_k|\}_{k\in\mathbb{N}}$ converges (even strongly) to 1, so it does not converge weakly* to $|\text{w*-}\lim_{k\to\infty}\xi_k| = 0$.

On the other hand, it is important to describe such a limit and to record fast oscillations of the "original" sequences. A useful tool for this task are *parametrized measures* which encode a much richer piece of information than a mere weak limit. Here we briefly describe two prominent classes. The first one are the so-called Young measures [410, 452, 537] which are tailored to treat just the oscillatory behavior of weakly converging sequences, the second class is a more general tool, called DiPerna-Majda measures in [452], see also [159, 296], which can deal simultaneously with oscillation and concentration phenomena.

D.1 Young measures

For $p \geq 0$ we define the following subspace of the space $C(\mathbb{R}^N)$ of all continuous functions on $\mathbb{R}^N$:

$$C_p(\mathbb{R}^N) = \{\psi \in C(\mathbb{R}^N);\ \lim_{|\xi|\to+\infty} \psi(\xi)/|\xi|^p = 0\}\,.$$

M. Kružík and T. Roubíček, *Mathematical Methods in Continuum Mechanics of Solids*, Interaction of Mechanics and Mathematics,
https://doi.org/10.1007/978-3-030-02065-1

The Young measures on a bounded domain $\Omega\subset\mathbb{R}^d$ are weakly* measurable mappings $x\mapsto\nu_x:\Omega\to\mathfrak{M}(\mathbb{R}^N)$ with values in probability measures.[1] Let us remind that, by the Riesz theorem the space $\mathfrak{M}(\mathbb{R}^N)$, normed by the total variation, is a Banach space which is isometrically isomorphic with $C_0(\mathbb{R}^N)^*$, where $C_0(\mathbb{R}^N)$ stands for the space of all continuous functions $\mathbb{R}^N\to\mathbb{R}$ vanishing at infinity. Let us denote the set of all Young measures by $\mathcal{Y}(\Omega;\mathbb{R}^N)$. It is known that $\mathcal{Y}(\Omega;\mathbb{R}^N)$ is a convex subset of $L^\infty_{\rm w}(\Omega;\mathfrak{M}(\mathbb{R}^N))\cong L^1(\Omega;C_0(\mathbb{R}^N))^*$, where the subscript "w" indicates the property "weakly* measurable". A classical result [537] is that, for every sequence $\{\xi_k\}_{k\in\mathbb{N}}$ bounded in $L^\infty(\Omega;\mathbb{R}^N)$, there exists its subsequence (denoted by the same indices for notational simplicity) and a Young measure $\nu=\{\nu_x\}_{x\in\Omega}\in\mathcal{Y}(\Omega;\mathbb{R}^N)$ such that, for every Carathéodory integrand $\tilde{\psi}:\Omega\times\mathbb{R}^N\to\mathbb{R}$, it holds

$$\lim_{k\to\infty}\int_\Omega\tilde{\psi}(x,\xi_k(x))\,\mathrm{d}x=\int_\Omega\int_{\mathbb{R}^N}\tilde{\psi}(x,s)\nu_x(\mathrm{d}s)\,\mathrm{d}x\,. \tag{D.1.1}$$

Let us denote by $\mathcal{Y}^\infty(\Omega;\mathbb{R}^N)$ the set of all Young measures which are created by this way, i.e. by taking all bounded sequences in $L^\infty(\Omega;\mathbb{R}^N)$.

It is easy to see that the Young measure corresponding to the sequence $\{\xi_k\}_{k\in\mathbb{N}}$ defined in Example D.0.1 is $\nu_x=\frac{1}{2}\delta_{-1}+\frac{1}{2}\delta_1$ for all $x\in(0,1)$. Hence $\nu:=\{\nu_x\}_{x\in\Omega}$ is even independent of x. Such a Young measure is called *homogeneous*.

A generalization of this result was formulated by Schonbek [478] (cf. also [28]): if $1\le p<+\infty$, then for every sequence $\{\xi_k\}_{k\in\mathbb{N}}$ bounded in $L^p(\Omega;\mathbb{R}^N)$ there exists its subsequence (denoted by the same indices) and a Young measure $\nu=\{\nu_x\}_{x\in\Omega}\in\mathcal{Y}(\Omega;\mathbb{R}^N)$ such that

$$\lim_{k\to\infty}\int_\Omega\tilde{\psi}(x,\xi_k(x))\,\mathrm{d}x=\int_\Omega\int_{\mathbb{R}^N}\tilde{\psi}(x,s)\nu_x(\mathrm{d}s)\,\mathrm{d}x \tag{D.1.2}$$

for every Carathéodory integrand $\tilde{\psi}:\Omega\times\mathbb{R}^N\to\mathbb{R}$ satisfying for almost all $x\in\Omega$ that $\tilde{\psi}(x,\cdot)\in C_p(\mathbb{R}^N)$. We say that $\{\xi_k\}$ generates ν if (D.1.1) or (D.1.2) hold.

Let us denote by $\mathcal{Y}^p(\Omega;\mathbb{R}^N)$ the set of all Young measures which are created by this way, i.e. by taking all bounded sequences in $L^p(\Omega;\mathbb{R}^N)$.

In fact, (D.1.2) can be generalized for nonnegative integrands. In this case, we only obtain the inequality sign between the left- and the right-hand side. We have the following results, see [410, Thm.6.11]

Theorem D.1.1 (Young measure representation). *Let $1\le p\le+\infty$ and let $\{\xi_k\}_{k\in\mathbb{N}}$ be bounded in $L^p(\Omega;\mathbb{R}^N)$. Then there exists a subsequence (denoted by the same indices) and a Young measure $\nu=\{\nu_x\}_{x\in\Omega}\in\mathcal{Y}(\Omega;\mathbb{R}^N)$ such that for every nonnegative Carathéodory integrand $\tilde{\psi}:\Omega\times\mathbb{R}^N\to[0;+\infty)$:*

$$\lim_{k\to\infty}\int_\Omega\tilde{\psi}(x,\xi_k(x))\,\mathrm{d}x\ge\int_\Omega\int_{\mathbb{R}^N}\tilde{\psi}(x,s)\nu_x(\mathrm{d}s)\,\mathrm{d}x\,. \tag{D.1.3}$$

The equality in (D.1.3) *occurs if $\{\tilde{\psi}(\cdot,\xi_k)\}$ is equiintegrable in $L^1(\Omega)$; cf.* (B.2.10).

---

[1] The adjective "weakly* measurable" means that, for any $\psi\in C_0(\mathbb{R}^N)$, the mapping $\Omega\to\mathbb{R}:x\mapsto\langle\nu_x,\psi\rangle=\int_{\mathbb{R}^N}\psi(s)\nu_x(\mathrm{d}s)$ is measurable in the usual sense. Cf. also Sect. B.4.

Indeed, the equality cannot be expected in general, as we can construct a sequence $\{\xi_k\}_{k\in\mathbb{N}} \subset L^p(0,2)$, $1 \le p < +\infty$, such that

$$\xi_k(x) := \begin{cases} k^{1/p} & \text{if } 1-1/k \le x \le 1, \\ 0 & \text{otherwise.} \end{cases}$$

Then $\nu_x = \delta_0$ for all $x \in (0,2)$. Taking $\tilde{\psi}(x,s) := |s|^p$ in (D.1.3), we easily see that the left-hand side is equal to one, while the right-hand side vanishes. The reason is that $\{\tilde{\psi}(\cdot,\xi_k)\}_{k\in\mathbb{N}} \subset L^1(0,2)$ is not uniformly integrable; see (B.2.10). Notice that the mentioned integrand is not admissible for (D.1.2) either. Nevertheless, it is well known that every $\nu \in \mathcal{Y}^p(\Omega;\mathbb{R}^N)$, $1 \le p < +\infty$, can be generated by a sequence $\{\xi_k\} \subset L^p(\Omega;\mathbb{R}^d)$ such that $\{|\xi_k|^p\}$ is uniformly integrable. In such case, we call $\{\xi_k\}$ *p-equiintegrable*. We refer to [295] for a proof.

An important subset of $\mathcal{Y}^p(\Omega;\mathbb{R}^N)$ for $N := d\times d$ is the set containing Young measures generated by gradients of $W^{1,p}(\Omega;\mathbb{R}^d)$-maps (the so-called *gradient Young measures*) which will be denoted by $\mathcal{GY}^p(\Omega;\mathbb{R}^{d\times d})$. Its characterization for $1 < p \le +\infty$ was performed by D. Kinderlehrer and P. Pedregal in [271, 272].

Theorem D.1.2 (Characterization of gradient Young measures). *Let $\Omega \subset \mathbb{R}^d$ be a bounded Lipschitz domain, $1 < p \le +\infty$, and $\nu \in \mathcal{Y}^p(\Omega;\mathbb{R}^{d\times d})$. Then there is a bounded sequence $\{y_k\}_{k\in\mathbb{N}} \subset W^{1,p}(\Omega;\mathbb{R}^d)$ such that $\{\nabla y_k\}_{k\in\mathbb{N}}$ generates ν, i.e., $\nu \in \mathcal{GY}^p(\Omega;\mathbb{R}^{d\times d})$, if and only if the following three conditions hold:*

$$\exists y \in W^{1,p}(\Omega;\mathbb{R}^d)\ \forall_{\text{a.a.}} x\in\Omega: \quad \nabla y(x) = \bar{\nu}(x) := \int_{\mathbb{R}^{d\times d}} F\nu_x(\mathrm{d}F)\,, \tag{D.1.4a}$$

for all $\psi : \mathbb{R}^{d\times d} \to \mathbb{R}$ quasiconvex (and if $p < +\infty$ satisfying $|\psi| \le K(1+|\cdot|^p)$ for some $K > 0$ depending on ψ) the following inequality is fulfilled

$$\forall_{\text{a.a.}} x\in\Omega: \qquad \psi(\nabla y(x)) \le \int_{\mathbb{R}^{d\times d}} \psi(F)\nu_x(\mathrm{d}F)\,, \tag{D.1.4b}$$

and (if $p < +\infty$)

$$\int_\Omega \int_{\mathbb{R}^{d\times d}} |F|^p \nu_x(\mathrm{d}F)\,\mathrm{d}x < +\infty\,, \tag{D.1.4c}$$

or (if $p = +\infty$) ν_x is for almost all $x \in \Omega$ supported on a bounded set in $\mathbb{R}^{d\times d}$ which is independent of $x \in \Omega$.

We wish to emphasize that a zero-measure subset of Ω on which (D.1.4b) possibly does not hold is independent of ψ. The mapping $\bar{\nu} \in L^p(\Omega;\mathbb{R}^{d\times d})$ is called the first moment (or expectation) of ν. If $\varphi : \Omega\times\mathbb{R}^{d\times d} \to \mathbb{R}$ is a Carathéodory integrand and, for some $K > 1$ and $1 < p < +\infty$, φ satisfies $|F|^p/K - K \le \varphi(x,F) \le K(|F|^p + 1)$, then a minimizing sequence of $y \mapsto \int_\Omega \varphi(x,\nabla y(x))\,\mathrm{d}x$ for $y \in W^{1,p}(\Omega)$ is generically bounded only in $W^{1,p}(\Omega;\mathbb{R}^d)$, which means that (D.1.2) is not applicable with $N := d\times d$, $\tilde{\psi} := \varphi$ and $\xi_k := \nabla y_k$. However, if we knew that $\{|\nabla y_k|^p\} \subset L^1(\Omega)$ is equiintegrable, we would be authorized to use Theorem D.1.1 to write (at least for a non-relabeled subsequence) that

$$\lim_{k\to\infty} \int_\Omega \varphi(x,\nabla y_k(x))\,\mathrm{d}x = \int_\Omega \int_{\mathbb{R}^{d\times d}} \varphi(x,F)\nu_x(\mathrm{d}F)\,\mathrm{d}x\,. \tag{D.1.5}$$

The following lemma shows that this is indeed the case.

Lemma D.1.3 (Decomposition lemma [193, 287]). *Let $1 < p < +\infty$, $\Omega \subset \mathbb{R}^d$ be an open bounded set and let $\{y_k\}_{k\in\mathbb{N}} \subset W^{1,p}(\Omega;\mathbb{R}^d)$ be bounded. Then there is a bounded sequence $\{\tilde{y}_j\}_{j\in\mathbb{N}} \subset W^{1,p}(\Omega;\mathbb{R}^d)$ such that*

$$\lim_{j\to\infty} \mathrm{meas}_d(\{x\in\Omega;\ \tilde{y}_j(x) \neq y_j(x) \text{ or } \nabla\tilde{y}_j(x) \neq \nabla y_j(x)\}) = 0 \tag{D.1.6}$$

and $\{|\nabla\tilde{y}_j|^p\}_{j\in\mathbb{N}}$ is equiintegrable in $L^1(\Omega)$. In particular, $\{\nabla y_k\}_{k\in\mathbb{N}}$ and $\{\nabla\tilde{y}_k\}_{k\in\mathbb{N}}$ generate the same Young measure. Moreover, if Ω is a Lipschitz domain then each $\tilde{y}_j$ can be chosen to be a Lipschitz map.

This lemma implies that if $\{y_k\}$ in (D.1.5) is a minimizing sequence of the integral functional $y \mapsto \varphi(x,\nabla y(x))\,\mathrm{d}x$ where $\varphi \geq 0$ then it can always be considered that $\{\nabla y_k\}$ is p-equiintegrable. Indeed, it easily follows from (D.1.3).

The following result is a generalization of a former result due to Zhang [541]. It roughly says that a Young measure supported on a convex compact set can be generated by a sequence of gradients of Sobolev functions taking values in an arbitrarily small neighborhood of this set. Before giving the precise statement of this result, we recall that $\mathrm{dist}(A,\mathcal{S}) := \inf_{F\in\mathcal{S}} |A - F|$ for $A \in \mathbb{R}^{d\times d}$, $\mathcal{S} \subset \mathbb{R}^{d\times d}$.

Proposition D.1.4 (Müller's version of Zhang's theorem [380]). *Suppose that $\Omega \subset \mathbb{R}^d$ is a bounded Lipschitz domain. Let $\nu \in \mathcal{GY}^p(\Omega;\mathbb{R}^{d\times d})$ and let there be a convex compact set $\mathcal{S} \subset \mathbb{R}^{d\times d}$ such that $\mathrm{supp}\,\nu_x \subset \mathcal{S}$ for almost all $x \in \Omega$. Then there is $\{y_k\}_{k\in\mathbb{N}} \subset W^{1,\infty}(\Omega;\mathbb{R}^d)$ such that $\{\nabla y_k\}_{k\in\mathbb{N}}$ generates ν and $\mathrm{dist}(\nabla y_k,\mathcal{S}) \to 0$ in $L^\infty(\Omega;\mathbb{R}^{d\times d})$ as $k \to \infty$.*

The following "homogenization" proposition can be found in [410, Th. 8.1]. It allows us to replace a general Young measure by a homogeneous one in the case that it is generated by a sequence of gradients of maps which all have the same affine boundary conditions. Before we state it we define the support of a Young measure $\nu = \{\nu_x\}_{x\in\Omega}$ as

$$\mathrm{supp}\,\nu := \bigcup_{x\in\Omega} \mathrm{supp}\,\nu_x\,. \tag{D.1.7}$$

Proposition D.1.5 (Homogenization). *Let $1 < p \leq +\infty$, Ω be a bounded Lipschitz domain in $\mathbb{R}^d$, $F \in \mathbb{R}^{d\times d}$ given, and $\{y_k\}_{k\in\mathbb{N}} \subset W^{1,p}(\Omega;\mathbb{R}^d)$ be a bounded sequence such that $y_k(x) = Fx$ for $x \in \partial\Omega$ and all $k \in \mathbb{N}$. Let the Young measure $\nu \in \mathcal{GY}^p(\Omega;\mathbb{R}^{d\times d})$ be generated by $\{\nabla y_k\}_{k\in\mathbb{N}}$. Then there is a another bounded sequence $\{\tilde{y}_k\}_{k\in\mathbb{N}} \subset W^{1,p}(\Omega;\mathbb{R}^d)$, $\tilde{y}_k(x) = Fx$ if $x \in \partial\Omega$ that generates a homogeneous (i.e. independent of x) measure μ defined by*

$$\int_{\mathbb{R}^{d\times d}} \psi(A)\mu(\mathrm{d}A) := \frac{1}{\mathrm{meas}_d(\Omega)} \int_\Omega \int_{\mathbb{R}^{d\times d}} \psi(A)\nu_x(\mathrm{d}A)\,\mathrm{d}x\,, \tag{D.1.8}$$

for all $\psi \in C(\mathbb{R}^{d\times d})$ such that $|\psi| \leq K(1 + |\cdot|^p)$, $K > 0$, if $p < +\infty$. Moreover, $\mu \in \mathcal{GY}^p(\Omega;\mathbb{R}^{d\times d})$ and $\bar{\mu} = F$ and if there is $\tilde{K} > 0$ such that $\|\nabla y_k\|_{L^\infty(\Omega;\mathbb{R}^{d\times d})} \leq \tilde{K}$ for all $k \in \mathbb{N}$ then also $\mathrm{supp}\,\nu \subset \overline{B(0,\tilde{K})}$.

Sometimes it is useful to be allowed to manipulate boundary conditions as in the proposition below whose proof can be found in [410, Lemma 8.3]. Heuristically said, it asserts that there is always a generating sequence of a gradient Young measure such that the corresponding maps share the same trace with the weak limit of the sequence in the Sobolev space.

Proposition D.1.6 (Inherited boundary conditions). *Let $\Omega \subset \mathbb{R}^d$ be a bounded Lipschitz domain and let $1 < p \le +\infty$. Let $\{y_k\}_{k\in\mathbb{N}} \subset W^{1,p}(\Omega;\mathbb{R}^d)$ be a bounded sequence such that $y_k \rightharpoonup y$ in $W^{1,p}(\Omega;\mathbb{R}^d)$ as $k \to \infty$. Assume that $\{\nabla y_k\}_{k\in\mathbb{N}}$ generates $\nu \in \mathcal{GY}^p(\Omega;\mathbb{R}^{d\times d})$. Then there exists another sequence $\{\tilde{y}_k\} \subset W^{1,p}(\Omega;\mathbb{R}^d)$ such that $\{\nabla\tilde{y}_k\}_{k\in\mathbb{N}}$ generates ν, $\tilde{y}_k \to y$ weakly in $W^{1,p}(\Omega;\mathbb{R}^d)$ as $k \to \infty$, and $\tilde{y}_k - y \in W^{1,p}_0(\Omega;\mathbb{R}^d)$ for all $k \in \mathbb{N}$. Moreover, if $p \ne +\infty$ then $\{\tilde{y}_k\}$ can be chosen in such a way that $\{|\nabla\tilde{y}_k|^p\}_{k\in\mathbb{N}}$ is equiintegrable.*

D.2 DiPerna-Majda measures

Sometimes we would like to analyze not only oscillation but also concentration effects of sequences bounded in L^p spaces, see for example Proposition 3.2.4. To this aim, Young measures must be replaced by a finer tool handling simultaneously oscillations and concentrations.

We consider the following commutative ring[2] of continuous bounded functions

$$\mathcal{S} := \Big\{ \psi \in C(\mathbb{R}^N);\ \exists \psi_0 \in C_0(\mathbb{R}^N),\ \psi_1 \in C(S^{N-1}),\ K \in \mathbb{R} : \\ \psi(s) := K + \psi_0(s) + \psi_1\Big(\frac{s}{|s|}\Big)\frac{|s|^p}{1+|s|^p} \text{ if } s \ne 0 \text{ and } \psi(0) := c + \psi_0(0) \Big\}, \tag{D.2.1}$$

where S^{N-1} denotes the $(N-1)$-dimensional unit sphere in $\mathbb{R}^N$. Then $\beta_{\mathcal{S}}\mathbb{R}^N$ is homeomorphic to the unit ball $\overline{B(0,1)} \subset \mathbb{R}^N$ via the mapping $a : \mathbb{R}^N \to B(0,1)$, $a(s) := s/(1+|s|)$ for all $s \in \mathbb{R}^N$. Note that $a(\mathbb{R}^N)$ is dense in $\overline{B(0,1)}$. We further denote by

$$\Upsilon^p_{\mathcal{S}}(\mathbb{R}^N) := \{\psi_p; \psi_p = \psi(1+|\cdot|^p)\ \&\ \psi \in \mathcal{S}\}\,.$$

For any $\psi_p \in \Upsilon^p_{\mathcal{S}}(\mathbb{R}^N)$ there exists a continuous and positively p-homogeneous function $\psi_\infty : \mathbb{R}^N \to \mathbb{R}$ (i.e. $\psi_\infty(\alpha s) = \alpha^p \psi_\infty(s)$ for all $\alpha \ge 0$ and $F \in \mathbb{R}^N$) such that

$$\lim_{|s|\to\infty} \frac{\psi_p(s) - \psi_\infty(s)}{|s|^p} = 0\,. \tag{D.2.2}$$

Indeed, if ψ is as in (D.2.1) and $\psi_p(\cdot) = \psi(\cdot)(1+|\cdot|^p)$, then set

[2] A commutative ring is a commutative group (with respect to addition-operation) equipped also with a multiplication-operation (another commutative group) compatible with the addition structure. Here it is also equipped with a multiplication by scalars, so that the addition group becomes a linear space.

$$\psi_\infty(s) := \Big(K + \psi_1\Big(\frac{s}{|s|}\Big)\Big)|s|^p \text{ for } s \in \mathbb{R}^N \setminus \{0\}.$$

By continuity we define $\psi_\infty(0) := 0$. It is easy to see that ψ_∞ satisfies (D.2.2). Such ψ_∞ is called the *recession function* of ψ.

Let $\sigma \in \mathfrak{M}(\bar{\Omega})$ be a positive Radon measure on a bounded domain $\Omega \subset \mathbb{R}^d$. A mapping $\hat{\nu} : x \mapsto \hat{\nu}_x$ belongs to the space $L^\infty_{\rm w}(\bar{\Omega},\sigma;\mathfrak{M}(\beta_{\mathcal{S}}\mathbb{R}^N))$ if it is weakly* σ-measurable (i.e., for any $\psi_0 \in C_0(\mathbb{R}^N)$, the mapping $\bar{\Omega} \to \mathbb{R} : x \mapsto \int_{\beta_{\mathcal{S}}\mathbb{R}^N} \psi_0(s)\hat{\nu}_x(\mathrm{d}s)$ is σ-measurable in the usual sense). If additionally $\hat{\nu}_x \in \mathfrak{M}_1^+(\beta_{\mathcal{S}}\mathbb{R}^N)$ for σ-a.a. $x \in \bar{\Omega}$ the collection $\{\hat{\nu}_x\}_{x\in\bar{\Omega}}$ is the so-called Young measure on $(\bar{\Omega},\sigma)$ [537], see also [28, 452, 505].

DiPerna and Majda [159] showed that having a bounded sequence in $L^p(\Omega;\mathbb{R}^N)$ with $1 \le p < +\infty$ and Ω an open domain in $\mathbb{R}^d$, there exists its subsequence (denoted by the same indices) a positive Radon measure $\sigma \in \mathfrak{M}(\bar{\Omega})$ and a Young measure $\hat{\nu} : x \mapsto \hat{\nu}_x$ on $(\bar{\Omega},\sigma)$ such that $(\sigma,\hat{\nu})$ is attainable by a sequence $\{z_k\}_{k\in\mathbb{N}} \subset L^p(\Omega;\mathbb{R}^N)$ in the sense that, for all $g \in C(\bar{\Omega})$ and $\psi \in \mathcal{S}$:

$$\begin{aligned}\lim_{k\to\infty}\int_\Omega g(x)\psi_p(z_k(x))\mathrm{d}x &= \int_{\bar{\Omega}}\int_{\beta_{\mathcal{S}}\mathbb{R}^N} g(x)\psi(s)\hat{\nu}_x(\mathrm{d}s)\sigma(\mathrm{d}x)\\ &= \int_\Omega\int_{\mathbb{R}^N} g(x)\psi(s)\hat{\nu}_x(\mathrm{d}s)d_\sigma(x)\mathrm{d}x\\ &\quad+\int_{\bar{\Omega}}\int_{\beta_{\mathcal{S}}\mathbb{R}^N\setminus\mathbb{R}^N} g(x)\psi(s)\hat{\nu}_x(\mathrm{d}s)\sigma(\mathrm{d}x)\end{aligned} \tag{D.2.3}$$

where $\psi_p \in \Upsilon^p_{\mathcal{S}}(\mathbb{R}^N) := \{\psi(1+|\cdot|^p);\ \psi \in \mathcal{S}\}$ and $d_\sigma \in L^1(\Omega)$ is the density of σ with respect to the Lebesgue measure on Ω. In particular, putting $\psi_0 = 1 \in \mathcal{S}$ in (D.2.3) we can see that

$$\lim_{k\to\infty}(1+|z_k|^p) = \sigma \quad \text{weakly* in } \mathfrak{M}(\bar{\Omega}). \tag{D.2.4}$$

If (D.2.3) holds, we say that $\{y_k\}_{k\in\mathbb{N}}$ generates $(\sigma,\hat{\nu})$. Let us denote by $\mathcal{DM}^p_{\mathcal{S}}(\Omega;\mathbb{R}^N)$ the set of all pairs $(\sigma,\hat{\nu}) \in \mathfrak{M}(\bar{\Omega})\times L^\infty_{\rm w}(\bar{\Omega},\sigma;\mathfrak{M}(\beta_{\mathcal{S}}\mathbb{R}^N))$ attainable by sequences from $L^p(\Omega;\mathbb{R}^N)$; note that, taking $\psi := 1$ in (D.2.3), one can see that these sequences must be inevitably bounded in $L^p(\Omega;\mathbb{R}^N)$. The explicit description of the elements from $\mathcal{DM}^p_{\mathcal{R}}(\Omega;\mathbb{R}^N)$, called *DiPerna-Majda measures* (DM measures), for unconstrained sequences was given in [296, Theorem 2].

Proposition D.2.1 (Characterization of DM measures). *Let $\Omega \subset \mathbb{R}^d$ be a bounded open domain such that $|\partial\Omega| = 0$, $\mathcal{S}$ be a separable complete subring of the ring of all continuous bounded functions on $\mathbb{R}^N$ and $(\sigma,\hat{\nu}) \in \mathfrak{M}(\bar{\Omega})\times L^\infty_{\rm w}(\bar{\Omega},\sigma;\mathfrak{M}(\beta_{\mathcal{S}}\mathbb{R}^N))$ and $1 \le p < +\infty$. Then the following two statements are equivalent with each other:*

(i) *The pair* $(\sigma,\hat\nu)$ *is the DiPerna-Majda measure, i.e.* $(\sigma,\hat\nu)\in\mathcal{DM}^p_{\mathcal{S}}(\Omega;\mathbb{R}^N)$.

(ii) *The following properties are satisfied simultaneously:*

1. σ *is positive,*
2. $\sigma_{\hat\nu}\in\mathfrak{M}(\bar\Omega)$ *defined by* $\sigma_{\hat\nu}(\mathrm{d}x)=(\int_{\mathbb{R}^N}\hat\nu_x(\mathrm{d}F))\sigma(\mathrm{d}x)$ *is absolutely continuous with respect to the Lebesgue measure* ($d_{\sigma_{\hat\nu}}$ *will denote its density),*
3. *for a.a.* $x\in\Omega$ *it holds*
$$\int_{\mathbb{R}^N}\hat\nu_x(\mathrm{d}s)>0,\qquad d_{\sigma_{\hat\nu}}(x)=\bigg(\int_{\mathbb{R}^N}\frac{\hat\nu_x(\mathrm{d}s)}{1+|s|^p}\bigg)^{-1}\int_{\mathbb{R}^N}\hat\nu_x(\mathrm{d}s)\,,$$
4. *for* σ*-a.a.* $x\in\bar\Omega$ *it holds* $\hat\nu_x\ge 0$ *and* $\int_{\beta_{\mathcal{S}}\mathbb{R}^N}\hat\nu_x(\mathrm{d}s)=1$.

We say that $\{z_k\}$ generates $(\sigma,\hat\nu)$ if (D.2.3) holds. Moreover, we denote $d_\sigma\in L^1(\Omega)$ the absolutely continuous (with respect to the Lebesgue measure) part of σ in the Lebesgue decomposition of σ.

Eventually, we sketch results about DiPerna-Majda measures generated by gradients, called *gradient DiPerna-Majda measures*. We will denote elements from $\mathcal{DM}^p_{\mathcal{R}}(\Omega;\mathbb{R}^{d\times d})$ which are generated by $\{\nabla y_k\}_{k\in\mathbb{N}}$ for some bounded sequence $\{y_k\}\subset W^{1,p}(\Omega;\mathbb{R}^d)$ by $\mathcal{GDM}^p_{\mathcal{S}}(\Omega;\mathbb{R}^{d\times d})$. The first result was proved in [270]. We recall that $\mathrm{Q}\psi_p$ denotes the quasiconvex envelope of ψ_p; cf. (4.4.25).

Theorem D.2.2 (Characterization of gradient DM measures I). *Let* $\Omega\subset\mathbb{R}^d$ *be a bounded domain with the extension property in* $W^{1,p}(\Omega;\mathbb{R}^d)$, $1<p<+\infty$ *and* $(\sigma,\hat\nu)\in\mathcal{DM}^p_{\mathcal{R}}(\Omega;\mathbb{R}^{d\times d})$. *Then then there is a bounded sequence* $\{y_k\}_{k\in\mathbb{N}}\subset W^{1,p}(\Omega;\mathbb{R}^d)$ *such that* $y_j-y_k\in W^{1,p}_0(\Omega;\mathbb{R}^d)$ *for any* $j,k\in\mathbb{N}$ *and* $\{\nabla y_k\}_{k\in\mathbb{N}}$ *generates* $(\sigma,\hat\nu)$ *if and only if the following three conditions hold*

$$\exists y\in W^{1,p}(\Omega;\mathbb{R}^d)\ \forall_{\mathrm{a.a.}}x\in\Omega:\ \ \nabla y(x)=d_\sigma(x)\int_{\beta_{\mathcal{S}}\mathbb{R}^{d\times d}}\frac{F}{1+|F|^p}\hat\nu_x(\mathrm{d}F)\,,\tag{D.2.5a}$$

for almost all $x\in\Omega$ *and for all* $\psi_p\in\Upsilon^p_{\mathcal{S}}(\mathbb{R}^{d\times d})$ *the following inequality is fulfilled*

$$\mathrm{Q}\psi_p(\nabla y(x))\le d_\sigma(x)\int_{\beta_{\mathcal{S}}\mathbb{R}^{m\times n}}\psi(F)\hat\nu_x(\mathrm{d}F)\,,\tag{D.2.5b}$$

for σ*-almost all* $x\in\bar\Omega$ *and all* $\psi_p\in\Upsilon^p_{\mathcal{R}}(\mathbb{R}^{d\times d})$ *with* $\mathrm{Q}\psi_p>-\infty$ *it holds that*

$$0\le\int_{\beta_{\mathcal{S}}\mathbb{R}^{d\times d}\setminus\mathbb{R}^{d\times d}}\psi(F)\hat\nu_x(\mathrm{d}F)\,.\tag{D.2.5c}$$

The next theorem states a complete characterization of gradient DiPerna-Majda measure up to the boundary of Ω. We refer to [292] for a proof. If $1<p<+\infty$ and $\vec n\in\mathbb{R}^d$ has a unit length we put for $\psi_\infty:\mathbb{R}^{d\times d}\to\mathbb{R}$ positively p-homogeneous

$$Q_{b,\vec n}\psi_\infty(0):=\inf_{y\in W^{1,p}(B(0,1);\mathbb{R}^d)}\int_{D_{\vec n}}\psi_\infty(\nabla y(x))\,\mathrm{d}x\,,\tag{D.2.6}$$

where $D_{\vec{n}} := B(0,1) \cap \{x \in \mathbb{R}^d;\ \vec{n} \cdot x < 0\}$. It is easy to see that either $Q_{b,\vec{n}}\psi_\infty(0) = 0$ or $Q_{b,\vec{n}}\psi_\infty(0) = -\infty$.

Theorem D.2.3 (Characterization of gradient DM measures II). *Let $\Omega \subset \mathbb{R}^d$ be a bounded domain with boundary of class C^1, $1 < p < +\infty$, and $(\sigma, \hat{\nu}) \in \mathcal{DM}^p_{\mathcal{S}}(\Omega; \mathbb{R}^{d\times d})$. Then there is a bounded sequence $\{y_k\}_{k\in\mathbb{N}} \subset W^{1,p}(\Omega; \mathbb{R}^d)$ such that $\{\nabla y_k\}_{k\in\mathbb{N}}$ generates $(\sigma, \hat{\nu})$ if and only if the following four conditions hold:*

$$\exists y \in W^{1,p}(\Omega; \mathbb{R}^d)\ \forall_{\text{a.a.}\, x\in\Omega}: \quad \nabla y(x) = d_\sigma(x) \int_{\beta_{\mathcal{S}} \mathbb{R}^{d\times d}} \frac{F}{1+|F|^p} \hat{\nu}_x(\mathrm{d}F)\,, \tag{D.2.7a}$$

for almost all $x \in \Omega$ and for all $\psi_p \in \Upsilon^p_{\mathcal{S}}(\mathbb{R}^{d\times d})$ the following inequality is fulfilled

$$\mathrm{Q}\psi_p(\nabla y(x)) \le d_\sigma(x) \int_{\beta_{\mathcal{S}} \mathbb{R}^{d\times d}} \psi(F)\hat{\nu}_x(\mathrm{d}F)\,, \tag{D.2.7b}$$

for σ-almost all $x \in \Omega$ and all $\psi_p \in \Upsilon^p_{\mathcal{S}}(\mathbb{R}^{d\times d})$ with $\mathrm{Q}\psi_p > -\infty$ it holds that

$$0 \le \int_{\beta_{\mathcal{S}} \mathbb{R}^{d\times d} \setminus \mathbb{R}^{d\times d}} \psi(F)\hat{\nu}_x(\mathrm{d}F)\,, \tag{D.2.7c}$$

and for σ-almost all $x \in \partial\Omega$ with the outer unit normal to the boundary $\vec{n}(x)$ and all $\psi \in \Upsilon^p_{\mathcal{S}}(\mathbb{R}^{d\times d})$ with $Q_{b,\vec{n}(x)}\psi_\infty(0) = 0$ it holds that

$$0 \le \int_{\beta_{\mathcal{S}} \mathbb{R}^{d\times d} \setminus \mathbb{R}^{d\times d}} \psi(F)\hat{\nu}_x(\mathrm{d}F)\,. \tag{D.2.7d}$$

Since polyconvexity implies quasiconvexity (D.2.5b) and (D.2.7b) hold also for polyconvex functions from $\Upsilon^p_{\mathcal{S}}(\mathbb{R}^{d\times d})$.

The following Lemma and Theorem D.2.3 were proven in [292].

Lemma D.2.4 (Equiintegrability condition). *Let $1 \le p < +\infty$, $0 \le \psi_p \in \Upsilon^p_{\mathcal{S}}(\mathbb{R}^{d\times d})$, let $\psi_p = \psi(\cdot)(1 + |\cdot|^p)$, and let $\{y_k\} \subset W^{1,p}(\Omega; \mathbb{R}^d)$ be a bounded sequence with $\{\nabla y_k\}_{k\in\mathbb{N}} \subset L^p(\Omega; \mathbb{R}^{d\times d})$ generating $(\sigma, \hat{\nu}) \in \mathcal{DM}^p_{\mathcal{S}}(\Omega; \mathbb{R}^{d\times d})$. Then*[3]

$$\{\psi_p(\nabla y_k)\}_{k\in\mathbb{N}} \text{ is weakly relatively compact in } L^1(\Omega)$$

if and only if

$$\int_{\bar{\Omega}} \int_{\beta_{\mathcal{S}} \mathbb{R}^{d\times d} \setminus \mathbb{R}^{d\times d}} \psi(F)\hat{\nu}_x(\mathrm{d}F)\sigma(\mathrm{d}x) = 0\,. \tag{D.2.8}$$

[3] For the assertion to hold true, it actually suffices to have that $\psi_p(\nabla y_k) \ge 0$ for every k. This can easily be seen by applying Lemma D.2.4 to $|\psi_p|$ instead of ψ_p.

Solutions or hints for selected exercises

The exercises occasionally included for educational purposes are here, in most cases, accompanied with some brief hints or solutions.

Exercises in Chapter 1

Exercise 1.1.8 on p. 9: Hint: If $v \in \mathbb{R}^d$ then $v \cdot G^\top G v = (Gv) \cdot (Gv) \geq 0$. As $G \in \mathrm{GL}(d)$ we have that $Gv = 0$ if and only if $v := 0$.

Exercise 1.1.14 on p. 13: Hint: We have $F = F^{-\top} \hat{F}^\top \hat{F}$, so we need to show that $F^{-\top} \hat{F}^\top$ belongs to $\mathrm{SO}(d)$. Notice that $F^{-\top} = (\hat{F}^\top \hat{F} F^{-1})^{-1} = F \hat{F}^{-1} \hat{F}^{-\top}$. This means that $(F^{-\top} \hat{F}^\top)^\top (F^{-\top} \hat{F}^\top) = \mathbb{I}$.

Exercise 1.1.15 on p. 13: Solution: In view of Exercise 1.1.14, we get that, for all $x \in \Omega$, $\nabla y(x) = R(x)$ for some $R(x) \in \mathrm{SO}(d)$. It means due to Theorem 1.1.12 that $\mathrm{dist}(\nabla y, \mathrm{SO}(d)) = 0$. Consequently, $\nabla y = R$ for some $R \in \mathrm{SO}(d)$. Another simpler proof[4] is as follows: We have that $\nabla y(x) = R(x)$ where $R(x) \in \mathrm{SO}(d)$ for almost all $x \in \Omega$. Therefore, by (1.1.19), one gets $0 = \mathrm{div}\,\mathrm{Cof}\,\nabla y = \mathrm{div}\,\nabla y = \Delta y$, i.e., y is a harmonic function and therefore it is smooth. We calculate

$$\frac{1}{2}\Delta(|\nabla y|^2) = |\nabla^2 y|^2 + \nabla y \cdot \nabla(\Delta y).$$

The second term on the right-hand side is zero. Also the left hand-side vanishes because $|\nabla y|^2 = |R|^2 = d$; here we exploit that $|\cdot|$ is the Frobenius norm and therefore constant on rotations. It follows that $|\nabla^2 y|^2 = 0$ and y is affine.

Exercise 1.2.3 on p. 17: Hint: Use a right triangle with catheti parallel to axes to the Cartesian coordinate system instead of the tetrahedron. Otherwise the proof is a direct adaptation of the three-dimensional version.

Exercise 1.2.5 on p. 21: Hint: The option A is correct. Indeed, formula (1.2.13) shows that density of volume forces in the reference and in the deformed configuration are parallel and have the same orientation. Assume that $g : \mathbb{R}^d \to \mathbb{R}^d$ is an inhomogeneous gravitational field. Then the total force acting on a part ω^y of the

[4] Personal communication and courtesy C. Kreisbeck.

M. Kružík and T. Roubíček, *Mathematical Methods in Continuum Mechanics of Solids*, Interaction of Mechanics and Mathematics,
https://doi.org/10.1007/978-3-030-02065-1

deformed specimen in the deformed configuration is $\int_{\omega^y} \varrho^y(x^y)g(x^y)\,\mathrm{d}x^y$, where ϱ^y is the mass density in the deformed configuration. We have

$$\int_{\omega^y} \varrho^y(x^y)g(x^y)\,\mathrm{d}x^y = \int_{\omega} \varrho^y(y(x))g(y(x))\det\nabla y(x)\,\mathrm{d}x\,,$$

so that, in view of (1.2.14), $f(x) := g(y(x))\varrho(x)$ is the corresponding volume density force in the reference configuration.

Exercises in Chapter 2

Exercise 2.5.7 on p. 46: Hint: Use formulas (2.5.8), (2.5.12), and $C = U^2$. From this, get that, for all $i,j,k,m \in \{1,\dots,d\}$, it holds $\partial_{x_m} C_{ij} = 2\partial_{x_m} U_{ik}U_{kj}$, where one sums over k.

Exercise 2.5.8 on p. 46: Hint: Use the Hölder continuity of $x \mapsto \det\nabla y(x)$ in the form $|\det\nabla y(x_1) - \det\nabla y(x_2)| \le K|x_1 - x_2|^\gamma$ for all $x_1, x_2 \in \bar{\Omega}$. A straightforward calculation shows that $\lim_{r\to 0} r^d h(Kr^\gamma) = +\infty$.

Exercises in Chapter 3

Exercise 3.1.5 on p. 55: Hint: Calculating the second derivative of φ, the Legendre-Hadamard condition (3.1.8) results in $\varphi(a\otimes b) \ge 0$ for every $a, b \in \mathbb{R}^d$.

Exercise 3.4.15 on p. 70: Hint: Notice that the nonlinear constraint $\det\nabla y = 1$ passes to the limit along a weakly converging minimizing sequence due to Theorem 3.2.2.

Exercise 3.4.16 on p. 70: Hint: Since $L : \mathrm{GL}^+(d) \to \mathbb{R}$ is polyconvex, there is a convex (and therefore continuous) function $\mathbb{L}$ such that $L(F) = \mathbb{L}(F, \mathrm{Cof}\,F, \det F)$ for all $F \in \mathrm{GL}^+(d)$. The a-priori estimates allow for extraction a minimizing sequence $\{y_k\} \subset \mathcal{A}$ satisfying

$$(y_k, \mathrm{Cof}\,\nabla y_k, \det\nabla y_k) \rightharpoonup (y, \mathrm{Cof}\,\nabla y, \det\nabla y) \text{ in } W^{1,p}(\Omega;\mathbb{R}^d)\times L^q(\Omega;\mathbb{R}^{d\times d})\times L^r(\Omega)\,.$$

Then apply Mazur's Lemma (A.1.7) to this sequence and exploit convexity of $\mathbb{L}$ to obtain

$$\begin{aligned} L(\nabla y) = \mathbb{L}(\nabla y, \mathrm{Cof}\,\nabla y, \det\nabla y) &= \lim_{k\to\infty} \mathbb{L}\Big(\sum_{j=k}^{N(k)} \lambda_k^j \nabla y_j, \sum_{j=k}^{N(k)} \lambda_k^j \mathrm{Cof}\,\nabla y_j, \sum_{j=k}^{N(k)} \lambda_k^j \det\nabla y_j\Big) \\ &\le \limsup_{k\to\infty} \sum_{j=k}^{N(k)} \lambda_k^j \mathbb{L}(\nabla y_j, \mathrm{Cof}\,\nabla y_j, \det\nabla y_j) \le 0\,. \end{aligned}$$

Exercise 3.4.18 on p. 70: Hint: Use the Hölder inequality to estimate the product $|\nabla y|^d(\det\nabla y)^{-1}$.

Exercise 3.6.5 on p. 78: Hint: Show that $(x,y)\mapsto x^m/y^{m-1}$ is convex for $x,y>0$.

Exercises in Chapter 4

Exercise 4.5.16 on p. 134: Hint: Once it is proved that the weak limit y of some minimizing sequence satisfies $\det\nabla y>0$ almost everywhere in Ω, the proof follows the lines of the proof of Theorem 4.5.7. Notice that, due to (4.5.10), $\varphi(x,F,\Delta_1,\Delta_2)\ge K|\det F|^{-r}$. Assume now that $\{y_k\}_{k\in\mathbb{N}}\subset\mathcal{A}$ is a minimizing sequence of $\mathcal{E}$ and that it weakly converges to y in $W^{1,p}(\Omega;\mathbb{R}^d)$ and $\det\nabla y=0$ on $\omega\subset\Omega$ where $\mathrm{meas}_d(\omega)>0$. Indeed, it cannot happen that $\det\nabla y<0$ on a set of a positive Lebesgue measure in $\mathbb{R}^d$ because it is a weak limit of positive functions. Therefore, we have

$$\lim_{k\to\infty}\int_\omega \det\nabla y_k(x)\,\mathrm{d}x=\lim_{k\to\infty}\int_\omega |\det\nabla y_k(x)|\,\mathrm{d}x=\int_\omega \det\nabla y(x)\,\mathrm{d}x=0\,.$$

Hence, $\det\nabla y_k\to 0$ strongly in $L^1(\omega)$ as $k\to\infty$ and there is a (non-relabeled) subsequence converging almost everywhere to zero. By the Fatou lemma B.2.3, it holds

$$\liminf_{k\to\infty}\int_\omega \varphi(x,\nabla y_k(x),\nabla\mathrm{Cof}\,\nabla y_k(x),\nabla\det\nabla y_k(x))\,\mathrm{d}x \ge\int_\omega \liminf_{k\to\infty}\frac{1}{(\det\nabla y_k(x))^r}\,\mathrm{d}x=+\infty,$$

which contradicts the assumption that $\{y_k\}$ is a minimizing sequence.

Exercise 4.5.18 on p. 134: Hint: The assumptions imply that $\{\mathrm{Cof}\,\nabla y_k\}_{k\in\mathbb{N}}$ and $\{\det\nabla y_k\}_{k\in\mathbb{N}}$ are bounded in $W^{1,1}(\Omega;\mathbb{R}^{d\times d})$ and $W^{1,1}(\Omega;\mathbb{R}^d)$, respectively. This implies strong convergence of these sequences in $L^1(\Omega:\mathbb{R}^{d\times d})$ and $L^1(\Omega:\mathbb{R}^d)$. Then proceed as in the proof of Theorem 4.5.7.

Exercise 4.5.21 on p. 135: Hint: We exploit the formula $(\nabla y)^{-1}=(\mathrm{Cof}\,\nabla y)^\top/\det\nabla y$. If $\{y_k\}_{k\in\mathbb{N}}\subset\mathcal{A}$ is a bounded minimizing sequence converging weakly to y, we have $\mathrm{Cof}\,\nabla y_k\to\mathrm{Cof}\,\nabla y$. Moreover, by the Vitali theorem B.2.4 $(\det\nabla y_k)^{-1}\to(\det\nabla y)^{-1}$ in $L^q(\Omega)$ for $1\le q<r$. Hence, $(d-1)/p+1/q=(q(d-1)+p)/pq$ and consequently, $\{(\nabla y_k)^{-1}\}_k$ is bounded in $L^{pq/(p+q(d-1))}(\Omega;\mathbb{R}^{d\times d})$ and converges weakly to $(\nabla y)^{-1}$.

Exercise 4.5.22 on p. 135: Hint: If $F\in W^{1,q}(\Omega;\mathbb{R}^{3\times 3})$ which is the Banach algebra then $F_{ij}F_{kl}\in W^{1,q}(\Omega)$ for all admissible i,j,k,l. This implies that $\mathrm{Cof}\,F\in W^{1,q}(\Omega;\mathbb{R}^{3\times 3})$. Note that we have not used the assumption on the determinant. To prove the opposite inequality we use (A.2.18) and (A.2.19) and proceed analogously as is Remark 4.5.11.

Exercise 4.6.4 on p. 149: Hint: Realize that in two dimensions matrices with the rank at most one are exactly those which are not invertible. Hence, $\det(R_2^\top R_1AB^{-1}-\mathbb{I})=0$. This is equivalent to finding $R\in\mathrm{SO}(2)$ such that $\det(AB^{-1}-R^\top)=0$.

Exercise 4.6.7 on p. 153: Hint: Use the definition of rank-one convexity and (4.6.19) together with (4.6.20).

Exercise 4.6.11 on p. 157: Hint: Add a convex term in $\nabla^2 y$, or $\nabla[\mathrm{Cof}\,\nabla y]$ and apply Theorem 4.5.1, or Theorem 4.5.8, respectively.

Exercises in Chapter 5

Exercise 5.3.3 on p. 168: Hint: The bilinear form B in (5.1.8) expands by the term $\int_\Gamma \mathbb{B}u{\cdot}v\,\mathrm{d}S$ and one needs to qualify $\mathbb{B}(x)$ positive semidefinite with $\mathbb{B}(x)u{\cdot}u \ge \varepsilon|u|^2$ with some $\varepsilon > 0$ for all $x{\in}\Gamma_{\mathrm{N}}$ and all $u{\in}\mathbb{R}^d$ to make it coercive.

Exercise 5.4.4 on p. 170: Hint: Consider the functional

$$u \mapsto \int_\Omega \frac{1}{2}\mathbb{C}e(u{+}u_{\mathrm{D}}){:}e(u{+}u_{\mathrm{D}}) + \frac{1}{2}\mathbb{H}\nabla e(u{+}u_{\mathrm{D}}) \vdots \nabla e(u{+}u_{\mathrm{D}}) - f{\cdot}u\,\mathrm{d}x - \int_{\Gamma_{\mathrm{N}}} g{\cdot}u\,\mathrm{d}S\,.$$

defined on the set $\{u \in H^2(\Omega;\mathbb{R}^d);\ u = 0 \text{ on } \Gamma_{\mathrm{D}}\}$. The existence of a unique solution follows by the direct method. Note that $\nabla^2 u$ is bounded due to (5.2.4) in $L^2(\Omega;\mathbb{R}^{d\times d\times d})$, and hence the Korn's inequality for the 2nd-grade materials (5.2.6) is to be used. Uniqueness of the solution is implied by strict convexity of this functional due to the assumed uniform positive definiteness of $\mathbb{C}$ and $\mathbb{H}{\in}\mathrm{SLin}(\Omega;\mathbb{R}^{d\times d}_{\mathrm{sym}})$. This solution then satisfies the corresponding boundary-value problem (5.4.6).

Exercise 5.4.5 on p. 171: Hint: Replace Γ by Γ_{N} in (5.4.6) and add the condition $\nabla v{\cdot}\vec{n} = 0$ on Γ_{D} into Definition 5.4.1 and realize that such a space of v's is closed in $H^2(\Omega;\mathbb{R}^d)$.

Exercise 5.6.4 on p. 176: Hint: The enhanced potential (5.6.4) now as a function of $(u,\phi,\vec{q})$ is to be minimized for $(u,\vec{q})$ and maximized for ϕ. The system (5.6.5) is enhanced by the equation $\mathrm{div}(\kappa\nabla\vec{q} + \mathbb{P}e(u)) = \phi'(\vec{q})$ and an additional stress $\mathbb{P}\cdot\vec{q}$ occurs together with $\mathbb{C}e(u)$ in the left-hand side of (5.6.5a).

Exercises in Chapter 6

Exercise 6.3.9 on p. 209: Hint: First consider the test-functions for (6.2.7) or (6.2.8) with a compact support in Q and make the by-part integration in time once or twice, respectively, and then (knowing that the integral over Q already vanishes) choose test functions with arbitrary nonzero traces on Γ_{N} or with arbitrary non-zero values at $t = 0$ or arbitrary non-zero time derivatives at $t = 0$.

Exercise 6.3.10 on p. 210: Solution: Consider $\sigma = \mathbb{C}e(u)$ in (6.2.7) and a time instance $0 \le s < T$, and take, e.g., the test function

$$v_\varepsilon(t)=\begin{cases}\dot u(t) & \text{for } t\le s,\\ (s{-}t{+}\varepsilon)\dot u(s)/\varepsilon & \text{for } s<t<s+\varepsilon,\\ 0 & \text{for } t\ge s+\varepsilon.\end{cases}$$

Note that v_ε is well defined for any $0<\varepsilon<T-s$ and that $v_\varepsilon|_{t=T}=0$ as required in (6.2.7). Then (6.2.7) results to the identity

$$\int_0^s\!\!\int_\Omega \mathbb{C}e(u){:}e(\dot u)-\varrho\dot u\cdot\ddot u\,\mathrm{d}x\mathrm{d}t$$
$$=\int_\Omega \varrho|v_0|^2\,\mathrm{d}x+\int_0^s\!\!\int_\Omega f\cdot\dot u\,\mathrm{d}x\mathrm{d}t+\int_0^s\!\!\int_{\Gamma_\mathrm{N}} g\cdot\dot u\,\mathrm{d}S\,\mathrm{d}t+I_\varepsilon(s)$$

$$\text{with}\quad I_\varepsilon(s)=\int_s^{s+\varepsilon}\bigg(\int_\Omega\frac{\varrho}{\varepsilon}\dot u(t)\cdot\dot u(s)+\frac{s{-}t{+}\varepsilon}{\varepsilon}\big(f(t)\cdot\dot u(s)-\mathbb{C}e(u(t)){:}e(\dot u(s))\big)\mathrm{d}x$$
$$+\int_{\Gamma_\mathrm{N}}\frac{s{-}t{+}\varepsilon}{\varepsilon}g(t)\cdot\dot u(s)\,\mathrm{d}S\bigg)\,\mathrm{d}t.$$

Relying on the assumed regularity $u\in C^1_\mathrm{weak}(I;L^2(\Omega;\mathbb{R}^d))\cap C_\mathrm{weak}(I;H^1(\Omega;\mathbb{R}^d))$, realize that

$$\lim_{\varepsilon\to0}I_\varepsilon(s)=\int_\Omega\varrho|\dot u(s)|^2\mathrm{d}x$$

at every $s\in I$ where $\dot u(s)\in H^1(\Omega;\mathbb{R}^d)$, i.e. for a.a. $s\in I$. Further, relying that $\sqrt{\varrho}\ddot u\in L^2(I;H^1(\Omega;\mathbb{R}^d)^*)$ is in duality with $\sqrt{\varrho}\dot u\in L^2(I;H^1(\Omega;\mathbb{R}^d))$, use

$$\int_0^s\!\!\int_\Omega\varrho\dot u\cdot\ddot u\,\mathrm{d}x\mathrm{d}t=\int_\Omega\frac{\varrho}{2}|\dot u(s)|^2-\frac{\varrho}{2}|v_0|^2\,\mathrm{d}x,$$

cf. Lemma C.2.2 on p. 535. Moreover, use the analogous argument to claim that

$$\int_0^s\!\!\int_\Omega\mathbb{C}e(u){:}e(\dot u)\,\mathrm{d}x\mathrm{d}t=\int_\Omega\frac12\mathbb{C}e(u(s)){:}e(u(s))-\frac12\mathbb{C}e(u_0){:}e(u_0)\,\mathrm{d}x.$$

Exercise 6.3.11 on p. 210: Hint: Test the second equation in (6.3.25) by $v_\tau^k+v_\tau^{k-1}$ (i.e. multiply it by $v_\tau^k+v_\tau^{k-1}$, integrate it over Ω, use the Green formula and the boundary condition (6.3.25b)), and substitute $v_\tau^k{+}v_\tau^{k-1}=\frac{2}{\tau}(u_\tau^k{-}u_\tau^{k-1})$ from the first equation in (6.3.25), and further use the bi-nomial formulas

$$\varrho\frac{v_\tau^k{-}v_\tau^{k-1}}{\tau}\cdot\frac{v_\tau^k{+}v_\tau^{k-1}}{2}=\frac{\varrho}{2}\frac{|v_\tau^k|^2-|v_\tau^{k-1}|^2}{\tau}$$

and, exploiting also the symmetry of $\mathbb{C}\in\mathrm{SLin}(\mathbb{R}^{d\times d}_\mathrm{sym})$,

$$\mathbb{C}e\Big(\frac{u_\tau^k{+}u_\tau^{k-1}}{2}\Big){:}e\Big(\frac{v_\tau^k{+}v_\tau^{k-1}}{2}\Big)=\mathbb{C}e\Big(\frac{u_\tau^k{+}u_\tau^{k-1}}{2}\Big){:}e\Big(\frac{u_\tau^k{-}u_\tau^{k-1}}{\tau}\Big)$$
$$=\frac{\frac12\mathbb{C}e(u_\tau^k){:}e(u_\tau^k)-\frac12\mathbb{C}e(u_\tau^{k-1}){:}e(u_\tau^{k-1})}{\tau}.$$

The constant in (6.3.26) is the initial energy $\int_\Omega \frac{\varrho}{2}|v_0|^2 + \frac{1}{2}\mathbb{C}e(u_0){:}e(u_0)\,\mathrm{d}x$.

Exercise 6.3.12 on p. 210: Hint: Make the energetic test of (6.3.21) to obtain

$$\begin{aligned}\|\dot u(t)\|^2_{L^2(\Omega)} + C\|\partial_x u(t)\|^2_{L^2(\Omega)} + H_2\|\partial^3_{xxx}u(t)\|^2_{L^2(\Omega)} &= -H_2\|\partial^2_{xx}u(t)\|^2_{L^2(\Omega)}\\ &\le C_{\rm GN}|H_2|\Big(\varepsilon\|\|\|\partial^3_{xxx}u(t)\|^2_{L^2(\Omega)} + \frac{1}{\varepsilon}\|u(t)\|^2_{L^2(\Omega)}\Big)\\ &= C_{\rm GN}|H_2|\Big(\varepsilon\|\|\|\partial^3_{xxx}u\|^2_{L^2(\Omega)} + \frac{1}{\varepsilon}\Big\|\int_0^t \dot u\,\mathrm{d}t\Big\|^2_{L^2(\Omega)}\Big)\end{aligned}$$

when using also an interpolation by the Gagliardo-Nirenberg inequality (B.3.20) with $k = 2$, $p = q = r = 2$, $\beta = 1$, and $\lambda = 1/2$ and the Young inequlality. For $\varepsilon > 0$ small enough, the only term to be estimate is $\|\int_0^t \dot u\,\mathrm{d}t\|^2_{L^2(\Omega)}$ for which the Gronwall inequality is to be used when estimating this terms still by Hölder's inequality as $t\int_0^t \|\dot u\|^2_{L^2(\Omega)}\,\mathrm{d}t$. For $d > 1$, $\partial_x u$ is to be replaced by $e(u)$ and similarly $\partial^2_{xx}u$ by $\nabla e(u)$ and $\partial^3_{xxx}u$ by $\nabla^2 e(u)$, and one must still use Korn's inequality like (5.2.5).

Exercise 6.3.13 on p. 210: Hint: Integrate (6.3.15) over $[0,t]$ and estimate $\int_\Omega \nabla\phi\cdot u\,\mathrm{d}x \le \epsilon\|\nabla\phi\|^2_{L^2(\Omega;\mathbb{R}^d)} + C_\epsilon(\int_0^t \|\dot u\|^2_{L^2(\Omega;\mathbb{R}^d)}\,\mathrm{d}t + \|u_0\|^2_{L^2(\Omega;\mathbb{R}^d)})$ and use Gronwall inequality to obtain the a-priori estimate of $u \in W^{1,\infty}(I;L^2(\Omega;\mathbb{R}^d)) \cap L^\infty(I;H^1(\Omega;\mathbb{R}^d))$ and $\phi \in L^\infty(I;H^1(\mathbb{R}^d))$, provided $u_0 \in H^1(\Omega;\mathbb{R}^d)$ and $v_0 \in L^2(\Omega;\mathbb{R}^d)$. Also use the estimate $\|\phi\|_{L^2(\mathbb{R}^d)} \le C\|\nabla\phi\|_{L^2(\mathbb{R}^d;\mathbb{R}^d)}$ provided $\phi(\infty) = 0$, which can (and should) be added as a "boundary" condition for the gravitational potential. The involvement of the Coriolis force does not influence the a-priori estimates because $\omega\times\dot u$ is always orthogonal to $\dot u$ so that $(\omega\times\dot u)\cdot\dot u = 0$ while the centrifugal force $\omega\times(\omega\times(x+u))$ has a linear growth in u and can be easily treated by the Gronwall inequality. The system is (at least formally) still conservative but the Hamilton-principle (6.1.3) cannot be directly applied because the Coriolis force does not have a potential. On the other hand, for qualified loading, the regularity like in Remark 6.4.4 holds (note that the orthogonality $(\omega\times\ddot u)\cdot\ddot u = 0$ is again at disposal), and then the uniqueness of the solution and energy conservation can rigorously be proved as in Proposition 6.3.2.

Exercise 6.3.14 on p. 210: Hint: considering smooth test function v in (6.2.8) make the limit passage in the term $\int_Q \varrho u\cdot\ddot v\,\mathrm{d}x\mathrm{d}t$ using the weak convergence in ϱ e.g. in $L^2(Q)$ when considering $\varrho(t,x) \equiv \varrho(x)$ as a function $Q \to \mathbb{R}^+$ constant in time and the strong convergence in u due to the Aubin-Lions compact embedding in $u \in L^\infty(I;H^1(\Omega;\mathbb{R}^d)) \cap W^{1,\infty}(I;L^2(\Omega;\mathbb{R}^d)) \subset L^2(Q)$. Prove existence of solutions for (6.3.27) by the direct method, taking a minimizing sequence and using weak* compactness of the set of admissible ϱ's in $L^\infty(\Omega)$. For the modification involving the trace of the velocity, considering smooth test function v in (6.2.7), make a limit passage in the term $\int_Q \varrho\dot u\cdot\dot v\,\mathrm{d}x\mathrm{d}t$ using again the weak convergence in ϱ's in $L^2(Q)$ but applying the Aubin-Lions compactness theorem on $\dot u$ instead of u, using the regularity from Proposition 6.3.2. For the modification involving the trace of the acceleration, use still more regular data to be able to make twice differentiation in time and then test by $\dddot u$.

Exercise 6.4.19 on p. 222: Hint: Consider $\{V_k\}_{k\in\mathbb{N}}\subset H^1_{\rm D}(\Omega;\mathbb{R}^d)$ the nested finite-dimensional spaces used for Galerkin method and $\tilde u_k$ valued in V_k such that $\tilde u_k\to u$ strongly in $H^1(I;H^1_{\rm D}(\Omega;\mathbb{R}^d))\cap C(I;L^2(\Omega;\mathbb{R}^d))$ with $\tilde u_k(0)=u_0$ and $\dot{\tilde u}_k(0)=v_0$. Then estimate

$$\begin{aligned}\frac12\int_Q\mathbb{D}e(\dot u_k-\dot u){:}e(\dot u_k-\dot u)\,\mathrm{d}x\mathrm{d}t &\le \int_Q\mathbb{D}e(\dot u_k-\dot{\tilde u}_k){:}e(\dot u_k-\dot{\tilde u}_k)+\mathbb{D}e(\dot{\tilde u}_k-u){:}e(\dot{\tilde u}_k-u)\,\mathrm{d}x\mathrm{d}t\\
\le \int_Q\mathbb{D}e(\dot u_k-\dot{\tilde u}_k){:}e(\dot u_k-\dot{\tilde u}_k)&+\mathbb{D}e(\dot{\tilde u}_k-u){:}e(\dot{\tilde u}_k-u)\,\mathrm{d}x\mathrm{d}t\\
+\frac12\int_\Omega\mathbb{C}e(u_k(T)-\tilde u_k(T)){:}e(u_k(T)&-\tilde u_k(T))+\varrho|(u_k(T)-\tilde u_k(T)|^2\,\mathrm{d}x\\
=\int_Q f\cdot(u_k(T)-\tilde u_k(T))+\mathbb{D}e(\dot{\tilde u}_k-u){:}e(\dot{\tilde u}_k-u)\,\mathrm{d}x\mathrm{d}t&+\int_{\Sigma_{\rm N}}g\cdot(u_k(T)-\tilde u_k(T))\,\mathrm{d}S\,\mathrm{d}t\to0.\end{aligned}$$

Exercise 6.4.20 on p. 222: Hint: Abbreviating $v^k_\tau=(u^k_\tau-u^{k-1}_\tau)/\tau$, realize that the left- and the right-hand sides of (C.2.51) expand by $|v^k_\tau-v^{k-1}_\tau|^2$ and $|v^{k-1}_\tau|^2$, respectively. Via an elementary algebra, it leads to an estimate $\frac12|v^k_\tau|^2\le2|v^{k-1}_\tau|^2$ added to (C.2.51) but it does not give an bound uniform in $\tau>0$.

Exercise 6.5.7 on p. 228: Solution: Comparing the Zener model (6.5.13) with the Poynting-Thomson model (6.5.3), one can see that the desired relations are

$$\mathbb{C}_1\mathbb{D}^{-1}(\mathbb{I}{+}\mathbb{C}_2\mathbb{C}_1^{-1})=\tilde{\mathbb{C}}_1\tilde{\mathbb{D}}^{-1},\quad \mathbb{C}_1\mathbb{D}^{-1}\mathbb{C}_2=\tilde{\mathbb{C}}_1\tilde{\mathbb{D}}^{-1}\tilde{\mathbb{C}}_2,\ \text{ and }\ \mathbb{C}_1=\tilde{\mathbb{C}}_1{+}\tilde{\mathbb{C}}_2.$$

The last relation fits the response under very fast loading (when both dampers are very rigid) while for a very slow loading (when both dampers "vanish") all time-derivatives in (6.5.13) and (6.5.3) vanish so that they take the form $\mathbb{C}_1\mathbb{D}^{-1}(\mathbb{I}+\mathbb{C}_2\mathbb{C}_1^{-1})\sigma=\mathbb{C}_1\mathbb{D}^{-1}\mathbb{C}_2e$ and $\tilde{\mathbb{C}}_1\tilde{\mathbb{D}}^{-1}\sigma=\tilde{\mathbb{C}}_1\tilde{\mathbb{D}}^{-1}\tilde{\mathbb{C}}_2e$, respectively. Hence $\sigma=(\mathbb{I}+\mathbb{C}_2\mathbb{C}_1^{-1})^{-1}\mathbb{C}_2e$ and $\sigma=\tilde{\mathbb{C}}_2e$, which reveals that

$$\tilde{\mathbb{C}}_2=(\mathbb{I}+\mathbb{C}_2\mathbb{C}_1^{-1})^{-1}\mathbb{C}_2=\mathbb{C}_1(\mathbb{C}_1{+}\mathbb{C}_2)^{-1}\mathbb{C}_2.$$

Substituting it into $\mathbb{C}_1=\tilde{\mathbb{C}}_1{+}\tilde{\mathbb{C}}_2$, we obtain

$$\tilde{\mathbb{C}}_1=\mathbb{C}_1-\mathbb{C}_1(\mathbb{C}_1{+}\mathbb{C}_2)^{-1}\mathbb{C}_2=\mathbb{C}_1(\mathbb{C}_1{+}\mathbb{C}_2)^{-1}\mathbb{C}_1.$$

Substituting this $\tilde{\mathbb{C}}_1$ into the desired equation $\mathbb{C}_1\mathbb{D}^{-1}(\mathbb{I}{+}\mathbb{C}_2\mathbb{C}_1^{-1})=\tilde{\mathbb{C}}_1\tilde{\mathbb{D}}^{-1}$, one gets

$$\tilde{\mathbb{D}}=(\mathbb{I}+\mathbb{C}_2\mathbb{C}_1^{-1})^{-1}\mathbb{D}(\mathbb{C}_1{+}\mathbb{C}_2)^{-1}\mathbb{C}_1.$$

Eventually, it is important that the last desired equation $\mathbb{C}_1\mathbb{D}^{-1}\mathbb{C}_2=\tilde{\mathbb{C}}_1\tilde{\mathbb{D}}^{-1}\tilde{\mathbb{C}}_2$ is satisfied, too. Conversely, it is not difficult to see that, given the Zener model $(\tilde{\mathbb{C}}_1,\tilde{\mathbb{C}}_2,\tilde{\mathbb{D}})$, we can evaluate the moduli in the Poynting-Thomson model: beside $\mathbb{C}_1=\tilde{\mathbb{C}}_1{+}\tilde{\mathbb{C}}_2$ already mentioned, we have also $\mathbb{C}_2=\mathbb{C}_1\tilde{\mathbb{C}}_1^{-1}\tilde{\mathbb{C}}_2=(\tilde{\mathbb{C}}_1{+}\tilde{\mathbb{C}}_2)\tilde{\mathbb{C}}_1^{-1}\tilde{\mathbb{C}}_2$, and also $\mathbb{D}=(\mathbb{I}+\mathbb{C}_2\mathbb{C}_1^{-1})\tilde{\mathbb{D}}\mathbb{C}_1^{-1}(\mathbb{C}_1{+}\mathbb{C}_2)=(\mathbb{I}+(\tilde{\mathbb{C}}_1{+}\tilde{\mathbb{C}}_2)\tilde{\mathbb{C}}_1^{-1}\tilde{\mathbb{C}}_2(\tilde{\mathbb{C}}_1{+}\tilde{\mathbb{C}}_2)^{-1})\tilde{\mathbb{D}}(\mathbb{I}+\tilde{\mathbb{C}}_1^{-1}\tilde{\mathbb{C}}_2)$. Note that these transformations preserve positive definiteness. Also note that commutativity of the visco/elastic tensors is not assumed, otherwise some of these formulas would simplify.

Exercise 6.5.8 on p. 228: Hint: Use the equations in (6.5.12) and calculate

$$\sigma{:}\dot{e} = (\tilde{\sigma}_1{+}\tilde{\sigma}_2){:}\dot{e} = \tilde{\sigma}_1{:}(\dot{e}_1{+}\dot{e}_2) + \tilde{\sigma}_2{:}\dot{e} = \tilde{\sigma}_1{:}\tilde{\mathbb{C}}_1^{-1}\dot{\tilde{\sigma}}_1 + \tilde{\sigma}_1{:}\tilde{\mathbb{D}}^{-1}\tilde{\sigma}_1 + e{:}\tilde{\mathbb{C}}_2\dot{e}$$

to reveal the stored energy $\varphi(e,\tilde{\sigma}_1) = \frac{1}{2}\tilde{\mathbb{C}}_1^{-1}\tilde{\sigma}_1{:}\tilde{\sigma}_1 + \frac{1}{2}\tilde{\mathbb{C}}_2 e{:}e$ and the dissipation rate $\xi(\tilde{\sigma}_1) = \tilde{\mathbb{D}}^{-1}\tilde{\sigma}_1{:}\tilde{\sigma}_1$, or alternatively written also as $\varphi(e,\tilde{e}_1) = \frac{1}{2}\tilde{\mathbb{C}}_1\tilde{e}_1{:}\tilde{e}_1 + \frac{1}{2}\tilde{\mathbb{C}}_2 e{:}e$ and $\xi(\dot{\tilde{e}}_2) = \tilde{\mathbb{D}}\dot{\tilde{e}}_2{:}\dot{\tilde{e}}_2$.

Exercise 6.6.8 on p. 237: Solution: Multiply (6.6.30) by $\sigma - \tilde{\mathbb{C}}(e(u){-}\tilde{e}_2)$ and use $\tilde{\sigma}_2 = \mathbb{D}_1(e(\dot{u}) - \dot{\tilde{e}}_2)$ and $\sigma = \tilde{\mathbb{D}}_2\dot{\tilde{e}}_2$ to calculate:

$$\begin{aligned}
\sigma{:}e(\dot{u}) &= \tilde{\mathbb{D}}_1^{-1}(\sigma - \tilde{\mathbb{C}}(e(u){-}\tilde{e}_2)){:}(\sigma - \tilde{\mathbb{C}}(e(u){-}\tilde{e}_2)) \\
&\quad + \tilde{\mathbb{D}}_2^{-1}\sigma{:}(\sigma - \tilde{\mathbb{C}}(e(u){-}\tilde{e}_2)) + \tilde{\mathbb{C}}(e(u){-}\tilde{e}_2)){:}e(\dot{u}) \\
&= \tilde{\mathbb{D}}_1^{-1}\tilde{\sigma}_2{:}\tilde{\sigma}_2 + \tilde{\mathbb{D}}_2^{-1}\sigma{:}\sigma + \tilde{\mathbb{C}}(e(u){-}\tilde{e}_2)){:}(e(\dot{u}) - \tilde{\mathbb{D}}_2^{-1}\sigma) \\
&= \tilde{\sigma}_2{:}(e(\dot{u}) - \dot{\tilde{e}}_2) + \sigma{:}\dot{\tilde{e}}_2 + \tilde{\mathbb{C}}(e(u){-}\tilde{e}_2)){:}(e(\dot{u}) - \dot{\tilde{e}}_2) \\
&= \tilde{\mathbb{D}}_1(\dot{e}{-}\dot{\tilde{e}}_2){:}(\dot{e}{-}\dot{\tilde{e}}_2) + \tilde{\mathbb{D}}_2\dot{\tilde{e}}_2{:}\dot{\tilde{e}}_2 + \frac{\partial}{\partial t}\frac{1}{2}\tilde{\mathbb{C}}(e{-}\tilde{e}_2){:}(e{-}\tilde{e}_2).
\end{aligned}$$

The qualification of the data needed for the a-priori estimates are the same as for the Kelvin-Voigt model due to the ϱ- and the $\mathbb{D}$-terms. The manipulation through the Korn, Hölder, Young, and Gronwall's inequalities is similar.

Exercise 6.6.9 on p. 237: Hint: Considering $\mathbb{C} = \mathbb{C}_1/\varepsilon$ and denoting by $(u_\varepsilon,\sigma_\varepsilon)$ the solution to the corresponding Maxwell model, use (6.6.4)–(6.6.5) to derive the a-priori estimates $\|u_\varepsilon\|_{L^\infty(I;L^2(\Omega;\mathbb{R}^d))} \le C$, $\|\sigma_\varepsilon\|_{L^2(Q;\mathbb{R}^{d\times d})} \le C$, and $\|\sigma_\varepsilon\|_{L^\infty(I;L^2(\Omega;\mathbb{R}^{d\times d}))} \le C/\sqrt{\varepsilon}$ under the same data qualification (6.3.3). Realize how (6.6.12) works for $\mathbb{C}^{-1} = \varepsilon\mathbb{C}_1^{-1} \to 0$. Then, using the a-priori estimates which are uniform with respect to ε and resulting convergences

$$\begin{aligned}
u_\varepsilon &\to u && \text{weakly* in } L^\infty(I;L^2(\Omega;\mathbb{R}^d)) \text{ and} \\
\sigma_\varepsilon &\to \sigma && \text{weakly in } L^2(Q;\mathbb{R}^{d\times d}),
\end{aligned}$$

make the limit passage for $\varepsilon \to 0$ in the weak formulation (6.2.7) or (6.2.8) and in the constitutive relation written in the form (6.6.1).

Exercise 6.6.10 on p. 237: Solution: Putting $\mathbb{D}_1 = 0$ into (6.6.28) is just a trivial algebra yielding $\dot{\sigma} + \mathbb{C}\mathbb{D}_2^{-1}\sigma = \mathbb{C}\dot{e}$ which is the Maxwell model, cf. (6.6.2). For the convergence with $\mathbb{D}_1 \to 0$, take $\mathbb{D}_1 = \varepsilon\mathbb{D}$ and denote by u_ε and $e_{2\varepsilon}$ the corresponding solution. From the energetic estimate based on (6.6.29), $\|e(\dot{u}_\varepsilon)\|_{L^2(Q;\mathbb{R}^{d\times d})} \le C/\sqrt{\varepsilon}$. The weak convergence uses $\|\varepsilon\mathbb{D}e(\dot{u}_\varepsilon)\| = \mathscr{O}(\sqrt{\varepsilon}) \to 0$. For the strong convergence, estimate the difference of the momentum equilibrium $\varrho\ddot{u}_\varepsilon - \mathrm{div}(\varepsilon\mathbb{D}e(\dot{u}_\varepsilon) + \mathbb{C}(e(u_\varepsilon){-}e_{2\varepsilon})) = f$ from the limit $\varrho\ddot{u} - \mathrm{div}(\mathbb{C}(e(u){-}e_2)) = f$ tested by $\dot{u}_\varepsilon{-}\dot{u}$ together with the differences $\mathbb{D}_2\dot{e}_{2\varepsilon}+\mathbb{C}e_{2\varepsilon}=\mathbb{C}e(u_\varepsilon)$ between $\mathbb{D}_2\dot{e}_2+\mathbb{C}e_2=\mathbb{C}e(u)$ tested by $\dot{e}_{2\varepsilon}-\dot{e}_2$; these equations express the equilibrium of the viscous stress $\mathbb{D}_2\dot{e}_{2\varepsilon}$ with the elastic stress $\mathbb{C}e_{1\varepsilon}=\mathbb{C}(e(u_\varepsilon)-e_{2\varepsilon}$, cf. Fig. 6.5. Altogether, it is to obtain

$$\int_\Omega \frac{1}{2}\mathbb{C}(e(u_\varepsilon(T)-u(T))-e_{2\varepsilon}(T)+e_2(T)):(e(u_\varepsilon(T)-u(T))-e_{2\varepsilon}(T)+e_2(T))$$
$$+\frac{\varrho}{2}|\dot{u}_\varepsilon(T)-\dot{u}(T)|^2\mathrm{d}x=\int_Q -\varepsilon\mathbb{D}e(\dot{u}_\varepsilon):e(\dot{u}_\varepsilon-\dot{u})\mathrm{d}x\mathrm{d}t\le\int_Q \varepsilon\mathbb{D}e(\dot{u}_\varepsilon):e(\dot{u})\mathrm{d}x\mathrm{d}t\to 0,$$

where we used that assumed regularity $e(\dot{u})\in L^2(Q;\mathbb{R}^{d\times d})$. This gives strong convergence $u_\varepsilon\to u$ in $L^\infty(I;H^1(\Omega;\mathbb{R}^d))$ and $e_{2\varepsilon}\to e_2$ in $L^2(Q;\mathbb{R}^{d\times d})$.

Exercise 6.6.11 on p. 238: Solution: Sending $\mathbb{D}_2\to\infty$, one can see the limit from (6.6.28) when putting $\mathbb{D}_2^{-1}=0$, which yields $\dot{\sigma}=\mathbb{D}_1\ddot{e}+\mathbb{C}\dot{e}$ and then, after integration in time, the Kelvin-Voigt model $\sigma=\mathbb{D}_1\dot{e}+\mathbb{C}e$. For the strong convergence with $\mathbb{D}_2\to\infty$, take $\mathbb{D}_2=\mathbb{D}/\varepsilon$ and denote by u_ε and $e_{2\varepsilon}$ the corresponding solution. From the energetic estimate based on (6.6.29), $\|\dot{e}_{2\varepsilon}\|_{L^2(Q;\mathbb{R}^{d\times d})}\le C\sqrt{\varepsilon}$. Fixing the initial condition $e_{2\varepsilon}(0)=0$, we have also $\|e_{2\varepsilon}\|_{L^2(Q;\mathbb{R}^{d\times d})}\le C\sqrt{\varepsilon}$. Then estimate the difference of the momentum equilibrium $\varrho\ddot{u}_\varepsilon-\mathrm{div}(\mathbb{D}_1e(\dot{u}_\varepsilon)+\mathbb{C}(e(u_\varepsilon)-e_{2\varepsilon}))=f$ from the limit $\varrho\ddot{u}-\mathrm{div}(\mathbb{D}_1e(\dot{u})+\mathbb{C}e(u))=f$ tested by $\dot{u}_\varepsilon-\dot{u}$ to obtain

$$\int_\Omega \frac{\varrho}{2}|\dot{u}_\varepsilon(T)-\dot{u}(T)|^2+\frac{1}{2}\mathbb{C}e(u_\varepsilon-u):e(u_\varepsilon-u)\mathrm{d}x+\int_Q \mathbb{D}_1e(\dot{u}_\varepsilon-\dot{u}):e(\dot{u}_\varepsilon-\dot{u})\,\mathrm{d}x\mathrm{d}t$$
$$=\int_Q f\cdot(\dot{u}_\varepsilon-\dot{u})-\mathbb{C}e_{2\varepsilon}:e(\dot{u}_\varepsilon-\dot{u})\,\mathrm{d}x\mathrm{d}t\to 0$$

because $e_{2\varepsilon}\to 0$ strongly in $L^2(Q;\mathbb{R}^{d\times d})$ while $e(\dot{u}_\varepsilon-\dot{u})$ is bounded (or even weakly converges to 0) in $L^2(Q;\mathbb{R}^{d\times d})$. Moreover, writing the energy balance based on (6.6.29), we can estimate the overall dissipated energy on the damper $\mathbb{D}_2=\mathbb{D}/\varepsilon$ as

$$\limsup_{\varepsilon\to 0}\int_Q \frac{1}{\varepsilon}\mathbb{D}\dot{e}_{2\varepsilon}:\dot{e}_{2\varepsilon}\,\mathrm{d}x\mathrm{d}t=\int_\Omega \frac{\varrho}{2}|v_0|^2+\frac{1}{2}\mathbb{C}e(u_0):e(u_0)\,\mathrm{d}x$$
$$-\liminf_{\varepsilon\to 0}\Big(\int_\Omega \frac{\varrho}{2}|\dot{u}_\varepsilon(T)|^2+\frac{1}{2}\mathbb{C}(e(u_\varepsilon(T))-e_{2\varepsilon}(T)):(e(u_\varepsilon(T))-e_{2\varepsilon}(T))\,\mathrm{d}x$$
$$+\int_Q \mathbb{D}_1e(\dot{u}_\varepsilon):e(\dot{u}_\varepsilon)\,\mathrm{d}x\mathrm{d}t\Big)+\lim_{\varepsilon\to 0}\int_Q f\cdot\dot{u}_\varepsilon\,\mathrm{d}x\mathrm{d}t$$
$$\le\int_\Omega \frac{\varrho}{2}|v_0|^2+\frac{1}{2}\mathbb{C}e(u_0):e(u_0)-\frac{\varrho}{2}|\dot{u}(T)|^2-\frac{1}{2}\mathbb{C}(e(u(T))):(e(u(T))\,\mathrm{d}x$$
$$+\int_Q f\cdot\dot{u}-\mathbb{D}_1e(\dot{u}):e(\dot{u})\,\mathrm{d}x\mathrm{d}t=0,$$

where the last equation expresses the energy conservation in the limit system, cf. Proposition 6.4.1. The boundary loading is not considered.

Exercise 6.6.13 on p. 238: Hint: Modify (6.3.25) by employing also (6.6.2) suitably discretized by the Crank-Nicolson scheme:

$$\frac{u_\tau^k-u_\tau^{k-1}}{\tau}=\frac{v_\tau^k+v_\tau^{k-1}}{2},\qquad \varrho\frac{v_\tau^k-v_\tau^{k-1}}{\tau}-\mathrm{div}\Big(\frac{\sigma_\tau^k+\sigma_\tau^{k-1}}{2}\Big)=f_\tau^k\qquad\text{on }\Omega,$$

$$\mathbb{C}^{-1}\frac{\sigma_\tau^k-\sigma_\tau^{k-1}}{\tau}+\mathbb{D}^{-1}\frac{\sigma_\tau^k+\sigma_\tau^{k-1}}{2}=e\Big(\frac{u_\tau^k-u_\tau^{k-1}}{\tau}\Big)\qquad\text{on }\Omega,$$

$$u_\tau^k=u_{\mathrm{D},\tau}^k\quad\text{on }\Gamma_{\mathrm{D}}\qquad\text{and}\qquad \vec{n}^\top\sigma_\tau^k=g_\tau^k\qquad\text{on }\Gamma_{\mathrm{N}},$$

to be solved successively for $k=1,2,...,$ starting with $u_0^\tau=u_0$ and $v_\tau^0=v_0$, and $\sigma_\tau^k=\sigma_0$. Treat the first equation as in Exercise 6.3.11 together with testing the second one by $\mathbb{C}(\sigma_\tau^k+\sigma_\tau^{k-1})$ to imitate (6.6.3) again as an equality.

Exercise 6.7.5 on p. 244: Hint: Realize all the constitutive relations $\sigma_1=\mathbb{C}_2e_3$, $\sigma_2=\mathbb{D}_2\dot{e}_3$, $\sigma=\sigma_1+\sigma_2=\mathbb{C}_1e_2=\mathbb{D}_1\dot{e}_1$, and $e=e_1+e_2+e_2$.

Exercise 6.7.6 on p. 244: Hint: Use the arguments from Exercise 6.6.11 for the volumetric part.

Exercise 6.7.7 on p. 245: Solution: The bulk loading f is to be treated simply by the Young and the Gronwall inequalities. One can estimate

$$\int_\Omega\frac{\varrho}{2}|\dot{u}_\varepsilon(t)|^2+K_{\mathrm{E}}|\mathrm{sph}\,e(u_\varepsilon(t))|^2\,\mathrm{d}x+\varepsilon\int_0^t\!\!\int_\Omega K_{\mathrm{V}}(\mathrm{sph}\,e(\dot{u}_\varepsilon))^2+G_{\mathrm{V}}|\mathrm{dev}\,e(\dot{u}_\varepsilon)|^2\,\mathrm{d}x\mathrm{d}t$$
$$=\int_0^t\!\!\int_\Omega f\cdot\dot{u}_\varepsilon\,\mathrm{d}x\mathrm{d}t+\int_\Omega\frac{\varrho}{2}|v_0|^2+K_{\mathrm{E}}|\mathrm{sph}\,e(u_0)|^2\,\mathrm{d}x$$
$$\le\int_0^t\|f\|_{L^2(\Omega)}\Big(1+\|\dot{u}_\varepsilon\|_{L^2(\Omega)}^2\Big)+C_{\delta,u_0,v_0}$$

and, choosing $\delta>0$ small and using Gronwall's inequality, obtain uniform bounds for u_ε in $W^{1,\infty}(I;L^2(\Omega;\mathbb{R}^d))$, and $\mathrm{div}\,u_\varepsilon$ in $L^\infty(I;L^2(\Omega))$, and also the estimate $\|e(\dot{u}_\varepsilon)\|_{L^2(Q;\mathbb{R}^{d\times d})}\le C/\sqrt{\varepsilon}$ and use the last estimate for $\mathrm{dev}\,\sigma_\varepsilon=\varepsilon G_{\mathrm{V}}\mathrm{dev}\,e(\dot{u}_\varepsilon)\to0$ in $L^2(Q;\mathbb{R}^{d\times d})$. The weak formulation for the limit relies on the orthogonality $\sigma:e=\mathrm{sph}\,\sigma:\mathrm{sph}\,e+\mathrm{dev}\,\sigma:\mathrm{dev}\,e=\mathrm{sph}\,\sigma:\mathrm{sph}\,e$ if $\mathrm{dev}\,e=0$. The weak solution $(u,\sigma)\in W^{1,\infty}(I;L^2(\Omega;\mathbb{R}^d))\times L^2(Q;\mathbb{R}^{d\times d}_{\mathrm{sym}})$ with $\mathrm{div}\,u\in L^2(Q)$ and $\mathrm{dev}\,\sigma=0$ is to satisfy $\sigma=K_{\mathrm{E}}(\mathrm{div}\,u)\mathbb{I}$ and the integral identity (6.2.7) modified as[5]

$$\int_Q dK_{\mathrm{E}}(\mathrm{div}\,u)(\mathrm{div}\,v)-\varrho\dot{u}\cdot\dot{v}\,\mathrm{d}x\mathrm{d}t=\int_\Omega\varrho v_0\cdot v(0,\cdot)\,\mathrm{d}x+\int_Q f\cdot v\,\mathrm{d}x\mathrm{d}t$$

while replacing $H^1_{\mathrm{D}}(\Omega;\mathbb{R}^d)$ in (6.2.7) with the Banach space of deviatoric-free functions $H^1_{\mathrm{dev}}(\Omega;\mathbb{R}^d):=\{v\in H^1(\Omega;\mathbb{R}^d);\ \mathrm{dev}\,e(v)=0\}$. Then, for $\varepsilon\to0$, make a limit passage in (6.2.7) written for the viscous Boger model towards this limit weak formulation. Moreover, as $\|\mathrm{dev}\,e(\dot{u}_\varepsilon)\|_{L^2(Q;\mathbb{R}^{d\times d})}=\mathscr{O}(1/\sqrt{\varepsilon})$, it holds $\|\mathrm{dev}\,\sigma_\varepsilon\|_{L^2(Q;\mathbb{R}^{d\times d})}=$

[5] Note $(\mathrm{sph}\,e(u)):(\mathrm{sph}\,e(v))=(\mathrm{div}\,u)\mathbb{I}:(\mathrm{div}\,v)\mathbb{I}=(\mathrm{div}\,u)(\mathrm{div}\,v)(\mathbb{I}:\mathbb{I})=d(\mathrm{div}\,u)(\mathrm{div}\,v)$, which is where the factor d came from.

$\|\varepsilon G_{\rm V}{\rm dev}\,e(\dot u_\varepsilon)\|_{L^2(Q;\mathbb{R}^{d\times d})} = \mathscr{O}(\sqrt{\varepsilon}) \to 0$. For the uniqueness, use the test function (6.3.8) to modify (6.6.16) for $\int_\Omega \frac{\varrho}{2}|u(s)|^2 + \frac{1}{2}K_{\rm E}|{\rm div}\,v(0)|^2\,{\rm d}x = 0$.

Exercise 6.7.8 on p. 245: Solution: Make differentiation of the system in time and test it by $\ddot u$. It puts, in particular, $\sqrt{\varrho}\ddot u \in L^2([0,t];L^2(\Omega;\mathbb{R}^d))$ into duality with $\sqrt{\varrho}\dot u$, so that the by-part integration in time used within the "energetic" test by $\dot u$ is legitimate. This yields the energy conservation of the limit solution on any time interval $[0,t]$. For the strong convergence (6.7.16), still use

$$\begin{aligned}
&\limsup_{\varepsilon\to0}\int_\Omega \frac{\varrho}{2}|\dot u_\varepsilon(t)|^2 + \frac{d}{2}K_{\rm E}|{\rm sph}\,e(u_\varepsilon(t))|^2\,{\rm d}x\\
&\le \limsup_{\varepsilon\to0}\bigg(\int_\Omega \frac{\varrho}{2}|\dot u_\varepsilon(t)|^2 + \frac{d}{2}K_{\rm E}|{\rm sph}\,e(u_\varepsilon(t))|^2\,{\rm d}x\\
&\qquad\qquad + \varepsilon\int_0^t\!\!\int_\Omega dK_{\rm V}({\rm sph}\,e(\dot u_\varepsilon))^2 + 2G_{\rm V}|{\rm dev}\,e(\dot u_\varepsilon)|^2\,{\rm d}x{\rm d}t\bigg)\\
&= \lim_{\varepsilon\to0}\int_0^t\!\!\int_\Omega f\cdot\dot u_\varepsilon\,{\rm d}x{\rm d}t + \int_\Omega \frac{\varrho}{2}|v_0|^2 + \frac{d}{2}K_{\rm E}|{\rm sph}\,e(u_0)|^2\,{\rm d}x\\
&= \int_0^t\!\!\int_\Omega f\cdot\dot u\,{\rm d}x{\rm d}t + \int_\Omega \frac{\varrho}{2}|v_0|^2 + \frac{d}{2}K_{\rm E}|{\rm sph}\,e(u_0)|^2\,{\rm d}x\\
&\qquad\qquad\qquad = \int_\Omega \frac{\varrho}{2}|\dot u(t)|^2 + \frac{d}{2}K_{\rm E}|{\rm sph}\,e(u(t))|^2\,{\rm d}x,
\end{aligned}$$

where the last equality is the proved energy conservation on the time interval $[0,t]$. Since also $\dot u_\varepsilon(t) \to \dot u(t)$ and ${\rm sph}\,e(u_\varepsilon(t)) \to {\rm sph}\,e(u(t))$ $L^2(\Omega)$-weakly and the opposite inequality holds always by the weak lower semicontinuity, the strong convergence (6.7.16) follows by the uniform convexity of the L^2-space, cf. Theorem A.3.8. Putting (6.7.16) again to the above calculations for $t = T$,

$$\begin{aligned}
&\limsup_{\varepsilon\to0}\varepsilon\int_Q dK_{\rm V}({\rm sph}\,e(\dot u_\varepsilon))^2 + 2G_{\rm V}|{\rm dev}\,e(\dot u_\varepsilon)|^2\,{\rm d}x{\rm d}t\\
&= \limsup_{\varepsilon\to0}\bigg(\int_\Omega \frac{\varrho}{2}|\dot u_\varepsilon(T)|^2 + \frac{d}{2}K_{\rm E}|{\rm sph}\,e(u_\varepsilon(T))|^2\,{\rm d}x\\
&\qquad\qquad + \varepsilon\int_Q dK_{\rm V}({\rm sph}\,e(\dot u_\varepsilon))^2 + 2G_{\rm V}|{\rm dev}\,e(\dot u_\varepsilon)|^2\,{\rm d}x{\rm d}t\bigg)\\
&\quad - \lim_{\varepsilon\to0}\int_\Omega \frac{\varrho}{2}|\dot u_\varepsilon(T)|^2 + \frac{d}{2}K_{\rm E}|{\rm sph}\,e(u_\varepsilon(T))|^2\,{\rm d}x = \lim_{\varepsilon\to0}\int_Q f\cdot\dot u_\varepsilon\,{\rm d}x{\rm d}t\\
&\quad + \int_\Omega \frac{\varrho}{2}|v_0|^2 + \frac{d}{2}K_{\rm E}|{\rm sph}\,e(u_0)|^2 - \frac{\varrho}{2}|\dot u_\varepsilon(T)|^2 - \frac{d}{2}K_{\rm E}|{\rm sph}\,e(u_\varepsilon(T))|^2\,{\rm d}x = 0.
\end{aligned}$$

Thus (6.7.17) follows.

Exercises in Chapter 7

Exercise 7.1.2 on p. 250: Hint: In terms of Figure 6.7(right), take $z = (e_1, e_3)$ and then $\varphi(e,z) \equiv \varphi(e,e_1,e_3) = \frac{1}{2}\mathbb{C}_1(e-e_1-e_3):(e-e_1-e_3)+\frac{1}{2}\mathbb{C}_2e_3:e_3$ and $\zeta_2(\dot z) \equiv \zeta_2(\dot e_1,\dot e_3) = \frac{1}{2}\mathbb{D}_1\dot e_1:\dot e_1+\frac{1}{2}\mathbb{D}_2\dot e_3:\dot e_3$.

Exercise 7.3.5 on p. 262: Hint: transform the 2nd-order system into the 1-st order one as in (C.2.3) on p. 534, and further realize that the stored energy $\mathcal{E}$ is quadratic in (u,z), employ the Crank-Nicolson time discretization (C.2.43) to both u and z simultaneously, without making the fractional-step splitting a'la (C.2.49).

Exercise 7.3.7 on p. 267: Hint: Make the limit passage by only weak convergence, using the Minty trick. More specifically, consider (7.3.24) with $c = 0$ and an arbitrary v instead of u, then substitute $v = u \pm \epsilon w$, cancel $\epsilon > 0$, and push the remaining ϵ to zero.

Exercise 7.3.13 on p. 275: Hint: Use the arguments like (7.3.24) to show strong convergence of strains. For the gradient theory in z, use the compact embedding and strong convergence also of the internal variable z.

Exercise 7.4.10 on p. 284: Hint: The energy-based a-priori estimates are as (7.4.12a-c) together with the additional bound of $e(\dot u)-\dot\pi$ in $L^2(Q;\mathbb{R}^{d\times d}_{\rm sym})$ ensured by positive definiteness of $\mathbb{D}$ which is to be assumed, of course. Combining it with (7.4.12c) the gives also a bound for $e(\dot u)$ in $L^1(I;L^2(\Omega;\mathbb{R}^{d\times d}_{\rm sym}))$, and by (7.4.12a) and Korn's inequality, also $\dot u \in L^1(I;H^1(\Omega;\mathbb{R}^d))$. The regularity copies the proof of Proposition 7.4.6. Now both $\varrho\ddot u$ and ${\rm div}\,\sigma$ with $\sigma = \mathbb{D}(e(\dot u)-\dot\pi)+\mathbb{C}(e(\dot u)-\dot\pi)$ are in duality with $\varrho\dot u$ as used in Proposition 7.3.9 so that the energy conservation and also the uniqueness can be proved.

Exercise 7.5.9 on p. 301: Hint: Heuristically, let us consider $\varphi(e,\alpha) = G(\alpha)|e|^2/\sqrt{1+|e|^2}$. Then the Hessian of $(e,\alpha) \mapsto \varphi(e,\alpha)+\frac{1}{2}K(|e|^2+\alpha^2)$ is $\begin{pmatrix} G(\alpha)\sigma'(e)+K, & G'(\alpha)\sigma(e) \\ G'(\alpha)\sigma(e) & ,G''(\alpha)|e|^2/\sqrt{1+|e|^2}+K \end{pmatrix}$ with $\sigma(e) = 2e/\sqrt{1+|e|^2} - |e|^2e/\sqrt[3/2]{1+|e|^2}$ bounded with $\sigma'(\cdot)$ bounded too. Hence, for K sufficiently large, the Hessian in question is positive definite. Analogously, it holds for ${\rm dev},e$ instead of e as used in (7.5.25).

Exercise 7.5.23 on p. 311: Hint: Let us consider heuristically $d = 1$ and $\mathbb{C} = 1$. The Hessian of $\frac{1}{2}\alpha e^2+\frac{1}{2}K(|e|^2+\alpha^2)$ is then $\begin{pmatrix} \alpha+K, & e \\ e & ,K \end{pmatrix}$. Its determinant, being $K^2+\alpha K-e^2$, cannot be made non-negative for all $(e,\alpha)\in\mathbb{R}^{d\times d}_{\rm sym}\times[0,1]$ no matter how K is large.

Exercise 7.5.24 on p. 311: Hint: Since $[0,1]$ is compact and $g(\cdot) > 0$, also $\min_{[0,1]} g(\cdot) > 0$. The Hessian, which is

$$(e,\alpha) \mapsto \begin{pmatrix} g(\alpha)\mathbb{C}+K\mathbb{I}, & g'(\alpha)\mathbb{C}e \\ g'(\alpha)\mathbb{C}e & ,g''(\alpha)\mathbb{C}e{:}e \end{pmatrix},$$

is to be positive semidefinite for a sufficiently big K. Since $g(\alpha)\mathbb{C}$ is surely positive definite, we can forgot this term. Also the block $K\mathbb{I}$ is positive definite, and by Sylvester's criterion, it suffices to show the nonnegativity of the determinant of the whole resting matrix. To this goal, use the matrix algebra $\det\begin{pmatrix} A\ ,B\\ B^\top,C\end{pmatrix} = \det(A)\det(C-B^\top A^{-1}B)$ to calculate

$$\det\begin{pmatrix} K\mathbb{I} & , & g'(\alpha)\mathbb{C}e \\ g'(\alpha)\mathbb{C}e, & \tfrac{1}{2}g''(\alpha)\mathbb{C}e{:}e \end{pmatrix} = K^{d^2}\Big(\frac{1}{2}g''(\alpha)\mathbb{C}e{:}e - \frac{1}{K}|g'(\alpha)\mathbb{C}e|^2\Big)$$
$$\geq K^{d^2}\Big(\frac{1}{2}cg''(\alpha) - \frac{1}{K}g'(\alpha)^2|\mathbb{C}|^2\Big)|e|^2 \geq 0,$$

the diagonal terms $g(\alpha)\mathbb{C}+K\mathbb{I}$ and $g''(\alpha)\mathbb{C}e:e$ are surely positive (definite), by Sylvester's criterion, one has to show that $g(\alpha)g''(\alpha)(\det\mathbb{C}+K)\mathbb{C}e{:}e-2g'(\alpha)^2|\mathbb{C}e|^2 \geq (c(g(\alpha)g''(\alpha)(\det\mathbb{C}{+}K)-2g'(\alpha)^2|\mathbb{C}|^2)|e|^2 \geq 0$ where $c>0$ is the positive-definiteness constant of $\mathbb{C}$, i.e. $c := \min_{|e|=1}\mathbb{C}e{:}e$. It reveals that it suffices to take $K \geq 2\max_{0\leq\alpha\leq 1} g'(\alpha)^2|\mathbb{C}|^2/(c\min_{0\leq\alpha\leq 1} g''(\alpha))$.

Exercise 7.5.25 on p. 311: Solution: Use the new variable $w = u+\tau_{\rm R}v$ like in (7.5.31) and denote the piecewise affine and the piecewise constant interpolants respectively as

$$w_\tau = u_\tau + \tau_{\rm R}v_\tau \qquad \text{and} \qquad \overline{w}_\tau = \overline{u}_\tau + \tau_{\rm R}\dot{u}_\tau = \overline{u}_\tau + \tau_{\rm R}\underline{\overline{v}}_\tau.$$

Then write the time-discrete approximation of the force equilibrium (7.5.40a,b) equivalently as

$$\frac{\varrho}{\tau_{\rm R}}\dot{w}_\tau - {\rm div}(\mathbb{C}(\overline{\underline{\alpha}}_\tau)e(\overline{w}_\tau)) = \overline{f}_\tau + \frac{\varrho}{\tau_{\rm R}}\dot{u}_\tau + {\rm div}((\mathbb{C}(\overline{\underline{\alpha}}_\tau){-}\mathbb{C}(\underline{\alpha}_\tau))e(\dot{u}_\tau))$$

and modify (7.5.31) as

$$\int_\Omega \frac{\varrho}{2\tau_{\rm R}}\big|w_\tau(T){-}w(T)\big|^2 + \frac{1}{2}\mathbb{D}_0 e(u_\tau(t){-}u(t)){:}\ e(u_\tau(t){-}u(t))\,{\rm d}x$$
$$+\int_Q \tau_{\rm R}\mathbb{D}_0 e(\dot{u}_\tau{-}\dot{u}){:}\ (\dot{u}_\tau{-}\dot{u}) + \mathbb{C}(\overline{\underline{\alpha}}_\tau)e(\overline{w}_\tau{-}w){:}\ e(\overline{w}_\tau{-}w)\,{\rm d}x{\rm d}t$$
$$\leq \int_Q\Big(\Big(\overline{f}_\tau + \frac{\varrho}{\tau_{\rm R}}(\dot{u}_\tau{-}\dot{w})\Big)\cdot(\overline{w}_\tau{-}w) - (\mathbb{D}_0+\mathbb{C}(\overline{\underline{\alpha}}_\tau))e(w){:}\ e(\overline{w}_\tau{-}w)\Big){\rm d}x{\rm d}t$$
$$+\int_\Sigma \overline{g}_\tau\cdot(\overline{w}_\tau{-}w)\,{\rm d}S\,{\rm d}t + \int_Q(\mathbb{C}(\underline{\alpha}_\tau){-}\mathbb{C}(\overline{\underline{\alpha}}_\tau)e(\dot{u}_\tau){:}\ e(\overline{w}_\tau{-}w)\,{\rm d}x{\rm d}t.$$

The arguments like for (7.5.31) can be used, additionally with the estimate of the last term as

$$\Big|\int_Q(\mathbb{C}(\underline{\alpha}_\tau){-}\mathbb{C}(\overline{\underline{\alpha}}_\tau))e(\dot{u}_\tau){:}\ e(\overline{w}_\tau{-}w)\,{\rm d}x{\rm d}t\Big|$$
$$\leq \big\|\mathbb{C}(\underline{\alpha}_\tau){-}\mathbb{C}(\overline{\underline{\alpha}}_\tau)\big\|_{L^\infty(Q;\mathbb{R}^{d^4})}\big\|e(\dot{u}_\tau)\big\|_{L^2(Q;\mathbb{R}^{d\times d})}\big\|e(\overline{w}_\tau{-}w)\big\|_{L^2(Q;\mathbb{R}^{d\times d})} \to 0$$

the compact embedding of $L^\infty(I;W^{1,p}(\Omega))\cap H^1(I;L^2(\Omega))$ into $C(\overline{Q})$ for $p>d$.

Exercise 7.5.26 on p. 311: Solution: Using also (C.2.44), in terms of (u,α), these potentials are

$$(u,\alpha)\mapsto\int_\Omega\frac{\varrho}{2\tau}\Big|\frac{u-u_\tau^{k-1}}{\tau}-v_\tau^{k-1}\Big|^2+\frac{1}{2\tau}\mathbb{D}(\alpha_\tau^{k-1})e(u-u_\tau^{k-1}){:}e(u-u_\tau^{k-1})$$
$$+2\varphi\Big(e\big(\frac{u+u_\tau^{k-1}}{2}\big),\alpha\Big)+2\tau\zeta_{\rm IN}\Big(\frac{\alpha-\alpha_\tau^{k-1}}{\tau}\Big)+\kappa|\nabla\alpha|^2-f_\tau^k\cdot u\,\mathrm{d}x-\int_\Gamma g_\tau^k\cdot u\,\mathrm{d}S,$$

and

$$(u,\alpha)\mapsto\int_\Omega\frac{\varrho}{2\tau}\Big|\frac{u-u_\tau^{k-1}}{\tau}-v_\tau^{k-1}\Big|^2+\frac{1}{2\tau}\mathbb{D}(\alpha_\tau^{k-1})e(u-u_\tau^{k-1}){:}e(u-u_\tau^{k-1})$$
$$+2\varphi\Big(e\big(\frac{u+u_\tau^{k-1}}{2}\big),\frac{\alpha+\alpha_\tau^{k-1}}{2}\Big)+\tau\zeta_{\rm IN}\Big(\frac{\alpha-\alpha_\tau^{k-1}}{\tau}\Big)+\frac{\kappa}{2}|\nabla\alpha|^2-f_\tau^k\cdot u\,\mathrm{d}x-\int_\Gamma g_\tau^k\cdot u\,\mathrm{d}S,$$

respectively. The critical points of these potentials are to be used for the incremental problems in question when putting

$$u_\tau^k=u,\qquad v_\tau^k=\frac{2}{\tau}(u_\tau^k-u_\tau^{k-1})-v_\tau^{k-1},\qquad \alpha_\tau^k=\alpha.$$

Exercise 7.6.8 on p. 322: Hint: $\varphi(e,c)=\frac12\mathbb{C}_1e_1{:}e_1+\frac12\mathbb{C}_2e{:}e=\frac12\mathbb{C}_1(e-Ec){:}(e-Ec)+\frac12\mathbb{C}_2e{:}e$ and $\alpha(\dot e,\dot c)=\frac12\mathbb{C}_1\dot e_1{:}\dot e_1+\frac12\mathbb{C}_2\dot e{:}\dot e=\frac12\mathbb{C}_1(\dot e-E\dot c){:}(\dot e-E\dot c)+\frac12\mathbb{C}_2\dot e{:}\dot e$.

Exercise 7.6.13 on p. 325: Hint: Write $j=(\mathrm{id}-\nu\Delta)^{-1}M\nabla\mu$ where $(\mathrm{id}-\nu\Delta)^{-1}:J\mapsto j$ is the symmetric linear operator assigning F the unique weak solution to the boundary-value problem for $j-\nu\Delta j=F$ on Ω with $\nabla j\cdot\vec n=0$ on Γ. Then use the usual tests of (7.6.29a) successively by μ and $\dot c$ to obtain the energetics

$$\frac{\mathrm{d}}{\mathrm{d}t}\int_\Omega\frac{\varrho}{2}|\dot u|^2+\varphi(e(u),c)+\frac{\kappa}{2}|\nabla c|^2\,\mathrm{d}t+\int_\Omega\tau_{\rm R}|\dot c|^2+M\big|(\mathrm{id}-\nu\Delta)^{-1/2}\nabla\mu\big|^2\,\mathrm{d}x$$
$$+\int_\Gamma\alpha\mu^2\,\mathrm{d}S=\int_\Omega f\cdot\dot u\,\mathrm{d}x+\int_\Gamma g\cdot\dot u+\alpha\mu_{\rm ext}\mu\,\mathrm{d}S\,.$$

In particular, $(\mathrm{id}-\nu\Delta)^{-1/2}\nabla\mu$ is controlled in $L^2(Q)$. Then the weak formulation of (7.6.29a) is to be devised as

$$\int_Q M(\mathrm{id}-\nu\Delta)^{-1/2}\nabla\mu\cdot(\mathrm{id}-\nu\Delta)^{-1/2}\nabla v-c\dot v\,\mathrm{d}x\mathrm{d}t=\int_\Omega c_0v(0)\,\mathrm{d}x\mathrm{d}t$$

with $\int_Q(\partial_c\varphi(e(u),c)-\mu)v+\kappa\nabla c\cdot\nabla v-\tau_{\rm R}c\dot v\,\mathrm{d}x\mathrm{d}t=\int_\Omega\tau_{\rm R}c_0v(0)\,\mathrm{d}x\mathrm{d}t$ for all v smooth with $v(T)=0$.

Exercise 7.6.15 on p. 329: Hint: denote by (u_M,c_M) the solution to (7.6.1) with φ from (7.6.44) and $\zeta(\dot e)=\frac12\mathbb{D}\dot e{:}\dot e$, i.e.

$$\varrho\ddot u_M-\mathrm{div}\,\sigma_M=f-\nabla p_M\qquad\text{with }\ \sigma_M=\mathbb{C}e(u_M)+\mathbb{D}e(\dot u_M)$$
$$\text{and }\ p_M=M(\beta\mathrm{tr}\ e(u_M)-c_M),$$
$$\dot c_M-\mathrm{div}(\mathbb{M}(c_M)\nabla\mu_M)=0\qquad\text{with }\ \mu_M=M(c_M-\beta\mathrm{tr}\ e(u_M))/\beta.$$

Note that $\partial_e\varphi_M(e_M,c_M) = \sigma_M + p_M\mathbb{I}$ and $\partial_c\varphi_M(e_M,c_M) = \mu_M = -\beta p_M$ with $e_M = e(u_M)$. Testing the force equilibrium by $\dot{u}_M$ and the diffusion equation by μ_M, as in (7.6.4), we have the a-priori estimates

$$\|\dot{u}_M\|_{L^\infty(I;L^2(\Omega;\mathbb{R}^d))\cap H^1(I;H^1(\Omega;\mathbb{R}^d))} \le C, \quad \|\nabla\mu_M\|_{L^2(I;L^2(\Omega))} \le C, \quad \|p_M\|_{L^2(I;L^2(\Omega))} \le C.$$

Then, for a subsequence $u_M \to u$ and $p_M \to p_{\text{fluid}}$ weakly and make a limit passage in the system (7.6.1). Show in particular that $\beta\text{tr } e(u_M)) - c_M \to 0$ so that, in the limit, $\beta\text{tr } e(u) = c$ and thus $\beta\text{tr } e(\dot{u}) = \dot{c}$ so that, in the limit, $\dot{c} - \text{div}(\mathbb{M}(c)\nabla\mu) = 0$ which means $\beta\text{tr } e(\dot{u}) = \text{div}(\mathbb{M}(c)\nabla\mu) = -\beta\text{div}(\mathbb{M}(c)\nabla p_{\text{fluid}})$.

Exercise 7.7.8 on p. 346: Hint: The formula (7.7.27) follows by

$$\int_\Omega \Delta z(\dot{u}\cdot\nabla z)\,\mathrm{d}x = -\int_\Omega \nabla z\cdot\nabla(\dot{u}\cdot\nabla z)\,\mathrm{d}x = -\int_\Omega (\nabla z\otimes\nabla z){:}e(\dot{u}) + (\nabla z\otimes\dot{u}){:}\nabla^2 z\,\mathrm{d}x$$

which can be written as (7.7.27) when also use

$$\int_\Omega (\nabla z\otimes\dot{u}){:}\nabla^2 z\,\mathrm{d}x = -\int_\Omega \text{div}(\nabla z\otimes\dot{u})\cdot\nabla z\,\mathrm{d}x = -\int_\Omega |\nabla z|^2\text{div}\,\dot{u} + (\nabla z\otimes\dot{u}){:}\nabla^2 z\,\mathrm{d}x.$$

Exercise 7.7.9 on p. 346: Solution: Test the momentum equation by $\dot{u}$ and the transport equation (7.7.31) by $\frac{\mathrm{D}z}{\mathrm{D}t}$. Disregarding possible boundary terms, when using the calculus from Example 7.7.6 and taking $\sigma_{\text{K,int}}$ from (7.7.29), one arrives to

$$\int_\Omega \Big|\frac{\mathrm{D}z}{\mathrm{D}t}\Big|^2\mathrm{d}x + \frac{\mathrm{d}}{\mathrm{d}t}\int_\Omega \Big(\frac{\varrho}{2}|\dot{u}|^2 + \sigma{:}e(\dot{u}) + \varphi_{\mathrm{R}}(z) + \frac{\kappa}{2}|\nabla z|^2\Big)\mathrm{d}x = \int_\Omega f\cdot\dot{u} - \varphi_{\mathrm{R}}'(z)(\dot{u}\cdot\nabla z)\,\mathrm{d}x$$

and, assuming $|\varphi_{\mathrm{R}}'(\cdot)|$ bounded, one can estimate the last term by Hölder inequality and then handle it via Gronwall inequality by using the terms $\frac{\varrho}{2}|\dot{u}|^2$ and $\frac{\kappa}{2}|\nabla z|^2$ on the left-hand side to obtain the usual estimates of u (depending on a particular rheology hidden behind the term $\sigma{:}e(\dot{u})$ and now also of z through the coercivity of φ_{R} to be assumed, and then also on $\dot{z} = \frac{\mathrm{D}z}{\mathrm{D}t} - \dot{u}\cdot\nabla z$ by using the estimate $\frac{\mathrm{D}z}{\mathrm{D}t} \in L^2(Q)$ and the already obtained estimates on $\dot{u}$ and ∇z. Alternatively, taking $\sigma_{\text{K,int}}$ from (7.7.32), one arrives to

$$\int_\Omega \Big|\frac{\mathrm{D}z}{\mathrm{D}t}\Big|^2\mathrm{d}x + \frac{\mathrm{d}}{\mathrm{d}t}\int_\Omega \Big(\frac{\varrho}{2}|\dot{u}|^2 + \sigma{:}e(\dot{u}) + \varphi_{\mathrm{R}}(z) + \frac{\kappa}{2}|\nabla z|^2\Big)\mathrm{d}x = \int_\Omega f\cdot\dot{u}\,\mathrm{d}x$$

and the estimation is then even simpler due to the absence of the term $\varphi_{\mathrm{R}}'(z)(\dot{u}\cdot\nabla z)$.

Exercise 7.7.10 on p. 346: Solution: the transport equation $\frac{\mathrm{D}z}{\mathrm{D}t} = \kappa\text{div}(|\nabla z|^{r-2}\nabla z)$ is to be tested by the material time derivative $\frac{\mathrm{D}z}{\mathrm{D}t} = \dot{z} + \dot{u}\cdot\nabla z$:

$$\begin{aligned}\int_\Omega \Big|\frac{\mathrm{D}z}{\mathrm{D}t}\Big|^2 + \frac{\kappa}{r}\frac{\partial}{\partial t}|\nabla z|^r\,\mathrm{d}x &= -\int_\Omega \kappa(\text{div}|\nabla z|^{r-2}\nabla z)(\dot{u}\cdot\nabla z)\,\mathrm{d}x\\ &= \int_\Omega \kappa|\nabla z|^{r-2}(\nabla z\otimes\nabla z){:}e(\dot{u}) - \frac{\kappa}{2}|\nabla z|^r\text{div}\,\dot{u}\,\mathrm{d}x = \int_\Omega \sigma_{\text{K,int}}{:}e(\dot{u})\,\mathrm{d}x\end{aligned}$$

from which $\sigma_{\rm K,int} = \kappa|\nabla z|^{r-2}\nabla z\otimes\nabla z - \kappa|\nabla z|^r\mathbb{I}/2$ can be identified. Here use a modification of the formula (7.7.27), namely:

$$\int_\Omega \mathrm{div}(|\nabla z|^{r-2}\nabla z)(\dot u\cdot\nabla z)\,\mathrm{d}x = \int_\Omega \frac{1}{2}|\nabla z|^r \mathrm{div}\,\dot u - |\nabla z|^{r-2}(\nabla z\otimes\nabla z){:}e(\dot u)\,\mathrm{d}x$$

which follows by $\int_\Omega(\mathrm{div}|\nabla z|^{r-2}\nabla z)(\dot u\cdot\nabla z)\,\mathrm{d}x = -\int_\Omega|\nabla z|^{r-2}\nabla z\cdot\nabla(\dot u\cdot\nabla z)\,\mathrm{d}x = -\int_\Omega|\nabla z|^{r-2}(\nabla z\otimes\nabla z){:}e(\dot u) + |\nabla z|^{r-2}(\nabla z\otimes\dot u){:}\nabla^2 z\,\mathrm{d}x$ and also $\int_\Omega|\nabla z|^{r-2}(\nabla z\otimes\dot u){:}\nabla^2 z\,\mathrm{d}x = -\int_\Omega \mathrm{div}(|\nabla z|^{r-2}\nabla z\otimes\dot u)\cdot\nabla z\,\mathrm{d}x = -\int_\Omega|\nabla z|^r\mathrm{div}\,\dot u + (|\nabla z|^{r-2}\nabla z\otimes\dot u){:}\nabla^2 z\,\mathrm{d}x$.

Exercises in Chapter 8

Exercise 8.2.5 on p. 375: Hint: For (8.2.29) and (8.2.30), use

$$\frac{1}{p} = \frac{\lambda}{r^*} + \frac{1-\lambda}{\omega} \quad\text{and}\quad \frac{1}{p} = \frac{\lambda}{r^{**}} + \frac{1-\lambda}{\omega}$$

with $1/r^* = 1/r - 1/d$ and $1/r^{**} = 1/r^* - 1/d = 1/r - 2/d$, respectively. Realize that it gives just (B.3.19) only with (p,r,ω) instead of (r,p,q) with $\beta = 0$ for $k = 1$ and 2, respectively.

Exercise 8.3.9 on p. 385: Hint: Perform the estimate $\varepsilon\|e(\dot u_k-\dot u)\|^q_{L^q(I;\mathbb{R}^{d\times d})} \le \int_Q(\zeta'(e(\dot u_k))-\zeta'(e(\dot u))) : e(\dot u_k-\dot u)\,\mathrm{d}x\mathrm{d}t \to 0$ when using the Galerkin identity for u_k tested by $\dot u_k-\dot{\tilde u}_k$ with some $\tilde u_k$ valued in the respective finite-dimensional space and approximating u strongly in $W^{1,q}(I,W^{1,q}(\Omega;\mathbb{R}^d))$.

Exercise 8.4.8 on p. 394: Hint: One should consider $\mathbb{D}(\theta_k)$ in (8.4.15) while expanding it by the term $(\mathbb{D}(\theta_k)-\mathbb{D}(\theta))e(\dot u){:}e(u_k-u)$. This term converges to 0 weakly in $L^1(Q)$ provided $\mathbb{D} : \mathbb{R}\to\mathrm{Lin}(\mathbb{R}^{d\times d}_{\rm sym})$ is continuous and bounded, as one should naturally supposed together with uniform positive definiteness. The argumentation (8.4.16) is to be modified for the temperature-dependent dissipation rate, i.e. $\xi = \xi(\theta,\dot e,\dot z)$, relying on the mentioned generalization of Theorem B.2.9.

Exercise 8.4.9 on p. 394: Hint: The term $\partial_e\mathscr{C}_{\rm v}(e(u),z,\theta)\nabla e(u)$ is to be added in (8.4.14) and estimated. The argumentation (8.4.15) is not needed when replaced by Aubin-Lions compactness theorem. The $\mathbb{H}_{\rm v}$-term ensures that $\varrho\ddot u$ is in duality with $\dot u$ and with $\mathrm{div}^2(\mathbb{H}_{\rm v}e(\dot u)+\mathbb{H}e(u))$ needed for (8.4.16) to be written for $\xi_{\rm E}$ instead of ξ.

Exercise 8.6.5 on p. 405: Hint: Realize monotonicity of $\partial_e\varphi(\cdot,c)$ and use the Minty trick for (8.6.2a) like in (7.3.15) on p. 263.

Exercises in Chapter 9

Exercise 9.2.8 on p. 421: Solution: Using the arguments of Sect. 4.5, obtain a global minimizer of the nonconvex functional

$$\int_\Omega \frac{\varphi(\nabla y)}{\tau} + \sqrt{\tau}\Big|\nabla\frac{y-y_\tau^{k-1}}{\tau}\Big|^2 + \tau\varrho\Big|\frac{y-2y_\tau^{k-1}+y_\tau^{k-2}}{\tau^2}\Big|^2 + \frac{\mathscr{H}(\nabla^2 y)}{\tau} - f_\tau^k\cdot y\,\mathrm{d}x - \int_\Gamma g_\tau^k\cdot y\,\mathrm{d}S$$

to be minimized for $y\in H^2(\Omega;\mathbb{R}^d)$, cf. (C.2.36). Chose one of the (not uniquely defined) minimizers and denote it by y_τ^k. Writing (at least formally) the Euler-Lagrange equation for this minimizer (i.e. the boundary value problem (9.2.21)) and testing it by $y_\tau^k - y_\tau^{k-1}$, obtain[6]

$$\int_\Omega \frac{\varrho}{2}|\dot{y}_\tau(t)|^2 + \varphi(\nabla y_\tau(t)) + \frac{1}{2}\mathbb{H}\nabla^2 y_\tau(t)\vdots\nabla^2 y_\tau(t)\,\mathrm{d}x + \sqrt{\tau}\int_0^t\int_\Omega |\nabla\dot{y}_\tau|^2\,\mathrm{d}x\mathrm{d}t$$
$$\le \int_0^t\int_\Omega \bar{f}_\tau\cdot\dot{y}_\tau\,\mathrm{d}x\mathrm{d}t + \int_0^t\int_\Gamma \bar{g}_\tau\cdot\dot{y}_\tau\,\mathrm{d}S\,\mathrm{d}t + \int_\Omega \frac{\varrho}{2}|v_0|^2 + \varphi(\nabla y_0) + \frac{1}{2}\mathbb{H}\nabla^2 y_0\vdots\nabla^2 y_0\,\mathrm{d}x$$

for any $t = k\tau$ with $k = 1,...,T/\tau$ for τ sufficiently small. Then apply the discrete Gronwall inequality together with the by-part integration of the boundary term $\bar{g}_\tau\cdot\dot{y}_\tau$, cf. the estimation strategy (6.6.11)–(6.5.9).

Exercise 9.2.9 on p. 421: Hint: Comparing the global minimizer of (C.2.36) with the previous time level (as it is the ultimate choice to get the stored and the dissipative energies estimated, as in Remark C.2.3 on p. 540), the inertial term itself yields the estimate of the type $\varrho|v_\tau^k - v_\tau^{k-1}|^2 \le \varrho|v_\tau^{k-1}|^2$ with the velocity $v_\tau^k := (y_\tau^k - y_\tau^{k-1})/\tau$, i.e. $\varrho|v_\tau^k|^2 \le 2\varrho v_\tau^k\cdot v_\tau^{k-1} \le \varrho(\frac{1}{2}\varepsilon|v_\tau^k|^2 + 2|v_\tau^{k-1}|^2/\varepsilon)$ with $\varepsilon>0$ to be chosen, which gives $\varrho|v_\tau^k|^2 \le \frac{2}{2-\varepsilon}\varrho|v_\tau^{k-1}|^2$ but it does not lead to an expected uniform bound of $|v_\tau^k|^2$ because always $\frac{2}{2-\varepsilon} > 1$, however.

Exercise 9.2.15 on p. 428: Hint: instead of (2.4.3), consider $\varphi(F) = (\frac{\lambda}{2}(\mathrm{tr}\ E)^2 + G|E|^2)/(1+\varepsilon|E|^2)^{1/2}$ with $E = \frac{1}{2}(F^\top F - \mathbb{I})$. Further, using (A.2.22), it holds $\varphi''(F) + K\mathbb{I} = v(\det F)'' + K\mathbb{I} = (v'(\det F)\mathrm{Cof}\,F)' + K\mathbb{I} = v''(\det F)\mathrm{Cof}\,F\otimes\mathrm{Cof}\,F + v'(\det F)\mathrm{Cof}'\,F + K\mathbb{I}$, which is not positive definite even for big K in general.

Exercise 9.3.6 on p. 434: Hint: Use

---

[6] One is to use the calculation like (C.2.31) but with the viscosity coefficient $\sqrt{\tau}$ instead of 1:

$$\Big(\frac{F_\tau^k - F_\tau^{k-1}}{\sqrt{\tau}} + \varphi'(F_\tau^k)\Big):(F_\tau^k - F_\tau^{k-1}) = \Big(\frac{F_\tau^k}{\sqrt{\tau}} + \varphi'(F_\tau^k)\Big):(F_\tau^k - F_\tau^{k-1}) - F_\tau^{k-1}:\frac{F_\tau^k - F_\tau^{k-1}}{\sqrt{\tau}}$$
$$\ge \frac{1}{2\sqrt{\tau}}|F_\tau^k|^2 + \varphi(F_\tau^k) - \frac{1}{2\sqrt{\tau}}|F_\tau^{k-1}|^2 - \varphi(F_\tau^{k-1}) - F_\tau^{k-1}:\frac{F_\tau^k - F_\tau^{k-1}}{\sqrt{\tau}}$$
$$= \varphi(F_\tau^k) - \varphi(F_\tau^{k-1}) + \tau\frac{\sqrt{\tau}}{2}\Big|\frac{F_\tau^k - F_\tau^{k-1}}{\tau}\Big|^2.$$

$$0 \le \int_Q (S(\nabla y_k) - S(\nabla \tilde{y})) : \nabla (y_k - \tilde{y})\,dxdt$$
$$= \int_Q \dot{y}_k \cdot (\dot{y}_k - \dot{\tilde{y}}) - \mathbb{D}e(\dot{y}_k) : e(y_k - \tilde{y}) - S(\nabla \tilde{y}) : \nabla (y_k - \tilde{y})\,dxdt$$
$$+ \int_\Omega \dot{y}_0 \cdot (y_0 - \tilde{y}(0)) - \dot{y}_k(T) \cdot (\dot{y}_k(T) - \dot{\tilde{y}}(T))\,dx$$

and estimate its limit superior from above by weak upper semi-continuity and then substitute $\tilde{y} := y \pm \varepsilon w$, cancel $\varepsilon > 0$, and eventually pass $\varepsilon \to 0$.

Exercise 9.4.7 on p. 441: Hint: Realize that the partial Gâteaux differentials of the quadratic functional $(F,\Pi) \mapsto \int_\Omega \frac{1}{2}\mathbb{H}\nabla F_{\rm el} \vdots \nabla F_{\rm el}\,dx = \int_\Omega \frac{1}{2}\mathbb{H}\nabla(F\Pi^{-1}) \vdots \nabla(F\Pi^{-1})\,dx$ at (F,Π) are

$$\tilde{F} \mapsto \int_\Omega \mathbb{H}\nabla(F\Pi^{-1}) \vdots \nabla(\tilde{F}\Pi^{-1})\,dx = -\int_\Omega \Pi^{-\top}{\rm div}(\mathbb{H}\nabla(F\Pi^{-1})) : \tilde{F}\,dx \quad \text{and}$$

$$\tilde{\Pi} \mapsto \int_\Omega \mathbb{H}\nabla(F\Pi^{-1}) \vdots \nabla(F(\Pi^{-1})'\tilde{\Pi})\,dx = -\int_\Omega ((\Pi^{-1})'F)^\top {\rm div}(\mathbb{H}\nabla(F\Pi^{-1})) : \tilde{\Pi}\,dx.$$

Exercise 9.4.8 on p. 441: Hint: First, realize that

$$0 \le \int_Q (Z(\dot{\Pi}_k\Pi_k^{-1}) - Z(P_k)) : (\dot{\Pi}_k\Pi_k^{-1} - P_k)\,dxdt$$
$$= \int_Q \Big({\rm div}(\kappa|\nabla\Pi_k|^{p-2}\nabla\Pi_k)\Pi^\top - S_{{\rm in},k}\Pi^\top\Pi^\top - Z(P_k)\Big) : (\dot{\Pi}_k\Pi_k^{-1} - P_k)\,dxdt$$

with $S_{{\rm in},k} = (\nabla y_k)^\top \varphi'_{\rm el}((\nabla y_k)\Pi_k^{-1}) : (\Pi_k^{-1})' + \varphi'_{\rm hd}(\Pi_k)$ and with P_k valued in the corresponding finitely-dimensional space used for the Galerkin method. Here, the Galerkin approximation conformal with $W^{2,\infty}(\Omega;\mathbb{R}^{d\times d})$ is needed. Using also the estimate of ${\rm div}(\kappa|\nabla\Pi_k|^{p-2}\nabla\Pi_k)$ in $L^2(Q;\mathbb{R}^{d\times d})$, the convergence of ${\rm div}(\kappa|\nabla\Pi_k|^{p-2}\nabla\Pi_k)\Pi^\top : (\dot{\Pi}_k\Pi_k^{-1})$ is due to the arguments (9.4.21). Also the strong convergence $\nabla\Pi_k \to \nabla\Pi$ in $L^p(Q;\mathbb{R}^{d\times d})$ holds, cf. (9.4.17) and realize that $Z(\dot{\Pi}_k\Pi_k^{-1})$ is bounded in $L^2(Q;\mathbb{R}^{d\times d})$ while $\Pi_k - \tilde{\Pi}_k \to 0$ strongly even in $L^\infty(Q;\mathbb{R}^{d\times d})$. Use $P_k \to P$ strongly in $L^2(Q;\mathbb{R}^{d\times d})$ and then replace P with $\dot{\Pi}\Pi^{-1} \pm \varepsilon P$ and converge $\varepsilon \to 0+$.

Exercise 9.6.4 on p. 466: Hint: use the fractional-step splitting combined with the vanishing-viscosity regularization:

$$\varrho\frac{y_\tau^k - 2y_\tau^{k-1} + y_\tau^{k-2}}{\tau^2} - {\rm div}\,S_\tau^k = f_\tau^k(y_\tau^k, \nabla y_\tau^k)$$
$$\text{with}\quad S_\tau^k = \partial_F\varphi(\nabla y_\tau^k, c_\tau^{k-1}) - {\rm Div}(\mathbb{H}\nabla^2 y^k) + \frac{\nabla y_\tau^k - \nabla y_\tau^{k-1}}{\sqrt{\tau}}$$
$$\frac{c_\tau^k - c_\tau^{k-1}}{\tau} - {\rm div}(\mathscr{M}(\nabla y_\tau^k, c_\tau^{k-1})\nabla\mu_\tau^k) = 0 \quad \text{with}\quad \mu_\tau^k = \partial_c\phi(\nabla y_\tau^k, c_\tau^k)$$

on Q with the corresponding initial and boundary conditions. Assuming $\tau > 0$ small enough, realize the cancellation effect when tested by $(y_\tau^k - y_\tau^{k-1})/\tau$ and μ_τ^k, respectively.

List of symbols

$\mathrm{AC}(I;V)$	the space of absolutely continuous mappings $I \to V$, p. 522
$\mathrm{BD}(\Omega;\mathbb{R}^d)$	the Banach space of functions with bounded deformations, p. 285
$\mathrm{B}(I;V)$	space of bounded measurable mappings $z: I \to V$, p. 522
$\mathfrak{B}(S)$	σ-algebra of Borel subsets of a topological space S, p. 506
$\mathrm{BV}(I;V)$	Banach space of mappings $I \to V$ of bounded variation, p. 522
$\mathbb{C}$	4th-order tensor of elastic moduli, p. 162
$C = F^\top F$	right Cauchy-Green strain tensor, p. 12
$C^{0,\lambda}(\Omega)$	space of Lipschitz ($\lambda = 1$) or Hölder continuous functions, p. 485
$C_\mathrm{w}(I;V)$	Banach space of weakly continuous mappings $I \to V$, p. 523
$\mathrm{cl}(A)$	the closure of a set A in a metric space, p. 484
$\mathrm{Cof}\,A$	the cofactor matrix of A, p. 491
d	dimension of the space, usually considered $d = 2$ or 3, p. 3
$\mathscr{D}(\cdot)$	infinitely differentiable functions with a compact support, p. 513
$\mathrm{D}A(u,v)$	directional derivative of a mapping A at u in the direction v, p. 494
$\mathrm{dev}\,e$	the deviatoric strain, p. 240
$\mathrm{diam}(U)$	the diameter of the set U, p. 506
div	the divergence (as a scalar of vector), p. 500
div_S	surface divergence, p. 520
Div	the divergence (as a tensor), p. 40
$\mathrm{d}S$	the $(d{-}1)$-dimensional surface measure on Γ used in $\int_\Gamma \ldots \mathrm{d}S$, p. 516
$E = (C - \mathbb{I})/2$	Green-Lagrange (also called Green-St. Venant) strain tensor, p. 12
$\mathrm{epi}(f)$	the epigraph of a functional f, p. 497
ess inf	essential supremum with respect to Lebesgue measure, p. 508
ess sup	essential supremum with respect to Lebesgue measure, p. 508
$e = e(u)$	small-strain tensor $\mathrm{sym}(\nabla u) = \frac{1}{2}\nabla u + \frac{1}{2}(\nabla u)^\top$, p. 161
$F = \nabla y$	deformation gradient, p. 4
G	shear modulus or a placeholder for $\nabla^2 y$ or ∇e, p. 36, p. 39
$\mathfrak{g}_\mathrm{c}$	a fracture toughness (energy needed for damage), p. 297
$\mathrm{GL}(d)$	the group of $d \times d$ matrices with nonzero determinant, p. 492
$\mathrm{GL}^+(d)$	the subgroup of $\mathrm{GL}(d)$ of matrices with positive determinant, p. 492
$\mathrm{Gr}F$	graph of a (set-valued) mapping F, p. 506

M. Kružík and T. Roubíček, *Mathematical Methods in Continuum Mechanics of Solids*, Interaction of Mechanics and Mathematics,
https://doi.org/10.1007/978-3-030-02065-1

$\mathcal{GY}^p(\Omega;\mathbb{R}^{d\times d})$ the set of gradient Young measures, p. 551
$\mathcal{H}^d$ d-dimensional Hausdorff measure, p. 507
$H^k(\Omega;\mathbb{R}^m)$ the Sobolev space $W^{k,2}(\Omega;\mathbb{R}^m)$, p. 514
$H^k_0(\Omega)$ the subspace of $H^k(\Omega)$ with zero Dirichlet conditions on $\partial\Omega$, p. 516
$H^k_{\rm D}(\Omega)$ the subspace of $H^k(\Omega)$ with zero Dirichlet conditions on $\Gamma_{\rm D}$, p. 326
$H^{-k}(\Omega)$ the dual space to $H^k_0(\Omega)$, p. 516
$I:=[0,T]$ the time interval, T a fixed time horizon, p. 194, p. 522
$\mathbb{I}\in\mathbb{R}^{d\times d}$ the identity matrix, p. 489
$\mathrm{int}(A)$ the interior of a set A in a metric space, p. 484
$\mathrm{ISO}:\mathbb{R}^2\to\mathrm{SLin}(\mathbb{R}^{d\times d}_{\rm sym})$ the visco/elastic moduli , p. 242
$\mathbb{K}$ a heat-conductivity $(d\times d)$-matrix (tensor), p. 188
K bulk modulus, p. 36
$K_y:\Omega\to[0;+\infty]$ distortion of $y:\Omega\to\mathbb{R}^d$, p. 68
meas_d d-dimensional Lebesgue measure, p. 508
$\liminf/\limsup$ limit inferior/superior, p. 485
$\mathrm{Lin}(\mathbb{R}^{d\times d}_{\rm sym})$ the set 4th order tensors with minor symmetry, p. 198
$\mathrm{Lin}(V_1,V_2)$ the linear space of continuous linear operators $V_1\to V_2$, p. 486
$\mathrm{Lin}(V)$ the linear space of continuous linear operators $V\to V$, p. 486
$L^p(\Omega;\mathbb{R}^m)$ Lebesgue space of functions with p-powers integrable, p. 508
$L^p(I;V)$ Bochner measurable p-integrable functions $I\to V$, p. 524
$L^p_{\rm w*}(I;V)$ weakly* measurable p-integrable functions $I\to V$, p. 524
$\mathbb{M}$ a mobility $(d\times d)$-matrix (tensor), p. 75
$\mathsf{M}(F)$ the vector of all minors of the matrix F, p. 52
$\mathcal{M}(S)$ Banach space of measures on S, p. 505
$\mathfrak{M}(S)$ Banach space of Radon measures on a topological space S, p. 505
$\mathfrak{M}^+_1(S)$ the set of probability measures, p. 506
$\vec{n}$ unit outward normal to $\partial\Omega$, p. 516
p' conjugate exponent to p, see (B.2.6), p. 509
p^* Sobolev exponent to p, see (B.3.7), p. 515
$p^\sharp$ Sobolev trace exponent to p, see (B.3.12), p. 516
$\mathscr{P}_F$ a pullback of a vector or tensor, p. 6
Q the times/space cylinder $I\times\Omega$, p. 199
q electric charge, p. 173
$\mathbb{R}^{d\times d}$ the set of all $d\times d$ matrices, p. 3
$\mathbb{R}^{d\times d}_{\rm sym}$ symmetric $(d\times d)$-matrices, p. 490
$\mathbb{R}^{d\times d}_{\rm dev}$ the set of all symmetric trace-free $(d\times d)$-matrices, p. 276
$\mathrm{Riem}(u,\Pi)$ Riemann sum of $u:I\to\mathbb{R}$ on the partition Π of I, p. 523
$\mathfrak{s}$ entropy, p. 358
$S:\bar\Omega\to\mathbb{R}^{d\times d}$ 1st Piola-Kirchhoff stress tensor , p. 18
S^{d-1} the unit sphere in $\mathbb{R}^d$, p. 13
$S_{\rm EL}$ a convex closed set containing origin = an elasticity domain, p. 253
$\mathrm{skew}(A)$ skew symmetric part of the matrix A, p. 490
$\mathrm{SL}(d)$ $d\times d$ matrices , p. 492
$\mathrm{SLin}(\mathbb{R}^{d\times d}_{\rm sym})$ the set 4th order tensors with major symmetry, p. 45

$\mathrm{SLin}(V_1,V_2)$, $\mathrm{SLin}(V)$ the linear space of symmetric (self-adjoint) continuous linear operators, p. 486, p. 487
$\mathrm{SO}(d)$ special orthogonal group of $d\times d$ matrices, p. 492
$\mathrm{sph}\, e$ the spherical (volumetric) strain, p. 240
$\mathrm{sym}(A)$ symmetric part of the matrix A, p. 490
T^y Cauchy stress tensor, p. 16
$t^y : \bar{\Omega}^y \times S^{d-1} \to \mathbb{R}^d$ Cauchy stress vector, p. 13
w internal energy, p. 358
$W^{1,p}(I;V)$ Sobolev-Bochner space of functions $I \to V$, p. 525
$W^{k,p}(\Omega)$ Sobolev space of functions with all kth derivatives in $L^p(\Omega)$, p. 514
$W^{\alpha,p}(\Omega)$ Sobolev-Slobodeckiĭ space (α non-integer), p. 519
$W_0^{k,p}(\Omega)$ space of functions from $W^{k,p}(\Omega)$ with zero traces on Γ, p. 516
$W^{1,p}_{\Gamma_{\mathrm{D}}}(\Omega$ the set of functions from $W^{1,p}_{\Gamma_{\mathrm{D}}}$ with fixed traces on Γ_{D}, p. 77
$W^{-k,p}(\Omega)$ the dual space to $W_0^{k,p}(\Omega)$, p. 516
x^y the point x mapped by deformation y, i.e. $y(x)$, or space variable, p. 4
$y : \bar{\Omega} \to \mathbb{R}^d$ deformation of $\bar{\Omega}$, p. 4
$\mathcal{Y}(\Omega;\mathbb{R}^N)$ the set of Young measures, p. 550
z a general internal variable, p. 248
α a scalar-valued internal variable describing damage (or aging), p. 290
χ_A the characteristic function of a set A (valued in $\{0,1\}$), p. 508
δ_s Dirac measure supported at $s \in S$, p. 506
δ_K indicator function of the set K (valued in $\{0,+\infty\}$), p. 499
ϕ potential (magneto/electro-static or gravitational), p. 71, p. 158, p. 205
Γ the boundary $\partial\Omega$ of the domain Ω, p. 514
Γ_{D}, Γ_{N} parts of Γ for Dirichlet/Neumann conditions, p.23, p.199, p. 529
μ chemical potential, p.75, p. 315
$\Omega \subset \mathbb{R}^d$ a bounded domain, p.3, p. 513
$\bar{\Omega} := \mathrm{cl}(\Omega)$ the closure of the domain Ω, p. 514
ϱ mass density, p. 19
Σ 2nd Piola-Kirchhoff tensor or the boundary $I\times\Gamma$, p.18, p. 199
Σ_{D}, Σ_{N} the times/space boundary $I{\times}\Gamma_{\mathrm{D}}$ or $I{\times}\Gamma_{\mathrm{N}}$, p. 199
$\vec{\tau}_{\mathrm{D}}$ the traction vector on Γ_{D}, p. 202
$(\cdot)'$ Gâteaux differential, p. 494
$(\cdot)^*$ the dual space, p. 486, or the convex conjugate functional, p. 498
$(\cdot)^{\boldsymbol{\cdot}} = \frac{\partial}{\partial t}$ or $= \frac{\mathrm{d}}{\mathrm{d}t}$ the (partial or total) time derivative, p. 194, p. 525
$|\cdot|$ a variation of a measure, p.505, or an Euclidean norm, p. 508
∇ the gradient, p. 500
∇_{S} surface gradient, p. 520
$\langle\cdot,\cdot\rangle_{V^*\times V}$ the duality pairing $V^* \times V \to \mathbb{R}$, p. 486
$(\cdot)^*$ the dual space or the adjoint mapping, p. 486
$\rightrightarrows$ a set-valued map, p. 506
$\rightharpoonup$ weak convergence in a normed linear space, p. 486
$\overset{*}{\rightharpoonup}$ weak* convergence in the dual to a normed linear space, p. 486
$\overset{\Gamma}{\longrightarrow}$ Γ-convergence of functionals, p. 501

∂	a partial derivative, or the boundary of a set, p. 484, or the convex subdifferential, p. 497
$\#(\cdot)$	the counting measure, p. 507
$\{\cdot\}_{k\in\mathbb{N}} \subset X$	a sequence, p. 484

References

[1] H. Abels, S. Bosia, and M. Grasselli. Cahn-Hilliard equation with nonlocal singular free energies. *Annali Mat. Pura Applicata*, 194:1071–1106, 2015.

[2] E. Acerbi and N. Fusco. Semicontinuity problems in the calculus of variations. *Arch. Ration. Mech. Anal.*, 86:125–145, 1984.

[3] R.A. Adams and J.J.F. Fournier. *Sobolev Spaces*. Acad. Press/Elsevier, Amsterdam, 2 edition, 2003.

[4] E.C. Aifantis. On the microstructural origin of certain inelastic models. *ASME J. Eng. Mater. Technol.*, 106:326–330, 1984.

[5] H.-D. Alber. *Materials with Memory*. Springer, Berlin, 1998.

[6] J.J. Alibert and B. Dacorogna. An example of a quasiconvex function that is not polyconvex in dimension two. *Arch. Ration. Mech. Anal.*, 117:155–166, 1992.

[7] M. Ambati, R. Kruse, and L. De Lorenzis. A phase-field model for ductile fracture at finite strains and its experimental verification. *Comput. Mech.*, 57:149–167, 2016.

[8] L. Ambrosio, N. Fusco, and D. Pallara. *Functions of Bounded Variation and Free Discontinuity Problems*. Clarendon Press, Oxford, 2000.

[9] L. Ambrosio, N. Gigli, and G. Savare. *Gradient Flows: In Metric Spaces and in the Space of Probability Measures*. Birkhäuser, Basel, 2008.

[10] L. Ambrosio and V. M. Tortorelli. On the approximation of free discontinuity problems. *Bollettino Unione Mat. Italiana*, 7:105–123, 1992.

[11] L. Ambrosio and V.M. Tortorelli. Approximation of functional depending on jumps via by elliptic functionals via Γ-convergence. *Comm. Pure Appl. Math.*, 43:999–1036, 1990.

[12] E.N. da Costa Andrade. On the viscous flow in metals, and applied phenomena. *Proc. R. Soc. London Ser. A*, 84:1–12, 1910.

[13] S.S. Antman. Physically unacceptable viscous stresses. *Zeitschrift angew. Math. Physik*, 49:980–988, 1998.

[14] S.S. Antman. *Nonlinear Problems of Elasticity*. Springer, New York, 2nd edition, 2005.

M. Kružík and T. Roubíček, *Mathematical Methods in Continuum Mechanics of Solids*, Interaction of Mechanics and Mathematics,
https://doi.org/10.1007/978-3-030-02065-1

[15] O. Anza Hafsa and J.-P. Mandallena. Relaxation theorems in nonlinear elasticity. *Ann. Inst. Henri Poincaré, Anal. Non Linéaire*, 25:135–148, 2008.

[16] G. Anzellotti and S. Luckhaus. Dynamical evolution of elasto-perfectly plastic bodies. *Appl. Math. Optim.*, 15:121–140, 1987.

[17] N. Aronszajn. Boundary values of functions with finite dirichlet integral. *Tech. Report of Univ. of Kansas*, 14:77–94, 1955.

[18] R.B. Ash. *Real Analysis and Probablity*. Acad. Press, New York, 1972.

[19] T.M. Atanacković, S. Pilipović, B. Stanković, and D. Zorica. *Fractional Calculus with Applications in Mechanics*. ISTE/J.Wiley, London/Hoboken, 2014.

[20] J.-P. Aubin. Un théorème de compacité. *C.R. Acad. Sci.*, 256:5042–5044, 1963.

[21] S. Aubry, M. Fago, and M. Ortiz. A constrained sequential-lamination algorithm for the simulation of sub-grid microstructure in martensitic materials. *Computer Methods in Applied Mechanics and Engineering*, 192(26):2823–2843, 2003.

[22] R.J. Aumann and S. Hart. Bi-convexity and bi-martingales. *Israel J. Math.*, 54:159–180, 1986.

[23] R. Awi and W. Gangbo. A polyconvex integrand; Euler-Lagrange equations and uniqueness of equilibrium. *Arch. Ration. Mech. Anal.*, 214:143–182, 2014.

[24] Y. Bai and Z. Li. Numerical solution of nonlinear elasticity problems with Lavrentiev phenomenon. *Math. Models Meth. Appl. Sci.*, 17:1619–1640, 2007.

[25] E.J. Balder. A general approach to lower semicontinuity and lower closure in optimal control theory. *SIAM J. Control Optim.*, 22:570–598, 1984.

[26] J.M. Ball. Convexity conditions and existence theorems in nonlinear elasticity. *Arch. Ration. Mech. Anal.*, 63:337–403, 1977.

[27] J.M. Ball. Global invertibility of Sobolev functions and the interpenetration of matter. *Proc. R. Soc. Edinb., Sect. A*, 88:315–328, 1981.

[28] J.M. Ball. A version of the fundamental theorem for Young measures. In *PDEs and Continuum Models of Phase Transitions (Nice, 1988)*, pages 207–215. Springer, Berlin, 1989.

[29] J.M. Ball. Some open problems in elasticity. In P. Newton, P. Holmes, and A. Weinstein, editors, *Geometry, Mechanics, and Dynamics*, pages 3–59. Springer, New York, 2002.

[30] J.M. Ball. Progress and puzzles in nonlinear elasticity. In J. Schröder and P. Neff, editors, *Poly-, Quasi- and Rank-One Convexity in Applied Mechanics*, CISM Intl. Centre for Mech. Sci. 516, pages 1–15. Springer, Wien, 2010.

[31] J.M. Ball, J.C. Currie, and P.J. Olver. Null Lagrangians, weak continuity, and variational problems of arbitrary order. *J. Funct. Anal.*, 41:135–174, 1981.

[32] J.M. Ball and R.D. James. Fine phase mixtures as minimizers of energy. *Arch. Ration. Mech. Anal.*, 100:13–52, 1987.

[33] J.M. Ball and R.D. James. Proposed experimental tests of a theory of fine microstructure and the two-well problem. *Phil. Trans. Roy. Soc. London Ser. A*, 338:389–450, 1992.
[34] J.M. Ball, B. Kirchheim, and J. Kristensen. Regularity of quasiconvex envelopes. *Calc. Var. Partial Diff. Eqs.*, 11:333–359, 2000.
[35] J.M. Ball and G. Knowles. A numerical method for detecting singular minimizers. *Numer. Math.*, 51:181–197, 1987.
[36] J.M. Ball and K. Koumatos. Quasiconvexity at the boundary and the nucleation of austenite. *Archive Ration. Mech. Anal.*, 219:89–157, 2016.
[37] J.M. Ball and J. Marsden. Quasiconvexity at the boundary, positivity of the second variation and elastic stability. *Arch. Ration. Mech. Anal.*, 86:251–277, 1984.
[38] J.M. Ball and V.J. Mizel. Singular minimizers for regular one-dimensional problems in the calculus of variations. *Bull. Amer. Math. Soc. (N.S.)*, 11:143–146, 07 1984.
[39] J.M. Ball and V.J. Mizel. One-dimensional variational problems whose minimizers do not satisfy the Euler-Lagrange equation. *Arch. Ration. Mech. Anal.*, 90:325–388, 1985.
[40] J.M. Ball and F. Murat. $W^{1,p}$-quasiconvexity and variational problems for multiple integrals. *J. Funct. Anal.*, 58:225–253, 1984.
[41] S. Banach. Sur les opérations dans les ensembles abstraits et leur applications aux équations intégrales. *Fund. Math.*, 3:133–181, 1922.
[42] S. Banach. Sur les fonctionelles linéaires. *Studia Math.*, 1:211–216, 223–239, 1929.
[43] S. Banach. *Théorie des Opérations Linéaires.* M. Garasiński, Warszawa, 1932. Engl. transl. North-Holland, Amsterdam, 1987.
[44] V. Barbu and T Precupanu. *Convexity and Optimization in Banach Spaces.* D. Reidel Publishing Co., Dordrecht, 3 edition, 1986.
[45] M. Barchiesi, D. Henao, and C. Mora-Corral. Local invertibility in Sobolev spaces with applications to nematic elastomers and magnetoelasticity. *Archive Ration. Mech. Anal.*, 224:743–816, 2017.
[46] G.I. Barenblatt and D.D. Joseph. *Collected Papers of R.S. Rivlin: Volume I and II.* Springer, New York, 2013.
[47] S. Bartels and M. Kružík. An efficient approach to the numerical solution of rate-independent problems with nonconvex energies. *Multiscale Modeling & Simulation*, 9:1276–1300, 2011.
[48] Y. Basar and D. Weichert. *Nonlinear Continuum Mechanics of Solids.* Springer, Berlin, 2000.
[49] Z.P. Bažant. Why continuum damage is nonlocal: Justification by quasiperiodic microcrack array. *Mechanics Res. Comm.*, 14:407–419, 1987.
[50] Z.P. Bažant and M. Jirásek. Nonlocal integral formulations of plasticity and damage: Survey of progress. *J. Engr. Mech. ASCE*, 128:1119–1149, 2002.
[51] Z.P. Bažant and M. Jirásek. *Creep and Hygrothermal Effects in Concrete Structures.* Springer, 2018.

[52] J. Bear. *Modeling Phenomena of Flow and Transport in Porous Media*. Springer, Switzerland, 2018.
[53] E. Bécache, P. Joly, and C. Tsogka. A new family of mixed finite elements for the linear elastodynamic problem. *SIAM J. Numer. Anal.*, 39:2109–2132, 2002.
[54] A. Bedford. *Hamilton's Principle in Continuum Mechanics*. Pitman, Boston, 1985.
[55] R. Bellman. The stability of solutions of linear differential equations. *Duke Math. J.*, 10:643–647, 1943.
[56] M.P. Bendsoe. *Optimization of Structural Topology, Shape, And Material*. Springer, Berlin, 1995.
[57] B. Benešová. Global optimization numerical strategies for rate-independent processes. *Journal of Global Optimization*, 50:197–220, 2011.
[58] B. Benešová and M. Kampschulte. Gradient Young measures generated by quasiconformal maps the plane. *SIAM J. Math. Anal.*, 47:4404–4435, 2015.
[59] B. Benešová and M. Kružík. Characterization of gradient Young measures generated by homeomorphisms in the plane. *ESAIM: Control Optim. Calc. Var.*, 22:267–288, 2016.
[60] B. Benešová and M. Kružík. Weak lower semicontinuity of integral functionals and applications. *SIAM Review*, 59:703–766, 2017.
[61] B. Benešová, M. Kružík, and A. Schlömerkemper. A note on locking materials and gradient polyconvexity. *Math. Models Meth. Appl. Sci.*, 28:2367–2401, 2018.
[62] A. Bensoussan, J.L. Lions, and G. Papanicolaou. *Asymptotic Analysis for Periodic Structures*. AMS Chelsea Publ., Providence, 2011.
[63] A. Berezovski, J. Engelbrecht, and G.A. Maugin. *Numerical Simulation of Waves and Fronts in Inhomogeneous Solids*. World Scientific, Singapore, 2008.
[64] A. Berezovski and P. Ván. *Internal Variables in Thermoelasticity*. Springer, Switzerland, 2017.
[65] A. Bertram. *Elasticity and Plasticity of Large Deformations: An Introduction*. Springer, Berlin, 2011.
[66] A. Bertram and R. Glüge. *Solid Mechanics: Theory, Modeling, and Problems*. Springer, 2015.
[67] J.J. Bevan and C.I. Zeppieri. A simple sufficient condition for the quasiconvexity of elastic stored-energy functions in spaces which allow for cavitation. *Calc. Var. Partial Diff. Eqs.*, 55:Art.no.42 (25 pages), 2016.
[68] K. Bhattacharya. *Microstructure of Martensite: Why it Forms and how it Gives Rise to the Shape-memory Effect*. Oxford Univ. Press, Oxford, 2003.
[69] M.A. Biot. General theory of three-dimensional consolidation. *J. Appl. Phys.*, 12:155–164, 1941.
[70] M.A. Biot. Theory of deformation of a porous viscoelastic anisotropic solid. *J. Appl. Phys.*, 27:459–467, 1956.
[71] D. Blanchard and O. Guibé. Existence of a solution for a nonlinear system in thermoviscoelasticity. *Adv. Diff. Eq.*, 5:1221–1252, 2000.

[72] L. Boccardo, A. Dall'aglio, T. Gallouët, and L. Orsina. Nonlinear parabolic equations with measure data. *J. Funct. Anal.*, 147:237–258, 1997.
[73] L. Boccardo and T. Gallouët. Non-linear elliptic and parabolic equations involving measure data. *J. Funct. Anal.*, 87:149–169, 1989.
[74] S. Bochner. Integration von Funktionen, deren Werte die Elemente eines Vektorraumes sind. *Fund. Math.*, 20:262–276, 1933.
[75] D.V. Boger. A highly elastic constant-viscosity fluid. *J. Non-Newtonian Fluid Mechanics*, 3:87–91, 1977.
[76] B. Bojarski and T. Iwaniec. Analytical foundations of the theory of quasiconformal mappings in rn. *Ann. Acad. Sci. Fenn. Ser. AI Math*, 8:257–324, 1983.
[77] B. Bolzano. *Schriften Bd.I: Funktionenlehre.* 1930. After a manuscript from 30ties of 19th century by K.Rychlík in: Abh. Königl. Böhmischen Gesellschaft Wiss. XVI+183+24+IV S.
[78] J. Bonet. Large strain viscoelastic constitutive models. *Intl. J. Solids Structures*, 38:2953–2968, 2001.
[79] M.J. Borden. *Isogeometric analysis of phase-field models for dynamic brittle and ductile fracture.* PhD thesis, Univ. of Texas, Austin, 2012.
[80] G. Bouchitté, A. Mielke, and T. Roubíček. A complete damage problem at small strains. *Zeitschrift angew. Math. Phys.*, 60:205–236, 2009.
[81] B. Bourdin, G.A. Francfort, and J.-J. Marigo. The variational approach to fracture. *J. Elasticity*, 91:5–148, 2008.
[82] B. Bourdin, C.J. Larsen, and C.L. Richardson. A time-discrete model for dynamic fracture based on crack regularization. *Int. J. of Fracture*, 10:133–143, 2011.
[83] B. Bourdin, J.-J. Marigo, C. Maurini, and P. Sicsic. Morphogenesis and propagation of complex cracks induced by thermal shocks. *Phys. Rev. Lett.*, 112:014301, 2014.
[84] A. Braides. *Approximation of Free-Discontinuity Problems.* Springer, Berlin, 1998.
[85] A. Braides. *Gamma-convergence for Beginners.* Oxford Univ. Press, 2002.
[86] A. Braides and A. Defranceschi. *Homogenization of Multiple Integrals.* Clarendon Press, Oxford, 1998.
[87] S.C. Brenner and L.R. Scott. *The Mathematical Theory of Finite Element Methods.* Springer, New York, 3rd edition, 2008.
[88] D. Bresch, B. Desjardins, M. Gisclon, and R. Sart. Instability results related to compressible Korteweg system. *Ann. Univ. Ferrara*, 54:11–36, 2008.
[89] H. Brézis. Équations et inéquations non linéaires dans les espaces vectoriels en dualité. *Annales de l'institut Fourier*, 18:115–175, 1968.
[90] P.W. Bridgman. *The Nature of Thermodynamics.* Harward Univ. Press, Cambridge (MA), 1943.
[91] H.C. Brinkman. A calculation of the viscous force exerted by a flowing fluid on a dense swarm of particles. *Appl. Sci. Res. A*, 1:27–34, 1947.
[92] H.C. Brinkman. On the permeability of media consisting of closely packed porous particles. *Appl. Sci. Res. A*, 1:81–86, 1947.

[93] M. Brokate, C. Carstensen, and J. Valdman. A quasi-static boundary value problem in multi-surface elastoplasticity: part I - analysis. *Math. Meth. Appl. Sci.*, 27:1697–1710, 2004.

[94] L.E. J. Brouwer. Über Abbildungen von Mannigfaltigkeiten. *Math. Annalen*, 71:97–115, 1912.

[95] W.F. Brown, Jr. *Magnetoelastic Interactions*. Springer, Berlin, 1966.

[96] M. Bulíček, E. Feireisl, and J. Málek. Navier-Stokes-Fourier system for incompressible fluids with temperature dependent material coefficients. *Nonlinear Anal., Real World Appl.*, 10:992–1015, 2009.

[97] M. Bulíček, E. Feireisl, J. Málek, and R. Shvydkoy. On the motion of incompressible inhomogeneous Euler-Korteweg fluids. *Disc. Cont. Dynam. Syst. S*, 3:497–515, 2010.

[98] M. Bulíček, J. Málek, and K.R. Rajagopal. On Kelvin-Voigt model and its generalizations. *Evol. Eqs. & Control Theory*, 1:17–42, 2012.

[99] M. Bulíček, J. Málek, K.R. Rajagopal, and J.R. Walton. Existence of solutions for the anti-plane stress for a new class of "strain-limiting" elastic bodies. *Calc. Var. PDE*, 54:2115–2147, 2015.

[100] J.M. Burgers. *Second report on viscosity and plasticity*, chapter III. Amsterdam Acad. of Sci. Nordemann, 1938.

[101] G. Buttazzo, M. Giaquinta, and S. Hildebrandt. *One-dimensional Variational Problems*. Clarendon Press, Oxford, 1998.

[102] J.W. Cahn and J.E. Hilliard. Free energy of a uniform system I., Interfacial free energy. *J. Chem. Phys.*, 28:258–267, 1958.

[103] A. Capella and F. Otto. A rigidity result for a perturbation of the geometrically linear three-well problem. *Comm. Pure Appl. Math.*, 62:1632–1669, 2009.

[104] A. Capella and F. Otto. A quantitative rigidity result for the cubic-to-tetragonal phase transition in the geometrically linear theory with interfacial energy. *Proc. Roy. Soc. Edinburgh Sect. A*, 142:273–327, 2012.

[105] G. Capriz. Continua with latent microstructure. *Arch. Ration. Mech. Anal.*, 90:43–56, 1985.

[106] M. Caputo and F. Mainardi. Linear models of dissipation in anelastic solids. *La Rivista del Nuovo Cimento*, 1:161–198, 1971.

[107] C. Carstensen and C. Ortner. Analysis of a class of penalty methods for computing singular minimizers. *Computational Meth. Appl. Math.*, 10:137–163, 2010.

[108] E. Casadio-Tarabusi. An algebraic characterization of quasi-convex functions. *Ricerche Mat.*, 42:11–24, 1993.

[109] P.P. Castaneda, J.J. Telega, and B. Gambin. *Nonlinear Homogenization and its Applications to Composites, Polycrystals and Smart Materials*. Kluwer, Dordrecht, 2004.

[110] A. Chan and S. Conti. Energy scaling and branched microstructures in a model for shape-memory alloys with SO(2) invariance. *Math. Models Methods Appl. Sci.*, 25:1091–1124, 2015.

[111] K. Chełmiński. Perfect plasticity as a zero relaxation limit of plasticity with isotropic hardening. *Math. Math. Appl. Sci.*, 24:117–136, 2001.
[112] A.H.-D. Cheng. *Poroelasticity*. Springer, Switzerland, 2016.
[113] M. Chipot, C. Collins, and D. Kinderlehrer. Numerical analysis of oscillations in multiple well problems. *Numer. Math.*, 70:259–282, 1995.
[114] P.G. Ciarlet. *Mathematical Elasticity. Vol.I: Three-Dimensional Elasticity*. North-Holland, Amsterdam, 1988.
[115] P.G. Ciarlet. *Mathematical Elasticity. Vol. II: Theory of Plates*. North-Holland, Amsterdam, 1997.
[116] P.G. Ciarlet. *Mathematical Elasticity. Vol. III: Theory of Shells*. Elsevier Science, Amsterdam, 2000.
[117] P.G. Ciarlet. Korn's inequalities: The linear vs. the nonlinear case. *Discrete and Continuous Dynamical Systems - S*, 5:473–483, 2012.
[118] P.G. Ciarlet and J. Nečas. Unilateral problems in nonlinear, three-dimensional elasticity. *Archive Ration. Mech. Anal.*, 87:319–338, 1985.
[119] P.G. Ciarlet and J. Nečas. Injectivity and self-contact in nonlinear elasticity. *Arch. Ration. Mech. Anal.*, 97:171–188, 1987.
[120] D. Cioranescu and P. Donato. *An Introduction to homogenization*. Oxford Univ. Press, Oxford, 1999.
[121] J.A. Clarkson. Uniformly convex spaces. *Trans. Amer. Math. Soc.*, 40:396–414, 1936.
[122] B.D. Coleman and W. Noll. The thermodynamics of elastic materials with heat condution and viscosity. *Arch. Ration. Mech. Anal.*, 13:167–178, 1963.
[123] C. Collins, D. Kinderlehrer, and M. Luskin. Numerical approximation of the solution of a variational problem with a double well potential. *SIAM J. Numer. Anal.*, 28:321–332, 1991.
[124] C. Collins and M. Luskin. Optimal order estimates for the finite element approximation of the solution of a nonconvex variational problem. *Math. Comp.*, 57:621–637, 1991.
[125] S. Conti. A lower bound for a variational model for pattern formation in shape-memory alloys. *Contin. Mech. Thermodyn.*, 17:469–476, 2006.
[126] S. Conti and G. Dolzmann. On the theory of relaxation in nonlinear elasticity with constraints on the determinant. *Arch. Ration. Mech. Anal.*, 217:413–437, 2015.
[127] R.D. Cook, D.S. Malkus, M.E. Plesha, and R.J. Witt. *Concepts and Applications of Finite Element Analysis*. J. Wiley, New York, 4th edition, 2002.
[128] P. Cornetti, N. Pugno, A. Carpinteri, and D. Taylor. Finite fracture mechanics: a coupled stress and energy failure criterion. *Engineering Fracture Mechanics*, 73:2021–2033, 2006.
[129] R. Courant. Variational methods for the solution of problems of equilibrium and vibrations. *Bull. Amer. Math. Soc. (N.S).*, 49:1–23, 1943.
[130] R. Courant, K. Friedrichs, and H. Lewy. Über die partiellen Differenzengleichungen der mathematischen Physik. *Math. Annalen*, 100:32–74, 1928.
[131] O. Coussy. *Poromechanics*. J.Wiley, Chichester, 2004.

[132] S.C. Cowin. *Continuum Mechanics of Anisotropic Materials*. Springer, New York, 2013.

[133] J. Crank and P. Nicolson. A practical method for numerical evaluation of solutions of partial differential equations of the heat conduction type. *Proc. Camb. Phil. Soc.*, 43:50–67, 1947.

[134] B. Dacorogna. *Weak continuity and weak lower semicontinuity of nonlinear functionals*. Springer, Berlin, 1982. Lecture Notes in Math. Vol. 922.

[135] B. Dacorogna. *Direct Methods in the Calculus of Variations*. Springer, New York, 2007.

[136] C.M. Dafermos. Global smooth solutions to the initial boundary value problem for the equations of one-dimensional thermoviscoelasticity. *SIAM J. Math. Anal.*, 13:397–408, 1982.

[137] F.A. Dahlen and J. Tromp. *Theoretical Global Seismology*. Princeton Univ. Press, Princeton, N.J., 1998.

[138] G. Dal Maso. *An Introduction to Γ-Convergence*. Birkhäuser, Boston, 1993.

[139] G. Dal Maso, A. DeSimone, and M.G. Mora. Quasistatic evolution problems for linearly elastic-perfectly plastic materials. *Arch. Ration. Mech. Anal.*, 180:237–291, 2006.

[140] G. Dal Maso, G.A. Francfort, and R. Toader. Quasistatic crack growth in nonlinear elasticity. *Arch. Ration. Mech. Anal.*, 176:165–225, 2005.

[141] G. Dal Maso, M. Negri, and D. Percivale. Linearized elasticity as γ-limit of finite elasticity. *Set-Valued Analysis*, 10(2-3):165–183, 2002.

[142] H. Darcy. *Les Fontaines Publiques de La Ville de Dijon*. Victor Dalmont, Paris, 1856.

[143] R. Dautray and J.-L. Lions. *Mathematical Analysis and Numerical Methods for Science and Technology. Vol.5*. Springer, Berlin, 2000.

[144] E. Davoli, T. Roubíček, and U. Stefanelli. Dynamic perfect plasticity and damage in viscoelastic solids. *Zeit. angew. Math. Mech.*, to appear.

[145] S. de Boer. *Trends in Continuum Mechanics of Porous Media*. Springer, Dordrecht, 2005.

[146] R. de Borst, M.A. Crisfield, J.J.C. Remmers, and C.V. Verhoosel. *Non-linear finite element analysis of solids and structures*. J.Wiley, Chichester, 2nd edition, 2012.

[147] E. De Giorgi. *Γ-convergenza e G-convergenza*. *Boll. Unione Mat. Ital., V. Ser.*, A 14:213–220, 1977.

[148] E. De Giorgi and T. Franzoni. Su un tipo di convergenza variazionale. *Atti Accad. Naz. Lincei Rend. Cl. Sci. Fis. Mat. Natur. (8)*, 58:842–850, 1975.

[149] S.R. de Groot and P. Mazur. *Non-equilibrium thermodynamics*. North-Holland (reprinted Dover Publ.), Amsterdam (New York), 1962 (1984).

[150] F. Dell'Isola, G. Sciarra, and S. Vidoli. Generalized Hooke's law for isotropic second gradient materials. *Proc. Royal Soc. London A*, 465:2177–2196, 2009.

[151] F. Demengel and P. Suquet. On locking materials. *Acta Applicandae Mathematica*, 6:185–211, 1986.

[152] S. Demoulini. Weak solutions for a class of nonlinear systems of viscoelasticity. *Arch. Ration. Mech. Anal.*, 155:299–334, 2000.

[153] S. Demoulini, D. Stuart, and A. Tzavaras. A variational approximation scheme for three dimensional elastodynamics with polyconvex energy. *Arch. Ration. Mech. Anal.*, 157:325–344, 2001.

[154] A. DeSimone. Energy minimizers for large ferromagnetic bodies. *Arch. Ration. Mech. Anal.*, 125:99–143, 1993.

[155] A. DeSimone and R.D. James. A constrained theory of magnetoelasticity. *J. Mech. Phys. Solids*, 50:283–320, 2002.

[156] E. Di Nezza, G. Palatucci, and E. Valdinoci. Hitchhiker's guide to the fractional Sobolev spaces. *Bull. Sci. Math.*, 136:521–573, 2012.

[157] Y.I. Dimitrienko. *Nonlinear Continuum Mechanics and Large Inelastic Deformations*. Springer, Dordrecht, 2011.

[158] R.J. DiPerna. Measure-valued solutions to conservation laws. *Arch. Ration. Mech. Anal.*, 88:223–270, 1985.

[159] R.J. DiPerna and A.J. Majda. Oscillations and concentrations in weak solutions of the incompressible fluid equations. *Comm. Math. Phys.*, 108:667–689, 1987.

[160] G. Dolzmann. *Variational Methods for Crystalline Microstructure – Analysis and Computation*. Springer, Berlin, 2003.

[161] L. Dorfmann and R.W. Ogden. *Nonlinear Theory of Electroelastic and Magnetoelastic Interactions*. Springer, New York, 2014.

[162] W. Dreyer, J. Giesselmann, and C. Kraus. Modeling of compressible electrolytes with phase transition. *Preprint: arXiv:1405.6625*, 2014.

[163] W. Dreyer, J. Giesselmann, C. Kraus, and C. Rohde. Asymptotic analysis for Korteweg models. *Interfaces Free Bound.*, 14:105–143, 2012.

[164] F.P. Duda, A.C. Souza, and E. Fried. A theory for species migration in a finitely strained solid with application to polymer network swelling. *J. Mech. Phys. Solids*, 58:515–529, 2010.

[165] P. Duhem. Le potentiel thermodynamique et la pression hydrostatique. *Ann. Ecole Normale (Paris)*, 10:183–230, 1893.

[166] P. Duhem. *Traité dénergétique ou de thermodynamique générale*. Gauthier-Villars, Paris, 1911. [Facsimile reprint, Gabay, Paris, 1997].

[167] N. Dunford and J.T. Pettis. Linear operators on summable functions. *Trans. Amer. Math. Soc.*, 47:323–392, 1940.

[168] N. Dunford and J.T. Schwartz. *Linear Operators part I: General Theory*. Interscience Publ., New York, 1958.

[169] J.E. Dunn and J. Serrin. On the thermomechanics of interstitial working. *Arch. Ration. Mech. Anal.*, 88:95–133, 1985.

[170] G. Duvaut and J.-L. Lions. *Inequalities in mechanics and physics*. Springer, Berlin, 1976.

[171] C. Eckart. The thermodynamics of irreversible processes Part i: the simple fluid; Part II: fluid mixtures. *Phys. Rev.*, 58:267–269, 269–275, 1940.

[172] G. Eisen. A selection lemma for sequences of measurable sets, and lower semicontinuity of multiple integrals. *Manuscr. Math.*, 27:73–79, 1979.

[173] I. Ekeland and R. Temam. *Convex Analysis and Variational Problems*. North Holland, Amsterdam, 1976.

[174] K.G. Eptaimeros, C.C. Koutsoumaris, and G.J. Tsamasphyros. Nonlocal integral approach to the dynamical response of nanobeams. *Intl. J. Mech. Sci.*, 115–115:68–80, 2016.
[175] A.C. Eringen. Linear theory of nonlocal elasticity and dispersion of plane waves. *Intl. J. Engr. Sci.*, 10:425–435, 1972.
[176] A.C. Eringen. *Nonlocal Continuum Field Theories.* Springer, New York, 2002.
[177] A.C. Eringen (with C.B. Kafadar and D.G.B. Edelen). *Continuum physics IV – Polar and Nonlocal Field Theories.* Acad. Press, New York, 1976.
[178] L.C. Evans. *Weak Convergence Methods for Nonlinear Partial Differential Equations.* AMS, Providence, 1990.
[179] L.C. Evans. *Partial Differential Equations.* AMS, Providence, 1998.
[180] L.C. Evans and R.F. Gariepy. *Measure Theory and Fine Properties of Functions.* CRC Press, Boca Raton, FL, 2nd edition, 2015.
[181] S. Faedo. Un nuovo metodo per l'analysi esistenziale e qualitativa dei problemi di propagazione. *Annali Scuola Norm. Sup. Pisa, Sér.III*, 1:1–40, 1949.
[182] K. Fan. Fixed-point and minimax theorems in locally convex linear spaces. *Proc. Nat. Acad. Sci. USA*, 38:121–126, 1952.
[183] K. Fan and I.L. Glicksberg. Some geometric properties of the spheres in a normed linear space. *Duke Math. J.*, 25:553–568, 1958.
[184] P. Fatou. Séries trigonométriques et séries de Taylor. *Acta Math.*, 30:335–400, 1906.
[185] H. Federer. *Geometric Measure Theory.* Springer, Berlin, 1969.
[186] E. Feireisl and J. Málek. On the Navier-Stokes equations with temperature-dependent transport coefficients. *Diff. Equations Nonlin. Mech.*, pages 14pp.(electronic), Art.ID 90616, 2006.
[187] A.E. Fick. Über Diffusion. *Ann. Phys. Chemie*, 94:59–86, 1855.
[188] M. Focardi. On the variational approximation of free-discontinuity problems in the vectorial case. *Math. Models Methods Appl. Sci.*, 11:663–684, 2001.
[189] I. Fonseca. Interfacial energy and the Maxwell rule. *Arch. Ration. Mech. Anal.*, 106:63–95, 1989.
[190] I. Fonseca and W. Gangbo. *Degree Theory in Analysis and Applications.* Clarendon Press, Oxford, 1995.
[191] I. Fonseca, D. Kinderlehrer, and P. Pedregal. Energy functional depending on elastic strain and chemical composition. *Calc. Var. PDE*, 2:283–313, 1994.
[192] I. Fonseca and G. Leoni. *Modern Methods in the Calculus of Variations: L^p Spaces.* Springer, New York, 2007.
[193] I. Fonseca, S. Müller, and P. Pedregal. Analysis of concentration and oscillation effects generated by gradients. *SIAM J. Math. Anal.*, 29:736–756, 1998.
[194] Fortuné, D. and Vallée, C. Bianchi identities in the case of large deformations. *Intl. J. Engr. Sci.*, 39:113–123, 2001.
[195] R. Fosdick and G. Royer-Carfagni. The constraint of local injectivity in linear elasticity theory. *Proc. Royal Soc. London A*, 457(2013):2167–2187, 2001.

[196] R. Fosdick and G. Royer-Carfagni. The Lagrange multipliers and hyperstress constraint reactions in incompressible multipolar elasticity theory. *J. Mech. Phys. Solids*, 50:1627–1647, 2002.
[197] R.L. Fosdick and J. Serrin. On the impossibility of linear Cauchy and Piola-Kirchhoff constitutive theories for stress in solids. *J. Elasticity*, 9:83–89, 1979.
[198] M. Foss, W.J. Hrusa, and V.J. Mizel. The Lavrentiev gap phenomenon in nonlinear elasticity. *Arch. Ration. Mech. Anal.*, 167:337–365, 2003.
[199] G. Francfort and A. Giacomini. On periodic homogenization in perfect elasto-plasticity. *J. Eur. Math. Soc.*, 16:409–461, 2014.
[200] G. Francfort and A. Mielke. Existence results for a class of rate-independent material models with nonconvex elastic energies. *J. reine angew. Math.*, 595:55–91, 2006.
[201] M. Frémond. *Non-Smooth Thermomechanics*. Springer, Berlin, 2002.
[202] M. Frémond. *Phase Change in Mechanics*. Springer, Berlin, 2012.
[203] E. Fried and M.E. Gurtin. Tractions, balances, and boundary conditions for nonsimple materials with application to liquid flow at small-lenght scales. *Arch. Ration. Mech. Anal.*, 182:513–554, 2006.
[204] A. Friedman and J. Nečas. Systems of nonlinear wawe equations with nonlinear viscosity. *Pacific J. Math.*, 135:29–55, 1988.
[205] M. Friedrich and M. Kružík. On the passage from nonlinear to linearized viscoelasticity. *SIAM J. Math. Anal.*, 50:4426–4456, 2018.
[206] G. Friesecke, R.D. James, and S. Müller. A theorem on geometric rigidity and the derivation of nonlinear plate theory from three-dimensional elasticity. *Comm. Pure Appl. Math.*, 55:1461–1506, 2002.
[207] G. Friesecke, R.D. James, and S. Müller. A hierarchy of plate models derived from nonlinear elasticity by Gamma-convergence. *Arch. Ration. Mech. Anal.*, 180:183–236, 2006.
[208] G. Fubini. Sugli integrali multipli. *Rend. Accad. Lincei Roma*, 16:608–614, 1907.
[209] J. Fuhrmann. Mathematical and numerical modeling of flow, transport and reactions in porous structures of electrochemical devices. In P. Bastian et al., editor, *Simulation of Flow in Porous Media*, pages 139–162. DeGruyter, Berlin, 2013.
[210] E. Gagliardo. Proprietà di alcune classi di funzioni in più variabili. *Ricerche Mat.*, 7:102–137, 1958.
[211] H. Gajewski and K. Zacharias. On a nonlocal phase separation model. *J. Math. Anal. Appl.*, 286:11–31, 2003.
[212] B.G. Galerkin. Series development for some cases of equilibrium of plates and beams. (In Russian). *Vestnik Inzhinierov Teknik*, 19:897–908, 1915.
[213] I.G. García, B.J. Carter, A.R. Ingraffea, and V. Mantič. A numerical study of transverse cracking in cross-ply laminates by 3D finite fracture mechanics. *Composites Part B*, 95:475–487, 2016.
[214] H. Garcke. On Cahn-Hilliard system with elasticity. *Proc. Roy. Soc. Edinburgh*, 133A:307–331, 2003.

[215] C.W. Gear. *Numerical Initial Value Problems in Ordinary Differential Equations*. Prentice-Hall, Englewood Cliffs, 1971.

[216] B.R. Geldbaum and J.M.H. Olmsted. *Counterexamples in Analysis*. Dover Publications Inc., Mineola New York, 2003.

[217] P. Germain. Mathematics and mechanics: two very close neighbours. In *Proc. ICIAM'87*, Philadelphia, 1988. SIAM.

[218] G. Geymonat and P. Suquet. Functional spaces for Norton-Hoff materials. *Math. Meth. Appl. Sci.*, 8:206–222, 1986.

[219] S. Ghosh, V. Sundararaghavan, and A.M. Waas. Construction of multi-dimensional isotropic kernels for nonlocal elasticity based on phonon dispersion data. *Intl. J. Solids Structures*, 51:392–401, 2014.

[220] A. Giacomin and J.L. Lebowitz. Phase segregation dynamics in particle systems with long range interactions. I. Macroscopic limits. *J. Statistical Phys.*, 87:37–61, 1997.

[221] A. Giacomini. Ambrosio-Tortorelli approximation of quasi-static evolution of brittle fractures. *Calc. Var. Partial Diff. Eqs.*, 22:129–172, 2005.

[222] E. Giusti. *Direct Methods in Calculus of Variations*. World Scientific, Singapore, 2003.

[223] F. Golay and P. Seppecher. Locking materials and the topology of optimal shapes. *Europ. J. Mech. - A/Solids*, 20:631–644, 2001.

[224] S. Govindjee and J.C. Simo. Coupled stress-diffusion: case II. *J. Mech. Phys. Solids*, 41:863–887, 1993.

[225] Y. Grabovsky. From microstructure-independent formulas for composite materials to rank-one convex, non-quasiconvex functions. *Arch. Ration. Mech. Anal.*, 227:607–636, 2018.

[226] D. Grandi and U. Stefanelli. Finite plasticity in $P^{\top}P$. Part I: constitutive model. *Cont. Mech. Thermodyn*, 29:97–116, 2017.

[227] A. Green and P. Naghdi. A general theory of an elastic-plastic continuum. *Arch. Ration. Mech. Anal.*, 18:251–281, 1965.

[228] A.E. Green and R.S. Rivlin. Multipolar continuum mechanics. *Arch. Ration. Mech. Anal.*, 17:113–147, 1964.

[229] G. Green. An essay on the application of mathematical analysis to the theories on electricity and magnetism. Nottingham, 1828.

[230] T.H. Gronwall. Note on the derivatives with respect to a parameter of the solution of a system of differential equations. *Ann. Math.*, 20:292–296, 1919.

[231] M.E. Gurtin. *An Introduction to Continuum Mechanics*. Acad. Press, New York, 1981.

[232] M.E. Gurtin. *Topics in Finite Elasticity*. SIAM, Philadelphia, 1983.

[233] M.E. Gurtin. Generalized Ginzburg-Landau and Cahn-Hilliard equations based on a microforce balance. *Physica D*, 92:178–192, 1996.

[234] M.E. Gurtin, E. Fried, and L. Anand. *The Mechanics and Thermodynamics of Continua*. Cambridge Univ. Press, New York, 2010.

[235] K. Hackl. Generalized standard media and variational principles in classical and finite strain elastoplasticity. *J. Mech. Phys. Solids*, 45:667–688, 1997.

[236] K. Hackl and F. D. Fischer. On the relation between the principle of maximum dissipation and inelastic evolution given by dissipation potential. *Proc. Royal Soc. A*, 464:117–132, 2007.

[237] H. Hahn. Über eine Verallgemeinerung der Riemannschen Inetraldefinition. *Monatshefte Math. Physik*, 26:3–18, 1915.

[238] H. Hahn. Über lineare gleichungsysteme in linearen Räume. *J. Reine Angew. Math.*, 157:214–229, 1927.

[239] B. Halphen and Q. S. Nguyen. Sur les matériaux standards généralisés. *J. Mécanique*, 14:39–63, 1975.

[240] W.R. Hamilton. On a general method in dynamics, part II. *Phil. Trans. Royal Soc.*, pages 247–308, 1834.

[241] W. Han and B.D. Reddy. *Plasticity: Mathematical Theory and Numerical Analysis*. Springer, New York, 1999.

[242] J. Haslinger and P. Neittaanmäki. *Finite Element Approximation for Optimal Shape, Material and Topology Design*. J.Wiley, 1996.

[243] P. Haupt. *Continuum Mechanics and Theory of Materials*. Springer, Berlin, second edition, 2002.

[244] F. Hausdorff. Dimension und äusseres Mass. *Math. Annalen*, 79:157–179, 1918.

[245] T.J. Healey and S. Krömer. Injective weak solutions in second-gradient nonlinear elasticity. *ESAIM: Control, Optim. & Cal. Var.*, 15:863–871, 2009.

[246] M. Heida and J. Málek. On compressible Korteweg fluid-like materials. *Intl. J. Engr. Sci.*, 48:1313–1324, 2010.

[247] E. Helly. Über lineare Funktionaloperationen. *Sitzungsberichte der Math.-Natur. Klasse der Kaiserlichen Akademie der Wissenschaften*, 121:265–297, 1912.

[248] S. Hencl and P. Koskela. *Lectures on Mappings of Finite Distortion*. Springer, Cham, 2014.

[249] H.M. Hilber, T.J.R. Hughes, and R.L. Taylor. Improved numerical dissipation for time integration algorithms in structural dynamics. *Earthquake Eng. Struct. Dyn.*, 5:283–292, 1977.

[250] D. Hilbert. Mathematische Probleme. *Archiv d. Math. u. Physik*, 1:44–63, 213–237, 1901. Engl. transl.: *Bull. Amer. Math. Soc.* **8** (1902), 437–479.

[251] R. Hill. A variational principle of maximum plastic work in classical plasticity. *Q.J. Mech. Appl. Math.*, 1:18–28, 1948.

[252] I. Hlaváček, J. Haslinger, J. Nečas, and J. Lovíšek. *Solution of Variational Inequalities in Mechanics*. Springer, New York, 1988.

[253] O. Hölder. Über einen Mittelwertsatz. *Nachr. Ges. Wiss. Göttingen*, pages 38–47, 1889.

[254] W. Hong and X. Wang. A phase-field model for systems with coupled large deformation and mass transport. *J. Mech. Phys. Solids*, 61:1281–1294, 2013.

[255] G.T. Houlsby and A.M. Puzrin. *Principles of Hyperplasticity, An Approach to Plasticity Theory Based on Thermodynamic Principles*. Springer, London, 2006.

[256] M.M. Hrabok and T.M. Hrudley. A review and catalog of plate bending finite elements. *Computers and Structures*, 19:479–495, 1984.

[257] W.D. Iwan. On a class of models for the yielding behavior of continuous and composite systems. *J. Applied Mech*, 34:612–617, 1967.

[258] A.J. Izzo. Existence of continuous functions that are one-to-one almost everywhere. *Math. Scand.*, 118:269–276, 2016.

[259] R.D. James and D. Kinderlehrer. Frustration in ferromagnetic materials. *Cont. Mech. Thermodyn.*, 2:215–239, 1990.

[260] R.D. James and D. Kinderlehrer. Theory of magnetostriction with applications to $Tb_xDy_{1-x}Fe_2$. *Phil. Mag. B*, 68:237–274, 1993.

[261] R.D. James and D. Kinderlehrer. Theory of magnetostriction with application to terfenol d. *J. Appl. Phys.*, 76:7012–7014, 1994.

[262] J. Jani, M. Leary, A. Subic, and M. Gibson. A review of shape memory alloy research, applications and opportunities. *Materials & Design*, 56:1078–1113, 2014.

[263] H. Jeffreys. *The Earth*. Cambridge Univ. Press, Cambridge, 1929.

[264] V.V. Jikov, S.M. Kozlov, and O.A. Oleinik. *Homogenization of Differential Operators and Integral Functionals*. Springer, Berlin, 2012.

[265] M. Jirásek. Nonlocal theories in continuum mechanics. *Acta Polytechnica*, 44:16–34, 2004.

[266] M. Jirásek and Z.P. Bažant. *Inelastic Analysis of Structures*. J.Wiley, Chichester, 2002.

[267] L.M. Kachanov. Time of rupture process under creep conditions. *Izv. Akad. Nauk SSSR*, 8:26, 1958.

[268] L.M. Kachanov. *Introduction to Continuum Damage Mechanics*. Springer, 1986.

[269] A. Kałamajska, S. Krömer, and M. Kružík. Sequential weak continuity of null Lagrangians at the boundary. *Calc. Var.*, 49:1263–1278, 2014.

[270] A. Kałamajska and M. Kružík. Oscillations and concentrations in sequences of gradients. *ESAIM: Control, Optimisation and Calculus of Variations*, 14:71–104, 2008.

[271] D. Kinderlehrer and P. Pedregal. Characterizations of Young measures generated by gradients. *Arch. Ration. Mech. Anal.*, 115:329–365, 1991.

[272] D. Kinderlehrer and P. Pedregal. Gradient Young measures generated by sequences in Sobolev spaces. *J. Geom. Anal.*, 4:59–90, 1994.

[273] H.-A. Klei and M. Miyara. Une extension du lemme de Fatou. *Bull. Sci. Math.*, 115:211–221, 1991.

[274] P. Klouček, B. Li, and M. Luskin. Analysis of a class of nonconforming finite elements for crystalline microstructures. *Math. Comp.*, 65:1111–1135, 1996.

[275] H. Knüpfer and R.V. Kohn. Minimal energy for elastic inclusions. *Proc. R. Soc. Lond. Ser. A Math. Phys. Eng. Sci.*, 467:695–717, 2011.

[276] H. Knüpfer, R.V. Kohn, and F. Otto. Nucleation barriers for the cubic-to-tetragonal phase transformation. *Comm. Pure Appl. Math.*, 66:867–904, 2013.

[277] R. Kohn and G. Strang. Optimal design and relaxation of variational problems I, II, III. *Comm. Pure Appl. Math.*, 39:113–137, 139–182, 353–357, 1986.

[278] R. Kohn and R. Temam. Dual spaces of stresses and strains, with applications to Hencky plasticity. *Appl. Math. Optim.*, 10:1–35, 1983.

[279] R.V. Kohn and S. Müller. Branching of twins near an austenite/twinned-martensite interface. *Philosophical Magazine A*, 66:697–715, 1992.

[280] R.V. Kohn and S. Müller. Surface energy and microstructure in coherent phase transitions. *Comm. Pure Appl. Math.*, 47:405–435, 1994.

[281] V.I. Kondrachov. Sur certaines propriétés fonctions dans l'espace L^p. *C.R. (Doklady) Acad. Sci. USSR (N.S.)*, 48:535–538, 1945.

[282] V.A. Kondrat'ev and O.A. Oleinik. Boundary-value problems for the system of elasticity theory in unbounded domains. korn's inequalities. *Russian Mathematical Surveys*, 43:65–120, 1988.

[283] A. Korn. Die Eigenschwingungen eines elastischen Korpers mit ruhender Oberflache. *Akad. der Wissensch., München, Math.-phys.Kl,*, 36:351–401, 1906.

[284] D. J. Korteweg. Sur la forme que prennent les équations du mouvement des fuides si lón tient compte des forces capillaires causées par des variations de densité considérables mais continues et sur la théorie de la capillarité dans l'hypothèse d'une variation continue de la densité. *Arch. Néerl. Sci. Exactes Nat.*, 6:1–24, 1901.

[285] K. Koumatos, F. Rindler, and E. Wiedemann. Differential inclusions and Young measures involving prescribed Jacobians. *SIAM J. Math. Anal.*, 47:1169–1195, 2015.

[286] P. Krejčí, E. Rocca, and J. Sprekels. A nonlocal phase-field model with non-constant specific heat. *Interfaces & Free Boundaries*, 9:285–306, 2007.

[287] J. Kristensen. *Finite functionals and Young measures generated by gradients of Sobolev functions*. PhD thesis, Math. Inst., Tech. Univ. of Denmark, 1994.

[288] E. Kröner. Allgemeine Kontinuumstheorie der Versetzungen und Eigenspannungen. *Arch. Ration. Mech. Anal.*, 4:273–334, 1960.

[289] M. Kružík. On the composition of quasiconvex functions and the transposition. *Journal of Convex Analysis*, 6:207–213, 1999.

[290] M. Kružík. Numerical approach to double well problems. *SIAM J. Numer. Anal.*, 35:1833–1849, 1998.

[291] M. Kružík. Bauer's maximum principle and hulls of sets. *Calculus of Variations and Partial Differential Equations*, 11:321–332, 2000.

[292] M. Kružík. Quasiconvexity at the boundary and concentration effects generated by gradients. *ESAIM: Control, Optim. Calc. Var.*, 19:679–700, 2013.

[293] M. Kružík and M. Luskin. The computation of martensitic microstructures with piecewise laminates. *J. Sci. Comput.*, 19:293–308, 2003.

[294] M. Kružík, A. Mielke, and T. Roubíček. Modelling of microstructure and its evolution in shape-memory-alloy single-crystals, in particular in CuAlNi. *Meccanica*, 40:389–418, 2005.

[295] M. Kružík and T. Roubíček. Explicit characterization of L^p-Young measures. *J. Math. Anal. Appl.*, 198:830–843, 1996.
[296] M. Kružík and T. Roubíček. On the measures of DiPerna and Majda. *Mathematica Bohemica*, 122:383–399, 1997.
[297] M. Kružík, U. Stefanelli, and J. Zeman. Existence results for incompressible magnetoelasticity. *Disc. Cont. Dynam. Systems*, 35:2615–2623, 2015.
[298] A. Kufner, O. John, and S. Fučík. *Function spaces*. Academia & Nordhoff Int. Publ., Praha & Leyden, 1977.
[299] A. A. Kulikovsky. *Analytical Modelling of Fuel Cells*. Elsevier, Amsterdam, 2010.
[300] 0.A. Ladyzhenskaya. *The Boundary Value Problems of Mathematical Physics*. Springer, New York, 1985.
[301] R.S. Lakes. *Viscoelastic Materials*. Cambridge Univ. Press, 2009.
[302] D. C. C. Lam, F. Yang, A. C. M Chong, J. Wang, and P. Tong. Experiments and theory in strain gradient elasticity. *Journal of the Mechanics and Physics of Solids*, 51:1477–1508, 2003.
[303] F.C. Larché and J.W. Cahn. The effect of self-stress on diffusion in solids. *Acta Metall.*, 30:1835–1845, 1982.
[304] C.J. Larsen, C. Ortner, and E. Süli. Existence of solution to a regularized model of dynamic fracture. *Math. Models Meth. Appl. Sci.*, 20:1021–1048, 2010.
[305] M. Lavrentieff. Sur quelques problèmes du calcul des variations. *Annali Matem. Pura Appl.*, 4:7–28, 1927.
[306] P.D. Lax and A.N. Milgram. Parabolic equations. *Annals of Mathematics Studies*, 33:167–190, 1964.
[307] M. Lazar and G.A. Maugin. Nonsingular stress and strain fields of dislocations and disclinations in first strain gradient elasticity. *Intl. J. Engr. Sci.*, 43:1157–1184, 2005.
[308] M. Lazar, G.A. Maugin, and E.C. Aifantis. On a theory of nonlocal elasticity of bi-Helmholtz type and some applications. *Intl. J. Solids Structures*, 43:1404–1421, 2006.
[309] H. Le Dret and A. Raoult. Remarks on the quasiconvex envelope of stored energy functions in nonlinear elasticity. *Commun. Appl. Nonlinear Anal.*, 1:85–96, 1994.
[310] J. Le Roux. Etude géométrique de la torsion et de la flexion, dans les déformations infinitésimales d'un milieu continu. *Ann Ecole Norm. Sup.*, 28:523–579, 1911.
[311] H. Lebesgue. Sur les intégrales singuliéres. *Ann. Fac. Sci Univ. Toulouse, Math.-Phys.*, 1:25–117, 1909.
[312] E. Lee and D. Liu. Finite-strain elastic-plastic theory with application to plain-wave analysis. *J. Applied Phys.*, 38:19–27, 1967.
[313] D. Leguillon. Strength or toughness? A criterion for crack onset at a notch. *European J. of Mechanics A/Solids*, 21:61–72, 2002.

[314] M. Lewicka and P. B. Mucha. A local existence result for system of viscoelasticity with physical viscosity. *Evolution Equations & Control Theory*, 2:337–353, 2013.
[315] Z. Li. Numerical methods for mimizers and microstructures in nonlinear elasticity. *Math. Models Meth. Appl. Sci.*, 6:957–975, 1996.
[316] F.H. Lin and C. Liu. Nonparabolic dissipative systems modeling the flow of liquid crystals. *Comm. Pure Appl. Math.*, 48:501–537, 1995.
[317] J.L. Lions. *Quelques Méthodes de Résolution des Problémes aux Limites non linéaires*. Dunod, Paris, 1969.
[318] J. Lubliner. *Plasticity theory*. Macmillan Publ., New York, 1990.
[319] R. Lucchetti and F. Patrone. On Nemytskii's operator and its application to the lower semicontinuity of integral functionals. *Indiana Univ. Math. J.*, 29:703–713, 1980.
[320] M. Luskin. Numerical analysis of microstructure for crystals with nonconvex energy density. In M. Chipot and J. Saint Jean Paulin, editors, *The Metz Days Surveys 1989-90*, Pitman Res. Notes in Math, pages 156–165, 1991.
[321] M. Luskin. On the computation of crystalline microstructure. *Acta Numerica*, 5:191–257, 1996.
[322] M. Luskin. Approximation of a laminated microstructure for a rotationally invariant, double well energy density. *Numer. Math.*, 75:205–221, 1997.
[323] V. Lyakhovsky and Y. Ben-Zion. Scaling relations of earthquakes and aseismic deformation in a damage rheology model. *Geophys. J. Int.*, 172:651–662, 2008.
[324] V. Lyakhovsky and Y. Ben-Zion. A continuum damagebreakage faulting model and solid-granular transitions. *Pure Appl. Geophys.*, 171:3099–3123, 2014.
[325] V. Lyakhovsky, Y. Hamiel, and Y. Ben-Zion. A non-local visco-elastic damage model and dynamic fracturing. *J. Mech. Phys. Solids*, 59:1752–1776, 2011.
[326] E. Magenes and G. Stampacchia. I problemi al contorno per le equazioni differenziali di tipo ellittico. *Annali Scuola Normale Superiore Pisa - Cl. Sci.*, 12:247–358, 1958.
[327] F. Mainardi. *Fractional calculus and waves in linear viscoelasticity*. Imperial College Press, London, 2010.
[328] F. Mainardi and G. Spada. Creep, relaxation and viscosity properties for basic fractional models in rheology. *The Europ. Phys. J. Special Topics*, 193:133–160, 2011.
[329] J. Malý. Personal communication. Prague, 2015.
[330] J. Mandel. *Plasticité classique et viscoplasticité*. Springer, Berlin, 1972.
[331] V. Mantič. Interface crack onset at a circular cylindrical inclusion under a remote transverse tension. Application of a coupled stress and energy criterion. *Intl. J. Solids Structures*, 46:1287–1304, 2009.
[332] V. Mantič. Prediction of initiation and growth of cracks in composites. Coupled stress and energy criterion of the finite fracture mechanics. In *ECCM-16*th *Europ. Conf. on Composite Mater. 2014*, pages 1–16,

http://www.escm.eu.org/eccm16/assets/1252.pdf, 2014. Europ. Soc. Composite Mater. (ESCM).

[333] C. Maor. A simple example of the weak discontinuity of $f \mapsto \int \det\nabla f$. *Preprint arXiv:1802.03066*, 2018.

[334] M. Marcus and V.J. Mizel. Transformations by functions in Sobolev spaces and lower semicontinuity for parametric variational problems. *Bull. Amer. Math. Soc.*, 79:790–795, 1973.

[335] M. Marcus and V.J. Mizel. Complete characterization of functions which act, via superposition, on Sobolev spaces. *Trans. Amer. Math. Soc.*, 251:187–218, 1979.

[336] L.C. Martins and P. Podio-Guidugli. A variational approach to the polar decomposition theorem. *Rend. delle sedute dell'Accademia nazionale dei Lincei*, 66(6):487–493, 1979.

[337] O. Martio and W.P. Ziemer. Lusin's condition (N) and mappings with non-negative Jacobians. *Michigan Math. J.*, 39:495–508, 1992.

[338] J. Matoušek and P. Plecháč. On functional separately convex hulls. *Discrete & Computational Geometry*, 19:105–130, 1998.

[339] G.A. Maugin. Nonlocal theories or gradient-type theories - A matter of convenience. *Archives of Mechanics*, 31:15–26, 1979.

[340] G.A. Maugin. *The Thermomechanics of Plasticity and Fracture*. Cambridge Univ. Press, Cambridge, 1992.

[341] G.A. Maugin. *Nonlinear Waves in Elastic Crystals*. Oxford Univ. Press, Oxford, 1999.

[342] G.A. Maugin. *The Thermomechanics of Nonlinear Irreversible Behaviors*. World Scientific, Singapore, 1999.

[343] G.A. Maugin. *Continuum Mechanics Through the Twentieth Century*. Springer, Dordrecht, 2013.

[344] G.A. Maugin. The saga of internal variables of state in continuum thermomechanics (1893-2013). *Mechanics Research Communications*, 69:79–86, 2015.

[345] G.A. Maugin. Some remarks on generalized continuum mechanics. *Math. Mech. Solids*, 20:280–291, 2015.

[346] J.C. Maxwell. On the dynamical theory of gasses. *Phil. Trans. Roy. Soc. London*, 157:49–88, 1868.

[347] S. Mazur. Über konvexe Mengen in linearen normierten Räumen. *Studia Mathematica*, 4:70–84, 1933.

[348] V. Maz'ya, S. Nazarov, and B. Plamenevskij. *Asymptotic Theory of Elliptic Boundary Value Problems in Singularly Perturbed Domains, Vol.I+II*. Springer/Birkhäuser, Basel, 2000.

[349] N.G. Meyers. Quasi-convexity and lower semi-continuity of multiple variational integrals of any order. *Trans. Amer. Math. Soc.*, 119:125–149, 1965.

[350] C. Miehe, S. Mauthe, and H. Ulmer. Formulation and numerical exploitation of mixed variational principles for coupled problems of Cahn-Hilliard-type and standard diffusion in elastic solids. *Int. J. Numer. Meth. Engng.*, 99:737–762, 2014.

[351] C. Miehe, F. Welschinger, and M. Hofacker. Thermodynamically consistent phase-field models of fracture: Variational principles and multi-field FE implementations. *Intl. J. Numer. Meth. Engr.*, 83:1273–1311, 2010.

[352] A. Mielke. Finite elastoplasticity, Lie groups and geodesics on SL(d). In P. Newton, A. Weinstein, and P. J. Holmes, editors, *Geometry, Mechanics, and Dynamics*, pages 61–90. Springer–Verlag, New York, 2002.

[353] A. Mielke. Evolution in rate-independent systems (Ch. 6). In C.M. Dafermos and E. Feireisl, editors, *Handbook of Differential Equations, Evolutionary Equations, vol. 2*, pages 461–559. Elsevier B.V., Amsterdam, 2005.

[354] A. Mielke. Complete-damage evolution based on energies and stresses. *Disc. Cont. Dynam. Syst. S*, 4:423–439, 2011.

[355] A. Mielke. Formulation of thermoelastic dissipative material behavior using GENERIC. *Contin. Mech. Thermodyn.*, 23:233–256, 2011.

[356] A. Mielke. On evolutionary Γ-convergence for gradient systems. In A. Muntean, J. Rademacher, and A. Zagaris, editors, *Macroscopic and Large Scale Phenomena*, pages 187–249. Springer, 2016.

[357] A. Mielke, C. Ortner, and Y. Sengül. An approach to nonlinear viscoelasticity via metric gradient flows. *SIAM J. Math. Anal*, 46:1317–1347, 2013.

[358] A. Mielke, R. Rossi, and G. Savaré. Global existence results for viscoplasticity at finite strain. *Arch. Ration. Mech. Anal.*, 227, 2018.

[359] A. Mielke and T. Roubíček. A rate-independent model for inelastic behavior of shape-memory alloys. *Multiscale Model. Simul.*, 1:571–597, 2003.

[360] A. Mielke and T. Roubíček. *Rate-Independent Systems – Theory and Application*. Springer, New York, 2015.

[361] A. Mielke and T. Roubíček. Rate-independent elastoplasticity at finite strains and its numerical approximation. *Math. Models Meth. Appl. Sci.*, 6:2203–2236, 2016.

[362] A. Mielke and T. Roubíček. Thermoviscoelasticity in Kelvin-Voigt rheology at large strains. In preparation.

[363] A. Mielke, T. Roubíček, and U. Stefanelli. Γ-limits and relaxations for rate-independent evolutionary problems. *Calc. Var. Part. Diff. Eqns.*, 31:387–416, 2008.

[364] A. Mielke, T. Roubíček, and J. Zeman. Complete damage in elastic and viscoelastic media and its energetics. *Comput. Methods Appl. Mech. Engrg.*, 199:1242–1253, 2010.

[365] A. Mielke and U. Stefanelli. Linearized plasticity is the evolutionary Γ-limit of finite plasticity. *J. Europ. Math. Soc.*, 15:923–948, 2013.

[366] A. Mielke and F. Theil. On rate-independent hysteresis models. *Nonl. Diff. Eqns. Appl.*, 11:151–189, 2004.

[367] A. Mielke, F. Theil, and V.I. Levitas. A variational formulation of rate–independent phase transformations using an extremum principle. *Arch. Ration. Mech. Anal.*, 162:137–177, 2002.

[368] G.W. Milton. *The Theory of Composites*. Cambridge Univ. Press, Cambridge, 2002.

[369] R.D. Mindlin. Micro-structure in linear elasticity. *Archive Ration. Mech. Anal.*, 16:51–78, 1964.

[370] R.D. Mindlin and N.N. Eshel. On first strain-gradient theories in linear elasticity. *Intl. J. Solids Strucrures*, 4:109–124, 1968.

[371] A. Miroshnikov and A.E. Tzavaras. Convergence of variational approximation schemes for elastodynamics with polyconvex energy. *J. Anal. its Appl. (ZAA)*, 33:43–64, 2014.

[372] L. Modica and S. Mortola. Il limite nella Γ-convergenza di una famiglia di funzionali ellittici. *Boll. Unione Mat. Italiana A*, 14:526–529, 1977.

[373] M.D.P. Monteiro Marques. *Differential Inclusions in Nonsmooth Mechanical Problems. Shocks and Dry Friction.* Birkhäuser Verlag, Basel, 1993.

[374] M. Mooney. A theory of large elastic deformation. *J. Appl. Physics*, 11:582–592, 1940.

[375] C.B. Morrey. *Multiple Integrals in the Calculus of Variations*. Springer, 1966.

[376] C.B.Jr. Morrey. Quasi-convexity and the lower semicontinuity of multiple integrals. *Pac. J. Math.*, 2:25–53, 1952.

[377] I. Müller and W. Weiss. *Entropy and Energy: a Universal Competition.* Springer, Berlin, 2005.

[378] S. Müller. A surprising higher integrability property of mappings with positive determinant. *Bull. Am. Math. Soc., New Ser.*, 21:245–248, 1989.

[379] S. Müller. Higher integrability of determinants and weak convergence in L^1. *J. Reine Angew. Math.*, 412:20–34, 1990.

[380] S. Müller. A sharp version of Zhang's theorem on truncating sequences of gradients. *Trans. Amer. Math. Soc.*, 351:4585–4597, 1999.

[381] S. Müller. Variational models for microstructure and phase transisions. Lect. Notes in Math. 1713, Springer Berlin, 1999, pp. 85–210, 1999.

[382] S. Müller. Quasiconvexity is not invariant under transposition. *Proceedings of the Royal Society of Edinburgh Section A: Mathematics*, 130:389–395, 2000.

[383] D. Mumford and J. Shah. Optimal approximations by piecewise smooth functions and associated variational problems. *Comm. Pure Appl. Math.*, 42:577–685, 1989.

[384] I. Münch and P. Neff. Rotational invariance conditions in elasticity, gradient elasticity and its connection to isotropy. *Math. Mech. Solids*, 23:3–42, 2018.

[385] F. Murat. Compacité par compensation. *Bull. Soc. Math. Fr., Suppl., Mém.*, 60:125–127, 1979.

[386] J. Nečas. *Sur les normes équivalentes dans* $W_p^k(\Omega)$ *et sur la coercivité des formes formellement positives*, pages 102–128. Séminaire de Mathématiques Supérieures. 19. Les Presses de l'Univ. de Montréal, Montréal, 1966.

[387] J. Nečas. *Les Méthodes Directes en Théorie des Équations Elliptiques*. Masson, Paris, 1967.

[388] J. Nečas and I. Hlaváček. *Mathematical Theory of Elastic and Elasto-Plastic Bodies: An Introduction*. Elsevier, Amsterdam, 1981.

[389] J. Nečas and T. Roubíček. Approximation of a nonlinear thermoelastic problem with a moving boundary via a fixed-domain method. *Aplikace matematiky*, 35:361–372, 1990.

[390] J. Nečas and T. Roubíček. Buoyancy-driven viscous flow with L^1-data. *Nonlinear Anal.*, 46:737–755, 2001.

[391] P. Neff. On Korn's first inequality with non-constant coefficients. *Proc. Royal Soc. Edinburgh*, 132A:221–243, 2002.

[392] P.V. Negrón-Marrero. A numerical method for detecting singular minimizers of multidimensional problems in nonlinear elasticity. *Numer. Math.*, 58:135–144, 1990.

[393] J. von Neumann. Zur Theorie der Gesellschaftsspiele. *Math. Ann.*, 100:295–320, 1928.

[394] N.M. Newmark. A method of computation for structural dynamics. *J. Eng. Mech. Div.*, 85:67–94, 1959.

[395] Q.S. Nguyen. *Stability and nonlinear solid mechanics*. J. Wiley, Chichester, 2000.

[396] R.A. Nicolaides and N.J. Walkington. Computation of microstructure utilizing Young measure representations. In C.A. Rogers and R.A. Rogers, editors, *Rec. Adv. Adapt. Sensory Mater. Appl.*, pages 131–141. Technomic Publ., 1992.

[397] M. Niezgodka and I. Pawłow. Generalized Stefan problem in several space variables. *Appl. Math. Optim.*, 9:193–224, 1983.

[398] O. Nikodým. Sur une généralisation des intégrales de M.J. Radon. *Fund. Mathematicae*, 15:131–179, 1930.

[399] L. Nirenberg. On elliptic partial differential equations. *Ann. Scuola Norm. Sup. Pisa Cl. Sci.*, 13:115–162, 1959.

[400] L. Nirenberg. An extended interpolation inequality. *Ann. Scuola Norm. Sup. Pisa Cl. Sci.*, 20:733–737, 1966.

[401] L. Nirenberg. *Topics in Nonlinear Functional Analysis*. Lecture Notes. Courant Institute, 1974.

[402] J.A. Nitsche. On Korn's second inequality. *ESAIM Math. Model. Numer. Anal.*, 15:237–248, 1981.

[403] R.W. Ogden. Large deformation isotropic elasticity on the correlation of theory and experiment for incompressible rubberlike solids. *Proc. Royal Soc. London, Ser. A, Math. Phys. Sci.*, 326:565–584, 1972.

[404] D. Ornstein. A non-inequality for differential operators in the L^1-norm. *Arch. Ration. Mech. Anal.*, 11:40–49, 1962.

[405] H.C. Öttinger. *Beyond Equilibrium Thermodynamics*. J. Wiley, Hoboken, NJ, 2005.

[406] A.Z. Palmer and T.J. Healey. Injectivity and self-contact in second-gradient nonlinear elasticity. *Calc. Var.*, 56:Art. no.114, 2017.

[407] E. Paraskevopoulos, C. Panagiotopoulos, and D. Talaslidis. Rational derivation of conserving time integration schemes: the moving-mass case. In M. Papadrakakis et al., editor, *Comput. Struct. Dynamics Earthquake Engr.*, chapter 10, pages 149–164. CRC Press, Bocca Raton, 2009.

[408] G.P. Parry. On shear bands in unloaded crystals. *J. Mech. Phys. Solids*, 35:367–382, 1987.

[409] P. Pedregal. Laminates and microstructure. *Europ. J. Appl. Math.*, 4:121–149, 1993.
[410] P. Pedregal. *Parametrized Measures and Variational Principles*. Birkhäuser, Basel, 1997.
[411] R.H.J. Peerlings and N.A. Fleck. Computational evaluation of strain gradient elasticity constants. *Intl. J. Multiscale Comput. Engr.*, 2:599–619, 2004.
[412] B.J. Pettis. On integration in vector spaces. *Trans. Amer. Math. Soc.*, 44:277–304, 1938.
[413] A. Phillips. The theory of locking materials. *Transactions of The Society of Rheology*, 3:13–26, 1959.
[414] K. Piechór. Non-local Korteweg stresses from kinetic theory point of view. *Arch. Mech.*, 60:23–58, 2008.
[415] M. Pitteri and G. Zanzotto. *Continuum Models for Phase Transitions and Twinning in Crystals*. Chapman & Hall/CRC, Boca Raton, 2003.
[416] P. Podio-Guidugli. Contact interactions, stress, and material symmetry, for nonsimple elastic materials. *Theor. Appl. Mech.*, 28–29:261–276, 2002.
[417] P. Podio-Guidugli and G.V. Caffarelli. Surface interaction potentials in elasticity. *Arch. Ration. Mech. Anal.*, 109:343–383, 1990.
[418] P. Podio-Guidugli and M. Vianello. Hypertractions and hyperstresses convey the same mechanical information. *Continuum Mech. Thermodynam.*, 22:163–176, 2010.
[419] W. Pompe. Korn's First Inequality with variable coefficients and its generalization. *Comment. Math. Univ. Carolinae*, 44:57–70, 2003.
[420] J.H. Poynting and J.J. Thomson. *Properties of Matter*. C. Griffin and Co., London, 1902 (6th ed. 1913).
[421] W. Prager. On ideal locking materials. *Trans. Soc. Rheology*, 1:169–175, 1957.
[422] K. Promislow and B. Wetton. PEM fuel cells: a mathematical overview. *SIAM J. Appl. Math.*, 70:369–409, 2009.
[423] A.M. Puzrin and G.T. Houlsby. Fundamentals of kinematic hardening hyperplasticity. *Intl. J. Solids Structures*, 38:3771–3794, 2001.
[424] A. Quarteroni and A. Valli. *Numerical Approximation of Partial Differential Equations*. Springer, Berlin, 1994.
[425] Yu.N. Rabotnov. *Creep Problems in Structural Members*. North-Holland, Amsterdam, 1969.
[426] J. Radon. Theorie u. anwendungen der absolut additiven mengenfunktionen. *Sitzber. Natur-Wiss. Klasse der Kais. Akademie der Wiss., Wien*, 112 Abt. II a/2, 1913.
[427] K.R. Rajagopal. On implicit constitutive theories. *Appl. Math.*, 48:279–319, 2003.
[428] K.R. Rajagopal. On a hierarchy of approximate models for flows of incompressible fluids through porous solids. *Math. Models Meth. Appl. Sci.*, 17:215–252, 2007.
[429] K.R. Rajagopal and T. Roubíček. On the effect of dissipation in shape-memory alloys. *Nonlinear Anal., Real World Appl.*, 4:581–597, 2003.

[430] K.R. Rajagopal and A.R. Srinivasa. On a class of non-dissipative materials that are not hyperelastic. *Proc. Royal Soc. London A*, 465:493–500, 2009.

[431] J.N. Reddy. *Theory and Analysis of Elastic Plates and Shells*. CRC Press, Taylor and Francis, 2007.

[432] J.N. Reddy. *An Introduction to Continuum Mechanics*. Cambridge Univ. Press, 2013.

[433] M. Reiner. The Deborah number. *Physics Today*, 17:62, 1964.

[434] J. Reinoso, A. Arteiro, M. Paggi, and P.P. Camanho. Strength prediction of notched thin ply laminates using finite fracture mechanics and the phase field approach. *Composites Sci. Technol.*, 150:205–216, 2017.

[435] J. Reinoso, M. Paggi, and C. Linder. Phase field modeling of brittle fracture for enhanced assumed strain shells at large deformations: formulation and finite element implementation. *Computational Mechanics*, 59:981–1001, 2017.

[436] F. Rellich. Ein Satz über mittlere Konvergenz. *Nachr. Akad. Wiss. Göttingen*, pages 30–35, 1930.

[437] M. Renardy and R.C. Rogers. *An Introduction to Partial Differential Equations*. Springer, New York, 2 edition, 2004.

[438] Y.G. Reshetnyak. Weak convergence of completely additive vector functions on a set. *Siberian Math. J.*, 9:1039–1045, 1968.

[439] Y.G. Reshetnyak. *Space Mappings with Bounded Distortion*. Amer. Math. Soc., Providence, RI, 1989.

[440] F. Rindler. *Calculus of Variations*. Springer Nature, Cham, 2018.

[441] W. Ritz. Über eine neue Methode zur Lösung gewisser Variationsprobleme der mathematischen Physik. *J. für reine u. angew. Math.*, 135:1–61, 1908.

[442] R.S. Rivlin. Large elastic deformations of isotropic materials. IV. Further developments of the general theory. *Phil. Trans. Royal Soc. London. Ser. A, Math. Phys. Sci.*, 241(835):379–397, 1948.

[443] R.T. Rockafellar. *Convex Analysis*. Princeton Univ. Press, 1970.

[444] R.C. Rogers. Nonlocal variational problems in nonlinear electromagneto-elastostatics. *SIAM J. Math. Anal.*, 19:1329–1347, 1988.

[445] R.C. Rogers. A nonlocal model for the exchange energy in ferromagnetic materials. *J. Int. Eq. Appl.*, 33:85–127, 1991.

[446] R.C. Rogers. Some remarks on nonlocal interactions and hysteresis in phase transitions. *Continuum Mech. Thermodyn.*, 8:65–73, 1996.

[447] D. Rogula. Introduction to nonlocal theory of material media. In D. Rogula, editor, *Nonlocal Theory of Material Media*, CISM Courses and Lectures 268, pages 125–222. Springer, Wien, 1982.

[448] R. Rossi and M. Thomas. Coupling rate-independent and rate-dependent processes: existence results. *SIAM J. Math. Anal.*, 49:1419–1494, 2017.

[449] Y.A. Rossikhin and M.V. Shitikova. Application of Fractional Calculus for Dynamic Problems of Solid Mechanics: Novel Trends and Recent Results. *Appl. Mech. Rev.*, 63:Art. no. 010801, 2010.

[450] E. Rothe. Zweidimensionale parabolische Randwertaufgaben als Grenzfall eindimensionaler Randwertaufgaben. *Math. Ann.*, 102:650–670, 1930.

[451] T. Roubíček. Unconditional stability of difference formulas. *Apl. Mat.*, 28:81–90, 1983.

[452] T. Roubíček. *Relaxation in Optimization Theory and Variational Calculus.* W. de Gruyter, Berlin, 1997.

[453] T. Roubíček. Direct method for parabolic problems. *Adv. Math. Sci. Appl.*, 10:57–65, 2000.

[454] T. Roubíček. Rate independent processes in viscous solids at small strains. *Math. Methods Appl. Sci.*, 32:825–862, 2009. Erratum p. 2176.

[455] T. Roubíček. Thermo-visco-elasticity at small strains with L^1-data. *Quarterly Appl. Math.*, 67:47–71, 2009.

[456] T. Roubíček. *Nonlinear Partial Differential Equations with Applications.* Birkhäuser, Basel, 2nd edition, 2013.

[457] T. Roubíček. Nonlinearly coupled thermo-visco-elasticity. *Nonlin. Diff. Eq. Appl.*, 20:1243–1275, 2013.

[458] T. Roubíček. Thermodynamics of perfect plasticity. *Discrete and Cont. Dynam. Syst. - S*, 6:193–214, 2013.

[459] T. Roubíček. Calculus of variations. In M. Grinfeld, editor, *Mathematical Tools for Physicists*, chapter 17, pages 551–588. J. Wiley, Weinheim, 2014.

[460] T. Roubíček. An energy-conserving time-discretisation scheme for poroelastic media with phase-field fracture emitting waves and heat. *Disc. Cont. Dynam. Syst. S*, 10:867–893, 2017.

[461] T. Roubíček. Geophysical models of heat and fluid flow in damageable poro-elastic continua. *Cont. Mech. Thermodyn.*, 29:625–646, 2017.

[462] T. Roubíček. Variational methods for steady-state Darcy/Fick flow in swollen and poroelastic solids. *Zeit. angew. Math. Mech.*, 97:990–1002, 2017.

[463] T. Roubíček. Seismic waves and earthquakes in a global monolithic model. *Cont. Mech. Thermodynam.*, 30:709–729, 2018.

[464] T. Roubíček and K.-H. Hoffmann. About the concept of measure-valued solutions to distributed parameter systems. *Math. Meth. in the Appl. Sci.*, 18:671–685, 1995.

[465] T. Roubíček and U. Stefanelli. Finite thermoelastoplasticity and creep under small elastic strain. *Math. Mech. of Solids (printed on line)*, 2018.

[466] T. Roubíček and G. Tomassetti. Dynamics of charged elastic bodies at large strains. *Submitted.*

[467] T. Roubíček and G. Tomassetti. A thermodynamically consistent model of magneto-elastic materials under diffusion at large strains and its analysis. *Zeit. angew. Math. Phys.*, 64:1–28, 2013.

[468] T. Roubíček and J. Valdman. Rate-independent perfect plasticity with damage and healing at small strains, its modelling, analysis, and computer implementation. *SIAM J. Appl. Math.*, 76:314–340, 2016.

[469] T. Roubíček and R. Vodička. A monolithic model for seismic sources and seismic waves. *Submitted.*

[470] T. Roubíček, V. Mantič, and C.G. Panagiotopoulos. Quasistatic mixed-mode delamination model. *Disc. Cont. Dynam. Syst. - S*, 6:591–610, 2013.

[471] P. Rybka and M. Luskin. Existence of energy minimizers for magnetostrictive materials. *SIAM J. Math. Anal.*, 36:2004–2019, 2005.

[472] M. Saadoune and M. Valadier. Extraction of a “good” sequence from a bounded sequence of integrable functions. *J. Convex Anal.*, 2:345–357, 1994.

[473] M.H. Sadd. *Elasticity: Theory, Applications, and Numerics*. Elsevier, third edition, 2014.

[474] S. Saks. *Theory of the Integral*. Hafner Publ. Comp., New York, 2nd edition, 1937.

[475] E. Sanchez-Palencia. *Non-homogeneous Media and Vibration Theory*. Springer, Berlin, 1980.

[476] J.M. Sargado, E. Keilegavlen, I. Berre, and J.M. Nordbotten. High-accuracy phase-field models for brittle fracture based on a new family of degradation functions. *J. Mech. Phys. Solids*, 111:458–489, 2018.

[477] J. Schauder. Der Fixpunktsatz in Fuktionalräumen. *Studia Math.*, 2:171–180, 1930.

[478] M.E. Schonbek. Convergence of solutions to nonlinear dispersive equations. *Commun. Partial Differ. Equations*, 7:959–1000, 1982.

[479] F. Schuricht. Variational approach to contact problems in nonlinear elasticity. *Calc. Var.*, 15:433–449, 2002.

[480] F. Schuricht. Locking constraints for elastic rods and a curvature bound for spatial curves. *Calculus of Variations and Partial Differential Equations*, 24:377–402, 2005.

[481] R.E. Showalter and U. Stefanelli. Diffusion in poro-elastic media. *Math. Methods Appl. Sci.*, 27:2131–2151, 2004.

[482] M. Šilhavý. Phase transitions in non-simple bodies. *Arch. Ration. Mech. Anal.*, 88:135–161, 1985.

[483] M. Šilhavý. *The Mechanics and Thermodynamics of Continuous Media*. Springer, Berlin, 1997.

[484] M. Šilhavý. Phase transitions with interfacial energy: Interface null lagrangians, polyconvexity, and existence. In K. Hackl, editor, *IUTAM Symp. on Var. Concepts with Appl. to Mech. of Mater.*, pages 233–244. Springer, Netherlands, 2010.

[485] M. Šilhavý. Equilibrium of phases with interfacial energy: a variational approach. *J. Elasticity*, 105:271–303, 2011.

[486] J.C. Simo and J.R. Hughes. *Computational Inelasticity*. Springer, Berlin, 1998.

[487] M. Slemrod. Dynamic of measure-valued solutions to a backward-forward heat equation. *J. of Dynamics and Differential Equations*, 3:1–28, 1991.

[488] L.N. Slobodeckiĭ. Generalized Sobolev spaces and their applications to boundary value problems of partial differential equations. *Leningrad. Gos. Ped. Inst. Ucep. Zap.*, 197:54–112, 1958.

[489] D.R. Smith. *An Introduction to Continuum Mechanics*. Kluwer, Dordrecht, 1993.

[490] S.L. Sobolev. On some estimates relating to families of functions having derivatives that are square integrable. (In Russian). *Dokl. Akad. Nauk SSSR*, 1:267–270, 1936.
[491] M. Spitzer. *Digitale Demenz. Wie wir uns und unsere Kinder um den Verstand bringen.* Droemer Verlag, München, 2012.
[492] P. Sprenger. *Quasikonvexität am Rande und Null-Lagrange-Funktionen in der nichtkonvexen Variationsrechnung*. PhD thesis, Univ. Hannover, 1996.
[493] G. Stampacchia. Le problème de Dirichlet pour les équations elliptiques du second ordre à coefficients discontinus. *Ann. Inst. Fourier (Grenoble)*, 15:189–258, 1965.
[494] D.J. Steigmann. Equilibrium theory for magnetic elastomers and magnetoelastic membranes. *Intl. J. Non-Linear Mech.*, 39:1193–1216, 2004.
[495] I. Steinbach. Phase-field models in materials science. *Modelling Simul. Mater. Sci. Eng.*, 17:073001 (31pp), 2009.
[496] M. Struwe. *Variational methods. Applications to Nonlinear Partial Differential Equations and Hamiltonian Systems*. Springer, Berlin, 2010.
[497] P.-M. Suquet. Existence et régularité des solutions des équations de la plasticité parfaite. *C. R. Acad. Sci. Paris Sér.A*, 286:1201–1204, 1978.
[498] V. Šverák. Regularity properties of deformations with finite energy. *Arch. Ration. Mech. Anal.*, 100:105–127, 1988.
[499] V. Šverák. New examples of quasiconvex functions. *Archive Ration. Mech. Anal.*, 119:293–300, 1992.
[500] V. Šverák. Rank-one convexity does not imply quasiconvexity. *Proc. R. Soc. Edinb*, 120A:185–189, 1992.
[501] M.A. Sychev. A new approach to Young measure theory, relaxation and convergence in energy. *Ann. Inst. Henri Poincaré, Anal. Non Linéaire*, 16:773–812, 1999.
[502] E. Tanné, T. Li, B. Bourdin, J.-J. Marigo, and C. Maurini. Crack nucleation in variational phase-field models of brittle fracture. *J. Mech. Phys. Solids*, 110:80–99, 2018.
[503] L. Tartar. Compensated compactness and applications to partial differential equations. In R. J. Knops, editor, *Res. Notes in Math. 39, Nonlinear Analysis and Mechanics*, pages 136–212, Boston, MA, 1979. Pitman.
[504] L. Tartar. Some remarks on separately convex functions. In D. Kinderlehrer, R.D. James, M. Luskin, and J.L. Ericksen, editors, *Microstructure and Phase Transition*, pages 191–204. Springer, New York, 1993.
[505] L. Tartar. Beyond Young measures. *Meccanica*, 30:505–526, 1995.
[506] L. Tartar. *An Introduction to Sobolev Spaces and Interpolation Spaces*. Lect. Notes. of the Unione Mat. Italiana (Vol.3). Springer, Berlin, 2007.
[507] L. Tartar. *The General Theory of Homogenization*. Lect. Notes Unione Mat. Italiana (Vol.7). Springer, Heidelberg, 2009.
[508] J.J. Telega and S. Jemioło. Macroscopic behaviour of locking materials with microstructure Part 1. Primal and dual elastic-locking potential. Relaxation. *Bull. Polish Acad. of Sci., Tech. Sci.*, 46:265–276, 1998.

[509] R. Temam. *Mathematical Problems in Plasticity*. Gauthier-Villars (Engl. Transl. Bordas, Paris), 1985.
[510] R. Temam and A. Miranville. *Mathematical Modeling in Continuum Mechanics*. Cambridge Univ. Press, 2005.
[511] M. Thomas. *Rate-Independent Damage Processes in Nonlinearly Elastic Materials*. PhD thesis, Inst. f. Mathematik, Humboldt-Universität zu Berlin, 2010.
[512] M. Thomas and A. Mielke. Damage of nonlinearly elastic materials at small strain – Existence and regularity results –. *Zeitschrift angew. Math. Mech.*, 90:88–112, 2010.
[513] V. Thomée. *Galerkin Finite Element Methods for Parabolic Problems*. Springer, Berlin, 1997.
[514] H. Tietze. Über Funktionen, die auf einer abgeschlossenen Menge stetig sind. *Journal für die reine und angewandte Mathematik*, 145:9–14, 1915.
[515] L. Tonelli. Sur un méthode directe du calcul des variations. *Rend. Circ. Mat. Palermo*, 39, 1915.
[516] R.A. Toupin. Elastic materials with couple stresses. *Arch. Ration. Mech. Anal.*, 11:385–414, 1962.
[517] R.A. Toupin. Theories of elasticity with couple stress. *Archive Ration. Mech. Anal.*, 17:85–112, 1964.
[518] C. Truesdell. *A First Course in Rational Continuum Mechanics*. Acad. Press, San Diego, 2 edition, 1991.
[519] C. Truesdell and W. Noll. *The Non-Linear Field Theories of Mechanics*. Pringer, Berlin, 3rd edition, 2004.
[520] L. Truskinovsky and A. Vainchtein. Quasicontinuum modelling of short-wave instabilities in crystal lattices. *Phil. Mag.*, 85:4055–4065, 2005.
[521] J. Vignollet, S. May, R. de Borst, and C.V. Verhoosel. Phase-field models for brittle and cohesive fracture. *Meccanica*, 49:2587–2601, 2014.
[522] A. Visintin. Strong convergence results related to strict convexity. *Comm. Partial Diff. Equations*, 9:439–466, 1984.
[523] A. Visintin. *Models of Phase Transitions*. Birkhäuser, Boston, 1996.
[524] A. Visintin. Mathematical models of hysteresis. In I. Mayergoyz and G. Bertotti, editors, *The Science of Hysteresis Vol. 1*. Elsevier, 2005. ISBN: 0-1248-0874-3.
[525] G. Vitali. Sui gruppi di punti e sulle funzioni di variabili reali. *Atti Accad. Sci. Torino*, 43:75–92, 1908.
[526] G. Vitali. Sui gruppi di punti e sulle funzioni di variabili reali. *Atti. Accad. Sci. Torino*, 43:75–92, 1908.
[527] W. Voigt. *Lehrbuch der Kristallphysik: (mit Ausschlusch der Kristalloptik)*. Teubner, Leipzig, 1910.
[528] J. von Neumann. *Functional operators. Vol.I: Measures and integrals*. Princeton Univ. Press, Princeton, 1950.
[529] T. Waffenschmidt, C. Polindara, A. Menzel, and S. Blanco. A gradient-enhanced large-deformation continuum damage model for fibre-reinforced materials. *Comput. Methods Appl. Mech. Engrg.*, 268:801–842, 2014.

[530] P. Weißgraeber, D. Leguillon, and W. Becker. A review of Finite Fracture Mechanics: crack initiation at singular and non-singular stress raisers. *Arch. Appl. Mech.*, 86:375–401, 2016.

[531] L.C. Wellford Jr. and S.M. Hamdan. An analysis of the stability and convergence properties of a Crank-Nicholson algorithm for nonlinear elastodynamics problems. In W. Wunderlich, E. Stein, and K.-J. Bathe, editors, *Nonlinear Finite Element Analysis in Structural Mechanics*, pages 502–518, Berlin, 1981. Springer.

[532] K. Wilmanski. *Continuum Thermodynamics, Part 1: Foundations.* World Scientific, New Jersey, 2008.

[533] J.H. Woodhouse and A. Deuss. Theory and observations Earths free oscillations. In B. Romanowicz and A. Dziewonski, editors, *Seismology and Structure of the Earth*, chapter 1.02. Elsevier, 2007.

[534] J.-Y. Wu and V.P. Nguyen. A length scale insensitive phase-field damage model for brittle fracture. *J. Mech. Phys. Solids*, 119:20–42, 2018.

[535] N.N. Yanenko. *The Method of Fractional Steps.* Springer, Berlin, 1971. (Russian Original: Nauka, Novosibirsk 1967).

[536] K. Yosida and E. Hewitt. Finitely additive measures. *Trans. Amer. Math. Soc.*, 72:46–66, 1952.

[537] L.C. Young. Generalized curves and the existence of an attained absolute minimum in the calculus of variations. *C. R. Soc. Sci. Varsovie, Cl. III*, 30:212–234, 1937.

[538] S.M.J. Zaidi and T. Matsuura, editors. *Polymer Membranes for Fuel Cells.* Springer, New York, 2009.

[539] E. Zeidler. *Nonlinear Functional Analysis and its Applications. I–IV.* Springer, New York, 1985-90.

[540] C.M. Zener. *Elasticity and Anelasticity of Metals.* Chicago Univ. Press, 1948.

[541] K. Zhang. A construction of quasiconvex functions with linear growth at infinity. *Annali Scuola Normale Superiore di Pisa - Cl. Sci.*, 19:313–326, 1992.

[542] W.P. Ziemer. *Weakly Differentiable Functions.* Springer, New York, 1989.

[543] O.C. Zienkiewicz and R.L. Taylor. *The Finite Element Method (Vol 1: The Basis, Vol 2: Solid Mechanics).* McGraw-Hill, 1967. (5th ed.: Butterworth-Heinemann, Oxford, 2000).

[544] J. Zimmer. Stored energy functions for phase transitions in crystals. *Arch. Ration. Mech. Anal.*, 172:191–212, 2004.

[545] B. Zwicknagl. Microstructures in low-hysteresis shape memory alloys: scaling regimes and optimal needle shapes. *Arch. Ration. Mech. Anal.*, 213:355–421, 2014.

Index

M. Kružík and T. Roubíček, *Mathematical Methods in Continuum Mechanics of Solids*, Interaction of Mechanics and Mathematics,
https://doi.org/10.1007/978-3-030-02065-1

Printed in the United States
By Bookmasters